Biology

Concepts & Applications 10e

About the Cover Photo
Portuguese Man-of-War

Anyone unfamiliar with the biology of the venomous Portuguese man-of-war would likely mistake it for a jellyfish. Not only is it not a jellyfish, it's not even an "it," but a "they." The Portuguese man-of-war is a siphonophore, an animal made up of a colony of organisms working together.

The man-of-war comprises four separate polyps. It gets its name from the uppermost polyp, a gas-filled bladder, or pneumatophore, which sits above the water and somewhat resembles an old warship at full sail. A man-of-war is also known as a bluebottle for the purple-blue color of its pneumatophores.

The tentacles are the man-of-war's second organism. These long, thin tendrils can extend 165 feet (50 meters) in length below the surface, although 30 feet (10 meters) is more the average. They are covered in venom-filled nematocysts used to paralyze and kill fish and other small creatures. For humans, a man-of-war sting is excruciatingly painful, but rarely deadly. But beware—even a dead man-of-war washed up on shore can deliver a sting.

Muscles in the tentacles draw prey up to a polyp containing the gastrozooids or digestive organisms. A fourth polyp contains the reproductive organisms.

A man-of-war is found, sometimes in groups of 1,000 or more, floating in warm waters throughout the world's oceans. They have no independent means of propulsion and either drift on the currents or catch the wind with their pneumatophores. To avoid threats on the surface, they can deflate their air bags and briefly submerge.

Australia • Brazil • Japan • Korea • Mexico • Singapore • Spain • United Kingdom • United States

Biology

Concepts & Applications 10e

Cecie Starr
Christine A. Evers
Lisa Starr

Biology: Concepts & Applications,
Tenth Edition
Cecie Starr, Christine A. Evers, Lisa Starr

Product Director: Dawn Giovanniello

Product Manager: April Cognato

Senior Content Developer: Jake Warde

Product Assistant: Kristina Cannon

Executive Marketing Manager: Tom Ziolkowski

Content Project Manager: Hal Humphrey

Managing Art Director: Andrei Pasternak

Manufacturing Planner: Karen Hunt

Production Service: Grace Davidson & Associates

Photo Researcher: Cheryl DuBois, Lumina Datamatics

Text Researcher: Manula Subramanian, Lumina Datamatics

Intellectual Property Analyst: Christine Myaskovsky

Intellectual Property Project Manager: Betsy Hathaway

Copy Editor: Anita Wagner Hueftle

Illustrators: Lisa Starr, ScEYEnce Studios, Lachina, Gary Head

Cover and Title Page Image: © Matthew Smith, 2016

Compositor: Lachina

For product information and technology assistance, contact us at
Cengage Learning Customer & Sales Support, 1-800-354-9706

For permission to use material from this text or product,
submit all requests online at **www.cengage.com/permissions.**
Further permissions questions can be e-mailed to
permissionrequest@cengage.com.

Library of Congress Control Number: 2016945595

High School Edition

978-1-337-09482-5

Cengage Learning
20 Channel Center Street
Boston, MA 02210
USA

Cengage Learning is a leading provider of customized learning solutions with employees residing in nearly 40 different countries and sales in more than 125 countries around the world. Find your local representative at **www.cengage.com/global.**

Cengage Learning products are represented in Canada by Nelson Education, Ltd.

To learn more about Cengage Learning Solutions, visit **www.cengage.com.**

To find online supplements and other instructional support, please visit **www.cengagebrain.com.**

Printed in Canada
Print Number: 02 Print Year: 2017

Contents in Brief

Detailed Contents

Enric Sala/National Geographic Creative

3 MOLECULES OF LIFE

4 CELL STRUCTURE

Detailed Contents (continued)

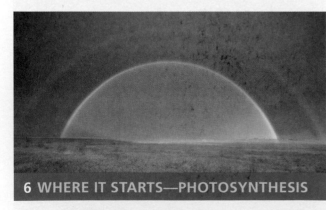

6 WHERE IT STARTS—PHOTOSYNTHESIS

7 RELEASING CHEMICAL ENERGY

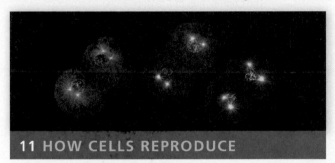

Detailed Contents (continued)

18 LIFE'S ORIGIN AND EARLY EVOLUTION

Detailed Contents (continued)

22 FUNGI

23 ANIMALS I: MAJOR INVERTEBRATE GROUPS

Detailed Contents (continued)

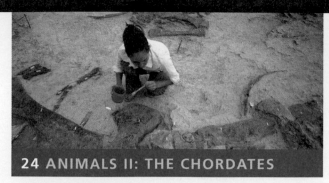

Detailed Contents (continued)

Detailed Contents (continued)

43 THE BIOSPHERE

44 HUMAN EFFECTS ON THE BIOSPHERE

APPENDICES

Preface

A revolution in the way information is shared has fundamentally changed the nature of biological inquiry. Interdisciplinary collaborations facilitated by instant, global access to data and ideas have fostered entirely new areas of research, both theoretical and practical. New discoveries and new technologies emerging from these collaborations are altering the way biologists think about their work—and the field in general.

Realizing that a traditional life science education would not adequately prepare students for the changing field, the American Association for the Advancement of Science and the National Science Foundation initiated a series of national conversations among leading life scientists, policy makers, educators, and students. The result was a document, *Vision and Change in Undergraduate Biology Education*, that calls for a fundamental change in the way life sciences are taught to all undergraduate students. A broad consensus recommends that science education become much more active, because personal experience with the process and limits of science better prepares students to evaluate scientific content and differentiate it from other information. A more concept-oriented approach that uses fundamental biological principles as a context for information (rather than the reverse) better prepares students to understand rapid changes in the field. Our future citizens and leaders will need this understanding to confront urgent societal problems such as climate change, threats to biodiversity, and the global spread of disease.

This book has been revised in alignment with "Vision and Change" recommendations. As always, recent discoveries are integrated in an accessible and appealing introduction to the study of life. In addition, this edition includes tools to explore basic biological concepts from a variety of perspectives (molecular, cellular, organismal, ecological, and so on).

Our quest to educate and engage is shared by the National Geographic Society, with whom we have partnered for this book. You will see the fruits of the partnership in spectacular photographs, informative illustrations, and text features that highlight the wide variety of work supported by the society.

FEATURES OF THIS EDITION

Setting the Stage
Each chapter opens with a dramatic two-page photo spread. A brief *Links to Earlier Concepts* paragraph reminds students of relevant information in previous chapters. A summary of chapter content is organized and presented in terms of three *Core Concepts*, each involving one of the following: evolution; information flow, exchange, and storage; structure and function; pathways and transformation of energy and matter; or systems.

Concept Spreads
The content of every chapter is organized as a series of concepts, each explored in a section that is two pages or less. *Learning Objectives* associated with each concept are phrased as activities that the student should be able to carry out after reading the text.

Engage
Our *Engage* feature associates chapter content with relevant research, while highlighting the diversity of the modern scientific community. Each illustrates one of six core competencies: the ability to apply the process of science (Process of Science), use quantitative reasoning (Quantitative Reasoning), use modeling and simulation (Complex Systems), tap into the interdisciplinary nature of science (Interdisciplinary Science), communicate and collaborate with other disciplines (Collaborative Science), or understand the relationship between science and society (Science & Society). Individuals whose work is spotlighted include well-established and newly minted scientists, as well as a few nonscientists; most are National Geographic Explorers or Grantees.

On-Page Glossary
An *On-Page Glossary* includes boldface key terms introduced in each section. This running glossary, which can be used as a convenient study aid, offers non-phonemic pronunciations and definitions in alternate wording. All glossary terms also appear in boldface in the Chapter Summary.

Emphasis on Relevance
We continue to focus on real world applications, including social issues arising from new research and developments—particularly the many ways in which human activities are continuing to alter the environment and threaten both human health and Earth's biodiversity. Each chapter ends with an *Application* section that explains a current topic in light of the chapter content, and also illustrates one of the core competencies listed above.

Self-Assessment Tools
Many figure captions include a *Figure It Out* question designed to engage students in an active learning process; an upside-down answer allows a quick check of understanding. At the end of each chapter, *Self-Assessment* and *Critical Thinking* questions provide additional self-assessment material. Another active-learning feature, the chapter-end *Data Analysis* class activity, sharpens analytical skills by asking students to interpret data presented in graphic or tabular form. The data is related to the chapter material, and is taken from a published scientific study in most cases.

Chapter-Specific Changes

This new edition contains 208 new photographs and 330 new or updated illustrations. In addition, the text of every chapter has been updated and revised for clarity. A complete section-by-section guide to new content and figures is available upon request, but the highlights are summarized here.

1 The Science of Biology Expanded material on the process of science includes the concept that research is typically nonlinear and unpredictable; text and table contrasting science with pseudoscience; and text detailing the way theories can be modified upon discovery of new data.

2 Life's Chemical Basis New table compares elemental composition of human body with Earth's crust, seawater, and the universe; effects of acid rain are now exemplified with dissolving shells of marine animals.

3 Molecules of Life New chapter opener illustrates formation of glycoaldehyde in interstellar gas; revised text and new art emphasizing protein structure–function relationship includes expanded discussion of amyloid fibrils and plaques.

4 Cell Structure Added nonmotile cilia and their newly discovered roles in cell signaling. New art shows ultrastructural details of cell junctions per recent discoveries. Expanded section on the nature of life now includes theory of living systems.

5 Ground Rules of Metabolism Coenzymes are now illustrated with ascorbic acid/scurvy example.

6 Where It Starts—Photosynthesis Chapter has been reorganized for a better introductory sequence. Expanded discussion of the cyclic pathway emphasizes the interplay between both versions of light reactions; expanded discussion of photorespiration incorporates new research on its adaptive value. New Application discusses biofuels in context of anthropogenic CO_2 and global warming.

7 Releasing Chemical Energy New art and tables emphasize the movement of energy and matter in aerobic respiration. Expanded Application includes mechanisms of mitochondrial malfunction, narrative of affected child, and three-person IVF. New Data Analysis concerns reprogramming of mitochondria in brown fat by dietary fat overload.

8 DNA Structure and Function New micrographs and revised art reinforce DNA structure and clarify mutations.

9 From DNA to Protein Expanded section on mutations includes material on a beneficial hemoglobin mutation (E6K, HbC) that offers resistance to malaria without the health consequences of HbS (E6V); and new art detailing the intron mutation that causes hairlessness in sphynx cats.

10 Control of Gene Expression Now includes RNA interference and microRNAs, and additional evidence for heritability of epigenetic modifications. Application section has expanded information about *BRCA* genes as tumor suppressors.

11 How Cells Reproduce New ultra-high resolution confocal live-cell images of mitosis by Dr. George von Dassow. Text and new art showing cytokinesis include ultrastructural details/processes per current research and paradigms. Expanded material on telomeres now includes telomere-associated triggering and consequences of cell senescence.

12 Meiosis and Sexual Reproduction New 3D structured illumination micrographs of meiosis in corn show synaptonemal complex detail in ultra-high resolution. Newly discovered mechanism of gene acquisition by individual rotifers added to Application essay.

13 Patterns in Inherited Traits Marfan syndrome discussion updated to reflect change in life expectancy due to increased awareness, accompanied by new photo of Baylor's Isaiah Austin. New material details environmentally induced hemoglobin production in *Daphnia*; new photo shows green-to-red phenotype that accompanies the switch.

14 Human Inheritance Molecular pathogenesis of CF, Huntington's, and DMD updated to reflect current research.

15 Biotechnology Heavily revised material includes mechanism, application, and social implications of CRISPR-Cas9 gene editing system.

16 Evidence of Evolution Cetacean evolutionary sequence updated to reflect current accepted narrative.

17 Processes of Evolution New photo of velvet walking worm shooting streams of glue from its head illustrates parapatric speciation. Expanded Application includes current statistics and research on generation and spread of antibiotic-resistant bacteria associated with factory farms, and example of cross-resistance to veterinary and human antibiotics.

18 Life's Origin and Early Evolution Improved descriptions of the Hadean, Archean, and Phanerozoic eons. Added information about Miller–Urey's experiments that simulated conditions around volcanoes. Updated discussion of protocells. Deleted coverage of Jeon's study of endosymbionts in amoebas. New subsection summarizes events of the Precambrian.

Chapter-Specific Changes (continued)

19 Viruses, Bacteria, and Archaea Improved art comparing viral structures. Updated information about Ebola, AIDS, West Nile virus. Added information about the Zika virus. New figure shows viral reassortment. Added figure illustrating the three common bacterial shapes. New table lists common bacterial diseases. New subsection discusses metagenomics, relationships among prokaryotes, and the relationship between prokaryotes and eukaryotes.

20 The Protists New table lists the various protist groups and their traits. Chapter now organized around the major eukaryotic supergroups. New overview of protist cell structure. New information about African trypanosomiasis and expanded coverage of malaria.

21 Plant Evolution Revised life cycle graphics throughout the chapter; improved figure illustrating generalized process of seed production.

22 Fungi Added text and table comparing traits of fungi, plants, and animals. New photos of athlete's foot and ringworm. Updated information about white nose syndrome. New information about medical use of psilocybin.

23 Animals I: Major Invertebrate Groups More on the similarities between choanoflagellates and animals. Coverage of placozoans deleted. New figure illustrates bivalve (clam) anatomy, and sea star anatomy figure has been updated. Expanded coverage of roundworms as a model organism. New information about overharvesting of krill and about copepods as reservoirs for cholera-causing bacteria.

24 Animals II: The Chordates Updated lancelet art. New subsection and figure devoted to declining amphibian diversity. New figure of avian skeleton. Updated discussion of fossil hominins and evidence of interbreeding among humans, Neandertals, and Denisovans.

25 Plant Tissues Many new photos. Expanded Application material includes carbon release by decomposition, relative stability of dead plant matter, and CO_2 and climate change.

26 Plant Nutrition and Transport Many new photos and updated art pieces. Mechanism of regulation of water flow through bordered pits updated per current research.

27 Reproduction and Development of Flowering Plants Revisions include addition/illustration of nastic movements, and explanation of *Phylloxera* resistance in American grapevines based on enhanced hypersensitive response involving resveratrol. New Application involves benefits of plant secondary metabolites using as an example the discovery of epicatechin in cocoa via the low incidence of hypertension among the Kuna tribe, with emphasis on similar signaling pathways in plants and humans.

28 Animal Tissues and Organ Systems New summary table describing tissue types. Improved graphic illustrates relative volumes of the fluid components of a human body. Added information about brown fat versus white fat and white matter and gray matter. New graphic illustrates properties of stem cells. Updated information about research on and clinical use of induced pluripotent stem cells (IPSCs).

29 Neural Control New overview of intracellular signaling. Revised/reorganized coverage of the peripheral nervous system. New photo of a paralyzed veteran using a robotic exoskeleton. New subsection covers tissues and fluid of the CNS. Updated Application with the latest findings about brain damage among professional football players.

30 Sensory Perception Simplified figure for human olfaction. Added frontal view to illustration of visual accommodation. New graphic depicts the anatomy of the retina and includes light-channeling neuroglia. Application now covers both cochlear and retinal implants.

31 Endocrine Control Added information about sites of human steroid hormone production (gonads/adrenals). Added information about hormone type (amino acid or steroid) to Table 31.1. Updated figure depicting feedback control of thyroid hormone.

32 Structural Support and Movement Improved figures depicting locomotion of fly and earthworm. Revised figure showing the structure of skeletal muscle.

33 Circulation Updated photo depicting measurement of blood pressure. Expanded coverage of venule function. New art depicts atherosclerosis. Added discussion of heart attack symptoms and of causes and symptoms of stroke. New Data Analysis exercise on hypertension and risk of stroke.

34 Immunity Text updates reflect current research on role of keratinocytes as immune cells. New Application essay that details vaccination and benefits of herd immunity features a narrative by a mother whose unvaccinated child has permanent health consequences of contracting measles.

35 Respiration Improved photo of insect tracheal system. Increased emphasis on evolutionary trends in discussion of vertebrate lungs. Updated information about risks for SIDS. Added information about risks of vaping.

36 Digestion and Human Nutrition Expanded discussion of sponge digestion with new graphic. New graphic shows diet-related variation in bird beaks. Improved photo of interior of small intestine. New graphics depict peristalsis and segmentation. Updated coverage of human nutrition and factors affecting weight.

37 Maintaining the Internal Environment Improved description of variations in types of nitrogenous wastes. New photo of donor/recipient in living donor kidney transplant. New information about how diseases affect composition of urine. Added information about human body hair as a temperature-related adaptation.

38 Reproduction and Development Improved description of apoptosis. Reorganized discussion of human reproduction; female now precedes male. New information about genetics of bird beak and human facial development, post-fertilization epigenetic reset, mechanism of Ru-486, developmental effects of Zika virus, and banking of sperm and eggs. New Data Analysis exercise about regional variations in male infertility.

39 Animal Behavior New information about oxytocin and human autism and about epigenetic effects in a variety of contexts. Improved honeybee dance language figure. New photo/information about tent caterpillars, a pre-social species. Application now covers effects of shipping noise on whale communication.

40 Population Ecology Introduction now explains applications of population ecology. Updated human population statistics. Improved graphic of intermediate disturbance model.

41 Community Ecology New figure shows sundew plants and spiders as competitors for insect prey. New discussion of study that showed character displacement in Galápagos finches.

42 Ecosystems Updated information about current level of atmospheric carbon dioxide; expanded discussion of sources of fossil fuels.

43 The Biosphere Improved description of how Earth's shape influences wind direction. More details about plant adaptations to coastal life. New coverage of ocean acidification and other threats to reefs.

44 Human Effects on the Biosphere Added information about habitat fragmentation, the Great Pacific Garbage Patch, ill effects of ammonium and mercury deposition, nitrous oxide's effect on the ozone layer, and the leakage of methane from pipelines.

Acknowledgments

Writing, revising, and illustrating a biology textbook is a major undertaking for two full-time authors, but our efforts constitute only a small part of what is required to produce and distribute this one. We are truly fortunate to be part of a huge team of very talented people who are just as committed to creating and disseminating this exceptional science education product.

Biology is not dogma; the fantastic amount of research in the field makes paradigm shifts common. Only with the ongoing input of our many academic reviewers and advisors (see following page) can we continue to tailor this book to the needs of instructors and students while integrating new information and models. We continue to learn from and be inspired by these dedicated educators.

This book benefits from a collaborative association with the National Geographic Society in Washington D.C. The Society has graciously allowed us to enhance our presentation with its extensive resources, including beautiful maps and images, Society explorer and grantee materials, and online videos.

At Cengage, Hal Humphrey managed all of the inhouse production operations. April Cognato's efforts, vision, and leadership brought this book forward with Vision and Change; Jake Warde provided insightful suggestions, kind support, and encouragement. Kellie Petruzzelli coordinated nearly all of the development of study, testing, and online active learning resources. Grace Davidson orchestrated a continuous flow of files, photos, and illustrations while managing schedules, budgets, and whatever else happened to be on fire. (Grace, thank you as always for your patience and dedication.) Cheryl DuBois at Lumina Datamatics, and Christine M. Myaskovsky and Betsy Hathaway at Cengage Learning, were responsible for photo research. Valuable suggestions from copyeditor Anita Wagner Hueftle and proofreader Diane Miller again kept our text clear and concise.

—*Lisa Starr, Chris Evers, and Cecie Starr 2016*

Reviewers for This Edition

Jessica Adams
Johnson & Wales University

Stanton Belford
Martin Methodist College

Joel Benington
St. Bonaventure University

Harriette Block
Prairie View A&M University

Rob Bogardus
Mid-Plains Community College

Ed Budde
Youngstown State University

Carolyn Bunde
Idaho State University

Wilbert Butler
Tallahassee Community College

Wayne Busch
Riverland Community College

Diomede Buzingo
Langston University

Nancy Cain
Colorado Mountain College

Ryan Chabarria
Lone Star College, Kingwood

Diane Cook
Louisburg College

John Dilustro
Chowan University

Stella Doyungan
Texas A&M University, Corpus Christi

Victor Fet
Marshall University

Solomon Gebru
Howard University

Joel Gluck
Johnson & Wales University

Marla Gomez
Nicholls State University

Nazanin Z. Hebel
Houston Community College, Northwest

Lauri Heintz
Hill College

Mark Hens
University of North Carolina, Greensboro

Sherry Hickman
Hillsborough Community College

Inigo Howlett
Rappahannock Community College

Daniel Husband
Florida State College at Jacksonville, Kent Campus

Rosalyn Hunter
Hill College

Stephen Kash
Oklahoma City Community College

Mijung Kim
Chicago State University

Patrick J. Krug
California State University, Los Angeles

Maureen Lemke
Texas State University

Paul Lepp
Minot State University

Abby L. Levitt
North Central State College

Todd Martin
Metropolitan Community College

Mitch McVey
Tufts University

Lisa Merritt
Rappahannock Community College

Kiran Misra
Edinboro University

Jonas Okeagu
Fayetteville State University

Jennifer O'Neil
Houston Community College, Northwest

Peter J. Park
Nyack College

Marc Perkins
Orange Coast College

Mary Poffenroth
San Jose State University

Sean Rice
Texas Tech University

Pele Eve Rich
North Hennepin Community College

Jeffrey Rousch
Elizabeth City State University

Christina Russin
Northwestern University

Brian Schmaefsky
Lone Star College, Kingwood

Amanda Schaetzel
University of Colorado, Boulder

Sheila Schreiner
Salem State University

Randal Snyder
Buffalo State College

Neeti Srivastava
Mountain View College

Linda Staffero
Yuba College

Bob Starkey
College of the Ouachitas

Patricia Valella
College of Southern Nevada

Phil Veillette
Johnson & Wales University

Jacqueline Washington
Nyack College

R. Douglas Watson
University of Alabama at Birmingham

Gordon Woolam
Ranger College

David Wessner
Davidson College

Ovid Wong
Benedictine University

Xiaoning Zhang
St. Bonaventure University

Student and Instructor Resources

MindTap Through personalized paths of dynamic assignments and applications, MindTap is a digital learning solution that turns cookie cutter into cutting edge, apathy into engagement, and memorizers into higher-level thinkers.

Nobody knows how to engage your students better than you do, so don't settle for static course content. MindTap's innovative personalization tools allow you to customize every element of your course—from rearranging the Table of Contents to inserting videos, comments, activities, and more into the Learning Path and MindTap Reader itself.

MindTap for Biology: Concepts & Applications 10e

The MindTap Learning Path includes these engaging learning opportunites in every chapter, and more!

MAKE IT RELEVANT

Motivate and Engage: Each chapter starts with an introductory reading based on an important application of chapter content. This includes assignable and auto-gradable questions and the popular **How Would You Vote?** student engagement feature. After reading the chapter, students can revisit their original vote in a new feature called **Make an Informed Decision.**

ASSIGN AND GRADE CONTENT THAT MATTERS

- **Data Analysis activities** are fully assignable in MindTap! Ensure your students are sharpening their analytical skills by assigning these engaging activities. The data is related to the chapter material and is taken from a published scientific study in most cases.

Other assignable and gradable activities in every chapter include:

- **Conceptual Learning Assignments**
- **Section-Based Concept Checks**
- **Critical Thinking Questions**
- **Post Learning Assessment**
- **Chapter Test**

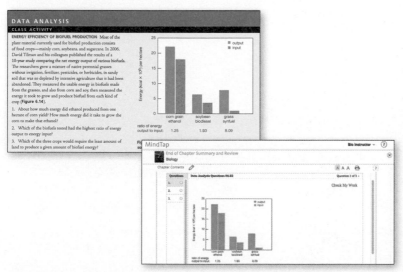

Data Analysis activities in every chapter... ...assigned in MindTap.

MindTap is fully customizable to meet your course goals. Easily assign students the content you want them to learn, in the order you want them to learn it.

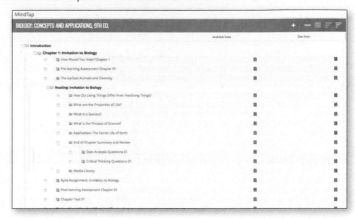

Make it your own. Insert your own materials—slides, videos, and lecture notes—wherever you want your students to see them.

Search for YouTube videos and insert the link directly into the chapter reading. And much more!

MindTap Course Development Effectively introducing digital solutions into your classroom—online or on-ground —is now easier than ever. We're with you every step of the way.

Our Digital Course Support Team When you adopt from Cengage Learning, you have a dedicated team of Digital Course Support professionals, who will provide hands-on start-to-finish support, making digital course delivery a success for you and your students.

"The technical support provided by Cengage staff in using MindTap was superb. The Digital Support Team was proactive in training and prompt in answering follow-up questions." —K. Sata Sathasivan, The University of Texas at Austin

Instructor Companion Site Everything you need for your course in one place! This collection of book-specific lecture and class tools is available online via www.cengagebrain.com/login. Access and download PowerPoint presentations, images, instructor's manuals, videos, test banks, and more.

Biology

Concepts & Applications 10e

1

The Science of Biology

Links to Earlier Concepts

Whether or not you have studied biology, you already have an intuitive understanding of life on Earth because you are part of it. Every one of your experiences with the natural world—from the warmth of the sun on your skin to the love of your pet—contributes to that understanding.

Core Concepts

Interactions among the components of a biological system give rise to complex properties.

We can understand life by studying it at increasingly inclusive levels, starting with atoms that compose matter, and extending to the biosphere. Each level is a biological system composed of interacting parts. Interactions among the components of a system give rise to complex properties not found in any of the components. Interactions among organisms and their environment result in the movement of matter and energy.

Evolution underlies the unity and diversity of life.

Shared core processes and features that are widely distributed among organisms provide evidence that all living things are linked by lines of descent from common ancestors. All biological systems are sustained by the exchange of matter and energy; all store, retrieve, transmit, and respond to information essential for life.

The field of biology relies upon experimentation and the collection and analysis of scientific evidence.

Science addresses only testable ideas about observable events and processes. Observation, experimentation, quantitative analysis, and critical thinking are key aspects of research in biology. Carefully designed experiments that yield objective data help researchers unravel cause-and-effect relationships in complex biological systems.

◀ Near a tent serving as a makeshift laboratory, herpetologist Paul Oliver records the call of a frog on an expedition to New Guinea's Foja Mountains cloud forest.

LEARNING OBJECTIVES

Describe the successive levels of life's organization.

Explain the idea of emergent properties and give an example.

EMERGENT PROPERTIES

Biologists study life. What, exactly, is "life"? We may never actually come up with a concise definition, because living things are too diverse, and they consist of the same basic components as nonliving things. When we try to define life, we end up with a long list of properties that differentiate living from nonliving things. These properties often emerge from the interactions of basic components. To understand how that works, take a look at the groups of squares in **Figure 1.1**. The property of "roundness" emerges when the component squares are organized one way, but not other ways. Another example is a complex behavior called swarming. When honeybees swarm, they fly en masse to establish a hive in a new location. Each bee is autonomous, but the new hive's location is decided collectively based on an integration of signals from hivemates. The swarm's collective intelligence is the emergent property in this example.

A characteristic of a system (a colony of bees swarming, for example) that does not appear in any of the system's components (individual bees) is called an **emergent property**. The idea that structures or systems with emergent properties can be assembled from the same components is a recurring theme in our world—and also in biology.

LIFE'S ORGANIZATION

Biologists view life as having nested levels of organization; interactions among the components of each level give rise to emergent properties (**Figure 1.2**). This organization begins with atoms. **Atoms** are the smallest units of a substance; they and the fundamental particles that compose them are the building blocks of all matter ❶. Atoms bond together to form **molecules** ❷. There are no atoms unique to living things, but there are unique molecules. A **cell** ❸, which is the smallest unit of life, consists of many of these "molecules of life."

Some cells live and reproduce independently. Others do so as part of a multicelled organism. An **organism** is an individual that consists of one or more cells ❼. In most multicelled organisms, cells are organized as tissues ❹. A **tissue** consists of specific types of cells in an arrangement that allows the cells to collectively perform some function—protection from injury (dermal tissue) or movement (muscle tissue), for example. An **organ** is a structure composed of tissues that collectively carry out a particular task or set of

tasks ❺. For example, a flower is an organ of reproduction in plants; a heart, an organ that pumps blood in animals. An **organ system** is a set of interacting organs and tissues that fulfill one or more body functions ❻. Examples of organ systems include the aboveground parts of a plant (the shoot system), and the heart and blood vessels of an animal (the circulatory system).

Unique types of organisms—California poppies, for example—are called **species**. A **population** is a group of interbreeding individuals of the same species living in a given area. For example, all California poppies growing in California's Antelope Valley Poppy Reserve form a population ❽. A **community** consists of all populations of all species in a given area. The Antelope Valley Reserve community includes California poppies and all other plants, as well as animals, microorganisms, and so on ❾. Communities may be large or small, depending on the area defined. The next level of organization is the **ecosystem**, which is a community interacting with its physical and chemical environment ❿. Earth's largest ecosystem is the **biosphere**, and it encompasses all regions of the planet's crust, waters, and atmosphere in which organisms live ⓫.

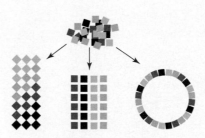

Figure 1.1 The same materials, assembled in different ways, form objects with different properties. The property of "roundness" emerges when the squares are assembled one way, but not the others.

atom Smallest unit of a substance; consists of subatomic particles.

biosphere (BY-oh-sfeer) All regions of Earth where organisms live.

cell Smallest unit of life.

community All populations of all species in a defined area.

ecosystem A community interacting with its environment.

emergent property (ee-MERGE-ent) A characteristic of a system that does not appear in any of the system's component parts.

molecule (MAUL-ick-yule) Two or more atoms bonded together.

organ In multicelled organisms, a structure that consists of tissues engaged in a collective task.

organism (ORG-uh-niz-um) An individual that consists of one or more cells.

organ system In multicelled organisms, set of interacting organs that carry out a particular body function.

population Group of interbreeding individuals of the same species living in a defined area.

species (SPEE-sheez) Unique type of organism.

tissue In multicelled organisms, specialized cells organized in a pattern that allows them to perform a collective function.

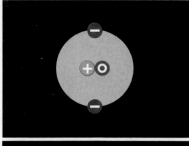

1 atom
Atoms and the particles that compose them make up all matter.

2 molecule
Atoms join other atoms in molecules. This is a model of a water molecule. The molecules special to life are much larger and more complex than water.

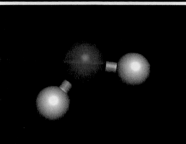

3 cell
The cell is the smallest unit of life. Some, like these plant cells, live and reproduce as part of a multicelled organism; others do so on their own.

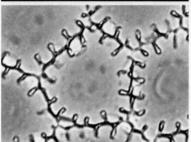

4 tissue
Organized array of cells that interact in a collective task. This is dermal tissue on the outer surface of a flower petal.

5 organ
Structural unit of interacting tissues. Flowers are the reproductive organs of some plants.

Figure 1.2 Levels of life's organization.
New emergent properties appear at each successive level.

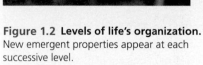

FIGURE IT OUT
At which level does the emergent property of "life" appear?
Answer: The cell

6 organ system
A set of interacting organs. The shoot system of this poppy plant includes its above-ground parts: leaves, flowers, and stems.

7 multicelled organism
Individual that consists of more than one cell. Cells of this California poppy plant make up its shoot system and root system.

8 population
Group of single-celled or multicelled individuals of a species in a given area. This population of California poppy plants is in California's Antelope Valley Poppy Reserve.

9 community
All populations of all species in a specified area. These plants are part of the community in the Antelope Valley Poppy Reserve.

10 ecosystem
A community interacting with its physical environment through the transfer of energy and materials. Sunlight and water sustain the community in the Antelope Valley.

11 biosphere
The sum of all ecosystems: every region of Earth's waters, crust, and atmosphere in which organisms live.

CREDITS: (2) 3, 4: © Umberto Salvagnin, www.flickr.com/photos/kaibara.; 5: California Poppy, © 2009, Christine M. Welter; 6: Lady Bird Johnson Wildflower Center; 7: Michael Szoenyi/Science Source; 8: James Randklev/Exactostock-1672/ SuperStock; 9: © Sergei Krupnov, www.flickr.com/photos/7969319@N03; 10: © Mark Koberg Photography; 11: NASA.

1.2 Life's Unity

Distinguish producers from consumers.

Define homeostasis and explain why it is important for sustaining life.

List some functions that are guided by an organism's DNA.

Explain how DNA is the basis of similarities and differences among organisms.

All living things share a particular set of key features. You already know one of these features: Because the cell is the smallest unit of life, all organisms consist of at least one cell. For now, we introduce three more: All living things require ongoing inputs of energy and raw materials; all sense and respond to change; and all use DNA as the carrier of genetic information (**Table 1.1**).

ENERGY AND NUTRIENTS

Not all living things eat, but all require nutrients on an ongoing basis. A **nutrient** is a substance that an organism acquires from the environment in order to support growth and survival. Both nutrients and energy are essential to maintain the organization of life, so organisms spend a lot of time acquiring them. However, the source of energy and the type of nutrients required differ among organisms. These differences allow us to classify all living things into two categories: producers and consumers (**Figure 1.3**). A **producer** makes its own food using energy and simple raw materials it obtains from nonbiological sources ❶. Plants are producers. By a process called **photosynthesis**, plants can use the energy of sunlight to make sugars from carbon dioxide (a gas in air) and water. Consumers, by contrast, cannot make their own food. A **consumer** obtains energy and nutrients by feeding on other organisms ❷. Animals are consumers. So are decomposers, which feed on the wastes or remains of other organisms. Leftovers from consumers' meals end up in the environment, where they serve as nutrients for producers. Said another way, nutrients cycle between producers and consumers ❸.

TABLE 1.1
Some Key Features of Life

Cellular basis	All living things consist of one or more cells.
Requirement for energy and nutrients	Life is sustained by ongoing inputs of energy and nutrients.
Homeostasis	Living things sense and respond appropriately to change.
DNA is hereditary material	Genetic information in the form of DNA is passed to offspring.

❶ producer acquiring energy and nutrients from the environment

❷ consumer acquiring energy and nutrients by eating a producer

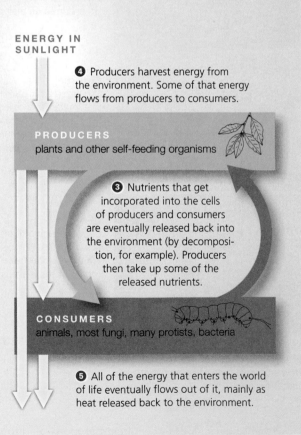

ENERGY IN SUNLIGHT

❹ Producers harvest energy from the environment. Some of that energy flows from producers to consumers.

PRODUCERS
plants and other self-feeding organisms

❸ Nutrients that get incorporated into the cells of producers and consumers are eventually released back into the environment (by decomposition, for example). Producers then take up some of the released nutrients.

CONSUMERS
animals, most fungi, many protists, bacteria

❺ All of the energy that enters the world of life eventually flows out of it, mainly as heat released back to the environment.

Figure 1.3 The one-way flow of energy and cycling of materials through the world of life.

Unlike nutrients, energy is not cycled. It flows in one direction: from the environment ❹, through organisms, and back to the environment ❺. This flow maintains the organization of every living cell and body, and it also influences how individuals interact with one another and their environment. The energy flow is one-way, because with each transfer, some energy escapes as heat, and cells cannot use heat as an energy source. Thus, energy that enters the world of life eventually leaves it (we return to this topic in Chapter 5).

HOMEOSTASIS

An organism cannot survive for very long unless it can respond appropriately to specific stimuli inside and outside of itself. For example, humans and some other animals normally perspire (sweat) when the body's internal temperature rises above a certain set point (**Figure 1.4**). The moisture cools the skin, which in turn helps cool the body.

All of the internal fluids that bathe the cells in your body are collectively called your internal environment. Temperature and many other conditions in that environment must be kept within certain ranges, or your cells will die (and so will you). By sensing and adjusting to change, all organisms keep condi-

Figure 1.4 Living things sense and respond to their environment. Sweating is a response to an internal body temperature that exceeds the normal set point. The response cools the skin, which in turn helps return the internal temperature to the set point.

tions in their internal environment within ranges that favor cell survival. **Homeostasis** is the name for this process, and it is one of the defining features of life.

DNA IS HEREDITARY MATERIAL

With little variation, the same types of molecules perform the same basic functions in every organism. For example, information in an organism's **DNA** (deoxyribonucleic acid) guides ongoing cellular activities that sustain the individual through its lifetime. Such functions include **development**: the process by which the first cell of a new individual gives rise to a multicelled adult; **growth**: increases in cell number, size, and volume; and **reproduction**: processes by which organisms produce offspring. **Inheritance**, the transmission of DNA to offspring, occurs during reproduction. All organisms inherit their DNA from one or more parents.

Individuals of every natural population are alike in most aspects of body form and behavior because their DNA is very similar: Humans look and act like humans and not like poppy plants because they inherited human DNA, which differs from poppy plant DNA in the information it carries. Individuals of almost every natural population also vary—just a bit—from one another: One human has blue eyes, the next has brown eyes, and so on. Such variation arises from small differences in the details of DNA molecules, and herein lies the source of life's diversity. As you will see in later chapters, differences among individuals of a species are the raw material of evolutionary processes.

consumer (kun-SUE-murr) Organism that gets energy and nutrients by feeding on tissues, wastes, or remains of other organisms.

development (dih-VELL-up-ment) Process by which the first cell of a multicelled organism gives rise to a multicelled adult.

DNA Deoxyribonucleic (dee-ox-ee-ribe-oh-nuke-LAY-ick) acid; molecule that carries hereditary information; guides development and other activities.

growth In multicelled species, an increase in the number, size, and volume of cells.

homeostasis (home-ee-oh-STAY-sis) Process in which organisms keep their internal conditions within tolerable ranges by sensing and responding appropriately to change.

inheritance (in-HAIR-ih-tunce) Transmission of DNA to offspring.

nutrient (NEW-tree-unt) Substance that an organism acquires from the environment to support growth and survival.

photosynthesis (foe-toe-SIN-thuh-sis) Process by which producers use light energy to make sugars from carbon dioxide and water.

producer Organism that makes its own food using energy and nonbiological raw materials from the environment.

reproduction (ree-pruh-DUCK-shun) Processes by which organisms produce offspring.

1.3 Life's Diversity

LEARNING OBJECTIVES

List two characteristics of prokaryotes.

Name the four main groups of eukaryotes.

Differences in the details of DNA molecules are the basis of a tremendous range of differences among types of organisms. Various classification schemes help us organize what we understand about the scope of this variation, which is an important aspect of Earth's biodiversity. For example, organisms can be grouped on the basis of whether they have a nucleus, which is a saclike structure containing a cell's DNA. **Bacteria** (singular, bacterium) and **archaea** (singular, archaeon) are organisms whose DNA is *not* contained within a nucleus. All bacteria and archaea are single-celled, which means each organism consists of one cell (**Figure 1.5A,B**). Collectively, these organisms are the most diverse representatives of life. Different kinds are producers or consumers in nearly all regions of Earth. Some inhabit such extreme environments as frozen desert rocks, boiling sulfurous lakes, and nuclear reactor waste. The first cells on Earth may have faced similarly hostile conditions.

Traditionally, organisms without a nucleus have been classified as **prokaryotes**, but the designation is now used only informally. This is because bacteria and archaea are less related to one another than we once thought, despite their similar appearance. Archaea turned out to be more closely related to **eukaryotes**, which are organisms whose DNA is contained within a nucleus. Some eukaryotes live as individual cells; others are multicelled (**Figure 1.5C**). Eukaryotic cells are typically larger and more complex than bacteria or archaea.

There are four main groups of eukaryotes: protists, fungi, plants, and animals. **Protist** is the common term for a collection of eukaryote groups that are not plants, animals, or fungi. Collectively, they vary dramatically, from single-celled consumers to giant, multicelled producers.

Fungi (singular, fungus) are eukaryotic consumers that secrete substances to break down food externally, then absorb nutrients released by this process. Many fungi are decomposers. Most fungi, including those that form mushrooms, are multicellular. Fungi that live as single cells are called yeasts.

Plants are multicelled eukaryotes, and the vast majority of them are photosynthetic producers that live on land. Besides feeding themselves, plants also serve as food for most other land-based organisms.

Animals are multicelled consumers that ingest other organisms or components of them. Unlike fungi, animals break down food inside their body. They also develop through a series of stages that lead to the adult form. All animals actively move about during at least part of their lives.

---

animal Multicelled consumer that breaks down food inside its body, develops through a series of stages, and moves about during part or all of its life.

archaea (are-KEY-uh) Group of single-celled organisms that lack a nucleus but are more closely related to eukaryotes than to bacteria.

bacteria The most diverse and well-known group of single-celled organisms that lack a nucleus.

eukaryote (you-CARE-ee-oat) Organism whose cells characteristically have a nucleus.

fungus Single-celled or multicelled eukaryotic consumer that breaks down material outside itself, then absorbs nutrients released from the breakdown.

plant A multicelled, typically photosynthetic producer.

prokaryote (pro-CARE-ee-oat) Single-celled organism without a nucleus.

protist Common term for a eukaryote that is not a plant, animal, or fungus.

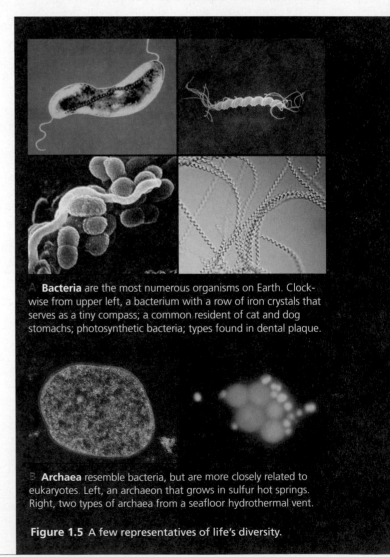

A **Bacteria** are the most numerous organisms on Earth. Clockwise from upper left, a bacterium with a row of iron crystals that serves as a tiny compass; a common resident of cat and dog stomachs; photosynthetic bacteria; types found in dental plaque.

B **Archaea** resemble bacteria, but are more closely related to eukaryotes. Left, an archaeon that grows in sulfur hot springs. Right, two types of archaea from a seafloor hydrothermal vent.

Figure 1.5 A few representatives of life's diversity.

CREDITS: (5A) top left, Dr. Richard Frankel; top right, Science Source; bottom left, www.zahnarzt-stuttgart.com; bottom right, © Susan Barnes; (5B) left, Eye of Science/Science Source; right, © Dr. Harald Huber, Dr. Michael Hohn, Prof. Dr. K.O. Stetter, University of Regensburg, Germany.

Protists are a group of extremely diverse eukaryotes that range from microscopic free-living cells (left) to giant multicelled seaweeds (right).

Fungi are eukaryotic consumers that secrete substances to break down food outside their body. Some are single-celled (left), but most are multicelled (right).

Plants are multicelled eukaryotes, most of which are photosynthetic Nearly all have roots, stems, and leaves.

Animals are multicelled eukaryotes that ingest other organisms or their parts, and they actively move about during part or all of their life cycle.

C **Eukaryotes** are single-celled or multicelled organisms whose DNA is contained within a nucleus. Eukaryotes include protists, plants, fungi, and animals.

Figure 1.6 On a survey in New Guinea, Kristofer Helgen finds a critically endangered long-beaked echidna. The only known living populations of this animal occur in New Guinea, and all are in rapid decline.

Kristofer Helgen

National Geographic Explorer Kristofer Helgen discovers new animals. Deep in a New Guinea rain forest (**Figure 1.6**). High on an Andean mountainside. Resting in a museum's specimen drawer. "Conventional wisdom would have it that we know all the mammals of the world," he notes. "In fact, we know so little. Unique species, profoundly different from anything ever discovered, are out there waiting to be found." His own efforts prove this. Helgen himself has discovered approximately 100 new species of mammals previously unknown to science. "Since I was three years old, I've been transfixed by animals," he recalls. "Even then, my excitement revolved around figuring out how many different kinds there were."

Helgen's search plunges him into the wild on almost every continent. Yet about three times as many new finds are made within the walls of museums. "An expert can go into any large natural history museum and identify kinds of animals no one knew existed," he explains. When only a few specimens of a species exist, and reside in museums scattered across the globe, sheer logistics often prevent researchers from pinpointing a new find. Helgen recently discovered a specimen of a long-beaked echidna in a London museum. It had been collected from Australia in 1901, misidentified, and left forgotten in the bottom of a drawer. Long-beaked echidnas were thought to be extinct in Australia for at least 5,000 years. Helgen's discovery that one was alive in 1901 means a population might still exist there, waiting to be discovered.

A ROSE BY ANY OTHER NAME . . .

Each time we discover a new species, we name and classify it, a practice called **taxonomy**. Taxonomy began thousands of years ago, but naming species in a consistent way did not become a priority until the eighteenth century. At the time, European explorers who were just discovering the scope of life's diversity started having more and more trouble communicating with one another because species often had multiple names. For example, the dog rose (a plant native to Europe, Africa, and Asia) was alternately known as briar rose, witch's briar, herb patience, sweet briar, wild briar, dog briar, dog berry, briar hip, eglantine gall, hep tree, hip fruit, hip rose, hip tree, hop fruit, and hogseed—and those are only the English names! Species often had multiple scientific names too, in Latin that was descriptive but often cumbersome. The scientific name of the dog rose was *Rosa sylvestris inodora seu canina* (odorless woodland dog rose), and also *Rosa sylvestris alba cum rubore, folio glabro* (pinkish white woodland rose with smooth leaves).

An eighteenth-century naturalist, Carolus Linnaeus, standardized a naming system that we still use. By the Linnaean system, each species is given a unique two-part scientific name. The first part of a scientific name is the **genus** (plural, genera), which is defined as a group of species that share a unique set of features. Combined with the second part of the name (the "specific epithet"), it designates one species. Thus, the dog rose now has one official name, *Rosa canina*, that is recognized worldwide.

The genus name is always capitalized, and the specific epithet is not. Both are always italicized. Consider *Panthera*, a genus of big cats. Lions belong to the species *Panthera leo*. Tigers belong to a different species in the same genus (*Panthera tigris*), and so do leopards (*P. pardus*). Note how the genus name may be abbreviated after it has been spelled out.

DISTINGUISHING SPECIES

The individuals of a species share a unique set of traits. For example, giraffes normally have very long necks, brown spots on white coats, and so on. These are morphological (structural) traits. Individuals of a species also share biochemical traits (they make and use the same molecules) and behavioral traits (they respond the same way to certain stimuli, as when hungry giraffes feed on tree leaves). We can rank a species into ever more inclusive categories based on some subset of traits it shares with other species. Each rank,

	wild carrot	marijuana	apple	prickly rose	dog rose
domain	Eukarya	Eukarya	Eukarya	Eukarya	Eukarya
kingdom	Plantae	Plantae	Plantae	Plantae	Plantae
phylum	Magnoliophyta	Magnoliophyta	Magnoliophyta	Magnoliophyta	Magnoliophyta
class	Magnoliopsida	Magnoliopsida	Magnoliopsida	Magnoliopsida	Magnoliopsida
order	Apiales	Rosales	Rosales	Rosales	Rosales
family	Apiaceae	Cannabaceae	Rosaceae	Rosaceae	Rosaceae
genus	*Daucus*	*Cannabis*	*Malus*	*Rosa*	*Rosa*
species	*carota*	*sativa*	*domestica*	*acicularis*	*canina*

Figure 1.7 Taxonomy of five species that are related at different levels. Each species has been assigned to ever more inclusive groups, or taxa: in this case, from genus to domain.

FIGURE IT OUT

Which of the plants shown here are in the same family?

Answer: Apple, prickly rose, and dog rose

or **taxon** (plural, taxa), is a group of organisms that share a unique set of inherited traits. Each category above species—genus, family, order, class, phylum (plural, phyla), kingdom, and domain—consists of a group of the next lower taxon (**Figure 1.7**). Using this system, we can sort all life into a few categories (**Figure 1.8** and **Table 1.2**). Later chapters return to details of these and other classification systems.

It is easy to tell that humans and dog roses are different species because they appear very different. Distinguishing other species that are more closely related may be much more challenging (**Figure 1.9**). In addition, traits shared by members of a species often vary a bit among individuals.

How do biologists decide whether similar-looking organisms belong to different species? The short answer to that question is that they rely on whatever information is available. Early naturalists studied anatomy and distribution—essentially the only methods available at the time—so species were named and classified according to what they looked like and where they lived. Today's biologists are able to compare traits that the early naturalists did not even know about, including biochemical ones such as DNA molecules.

Consider that the information in a molecule of DNA changes a bit each time it passes from parents to offspring, and it has done so since life began. Over long periods of time, these tiny changes have added up to big differences between species such as humans and dog roses. Thus, differences in DNA are one way to measure relative relatedness: The fewer differences between species, the closer the relationship. For example, we know that the DNA of humans is more similar to chimpanzee DNA than it is to cat DNA, so we can assume that chimpanzees and humans are closer relatives than cats and humans. Every living species known has DNA in common with every other species, so every living species is related to one extent or another. Unraveling these relationships has become a major focus of biology (later chapters return to this topic).

The discovery of new information sometimes changes the way we distinguish a particular species or how we group it with others. For example, Linnaeus grouped plants by the number and arrangement of reproductive parts, a scheme that resulted in odd pairings such as castor-oil plants with pine trees. Having more information today, we place these plants in separate phyla.

Evolutionary biologist Ernst Mayr defined a species as one or more groups of individuals that potentially can inter-

genus (JEE-nuss) A group of species that share a unique set of traits.

taxon (TAX-on) A rank of organisms that share a unique set of traits.

taxonomy (tax-ON-oh-me) Naming and classifying species in a systematic way.

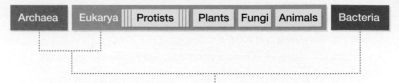

Figure 1.8 The big picture of life. This diagram summarizes one hypothesis about how all life is connected by shared ancestry. Lines indicate proposed evolutionary connections between the domains.

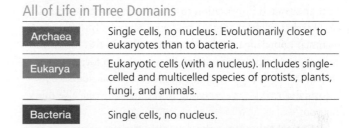

FIGURE IT OUT
Protists, plants, fungi, and animals are at which taxonomic level(s)?

Answer: All are kingdoms

TABLE 1.2

All of Life in Three Domains

Archaea	Single cells, no nucleus. Evolutionarily closer to eukaryotes than to bacteria.
Eukarya	Eukaryotic cells (with a nucleus). Includes single-celled and multicelled species of protists, plants, fungi, and animals.
Bacteria	Single cells, no nucleus.

Figure 1.9 Four butterflies, two species: Which are which?

The top row shows two forms of the species *Heliconius melpomene*; the bottom row, two forms of *H. erato*.

H. melpomene and *H. erato* never cross-breed; DNA comparisons confirmed that they are truly different species. The alternate but similar patterns of coloration evolved as a shared warning signal to predatory birds that these butterflies taste terrible.

breed, produce fertile offspring, and do not interbreed with other groups. This "biological species concept" is useful but not universally applicable. For example, we may never know whether populations of some species could interbreed because impassable geographical barriers keep them separated. As another example, populations often continue to interbreed even as they diverge, so the exact moment at which two populations become two species is often impossible to pinpoint. We return to speciation and how it occurs in Chapter 17, but for now it is important to remember that a "species" is a convenient but artificial construct of the human mind.

CREDITS: (9) From Meyer A., Repeating Patterns of Mimicry. *PLoS Biology* Vol. 4, No. 10, e341 doi:10.1371/journal.pbio.0040341.

LEARNING OBJECTIVES

Describe critical thinking and give some examples of how to do it.

Distinguish between inductive and deductive reasoning.

Use suitable examples to explain dependent and independent variables.

Explain how a control group is used in an experiment.

List the tasks that are part of the scientific method.

THINKING ABOUT THINKING

Most of us assume that we do our own thinking, but do we, really? You might be surprised to find out how often we let others think for us. Consider how a school's job (which is to impart as much information to students as quickly as possible) meshes perfectly with a student's job (which is to acquire as much knowledge as quickly as possible). In this rapid-fire exchange of information, it can be very easy to forget about the merit of what is being exchanged. Any time you accept information without evaluating it, you let someone else think for you.

Critical thinking is the deliberate process of judging the quality of information before accepting it. "Critical" comes from the Greek *kriticos* (discerning judgment). When you use critical thinking, you move beyond the content of new information to consider supporting evidence, bias, and alternative interpretations. How does the busy student manage this? Critical thinking does not necessarily require extra time, just

a bit of extra awareness. There are many ways to do it. For example, you might ask yourself some of the following questions while learning something new:

▶ What message am I being asked to accept?
▶ Is the message based on facts or opinion?
▶ Is there a different way to interpret the facts?
▶ What biases might the presenter have?
▶ How do my own biases affect what I'm learning?

Questions like these are a way of being conscious about learning. They can help you decide whether to allow new information to guide your beliefs and actions.

THE SCIENTIFIC METHOD

Critical thinking is a crucial part of **science**, the systematic study of the physical universe and how it works. **Biology**, the branch of science concerning past and present life, comprises hundreds of specializations (**Table 1.3** and **Figure 1.10**). A line of inquiry in biology typically begins with a researcher's curiosity about something observable in nature, such as, say, an unusual decrease in the number of birds occupying a particular area. The researcher reads about what others have discovered before making a **hypothesis**, a testable explanation for a natural phenomenon. An example of a hypothesis would be: The number of birds is decreasing because the number

TABLE 1.3

A Few Research Specializations in Biology

Field	Focus
Astrobiology	Potential life elsewhere in the universe
Biogeography	Distribution of life on Earth
Bioinformatics	Development of tools to analyze data
Botany	Plant structure and processes
Cell biology	Cell structure and processes
Ecology	Interactions among organisms, and among organisms and their environment
Ethology	Animal behavior
Genetics	Inheritance
Marine biology	Life in saltwater habitats
Medicine	Human health
Paleontology	Life in the ancient past
Structural biology	Architecture-dependent function of large biological molecules

Figure 1.10 Example of scientific research in biology. Left, National Geographic Explorer Tierney Thys travels the world's oceans to study the giant sunfish (mola). "When it comes to fishes, the mola really pushes the boundary of fish form," she says. "It seems a somewhat counterintuitive design for plying the waters of the open seas—a rather goofy design—and yet the more I learn about it, the more respect and admiration I have for it."

of cats is increasing. Making a hypothesis is an example of **inductive reasoning**, which means arriving at a conclusion based on one's observations. Inductive reasoning is the way we come up with new ideas about groups of objects or events. Note that a scientific hypothesis must be testable in the real world; consider how, for example, there would be no way we could test a hypothesis that the number of birds is decreasing because undetectable aliens are taking them to another planet (Section 1.8 returns to the topic of testability).

A **prediction**, or statement of some condition that should occur if the hypothesis is correct, comes next. Making predictions is called the if–then process, in which the "if" part is the hypothesis, and the "then" part is the prediction: *If the number of birds is decreasing because the number of cats is increasing, then removing cats from the area should stop the decline.* Using a hypothesis to make a prediction is a form of **deductive reasoning**, the logical process of using a general premise to draw a conclusion about a specific case.

Next, a researcher tests the "then" part of the prediction. Some predictions are tested by systematic observation; others require experimentation. **Experiments** are tests designed to determine whether a prediction is valid. In an investigation of our hypothetical bird–cat relationship, the researcher may remove all cats from the area. If working with an object or event directly is not possible, experiments may be performed on a **model**, or analogous system (for example, animal diseases are often used as models of similar human diseases).

A typical experiment explores a cause-and-effect relationship using variables. A **variable** is something that varies: a characteristic that differs among individuals, say, or an event that differs over time. An **independent variable** is defined or controlled by the person doing the experiment (in the bird–cat experiment, the independent variable is the presence or absence of cats). A **dependent variable** is presumed to be influenced by the independent variable (the dependent variable in the bird–cat experiment would be the number of birds).

Biological systems are complex because they involve many interdependent variables. It can be difficult to study one variable separately from the rest. Thus, biology researchers often test two groups of individuals simultaneously. An **experimental group** is a set of individuals that have a certain characteristic or receive a certain treatment. This group is tested side by side with a **control group**, which is identical to the experimental group except for one independent variable: the characteristic or the treatment being tested. Any differences in experimental results between the two groups is likely to be an effect of changing the variable.

Test results—**data**—offer a quantifiable way of evaluating the hypothesis. Data that validate the prediction are evidence in support of the hypothesis. Data that show the prediction is invalid are evidence that the hypothesis is flawed and should be revised. A necessary part of science is reporting one's data and conclusions in a standard way, such as in a peer-reviewed journal article. The communication gives other scientists an opportunity to evaluate the tested hypotheses for themselves, both by checking the conclusions drawn and by repeating the experiments.

Forming a hypothesis based on observation, and then systematically testing and evaluating the hypothesis, are collectively called the **scientific method** (**Table 1.4**).

TABLE 1.4

The Scientific Method

1. Observe some aspect of nature.

2. Think of an explanation for your observation (in other words, form a hypothesis).

3. Evaluate the hypothesis.
   a. Make a prediction based on the hypothesis (If . . . then).
   b. Test the prediction using experiments or surveys.
   c. Analyze the results (data).

4. Decide whether the results support your hypothesis or not (form a conclusion).

5. Report your experiment, data, and conclusion to the scientific community.

---

biology The scientific study of life.

control group Group of individuals identical to an experimental group except for the independent variable under investigation.

critical thinking Evaluating information before accepting it.

data (DAY-tuh) Experimental results.

deductive reasoning Using a general idea to make a conclusion about a specific case.

dependent variable In an experiment, a variable that is presumably affected by an independent variable being tested.

experiment A test designed to evaluate a prediction.

experimental group In an experiment, a group of individuals who have a certain characteristic or receive a certain treatment.

hypothesis (hi-POTH-uh-sis) Testable explanation of a natural phenomenon.

independent variable Variable that is controlled by an experimenter in order to explore its relationship to a dependent variable.

inductive reasoning Drawing a conclusion based on observation.

model Analogous system used for testing hypotheses.

prediction Statement, based on a hypothesis, about a condition that should occur if the hypothesis is correct.

science Systematic study of the physical universe.

scientific method Making, testing, and evaluating hypotheses.

variable (VAIR-ee-uh-bull) In an experiment, a characteristic or event that differs among individuals or over time.

1.6 Examples of Experiments in Biology

Give an example, real or hypothetical, of an experiment in which a dependent variable is affected by an independent variable.

RESEARCH IN THE REAL WORLD

Particularly in biology, scientific research rarely proceeds in a direct, start-to-finish fashion as Table 1.4 might suggest. Scientists often describe their work as a nonlinear process of exploration, asking questions, testing hypotheses, and changing directions. Hypotheses are often wrong and experimental results are often unpredictable. Research usually raises more question than it answers, so there never really is an endpoint. The unpredictability might be frustrating at times, but researchers typically say they enjoy their work because of the surprising twists and turns it takes.

To give you a sense of how biology experiments work, we summarize two published studies here.

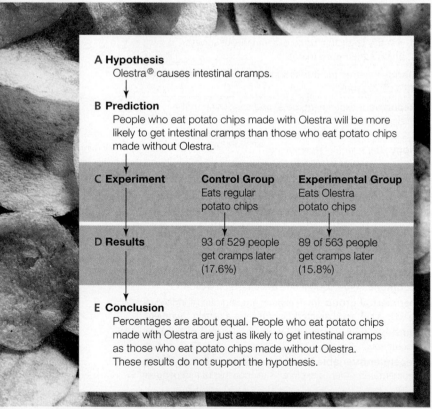

A Hypothesis
Olestra® causes intestinal cramps.

B Prediction
People who eat potato chips made with Olestra will be more likely to get intestinal cramps than those who eat potato chips made without Olestra.

C Experiment	Control Group	Experimental Group
	Eats regular potato chips	Eats Olestra potato chips
D Results	93 of 529 people get cramps later (17.6%)	89 of 563 people get cramps later (15.8%)

E Conclusion
Percentages are about equal. People who eat potato chips made with Olestra are just as likely to get intestinal cramps as those who eat potato chips made without Olestra. These results do not support the hypothesis.

Figure 1.11 The steps in a scientific experiment to determine if Olestra causes cramps. A report of this study was published in the *Journal of the American Medical Association* in January 1998.

FIGURE IT OUT
What was the dependent variable in this experiment?

Answer: Whether or not a person got cramps

POTATO CHIPS AND STOMACHACHES

In 1996 the U.S. Food and Drug Administration (the FDA) approved Olestra®, a fat replacement manufactured from sugar and vegetable oil, as a food additive. Potato chips were the first Olestra-containing food product to be sold in the United States. Controversy about the chip additive soon raged. Many people complained of intestinal problems after eating the chips, and thought that the Olestra was at fault. Two years later, researchers at the Johns Hopkins University School of Medicine designed an experiment to test whether Olestra causes cramps.

The researchers made the following prediction: *if* Olestra causes cramps, *then* people who eat Olestra should be more likely to get cramps than people who do not eat it. To evaluate their prediction, they used a Chicago theater as a "laboratory." They asked 1,100 people between the ages of thirteen and thirty-eight to watch a movie and eat their fill of potato chips. Each person received an unmarked bag that contained 13 ounces of chips.

In this experiment, the individuals who received Olestra-containing potato chips constituted the experimental group, and individuals who received regular chips were the control group. The independent variable was the presence or absence of Olestra in the chips.

A few days later, the researchers contacted everyone and collected reports of any post-movie gastrointestinal problems (the dependent variable). Of 563 people making up the experimental group, 89 (15.8 percent) complained about cramps. However, so did 93 of the 529 people (17.6 percent) making up the control group—who had eaten the regular chips. People were about as likely to get cramps whether or not they ate chips made with Olestra. These results showed that the prediction was invalid, so the researchers concluded that eating Olestra does not cause cramps (**Figure 1.11**).

BUTTERFLIES AND BIRDS

The peacock butterfly is a winged insect named for the large, colorful spots on its wings. In 2005, researchers reported the results of experiments investigating whether certain behaviors help these butterflies defend themselves against insect-eating birds. The study began with the observation that a resting peacock butterfly sits motionless with its wings folded (**Figure 1.12A**). The dark underside of the wings provides appropriate camouflage. However, when a peacock butterfly sees a predator approaching, it exposes its brilliant spots by repeatedly flicking its wings open and closed (**Figure 1.12B**). At the same time, it moves the hindwings in a way that produces a hissing sound and a series of clicks. A colorful, moving, noisy insect is usually very attractive to insect-eating birds, so the researchers were curious about why the peacock butterfly moves and makes noises only in the *presence* of predators. After they reviewed earlier studies, the scientists made two hypotheses that might explain the wing-flicking behavior:

Hypothesis 1: Peacock butterflies flick their wings in the presence of a predator because it exposes brilliant wing spots, thereby reducing predation. Peacock butterfly wing spots resemble owl eyes, and anything that looks like owl eyes is known to startle insect-eating birds.

Hypothesis 2: Peacock butterflies hiss and click when they flick their wings because these sounds reduce predation by birds. The sounds may be an additional defense that startles insect-eating birds.

The researchers then used their hypotheses to make the following predictions:

Prediction 1: If exposing brilliant wing spots by wing-flicking reduces predation by insect-eating birds, then removing the wingspots from peacock butterflies should make them more likely to get eaten.

Prediction 2: If the hissing and clicking sounds produced during wing-flicking reduce predation by insect-eating birds, then silencing peacock butterflies should make them more likely to get eaten.

Next came the experiments. The researchers used a black marker to cover up the wing spots of some butterflies. Other butterflies had the sound-making part of the wings removed with scissors. A third group had both treatments: Their spots were covered and wings were cut. Each butterfly was then put into a large cage with a hungry blue tit (**Figure 1.12C**), and the pair was watched for thirty minutes.

Figure 1.12D lists the results. All of the butterflies with unmodified wing spots survived, regardless of whether they made sounds. These results were consistent with the first hypothesis: By exposing brilliant spots, peacock butterfly wing-flicking behavior decreases predation by blue tits.

In contrast, a large proportion of butterflies without spots got eaten, whether or not they could make sounds. Experimental results were not consistent with the second hypothesis that peacock butterfly sounds reduce predation by birds. Other questions raised by these results offer an example of how research leads to more research: Do other predatory birds respond differently to the sounds than blue tits? If not, do the sounds reduce predation by other organisms (such as mice) that eat peacock butterflies? If sound-making is unrelated to predation, what is its function? Several additional experiments would be necessary to answer these questions.

A With wings folded, a peacock butterfly resembles a dead leaf, so it is appropriately camouflaged from predatory birds.

B When a predatory bird approaches, a butterfly flicks its wings open and closed, revealing brilliant spots and producing hissing and clicking sounds.

C Researchers tested whether the wing flicking and sound-making behaviors of peacock butterflies affected predation by blue tits (a type of songbird).

Experimental Treatment	Number of Butterflies Eaten
Spots painted out	5 of 10 (50%)
Wings cut	0 of 8 (0%)
Spots painted, wings cut	8 of 10 (80%)
None	0 of 9 (0%)

* Proceedings of the Royal Society of London, Series B (2005) 272: 1203–1207.

D The researchers painted out the spots of some butterflies, cut the sound-making part of the wings on others, and did both to a third group; then exposed each butterfly to a hungry blue tit for 30 minutes. Results support only the hypothesis that peacock butterfly spots deter predatory birds.

Figure 1.12 Testing the defensive value of two peacock butterfly behaviors.

FIGURE IT OUT
What was the dependent variable in this series of experiments?　　Answer: Getting eaten

WHAT SCIENCE IS

A scientific hypothesis, remember, is a testable explanation for a natural phenomenon, and testing a hypothesis may prove it false. Suppose a hypothesis survives after serious attempts to falsify it by years of rigorous testing. It is consistent with existing evidence, and scientists use it to make successful predictions about a wide range of other phenomena. A hypothesis that meets these criteria is a **scientific theory** (**Table 1.5**).

Consider the hypothesis that matter consists of atoms and their tiny components, which are called subatomic particles. Researchers no longer spend time testing this hypothesis for the compelling reason that, since we started looking 200 years ago, no one has discovered matter that consists of anything else. Thus, the hypothesis—now part of atomic theory—has been incorporated into our general understanding of matter. This understanding underpins research in many fields, including biology.

Scientific theories are our best objective descriptions of the natural world. However, they can never be proven under every possible circumstance. For example, proving that atomic theory holds under all circumstances would mean the composition of all matter in the universe would have to be

checked—an impossible task even if someone wanted to try. Thus, scientists avoid using the word "proven" to describe a theory. Instead, a theory is "accepted," along with the possibility—however remote—that new data inconsistent with it might be found.

What happens if someone discovers an exception to a theory? By our definition, a scientific theory has been tested rigorously and repeatedly. New results do not invalidate pre-existing results, but the interpretation of what the results mean can change. Thus, new data inconsistent with a theory may trigger its revision. Atomic theory has been modified many times since it was proposed thousands of years ago. Ancient Greeks defined atoms as fundamental, indivisible units of matter (*atomo* is the Greek word for "uncuttable"), but they had no way of knowing what atoms actually are. Their definition prevailed until 1897, when the newly invented cathode ray tube allowed Joseph J. Thomson to discover the existence of subatomic particles. Atoms were not indivisible after all! This new understanding did not falsify atomic theory, but it did prompt a revision that removed "indivisible" from the definition of atoms. Even today, the theory is revised periodically as improved equipment reveals new information about subatomic particles. If someone ever discovers matter that does not consist of atoms and subatomic particles, atomic theory would be revised to include the exception (all matter consists of atoms and subatomic particles except . . .). If many exceptions accumulate, the theory will be rewritten so it better accounts for the discrepancies.

The theory of evolution by natural selection, which states that environmental pressures can drive change in the inherited traits of a population, still stands after more than a century of concerted observations, testing, and challenges. Natural selection is not the only mechanism by which evolution occurs, but it is by far the most studied. Few other scientific theories have withstood as much scrutiny.

You may hear people apply the word "theory" to a speculative idea, as in the phrase "It's just a theory." This everyday usage of the word differs from the specific way it is used in science. A scientific theory also differs from a **law of nature**, which describes a phenomenon that always occurs under certain circumstances. Laws do not include mechanisms, so they do not necessarily have a complete scientific explanation. The laws of thermodynamics, which describe energy, are examples. We understand *how* energy behaves, but not exactly *why* it behaves the way it does (Chapter 5 returns to energy).

WHAT SCIENCE IS NOT

Not everything that uses scientific vocabulary is actually science. Claims, arguments, or methods that are presented as

TABLE 1.5

Examples of Scientific Theories

Atomic theory	All matter consists of atoms and their smaller subatomic parts; properties of matter arise from these components.
Big bang	The universe originated with an explosion of high-density matter.
Cell theory	All organisms consist of one or more cells, the cell is the basic unit of life, and all cells arise from preexisting cells.
Climate change	Earth's average temperature is rising; human activities are part of the cause.
Evolution by natural selection	Environmental pressures drive change in the inherited traits of a population.
Plate tectonics	Earth's lithosphere (crust and upper mantle) is cracked into pieces that move in relation to one another.

CREDITS: (in text) Alhovik/Shutterstock.com; (Table 1.5 photo) Raymond Gehman/Documentary Value/Corbis.

TABLE 1.6

How To Tell If a Theory Is Scientific or Not

1. Does the theory use a natural explanation?
Science addresses only the observable, natural world.

2. Is the explanation the simplest one?
All else being equal, the explanation that requires the fewest possible assumptions is most likely to be correct.

3. Is the theory consistent with other scientific evidence?
Scientific theories are consistent with all (or almost all) existing observations and experimental evidence in all scientific fields.

4. Has the theory been tested?
A scientific theory has survived serious efforts to falsify it. Having been tested and confirmed many times, it is supported by a large body of scientific evidence.

5. Is the theory useful for making predictions?
Scientific theories are consistently useful for making valid predictions about a wide range of natural phenomena.

scientific but do not follow scientific principles are called **pseudoscience** (*pseudo–* means false).

Consider how a scientific hypothesis is constructed to be testable in ways that could prove it wrong; testing a hypothesis is technically an attempt to falsify it. Pseudoscientific statements such as "Earth appears older than it is because it came into existence that way" are inherently impossible to falsify by testing—in this case, because no measurement or observation could possibly reveal Earth's true age. A statement that cannot be tested and potentially proven wrong by evidence is not part of science.

All scientific theories are supported by a large body of evidence, which, as you now know, consists of data derived from repeated experiments or surveys. Pseudoscientific claims may not be. Anecdotal evidence, not scientific data, is often used to support pseudoscience ("UFOs are real because my uncle saw one," for example).

The scientific method begins with observations and ends with conclusions. Many pseudoscientific methods begin with a conclusion ("eating bananas causes weight loss," for example), followed by a search for supporting evidence (collecting anecdotes from people who lost weight and ate bananas). Contrary evidence is ignored, downplayed, or dismissed.

Distinguishing pseudoscience from the real thing can be tricky, but your critical thinking skills will help (**Table 1.6**).

law of nature Generalization that describes a consistent natural phenomenon but does not propose a mechanism.
pseudoscience A claim, argument, or method that is presented as if it were scientific, but is not.
scientific theory Hypothesis that has not been falsified after many years of rigorous testing, and is useful for making predictions about a wide range of phenomena.

APPLICATION: PROCESS OF SCIENCE

Figure 1.15 The Pinocchio frog. Paul Oliver discovered this tiny tree frog perched on a sack of rice during the first survey of a cloud forest in New Guinea. It was named after the Disney character because the male frog's long nose inflates and points upward during times of excitement.

THE SECRET LIFE OF EARTH

In this era of cell phone GPS, could there possibly be any places left on Earth that humans have not yet explored? Actually, there are plenty. Consider a 2-million-acre cloud forest in the Foja Mountains of New Guinea that was not penetrated by humans until 2005. About forty new species were discovered living there, including a rhododendron plant with flowers the size of dinner plates, a rat the size of a cat, and a frog the size of a pea. Also among the forest's inhabitants were hundreds of species that are on the brink of extinction in other parts of the world, and some that supposedly had been extinct for decades. The animals had never learned to be afraid of humans, so they could easily be approached. A few were discovered casually wandering through campsites (**Figure 1.15**).

New species are discovered all the time, in places much more mundane than tropical cloud forests. A survey in 2015 revealed 30 new insect species living in urban Los Angeles, for example. Each discovery is a reminder that we do not yet know all of the organisms that share our planet. We don't even know how many to look for. Why does that matter? Understanding the scope of life on Earth gives us perspective on where we fit into it. For example, the current rate of extinctions is about 1,000 times faster than ever recorded, and we now know that human activities are responsible for the acceleration. At this rate, we will never know about most of the species that are alive today. Is that important? Biologists think so. Whether or not we are aware of it, humans are intimately connected with the world around us. Our activities are profoundly changing the entire fabric of life on Earth. The changes are, in turn, affecting us in ways we are only beginning to understand.

Section 1.1 Biologists think about life at different levels of organization, with **emergent properties** appearing at successive levels. All matter, living or not, consists of **atoms** and their subatomic components. Atoms combine as **molecules**. The unique properties of life emerge as certain kinds of molecules become organized into a **cell**. **Organisms** are individuals that consist of one or more cells. In larger multicelled organisms, cells are organized as **tissues**, **organs**, and **organ systems**. A **population** is a group of interbreeding individuals of a **species** in a given area; a **community** is all populations of all species in a given area. An **ecosystem** is a community interacting with its environment. The **biosphere** includes all regions of Earth that hold life.

Section 1.2 All living things share a set of characteristics that include a requirement for energy and **nutrients**. **Producers** harvest energy from the environment to make their own food by processes such as **photosynthesis**; **consumers** ingest other organisms, or their wastes or remains. Organisms sense and respond to change, making adjustments that keep conditions in their internal environment within tolerable ranges—a process called **homeostasis**. Information in an organism's **DNA** guides its **growth**, **development**, and **reproduction**. The passage of DNA from parents to offspring is called **inheritance**. DNA is the basis of similarities and differences among organisms.

Section 1.3 The many species alive today differ greatly in form and function. **Bacteria** and **archaea** are **prokaryotes**: single-celled organisms whose DNA is not contained within a nucleus. The DNA of single- or multicelled **eukaryotes** (**protists**, **plants**, **fungi**, and **animals**) is contained in a nucleus.

Section 1.4 Each species is given a two-part name. The first part is the **genus** name. When combined with the specific epithet, it designates the particular species. With **taxonomy**, species are named and ranked into more inclusive **taxa** on the basis of shared traits.

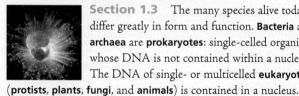

Section 1.5 **Critical thinking**, the act of judging the quality of information as one learns, is an important part of **science**. **Biology** is the scientific study of life. Generally, a researcher observes something in nature, uses **inductive reasoning** to form a **hypothesis** (testable explanation) for it, then uses **deductive reasoning** to make a **prediction** about what might occur if the hypothesis is correct. Predictions are evaluated with observations, **experiments**, or both. Experiments with **variables** are typically performed on an **experimental group** and compared with a **control group**, and sometimes on **model** systems. A researcher changes an **independent variable**, then observes the effects of the change on a **dependent variable**. Conclusions are drawn from the resulting **data**. The **scientific method** involves making, testing, and evaluating hypotheses, then sharing results.

Section 1.6 Science is a nonlinear process of exploration, asking questions, testing hypotheses, and changing directions. Hypotheses are often wrong; results are often unpredictable and raise more questions than they answer.

Section 1.7 Checks and balances inherent in the scientific process help researchers to be objective about their observations. Researchers minimize **sampling error** by using large sample sizes and by repeating their experiments. **Probability** calculations can show whether a result is **statistically significant** (very unlikely to have occurred by chance alone). Science is ideally self-correcting because it is carried out by a large community of people systematically checking one another's ideas. It helps us be objective because it is only concerned with testable ideas about observable aspects of nature. Opinion and belief are not addressed by science.

Section 1.8 A **scientific theory** is a long-standing hypothesis that is useful for making predictions about a wide range of other phenomena. It may be revised upon the discovery of new data, but it is our most objective way of describing the natural world. A **law of nature** describes a consistent natural phenomenon but not an explanation for it. **Pseudoscience** is anything that is not scientific but presented as if it were.

Application We know about only a fraction of the organisms that live on Earth, in part because we have explored only a fraction of it. Understanding the scope of Earth's biodiversity gives us perspective on where we fit into it.

SELF-ASSESSMENT
ANSWERS IN APPENDIX VII

1. _____ are the smallest units of all substances.
   - a. Atoms
   - b. Molecules
   - c. Cells
   - d. Organisms

2. The smallest unit of life is the _____ .
   - a. atom
   - b. molecule
   - c. cell
   - d. organism

3. Organisms require _____ and _____ to maintain themselves, grow, and reproduce.

4. By sensing and responding to change, organisms keep conditions in the internal environment within ranges that cells can tolerate. This process is called _____ .

5. DNA _____ .
   - a. guides form and function
   - b. is the basis of traits
   - c. is transmitted from parents to offspring
   - d. all of the above

6. Bacteria, Archaea, and Eukarya are three _____ .

PEACOCK BUTTERFLY PREDATOR DEFENSES The photographs below represent experimental and control groups used in the peacock butterfly experiment discussed in Section 1.6. Identify the experimental groups, and match them up with the relevant control group(s). *Hint*: Determine which variable is being tested in each group (each variable has a control).

A Wing spots painted out

B Wing spots visible; wings silenced

C Wing spots painted out; wings silenced

D Wings painted but spots visible

E Wings cut but not silenced

F Wings painted, spots visible; wings cut, not silenced

7. A process by which an organism produces offspring is called _____ .

8. _____ is the transmission of DNA to offspring.
   - a. Reproduction
   - b. Development
   - c. Homeostasis
   - d. Inheritance

9. A butterfly is a(n) _____ (choose all that apply).
   - a. organism
   - b. domain
   - c. species
   - d. eukaryote
   - e. consumer
   - f. producer
   - g. prokaryote
   - h. trait

10. A bacterium is _____ (choose all that apply).
    - a. an organism
    - b. single-celled
    - c. an animal
    - d. a eukaryote

11. A control group is _____ .
    - a. a set of individuals that have a certain characteristic or receive a certain treatment
    - b. the standard against which an experimental group is compared
    - c. the experiment that gives conclusive results

12. Fifteen randomly selected students are found to be taller than 6 feet. The researchers concluded that the average height of a student is greater than 6 feet. This is an example of _____ .
    - a. experimental error
    - b. sampling error
    - c. a subjective opinion
    - d. experimental bias

13. Science addresses only that which is _____ .
    - a. alive
    - b. observable
    - c. variable
    - d. indisputable

14. Which of the following statements can be falsified?
    - a. All of the fish in Lake Michigan are brown.
    - b. French cheese is the tastiest.
    - c. Homeopathic remedies work only if you believe in them.

15. Match the terms with the most suitable description.
    - ___ data
    - ___ probability
    - ___ species
    - ___ theory
    - ___ prediction
    - ___ producer
    - ___ hypothesis
    - ___ life
    - a. if–then statement
    - b. unique type of organism
    - c. emerges with cells
    - d. testable explanation
    - e. measure of chance
    - f. time-tested hypothesis
    - g. makes its own food
    - h. scientific results

1. A person is declared dead upon the irreversible ceasing of spontaneous body functions: brain activity, blood circulation, and respiration. Only about 1% of a body's cells have to die in order for all of these things to happen. How can a person be dead when 99% of his or her cells are alive?

2. What is the difference between a one-celled organism and a single cell of a multicelled organism?

3. Why would you think twice about ordering from a restaurant menu that lists the specific epithet but not the genus name of its offerings? *Hint*: Look up *Homarus americanus*, *Ursus americanus*, *Ceanothus americanus*, *Bufo americanus*, *Lepus americanus*, and *Nicrophorus americanus*.

4. Once there was a highly intelligent turkey that had nothing to do but reflect on the world's regularities. Morning always started out with the sky turning light, followed by the master's footsteps, which were always followed by the appearance of food. Other things varied, but food always followed footsteps. The sequence of events was so predictable that it eventually became the basis of the turkey's theory that footsteps bring food. One morning, after more than 100 confirmations of this theory, the turkey listened for the master's footsteps, heard them, and had its head chopped off.

   Scientific theories can be revised upon the discovery of inconsistent evidence. Suggest a modification to the turkey's theory that the remaining members of the flock would find more useful for making accurate predictions.

5. In 2005, researcher Woo-suk Hwang reported that he had made immortal stem cells from human patients. His research was hailed as a breakthrough for people affected by degenerative diseases, because stem cells may be used to repair a person's own damaged tissues. Hwang published his results in a peer-reviewed journal. In 2006, the journal retracted his paper after other scientists discovered that Hwang's group had faked their data. Does the incident show that results of scientific studies cannot be trusted? Or does it confirm the usefulness of a scientific approach, because other scientists discovered and exposed the fraud?

for additional quizzes, flashcards, and other study materials
▶ ACCESS MINDTAP AT WWW.CENGAGEBRAIN.COM

CREDIT: Adrian Vallin, Sven Jakobsson, Johan Lind, Christer Wiklund, Prey survival by predator intimidation: an experimental study of peacock butterfly defence against blue tits, Proceedings B, 2005, Vol 272, Issue 1569, 1203–1207 by permission of The Royal Society.

2 Life's Chemical Basis

Links to Earlier Concepts

In this chapter, you will explore the first level of life's organization—atoms—as you encounter the first example of how the same building blocks, arranged different ways, form different products (Section 1.1). You will also see one aspect of homeostasis, the process by which organisms keep themselves in a state that favors cell survival (1.2).

Core Concepts

Interactions among the components of each biological system give rise to complex properties.

When components of a biological system interact, they enable properties not found in the individual components alone. The behavior of atoms arises from the number and type of subatomic particles that compose them. Similarly, the properties of molecules arise from the types and arrangement of their component atoms; and the properties of substances, from the kinds and amounts of their component molecules. Unique properties that make water essential to life arise from the polarity of individual molecules.

Organisms exchange matter and energy with the environment in order to grow, maintain themselves, and reproduce.

All organisms harvest atomic and molecular building blocks from their environment. Electrons have energy that can be stored in and retrieved from atoms and molecules they compose. The movement of atoms and molecules is influenced by external factors such as temperature.

Living things sense and respond appropriately to their internal and external environments.

Homeostatic mechanisms allow living things to maintain themselves by responding dynamically to internal and external conditions, including shifts in pH and temperature. Organisms keep their internal environment stable by returning a changed condition back to its target set point, but extreme variation can overwhelm these mechanisms.

◀ Water is essential to life. Interactions among individual water molecules give rise to unique properties that all organisms depend on.

23

Photograph by Paul Nicklen, National Geographic Creative.

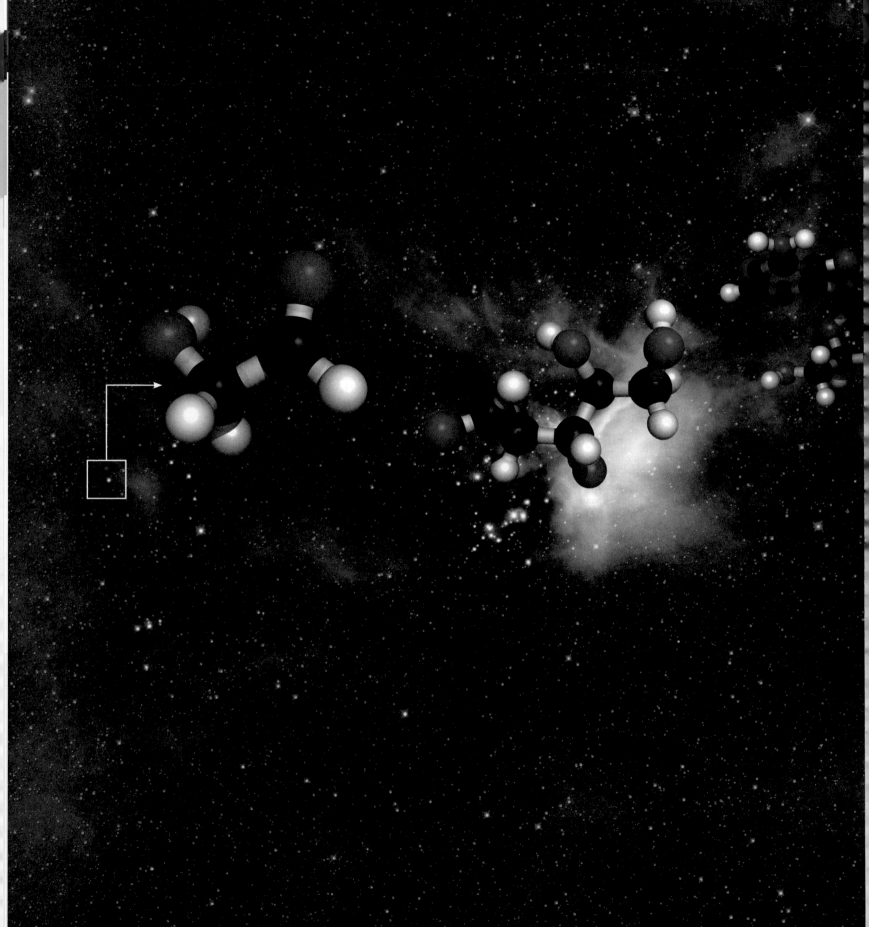

3 Molecules of Life

Links to Earlier Concepts

Having learned about atomic interactions (Section 2.3), you are now in a position to understand the structure of the molecules of life. Keep the big picture in mind by reviewing Section 1.1. You will be building on your knowledge of covalent bonding (2.3), acids and bases (2.5), and the effects of hydrogen bonds (2.4).

Core Concepts

Organisms exchange matter and energy with the environment in order to grow, maintain themselves, and reproduce.

All organisms take up carbon-containing compounds from the environment and use them to build the molecules of life: complex carbohydrates and lipids, proteins, and nucleic acids. Nitrogen is essential for building proteins and nucleic acids; phosphorus is incorporated into lipids and nucleic acids. All biological molecules are eventually broken down, and their components are cycled back to the environment in by-products, wastes, and remains.

Interactions among the components of each biological system give rise to complex properties.

The molecules of life are assembled from simpler organic subunits. The structure and function of a biological molecule arises from (and depends on) the order, orientation, and interactions of its component subunits. The dual hydrophilic and hydrophobic properties of individual phospholipids give rise to the lipid bilayer structure of cell membranes.

A molecule's function and ability to interact with other molecules depend on its three-dimensional structure.

The same sugar molecules, bonded together in slightly different ways, form carbohydrates with very different properties. A protein's function depends on its structure, which arises from interactions among its amino acid components. Biological information encoded in a nucleic acid consists of the sequence of nucleotide monomers that compose it.

◀ A sugar called glycolaldehyde has been spotted in space gas falling toward a young, sunlike star. This sugar can combine with other molecules in the gas to form ribose, a building block of RNA. RNA is one of the carbon-rich molecules common to all life.

NASA/JPL-Caltech/UCLA.

3.1 The Chemistry of Biology

A Carbon's versatile bonding behavior allows it to form a variety of structures, including rings.

B Carbon rings form the framework of many sugars, starches, and fats (including those that make up doughnuts).

Figure 3.1 Carbon rings.

CARBON, THE STUFF OF LIFE

The same elements that make up a living body also occur in nonliving things, but their proportions differ. For example, compared to sand or seawater, a human body has a much larger proportion of carbon atoms (Section 2.1). Why? Unlike sand or seawater, a body has a lot of the molecules of life—complex carbohydrates and lipids, proteins, and nucleic acids—which, in turn, consist of a high proportion of carbon atoms. Compounds that consist mainly of carbon and hydrogen are said to be **organic**. The term is a holdover from a time

when these molecules were thought to be made only by living things, as opposed to the "inorganic" molecules that formed by nonliving processes. We now know that organic compounds were present on Earth long before organisms were.

A carbon atom is unusual among elements because it can bond stably with many other elements. It also has four vacancies (Section 2.2), so it can form four covalent bonds with other atoms—including other carbon atoms. Many organic molecules have a chain of carbon atoms, and this backbone often forms rings (**Figure 3.1**).

The versatility of carbon atoms means that they can be assembled into a wide variety of organic compounds. A molecule that consists only of carbon and hydrogen atoms is called a **hydrocarbon**, and it is nonpolar. The molecules of life have other elements in addition to carbon and hydrogen. These other elements are often part of **functional groups**: small molecular groups covalently bonded to the carbon backbone (**Table 3.1**).

Each functional group imparts a particular chemical property to an organic compound. For example, carboxyl groups ($-COOH$) make amino acids and fatty acids acidic. A hydroxyl group ($-OH$) adds polar character, thus increasing solubility in water. Hydroxyl groups turn hydrocarbons into alcohols. A methyl group ($-CH_3$) adds nonpolar character. Methyl groups added to DNA act like an "off" switch for this molecule; acetyl groups act like an "on" switch (we return to this topic in Chapter 10). The chemical behavior of the molecules of life arises mainly from the number, kind, and arrangement of their functional groups.

The structure of biological molecules can be quite complex (**Figure 3.2A**). For clarity, we may omit some features when representing them: some of the bonds in a structural formula, for example, or hydrogen atoms bonded to a carbon backbone. Carbon rings such as the ones that occur in glucose and other sugars are often depicted as polygons (**Figure 3.2B**). If no atom is shown at a corner or at the end of a bond, a carbon is implied there. Ball-and-stick models are used to depict an organic molecule's three-dimensional arrangement of atoms (**Figure 3.2C**); space-filling models are used to show its overall shape (**Figure 3.2D**). Proteins

TABLE 3.1

Some Functional Groups in Biological Molecules

Group	Structure	Character	Formula	Found in:
methyl	H|—C—H|H	nonpolar	$-CH_3$	fatty acids, some amino acids
hydroxyl	—O—H	polar	$-OH$	alcohols, sugars
sulfhydryl	—S—H	forms rigid disulfide bonds	$-SH$	cysteine, many cofactors
amine	—N(H)(H)	very basic	$-NH_2$	nucleotide bases, amino acids
carbonyl	O‖—C—	polar, reactive	$-CO$	alcohols, other functional groups
carboxyl	O‖—C—C—OH	acidic, reactive	$-COOH$	fatty acids, amino acids
aldehyde	O‖—C—C—H	polar, reactive	$-CHO$	simple sugars
acetyl	O‖—C—CH	polar, acidic	$-COCH_3$	some proteins, coenzymes
amide	O‖—C—N	weakly basic, stable, rigid	$-CON-$	proteins, nucleotide bases
ketone	O‖—C—C—C—	polar, acidic	$-CO-$	simple sugars, nucleotide bases
phosphate	O‖—O—P—O—|O⁻	polar, reactive	$-PO_4$	nucleotides, DNA, RNA, phospholipids, proteins

CREDIT: (1B) © JupiterImages/Getty Images.

and nucleic acids are often represented as ribbon structures, which, as you will see in Section 3.4, show how the backbone of these molecules folds and twists.

METABOLIC REACTIONS

All biological systems are based on the same organic molecules, a similarity that is one of many legacies of life's common origin. However, the details of those molecules differ. Just as atoms bonded in different numbers and arrangements form different molecules, simple organic building blocks bonded in different numbers and arrangements form different versions of the molecules of life. The building blocks—sugars, fatty acids, amino acids, and nucleotides—are **monomers** when used as subunits of larger molecules. A molecule that consists of repeated monomers is a **polymer**.

Cells link monomers to form polymers, and break apart polymers to release monomers. These and any other processes of molecular change are called **reactions**. Cells constantly run reactions as they acquire and use energy to stay alive, grow, reproduce, and so on. Collectively, these reactions are called **metabolism (Figure 3.3)**. Metabolism requires **enzymes**, which are organic molecules (usually proteins) that speed up reactions without being changed by them. Enzymes remove monomers from polymers in a common metabolic reaction called **hydrolysis ❶**. Hydrolysis requires water (hence the name). The reverse of hydrolysis is a reaction called **condensation**, in which an enzyme joins one monomer to another ❷. Water forms during condensation, so the reaction is also called dehydration.

Chapter 5 returns to metabolism. The rest of this chapter introduces the different types of biological molecules and the monomers from which they are built.

---

condensation Chemical reaction in which a large molecule is assembled from smaller subunits; water also forms.

enzyme Organic molecule that speeds up a reaction without being changed by it.

functional group An atom (other than hydrogen) or a small molecular group bonded to a carbon of an organic compound; imparts a specific chemical property.

hydrocarbon Compound that consists only of carbon and hydrogen atoms.

hydrolysis (hy-DRAWL-uh-sis) Water-requiring chemical reaction that breaks a molecule into smaller subunits.

metabolism All of the enzyme-mediated reactions in a cell.

monomer Molecule that is a subunit of a polymer.

organic Describes a molecule that consists mainly of carbon and hydrogen atoms.

polymer Molecule that consists of repeated monomers.

reaction Process of molecular change.

A A structural formula for an organic molecule—even a simple one—can be very complicated. The overall structure is obscured by detail.

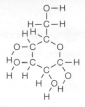

B Structural formulas of organic molecules are often simplified by using polygons as symbols for rings, and omitting some bonds and element labels. If no atom is shown at a corner or at the end of a bond, a carbon is implied there.

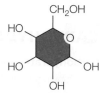

C A ball-and-stick model shows the arrangement of atoms and bonds in three dimensions.

D A space-filling model can be useful to show a molecule's overall shape. Individual atoms are visible in this model. Space-filling models of larger molecules often show only the surface contours.

Figure 3.2 Modeling an organic molecule. All of these models represent the same molecule: glucose.

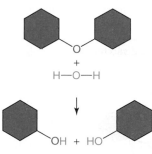

❶ **Hydrolysis.** Cells use this water-requiring reaction to split polymers into monomers. An enzyme attaches a hydroxyl group and a hydrogen atom (both from water) at the site of the split.

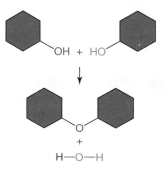

❷ **Condensation.** Cells use this reaction to build polymers from monomers. An enzyme removes a hydroxyl group from one molecule and a hydrogen atom from another. A covalent bond forms between the two molecules. Water forms, so this reaction is also called dehydration.

Figure 3.3 Examples of metabolic reactions. Common reactions by which cells build and break down organic molecules are shown.

3.5 The Importance of Protein Structure

LEARNING OBJECTIVES

Describe protein denaturation and its effects.

Using appropriate examples, explain why prions are dangerous.

Protein shape depends on hydrogen bonds. These bonds can be disrupted by shifts in pH or temperature, or by exposure to detergents or salts. Such disruption causes a protein to undergo **denaturation**, which means it loses its secondary, tertiary, and quaternary structure. Once a protein's shape unravels, so does its function.

You can see denaturation in action when you cook an egg. A protein called albumin is a major component of egg white. Cooking does not disrupt the covalent bonds of albumin's primary structure, but it does destroy the hydrogen bonds that maintain the protein's shape. When albumin denatures, the translucent egg white turns opaque. For a very few proteins, denaturation is reversible if normal conditions return, but albumin is not one of them. There is no way to uncook an egg.

Figure 3.15 Variant Creutzfeldt–Jakob disease (vCJD).

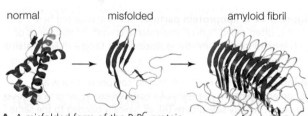

normal   misfolded   amyloid fibril

A A misfolded form of the PrP^C protein causes other PrP^C proteins to misfold and align tightly. Misfolded proteins accumulate in amyloid fibrils that form plaques in the brain as they grow.

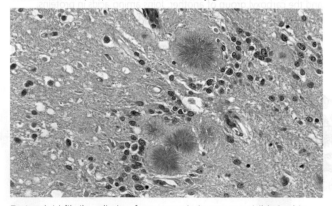

B Amyloid fibrils radiating from several plaques are visible in this slice of brain tissue from a person with vCJD.

FIGURE IT OUT
How does PrPC secondary structure change when it misfolds?

Answer: It loses helices and gains sheets.

Mad cow disease (bovine spongiform encephalitis, or BSE) in cattle, Creutzfeldt–Jakob disease in humans, and scrapie in sheep are the dire aftermath of a protein that changes shape. These diseases may be inherited, but more often they arise spontaneously; all are characterized by relentless deterioration of mental and physical abilities that eventually causes death. They begin with a glycoprotein called PrP^C, which occurs normally in cell membranes of the mammalian body. This protein is abundant in brain cells, but we still know little about its normal function. Sometimes, a PrP^C protein misfolds. One misfolded molecule should not pose a threat, but when this particular protein misfolds it becomes a **prion**, or infectious protein. The shape of the misfolded protein causes normally folded PrP^C proteins to misfold too (**Figure 3.15A**). Each protein that misfolds becomes infectious, so the number of prions increases exponentially.

Misfolded PrP^C proteins align into long fibers called amyloid fibrils. Amyloid fibrils resist cellular mechanisms that normally destroy misfolded proteins. They grow from their ends as newly misfolded proteins join the aggregation, forming characteristic patches in the brain called plaques (**Figure 3.15B**). The misfolded PrP^C protein is also toxic to brain cells, and holes form in the brain as its cells die. Progressive symptoms such as confusion, memory loss, and lack of coordination precede death.

In the mid-1980s, an epidemic of mad cow disease in Britain was followed by an outbreak of a new variant of Creutzfeldt–Jakob disease (vCJD) in humans. A prion similar to the one in scrapie-infected sheep was found in cows with BSE, and also in humans affected by vCJD. How did the prion get from sheep to cattle to people? Prions resist denaturation, so treatments such as cooking that inactivate other infectious agents have little effect on them. The cattle became infected by the prion after eating feed prepared from the remains of scrapie-infected sheep, and people became infected by eating beef from the infected cattle.

PrP^C is not the only protein that is harmful when misfolded. In the last decade, researchers discovered similar mechanisms that underlie several neurodegenerative disorders, including Alzheimer's, Parkinson's, and Huntington's diseases, atherosclerosis, and amyotrophic lateral sclerosis (ALS). Like vCJD, all involve proteins that are normally found in the body. When these proteins misfold, they cause normal versions of the protein to misfold too. The misfolded proteins are toxic, and they disrupt normal function of the brain, nerves, arteries, or other parts of the body.

denaturation (dee-nay-turr-AY-shun) Loss of a protein's three-dimensional shape, as by a shift in temperature.

prion (PREE-on) Infectious protein.

CREDITS: (15A) left, pdb 1QM2, Zahn, R., Liu, A., Luhrs, T., Riek, R., Von Schroetter, C., Garcia, F.L., Billeter, M., Calzolai, L., Wider, G., Wuthrich, K., "NMR solution structure of the human prion protein." Journal: (2000) Proc.Natl.Acad.Sci. USA 97: 145; middle and right, pdb 2RNM, Wasmer, C., Lange, A., Van Melckebeke, H., Siemer, A.B., Riek, R., Meier, B.H., "Amyloid fibrils of the HET-s(218–289) prion form a beta solenoid with a triangular hydrophobic core." Journal: (2008) Science 319: 1523–1526; (15B) Sherif Zaki, MD PhD, Wun-Ju Shieh, MD PhD; MPH/ CDC.

3.6 Nucleic Acids

Nucleotides are small organic molecules that function as energy carriers, enzyme helpers, chemical messengers, and subunits of DNA and RNA. Each consists of a monosaccharide ring bonded to a nitrogen-containing base and one, two, or three phosphate groups (**Figure 3.16A**). The monosaccharide is a five-carbon sugar, either ribose or deoxyribose, and the base is one of five compounds with a flat ring structure (more about this in Section 8.2). When the third phosphate group of a nucleotide is transferred to another molecule, energy is transferred along with it. As you will see in Section 5.5, the nucleotide **ATP** (adenosine triphosphate) serves an especially important role as an energy carrier in cells.

Nucleic acids are chains of nucleotides in which the sugar of one nucleotide is joined to the phosphate group of the next (**Figure 3.16B**). **RNA**, or ribonucleic acid, is named after the ribose sugar of its component nucleotides. An RNA is a single chain of nucleotides, one of which is ATP. Different types of RNA interact with DNA during protein synthesis. **DNA**, or deoxyribonucleic acid, is a nucleic acid named after the deoxyribose sugar of its component nucleotides. DNA has two chains of nucleotides twisted into a double helix (**Figure 3.16C**). The order of nucleotides composing a DNA molecule is biological information: an idea so important that we devote an entire unit to it (Unit 2, Genetics).

**Figure 3.16
Nucleic acid structure.**

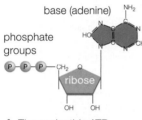

A The nucleotide ATP.

B The nucleic acid RNA.

C The nucleic acid DNA.

ATP Adenosine triphosphate. (uh-DEN-uh-seen) Nucleotide monomer of RNA; also is an important energy carrier in cells.

DNA Deoxyribonucleic acid. (dee-ox-ee-rye-bo-new-CLAY-ick) Nucleic acid with two nucleotide chains twisted into a double helix; carries hereditary information.

nucleic acid (new-CLAY-ick) Polymer of nucleotides; DNA or RNA.

nucleotide (NEW-klee-uh-tide) Small molecule with a five-carbon sugar, a nitrogen-containing base, and phosphate groups. Monomer of nucleic acids; some have additional metabolic roles.

RNA Ribonucleic acid. (rye-bo-new-CLAY-ick) Nucleic acid that consists of a chain of nucleotides. Carries out protein synthesis.

APPLICATION: SCIENCE & SOCIETY

FEAR OF FRYING

Seemingly small differences in the structure of a biological molecule can have very big effects in an organism. Consider how fatty acid tails vary, for example in saturation (Section 3.3). Another variation occurs in the architecture of double bonds in the tails of unsaturated fats. In most naturally occurring fats, the two hydrogen atoms flanking these bonds are *cis*, which means they are on the same side of the carbon backbone (right). Fats with hydrogen atoms on opposite sides of the tail are called *trans* fats.

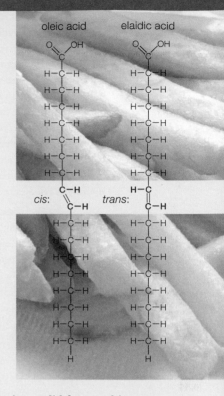

In the Western diet, the main source of *trans* fats is an artificial food product called partially hydrogenated vegetable oil. Hydrogenation is a manufacturing process that adds hydrogen atoms to oils in order to change them into solid fats, and it creates abundant *trans* bonds in fatty acid tails. In 1908, Procter & Gamble Co. developed partially hydrogenated soybean oil as a substitute for the more expensive solid animal fats they had been using to make candles. However, the demand for candles began to wane as more households in the United States became wired for electricity, and P&G looked for another way to sell its proprietary fat. Partially hydrogenated vegetable oil looks like lard, so the company began aggressively marketing it as a revolutionary new food: a solid cooking fat with a long shelf life, mild flavor, and lower cost than lard or butter. By the mid-1950s, hydrogenated vegetable oil had become a major part of the American diet, and today it is still found in many manufactured and fast foods, including french fries.

Partially hydrogenated vegetable oil was once thought to be healthier than animal fats, but we have known otherwise since 2002. *Trans* fats raise the level of cholesterol in blood more than any other fat, and they directly alter the function of arteries and veins. The effects of such changes are quite serious. Eating as little as 2 grams a day (about 0.4 teaspoon) of hydrogenated vegetable oil measurably increases a person's risk of atherosclerosis (hardening of the arteries), heart attack, and diabetes. A small serving of french fries made with hydrogenated vegetable oil contains about 5 grams of *trans* fat.

The use of *trans* fats in the U.S. food supply is now declining, and will be banned in manufactured foods in 2018. Until then, packaged foods in the United States are required to list *trans* fat content. However, they may be marked "zero grams of *trans* fats" even when a single serving contains up to half a gram of *trans* fats.

Section 3.1 Molecules that consist mainly of carbon and hydrogen atoms are **organic**. **Hydrocarbons** have only carbon and hydrogen atoms. The structure of the molecules of life—complex carbohydrates and lipids, proteins, and nucleic acids—starts with a chain of carbon atoms (the backbone) that may form rings. **Functional groups** attached to the backbone influence the molecule's chemical character. Different molecular models reveal different aspects of structure.

Metabolism includes all enzyme-mediated **reactions** in a cell. In **condensation** reactions, **enzymes** build **polymers** from **monomers** of simple sugars, fatty acids, amino acids, and nucleotides. **Hydrolysis** releases monomers by breaking apart polymers.

Section 3.2 Cells use simple **carbohydrates** (sugars) for energy and to build other molecules. **Monosaccharides** are bonded together to form **disaccharides** (two sugars), oligosaccharides (a few sugars), and **polysaccharides** (many sugars). **Cellulose**, **starch**, and **glycogen** are polysaccharides that consist of the same glucose monomers, bonded different ways. **Chitin** is a polysaccharide of nitrogen-containing sugar monomers.

Section 3.3 **Lipids** in biological systems are partially or entirely nonpolar. A **fatty acid** has an acidic head and a long hydrocarbon tail. Only single bonds link the carbons in the tail of a **saturated fatty acid**; the tail of an **unsaturated fatty acid** has one or more double bonds. **Fats** are substances that consist primarily of **triglycerides**, which have three fatty acid tails. Triglycerides are called unsaturated fats if there are double bonds in one or more of their fatty acid tails, and saturated fats if there are none. The basic structure of cell membranes is a **lipid bilayer** of mainly **phospholipids**. **Waxes** are water-repellent substances that consist of varying mixtures of lipids. **Steroids**, with four carbon rings and no fatty acid tails, serve important physiological roles such as starting materials for sex hormone synthesis.

Section 3.4 **Peptides** and **polypeptides** are (short and long) chains of **amino acids** linked by **peptide bonds**. A **protein** consists of one or more polypeptides. The order of amino acids making up a polypeptide (primary structure) dictates the type of protein and its shape. A protein's shape is the source of its function. Polypeptides twist into helices, sheets, loops, and turns (secondary structure) that fold into functional domains (tertiary structure). Most enzymes consist of two or more polypeptides (quaternary structure). A protein that can bind to lipids is a lipoprotein; a protein with attached oligosaccharides is a glycoprotein.

Section 3.5 Changes in a protein's structure may also alter its function. Hydrogen bonds that stabilize protein shape may be disrupted by shifts in pH or temperature, or exposure to detergent or some salts. This causes **denaturation**, which means the protein loses its shape, and so loses its function. **Prion** diseases are a fatal consequence of misfolded proteins.

Section 3.6 **Nucleotides** are monomers of **DNA** and **RNA**, which are **nucleic acids**. Some nucleotides have additional roles. **ATP**, for example, is an important energy carrier in cells. The order of nucleotides in a DNA molecule is biological information. Different types of RNAs interact with DNA in protein synthesis.

Application All life consists of the same kinds of molecules. Seemingly small differences in the way those molecules are put together can have big effects inside a living organism. *Cis* and *trans* bonds in fatty acid tails are examples.

SELF-ASSESSMENT
ANSWERS IN APPENDIX VII

1. Organic molecules consist mainly of _____ atoms.
   a. carbon   c. carbon and hydrogen
   b. carbon and oxygen   d. carbon and nitrogen

2. Each carbon atom can bond with as many as _____ other atom(s).

3. _____ groups are the "acid" part of amino acids and fatty acids.
   a. Hydroxyl (—OH)   c. Methyl (—CH$_3$)
   b. Carboxyl (—COOH)   d. Phosphate (—PO$_4$)

4. _____ is a simple sugar (a monosaccharide).
   a. Ribose   c. Starch
   b. Sucrose   d. all are monosaccharides

5. Unlike saturated fats, the fatty acid tails of unsaturated fats incorporate one or more _____ .
   a. phosphate groups   c. double bonds
   b. glycerols   d. single bonds

6. Is this statement true or false? Unlike saturated fats, all unsaturated fats are beneficial to health because their fatty acid tails kink and do not pack together.

7. Steroids are among the lipids with no _____ .
   a. double bonds   c. hydrogens
   b. fatty acid tails   d. carbons

8. Name three kinds of carbohydrates that can be built using only glucose monomers.

9. Which of the following is a class of molecules that encompasses all of the other molecules listed?
   a. triglycerides   c. waxes   e. lipids
   b. fatty acids   d. steroids   f. phospholipids

10. _____ are to proteins as _____ are to nucleic acids.
    a. Sugars; lipids   c. Amino acids; hydrogen bonds
    b. Sugars; proteins   d. Amino acids; nucleotides

11. A denatured protein has lost its _____ .
    a. hydrogen bonds   c. function
    b. shape   d. all of the above

12. A _____ is an example of protein secondary structure.
    a. barrel   c. domain
    b. polypeptide   d. helix

13. In the following list, identify the carbohydrate, the fatty acid, the amino acid, and the polypeptide:
    a. methionine-valine-proline-leucine-serine
    b. $C_6H_{12}O_6$
    c. $NH_2—CHR—COOH$
    d. $CH_3(CH_2)_{16}COOH$

14. Match the molecules with the best description.
    ___ wax   a. sugar storage in plants
    ___ starch   b. richest energy source
    ___ triglyceride   c. water-repellent secretions
    ___ sucrose   d. animal muscle energy source
    ___ glycogen   e. a disaccharide

15. Match each polymer with the components.
    ___ protein   a. phosphate, fatty acids
    ___ phospholipid   b. amino acids, sugars
    ___ glycoprotein   c. glycerol, fatty acids
    ___ triglyceride   d. nucleotides
    ___ nucleic acid   e. glucose only
    ___ lipoprotein   f. lipids, amino acids
    ___ cellulose   g. amino acids

CLASS ACTIVITY
CRITICAL THINKING

1. Lipoprotein particles are relatively large, spherical clumps of protein and lipid molecules (see **Figure 3.14**) that circulate in the blood of mammals. They are like suitcases that move cholesterol, fatty acid remnants, triglycerides, and phospholipids from one place to another in the body. Given what you know about lipids and their solubility in water, which of the four kinds of lipids would you predict to be on the outside of a lipoprotein clump, bathed in the water-based fluid portion of blood?

2. In 1976, a team of chemists in the United Kingdom was developing new insecticides by modifying sugars with chlorine (Cl_2), phosgene (Cl_2CO), and other toxic gases. One young member of the team misunderstood his verbal instructions to "test" a newly made candidate substance. He thought he had been told to "taste" it. Luckily for him, the substance was not toxic, but it was very sweet. It became the food additive sucralose.

Sucralose has three chlorine atoms substituted for three hydroxyl groups of sucrose (table sugar). It binds so strongly to the sweet-taste receptors on the tongue that the human brain perceives it as 600 times sweeter than sucrose. Sucralose was originally marketed as an artificial sweetener called Splenda®, but it is now available under several other brand names.

EFFECTS OF DIETARY FATS ON LIPOPROTEIN LEVELS Cholesterol that is made by the liver or that enters the body from food cannot dissolve in blood, so it is carried through the bloodstream in lipoprotein particles. Low-density lipoprotein (LDL) particles carry cholesterol to body tissues such as artery walls, where they can form deposits associated with cardiovascular disease. Thus, LDL is often called "bad" cholesterol. High-density lipoprotein (HDL) particles (**Figure 3.14**) carry cholesterol away from tissues to the liver for disposal, so HDL is often called "good" cholesterol. In 1990, Ronald Mensink and Martijn Katan published a study that tested the effects of different dietary fats on blood lipoprotein levels. Their results are shown in **Figure 3.17**.

1. In which group was the level of LDL ("bad" cholesterol) highest?

2. In which group was the level of HDL ("good" cholesterol) lowest?

3. An elevated risk of heart disease has been correlated with increasing LDL-to-HDL ratios. Which group had the highest LDL-to-HDL ratio?

4. Rank the three diets from best to worst according to their potential effect on heart disease.

	Main Dietary Fats			
	cis fatty acids	*trans* fatty acids	saturated fats	optimal level
LDL	103	117	121	<100
HDL	55	48	55	>40
ratio	1.87	2.44	2.2	<2

Figure 3.17 Effect of diet on lipoprotein levels. Researchers placed 59 men and women on a diet in which 10 percent of their daily energy intake consisted of *cis* fatty acids, *trans* fatty acids, or saturated fats. Blood LDL and HDL levels were measured after three weeks on the diet; averaged results are shown in mg/dL (milligrams per deciliter of blood). All subjects were tested on each of the diets. The ratio of LDL to HDL is also shown.

sucrose   sucralose

Researchers investigated whether the body recognizes sucralose as a carbohydrate. They began by feeding sucralose labeled with ^{14}C to volunteers. Analysis of the radioactive molecules in the volunteers' urine and feces showed that 92.8 percent of the sucralose passed through the body without being altered. Many people are worried that the chlorine atoms impart toxicity to sucralose. How would you respond to that concern?

for additional quizzes, flashcards, and other study materials
► ACCESS MINDTAP AT WWW.CENGAGEBRAIN.COM

CREDITS: (18) Source, Mensink RP, Katan MB., "Effect of dietary trans fatty acids on high-density and low-density lipoprotein cholesterol levels in healthy subjects." NEJM 323(7):439–45, 1990.

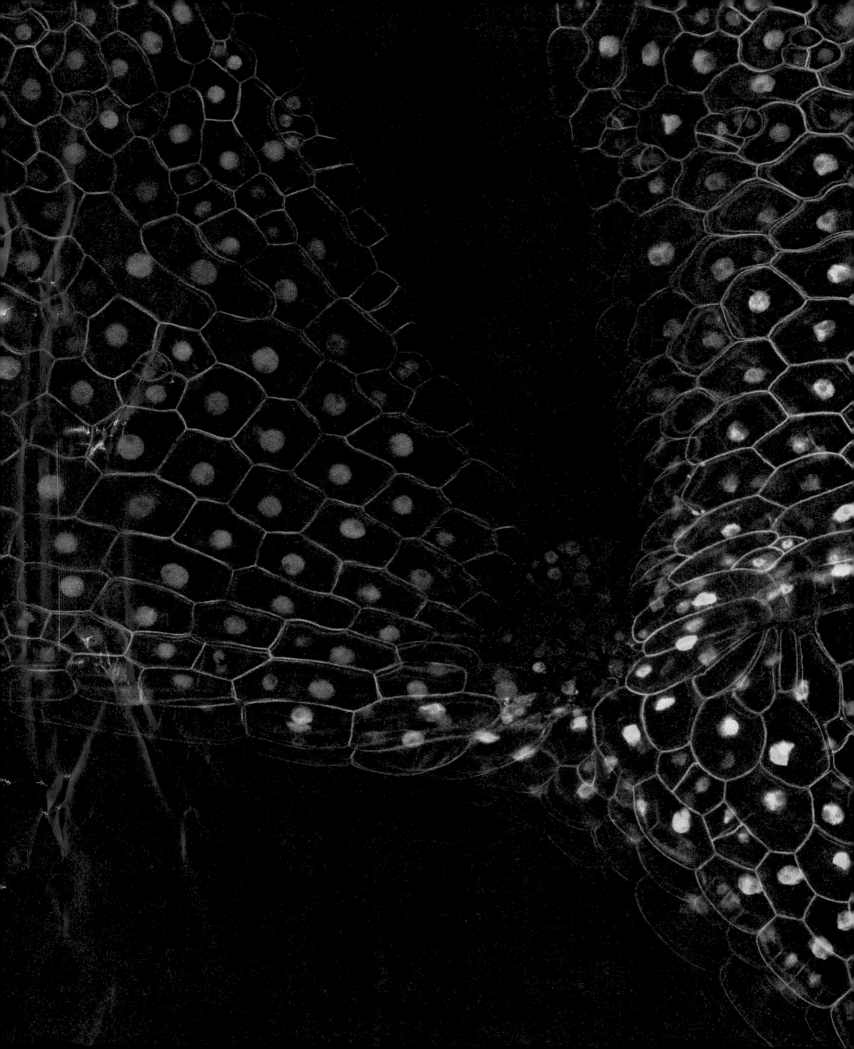

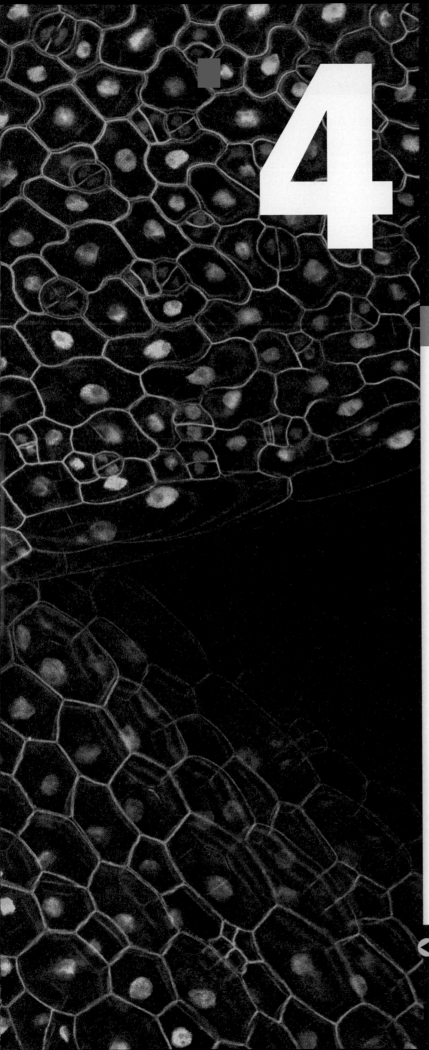

Cell Structure

4

Links to Earlier Concepts

Reflect on the overview of life's levels of organization in Section 1.1. In this chapter, you will see how the properties of lipids (3.3) give rise to cell membranes; consider the location of DNA (3.6) and the sites where carbohydrates are built and broken apart (3.1, 3.2); and expand your understanding of the vital roles of proteins in cell function (3.4, 3.5). You will also revisit the philosophy of science (1.5, 1.8) and tracers (2.1).

Core Concepts

Organisms exchange matter and energy with the environment in order to grow, maintain themselves, and reproduce.

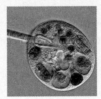

All cells acquire water and nutrients for building new molecules, and all eliminate metabolic wastes to maintain their internal environment. The behavior of matter and energy have shaped the way such tasks are carried out. All organisms have areas or compartments in which functions related to energy and matter occur.

Interactions among biological systems give rise to complex properties.

When components of a biological system interact, they enable properties not found in the individual components alone. Cellular functioning arises from and depends on interactions among cell parts specialized for different tasks. Some cells have structures that allow them to carry out special processes such as photosynthesis.

Cells create and maintain internal environments that differ from external environments.

A membrane separates a cell from the external environment and allows it to maintain internal conditions that support life. Internal membranes that partition the cytoplasm into specialized regions maximize cellular efficiency. Selective permeability required for such separation arises from membrane structure. Other membrane features allow cells to sense and respond appropriately to their environment.

◀ Each cell making up this seedling contains a nucleus (orange spots), which is the defining characteristic of eukaryotes. A plasma membrane (blue-green) surrounds each cell.

LEARNING OBJECTIVES

List the four generalizations that constitute cell theory.

Describe the three components that all cells have.

Explain how the surface-to-volume ratio limits cell size.

CELL THEORY

Before microscopes were invented, no one knew that cells existed because nearly all are invisible to the naked eye. By 1665, Antoni van Leeuwenhoek had constructed an early microscope that revealed tiny organisms in rainwater, insects, fabric, sperm, feces, and other samples. In scrapings of tartar from his teeth, Leeuwenhoek saw "many very small animalcules, the motions of which were very pleasing to behold." He (incorrectly) assumed that movement defined life, and (correctly) concluded that the moving "beasties" he saw were alive. Leeuwenhoek might have been less pleased to behold his animalcules if he had grasped the implications of what he saw: Our world, and our bodies, teem with microbial life.

Today we know that a cell carries out metabolism and homeostasis, and reproduces either on its own or as part of a larger organism. By this definition, each cell is alive even if it is part of a multicelled body, and all living organisms consist of one or more cells. We also know that cells reproduce by dividing, so it follows that all existing cells must have arisen by division of other cells (later chapters discuss processes by which cells reproduce). As a cell divides, it passes its hereditary material—DNA—to offspring. Taken together, these generalizations constitute the **cell theory**, which is one of the foundations of modern biology (**Table 4.1**).

TABLE 4.1

Cell Theory

1. Each organism consists of one or more cells.

2. The cell is the structural and functional unit of all organisms. A cell is the smallest unit of life, individually alive even as part of a multicelled organism.

3. All living cells arise by division of preexisting cells.

4. Cells contain hereditary material (DNA), which they pass to their offspring when they divide.

COMPONENTS OF ALL CELLS

Cells vary in shape and in what they do, but all share certain organizational and functional features: a plasma membrane, cytoplasm, and DNA (**Figure 4.1**). A **plasma membrane** separates the cell from the external environment (**Figure 4.2**). Like all other cell membranes, it consists mainly of a lipid bilayer, and it is selectively permeable, which means that only certain materials can cross it. Thus, a plasma membrane controls exchanges between the cell and its environment. Many different proteins embedded in a lipid bilayer or attached to one of its surfaces carry out particular membrane functions.

The plasma membrane encloses **cytoplasm**, a jellylike mixture of water, sugars, ions, and proteins. Most of a cell's metabolism occurs in cytoplasm, and the cell's internal components, including organelles, are suspended in it. **Organelles** are structures that carry out special metabolic functions inside a cell. Those with membranes compartmentalize substances and activities.

Every cell starts out life with DNA. In nearly all bacteria and archaea, that DNA is suspended directly in cytoplasm.

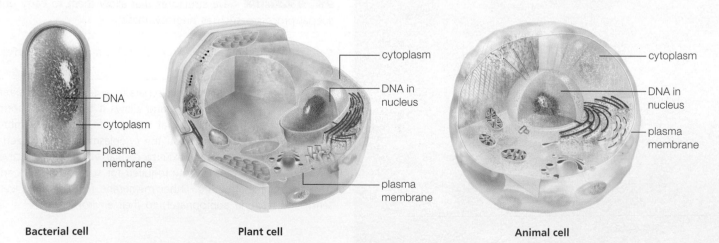

Bacterial cell

Plant cell

Animal cell

Figure 4.1 Overview of the general organization of a cell.
All cells start out life with a plasma membrane, cytoplasm, and DNA. Archaea are similar to bacteria in overall structure; both are typically much smaller than eukaryotic cells. If the cells depicted here had been drawn to the same scale, the bacterium would be about this big:

By contrast, the DNA of a eukaryotic cell is contained in a **nucleus** (plural, nuclei), an organelle with a double membrane. All protists, fungi, plants, and animals are eukaryotes. Some of these organisms are independent, free-living cells; others consist of many cells working together as a body.

CONSTRAINTS ON CELL SIZE

A living cell must exchange substances with its environment at a rate that keeps pace with its metabolism. These exchanges occur across the plasma membrane, which can handle only so many exchanges at a time. The rate of exchange across a plasma membrane depends on its surface area: The bigger it is, the more substances can cross it during a given interval. Thus, cell size is limited by a physical relationship called the **surface-to-volume ratio**. By this ratio, an object's volume increases with the cube of its diameter, but its surface area increases only with the square.

Let's apply the surface-to-volume ratio to a cell. As **Figure 4.3** shows, when a round cell expands in diameter, its volume increases faster than its surface area does. Imagine that the cell expands until it is four times its original diameter. The cell's volume has increased 64 times (4^3), but its surface area has increased only 16 times (4^2). Each unit of plasma membrane must now handle exchanges with four times as much cytoplasm ($64 \div 16 = 4$). If the cell gets too big, the inward flow of nutrients and the outward flow of wastes across that membrane will not be fast enough to keep the cell alive.

The surface-to-volume ratio also constrains cell form in colonial and multicelled organisms. For example, cells of some colonial algae attach end to end to form a strand (**Figure 4.4**). This arrangement allows each cell to interact directly with the environment. As another example, consider that some muscle cells run the length of your thigh. Although surprisingly long, each of these cells is very thin, so it exchanges substances efficiently with fluids in the surrounding tissue.

---

cell theory Theory that all organisms consist of one or more cells, which are the basic unit of life; all cells come from division of preexisting cells; and all cells pass hereditary material to offspring.

cytoplasm (SITE-uh-plaz-um) Semifluid substance enclosed by a cell's plasma membrane.

nucleus (NEW-klee-us) Of a eukaryotic cell, organelle with a double membrane that holds the cell's DNA.

organelle (or-guh-NEL) Structure that carries out a specialized metabolic function inside a cell.

plasma membrane (PLAZ-muh) Membrane that encloses a cell and separates it from the external environment.

surface-to-volume ratio A relationship in which the volume of an object increases with the cube of the diameter, and the surface area increases with the square.

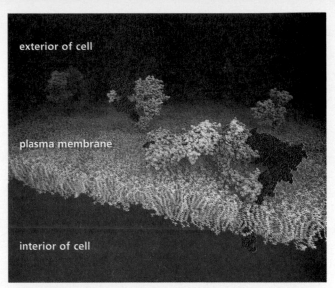

Figure 4.2 The plasma membrane. A plasma membrane separates a cell from its external environment. Proteins (in color) embedded in the lipid bilayer (gray) carry out special membrane functions.

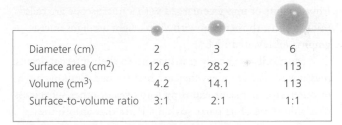

Diameter (cm)	2	3	6
Surface area (cm²)	12.6	28.2	113
Volume (cm³)	4.2	14.1	113
Surface-to-volume ratio	3:1	2:1	1:1

Figure 4.3 Examples of surface-to-volume ratio. This physical relationship between volume and surface area limits the size of cells and influences their shape.

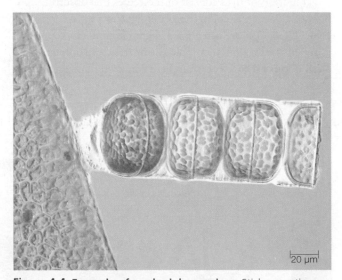

20 µm

Figure 4.4 Example of a colonial organism. Sticky secretions hold these pill-shaped algal cells together end to end, forming a long strand. The arrangement allows each cell to exchange substances directly with the surrounding water. Secretions also anchor the strand to a solid surface (such as the plant on the left).

CREDIT: (2) Courtesy of © Johannes Kästner, Universität Stuttgart; (4) © Frank Fox Shaw & Todd, Inc.

4.2 How We See Cells

Explain how a light microscope works.

Describe the use of fluorescent dyes in microscopy.

Explain the difference between a transmission electron micrograph (TEM) and a scanning electron micrograph (SEM).

TABLE 4.2

Equivalent Units of Length

Unit	Equivalent Meter(s)	Inch(es)
kilometer	1,000	39,370
meter (m)	1	39.37
centimeter (cm)	1/100	0.4
millimeter (mm)	1/1000	0.04
micrometer (µm)	1/1,000,000	0.00004
nanometer (nm)	1/1,000,000,000	0.00000004

Most cells are between 1–100 micrometers in diameter, much smaller than the unaided human eye can perceive (**Figure 4.5** and **Table 4.2**). We use microscopes to observe cells and other objects in the micrometer range of size.

Visible light illuminates a sample in a light microscope. As you will learn in Chapter 6, all light travels in waves. This property of light causes it to bend when passing through a curved glass lens. Inside a light microscope, such lenses focus light that passes through a specimen, or bounces off of one, into a magnified image (**Figure 4.6A**). Microscopes that use polarized light can yield images in which the edges of some structures appear in three-dimensional relief (**Figure 4.6B**). Photographs of images enlarged with a microscope are called micrographs; those taken with visible light are light micrographs, abbreviated LM.

Most cells are nearly transparent, so their internal details may not be visible unless they are first stained. Staining a cell or other sample means exposing it to dyes or other substances that only some of its parts soak up. Parts that absorb the most dye appear darkest. Staining results in an increase in contrast (the difference between light and dark) that allows us to see a greater range of detail.

Fluorescent dyes consist of molecules that absorb light of a particular color, then emit light of a different color. These dyes are often used as tracers in microscopy to pinpoint the location of a molecule or structure of interest in a cell (**Figure 4.6C**). For example, a common fluorescent dye

called DAPI binds preferentially to DNA. This dye emits blue light after absorbing ultraviolet (UV) light. When cells are stained with DAPI and then illuminated with UV light, their DNA glows blue. A magnified image of the emitted light reveals the location of the DNA in each cell.

Structures smaller than about 200 nanometers across appear blurry under light microscopes. To observe objects of this size range clearly, we would have to switch to an electron microscope. There are two types of electron microscope; both use magnetic fields as lenses to focus a beam of electrons onto a sample. A transmission electron microscope directs electrons through a thin specimen. The specimen's internal details appear as shadows in the resulting image, which is called a transmission electron micrograph, or TEM (**Figure 4.6D**). A scanning electron microscope directs electrons back and forth across the surface of a specimen that has been coated with a thin layer of gold or other metal. The irradiated metal emits electrons and x-rays, which are converted into an image (a scanning electron micrograph, or SEM) of the surface (**Figure 4.6E**). SEMs and TEMs are always black and white; colored versions have been digitally altered to highlight specific details.

Figure 4.5 Relative sizes. The diameter of most cells is between 1 and 100 micrometers. See also Table 4.2 and Appendix VI, Units of Measure.

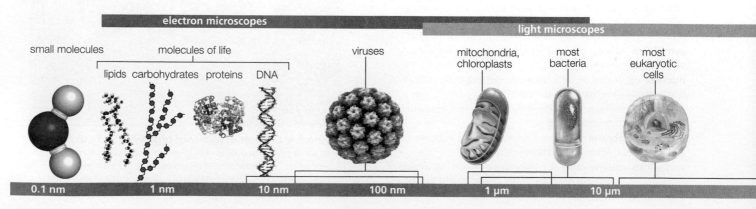

electron microscopes | light microscopes

small molecules | molecules of life | viruses | mitochondria, chloroplasts | most bacteria | most eukaryotic cells

lipids carbohydrates proteins DNA

0.1 nm | 1 nm | 10 nm | 100 nm | 1 µm | 10 µm

CREDIT: (5) Virus, CDC.

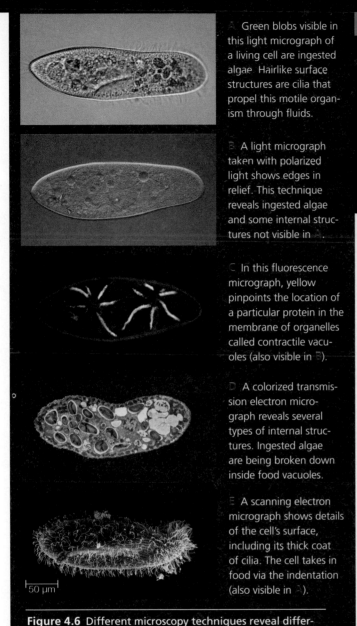

A Green blobs visible in this light micrograph of a living cell are ingested algae. Hairlike surface structures are cilia that propel this motile organism through fluids.

B A light micrograph taken with polarized light shows edges in relief. This technique reveals ingested algae and some internal structures not visible in A.

C In this fluorescence micrograph, yellow pinpoints the location of a particular protein in the membrane of organelles called contractile vacuoles (also visible in B).

D A colorized transmission electron micrograph reveals several types of internal structures. Ingested algae are being broken down inside food vacuoles.

E A scanning electron micrograph shows details of the cell's surface, including its thick coat of cilia. The cell takes in food via the indentation (also visible in A).

50 μm

Figure 4.6 Different microscopy techniques reveal different characteristics. All of these micrographs show the same organism, a protist called *Paramecium* that is about 250 μm long.

Dr. Aydogan Ozcan

National Geographic Explorer Aydogan Ozcan grew up in Turkey, where many rural areas had no access to medicine. Today, his UCLA team pioneers global health solutions using one of the most common forms of technology available—the smart phone. "Here's this extremely inexpensive, yet advanced, technology that's in use almost everywhere," he says. Using readily available parts that cost about $10, Ozcan builds adapters that transform a smart phone into a mobile medical lab with the capability to test and diagnose diseases remotely.

Conventional microscopes, the mainstay of diagnosis for centuries, are impractical on a global level. What's more, because technicians in remote areas may be poorly trained, they often interpret images inaccurately. "You can't even assume there will be consistent electricity in rural clinics or villages," he says. Ozcan's invention solves all of these problems by arming cell phones with sophisticated algorithms that do the interpreting. To tackle the most expensive part of microscopes—lenses—his team simply eliminated them. Ozcan's attachment uses the phone's camera to capture an image of a patient's blood or saliva, essentially turning the phone into a lens-free microscope able to resolve structures smaller than one micron.

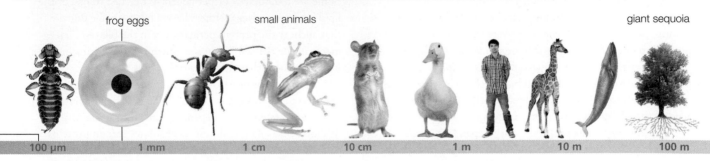

human eye (no microscope)

frog eggs · small animals · giant sequoia

100 μm · 1 mm · 1 cm · 10 cm · 1 m · 10 m · 100 m

4.3 Cell Membrane Structure

A plasma membrane physically separates a cell's external and internal environments, but that is not its only function. For example, Section 4.1 mentioned that a plasma membrane allows some substances, but not others, to cross it. Membranes around organelles do this too. We return to membrane functions in Chapter 5; here, we explore the structure that gives rise to these functions.

THE FLUID MOSAIC MODEL

The foundation of all cell membranes is a lipid bilayer that consists mainly of phospholipids. Remember from Section 3.3 that a phospholipid has a phosphate-containing head and two fatty acid tails. The head is polar and hydrophilic, so it interacts with water molecules. The two long hydrocarbon tails are nonpolar and hydrophobic, so they do not interact with water molecules. As a result of these opposing properties, phospholipids mixed with water will spontaneously organize themselves into lipid bilayer sheets or bubbles, with hydrophobic tails together, hydrophilic heads facing the watery surroundings (**Figure 4.7A**). A cell's basic structure is essentially a lipid bilayer bubble filled with fluid (left).

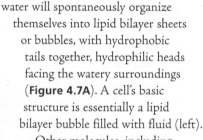

fluid

Other molecules, including cholesterol, proteins, glycoproteins, and glycolipids, are embedded in or attached to the lipid bilayer of a cell membrane. Many of these molecules move around the membrane more or less freely. The **fluid mosaic model** describes a membrane as a two-dimensional liquid of mixed composition. The "mosaic" part of the name comes from the many molecules, and many different types of molecules, in the membrane. Membrane fluidity occurs because phospholipids in the bilayer are not chemically bonded to one another; they stay organized as a result of collective hydrophobic and hydrophilic attractions. These interactions are, on an individual basis, relatively weak. Thus, individual phospholipids in the bilayer drift sideways and spin around their long axis, and their tails wiggle.

A cell membrane's properties vary depending on the types and proportions of molecules composing it. For exam-

A In a watery fluid, phospholipids spontaneously line up into two layers: the hydrophobic tails cluster together, and the hydrophilic heads face outward, toward the fluid. This lipid bilayer forms the framework of all cell membranes. Many types of proteins intermingle among the lipids; a few that are typical of plasma membranes are shown on the opposite page.

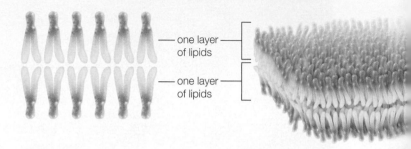

— one layer of lipids

— one layer of lipids

Figure 4.7 Cell membrane structure.

A Organization of phospholipids in cell membranes.

B–E Examples of common membrane proteins. Many membrane proteins have an extremely complex structure; for clarity, they are often modeled as blobs or geometric shapes.

ple, membrane fluidity decreases with increasing cholesterol content. A membrane's fluidity also depends on the length and saturation of its phospholipids' fatty acid tails. Archaea do not even use fatty acids to build their phospholipids. Instead, they use molecules with reactive side chains, so the tails of archaeal phospholipids form covalent bonds with one another. As a result of this rigid crosslinking, archaeal phospholipids do not drift, spin, or wiggle in a bilayer. This makes the membranes of archaea stiffer than those of bacteria or eukaryotes, a characteristic that may help these cells survive in extreme habitats.

PROTEINS ADD FUNCTION

Many types of proteins are associated with a cell membrane. Some are temporarily or permanently attached to one of the lipid bilayer's surfaces. Others have a hydrophobic domain that anchors the protein permanently in the bilayer. Filaments inside the cell fasten some membrane proteins in place, including those that cluster as rigid pores.

Each type of protein in a membrane imparts a specific function to it (**Table 4.3**). Thus, different cell membranes can carry out different tasks depending on which proteins they include. A plasma membrane incorporates certain proteins that no internal cell membrane has, so it has functions that

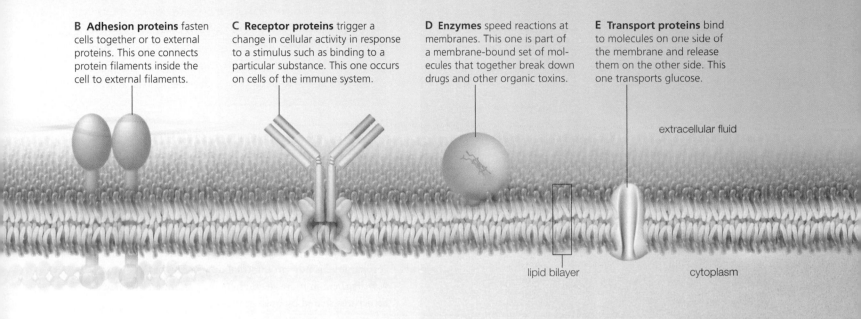

B Adhesion proteins fasten cells together or to external proteins. This one connects protein filaments inside the cell to external filaments.

C Receptor proteins trigger a change in cellular activity in response to a stimulus such as binding to a particular substance. This one occurs on cells of the immune system.

D Enzymes speed reactions at membranes. This one is part of a membrane-bound set of molecules that together break down drugs and other organic toxins.

E Transport proteins bind to molecules on one side of the membrane and release them on the other side. This one transports glucose.

extracellular fluid

lipid bilayer

cytoplasm

no other membrane does. For example, cells stay organized in animal tissues because **adhesion proteins** in their plasma membranes fasten them together and hold them in place (**Figure 4.7B**). Adhesion proteins are more than just sticky molecules, however; they also provide a cell with information about its position relative to other cells or structures.

Plasma membranes and some internal membranes incorporate **receptor proteins**, which trigger a change in the cell's activities in response to a stimulus (**Figure 4.7B**). Each type of receptor protein receives a particular stimulus such as a hormone binding to it. The response triggered by a receptor may involve (for example) metabolism, movement, division, or even cell death.

All cell membranes incorporate enzymes (**Figure 4.7D**). Some of these enzymes act on other proteins or lipids that are part of the membrane; others work in series and use the membrane as a scaffold. All membranes also have **transport proteins**, which move specific substances across the bilayer (**Figure 4.7E**). Transport proteins are important because lipid bilayers are impermeable to most substances, including the ions and polar molecules that cells must take in and expel on a regular basis (Chapter 5 returns to this topic).

TABLE 4.3

Common Membrane Proteins

Category	Function	Examples
Adhesion protein	Holds cells together in animal tissues and helps them sense their position.	Integrins; cadherins
Receptor protein	Initiates change in a cell activity by responding to an outside signal (e.g., by binding to a hormone or absorbing light energy).	Insulin receptor; B cell receptor
Enzyme	Speeds a specific reaction. Membranes provide a relatively stable reaction site for enzymes that work in series with other molecules.	Cytochrome P450
Transport protein	Moves or allows specific ions or molecules across a membrane. Some types require an energy input, as from ATP.	Calcium pump; glucose transporter

adhesion protein Plasma membrane protein that helps cells stick together in animal tissues.

fluid mosaic model Model of a cell membrane as a two-dimensional fluid of mixed composition.

receptor protein Membrane protein that triggers a change in cell activity in response to a stimulus such as binding a hormone.

transport protein Protein that moves specific ions or molecules across a membrane.

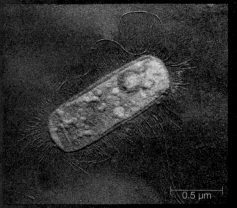

Escherichia coli is a common bacterial inhabitant of human intestines. Short, hair-like structures are pili; longer ones are flagella. This one is harmless; others can cause disease in humans.

0.5 μm

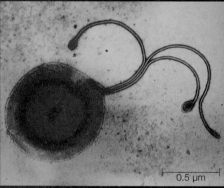

Dermocarpa is a type of cyanobacteria, an ancient lineage of photosynthetic bacteria. Photosynthesis occurs at internal membranes (green). The dark, multisided structures are carboxysomes, protein-enclosed organelles that assist photosynthesis.

1 μm

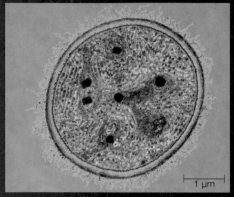

Helicobacter pylori is a spiral-shaped bacterium that can cause ulcers and cancer when it infects the lining of the stomach. It can take on a ball-shaped form (shown) that offers protection from environmental challenges such as exposure to an antibiotic.

0.5 μm

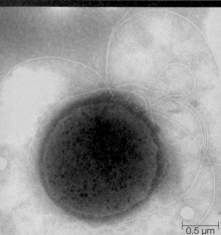

The archaeon *Thermococcus gammatolerans* lives under extreme conditions of salt, temperature, and pressure. It is by far the most radiation-resistant organism ever discovered, capable of withstanding thousands of times more radiation than humans can.

0.5 μm

Figure 4.8 Some representatives of bacteria (–) and an archaeon ( ).

All bacteria and archaea are single-celled (**Figure 4.8**), but individual cells of many species cluster in filaments or colonies. Outwardly, cells of the two groups appear so similar that archaea were once presumed to be an unusual group of bacteria. Both were classified as prokaryotes, a word that means "before the nucleus." By 1977, it had become clear that archaea are more closely related to eukaryotes than to bacteria, so they were given their own separate domain. The term "prokaryote" is now an informal designation only. Chapter 19 revisits them in more detail; here we present an overview of structures shared by both groups.

STRUCTURAL FEATURES

Prokaryotic cells have a less elaborate internal framework than eukaryotic cells (**Figure 4.9**). Protein filaments under the plasma membrane reinforce the cell's shape and provide a scaffolding for internal structures. The cytoplasm ❶ contains many **ribosomes**—organelles upon which proteins are assembled—and in some species, additional organelles. Cytoplasm also contains **plasmids** ❷, which are small circles of DNA that carry a few genes (units of inheritance). Plasmid genes can provide advantages such as resistance to antibiotics; the cell's essential genetic information occurs on one or two circular DNA molecules located in an irregularly shaped region of cytoplasm called the **nucleoid** ❸. In a few species, the nucleoid is enclosed by a membrane. Other internal membranes of some prokaryotes carry out special processes such as photosynthesis.

Like all cells, bacteria and archaea have a plasma membrane ❹. Almost all prokaryotes have a permeable layer of secreted material called a **cell wall** ❺ that encloses the plasma membrane. A wall protects the cell and supports its shape. Cell walls are not membranes and they do not consist of lipids; rather, they are crystalline arrays of other molecules. Ions and other solutes cross them easily (unlike lipid bilayers). Tightly packed proteins make up the durable cell wall of most archaea. By contrast, bacterial cell walls consist of peptidoglycan, a polymer of peptides and sugars that is unique to bacteria. In many species of bacteria, including the ones shown in **Figure 4.8**, a second membrane surrounds the cell wall. Enclosing this second membrane (or the cell wall)

CREDITS: (8A) Biophoto Associates/Science Photo Library; (8B) Dr. Dennis Kunkel/Visuals Unlimited, Inc.; (8C) Biomedical Imaging Unit, Southampton General Hospital/Science Photo Library; (8D) Archivo Angels Tapias y Fabrice Confalonieri.

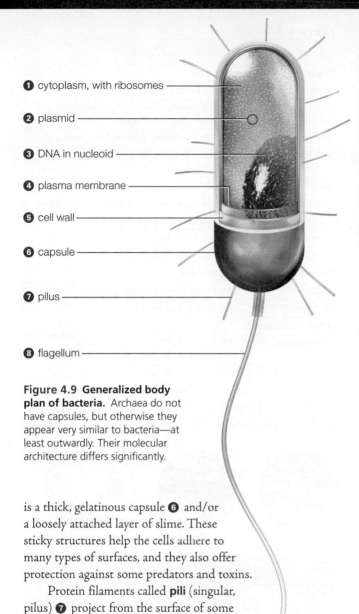

① cytoplasm, with ribosomes

② plasmid

③ DNA in nucleoid

④ plasma membrane

⑤ cell wall

⑥ capsule

⑦ pilus

⑧ flagellum

Figure 4.9 Generalized body plan of bacteria. Archaea do not have capsules, but otherwise they appear very similar to bacteria—at least outwardly. Their molecular architecture differs significantly.

is a thick, gelatinous capsule ⑥ and/or a loosely attached layer of slime. These sticky structures help the cells adhere to many types of surfaces, and they also offer protection against some predators and toxins.

Protein filaments called **pili** (singular, pilus) ⑦ project from the surface of some prokaryotes. Pili help these cells move across or cling to surfaces. One kind, a "sex" pilus,

---

biofilm Community of microorganisms living within a shared mass of secreted slime.

cell wall Rigid but permeable structure that surrounds the plasma membrane of some cells.

flagellum (fluh-JEL-um) Long, slender cellular structure used for motility.

nucleoid (NEW-klee-oyd) Of a bacterium or archaeon, region of cytoplasm where the DNA is concentrated.

pilus (PIE-luss) A protein filament that projects from the surface of some prokaryotic cells.

plasmid (PLAZ-mid) Small circle of DNA in some bacteria and archaea.

ribosome (RYE-buh-sohm) Organelle of protein synthesis.

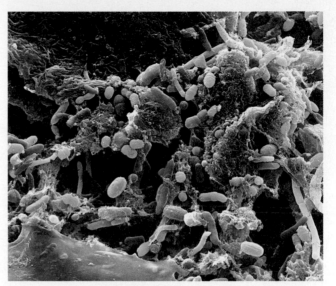

Figure 4.10 Bacteria of the human large intestine. An SEM of human feces reveals a few of the more than 30,000 species of bacteria that normally inhabit our large intestine. These organisms form a biofilm that protects intestinal surfaces from colonization by harmful species, while producing needed vitamins and helping us digest our food.

attaches to another cell and then shortens. The attached cell is reeled in, and DNA is transferred from one cell to the other through the pilus. Many prokaryotes also have one or more flagella projecting from their surface. **Flagella** (singular, flagellum) are long, slender cellular structures used for motion ⑧. A prokaryotic flagellum rotates like a propeller that drives the cell through fluid habitats.

BIOFILMS

Bacterial cells often live so close together that an entire community shares a layer of secreted slime. A communal living arrangement in which single-celled organisms occupy a shared mass of slime is called a **biofilm**. A biofilm is often attached to a solid surface, and may include bacteria, algae, fungi, protists, and/or archaea. Participating in a biofilm allows the cells to linger in a favorable spot rather than be swept away by fluid currents, and to reap the benefits of living communally. For example, rigid or netlike secretions of some species serve as permanent scaffolding for others; species that break down toxic chemicals allow more sensitive ones to thrive in habitats that they could not withstand on their own; and waste products of some serve as raw materials for others. The human gastrointestinal tract hosts a large and beneficial biofilm (**Figure 4.10**). Other biofilms, including the dental plaque that forms on our teeth, can be harmful to human health.

Identify some cellular components unique to eukaryotes.

Explain the function of membranes that enclose organelles.

List the functions of the cell nucleus and explain how these functions arise from its structure.

GENERAL FEATURES OF EUKARYOTIC CELLS

Every eukaryotic cell starts out life with a nucleus, and a typical one has many additional organelles. Ribosomes are found in all cells. However, organelles enclosed by lipid bilayer membranes, including endoplasmic reticulum, Golgi bodies, chloroplasts, and mitochondria, are characteristic of eukaryotes (**Table 4.4** and **Figure 4.11**). An enclosing membrane allows these structures to regulate the types and amounts of substances that enter and exit. Through this type of control, an organelle can maintain a special internal environment that allows it to carry out a particular function—for example, isolating toxic or sensitive substances from the rest of the cell, storing large amounts of water, transporting substances through cytoplasm, maintaining fluid balance, or providing a favorable environment for a special process.

TABLE 4.4

Some Components of Eukaryotic Cells

Organelles with membranes	
Nucleus	Protects and controls access to DNA
Endoplasmic reticulum (ER)	Makes and modifies new polypeptides and lipids, among other tasks
Golgi body	Modifies and sorts polypeptides and lipids
Vesicle	Transports, stores, or breaks down substances
Mitochondrion	Makes ATP by glucose breakdown
Chloroplast	Makes sugars (in plants and some protists)
Lysosome	Intracellular digestion
Peroxisome	Breaks down fatty acids, amino acids, toxins
Vacuole	Stores, breaks down substances
Organelles without membranes	
Ribosome	Assembles polypeptides
Centriole	Anchors cytoskeleton
Other components	
Cytoskeleton	Contributes to cell shape, internal organization, movement

In this section, we detail the nucleus, the defining characteristic of eukaryotes. The remaining sections of the chapter discuss the functions of other eukaryotic organelles.

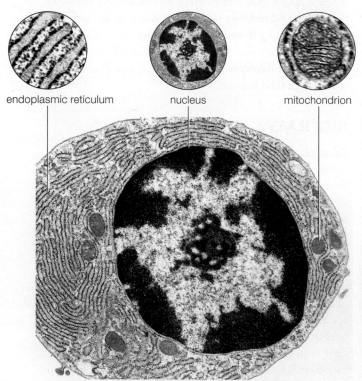

endoplasmic reticulum   nucleus   mitochondrion

An animal cell. This is a white blood cell from a guinea pig.

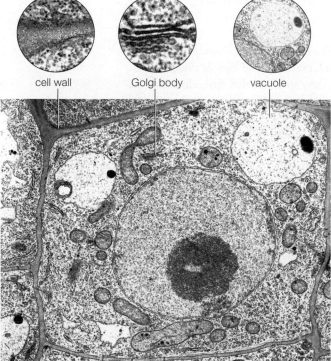

cell wall   Golgi body   vacuole

A plant cell. This is a cell from the root of a thale cress plant.

Figure 4.11 Some components of typical eukaryotic cells.

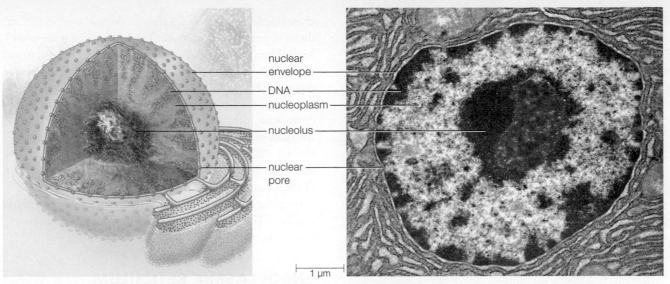

Figure 4.12 **The cell nucleus.** The TEM shows the nucleus of a pancreas cell from a mouse.

THE NUCLEUS

The cell nucleus (**Figure 4.12**) serves two important functions. First, it keeps the cell's genetic material—its one and only copy of DNA—safe from metabolic processes that might damage it. The DNA is suspended in **nucleoplasm**, a viscous fluid similar to cytoplasm, that fills the nucleus. Isolated in its own compartment, DNA stays separated from the bustling activity of the cytoplasm. Second, a nucleus controls the passage of certain molecules across its membrane. This function is carried out by the **nuclear envelope**, a special membrane that encloses the nucleus. A nuclear envelope consists of two lipid bilayers folded together (**Figure 4.13**). Proteins embedded in the two lipid bilayers assemble into thousands of tiny nuclear pores that span the envelope and form holes in it. (Some bacteria have a membrane enclosing their DNA, but this structure is not a nucleus because the membrane has no pores.) Proteins making up nuclear pores are anchored by the nuclear lamina, a dense mesh of fibrous proteins that supports the inner surface of the membrane.

As you will see in Chapter 5, large molecules, including RNA and proteins, cannot cross lipid bilayers on their own. Nuclear pores function as gateways for these molecules to enter and exit a nucleus. Protein synthesis offers an example

nuclear envelope A double membrane that constitutes the outer boundary of the nucleus. Pores in the membrane control which molecules can cross it.
nucleolus (new-KLEE-oh-luss) In a cell nucleus, a dense, irregularly shaped region where ribosomal subunits are assembled.
nucleoplasm (NEW-klee-oh-plaz-um) Viscous fluid enclosed by the nuclear envelope.

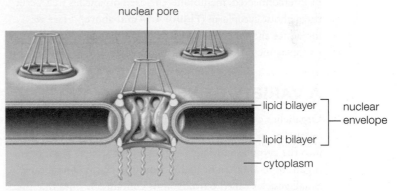

Figure 4.13 **The nuclear envelope.** The nuclear envelope consists of two lipid bilayers folded together as a single membrane. The envelope is studded with thousands of nuclear pores, each an organized cluster of membrane proteins. Nuclear pores selectively allow certain substances to cross the membrane.

of why this movement is important. Protein synthesis occurs in cytoplasm, and it requires the participation of many molecules of RNA. RNA is produced in the nucleus. Thus, RNA molecules must move from nucleus to cytoplasm, and they do so through nuclear pores. Proteins that participate in RNA synthesis must move in the opposite direction, because this process occurs in the nucleus. A cell can regulate the amounts and types of proteins it makes at a given time by selectively restricting the passage of certain molecules through nuclear pores. (Later chapters return to details of protein synthesis and controls over it.)

Depending on a cell's metabolic state, its nucleus contains one or more nucleoli. A **nucleolus** (plural, nucleoli) is a dense, irregularly shaped region of proteins and nucleic acid where subunits of ribosomes are produced.

4.6 The Endomembrane System

LEARNING OBJECTIVES

List the main components of the endomembrane system.

Name some different types of vesicles and describe their functions.

Explain the differences between rough and smooth endoplasmic reticulum.

Describe the function of Golgi bodies.

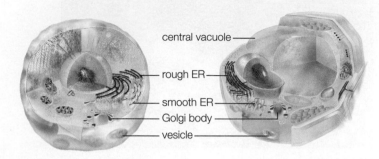

central vacuole

rough ER

smooth ER
Golgi body
vesicle

The **endomembrane system** is a multifunctional network of interconnected, membrane-enclosed organelles that occur throughout cytoplasm (**Figure 4.14** and above). Here we introduce the functions of its main components: vesicles, endoplasmic reticulum, and Golgi bodies.

A VARIETY OF VESICLES

Organelles called **vesicles** ❶ form by budding from other organelles or when a patch of plasma membrane sinks into the cytoplasm. Many types carry substances from one organelle to another, or to and from the plasma membrane. Small vesicles called **peroxisomes** contain enzymes that break down amino acids and long hydrocarbon chains of fatty acids, as well as toxic hydrogen peroxide and ammonia produced by

these reactions. Other toxins such as alcohols are also broken down in peroxisomes. Enzymes in vesicles called **lysosomes** break down cellular debris and wastes. Bacteria, cell parts, and other particles taken in by a cell are delivered to lysosomes for digestion (**Figure 4.15**).

Large, fluid-filled vesicles called **vacuoles** store or break down waste, debris, toxins, or food. Some cells have contractile vacuoles that expel excess water (several contractile vacuoles are visible in **Figure 4.6**). In plants, lysosome-like vesicles fuse to form a very large **central vacuole** that makes up most of the volume of the cell. Fluid pressure in a central vacuole keeps plant cells plump, so stems, leaves, and other plant parts stay firm. The central vacuole has additional functions in some cells.

ENDOPLASMIC RETICULUM

The nuclear envelope is often considered to be part of the endomembrane system. This is because the system of sacs and tubes that make up **endoplasmic reticulum** (**ER**) extends from the outer lipid bilayer of the nuclear envelope. The space between the nuclear envelope's two lipid bilayers is continuous with the space enclosed by the ER membrane.

Two kinds of ER, rough and smooth, are named for their appearance in electron micrographs. The membrane of rough ER is typically folded into flattened sacs, and has thousands of attached ribosomes that give it a "rough" appearance. These ribosomes make polypeptides that thread into the ER's interior as they are assembled ❷. The polypeptides take on their tertiary structure in the interior of rough ER, and many assemble with other polypeptides (Section 3.4). Cells that make, store, and secrete proteins have a lot of rough ER.

Figure 4.14 The endomembrane system.

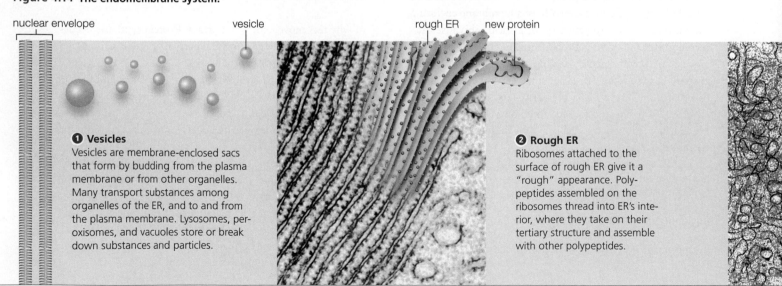

nuclear envelope

vesicle

rough ER   new protein

❶ Vesicles
Vesicles are membrane-enclosed sacs that form by budding from the plasma membrane or from other organelles. Many transport substances among organelles of the ER, and to and from the plasma membrane. Lysosomes, peroxisomes, and vacuoles store or break down substances and particles.

❷ Rough ER
Ribosomes attached to the surface of rough ER give it a "rough" appearance. Polypeptides assembled on the ribosomes thread into ER's interior, where they take on their tertiary structure and assemble with other polypeptides.

CREDIT: (14-2) Don W. Fawcett/Science Source.

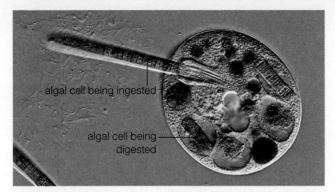

Figure 4.15 An example of vesicle function. *Nassula* (round cell) is a protist that uses a distinctive oral basket to feed on strands of photosynthetic algae (green). Ingested cells are delivered to lysosomes for digestion; they change color (green to purple to brown to gold) as chlorophyll molecules inside them are broken down.

algal cell being ingested

algal cell being digested

central vacuole Very large fluid-filled vesicle of plant cells.

endomembrane system Multifunctional network of membrane-enclosed organelles (endoplasmic reticulum, Golgi bodies, vesicles).

endoplasmic reticulum (ER) (en-doh-PLAZ-mick ruh-TICK-you-lum) Membrane-enclosed organelle that is a system of sacs and tubes extending from the nuclear envelope. Smooth ER makes phospholipids, stores calcium, and has additional functions in some cells; ribosomes on the surface of rough ER make proteins.

Golgi body (GOAL-jee) Organelle that modifies proteins and lipids, then packages the finished products into vesicles.

lysosome (LICE-uh-sohm) Enzyme-filled vesicle that breaks down cellular wastes and debris.

peroxisome (purr-OX-uh-sohm) Enzyme-filled vesicle that breaks down fatty acids, amino acids, and toxic substances.

vacuole (VAK-you-ole) Large, fluid-filled vesicle that isolates or breaks down waste, debris, toxins, or food.

vesicle (VESS-ih-cull) Membrane-enclosed organelle; different kinds store, transport, or break down their contents.

For example, rough ER-rich cells in the pancreas make digestive enzymes that they secrete into the small intestine.

Some proteins made in rough ER become part of its membrane. Others migrate through the ER compartment to the network of tubules that make up smooth ER. Smooth ER has no ribosomes on its surface, so it does not make its own proteins ❸. Some proteins that arrive in smooth ER are packaged into vesicles for delivery elsewhere. Others are enzymes that stay and carry out various functions of the smooth ER, including synthesis of phospholipids for the cell's membranes. Smooth ER also stores calcium ions in its interior. As you will see in later chapters, these ions trigger various cellular processes, so their concentration in cytoplasm is normally kept very low. Some cells have a lot of smooth ER, and in these cells it has special functions. For example, the abundant smooth ER in liver cells releases glucose from glycogen (Section 3.2) and breaks down hydrophobic toxins.

GOLGI BODIES

A **Golgi body** (also called a Golgi apparatus) has a folded membrane that often looks like a stack of pancakes ❹. Enzymes inside Golgi bodies put finishing touches on proteins and lipids that have been delivered from ER, for example by attaching sugars, oligosaccharides, or phosphate groups. The finished products—membrane proteins and lipids, proteins for secretion, and enzymes—are sorted and packaged in new vesicles. Some of the new vesicles deliver their cargo to the plasma membrane; others become lysosomes. In plant cells, Golgi bodies have an additional, major function: They make pectin and other complex, branched polysaccharides for the cell wall.

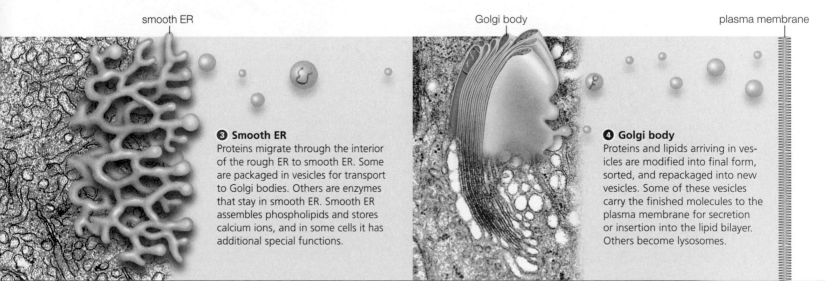

smooth ER

Golgi body

plasma membrane

❸ **Smooth ER**
Proteins migrate through the interior of the rough ER to smooth ER. Some are packaged in vesicles for transport to Golgi bodies. Others are enzymes that stay in smooth ER. Smooth ER assembles phospholipids and stores calcium ions, and in some cells it has additional special functions.

❹ **Golgi body**
Proteins and lipids arriving in vesicles are modified into final form, sorted, and repackaged into new vesicles. Some of these vesicles carry the finished molecules to the plasma membrane for secretion or insertion into the lipid bilayer. Others become lysosomes.

CREDITS: (14-3, 14-4) Don W. Fawcett/Science Source; (15) © Michael Plewka, plingfactory.de.

LEARNING OBJECTIVES

Explain the function of mitochondria in eukaryotic cells.

Describe the structure of a mitochondrion.

mitochondrion

As you will see in Chapter 5, biologists think of the nucleotide ATP as a type of cellular currency because it carries energy between reactions. Cells require a lot of ATP. The most efficient way they can produce it is by aerobic respiration, a series of oxygen-requiring reactions that harvests energy from sugars by breaking their bonds. In eukaryotes, aerobic respiration occurs inside organelles called **mitochondria** (singular, mitochondrion).

The structure of a mitochondrion is specialized for carrying out reactions of aerobic respiration. Each mitochondrion has two membranes, one highly folded inside the other (**Figure 4.16**). This arrangement creates two compartments: an outer one (between the two membranes) and an inner one (inside the inner membrane). Hydrogen ions accumulate in the outer compartment. The buildup pushes the ions across the inner membrane, into the inner compartment, and this flow drives ATP formation (Chapter 7 returns to details of aerobic respiration).

Nearly all eukaryotic cells (including plant cells) have mitochondria, but the number varies by the type of cell and by the organism. For example, single-celled eukaryotes such as yeast often have only one mitochondrion, but human skeletal muscle cells have a thousand or more. In general, cells that have the highest demand for energy tend to have the most mitochondria.

Typical mitochondria are between 1 and 4 micrometers in length. They resemble bacteria in size, form, and biochemistry. For example, they have their own DNA, which is circular and otherwise similar to bacterial DNA. They also divide independently of the cell, and have their own ribosomes. These features led to the theory that mitochondria evolved from bacteria that took up permanent residence inside a host cell (we return to this topic in Section 18.4).

Some eukaryotes that live in oxygen-free environments have modified mitochondria that produce hydrogen in addition to ATP. Like mitochondria, these organelles have two membranes. Unlike mitochondria, they have no DNA, so they cannot divide independently of the cell.

mitochondrion (my-tuh-CON-dree-un) Double-membraned organelle that produces ATP by aerobic respiration in eukaryotes.

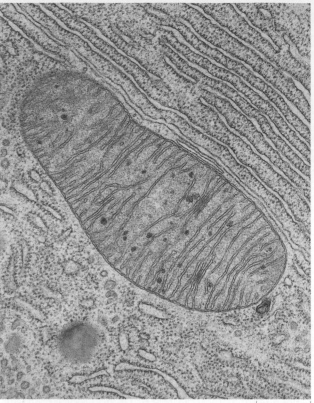

A Mitochondrion in a cell from bat pancreas.

0.5 µm

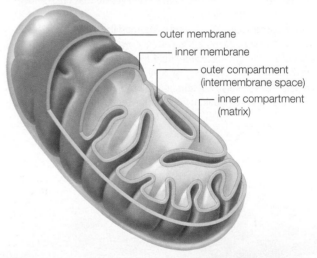

outer membrane
inner membrane
outer compartment (intermembrane space)
inner compartment (matrix)

B Each mitochondrion has two membranes, one highly folded inside the other. The inner compartment formed by these membranes is called the mitochondrial matrix; the outer compartment is called the intermembrane space.

Figure 4.16 The mitochondrion. This eukaryotic organelle specializes in producing ATP.

FIGURE IT OUT

What kind of microscope was used to make the photo?

Answer: A transmission electron microscope

4.8 Chloroplasts and Other Plastids

LEARNING OBJECTIVES

Describe the structure of chloroplasts.

Explain the function of chloroplasts and amyloplasts.

chloroplast

Plastids are double-membraned organelles that function in photosynthesis, storage, or pigmentation in plant and algal cells. Photosynthetic cells of plants and many protists contain **chloroplasts**, which are plastids specialized for photosynthesis (**Figure 4.17**). Plant chloroplasts are oval or disk-shaped. Each has two outer membranes enclosing a semifluid interior, the stroma, that contains enzymes and the chloroplast's own DNA. In the stroma, a third, highly folded membrane forms a single, continuous compartment. Photosynthesis occurs at this inner membrane, which is called the thylakoid membrane.

A thylakoid membrane incorporates many pigments, including a green one called chlorophyll (the abundance of chlorophyll in plant cell chloroplasts is the reason most plants are green). During photosynthesis, these pigments capture energy from sunlight, then pass it to other molecules that use the energy to make ATP. The ATP produced by this process is used to power reactions in the stroma that build sugars from carbon dioxide and water. (Chapter 6 returns to details of photosynthesis.) In many ways, chloroplasts resemble the photosynthetic bacteria from which they evolved.

Chromoplasts are plastids that make and store pigments other than chlorophylls. They often contain red or orange pigments called carotenoids that color flowers, leaves, roots, and fruits (**Figure 4.18**). Chromoplasts are related to chloroplasts, and the two types of plastids are interconvertible. For example, as fruits such as tomatoes ripen, green chloroplasts in their cells are converted to red chromoplasts, so the color of the fruit changes (Section 27.8 details fruit ripening).

Amyloplasts are unpigmented plastids that make and store starch grains. They are notably abundant in cells of stems, tubers (underground stems), fruits, and seeds. Like chromoplasts, amyloplasts are related to chloroplasts, and one type can change into the other. Starch-packed amyloplasts are dense and heavy compared to cytoplasm; in some plant cells, they function as gravity-sensing organelles (we return to this topic in Section 27.10).

---

chloroplast (KLOR-uh-plast) Organelle of photosynthesis in the cells of plants and photosynthetic protists.

plastid One of several types of double-membraned organelles in plants and algal cells; for example, a chloroplast or amyloplast.

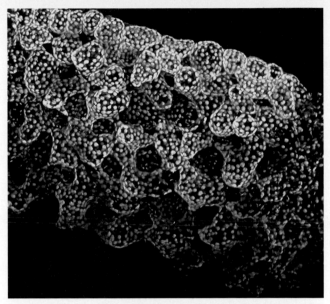

A Chloroplast-packed cells make up a leaf of a flowering plant.

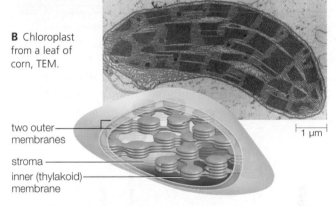

B Chloroplast from a leaf of corn, TEM.

1 μm

two outer membranes

stroma

inner (thylakoid) membrane

C Each chloroplast has two outer membranes. Photosynthesis occurs at a third, highly folded inner membrane that is called the thylakoid membrane.

Figure 4.17 The chloroplast.

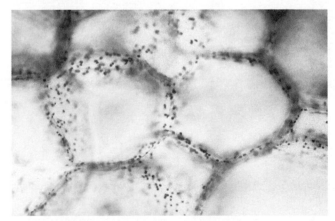

Figure 4.18 Chromoplasts. The color of a red bell pepper (a fruit) arises from carotenoid-containing chromoplasts in its cells.

4.9 The Cytoskeleton

CYTOSKELETAL ELEMENTS

Between the nucleus and plasma membrane of all eukaryotic cells is a system of interconnected protein filaments collectively called the **cytoskeleton**. Elements of the cytoskeleton reinforce, organize, and move cell structures, and often the whole cell. Some are permanent; others form only at certain times.

Microtubules Long cylinders called **microtubules** consist of subunits of the protein tubulin (**Figure 4.19A**). These cytoskeletal elements form a dynamic scaffolding for many cellular processes, rapidly assembling when they are needed, disassembling when they are not. For example, microtubules assemble before a eukaryotic cell divides, separate the cell's duplicated DNA molecules, then disassemble.

Microfilaments Fine fibers called **microfilaments** consist primarily of subunits of the protein actin (**Figure 4.19B**). These fibers strengthen or change the shape of eukaryotic cells, and have a critical function in cell movement, contraction, and migration (**Figure 4.20**). Crosslinked, bundled, or gel-like arrays of them make up the **cell cortex**, a reinforcing mesh under the plasma membrane. Microfilaments also connect plasma membrane proteins to other proteins in the cell.

Intermediate Filaments Animals and some protists have an additional type of cytoskeletal element: the intermediate filament. **Intermediate filaments** form a stable framework that lends structure and resilience to cells and tissues. Several types are assembled from different proteins (**Figure 4.19C**). For example, intermediate filaments that make up your hair consist of keratin, a fibrous protein (Section 3.4). Intermediate filaments made of lamins (another fibrous protein) form the nuclear lamina of animal cells, and also help regulate processes inside the nucleus such as DNA replication.

CELLULAR MOVEMENT

Motor proteins that associate with cytoskeletal elements move cell parts when energized by a phosphate-group transfer from ATP (Section 3.6). A cell is like a bustling train station, with molecules and structures moving continuously throughout its interior. Motor proteins are like freight trains, dragging cellular cargo along tracks of microtubules and microfilaments (**Figure 4.21**).

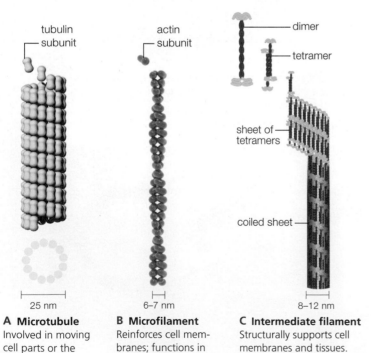

A Microtubule
Involved in moving cell parts or the whole cell.

B Microfilament
Reinforces cell membranes; functions in muscle contraction.

C Intermediate filament
Structurally supports cell membranes and tissues. Most stable element.

Figure 4.19 Cytoskeletal elements.

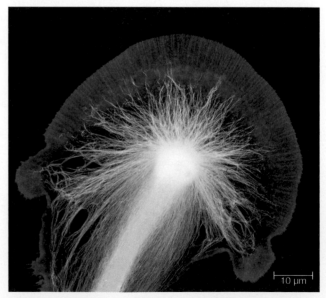

Figure 4.20 Cytoskeletal elements in a nerve cell.

This fluorescence micrograph shows microtubules (yellow) and microfilaments (blue) in the growing end of a nerve cell. These cytoskeletal elements support and guide the cell's lengthening in a particular direction.

A motor protein called dynein interacts with microtubules to bring about movement of eukaryotic flagella and cilia. **Cilia** (singular, cilium) are similar to flagella except they are shorter, and they often occur in clumps that beat in unison. Running lengthwise through each cilium or flagellum is a ring of nine microtubule pairs surrounding a central pair (**Figure 4.22A**). The microtubules grow from a barrel-shaped organelle called a **centriole**, which remains below the finished array as a basal body (**Figure 4.22B**). ATP-energized dynein causes the outer microtubule pairs to slide past one another, so the whole structure bends (**Figure 4.22C**). The whiplike motion that arises from this interaction—a power stroke that begins at the base and propagates to the tip—differs from the propeller-like motion of a prokaryotic flagellum, but it can propel motile cells through fluid (**Figure 4.22D**).

Almost all cells of mammals have a single cilium that lacks one or more of the internal components required for movement. Although these cilia cannot move, they serve important roles as sensory structures in the perception of (for example) light, sound, chemicals, or mechanical force.

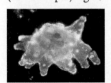

Some eukaryotic cells, including the amoeba at left, form **pseudopods**, or "false feet." As these temporary, irregular lobes bulge outward, they can move the cell or engulf a target such as prey. Elongating microfilaments inside the cell force the lobe to advance in a steady direction. Motor proteins attached to the microfilaments drag the plasma membrane along with them.

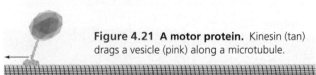

Figure 4.21 A motor protein. Kinesin (tan) drags a vesicle (pink) along a microtubule.

cell cortex Reinforcing mesh of cytoskeletal elements under a plasma membrane.

centriole (SEN-tree-ole) Barrel-shaped organelle from which microtubules grow.

cilia (SILL-ee-uh) Short, hairlike structures that project from the plasma membrane of some eukaryotic cells.

cytoskeleton (sigh-toe-SKEL-uh-ton) Network of protein filaments that support, organize, and move eukaryotic cells and their internal structures.

intermediate filament Stable cytoskeletal element of animals and some protists; structurally supports cell membranes and tissues.

microfilament Reinforcing cytoskeletal element that functions in cell movement; a fiber of actin subunits.

microtubule (my-crow-TUBE-yule) Cytoskeletal element involved in movement; hollow filament of tubulin subunits.

motor protein Type of energy-using protein that interacts with cytoskeletal elements to move the cell's parts or the whole cell.

pseudopod (SUE-doh-pod) A temporary protrusion that helps some eukaryotic cells move and engulf prey.

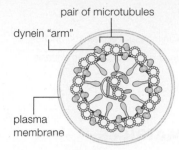

A A 9+2 array consists of a ring of nine pairs of microtubules plus one pair at their core. Stabilizing spokes and linking elements connect the microtubules and keep them aligned in this pattern. Projecting from each pair of microtubules are "arms" of the motor protein dynein.

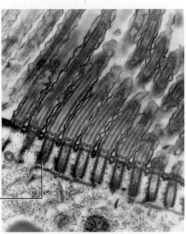

B Microtubules of a developing 9+2 array grow from a centriole, which remains under the finished array as a basal body. The micrograph below shows basal bodies underlying cilia of the protist pictured in Figure 4.6.

basal body

C Phosphate-group transfers from ATP cause the dynein arms in a 9+2 array to repeatedly bind the adjacent pair of microtubules, bend, and then disengage. The dynein arms "walk" along the microtubules, so adjacent microtubule pairs slide past one another. The short, sliding strokes of the dynein arms occur in a coordinated sequence around the ring, down the length of the microtubules. The movement causes the entire structure to bend.

flagellum

D A 9+2 array gives rise to a whiplike motion that can propel mobile eukaryotic cells such as sperm through fluid surroundings.

Figure 4.22 How eukaryotic flagella and cilia move.

CREDITS: (in text) Astrid & Hanns-Frieder Michler/Science Source; (22B right) Dennis Kunkel Microscopy, Inc./Visuals Unlimited, Inc.

4.10 Cell Surface Specializations

LEARNING OBJECTIVES

Describe an extracellular matrix of plants.

Give an example of extracellular matrix in animals and explain its function.

Explain the functions of the three major types of cell junctions in animals.

Describe a plasmodesma.

EXTRACELLULAR MATRICES

Many cells secrete an **extracellular matrix (ECM)**, a complex mixture of molecules that often includes polysaccharides and fibrous proteins. The composition and function of ECM vary by cell type.

A cell wall is an example of ECM. You learned in Section 4.4 that many prokaryotes have cell walls. Plants have them too, as do fungi and some protists. The composition of the wall differs among these groups, but in all cases it supports and protects the cell. Like a prokaryotic cell wall, a eukaryotic cell wall is porous: Water and solutes easily cross it on the way to and from the plasma membrane.

In plants, the cell wall forms as a young cell secretes pectin and other polysaccharides onto the outer surface of its plasma membrane. The sticky coating glues adjacent cells together. Each cell then forms a **primary wall** by secreting strands of cellulose into the coating. Some pectin remains as the middle lamella, a sticky layer in between the primary walls of abutting plant cells (**Figure 4.23A**). Being thin and pliable, a primary wall allows a growing plant cell to enlarge and change shape. In some plants, mature cells secrete material onto the primary wall's inner surface. These deposits form a firm **secondary wall** (**Figure 4.23B**). One of the materials deposited is **lignin**, a complex organic polymer that makes up as much as 25 percent of the secondary wall of cells in older stems and roots. Lignified plant parts are stronger, more waterproof, and less susceptible to plant-attacking organisms than younger tissues.

Animal cells have no walls, but some types secrete an extracellular matrix called basement membrane (**Figure 4.23C**). Despite the name, basement membrane is not a cell membrane because it does not consist of a lipid bilayer. Rather, it is a sheet of fibrous material that structurally supports and organizes tissues, and it has roles in cell signaling. Bone is mainly an ECM of the fibrous protein collagen hardened into a mineral by bonding with calcium, magnesium, and phosphate ions.

A covering called **cuticle** is a type of ECM secreted by cells at a body surface. In plants, a cuticle of waxes and proteins helps stems and leaves fend off insects and retain water (**Figure 4.23D**). Crabs, spiders, insects, and other arthropods have a cuticle that consists mainly of chitin (Section 3.2).

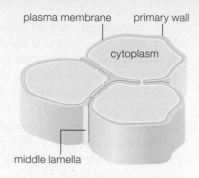

plasma membrane   primary wall

cytoplasm

A Plant cell secretions form a primary wall. The middle lamella cements adjoining cells together.

middle lamella

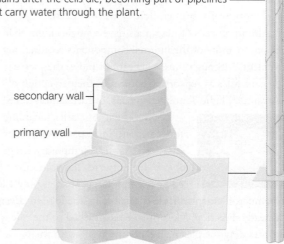

B Plant cells also secrete materials in layers on the inner surface of their primary wall. These layers form a sturdy secondary wall. In some tissues, the wall remains after the cells die, becoming part of pipelines that carry water through the plant.

secondary wall

primary wall

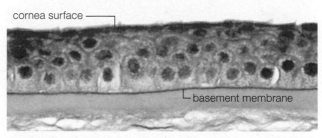

cornea surface

basement membrane

C In this section of human cornea (a part of the eye), basement membrane is visible as a dark line underlying the cells.

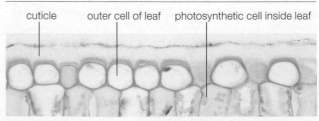

cuticle   outer cell of leaf   photosynthetic cell inside leaf

D Section through a plant leaf showing cuticle, a protective covering secreted by living cells.

Figure 4.23 Examples of extracellular matrix.

CELL JUNCTIONS

In multicelled species, cells can interact with one another and their surroundings by way of cell junctions. **Cell junctions** are structures that connect a cell directly to other cells or to ECM. Cells send and receive substances and signals through some junctions. Other junctions help cells recognize and stick to each other in tissues.

Three types of cell junctions are common in animals (**Figure 4.24**). In tissues that line body surfaces and internal cavities, rows of **tight junctions** fasten the plasma membranes of adjacent cells together and prevent body fluids from seeping between them. For example, the lining of the stomach is leak-proof because the cells that make it up are joined by tight junctions.

Adhesion proteins assemble into various types of **adhering junctions**, which fasten cells to one another and to basement membrane. These junctions make a tissue quite strong because they connect to cytoskeletal elements inside the cells. Contractile tissues (such as heart muscle) have a lot of adhering junctions, as do tissues subject to abrasion or stretching (such as skin).

Gap junctions are closable channels that connect the cytoplasm of adjoining animal cells. When open, they permit water, ions, and small molecules to pass directly from the cytoplasm of one cell to another. These channels allow entire regions of cells to respond to a single stimulus. Heart muscle and other tissues in which the cells perform a coordinated action have many gap junctions.

In plants, open channels called **plasmodesmata** (singular, plasmodesma) connect the cytoplasm of adjoining cells (**Figure 4.25**). Plasmodesmata also allow substances in cytoplasm to move from cell to cell.

---

adhering junction Cell junction that fastens an animal cell to another cell, or to basement membrane. Connects to cytoskeletal elements inside the cell; composed of adhesion proteins.

cell junction Structure that connects a cell to another cell or to extracellular matrix.

cuticle (KEW-tih-cull) Secreted covering at a body surface.

extracellular matrix (ECM) Complex mixture of cell secretions; its composition and function vary by cell type.

gap junction Cell junction that forms a closable channel across the plasma membranes of adjoining animal cells.

lignin (LIG-nin) Material that strengthens cell walls of some plants.

plasmodesma (plaz-mow-DEZ-muh) Cell junction that forms an open channel between the cytoplasm of adjacent plant cells.

primary wall The first cell wall of young plant cells.

secondary wall Lignin-reinforced wall that forms inside the primary wall of a mature plant cell.

tight junction Cell junction that fastens together the plasma membrane of adjacent animal cells; collectively prevent fluids from leaking between the cells.

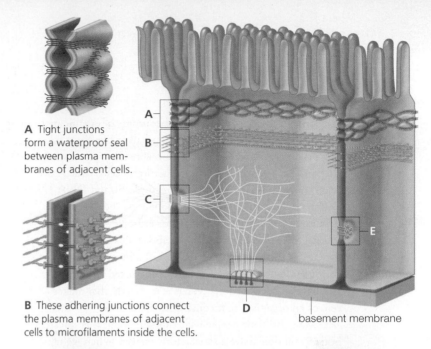

A Tight junctions form a waterproof seal between plasma membranes of adjacent cells.

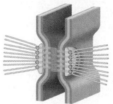

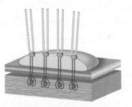

B These adhering junctions connect the plasma membranes of adjacent cells to microfilaments inside the cells.

basement membrane

C These adhering junctions connect the plasma membranes of adjacent cells to intermediate filaments inside the cells.

D These adhering junctions attach the plasma membrane of a cell to basement membrane and to intermediate filaments inside the cell.

E Gap junctions are closable channels that connect the cytoplasm of adjacent cells.

Figure 4.24 Cell junctions in animal tissues. Three types of cell junctions are shown: tight junctions (**A**), adhering junctions (**B**, **C**, and **D**), and gap junctions (**E**).

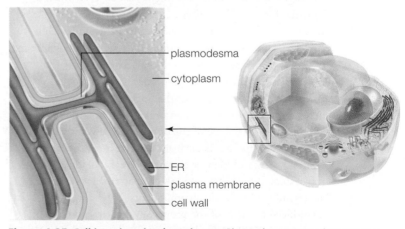

plasmodesma

cytoplasm

ER

plasma membrane

cell wall

Figure 4.25 Cell junctions in plant tissues. Plasmodesmata are channels that connect the cytoplasm and ER of adjacent plant cells.

LEARNING OBJECTIVE
List some of the properties that are associated with living systems.

TABLE 4.5

Collective Properties of Living Systems

A living system:

Consists of one or more cells

Harvests matter and energy from the environment

Requires water

Makes and uses complex carbon-containing molecules

Engages in self-sustaining biological processes such as homeostasis and metabolism

Has the capacity to respond to internal and external stimuli

Has the capacity for growth

Changes over its lifetime, for example by maturing and aging

Passes hereditary material (DNA) to offspring

Has the capacity to adapt to environmental pressures over successive generations

LIFE IS SQUISHY

Carbon, hydrogen, oxygen, and other atoms of organic molecules are the stuff of you, and us, and all of life. Yet it takes more than organic molecules to complete the picture. In this chapter, you learned about the structure of cells, which have at minimum a plasma membrane, cytoplasm, and DNA. Differences in other cellular components—the presence or absence of a particular organelle, for example—are often used to categorize life's diversity. What about life's commonality? The cell is the smallest unit of life, but a living cell does not spring to life from cellular components mixed in the proper amounts and proportions. What is it, exactly, that makes a cell, or an organism that consists of them, alive?

Many brilliant people have thought about that question, but we still don't have a satisfactory answer. When we try to define life, we end up with a very long list of properties that collectively set the living apart from the nonliving (Section 1.1). However, even that list can be tricky. For example, living things have a high proportion of the organic molecules of life, but so do the remains of dead organisms in seams of coal. Living things use energy to reproduce themselves, but computer viruses, which are arguably not alive, can do that too. Living things can pass hereditary material to offspring, but a single living rabbit (for example) cannot. No list of properties reliably and unambiguously unites all things that we would consider alive, and excludes all things we would consider not alive.

So how do biologists, who study life as a profession, define it? According to evolutionary biologist Gerald Joyce, the simplest definition of life might well be "that which is squishy." He says, "Life, after all, is protoplasmic and cellular. It is made up of cells and organic stuff and is undeniably squishy." The point is that defining life may be impossible and pointless from a scientific standpoint. The property that we call "life" is not the same as properties (such as homeostasis, reproduction, and metabolism) associated with the state of being alive; a description of an organism that is alive is not a description of the life of the organism. However, that very long list of properties is far from useless: It constitutes a working theory of living systems, a way of looking at a biological system as part of a network of relationships rather than an assembly of individual components. You have already been introduced to many of the properties associated with biological systems (**Table 4.5**). The remaining units of this book explore biologists' current understanding of these properties.

Dr. Kevin Peter Hand

Today's weather forecast for Europa, Jupiter's fourth-largest moon, is –280°F. A layer of ice several miles thick coats its fractured surface, with 1,000-foot ice cliffs piercing a pitch-black sky. It is devoid of atmosphere, bombarded by fierce radiation—and National Geographic Explorer Kevin Hand can hardly wait to get there.

Hand works at the Jet Propulsion Laboratory (JPL), where he is helping NASA plan a mission to Jupiter's moons—an orbiting probe that will give Earthlings a closer look at Europa. Beneath Europa's icy shell lies a vast global liquid water ocean that Hand thinks could be a great place for life. "I want to know if DNA is the only game in town. Are there different biochemical pathways that could lead to other kinds of life? That's at the heart of why I want to go to Europa—to find something living in that ocean we can poke at and use to understand and define life in a much more comprehensive way."

For the first time, we have the technological capability of taking our search for life to distant worlds. "Nevertheless, our understanding of life as a phenomenon remains largely qualitative and poorly constrained," Hand says. In other words, without an exact definition of "life," how do we determine whether it exists on Europa? "Biology preferentially uses specific organic subunits to build larger compounds while abiotic organic chemistry proceeds randomly," says Hand. "The structures of life arise from a relatively small set of universal building blocks; thus, when we search for life we look for patterns indicative of life's structural biases."

FOOD FOR THOUGHT

There are about as many microorganisms living in and on a human body as there are cells making up that body. Most are bacteria in the digestive tract, where they help with digestion, make vitamins that mammals cannot, prevent the growth of dangerous germs, and shape the immune system. One of the most common intestinal bacteria is *Escherichia coli*. Most of the hundreds of types, or strains, of *E. coli* are helpful, but some make a toxic protein that can severely damage the lining of the intestine. After ingesting as few as ten cells of a toxic strain, a person may become ill with severe cramps and bloody diarrhea that lasts up to ten days. In some people, complications of infection result in kidney failure, blindness, paralysis, and death. Each year, about 265,000 people in the United States become infected with toxin-producing *E. coli*.

Strains of *E. coli* that are toxic to people live in the intestines of other animals without sickening them. Humans are exposed to the bacteria when they come into contact with animals that harbor them, for example, by eating fresh fruits and vegetables that have contacted animal feces (**Figure 4.26**). People also become infected by eating contaminated ground meat. An animal's feces can contaminate its meat during slaughter. Bacteria in the feces stick to the meat, then get thoroughly mixed into it during the grinding process. Unless the contaminated meat is cooked to at least 71°C (160°F), live bacteria will enter the digestive tract of anyone who eats it.

The United States Department of Agriculture (USDA) recalls food products in which toxic bacteria are discovered. Recalled meat is not necessarily discarded; often, it is cooked and processed into ready-to-eat products such as canned chili and frozen dinners. Sterilization by cooking or other means kills bacteria, and it is one way to ensure food safety. Raw beef trimmings, which have a high risk of contact with fecal matter during the butchering process, are effectively sterilized when sprayed with ammonia; ground to a paste and formed into pellets or blocks, the resulting product is termed "lean finely textured beef" or "boneless lean beef trimmings." This product is routinely used as a filler in prepared food products such as hamburger patties, fresh ground beef, hot dogs, lunch meats, sausages, frozen entrees, canned foods, and other items sold to quick-service restaurants, hotel and restaurant chains, institutions, and school lunch programs. In 2012, a series of news reports that nicknamed the product "pink slime" provoked public outrage at its widespread use. Meat industry organizations and the USDA agree that lean finely textured beef, appetizing or not, is perfectly safe to eat because it has been sterilized.

Figure 4.26 *Escherichia coli* cells sticking to the surface of a lettuce leaf.

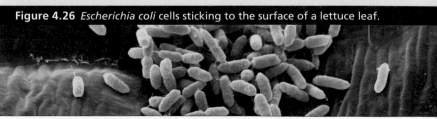

Section 4.1 Observations of cells led to the **cell theory**: All organisms consist of one or more cells; the cell is the smallest unit of life; each new cell arises from another cell; and a cell passes hereditary material to its offspring.

All cells start out life with **cytoplasm**, DNA, and a **plasma membrane** that controls the types and kinds of substances that cross it. Most have many additional components (**Table 4.6** and **Figure 4.27**). A eukaryotic cell's DNA is contained within a **nucleus**, which is a membrane-enclosed **organelle**.

A cell's surface area increases with the square of its diameter, while its volume increases with the cube. This **surface-to-volume ratio** limits cell size and influences cell (and body) shape.

Section 4.2 Almost all cells are far too small to see with the naked eye, so we use microscopes to observe them. Different types of microscopes and techniques reveal different internal and external details of cells.

Section 4.3 The foundation of almost all cell membranes is the lipid bilayer—two layers of lipids (mainly phospholipids), with tails sandwiched between heads. A bacterial or eukaryotic cell membrane can be described as a **fluid mosaic**; archaeal membranes are not fluid. Proteins embedded in or attached to a lipid bilayer add specific functions to each type of cell membrane. All cell membranes have enzymes, and all have **transport proteins** that help substances cross the lipid bilayer. Plasma membranes also incorporate **adhesion proteins** that lock cells together in tissues. Plasma membranes and some internal membranes have **receptor proteins** that trigger a change in cell activities in response to a stimulus.

Section 4.4 Bacteria and archaea (prokaryotes) are single-celled organisms with no nucleus. All have **ribosomes**. Some have **plasmids** in addition to one or two circular molecules of DNA in the **nucleoid**. Many types also have a **cell wall**; some have motile structures (**flagella**) and other projections (**pili**). Bacteria and other microorganisms may live together in a shared mass of slime as a **biofilm**.

Section 4.5 Most eukaryotes have many membrane-enclosed organelles, which compartmentalize tasks and substances that may be sensitive or dangerous to the rest of the cell. The nucleus protects and controls access to the cell's DNA. The nucleus has a **nuclear envelope**, a double lipid bilayer studded with special pores that allow some molecules, but not others, to pass into and out of the nucleus. Inside the nuclear envelope, the cell's DNA is suspended in viscous **nucleoplasm**. Also inside the nucleus, ribosome subunits are assembled in dense, irregularly shaped areas called **nucleoli**.

TABLE 4.6

Summary of Typical Components of Cells

Cell Component	Example(s) of Function	Prokaryotes	Eukaryotes			
			Protists	Fungi	Plants	Animals
Cell wall	Protection, structural support	+	+	+	+	−
Plasma membrane	Control of substances moving into and out of cell	+	+	+	+	+
Nucleus	Physical separation of DNA from cytoplasm	−	+	+	+	+
Nucleolus	Assembly of ribosome subunits	−	+	+	+	+
Cytoplasm	Fills the cell and suspends organelles	+	+	+	+	+
DNA	Encoding of hereditary information	+	+	+	+	+
RNA	Protein synthesis	+	+	+	+	+
Ribosome	Protein synthesis	+	+	+	+	+
Endoplasmic reticulum	Protein, lipid synthesis; carbohydrate, fatty acid breakdown	−	+	+	+	+
Golgi body	Final modification of proteins, lipids	−	+	+	+	+
Lysosome	Intracellular digestion	−	+	+	+	+
Peroxisome	Breakdown of fatty acids, amino acids, and toxins	−	+	+	+	+
Mitochondrion	Production of ATP by aerobic respiration	−	+	+	+	+
Chloroplast	Photosynthesis; starch storage	−	+	−	+	−
Central vacuole	Increasing cell surface area; storage	−	−	+	+	−
Flagellum	Locomotion through fluid surroundings	+	+	+	+	+
Cilium	Movement through (and of) fluid	+	+	−	+	+
Cytoskeleton	Physical reinforcement; internal organization; movement	+	+	+	+	+

+ Exists in some or all groups; **−** exists in no groups.

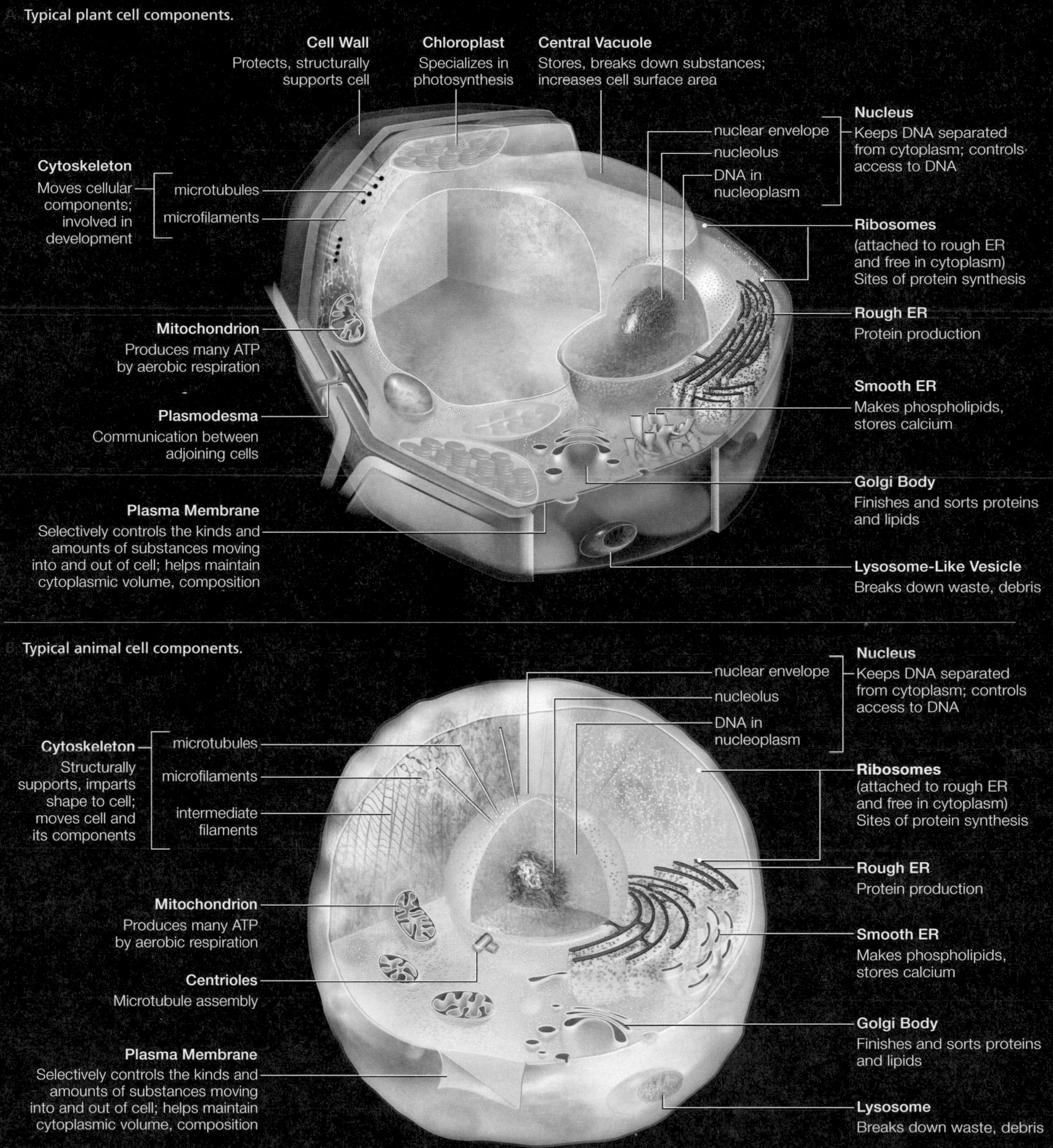

A Typical plant cell components.

Cell Wall
Protects, structurally supports cell

Chloroplast
Specializes in photosynthesis

Central Vacuole
Stores, breaks down substances; increases cell surface area

nuclear envelope
nucleolus
DNA in nucleoplasm

Nucleus
Keeps DNA separated from cytoplasm; controls access to DNA

Cytoskeleton
Moves cellular components; involved in development
microtubules
microfilaments

Ribosomes
(attached to rough ER and free in cytoplasm)
Sites of protein synthesis

Mitochondrion
Produces many ATP by aerobic respiration

Rough ER
Protein production

Plasmodesma
Communication between adjoining cells

Smooth ER
Makes phospholipids, stores calcium

Golgi Body
Finishes and sorts proteins and lipids

Plasma Membrane
Selectively controls the kinds and amounts of substances moving into and out of cell; helps maintain cytoplasmic volume, composition

Lysosome-Like Vesicle
Breaks down waste, debris

B Typical animal cell components.

nuclear envelope
nucleolus
DNA in nucleoplasm

Nucleus
Keeps DNA separated from cytoplasm; controls access to DNA

Cytoskeleton
Structurally supports, imparts shape to cell; moves cell and its components
microtubules
microfilaments
intermediate filaments

Ribosomes
(attached to rough ER and free in cytoplasm)
Sites of protein synthesis

Mitochondrion
Produces many ATP by aerobic respiration

Rough ER
Protein production

Centrioles
Microtubule assembly

Smooth ER
Makes phospholipids, stores calcium

Golgi Body
Finishes and sorts proteins and lipids

Plasma Membrane
Selectively controls the kinds and amounts of substances moving into and out of cell; helps maintain cytoplasmic volume, composition

Lysosome
Breaks down waste, debris

Figure 4.27 Organelles and structures typical of A plant cells and B animal cells.

Section 4.6 The **endomembrane system** is a series of organelles (endoplasmic reticulum, Golgi bodies, and various types of vesicles) that interact mainly to make lipids, enzymes, and proteins for insertion into membranes or secretion. Different types of **vesicles** store, break down, or transport substances through the cell. Enzymes in **peroxisomes** break down amino acids, fatty acids, and toxins. Enzymes inside **lysosomes** break down cellular wastes and debris. Fluid-filled **vacuoles** store or break down waste, food, and toxins. Fluid pressure inside a **central vacuole** keeps plant cells plump, thus keeping plant parts firm. **Endoplasmic reticulum (ER)** is a continuous system of sacs and tubes extending from the nuclear envelope. Polypeptides made by ribosomes on rough ER take on tertiary structure in the ER compartment. Smooth ER makes phospholipids and stores calcium ions. **Golgi bodies** modify proteins and lipids before sorting them into vesicles.

Section 4.7 **Mitochondria** are organelles with two membranes, one folded inside the other. They are specialized to produce ATP by aerobic respiration, an oxygen-requiring series of reactions that breaks down carbohydrates.

Section 4.8 **Plastids** of plants and some protists function in photosynthesis, storage, and pigmentation. In eukaryotes, photosynthesis takes place at the inner (thylakoid) membrane of **chloroplasts**. Pigment-filled chromoplasts and starch-filled amyloplasts are used for storage and other roles.

Section 4.9 A **cytoskeleton** of protein filaments is the basis of eukaryotic cell shape, internal structure, and movement. **Microtubules** help move cell parts. Networks of **microfilaments** reinforce cell shape, for example by forming a mesh called the **cell cortex** that reinforces plasma membranes. Microfilaments also function in movement, for example, of **pseudopods**. **Intermediate filaments** strengthen and maintain the shape of animal cell membranes and tissues, and form external structures such as hair. The nuclear lamina consists of intermediate filaments. Interactions between ATP-driven **motor proteins** and cytoskeletal elements bring about the movement of cell parts. For example, motor proteins interact with a special 9+2 array of microtubules inside **cilia** and eukaryotic flagella to move these structures. The 9+2 array arises from **centrioles**.

Section 4.10 Many cells secrete **extracellular matrix (ECM)**. ECM varies in composition and function depending on cell type. In animals, basement membrane supports and organizes cells in tissues. In plants, fungi, and some protists, a cell wall surrounds the plasma membrane. Animal cells are unwalled. Older plant cells secrete a rigid, **lignin**-containing **secondary wall** inside their pliable **primary wall**. Many eukaryotic cell types also secrete a protective **cuticle**.

Cell junctions structurally and functionally connect cells in tissues. **Plasmodesmata** (in plants) and **gap junctions** (in animals) connect the cytoplasm of adjacent cells. Also in animal cells, **adhering junctions** that connect to cytoskeletal elements fasten cells to one another and to basement membrane. **Tight junctions** form a waterproof seal between cells.

Section 4.11 Living systems consist of cells that harvest energy and matter from the environment. They also make and use complex carbon-based molecules; require water; grow; engage in self-sustaining biological processes; respond to stimuli; change over their lifetime, and over generations; and pass hereditary material (DNA) to offspring.

Application A huge number of bacteria live in and on the human body. Most of them are helpful; a few types can cause disease. Contamination of food with disease-causing bacteria can result in illness that is sometimes fatal.

SELF-ASSESSMENT
ANSWERS IN APPENDIX VII

1. All cells have these three things in common: _____ .
   a. cytoplasm, DNA, and organelles with membranes
   b. a plasma membrane, DNA, and a nuclear envelope
   c. cytoplasm, DNA, and a plasma membrane
   d. a cell wall, cytoplasm, and DNA

2. Every cell is descended from another cell. This idea is part of _____ .
   a. evolution   c. the cell theory
   b. the theory of heredity   d. cell biology

3. The surface-to-volume ratio _____ .
   a. does not apply to prokaryotic cells
   b. constrains cell size
   c. is part of the cell theory

4. Cell membranes consist mainly of _____ and _____ .
   a. lipids; carbohydrates   c. lipids; phospholipids
   b. phospholipids; protein   d. phospholipids; ECM

5. In a lipid bilayer, the _____ of all the lipid molecules are sandwiched between all of the _____ .
   a. hydrophilic tails; hydrophobic heads
   b. hydrophilic heads; hydrophilic tails
   c. hydrophobic tails; hydrophilic heads
   d. hydrophobic heads; hydrophilic tails

6. What controls the passage of molecules into and out of the nucleus?
   a. endoplasmic reticulum, an extension of the nucleus
   b. nuclear pores, which consist of membrane proteins
   c. nucleoli, in which ribosome subunits are made
   d. dynamically assembled microtubules
   e. tight junctions

KARTAGENER SYNDROME An abnormal form of the motor protein dynein causes Kartagener syndrome, a genetic disorder characterized by chronic sinus and lung infections. Biofilms form in the thick mucus that collects in the airways, and the resulting bacterial activities and inflammation damage tissues.

Affected men produce sperm but are infertile (**Figure 4.28**). Some have become fathers after a doctor injected their sperm cells directly into eggs.

Review **Figure 4.22**, then explain how abnormal dynein could cause these observed effects.

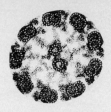

Figure 4.28 Cross-section of the flagellum of a sperm cell from a man affected by Kartagener syndrome (left) and an unaffected man (right).

7. Most of a membrane's diverse functions are carried out by _____ .
   a. proteins
   b. phospholipids
   c. nucleic acids
   d. hormones

8. The main function of the endomembrane system is _____ .
   a. building and modifying proteins and lipids
   b. isolating DNA from toxic substances
   c. secreting extracellular matrix onto the cell surface
   d. producing ATP by aerobic respiration

9. Which of the following statements is correct?
   a. Ribosomes are only found in bacteria and archaea.
   b. Some animal cells are prokaryotic.
   c. Only eukaryotic cells have mitochondria.
   d. The plasma membrane is the outermost boundary of all cells.
   e. Some cell membranes do not consist of lipids.

10. Enzymes contained in _____ break down worn-out organelles, bacteria, and other particles.
   a. lysosomes
   b. mitochondria
   c. endoplasmic reticulum
   d. peroxisomes

11. Put the following structures in order according to the pathway of a secreted protein:
   a. plasma membrane
   b. Golgi bodies
   c. endoplasmic reticulum
   d. post-Golgi vesicles

12. No animal cell has a _____ .
   a. plasma membrane
   b. flagellum
   c. lysosome
   d. cell wall

13. Which of the following organelles contains no DNA? Choose all that are correct.
   a. nucleus
   b. Golgi body
   c. mitochondrion
   d. chloroplast

14. _____ connect the cytoplasm of plant cells.
   a. Plasmodesmata
   b. Adhering junctions
   c. Tight junctions
   d. Adhesion proteins
   e. Chloroplasts
   f. Gap junctions

15. Match each cell component with its function.
   ___ mitochondrion   a. connection
   ___ chloroplast   b. protective covering
   ___ ribosome   c. ATP production
   ___ nucleus   d. protects DNA
   ___ cell junction   e. protein synthesis
   ___ flagellum   f. photosynthesis
   ___ cuticle   g. movement

CLASS ACTIVITY
CRITICAL THINKING

1. In a classic episode of *Star Trek*, a gigantic amoeba engulfs an entire starship. Spock blows the cell to bits before it can reproduce. Think of at least one inaccuracy that a biologist would identify in this scenario.

2. In plants, the cell wall forms as a young plant cell secretes polysaccharides onto the outer surface of its plasma membrane. Being thin and pliable, this primary wall allows the cell to enlarge and change shape. At maturity, cells in some plant tissues deposit material onto the primary wall's inner surface. Why doesn't this secondary wall form on the outer surface of the primary wall?

3. What type of micrograph is shown below? Is the organism pictured prokaryotic or eukaryotic? How can you tell? Which structures can you identify?

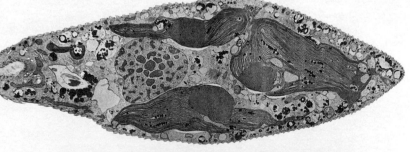

for additional quizzes, flashcards, and other study materials
▶ ACCESS MINDTAP AT WWW.CENGAGEBRAIN.COM

CREDITS: (27) From "Tissue & Cell," Vol. 27, pp.421–427, Courtesy of Bjorn Afzelius, Stockholm University; (in text) P.L. Walne and J. H. Arnott, *Planta*, 77:325–354, 1967.

CHAPTER 4 CELL STRUCTURE   **75**

5 Ground Rules of Metabolism

Links to Earlier Concepts

In this chapter, you will gain insight into the one-way flow of energy through the world of life (Sections 1.1, 1.2) as you learn more about specific types of energy (2.4) and the laws of nature (1.8) that describe it. The chapter also revisits the structure and function of atoms (2.2), molecules (2.3, 3.1–3.6), and cells (4.3, 4.6, 4.7, 4.9).

Core Concepts

Organisms exchange matter and energy with the environment in order to grow, maintain themselves, and reproduce.

Energy cannot be created or destroyed, but it can be transferred and converted to other forms. With each transfer or conversion, some energy disperses. Maintaining the organization of living systems requires ongoing inputs of energy to offset this loss.

Organisms have various strategies to capture and store energy for use in biological processes.

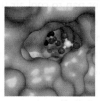

Cells store energy by building organic molecules, and retrieve energy by breaking them down. Many metabolic pathways couple energy-releasing reactions with energy-requiring reactions. The coupled reactions of electron transfer chains allow cells to harvest energy from electrons in small, manageable increments.

Cellular homeostasis requires maintenance of an internal environment that differs from the external environment.

Homeostasis depends on constant movement of substances across cell membranes, which are not permeable to ions or most polar molecules. Transport proteins and other membrane-crossing mechanisms help cells maintain control over their internal environment. Controls that govern steps in metabolic pathways quickly shift cell activities in response to changes in internal and external conditions.

◀ A single-celled protist with the common name Sea Sparkle glows blue when agitated—for example, by waves breaking on this seashore in Maldives. The light is energy released from chemical reactions that run inside these cells.

Figure 5.31 A tailgate party at a Notre Dame–Alabama football game.

A TOAST TO ALCOHOL DEHYDROGENASE

Most college students are under the legal drinking age, but alcohol abuse continues to be the most serious drug problem on college campuses in the United States (**Figure 5.31**). Recent surveys polled tens of thousands of undergraduates about their drinking habits, and more than half of them reported regularly consuming five or more alcoholic beverages within a two-hour period—a self-destructive behavior called binge drinking. Drinking large amounts of alcohol in a brief period of time is an extremely risky behavior, both for the drinkers and people around them. Every year, around 600,000 students injure themselves while under the influence of alcohol; intoxicated students physically assault 690,000 people and sexually assault 97,000 others. Binge drinking is responsible for killing or causing the death of 1,825 students per year.

Before you drink, consider what you are consuming. All alcoholic drinks—beer, wine, hard liquor—contain the same psychoactive ingredient: ethanol. Almost all ingested ethanol ends up in the liver, a large organ in the abdomen with many important functions. Liver cells make alcohol dehydrogenase

(ADH), an enzyme essential to a metabolic pathway that detoxifies alcohols.

ADH converts ethanol to acetaldehyde, an organic compound even more toxic than ethanol and the most likely source of various hangover symptoms. A second enzyme, ALDH, converts the toxic acetaldehyde to acetate, which is a nontoxic salt. Both ADH and ALDH use the coenzyme NAD^+ to accept electrons and hydrogen atoms. Thus, the overall pathway of ethanol metabolism in humans is:

$$\text{ethanol} \xrightarrow[\underset{NAD^+ \quad NADH}{}]{ADH} \text{acetaldehyde} \xrightarrow[\underset{NAD^+ \quad NADH}{}]{ALDH} \text{acetate}$$

In the average healthy adult, this metabolic pathway can detoxify between 7 and 14 grams of ethanol per hour. The average alcoholic beverage contains between 10 and 20 grams of ethanol, which is why having more than one drink in any two-hour interval may result in a hangover.

Putting more alcohol into your body than your enzymes can detoxify damages it more permanently than a hangover, however. Ethanol breakdown harms liver cells, so the more a person drinks, the fewer liver cells are left to do the breaking down. Ethanol also interferes with normal processes of metabolism. For example, oxygen that would ordinarily take part in breaking down fatty acids is diverted to breaking down the ethanol. As a result, fats tend to accumulate as large globules in the tissues of heavy drinkers.

Long-term heavy drinking causes alcoholic hepatitis, a disease characterized by inflammation and destruction of liver tissue; and cirrhosis, a condition in which the liver becomes so scarred, hardened, and filled with fat that it loses its function. (The term cirrhosis is from the Greek *kirros*, meaning orange-colored, after the abnormal skin color of people with the disease.) A cirrhotic liver stops making albumin, a protein that maintains normal tonicity of blood and tissue fluids. Without sufficient albumin in circulation, body tissues swell with watery fluid, especially in the legs and abdomen. Drugs and other toxins can no longer be removed from the blood, so they accumulate in the brain—which impairs mental functioning and alters personality. Restricted blood flow through the liver causes veins to enlarge and rupture, so internal bleeding is a risk. The damage to the body results in a heightened susceptibility to diabetes and liver cancer. Once cirrhosis has been diagnosed, a person has about a 50 percent chance of dying within 10 years (**Figure 5.32**).

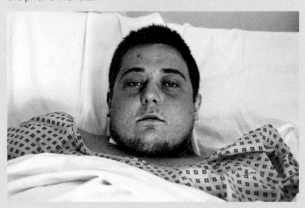

Figure 5.32 Gary Reinbach. The 22-year-old died from alcoholic liver disease shortly after this photograph was taken. The odd color of his skin is a symptom of cirrhosis.

Transplantation is a last-resort treatment for a failed liver, but there are not enough donors for everyone who needs a transplant. Reinbach was refused a transplant that may have saved his life because he had not abstained from drinking for the prior 6 months.

CHAPTER SUMMARY

Section 5.1 **Energy**, which is the capacity to do work, cannot be created or destroyed (**first law of thermodynamics**), and it tends to disperse spontaneously (**second law of thermodynamics**). **Entropy** is a measure of how much the energy of a system is dispersed. Energy can be transferred between systems or converted from one form to another (for example, **potential energy** can be converted to **kinetic energy**), but some is lost (often as heat) during every such exchange. Sustaining life's organization requires ongoing energy inputs to counter energy loss. Organisms stay alive by replenishing themselves with energy they harvest from someplace else.

Section 5.2 In chemical reactions, **reactants** are converted to **products**. Cells store energy in chemical bonds by running **endergonic** reactions that build organic compounds. To release this stored energy, they run **exergonic** reactions that break the bonds. Both endergonic and exergonic reactions require an input of **activation energy** to begin.

Section 5.3 Enzymes greatly enhance the rate of reactions without being changed by them, a process called **catalysis**. Interacting with an enzyme's **active site** causes a **substrate** to reach its transition state. Enzymes lower a reaction's activation energy, for example improving the fit between substrate and active site (**induced-fit model**), or by forcing substrates together. Each enzyme works best within a certain range of environmental conditions that include temperature, pH, pressure, and salt concentration.

Section 5.4 A **metabolic pathway** is a stepwise series of enzyme-mediated reactions that collectively build, remodel, or break down an organic molecule. Cells conserve energy and resources by producing only what they need at a given time. Such metabolic control can arise from mechanisms (such as a regulatory molecule binding to an enzyme) that start, stop, or alter the rate of a single reaction. **Allosteric regulation** occurs when a regulatory molecule binds an enzyme in a region other than the active site. Other mechanisms (such as **feedback inhibition**) influence an entire pathway. **Redox** (oxidation–reduction) **reactions** in **electron transfer chains** allow cells to harvest energy stepwise, in small, manageable amounts.

Section 5.5 **Cofactors** associate with enzymes and assist their function. Many **coenzymes** carry chemical groups, atoms, or electrons from one reaction to another. ATP is often used as a coenzyme that carries energy between reactions. When a phosphate group is transferred from ATP to another molecule, energy is transferred along with it. Such phosphate-group transfers (**phosphorylations**) to and from ATP couple exergonic with

endergonic reactions. Cells regenerate ATP in the **ATP/ADP cycle**. **Antioxidants** interfere with the oxidation of other molecules.

Section 5.6 **Diffusion** is influenced by concentration gradients, temperature, molecular size, charge, and pressure. Nonpolar molecules as well as gases and small uncharged polar molecules such as water can diffuse across a lipid bilayer. Most other substances—ions and charged molecules, in particular—cannot. Solutes tend to diffuse into an adjoining region of fluid in which they are not as concentrated. When two fluids of different solute concentration are separated by a selectively permeable membrane such as a lipid bilayer, water diffuses across the membrane from the **hypotonic** to the **hypertonic** fluid (there is no net movement of water between **isotonic** solutions). This movement, **osmosis**, is opposed by **turgor** (fluid pressure against a cell membrane or wall). **Osmotic pressure** is the amount of turgor sufficient to halt osmosis.

Section 5.7 Transport proteins move specific ions or molecules across a cell membrane. In **facilitated diffusion** (a type of **passive transport**), a solute binds to a transport protein that releases it on the opposite side of the membrane. The movement is driven by the solute's concentration gradient. In **active transport**, a transport protein uses energy to pump a solute across a membrane against its concentration gradient. The movement requires energy, as from ATP.

Section 5.8 Exocytosis and endocytosis move particles and substances in bulk across plasma membranes. In **exocytosis**, a cytoplasmic vesicle fuses with the plasma membrane, and its contents are released to the outside of the cell. In **endocytosis**, a patch of plasma membrane balloons into the cell, taking with it a drop of extracellular fluid. The balloon forms a vesicle that sinks into the cytoplasm. Bulk-phase endocytosis is not specific about what it takes in; receptor-mediated endocytosis targets specific molecules. Some cells engulf large solid particles such as other cells by the endocytic pathway of **phagocytosis**.

Application Alcohol abuse continues to be the most serious drug problem on college campuses. Drinking more alcohol than the body's enzymes can detoxify can be lethal in both the short term and the long term.

SELF-ASSESSMENT
ANSWERS IN APPENDIX VII

1. Which of the following statements is not correct?
   a. Energy cannot be created or destroyed.
   b. Energy cannot change from one form to another.
   c. Energy tends to disperse spontaneously.
   d. Energy can be transferred.

2. _____ is life's primary source of energy.
   a. Food   b. Water   c. Sunlight   d. ATP

3. Entropy _____ .
   a. tends to disperse   c. tends to decrease, overall
   b. is a measure of disorder d. is free energy

4. In an endergonic reaction, activation energy is a bit like _____ .
   a. a burst of speed
   b. coasting downhill
   c. an energy hill
   d. putting on the brakes

5. _____ are always changed by participating in a reaction.
   a. Enzymes   c. Reactants
   b. Cofactors   d. Coenzymes

6. Name one environmental factor that typically influences enzyme function.

7. A metabolic pathway _____ .
   a. may build or break down molecules
   b. generates heat
   c. can include redox reactions
   d. all of the above

8. Which of the following statements is incorrect?
   a. Some metabolic pathways are cyclic.
   b. Glucose can diffuse directly through a lipid bilayer.
   c. Feedback inhibition regulates some metabolic pathways.
   d. All coenzymes are cofactors.
   e. Osmosis is a type of diffusion.

9. Antioxidants _____ .
   a. prevent other molecules from being oxidized
   b. are coenzymes
   c. balance charge
   d. deoxidize free radicals

10. A molecule that donates electrons becomes _____ , and the one that accepts the electrons becomes _____ .
    a. reduced; oxidized   c. oxidized; reduced
    b. ionic; electrified   d. electrified; ionic

11. Solutes tend to diffuse from a region where they are _____ (more/less) concentrated to an adjacent region where they are _____ (more/less) concentrated.

12. _____ do not easily diffuse across a lipid bilayer.
    a. Water molecules   c. Ions
    b. Gases   d. Nonpolar molecules

13. A transport protein requires ATP to pump sodium ions across a membrane. This is a case of _____ .
    a. passive transport   c. osmosis
    b. active transport   d. facilitated diffusion

14. Immerse a human blood cell in a hypotonic solution, and water _____ .
    a. diffuses into the cell   c. shows no net movement
    b. diffuses out of the cell d. moves in by endocytosis

ONE TOUGH BUG The genus *Ferroplasma* consists of a few species of acid-loving archaea. One species, *F. acidarmanus*, was discovered to be the main constituent of slime streamers (a type of biofilm) deep inside an abandoned California copper mine (**Figure 5.33A**). These cells have an ancient energy-harvesting pathway that uses electrons pulled from iron–sulfur compounds in minerals such as pyrite. Oxidizing these minerals dissolves them, so groundwater in the mine ends up with extremely high concentrations of solutes, including metal ions such as copper, zinc, cadmium, and arsenic. The reaction also produces sulfuric acid, which lowers the pH of the water around the cells to zero.

F. acidarmanus cells keep their internal pH at a cozy 5.0 despite living in an environment similar to hot battery acid. However, most of the cells' enzymes function best at much lower pH (**Figure 5.33B**). Thus, researchers think *F. acidarmanus* may have an unknown type of internal compartment that keeps their enzymes in a low-pH environment.

1. What do the dashed lines in the graphs signify?

2. Of the four enzymes profiled in the graph, how many function optimally at a pH lower than 5? How many retain significant function at pH 5?

3. What is the optimal pH for the carboxylesterase?

A Deep inside one of the most toxic sites in the United States: Iron Mountain Mine, in California. The water in this stream, which is about 1 meter (3 feet) wide in this view, is hot (around 40°C, or 104°F), heavily laden with arsenic and other toxic metals, and has a pH of zero. The slime streamers growing in it are a biofilm dominated by a species of archaea, *Ferroplasma acidarmanus*.

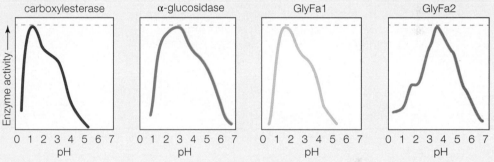

B pH activity profiles of four enzymes isolated from *Ferroplasma*. Researchers had expected these enzymes to function best at the cells' cytoplasmic pH (5.0).

Figure 5.33 pH anomaly of *Ferroplasma acidarmanus* enzymes.

---

15. Match each term with its most suitable description.

___ reactant	a. assists enzymes
___ phagocytosis	b. forms at reaction's end
___ first law of thermodynamics	c. energy cannot be created or destroyed
___ product	d. cell eats cell
___ cofactor	e. enters a reaction
___ concentration gradient	f. requires energy input
___ passive transport	g. influences diffusion
___ active transport	h. no energy input required
___ electron transfer chain	i. internal pressure
___ cyclic pathway	j. goes in circles
___ turgor	k. series of redox reactions

for additional quizzes, flashcards, and other study materials
▶ **ACCESS MINDTAP AT WWW.CENGAGEBRAIN.COM**

1. Beginning physics students are often taught the basic concepts of thermodynamics with two phrases: First, you can never win. Second, you can never break even. Explain.

2. Describe diffusion in terms of entropy.

3. How do you think a cell regulates the amount of glucose it brings into its cytoplasm from the extracellular environment?

4. The enzyme trypsin is sold as a dietary enzyme supplement. Explain what happens to trypsin that is taken with food.

5. Catalase combines two hydrogen peroxide molecules ($H_2O_2 + H_2O_2$) to make two molecules of water. A gas also forms. What is the gas?

6

Where It Starts— Photosynthesis

Links to Earlier Concepts

This chapter explores the main metabolic pathways (Section 5.4) by which organisms harvest energy from the sun (5.1). We revisit experimental design (1.6), electrons and energy levels (2.2), bonds (2.3), carbohydrates (3.2), membrane proteins (4.3), chloroplasts (4.8), antioxidants (5.5), and concentration gradients (5.6).

Core Concepts

Organisms exchange matter and energy with the environment in order to grow, maintain themselves, and reproduce.

The main flow of energy through the biosphere starts with photosynthesis. In this pathway, light energy is captured from the environment and converted to chemical energy that supports photosynthetic organisms as well as most consumers. During photosynthesis, carbon (CO_2) moves from the environment to organisms that incorporate it into sugars. The main pathway uses water and produces O_2.

Interactions among the components of each biological system give rise to complex properties.

A particular pattern of molecular structure allows pigments to capture light energy. In eukaryotes, photosynthesis occurs in chloroplasts. In a series of coordinated reaction pathways, molecules in the chloroplast's thylakoid membrane use light energy to make ATP, which is used to power synthesis of sugars from carbon dioxide in the stroma.

Adaptation of organisms to a variety of environments has resulted in diverse structures and physiological processes.

Variations in photosynthetic pathways are evolutionary adaptations that allow photosynthetic organisms to thrive in a variety of environments. Like building blocks, molecules and processes in complex metabolic pathways are often evolutionarily reused in improved pathways that offer enhanced benefit.

◀ Raindrops can separate light from the sun into its different component wavelengths, which we see as different colors in a rainbow.

99

Explain why photosynthesis feeds most life on Earth.

Describe the structure of the thylakoid membrane.

Write an equation that summarizes the overall pathway of photosynthesis.

AUTOTROPHS AND HETEROTROPHS

All life is sustained by inputs of energy, but not all forms of energy can sustain life. Sunlight, for example, is abundant here on Earth, but it cannot directly power protein synthesis or other energy-requiring reactions that all organisms run in order to stay alive. For this purpose, sunlight must first be converted to chemical bond energy (Section 5.2). Unlike light, chemical energy can power the reactions of life, and it can be stored for later use.

Energy flow through most ecosystems on Earth begins with **autotrophs**: organisms that make their own food by harvesting energy directly from the environment (*auto-* means self; *-troph* refers to nourishment). All organisms need carbon to build the molecules of life; autotrophs obtain it from inorganic molecules such as carbon dioxide (CO_2). Plants and most other autotrophs harvest energy from the environment by photosynthesis (Section 1.2). **Photosynthesis** is a metabolic pathway that uses the energy of sunlight to drive the assembly of carbohydrates—sugars—from carbon dioxide and water. The sugars can be stored as polysaccharides for later use, remodeled into other compounds, or broken down to release energy in their bonds (a topic of the next chapter).

Autotrophs are an ecosystem's producers. **Heterotrophs** are consumers; they get their carbon by breaking down organic molecules acquired from other organisms (*hetero-* means other). Humans and almost all other heterotrophs obtain carbon and energy from organic molecules originally assembled by photosynthesizers. Thus, directly or indirectly, photosynthesis feeds most life on Earth (**Figure 6.1**).

TWO STAGES OF REACTIONS

Photosynthesis is often summarized by an equation:

$$CO_2 + H_2O \xrightarrow{\text{light energy}} \text{sugars} + O_2$$

CO_2 is carbon dioxide, and O_2 is oxygen; both are gases abundant in the atmosphere. The equation means photosynthesis converts carbon dioxide and water to sugars and oxygen. However, photosynthesis is not a single reaction. Rather, it is a metabolic pathway with many reactions that occur in two stages. The reactions of the first stage are driven by light and thus called the **light-dependent reactions**. The "photo" in photosynthesis means light, and it refers to the conversion

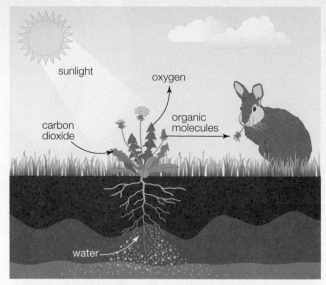

Figure 6.1 How photosynthesis sustains life. Photosynthetic autotrophs get energy from sunlight, and carbon from carbon dioxide. Most heterotrophs get their energy and carbon from organic molecules assembled by photosynthetic autotrophs.

of light energy to the chemical bond energy of ATP during this stage.

In addition to making ATP, the main light-dependent pathway splits water molecules and releases O_2. Hydrogen ions and electrons from the water molecules are loaded onto the coenzyme $NADP^+$, so NADPH forms (**Figure 6.2A**).

The "synthesis" part of photosynthesis refers to the reactions of the second stage, which build sugars from CO_2 and water (**Figure 6.2B**). They are collectively called the **light-independent reactions** because light energy does not power them. Instead, they run on energy delivered by NADPH and ATP that formed during the first stage. At the end of the second stage, $NADP^+$ and ADP are recycled to work again in the reactions of the first stage (**Figure 6.2C**).

autotroph (AH-toe-trof) Organism that makes its own food using energy from the environment and carbon from inorganic molecules such as CO_2.

heterotroph (HET-er-oh-trof) Organism that obtains carbon from organic compounds assembled by other organisms.

light-dependent reactions First stage of photosynthesis; convert light energy to chemical energy.

light-independent reactions Second stage of photosynthesis; use ATP and NADPH to assemble sugars from water and CO_2.

photosynthesis Metabolic pathway by which most autotrophs use sunlight to make sugars from carbon dioxide and water. Converts light energy into chemical energy.

stroma (STROH-muh) Thick, cytoplasm-like fluid between the thylakoid membrane and the two outer membranes of a chloroplast.

thylakoid membrane Inner membrane system that carries out light-dependent reactions in chloroplasts and cyanobacteria.

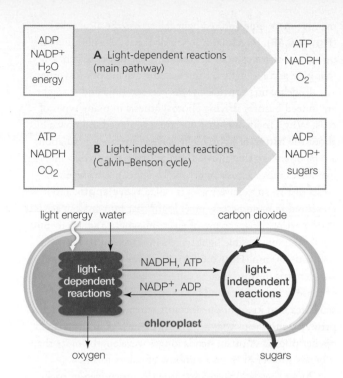

ADP NADP+ H₂O energy	**A** Light-dependent reactions (main pathway)	ATP NADPH O₂

ATP NADPH CO₂	**B** Light-independent reactions (Calvin–Benson cycle)	ADP NADP+ sugars

C Light energy drives the production of ATP in the light-dependent reactions; the main pathway also produces NADPH and O₂. NADPH and ATP drive sugar production in the light-independent reactions.

Figure 6.2 How coenzymes connect the first- and second-stage reactions of photosynthesis. Substrates and products of the main pathways are shown as they occur in chloroplasts. With some variation, cyanobacteria use the same pathways.

WHERE PHOTOSYNTHESIS OCCURS

In plants, photosynthetic protists, and cyanobacteria, the light-dependent reactions are carried out by molecules embedded in a **thylakoid membrane** (Section 4.8). Folds of this membrane form disks called thylakoids. In eukaryotes, the thylakoid membrane occurs inside chloroplasts (**Figure 6.3A**). A chloroplast's thylakoid membrane is highly folded into stacks of interconnected thylakoids (**Figure 6.3B**), and it encloses a single, continuous internal space (compartment). The light-independent reactions run in **stroma**, a thick, cytoplasm-like fluid that fills the chloroplast. The thylakoid membrane is suspended in the stroma, as are the chloroplast's own DNA and ribosomes.

Chloroplasts are descendants of ancient cyanobacteria, which is why photosynthesis in eukaryotes is similar to cyanobacterial photosynthesis. Modern cyanobacteria have thylakoid membranes (as shown in Figure 4.8B) that carry out the light-dependent reactions; the light-independent reactions run in cytoplasm of these cells.

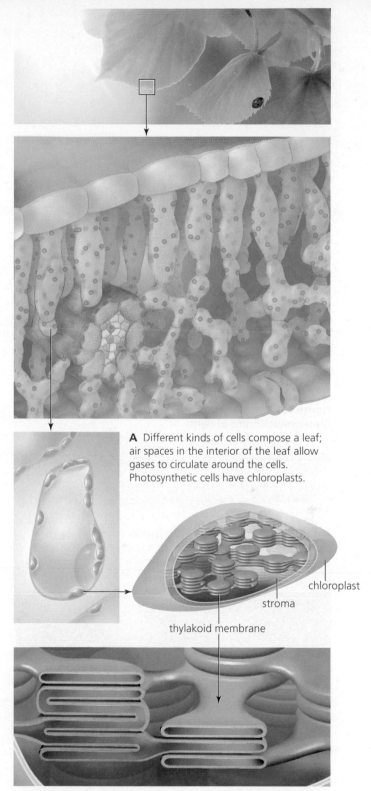

A Different kinds of cells compose a leaf; air spaces in the interior of the leaf allow gases to circulate around the cells. Photosynthetic cells have chloroplasts.

B Inside a chloroplast, stroma surrouds a highly folded thylakoid membrane that encloses one continuous space (compartment).

Figure 6.3 Zooming in on the site of photosynthesis in a leaf.

6.2 Sunlight as an Energy Source

VISIBLE LIGHT DRIVES PHOTOSYNTHESIS

In 1882, botanist Theodor Engelmann designed a set of experiments to test his hypothesis that the color of light affects the rate of photosynthesis. It had long been known that photosynthesis releases oxygen, so Engelmann used oxygen emission as an indirect measurement of photosynthetic activity. He directed a spectrum of light across individual strands of green algae suspended in water (**Figure 6.4A**). Oxygen-sensing equipment had not yet been invented, so Engelmann used motile, oxygen-requiring bacteria to show him where the oxygen concentration in the water was highest. The bacteria moved through the water and gathered

Figure 6.4 Discovery that photosynthesis is driven by particular wavelengths of light.

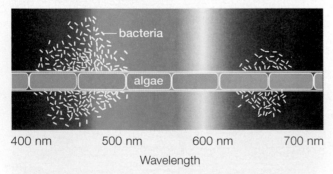

A Each cell in a strand of *Cladophora* is filled with a single chloroplast. Theodor Engelmann used this and other species of green algae in a series of experiments to determine whether some colors of light are better for photosynthesis than others.

400 nm　　　500 nm　　　600 nm　　　700 nm
Wavelength

B Engelmann directed light through a prism so that bands of colors crossed a water droplet on a microscope slide. The water held a strand of photosynthetic algae, and also oxygen-requiring bacteria. The bacteria swarmed around the algal cells that were releasing the most oxygen—the ones most actively engaged in photosynthesis. Those cells were under blue and red light.

mainly where blue and red light fell across the algal cells (**Figure 6.4B**). Engelmann concluded that cells illuminated by light of these colors were releasing the most oxygen, a sign that blue and red light are the best for driving photosynthesis.

Engelmann's conclusion was correct: Blue and red light are indeed best for driving photosynthesis in many types of cells. Why? To answer that question, you need to know a bit about the nature of light. Light is electromagnetic radiation, a type of energy that moves through space in waves, a bit like waves move across an ocean. The distance between the crests of two successive waves is called **wavelength**, and it is measured in nanometers (nm). Light that humans can see is a small part of the spectrum of electromagnetic radiation emitted by the sun. Visible light travels in wavelengths between 380 and 750 nm (**Figure 6.5A**). Our eyes perceive particular wavelengths in this range as different colors; light with all of these wavelengths combined appears white. White light separates into its component colors when it passes through a prism (or raindrops that act as tiny prisms, as in the chapter opener photo). A prism bends longer wavelengths more than it bends shorter ones, so a rainbow of colors forms.

Light travels in waves, but it is also organized in packets of energy called photons. A photon's wavelength and its energy are related, so all photons traveling at the same wavelength carry the same amount of energy. Photons with the most energy travel in shorter wavelengths; those with the least energy travel in longer wavelengths (**Figure 6.5B**).

TO CATCH A RAINBOW

Photosynthetic pigments can absorb photons of visible light. A **pigment** is an organic molecule that selectively absorbs light of specific wavelengths. The molecule has a light-trapping region, a carbon chain or ring in which single bonds alternate with double bonds. Electrons (Section 2.2) can move freely among all the atoms of this region. These electrons can easily absorb a photon—but not just any photon. Only photons with exactly enough energy to boost the electrons to a higher energy level (shell) are absorbed. This is why a pigment absorbs light of only certain wavelengths.

Wavelengths of light that are not absorbed are reflected, and that reflected light gives each pigment its characteristic color. The most common photosynthetic pigment in plants, cyanobacteria, and photosynthetic protists is **chlorophyll a** (**Figure 6.6**). Chlorophyll *a* absorbs violet, red, and orange light, and it reflects green light, so it appears green to us. Most photosynthetic cells also have accessory pigments, including other chlorophylls. Together with chlorophyll *a*, these pigments maximize the range of wavelengths that an organism can use for photosynthesis. Accessory pigments

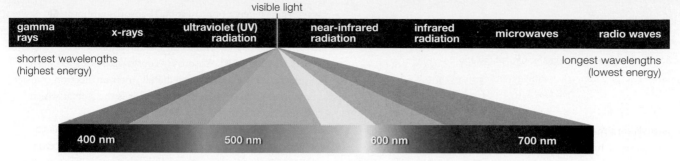

visible light

| gamma rays | x-rays | ultraviolet (UV) radiation | near-infrared radiation | infrared radiation | microwaves | radio waves |

shortest wavelengths
(highest energy)

longest wavelengths
(lowest energy)

400 nm   500 nm   600 nm   700 nm

A Electromagnetic radiation moves through space in waves that we measure in nanometers (nm). Visible light makes up a very small part of this energy. Raindrops or a prism can separate visible light's different wavelengths, which we see as different colors. About 25 million nanometers are equal to 1 inch.

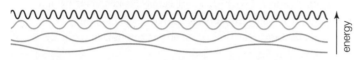

energy

B Light is organized as packets of energy called photons. The shorter a photon's wavelength, the greater its energy. Thus, photons of violet light carry more energy than photons of blue light, and so on.

Figure 6.5 Properties of light.

have other important roles in plants. Appealing colors, for example, attract pollinators to flowers and animals to fruits. You are already familiar with accessory pigments: Carrots, for example, are orange because they contain beta-carotene (β-carotene), which functions as an antioxidant and also in signaling pathways between chloroplast and nucleus. Roses are red and violets are blue because their cells make anthocyanin, another antioxidant and also a natural sunscreen.

Different photosynthetic species use different combinations of accessory pigments during photosynthesis. Why? Like all organisms, photosynthesizers are adapted to the environment in which they evolved, and light that reaches different environments varies in its proportions of wavelengths. Consider that seawater absorbs green and blue-green light less efficiently than other colors. Thus, more green and blue-green light penetrates deep ocean water. Algae that can live far below the sea surface tend to be rich in accessory pigments that absorb green and blue-green light.

In green plants, chlorophylls are usually so abundant that they mask the colors of the accessory pigments. Plants that change color during autumn are preparing for a period of dormancy. These plants conserve resources by moving nutrients from tender parts (such as leaves) that would be damaged by winter cold to protected parts (such as roots). Chlorophylls are not needed during dormancy, so they are disassembled and their components recycled. Yellow and

orange accessory pigments are also recycled, but not as quickly as chlorophylls. Their colors begin to show as the chlorophyll content declines in leaves. Anthocyanin synthesis also increases in some plants, adding red and purple tones to turning leaf colors. (Chapter 27 returns to dormancy in plants.)

Figure 6.6 Photosynthetic pigments.

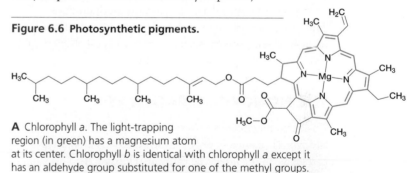

A Chlorophyll *a*. The light-trapping region (in green) has a magnesium atom at its center. Chlorophyll *b* is identical with chlorophyll *a* except it has an aldehyde group substituted for one of the methyl groups.

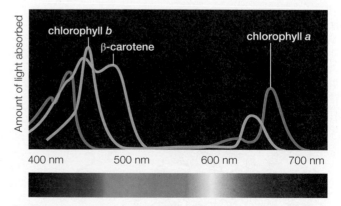

Amount of light absorbed

chlorophyll *b*

β-carotene

chlorophyll *a*

400 nm   500 nm   600 nm   700 nm

B Absorption spectra of three photosynthetic pigments reveal the efficiency with which each absorbs different wavelengths of visible light. Line color indicates the characteristic color of each pigment. Compare Figure 6.4B.

chlorophyll *a* (KLOR-uh-fil) Main photosynthetic pigment in plants.
pigment An organic molecule that can absorb light of certain wavelengths.
wavelength Distance between the crests of two successive waves.

LEARNING OBJECTIVES

Describe what happens to a chlorophyll molecule when it absorbs light.

Follow the path of electrons through the noncyclic pathway.

Explain the process of electron transfer phosphorylation.

Compare the cyclic and noncyclic pathways of photosynthesis.

A thylakoid membrane contains millions of light-harvesting complexes, which are circular arrays of chlorophylls and other pigments, lipids, and proteins (**Figure 6.7A**). When a pigment in a light-harvesting complex absorbs a photon, one of its electrons jumps to a higher energy level, or shell (Section 2.2). The electron quickly drops back down to a lower shell by emitting its extra energy as a photon. Light-harvesting complexes hold on to that photon by passing it back and forth, a bit like volleyball players pass a ball among a circle of team members. Photosynthesis begins when a photon being passed around the thylakoid membrane reaches and becomes absorbed by a photosystem. A **photosystem** is an extremely large complex of molecules in the thylakoid membrane; it consists of a reaction center—a core group of proteins, pigments, and cofactors—surrounded by a ring of light-harvesting complexes.

There are two versions of light-dependent reactions, a noncyclic pathway and a cyclic pathway. Both produce ATP. The noncyclic pathway, which is the primary one in plants, photosynthetic protists (such as algae), and cyanobacteria, also produces oxygen (O_2) and NADPH.

THE NONCYCLIC PATHWAY

Thylakoid membranes have two kinds of photosystems, type I and type II (named in order of their discovery). Both types participate in the noncyclic pathway (**Figure 6.8**). This path-

way begins when a photosystem II absorbs a photon **❶**. The photon's energy is absorbed by a "special pair" of closely associated chlorophyll molecules in the photosystem's reaction center (**Figure 6.7B**). Absorbing energy causes a special pair to release electrons that immediately enter an electron transfer chain in the thylakoid membrane. This is the step at which light energy is converted to chemical energy.

A photosystem that loses electrons must immediately replace them. Photosystem II does this by pulling electrons off of water molecules in the thylakoid compartment. A water molecule does not give up electrons easily; doing so breaks it apart into hydrogen ions and oxygen atoms **❷**. This and any other process by which a molecule is broken apart by light energy is called **photolysis**. The hydrogen ions stay in the thylakoid compartment; oxygen atoms combine and diffuse out of the cell as oxygen gas (O_2).

Meanwhile, the electrons released by the photosystem move through the electron transfer chain in the thylakoid membrane **❸**. Electron transfer chains release the energy of electrons little by little in a series of redox reactions (Section 5.4). In this case, molecules of the electron transfer chain use the released energy to actively transport hydrogen ions (H^+) across the membrane, from the stroma to the thylakoid compartment **❹**. Thus, the movement of electrons through the electron transfer chain sets up and maintains a hydrogen ion gradient across the thylakoid membrane.

Electrons that have reached the end of the electron transfer chain are accepted by a photosystem I. When this photosystem absorbs a photon, its special pair of chlorophylls releases electrons **❺** that immediately enter a second, different electron transfer chain. At the end of this second chain, the coenzyme $NADP^+$ accepts the electrons along with H^+, so NADPH forms **❻**:

$$NADP^+ + electrons + H^+ \longrightarrow \boxed{\textbf{NADPH}}$$

NADPH is a powerful reducing agent (electron donor) that will be used to make sugars in the second-stage reactions.

The hydrogen ion gradient that forms across the thylakoid membrane is a type of potential energy that can be tapped to make ATP. The ions want to follow their gradient by moving back into the stroma, but ions cannot diffuse through a lipid bilayer (Section 5.6). Hydrogen ions leave the thylakoid compartment only by flowing through proteins called ATP synthases embedded in the thylakoid membrane **❼**. An ATP synthase is a small molecular motor that is both a transport protein and an enzyme. When hydrogen ions flow through its interior, the protein phosphorylates ADP, so ATP forms in the stroma **❽**. Any process in which the flow of electrons through electron transfer chains drives ATP formation is called **electron transfer phosphorylation**.

Figure 6.7 Special arrangements of pigments in the thylakoid membrane. Chlorophylls are shown in green; beta-carotene and other pigments, in orange and yellow; lipids, in pink.

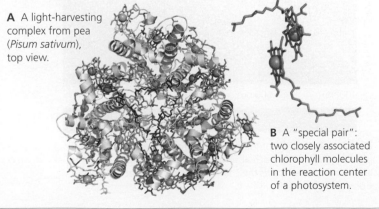

A A light-harvesting complex from pea (*Pisum sativum*), top view.

B A "special pair": two closely associated chlorophyll molecules in the reaction center of a photosystem.

CREDITS: (7A) Mechanisms of photoprotection and nonphotochemical quenching in pea light-harvesting complex at 2.5 Å resolution. Stanfuss, J., Terwisscha, Van Scheltinga, A.C., Lamborghini, M., Kuchlbrandt, W. Journal: (2005) *Embo J.* 24:919; Pub Med:15719016; PubMedCentral: PMC554132 DOI:10.1038/sj.embo, 7600585 made by Lisa Starr using pdb 2BHW; (7B) Architecture of the photosynthetic oxygen-evolving center. Ferreira, K.N., Iverson, T.M., Maghlaoui, K., Barber, J., Iwata, S. Journal: (2004) *Science* 303:1831–1838. PubMed:14764885 DOI:10.1126/science.1093087 made by Lisa Starr using pdb IS5L.

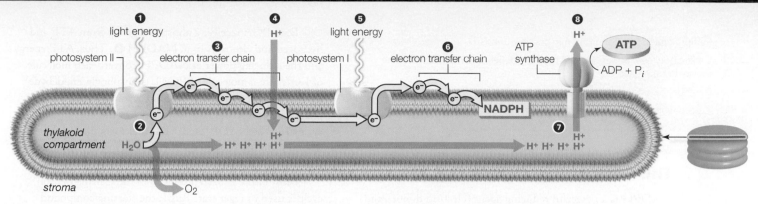

1 A photosystem II absorbs a photon and emits electrons.

2 The photosystem pulls replacement electrons from water molecules, which then break apart into hydrogen ions and oxygen. The oxygen leaves the cell as O_2 gas.

3 The electrons enter an electron transfer chain in the thylakoid membrane.

4 Energy released by the electrons as they move through the chain is used to actively transport hydrogen ions from the stroma into the thylakoid compartment. A hydrogen ion gradient forms across the thylakoid membrane.

5 A photosystem I absorbs a photon and emits electrons. Replacement electrons come from photosystem II via an electron transfer chain.

6 Electrons emitted from photosystem I move through a different electron transfer chain, then combine with $NADP^+$ and H^+ to form NADPH.

7 Hydrogen ions in the thylakoid compartment follow their gradient across the thylakoid membrane by flowing through the interior of ATP synthases.

8 Hydrogen ion flow causes ATP synthases to phosphorylate ADP, so ATP forms in the stroma.

Figure 6.8 Light-dependent reactions, noncyclic pathway. ATP and oxygen gas are produced in this pathway. Electrons that travel through two different electron transfer chains end up in NADPH. P_i is an abbreviation for phosphate group.

THE CYCLIC PATHWAY

Light-dependent reactions that occur in the cyclic pathway use only photosystem I (**Figure 6.9**). The pathway begins when a photosystem I absorbs a photon and emits electrons, which then enter an electron transfer chain. As in the noncyclic pathway, the electron transfer chain uses energy released from the electrons to move hydrogen ions into the thylakoid compartment, and the resulting hydrogen ion gradient drives ATP formation. However, after electrons pass through the electron transfer chain, they are not used to reduce $NADP^+$, so NADPH does not form. Instead, the electrons return to photosystem I. This photosystem does not replace lost electrons by pulling them off of water, so O_2 does not form.

The cyclic pathway evolved before the noncyclic pathway. Organisms that used the older reactions produced ATP photosynthetically, but not the NADPH required to make sugars. Later, the molecular machinery of the cyclic pathway evolved to include additional reactions. Photosystem I became remodeled into photosystem II, and the new photosystem was added to the existing reactions. The new noncyclic pathway that

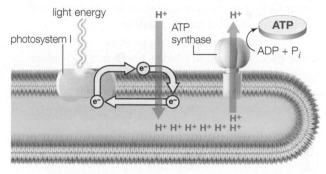

Figure 6.9 Light-dependent reactions, cyclic pathway. Electrons released from photosystem I enter an electron transfer chain, then cycle back to the photosystem. ATP is produced, but not oxygen or NADPH.

resulted offered an essentially unlimited supply of electrons (from water). Stored in NADPH, these electrons could now be used to produce sugars from CO_2, as you will see shortly.

The cyclic pathway still operates in nearly all photosynthesizers. On its own, the noncyclic pathway does not yield enough ATP to balance NADPH use in sugar production. The cyclic pathway provides extra ATP for this purpose. It also allows light-dependent reactions to continue when the noncyclic pathway stalls, for example as occurs under intense illumination. Light energy in excess of what can be used for photosynthesis can cause dangerous free radicals to form (Section 2.2). Photosystem II minimizes this effect. In strong light, it stops releasing electrons to electron transfer chains; instead, it emits absorbed photon energy as heat. The cyclic pathway becomes the dominant one in this circumstance.

electron transfer phosphorylation Process in which electron flow through electron transfer chains sets up a hydrogen ion gradient that drives ATP formation.

photolysis (foe-TALL-ih-sis) Process by which light energy breaks down a molecule.

photosystem Large protein complex in the thylakoid membrane; consists of pigments and other molecules that collectively convert light energy to chemical energy in photosynthesis.

State the role of carbon fixation in photosynthesis.

Explain how coenzymes connect the light-dependent reactions with the Calvin–Benson cycle.

Describe stomata and their function.

Explain why photorespiration makes sugar production inefficient.

THE CALVIN–BENSON CYCLE

NADPH is a powerful reducing agent (electron donor), and a lot of it is required to run the second stage of photosynthesis. You learned in Section 6.1 that reactions of this stage are light-independent because light energy does not power them. Energy that drives these reactions is provided by phosphate-group transfers from ATP, and electrons from NADPH.

The light-independent reactions, which are collectively called the **Calvin–Benson cycle**, produce sugars in the stroma of chloroplasts (**Figure 6.10**). This cyclic pathway uses carbon atoms from CO_2 to build carbon backbones of the sugar molecules. Extracting carbon atoms from an inorganic source (such as CO_2) and incorporating them into an organic molecule is a process called **carbon fixation**.

The Calvin–Benson cycle begins when the enzyme **rubisco** fixes carbon by attaching CO_2 ❶ to a five-carbon organic compound, RuBP (ribulose bisphosphate). The six-carbon intermediate that forms is unstable, and it splits into two three-carbon molecules of PGA (phosphoglycerate).

Each PGA receives a phosphate group from ATP, and hydrogen and electrons from NADPH ❷. Thus, ATP energy and the reducing power of NADPH convert each molecule of PGA into a molecule of PGAL (phosphoglyceraldehyde). PGAL is a three-carbon simple sugar with an attached phosphate group, and it is the product of the Calvin–Benson cycle ❸. (The $NADP^+$ and ADP that form during the reactions diffuse back to the thylakoid for reuse in the light-dependent reactions.)

Most of the PGAL produced in the Calvin–Benson cycle is used to regenerate RuBP, the starting compound of the cycle ❹. In plants, the remaining PGALs are usually exported from the chloroplast to the cell's cytoplasm. PGAL is an intermediate in several metabolic pathways, and it can be assembled into a variety of carbohydrates. Most of the PGAL molecules that enter plant cell cytoplasm are combined to make sucrose ❺, the main sugar in plants (Section 3.2). Sucrose produced by photosynthetic cells is loaded into vascular tissues for transport to other parts of the plant. (Section 26.5 returns to the transport of organic molecules through the plant body.)

When sucrose production is high (for example, during the day when the light-dependent reactions are running), some of the PGAL molecules that stay in chloroplasts are assembled into starch. The starch is disassembled at night, and its monosaccharide monomers are used to produce sucrose. An uninterrupted supply of sucrose sustains the plant's metabolism and growth even in the dark.

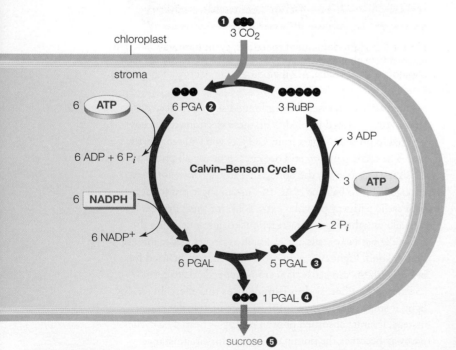

Figure 6.10 The Calvin–Benson cycle in a chloroplast.

This is a cross-section of a chloroplast with Calvin–Benson reactions cycling in the stroma. The steps are a summary of three cycles of reactions. Water is also a substrate, but not shown for clarity (Appendix III details the reactions). Black balls signify carbon atoms.

❶ Three CO_2 molecules diffuse into a photosynthetic cell, and then into a chloroplast. Rubisco attaches each to a five-carbon RuBP molecule. The resulting intermediates split, so six molecules of three-carbon PGA form.

❷ Each PGA gets a phosphate group from ATP, plus hydrogen and electrons from NADPH, so six PGAL form. This phosphorylated sugar is the product of the Calvin–Benson cycle.

❸ Five PGAL continue in reactions that regenerate 3 RuBP.

❹ The remaining PGAL is usually exported from the chloroplast. In plant cell cytoplasm, PGAL molecules are combined to form sucrose ❺.

FIGURE IT OUT For every molecule of carbon dioxide that enters the Calvin–Benson cycle, how many NADPH are reduced?

Answer: Two

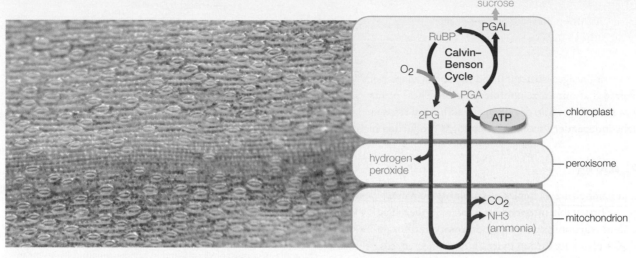

A Stomata on the surface of a leaf. When these tiny pores are open, they allow gas exchange between the plant's internal tissues and air.

Stomata close to conserve water on hot, dry days, and then gas exchange stops. Oxygen produced by the light-dependent reactions cannot exit the plant, and CO_2 required for the light-independent reactions cannot enter it. Thus, the amount of oxygen rises and the amount of CO_2 declines in photosynthetic tissues.

Figure 6.11 Photorespiration.

B Photorespiration occurs when the oxygen/CO_2 ratio increases in photosynthetic tissues, causing rubisco to use oxygen as a substrate. Several reactions are necessary to convert the product of this reaction to a molecule that can reenter the Calvin–Benson cycle. Intermediates are transported from the chloroplast to a peroxisome, then a mitochondrion, a peroxisome again, and back to a chloroplast. The energy required to carry out the extra steps make photorespiration an inefficient way to produce sugars.

PHOTORESPIRATION

Most plants have a thin, waterproof cuticle that limits evaporative water loss from their aboveground parts. Gases cannot diffuse across the cuticle, but oxygen produced by the light-dependent reactions must escape the plant, and carbon dioxide needed for the Calvin–Benson cycle must enter it. Thus, the surfaces of leaves and stems are studded with tiny, closable gaps called **stomata** (singular, stoma). When stomata are open, CO_2 diffuses from the air into photosynthetic tissues, and O_2 diffuses out of the tissues into the air (**Figure 6.11A**). Stomata close to conserve water on hot, dry days. When that happens, gas exchange comes to a halt.

Plants that fix carbon only by the Calvin–Benson cycle are called **C3 plants** because a three-carbon molecule (PGA) is the first stable intermediate to form in their light-independent reactions. In C3 plants, both stages of photosynthesis run during the day. With stomata closed, the O_2 level in the plant's tissues rises, and the CO_2 level declines. This outcome can reduce the efficiency of sugar production because both gases are substrates of rubisco, and they compete for its active site. Rubisco initiates the Calvin–Benson cycle by attaching CO_2 to RuBP. It also initiates a pathway called **photorespiration** by attaching O_2 to RuBP. The remainder of the photorespiration pathway converts the product of this reaction to a substrate of the Calvin–Benson cycle. ATP is required, and intermediates must be transported

among three organelles (**Figure 6.11B**). Photorespiration produces CO_2 (so carbon is lost instead of being fixed), and also ammonia that must be detoxified (requiring additional ATP). These extra steps and energy requirements make photorespiration an extremely inefficient way to produce sugars. C3 plants compensate for the inefficiency by making a lot of rubisco: It is the most abundant protein on Earth.

Despite being wasteful in terms of sugar production, photorespiration is an essential part of a wide range of processes in the plant. For example, it is a major source of hydrogen peroxide in photosynthetic cells, and as such it is required for several signaling pathways that govern growth and defense responses. (Chapter 27 returns to this topic.)

C3 plant Type of plant that uses only the Calvin–Benson cycle to fix carbon.
Calvin–Benson cycle Cyclic carbon-fixing pathway that builds sugars from CO_2; the light-independent reactions of photosynthesis.
carbon fixation Process by which carbon from an inorganic source such as carbon dioxide is incorporated into an organic molecule.
photorespiration Pathway initiated by rubisco when it attaches oxygen instead of carbon dioxide to RuBP (ribulose bisphosphate).
rubisco (roo-BIS-co) Ribulose bisphosphate carboxylase. Carbon-fixing enzyme of the Calvin–Benson cycle.
stomata (stow-MA-tuh) Gaps that open on plant surfaces; allow water vapor and gases to diffuse into and out of plant tissues.

6.5 Carbon-Fixing Adaptations of Plants

Explain how C4 and CAM plants minimize photorespiration.

In some plant lineages, structural and metabolic adaptations have evolved that minimize photorespiration. These plants also close stomata on hot, dry days, but additional steps in their light-independent reactions keep sugar production high.

C4 PLANTS

Corn and bamboo are examples of **C4 plants**, so named because the first stable intermediate to form in their light-independent reactions is a four-carbon compound (oxalo-acetate). C4 plants fix carbon twice, in two kinds of cells (**Figure 6.12A**). The reactions begin in mesophyll cells, where carbon is fixed by an enzyme that does not use oxygen even when the CO_2 level is low (**Figure 6.12B**). An intermediate (malate) is moved into bundle-sheath cells, where it is converted back to CO_2. Rubisco fixes carbon for the second time as the CO_2 enters the Calvin–Benson cycle. Bundle-sheath cells in C4 plants have chloroplasts that carry out light-dependent reactions, but only in the cyclic pathway. No oxygen is released, so the oxygen level near rubisco stays low. This, along with the high CO_2 level provided by the C4 reactions, minimizes photorespiration. The pathway uses an extra ATP per PGAL, but efficient sugar production in hot, dry weather more than compensates for the loss (**Figure 6.12C**).

CAM PLANTS

Succulents, cacti, and other **CAM plants** use a carbon-fixing pathway that allows them to conserve water even in desert regions with extremely high daytime temperatures. CAM stands for crassulacean acid metabolism, after the Crassula-ceae family of plants in which this pathway was first studied. Like C4 plants, CAM plants fix carbon twice, but the reactions occur at different times rather than in different cells. Stomata on a CAM plant open at night, when typically lower temperatures minimize evaporative water loss. The plants fix carbon from CO_2 in the air at this time. An intermediate (malate) is stored in the cell's central vacuole. When stomata close the next day, the malate is moved out of the vacuole and converted back to CO_2. Rubisco then fixes carbon for the second time as the CO_2 enters the Calvin–Benson cycle.

C4 plant Type of plant that minimizes photorespiration by fixing carbon twice, in two cell types.

CAM plant Type of plant that conserves water by fixing carbon twice, at different times of day.

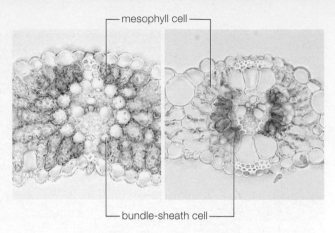

A In a C3 plant (barley, left), chloroplasts—the sites of carbon fixation—occur mainly in mesophyll cells. A C4 plant (millet, right) has chloroplasts in mesophyll cells and in bundle-sheath cells, so it can carry out photosynthesis in both types of cells. C4 plants minimize photorespiration by fixing carbon twice, in the two cell types.

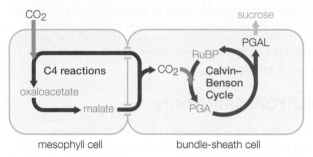

B In C4 plants, as in C3 plants, oxygen builds up when stomata close during photosynthesis. However, C4 plants fix carbon twice, first in mesophyll cells and then in bundle-sheath cells. This strategy minimizes photorespiration because it keeps the CO_2 concentration high and the O_2 concentration low near rubisco.

C Crabgrass "weeds" overgrowing a lawn. Crabgrasses, which are C4 plants, thrive in hot, dry summers, when they easily outcompete Kentucky bluegrass and other fine-leaved C3 grasses commonly planted in residential lawns.

Figure 6.12 C4 plant adaptations minimize photorespiration.

CREDITS: (12A) Eri Maai, Shouu Shimada, Masahiro Yamada, Tatsuo Sugiyama, Hiroshi Miyake, Mitsutaka Taniguchi; The avoidance and aggregative movements of mesophyll chloroplasts in C4 monocots in response to blue light and abscisic acid; Journal of Experimental Botany, May 2011, Volume 62, issue number 9, 3213–3221, by permission of Oxford University Press; (12C) Image courtesy msuturfweeds.net.

Figure 6.13 Air pollution—smog—from fossil fuel use blankets Lianyungang, China, in December 2013. The brown color of smog comes from a gas (nitric oxide) that is toxic in large amounts. Invisible components include other nitrogen compounds, sulfur compounds, organic molecules, mercury and other heavy metals, and carbon dioxide.

A BURNING CONCERN

Photosynthesis removes carbon dioxide from the atmosphere, and fixes its carbon atoms in organic compounds. The abundant energy locked up in the chemical bonds of those compounds can fuel heterotrophs, as when an animal cell makes ATP by aerobic respiration (the topic of the next chapter). It can also fuel our cars. Both processes are fundamentally the same: They release energy by breaking the bonds of organic molecules. Both use oxygen to break those bonds, and both produce CO_2.

As early as 8,000 years ago, humans began burning forests to clear land for agriculture. Burning breaks the bonds of cellulose and other organic compounds in plant tissues, and releases their carbon mainly as CO_2. Today, we are burning fossil fuels—coal, petroleum, and natural gas—to satisfy our increasing demands for energy. Fossil fuels are the organic remains of ancient organisms. When we burn these fuels, carbon that has been locked in organic molecules for hundreds of millions of years is released in CO_2 that reenters the atmosphere.

Human activities are adding far more CO_2 to the atmosphere than photosynthetic organisms are removing from it, and the resulting imbalance is the major cause of global climate change (Chapter 44 returns to this topic). In 2014 alone, humans released 36 billion tons of CO_2 into the atmosphere—almost twice as much as photosynthesis removed from the atmosphere during the same year. By far, most of the CO_2 that humans release comes from burning fossil fuels (**Figure 6.13**).

Burning biofuels also releases CO_2 into the atmosphere, but it contributes less to global climate change. Biofuels are oils, gases, or alcohols made from organic

Sanga Moses

National Geographic Explorer Sanga Moses has a vision: to provide clean, inexpensive cooking energy to all Africans while improving socioeconomic outcomes and reversing deforestation. He grew up barefoot in a small Ugandan village of thatched roof dwellings that lacked electricity. Yet he became his clan's first college graduate and took a bank job in a big city. Returning home for a visit in 2009, he met his 12-year-old sister on the road. "She stood there crying, with a heavy bundle of wood on her head," Moses remembers. "She was upset because, like most rural girls, she missed days of school each week searching for fuelwood." This troubled him. "My sister was losing the only opportunity she had to make her life better: education."

The problems that wood burning created for Moses's family and in his hometown can be seen across sub-Saharan Africa. Eight in ten people in the region depend on wood to cook and to heat their homes. As more forests are destroyed to feed that demand—in Uganda, 70 percent of protected forests are gone—families must walk more miles every day to buy increasingly scarce and costly wood. "Burning fuelwood not only destroys Uganda's trees," Moses says, but it also affects "the health and educational opportunities of our poorest people."

Moses immediately quit his job and began working with engineering students to find a sustainable solution. Now, 2,500 farmers use his kilns to burn farm waste—coffee husks and waste from sugarcane and rice. A company that Moses founded buys the resulting "char" and turns it into briquettes for cooking that burn cleaner and cost less than wood. The company takes those briquettes to market, providing fuel for more than 19,000 Ugandan families.

matter that is not fossilized. Most biofuels are made from plants, and growing plants for fuel recycles carbon that is already in the atmosphere. Unlike fossil fuels, biofuels are a renewable source of energy: We can always make more simply by growing more plants. A lot of corn is grown to make ethanol, for example. The starch in corn kernels is enzymatically broken down to glucose, which is converted to ethanol by bacteria or yeast. Biofuels are also made from weeds, wood chips, or agricultural wastes. These materials have a higher proportion of cellulose, and making biofuels from them requires additional steps to break down the tough carbohydrate to its glucose monomers.

Section 6.1   Plants and other **autotrophs** make their own food using energy from the environment and carbon from inorganic sources such as carbon dioxide. By metabolic pathways of **photosynthesis**, plants and most other autotrophs capture the energy of light and use it to build sugars from water and CO_2. Humans and almost all other **heterotrophs** obtain carbon and energy from organic molecules originally assembled by photosynthetic organisms.

In eukaryotes, photosynthesis occurs in chloroplasts: the **light-dependent reactions** at the **thylakoid membrane**, and the **light-independent reactions** in **stroma**. In the light-dependent reactions, light energy drives the formation of NADPH and ATP. In turn, these coenzymes drive the synthesis of sugars from water and carbon dioxide in the light-independent reactions:

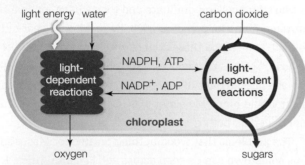

Section 6.2   Visible light is a small part of the electromagnetic energy radiating from the sun. That energy travels in waves, and it is organized as photons. We see different **wavelengths** of visible light as different colors. Visible light powers photosynthesis, which begins when photons are absorbed by photosynthetic **pigments**. The main photosynthetic pigment, **chlorophyll a**, absorbs violet and red light, so it appears green. Most photosynthetic organisms use a combination of pigments to capture light for photosynthesis.

Section 6.3   Photosynthetic pigments are part of light-harvesting complexes in the thylakoid membrane. In the light-dependent reactions, the pigments transfer the energy of light to **photosystems**. Receiving energy causes photosystems to release electrons that enter electron transfer chains. Electron flow through electron transfer chains sets up a hydrogen ion gradient that drives ATP formation, a process called **electron transfer phosphorylation**. The noncyclic pathway uses two photosystems. Water molecules are split (**photolysis**), oxygen is released, and electrons end up in NADPH. The cyclic pathway uses only photosystem I. No NADPH forms, and no oxygen is released.

Section 6.4   NADPH and ATP produced by the light-dependent reactions power the light-independent reactions of the **Calvin–Benson cycle**, which builds sugars from CO_2. The reactions begin when the enzyme **rubisco** carries out

carbon fixation by attaching CO_2 to an organic molecule. When **stomata** close on hot, dry days, plant tissues cannot exchange gases with air. The resulting high O_2/CO_2 ratio causes rubisco to initiate **photorespiration**, a pathway that reduces the efficiency of sugar production in **C3 plants**.

Section 6.5   Some plants limit photorespiration by fixing carbon twice. **C4 plants** separate the two sets of reactions by carrying them out in different cell types; **CAM plants** separate the reactions in time.

Application   Autotrophs remove CO_2 from the atmosphere, and the metabolic activity of most organisms puts it back. Humans disrupt this cycle by burning fossil fuels, which adds extra CO_2 to the atmosphere. The resulting imbalance is the major cause of global climate change.

SELF-ASSESSMENT
ANSWERS IN APPENDIX VII

1. A cat eats a bird, which ate a caterpillar that chewed on a weed. Which of these organisms are autotrophs? Which ones are heterotrophs?

2. Plants use _____ as an energy source to drive photosynthesis.
    a. sunlight   b. hydrogen ions   c. O_2 d. CO_2

3. In plants, photosynthetic eukaryotes, and cyanobacteria, the light-dependent reactions proceed in/at the _____ .
    a. thylakoid membrane   c. stroma
    b. plasma membrane   d. cytoplasm

4. Most of the carbon that land plants use for photosynthesis comes from _____ .
    a. sugars   c. water
    b. the atmosphere   d. soil

5. Which of the following statements is incorrect?
    a. Pigments absorb light of certain wavelengths only.
    b. Some accessory pigments are antioxidants.
    c. Chlorophyll is green because it absorbs green light.

6. When a photosystem absorbs light energy, _____ .
    a. sugar phosphates are produced
    b. electrons are transferred to ATP
    c. electrons are ejected from its special pair

7. In the light-dependent reactions, _____ .
    a. carbon dioxide is fixed   c. CO_2 accepts electrons
    b. ATP forms   d. sugars form

8. In the light-dependent reactions, what accumulates in the thylakoid compartment of chloroplasts?
    a. sugars   b. hydrogen ions   c. O_2 d. CO_2

9. The atoms in the oxygen molecules released during photosynthesis come from _____ .
    a. sugars   b. CO_2 c. water   d. O_2

ENERGY EFFICIENCY OF BIOFUEL PRODUCTION Most of the plant material currently used for biofuel production consists of food crops—mainly corn, soybeans, and sugarcane. In 2006, David Tilman and his colleagues published the results of a 10-year study comparing the net energy output of various biofuels. The researchers grew a mixture of native perennial grasses without irrigation, fertilizer, pesticides, or herbicides, in sandy soil that was so depleted by intensive agriculture that it had been abandoned. They measured the usable energy in biofuels made from the grasses, and also from corn and soy, then measured the energy it took to grow and produce biofuel from each kind of crop (**Figure 6.14**).

1. About how much energy did ethanol produced from one hectare of corn yield? How much energy did it take to grow the corn to make that ethanol?

2. Which of the biofuels tested had the highest ratio of energy output to energy input?

3. Which of the three crops would require the least amount of land to produce a given amount of biofuel energy?

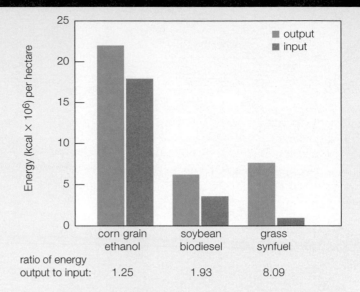

| ratio of energy output to input: | 1.25 | 1.93 | 8.09 |

Figure 6.14 Energy inputs and outputs of biofuels from different sources. One hectare is about 2.5 acres.

10. In chloroplasts, the light-independent reactions proceed at/in the _____ .
    a. thylakoid membrane   c. stroma
    b. plasma membrane   d. cytoplasm

11. The Calvin–Benson cycle starts with _____ .
    a. the absorption of photon energy
    b. carbon fixation
    c. the release of electrons from photosystem II
    d. $NADP^+$ formation

12. Which of the following substances does *not* participate in the Calvin–Benson cycle?
    a. ATP   c. NADPH   e. PGAL
    b. O_2 d. RuBP   f. CO_2

13. Closed stomata _____ .
    a. limit gas exchange   c. prevent photosynthesis
    b. permit water loss   d. absorb light

14. In C3 plants, _____ makes sugar production inefficient when stomata close on hot, dry days.
    a. photosynthesis   c. photorespiration
    b. photolysis   d. carbon fixation

15. Match each term with its most suitable description.

    ___ PGAL formation   a. absorbs light
    ___ CO_2 fixation   b. converts light to chemical energy
    ___ autotroph
    ___ ATP forms;   c. self-feeder
        NADPH does not   d. electrons cycle back to photosystem I
    ___ photorespiration
    ___ photosynthesis   e. problem in C3 plants
    ___ pigment   f. Calvin–Benson cycle product
    ___ photolysis in   g. water molecules split
        photosynthesis   h. rubisco function

CLASS ACTIVITY

CRITICAL THINKING

1. About 200 years ago, Jan Baptista van Helmont wanted to know where growing plants get the materials necessary for increases in size. He planted a tree seedling weighing 2.2 kilograms (5 pounds) in a barrel filled with 90 kilograms (200 pounds) of soil and then watered the tree regularly. After five years, the tree had gained almost 75 kilograms (164 pounds), and the soil's weight was unchanged. He incorrectly concluded that the tree had gained all of its additional weight by absorbing water. How did the tree really gain most of its weight?

2. While gazing into an aquarium, you observe bubbles coming from an aquatic plant (left). What are the bubbles?

3. A C3 plant absorbs a carbon radio-isotope (as part of $^{14}CO_2$). In which stable organic compound does the labeled carbon first appear?

4. In 2005, a new species of green sulfur bacteria was discovered near a geothermal vent at the bottom of the ocean. An exceptionally efficient light-harvesting mechanism allows these bacteria to carry out photosynthesis where the only illumination is a dim volcanic glow emanating from the vent. Cell membranes are required for electron transfer chains in the light-dependent reactions, but these bacteria have no internal membranes. How do you think they carry out the light-dependent reactions?

7 Releasing Chemical Energy

Links to Earlier Concepts

This chapter focuses on metabolic pathways (Section 5.4) that harvest energy (5.1) stored in the chemical bonds of sugars (3.2). Some reactions (3.1) of these pathways occur in mitochondria (4.7). You will revisit free radicals (2.2), lipids (3.3), proteins (3.4), electron transfer chains (5.4), coenzymes (5.5), membrane transport (5.6, 5.7), and photosynthesis (6.1).

Core Concepts

Evolution underlies the unity and diversity of life.

 Photosynthesis and respiration are linked by reactants, products, and energy flow. The evolution of photosynthesis in bacteria permanently changed Earth's atmosphere by adding oxygen gas, which wiped out most life on Earth. Aerobic respiration evolved in some of the survivors; molecules and processes that already existed were reused in this new pathway.

Organisms exchange matter and energy with the environment in order to grow, maintain themselves, and reproduce.

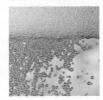 All organisms require ongoing inputs of energy. Cells store energy by building organic molecules. They retrieve energy stored in organic molecules by breaking apart the carbon backbones. Redox reactions that capture electrons in coenzymes are typically part of energy-harvesting pathways.

Organisms have various strategies to capture and store energy for use in biological processes.

 Most organisms release free energy in carbon compounds by breaking them down in either aerobic respiration or anaerobic fermentation. Both pathways store the released energy in ATP, and both begin with the same reactions in cytoplasm, but their concluding reactions use different electron acceptors. Aerobic respiration, the more efficient way of making ATP, uses oxygen to accept electrons.

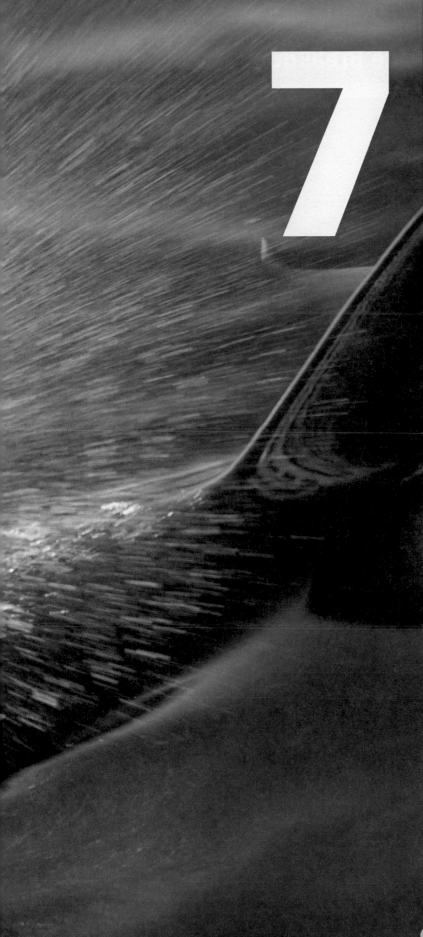

Like you, a whale breathes air to provide its cells with a fresh supply of oxygen for aerobic respiration. Carbon dioxide released from aerobically respiring cells leaves the body in each exhalation.

Photograph by Ralph Lee Hopkins, National Geographic Creative.

Explain how oxygen became a major component of Earth's atmosphere.

Outline the events that led to the evolution of aerobic respiration.

Compare the processes of cellular respiration and fermentation.

Figure 7.1 Artist's view of Earth's early atmosphere. Earth's early atmosphere may have been abundant in gases such as methane, hydrogen sulfide, and ammonia. There was no oxygen in the air before the noncyclic pathway of photosynthesis evolved, about 2.5 billion years ago.

Autotrophs harvest energy directly from the environment and store it in the form of sugars and other carbohydrates. They and all other organisms use energy stored in sugars to power various endergonic reactions that sustain life (Section 5.2). However, in order to use the energy stored in sugars, cells must first transfer it to molecules—especially ATP—that can participate directly in these reactions.

Remember, organic molecules hold energy that can be released when their carbon–carbon bonds are broken (Section 5.2). Cells harvest energy from an organic molecule by breaking its carbon backbone, one bond at a time. This releases the energy of the molecule stepwise, in small increments that can be captured for cellular work. A number of different pathways drive ATP synthesis by breaking the bonds of sugar molecules this way. All of them are ancient.

The first cells we know of appeared on Earth about 3.4 billion years ago, and they lived in a world we probably would not recognize (**Figure 7.1**). The atmosphere contained no oxygen, and it would stay that way for another billion years. Organisms at the time were necessarily **anaerobic**, which means they lived in the absence of oxygen. Current research suggests these cells released energy from sugars by cellular respiration. **Cellular respiration** refers to any pathway that uses an electron transfer chain (Section 5.4) to harvest energy from an organic molecule and make ATP. You already learned about one pathway that couples ATP synthesis with electron transfer chains: electron transfer phosphorylation, which is part of photosynthesis (Section 6.3). Electron transfer phosphorylation is also part of cellular respiration, but the electrons come from organic molecules, not from water as occurs in photosynthesis.

When the noncyclic pathway of photosynthesis evolved (Section 6.3), oxygen gas (O_2) released from uncountable numbers of water molecules began seeping out of photosynthetic prokaryotes. In what may have been the earliest case of catastrophic air pollution, the new abundance of oxygen gas exerted tremendous pressure on all life at the time. Why? Molecules that carry electrons are a critical part of metabolism. Oxygen gas easily removes the electrons from (oxidizes) these molecules, interfering with their function and also producing dangerous free radicals in the process (Section 2.2). Most cells had no way to counter these effects of oxygen and so were wiped out everywhere except deep water, muddy sediments, and other habitats that remained anaerobic.

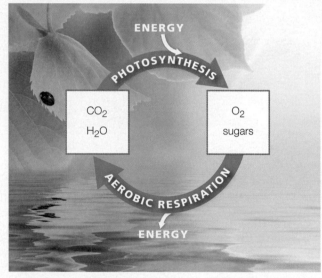

Figure 7.2 Substrates and products link photosynthesis with aerobic respiration. Photosynthesis produces sugars and, in most organisms, it also releases oxygen. The breakdown of sugars in aerobic respiration requires oxygen, and it produces carbon dioxide and water—the raw materials from which photosynthetic organisms make the sugars in the first place.

By lucky circumstance, a few types of cells were already making antioxidants (Section 5.5) that could prevent oxidative damage caused by O_2 gas. They were the first **aerobic** organisms—they could live in the presence of oxygen. As these organisms evolved in environments with abundant oxygen, so did their metabolic pathways. Oxygen, the reactive molecule that had poisoned most life on Earth, turned out to be a useful addition to cellular respiration, and many lineages incorporated it into their pathways.

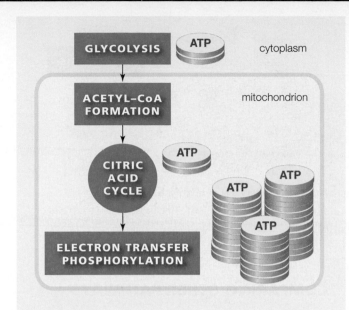

TABLE 7.1

Inputs and Outputs Linking Steps in Aerobic Respiration

Pathway	Main Reactants	Net Yield
Glycolysis	1 six-carbon sugar (such as glucose)	2 pyruvate 2 NADH 2 ATP
Acetyl–CoA formation	2 pyruvate	2 acetyl–CoA 2 CO_2 2 NADH
Citric acid cycle	2 acetyl–CoA	4 CO_2 6 NADH 2 $FADH_2$ 2 ATP
Electron transfer phosphorylation	10 NADH 2 $FADH_2$ O_2	32 ATP (approx.) H_2O
Overall	1 six-carbon sugar O_2	6 CO_2 H_2O 36 ATP (approx.)

Figure 7.3 Overview of aerobic respiration. Left, aerobic respiration consists of four pathways: glycolysis, acetyl–CoA formation, the citric acid cycle, and electron transfer phosphorylation. In eukaryotes only, the last three steps occur in mitochondria. Right, Table 7.1 summarizes the reactants and products of each step.

Today, most cellular respiration uses oxygen. Oxygen-requiring cellular respiration is called **aerobic respiration**. Aerobic respiration means "breathing air," and the name is a reference to this pathway's requirement for oxygen. With each breath, you take in oxygen for your trillions of aerobically respiring cells. You exhale the pathway's products: carbon dioxide and water. Carbon dioxide and water are the raw materials of photosynthesis, which produces sugars and, in most organisms, releases oxygen (Section 6.1). Thus, raw materials and products link the two pathways in most of the biosphere (**Figure 7.2**).

REACTION PATHWAYS

Aerobic respiration harvests energy from an organic molecule—glucose, in particular—by completely breaking apart its carbon backbone, bond by bond. The following equation summarizes the overall pathway:

$$C_6H_{12}O_6 \ + \ O_2 \longrightarrow CO_2 \ + \ H_2O \ + \ \text{ATP}$$

The equation means that glucose and oxygen are converted to carbon dioxide and water, for a yield of ATP. However, aerobic respiration is not a single reaction. Rather, it consists of four separate metabolic pathways: glycolysis, acetyl–CoA formation, the citric acid cycle, and electron transfer phosphorylation (**Figure 7.3**). These four pathways are linked by products and substrates (**Table 7.1**), and all involve electron transfers (redox reactions, Section 5.4).

Fermentation refers to glucose-breakdown pathways that make ATP without the use of oxygen or electron transfer chains. Many organisms supplement cellular respiration with fermentation; a few species of bacteria that have lost the ability to carry out cellular respiration use fermentation exclusively. Fermentation pathways do not break all of the carbon–carbon bonds in glucose, so their products include two-carbon or three-carbon organic molecules. Like aerobic respiration, fermentation begins with the pathway of glycolysis. Unlike aerobic respiration, fermentation does not include electron transfer phosphorylation, and it does not produce many ATP. Fermentation is efficient enough to sustain some single-celled species. It also helps cells of multicelled species produce ATP, especially under anaerobic conditions. However, aerobic respiration is a much more efficient way of harvesting energy from glucose. You and other large, multicelled organisms could not live without its higher yield.

aerobic (air-OH-bick) Involving or occurring in the presence of oxygen.
aerobic respiration Oxygen-requiring cellular respiration; breaks down organic molecules (particularly glucose) and produces ATP, carbon dioxide, and water.
anaerobic (an-air-OH-bick) Occurring in the absence of oxygen.
cellular respiration Pathway that breaks down an organic molecule to form ATP and includes an electron transfer chain.
fermentation Anaerobic glucose-breakdown pathway that produces ATP without use of an electron transfer chain.

7.2 Glycolysis—Sugar Breakdown Begins

LEARNING OBJECTIVES

Write an equation that summarizes glycolysis.

Explain the difference between substrate-level phosphorylation and electron transfer phosphorylation.

Describe the transfer of energy in glycolysis.

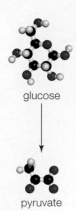

glucose

pyruvate

Glycolysis is a series of reactions that produce ATP by converting glucose to **pyruvate**, an organic compound with a three-carbon backbone (left). The pathway occurs in the cytoplasm of all cells (**Figure 7.4**), and it is the first step in both aerobic respiration and fermentation. Its name (which comes from the Greek words *glyk-*, sweet, and *-lysis*, loosening) refers to the release of chemical energy from sugars. Glycolysis breaks one carbon–carbon bond of a glucose molecule. The energy released when that bond breaks is captured in electrons carried by NADH, and in high-energy phosphate bonds of ATP. The reactions use two ATP and produce four, so we say the net yield of glycolysis is two ATP.

GLYCOLYSIS REACTIONS

Glycolysis begins when a molecule of glucose enters a cell through a glucose transporter, a passive transport protein that you encountered in Section 5.7. The cell invests two ATP in the endergonic reactions that begin the pathway (**Figure 7.5**). In the first reaction, a phosphate group is transferred from ATP to the glucose, thus forming glucose-6-phosphate ❶ (the enzyme hexokinase is pictured catalyzing this reaction in Figure 5.9D). A phosphate group from a second ATP is transferred to an intermediate ❷, so a six-carbon molecule with two phosphate groups forms. This molecule is split in half to form two molecules of PGAL ❸, a phosphorylated three-carbon sugar. Both PGAL molecules continue in glycolysis, so the remaining reactions are carried out in duplicate.

At this point in glycolysis, two ATP have been invested to break one carbon–carbon bond, but no energy has been harvested. Energy-harvesting reactions begin with a redox reaction that transfers electrons and a hydrogen ion from PGAL to the coenzyme NAD⁺, which is thereby reduced to NADH ❹. (You will see in Section 7.4 how the energy of electrons carried by this NADH helps power the final stage of aerobic respiration.) The reaction also attaches a second phosphate group to the three-carbon intermediate.

The next reaction transfers one of the phosphate groups to ADP, so ATP forms ❺. The remaining phosphate group is transferred from one carbon to another ❻, and then to another ADP, so ATP forms again ❼. Pyruvate is the product of the final reaction.

glucose

GLYCOLYSIS

2 pyruvate

2 ATP

2 NADH

Figure 7.4 Overview of glycolysis.

Glycolysis converts one molecule of glucose to two molecules of pyruvate, for a net yield of two ATP and two NADH. An animal cell is shown, but the pathway occurs in the cytoplasm of all cells, prokaryotic and eukaryotic.

COMPARING OTHER PATHWAYS

Aerobic respiration has more than a few molecules and processes in common with photosynthesis, an outcome of evolutionary repurposing (later chapters return to this topic). For example, glycolysis uses some of the same enzymes as the Calvin–Benson cycle (Section 6.4), and some of the same intermediates form (such as PGAL). Also, ATP forms by substrate-level phosphorylation in both pathways. In **substrate-level phosphorylation**, a phosphate group is transferred directly from a phosphorylated molecule to ADP, so ATP forms. Note that substrate-level phosphorylation differs from electron transfer phosphorylation, which is the way ATP forms during the light-dependent reactions of photosynthesis (Section 6.3). In electron transfer phosphorylation, the movement of electrons through electron transfer chains sets up a hydrogen ion gradient, and the resulting flow of hydrogen ions through ATP synthases causes these proteins to attach a free phosphate group to ADP.

A pathway that builds glucose from pyruvate is almost the reverse of glycolysis, and the two pathways must be tightly regulated or the cell would make glucose and break it down at the same time. Consider how the two-ATP investment during the first reactions of glycolysis prevents the pathway from running in reverse. Remember from Section 5.4 that most reactions can run backward as well as forward. After two inputs of ATP energy, the reverse pathway would require too much energy to run spontaneously.

---

glycolysis (gly-COLL-ih-sis) First stage of aerobic respiration and fermentation; set of reactions that convert glucose to two pyruvate for a net yield of two ATP and two NADH.

pyruvate (pie-ROO-vait) Three-carbon product of glycolysis.

substrate-level phosphorylation ATP formation by the transfer of a phosphate group from a phosphorylated molecule to ADP.

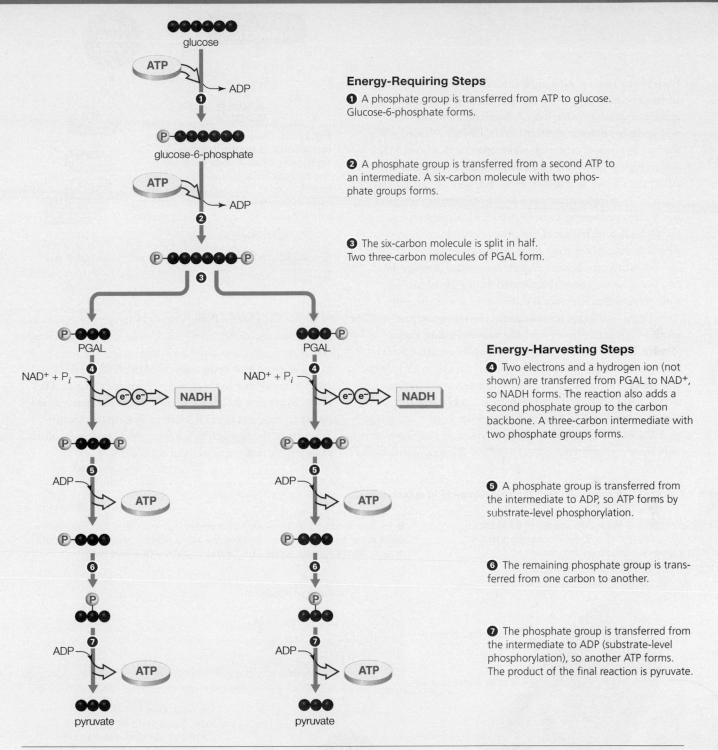

glucose

Energy-Requiring Steps

❶ A phosphate group is transferred from ATP to glucose. Glucose-6-phosphate forms.

glucose-6-phosphate

❷ A phosphate group is transferred from a second ATP to an intermediate. A six-carbon molecule with two phosphate groups forms.

❸ The six-carbon molecule is split in half. Two three-carbon molecules of PGAL form.

PGAL   **PGAL**

NAD$^+$ + P$_i$ NADH   NAD$^+$ + P$_i$ NADH

Energy-Harvesting Steps

❹ Two electrons and a hydrogen ion (not shown) are transferred from PGAL to NAD$^+$, so NADH forms. The reaction also adds a second phosphate group to the carbon backbone. A three-carbon intermediate with two phosphate groups forms.

❺ A phosphate group is transferred from the intermediate to ADP, so ATP forms by substrate-level phosphorylation.

❻ The remaining phosphate group is transferred from one carbon to another.

❼ The phosphate group is transferred from the intermediate to ADP (substrate-level phosphorylation), so another ATP forms. The product of the final reaction is pyruvate.

pyruvate   **pyruvate**

Figure 7.5 Reactions of glycolysis. This first stage of sugar breakdown starts and ends in the cytoplasm of all cells. All reactions are mediated by enzymes. For clarity, we track only the six carbon atoms (black balls) of the glucose molecule that enters the pathway.

Cells invest two ATP to start glycolysis, and two ATP form, so the net yield is two ATP per molecule of glucose. Electrons are also accepted by NADH. Depending on the type of cell and environmental conditions, the pyruvate product of glycolysis may enter the second stage of aerobic respiration or it may be used in other ways, such as in fermentation. Appendix III has more details for interested students.

7.3 Acetyl–CoA Formation and the Citric Acid Cycle

Describe the movement of energy and matter during acetyl–CoA formation and the citric acid cycle.

Each of the two pyruvate molecules that formed in glycolysis has two carbon–carbon bonds. The next two steps of aerobic respiration, acetyl–CoA formation and the citric acid cycle, break both of these bonds. Energy released when the bonds break is captured in electrons carried by NADH, and in high-energy phosphate bonds of ATP (**Figure 7.6**). All of the carbon atoms that were once part of glucose end up in CO_2, which diffuses out of the cell (remember, gases freely cross cell membranes, Section 5.6). In prokaryotes, acetyl–CoA formation and the citric acid cycle occur in cytoplasm; in eukaryotes, they occur in mitochondria.

In eukaryotes, aerobic respiration continues when the two pyruvate molecules that formed during glycolysis enter a mitochondrion. Pyruvate is transported across the mitochondrion's two membranes and into the inner compartment, which is filled with a gel-like material called matrix (**Figure 7.7A**). There, a redox reaction splits a carbon from pyruvate, and this carbon diffuses out of the cell in CO_2 (**Figure 7.8 ❶**). The reaction transfers electrons and a hydrogen ion to NAD^+, which is thereby reduced to NADH. The remaining two-carbon fragment of pyruvate is an acetyl group ($—COCH_3$). This group becomes attached to a coenzyme called CoA (an abbreviation of coenzyme A), so a molecule

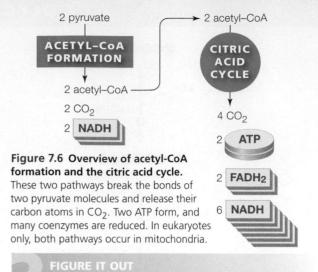

Figure 7.6 Overview of acetyl-CoA formation and the citric acid cycle.
These two pathways break the bonds of two pyruvate molecules and release their carbon atoms in CO_2. Two ATP form, and many coenzymes are reduced. In eukaryotes only, both pathways occur in mitochondria.

FIGURE IT OUT
Which of these two pathways produces more ATP?

Answer: The citric acid cycle

of acetyl–CoA forms. Both pyruvates from glycolysis undergo these reactions, so two acetyl–CoA molecules form; each now carries two carbons into the citric acid cycle.

The **citric acid cycle** (also called the Krebs cycle) is a cyclic pathway that harvests energy from acetyl–CoA. The reactions capture the energy in the form of electrons carried by coenzymes, and in ATP. It is a cyclic pathway because a substrate of the first reaction, a four-carbon molecule called oxaloacetate, is also a product of the last.

Figure 7.7 Acetyl–CoA formation and the citric acid cycle in mitochondria.

A An inner membrane divides the interior of a mitochondrion into two compartments. The inner compartment is filled with a gel-like substance called matrix.

The location of this mitochondrion is shown in an animal cell, but all eukaryotic cells have these organelles.

B Pyruvate produced by glycolysis is transported from cytoplasm into a mitochondrion, across both membranes, and into the matrix. Aerobic respiration continues here, in the matrix, with acetyl–CoA formation and the citric acid cycle.

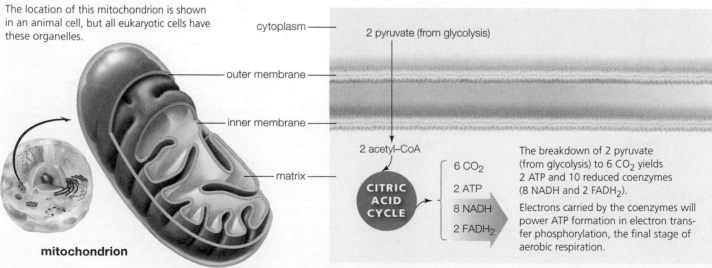

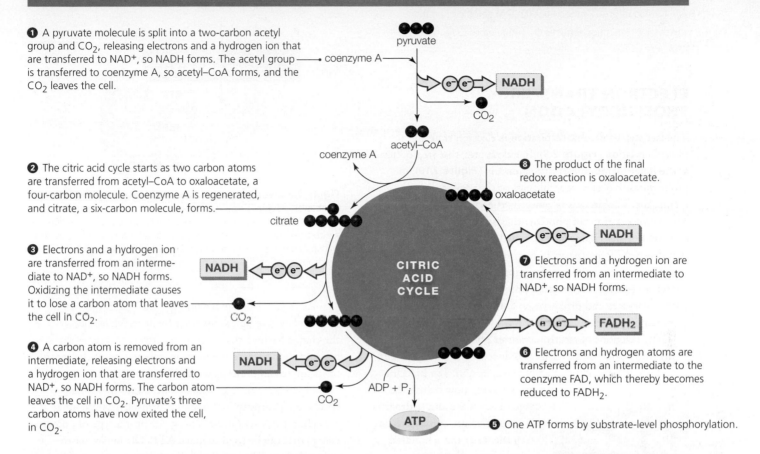

❶ A pyruvate molecule is split into a two-carbon acetyl group and CO_2, releasing electrons and a hydrogen ion that are transferred to NAD^+, so NADH forms. The acetyl group is transferred to coenzyme A, so acetyl–CoA forms, and the CO_2 leaves the cell.

❷ The citric acid cycle starts as two carbon atoms are transferred from acetyl–CoA to oxaloacetate, a four-carbon molecule. Coenzyme A is regenerated, and citrate, a six-carbon molecule, forms.

❸ Electrons and a hydrogen ion are transferred from an intermediate to NAD^+, so NADH forms. Oxidizing the intermediate causes it to lose a carbon atom that leaves the cell in CO_2.

❹ A carbon atom is removed from an intermediate, releasing electrons and a hydrogen ion that are transferred to NAD^+, so NADH forms. The carbon atom leaves the cell in CO_2. Pyruvate's three carbon atoms have now exited the cell, in CO_2.

❺ One ATP forms by substrate-level phosphorylation.

❻ Electrons and hydrogen atoms are transferred from an intermediate to the coenzyme FAD, which thereby becomes reduced to $FADH_2$.

❼ Electrons and a hydrogen ion are transferred from an intermediate to NAD^+, so NADH forms.

❽ The product of the final redox reaction is oxaloacetate.

Figure 7.8 Acetyl–CoA formation and the citric acid cycle. Follow the carbons (black balls) to understand the movement of matter; follow the electrons to see the movement of energy. Each time these reactions run, they break down one pyruvate molecule. Thus, it takes two sets of reactions to break down the two pyruvate molecules that formed during glycolysis. Then, all six carbons that entered glycolysis in one glucose molecule have exited the cell, in six CO_2. Electrons and hydrogen ions are released as each carbon is removed from the backbone of intermediate molecules; these combine with ten coenzymes (thus reducing them). Not all reactions are shown; see Appendix III for details.

In the first reaction, two carbon atoms of an acetyl–CoA molecule are transferred to oxaloacetate, forming citrate, the ionized form of citric acid ❷. The citric acid cycle is named after this first intermediate.

A series of redox reactions follows. Electrons and hydrogen ions are transferred to NAD^+, so NADH forms ❸ and ❹. The reactions remove two carbon atoms from intermediates, and these depart the cell in CO_2. No more carbons are split from the backbone of the four-carbon intermediate that forms; the remaining reactions of the citric acid cycle harvest energy from this molecule by rearranging its bonds.

In the next reaction, ATP forms by substrate-level phosphorylation ❺. Then, electrons and hydrogen atoms are transferred from an intermediate to the coenzyme FAD (flavin adenine dinucleotide), which becomes reduced to $FADH_2$ ❻. A final reaction transfers electrons and a hydrogen to NAD^+, so another NADH forms ❼. The product of this final redox reaction is oxaloacetate ❽.

It takes two rounds of citric acid cycle reactions to harvest the energy from the two acetyl–CoA molecules that formed in the previous step. At this point in aerobic respiration, the carbon backbone of one glucose molecule has been broken apart completely, its six carbon atoms having exited the cell in six molecules of CO_2. Two ATP that form during the citric acid cycle add to the small net yield of 2 ATP from glycolysis. However, acetyl–CoA formation reduced two coenzymes, and the citric acid cycle reduced eight more (**Figure 7.7B**). Add in the two coenzymes reduced in glycolysis, and the full breakdown of each glucose molecule has a big potential payoff: Twelve coenzymes will deliver electrons—and the energy they carry—to the final stage of aerobic respiration.

citric acid cycle Cyclic pathway that harvests energy from acetyl–CoA; part of aerobic respiration.

LEARNING OBJECTIVES

Describe the formation of ATP in the final stage of aerobic respiration.

Explain why aerobic respiration requires O_2 and releases CO_2.

ELECTRON TRANSFER PHOSPHORYLATION

The last step of aerobic respiration is electron transfer phosphorylation (**Figure 7.9**). In eukaryotes, this step occurs at the inner membrane of mitochondria (**Figure 7.10**); in prokaryotes, it occurs at infoldings of the inner (plasma) membrane. The process is generally the same as in the light-dependent reactions of photosynthesis: Electron flow through electron transfer chains sets up a hydrogen ion (H^+) gradient that drives ATP synthesis. In fact, several of the molecules that carry out electron transfer phosphorylation are highly conserved (similar or identical) among the two pathways in both eukaryotes and prokaryotes, a molecular legacy of life's common origin.

The reactions of electron transfer phosphorylation begin with coenzymes that were reduced during the previous steps of aerobic respiration. Ten NADH and two $FADH_2$ now deliver their cargo of electrons and hydrogen ions to electron transfer chains in the inner mitochondrial membrane **❶**.

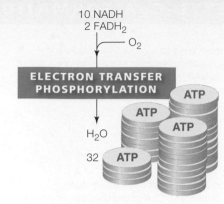

Figure 7.9 Overview of electron transfer phosphorylation in aerobic respiration. In eukaryotes only, this pathway occurs in mitochondria. Coenzymes reduced during the previous stages of aerobic respiration deliver electrons and hydrogen ions to electron transfer chains in the inner mitochondrial membrane. The flow of electrons through these chains sets up a hydrogen ion gradient that powers ATP formation.

As the electrons move through the electron transfer chains, they give up energy little by little. Molecules in the chains harness that energy to actively transport the hydrogen ions across the inner membrane, from the matrix to the intermembrane space **❷**.

Hydrogen ions accumulate in the intermembrane space, so a hydrogen ion gradient forms across the inner mitochondrial membrane. This gradient is a type of potential energy that can be used to make ATP. The ions want to follow their gradient by moving back into the matrix, but

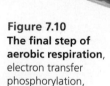

**Figure 7.10
The final step of aerobic respiration**, electron transfer phosphorylation, in a mitochondrion.

❶ NADH and $FADH_2$ deliver electrons and hydrogen ions to electron transfer chains in the inner membrane.

❷ Energy lost by electrons moving through electron transfer chains fuels the active transport of hydrogen ions (H^+) from the matrix to the intermembrane space. A hydrogen ion gradient forms across the inner membrane.

❸ Hydrogen ion flow back to the matrix through ATP synthases drives the formation of ATP from ADP and phosphate (P_i).

❹ Oxygen accepts electrons and hydrogen ions at the end of the electron transfer chains, so water forms.

ELECTRON TRANSFER PHOSPHORYLATION

10 NADH

2 FADH₂

H^+

e^- e^-

❶

electron transfer chain

ATP synthase **❸**

H^+

ATP

ADP + P_i

matrix

inner membrane

e^-

e^-

e^-

H^+ H^+ H^+ H^+
H^+ H^+ H^+ H^+

H^+
❷ H^+ H^+ H^+

O_2

intermembrane space

❹ H_2O

outer membrane

cell cytoplasm

lipid bilayers are impermeable to ions (Section 5.6). Hydrogen ions leave the intermembrane space only by flowing through ATP synthases in the inner mitochondrial membrane. An ATP synthase is a small molecular motor that is both a transport protein and an enzyme (Section 6.3). When hydrogen ions flow through its interior, the protein phosphorylates ADP, so ATP forms in the matrix ❸.

Oxygen accepts electrons at the end of mitochondrial electron transfer chains ❹. An oxygen molecule combines with electrons and hydrogen ions to form water:

$$O_2 + 4e^- + 4H^+ \longrightarrow 2H_2O$$

This equation is the reverse of photolysis during the noncyclic, light-dependent reactions.

SUMMING UP

The twelve coenzymes reduced in the previous steps of aerobic respiration deliver enough electrons to fuel synthesis of about thirty-two ATP in electron transfer phosphorylation. Thus, the full breakdown of one glucose molecule typically yields thirty-six ATP (**Figure 7.11**), but the yield varies by cell type. For example, aerobic respiration in brain and skeletal muscle cells yields thirty-eight ATP per glucose.

Aerobic respiration is the most efficient way that cells make ATP by carbohydrate breakdown. However, a lot of energy transfers occur in aerobic respiration, and some energy is lost with every transfer (Section 5.1). About 60 percent of the energy stored in glucose disperses as metabolic heat during the pathway.

Figure 7.11 Summary of aerobic respiration in eukaryotes.

Glycolysis
Glycolysis (in cytoplasm) splits a glucose molecule into 2 pyruvate. Two ATP form, and 2 NAD^+ accept electrons and hydrogen ions, so 2 NADH form. The pyruvate and NADH are transported into the mitochondrial matrix.

Acetyl–CoA Formation
In the mitochondrial matrix, enzymes remove a carbon from both pyruvates. The carbon exits the cell in 2 CO_2. Two acetyl–CoA and 2 NADH also form.

Citric Acid Cycle
In the mitochondrial matrix, the 2 acetyl–CoA are broken down to 4 CO_2 (which leave the cell). Two ATP form. The reactions also produce 6 NADH and 2 $FADH_2$.

Electron Transfer Phosphorylation
At the inner membrane, electrons from the 12 coenzymes reduced in the previous steps power the synthesis of approximately 32 ATP. Water forms when oxygen accepts electrons at the end of electron transfer chains.

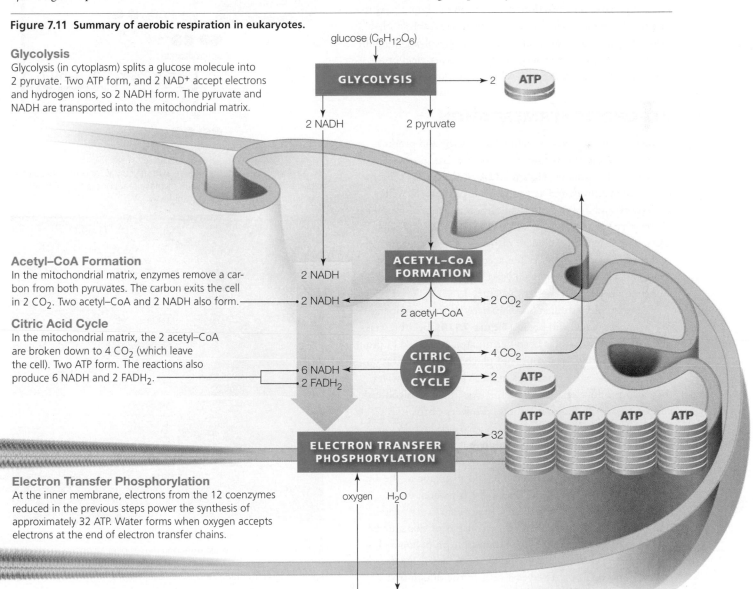

glucose ($C_6H_{12}O_6$)

GLYCOLYSIS → 2 ATP

2 NADH   2 pyruvate

ACETYL–CoA FORMATION

2 NADH   2 CO_2

2 acetyl–CoA

CITRIC ACID CYCLE → 4 CO_2

6 NADH   → 2 ATP
2 $FADH_2$

ATP   ATP   ATP   ATP

ELECTRON TRANSFER PHOSPHORYLATION → 32

oxygen H_2O

Almost all cells can carry out fermentation, and many can switch between aerobic respiration and fermentation as needed. Fermentation reactions vary greatly, but all occur in cytoplasm. They require no oxygen because the final acceptor of electrons is an organic molecule (not oxygen). Here we discuss two common fermentation pathways that, like aerobic respiration, begin with glycolysis. Unlike aerobic respiration, neither pathway fully breaks down the carbon backbone of glucose. The reactions after glycolysis simply serve to remove electrons and hydrogen ions from NADH, so NAD^+ forms. Regenerating this coenzyme allows glycolysis—and the ATP yield it offers—to continue. Thus, the net yield of fermentation consists of the two ATP that form in glycolysis.

ALCOHOLIC FERMENTATION

Fermentation pathways are named after their end product. **Alcoholic fermentation**, for example, converts glucose to ethyl alcohol—ethanol (**Figure 7.12A**). Glycolysis is the first part of the pathway, and as you know it produces 2 ATP, 2 NADH, and 2 pyruvates. One carbon is removed from each of the pyruvate molecules, and this carbon leaves the cell in CO_2. The remaining fragment of pyruvate is an organic molecule called acetaldehyde. When acetaldehyde accepts electrons and a hydrogen ion from NADH, it becomes ethanol. This final reaction regenerates NAD^+.

A yeast species called *Saccharomyces cerevisiae* helps us produce beer, wine, and bread (**Figure 7.12B**). Beer brewers often use barley that has been germinated and dried (a process called malting) as a source of glucose for fermentation by this yeast. As the yeast cells make ATP for themselves, they also produce ethanol (which makes the beer alcoholic) and CO_2 (which makes it bubbly). Flowers of the hop plant add flavor and help preserve the finished product. Winemakers use crushed grapes as a source of sugars for yeast fermentation. The ethanol produced by the cells makes the wine alcoholic, and the CO_2 is allowed to escape to the air.

To make bread, flour is kneaded with water, yeast, and sometimes other ingredients. Flour contains a protein (gluten) and a sugar (maltose) that consists of two glucose subunits. Kneading causes the gluten to form polymers in long, interconnected strands that make the resulting dough stretchy and resilient. The yeast cells in the dough first break down the maltose, then use the released glucose for

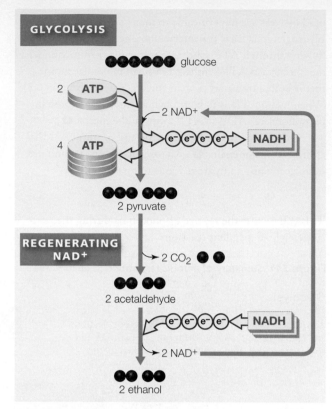

A Alcoholic fermentation begins with glycolysis, and the final steps regenerate NAD^+. The net yield of these reactions is two ATP per molecule of glucose (from glycolysis).

B *Saccharomyces cerevisiae* cells (top). One product of alcoholic fermentation by this yeast (ethanol) makes beer alcoholic; another (CO_2) makes it bubbly. Holes in bread are pockets where CO_2 released by fermenting yeast cells accumulated in the dough.

Figure 7.12 Alcoholic fermentation.

alcoholic fermentation. The CO_2 they produce accumulates in bubbles that are trapped by the mesh of gluten strands. As the bubbles expand, they cause the dough to rise. The ethanol product of fermentation evaporates during baking.

LACTATE FERMENTATION

Lactate fermentation converts glucose to lactate (**Figure 7.13A**). Lactate is the ionized form of lactic acid, so this pathway is also called lactic acid fermentation. Glycolysis is the first step of the pathway, and produces 2 ATP, 2 NADH, and 2 pyruvates. The second step is a reaction that transfers electrons and hydrogen ions from NADH directly to the pyruvates. The reaction regenerates NAD^+ and also converts pyruvate to lactate. No carbons are lost, so lactate fermentation produces no CO_2.

We use lactate fermentation by beneficial bacteria to prepare many foods. Yogurt, for example, is made by allowing bacteria such as *Lactobacillus bulgaricus* and *Streptococcus thermophilus* to grow in milk (**Figure 7.13B**). Milk contains a disaccharide (lactose) and a protein (casein). The cells first break down the lactose into its monosaccharide subunits, then use the sugars for lactate fermentation. The lactate they produce reduces the pH of the milk, which imparts tartness and causes the casein to form a gel.

Cells in animal skeletal muscles are fused as long fibers that carry out aerobic respiration, lactate fermentation, or both. Aerobic respiration predominates under most circumstances. However, there are times when fermentation is required, for example when intense exercise depletes oxygen in muscles faster than it can be replenished. Under the resulting anaerobic conditions, muscle cells produce ATP mainly by lactate fermentation. This pathway makes ATP quickly, so it is useful for strenuous bursts of activity, but the low ATP yield does not support prolonged exertion.

Contrary to popular opinion, the buildup of lactate in muscles after exercise does not cause muscle soreness the following day. Lactate very quickly leaves muscles and enters the bloodstream, where it is taken up by cells that are not oxygen-depleted. These cells convert the lactate back to pyruvate for use in aerobic respiration.

Lactate fermentation sustains a few animals that hibernate without oxygen for long periods of time. Some freshwater turtles, for example, are adapted to spend the winter months unable to breathe because they are buried in mud or trapped under ice. The normal metabolic rate of a turtle is much lower than a warm-blooded animal in the first place, but during these times it drops even lower, to about 1/10,000 the rate of a resting mammal. Very little ATP is required to maintain such a low metabolism, so the indi-

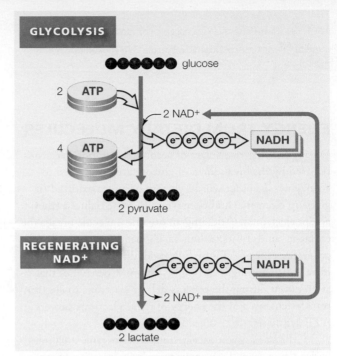

A Lactate fermentation begins with glycolysis, and the final steps regenerate NAD^+. The net yield of these reactions is two ATP per molecule of glucose (from glycolysis).

B Yogurt is a product of lactate fermentation by bacteria in milk. The micrograph shows *Lactobacillus bulgaricus* (red) and *Streptococcus thermophilus* (purple) in yogurt.

Figure 7.13 Lactate fermentation.

vidual can meet its energy demands using only lactate fermentation. Excess lactate produced by the pathway is taken up by the animal's bones and shell.

alcoholic fermentation Anaerobic sugar breakdown pathway that produces ATP, CO_2, and ethanol.
lactate fermentation Anaerobic sugar breakdown pathway that produces ATP and lactate.

LEARNING OBJECTIVES

Explain how cells convert energy in an organic compound to energy in ATP.

Describe how different kinds of organic molecules are broken down in aerobic respiration.

ENERGY FROM DIETARY MOLECULES

During the first two stages of aerobic respiration, electrons released by the breakdown of glucose are transferred to coenzymes. In other words, glucose becomes oxidized (it gives up electrons) and coenzymes become reduced (they accept electrons). Oxidizing an organic molecule breaks the covalent bonds of its carbon backbone. Aerobic respiration generates a lot of ATP by fully oxidizing glucose, completely dismantling it carbon by carbon. Coenzymes that are reduced during the process deliver electrons to electron transfer chains, and the energy of these electrons powers ATP synthesis.

Cells also harvest energy from other organic molecules by oxidizing them. Fats, complex carbohydrates, and proteins in food can be converted to molecules that enter aerobic respiration at various stages (**Figure 7.14**). As in glucose breakdown, many coenzymes are reduced, and the energy of the electrons they carry ultimately drives the synthesis of ATP in electron transfer phosphorylation.

Fats A triglyceride molecule has three fatty acid tails attached to a glycerol (Section 3.3). Cells dismantle triglycerides in fats by first breaking the bonds that connect fatty acid tails to the glycerol (**Figure 7.14A**). Nearly all cells in the body can oxidize the released fatty acids by splitting their long backbones into two-carbon fragments. These fragments are converted to acetyl–CoA, which can enter the citric acid cycle ❶. Enzymes in liver cells convert the glycerol to PGAL ❷, an intermediate of glycolysis.

On a per carbon basis, fats are a richer source of energy than carbohydrates. Carbohydrate backbones have many oxygen atoms, so they are partially oxidized. A fatty acid's long tails are hydrocarbon chains that typically have no oxygen atoms bonded to them, so they have a longer way to go to become oxidized—more reactions are required to fully break them down. Coenzymes accept electrons in these oxidation reactions. The more reduced coenzymes that form, the more electrons can be delivered to the ATP-forming machinery of electron transfer phosphorylation.

Complex Carbohydrates In humans and other mammals, the digestive system breaks down starch and other complex carbohydrates to monosaccharides (**Figure 7.14B**).

Cells quickly take up these simple sugars. Glucose is immediately converted to glucose-6-phosphate, which can continue in glycolysis ❸. Six-carbon sugars other than glucose also enter glycolysis. Fructose, for example, can be phosphorylated by hexokinase, the same enzyme that phosphorylates glucose. The product of this reaction, fructose-6-phosphate, is an intermediate of glycolysis.

When a cell produces more ATP than it uses, the concentration of ATP rises in the cytoplasm. The high concentration of ATP causes glucose-6-phosphate to be diverted away from glycolysis and into a pathway that builds glycogen (Section 3.2). Liver and muscle cells especially favor the conversion of glucose to glycogen, and these cells contain the body's largest stores of it. Between meals, the liver maintains the glucose level in blood by converting the stored glycogen back to glucose.

What happens if you eat too many carbohydrates? When the blood level of glucose gets too high, acetyl–CoA is diverted away from the citric acid cycle and into a pathway that builds fatty acids. This is why excess dietary carbohydrate ends up as fat.

Proteins Enzymes in the digestive system split dietary proteins into their amino acid subunits (**Figure 7.14C**), which are absorbed into the bloodstream. Cells use these free amino acids to build proteins or other molecules. When you eat more protein than your body needs for this purpose, the amino acids are broken down ❹. The amino ($-NH_2$) group is removed, and the carbon backbone is split. The removed amino group is converted to ammonia (NH_3), a waste product that is eliminated in urine. The carbon-containing fragments are converted to pyruvate, acetyl–CoA, or an intermediate of the citric acid cycle, depending on the amino acid. These molecules enter aerobic respiration at the acetyl–CoA formation stage or the citric acid cycle.

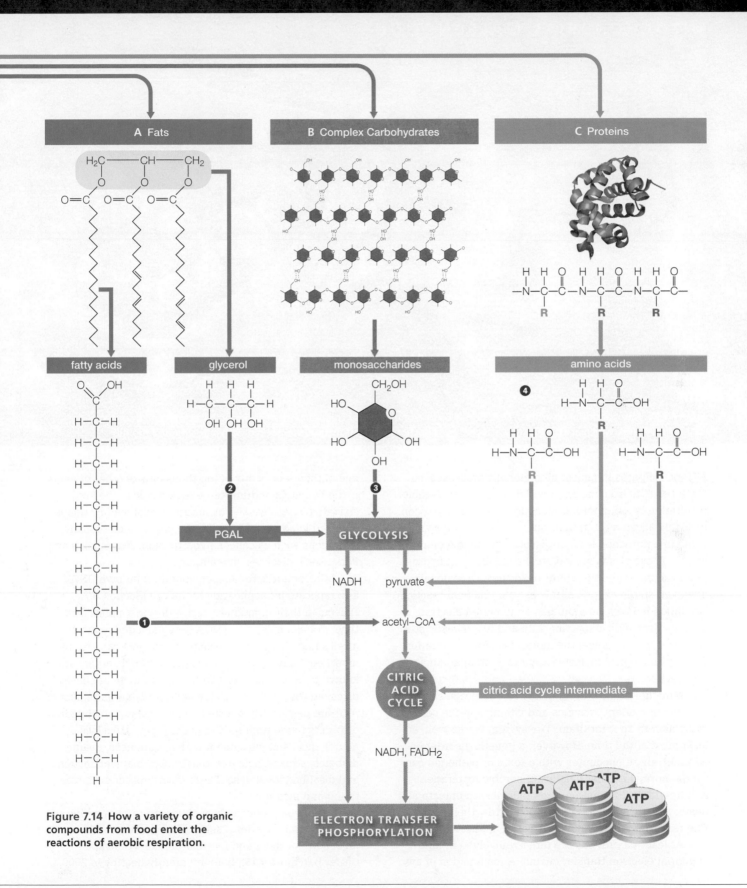

A Fats

H₂C — CH — CH₂

O=C   O=C   O=C

fatty acids

glycerol

B Complex Carbohydrates

monosaccharides

CH₂OH

C Proteins

amino acids

② PGAL → **GLYCOLYSIS** ← **③**

NADH   pyruvate

① → acetyl–CoA ←

CITRIC ACID CYCLE ← citric acid cycle intermediate

NADH, FADH₂

ATP ATP ATP ATP

ELECTRON TRANSFER PHOSPHORYLATION →

Figure 7.14 How a variety of organic compounds from food enter the reactions of aerobic respiration.

Figure 7.15 Mitochondrial disorders. Left, Carter B., one of millions of children with devastating mitochondrial disorders. Right, nerve cells, which are particularly affected by mitochondrial malfunction, are packed with mitochondria (gold).

ATP participates in almost all metabolic processes, so a cell benefits from making a lot of it. However, aerobic respiration is a dangerous occupation. When an oxygen molecule accepts electrons from an electron transfer chain in a mitochondrion, it dissociates into oxygen atoms. These atoms immediately combine with hydrogen ions and end up in water molecules. Occasionally, however, single oxygen atoms escape this final reaction. An unbonded oxygen atom has an unpaired electron, so it is a free radical (Section 2.2), and it can easily strip electrons from (oxidize) the molecules of life. Oxidation destroys biological molecules by breaking apart their carbon backbones.

Mitochondria cannot detoxify free radicals, so they rely on antioxidant enzymes and vitamins in the cell's cytoplasm to do it for them. The system works well, at least most of the time. However, a genetic disorder or an unfortunate encounter with a toxin or pathogen can tip the normal cellular balance of aerobic respiration and free radical formation. Free radicals accumulate and destroy first the function of mitochondria, then the cell. The resulting tissue damage is called oxidative stress.

At least 83 proteins are directly involved in mitochondrial electron transfer chains. A malfunction of any one of them—or in any of the thousands of other proteins made by mitochondria—can wreak havoc in the body. Oxidative stress caused by mitochondrial malfunction is involved in aging, and also in the progression of many conditions such as cancer, hypertension, Alzheimer's and Parkinson's diseases, and autism.

Mitochondria, remember, contain their own DNA and replicate independently of the cell (Section 4.7). Defects in their function can arise during a person's lifetime, (for example) if the DNA gets damaged in most of a cell's mitochondria. Mitochondrial defects can also be inherited: Hundreds of genetic (heritable) disorders are known to be associated with them, and more are being discovered all the time. Nerve cells, which require a lot of ATP, are particularly affected. Symptoms of mitochondrial disorders range from mild to major progressive loss of neurological and muscular function, blindness, deafness, diabetes, strokes, seizures, gastrointestinal malfunction, and disabling weakness. These disorders can be devastating and incurable.

Every year, thousands of children are born with a mitochondrial disorder. Their suffering can be unbearable, both for them and for their families. Consider Carter B. (**Figure 7.15**), born apparently healthy in 2009.

Minor persistent problems led to a diagnosis of a mitochondrial disorder when he was 18 months old. Testing revealed genetic defects in two of the four major components of his mitochondrial electron transfer chains. The first defect impairs the enzyme that removes electrons from NADH; the second affects a protein complex that moves hydrogen ions from the mitochondrial matrix to the intermembrane space. Too few electrons enter mitochondrial electron transfer chains, and the ones that do cannot efficiently maintain a hydrogen ion gradient. The result is that Carter's mitochondria cannot make enough ATP to support normal cellular functioning, and his whole body is affected. Countless surgeries, visits to the emergency room, and hospital admissions have filled his short life. Today, Carter has symptoms of cerebral palsy, epilepsy, and cognitive impairment; he cannot walk, talk, eat most foods, crawl, or respond appropriately to his environment. For now, his survival depends on 22 injections per day of 14 different drugs. Like most children born with a mitochondrial disorder, he is not expected to reach his teens.

A child inherits mitochondria from one parent only: the mother. For a variety of reasons, some women's eggs contain a lot of defective mitochondria. Children born to these women have a higher than normal risk of mitochondrial disorders.

A reproductive technology called mitochondrial donation, or three-person IVF (*in vitro* fertilization), can help women with defective mitochondria in their eggs have healthy babies. For example, nuclear DNA from the mother and father can be inserted into a donor egg from a woman with healthy mitochondria. The embryo that forms from the hybrid egg is implanted into the mother, much as normal IVF works. Babies born from this procedure have three parents: Their cells have nuclear DNA inherited from the mother and father, and mitochondria inherited from the donor.

Three-person IVF was approved for general use in the United Kingdom in 2015, and for conditional use in the United States in 2016. Opponents point out that female offspring will pass donated mitochondria to future generations. Thus, even if the procedure works as expected, using it puts us on a slippery slope toward genetically modifying babies (we return to genetic engineering in Chapter 15).

CHAPTER SUMMARY

 Section 7.1 When photosynthesis evolved, oxygen released by the pathway permanently changed the atmosphere, with profound effects on life's evolution. Organisms that could not tolerate the increased atmospheric oxygen persisted only in **anaerobic** habitats. Antioxidants allowed other organisms to thrive in **aerobic** conditions. Most modern organisms convert the chemical energy of glucose to the chemical energy of ATP by a **cellular respiration** pathway called **aerobic respiration**, which requires oxygen and includes glycolysis, acetyl–CoA formation, the citric acid cycle, and electron transfer phosphorylation. **Fermentation** pathways begin with glycolysis, do not require oxygen, and yield fewer ATP than aerobic respiration.

 Section 7.2 **Glycolysis**, which occurs in cytoplasm, converts one molecule of glucose to two molecules of **pyruvate**. Energy released when the carbon–carbon backbone breaks is captured in electrons carried by NADH, and two ATP form by **substrate-level phosphorylation**.

 Section 7.3 In eukaryotes, aerobic respiration continues in the matrix that fills the inner compartment of a mitochondrion. The first steps convert the two pyruvate from glycolysis to two acetyl–CoA, two CO_2, and two NADH. Reactions of the **citric acid cycle** break down the two acetyl–CoA to four CO_2, and yield two ATP and eight reduced coenzymes (six NADH, two $FADH_2$). At this point in aerobic respiration, the glucose molecule that entered glycolysis has been dismantled completely; energy released during the process has been captured in high-energy phosphate bonds of four ATP, and in electrons carried by twelve reduced coenzymes.

 Section 7.4 The final step of aerobic respiration is electron transfer phosphorylation. Coenzymes that were reduced in the previous steps deliver electrons and hydrogen ions to electron transfer chains in the inner mitochondrial membrane. Energy released by electrons as they pass through electron transfer chains is used to pump H^+ from the matrix to the intermembrane space. The H^+ gradient that forms across the inner mitochondrial membrane drives the flow of hydrogen ions through ATP synthases, which results in ATP formation. Oxygen accepts electrons at the end of the chains and combines with hydrogen ions, so water forms.

About thirty-two ATP form during the third-stage reactions, so a typical net yield of all three stages is about thirty-six ATP per glucose. A summary of all steps of aerobic respiration is shown on the next page, in **Figure 7.16**.

 Section 7.5 Prokaryotes and eukaryotes use fermentation pathways, which are anaerobic, to produce ATP by breaking down carbohydrates in cytoplasm. An organic molecule, rather than oxygen, accepts electrons at the end of

these pathways. The end product of **alcoholic fermentation** is ethyl alcohol, or ethanol. The end product of **lactate fermentation** is lactate. Both pathways begin with glycolysis. Their final steps regenerate NAD⁺, which is required for glycolysis to continue. Fermentation's small ATP yield (two per molecule of glucose) is from glycolysis.

Section 7.6 Oxidizing organic molecules breaks their carbon backbones, releasing electrons whose energy can be harnessed to drive ATP formation in aerobic respiration. In humans and other mammals, first the digestive system and then individual cells convert molecules in food (fats, complex carbohydrates, and proteins) into substrates of glycolysis or aerobic respiration's second-stage reactions.

Application Free radicals that form during aerobic respiration are detoxified by antioxidant molecules in the cell's cytoplasm. Defects in mitochondrial electron transfer chain components can cause a buildup of free radicals that damage the cell—and ultimately, the individual. Symptoms can be lethal. Oxidative stress due to mitochondrial malfunction plays a role in many illnesses and genetic disorders.

SELF-ASSESSMENT
ANSWERS IN APPENDIX VII

1. Is the following statement true or false? Unlike animals, which make many ATP by aerobic respiration, plants make all of their ATP by photosynthesis.

2. Glycolysis starts and ends in the _____ .
   a. nucleus   c. plasma membrane
   b. mitochondrion   d. cytoplasm

3. Which of the following pathways require(s) molecular oxygen (O_2)?
   a. aerobic respiration
   b. lactate fermentation
   c. alcoholic fermentation
   d. photosynthesis

4. Which molecule does not form during glycolysis?
   a. NADH   c. oxygen (O_2)
   b. pyruvate   d. ATP

5. In eukaryotes, the final reactions of aerobic respiration are completed in _____ .
   a. the nucleus   c. the plasma membrane
   b. mitochondria   d. cytoplasm

6. In eukaryotes, the final reactions of fermentation are completed in _____ .
   a. the nucleus   c. the plasma membrane
   b. mitochondria   d. cytoplasm

7. After the citric acid cycle reactions run _____ , one six-carbon glucose molecule has been completely broken down to CO_2.
   a. once   b. twice   c. six times   d. twelve times

8. In the final stage of aerobic respiration, _____ accepts electrons.
   a. water   c. oxygen (O_2)
   b. hydrogen   d. NADH

9. Most of the energy that aerobic respiration releases from glucose ends up in _____ .
   a. NADH   c. heat
   b. ATP   d. electrons

10. _____ accepts electrons in lactate fermentation.
    a. Oxygen   c. Acetaldehyde
    b. Pyruvate   d. Ethanol

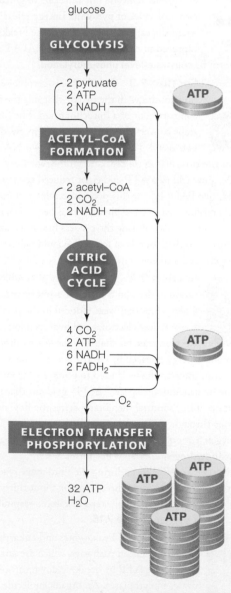

Figure 7.16 Reviewing the steps of aerobic respiration.

DATA ANALYSIS

CLASS ACTIVITY

DIETARY FAT OVERLOAD REPROGRAMS MITOCHONDRIA

The bodies of humans and other mammals store triglycerides in two kinds of tissue called white and brown fat. The main function of white fat is to store triglycerides for energy. Brown tissue has more mitochondria than white tissue, and these are used to generate body heat.

Mitochondria in brown fat make thermogenin, a protein that uncouples ATP synthesis from mitochondrial electron transfer chains. Thermogenin is a transport protein that allows hydrogen ions to move across the inner mitochondrial membrane. Hydrogen ions that move through this protein instead of ATP synthase power no ATP formation.

In cells of brown fat, organic compounds are rapidly broken down by aerobic respiration. Electrons pass through mitochondrial electron transfer chains, which actively transport hydrogen ions from the matrix to the intermembrane space. However, hydrogen ion gradient formation is inefficient across the "leaky" inner membrane, so not much ATP forms. Thus, the major product of aerobic respiration in these cells is heat lost from the ongoing electron transfers.

In 2015, Daniele Barbato and his colleagues investigated the effect of a high-fat diet on brown fat in mice. They maintained an experimental group of mice on a high-fat diet (60 percent fat), and a control group of mice on a normal diet (12 percent fat). Results are shown in **Figure 7.17**.

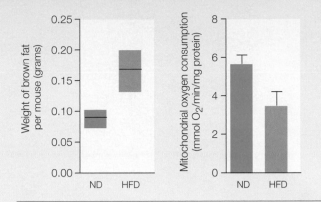

Figure 7.17 Effects of a high-fat diet on brown fat.
ND, normal diet group. HFD, high-fat diet group.
Oxygen consumption was normalized for protein concentration.

1. The number of mitochondria per gram of brown fat tissue was measured and found to be identical in both groups. Which group had the highest total number of brown fat mitochondria per mouse?

2. In which group was the oxygen consumption per mitochondrion greatest?

3. How did a high-fat diet affect brown fat?

---

11. Put the following pathways in the order they occur during aerobic respiration (use **Figure 7.16** to check your answers).
    a. electron transfer phosphorylation
    b. acetyl–CoA formation
    c. citric acid cycle
    d. glycolysis

12. Your body cells can break down _____ as a source of energy to fuel ATP production.
    a. fatty acids   c. amino acids
    b. glycerol   d. all of the above

13. Which of the following is *not* produced by an animal muscle cell operating under anaerobic conditions?
    a. heat   d. ATP
    b. pyruvate   e. lactate
    c. PGAL   f. oxygen

14. Hydrogen ion flow drives ATP synthesis during _____ .
    a. glycolysis
    b. the citric acid cycle
    c. aerobic respiration
    d. fermentation

15. Match the term with the best description.
    ___ mitochondrial matrix   a. needed for glycolysis
    ___ pyruvate   b. inner gel
    ___ NAD$^+$ c. makes many ATP
    ___ mitochondrion   d. product of glycolysis
    ___ NADH   e. reduced coenzyme
    ___ anaerobic   f. no oxygen required

CLASS ACTIVITY

CRITICAL THINKING

1. The higher the altitude, the lower the oxygen level in air. Climbers of very tall mountains risk altitude sickness, which is an illness characterized by shortness of breath, weakness, dizziness, and confusion. The early symptoms of cyanide poisoning are the same as those for altitude sickness. Cyanide binds tightly to cytochrome *c* oxidase, the protein that reduces oxygen molecules in the final step of mitochondrial electron transfer chains. Cytochrome *c* oxidase with bound cyanide can no longer transfer electrons. Explain why cyanide poisoning starts with the same symptoms as altitude sickness.

2. The bar-tailed godwit is a type of shorebird that makes an annual migration from Alaska to New Zealand and back. The birds make each 11,500-kilometer (7,145-mile) trip by flying over the Pacific Ocean in about nine days, depending on weather, wind speed, and direction of travel. One bird was observed to make the entire journey uninterrupted, a feat that is comparable to a human running a nonstop seven-day marathon at 70 kilometers per hour (43.5 miles per hour). Would you expect the flight (breast) muscles of bar-tailed godwits to use mainly aerobic respiration or fermentation? Explain your answer.

8 DNA Structure and Function

Links to Earlier Concepts

Radioisotope tracers (Section 2.1) were used in research that led to the discovery that DNA (3.6), not protein (3.4, 3.5), is the hereditary material of all organisms (1.2). This chapter revisits free radicals (2.2), the cell nucleus (4.5), and metabolism (5.3–5.5). Your knowledge of carbohydrate ring numbering (3.2) will help you understand DNA replication.

Core Concepts

All organisms alive today are linked by lines of descent from shared ancestors.

Shared core processes and features provide evidence that all living things are related. A series of historical experiments led to the discovery that DNA is the primary source of heritable information for all life. This information passes from parent to offspring during processes of reproduction. DNA replication ensures the continuity of hereditary information, and so it maintains the continuity of life.

Living systems store, retrieve, transmit, and respond to information essential for life.

DNA in each cell encodes all of the instructions necessary for growth, survival, and reproduction of the cell and (in multicelled organisms) the individual. The double-stranded structure of DNA is a simple, efficient way of storing this gigantic amount of information; it also allows a DNA molecule to be copied with high fidelity for transmission to offspring.

Interactions among the components of each biological system give rise to complex properties.

Random changes in DNA that are not repaired become mutations, which can alter the structure and function of cells, and also of multicelled organisms. DNA replication is a source of mutations because it is an imperfect process. Exposure to some environmental factors also results in mutations.

Information encoded in DNA is the basis of visible traits that define species and distinguish individuals. Identical twins appear identical because they inherited identical DNA.

131

8.1 The Discovery of DNA's Function

Describe the four properties required of any genetic material.

Summarize the classic experiments of Griffith, Avery, and Hershey and Chase that demonstrated DNA is genetic material.

EARLY CLUES

DNA, the substance (**Figure 8.1**), was first described in 1869 by Johannes Miescher, a chemist who extracted it from cell nuclei. Miescher determined that DNA is not a protein, and that it is rich in nitrogen and phosphorus, but he never learned its function. Sixty years later, Frederick Griffith unexpectedly found a clue to that function. Griffith was studying pneumonia-causing bacteria in the hope of making a vaccine. He isolated two strains (types) of the bacteria, one harmless (R), the other lethal (S). Griffith used R and S cells in a series of experiments testing their ability to cause pneumonia in mice (**Figure 8.2**). He discovered that heat destroyed the ability of lethal S bacteria to cause pneumonia, but it did not destroy their hereditary material, including whatever specified "kill mice." That material could be transferred from dead S cells to live R cells, which put it to use. The transformation was permanent and heritable: Even after hundreds of generations, descendants of transformed R cells retained the ability to kill mice.

A SURPRISING RESULT

What was the substance that transformed harmless R cells into killers? Oswald Avery, Colin MacLeod, and Maclyn McCarty decided to identify this "transforming principle." They extracted lipid, protein, and nucleic acids from S cells, then used a process of elimination to determine which transformed bacteria. Treating the extract with enzymes that destroy lipids and proteins did not destroy the transforming principle. The researchers realized that the substance they were seeking must be nucleic acid—DNA or RNA. DNA-degrading enzymes destroyed the extract's ability to transform cells, but RNA-degrading enzymes did not. Thus, DNA had to be the transforming principle.

The result surprised the researchers, who, along with most other scientists, had assumed that proteins were the material of heredity. After all, traits are diverse, and proteins are the most diverse of all biological molecules. DNA was widely assumed to be too simple (experts of the era called it "a stupid molecule") because it has only four nucleotide components (Section 8.2 returns to DNA structure). Avery's team was so skeptical that they spent a decade confirming their result. In 1944, they finally published their work, along with the conclusion that DNA must be hereditary material.

Figure 8.1 DNA, the substance, extracted from human cells.

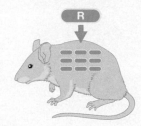

A Griffith's first experiment showed that R cells were harmless. When injected into mice, the bacteria multiplied, but the mice remained healthy.

B The second experiment showed that an injection of S cells caused mice to develop fatal pneumonia. Their blood contained live S cells.

C For a third experiment, Griffith killed S cells with heat before injecting them into mice. The mice remained healthy, indicating that the heat-killed S cells were harmless.

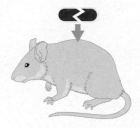

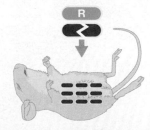

D In his fourth experiment, Griffith injected a mixture of heat-killed S cells and live R cells. To his surprise, the mice became fatally ill, and their blood contained live S cells.

Figure 8.2 Fred Griffith's experiments.
Griffith discovered that hereditary information passes between two strains of bacteria (R and S).

FIGURE IT OUT
In the fourth experiment, did hereditary material pass from R cells to S cells, or from S cells to R cells?

Answer: S cells to R cells

CREDIT: (1) Patrick Landmann/Science Source.

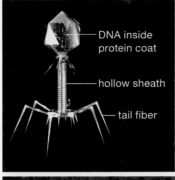

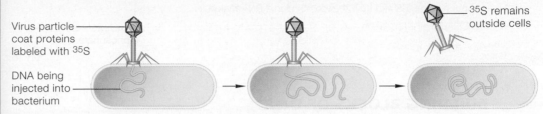

A Top left, model of a bacteriophage. Bottom left, micrograph of bacteriophage injecting DNA into a bacterium.

DNA inside protein coat

hollow sheath

tail fiber

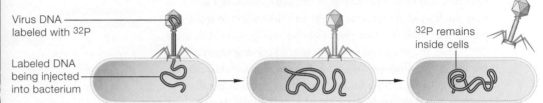

Virus particle coat proteins labeled with ^{35}S

DNA being injected into bacterium

^{35}S remains outside cells

B In one experiment, bacteriophage were labeled with a radioisotope of sulfur (^{35}S), a process that makes their protein components radioactive. The labeled viruses were mixed with bacteria long enough for infection to occur, and then the mixture was whirled in a kitchen blender. Blending dislodged viral parts that remained on the outside of the bacteria. Afterward, most of the radioactive sulfur was detected outside the bacterial cells. The viruses had not injected protein into the bacteria.

Virus DNA labeled with ^{32}P

Labeled DNA being injected into bacterium

^{32}P remains inside cells

C In another experiment, bacteriophage were labeled with a radioisotope of phosphorus (^{32}P), a process that makes their DNA radioactive. The labeled viruses were allowed to infect bacteria. After the external viral parts were dislodged from the bacteria, the radioactive phosphorus was detected mainly inside the bacterial cells. The viruses had injected DNA into the cells—evidence that DNA is the genetic material of this virus.

Figure 8.3 The Hershey–Chase experiments. Alfred Hershey and Martha Chase carried out experiments to determine the composition of the hereditary material that bacteriophage inject into bacteria. The experiments were based on the knowledge that proteins contain more sulfur (S) than phosphorus (P), and DNA contains more phosphorus than sulfur.

FINAL PIECES OF EVIDENCE

Avery, MacLeod, and McCarty's surprising result prompted a stampede of other scientists into the field of DNA research. The resulting explosion of discovery confirmed the molecule's role as carrier of hereditary (genetic) information. Key in this advance was the realization that any molecule, DNA or otherwise, had to have a certain set of properties in order to function as the sole repository of hereditary information. First, a full complement of hereditary information must be transmitted along with the molecule from one generation to the next; second, cells of a given species should contain the same amount of it; third, it must be exempt from major change in order to function as a genetic bridge between generations; and fourth, it must be capable of encoding the unimaginably huge amount of information required to build a new individual.

In the late 1940s, Alfred Hershey and Martha Chase proved that DNA, and not protein, satisfies the first expected property of a hereditary molecule: It transmits a full comple-

ment of hereditary information. Hershey and Chase worked with **bacteriophage**, a type of virus that infects bacteria (**Figure 8.3A**). Like all viruses, these infectious particles carry hereditary material that specifies how to make new viruses. After a virus injects a cell with this material, the cell starts making new virus particles. Hershey and Chase carried out an elegant series of experiments proving that the material a bacteriophage injects into bacteria is DNA, not protein (**Figure 8.3B,C**).

In 1948, the second property expected of a hereditary molecule was pinned on DNA. By meticulously measuring the amount of DNA in cell nuclei from a number of species, André Boivin and Roger Vendrely proved that body cells of any individual of a species contain precisely the same amount of DNA. Proof that DNA has the third property expected of a hereditary molecule came from a demonstration that DNA is not involved in metabolism: Daniel Mazia's laboratory discovered that the protein and RNA content of cells varies over time, but the DNA content does not. The fourth property—that a hereditary molecule must somehow encode a huge amount of information—would be proven along with the discovery of DNA's structure.

---

bacteriophage (bac-TEER-ee-oh-fayj) Virus that infects bacteria.

LEARNING OBJECTIVES

Summarize the events that led to the discovery of DNA's structure.

Identify the subunits of DNA and how they combine in a DNA molecule.

Explain how DNA holds information.

Describe base pairing.

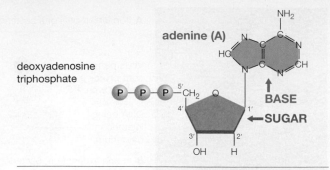

BUILDING BLOCKS OF DNA

DNA is a polymer of four types of nucleotides (**Figure 8.4**). Each has a deoxyribose sugar, three phosphate groups, and a nitrogen-containing base after which it is named: adenine (A), guanine (G), thymine (T), or cytosine (C). Nucleotide structure had been worked out in the early 1900s, but just how the four kinds are arranged in a DNA molecule was a puzzle that took 50 more years to solve.

Clues about DNA's structure started coming together around 1950, when Erwin Chargaff (one of many researchers investigating DNA's function) made two important discoveries about the molecule. First, the amounts of thymine and adenine are identical in any DNA molecule, as are the amounts of cytosine and guanine (A = T and G = C). We call this discovery Chargaff's first rule. Chargaff's second discovery, or rule, is that the DNA of different species differs in its proportions of adenine and guanine.

Meanwhile, biologist James Watson and biophysicist Francis Crick had been sharing ideas about the structure of DNA. The helical (coiled) pattern of secondary structure that occurs in many proteins (Section 3.4) had just been discovered, and Watson and Crick suspected that the DNA molecule was also a helix. The two spent many hours arguing about the size, shape, and bonding requirements of the four DNA nucleotides. They pestered chemists to help them identify bonds they might have overlooked, fiddled with cardboard cutouts, and made models from scraps of metal connected by suitably angled "bonds" of wire.

Biochemist Rosalind Franklin had also been working on the structure of DNA. Like Crick, Franklin specialized in x-ray crystallography, a technique in which x-rays are directed through a purified and crystallized substance. Atoms in the substance's molecules diffract (scatter) the x-rays in a pattern that can be captured as an image. Researchers use the pattern to calculate the size, shape, and spacing between any repeating elements of the molecules—all of which are details of molecular structure.

 As molecules go, DNA is gigantic, and it was difficult to crystallize given the techniques of the time. Franklin made the first clear x-ray diffraction image (left) of DNA as it occurs in cells. She used the image

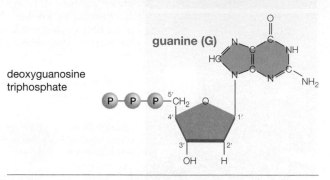

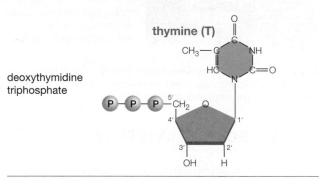

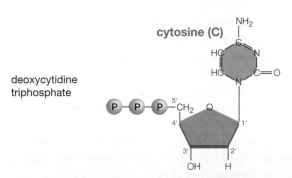

Figure 8.4 The four nucleotides that make up DNA.
Each of the four DNA nucleotides has three phosphate groups, a deoxyribose sugar (orange), and a nitrogen-containing base (blue) after which it is named. Biochemist Phoebus Levene identified the structure of these bases and how they are connected in nucleotides in the early 1900s.

Adenine and guanine bases are purines; thymine and cytosine, pyrimidines. Numbering the carbons in the sugars allows us to keep track of the orientation of nucleotide chains (compare Figure 8.6).

CREDIT: (in text) National Library of Medicine.

to calculate that a DNA molecule is very long compared to its diameter of 2 nanometers. She also identified a repeating pattern every 0.34 nanometer along its length, and another every 3.4 nanometers.

Franklin's image and data came to the attention of Watson and Crick, who now had all the information they needed to build the first accurate model of DNA (**Figure 8.5**). A DNA molecule has two chains (strands) of nucleotides running in opposite directions and coiled into a double helix (**Figure 8.6**). Covalent bonds link the deoxyribose of one nucleotide to a phosphate group of the next, forming the sugar–phosphate backbone of each chain ❶. Hydrogen bonds between the internally positioned bases hold the two strands together ❷. Only two kinds of base pairings form: A to T, and G to C, and this explains the first of Chargaff's rules. Most scientists had assumed (incorrectly) that the bases had to be on the outside of the helix because they would be more accessible to DNA-copying enzymes that way. You will see in Section 8.4 how DNA replication enzymes access the bases on the inside of the double helix.

DNA SEQUENCE

A small piece of DNA from a tulip, a human, or any other organism might be:

TGAGGACTCCTC
ACTCCTGAGGAG
one base pair

Notice how the two strands of DNA fit together. They are complementary—the base of each nucleotide on one strand pairs with a suitable partner base on the other. This base-pairing pattern (A to T, G to C) is the same in all molecules of DNA. How can just two kinds of base pairings give rise to the incredible diversity of traits we see among living things? Even though DNA is composed of only four nucleotides, the *order* in which the nucleotide bases occur along a strand—the **DNA sequence**—varies tremendously among species (which explains Chargaff's second rule). DNA molecules can be hundreds of millions of nucleotides long, so their sequence can encode a massive amount of information (we return to the nature of that information in the next chapter). DNA sequence variation is the basis of traits that define species and distinguish individuals. Thus DNA, the molecule of inheritance in every cell, is the basis of life's unity. Variations in its sequence are the foundation of life's diversity.

DNA sequence Order of nucleotide bases in a strand of DNA.

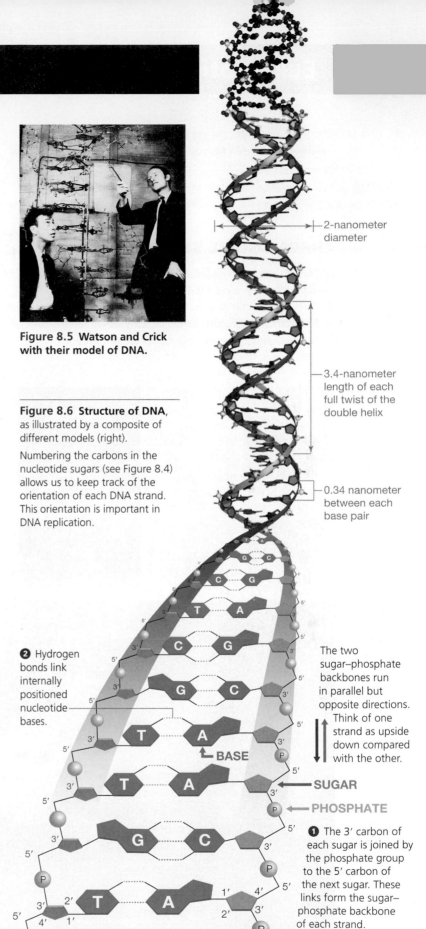

Figure 8.5 Watson and Crick with their model of DNA.

Figure 8.6 Structure of DNA, as illustrated by a composite of different models (right).

Numbering the carbons in the nucleotide sugars (see Figure 8.4) allows us to keep track of the orientation of each DNA strand. This orientation is important in DNA replication.

2-nanometer diameter

3.4-nanometer length of each full twist of the double helix

0.34 nanometer between each base pair

❷ Hydrogen bonds link internally positioned nucleotide bases.

BASE

The two sugar–phosphate backbones run in parallel but opposite directions. Think of one strand as upside down compared with the other.

SUGAR

PHOSPHATE

❶ The 3′ carbon of each sugar is joined by the phosphate group to the 5′ carbon of the next sugar. These links form the sugar–phosphate backbone of each strand.

Describe the way DNA is organized in a chromosome.

Explain how a eukaryotic cell's chromosomes carry its genetic information.

Explain the meaning of diploid.

Distinguish between autosomes and sex chromosomes.

CHROMOSOME STRUCTURE

Stretched out end to end, the DNA in a single human cell would be about 2 meters (6.5 feet) long. How can that much DNA cram into a nucleus less than 10 micrometers in diameter? Proteins that associate with DNA allow it to pack very tightly by organizing the molecule into a structure

called a **chromosome** (**Figure 8.7**). In eukaryotes, a DNA molecule ❶ wraps at regular intervals around proteins called **histones**. These DNA–histone spools are **nucleosomes** ❷, and they look like beads on a string in micrographs. Interactions among histones coil the DNA into a tight fiber ❸. This fiber coils again into a hollow cylinder, a bit like the cord of an old-style land line phone ❹.

Most of the time, a eukaryotic chromosome consists of one DNA molecule (and its associated proteins). A cell preparing to divide will duplicate its DNA. After duplication, a eukaryotic chromosome consists of two identical DNA molecules attached to one another at a constricted region called the **centromere**. The two identical halves of a duplicated eukaryotic chromosome are **sister chromatids**:

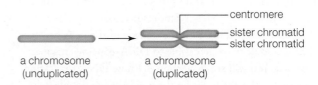

The duplicated chromosomes condense into their familiar "X" shapes ❺ just before the cell divides.

CHROMOSOME NUMBER

Most prokaryotes have a single circular chromosome. By contrast, eukaryotic cells have a number of linear chromosomes that differ in length and shape ❻.

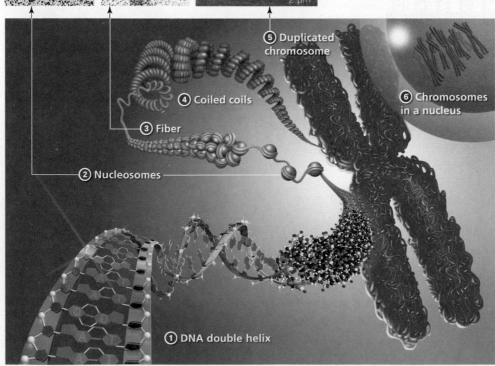

Figure 8.7 DNA organization in a eukaryotic chromosome.

❶ The two strands of a DNA molecule form a double helix.

❷ At regular intervals, the DNA molecule (blue) wraps around a core of histone proteins (purple), forming a nucleosome.

❸ The DNA and proteins associated with it twist tightly into a fiber.

❹ The fiber coils and then coils again to form a hollow cylinder.

❺ At its most condensed, a duplicated eukaryotic chromosome has an X shape. This structure consists of two identical DNA molecules (sister chromatids) attached at a centromere.

❻ The nucleus of a eukaryotic cell contains a characteristic number of chromosomes.

CREDITS: (7) top left, O. L. Miller, Jr., Steve L. McKnight; top middle, B. Hamkalo; top right, Andreew Syred/Science Source.

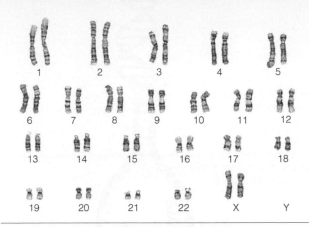

Figure 8.8 A karyotype of a human female, showing 22 pairs of autosomes and a pair of X chromosomes (XX).

The number of chromosomes in a cell of a given species is called the **chromosome number**. The chromosome number of humans is 46, so our cells have 46 chromosomes.

Actually, human body cells have two sets of 23 chromosomes: two of each type. Cells with two sets of chromosomes are **diploid**, or 2n (n stands for "number"). A **karyotype** is an image of an individual cell's chromosomes (**Figure 8.8**). To create a karyotype, cells taken from the individual are treated to make the chromosomes condense, and then stained so the chromosomes can be distinguished under a microscope. A micrograph of a single cell is digitally rearranged so the images of the chromosomes are lined up by centromere location, and arranged according to size, shape, and length.

In a human body cell, 22 of the 23 pairs of chromosomes are autosomes. The two **autosomes** of each pair are the same in both females and males, and they have the same length, shape, and centromere location. They also hold information about the same traits. Think of them as two sets of books on how to build a house. Your father gave you one set. Your mother had her own ideas about wiring, plumbing, and so on. She gave you an alternate set that says slightly different things about many of those tasks.

Members of a pair of **sex chromosomes** differ between females and males, and the difference determines an individual's sex. The sex chromosomes of humans are called X and Y. The body cells of typical human females have two X chromosomes (XX); those of typical human males have one X and one Y chromosome (XY). This pattern—XX females and XY males—is the rule among fruit flies, mammals, and many other animals, but there are other patterns. Female butterflies, moths, birds, and certain fishes have two nonidentical sex chromosomes; the two sex chromosomes of males are identical. In some species of invertebrates, frogs, and turtles, environmental factors (not chromosomes) determine sex.

Mariana Fuentes

A host of natural and human-caused challenges threaten sea turtles; only one in a thousand survives to sexual maturity. Climate change adds an immediate, serious threat, in part because the temperature of the sand in which their eggs are buried—not sex chromosomes—determines the sex of the hatchlings. National Geographic Explorer Mariana Fuentes predicts that rising global temperatures will soon skew the gender ratio of hatchlings toward all female, with disastrous results for sea turtle populations.

autosome Chromosome of a pair that is the same in males and females; a chromosome that is not a sex chromosome.

centromere (SEN-truh-meer) Of a duplicated eukaryotic chromosome, constricted region where sister chromatids attach to each other.

chromosome A structure that consists of DNA together with associated proteins; carries part or all of a cell's genetic information.

chromosome number The total number of chromosomes in a cell of a given species.

diploid (DIP-loyd) Having two of each type of chromosome characteristic of the species (2n).

histone (HISS-tone) Type of protein that associates with DNA and structurally organizes eukaryotic chromosomes.

karyotype (CARE-ee-uh-type) Image of a cell's chromosomes arranged by size, length, shape, and centromere location.

nucleosome (NEW-klee-uh-sohm) A length of DNA wound twice around histone proteins.

sex chromosomes Pair of chromosomes that differs between males and females, and the difference determines sex.

sister chromatids (KROME-uh-tid) The two attached DNA molecules of a duplicated eukaryotic chromosome.

When a cell reproduces, it divides. The two descendant cells must inherit a complete and accurate copy of their parent's genetic information, or they will not function like the parent cell. Thus, in preparation for division, the cell copies its chromosomes so that it contains two sets: one for each of its future offspring. The copying process, which is called **DNA replication**, ensures the continuity of genetic information from one generation to the next.

SEMICONSERVATIVE REPLICATION

Before DNA replication, a chromosome consists of one molecule of DNA—one double helix (**Figure 8.9**). As replication begins, enzymes unwind the double helix ❶ and break the hydrogen bonds that hold the two DNA strands together ❷. Separating the DNA strands exposes their internally positioned bases. Another enzyme starts making primers. A **primer** is a short, single strand of DNA or RNA that serves as an attachment point for the enzyme that carries out DNA synthesis. The nucleotide bases of a primer can form hydrogen bonds with the exposed bases of a single strand of DNA ❸. Thus, a primer can base-pair with a complementary strand of DNA (left). The establishment of base pairing between two strands of DNA (or DNA and RNA) is called **hybridization**. Hybridization is spontaneous and is driven entirely by hydrogen bonding.

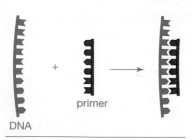

DNA + primer → hybridization

The enzyme that carries out DNA synthesis is called **DNA polymerase.** There are several DNA polymerases, but all can attach to a hybridized primer and begin DNA synthesis. A DNA polymerase moves along a DNA strand, using the sequence of exposed nucleotide bases as a template (guide) to assemble a new strand of DNA from free nucleotides ❹. Two of a nucleotide's three phosphate groups are removed when it is added to a DNA strand. Breaking those bonds releases enough energy to drive the attachment.

A DNA polymerase follows base-pairing rules: It adds a T to the end of the new DNA strand when

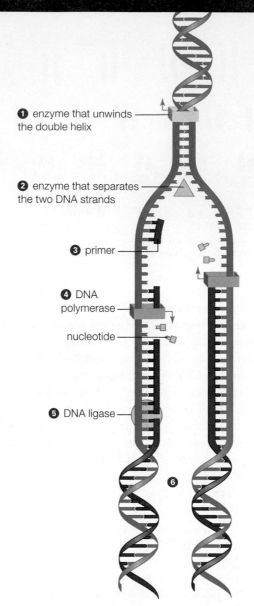

❶ enzyme that unwinds the double helix

❷ enzyme that separates the two DNA strands

❸ primer

❹ DNA polymerase

nucleotide

❺ DNA ligase

❻

Figure 8.9 DNA replication. Green arrows show the direction of synthesis for each strand. The Y-shaped structure of a DNA molecule undergoing replication is called a replication fork.

As replication begins, enzymes begin to ❶ unwind and ❷ separate the two strands of DNA.

❸ Primers base-pair with the exposed single DNA strands.

❹ Starting at primers, DNA polymerases (green boxes) assemble new strands of DNA from nucleotides, using the parent strands as templates.

❺ DNA ligase seals any gaps that remain between bases of the "new" DNA.

❻ Each parental DNA strand (blue) serves as a template for assembly of a new strand of DNA (magenta). The new DNA strand winds up with its template, so two double-stranded DNA molecules result. One strand of each is parental (conserved), and the other is new, so DNA replication is said to be semiconservative.

it reaches an A in the template strand; it adds a G when it reaches a C; and so on. Thus, the DNA sequence of each new strand is complementary to the template (parental) strand. The enzyme **DNA ligase** seals any gaps, so the new DNA strands are continuous.

Both strands of the parent molecule are copied at the same time. As each new DNA strand lengthens, it winds up with the template strand into a double helix. So, after replication, two double-stranded molecules of DNA have formed ❻. One strand of each molecule is conserved (parental), and the other is new; hence the name of the process, **semiconservative replication**. Both double-stranded molecules produced by DNA replication are duplicates of the parent molecule. In eukaryotes, these molecules are sister chromatids that remain attached at the centromere until cell division occurs.

DIRECTIONAL SYNTHESIS

Numbering the carbons of the deoxyribose sugars in nucleotides allows us to keep track of the orientation of strands in a DNA double helix (see Figures 8.4 and 8.6). Each strand has two ends. The last carbon atom on one end of the strand is a 5′ (5 prime) carbon of a sugar; the last carbon atom on the other end is a 3′ (three prime) carbon of a sugar:

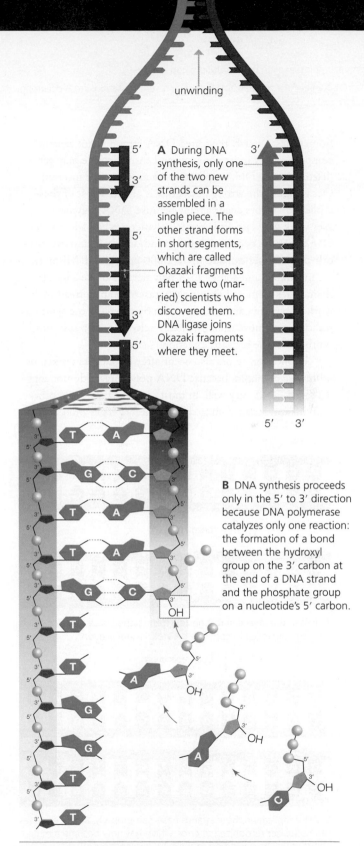

DNA polymerase can attach a nucleotide only to a 3′ end. Thus, during DNA replication, only one of two new strands of DNA can be constructed in a single piece (**Figure 8.10**). Synthesis of the other strand occurs in segments that must be joined by DNA ligase where they meet up. This is why we say that DNA synthesis proceeds only in the 5′ to 3′ direction.

DNA ligase (LIE-gayce) Enzyme that seals gaps or breaks in double-stranded DNA.

DNA polymerase Enzyme that carries out DNA replication. Uses a DNA template to assemble a complementary strand of DNA.

DNA replication Process by which a cell duplicates its DNA before it divides.

hybridization (hi-brih-die-ZAY-shun) Spontaneous establishment of base pairing between two nucleic acid strands.

primer Short, single strand of DNA or RNA that serves as an attachment point for DNA polymerase.

semiconservative replication Describes the process of DNA replication in which one strand of each copy of a DNA molecule is new, and the other is a strand of the original DNA.

A During DNA synthesis, only one of the two new strands can be assembled in a single piece. The other strand forms in short segments, which are called Okazaki fragments after the two (married) scientists who discovered them. DNA ligase joins Okazaki fragments where they meet.

B DNA synthesis proceeds only in the 5′ to 3′ direction because DNA polymerase catalyzes only one reaction: the formation of a bond between the hydroxyl group on the 3′ carbon at the end of a DNA strand and the phosphate group on a nucleotide's 5′ carbon.

Figure 8.10 Discontinuous synthesis of DNA.
This close-up of a replication fork shows that only one of the two new DNA strands is assembled in a single piece.

8.5 Mutations and Their Causes

LEARNING OBJECTIVES

Using examples, explain how mutations can arise.

Describe two cellular mechanisms that can prevent mutations from occurring.

Sometimes, a newly replicated DNA strand is not exactly complementary to its parent strand. A nucleotide may get deleted during DNA replication, or an extra one inserted. Occasionally, the wrong nucleotide is added. Most of these replication errors occur simply because DNA polymerases work very fast. Eukaryotic polymerases can add about 50 nucleotides per second to a strand of DNA, and prokaryotic polymerases can add up to 1,000 per second. Mistakes are inevitable, and some DNA polymerases make a lot of them. Luckily, most DNA polymerases also proofread their work. They can correct a mismatch by reversing the synthesis reaction to remove the mispaired nucleotide, then resuming synthesis in the forward direction.

Replication errors also occur after DNA gets broken or otherwise damaged, because DNA polymerases do not copy damaged DNA very well. In most cases, enzymes and other proteins can repair damaged DNA before replication begins.

When proofreading and repair mechanisms fail, an error becomes a **mutation**—a permanent change in the DNA sequence of a chromosome. Repair enzymes cannot fix a mutation after the altered DNA strand has been replicated, because they do not recognize correctly paired bases (**Figure 8.11**). Thus, a mutation is passed to the cell's descendants, their descendants, and so on.

Ultraviolet (UV) light in the range of 320 to 380 nanometers has enough energy to open up a double bond in the ring of a cytosine or thymine base. Then, the ring can form a covalent bond with the ring of an adjacent cytosine or thymine to form what is called a dimer. Dimers kink a DNA strand (right). A specific set of proteins can recognize, remove, and replace dimers before replication begins. If these proteins fail to fix a dimer, mutations can occur because DNA polymerases tend to copy the kinked DNA incorrectly. Mutations that arise as a result of dimers are the cause of most skin cancers: Exposing unprotected skin to sunlight increases the risk of cancer because UV wavelengths in the light cause dimers to form. For every second a skin cell spends in the sun, 50 to 100 dimers form in its DNA.

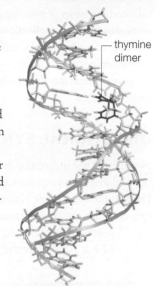

thymine dimer

Electromagnetic energy with a wavelength shorter than 320 nanometers (gamma rays, x-rays, and short-wave UV light) has enough energy to knock electrons out of atoms. Exposure to this ionizing radiation damages DNA, breaking it into pieces that get lost during replication (**Figure 8.12A**). Ionizing radiation can also cause covalent bonds to form between bases on opposite strands of the double helix, an outcome that permanently blocks DNA replication (we consider effects of such cell cycle interruptions in Chapter 11). The nucleotide bases themselves can be irreparably damaged by ionizing radiation. Repair enzymes remove bases damaged in this way, but they leave an empty space in the double helix or a strand break that requires further repair. Any of these events can result in mutations (**Figure 8.12B**).

Exposure to some chemicals also causes mutations. For instance, several of the cancer-causing chemicals in tobacco smoke transfer methyl groups ($—CH_3$) to the nucleotide bases in DNA. The added methyl groups can interfere with base-pairing. The body converts other chemicals in the smoke to compounds that bind irreversibly to DNA. In both cases,

Figure 8.11

| G | G | A | C | T | C | C | T | C | T | T | C | A | G |
| C | C | T | G | A | G | G | A | G | A | A | G | T | C |

↓ base change

| G | G | A | C | A | C | C | T | C | T | T | C | A | G |
| C | C | T | G | A | G | G | A | G | A | A | G | T | C |

A Repair enzymes can recognize a mismatched base (magenta), but sometimes fail to correct it before DNA replication.

↓ DNA replication

| G | G | A | C | A | C | C | T | C | T | T | C | A | G |
| C | C | T | G | T | G | G | A | G | A | A | G | T | C |

| G | G | A | C | T | C | C | T | C | T | T | C | A | G |
| C | C | T | G | A | G | G | A | G | A | A | G | T | C |

B After replication, both strands base-pair properly. Repair enzymes can no longer recognize the error, which has now become a mutation that will be passed on to the cell's descendants.

Figure 8.11 How a replication error becomes a mutation.

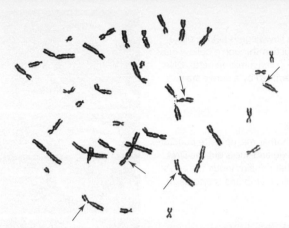

A Major breaks (red arrows) in chromosomes of a human white blood cell after it was exposed to ionizing radiation. Pieces of broken chromosomes often become lost during DNA replication.

normal flower

B These *Ranunculus* flowers were grown from plants harvested around Chernobyl, Ukraine, where in 1986 an accident at a nuclear power plant released a massive amount of radiation. All except one are abnormal—an effect of mutations caused by radiation exposure.

Figure 8.12 Exposure to ionizing radiation causes mutations.

the resulting replication errors can lead to mutation. Cigarette smoke also contains free radicals, which cause the same types of DNA damage as ionizing radiation.

Mutations can form in any type of cell, and they can alter any part of the cell's DNA. Those that arise during egg or sperm formation can be passed to offspring, and in fact each human child is born with an average of 70 mutations that neither parent has. Mutations that alter DNA's instructions may have a harmful or lethal outcome (Sections 9.5 and 11.5 return to this topic). However, not all mutations are dangerous: As you will see in Chapter 17, they give rise to the variation in traits that is the raw material of evolution.

mutation Permanent change in a chromosome's DNA sequence.

Rosalind Franklin

Rosalind Franklin arrived at King's College, London, in 1951. The expert x-ray crystallographer had been told she would be the only one in her department working on the structure of DNA, so she did not know that Maurice Wilkins was already doing the same thing just down the hall. No one had told Wilkins about Franklin's assignment; he assumed she was a technician hired to do his x-ray crystallography work. And so a clash began. Wilkins thought Franklin displayed an appalling lack of deference that technicians of the era usually accorded researchers. To Franklin, Wilkins seemed prickly and oddly overinterested in her work.

Wilkins and Franklin had been given identical samples of DNA. Franklin's meticulous work with hers yielded the first clear x-ray diffraction image of DNA as it occurs in cells, and she gave a presentation about it in 1952. DNA, she said, had two chains twisted into a double helix, with a backbone of phosphate groups on the outside, and bases arranged in an unknown way on the inside. She had calculated DNA's diameter, the distance between its chains and between its bases, the angle of the helix, and the number of bases in each coil. Crick, with his crystallography background, would have recognized the significance of the work—if he had been there. Watson was in the audience but he was not a crystallographer, and he did not understand the implications of Franklin's data.

Franklin started to write a research paper on her findings. Meanwhile, and perhaps without her knowledge, Wilkins reviewed Franklin's x-ray diffraction image with Watson, and Watson and Crick read a report detailing Franklin's unpublished data. Crick, who had more experience with molecular modeling than Franklin, immediately understood what the image and the data meant. Watson and Crick used that information to build their model of DNA. On April 25, 1953, Franklin's paper appeared last in a series of articles about the structure of DNA in the journal *Nature*. It supported with solid experimental evidence Watson and Crick's theoretical model, which appeared in the first article of the series.

Rosalind Franklin died in 1958, at the age of 37, of ovarian cancer probably caused by extensive exposure to x-rays during her work. At the time, the link between x-rays, mutations, and cancer was not understood. Because the Nobel Prize is not given posthumously, Franklin did not share in the 1962 honor that went to Watson, Crick, and Wilkins for the discovery of the structure of DNA.

CREDITS: (12A) Olga Shovman, Andrew C. Riches, Douglas Adamson, and Peter E. Bryant. An improved assay for radiation-induced chromatid breaks using a colcemid block and calyculin-induced PCC combination. *Mutagenesis* (2008) 23(4): 267–270 first published online March 6, 2008 doi:10.1093/mutage/gen009, by permission of Oxford University Press; (12B) Courtesy of Taavi Tuulik; (in text) National Library of Medicine.

LEARNING OBJECTIVES

Using a suitable example, explain the process of cloning an adult animal.

Explain why clones can be produced from a single body cell of an adult.

Describe the process of differentiation.

Barring mutations, all cells descended from a fertilized egg inherit the same DNA. That DNA is like a master blueprint for directing the development of an individual's body. Consider how identical twins have identical DNA, so their bodies develop the same way (which is why they look identical, as shown in the chapter opening photo). Identical twins are the product of a natural process called embryo splitting. The first few divisions of a fertilized egg form a tiny ball of cells that sometimes splits spontaneously. If both halves of the ball develop independently, identical twins result. Animal breeders have long exploited this phenomenon with a technique called artificial embryo splitting. A tiny ball of cells grown from a fertilized egg is teased apart into two halves that develop as separate embryos. The embryos are implanted in surrogate mothers, who give birth to identical twins.

Twins produced by embryo splitting are genetically identical to one another, but they are not identical to either parent. This is because, in humans and other animals, twins get their DNA from two parents that typically differ in their DNA sequence (Chapter 12 returns to this topic). Animal breeders who want an exact copy of a specific individual use **somatic cell nuclear transfer** (**SCNT**), a laboratory procedure in which an unfertilized egg's nucleus is replaced with the nucleus of a donor's somatic cell (**Figure 8.13**). A somatic cell is a body cell, as opposed to a reproductive cell (*soma* is a Greek word that means body). If all goes well, the transplanted DNA directs the development of an embryo, which is then implanted into a surrogate mother. The animal born to the surrogate is a **clone**—a genetic copy—of the donor. Clones produced by SCNT have the same features as their adult donors (**Figure 8.14**), and there are other benefits. SCNT can yield many more offspring in a given time frame than traditional breeding, and offspring can be produced from a donor animal that is castrated or even dead.

"Cloning" means making an identical copy of something, and it can refer to deliberate interventions intended to produce an exact genetic copy of an organism. **Reproductive cloning** refers to SCNT and any other technology that produces clones of an animal from a single cell. SCNT has been used since 1997, when a lamb called Dolly was cloned from a mammary cell of an adult sheep (the clone was named after voluptuous performer Dolly Parton). Dolly looked and acted like a normal sheep, but she died early, most likely because she was a clone (Section 11.4 returns to this topic).

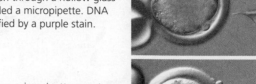

A A cow's egg is held in place by suction through a hollow glass tube called a micropipette. DNA is identified by a purple stain.

B Another micropipette punctures the egg and sucks out the DNA. All that remains inside the egg's plasma membrane is cytoplasm.

C A new micropipette prepares to enter the egg at the puncture site. The pipette contains a cell grown from the skin of a donor animal.

D The micropipette enters the egg and delivers the skin cell to a region between the cytoplasm and the plasma membrane.

E After the pipette is withdrawn, the donor's skin cell is visible next to the cytoplasm of the egg. The transfer is complete.

F An electric current causes the skin cell to empty its nucleus into the cytoplasm of the egg. If all goes well, the egg begins to divide, and an embryo forms.

Figure 8.13 Somatic cell nuclear transfer (SCNT) with cells from cattle. This series of micrographs was taken by scientists at Cyagra, a company that specializes in cloning livestock.

The outcome of SCNT is still unpredictable. Depending on the species, few implanted embryos may survive until birth. Until recently, most of the clones that did survive had serious health problems such as enlarged organs and obesity. Cloned mice developed lung and liver problems, and almost all died prematurely. Cloned pigs tended to limp and have heart problems; some developed without a tail or, even worse, an anus.

Figure 8.14 Champion dairy cow Liz (right) **and her clone, Liz II** (left). Liz II, who was produced by SCNT, had already begun to win championships by the time she was one year old.

Why the problems? SCNT works because the DNA in a somatic cell contains all the information necessary to build the individual all over again. However, a somatic cell does not automatically start dividing and form an embryo. This is because, during early development, an embryo's cells start using different subsets of their DNA. As they do, the cells become different in form and function, a process called **differentiation**. Differentiation in animals is usually a one-way path, which means that once a cell has become specialized, all of its descendant cells will be specialized the same way. By the time a liver cell, muscle cell, or other differentiated cell forms, most of its DNA has been turned off, and is no longer used (Chapter 10 returns to this topic). To clone an adult, the DNA of a somatic cell must be reprogrammed so the parts that trigger embryonic development are turned back on. We still have a lot to learn about that, but SCNT techniques have improved so much that health problems are much less common in animals cloned today.

As the techniques become routine, cloning humans is no longer only within the realm of science fiction. SCNT is already being used to produce human embryos for research purposes. Human cloning is not the intent of the research, but if it were, SCNT would indeed be the first step toward that end.

---

clone Genetic copy of an organism.
differentiation Process by which cells become specialized during development; occurs as different cells in an embryo begin to use different subsets of their DNA.
reproductive cloning Laboratory procedure such as SCNT that produces animal clones from a single cell.
somatic cell nuclear transfer (SCNT) Reproductive cloning method in which the DNA of a donor's body cell is transferred into an enucleated egg.

Figure 8.15 James Symington and his dog Trakr assisting in the search for victims at Ground Zero, September 2001.

A HERO DOG'S GOLDEN CLONES

Why clone animals? Consider the story of Canadian police officer James Symington and his search dog Trakr. On September 11, 2001, Constable James Symington drove his search dog Trakr from Nova Scotia to Manhattan. Within hours of arriving, the dog led rescuers to the area where the final survivor of the World Trade Center attacks was buried. She had been clinging to life, pinned under rubble from the building where she had worked. Symington and Trakr helped with the search and rescue efforts for three days nonstop, until Trakr collapsed from smoke and chemical inhalation, burns, and exhaustion (**Figure 8.15**).

Trakr survived the ordeal, but later lost the use of his limbs from a degenerative neurological disease probably linked to toxic smoke exposure at Ground Zero. The hero dog died in April 2009, but his DNA lives on—in his clones. Symington's essay about Trakr's superior nature and abilities as a search and rescue dog won the Golden Clone Giveaway, a contest to find the world's most clone-worthy dog. Trakr's DNA was shipped to Korea, where it was inserted into donor dog eggs, which were then implanted into surrogate mother dogs. Five puppies, all clones of Trakr, were delivered to Symington in July 2009. Symington trained the clones to be search and rescue dogs for his humanitarian organization, Team Trakr Foundation.

Cloning animals brings us closer to the possibility of cloning humans, both technically and ethically. The idea raises uncomfortable questions. For example, if cloning a lost animal for a grieving owner is acceptable, why would it not be acceptable to clone a lost child for a grieving parent? Different people have very different answers to such questions, so controversy over cloning continues to rage even as the technique improves.

Section 8.1 It took many scientists and decades of research using cells and viruses called **bacteriophage** to determine that deoxyribonucleic acid (DNA), not protein, is the hereditary material of life.

Section 8.2 Each DNA nucleotide has a five-carbon sugar, three phosphate groups, and one of four nitrogen-containing bases after which the nucleotide is named: adenine, thymine, guanine, or cytosine. A DNA molecule consists of two strands of these nucleotides coiled into a double helix. Hydrogen bonding between the internally positioned bases holds the strands together. The bases pair in a consistent way: adenine with thymine (A–T), and guanine with cytosine (G–C). The order of bases along a strand of DNA (the **DNA sequence**) varies among species and among individuals, and this variation is the basis of life's diversity.

2

Section 8.3 The DNA of eukaryotes is divided among a characteristic number of **chromosomes** that differ in length and shape. In eukaryotic chromosomes, the DNA wraps around **histone** proteins to form **nucleosomes**. When duplicated, a eukaryotic chromosome consists of two **sister chromatids** attached at a **centromere**. **Diploid** (2n) cells have two sets of chromosomes. **Chromosome number** is the total number of chromosomes in a cell of a given species. A human body cell has twenty-three pairs of chromosomes. Members of a pair of **sex chromosomes** differ among males and females. Members of a pair of **autosomes** are the same in males and females. Autosomes of a pair have the same length, shape, and centromere location. A **karyotype** reveals an individual's complete set of chromosomes.

Section 8.4 Before a cell divides, it copies all of its DNA (by **DNA replication**). For each molecule of DNA that is copied, two DNA molecules are produced; each is a duplicate of the parent molecule. One strand of each molecule is new, and the other is parental, so the process is called **semiconservative replication**. During DNA replication, enzymes unwind and separate the two strands of the double helix. **Primers** form and base-pair with the exposed nucleotides, a process called **hybridization**. Starting at the primers, **DNA polymerase** enzymes use each strand as a template to assemble new, complementary strands of DNA from nucleotides. **DNA ligase** seals any gaps.

Section 8.5 Proofreading by DNA polymerases corrects most DNA replication errors as they occur. DNA damage by environmental agents, including high-energy electromagnetic radiation, free radicals, and some chemicals, can lead to replication errors because DNA polymerase does not copy damaged DNA very well. Most DNA damage is repaired before replication begins. Uncorrected replication errors become **mutations**, which are permanent changes in the DNA sequence of a chromosome. Cancer begins with mutations, but not all mutations are harmful.

Section 8.6 **Somatic cell nuclear transfer (SCNT)** and other **reproductive cloning** technologies produce **clones**—genetic copies—from a single cell. SCNT works because the DNA in each body cell of an animal contains all the information necessary to build the individual all over again. The outcome of SCNT is unpredictable. This is because **differentiation** is usually a one-way path in animals. We are still learning how to reprogram the DNA of a differentiated cell to function like the DNA of an embryo.

Application Animal cloning continues to raise ethical questions, particularly about cloning humans.

SELF-ASSESSMENT
ANSWERS IN APPENDIX VII

1. Which is not a nucleotide base in DNA?
   a. adenine   c. glutamine   e. cytosine
   b. guanine   d. thymine   f. All are in DNA.

2. What are the base-pairing rules for DNA?
   a. A–G, T–C   c. A–T, G–C
   b. A–C, T–G   d. A–A, G–G, C–C, T–T

3. Similarities in _____ are the basis of similarities in traits.
   a. karyotype   c. the double helix
   b. DNA sequence   d. chromosome number

4. One species' DNA differs from others in its _____ .
   a. nucleotides   c. double helix
   b. DNA sequence   d. sugar–phosphate backbone

5. In eukaryotic chromosomes, DNA wraps around _____ .
   a. histone proteins   c. centromeres
   b. sister chromatids   d. nucleosomes

6. The chromosome number _____ .
   a. refers to a particular chromosome in a cell
   b. is a characteristic feature of a species
   c. is the number of autosomes in cells of a given type
   d. is the same in all species

7. Human body cells are diploid, which means they _____ .
   a. divide to form two cells
   b. have two full sets of chromosomes
   c. contain two sex chromosomes

8. DNA replication requires _____ .
   a. DNA polymerase   c. primers
   b. nucleotides   d. all are required

9. Energy that drives the attachment of a nucleotide to the end of a growing strand of DNA comes from _____ .
   a. the nucleotide   b. DNA polymerase
   c. phosphate-group transfers from ATP

HERSHEY–CHASE EXPERIMENTS The graph in **Figure 8.16** is reproduced from an original publication by Hershey and Chase. The data are from experiments described in Section 8.1, in which bacteriophage were labeled with radioactive tracers and allowed to infect bacteria. The virus–bacteria mixtures were then whirled in a blender to dislodge any viral components attached to the exterior of the bacteria. Afterward, radioactivity from the tracers was measured.

1. Before blending, what percentage of each isotope, ^{35}S and ^{32}P, was extracellular (outside the bacteria)?

2. After 4 minutes in the blender, what percentage of each isotope was outside the bacteria?

3. How did the researchers know that the radioisotopes in the fluid came from outside of the bacterial cells and not from bacteria that had been broken apart by whirling in the blender?

4. The extracellular concentration of which isotope increased the most with blending?

5. Do these results imply that viruses inject DNA or protein into bacteria? Why or why not?

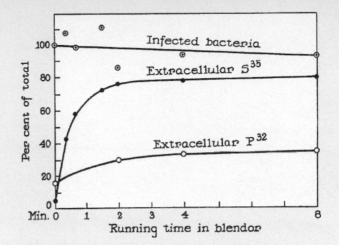

Figure 8.16 Detail of Alfred Hershey and Martha Chase's 1952 publication describing their experiments with bacteriophage. "Infected bacteria" refers to the percentage of bacteria that survived the blender.

10. When DNA replication begins, _____ .
    a. the two DNA strands unwind from each other
    b. the two DNA strands condense for base transfers
    c. old strands move to find new strands

11. The phrase 5′ to 3′ refers to the _____ .
    a. timing of DNA replication
    b. directionality of DNA synthesis
    c. number of phosphate groups

12. After DNA replication, a eukaryotic chromosome _____ .
    a. consists of two sister chromatids
    b. has a characteristic X shape
    c. is constricted at the centromere
    d. all of the above

13. All mutations _____ .
    a. result from radiation   c. are caused by DNA damage
    b. lead to evolution   d. change the DNA sequence

14. _____ is an example of reproductive cloning.
    a. Somatic cell nuclear transfer (SCNT)
    b. Multiple offspring from the same pregnancy
    c. Artificial embryo splitting
    d. a and c

15. Match the terms appropriately.
    ___ nucleotide
    ___ clone
    ___ autosome
    ___ DNA polymerase
    ___ mutation
    ___ bacteriophage
    ___ semiconservative replication
    ___ DNA ligase

    a. nitrogen-containing base, sugar, phosphate group(s)
    b. copy of an organism
    c. does not determine sex
    d. injects DNA
    e. replication enzyme
    f. can cause cancer
    g. something old, something new
    h. seals gaps and breaks

CLASS ACTIVITY
CRITICAL THINKING

1. Determine the complementary strand of DNA that forms on this template DNA fragment during replication:

    5′—GGTTTCTTCAAGAGA—3′

2. Woolly mammoths have been extinct for about 4,000 years, but we often find their well-preserved remains in Siberian permafrost. Research groups are now planning to use SCNT to resurrect these huge elephant-like mammals. No mammoth eggs have been recovered yet, so elephant eggs would be used instead. An elephant would also be the surrogate mother for the resulting embryo. The researchers may try a modified SCNT technique used to clone a mouse that had been dead and frozen for sixteen years. Ice crystals that form during freezing break up cell membranes, so cells from the frozen mouse were in bad shape. Their DNA was transferred into donor mouse eggs, and cells from the resulting embryos were fused with undifferentiated mouse cells. Four healthy clones were born from the hybrid embryos. What are some of the pros and cons of cloning an extinct animal?

3. Xeroderma pigmentosum is an inherited disorder characterized by rapid formation of many skin sores that develop into cancers. All forms of radiation trigger these symptoms, including fluorescent light, which contains UV light in the range of 320 to 400 nm. In most affected individuals, at least one of nine particular proteins is missing or defective. What is the collective function of these proteins?

for additional quizzes, flashcards, and other study materials
▶ **ACCESS MINDTAP AT WWW.CENGAGEBRAIN.COM**

CREDIT: (16) *Journal of General Physiology*, 36(1), Sept. 20, 1952.

CHAPTER 8 DNA STRUCTURE AND FUNCTION   **145**

9

From DNA to Protein

Links to Earlier Concepts

Your knowledge of base pairing (Section 8.2) and chromosomes (8.3) will help you understand the role of nucleic acids (3.6) in protein synthesis (3.4). This chapter revisits membrane proteins (4.3), the nucleus (4.5) and endomembrane system (4.6); hydrophobicity (2.4); cofactors (5.5); enzymes (5.3); DNA replication (8.4), and mutation (8.5).

Core Concepts

Living systems store, retrieve, transmit, and respond to information essential for life.

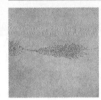

An organism's growth, survival, and reproduction depend on the genetic information carried by the DNA in its cells. In order for cells to use this information, it must be expressed: transcribed into RNA (DNA→RNA), and in most cases translated into protein (RNA→protein). The products of gene expression carry out metabolism, and they also give rise to traits of the individual.

A molecule's function and ability to interact with other molecules depend on its three-dimensional structure and its location in a cell.

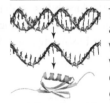

The chemical structures of DNA and RNA allow a cell to preserve its genetic information even while using it. Both nucleic acids are polymers of very similar nucleotides, but important structural differences between the two types of monomers give rise to major differences in molecular stability and function. Three types of RNA interact during protein synthesis.

The continuity of life arises from and depends on genetic information in DNA.

The organizational basis of cells and multicelled organisms is heritable information. A mutation that changes a gene's expression or its product can result in observable changes in anatomy and physiological function. Depending on the mutation and its location in DNA, these effects may be harmful, neutral, or even advantageous.

◀ The hairless appearance of a sphynx cat arises from a single base-pair mutation in its DNA. The change results in an altered form of the keratin protein that makes up cat fur.

147

Photograph by Glennis Siverson.

9.4 Translation: RNA to Protein

Translation occurs in the cytoplasm of all cells. Cytoplasm has many free amino acids, tRNAs, and ribosomal subunits available to participate in the process.

In a eukaryotic cell, RNAs involved in translation are produced in the nucleus (**Figure 9.10 ❶**), then transported through nuclear pores into the cytoplasm ❷. Translation begins when a small ribosomal subunit binds to an mRNA, and the anticodon of a special tRNA called an initiator base-pairs with the mRNA's first AUG codon. Then, a large ribosomal subunit joins the small subunit ❸.

The complex of molecules is now ready to carry out protein synthesis: The intact ribosome moves along the mRNA and assembles a polypeptide ❹. **Figure 9.11** zooms in on this process. Most initiator tRNAs carry methionine, so the first amino acid of most new polypeptide chains is methionine. Another tRNA joins the complex when its anticodon base-pairs with the second codon in the mRNA ❶. This tRNA brings with it the second amino acid. The ribosome then catalyzes formation of a peptide bond (Section 3.4) between the first two amino acids ❷.

As the ribosome moves to the next codon, it releases the first tRNA. Another tRNA brings the third amino acid to the complex as its anticodon base-pairs with the third codon of the mRNA ❸. A peptide bond forms between the second and third amino acids ❹, so a peptide forms.

Figure 9.10 Overview of translation.
In eukaryotes, RNA transcribed in the nucleus moves into the cytoplasm through nuclear pores. Translation occurs in the cytoplasm. Ribosomes simultaneously translating the same mRNA are called polysomes.

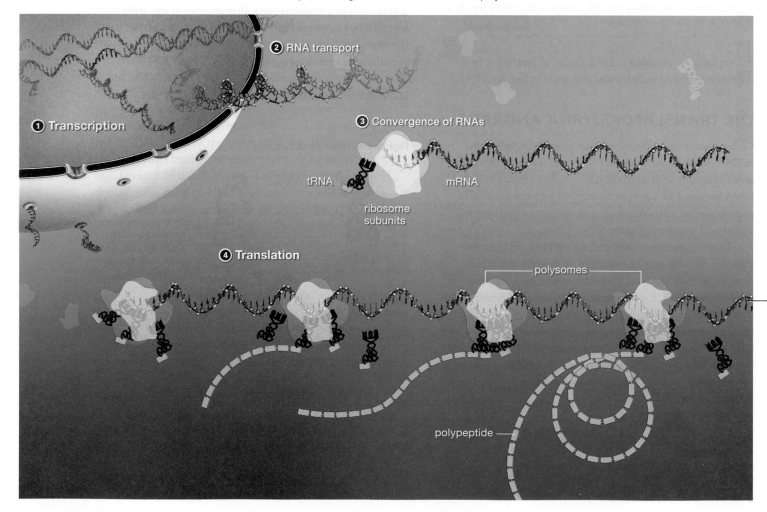

The second tRNA is released and the ribosome moves to the next codon. Another tRNA brings the fourth amino acid to the complex as its anticodon base-pairs with the fourth codon of the mRNA ❺. The ribosome then catalyzes the formation of a peptide bond between the third and fourth amino acids.

The amino acid chain continues to elongate as amino acids delivered by successive tRNAs are added. In eukaryotes, many ribosomes associate with rough endoplasmic reticulum (Section 4.6). Polypeptides made by these ribosomes thread into the interior of the ER as they lengthen.

Translation terminates when the ribosome reaches a stop codon in the mRNA ❻. The mRNA and the polypeptide detach from the ribosome, and the ribosomal subunits separate from each other. Translation is now complete.

Many ribosomes may simultaneously translate the same mRNA, in which case they are called polysomes (**Figure 9.12**). In bacteria and archaea, transcription and translation both occur in the cytoplasm, and these processes are closely linked in time and space. Translation begins before transcription ends, so a transcription "Christmas tree" is often decorated with polysome "balls."

Given that multiple polypeptides can be translated from one mRNA, why would a cell also make multiple copies of the mRNA? Compared with DNA, RNA is not very stable. An mRNA may last only a few minutes in cytoplasm before enzymes disassemble it. The fast turnover allows cells to adjust protein synthesis very quickly in response to changing needs.

Translation is fast—each ribosome adds about five to twenty amino acids per second—and energy intensive. Most of that energy is provided in the form of phosphate-group transfers from the RNA nucleotide GTP to molecules involved in the process.

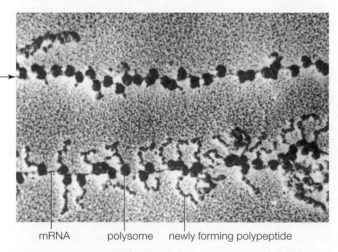

Figure 9.12 Polysomes. Protein synthesis proceeds quickly when multiple ribosomes (polysomes) simultaneously translate an mRNA.

mRNA   polysome   newly forming polypeptide

Figure 9.11 Zooming in on translation. Ribosomal subunits and an initiator tRNA converge on an mRNA. Then, tRNAs deliver amino acids in the order dictated by successive codons in the mRNA. As the ribosome moves along the mRNA, it links the amino acids together via peptide bonds, so a polypeptide forms and elongates. Translation ends when the ribosome reaches a stop codon.

❶ Ribosomal subunits and an initiator tRNA converge on an mRNA. A second tRNA binds to the second codon.

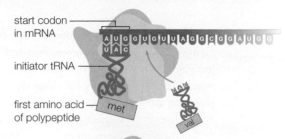

start codon in mRNA

initiator tRNA

first amino acid of polypeptide

❷ A peptide bond forms between the first two amino acids.

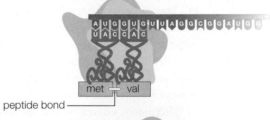

peptide bond

❸ The first tRNA is released and the ribosome moves to the next codon. A third tRNA binds to the third codon.

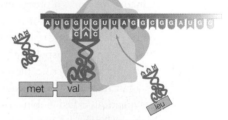

❹ A peptide bond forms between the second and third amino acids.

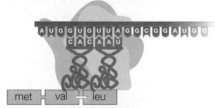

❺ The second tRNA is released and the ribosome moves to the next codon. A fourth tRNA brings the next amino acid to be added to the polypeptide chain.

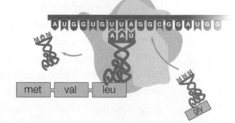

❻ The process repeats until the ribosome encounters a stop codon. Then, the new polypeptide is released and the ribosome subunits separate.

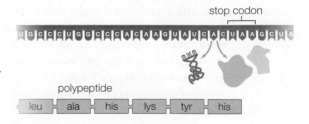

stop codon

polypeptide

leu | ala | his | lys | tyr | his

CREDIT: (12) Kiseleva and Donald Fawcett/Visuals Unlimited.

9.5 Consequences of Mutations

Describe three types of mutations.

Explain how mutations can affect protein structure.

Using appropriate examples, explain why some mutations are not harmful.

Mutations are relatively uncommon events in a normal cell. Consider that the chromosomes in a diploid human cell collectively consist of about 6.5 billion nucleotides, any of which may be copied incorrectly each time that cell divides. The mutation rate in human somatic cells has been measured: About five nucleotides change every time DNA replication occurs. Less than 2 percent of human DNA encodes products, however, so there is a low probability that any mutation will occur in a coding region. When a nucleotide in a protein-coding region does change, the redundancy of the genetic code offers the cell a margin of safety. For example, a mutation that changes a codon from CCU to CCC may have no further effect, because both of these codons specify proline.

Very rarely, a mutation changes an amino acid in a protein, or introduces a premature stop codon. Such mutations can have drastic effects in an organism, particularly if they occur during egg and sperm formation. Consider hemoglobin, an oxygen-transporting protein in your red blood cells. Hemoglobin's structure allows it to bind and release oxygen. In adult humans, a hemoglobin molecule consists of four polypeptides called globins: two alpha globins and two beta globins (**Figure 9.13A**). Each globin folds around a heme cofactor (Section 5.5). Oxygen molecules bind to hemoglobin at those hemes.

As red blood cells circulate through the lungs, the hemoglobin inside of them binds to oxygen molecules. The cells then travel in the bloodstream to other regions of the body, and the hemoglobin releases its oxygen cargo wherever the oxygen level is low. When the red blood cells return to the lungs, the hemoglobin binds to more oxygen.

Mutations that alter hemoglobin's structure can greatly affect health. For example, mutations in either the alpha or beta globin genes can cause a condition called anemia, in which a person's blood is deficient in red blood cells or in hemoglobin. Both outcomes limit the blood's ability to carry oxygen, and the resulting symptoms can range from mild to life-threatening. Sickle-cell anemia arises because of a particular mutation in the beta globin gene. The mutation changes one base pair to another, so it is called a **base-pair substitution**. In this case, the substitution results in a version of beta globin that has valine instead of glutamic acid as its sixth amino acid (**Figure 9.13B,C**). Hemoglobin assembled with this altered beta globin chain is called sickle hemoglobin, or HbS.

A Hemoglobin, an oxygen-binding protein in red blood cells. A working molecule of hemoglobin has four polypeptides: two alpha globins (blue) and two beta globins (green). Each globin has a pocket that cradles a heme (red). Oxygen molecules bind to the iron atom at the center of each heme.

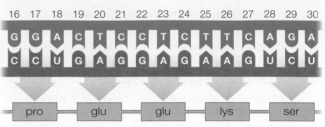

B Part of the DNA (blue), mRNA (brown), and amino acid sequence (green) of human beta globin. Numbers indicate the position of the nucleotide in the mRNA.

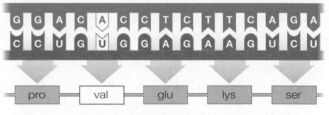

C A base-pair substitution replaces a thymine with an adenine in the beta globin gene. When the altered mRNA is translated, valine replaces glutamic acid as the sixth amino acid. Hemoglobin with this form of beta globin is called sickle hemoglobin, or HbS.

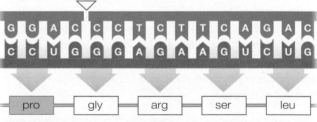

D A deletion of one nucleotide shifts the reading frame for the rest of the mRNA, so a completely different protein product forms. The mutation shown results in a defective beta globin. The outcome is beta thalassemia, a genetic disorder in which a person has an abnormally low amount of hemoglobin.

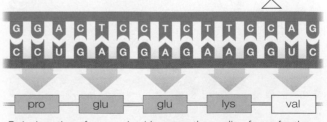

E An insertion of one nucleotide causes the reading frame for the rest of the mRNA to shift. The protein translated from this mRNA is too short and does not assemble correctly into hemoglobin molecules. As in **D**, the outcome is beta thalassemia.

Figure 9.13 Examples of mutations.

CREDIT: (13A) Image of PDB ID: 1GZX, Saito M., Okazaki I., A 45-ns molecular dynamics simulation of hemoglobin in water by vectorizing and parallelizing COSMOS90 on the earth simulator: dynamics of tertiary and quaternary structures, J Comput Chem. 2007 Apr 30;28(6):1129–36, created with PyMOL software from Schrödinger.

Unlike glutamic acid, which carries a negative charge, valine carries no charge. As a result of that one base-pair substitution, a tiny patch of the beta globin polypeptide that is normally hydrophilic becomes hydrophobic. This change slightly alters the behavior of hemoglobin. Under certain conditions, HbS molecules stick together and form large, rod-like clumps. Red blood cells that contain the clumps become distorted into a crescent (sickle) shape (**Figure 9.14**). Sickled cells clog tiny blood vessels, thus disrupting blood circulation throughout the body. Over time, repeated episodes of sickling can damage organs and eventually cause death.

A different type of anemia called beta thalassemia is caused by a **deletion**, which is a mutation in which one or more nucleotides is lost from the DNA. In this case, the twentieth nucleotide in the coding region of the beta globin gene is lost (**Figure 9.13D**). Like most other deletions, this one causes the reading frame of the mRNA codons to shift. A frameshift usually has drastic consequences because it garbles the genetic message, just as incorrectly grouping a series of letters garbles the meaning of a sentence:

> *The fat cat ate the sad rat.*
> *Th efa tca tat eth esa dra t.*

The frameshift caused by the beta globin deletion results in a polypeptide that is very different from normal beta globin in amino acid sequence and in length. This outcome is the source of the anemia. Beta thalassemia can also be caused by an **insertion**, which is a mutation in which nucleotides are added to the DNA. Insertions, like deletions, often cause frameshifts (**Figure 9.13E**).

Not all mutations that change proteins are harmful. Consider the sickle-cell anemia mutation shown in **Figure 9.13C**. A different mutation in the same codon, a base-pair substitution (GAG to AAG) replaces the sixth amino acid with a lysine. Hemoglobin assembled from the resulting beta-globin protein is called HbC. Unlike HbS, HbC does not clump or distort red blood cells. It can cause a mild anemia, but most people who carry the mutation have no symptoms at all. These people are particularly resistant to infection by the parasite that causes malaria, a trait that is quite helpful in regions of the world where malaria is common (we return to this topic in Chapter 17).

A mutation can affect a gene's expression without changing any of its codons, for example if it occurs in a promoter

---

base-pair substitution Type of mutation in which a single base pair changes.
deletion Mutation in which one or more nucleotides are lost.
insertion Mutation in which one or more nucleotides become inserted into DNA.

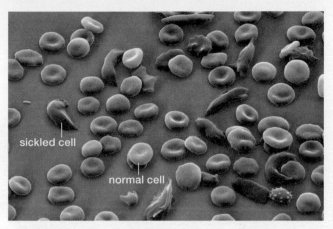

Figure 9.14 Sickled red blood cells. A base-pair substitution results in the abnormal beta globin of sickle hemoglobin (HbS). HbS molecules form rod-shaped clumps that distort normally round blood cells into sickle shapes that get stuck in small blood vessels.

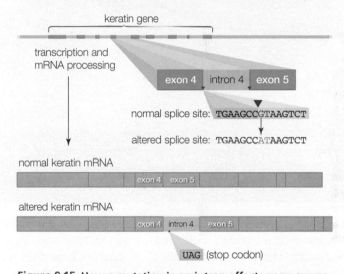

Figure 9.15 How a mutation in an intron affects gene expression. A mutation that causes hairlessness in sphynx cats is a base-pair substitution that changes a G to A in an intron–exon splice site. The altered site is no longer recognized during RNA processing, so the finished mRNA ends up with an intron in it. A stop codon in the intron sequence cuts short the protein translated from the mRNA.

or other special nucleotide sequence in the DNA. Consider a mutation that causes the hairless appearance of the sphynx cat (as shown in the chapter opener photo). In this case, a base-pair substitution disrupts an intron–exon splice site in a gene for keratin, a fibrous protein. The intron is not removed during post-transcriptional processing, and it introduces a premature stop codon in the finished mRNA (**Figure 9.15**). The altered protein translated from the resulting mRNA is too short and cannot properly assemble into filaments that make up hair. Cats that have this mutation still make hair, but it falls out before it gets very long.

Figure 9.16 Bulgarian spy's weapon: an umbrella modified to fire a tiny pellet of ricin into a victim. An umbrella like this one was used to assassinate Georgi Markov on the streets of London in 1978.

THE APTLY ACRONYMED RIPs

Castor-oil plants grow wild in tropical regions, and they are widely cultivated for their seeds. The seeds of the plant are rich in castor oil, which is used as an ingredient in plastics, cosmetics, paints, soaps, and many other items. They are also rich in ricin, a toxic protein that effectively deters beetles, birds, mammals, and other animals from eating them. After castor oil is extracted from the seeds, the leftover ricin-containing seed pulp is usually discarded.

Inhaled, ingested, or injected ricin is extremely toxic to humans. Most cases of ricin poisoning occur as a result of eating castor-oil seeds, and just a few of these are enough to kill an adult. Purified, a dose of ricin as small as a few grains of salt is lethal, and there is no known antidote.

Ricin's toxicity was known as long ago as 1888, and since then several countries have tried (unsuccessfully) to weaponize it. Most countries have since outlawed production, stockpiling, possession, and use of toxic chemicals as weapons. Ricin falls into this category. However, it is impossible to control production of the toxin because no special skills or equipment are required to manufacture it from easily obtained raw materials. Thus, ricin appears periodically in news, mostly in reports of amateur criminals getting caught extracting it or trying to poison someone with the purified material.

Perhaps the most famous ricin poisoning occurred in 1978 at the height of the Cold War, when defectors from countries under Russian control were

targets for assassination. Bulgarian journalist Georgi Markov had defected to England and was working for the BBC. As he made his way to a bus stop on a London street, an assassin used a modified umbrella (**Figure 9.16**) to fire a tiny, ricin-laced ball into Markov's leg. Markov died in agony three days later.

Ricin is a ribosome-inactivating protein (RIP). These proteins interfere with ribosome function, so they prevent the assembly of amino acids into proteins. Other RIPs are made by some bacteria, mushrooms, algae, and many plants (including food crops such as tomatoes, barley, and spinach), but most of these proteins are not particularly toxic to humans because they do not cross intact cell membranes very well. By contrast, ricin and other toxic

Castor-oil seeds

Ricin

Jequirity beans

Abrin

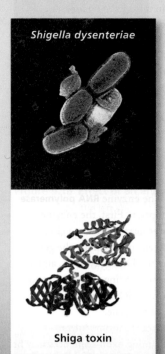

Shigella dysenteriae

Shiga toxin

Escherichia coli O157:H7

Shiga-like toxin

Figure 9.17 Lethal lineup: a few toxic ribosome-inactivating proteins (RIPs) and their sources. One polypeptide chain (red) of a toxic RIP helps the molecule cross a cell's plasma membrane; the other chain (gold) is an enzyme that inactivates ribosomes.

RIPs can enter cells. These proteins have a domain that binds tightly to plasma membrane glycolipids or glycoproteins (**Figure 9.17**). Binding of an RIP to these molecules triggers endocytosis (Section 5.8), so the cell takes in the RIP.

When an RIP enters a cell, its second domain—an enzyme—begins to inactivate ribosomes. The enzyme removes a particular adenine base from one of the rRNAs composing the large ribosomal subunit. The adenine is part of a binding site for proteins involved in GTP-requiring steps of elongation; after the base has been removed, the ribosome can no longer bind to these proteins—or to GTP. A ribosome requires an energy input from GTP in order to catalyze formation of a peptide bond. Without a source of energy, the ribosome stops working.

One molecule of ricin can inactivate more than 1,000 ribosomes per minute. If enough ribosomes are affected, protein synthesis grinds to a halt. Proteins are critical to all life processes, so cells that cannot make them die quickly.

Fortunately, few people actually encounter ricin. Contact with other toxic RIPs is much more common. Bracelets made from beautiful seeds were recalled from stores in 2011 after a botanist recognized them as jequirity beans. These beans contain abrin, an RIP even more toxic than ricin. Shiga toxin, an RIP made by *Shigella dysenteriae* bacteria, causes a severe bloody diarrhea (dysentery) that can be lethal. Some strains of *E. coli* bacteria make Shiga-like toxin, an RIP that is the source of intestinal illness (see Chapter 4 Application).

Despite their toxicity, the main function of RIPs may not be destroying ribosomes. Many are part of plant immunity, and they have antiviral and anticancer activity. Plants that make toxic RIPs have been used as traditional medicines for many centuries; now, Western scientists are exploiting the unique properties of these proteins to combat HIV and cancer. For example, researchers have modified ricin's glycolipid-binding domain to recognize plasma membrane proteins (Section 4.3) especially abundant in cancer cells. The modified ricin preferentially enters—and kills—cancer cells. Ricin's toxic enzyme has also been attached to an antibody that can find cancer cells in a person's body. The intent of both strategies: to assassinate the cancer cells without harming normal ones.

CREDITS: (17) top from left, Amawasri Pakdara/Shutterstock.com; Steve Hurst/USDA-NRCS PLANTS Database; Kwangshin Kim/Science Source; CDC/E.H. White/BSIP/SuperStock.

10

Control of Gene Expression

Links to Earlier Concepts

This chapter explores gene expression (9.1) in the context of metabolic pathways (Sections 5.3, 5.4, 5.5). You will apply what you know about functional groups (3.1), carbohydrates (3.2, 7.1, 7.6), active transport (5.7), photosynthesis (6.3–6.5), DNA structure (8.2, 8.3) and function (8.4, 8.6), mutations (8.5, 9.5), transcription (9.2), and translation (9.4).

Core Concepts

Living things sense and respond appropriately to their internal and external environments.

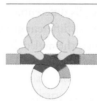

All cells respond to internal and external change, and their responses often involve adjustments to gene expression. In multicelled eukaryotes, normal embryonic development depends on appropriate cellular responses to molecules that govern gene expression. Environmental factors that influence gene expression during an individual's lifetime can have multi-generational effects.

Biological processes include and depend on proper timing and coordination of specific events.

The timing and coordination of specific molecular and cellular events are regulated by mechanisms that govern gene expression. Growth, reproduction, and homeostasis depend on expression of appropriate genetic information at the appropriate times, so every step of gene expression is regulated.

Interactions among biological systems give rise to complex properties.

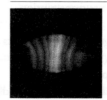

All cells of a multicelled organism start out with the same DNA. Differentiation occurs during development: The set of genes a cell expresses determine the products it makes, and these in turn determine the type of cell it will be. Cascades of master gene expression during embryonic development give rise to complex multicelled bodies. Many of these genes are evolutionarily conserved across species.

◀ National Geographic Explorer Jane Goodall observes two chimpanzees. About 97 percent of chimpanzee DNA is identical with human DNA, but how that DNA is used differs between the two species.

163

Photograph by Michael Nichols, National Geographic Creative.

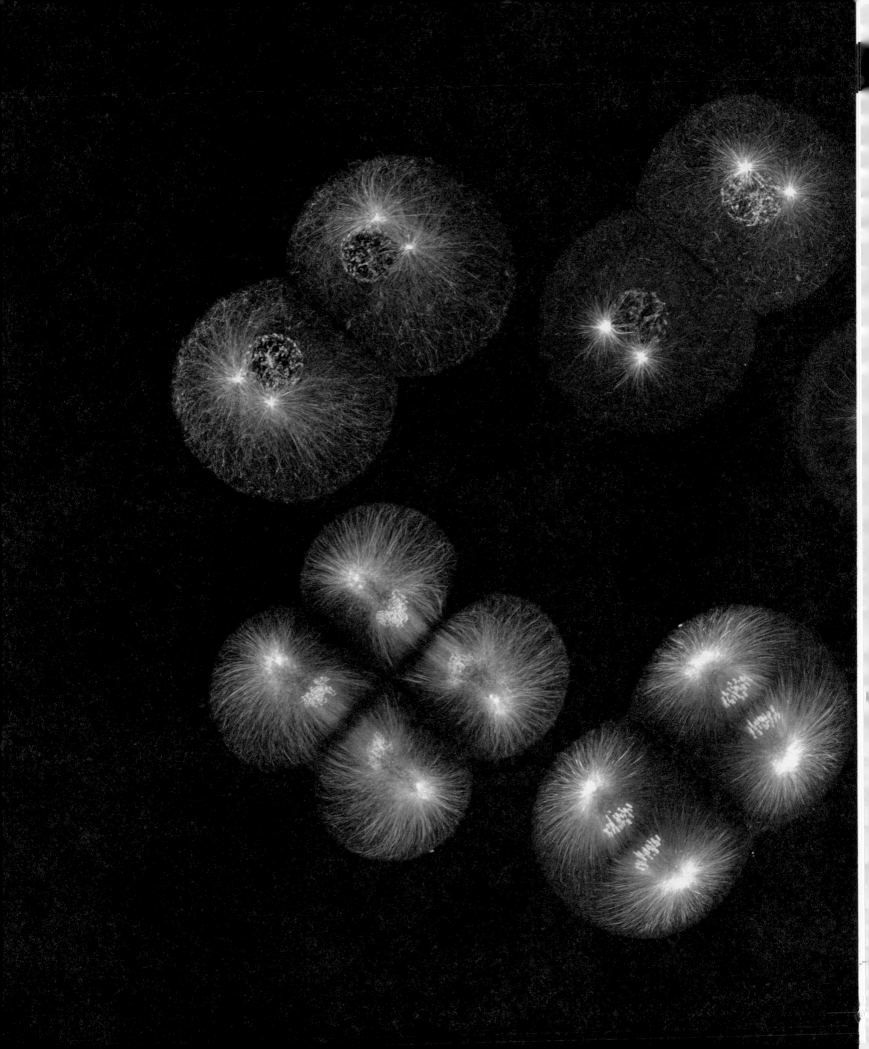

11

How Cells Reproduce

Links to Earlier Concepts

Before beginning this chapter, be sure you understand cell structure (Sections 4.1, 4.5, 4.9, 4.10); chromosomes (8.3); and DNA replication (8.4) and repair (8.5). What you know about receptors and recognition proteins (4.3), free radicals (5.5) and mutations (9.5), fermentation (7.5), and eukaryotic gene control (10.1) will help you understand how cancer develops.

Core Concepts

All organisms alive today are linked by lines of descent from shared ancestors.

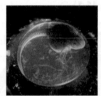
Cells of all multicelled eukaryotes reproduce by mitosis and cytoplasmic division. Together, these processes are the basis of growth, tissue repair, and asexual reproduction. Because mitosis links one generation of cells to the next, it is a mechanism by which the continuity of life occurs. All multicelled organisms use similar molecules to drive and regulate their cell cycle.

Living systems store, retrieve, transmit, and respond to information essential for life.

Cells transmit essential genetic information to their offspring in the form of DNA. When a cell divides by mitosis, it passes a complete set of chromosomes to each of its two descendant cells. The descendant cells are identical to one another and to the parent cell.

Growth, reproduction, and homeostasis depend on proper timing and coordination of specific molecular events.

Mechanisms that control gene expression orchestrate the timing and coordination of molecular and cellular events required for proper cellular function. Mutations that alter the molecules involved in these mechanisms may have complex effects on the cell, tissue, or whole organism. Cancer arises from disruptions in any of several molecular interactions governing the cell cycle.

Stages of mitosis in the two-cell embryo of a ribbon worm (*Cerebratulus*). Many more of these divisions will produce many more cells that will eventually form tissues and body parts of the new individual.

177

Photograph by George von Dassow, University of Oregon.

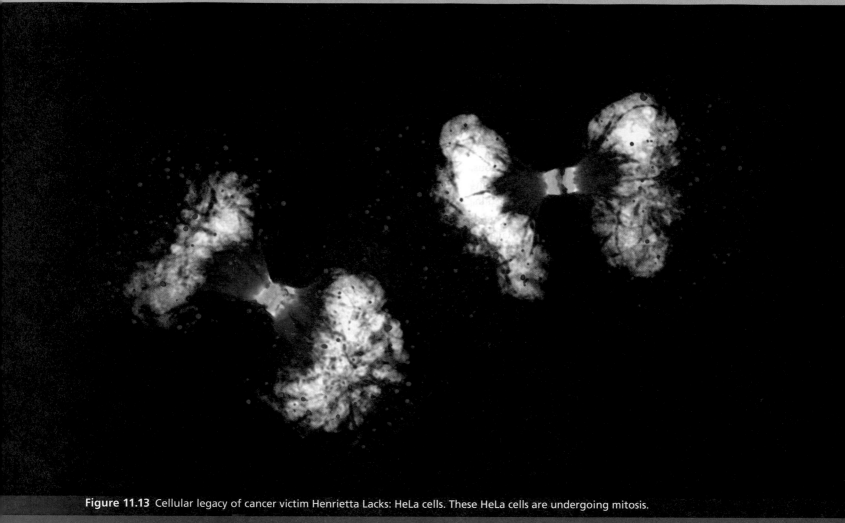

Figure 11.13 Cellular legacy of cancer victim Henrietta Lacks: HeLa cells. These HeLa cells are undergoing mitosis.

HENRIETTA'S IMMORTAL CELLS

Each human starts out as a fertilized egg. By the time of birth, that single cell has given rise to about a trillion other cells, all organized as a human body. Even in the adult, billions of cells divide every day as new cells replace worn-out ones. However, human cells grown in the laboratory tend to divide a limited number of times and die within weeks; a cell division limit caps the number of times that body cells can divide. As early as the mid-1800s, researchers were trying to coax human cells to keep dividing outside of the body because they realized immortal cell lineages— cell lines—would allow them to study human diseases (and potential cures) without experimenting on people.

The quest to create a human cell line continued unsuccessfully for over one hundred years. In 1951, George and Margaret Gey had been trying to create a human cell line for nearly thirty years when their assistant, Mary Kubicek, prepared a new sample of human cancer cells. Mary named the cells HeLa, after the first and last names of the patient from whom the cells had been taken.

The HeLa cells began to divide, again and again. The cells were astonishingly vigorous, quickly coating the inside of their test tube and consuming their nutrient broth. Four days later, there were so many cells that the researchers had to transfer them to more test tubes. The cells were dividing

every twenty-four hours and coating the inside of the tubes within days. Sadly, cancer cells in the patient were dividing just as fast. Only six months after she had been diagnosed with cervical cancer, malignant cells had invaded tissues throughout her body. Two months after that, Henrietta Lacks, a young woman from Baltimore, was dead.

Even after Henrietta had passed away, her cells lived on in the Geys' laboratory. The Geys discovered how to grow poliovirus in HeLa cells, a practice that enabled them to determine which strains of the virus cause polio. That work was a critical step in the development of polio vaccines, which have since saved millions of lives.

Henrietta Lacks

Henrietta Lacks (left) was just thirty-one, a wife and mother of five, when runaway cell divisions of cancer killed her. Her cells, however, are still dividing, again and again, more than sixty years after she died. Frozen away in tiny tubes packed in Styrofoam boxes, HeLa cells continue to be shipped among laboratories all over the world. They are still widely used to investigate cancer, viral growth, protein synthesis, the effects of radiation, and countless other processes important in medical research. HeLa cells helped several researchers win Nobel Prizes, and they even traveled into space for experiments on satellites.

HeLa cells were used in early tests of paclitaxel, a drug that keeps microtubules from disassembling. Spindle microtubules that cannot shrink cannot properly position the cell's chromosomes during metaphase, and this triggers a checkpoint that stops the cell cycle. Shortly thereafter, the cell either exits mitosis or dies. Frequent divisions make cancer cells more vulnerable to this microtubule poison than normal cells.

A more recent example of cancer research is shown in **Figure 11.13**. In this micrograph of mitotic HeLa cells, chromosomes appear white and the spindle is red. Blue dots pinpoint a protein that helps sister chromatids stay attached to one another at the centromere. Green identifies an enzyme called Aurora B kinase that helps attach spindle microtubules to centromeres. At the stage of telophase shown, the two proteins should be closely associated midway between the two clusters of chromosomes. The abnormal distribution means that the chromosomes are not properly attached to the spindle.

Defects in Aurora B kinase or its expression result in unequal distribution of chromosomes into descendant cells. Researchers recently correlated overexpression of Aurora B in cancer cells with shortened patient survival rates. Thus, drugs that inhibit Aurora B function are now being tested as potential cancer therapies.

Ongoing research with HeLa cells may one day allow researchers to identify drugs that target and destroy malignant cells or stop them from dividing. The research is far too late to have saved Henrietta Lacks, but it may one day yield drugs that put the brakes on cancer.

Dr. Iain Couzin

Can a million migrating wildebeests help explain why cancer spreads? Ask National Geographic Explorer Iain Couzin. He is investigating how collective behavior in animals can be quantified and analyzed to give us new insights into the patterns of nature—and ourselves.

"We're realizing that animals have highly coordinated social systems and make decisions together," Couzin says. "They can do things collectively that no individual could do alone. It's still a very unexplored area of animal behavior." Couzin blends fieldwork, lab experiments, computer simulations, and complex mathematical models to test theories about how and why cells, animals, and humans organize and work together.

Cancer cells migrating during metastasis may have some parallels to animal swarms. As animals do in swarms, metastatic cells collectively sense the environment and change their behavior in response to it. For example, some brain tumors send out tendrils into surrounding tissue. Couzin compares the tendrils to exploratory ant trails, and thinks they involve a division of labor not unlike that of ants: A few modified cells determine an accommodating path, and others follow. A marching column of cancer cells may be more effective at invading healthy tissue than individual cells. If researchers can find out what is guiding them, it might be possible to use drugs to control how these tumors spread, perhaps confining them or herding the cells into areas where they can be treated without damaging vital tissue.

Section 11.1 A eukaryotic cell reproduces by division: nucleus first, then cytoplasm. The **cell cycle** is a series of events and stages through which a cell passes during its lifetime. DNA replication occurs during **interphase**. **Mitosis** is a nuclear division mechanism that maintains the chromosome number. For example, diploid cells (with **homologous chromosomes**) produce diploid offspring by mitosis. Mitosis is the basis of development, cell replacements, and tissue repair in multicelled species, and of **asexual reproduction** in many species.

Section 11.2 DNA replication occurs before mitosis, so each chromosome consists of two DNA molecules attached as sister chromatids. During mitosis, microtubules form a **spindle** that moves the chromosomes.

Mitosis proceeds in four stages. During **prophase**, the chromosomes condense and the spindle forms. As the nuclear envelope breaks up, spindle microtubules attach to the chromosomes and move them toward the center of the cell.

At **metaphase**, the spindle has aligned all of the (still duplicated) chromosomes midway between spindle poles.

During **anaphase**, the sister chromatids of each chromosome separate, and the spindle moves them toward opposite spindle poles. Each DNA molecule is now an individual chromosome.

During **telophase**, a complete set of chromosomes reaches each spindle pole. A nuclear envelope forms around each cluster. Two new nuclei, each with the parental chromosome number, are the result. The two nuclei are identical to one another (and to the nucleus of the parent cell).

Section 11.3 In most cases, **cytokinesis** follows nuclear division. Vesicles guided by microtubules to the future plane of division merge to fill the space between the two descendant cells. In animal cells, a contractile ring pulls the plasma membrane inward, forming a **cleavage furrow**. In plant cells, vesicles merge as a **cell plate** that expands until it fuses with the plasma membrane, thus partitioning the cytoplasm.

Section 11.4 **Telomeres** (regions of noncoding DNA at the end of eukaryotic chromosomes) shorten with every cell division. Normal body cells can divide only a certain number of times before their telomeres get too short and they become **senescent**. This cell division limit is a fail-safe mechanism in case the cell loses control over its cell cycle.

Section 11.5 The products of checkpoint genes work together to control the cell cycle. These molecules monitor the integrity of the cell's DNA, and can pause the cycle until breaks or other problems are fixed. When checkpoint mechanisms fail, a cell loses control over its cell cycle, and its abnormally dividing descendants form a **neoplasm.** Neoplasms may form lumps called **tumors.**

Genes encoding **growth factor** receptors are examples of **proto-oncogenes**, which means mutations can turn them into tumor-causing **oncogenes**. Mutations in multiple checkpoint genes can give rise to a malignant neoplasm that gets progressively worse. Cells of malignant neoplasms can break loose from their home tissues and colonize other parts of the body, a process called **metastasis**. **Cancer** occurs when malignant neoplasms physically and metabolically disrupt normal body tissues.

Application An immortal line of human cells (HeLa) is a legacy of cancer victim Henrietta Lacks. Researchers all over the world continue to work with these cells in their efforts to unravel the mechanisms of cancer.

SELF-ASSESSMENT
ANSWERS IN APPENDIX VII

1. Mitosis and cytoplasmic division are the basis of _____ .
   a. asexual reproduction of single-celled prokaryotes
   b. development and tissue repair in multicelled species
   c. sexual reproduction in plants and animals

2. A duplicated chromosome has _____ chromatid(s).
   a. one   c. three
   b. two   d. four

3. Homologous chromosomes _____ .
   a. are inherited from two parents
   b. are sister chromatids
   c. are different in size and in length

4. Most cells spend the majority of their lives in _____ .
   a. prophase   d. telophase
   b. metaphase   e. interphase
   c. anaphase   f. senescence

5. The spindle attaches to chromosomes at the _____ .
   a. centriole   c. centromere
   b. contractile ring d. centrosome

6. Which is not a stage of mitosis?
   a. prophase   c. interphase
   b. metaphase   d. anaphase

7. In intervals of interphase, G stands for _____ .
   a. gap   c. Gey
   b. growth   d. gene

8. Interphase is the part of the cell cycle when _____ .
   a. a cell ceases to function
   b. a cell forms its spindle apparatus
   c. a cell grows and duplicates its DNA
   d. mitosis proceeds

9. After mitosis, the chromosome number of descendant cells is _____ the parent cell's.
   a. the same as   c. rearranged compared to
   b. one-half of   d. doubled compared to

HELA CELLS ARE A GENETIC MESS

HeLa cells can vary in chromosome number. Defects in proteins that orchestrate cell division result in descendant cells with too many or too few chromosomes, an outcome that is one of the hallmarks of cancer cells. The panel of chromosomes in **Figure 11.14**, originally published in 1989, shows all of the chromosomes in a single metaphase HeLa cell.

1. What is the chromosome number of this HeLa cell?

2. How many extra chromosomes does this cell have, compared to a normal human body cell?

3. Can you tell that this cell came from a female? How?

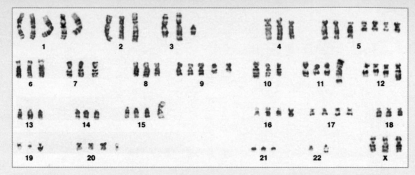

Figure 11.14 Chromosomes in one HeLa cell.

10. In the diagram of the nucleus below, fill in the blanks with the name of each interval.

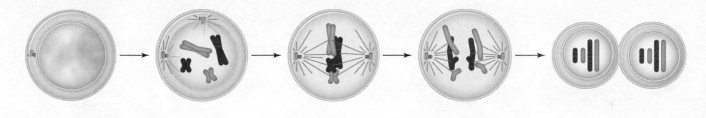

_____   _____   _____   _____

11. Cytoplasm of a plant cell divides by the process of _____ .
   a. telekinesis   c. fission
   b. nuclear division d. cytokinesis

12. *BRCA1* and *BRCA2* _____ .
   a. are checkpoint genes   c. encode tumor suppressors
   b. are proto-oncogenes   d. all of the above

13. _____ are characteristic of cancer.
   a. Malignant cells   b. Neoplasms   c. Tumors

14. Match each term with its best description.
   ___ cell plate   a. lump of cells
   ___ spindle   b. made of microfilaments
   ___ tumor   c. divides plant cells
   ___ cleavage furrow   d. spindle originates here
   ___ contractile ring   e. dangerous metastatic cells
   ___ cancer   f. made of microtubules
   ___ centrosomes   g. indentation
   ___ telomere   h. shortens with age

15. Match each stage with the events listed.
   ___ metaphase   a. sister chromatids move apart
   ___ prophase   b. chromosomes condense
   ___ telophase   c. new nuclei form
   ___ interphase   d. chromosomes aligned midway
   ___ anaphase   between spindle poles
   ___ cytokinesis   e. G1, S, G2
                       f. cytoplasmic division

1. When a cell reproduces by mitosis and cytoplasmic division, does its life end?

2. How is the human life cycle related to the cell cycle of the cells that make up our bodies?

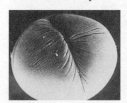

3. The eukaryotic cell in the photo on the left is in the process of cytoplasmic division. Is this cell from a plant or an animal? How do you know?

4. Exposure to radioisotopes or other sources of radiation can damage DNA. Humans exposed to high levels of radiation face a condition called radiation poisoning. Why do you think that exposure to radiation is used as a therapy to treat some kinds of cancers?

5. Suppose you have a way to measure the amount of DNA in one cell during the cell cycle. You first measure the amount at the G1 phase. At what points in the rest of the cycle will you see a change in the amount of DNA per cell?

for additional quizzes, flashcards, and other study materials
▶ **ACCESS MINDTAP AT WWW.CENGAGEBRAIN.COM**

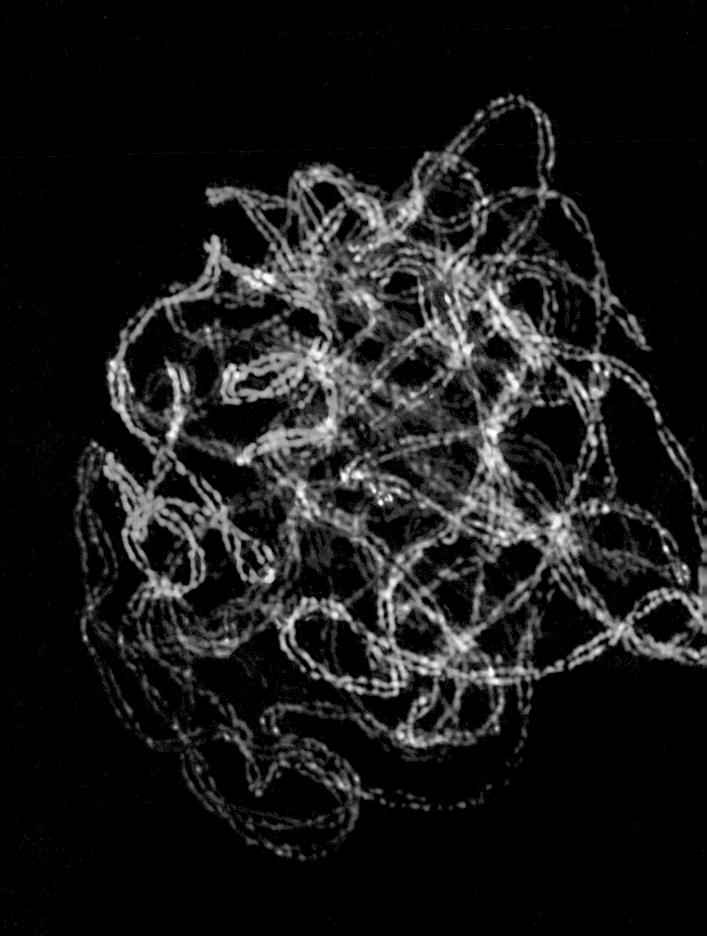

12 Meiosis and Sexual Reproduction

Links to Earlier Concepts

Before reading this chapter, be sure you understand the organization of eukaryotic chromosomes (Section 8.3) and how genes work (9.1). You will draw on your knowledge of DNA replication (8.4), cytoplasmic division (11.3), and cell cycle controls (11.5) as we compare meiosis with mitosis (11.2). This chapter also revisits clones (8.6) and effects of mutations (9.5).

Core Concepts

Interactions among the components of each biological system give rise to complex properties.

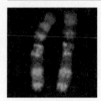

In general, the most resilient biological systems have many different components. Individuals of sexually reproducing species vary in the details of their DNA. Such variation is almost always advantageous, because genetically diverse populations are more resilient to environmental change than populations with low diversity. Asexual reproduction produces genetically identical individuals, so populations of asexual reproducers typically have low genetic diversity.

Living systems store, retrieve, transmit, and respond to information essential for life.

All organisms transmit genetic information to their offspring in the form of DNA. Sexual reproduction is one way in which this occurs, and it involves meiosis (which reduces the chromosome number for forthcoming gametes) and fertilization. Both processes mix genetic information from two parents who differ in the details of their DNA, so sexual reproduction produces offspring genetically different from one another and from the parents.

Evolution underlies the unity and diversity of life.

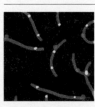

Like building blocks, molecules and processes are often evolutionarily reused in improved pathways that offer enhanced benefit. The process of meiosis is similar to mitosis, and both use some of the same molecules that are also involved in DNA repair.

◄ Meiosis mixes up genetic material of two parents. In this micrograph of a plant cell nucleus undergoing meiosis, tightly aligned homologous chromosomes (colored threads) are exchanging corresponding sections of DNA.

Explain why homologous chromosomes may carry different alleles.

List the differences between sexual reproduction and asexual reproduction.

Describe how genetic diversity makes a population more resilient to environmental challenges.

INTRODUCING ALLELES

Remember from Section 11.1 that your body cells contain pairs of homologous chromosomes. The two chromosomes of each pair have the same genes, but their DNA sequence is not identical. This is because you inherited your chromosomes from two parents who differ genetically—unique mutations accumulated in your parents' separate lines of descent over time. Thus, the DNA sequence of any of your genes may differ a bit from a corresponding gene on a homologous chromosome (**Figure 12.1**). Different forms of the same gene are called **alleles**.

New alleles arise by mutation. They may encode slightly different forms of a gene's product, and such differences influence the details of shared, inherited traits. Members of a species have the same traits because they have the same genes, but almost every shared trait varies a bit among individuals of a sexually reproducing species. Alleles of the shared genes are the basis of this variation. Consider just one human gene, *HBB*, which encodes beta globin (Section 9.5). *HBB* has more than 700 alleles; a few cause sickle-cell anemia, several cause beta thalassemia, and so on. There are more than 20,000 human genes, and most of them have multiple alleles.

ON THE ADVANTAGES OF SEX

You learned in Chapter 11 that mitosis and cytoplasmic division are part of asexual reproduction in eukaryotes, but not all species can use this reproductive mode. Many eukaryotes reproduce sexually, either exclusively or most of the time (**Figure 12.2**). **Sexual reproduction** is the process in which offspring arise from two parents and inherit genes from both. **Table 12.1** introduces some other differences between sexual and asexual reproduction.

If the function of reproduction is the perpetuation of one's genes, then an asexual reproducer would seem to win the evolutionary race. When it reproduces, it passes all of its genes to every one of its offspring. Only about half of a sexual reproducer's genes are passed to each offspring. So why is sex so widespread?

Think about how all offspring of an asexual reproducer are clones: Barring new mutations, they have the same alleles as their one parent. Consistency is a good evolutionary strat-

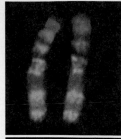

A Corresponding colored patches in this fluorescence micrograph indicate corresponding DNA sequences in a homologous chromosome pair.

Homologous chromosomes carry the same set of genes.

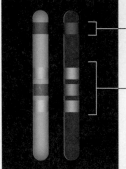

B Genes occur in pairs on homologous chromosomes.

The members of each pair of genes may be identical in DNA sequence, or they may differ slightly, as alleles (color variations represent DNA sequence differences).

Figure 12.1 Genes on homologous chromosomes.
Different forms of a gene are called alleles.

sexual   asexual

Figure 12.2 Different species use different modes of reproduction. Left, offspring of sexual reproducers differ from one another and from their parents. Right, asexual reproduction, which is not unusual in plants, gives rise to genetically identical offspring.

TABLE 12.1

Comparing Asexual and Sexual Reproduction

Mode:	Asexual	Sexual
Division	mitosis	meiosis and mitosis
Parents	1	2
Parental cell(s)	diploid cell	haploid gametes
Parental genes	100% of one parent's genes are passed to offspring	50% of each parent's genes are passed to offspring
Offspring	offspring are genetically identical	offspring are not genetically identical
Advantages	does not require a partner	high genetic diversity
Disadvantages	no genetic diversity	requires a partner

CREDITS: (1) Image courtesy of Carl Zeiss MicroImaging, Thornwood, NY; (2) left, cynoclub/Shutterstock.com; right, KYTan/Shutterstock.com.

egy in a favorable, unchanging environment; alleles that help an organism survive and reproduce in the environment do the same for its descendants. However, most environments are constantly changing, and change is not always favorable. Individuals that are identical are equally vulnerable to challenges. In a changing environment, genetic diversity gives sexual reproducers the evolutionary edge. Sexual reproduction randomly mixes up the genetic information of two parents who have different alleles. The offspring of sexual reproducers inherit different combinations of parental alleles, so they differ from one another and from their parents. Some of the offspring may be perfectly suited to a new environmental challenge. Thus, as a group, they have a better chance of surviving environmental change than clones.

To understand why environments constantly change, think about one example: interactions between a predatory species and its prey. To the prey species, the predators are an environmental challenge. Prey individuals that are best at escaping predation tend to leave more offspring. Thus, alleles that enhance an individual's ability to escape a predator tend to become more common in a prey species over generations. At the same time, prey species that become better at escaping predation are an environmental challenge to a predator species. Predator individuals that are best at capturing prey tend to leave more offspring. Thus, alleles that enhance an individual's ability to capture prey tend to become more common in the predator species over generations (Chapter 17 returns to evolutionary processes). The two species are locked in a constant race, with each genetic improvement in one countered by a genetic improvement in the other. This idea is called the Red Queen hypothesis, a reference to Lewis Carroll's book *Through the Looking Glass*. In the book, the Queen of Hearts tells Alice, "It takes all the running you can do, to keep in the same place."

Another advantage of sexual reproduction involves the inevitable occurrence of harmful mutations that can be passed to offspring. A population of sexual reproducers has a better chance of weathering the evolutionary effects of such mutations. With asexual reproduction, individuals bearing a moderately harmful mutation necessarily pass it to all of their offspring. This outcome is relatively rare with sexual reproduction, because each offspring of a sexual union has a 50 percent chance of inheriting a parent's mutation. Thus, a moderately harmful mutation tends to spread more slowly through a population of sexual reproducers.

---

alleles (uh-LEELZ) Forms of a gene with slightly different DNA sequences; may encode different versions of the gene's product.
sexual reproduction Reproductive mode by which offspring arise from two parents and inherit genes from both.

Maurine Neiman

Animals that reproduce solely by asexual means are very rare. However, some populations of *Potamopyrgus antipodarum*, a New Zealand freshwater snail (left), do just that. National Geographic Explorer Maurine Neiman compares sexually reproducing and asexual populations of these tiny snails to determine the costs and benefits of sex. The asexual populations consist of females who produce female offspring. In these clonal populations, all offspring can produce more offspring. By contrast, half of the offspring of sexual reproducers are male, and cannot produce offspring. Thus, all else being equal, a clonal population of females will expand much more quickly than a population of sexual reproducers. So why do the sexual snails persist?

Neiman and her collaborators Amy Krist and Adam Kay have discovered that sexual and asexual snails might differ in their need for phosphorus. Like most sexual organisms, the sexual snails have two chromosome sets; like most animals that cannot reproduce sexually, the asexual snails have at least three. In low-phosphorus environments, the extra chromosome sets pose a disadvantage. DNA has a high phosphorus content; having the extra sets of chromosomes multiplies each organism's requirement for this nutrient. Thus, sexual snails—with the fewest chromosome sets of all—are especially likely to beat out asexuals in regions where phosphorus is scarce.

Follow-up experiments will extend beyond explaining the predominance of sex. These tiny snails have spread far beyond their native New Zealand, and huge populations of them are now disrupting ecosystems all over the world. The invasive populations are always asexual (and in fact many other invasive species have three or more sets of chromosomes). Fertilizers and detergents contain phosphorus, so agricultural runoff and other types of water pollution could be fueling the gigantic populations of asexual snails currently invading ecosystems worldwide.

LEARNING OBJECTIVES

Describe the relationship between germ cells and gametes.

Explain how meiosis reduces the chromosome number, and why this is a necessary part of sexual reproduction.

MEIOSIS HALVES THE CHROMOSOME NUMBER

Sexual reproduction involves the fusion of mature reproductive cells—**gametes**—from two parents. Gametes have a single set of chromosomes, so they are **haploid** (n): Their chromosome number is half of the diploid ($2n$) number (Section 8.3). **Meiosis**, a nuclear division mechanism that halves the chromosome number, is a necessary part of gamete formation and sexual reproduction. As you will see in Section 12.4, meiosis also gives rise to new combinations of parental alleles.

In multicelled eukaryotes, gametes arise by division of immature reproductive cells called **germ cells**. In plants and animals, germ cells form in organs set aside for reproduction (**Figure 12.3**), but the two groups make gametes somewhat differently. In plants, haploid germ cells (spores) form by meiosis. These cells divide by mitosis to form structures that produce or contain gametes. In animals, germ cells are part of the germline, which is a lineage of cells dedicated to producing gametes. Meiosis in diploid germ cells gives rise to eggs (female gametes) or sperm (male gametes).

HOW MEIOSIS WORKS

The first part of meiosis is similar to mitosis. A cell duplicates its DNA before either nuclear division process begins. As in mitosis, a spindle forms, and its microtubules move the chromosomes. However, meiosis sorts the chromosomes into new nuclei not once but two times, so it results in the formation of four haploid nuclei. The two consecutive nuclear divisions are called meiosis I and meiosis II:

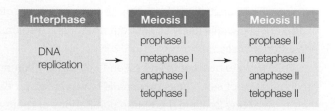

Interphase	Meiosis I	Meiosis II
DNA replication	prophase I metaphase I anaphase I telophase I	prophase II metaphase II anaphase II telophase II

During meiosis I, every duplicated chromosome aligns with its homologous partner (**Figure 12.4 ❶**). Then the homologous chromosomes are pulled away from one another and packaged into separate nuclei ❷. Each of the two new nuclei is haploid (n)—it has one copy of each chromosome—so its chromosome number is half that of the diploid parent cell.

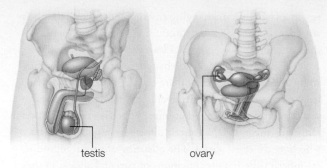

testis   ovary

A Reproductive organs of humans. Meiosis in germ cells inside testes and ovaries produces gametes (sperm and eggs).

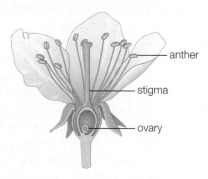

anther

stigma

ovary

B A flower is an organ used for sexual reproduction in many plants. Meiosis produces haploid male germ cells inside anthers, and haploid female germ cells in ovaries. Male and female gametes form when germ cells divide by mitosis.

Figure 12.3 Examples of reproductive organs in A animals and B plants.

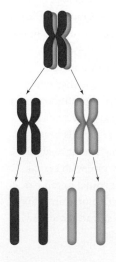

❶ During meiosis I, each chromosome in the nucleus pairs with its homologous partner. The nucleus contains two of each chromosome, so it is diploid ($2n$).

❷ Meiosis I separates homologous chromosomes and packages them in separate nuclei. Each new nucleus contains one of each chromosome, so it is haploid (n). The chromosomes are still duplicated.

❸ Meiosis II separates sister chromatids and packages them into four new nuclei. Each new nucleus contains one of each chromosome, so it is haploid (n). The chromosomes are now unduplicated.

Figure 12.4 How meiosis halves the chromosome number. DNA replication occurs before meiosis begins, so each chromosome consists of two identical molecules of DNA (sister chromatids). After meiosis, each chromosome consists of one molecule of DNA.

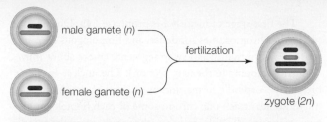

A At fertilization, two gametes meet to form a zygote.

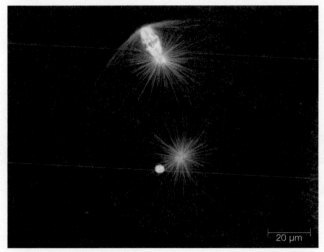

B In germ cells of female animals, meiosis pauses at metaphase II, then resumes and finishes after fertilization. This is a freshly fertilized egg of a *Cerebratulus* worm. Its chromosomes are completing meiosis at the top of the cell; the blue spot is the nucleus of the sperm that fertilized it. Orange shows spindle microtubules; blue, DNA.

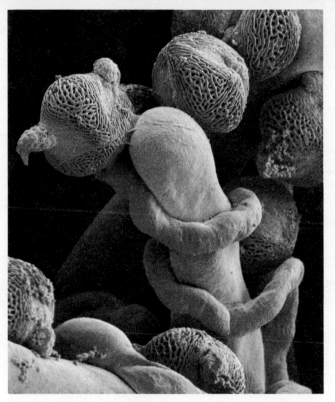

C In plants, meiosis completes before fertilization. These pollen grains (orange) are germinating on a stigma (yellow) of a flower. (Germination is growth after a period of metabolic rest.) Pollen tubes with male gametes inside are growing from the grains down into tissues of the ovary, which houses female gametes.

Figure 12.5 Sexual reproduction involves the fusion of two gametes.

After meiosis I, the chromosomes are still duplicated (sister chromatids remain attached). During meiosis II, the sister chromatids are pulled apart, and each becomes an individual, unduplicated chromosome ❸. The chromosomes are sorted into four new nuclei. Each new nucleus has one copy of each chromosome, so it is haploid (*n*).

In summary, meiosis partitions the duplicated chromosomes of one diploid nucleus (2*n*) into four haploid (*n*) nuclei. The next section zooms in on the details of the process.

---

fertilization Fusion of two gametes to form a zygote.

gamete (GAM-eat) Mature, haploid reproductive cell; e.g., a sperm.

germ cell Immature reproductive cell that gives rise to haploid gametes when it divides.

haploid (HAP-loyd) Having one of each type of chromosome characteristic of the species.

meiosis (my-OH-sis) Nuclear division process that halves the chromosome number. Required for sexual reproduction.

zygote (ZYE-goat) Diploid cell that forms when two gametes fuse; the first cell of a new individual.

FERTILIZATION RESTORES THE CHROMOSOME NUMBER

We leave details of sexual reproduction for later chapters, but you will need to know a few concepts before you get there. You already know that meiosis is required for the formation of haploid gametes. A male gamete fuses with a female gamete at **fertilization** (**Figure 12.5**). The fusion of haploid gametes at fertilization produces a diploid cell called a **zygote**. A zygote is the first cell of a new individual. Thus, meiosis halves the chromosome number, and fertilization restores it.

If meiosis did not precede fertilization, the chromosome number would double with every generation. If the chromosome number changes, so does the individual's genetic instructions. An individual's chromosomes are like a fine-tuned blueprint that must be followed exactly, in order to build a body that functions normally. As you will see in Chapter 14, chromosome number changes can have drastic consequences for health, particularly in animals.

12.3 A Visual Tour of Meiosis

LEARNING OBJECTIVES

Explain the steps of meiosis in a diploid (2*n*) cell.

Describe the major differences between meiosis I and meiosis II.

Figure 12.6 shows the steps of meiosis I and II in a diploid (2*n*) cell. DNA replication occurs before meiosis I, so each chromosome starts out with two sister chromatids.

Figure 12.6 Meiosis. Illustrations show two pairs of chromosomes in a diploid (2*n*) animal cell; homologous chromosomes are indicated in blue and pink. Micrographs show meiosis in a lily cell.

? FIGURE IT OUT
During which stage of meiosis does the chromosome number become reduced?

Answer: Anaphase I

The first stage of meiosis I is prophase I ❶. During this phase, the chromosomes condense, and homologous chromosomes align tightly and swap segments (more about this segment-swapping in the next section). The nuclear envelope breaks up. A spindle forms, and by the end of prophase I, microtubules attach one chromosome of each homologous pair to one spindle pole, and the other to the opposite spindle pole. These microtubules grow and shrink, pushing and pulling the chromosomes as they do.

At metaphase I ❷, all of the microtubules are the same length, and the chromosomes are aligned midway between the spindle poles.

During anaphase I ❸, microtubules of the spindle move the homologous chromosomes of each pair away from one

MEIOSIS I: ONE DIPLOID NUCLEUS TO TWO HAPLOID NUCLEI

❶ Prophase I
Homologous chromosomes condense, pair up, and swap segments. Spindle microtubules attach to homologous chromosomes as the nuclear envelope breaks up.

❷ Metaphase I
Homologous chromosome pairs are aligned midway between spindle poles. Spindle microtubules attach the two chromosomes of each pair to opposite spindle poles.

❸ Anaphase I
Microtubules of the spindle separate all of the homologous chromosomes and move them toward opposite spindle poles.

❹ Telophase I
A complete set of chromosomes clusters at both ends of the cell and loosen up. A nuclear envelope forms around each set to produce two haploid (*n*) nuclei. The cytoplasm may divide.

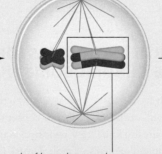
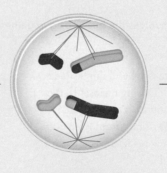

plasma membrane   spindle

nuclear envelope breaking up   pair of homologous chromosomes

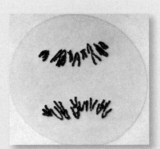

CREDITS: (6) Bottom photos, With thanks to the John Innes Foundation Trustees, computer enhanced by Gary Head.

another and toward opposite spindle poles. During telophase I, two sets of chromosomes reach the spindle poles, and a new nuclear envelope forms around each set as the DNA loosens up ❹. The two new nuclei are haploid (*n*); each contains one complete set of chromosomes. Each chromosome is still duplicated (it consists of two sister chromatids). The cytoplasm often divides at this point.

In some cells, meiosis II occurs immediately after meiosis I. In other cells, a period of protein synthesis—but no DNA replication—intervenes between the divisions.

Meiosis II proceeds simultaneously in both nuclei that formed in meiosis I. During prophase II ❺, the chromosomes condense and the nuclear envelope breaks up. A new spindle forms. By the end of prophase II, spindle microtubules attach

each chromatid to one spindle pole, and its sister chromatid to the opposite spindle pole. These microtubules push and pull the chromosomes, aligning them in the middle of the cell at metaphase II ❻.

During anaphase II ❼, the spindle microtubules move the sister chromatids apart and toward opposite spindle poles. Each chromosome is now unduplicated (it consists of one molecule of DNA).

During telophase II ❽, the chromosomes reach the spindle poles. New nuclear envelopes form around the four clusters of chromosomes as the DNA loosens up. Each of the four nuclei that form are haploid (*n*), with one set of (unduplicated) chromosomes. The cytoplasm often divides at this point.

MEIOSIS II: TWO HAPLOID NUCLEI TO FOUR HAPLOID NUCLEI

❺ Prophase II
The chromosomes condense. Spindle microtubules attach to each sister chromatid as the nuclear envelope breaks up.

❻ Metaphase II
The (still duplicated) chromosomes are aligned midway between spindle poles.

❼ Anaphase II
Sister chromatids separate. The (now unduplicated) chromosomes move toward the spindle poles.

❽ Telophase II
A complete set of chromosomes clusters at both ends of the cell. A nuclear envelope forms around each set to produce four haploid (*n*) nuclei.

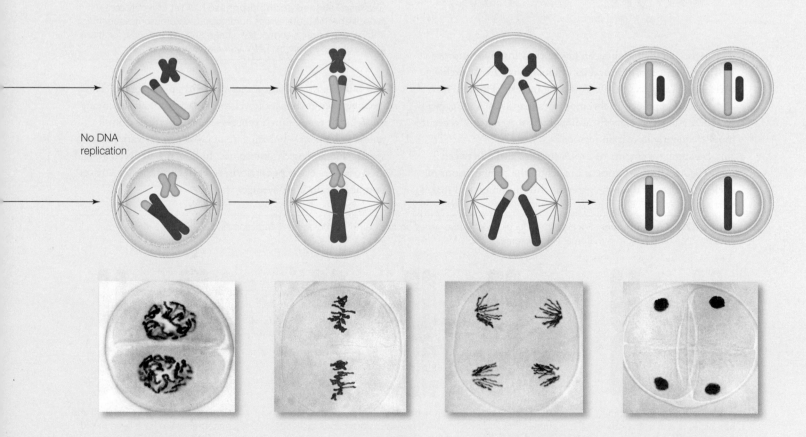

No DNA replication

LEARNING OBJECTIVES

Describe crossing over and how it introduces variation in traits among the offspring of sexual reproducers.

Explain the random nature of chromosome segregation during gamete formation, and its significance in terms of genetic variation.

The previous section mentioned briefly that homologous chromosomes swap segments in prophase I. It also showed how homologous chromosomes separate in anaphase I. Together with fertilization, these processes give rise to offspring that have different combinations of alleles. This genetic variation is the basis of variation in shared traits among the individuals of a sexually reproducing species.

CROSSING OVER

Early in prophase I of meiosis, all chromosomes in the cell condense. When they do, each is drawn close to its homologous partner, so that the chromatids align along their length:

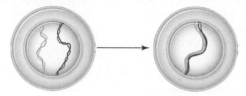

This tight, parallel orientation favors **crossing over**, a process in which a chromosome and its homologous partner exchange corresponding pieces of DNA during meiosis (**Figures 12.7** and **12.8**). Homologous chromosomes may swap any segment of DNA along their length, although crossovers tend to occur more frequently in certain regions.

Swapping segments of DNA shuffles alleles between homologous chromosomes. It breaks up the combinations of alleles that occurred on the parental chromosomes, and makes

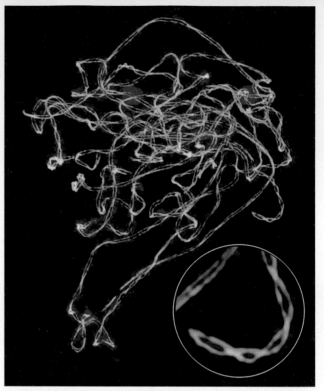

Figure 12.8 Homologous chromosomes in a cell of corn (*Zea mays*) **aligned during prophase I.** A set of highly conserved, meiosis-specific proteins zips homologous chromosomes together until crossing over is completed. Green and red show the locations of two of these proteins; DNA appears blue.

new ones on the chromosomes that end up in gametes. Thus, crossing over introduces novel combinations of alleles among offspring. It is a required process—meiosis will not finish unless it happens—but the rate of crossing over varies among species and among chromosomes. In humans, between 46 and 95 crossovers occur per meiosis, so each chromosome crosses over at least once, on average.

Figure 12.7 Crossing over. For clarity, we focus on only two genes on one pair of homologous chromosomes. Blue signifies the paternal chromosome, and pink, the maternal chromosome.

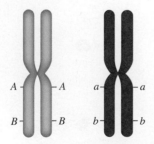

A In this example, one gene has alleles *A* and *a*; the other gene has alleles *B* and *b*.

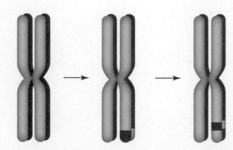

B Close contact between homologous chromosomes promotes crossing over, in which nonsister chromatids exchange corresponding pieces. Multiple crossovers are not uncommon.

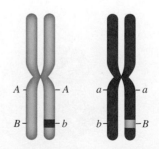

C After crossing over, paternal and maternal alleles have been mixed up on homologous chromosomes.

CREDIT: (8) Courtesy of Ding-Hua Lee and Chung-Ju Rachel Wang (Academia Sinica).

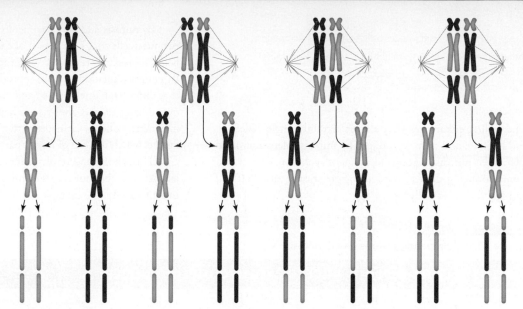

① The homologous chromosomes of three pairs can be divided between two spindle poles in four possible ways.

② There are eight possible combinations of maternal and paternal chromosomes in the two nuclei that form at telophase I.

③ There are eight possible combinations of maternal and paternal chromosomes in the four nuclei that form at telophase II.

Figure 12.9 Segregation of three chromosome pairs during meiosis. Maternal chromosomes are pink in this hypothetical example; paternal chromosomes are blue. For simplicity, we show no crossing over, so all sister chromatids are identical.

CHROMOSOME SEGREGATION

Normally, all of the new nuclei that form in meiosis I receive a complete set of chromosomes. However, whether a new nucleus ends up with the maternal or paternal version of a chromosome is entirely random. The chance that the maternal or the paternal version of any chromosome will end up in a particular nucleus is 50 percent. Why? The answer has to do with the way the spindle segregates homologous chromosomes during meiosis I.

The process of chromosome segregation begins in prophase I. Imagine a germ cell undergoing meiosis. Crossovers have already made genetic mosaics of its chromosomes, but for simplicity let's put crossing over aside for a moment and say the homologous chromosomes are maternal and paternal in origin.

During prophase I, microtubules fasten homologous chromosomes of each pair to opposite spindle poles. However, there is no pattern to the attachment of the maternal or paternal chromosome to a particular pole. Microtubules extending from a spindle pole attach to the centromere of the first chromosome they contact, regardless of whether it is maternal or paternal.

Now imagine that the germ cell has three pairs of chromosomes (**Figure 12.9**). By metaphase I, those three pairs of maternal and paternal chromosomes have been divided up between the two spindle poles in one of four ways **①**. In anaphase I, homologous chromosomes separate and are pulled toward opposite spindle poles. In telophase I, a new nucleus

forms around the chromosomes that cluster at each spindle pole. Each nucleus contains one of eight possible combinations of maternal and paternal chromosomes **②**.

In telophase II, each of the two nuclei divides and gives rise to two new haploid nuclei. The two new nuclei are identical because no crossing over occurred in our hypothetical example, so all of the sister chromatids were identical. Thus, at the end of meiosis in this cell, two (2) spindle poles have one homologous chromosomes from each of three (3) pairs. The resulting four nuclei have one of eight (2^3) possible combinations of maternal and paternal chromosomes **③**.

Cells that give rise to human gametes have twenty-three pairs of homologous chromosomes, not three. Each time a human germ cell undergoes meiosis, the four gametes that form end up with one of 8,388,608 (or 2^{23}) possible combinations of homologous chromosomes. In addition, any number of genes may occur as different alleles on the maternal and paternal chromosomes, and crossing over makes mosaics of that genetic information. Then, out of all the male and female gametes that form, which two actually get together at fertilization is a matter of chance. Are you getting an idea of why such fascinating combinations of traits show up among the generations of your own family tree?

crossing over Process by which homologous chromosomes exchange corresponding segments of DNA during prophase I of meiosis.

12.5 Mitosis and Meiosis—An Ancestral Connection?

LEARNING OBJECTIVES

Describe the similarities and differences between mitosis and meiosis II.

Support the argument that meiosis might have evolved from mutations in the process of mitosis.

The earliest eukaryotes were almost certainly single-celled and haploid, and they reproduced by mitosis. Meiosis evolved later, from mitosis. It seems like a giant evolutionary step from producing clones to producing genetically varied offspring, but was it really?

By mitosis and cytoplasmic division, one cell becomes two new cells that have copies of the parental chromosomes. Mitotic (asexual) reproduction produces clones of the parent. By contrast, meiotic (sexual) reproduction produces offspring that differ from one another, and also from parents. Crossing over during meiosis gives rise to haploid gametes that vary in genetic makeup. Gametes of two parents fuse to form a zygote, which is a cell of mixed parentage.

Though the end results differ, there are striking parallels between the four stages of mitosis and meiosis II (**Figure 12.10**). As one example, a spindle forms and separates

Figure 12.10 Comparing meiosis II with mitosis.

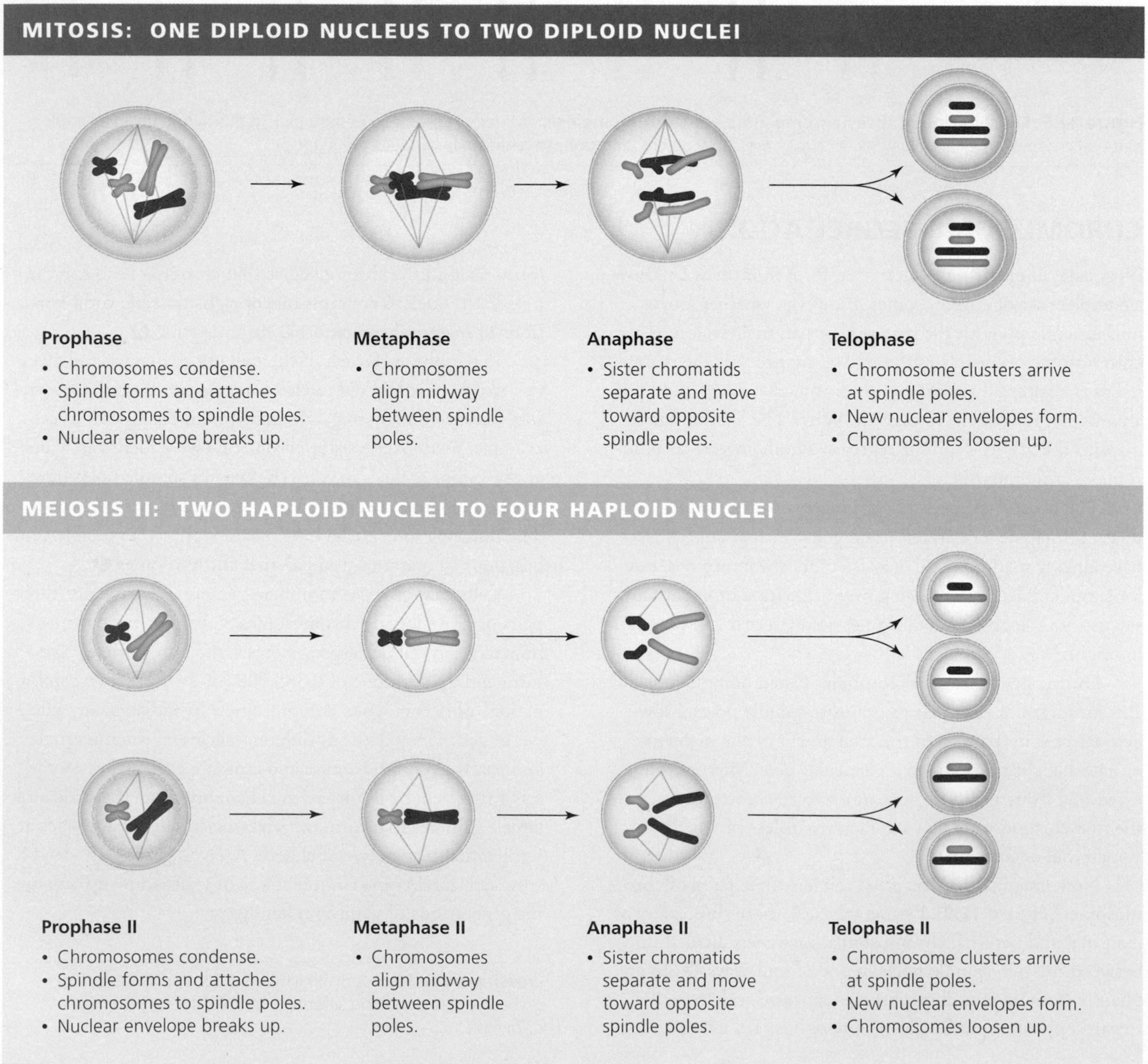
MITOSIS: ONE DIPLOID NUCLEUS TO TWO DIPLOID NUCLEI
Prophase
• Chromosomes condense.
• Spindle forms and attaches chromosomes to spindle poles.
• Nuclear envelope breaks up.
Metaphase
• Chromosomes align midway between spindle poles.
Anaphase
• Sister chromatids separate and move toward opposite spindle poles.
Telophase
• Chromosome clusters arrive at spindle poles.
• New nuclear envelopes form.
• Chromosomes loosen up.
MEIOSIS II: TWO HAPLOID NUCLEI TO FOUR HAPLOID NUCLEI
Prophase II
• Chromosomes condense.
• Spindle forms and attaches chromosomes to spindle poles.
• Nuclear envelope breaks up.
Metaphase II
• Chromosomes align midway between spindle poles.
Anaphase II
• Sister chromatids separate and move toward opposite spindle poles.
Telophase II
• Chromosome clusters arrive at spindle poles.
• New nuclear envelopes form.
• Chromosomes loosen up.

sister chromatids during both processes. There are many more similarities at the molecular level.

Long ago, the molecular machinery of mitosis became remodeled into meiosis. Evidence for this hypothesis includes a host of shared molecules, including the products of the *BRCA* genes (Chapter 10 Application and Section 11.5) that all modern eukaryotes have. By monitoring and fixing problems with the DNA—such as damaged or mismatched bases (Section 8.5)—these molecules actively maintain the integrity of a cell's chromosomes, particularly during DNA replication and mitosis. Many of the same molecules help homologous chromosomes cross over in prophase I of meiosis (**Figure 12.11**). Another example of shared molecules: The same regulatory molecules involved in checkpoints of mitosis are also involved in checkpoints of meiosis.

During anaphase of mitosis, sister chromatids are pulled apart. What would happen if the connections between the sisters did not break? Each duplicated chromosome would be pulled to one or the other spindle pole—which is exactly what happens during anaphase I of meiosis. The shared molecules and mechanisms imply a shared evolutionary history; sexual reproduction probably originated with mutations that affected processes of mitosis. As you will see in later chapters, the remodeling of existing processes into new ones is a common evolutionary theme.

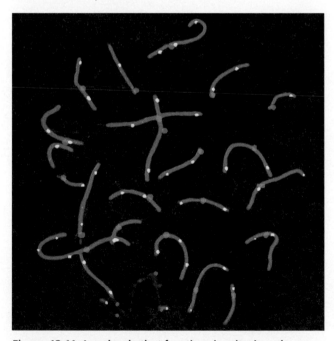

Figure 12.11 A molecule that functions in mitosis and meiosis. This fluorescence micrograph shows paired homologous chromosomes (red) in the nucleus of a human cell in prophase I of meiosis. Yellow pinpoints the location of a protein called MLH1 assisting with crossovers. MLH1 also helps repair DNA replication errors, and it prevents crossing over in mitosis. Centromeres are blue.

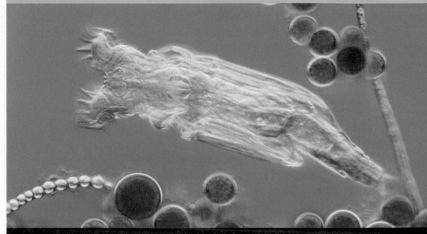

Figure 12.12 A bdelloid rotifer. All of these tiny animals are female.

HOW TO SURVIVE FOR 80 MILLION YEARS WITHOUT SEX

Why do males exist? No male has ever been found among the tiny freshwater creatures called bdelloid rotifers (**Figure 12.12**). Females have been reproducing for 80 million years solely through cloning themselves. Bdelloids are one of the few groups of animals to have completely abandoned sex.

Compared to sexual reproduction, asexual reproduction is often seen as a poor long-term strategy because it does not include crossing over or random segregation of chromosomes into gametes—the chromosomal shufflings that bring about genetic diversity thought to give species an adaptive edge in the face of new challenges. Bdelloids have contradicted this theory by being very successful; over 400 species are thriving today.

An unusual ability may help to explain the success of the bdelloids despite their rejection of sex. These rotifers can apparently import genes from bacteria, fungi, protists, and even plants. The direct swapping of genetic material was thought to be incredibly rare in complex organisms, but bdelloids are bringing in external genes to an extent unheard of in animals. Each rotifer is a genetic mosaic whose DNA spans almost all the major kingdoms of life: About 10 percent of its active genes have been pilfered from other organisms. If the main advantage of sex is that it promotes genetic diversity, why worry about it when you have the genes of entire kingdoms available to you?

Bdelloid rotifers can dry out and survive for decades in a dormant state, reanimating when they encounter liquid again. Researchers suspect that foreign genes slip into the animals during the reanimation process. When a rotifer dries out, its chromosomes break into pieces and its cell membranes become leaky. Fragments of foreign DNA in the environment easily cross the leaky membranes. When the individual reanimates, DNA repair enzymes (Section 8.5) reassemble its shattered chromosomes. As the reassembly occurs, any foreign DNA fragments that entered the rotifer when it was dried out can become incorporated into its chromosomes.

CREDIT: (11) Reprinted from Fertility and Sterility, Vol 87/edition 3, Fei Sun, Paul Turek, Calvin Greene, Evelyn Ko, Alfred Rademaker, Renée H. Martin, Abnormal progression through meiosis in men with nonobstructive azoospermia, 565–571, © 2007 Elsevier. Image supplied by Renée H. Martin, Ph.D.; (12) Wim van Egmond/Visuals Unlimited, Inc.

Section 12.1 **Sexual reproduction** mixes up the genetic information of two parents. The offspring of sexual reproducers typically vary in the details of shared, inherited traits. Particularly in a changing environment, this variation can offer an evolutionary advantage over genetically identical offspring produced by asexual reproduction.

Sexual reproduction produces offspring with pairs of chromosomes. One chromosome of each homologous pair was inherited from the mother; the other, from the father. The two chromosomes of a homologous pair carry the same genes. Paired genes on homologous chromosomes may vary in DNA sequence, in which case they are called **alleles**. Alleles are the basis of differences in shared, heritable traits. They arise by mutation.

Section 12.2 **Meiosis**, the basis of sexual reproduction in eukaryotes, is a nuclear division mechanism that halves the chromosome number. It occurs in immature reproductive cells called **germ cells**, and is required for production of mature reproductive cells called **gametes**. The fusion of two **haploid** (*n*) gametes during **fertilization** restores the diploid parental chromosome number in the **zygote**, the first cell of the new individual.

Section 12.3 DNA replication occurs before meiosis begins, so each chromosome consists of two molecules of DNA (sister chromatids). Two nuclear divisions occur during meiosis. In the first nuclear division (meiosis I), the chromosomes condense and align tightly with their homologous partners during prophase I. Microtubules extend from the spindle poles, penetrate the nuclear region, and attach to one or the other chromosome of each homologous pair. At metaphase I, all chromosomes are lined up midway between the spindle poles. During anaphase I, homologous chromosomes separate and move to opposite spindle poles. During telophase I, a nuclear envelope forms around each of the two sets of chromosomes. The cytoplasm may divide at this point. The cell may rest before meiosis resumes, but DNA replication does not occur.

The second nuclear division (meiosis II) occurs simultaneously in both haploid nuclei that formed during meiosis I. Each chromosome still consists of two sister chromatids. The chromosomes condense during prophase II, and align in metaphase II. The sister chromatids of each chromosome separate during anaphase II, so at the end of meiosis each chromosome consists of one molecule of DNA. By the end of telophase II, four haploid nuclei have typically formed, each with a complete set of (unduplicated) chromosomes.

Section 12.4 Meiosis shuffles parental DNA, so offspring inherit nonparental combinations of alleles. During prophase I, homologous chromosomes exchange corresponding segments. This **crossing over** mixes up the alleles on maternal and paternal chromosomes, thus giving rise to combinations of alleles not present in the chromosomes of either parent. Meiosis also contributes to variation in traits by randomly segregating homologous chromosomes into gametes. Microtubules can attach the maternal or the paternal chromosome of each pair to one or the other spindle pole. Either chromosome may end up in any new nucleus, and in any gamete. Novel combinations of alleles are the basis of novel combinations of traits among offspring of sexual reproducing organisms.

Section 12.5 The process of meiosis resembles that of mitosis, and probably evolved from it. The two processes have many similarities, including shared molecules. A spindle forms, separates sister chromatids, and disassembles during both mitosis and meiosis.

Application A few groups of animals have survived for millions of years by reproducing only asexually, despite the lack of chromosome shufflings that bring about genetic diversity among offspring. Bdelloid rotifers may have offset this disadvantage by picking up new genes from organisms in other kingdoms.

SELF-ASSESSMENT
ANSWERS IN APPENDIX VII

1. One evolutionary advantage of sexual over asexual reproduction may be that it produces _____ .
   a. more offspring per individual
   b. more variation among offspring
   c. healthier offspring

2. Alternative forms of the same gene are called _____ .
   a. gametes   c. alleles
   b. homologous   d. sister chromatids

3. Meiosis is a necessary part of sexual reproduction because it _____ .
   a. divides two nuclei into four new nuclei
   b. reduces the chromosome number
   c. produces clones that can cross over

4. Meiosis _____ .
   a. occurs only in germ cells of animals
   b. supports growth and tissue repair in multicelled species
   c. gives rise to genetic diversity among offspring
   d. is part of the life cycle of all cells

5. Sexual reproduction in animals requires _____ .
   a. meiosis   c. gametes
   b. fertilization   d. all of the above

6. Dogs have a diploid chromosome number of 78. How many chromosomes do their gametes have?
   a. 39   c. 156
   b. 78   d. 234

BPA AND ABNORMAL MEIOSIS In 1998, researchers at Case Western University were studying meiosis in mouse oocytes when they saw an unexpected and dramatic increase of abnormal meiosis events (**Figure 12.13**). The improper segregation of chromosomes during meiosis is one of the main causes of human genetic disorders, which we will discuss in Chapter 14.

The researchers discovered that the spike in abnormal meiosis events began immediately after the mouse facility started washing the animals' plastic cages and water bottles in a new, alkaline detergent. The detergent had damaged the plastic, which as a result was leaching bisphenol A (BPA). BPA is a synthetic chemical that mimics estrogen, the main female sex hormone in animals. Though it has since been banned for use in baby bottles, BPA is still widely used to manufacture other plastic items and epoxies (such as the coating on the inside of metal cans of food). BPA-free plastics are often manufactured with a related compound, bisphenol S (BPS), that has effects similar to BPA.

1. What percentage of mouse oocytes displayed abnormalities of meiosis with no exposure to damaged caging?

2. Which group of mice showed the most meiotic abnormalities?

3. What is abnormal about metaphase I as it is occurring in the oocytes shown in the micrographs in **Figure 12.13B, C**, and **D**?

Caging materials	Total number of oocytes	Abnormalities
Control: New cages with glass bottles	271	5 (1.8%)
Damaged cages with glass bottles		
Mild damage	401	35 (8.7%)
Severe damage	149	30 (20.1%)
Damaged bottles	197	53 (26.9%)
Damaged cages with damaged bottles	58	24 (41.4%)

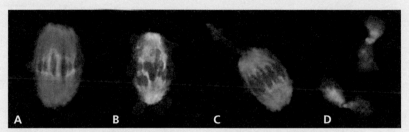

Figure 12.13 Meiotic abnormalities associated with exposure to BPA.

Top, the most abnormal meiosis events occurred in mice that were housed in damaged plastic caging with damaged plastic bottles. Damaged plastic releases BPA.

Bottom, fluorescent micrographs show nuclei of single mouse oocytes in metaphase I. **A** Normal metaphase; **B–D** examples of abnormal metaphase. Chromosomes are stained red; spindle fibers, green.

---

7. The cell in the diagram to the right is in anaphase I, not anaphase II. I know this because _____ .

8. The cell pictured to the right is in which stage of nuclear division?
   a. metaphase II   c. anaphase II
   b. anaphase I   d. interphase

9. Crossing over mixes up _____ .
   a. chromosomes   c. zygotes
   b. alleles   d. gametes

10. Crossing over happens during which phase of meiosis?
    a. prophase I   c. anaphase I
    b. prophase II   d. anaphase II

11. _____ contributes to variation in traits among the offspring of sexual reproducers.
    a. Crossing over   c. Fertilization
    b. Chromosome   d. all of the above
       segregation

12. Which of the following is one of the very important differences between mitosis and meiosis?
    a. Chromosomes align midway between spindle poles only in meiosis.
    b. Homologous chromosomes separate only in meiosis.
    c. DNA is replicated before mitosis only.
    d. Sister chromatids separate only in meiosis.
    e. Interphase occurs only in mitosis.

13. Match each term with its description.
    ___ interphase
    ___ metaphase I
    ___ alleles
    ___ zygotes
    ___ gametes
    ___ males
    ___ prophase I

    a. different forms of a gene
    b. useful for varied offspring
    c. none between meiosis I and meiosis II
    d. chromosome lineup
    e. haploid
    f. form at fertilization
    g. mash-up time

CLASS ACTIVITY
CRITICAL THINKING

1. In your own words, explain why sexual reproduction tends to give rise to greater genetic diversity among offspring in fewer generations than asexual reproduction.

2. Make a simple sketch of meiosis in a diploid germ cell that has one pair of homologous chromosomes. Now try it when the cell has three homologous chromosomes (it is triploid, with three copies of one chromosome).

3. The diploid chromosome number for the body cells of a frog is 26. What would the frog chromosome number be after three generations if meiosis did not occur before gamete formation?

for additional quizzes, flashcards, and other study materials
▶ **ACCESS MINDTAP AT WWW.CENGAGEBRAIN.COM**

CREDITS: (13) Reprinted from *Current Biology*, Vol 13, (Apr 03), Authors Hunt, Koehler, Susiarjo, Hodges, Ilagan, Voigt, Thomas, Thomas and Hassold, Bisphenol A Exposure Causes Meiotic Aneuploidy in the Female Mouse, pp. 546–553, © 2003 Cell Press. Published by Elsevier Ltd. With permission from Elsevier; (in text S-A 8) Michael Clayton/University of Wisconsin, Department of Botany.

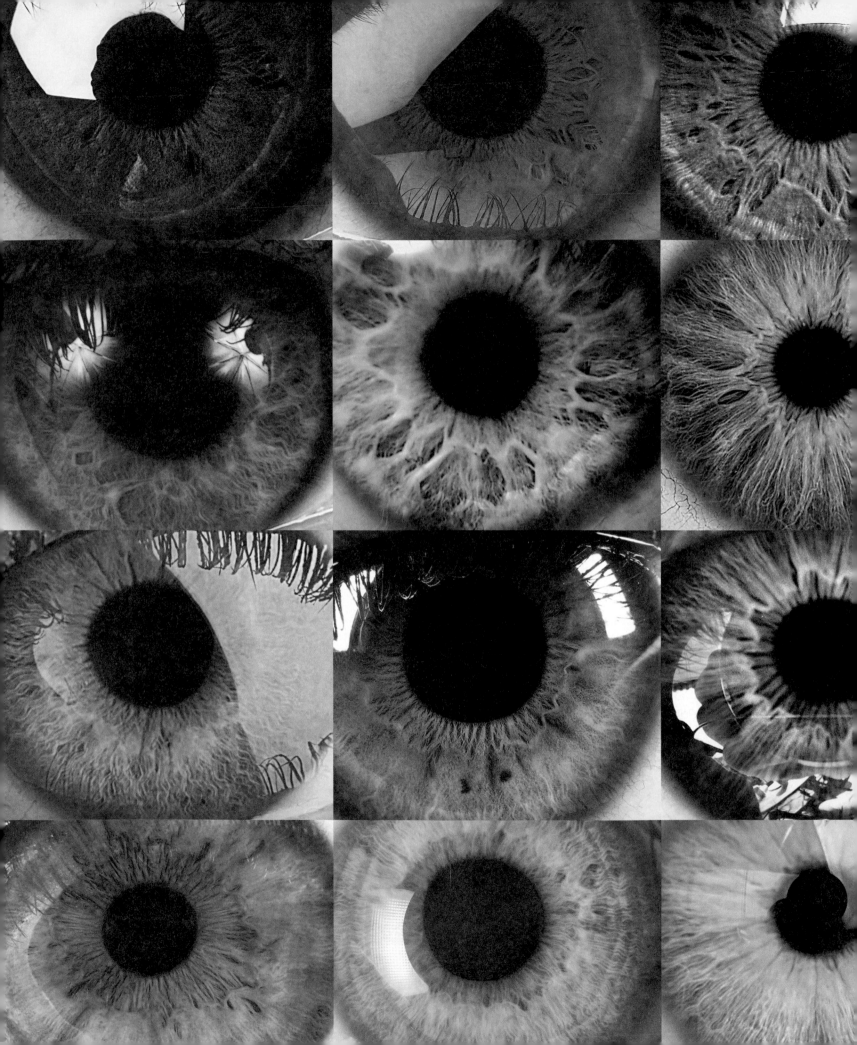

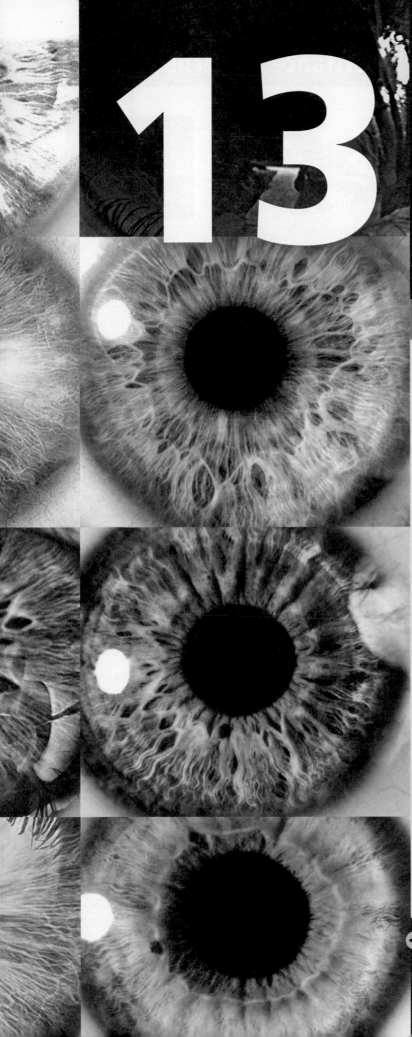

13 Patterns in Inherited Traits

Links to Earlier Concepts

You may want to review what you know about traits (Section 1.4), chromosomes (8.3), genes and gene expression (9.1), mutation (9.5), sexual reproduction and alleles (12.1), and meiosis (12.2–12.4). You will revisit probability and sampling error (1.7), laws of nature (1.8), protein structure (3.4, 3.5), pigments (6.2), clones (8.6), gene control (10.1, 10.2, 11.5), and epigenetics (10.5).

Core Concepts

Living systems store, retrieve, transmit, and respond to information essential for life.

The chromosomal basis of inheritance explains patterns in which traits appear among generations of offspring. For example, alleles of some genes are usually inherited together; alleles of others are inherited independently. Meiosis and fertilization produce a spectrum of variation in shared traits among offspring of sexual reproducers.

The field of biology consists of and relies upon experimentation and the collection and analysis of scientific evidence.

By applying mathematical reasoning to a biological phenomenon, Gregor Mendel discovered some simple patterns of inheritance in garden pea plants. Quantitative analysis of the distribution of traits among generations of offspring can reveal Mendelian patterns of inheritance, and can also confirm a non-Mendelian pattern.

Interactions among the components of a biological system, and among biological systems, give rise to complex properties.

Most traits are not inherited in simple Mendelian patterns. For example, one trait may be affected by multiple genes, or one gene may give rise to multiple traits. Many traits are also influenced by the environment. Changes in genetic information, such as occur by mutation, may result in observable changes in traits. Outcomes can include abnormal development and/or function.

◀ The continuous range of variation in human eye color is the result of interactions among several genes involved in making and distributing pigments.

LEARNING OBJECTIVES

Explain the contribution of Gregor Mendel to the study of inheritance.

Describe the difference between a homozygous and heterozygous genotype, and represent each symbolically with an example.

Distinguish between genotype and phenotype with an example.

Use an example to describe dominant and recessive alleles.

carpel
anther

❶ In garden pea plants (left), pollen grains that form in anthers of the flowers (right) produce male gametes. Female gametes form in carpels.

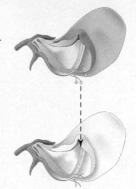

❷ Experimenters control the transfer of hereditary material from one pea plant to another by cutting off a flower's pollen-producing anthers (to prevent it from self-fertilizing), then brushing pollen from another flower onto its egg-producing carpel.

In this example, pollen from a plant with purple flowers is brushed onto the carpel of a white-flowered plant.

❸ Later, seeds develop inside pods of the cross-fertilized plant. When the seeds are planted, the embryo in each develops into a pea plant.

❹ In this example, every plant that arises from the cross has purple flowers. Predictable patterns such as this are evidence of how inheritance works.

Figure 13.1 Breeding experiments with the garden pea.

In the nineteenth century, people thought that hereditary material must be some type of fluid, with fluids from both parents blending at fertilization like milk into coffee. However, the idea of "blending inheritance" failed to explain what people could see with their own eyes. Children sometimes have traits such as freckles that do not appear in either parent. A cross between a black horse and a white one does not produce gray offspring.

The naturalist Charles Darwin had no hypothesis to explain such phenomena, even though inheritance was central to his theory of natural selection (we return to Darwin and natural selection in Chapter 16). At the time, no one knew that hereditary information (DNA) is divided into discrete  units (genes), an insight that is critical to understanding how traits are inherited. However, even before Darwin presented his theory, someone had been gathering evidence that would support it. Gregor Mendel (left), an Austrian monk, had been carefully breeding thousands of pea plants. By keeping detailed records of how traits passed from one generation to the next, Mendel had been collecting evidence of how inheritance works.

MENDEL'S EXPERIMENTS

Mendel cultivated the garden pea plant (**Figure 13.1**). This species is naturally self-fertilizing, which means each plant's flowers produce male and female gametes ❶ that form viable embryos when they meet up. In order to study inheritance, Mendel had to carry out controlled matings between individuals with specific traits. Mendel kept individual pea plants from self-fertilizing by removing the pollen-bearing parts (anthers) from their flowers. He then cross-fertilized the plants by brushing the egg-bearing parts (carpels) of their flowers with pollen from other plants ❷. He collected and planted seeds ❸ from the cross-fertilized individuals, and recorded the traits of the resulting pea plant offspring ❹.

Many of Mendel's experiments, which are called crosses, started with plants that "breed true" for particular traits such as white flowers or purple flowers. Breeding true for a trait means that, new mutations aside, all offspring have the same form of the trait as the parent(s), generation after generation. For example, all offspring of pea plants that breed true for white flowers also have white flowers. As you will see in Section 13.2, Mendel cross-fertilized pea plants that breed true for different forms of a trait, and discovered that the traits of the offspring often appear in predictable patterns. Mendel's meticulous work tracking pea plant traits led him to conclude (correctly) that hereditary information passes from one generation to the next in discrete units.

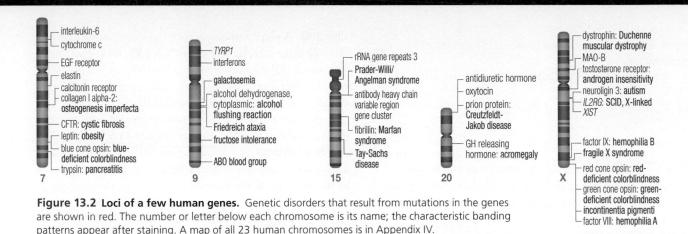

Figure 13.2 **Loci of a few human genes.** Genetic disorders that result from mutations in the genes are shown in red. The number or letter below each chromosome is its name; the characteristic banding patterns appear after staining. A map of all 23 human chromosomes is in Appendix IV.

INHERITANCE IN MODERN TERMS

Mendel discovered hereditary units, which we now call genes, almost a century before the discovery of DNA (Section 8.1). Today, we know that individuals of a species share certain traits because their chromosomes carry the same genes.

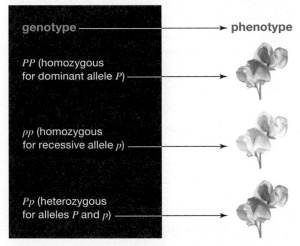

Figure 13.3 **Genotype gives rise to phenotype.** In this example, the dominant allele *P* specifies purple flowers; the recessive allele *p*, white flowers.

---

dominant Refers to an allele that masks the effect of a recessive allele paired with it in heterozygous individuals.

genotype (JEEN-oh-type) The particular set of alleles that is carried by an individual's chromosomes.

heterozygous (het-er-uh-ZYE-guss) Having two different alleles at the same locus on homologous chromosomes.

homozygous (ho-mo-ZYE-guss) Having two identical alleles at the same locus on homologous chromosomes.

hybrid (HI-brid) A heterozygous individual.

locus (LOW-cuss) A particular location on a chromosome.

phenotype (FEEN-oh-type) An individual's observable traits.

recessive Refers to an allele with an effect that is masked by a dominant allele on the homologous chromosome.

The location of a gene on a chromosome is called its **locus** (plural, loci, **Figure 13.2**). Diploid cells have pairs of homologous chromosomes (Section 11.1), so they have two copies of each gene; in most cases, both copies are expressed at the same level (Section 10.1). The two copies of any gene may be identical, or they may vary as alleles (Section 12.1). An individual with the same allele on both homologous chromosomes is **homozygous** for the allele (*homo-* means the same). Organisms breed true for a trait because they are homozygous for alleles governing the trait. By contrast, an individual with different alleles at a particular locus is **heterozygous** (*hetero-* means mixed). A **hybrid** is a heterozygous individual produced by a cross or mating between parents that breed true for different forms of a trait.

Homozygous and heterozygous are examples of **genotype**, the particular set of alleles an individual carries. Genotype is the basis of **phenotype**, which refers to the individual's observable traits. "White-flowered" and "purple-flowered" are examples of pea plant phenotypes that arise from differences in genotype.

The phenotype of a heterozygous individual depends on how the products of its two different alleles interact. In many cases, the effect of one allele influences the effect of the other, and the outcome of this interaction is reflected in the individual's phenotype. An allele is **dominant** when its effect masks that of a **recessive** allele paired with it. A dominant allele is often represented by an italic capital letter such as *A*; a recessive allele, with a lowercase italic letter such as *a*. Consider the purple-flowered and white-flowered pea plants that Mendel studied. In these plants, the allele that specifies purple flowers (let's call it *P*) is dominant over the allele that specifies white flowers (*p*). Thus, a pea plant homozygous for the dominant allele (*PP*) has purple flowers; one homozygous for the recessive allele (*pp*) has white flowers (**Figure 13.3**). A heterozygous plant (*Pp*) has purple flowers.

Describe the way a Punnett square is used to predict the outcome of a monohybrid cross.

Explain how a testcross can reveal the genotype of an individual with a dominant trait.

State the law of segregation in modern terms.

When homologous chromosomes separate during meiosis (Section 12.3), the gene pairs on those chromosomes separate too. Let's use our pea plant alleles for purple and white flowers in an example (**Figure 13.4**). A plant homozygous for the dominant allele (*PP*) can only make gametes that carry

allele *P* ❶. A plant homozygous for the recessive allele (*pp*) can only make gametes that carry allele *p* ❷. If the two plants are crossed (*PP* × *pp*), only one outcome is possible: A gamete carrying allele *P* meets with a gamete carrying allele *p* ❸. All offspring of this cross will have both alleles—they will be heterozygous (*Pp*). A grid called a **Punnett square** is helpful for predicting the outcomes of crosses (**Figure 13.5**).

In a **testcross**, an individual with a dominant trait (but unknown genotype) is crossed with an individual known to be homozygous for the recessive allele. The pattern of traits among the offspring of the cross can reveal whether the tested individual is heterozygous or homozygous. If all offspring of the testcross have the dominant trait (as occurred in our example above), then the parent with the unknown genotype is homozygous for the dominant allele. If some of the offspring have the recessive trait, then the parent is heterozygous.

Dominance relationships between alleles determine the phenotypic outcome of a **monohybrid cross**, in which individuals that are identically heterozygous at one gene locus are crossed (*Pp* × *Pp*, for example). The frequency at which traits associated with the alleles appear among the offspring depends on whether one of the alleles is dominant over the other.

Making a monohybrid cross starts with two individuals that breed true for different forms of a trait. In pea plants,

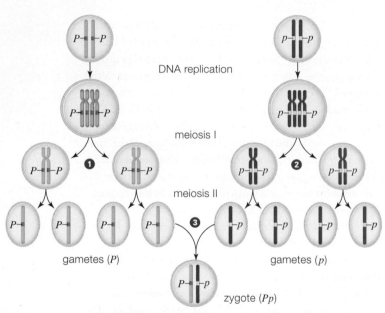

Figure 13.4 Segregation of genes on homologous chromosomes into gametes. Homologous chromosomes separate during meiosis, so the genes on them separate too. Each of the resulting gametes carries one of the two members of every gene pair. For clarity, only one set of chromosomes is illustrated.

❶ All gametes made by a parent homozygous for a dominant allele carry that allele.

❷ All gametes made by a parent homozygous for a recessive allele carry that allele.

❸ If these two parents are crossed, the union of any of their gametes at fertilization produces a zygote (Section 12.2) with both alleles. All offspring will be heterozygous.

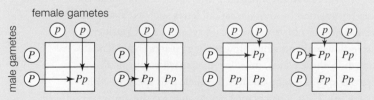

Figure 13.5 Making a Punnett square. Parental gametes are circled on the top and left sides of a grid. Each square is filled with the combination of alleles that would result if the gametes in the corresponding row and column met up.

TABLE 13.1

Mendel's Seven Pea Plant Traits

Trait	Dominant Form	Recessive Form
Seed Shape	Round	Wrinkled
Seed Color	Yellow	Green
Pod Texture	Smooth	Wrinkled
Pod Color	Green	Yellow
Flower Color	Purple	White
Flower Position	Along Stem	At Tip
Stem Length	Tall	Short

flower color (purple and white) is one example of a trait with two distinct forms, but there are many others. Mendel investigated seven of them: stem length (tall and short), seed color (yellow and green), pod texture (smooth and wrinkled), and so on (**Table 13.1**). A cross between individuals that breed true for different forms of a trait yields hybrid offspring, all with the same set of alleles governing the trait (**Figure 13.6A**). A cross between two of these F_1 (first generation) hybrids is the monohybrid cross. The frequency at which the two traits appear in the F_2 (second generation) offspring offers information about a dominance relationship between the two alleles. (F is an abbreviation for filial, which means offspring.)

A cross between two purple-flowered heterozygous pea plants (Pp) is an example of a monohybrid cross. Each individual can make two types of gametes: ones that carry allele P, and ones that carry allele p (**Figure 13.6B**). So, in a monohybrid cross between two Pp plants ($Pp \times Pp$), the two types of gametes can meet up in four possible ways at fertilization:

Possible Event	Probable Outcome
Sperm P meets egg P →	zygote genotype is PP → purple flowers
Sperm P meets egg p →	zygote genotype is Pp → purple flowers
Sperm p meets egg P →	zygote genotype is Pp → purple flowers
Sperm p meets egg p →	zygote genotype is pp → white flowers

Three out of four possible outcomes of this cross include at least one copy of the dominant allele P. In other words, each time fertilization occurs, there are 3 chances in 4 that the resulting offspring will have a P allele (and the individual will make purple flowers). There is 1 chance in 4 that the zygote will have two p alleles (and the individual will make white flowers). Thus, the probability that a particular offspring of this cross will have purple or white flowers is 3 purple to 1 white—a ratio of 3:1 (**Figure 13.6C**). The 3:1 pattern is evidence that purple and white flower color are specified by alleles with a clear dominance relationship: Purple is dominant; white, recessive. If the probability of one individual inheriting a particular genotype is difficult to imagine, think about the probability in terms of many offspring. In this example, there will be roughly three purple-flowered plants for every white-flowered one.

Our example illustrated a pattern so predictable that it can be used as evidence of a dominance relationship between alleles. The phenotype ratios in the F_2 offspring of Mendel's monohybrid crosses were all close to 3:1. These results became the basis of his **law of segregation**, which we state here in modern terms: A diploid cell has two copies of every gene that occurs on its homologous chromosomes, and the two copies may vary as alleles. Two alleles at any locus are distributed into separate gametes during meiosis.

A A monohybrid cross starts with two individuals that breed true for two forms of a trait. Each is homozygous for one allele and only makes gametes with that allele.

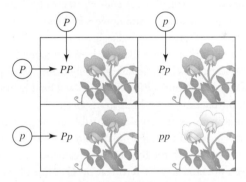

parent plant homozygous for purple flowers (PP)   parent plant homozygous for white flowers (pp)

one type of gamete P × p one type of gamete

B All of the F_1 offspring of a cross between two plants that breed true for different forms of a trait are identically heterozygous (Pp). These offspring make two types of gametes: P and p.

Pp hybrid

P p two types of gametes

C A cross between these F_1 offspring ($Pp \times Pp$) is a monohybrid cross. The phenotype ratio in F_2 offspring is 3:1 (3 purple to 1 white) in this example.

Figure 13.6 Example of a monohybrid cross.

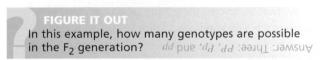

FIGURE IT OUT
In this example, how many genotypes are possible in the F_2 generation?

Answer: Three: PP, Pp, and pp

law of segregation A diploid cell has two copies of every gene that occurs on its homologous chromosomes. Two alleles at any locus are distributed into separate gametes during meiosis.

monohybrid cross Cross between two individuals identically heterozygous for one gene; for example $Aa \times Aa$.

Punnett square (PUN-it) Diagram used to predict the genotypic and phenotypic outcome of a cross.

testcross Method of determining the genotype of an individual with a dominant phenotype: a cross between the individual and another individual known to be homozygous recessive.

13.3 Mendel's Law of Independent Assortment

LEARNING OBJECTIVES

With a suitable example, explain how to make a dihybrid cross.

State the law of independent assortment in modern terms.

Explain why the relative location of two genes on a chromosome can affect the way their alleles are distributed into gametes.

A monohybrid cross allows us to study a dominance relationship between alleles of one gene. What about alleles of two genes? An individual heterozygous for two alleles at two loci (*AaBb*, for example) is called a dihybrid, and a cross between two such individuals is a **dihybrid cross**. As with a monohybrid cross, the frequency of traits appearing among the offspring of a dihybrid cross depends on the dominance relationships between the alleles.

To make a dihybrid cross, we would start with individuals that breed true for two different traits. Let's use a gene for flower color (*P*, purple; *p*, white) and one for plant height (*T*, tall; *t*, short) in an example (**Figure 13.7**). Making a dihybrid cross using these genes begins with one parent plant that breeds true for purple flowers and tall stems (*PPTT*), and one that breeds true for white flowers and short stems (*pptt*) ❶. The *PPTT* plant only makes gametes with the dominant alleles (*PT*); the *pptt* plant only makes gametes with the recessive alleles (*pt*). So, all offspring from a cross between these parent plants (*PPTT* × *pptt*) will be dihybrids (*PpTt*) with purple flowers and tall stems ❷.

Four combinations of alleles are possible in the gametes of *PpTt* dihybrids ❸. If two *PpTt* plants are crossed (a dihybrid cross, *PpTt* × *PpTt*), the four types of gametes can combine in sixteen possible ways at fertilization ❹. Nine of the sixteen genotypes would give rise to tall plants with purple flowers; three, to short plants with purple flowers; three, to tall plants with white flowers; and one, to short plants with white flowers. Thus, the ratio of phenotypes among the offspring of this dihybrid cross would be 9:3:3:1.

Mendel discovered the 9:3:3:1 ratio of phenotypes among offspring of his dihybrid crosses, but he had no idea what it meant. He could only say that "units" specifying one trait (such as flower color) are inherited independently of "units" specifying other traits (such as plant height). In time, Mendel's hypothesis became known as the **law of independent assortment**, which we state here in modern terms: During meiosis, two alleles at one gene locus tend to be sorted into gametes independently of alleles at other loci.

Mendel published his results in 1866, but his work was read by few and understood by no one at the time. In 1871 he was promoted, and his pioneering experiments ended. When he died in 1884, he did not know that his work with pea plants would be the starting point for modern genetics.

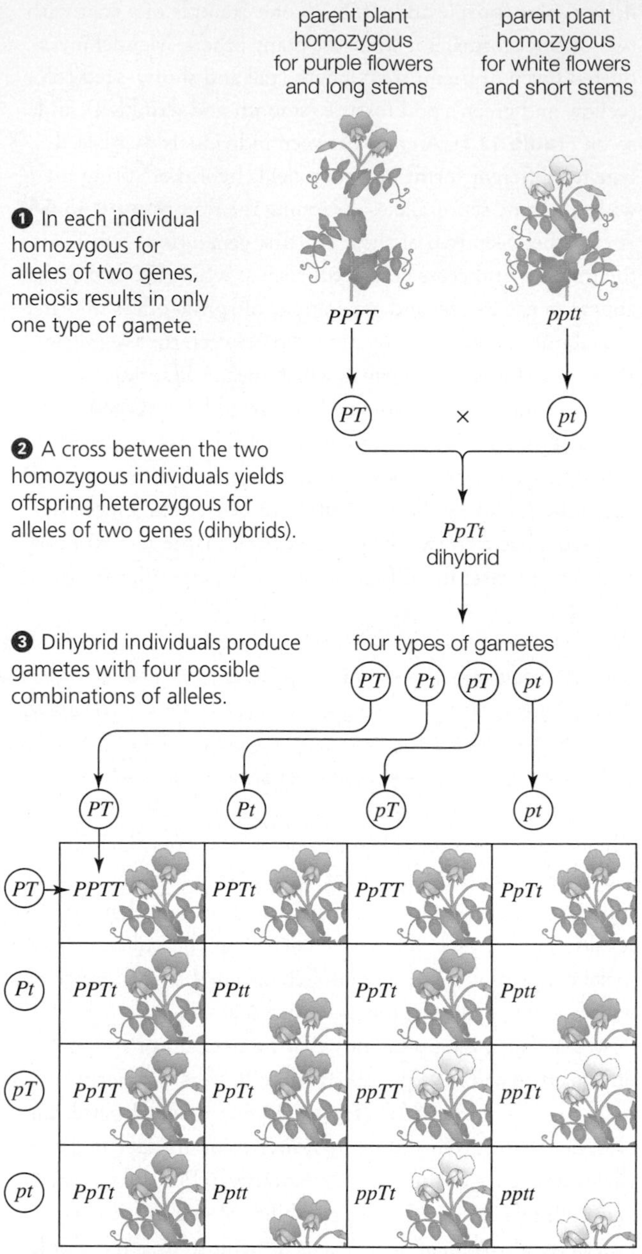

parent plant homozygous for purple flowers and long stems

parent plant homozygous for white flowers and short stems

❶ In each individual homozygous for alleles of two genes, meiosis results in only one type of gamete.

PPTT

pptt

PT × *pt*

❷ A cross between the two homozygous individuals yields offspring heterozygous for alleles of two genes (dihybrids).

PpTt dihybrid

❸ Dihybrid individuals produce gametes with four possible combinations of alleles.

four types of gametes

PT *Pt* *pT* *pt*

	PT	*Pt*	*pT*	*pt*
PT	*PPTT*	*PPTt*	*PpTT*	*PpTt*
Pt	*PPTt*	*PPtt*	*PpTt*	*Pptt*
pT	*PpTT*	*PpTt*	*ppTT*	*ppTt*
pt	*PpTt*	*Pptt*	*ppTt*	*pptt*

❹ If two of the dihybrid individuals are crossed, the four types of gametes can meet up in 16 possible ways. Of 16 possible offspring genotypes, 9 will result in plants that are purple-flowered and tall; 3, purple-flowered and short; 3, white-flowered and tall; and 1, white-flowered and short. Thus, the ratio of phenotypes is 9:3:3:1.

Figure 13.7 A dihybrid cross between plants that differ in flower color and plant height. In this example, *P* and *p* are dominant and recessive alleles for purple and white flower color; *T* and *t* are dominant and recessive alleles for tall and short plant height.

FIGURE IT OUT
What do the flowers inside the boxes represent?

Answer: Phenotypes of the F₂ offspring

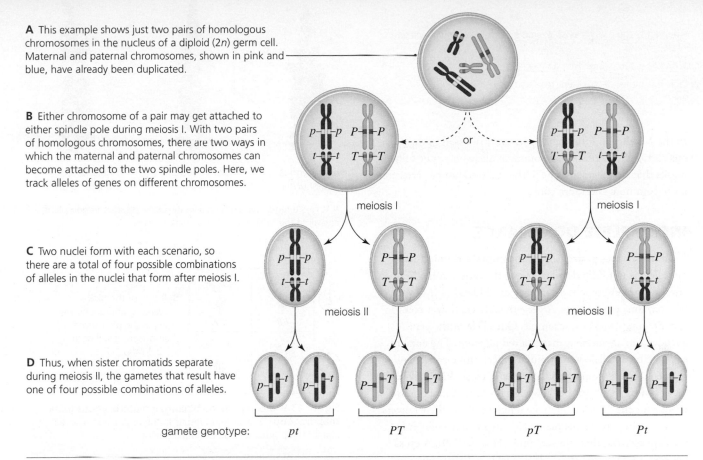

A This example shows just two pairs of homologous chromosomes in the nucleus of a diploid (2*n*) germ cell. Maternal and paternal chromosomes, shown in pink and blue, have already been duplicated.

B Either chromosome of a pair may get attached to either spindle pole during meiosis I. With two pairs of homologous chromosomes, there are two ways in which the maternal and paternal chromosomes can become attached to the two spindle poles. Here, we track alleles of genes on different chromosomes.

C Two nuclei form with each scenario, so there are a total of four possible combinations of alleles in the nuclei that form after meiosis I.

D Thus, when sister chromatids separate during meiosis II, the gametes that result have one of four possible combinations of alleles.

or

meiosis I meiosis I

meiosis II meiosis II

gamete genotype: *pt* *PT* *pT* *Pt*

Figure 13.8 Independent assortment. Alleles of genes on different chromosomes assort independently into gametes. Alleles of genes that are far apart on the same chromosome usually assort independently too, because crossovers typically separate them.

THE CONTRIBUTION OF CROSSOVERS

How alleles at two gene loci get sorted into gametes depends partly on whether the two genes are on the same chromosome. When the members of a pair of homologous chromosomes separate during meiosis, either can end up in either of the two new nuclei that form (Section 12.4). This random assortment happens independently for each pair of homologous chromosomes in the cell. Thus, alleles of genes on one chromosome pair assort into gametes independently of alleles of genes on other chromosome pairs (**Figure 13.8**).

What about genes on the same chromosome? Pea plants have seven chromosomes. Mendel studied seven pea genes,

and alleles of all of them assorted into gametes independently of one another. Was he lucky enough to choose one gene on each of those chromosomes? As it turns out, some of the genes Mendel studied *are* on the same chromosome. These genes are far enough apart that crossing over occurs between them very frequently—so frequently that their alleles tend to assort into gametes independently, just as if the genes were on different chromosomes. By contrast, alleles of genes that are very close together on a chromosome usually do not assort independently into gametes, because crossing over does not happen between them very often. Thus, gametes usually end up with parental combinations of alleles of these genes.

Genes whose alleles do not assort independently into gametes are said to be linked. Linked genes in humans were identified by tracking inheritance in families over several generations. A **linkage group** comprises all of the genes on a chromosome. Peas have 7 different chromosomes, so they have 7 linkage groups. Humans have 23 different chromosomes, so they have 23 linkage groups.

---

dihybrid cross Cross between two individuals identically heterozygous for two genes; for example *AaBb* × *AaBb*.
law of independent assortment During meiosis, alleles at one gene locus on homologous chromosomes tend to be distributed into gametes independently of alleles at other loci.
linkage group All of the genes on a chromosome.

13.4 Beyond Simple Dominance

LEARNING OBJECTIVES

Using the human blood groups as an example, explain codominance and multiple allele systems.

Explain incomplete dominance with an example.

With suitable examples, describe epistasis and pleiotropy.

In the Mendelian inheritance patterns discussed in the previous sections, the effect of a dominant allele on a trait fully masks that of a recessive one. Other inheritance patterns are more common, and more complex.

INCOMPLETE DOMINANCE

In an inheritance pattern called **incomplete dominance**, one allele is not fully dominant over the other, so the heterozygous phenotype is an intermediate blend of the two homozygous phenotypes. A gene that affects flower color in snapdragon plants is an example. One allele of the gene (*R*) encodes an enzyme that makes a red pigment. The enzyme encoded by an allele with a mutation (*r*) cannot make any pigment. Plants homozygous for the *R* allele (*RR*) make a lot of red pigment, so they have red flowers. Plants homozygous for the *r* allele (*rr*) make no pigment, so their flowers are white. Heterozygous plants (*Rr*) make only enough red pigment to color their flowers pink (**Figure 13.9**). A cross between two heterozygous plants yields red-flowered, pink-flowered, and white-flowered offspring in a 1:2:1 ratio.

CODOMINANCE

In an inheritance pattern called **codominance**, traits associated with two nonidentical alleles of a gene are fully and equally apparent in heterozygous individuals; neither allele is dominant or recessive. Alleles of the *ABO* gene offer an example. This gene encodes an enzyme that modifies a carbohydrate on the surface of human red blood cells. Two alleles, *A* and *B*, encode slightly different versions of the enzyme, which in turn modify the carbohydrate differently. A third allele, *O*, has a mutation that causes a frameshift (Section 9.5). The protein encoded by this allele has no enzymatic activity, so the carbohydrate remains unmodified.

The alleles you carry for the *ABO* gene determine the form of the carbohydrate on your red blood cells, so they are the basis of your ABO blood type (**Figure 13.10**). Alleles *A* and *B* are codominant when paired. If your genotype is *AB*, then you have both versions of the carbohydrate, and your blood type is AB. The *O* allele is recessive when paired with either the *A* or *B* allele. If your genotype is *AA* or *AO*, your blood type is A. If your genotype is *BB* or *BO*, it is type B.

homozygous parent (*RR*) × homozygous parent (*rr*) → heterozygous offspring (*Rr*)

A Cross a red-flowered with a white-flowered snapdragon plant, and all of the offspring will have pink flowers.

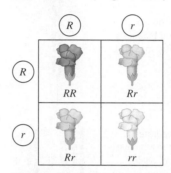

B If two of the pink-flowered snapdragons are crossed, the phenotypes of their offspring will occur in a 1:2:1 ratio.

Figure 13.9 Incomplete dominance in heterozygous (pink) snapdragons. One allele (*R*) results in the production of a red pigment; the other (*r*) results in no pigment.

FIGURE IT OUT Is the experiment in B a monohybrid cross or a dihybrid cross?

Answer: A monohybrid cross

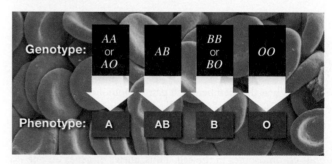

Genotype: *AA* or *AO* | *AB* | *BB* or *BO* | *OO*

Phenotype: A | AB | B | O

Figure 13.10 Combinations of alleles (genotype) **that are the basis of human blood type** (phenotype).

If you are *OO*, it is type O (**Figure 13.10**). A gene such as *ABO*, with three or more alleles persisting at relatively high frequency in a population, is called a **multiple allele system**.

The immune system attacks any cell bearing molecules that do not occur in one's own body, so receiving incompatible blood in a transfusion can be dangerous. An immune attack causes red blood cells to clump or burst, with potentially fatal results. Almost everyone makes the red blood cell carbohy-

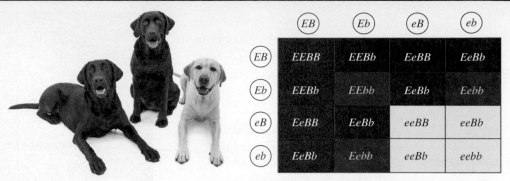

Figure 13.11 An example of epistasis. Interactions among products of two genes affect fur color in Labrador retrievers. Dogs with alleles *E* and *B* have black fur. Those with an *E* and two recessive *b* alleles have brown fur. Dogs homozygous for the recessive *e* allele have yellow fur.

	EB	*Eb*	*eB*	*eb*
EB	*EEBB*	*EEBb*	*EeBB*	*EeBb*
Eb	*EEBb*	*EEbb*	*EeBb*	*Eebb*
eB	*EeBB*	*EeBb*	*eeBB*	*eeBb*
eb	*EeBb*	*Eebb*	*eeBb*	*eebb*

drate (which is later modified in people with an *A* or *B* allele), so type O blood does not usually trigger an immune response in transfusion recipients. People with type O blood are called universal donors because they can donate blood to anyone. However, because their body is unfamiliar with the modified form of the carbohydrate made by people with type A or B blood, they can receive type O blood only. People with type AB blood can receive a transfusion of any *ABO* blood type, so they are called universal recipients.

EPISTASIS

In a common inheritance pattern called polygenic inheritance or **epistasis**, one trait is affected by multiple gene products. Consider how fur color in dogs and other animals arises from pigments called melanins. The color of brown or black fur arises from a dark brown form of melanin; a reddish melanin colors fur yellow or tan. The products of several genes interact to make these melanins and deposit them in fur. In Labrador retriever dogs, alleles of two of these genes determine whether the individual has black, brown, or yellow fur (**Figure 13.11**). The product of one gene (*TYRP1*) helps make the brown melanin. A dominant allele (*B*) of this gene results in a higher production of brown melanin than the recessive allele (*b*). A different gene (*MC1R*) affects which type of melanin is produced. A dominant allele (*E*) of this gene triggers production of the brown melanin; its recessive partner (*e*) carries a mutation that results in production of the reddish form. Dogs homozygous for the *e* allele are yellow because they make only the reddish melanin.

PLEIOTROPY

A **pleiotropic** gene influences multiple traits, so mutations that affect its expression or its product affect all of the traits. Many complex genetic disorders, including sickle-cell anemia (Section 9.5) and cystic fibrosis, are caused by mutations in single genes. Marfan syndrome, another example, is a result of mutations in the gene for fibrillin. Long fibers of this protein are part of elastic tissues that make up the heart, skin, blood

vessels, tendons, and other body parts. When the fibrillin gene carries a mutation that alters its product, body tissues form with defective fibrillin or none at all. The largest blood vessel leading from the heart, the aorta, is particularly affected. Without a proper scaffold of fibrillin, the aorta's thick wall is not as elastic as it should be, and it eventually stretches and becomes leaky. Thinned and weakened, the aorta can rupture during exercise—an abruptly fatal outcome. About 1 in 5,000 people have Marfan syndrome, and there is no cure. However, recently increased awareness of the symptoms of Marfan has undoubtedly saved many lives (**Figure 13.12**).

Figure 13.12 A heartbreaker: Marfan syndrome. Isaiah Austin was diagnosed with Marfan syndrome just days before the 2014 NBA draft. He was considered a first-round prospect, but learning that his heart could rupture unexpectedly during a game ended his dream of becoming a professional basketball player. The diagnosis probably saved Isaiah's life.

codominance Effect in which the full and separate phenotypic effects of two alleles are apparent in heterozygous individuals.

epistasis (epp-ih-STAY-sis) Polygenic inheritance, in which a trait is influenced by multiple genes.

incomplete dominance Inheritance pattern in which one allele is not fully dominant over another, so the heterozygous phenotype is an intermediate blend between the two homozygous phenotypes.

multiple allele system Gene for which three or more alleles persist in a population at relatively high frequency.

pleiotropic (ply-uh-TROH-pick) Refers to a gene that affects multiple traits.

CREDITS: (11) right, Susan Schmitz/Shutterstock; (12) Photo by Andy Lyons/Getty Images.

CHAPTER 13 PATTERNS IN INHERITED TRAITS   **213**

13.5 Nature and Nurture

LEARNING OBJECTIVES

With suitable examples, describe some environmental effects on phenotype.

Discuss the influence of an individual's environment on expression of genes associated with behavior.

The phrase "nature versus nurture" refers to a centuries-old debate about whether human behavioral traits arise from one's genetics (nature) or from environmental factors (nurture). Today, we know that both play a substantial role. The environment affects the expression of many genes, which in turn affects phenotype—including behavioral traits. We can summarize this thinking with an equation:

$$genotype \ + \ environment \longrightarrow phenotype$$

Epigenetics research is revealing that the environment makes an even greater contribution to this equation than most biologists had ever suspected (Section 10.5).

Environmental cues initiate cell-signaling pathways that in turn trigger changes in gene expression (you will learn more about such pathways in later chapters). Some of these cell-signaling pathways methylate particular regions of DNA, so they suppress gene expression in those regions (Section 10.1). In humans and other animals, DNA methylation patterns can be permanently and heritably affected by diet, stress, and exercise, and also by exposure to drugs and toxins such as tobacco, alcohol, arsenic, and asbestos. These mechanisms adjust phenotype in response to external cues; they are part of an individual's normal ability to adapt to its environment, as the following examples illustrate.

Water fleas (*Daphnia*) are tiny aquatic animals that inhabit seasonal ponds and other standing pools of fresh water. Conditions such as temperature, oxygen content, and salinity vary dramatically over time and between different areas of these pools. For example, water at the top of a still pond is typically warmer than water at the bottom, and it also contains more light and dissolved oxygen. Individual water fleas acclimate to such differences in environmental conditions by adjusting their gene expression. The adjustment provides an appropriate set of proteins to maintain cellular function.

Daphnia has a lot of genes, and the abundance offers a striking flexibility of phenotype. Environmental cues trigger adjustments in gene expression that change an individual's form and function to suit its current environment. For example, a water flea that swims to the bottom of a pond can survive the low oxygen conditions there by turning on expression of genes involved in the production of hemoglobin—and turning red (**Figure 13.13A**). Hemoglobin is a protein that carries oxygen (Section 9.5), and it enhances the individual's ability to absorb oxygen from the water.

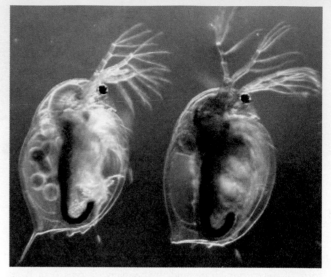

A Under low-oxygen conditions, a water flea switches on genes involved in producing hemoglobin. Making this red protein enhances the individual's ability to take up oxygen from water. The flea on the left has been living in water with a normal oxygen content; the one on the right, in water with a low oxygen content.

B The color of the snowshoe hare's fur varies by season. In summer, the fur is brown (left); in winter, white (right). Both forms offer seasonally appropriate camouflage from predators.

C The height of a mature yarrow plant depends on the elevation at which it grows.

Figure 13.13 Some environmental effects on phenotype.

CREDITS: (13A) From The Ecoresponsive Genome of Daphnia pulex, *Science* 04 Feb 2011: Vol 331 Issue 6017 pp.555–561, Reprinted with permission from AAAS; (13B) left, JupiterImages Corporation; right, © age fotostock/SuperStock; (13C) photo, Igor Sokolov (breeze)/Shutterstock.com.

Other environmental factors also affect water flea phenotype. The presence of insect predators causes water fleas to form a protective pointy helmet and lengthened tail spine, for example. Individual water fleas also switch between asexual and sexual modes of reproduction depending on the season.

In many mammals, seasonal changes in temperature and length of day affect production of melanin and other pigments that color skin and fur. These species have different color phases in different seasons (**Figure 13.13B**). Hormonal signals triggered by seasonal changes cause fur to be shed, and new fur grows back with different types and amounts of pigments deposited in it. The resulting change in phenotype provides the animals with seasonally appropriate camouflage from predators.

In plants, a flexible phenotype gives immobile individuals an ability to thrive in diverse habitats. For example, genetically identical yarrow plants grow to different heights at different altitudes (**Figure 13.13C**). More challenging temperature, soil, and water conditions are typically encountered at higher altitudes. Differences in altitude are also correlated with changes in the reproductive mode of yarrow: Plants at higher altitude tend to reproduce asexually, and those at lower altitude tend to reproduce sexually.

Researchers recently discovered several mutations associated with five human psychiatric disorders: autism, depression, schizophrenia, bipolar disorder, and attention deficit/hyperactivity disorder (ADHD). However, there must be environmental components to these disorders too, because the majority of people who have the mutations never end up with a psychiatric disorder. Moreover, one person with the mutations might get one type of disorder, while a relative with the same mutations might get another: two different phenotypic results from the same genetic underpinnings.

Animal models are helping us unravel some of the mechanisms by which environment can influence mental state. For example, we have discovered that learning and memory are associated with dynamic and rapid DNA modifications in animal brain cells. Mood is, too. Stress-induced depression causes methylation-based silencing of a particular nerve growth factor gene; some antidepressants work by reversing this methylation. As another example, rats whose mothers are not very nurturing end up anxious and having a reduced resilience for stress as adults. The difference between these rats and ones who had nurturing maternal care is traceable to epigenetic DNA modifications that result in a lower-than-normal level of another nerve growth factor. Drugs can reverse these modifications—and their effects. We do not yet know all of the genes that influence human mental state, but the implication of such research is that future treatments for many psychiatric disorders will involve deliberate modification of methylation patterns in an individual's DNA.

Dr. Gay Bradshaw

The air explodes with the sound of high-powered rifles and the startled infant watches his family fall to the ground, the image seared into his memory. He and other orphans are then transported to distant locales to start new lives. Ten years later, the teenaged orphans begin a raping and killing rampage, leaving more than a hundred victims.

A scene describing post-traumatic stress disorder (PTSD) in Kosovo or Rwanda? The similarities are striking—but the teenagers are young elephants, and the victims, rhinoceroses.

Gay Bradshaw, a psychologist and the director of the Kerulos Center in Oregon, has brought the latest insights from human neuroscience and psychology to bear on startling field observations of elephant behavior. She suspects that some threatened elephant populations might be suffering from chronic stress and trauma brought on by human encroachment and killing. "The loss of older elephants," says Bradshaw, "and the extreme psychological and physical trauma of witnessing the massacres of their family members interferes with a young elephant's normal development."

Under normal conditions, an early and healthy emotional relationship between an infant and its mother fosters the development of self-regulatory structures in the brain's right hemisphere. All mammals share this developmental attachment mechanism. With trauma, a malfunction can develop that makes the individual vulnerable to PTSD and predisposed to violence as an adult. Individuals who survive trauma often face a lifelong struggle with depression, suicide, or behavioral dysfunctions. In addition, children of trauma survivors can exhibit similar symptoms, an effect that is likely to be epigenetic at least in part.

All across Africa, India, and parts of Southeast Asia, from within and around whatever patches and corridors of their natural habitat remain, elephants have been striking out, destroying villages and crops, attacking and killing human beings. "Everybody pretty much agrees that the relationship between elephants and people has dramatically changed," Bradshaw says. "What we are seeing today is extraordinary. Where for centuries humans and elephants lived in relative peaceful coexistence, there is now hostility and violence. Now, I use the term 'violence' because of the intentionality associated with it, both in the aggression of humans and, at times, the recently observed behavior of elephants."

LEARNING OBJECTIVE
Using appropriate examples, explain continuous variation and its causes.

The pea plant phenotypes that Mendel studied appeared in two distinct forms, which made them easy to track through generations. However, many other traits do not appear in distinct forms. Such traits are often the result of complex genetic interactions—multiple genes, multiple alleles, or both—with added environmental influences (we return to this topic in Chapter 17, as we consider some evolutionary consequences of variation in shared traits). The complexity often makes these traits difficult to study, which is why the genetic basis of many of them has not yet been completely unraveled.

Some traits occur in a range of small differences that is called **continuous variation**. Continuous variation can be an outcome of epistasis, in which multiple genes affect a single trait. The more genes and environmental factors that influence a trait, the more continuous is its variation.

Traits that arise from genes with a lot of alleles may vary continuously. Consider that, in some genes, a series of 2 to 6 nucleotides is repeated many times in a row. These **short tandem repeats** can spontaneously expand or contract during DNA replication or repair, at a rate that is much faster than other mutations. The resulting changes in the gene's DNA sequence may be preserved as alleles. For example, short tandem repeats have given rise to 12 alleles of a homeotic gene that influences the length of the face in dogs, with longer repeats associated with longer faces (**Figure 13.14**).

Human skin color varies continuously (a topic that we return to in Chapter 14), as does human eye color (shown in the chapter opener). The colored part of the eye is a doughnut-shaped structure called the iris. Iris color, like skin color, is the result of interactions among gene products that make and distribute melanins. The more melanin deposited in the iris, the less light is reflected from it and the darker it appears.

How do we know if a trait varies continuously? First, the total range of phenotypes is divided into measurable categories (**Figure 13.15A**). The number of individuals in each category reveals the relative frequencies of phenotypes across the range of values. When the data are plotted as a bar chart, a graph line around the top of the bars shows the distribution of values for the trait (**Figure 13.15B**). If the line is a bell-shaped curve, or **bell curve**, the trait varies continuously.

---

bell curve Bell-shaped curve; typically results from graphing frequency versus distribution for a trait that varies continuously.
continuous variation Range of small differences in a trait.
short tandem repeat In chromosomal DNA, sequences of a few nucleotides repeated multiple times in a row.

Figure 13.14 Face length varies continuously in dogs.
A gene with 12 alleles influences this trait. All of the alleles arose by spontaneous expansion or contraction of short tandem repeats. The longer the allele, the longer the face.

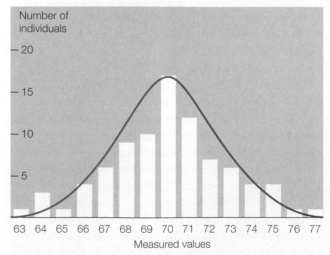

63 64 65 66 67 68 69 70 71 72 73 74 75 76 77

A To see if human height varies continuously, male biology students at the University of Florida were divided into categories of one-inch increments in height and counted.

B Graphing the data that resulted from the experiment in **A** produces a bell-shaped curve, an indication that height does vary continuously in humans.

Figure 13.15 How to determine whether a particular trait varies continuously.

CREDITS: (14) WilleeCole/Shutterstock.com; (15A) Courtesy of Ray Carson, University of Florida News and Public Affairs.

Lindsay, 22   Savannah, 19   Ben, 23   Jeff, 21   Brandon, 18   Cody, 23

Figure 13.16 A few of the many victims of cystic fibrosis. At least one young person dies every day in the United States from complications of this disease, which occurs most often in people of northern European ancestry.

MENACING MUCUS

Cystic fibrosis (CF), the most common fatal genetic disorder in the United States, occurs in people homozygous for a mutated allele of the *CFTR* gene. This gene encodes an active transport protein that moves chloride ions out of epithelial cells. Sheets of these cells line the passageways and ducts of the lungs, liver, pancreas, intestines, and reproductive system. When the CFTR protein pumps chloride ions out of these cells, water follows the solute by osmosis, a two-step process that maintains a thin, watery film on the surface of epithelial cell sheets. Mucus slides easily over the wet sheets of cells.

The allele most commonly associated with CF has a three-base-pair deletion. It is called *ΔF508* because it encodes a CFTR protein missing the phenylalanine (F) that is normally the 508th amino acid (Δ means deleted). A CFTR protein missing this amino acid misfolds in a tiny region. This small defect interferes with cellular processes that would otherwise finish the protein and install it in the plasma membrane. Normally, a newly translated CFTR polypeptide is modified by the endoplasmic reticulum (ER) and exported to a Golgi body, which attaches carbohydrates to it. The finished protein is then packaged in vesicles routed to the plasma membrane. CFTR polypeptides with the missing amino acid are produced properly, but a cellular quality control mechanism recognizes the misfolded region and destroys most of them before they leave the ER. A few misfolded proteins that reach the plasma membrane are quickly taken back into the cell by endocytosis and destroyed.

Epithelial cell membranes that lack the CFTR protein cannot transport chloride ions. Too few chloride ions leave these cells. Not enough water leaves them either, so the surfaces of epithelial cell sheets are too dry. Mucus that normally slips through the body's tubes sticks to the walls of the tubes instead. This outcome has pleiotropic effects because thick globs of mucus accumulate and clog passageways and ducts throughout the body. Breathing becomes difficult as the mucus obstructs the smaller airways of the lungs. Digestive problems arise as ducts that lead to the gut become clogged. Males are typically infertile because their sperm flow is hampered.

In addition to its role in chloride ion transport, the CFTR protein also helps alert the immune system to the presence of disease-causing bacteria in the lungs. It functions as a receptor by binding directly to these bacteria and causing them to be taken into the cell by endocytosis. In epithelial cells lining the respiratory tract, endocytosis of bacteria triggers an immune response. Bacteria-fighting molecules that are produced in this response keep microbial populations at bay. When the cells lack CFTR, the early alert system fails, so bacteria have time to multiply before being detected by the immune system. Thus, chronic bacterial infections of the lungs are a hallmark of cystic fibrosis. Antibiotics help control infections, but there is no cure for the disorder. Most affected people die before age thirty, when their tormented lungs fail (**Figure 13.16**).

The *ΔF508* allele is at least 50,000 years old and very common: In some populations, 1 in 25 people are heterozygous for it. Why does the allele persist if it is so harmful? *ΔF508* is codominant with the normal allele. Heterozygous individuals typically have no symptoms of cystic fibrosis because their cells have plasma membranes with enough CFTR to transport chloride ions normally. The *ΔF508* allele may offer these individuals an advantage in surviving certain deadly infectious diseases. CFTR's receptor function is an essential part of the immune response to bacteria in the respiratory tract. However, the same function allows bacteria to enter cells of the gastrointestinal tract, where they can be deadly. Cells lacking CFTR do not take up these bacteria. Thus, people who carry *ΔF508* are probably less susceptible to bacterial diseases that begin in the intestinal tract.

CREDITS: (16) from the left, Courtesy of ©Steve & Ellison Widener and Breathe Hope, http//breathehope.tamu.edu; Courtesy of ©The Family of Savannah Brooke Snider; Courtesy of the Family of Benjamin Hill, reprinted with permission of © Chappell/Marathonfoto; Courtesy of © Bobby Brooks and the Family of Jeff Baird; Courtesy of © The Family of Brandon Herriott; Courtesy of © The Cody Dieruf Benefit Foundation, www.breathinisbelievin.org.

Section 13.1 Gregor Mendel indirectly discovered the role of genes in inheritance by breeding and tracking traits of pea plants. **Genotype** (an individual's alleles) is the basis of **phenotype** (observable traits). Each gene occurs at a **locus** on a chromosome. A **homozygous** individual has the same allele of a gene on both homologous chromosomes. A **heterozygous** individual, or **hybrid**, has two different alleles. A **dominant** allele masks the effect of a **recessive** allele in a heterozygous individual.

Section 13.2 Crossing two individuals that breed true for different forms of a trait yields identically heterozygous offspring. A cross between such offspring is called a **monohybrid cross**. The frequency at which the traits appear in offspring of a **testcross** can reveal the genotype of an individual with a dominant phenotype. **Punnett squares** are useful for determining the probability of offspring genotype and phenotype. Mendel's monohybrid cross results led to his **law of segregation**, which we state here in modern terms: A diploid cell has two copies of every gene that occurs on its homologous chromosomes. Two alleles at any locus separate from each other during meiosis, so they end up in different gametes.

Section 13.3 Crossing individuals that breed true for two forms of two traits yields F_1 offspring identically heterozygous for alleles governing those traits. A cross between such offspring is a **dihybrid cross**. The frequency at which the two traits appear in F_2 offspring can reveal dominance relationships of the alleles. Mendel's dihybrid cross results led to his **law of independent assortment**, which we state here in modern terms: Alleles at one gene locus tend to sort into gametes independently of alleles at other loci. Crossovers can break up parental combinations of alleles in a **linkage group**.

Section 13.4 With **incomplete dominance**, the phenotype of heterozygous individuals is an intermediate blend of the two homozygous phenotypes. With **codominant** alleles, heterozygous individuals have both homozygous phenotypes. Codominance may occur in **multiple allele systems** such as the one underlying ABO blood type. With **epistasis**, two or more genes affect the same trait. A **pleiotropic** gene affects multiple traits.

Section 13.5 Changes in phenotype are part of an individual's ability to adapt to its environment. Such changes are an effect of environmental cues that alter gene expression (for example by methylating DNA).

Section 13.6 A trait that is influenced by multiple genes often occurs in a range of small increments of phenotype called **continuous variation**. Continuous variation typically occurs as a **bell curve** in the range of values. Multiple alleles such as those that arise in regions of **short tandem repeats** can give rise to continuous variation.

Application Symptoms of cystic fibrosis are pleiotropic effects of mutations in the *CFTR* gene. The allele associated with most cases persists at high frequency despite its devastating effects in homozygous people.

SELF-ASSESSMENT
ANSWERS IN APPENDIX VII

1. A heterozygous individual has _____ for a trait being studied.
   - a. the same allele on both homologous chromosomes
   - b. two different alleles of a gene
   - c. a haploid condition, in genetic terms

2. An organism's observable traits constitute its _____ .
   - a. phenotype   c. genotype
   - b. variation   d. pedigree

3. In genetics, F stands for filial, which means _____ .
   - a. friendly   c. final
   - b. offspring   d. hairlike

4. The offspring of the cross $AA \times aa$ are _____ .
   - a. all AA c. all Aa
   - b. all aa d. half are AA and half are aa

5. The second-generation offspring of a cross between individuals who are homozygous for different alleles of a gene are called the _____ .
   - a. F_1 generation   c. hybrid generation
   - b. F_2 generation   d. none of the above

6. Refer to question 4. Assuming complete dominance, the F_2 generation will show a phenotypic ratio of _____ .
   - a. 3:1   b. 9:1   c. 1:2:1   d. 9:3:3:1

7. Independent assortment means _____ .
   - a. alleles at one locus assort into different gametes
   - b. alleles at different loci tend to assort into gametes independently of each other

8. A testcross is a way to determine _____ .
   - a. phenotype   b. genotype   c. both a and b

9. Assuming complete dominance, crosses between two dihybrid F_1 pea plants, which are offspring from a cross $AABB \times aabb$, result in F_2 phenotype ratios of _____ .
   - a. 1:2:1   b. 3:1   c. 1:1:1:1   d. 9:3:3:1

10. The probability of a crossover occurring between two genes on the same chromosome _____ .
    - a. is unrelated to the distance between them
    - b. decreases with the distance between them
    - c. increases with the distance between them

11. True or false? All traits are inherited in a Mendelian pattern.

12. One gene that affects three traits is an example of _____ .
    - a. dominance   c. pleiotropy
    - b. codominance   d. epistasis

13. _____ in a trait is indicated by a bell curve.

CARRYING THE CYSTIC FIBROSIS ALLELE OFFERS PROTECTION FROM TYPHOID FEVER Epithelial cells that lack the CFTR protein cannot take up bacteria by endocytosis. Endocytosis is an important part of the respiratory tract's immune defenses against common *Pseudomonas* bacteria, which is why *Pseudomonas* infections of the lungs are a chronic problem in cystic fibrosis patients. Endocytosis is also the way that *Salmonella typhi* enter cells of the gastrointestinal tract, where internalization of this bacteria can result in typhoid fever.

Typhoid fever is a common worldwide disease. Its symptoms include extreme fever and diarrhea, and the resulting dehydration causes delirium that may last several weeks. If untreated, it kills up to 30 percent of those infected. Around 600,000 people, most of whom are children, die annually from typhoid fever.

In 1998, Gerald Pier and his colleagues compared the uptake of *S. typhi* (pictured at right) by different types of epithelial cells: those homozygous for the normal allele, and those heterozygous for the ΔF508 allele associated with CF. (Cells that are homozygous for the mutation do not take up any *S. typhi* bacteria.) Some of the results are shown in **Figure 13.17**.

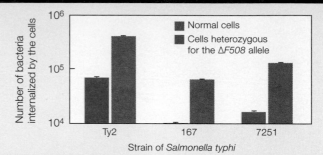

Figure 13.17 Effect of the ΔF508 mutation on the uptake of three different strains of *Salmonella typhi* bacteria by epithelial cells.

1. Regarding the Ty2 strain of *S. typhi*, about how many more bacteria were able to enter normal cells than cells heterozygous for the ΔF508 allele?

2. Which strain of bacteria entered normal epithelial cells most easily?

3. Entry of all three *S. typhi* strains into the heterozygous epithelial cells was inhibited. Is it possible to tell which strain was most inhibited?

---

14. The phenotype of individuals heterozygous for _____ alleles comprises both homozygous phenotypes.
    a. epistatic   c. pleiotropic
    b. codominant   d. hybrid

15. Match the terms with the best description.
    ___ dihybrid cross   a. *bb*
    ___ monohybrid cross   b. *AaBb* × *AaBb*
    ___ homozygous condition   c. *Aa*
    ___ heterozygous condition d. *Aa* × *Aa*

GENETICS PROBLEMS
ANSWERS IN APPENDIX VII

1. Mendel crossed a true-breeding pea plant with green pods and a true-breeding pea plant with yellow pods. All offspring had green pods. Which color is recessive?

2. Assuming that independent assortment occurs during meiosis, what type(s) of gametes will form in individuals with the following genotypes?
    a. *AABB*   b. *AaBB*   c. *Aabb*   d. *AaBb*

3. Determine the predicted genotype frequencies among the offspring of an *AABB* × *aaBB* cross.

4. Heterozygous individuals perpetuate some alleles that have lethal effects in homozygous individuals. A mutated allele (M^L) associated with taillessness in Manx cats (left) is an example. Cats homozygous for this allele (M^LM^L) typically die before birth due to severe spinal cord defects. In a case of incomplete dominance, cats heterozygous for the M^L allele and the normal, unmutated allele (M) have a short, stumpy tail or none at all. Two M^LM cats mate. What is the probability that any one of their surviving kittens will be heterozygous?

5. A single allele gives rise to the HbS form of hemoglobin. Individuals who are homozygous for the allele (*HbS/HbS*) develop sickle-cell anemia (Section 9.5). Heterozygous individuals (*HbA/HbS*) have few symptoms. A couple who are both heterozygous for the *HbS* allele plan to have children. For each of the pregnancies, determine the probability that their child will be:
    a. homozygous for the *HbS* allele
    b. homozygous for the normal allele (*HbA*)
    c. heterozygous: *HbA/HbS*

6. In sweet pea plants, an allele for purple flowers (*P*) is dominant when paired with a recessive allele for red flowers (*p*). An allele for long pollen grains (*L*) is dominant when paired with a recessive allele for round pollen grains (*l*). Bateson and Punnett crossed a plant having purple flowers/long pollen grains with one having white flowers/round pollen grains. All F₁ offspring had purple flowers and long pollen grains. Among the F₂ generation, the researchers observed the following phenotypes:
    296 purple flowers/long pollen grains
    19 purple flowers/round pollen grains
    27 red flowers/long pollen grains
    85 red flowers/round pollen grains
What is the best explanation for these results?

for additional quizzes, flashcards, and other study materials
▶ ACCESS MINDTAP AT WWW.CENGAGEBRAIN.COM

CREDITS: (in text) Genetics Problems #3, © Leslie Faltheisek/Clacritter Manx; Data Analysis inset, © Gary Gaugler/The Medical File/Peter Arnold, Inc.

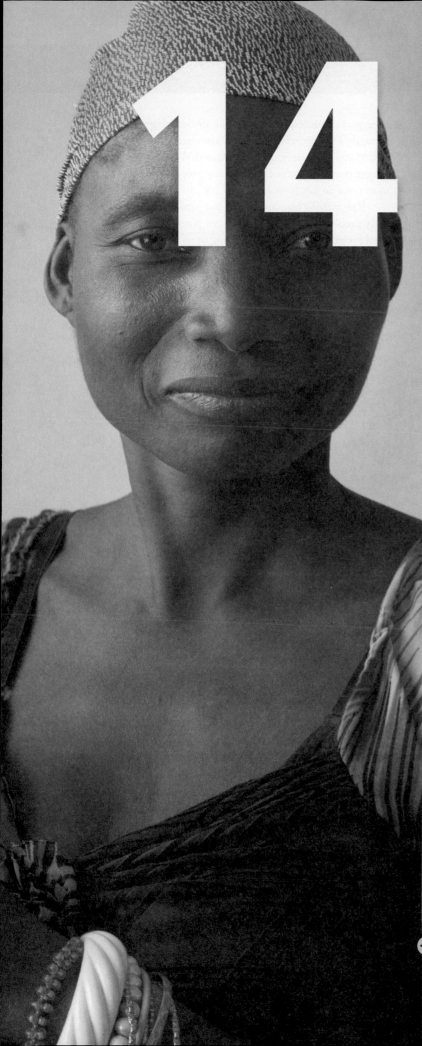

14

Human Inheritance

Links to Earlier Concepts

Be sure you understand dominance relationships (Sections 13.1, 13.4, and 13.5), gene expression (9.1, 9.2), and mutations (9.5). You will use your knowledge of chromosomes (8.3), DNA replication and repair (8.4, 8.5), meiosis (12.2, 12.3), and sex determination (10.3). Sampling error (1.7), proteins (3.4), cell components (4.5, 4.9, 4.10), transport proteins (5.7), pigments (6.2), telomeres (11.4), and oncogenes (11.5) will turn up again.

Core Concepts

Living systems store, retrieve, transmit, and respond to information essential for life.

A cell's DNA encodes all of the instructions necessary for growth, survival, and reproduction of the cell and (in multicelled organisms) the individual. The chromosomal basis of inheritance explains patterns in which genetic information is transmitted to offspring. Some human traits arise from single genes. The inheritance pattern of a single-gene trait can be revealed by tracking the appearance of the trait's alternative forms in individuals through generations of family trees.

Interactions among the components of each biological system give rise to complex properties.

Changes in genotype can affect human phenotype. Mutations in single genes give rise to many genetic disorders, and errors in mitosis or meiosis that change chromosome number or structure often result in developmental disorders. Growth, reproduction, and homeostasis depend on proper timing and coordination of specific molecular events. Mutations that alter expression of genes involved in these events may have complex effects on the cell, tissue, or whole organism.

Evolution underlies the unity and diversity of life.

Individuals of sexually reproducing species vary in the details of their DNA. Such variation is almost always advantageous, because genetically diverse populations are more resilient than populations with low diversity. Genetic diversity is fostered by processes inherent in sexual reproduction.

◀ This family lives in Tanzania, where exposure to intense sunlight is responsible for the skin cancer that kills almost everyone with the albino phenotype. An abnormally low amount of melanin leaves people with this trait defenseless against UV radiation in sunlight.

221

AP Photo/Jacquelyn Martin.

14.1 Human Chromosomes

Relatively few human traits follow a Mendelian inheritance pattern. Like the flower color of Mendel's pea plants, such traits arise from a single gene with alleles that have a clear dominance relationship. Consider the *MC1R* gene that you learned about in Section 13.4. In humans, dogs, and other animals, *MC1R* encodes a protein that triggers production of the brownish melanin. Mutations can result in a defective protein; an allele with one of these mutations is recessive when paired with an unmutated allele. A person who has two mutated *MC1R* alleles does not make the brownish melanin—only the reddish type—so this individual has red hair.

Red hair is fairly common in human populations, but other single-gene traits (one gene → one trait) are not. Most human traits are inherited in patterns much more complex than the Mendelian model (Section 13.4). Human skin color, for example, varies continuously (below) because it is an outcome of interactions among the products of about 350 genes.

their genetic underpinnings remain unclear despite decades of research. For example, mutations associated with an increased risk of autism (a developmental disorder) have been found on almost every chromosome, but most people who carry these mutations do not have autism.

Researching human inheritance patterns is also inherently problematic. Consider how pea plants and fruit flies are ideal for genetics research. Breeding them in a controlled manner poses few ethical problems. They reproduce quickly, so it does not take long to follow a trait through many generations. Humans, however, live under variable conditions, in different places, and we live as long as the geneticists who study our inheritance patterns. Most of us select our own mates and reproduce if and when we want to. Human families are relatively small, so sampling error (Section 1.7) is unavoidable.

Because of these challenges, geneticists often use historical records to track inherited disorders through generations of a family. They make and study **pedigrees**, which are standardized charts that illustrate the phenotypes of family members and genetic connections among them (**Figure 14.1**). Analysis of a pedigree can reveal whether a single-gene disorder is associated with a dominant or recessive allele, and

Our understanding of inheritance patterns in humans comes mainly from research involving genetic disorders, because this information helps us develop treatments for affected people. Note that a rare or uncommon version of a trait, such as having six fingers on a hand, or a web between two toes, is called a genetic abnormality. Heritable abnormalities are not dangerous, and how you view them is a matter of opinion. By contrast, genetic disorders sooner or later cause medical problems that may be quite severe.

Even though single-gene traits are the least common kind in humans, we actually know most about inheritance of single-gene disorders. This is partly an outcome of relative complexity. Genetic disorders usually have multiple symptoms (a syndrome) that may vary among affected individuals. Environmental effects on phenotype (Section 13.5) further complicate inheritance patterns. Diabetes, asthma, obesity, cancers, heart disease, multiple sclerosis, and many other disorders can be inherited, but in patterns so complex that

whether the allele is on an autosome or a sex chromosome (Section 8.3). Pedigree analysis also allows geneticists to estimate the probability that a disorder will reappear in future generations of a family or a population.

Alleles that give rise to severe genetic disorders are generally rare in populations because they compromise the health and reproductive ability of their bearers. Why do they persist? Inevitable mutations periodically reintroduce them. In some cases, a codominant allele persists because it offers a survival advantage in a particular environment. You learned about one example, the ΔF508 allele that causes cystic fibrosis, in Chapter 13: People heterozygous for this allele are protected from infection by bacteria that cause typhoid fever. You will encounter additional examples in later chapters.

Single-gene disorders collectively affect about 1 in 100 people; **Table 14.1** introduces some that arise from genes on autosomes. Appendix IV shows the loci of several genes known to be associated with disorders and abnormalities.

CREDITS: (in text) Sarah Leen/National Geographic Creative.

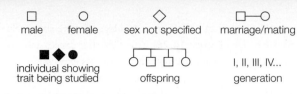

male female sex not specified marriage/mating

individual showing trait being studied   offspring   I, II, III, IV... generation

A Standard symbols used in pedigrees.

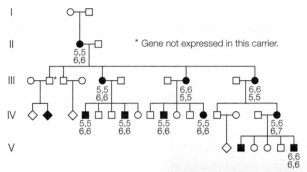

I

II
5,5
6,6
* Gene not expressed in this carrier.

III
5,5
6,6
6,6
5,5
6,6
5,5

IV
5,5
6,6
5,5
6,6
5,5
6,6
5,5
6,6
5,6
6,7

V
6,6
6,6

B Above, a pedigree for poly-dactyly, a genetic abnormality characterized by extra fingers (right), toes, or both. The number of fingers on each hand is indicated in black; toes on each foot, in red. Polydactyly on its own is often inherited in an autosomal dominant pattern. When part of a syndrome (such as Ellis–van Creveld syndrome), it can be inherited in an autosomal recessive pattern (more about these patterns in the next section).

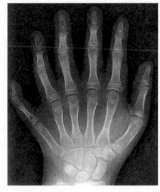

Figure 14.1 Pedigrees.

TABLE 14.1

Some Autosomal Abnormalities and Disorders

Disorder/Abnormality	Main Symptoms
Autosomal dominant inheritance pattern	
Achondroplasia	One form of dwarfism
Aniridia	Defects of the eyes
Huntington's disease	Degeneration of the nervous system
Marfan syndrome	Abnormal or missing connective tissue
Progeria	Drastic premature aging
Autosomal recessive inheritance pattern	
Albinism	Absence of pigmentation
Cystic fibrosis	Difficulty breathing; lung infections
Ellis–van Creveld syndrome	Dwarfism, heart defects, polydactyly
Phenylketonuria (PKU)	Mental impairment
Sickle-cell anemia	Anemia, swelling, frequent infections
Tay–Sachs disease	Deterioration of mental and physical abilities; early death

pedigree Chart of family connections that shows the appearance of a phenotype through generations.

Dr. Nancy Wexler

The village of Barranquitas, Venezuela, has the highest incidence of Huntington's disease in the world. Huntington's is an incurable, fatal hereditary disorder in which voluntary muscle control gradually gives way to involuntary jerking, twitching, and writhing movements. Eventually, serious problems with swallowing cause many patients to die from choking or malnutrition. Beyond the physical symptoms, deep depression can often take hold.

Huntington's affects 1 in 10,000 people worldwide, but in Barranquitas the rate is more like 1 in 10. Some 1,000 villagers already have full-blown Huntington's; many more carry the gene. Such a high concentration of Huntington's patients made this region the backbone of Nancy Wexler's research. Wexler has been coming here for more than 30 years to study the genetics behind the disorder.

This is more than an academic pursuit or a career goal for her. "My mother died of Huntington's and she was a scientist. My father was a scientist too, and so we said, 'Let's find a cure.'" Wexler and her colleagues collected DNA from and compiled an extended pedigree for nearly 10,000 Venezuelans. Her research was critical to the discovery that a dominant allele on human chromosome 4 causes Huntington's. As the daughter of a Huntington's sufferer, she has a one-in-two chance of carrying the fatal genetic flaw herself.

In 1999, Wexler cofounded a care home for Huntington's sufferers near Lake Maracaibo in Venezuela. The home, Casa Hogar, is a haven for more than 50 people whose families can no longer cope. Casa Hogar is facing a chronic lack of funding, possibly even closure. Still, Wexler remains confident that one day it won't be needed. "We never know when some miraculous discovery is going to be made," she said. "There are science breakthroughs on the horizon and happening now so I am very hopeful about the cure in the near future."

LEARNING OBJECTIVE

Using appropriate examples, explain and diagram the autosomal dominant and autosomal recessive inheritance patterns.

AUTOSOMAL DOMINANCE

A trait associated with a dominant allele on an autosome appears in people who are heterozygous for it as well as those who are homozygous. Such traits appear in every generation of a family, and they occur with equal frequency in both sexes. When one parent is heterozygous for a dominant allele, and the other is homozygous for the recessive allele, each of their children has a 50 percent chance of inheriting the dominant allele and having the associated trait (**Figure 14.2A**).

Achondroplasia A form of hereditary dwarfism called achondroplasia offers an example of an autosomal dominant disorder (one caused by a dominant allele on an autosome). Mutations associated with achondroplasia occur in a gene for a growth factor receptor (Section 11.5), the activity of which inhibits growth and differentiation of cells that give rise to bone. The mutations interfere with the normal cellular process that recycles the receptor after it is no longer needed. Cells end up with receptors when they should not have them, so bone growth is inhibited when it should not be. About 1 in 10,000 people is heterozygous for one of these mutations. As adults, affected people are, on average, about four feet, four inches (1.3 meters) tall, with arms and legs that are short relative to torso size (**Figure 14.2B**). An allele that causes achondroplasia can be passed to children because its expression does not interfere with reproduction, at least in heterozygous people. The homozygous condition results in severe skeletal malformations that are lethal before birth or shortly after.

Huntington's Disease Alleles that cause Huntington's disease are also inherited in an autosomal dominant pattern. The disease is caused by expansions of a short tandem repeat (Section 13.6) in a gene for a cytoplasmic protein (the many functions of this protein are still being discovered). The protein encoded by the altered gene has a section in which the amino acid glutamine is repeated thirty or more times. This region causes the protein to misfold, and also to resist cellular processes that normally dismantle and recycle defective proteins. The misfolded protein accumulates to high levels, particularly in brain cells involved in movement, thinking, and emotion. Cellular functioning is hampered, and a stress response is triggered that eventually causes the cell to die. Glutamine-rich protein fragments accumulate in amyloid fibrils, so amyloid plaques form in the brain (Section 3.5). In the most common form of Huntington's, symptoms do not start until after the age of thirty, and affected people die during their forties or fifties. With this and other late-onset disorders, people tend to reproduce before symptoms appear, so the allele is often passed unknowingly to children.

Figure 14.2 Autosomal dominant inheritance.

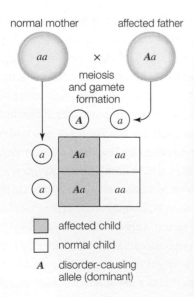

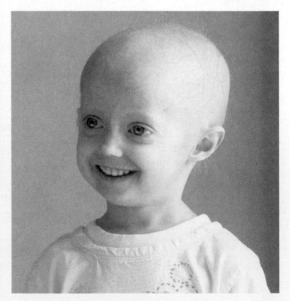

A A dominant allele (red) on an autosome affects heterozygous people.

B Achondroplasia affects Ivy Broadhead (left), as well as her brother, father, and grandfather.

C Symptoms of Hutchinson–Gilford progeria are already apparent in Megan, age 5.

Hutchinson–Gilford Progeria Drastically accelerated aging characterizes an autosomal dominant disorder called Hutchinson–Gilford progeria. A mutation in the gene for lamin A is the cause. Lamins are protein subunits of intermediate filaments that support the nuclear envelope (Section 4.9). They also act as a bridge between the inner nuclear membrane and chromosomes, with roles in mitosis, DNA synthesis and repair, and transcriptional regulation. The progeria mutation is a single base-pair substitution that adds a signal for a splice site (Section 9.2), resulting in a polypeptide that is too short and cannot be processed correctly after translation. The defective protein binds abnormally to other molecules that usually interact with lamin A, and interferes with their function. Effects are pleiotropic. Cells that carry the mutation have a nucleus that is grossly abnormal, with nuclear pore complexes that do not assemble properly and membrane proteins localized to the wrong side of the nuclear envelope. The function of the nucleus as protector of chromosomes is severely impaired, and DNA damage accumulates quickly. Abnormal DNA methylation patterns and unusually short telomeres result in early senescence of the individual's cells (Section 11.4).

Outward symptoms begin to appear before age two, as skin that should be plump and resilient starts to thin, muscles weaken, and bones soften. Premature baldness is inevitable (**Figure 14.2C**). Most people with the disorder die in their early teens as a result of a stroke or heart attack brought on by hardened arteries, a condition typical of advanced age. Progeria does not run in families because affected people do not live long enough to reproduce.

THE AUTOSOMAL RECESSIVE PATTERN

A recessive allele on an autosome is expressed only in homozygous individuals; heterozygous individuals are called carriers because they have the allele but not the associated trait. These traits appear in both sexes at equal frequency, and they tend to skip generations. Any child of two carriers has a 25 percent chance of inheriting the allele from both parents and developing the trait (**Figure 14.3A**).

Tay–Sachs Disease Alleles associated with Tay–Sachs disease are inherited in an autosomal recessive pattern. In the general population, about 1 in 300 people is a carrier for one of these alleles, but the incidence is ten times higher in Jews of eastern European descent and some other groups. The gene altered in Tay–Sachs encodes a lysosomal enzyme responsible for breaking down a particular type of lipid. Mutations cause this enzyme to misfold and become destroyed, so cells make the lipid but cannot break it down.

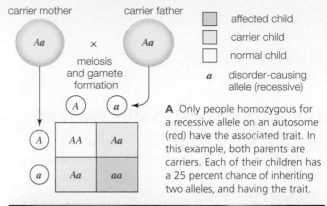

A Only people homozygous for a recessive allele on an autosome (red) have the associated trait. In this example, both parents are carriers. Each of their children has a 25 percent chance of inheriting two alleles, and having the trait.

B Conner Hopf was diagnosed with Tay–Sachs disease at age 7½ months. He died before his second birthday.

Figure 14.3 Autosomal recessive inheritance.

Typically, newborns homozygous for a Tay–Sachs allele seem normal, but within three to six months they become irritable, listless, and may have seizures as the lipid accumulates in their nerve cells. Blindness, deafness, and paralysis follow. Affected children usually die by age five (**Figure 14.3B**).

Albinism Albinism, a phenotype characterized by an abnormally low level of the pigment melanin, is inherited in an autosomal recessive pattern. Mutations associated with albinism affect proteins involved in melanin synthesis. Skin, hair, or eye pigmentation may be reduced or missing. In the most dramatic form, the skin is very white and does not tan, and the hair is white (as shown in the chapter opening photo). The irises of the eyes appear red because the lack of pigment allows underlying blood vessels to show through. Melanin also plays a role in the retina, so vision problems are typical. In skin, melanin acts as a sunscreen; without it, the skin is defenseless against UV radiation. Thus, people with the albino phenotype have a very high risk of skin cancer.

Using appropriate examples, explain and diagram the X-linked recessive inheritance pattern.

Explain why genetic disorders associated with an X chromosome allele are more common in males than in females.

THE X-LINKED RECESSIVE PATTERN

Traits inherited in an X-linked pattern (**Table 14.2**) arise from genes on the X chromosome. In most cases, X chromosome alleles that cause genetic disorders are recessive, and these leave two inheritance clues. First, an affected father never passes an X-linked recessive allele to a son, because all children who inherit their father's X chromosome are female. Only mothers can pass the allele to a son (**Figure 14.4A**). Second, the disorder appears more often in males than females. Having only one X chromosome, a male must inherit only one allele to be affected by the disorder; a female must inherit two, and inheriting two disorder-causing alleles is statistically less common than inheriting one.

An X-linked allele can have uneven effects in heterozygous females because of X chromosome inactivation (Section 10.3). About half of the cells making up the body of a heterozygous female express a recessive X-linked allele on one X chromosome; the other half express the dominant allele on the other X chromosome. Mild symptoms may appear if the

TABLE 14.2

Some X-Linked Traits

Disorder/Abnormality	Main Symptoms
X-linked recessive inheritance pattern	
Androgen insensitivity syndrome	XY individual but having some female traits; sterility
Red–green color blindness	Inability to distinguish red from green
Hemophilia	Impaired blood clotting ability
Muscular dystrophies	Progressive loss of muscle function
X-linked anhidrotic dysplasia	Mosaic skin (patches with or without sweat glands); other ill effects
X-linked dominant inheritance pattern	
Fragile X syndrome	Intellectual, emotional disability
Incontinentia pigmenti	Abnormalities of skin, hair, teeth, nails, eyes; neurological problems

effect of the recessive allele on her body is not fully masked by that of the dominant allele.

Red–Green Color Blindness Color blindness refers to a range of conditions in which an individual cannot distinguish among some or all colors of visible light. These conditions are typically inherited in an X-linked recessive pattern, because most of the genes involved in color vision are on the X chromosome. Humans normally sense the differences among 150 colors, and this perception depends on pigment-

Figure 14.4 X-linked recessive inheritance.

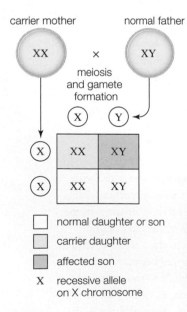

carrier mother   normal father

XX   ×   XY

meiosis and gamete formation

X   Y

	XX	XY
X	XX	XY
X	XX	XY

☐ normal daughter or son
▢ carrier daughter
▦ affected son
X recessive allele on X chromosome

A In this example, the mother carries a recessive allele on one of her two X chromosomes (red).

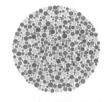

You may have one form of red–green color blindness if you see a 7 in this circle instead of a 29.

You may have another form of red–green color blindness if you see a 3 instead of an 8 in this circle.

B A view of color blindness, a trait inherited in an X-linked recessive pattern. The photo on the left shows how a person with red–green color blindness sees the photo on the right. The perception of blues and yellows is normal; red and green appear similar. The circle diagrams are part of a standardized test for color blindness. A set of 38 of these diagrams is commonly used to diagnose deficiencies in color perception.

containing receptors in the eyes. Mutations that result in altered or missing receptors affect color vision. For example, people who have red–green color blindness see fewer than 25 colors because receptors that respond to red and green light are weakened or absent (**Figure 14.4B**).

Duchenne Muscular Dystrophy

Progressive muscle degeneration characterizes a severe genetic disorder called Duchenne muscular dystrophy (DMD). The disorder is caused by mutations in the X chromosome gene for dystrophin, a rod-shaped, flexible protein found mainly in skeletal and heart (cardiac) muscle. Dystrophin connects contractile units in the cytoplasm of muscle cells to proteins that anchor the cell in extracellular matrix (Section 4.10). The flexible link imparts strength to muscle tissue and protects muscle cell membranes from mechanical damage during contraction.

Mutations associated with DMD result in defective or missing dystrophin. Without dystrophin, muscle cell plasma membranes are easily damaged during contraction, and the cells become flooded with calcium ions. Calcium ions act as potent messengers in cells, so their concentration in cytoplasm is normally kept very low (Section 5.7). Among other negative effects, a chronic calcium ion overload causes mitochondrial malfunction (Chapter 7 Application). Muscle cell mitochondria produce too little ATP to support normal cellular function, so muscles are abnormally weak. Free radicals accumulate and kill the cells. Eventually, the tissue's capacity to regenerate is overwhelmed by rapid cell turnover, and it becomes replaced by fat and connective tissue.

DMD affects about 1 in 3,500 people, almost all of them boys. Symptoms begin around age four and progress very quickly. Anti-inflammatory drugs can slow the progression of DMD, but there is no cure. When an affected boy is about ten years old, he will begin to need a wheelchair and his heart will start to fail. Even with the best care, he will probably die before the age of thirty, from a heart disorder or respiratory failure.

Hemophilia

Hemophilias are genetic disorders in which the blood does not clot properly. Most people have a blood clotting mechanism that quickly stops bleeding from minor injuries. That mechanism involves two proteins, clotting factors VIII and IX, both products of X chromosome genes. Mutations in these two genes cause two types of hemophilia (A and B, respectively). Males who carry one of these mutations have prolonged bleeding, as do homozygous females (heterozygous females make about half the normal amount of the clotting factor, but this is generally enough to sustain normal clotting). Affected people bruise easily, but internal bleeding is their most serious problem. Repeated bleeding inside the joints disfigures them and causes chronic arthritis.

In the nineteenth century, the incidence of hemophilia was relatively high among European and Russian royals (**Figure 14.5**), in part because a centuries-old practice of inbreeding among close relatives kept the allele circulating in royal families. Today, about 1 in 7,500 people in the general population is affected. That number may be rising because the disorder is now treatable, so more affected people are living long enough to transmit a mutated allele to children.

Figure 14.5 A classic case of X-linked recessive inheritance. Hemophilia B afflicted many descendants of Queen Victoria of England. The disease was caused by a rare X-linked allele that most likely arose from a spontaneous germline mutation.

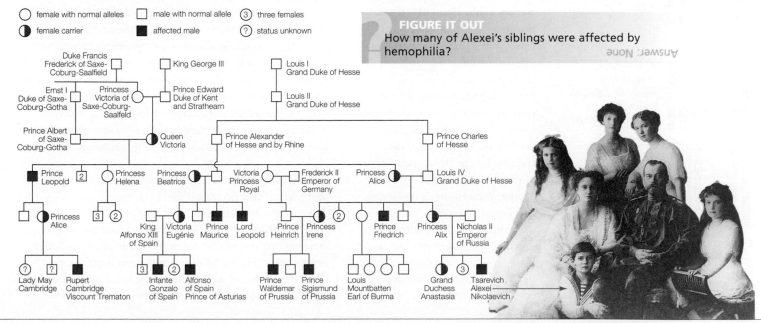

FIGURE IT OUT
How many of Alexei's siblings were affected by hemophilia?
Answer: None

14.4 Changes in Chromosome Structure

List some causes of large-scale structural changes in chromosomes.

Describe the main types of large-scale structural changes in chromosomes, and explain their potential consequences.

Give an example of structural change that has occurred in chromosomes during evolution.

Mutation is a term that usually refers to small-scale changes in DNA sequence—one or a few nucleotides. Chromosome changes on a larger scale also occur. Like mutations, some of these changes are induced by exposure to chemicals or radiation. Others are an outcome of faulty crossing over during prophase I of meiosis. For example, nonhomologous chromosomes sometimes swap segments at spots where the DNA sequence is similar. Homologous chromosomes also may misalign along their length. In both cases, crossing over results in the exchange of segments that are not equivalent.

Figure 14.6 Some types of major change in chromosome structure.

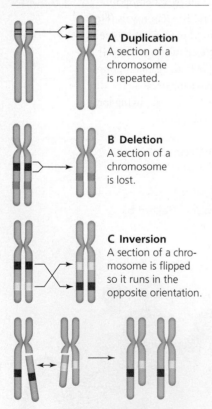

A Duplication
A section of a chromosome is repeated.

B Deletion
A section of a chromosome is lost.

C Inversion
A section of a chromosome is flipped so it runs in the opposite orientation.

D Translocation A piece of a broken chromosome is reattached in the wrong place. This example shows a reciprocal translocation, in which two nonhomologous chromosomes exchange chunks.

TYPES OF CHANGE

Large-scale changes in chromosome structure are categorized by the type of change. Any type can have a drastic effect on health (**Table 14.3**); about half of all miscarriages are due to chromosome abnormalities of the developing embryo.

Insertions Large insertions often result from the activity of **transposable elements** (**transposons**), which are segments of DNA hundreds or thousands of nucleotides long that can move spontaneously ("jump") within or between chromosomes. Transposons are common in the chromosomes of all species. The jumping mechanism involves enzymes encoded by the DNA sequence of the transposon itself, or by another transposon. A transposon that jumps into a gene becomes a major insertion that destroys the function of the gene product. In most cases the location of insertion is random; the effect depends on the gene interrupted.

TABLE 14.3

A Few Effects of Human Chromosome Structure Change

Cause	Chromosome	Disorder/Abnormality
Duplication	X	Fragile X syndrome
Duplication	4	Huntington's disease
Deletion	5	Cri-du-chat syndrome
Deletion	15	Prader–Willi syndrome
Inversion	X	Hemophilia A
Translocation	8⟷14	Burkitt lymphoma
Translocation	9⟷22	Chronic myelogenous leukemia (CML)

Duplications Even normal chromosomes have DNA sequences that are repeated two or more times. These repetitions are called **duplications** (**Figure 14.6A**). Some newly occurring duplications, such as the expansion mutations that cause Huntington's disease, cause genetic abnormalities or disorders. Others, as you will see shortly, have been evolutionarily important.

Deletions Large-scale deletions (**Figure 14.6B**) have severe consequences. Duchenne muscular dystrophy usually arises from deletions in the X chromosome. A deletion in chromosome 5 causes cri-du-chat syndrome, named after the sounds that affected infants make as a result of an abnormally shaped larynx (cri-du-chat is French for "cat's cry"). Impaired mental functioning and a short life span are other symptoms.

Inversions Sometimes a segment of a chromosome breaks off and gets reattached backwards, with no loss of DNA (**Figure 14.6C**). This type of structural change, an **inversion**, may not affect a carrier's health if the break point does not occur in a gene or control region, because the individual's cells still contain their full complement of genetic material. However, fertility may be compromised because a chromosome with an inversion does not pair properly with its homologous partner during meiosis. Crossovers between the mispaired chromosomes produce other chromosome abnormalities that reduce the viability of forthcoming embryos. Some people learn that they carry an inversion when their karyotype is checked because of fertility problems.

Translocations If a chromosome breaks, the broken part may get attached to a different chromosome, or to a different part of the same one. This type of structural change is called a **translocation**. Most translocations are reciprocal, meaning that two nonhomologous chromosomes exchange broken parts (**Figure 14.6D**). A reciprocal translocation between chromosomes 8 and 14 is the usual cause of Burkitt lymphoma, an aggressive cancer of the immune system.

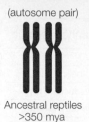 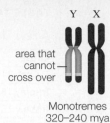 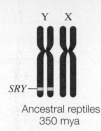 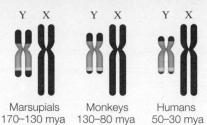

Ancestral reptiles >350 mya

Ancestral reptiles 350 mya

SRY

area that cannot cross over

Monotremes 320–240 mya

Marsupials 170–130 mya

Monkeys 130–80 mya

Humans 50–30 mya

A Before 350 mya, sex was determined by temperature, not by chromosome differences.

B The *SRY* gene begins to evolve 350 mya. It hampers crossing over, and the homologous chromosomes diverge as other mutations accumulate.

C By 320–240 mya, the DNA sequences of the chromosomes are so different that the pair can no longer cross over in one region. The Y chromosome begins to shorten.

D Three more times, the pair stops crossing over in yet another region. Each time, the DNA sequences of the chromosomes diverge, and the Y chromosome shortens. Today, the X and Y can cross over only at a small region near the ends.

Figure 14.7 Evolution of the Y chromosome. Today, the *SRY* gene on the Y chromosome determines male sex, but this was not always the case. Homologous regions are pink; mya, million years ago. Monotremes are egg-laying mammals; marsupials are pouched mammals.

This translocation moves a proto-oncogene to a region that is vigorously transcribed in white blood cells, resulting in uncontrolled cell divisions that are characteristic of cancer (Section 11.5). Many other reciprocal translocations have no adverse effects on health, but, like inversions, they can compromise fertility. During meiosis, translocated chromosomes pair abnormally and segregate improperly; about half of the resulting gametes carry major duplications or deletions. If one of these gametes unites with a normal gamete at fertilization, the resulting embryo almost always dies. As with inversions, people who carry a translocation may not find out about it until they have difficulty with fertility.

CHROMOSOMES AND EVOLUTION

There is evidence of major structural alterations in the chromosomes of all known species. For example, duplications have often allowed a copy of a gene to mutate while the original carried out its unaltered function. The multiple and strikingly similar globin chain genes of mammals apparently evolved this way. Globin chains, remember, associate to form molecules of hemoglobin (Section 9.5). Two identical genes for the alpha chain—and five other slightly different versions of it—form a cluster on chromosome 16. Five slightly different forms of the beta chain gene cluster on chromosome 11.

Consider the human X and Y chromosomes, which were once homologous autosomes in ancient, reptile-like ancestors of mammals (**Figure 14.7**). Ambient temperature probably determined the gender of those organisms, as it still does in turtles and some other modern reptiles. About 350 million years ago, a gene on one of the two homologous chromosomes mutated. The mutation interfered with crossing over during meiosis, and it was the beginning of the male sex determination gene *SRY* (Section 10.3). A reduced frequency of crossing over allowed mutations to accumulate separately

in the two chromosomes, so they began to diverge around the changed region. Over evolutionary time, the chromosomes became so different that they no longer crossed over at all in the changed region, so they diverged even more. Today, the Y chromosome is much smaller than the X, and is homologous with it only in a tiny part. The Y completes meiosis by crossing over with itself, translocating duplicated regions of its own DNA.

Some chromosome structure changes contributed to differences among closely related organisms, such as apes and humans. Human somatic cells have twenty-three pairs of chromosomes; cells of chimpanzees, gorillas, and orangutans have twenty-four. Thirteen human chromosomes are almost identical with chimpanzee chromosomes. Nine more are similar, except for some inversions. One human chromosome matches up with two in chimpanzees and the other great apes (**Figure 14.8**). During human evolution, two chromosomes evidently fused end to end and formed our chromosome 2. How do we know? The region where the fusion occurred contains remnants of a telomere (Section 11.4).

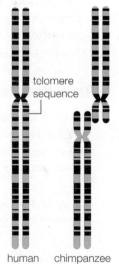

telomere sequence

human   chimpanzee

Figure 14.8 Human chromosome 2 compared with chimpanzee chromosomes 2A and 2B.

duplication Repeated section of a chromosome.

inversion Structural rearrangement of a chromosome in which part of the DNA has become oriented in the reverse direction.

translocation Structural rearrangement of a chromosome in which a broken piece has become reattached in the wrong location.

transposable element (**transposon**) Segment of DNA that can move spontaneously within or between chromosomes.

14.5 Changes in Chromosome Number

Distinguish between polyploidy and aneuploidy.

Explain how Down syndrome arises by nondisjunction during meiosis.

Using appropriate examples, describe some health effects of changes in chromosome number in humans.

TABLE 14.4

A Few Effects of Human Chromosome Number Change

Cause	Syndrome	Main Symptoms
Trisomy 21	Down	Mental impairment; heart defects
Trisomy 18	Edwards	Severe disability; low survival rate
Trisomy 13	Patau	Severe disability; low survival rate
XO	Turner	Abnormal ovaries and sexual traits
XXY	Klinefelter	Sterility; mild mental impairment
XXX	Trisomy X	Minimal abnormalities
XYY	Jacob's	Mild mental impairment or no effect

Individuals of some species have three or more complete sets of chromosomes, a condition called **polyploidy**. About 70 percent of flowering plant species are polyploid, as are some insects, fishes, and other animals—but not humans. In our species, inheriting more than two full sets of chromosomes is invariably fatal. A few babies survive with too many or too few copies of a particular chromosome, however, and this condition is called **aneuploidy** (**Table 14.4**). Aneuploidy is usually an outcome of **nondisjunction**, the failure of chromosomes to separate properly during mitosis or meiosis. Nondisjunction during meiosis (**Figure 14.9**) can affect the chromosome number at fertilization. For example, fusion of a normal gamete (n) with a gamete that has an extra chromosome ($n+1$) gives rise to a zygote with three copies of one type of chromosome and two of every other type ($2n+1$), an aneuploid condition called trisomy. Fusion of a normal gamete (n) with a gamete missing a chromosome ($n-1$) gives rise to a zygote with one copy of one chromosome and two of every other type ($2n-1$), an aneuploid condition called monosomy.

DOWN SYNDROME

In most cases, inheriting the wrong number of autosomes is fatal in humans before birth or shortly thereafter. An important exception is trisomy 21. A person born with three chromosomes 21 will have Down syndrome and a high likeli-hood of surviving infancy. Mild to moderate mental impairment and health problems such as heart disease are hallmarks of this syndrome. Other phenotypic effects may include a somewhat flattened facial profile, a fold of skin that starts at the inner corner of each eyelid, white spots on the iris (**Figure 14.10**), and one deep crease (instead of two shallow creases) across each palm. The skeleton develops abnormally, so older children have short body parts, loose joints, and misaligned bones of the fingers, toes, and hips. Muscles and reflexes are weak, and motor skills such as speech develop slowly. Early training can help affected individuals learn to care for themselves and to take part in normal activities.

Alzheimer's disease, a condition of progressive mental deterioration, is associated with Down syndrome. A gene on chromosome 21 is the culprit. Having three of these chromosomes, a person with Down syndrome makes an excess of the gene's product, a protein that forms the main component of amyloid fibrils (Section 3.5) characteristic of Alzheimer's. By age 40, all persons with Down syndrome will have these fibrils in their brain. Down syndrome occurs in about 1 of 700 live births, and the risk increases with maternal age.

Figure 14.9 An example of nondisjunction. Of the two pairs of homologous chromosomes shown here, one fails to separate properly during anaphase I of meiosis. The chromosome number is altered in the resulting gametes.

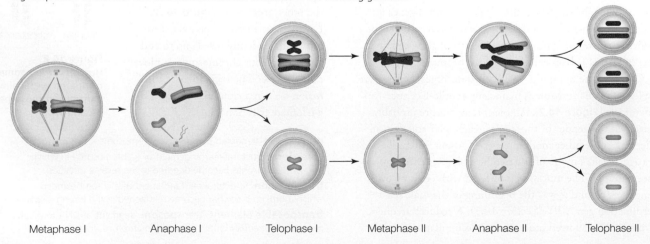

| Metaphase I | Anaphase I | Telophase I | Metaphase II | Anaphase II | Telophase II |

SEX CHROMOSOME CHANGES

About 1 in 400 human babies is born with an atypical number of sex chromosomes. Most often, such alterations lead to mild difficulties in learning and impairment in motor skills such as speech, but these problems may be very subtle.

Turner Syndrome Individuals with an X chromosome and no corresponding X or Y chromosome (XO) have a type of monosomy that results in Turner syndrome. The syndrome is thought to arise most frequently as an outcome of inheriting an unstable Y chromosome from the father. The zygote starts out being genetically male, with an X and a Y chromosome. Sometime during early development, the Y chromosome breaks up and is lost, so the embryo continues to develop as a female.

Fewer people are affected by Turner syndrome than by other chromosome abnormalities: Only about 1 in 2,500 newborns has it. XO individuals grow up well proportioned but short, with an average height of four feet, eight inches (1.4 meters). Their ovaries do not develop properly, so they do not make enough sex hormones to become sexually mature or to develop secondary sexual traits such as enlarged breasts.

XXX Syndrome A female may inherit an extra X chromosome, a condition called triple X syndrome or trisomy X. This syndrome occurs in about 1 of 1,000 births. Because of X chromosome inactivation (Section 10.3), only one X chromosome is typically active in female cells. Thus, having extra X chromosomes usually does not cause physical or medical problems, but mild mental impairment may occur.

Klinefelter Syndrome About 1 out of every 500 males has two or more X chromosomes (XXY, XXXY, and so on). The result is Klinefelter syndrome that develops at puberty. As adults, affected males tend to be overweight and tall, with small testes. Underproduction of the hormone testosterone interferes with sexual development and can result in sparse facial and body hair, a high-pitched voice, enlarged breasts, and infertility. Testosterone injections during puberty can minimize this feminization.

XYY Syndrome About 1 in 1,000 males is born with an extra Y chromosome (XYY), a result of nondisjunction of the Y chromosome during sperm formation. Adults with the resulting Jacob's syndrome tend to be taller than average and have mild mental impairment, but most are otherwise normal. XYY men were once thought to be predisposed to a life of crime. This misguided view was based on sampling error (too few cases in narrowly chosen groups such as prison inmates) and bias (the researchers who gathered the

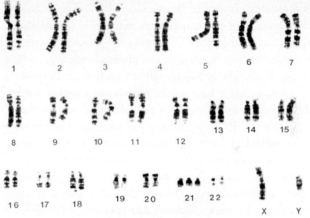

Figure 14.10 Down syndrome. Top, example of a Down syndrome phenotype: a somewhat flattened facial profile, and a fold of skin that starts at the inner corner of each eyelid. Excess tissue deposits on the iris also give rise to a ring of starlike white speckles, a lovely effect of the chromosome number change. Bottom, example of a Down syndrome genotype.

FIGURE IT OUT
Is the karyotype from an individual who is male or female?
Answer: Male

karyotypes also took the personal histories of the participants). That view has since been disproven: Men with XYY syndrome are only slightly more likely to be convicted for crimes than unaffected men. Researchers now believe this slight increase can be explained by poor socioeconomic conditions related to the effects of the syndrome.

---

aneuploidy (AN-you-ploy-dee) Condition of having too many or too few copies of a particular chromosome.
nondisjunction Failure of chromosomes to separate properly during nuclear division.
polyploidy (PALL-ee-ploy-dee) Condition of having three or more of each type of chromosome characteristic of the species.

CREDITS: (10) top, Ciarra, photo by © Michelle Harmon; bottom, L. Willatt, East Anglian Regional Genetics Service/ Science Source.

LEARNING OBJECTIVES

Explain how early screening for a genetic disorder can help a baby.

Describe three prenatal diagnosis methods.

Discuss the relative benefits and risks of prenatal diagnosis procedures.

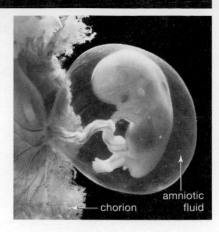

Figure 14.11 Tissues tested in prenatal diagnosis. With amniocentesis, fetal cells that have been shed into amniotic fluid are tested for genetic disorders. Chorionic villus sampling tests cells of the chorion, which is part of the placenta.

chorion   amniotic fluid

Studying human inheritance patterns has given us many insights into how genetic disorders arise and progress, and how to treat them. Surgery, prescription drugs, hormone replacement therapy, and dietary controls can minimize and in some cases eliminate the symptoms of a genetic disorder. Some disorders can be detected early enough to start countermeasures before symptoms develop; thus, most hospitals in the United States now screen newborns for mutations such as those that cause phenylketonuria, or PKU. The mutations affect an enzyme that converts one amino acid (phenylalanine) to another (tyrosine). Without this enzyme, the body becomes deficient in tyrosine, and phenylalanine accumulates to high levels. The imbalance inhibits protein synthesis in the brain, which in turn results in permanent intellectual disability. Restricting all intake of phenylalanine can slow the progression of PKU, so routine early screening has resulted in fewer individuals suffering from the symptoms of the disorder.

Genetic screening can also benefit people who are planning to have a child. The probability that a future child will inherit a genetic disorder can be estimated by testing prospective parents for certain alleles. Karyotypes and pedigrees are also useful in this type of screening, which can help people make informed decisions about family planning.

Genetic screening after conception but before birth is called prenatal diagnosis (prenatal means before birth). Prenatal diagnosis checks an embryo or fetus for physical and genetic abnormalities. Dozens of conditions are detectable prenatally, including aneuploidy, hemophilia, Tay–Sachs disease, sickle-cell anemia, muscular dystrophy, and cystic fibrosis. Early diagnosis gives parents time to prepare for the birth of an affected child, and an opportunity to decide whether to continue with the pregnancy. If a disorder is treatable, early detection may allow the newborn to receive prompt and appropriate treatment. A few conditions are even surgically correctable before birth.

As an example of how prenatal diagnosis works, consider a woman who becomes pregnant at age thirty-five. Her doctor will probably perform a procedure called obstetric sonography, in which ultrasound waves directed across the woman's abdomen form images of the fetus's limbs and internal organs. If the images reveal a physical defect that may be the result of a genetic disorder, a more invasive technique would be recommended. With fetoscopy, sound waves pulsed from inside the mother's uterus yield images much higher in resolution than

ultrasound. Samples of tissue or blood are often taken at the same time, and some corrective surgeries can be performed.

Human genetics studies show that our thirty-five-year-old woman has about a 1 in 80 chance that her baby will be born with a chromosomal abnormality, a risk more than six times greater than when she was twenty years old. Thus, even if no abnormalities are detected by ultrasound, she probably will be offered an additional diagnostic procedure, amniocentesis, in which a small sample of fluid is drawn from the amniotic sac enclosing the fetus (**Figure 14.11**). The fluid contains cells shed by the fetus, and those cells can be tested for genetic disorders. Chorionic villus sampling (CVS) can be performed earlier than amniocentesis. With this technique, a few cells from the chorion are removed and tested (the chorion is a membrane that surrounds the amniotic sac).

An invasive procedure often carries a risk to the fetus. The risks vary by procedure. Amniocentesis has improved so much that, in the hands of a skilled physician, it no longer increases the risk of miscarriage. CVS occasionally disrupts the placenta's development, causing underdeveloped or missing fingers and toes in 0.3 percent of newborns. Fetoscopy raises the miscarriage risk by a whopping 2 to 10 percent, so it is rarely performed unless surgery or another medical procedure is required before the baby is born.

Couples who discover they are at high risk of having a child with a genetic disorder may opt for reproductive interventions such as *in vitro* fertilization. With this procedure, sperm and eggs taken from the prospective parents are mixed in a test tube. If an egg becomes fertilized, the resulting zygote will begin to divide. In about forty-eight hours, it will have become an embryo that consists of a ball of eight undifferentiated cells. One cell can be removed and its genes analyzed, a procedure called preimplantation diagnosis. The withdrawn cell will not be missed. If the embryo has no detectable genetic defects, it is inserted into the woman's uterus to develop. Most of the resulting "test-tube babies" are born in good health.

Figure 14.12 Fraternal twins Kian and Remee with their parents. Both of the children's grandmothers are of European descent, and have pale skin. Both of their grandfathers are of African descent, and have dark skin. The twins inherited different alleles of some genes that affect skin color from their parents, who, given the appearance of their children, must be heterozygous for those alleles.

SHADES OF SKIN

The color of human skin begins with melanosomes, which are organelles that make melanin pigments. Most people have about the same number of melanosomes in their skin cells. Variations in skin color arise from differences in the size, shape, and cellular distribution of melanosomes in the skin, as well as in the kinds and amounts of melanins they make.

Variation in human skin color may have evolved as a balance between vitamin production and protection against harmful ultraviolet (UV) radiation in the sun's rays. Dark skin would have been (and still is) beneficial under the intense sunlight of African savannas where humans first evolved. Melanin is a natural sunscreen: It prevents ultraviolet radiation from breaking down folate, a vitamin essential for normal sperm formation and embryonic development.

Early human groups that migrated to regions with cold climates were exposed to less sunlight. In these regions, lighter skin color is beneficial. Why? Exposure to UV radiation stimulates skin cells to make a molecule the body converts to vitamin D. Where sunlight exposure is minimal, UV radiation is less of a risk than vitamin D deficiency, which has serious health consequences for children and developing fetuses. In regions with long, dark winters, people with dark, UV-shielding skin have a high risk of this deficiency.

The evolution of regional variations in human skin color began with mutations in genes involved in pigmentation. Consider a gene on chromosome 15 that encodes a transport protein in melanosome membranes. Nearly all people of African, Native American, or east Asian descent carry the same allele of this gene. Between 6,000

and 10,000 years ago, a mutation gave rise to a different allele. The mutation, a single base-pair substitution (Section 9.5), changed the 111th amino acid of the transport protein from alanine to threonine. The change results in less melanin—and lighter skin color—than the original African allele does. Today, nearly all people of European descent are homozygous for the mutated allele.

A person of mixed ethnicity makes gametes with different combinations of thousands of alleles for dark and light skin. It is rare for all of the alleles for dark skin (or for light skin) to end up in the same gamete, but this does occur (**Figure 14.12**). Skin color is one of many traits that vary as a result of single nucleotide mutations. The small scale of such changes offers a reminder that we all share the genetic legacy of common ancestry.

CREDIT: (12) Gary Roberts/worldwidefeatures.com.

Section 14.1 Relatively few human traits are governed by single genes inherited in a Mendelian pattern; most are polygenic with environmental contributions. Geneticists use **pedigrees** to study inheritance patterns in humans. Most of what we know about human inheritance comes from tracking genetic disorders in families, because this information helps us develop treatments for affected people. A genetic abnormality is an uncommon but harmless version of a heritable trait. By contrast, a genetic disorder sooner or later causes medical problems. Genetic disorders typically arise from mutations that affect pleiotropic genes, so their symptoms usually occur in a syndrome.

Section 14.2 A trait is inherited in an autosomal dominant pattern when the associated allele occurs on an autosome, and everyone who has it (homozygous or heterozygous) has the trait. The trait typically appears in every generation of a family, and both sexes are affected with equal frequency. A trait is inherited in an autosomal recessive pattern when the associated allele occurs on an autosome, and the trait it specifies appears only in homozygous people. The trait also appears in both sexes equally, but it can skip generations.

Section 14.3 An trait is inherited in an X-linked pattern when the associated allele occurs on the X chromosome. Most X-linked disorders are inherited in a recessive pattern, and these appear in men more often than in women. Heterozygous women have a dominant, normal allele that can mask the effects of the recessive one; men, with one X chromosome, do not have this benefit. Men can transmit an X-linked allele to their daughters, but not to their sons. Only a woman can pass an X-linked allele to a son.

Section 14.4 Faulty crossovers and the activity of **transposable elements** can give rise to major changes in chromosome structure, including insertions, deletions, **duplications**, **inversions**, and **translocations**. Some of these changes are harmful or lethal in humans; others affect fertility. Even so, major structural changes have accumulated in the chromosomes of all species over evolutionary time.

Section 14.5 Occasionally, new individuals end up with the wrong chromosome number. Consequences of such changes range from minor to lethal alterations in form and function. Chromosome number change is usually an outcome of **nondisjunction**. Having three or more of each type of chromosome is called **polyploidy**. Polyploidy is common in flowering plants and also occurs in some animals, but it is lethal in humans. In humans, autosomal **aneuploidy** is often lethal. Trisomy 21, which causes Down syndrome, is an exception. Sex chromosome aneuploidy can result in an impairment in learning and motor skills.

Section 14.6 Prospective parents can use genetic screening to estimate their risk of transmitting a harmful allele to offspring. The procedure involves analysis of parental pedigrees, karyotypes, and genotypes by a genetic counselor. Amniocentesis and other methods of prenatal testing can reveal a genetic disorder before birth.

Application Like most other human traits, skin color has a genetic basis. Minor differences in alleles that govern melanin production and deposition of melanosomes affect skin color. The differences probably evolved as a balance between vitamin production and protection against UV radiation.

SELF-ASSESSMENT
ANSWERS IN APPENDIX VII

1. Constructing a family pedigree is particularly useful when studying inheritance patterns in organisms that _____ .
   a. produce many offspring per generation
   b. produce few offspring per generation
   c. have a small chromosome number
   d. reproduce asexually
   e. have a fast life cycle

2. Pedigree analysis is necessary when studying human inheritance patterns because _____ .
   a. humans have more than 20,000 genes
   b. there are ethical issues with human experimentation
   c. human genes are more complicated than genes of other organisms
   d. genetic disorders occur in humans

3. A recognized set of symptoms that characterize a genetic disorder is a(n) _____ .
   a. syndrome   b. disease   c. abnormality

4. True or false? All traits are inherited in a Mendelian pattern.

5. One parent is heterozygous for an allele inherited in an autosomal dominant pattern; the other parent does not carry the allele. Any child of these two parents has a _____ chance of having the trait associated with the allele.
   a. 25 percent   c. 75 percent
   b. 50 percent   d. 100 percent

6. A trait that is present in a male child but not in either of his parents is characteristic of _____ inheritance.
   a. autosomal dominant   d. It is not possible to answer
   b. autosomal recessive   this question without more
   c. X-linked recessive   information.

7. Color blindness is a case of _____ inheritance.
   a. autosomal dominant   c. X-linked dominant
   b. autosomal recessive   d. X-linked recessive

8. A female child inherits one X chromosome from her mother and one from her father. What sex chromosome does a male child inherit from each of his parents?

SKIN COLOR SURVEY OF NATIVE PEOPLES In 2000, researchers measured the average amount of UV radiation received in more than fifty regions of the world, and correlated it with the average skin reflectance of people native to those regions (reflectance is a way to measure the amount of melanin pigment in skin: Skin with more melanin is darker, so it reflects less light). Some of the results of this study are shown in **Figure 14.13**.

1. Which country receives the most UV radiation?

2. Which country receives the least UV radiation?

3. People native to which country have the darkest skin?

4. People native to which country have the lightest skin?

5. According to these data, how does the skin color of indigenous peoples correlate with the amount of UV radiation incident in their native regions?

Country	Skin Reflectance	UVMED
Australia	19.30	335.55
Kenya	32.40	354.21
India	44.60	219.65
Cambodia	54.00	310.28
Japan	55.42	130.87
Afghanistan	55.70	249.98
China	59.17	204.57
Ireland	65.00	52.92
Germany	66.90	69.29
Netherlands	67.37	62.58

Figure 14.13 Skin color of indigenous peoples and regional incident UV radiation. Skin reflectance measures how much light of 685-nanometer wavelength is reflected from skin; UVMED is the annual average UV radiation received at Earth's surface.

9. Alleles for Tay–Sachs disease are inherited in an autosomal recessive pattern. Why would two parents with a normal phenotype have a child with Tay–Sachs?
   a. Both parents are homozygous for a Tay–Sachs allele.
   b. Both parents are heterozygous for a Tay–Sachs allele.
   c. New mutations gave rise to Tay–Sachs in the child.
   d. b or c

10. The *SRY* gene gives rise to the male phenotype in humans (Sections 10.3 and 14.4). What do you think the inheritance pattern of *SRY* alleles is called?

11. True or false? Transposable elements are common in the DNA of all species.

12. Nondisjunction may occur during _____ .
   a. mitosis
   b. meiosis
   c. mitosis and meiosis
   d. fertilization

13. Nondisjunction at meiosis can result in _____ .
   a. base-pair substitutions
   b. aneuploidy
   c. crossing over
   d. pleiotropy

14. Klinefelter syndrome (XXY) can be easily diagnosed by _____ .
   a. pedigree analysis
   b. aneuploidy
   c. karyotyping
   d. phenotypic treatment

15. Match the chromosome terms appropriately.
   ___ polyploid
   ___ deletion
   ___ aneuploidy
   ___ translocation
   ___ syndrome
   ___ transposable element
   ___ X-linked
   a. symptoms of a genetic disorder
   b. chromosomal mashup
   c. extra sets of chromosomes
   d. gets around
   e. a chromosome segment lost
   f. one extra chromosome
   g. allele on the X chromosome

GENETICS PROBLEMS
ANSWERS IN APPENDIX VII

1. Does the phenotype indicated by the red circles and squares in this pedigree show an inheritance pattern that is autosomal dominant, autosomal recessive, or X-linked?

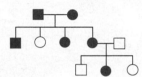

2. Human females have two X chromosomes (XX); males have one X and one Y chromosome (XY).
   a. With respect to X-linked alleles, how many different types of gametes can a male produce?
   b. A female homozygous for an X-linked allele can produce how many types of gametes with respect to that allele?
   c. A female heterozygous for an X-linked allele can produce how many types of gametes with respect to that allele?

3. Somatic cells of individuals with Down syndrome usually have an extra chromosome 21, so they contain forty-seven chromosomes. A few individuals with Down syndrome have forty-six chromosomes: two normal-appearing chromosomes 21, and a longer-than-normal chromosome 14. Speculate on how this chromosome abnormality arises.

4. An allele responsible for Marfan syndrome (Section 13.4) is inherited in an autosomal dominant pattern. What is the chance that a child will inherit the allele if one parent does not carry it and the other is heterozygous?

5. Both Duchenne muscular dystrophy and color blindness are caused by recessive alleles. DMD, unlike color blindness, nearly always occurs in males. Explain why.

CREDITS: (13) Based on *Journal of Human Evolution* (2000) 39, 57106 doi:10.1006/jhev. 2000-0403 2000 Academic Press.

15 Biotechnology

Links to Earlier Concepts

This chapter builds on your understanding of DNA (Sections 8.2, 8.3, 13.1, 14.6) and DNA replication (8.4). Clones (8.6), gene expression (9.1, 9.2), and knockouts (10.2) are important in genetic engineering, particularly in research on human genetic disorders (Chapter 14). You will revisit tracers (2.1), triglycerides (3.3), denaturation (3.5), bacteria (4.4), β-carotene (6.2), mutations (9.5), the *lac* operon (10.4), cancer (11.5), alleles (12.1), and short tandem repeats (13.6).

Core Concepts

Living systems store, retrieve, transmit, and respond to information essential for life.

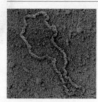

An organism's form and function arise from and depend on expression of genetic information in its DNA. Differences in genes are the basis of differences between species; differences in alleles are the basis of variation in shared traits among individuals of a species. Researchers can change an organism's phenotype in a specific way by altering its genes, or by transferring genes from another species into it.

All organisms alive today are linked by lines of descent from shared ancestors.

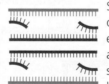

Shared core processes and features widely distributed among organisms alive today provide evidence that all living things are descended from a common ancestor. Genetic and biochemical similarities among organisms of different taxa help us understand our own genes, and also allow us to manipulate heritable information by techniques of genetic engineering.

Interactions among the components of each biological system give rise to complex properties.

Genetically modified organisms are used for research, medicine, or food. Gene editing raises social and ethical issues. Biological systems are too complex to accurately predict long-term effects of genetic manipulation on humans, or on the environment.

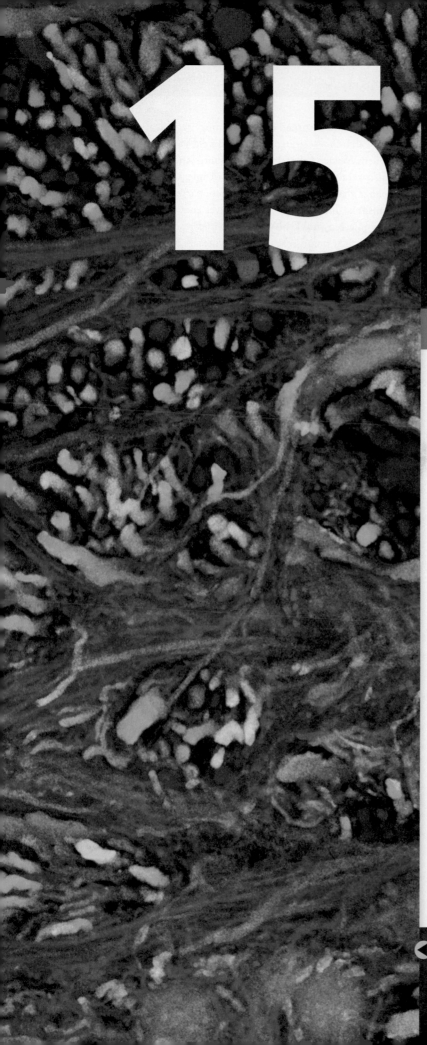

Mice transgenic for multiple pigments ("brainbow mice") are allowing researchers to map the complex neural circuitry of the brain. Individual nerve cells in the brain stem of a brainbow mouse are visible in this fluorescence micrograph.

Photograph courtesy of © Dr. Jean Levit. The Brainbow technique was developed in the laboratories of Jeff W. Lichtman and Joshua R. Sanes at Harvard University. This image has received the Bioscape imaging competition 2007 prize.

DNA CLONING

In the 1950s, excitement over the discovery of DNA's structure (Section 8.2) gave way to frustration because no one could determine the order of nucleotides in a molecule of DNA. Identifying a single base among thousands or mil-

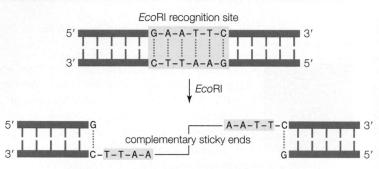

❶ The restriction enzyme *Eco*RI recognizes a specific base sequence in DNA and cuts it. *Eco*RI leaves single-stranded tails ("sticky ends").

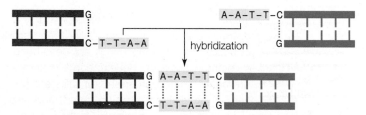

❷ When DNA fragments from two sources are cut with *Eco*RI and mixed together, matching sticky ends base-pair (hybridization, Section 8.4).

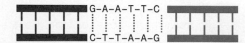

❸ DNA ligase seals the gaps between hybridized DNA fragments. The resulting hybrid molecules are called recombinant DNA.

Figure 15.1 Restriction enzymes and recombinant DNA.

FIGURE IT OUT
Why does the enzyme cut both strands of DNA?

Answer: Because the recognition site occurs on both strands

lions of others turned out to be a huge technical challenge. Research in a seemingly unrelated field yielded a solution when Werner Arber, Hamilton Smith, and their coworkers discovered how some bacteria resist infection by bacteriophage viruses (Section 8.1). In these bacteria, a defense mechanism against viral infection involves enzymes that chop up any injected viral DNA. The enzymes restrict viral growth; hence their name, restriction enzymes. A **restriction enzyme** cuts DNA wherever a specific nucleotide sequence occurs (**Figure 15.1**). For example, the enzyme *Eco*RI (named after *E. coli*, the bacteria from which it was isolated) cuts DNA at the nucleotide sequence G-A-A-T-T-C ❶. Other restriction enzymes cut at different sequences.

The discovery of restriction enzymes allowed researchers to cut chromosomal DNA into manageable chunks. It also allowed them to combine DNA fragments from different organisms. How? Many restriction enzymes, including *Eco*RI, leave single-stranded tails on DNA fragments. Researchers realized that, because the chemical structure of DNA is the same in all organisms, complementary tails will base-pair regardless of the source of DNA ❷. The tails are called "sticky ends," because two DNA fragments stick together when their matching tails base-pair. DNA ligase (Section 8.4) can be used to seal the gaps between hybridized sticky ends, so continuous DNA strands form ❸. Thus, using appropriate restriction enzymes and DNA ligase, researchers can cut and paste DNA from different sources. The result, a hybrid molecule that consists of genetic material from two or more organisms, is called **recombinant DNA**.

Making recombinant DNA is the first step in **DNA cloning**, a set of laboratory methods that uses living cells to mass-produce specific DNA fragments. Researchers clone a fragment of DNA by inserting it into a **vector**, which in this context is a molecule that can carry foreign DNA into host cells. Plasmids may be used as cloning vectors in cells that normally carry them—bacteria, for example (Section 4.4). When a bacterium reproduces, its offspring inherit a full complement of genetic information—one chromosome plus

cDNA Complementary strand of DNA synthesized from an RNA template by the enzyme reverse transcriptase.

DNA cloning Set of methods that uses living cells to mass-produce targeted DNA fragments.

recombinant DNA (ree-COM-bih-nent) A DNA molecule that contains genetic material from more than one organism.

restriction enzyme Type of enzyme that cuts DNA at a specific nucleotide sequence.

reverse transcriptase (trance-CRYPT-ace) An enzyme that uses mRNA as a template to make a strand of cDNA.

vector A DNA molecule that can accept foreign DNA and be replicated inside a host cell.

plasmids. A recombinant plasmid gets replicated and distributed to descendant cells just like other plasmids.

Researchers insert recombinant vectors into cells that can be easily grown in the laboratory (cultured) to yield huge populations of genetically identical descendants. Each of these clones (Section 8.6) contains a copy of the recombinant vector. The vector can be harvested in quantity from the clones, and the hosted DNA fragment excised from the vector using the same restriction enzyme used for cloning.

WHY CLONE DNA?

Learning how to clone DNA was a key step in the quest to understand the information it encodes (which is why Arber and Smith won a 1978 Nobel Prize for the discovery of restriction enzymes). The sequence of nucleotides in a cloned fragment of DNA could now be determined, and sequence information from many organisms started accumulating (Sections 15.3 and 15.4 return to this topic). The next step, finding the genes, became a major topic of research. Eukaryotic genes occur discontinuously on a chromosome—their coding sequences are interrupted by long introns—so finding and studying them posed a particular challenge at the time. Researchers realized that post-transcriptional processing removes introns (Section 9.2), so they developed a method for cloning mRNA.

An mRNA cannot be cut with restriction enzymes or pasted with DNA ligase, because these enzymes work only on double-stranded DNA. Thus, cloning mRNA requires reverse transcriptase, a replication enzyme made by some viruses. **Reverse transcriptase** uses an RNA template to assemble a strand of complementary DNA (**Figure 15.2A**); DNA made this way is called **cDNA**. Next, DNA polymerase is used to make a second strand of DNA (**Figure 15.2B**). The outcome is a double-stranded cDNA version of the original mRNA.

A cDNA contains all of the protein-building information encoded in a eukaryotic mRNA, but in DNA form. Inserted into an appropriate vector and host (**Figure 15.3**), it will direct synthesis of an mRNA, which in turn will direct the synthesis of a protein product. This method is useful for many reasons. First, it allows researchers to manipulate eukaryotic genes, for example by changing them, knocking them out, or inserting them into other species (more about this in Section 15.5). Second, the method allows genes from complex eukaryotes to be expressed in simpler organisms such as bacteria and yeast. This bypasses normal gene expression controls that would otherwise limit the protein's production in its eukaryotic cell of origin. Gene expression is also much more easily adjusted in these single-celled organisms than it is in complex eukaryotes. Particularly in bacteria,

A The enzyme reverse transcriptase uses an RNA template to assemble a complementary strand of DNA (which is called cDNA).

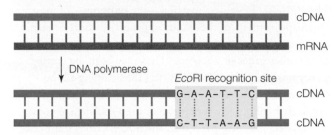

B DNA polymerase uses the cDNA as a template to assemble a complementary strand of DNA (the enzyme also removes the mRNA as it carries out synthesis). Like any other double-stranded DNA, the resulting double-stranded cDNA may be cut with restriction enzymes and pasted into a cloning vector using DNA ligase.

Figure 15.2 Making cDNA for cloning.

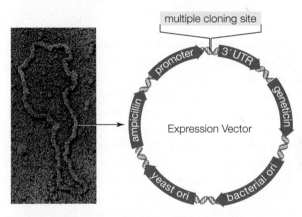

Figure 15.3 A plasmid cloning and expression vector.

A cDNA is inserted into the multiple cloning site, which has several restriction enzyme recognition sites useful for this purpose. Bacteria or yeast cells are then induced to take up the recombinant plasmid. Antibiotic resistance genes (here, ampicillin for bacteria, and geneticin for yeast) allow only cells that take up the plasmid to survive treatment with the antibiotic.

Yeast and bacterial origins of replication (ori) are sites where DNA replication of the plasmid begins in these cells. A eukaryotic promoter is necessary for yeast cells to transcribe RNA from the inserted DNA, as is a 3' untranslated region (UTR) that includes required regulatory signals.

expression of foreign genes can be adjusted to extremely high levels. Finally, bacteria and yeast can be easily grown on an industrial scale, for example to produce medically relevant proteins (Section 15.5 returns to this topic).

CREDIT: (3) left, Professor Stanley Cohen/Science Source.

DNA LIBRARIES

The entire set of genetic material—the **genome**—of most organisms consists of thousands of genes. To study or manipulate a single gene, researchers must first separate it from all of the other genes in a genome. They may begin by cutting an organism's DNA into pieces, and then cloning all the pieces simultaneously. The result is a genomic library, a set of clones that collectively contain all of the DNA in a genome. Researchers may also harvest all the mRNA from some cells, make cDNA from the mRNA, and then clone the cDNA. The resulting cDNA library represents only those genes being expressed at the time the mRNA was harvested.

Genomic and cDNA libraries are **DNA libraries**, sets of cells that host various cloned DNA fragments. In such libraries, a cell that contains a particular DNA fragment of interest is mixed up with thousands or millions of others that do not—a needle in a genetic haystack. One way to find that cell among the others involves the use of a **probe**, which is a fragment of DNA or RNA labeled with a tracer (Section 2.1). For example, to find a cell hosting a gene, researchers may assemble radioactive nucleotides into a short strand of DNA complementary in sequence to a similar gene. Because the nucleotide sequences of the probe and the gene are complementary, the two can hybridize. When the probe is mixed with DNA from a library, it will hybridize with the gene, but not with other DNA (**Figure 15.4**). Researchers can pinpoint a cell that hosts the gene by detecting the label on the probe. That cell is isolated and cultured, and DNA can be extracted in bulk from the cultured cells for research or other purposes.

PCR

The **polymerase chain reaction** (**PCR**) is a technique used to mass-produce copies of a particular section of DNA without having to clone it in living cells (**Figure 15.5**). The reaction can transform a needle in a haystack—that one-in-a-million fragment of DNA—into a huge stack of needles with a little hay in it.

The starting material for PCR is any sample of DNA with at least one molecule containing a targeted sequence. The source of the DNA might be a mixture of 10 million different clones, a sperm, a hair left at a crime scene, or a mummy: essentially any sample that has DNA in it.

A Individual bacterial cells from a DNA library are spread over the surface of a solid growth medium. The cells divide repeatedly and form colonies—clusters of millions of genetically identical descendant cells.

B Special paper is pressed onto the surface of the growth medium. Some cells from each colony stick to the paper.

C The paper is soaked in a solution that ruptures the cells and makes the released DNA single-stranded. The DNA clings to the paper in spots mirroring the distribution of colonies.

D A radioactive probe is added to the liquid bathing the paper. The probe hybridizes with any spot of DNA that contains a complementary sequence.

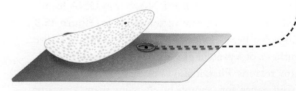

E The paper is pressed against x-ray film. The radioactive probe darkens the film in a spot where it has hybridized. The spot's position is compared to the positions of the original bacterial colonies. Cells from the colony that corresponds to the spot are cultured, and their DNA is harvested.

Figure 15.4 Using a probe to screen a DNA library.
In this example, a radioactive probe helps identify a bacterial colony that contains a targeted fragment of DNA.

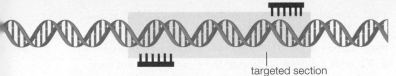

① DNA template (blue) with a targeted sequence is mixed with primers (pink), nucleotides, and heat-tolerant *Taq* DNA polymerase.

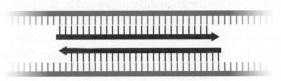

② When the mixture is heated, the double-stranded DNA separates into single strands. When the mixture is cooled, some of the primers base-pair with the DNA at opposite ends of the targeted sequence.

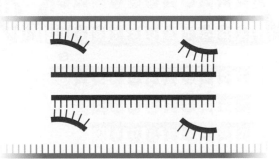

③ *Taq* polymerase begins DNA synthesis at the primers, so it produces complementary strands of the targeted DNA sequence.

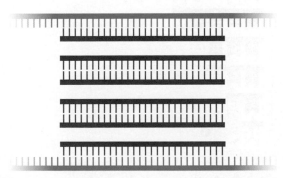

④ The mixture is heated again, so all double-stranded DNA separates into single strands. When it is cooled, primers base-pair with the targeted sequence in the original template DNA and in the new DNA strands.

⑤ Each cycle of heating and cooling can double the number of copies of the targeted DNA section.

Figure 15.5 Two rounds of PCR.

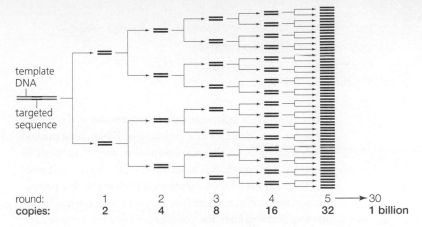

round:	1	2	3	4	5 → 30
copies:	2	4	8	16	32 ... 1 billion

Figure 15.6 Exponential amplification by PCR.

The PCR reaction is similar to DNA replication (Section 8.4). It requires two synthetic DNA primers designed to base-pair at opposite ends of the section of DNA to be amplified, or mass-produced **①**. Researchers mix these primers with the starting (template) DNA, nucleotides, and DNA polymerase, then expose the reaction mixture to repeated cycles of high and low temperatures. A few seconds at high temperature disrupts the hydrogen bonds that hold the two strands of a DNA double helix together (Section 8.2), so every molecule of DNA unwinds and becomes single-stranded. As the temperature of the reaction mixture is lowered, the single DNA strands hybridize with the primers **②**.

The DNA polymerases of most organisms denature at the high temperature required to separate DNA strands. PCR reactions make use of *Taq* polymerase, which is from *Thermus aquaticus*. These bacteria live in hot springs and hydrothermal vents, so their polymerase necessarily tolerates heat. Like other DNA polymerases, *Taq* polymerase recognizes hybridized primers as places to start DNA synthesis **③**. Synthesis proceeds along the template strand until the temperature rises and the DNA separates into single strands **④**. The newly synthesized DNA is a copy of the targeted section. When the mixture is cooled, the primers rehybridize, and DNA synthesis begins again. Each cycle of heating and cooling takes only a few minutes, but it can double the number of copies of the targeted section of DNA **⑤**. Thirty PCR cycles may amplify that number a billionfold (**Figure 15.6**).

DNA library Collection of cells that host different fragments of foreign DNA, often representing an organism's entire genome.

genome An organism's complete set of genetic material.

polymerase chain reaction (PCR) Method that rapidly generates many copies of a specific section of DNA.

probe Short fragment of DNA designed to hybridize with a nucleotide sequence of interest, and labeled with a tracer.

LEARNING OBJECTIVES

Describe how the sequence of a fragment of DNA is determined.

Explain the concept of electrophoresis.

Describe the Human Genome Project.

Researchers use a technique called **sequencing** to determine the order of nucleotides in a molecule of DNA. One method of sequencing DNA that was invented in the 1970s is still in use today. This method is based on the DNA synthesis reaction (Section 8.4). DNA polymerase is mixed with a primer, nucleotides, and a template (the DNA to be sequenced). Starting at the primer, the polymerase joins the nucleotides into a new strand of DNA, in the order dictated by the sequence of the template (**Figure 15.7**).

The sequencing reaction mixture includes all four kinds of nucleotides, and also all four kinds of dideoxynucleotides, which are nucleotides with no hydroxyl group on the 3′ carbon ❶. Each dideoxynucleotide base (A, C, G, or T) is labeled with a different colored pigment. During the reaction, the polymerase randomly adds either a regular nucleotide or a dideoxynucleotide to the 3′ end of a growing DNA strand. If it adds a regular nucleotide, synthesis can continue because the 3′ carbon of the strand will have a hydroxyl group on it. If it adds a dideoxynucleotide, the 3′ carbon will not have a hydroxyl group, so synthesis of the strand ends there ❷. (Remember, DNA synthesis proceeds only in the 5′ to 3′ direction.) Millions of DNA fragments of different lengths are produced, each a partial copy of the template DNA ❸. Every fragment of a given length ends with the same dideoxynucleotide. For example, if the template strand's tenth base was guanine, then any newly synthesized fragment that is 10 nucleotides long ends with a cytosine.

The DNA fragments are then separated by length. With a technique called **electrophoresis**, an electric field pulls the fragments through a semisolid gel. Fragments of different sizes move through the gel at different rates. The shorter the fragment, the faster it moves, because shorter fragments slip through the tangled molecules of the gel faster than longer fragments do. All fragments of the same length move through the gel at the same speed, so they gather into bands. All fragments in a given band have the same modified nucleotide

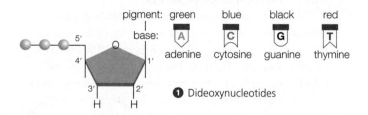
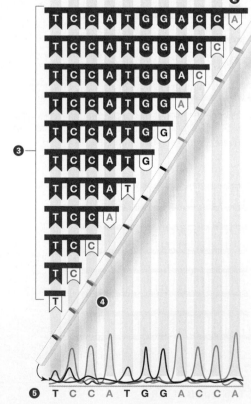

Figure 15.7 DNA sequencing.
In this method, DNA polymerase is used to incompletely replicate a section of DNA.

❶ This sequencing method depends on dideoxynucleotides, which are nucleotides that have a hydrogen atom instead of a hydroxyl group on their 3′ carbon (compare the structure with those in Figure 8.4). Each base (G, A, T, or C) is labeled with a different colored pigment.

❷ DNA polymerase uses a section of DNA as a template to synthesize new strands of DNA. Synthesis of each new strand stops when a dideoxynucleotide is added.

❸ At the end of the reaction, there are many incomplete copies of the template DNA in the mixture.

❹ Electrophoresis separates the copied DNA fragments into bands according to their length. All of the DNA strands in each band end with the same base, so each band is the color of the base's tracer pigment.

❺ A computer detects and records the color of successive bands on the gel (see Figure 15.8 for an example). The order of colors of the bands reflects the sequence of the DNA.

at their ends, and the pigment labels now impart distinct colors to the bands ❹. Each color designates one of the four modified nucleotides, so the order of colored bands in the gel reflects the sequence of the template DNA ❺.

THE HUMAN GENOME PROJECT

Ten years after DNA sequencing had been invented, it had become so routine that scientists began to consider sequencing the entire human genome—all 3 billion nucleotides of it. Proponents of the idea said it could provide huge payoffs for medicine and research. Opponents said this daunting task would divert attention and funding from more urgent research. It would take 50 years to sequence the human genome given the techniques of the time. However, the techniques continued to improve, and with each improvement more nucleotides could be sequenced in less time. Automated (robotic) DNA sequencing and PCR had just been invented. Both were still too cumbersome and expensive to be useful in routine applications, but they would not be so for long. Waiting for faster, cheaper technologies seemed the most efficient way to sequence the genome, but just how fast and cheap did they need to be before the project should begin?

A few privately owned companies decided not to wait, and started sequencing. One of them intended to determine the genome sequence in order to patent it. The idea of patenting the human genome provoked widespread outrage, but it also spurred commitments in the public sector. In 1988, the National Institutes of Health (NIH) essentially took over the project by hiring James Watson (of DNA structure fame) to head an official Human Genome Project, and providing $200 million per year to fund it. A partnership formed between the NIH and international institutions that were sequencing different parts of the genome. Watson set aside 3 percent of the funding for studies of ethical and social issues arising from the work. He later resigned over a patent disagreement, and geneticist Francis Collins took his place.

Amid ongoing squabbles over patent issues, Celera Genomics formed in 1998. With biologist Craig Venter at its helm, the company intended to commercialize human genetic information. Celera invented faster techniques for sequencing genomic DNA because the first to have the complete sequence had a legal basis for patenting it. The competition motivated the international partnership to accelerate its efforts. Then, in 2000, U.S. President Bill Clinton and

electrophoresis (eh-lek-troh-fur-EE-sis) Technique that separates DNA fragments by size.
sequencing Method of determining the order of nucleotides in DNA.

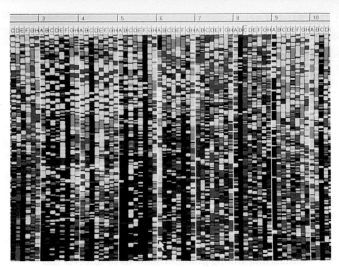

Figure 15.8 Part of the human genome sequence, raw data. The order of colored bands in each vertical lane reveals the sequence of about 100 nucleotides of the human genome.

TABLE 15.1

Some Human Genome Statistics

Chromosomes (number of)*	24
Nucleotides (number of)*	3,547,121,844
Genes	
Protein coding genes (number of)*	20,313
Non-protein-coding genes (number of)*	25,180
Pseudogenes (gene duplications, number of)*	14,453
Exons (number of)	381,122
Average exon length (base pairs)	163
Introns (number of)	378,089
Average intron length (base pairs)	5,849
Variation	
Short tandem repeats (number of)	741,587
Single-nucleotide polymorphisms (number of)‡	87,339,846

*Ensembl release 84.38; ‡NCBI dbSNP build 146

British Prime Minister Tony Blair jointly declared that the sequence of the human genome could not be patented. Celera kept sequencing anyway, and, in 2001, the competing governmental and corporate teams published about 90 percent of the sequence. In 2003, fifty years after the discovery of the structure of DNA, the sequence of the human genome was officially completed (**Figure 15.8**). The sequence is freely accessible worldwide (see www.ncbi.nlm.nih.gov/genome). Anyone can search it and see current statistics (**Table 15.1**).

From start to finish, the human genome project required about 16 years and $2.7 billion to complete, but it spurred development of much more efficient technologies for sequencing and data analysis. Today, an individual's genome can be sequenced in a few hours, for about $1,000.

Section 15.1 In **DNA cloning**, researchers use **restriction enzymes** to cut a sample of DNA into pieces, and then use DNA ligase to insert the fragments into plasmids or other **vectors**. The resulting molecules of **recombinant DNA** are delivered into host cells such as bacteria or yeast. Division of host cells produces huge populations of genetically identical descendant cells, each with a copy of the inserted DNA. Researchers can study eukaryotic gene expression by using **reverse transcriptase** to transcribe mRNA into **cDNA**. Inserted into an appropriate expression vector, a cDNA will direct the production of a protein in host cells.

Section 15.2 A **DNA library** is a collection of cells that host different fragments of DNA, often representing an organism's entire **genome**. Researchers can use **probes** to identify cells in a library that carry a specific fragment of DNA. The **polymerase chain reaction (PCR)** uses primers and a heat-resistant DNA polymerase to rapidly increase the number of copies of a targeted section of DNA.

Section 15.3 Advances in **sequencing**, which reveals the order of nucleotides in a sample of DNA, allowed the DNA sequence of the entire human genome to be determined. The technique uses DNA polymerase to partially replicate a DNA template, and it produces a mixture of DNA fragments of different lengths. **Electrophoresis** separates the fragments by length into bands.

Section 15.4 **Genomics** gives us insights into the function of the human genome. Similarities between genomes of different organisms are evidence of evolutionary relationships, and can be used as a predictive tool in research.

DNA profiling identifies a person by the unique parts of his or her DNA. DNA profiling techniques reveal an individual's array of short tandem repeats or **single-nucleotide polymorphisms (SNPs)**. Within the context of a criminal investigation, a DNA profile is called a DNA fingerprint.

Section 15.5 Recombinant DNA technology is the basis of **genetic engineering**, the directed modification of an organism's genetic makeup with the intent to modify its phenotype. A gene from one species can be inserted into an individual of a different species to make a **transgenic** organism, or a modified gene may be reinserted into the same individual. The result of either process is a **genetically modified organism (GMO)**. Bacteria and yeast, the most common genetically engineered organisms, produce proteins that have medical value. Most transgenic crop plants, which are now in widespread use worldwide, were created to help farmers produce food more efficiently. Animals being produced by genetic engineering are used for medical applications, research, and food.

Section 15.6 With **gene therapy**, a gene is transferred into a person's body cells to correct a genetic defect or treat a disease. Potential benefits of gene editing must be weighed against potential risks. Recent advances in gene editing technologies are accelerating research.

Application Personal genetic testing, which reveals an individual's unique array of SNPs, is revolutionizing medicine. Some SNPs are associated with health risks. However, it is not yet possible to accurately predict the effect of an individual's SNPs on his or her health.

SELF-ASSESSMENT
ANSWERS IN APPENDIX VII

1. _____ cut(s) DNA molecules at specific sites.
   a. DNA polymerase   c. Restriction enzymes
   b. DNA probes   d. DNA ligase

2. A _____ is a molecule that can be used to carry a fragment of DNA into a host organism.
   a. cloning vector   c. GMO
   b. chromosome   d. cDNA

3. Reverse transcriptase assembles a(n) _____ on a(n) _____ template.
   a. mRNA; DNA   c. DNA; ribosome
   b. cDNA; mRNA   d. protein; mRNA

4. For each species, all _____ in the complete set of chromosomes is/are the _____ .
   a. genomes; phenotype   c. mRNA; start of cDNA
   b. DNA; genome   d. cDNA; start of mRNA

5. A set of cells that host various DNA fragments collectively representing an organism's entire set of genetic information is called a _____ .
   a. genome   c. genomic library
   b. clone   d. GMO

6. _____ is a technique to determine the order of nucleotides in a sample of DNA.
   a. PCR   c. Electrophoresis
   b. Sequencing   d. Nucleic acid hybridization

7. Electrophoresis separates fragments of DNA according to _____ .
   a. sequence   b. length   c. species

8. PCR can be used _____ .
   a. as a cloning vector
   b. in DNA profiling
   c. to modify a human genome

9. An individual's set of unique _____ can be used as a DNA profile.
   a. DNA sequences   c. short tandem repeats
   b. SNPs   d. all of the above

ENHANCED SPATIAL LEARNING ABILITY IN MICE WITH AN AUTISM MUTATION Autism is a neurobiological disorder with symptoms that include impaired social interactions and stereotyped patterns of behavior; around 10 percent of autistic people also have an extraordinary skill or talent such as greatly enhanced memory. Mutations in the gene for neuroligin 3, an adhesion protein that connects brain cells to one another, have been associated with autism. One of these mutations is called *R451C* because the altered gene encodes a protein with an amino acid substitution: a cysteine (C) instead of an arginine (R) in position 451.

In 2007, Katsuhiko Tabuchi and his colleagues introduced the *R451C* mutation into the neuroligin 3 gene of mice. The researchers discovered that the genetically modified mice had impaired social behavior, and also enhanced spatial learning ability.

Spatial learning in mice is tested with a water maze, which consists of a small platform submerged a bit below the surface of a pool of water so it is invisible to swimming mice. Mice do not particularly enjoy swimming, so they try to locate the hidden platform as quickly as they can. When tested again later, they remember the platform's location by checking visual cues around the edge of the pool. How quickly they remember is a measure of their spatial learning ability. **Figure 15.16** shows some of Tabuchi's results.

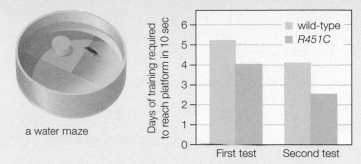

a water maze

Figure 15.16 Spatial learning ability in mice with a mutation in neuroligin 3 (*R451C*), compared to unmodified (wild-type) mice.

1. In the first test, how many days did it take unmodified mice to learn to find the location of a hidden platform in 10 seconds?

2. Did the modified or the unmodified mice learn the location of the platform faster in the first test?

3. Which mice learned faster the second time around?

4. Which mice had the greatest improvement in memory?

10. *Taq* polymerase is used for PCR because it _____ .
    a. tolerates the high temperature needed to separate DNA strands
    b. is an enzyme from a bacterium
    c. does not require primers
    d. is genetically modified

11. Put the following tasks in the order they would occur during a DNA cloning experiment.
    a. using DNA ligase to seal DNA fragments into vectors
    b. using a probe to identify a clone in the library
    c. sequencing the DNA of the clone
    d. making a DNA library of clones
    e. cutting genomic DNA with restriction enzymes

12. True or false? Some transgenic organisms can pass their foreign genes to offspring.

13. _____ can correct a genetic defect in an individual.
    a. Cloning vectors   c. Sequencing
    b. Gene therapy   d. Electrophoresis

14. True or false? Some humans are genetically modified.

15. Match each term with the most suitable description.
    ___ DNA profile
    ___ Ti plasmid
    ___ cDNA synthesis
    ___ SNP
    ___ transgenic
    ___ restriction enzyme

    a. GMO with a foreign gene
    b. alleles commonly contain them
    c. a person's unique collection of short tandem repeats
    d. requires reverse transcriptase
    e. cuts DNA
    f. used in plant gene transfers

CLASS ACTIVITY

CRITICAL THINKING

1. Restriction enzymes in bacterial cytoplasm cut injected bacteriophage DNA wherever certain sequences occur. Why do you think these enzymes do not cut the bacterial chromosome, which is exposed to the enzymes in cytoplasm?

2. The results of a paternity test are shown in the table below. Numbers indicate the number of short tandem repeats for loci tested. Who's the daddy? How sure are you?

Locus	Mother	Baby	Alleged Father #1	Alleged Father #2
CSF1PO	15, 17	17, 23	23, 27	17, 15
FGA	9, 9	9, 9	9, 12	9, 12
THO1	29, 29	29, 27	27, 28	29, 28
TPOX	14, 18	18, 20	15, 20	17, 22
VWA	14, 14	14, 14	14, 14	14, 16
D3S1358	11, 14	14, 16	12, 16	14, 20
D5S818	11, 13	10, 13	8, 10	18, 18
D7S820	7, 13	13, 13	13, 19	13, 13
D8S1179	13, 13	13, 15	12, 15	10, 12
D13S317	12, 12	10, 12	8, 10	12, 17
D16S539	12, 14	14, 12	14, 14	18, 25
D18S51	5, 6	6, 22	22, 6	5, 22

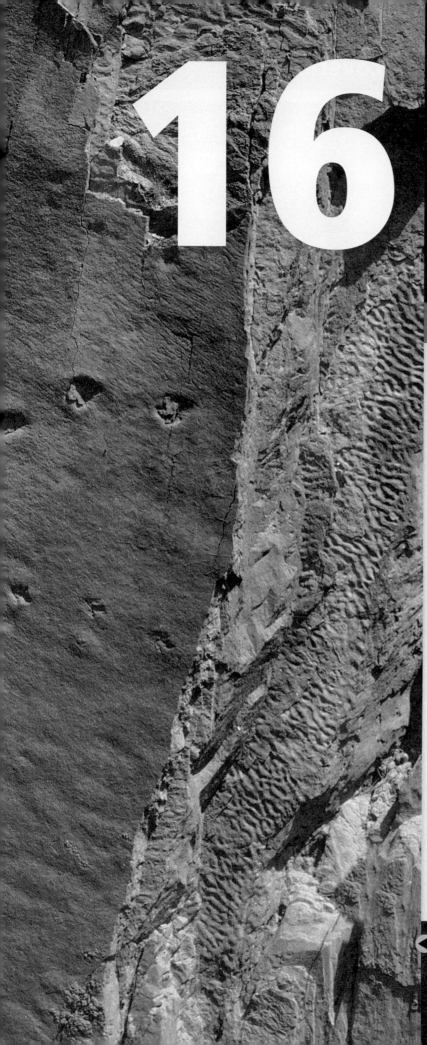

16 Evidence of Evolution

Links to Earlier Concepts

You may wish to review critical thinking (Section 1.5) before reading this chapter, which discusses a clash between belief and science (1.7). What you know about alleles (12.1) will help you understand natural selection. The chapter revisits radioisotopes (2.1), the effect of photosynthesis on Earth's early atmosphere (7.1), the genetic code and mutations (9.3, 9.5), master genes (10.2, 10.3), and gene duplications (14.4).

Core Concepts

The field of biology consists of and relies upon experimentation and the collection and analysis of scientific evidence.

We often reconstruct history by studying physical evidence of events that took place in the past. Reconstructing the history of life relies on a premise that is supported by a vast amount of data from many scientific disciplines: Natural phenomena that occurred in the past can be explained by the same physical, chemical, and biological processes operating today.

Natural selection leads to the evolution of structures and processes that increase fitness in a specific environment.

Members of a population must compete for resources that are limited. Individuals vary in the details of their shared traits, and those with forms of a trait that confer some advantage in this competition tend to leave more offspring (a process called natural selection). Thus, advantageous traits tend to become more common among members of a population over time. Change in a line of descent is called evolution.

All organisms alive today are linked by lines of descent from shared ancestors.

A record of ancient life persists in fossils, and also in the details of structure and function in modern organisms. Both provide evidence that evolution has occurred throughout the history of life, and also that all modern organisms are connected by shared ancestry.

High in the Andes, a scientist infers the stride of an extinct dinosaur by measuring the distance between its fossilized footprints on an ancient shoreline.

Photograph, © Louie Psihoyos/Corbis.

About 2,300 years ago, the Greek philosopher Aristotle described nature as a continuum of organization, from lifeless matter through complex plants and animals. Aristotle's work greatly influenced later European thinkers, who adopted his view of nature and modified it in light of their own beliefs. By the fourteenth century, Europeans generally believed that a "great chain of being" extended from the lowest form of life (plants), up through animals, humans, and spiritual beings. Each link in the chain was a species, and each was said to have been forged at the same time, in one place, and in a perfect state. The chain was complete. Because everything that needed to exist already did, there was no room for change.

In the 1800s, European naturalists embarked on globe-spanning survey expeditions and brought back tens of thousands of plants and animals from Asia, Africa, North and South America, and the Pacific Islands. Each newly discovered species was carefully catalogued as another link in the chain of being. The explorers began to see patterns in where species live and similarities in body plans, and they started to think about natural forces that shape life. These explorers were pioneers in **biogeography**, the study of patterns in the geographic distribution of species and communities. Some of the patterns raised questions that could not be answered within the framework of existing belief systems. For example, the European explorers had discovered plants and animals living in extremely isolated places. The isolated species looked suspiciously similar to species living on the other side of impassable mountain ranges, or across vast expanses of open ocean. Consider the emu, rhea, and ostrich, three types of bird native to three different continents. These birds share a set of unusual features (**Figure 16.1**). Alfred Wallace, an explorer particularly interested in the geographical distribution of animals, thought that the shared traits might mean that the birds descended from a common ancestor (and he was correct), but how could they have ended up on widely separated continents?

Naturalists of the time also had trouble classifying organisms that are very similar in some features but different in others. For example, both plants shown in **Figure 16.2** live in hot deserts where water is seasonally scarce. Both have rows of sharp spines that deter herbivores, and both store water in their thick, fleshy stems. However, their reproductive parts are very different, so these plants cannot be (and are not) as closely related as their outward appearance might otherwise suggest. Observations such as these are examples of

A Ostrich, native to Africa.

B Rhea, native to South America.

C Emu, native to Australia.

Figure 16.1 Similar-looking, related species native to distant geographic realms. These birds are unlike most others in several unusual features, including long, muscular legs and an inability to fly. All are native to open grassland regions about the same distance from the equator.

comparative morphology, the study of anatomical patterns: similarities and differences among the body plans of organisms. Today, comparative morphology is only one of several aspects of taxonomy (Section 1.4), but in the nineteenth century it was the only way to distinguish species. In some cases,

Figure 16.2 Similar-looking, unrelated species.
Left, saguaro cactus (*Carnegiea gigantea*), native to the Sonoran Desert of Arizona. Right, an African milk barrel plant (*Euphorbia horrida*), native to the Great Karoo desert of South Africa.

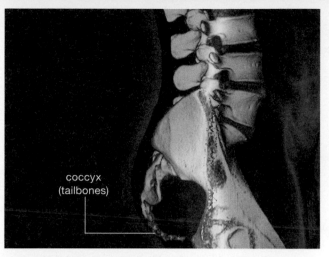

coccyx
(tailbones)

Figure 16.3 Human tailbones.
Nineteenth-century naturalists were well aware of—but had trouble explaining—body structures such as human tailbones that had apparently lost most or all function.

comparative morphology revealed anatomical details (body parts with no apparent function, for example) that added to the mounting confusion. If every species had been created in a perfect state, then why were there useless parts such as wings in birds that do not fly, eyes in moles that are blind, or remnants of a tail in humans (**Figure 16.3**)?

Fossils were puzzling too. A **fossil** is physical evidence—remains or traces—of an organism that lived in the ancient past. Fossilized skeletons of animals (such as gigantic dinosaurs) unlike any living ones were being unearthed. If these animals had been perfect at the time of creation, then why did they perish?

Geologists mapping rock formations exposed by erosion or quarrying had also discovered identical sequences of rock layers in different parts of the world. These layers held fossils of simple marine organisms that were clearly related to one another, but their form changed slightly from one layer to the next (**Figure 16.4**). Those in the uppermost layers resembled modern species. What did these sequences mean?

Taken as a whole, the accumulating discoveries from biogeography, comparative morphology, and geology challenged traditional beliefs of the nineteenth century. If species had not been created in a perfect state (and extinct species, fossil sequences, and "useless" body parts implied that they had not), then perhaps species had indeed changed over time.

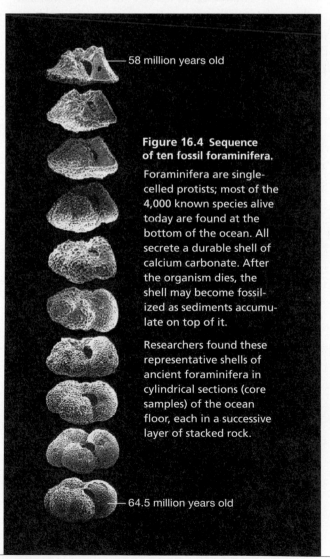

58 million years old

Figure 16.4 Sequence of ten fossil foraminifera.

Foraminifera are single-celled protists; most of the 4,000 known species alive today are found at the bottom of the ocean. All secrete a durable shell of calcium carbonate. After the organism dies, the shell may become fossilized as sediments accumulate on top of it.

Researchers found these representative shells of ancient foraminifera in cylindrical sections (core samples) of the ocean floor, each in a successive layer of stacked rock.

64.5 million years old

---

biogeography Study of patterns in the geographic distribution of species and communities.

comparative morphology (more-FALL-uh-jee) Scientific study of similarities and differences in body plans.

fossil Physical evidence of an organism that lived in the ancient past.

CREDITS: (2) left, © Marka/SuperStock; right, © Richard J. Hodgkiss, www.succulent-plant.com; (3) © Zephyr/Science Photo Library/Science Source; (4) Courtesy of Daniel C. Kelley, Anthony J. Arnold, and William C. Parker, Florida State University Department of Geological Science.

In the nineteenth century, naturalists were faced with increasing evidence that life on Earth, and even Earth itself, had changed. Around 1800, Georges Cuvier (left) was trying to make sense of the new information. He knew that many fossil species seemed to have no living counterparts. Given this evidence, he proposed an idea startling for the time: Many species that had once existed were now extinct. Cuvier also knew about evidence that Earth's surface had changed. For example, he had seen fossilized seashells in rocks at the tops of mountains far from modern seas. Like most others of his time, he assumed Earth's age to be in the thousands, not billions, of years. He reasoned that geologic forces unlike any that had ever been experienced would have been necessary to raise seafloors to mountaintops in such a short time span. Catastrophic geological events would have caused extinctions, after which surviving species repopulated the planet. We now know that this idea, which was called catastrophism, is incorrect: Geologic processes have not changed.

Another naturalist, Jean-Baptiste Lamarck, was thinking about processes that might drive **evolution**, or change in a

line of descent. A line of descent is also called a **lineage**. Lamarck (left) thought that a species gradually improved over generations because of an inherent drive toward perfection, up the chain of being. The drive directed an unknown "fluida" into body parts needing change. By Lamarck's hypothesis, environmental pressures cause an internal requirement for change in an individual's body, and the resulting change is inherited by offspring. Indeed, the environment does affect traits in a heritable way, but Lamarck's hypothesis about the mechanism of change was incorrect.

Charles Darwin (left) had earned a theology degree from Cambridge after an attempt to study medicine. All through school, however, he had spent most of his time with faculty members and other students who embraced natural history. In 1831, when he was 22, Darwin joined a 5-year survey expedition to South America on the ship *Beagle*, and he quickly became an enthusiastic naturalist. During the *Beagle*'s voyage, Darwin found many unusual fossils, and saw diverse species living in environments that ranged from the sandy shores of remote islands to plains high in the Andes. Along the way, he read *Principles of Geology*, a new and popular book

A Fossil of a glyptodon, an automobile-sized mammal that existed from 2 million to 15,000 years ago.

B A modern armadillo, about a foot long.

Figure 16.5 Ancient relatives: glyptodon and armadillo. Though widely separated in time, these animals share a restricted distribution and unusual traits, including a shell and helmet of keratin-covered bony plates similar to crocodile and lizard skin. (The fossil in **A** is missing its helmet.) Their similarities were a clue that helped Darwin develop the theory of natural selection.

by Charles Lyell (left). Lyell was a proponent of what became known as the theory of uniformity, the idea that gradual, everyday geologic processes such as erosion could have sculpted Earth's current landscape over great spans of time. The theory challenged the prevailing belief that Earth was 6,000 years old. By Lyell's calculations, it must have taken millions of years to sculpt Earth's surface. Darwin's exposure to Lyell's ideas gave him insights into the history of the regions he would encounter on his journey.

Among the thousands of specimens Darwin collected during the *Beagle*'s voyage were fossil glyptodons. These armored mammals are extinct, but they have many unusual traits in common with modern armadillos (**Figure 16.5**). Armadillos also live only in places where glyptodons once lived. Could the odd shared traits and restricted distribution mean that glyptodons were ancient relatives of armadillos? If so, perhaps traits of their common ancestor had changed in the line of descent that led to armadillos. But why would such changes have occurred?

After Darwin returned to England, he pondered his notes and fossils, and read an essay by one of his contemporaries, economist Thomas Malthus. Malthus (left) proposed the idea that famine, disease, and war limit the size of human populations. When people reproduce beyond the capacity of their environment to sustain them, they run out of food. Then, individuals compete with one another for limited resources. Some survive this struggle for existence, and some do not. Darwin realized that Malthus's ideas had wider application: Individuals of all populations, not just human ones, struggle for existence by competing for limited resources.

EVOLUTION BY NATURAL SELECTION

Reflecting on his journey, Darwin started thinking about how individuals of a species often vary a bit in the details of shared traits such as size, coloration, and so on. He saw such variation among finch species on isolated islands of the Galápagos archipelago. This island chain is separated from South America by 900 kilometers (550 miles) of open ocean, so most species living on them had no opportunity for interbreeding with mainland populations. The Galápagos island birds resembled finch species in South America, but unique traits suited many of them to their particular island habitats.

Darwin was familiar with dramatic variations in traits that selective breeding could produce in pigeons, dogs, and horses, and reasoned that natural environments could similarly "select" certain traits. In any natural population, some individuals have forms of shared traits that make them better suited to their environment than others. In other words, individuals of a population vary in fitness. Today, we define **fitness** as the degree of adaptation to a specific environment, and measure it by the individual's relative genetic contribution to future generations. An evolutionary **adaptation**, or **adaptive trait**, is a form of a heritable trait that enhances fitness.

Darwin realized that individuals with adaptive traits would tend to leave more offspring than their less fit rivals, a process he named **natural selection**. He understood that natural selection causes a population to change—in other words, it drives evolution. **Table 16.1** summarizes this reasoning. If an individual has an adaptive trait that makes it better suited to an environment, then it is better able to survive. If an individual is

---

adaptive trait (adaptation) A form of a heritable trait that enhances an individual's fitness.
evolution Change in a line of descent.
fitness Degree of adaptation to an environment, as measured by the individual's relative genetic contribution to future generations.
lineage (LINN-edge) Line of descent.
natural selection Differential survival and reproduction of individuals of a population based on differences in shared, heritable traits.

TABLE 16.1

Evolution by Natural Selection

Observations About Populations

> Natural populations have an inherent capacity to increase in size over time.

> As a population expands, resources that are used by its members (such as food and living space) eventually become limited.

> When resources are limited, members of a population compete for them.

Observations About Genetics

> Individuals of a species share certain traits.

> Individuals of a natural population vary in the details of those shared traits.

> Shared traits have a heritable basis, in genes. Slightly different forms of those genes (alleles) give rise to variation in shared traits.

Inferences

> A certain form of a shared trait may make its bearer better able to survive.

> Individuals of a population that are better able to survive tend to leave more offspring.

> Thus, an allele associated with an adaptive trait tends to become more common in a population over time.

---

better able to survive, then it has a better chance of living long enough to produce offspring. If individuals with an adaptive trait produce more offspring than other individuals, then the frequency of the adaptive trait in the population will increase over successive generations (the population will evolve).

Darwin wrote out his hypothesis of evolution by natural selection in the late 1830s, but was so conflicted about its implications that he collected evidence for twenty years without sharing it. Meanwhile, Alfred Wallace (left), who was studying wildlife in the Amazon basin and the Malay Archipelago, wrote to Darwin and Lyell about patterns in the geographic distribution of species. When he was ready to publish his own ideas about evolution, he sent them to Darwin for advice. To Darwin's shock, Wallace had come up with the same hypothesis: that evolution occurs by natural selection.

In 1858, the hypothesis of evolution by natural selection was presented at a scientific meeting. Darwin and Wallace were credited as authors, but neither attended the meeting. The next year, Darwin published *On the Origin of Species*, which laid out detailed evidence in support of the hypothesis. Many people had already accepted the idea of descent with modification (evolution). However, there was a fierce debate over the idea that evolution occurs by natural selection. Decades would pass before experimental evidence from the field of genetics led to its widespread acceptance as a theory by the scientific community. As with all scientific theories, the theory of evolution by natural selection has not been falsified after years of rigorous testing, and it has proved useful for making predictions about a wide range of other phenomena.

LEARNING OBJECTIVES

Give examples of three types of fossils and what kind of information each can provide about past life.

Explain the process of fossilization and why fossils are relatively rare.

Describe the kinds of evidence about the history of life that can be found in sedimentary rock formations.

Even before Darwin's time, fossils were recognized as stone-hard evidence of earlier forms of life (**Figure 16.6**). Most fossils consist of mineralized bones, teeth, shells, seeds, spores, or other durable parts of an ancient organism. Trace fossils such as footprints and other impressions, nests, burrows, trails, eggshells, or feces are evidence of activities.

The process of fossilization typically begins when an organism or its traces become covered by sediments, mud, or ash. Groundwater then seeps into the remains, filling spaces around and inside of them. Minerals dissolved in the water gradually replace minerals in bones and other hard tissues. Mineral particles that settle out of the groundwater and crystallize inside cavities and impressions form detailed imprints of internal and external structures. Sediments that slowly accumulate on top of the site exert increasing pressure, and, after a very long time, extreme pressure transforms the mineralized remains into rock.

EVIDENCE IN ROCK FORMATIONS

Most fossils are found in layers of sedimentary rock that forms as rivers wash silt, sand, volcanic ash, and other materials from land to sea. Mineral particles in the materials settle on the seafloor in horizontal layers that vary in thickness and composition. After many millions of years, the layers of sediments become buried and compacted into layered sedimentary rock. Geologic processes can tilt the rock and lift it far above sea level, where the layers may become exposed by the erosive forces of water and wind (the chapter opening photo shows an example).

Biologists study sedimentary rock formations in order to understand the historical context of ancient life. Usually, the deepest layers in a particular formation of rock were the first to form, and those closest to the surface formed most recently. Thus, in general, the deeper the layer, the older the fossils it contains. The layers themselves contain information about environmental conditions and events that occurred as they

formed. Consider banded iron, a unique formation named after its distinctive striped appearance (left). Huge deposits of this sedimentary rock are the source of most iron we mine for steel today, but

A Fossil skeleton of an ichthyosaur that lived about 200 million years ago. These marine reptiles were about the same size as modern porpoises, breathed air like them, and probably swam as fast, but the two groups are not closely related.

B Extinct wasp encased in amber, which is ancient tree sap. This 9-mm-long insect lived about 20 million years ago.

C Fossilized imprint of a leaf from a 260-million-year-old *Glossopteris,* a type of plant called a seed fern.

D Fossilized footprint of a theropod, a name that means "beast foot." This group of carnivorous dinosaurs, which includes the familiar *Tyrannosaurus rex*, arose about 250 million years ago.

E Coprolite (fossilized feces). Fossilized food remains and parasitic worms inside coprolites offer clues about the diet and health of extinct species. A foxlike animal excreted this one.

Figure 16.6 Examples of fossils.

CREDITS: (in text) Natural History Museum, London/Science Photo Library/Science Source; (6A) © Jonathan Blair; (6B) © Dr. Michael Engel, University of Kansas; (6C) © Martin Land/Science Source; (6D) © Pixtal/SuperStock; (6E) Courtesy of Stan Celestian/Glendale Community College Earth Science Image Archive.

they also provide a record of how the evolution of the noncyclic pathway of photosynthesis changed the chemistry of Earth. Banded iron became abundant about 2.4 billion years ago, after this pathway evolved (Section 7.1). Before then, Earth's atmosphere and ocean contained very little oxygen, so almost all of the iron on Earth was in a reduced form (Section 5.4). Reduced iron dissolves in water, and ocean water contained a lot of it. Oxygen released into the ocean by early photosynthetic bacteria quickly combined with the dissolved iron. The resulting oxidized iron compounds are completely insoluble in water, and they began to rain down on the ocean floor in massive quantities. These compounds accumulated in sediments that would eventually become compacted into banded iron formations. This process continued for about 600 million years. After that, ocean water no longer contained very much dissolved iron, and oxygen gas bubbling out of it had oxidized the iron in rocks exposed to the atmosphere.

THE FOSSIL RECORD

We have fossils for more than 250,000 known species. Considering the current range of biodiversity, there must have been many millions more, but we will never know all of them. Why not? The odds are against finding evidence of an extinct species. Few individuals of any species become fossilized in the first place. Typically, when an organism dies, its remains are obliterated quickly by scavengers. Organic materials decompose in the presence of moisture and oxygen, so remains that escape scavenging endure only if they dry out, freeze, or become encased in an air-excluding material such as sap, tar, or mud. Most remains that do become fossilized are crushed or scattered by erosion and other geologic assaults.

In order for us to know about an extinct species that existed long ago, we have to find a fossil of it. At least one specimen had to be buried before it decomposed or something ate it. The burial site had to escape destructive geologic events, and end up in a place that we can find today.

Most ancient species had no hard parts to fossilize, so we do not find much evidence of them. For example, there are many fossils of bony fishes and mollusks with hard shells, but few fossils of the jellyfishes and soft worms that were probably much more common. Also think about relative numbers of organisms. Fungal spores and pollen grains are typically released by the billions. By contrast, the earliest humans lived in small bands and few of their offspring survived. The odds of finding even one fossilized human bone are much smaller than the odds of finding a fossilized fungal spore. Finally, imagine two species, one that existed only briefly and the other for a hundred million years. Which is more likely to be represented in the fossil record?

Dr. Paul Sereno

A real-life Indiana Jones, National Geographic Explorer Paul Sereno blends his background as an artist with a love for science and history. Sereno's passion carries him to the remote corners of the world to discover new species under the harshest of conditions.

Sereno's fieldwork began in 1988 in the Andes, where his team discovered the first dinosaurs to roam the Earth, including the most primitive of all: *Eoraptor*. This work culminated in the most complete picture yet of the dawn of the dinosaur era, some 225 million years ago. In the 1990s, Sereno's expeditions shifted to the Sahara. There, his teams have since excavated more than 70 tons of fossils representing organisms such as the huge-clawed fish-eater *Suchomimus*, the gigantic *Carcharodontosaurus* (its jaws are pictured above, with Sereno), and a series of crocs including the 40-foot-long "SuperCroc" *Sarcosuchus*, the world's largest crocodile. In 2001, a trip to India yielded the Asian continent's first dinosaur skull—a new species of predator, *Rajasaurus*. Also in 2001, Sereno began an ongoing series of expeditions to China, first exploring remote areas of the Gobi in Inner Mongolia and discovering a herd of more than 20 dinosaurs that had died in their tracks. In 2012, he reported the discovery of *Pegomastax*, a bizarre, cat-sized dinosaur with a parrotlike beak and sharp fangs.

LEARNING OBJECTIVE

Explain how radioisotope decay can be used to determine the age of fossils.

Researchers use the predictability of radioactive decay (Section 2.1) to reveal the age of a fossil or other ancient object. Like the ticking of a perfect clock, each type of radioisotope decays at a constant rate into predictable products (daughter elements). The time it takes for half of the atoms in a sample of a radioisotope to decay is called its **half-life** (**Figure 16.7**).

Almost all carbon on Earth is in the form of ^{12}C; a tiny but constant amount of ^{14}C forms in the upper atmosphere and becomes incorporated into molecules of carbon dioxide. The ratio of ^{14}C to ^{12}C in atmospheric CO_2 is relatively stable; because carbon dioxide is life's main source of carbon, the same ratio occurs in the body of a living organism. As long as the organism lives, it continues to assimilate both isotopes in the same proportions. After it dies, no more carbon is assimilated, and the ratio of ^{14}C to ^{12}C in its remains declines over time as the ^{14}C decays. The half life of ^{14}C is known (5,730 years) so the ratio of ^{14}C to ^{12}C in the organism's remains can be used to calculate how long ago it died (**Figure 16.8**).

We have just described **radiometric dating**, a method that can reveal the age of a material by measuring its radioisotope content. Carbon isotope dating can be used to find the age of a biological material less than 60,000 years old; in older material, all of the ^{14}C has decayed. The age of older fossils can be estimated by dating volcanic rocks in lava flows above and below the fossil-containing layer of sedimentary rock.

How is rock dated? The original source of most rock on Earth is magma, a hot, molten material deep under Earth's surface. Atoms swirl and mix in this material. When magma cools, for example after reaching the surface as lava, it hardens and becomes rock. As this occurs, the atoms in it join and crystallize as different kinds of minerals, each with a characteristic structure and composition. Consider zircon, a mineral that consists mainly of zirconium silicate molecules ($ZrSiO_4$). Some of the molecules in a newly formed zircon crystal have uranium atoms substituted for zirconium atoms, but never lead atoms. Uranium is a radioactive element with a half-life of 4.5 billion years. Its daughter element, thorium, is another radioactive element, and it decays at a predictable rate into another radioactive element, and so on until the atom becomes lead, a stable element. Over time, uranium atoms disappear from a zircon crystal, and lead atoms accumulate in it. The ratio of uranium atoms to lead atoms in a zircon crystal can be measured and used to calculate how long ago the crystal formed (its age). Using this technique, we know that the oldest known terrestrial rock, a tiny zircon crystal from the Jack Hills in Western Australia, is 4.404 billion years old.

zircon

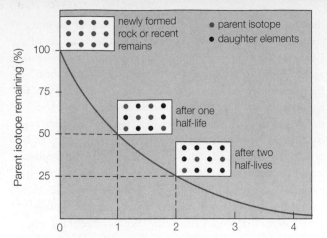

Figure 16.7 Half-life.

? FIGURE IT OUT
How much of any radioisotope remains after two of its half-lives have passed?   Answer: 25 percent

A Long ago, ^{14}C and ^{12}C were incorporated into the tissues of a nautilus. The carbon atoms were part of organic molecules in the animal's food. ^{12}C is stable and ^{14}C decays, but the proportion of the two isotopes in the nautilus's tissues remained the same. Why? The nautilus continued to gain both types of carbon atoms in the same proportions from its food.

B The nautilus stopped eating when it died, so its body stopped gaining carbon. The ^{12}C atoms in its tissues were stable, but the ^{14}C atoms (represented as red dots) were decaying into nitrogen atoms. Thus, over time, the amount of ^{14}C decreased relative to the amount of ^{12}C. After 5,730 years, half of the ^{14}C had decayed; after another 5,730 years, half of what was left had decayed, and so on.

C Fossil hunters discover the fossil and measure its content of ^{14}C and ^{12}C. They use the ratio of these isotopes to calculate how many half-lives passed since the organism died. For example, if its ^{14}C to ^{12}C ratio is one-eighth of the ratio in living organisms, then three half-lives $(1/2)^3$ must have passed since it died. Three half-lives of ^{14}C is 17,190 years.

Figure 16.8 Example of how radiometric dating is used to find the age of a carbon-containing fossil. Carbon 14 (^{14}C) is a radioisotope of carbon that decays into nitrogen. It forms in the atmosphere and combines with oxygen to become CO_2, which enters food chains by way of photosynthesis.

MISSING LINKS

The discovery of intermediate forms of cetaceans (an order of animals that includes whales, dolphins, and porpoises) offers an example of how fossil finds and radiometric dating can be used to reconstruct evolutionary history.

Modern cetacean skeletons have remnants of a pelvis and hindlimbs (**Figure 16.9A**), so evolutionary biologists had long thought that the ancestors of this group walked on land. However, no one had found fossils demonstrating a clear link between cetaceans and ancient land animals, so the rest of the story remained speculative. Intact fossil skeletons of extinct whale-like *Dorudon atrox* had been discovered, but these animals were clearly cetaceans, with little resemblance to land animals. *Dorudon* lived about 37 million years ago, and like modern whales it was a tail-swimmer and fully aquatic: The tiny hindlimbs could not have supported the animal's huge body out of water (**Figure 16.9B**).

In the early 1990s, DNA sequence comparisons suggested that modern cetaceans are more related to artiodactyls than to other groups. Artiodactyls are hooved mammals with an even number of toes (two or four) on each foot; modern representatives of the lineage include camels, hippopotamuses, pigs, antelopes, and sheep. The DNA findings were controversial. Some scientists thought the ancestors of cetaceans were Mesonychids, a lineage of carnivorous mammals; the ancestors of artiodactyls were a lineage of herbivores that looked a bit like tiny deer with long tails. Either way, a seemingly unimaginable number of skeletal and physiological changes would have been required for small, land-based animals to evolve into whales with gigantic bodies uniquely adapted to deep ocean swimming.

Then, in 2000, Philip Gingerich and his colleagues unearthed complete fossilized skeletons of two cetaceans in a 47-million-year-old rock formation in Pakistan. Later named *Artiocetus clavis* and *Rodhocetus balochistanensis*, both animals had whalelike skulls and robust hindlimbs (**Figure 16.9C**). Their bodies were built to swim with their feet (not their tails as whales do), and their ankle bones had clearly distinctive features of artiodactyls—settling the debate about cetacean ancestry (**Figure 16.9D**). *Rodhocetus* and *Artiocetus* were not direct ancestors of modern whales, but their telltale ankle bones mean they were long-lost relatives. Both ancient cetaceans were offshoots of the artiodactyl-to-modern-whale lineage as it transitioned from life on land to life in water.

half-life Characteristic time it takes for half of a quantity of a radioisotope to decay.

radiometric dating Method of estimating the age of a rock or fossil by measuring its radioisotope content.

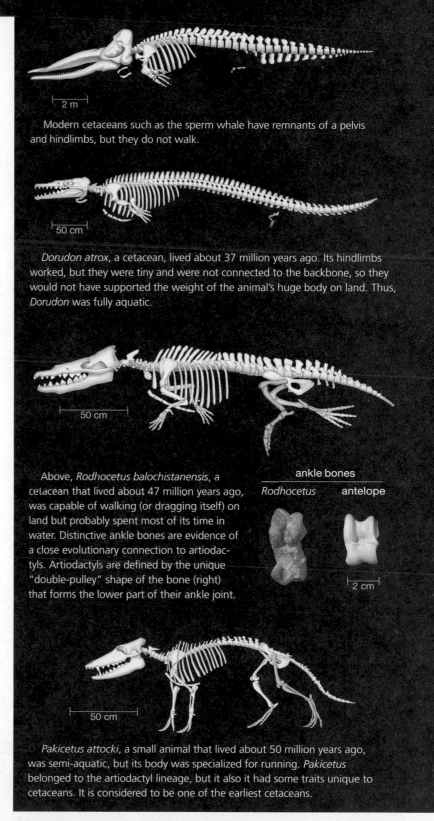

Modern cetaceans such as the sperm whale have remnants of a pelvis and hindlimbs, but they do not walk.

Dorudon atrox, a cetacean, lived about 37 million years ago. Its hindlimbs worked, but they were tiny and were not connected to the backbone, so they would not have supported the weight of the animal's huge body on land. Thus, *Dorudon* was fully aquatic.

Above, *Rodhocetus balochistanensis*, a cetacean that lived about 47 million years ago, was capable of walking (or dragging itself) on land but probably spent most of its time in water. Distinctive ankle bones are evidence of a close evolutionary connection to artiodactyls. Artiodactyls are defined by the unique "double-pulley" shape of the bone (right) that forms the lower part of their ankle joint.

ankle bones

Rodhocetus   antelope

Pakicetus attocki, a small animal that lived about 50 million years ago, was semi-aquatic, but its body was specialized for running. *Pakicetus* belonged to the artiodactyl lineage, but it also it had some traits unique to cetaceans. It is considered to be one of the earliest cetaceans.

Figure 16.9 Comparison of cetacean skeletons.
The ancestor of whales was an artiodactyl that walked on land. Over millions of years, the lineage transitioned from life on land to life in water, and as it did, bones of the hindlimb (highlighted in blue) became smaller.

CREDITS: (9A) W. B. Scott (1894); (9B) top, Doug Boyer in P. D. Gingerich et al. (2001) © American Association for Advancement of Science; bottom left and right, © Philip Gingerich/University of Michigan; (9C) © P. D. Gingerich and M. D. Uhen (1996), © University of Michigan. Museum of Paleontology.

16.5 Drifting Continents, Changing Seas

LEARNING OBJECTIVES

Explain how continents move around Earth.

Describe some geological features that form as a result of plate tectonics, and explain how these features arise.

Wind, water, and other forces continuously sculpt Earth's surface, but they are only part of a much bigger picture of geological change. For example, all continents on Earth today were once part of a bigger supercontinent—**Pangea**—that split into fragments and drifted apart. The idea that continents move around, originally called continental drift, was proposed in the early 1900s to explain why the Atlantic coasts of South America and Africa seem to "fit" like jigsaw puzzle pieces, and why the same types of fossils occur in identical rock formations on both sides of the Atlantic Ocean. The concept of continental drift also explained why the magnetic poles of gigantic rock formations point in different directions on different continents. As magma solidifies into rock, some iron-rich minerals in it become magnetic, and their magnetic poles align with Earth's poles when they do. If the continents never moved, then all of these ancient rocky magnets should be aligned north-to-south, like compass needles. Indeed, the magnetic poles of rocks in each formation are aligned with one another, but the alignment is not always north-to-south. Either Earth's magnetic poles have veered dramatically from their north–south axis, or the continents have wandered.

Continental drift was initially greeted with skepticism because there was no known mechanism for continents to move. Then, in the late 1950s, deep-sea explorers found huge ridges and trenches stretching thousands of kilometers across the seafloor. The discovery led to an explanation of continental drift (**Figure 16.10**). By the **plate tectonics theory**, Earth's outer layer of rock is cracked into huge plates, like a gigantic cracked eggshell. Magma welling up at an undersea ridge ❶ or continental rift at one edge of a plate pushes old rock at the opposite edge into a trench ❷. The movement is like that of a colossal conveyor belt that slowly transports continents on top of it to new locations. The plates move no more than 10 centimeters (4 inches) a year—about half as fast as your toenails grow—but it is enough to carry a continent all the way around the world after 40 million years or so.

The San Andreas Fault, which extends 800 miles through California, marks the boundary between two tectonic plates.

Figure 16.10 Plate tectonics. Slowly moving pieces of Earth's outer layer of rock convey continents around the globe.

❶ At oceanic ridges, magma (red) welling up from Earth's interior drives the movement of tectonic plates. New crust spreads outward as it forms on the surface, forcing adjacent tectonic plates away from the ridge and into trenches elsewhere.

❷ At trenches, the advancing edge of one plate plows under an adjacent plate and buckles it.

❸ Faults (ruptures in Earth's crust) often occur where plates meet.

❹ Magma ruptures Earth's crust at what are called "hot spots."

❺ An archipelago forms as a tectonic plate moves over a hot spot. The Hawaiian Islands have been forming from magma that continues to erupt from a hot spot under the Pacific Plate. This and other tectonic plates are shown in Appendix V.

FIGURE IT OUT

Is the tectonic plate over the hot spot moving left to right, or right to left?

Answer: Left to right

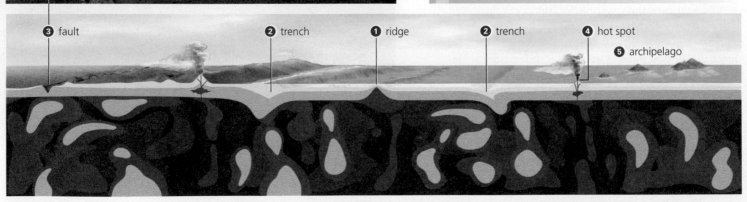

❸ fault   ❷ trench   ❶ ridge   ❷ trench   ❹ hot spot   ❺ archipelago

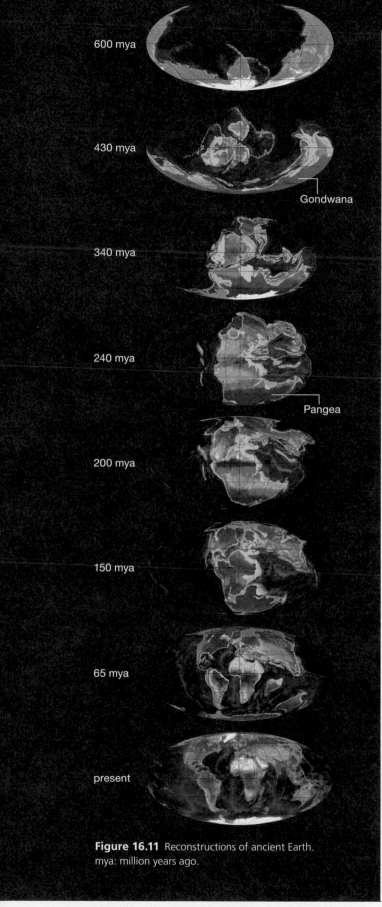

Figure 16.11 Reconstructions of ancient Earth. mya: million years ago.

600 mya

430 mya

Gondwana

340 mya

240 mya

Pangea

200 mya

150 mya

65 mya

present

Evidence of tectonic plate movement is all around us, in geological features such as faults. Faults are gigantic cracks in Earth's crust, and they often occur where plates meet ❸. As another example, consider volcanic hot spots, which are places where plumes of molten rock well up from deep inside Earth and rupture its crust ❹. Volcanic island chains (archipelagos) form as a plate moves across an undersea hot spot ❺. We can see this process occurring in archipelagos such as the Hawaiian Islands, where the hot spot is still active.

The fossil record also provides evidence in support of plate tectonics. Consider an unusual geological formation that occurs in a belt across Africa. The sequence of rock layers in this formation is so complex that it is quite unlikely to have formed more than once, but identical sequences also occur in huge belts that span India, South America, Madagascar, Australia, and Antarctica. Across all of these continents, the layers are the same ages. They also hold fossils found nowhere else, including remains of the seed fern *Glossopteris* (pictured in Figure 16.6C), which lived 299–252 million years ago, and an early reptile called *Lystrosaurus* that existed 270–225 million years ago. The unusual layered rock formation that contains fossils of these organisms must have formed in one long belt on a single continent that later broke up.

We have evidence of at least five supercontinents that formed and split up again since Earth's outer layer of rock solidified 4.55 billion years ago. One of them, a supercontinent called **Gondwana**, existed around 540 million years ago. Over the next 260 million years, Gondwana wandered across the South Pole, then drifted north until it merged with another supercontinent to form Pangea about 300 million years ago (**Figure 16.11**). Most of the landmasses currently in the Southern Hemisphere as well as India and Arabia were once part of Gondwana. Some modern species, including the birds pictured in Figure 16.1, live only in these places.

When two continents collided into one, they brought together organisms that had been living apart on the separate landmasses, and they physically separated organisms living in an ocean. When one continent broke up, organisms living on it would have been separated, and those living in different parts of the ocean would have come together. Events like these have been a major driving force of evolution, as you will see in the next chapter.

Gondwana (gond-WAN-uh) Supercontinent that existed before Pangea, more than 500 million years ago.

Pangea (pan-JEE-uh) Supercontinent that began to form about 300 million years ago; broke up 100 million years later.

plate tectonics theory Theory that Earth's outermost layer of rock is cracked into plates, the slow movement of which conveys continents to new locations over geologic time.

LEARNING OBJECTIVE

Explain how layers of sedimentary rock can be correlated with evolutionary events in the geologic time scale.

Similar sequences of sedimentary rock layers occur around the world. The **geologic time scale** is a chronology of Earth's history that correlates these sequences with time. Transi-

tions between layers of rock mark boundaries between time intervals (**Figure 16.12**). Each layer's composition offers clues about conditions that prevailed on Earth when it was deposited. For example, fossils are a record of life during the period of time that sediments in a layer accumulated.

geologic time scale Chronology of Earth's history.

Figure 16.12 The geologic time scale (below) **correlated with sedimentary rock exposed by erosion in the Grand Canyon** (opposite). Red triangles mark great mass extinctions. "First appearance" refers to appearance in the fossil record, not necessarily first appearance on Earth. * mya: million years ago. Dates are from the International Commission on Stratigraphy, 2014.

Eon	Era	Period	Epoch	mya*	Major Geologic and Biological Events
Phanerozoic	Cenozoic	Quaternary	Holocene	0.01	Modern humans evolve. Major extinction event is now under way.
			Pleistocene	2.6	
		Neogene	Pliocene	5.3	Tropics, subtropics extend poleward. Climate cools; dry woodlands and grasslands emerge. Adaptive radiations of mammals, insects, birds.
			Miocene	23.0	
		Paleogene	Oligocene	33.9	
			Eocene	56.0	
			Paleocene	66.0 ◄	**Major extinction event**
	Mesozoic	Cretaceous	Upper		Flowering plants diversify; sharks evolve. All dinosaurs and many marine organisms disappear at the end of this epoch.
				100.5	
			Lower		Climate very warm. Dinosaurs continue to dominate. Important modern insect groups appear (bees, butterflies, termites, ants, and herbivorous insects including aphids and grasshoppers). Flowering plants originate and become dominant land plants.
				145.0	
		Jurassic			Age of dinosaurs. Lush vegetation; abundant gymnosperms and ferns. Birds appear. Pangea breaks up.
				201.3 ◄	**Major extinction event**
		Triassic			Recovery from the major extinction at end of Permian. Many new groups appear, including turtles, dinosaurs, pterosaurs, and mammals.
				252 ◄	**Major extinction event**
	Paleozoic	Permian			Supercontinent Pangea and world ocean form. Adaptive radiation of conifers. Cycads and ginkgos appear. Relatively dry climate leads to drought-adapted gymnosperms and insects such as beetles and flies.
				299	
		Carboniferous			High atmospheric oxygen level fosters giant arthropods. Spore-releasing plants dominate. Age of great lycophyte trees; vast coal forests form. Ears evolve in amphibians; penises evolve in early reptiles (vaginas evolve later, in mammals only).
				359 ◄	**Major extinction event**
		Devonian			Land tetrapods appear. Explosion of plant diversity leads to tree forms, forests, and many new plant groups including lycophytes, ferns with complex leaves, seed plants.
				419	
		Silurian			Radiations of marine invertebrates. First appearances of land fungi, vascular plants, bony fishes, and perhaps terrestrial animals (millipedes, spiders).
				443 ◄	**Major extinction event**
		Ordovician			Major period for first appearances. The first land plants, fishes, and reef-forming corals appear. Gondwana moves toward the South Pole and becomes frigid.
				485	
		Cambrian			Earth thaws. Explosion of animal diversity. Most major groups of animals appear (in the oceans). Trilobites and shelled organisms evolve.
				541	
Precambrian	Proterozoic				Oxygen accumulates in atmosphere. Origin of aerobic metabolism. Origin of eukaryotic cells, then protists, fungi, plants, animals. Evidence that Earth mostly freezes over in a series of global ice ages between 750 and 600 mya.
				2,500	
	Archean and earlier				3,800–2,500 mya. Origin of bacteria and archaea.
					4,600–3,800 mya. Origin of Earth's crust, first atmosphere, first seas. Chemical, molecular evolution leads to origin of life (from protocells to anaerobic single cells).

Permian

(Ka) Kaibab Limestone

(To) Toroweap Formation

(Co) Coconino Sandstone

(He) Hermit Shale

(Es) Esplanade Sandstone

(We) Wescogame Formation

Carboniferous

(Ma) Manakacha Formation

(Wa) Watahomigi Formation

(Re) Redwall Limestone

(Te) Temple Butte Formation

(Mu) Muav Limestone

Cambrian

(Br) Bright Angel Shale

(Ta) Tapeats Sandstone

Proterozoic

Chuar Group

Nankoweap Formation

Unkar Group

(Vi) Vishnu Basement Rocks

* Layers not visible in this view of the Grand Canyon

Each rock layer has a composition and set of fossils that reflect events during its deposition. For example, Coconino Sandstone, which stretches from California to Montana, is mainly weathered sand. Ripple marks and reptile tracks are the only fossils in it. Many think it is the remains of a vast sand desert, similar to the modern Sahara.

FIGURE IT OUT
Which formation is marked with the **?**

Answer: Tapeats Sandstone

Evolutionary biologists are a bit like detectives, using clues to piece together history that no human witnessed. Fossils provide some clues. The body form and function of modern organisms provide others.

MORPHOLOGICAL DIVERGENCE

Comparative morphology is often used to unravel evolutionary relationships. Body parts that appear similar in separate lineages because they evolved in a common ancestor are called **homologous structures** (*hom–* means "the same"). Homologous structures may be used for different purposes in different groups, but the same genes direct their development.

A body part that outwardly appears very different in separate lineages may be homologous in underlying form. Vertebrate forelimbs, for instance, vary in size, shape, and function. However, they are alike in the structure and positioning of internal elements such as bones, nerves, blood vessels, and muscles.

Populations that are not interbreeding tend to diverge genetically, and in time these genetic divergences give rise to changes in body form. Change from the body form of a common ancestor is an evolutionary pattern called **morphological divergence**. Consider the limbs of modern vertebrate animals. Fossil evidence suggests that many vertebrates are descended from a family of ancient "stem reptiles" that crouched low to the ground on five-toed limbs. Descendants of this ancestral group diversified over millions of years, and eventually gave rise to modern reptiles, birds, and mammals. As you learned in Section 16.4, a few lineages even returned to life in the seas. As these lineages diversified, their five-toed limbs became adapted for appropriate purposes (**Figure 16.13**). Forelimbs became modified for flight in extinct reptiles called pterosaurs and in bats and most birds. In penguins and cetaceans, they are now flippers useful for swimming. Human forelimbs are arms and hands with four fingers and an opposable thumb. Elephant limbs are strong and pillarlike, capable of supporting a great deal of weight. The five-toed limb degenerated to nubs in pythons and boa constrictors, and disappeared entirely in other snakes.

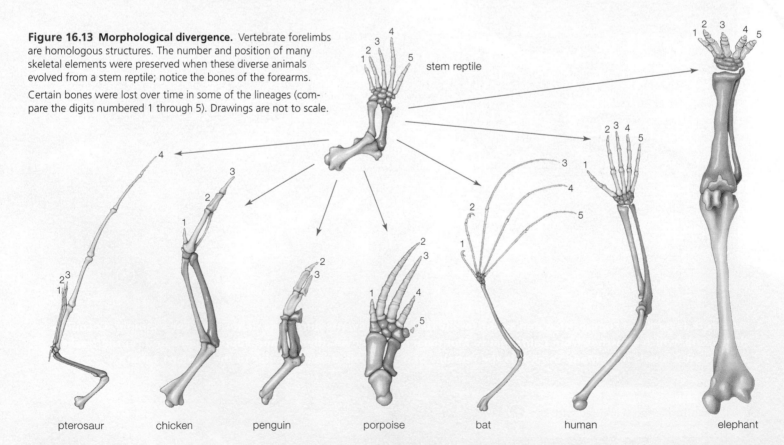

Figure 16.13 Morphological divergence. Vertebrate forelimbs are homologous structures. The number and position of many skeletal elements were preserved when these diverse animals evolved from a stem reptile; notice the bones of the forearms.

Certain bones were lost over time in some of the lineages (compare the digits numbered 1 through 5). Drawings are not to scale.

stem reptile

pterosaur   chicken   penguin   porpoise   bat   human   elephant

MORPHOLOGICAL CONVERGENCE

Body parts that appear similar in different species are not always homologous; they sometimes evolve independently in lineages subjected to the same environmental pressures. The independent evolution of similar body parts in different lineages is an evolutionary pattern called **morphological convergence**. Body parts that look alike but evolved independently after lineages diverged are **analogous structures**. Consider how bird, bat, and insect wings all perform the same function, which is flight. However, several clues tell us that the wing surfaces are not homologous. All of the wings are adapted to the same physical constraints that govern flight, but each is adapted in a different way. In the case of birds and bats, the limbs themselves are homologous, but the adaptations that make those limbs useful for flight differ. The surface of a bat wing is a thin, membranous extension of skin. By contrast, the surface of a bird wing is a sweep of feathers, which are specialized structures derived from skin. Insect wings differ even more. An insect wing forms as a saclike extension of the body wall. Except at forked veins, the sac flattens and fuses into a thin membrane. The sturdy, chitin-reinforced veins structurally support the wing. Unique adaptations for flight are evidence that wing surfaces of birds, bats, and insects are analogous structures that evolved after the ancestors of these modern groups diverged (**Figure 16.14A**).

As another example of morphological convergence, consider the similar appearance of the African euphorbia and American cactus in Figure 16.2. Both plants are adapted to similarly harsh desert environments where rain is scarce. Accordion-like pleats allow the plant body to swell with water when rain does come; water stored in the plants' tissues allows them to survive long dry periods. As the stored water is used, the plant body shrinks, and the folded pleats provide some shade in an environment that typically has none. Despite these similarities, a closer look reveals many differences that indicate the two plants are not closely related (**Figure 16.14B**). For example, cactus spines have a simple fibrous structure; they are modified leaves that arise from dimples on the plant's surface. Euphorbia spines project smoothly from the plant surface, and they are not modified leaves: In many species the spines are dried flower stalks.

analogous structures Similar body structures that evolved separately in different lineages.
homologous structures Body structures that are similar in different lineages because they evolved in a common ancestor.
morphological convergence Evolutionary pattern in which similar body parts evolve separately in different lineages.
morphological divergence Evolutionary pattern in which a body part of an ancestor changes differently in its different descendants.

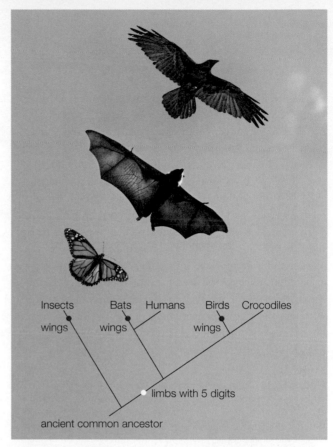

A The flight surfaces of an insect wing, a bat wing, and a bird wing are analogous structures. The diagram shows how the evolution of wings (red dots) occurred independently in the three separate lineages. You will read more about diagrams that show evolutionary relationships in Section 17.11.

B Spines of a saguaro cactus (left) differ from spines of an African milk barrel plant (a type of *Euphorbia*, right). This and some other differences indicate the two plants are not closely related despite their similar appearances (see Figure 16.2). Many similarities between these species are an outcome of morphological convergence.

Figure 16.14 Examples of morphological convergence.

CREDITS: (14A) top, © iStockphoto.com/DanCardiff; middle, © Taro Taylor, www.flickr.com/photos/tjt195; bottom, © Alberto J. Espiñeira Francés - Alesfra/Getty Images; (14B) left, George Burba/Shutterstock.com; right, © James C. Gaither, www.flickr.com/photos/jim-sf.

LEARNING OBJECTIVE

Explain why similarities between protein sequences (or between DNA sequences) can be used as a measure of relative relatedness.

Over time, inevitable mutations change a genome's DNA sequence. Most of these mutations are neutral—they have no effect on an individual's survival or reproduction—and they alter the DNA of each lineage independently of all other lineages. The more recently two lineages diverged, the less time there has been for unique mutations to accumulate in the DNA of each one. That is why the genomes of closely related species tend to be more similar than those of distantly related ones—a general rule that can be used to estimate relative times of divergence. Thus, similarities in the nucleotide sequence of a shared gene (or in the amino acid sequence of a shared protein) can be used to study evolutionary relationships. Biochemical comparisons like these may be combined with morphological comparisons, in order to provide data for hypotheses about shared ancestry.

Two species with a distant evolutionary relationship have few similarities between their proteins because many mutations have accumulated in the DNA of their separate lineages. Evolutionary biologists often compare a protein's sequence among several species, and use the number of amino acid differences as a measure of relative relatedness (**Figure 16.15**).

Among lineages that diverged relatively recently, many proteins have identical amino acid sequences. Nucleotide sequence differences may be instructive in such cases. Even if the amino acid sequence of a protein is identical among species, the nucleotide sequence of the gene that encodes the protein may differ because of redundancies in the genetic code. The DNA from nuclei, mitochondria, or chloroplasts can be used in nucleotide comparisons. In most animals, mitochondria are inherited intact from only one of the parents (the mother, in humans). Mitochondrial chromosomes do not undergo crossing over, so any changes in their DNA sequence occur by mutation. Thus, in most cases, differences in mitochondrial DNA sequences between maternally related people are due to new mutations.

PATTERNS IN ANIMAL DEVELOPMENT

In general, the more closely related animals are, the more similar is their development. For example, all vertebrates go through a stage during which a developing embryo has four limb buds, a tail, and a series of somites—divisions of the body that give rise to the backbone and associated skin and muscle (**Figure 16.16**). Animals have similar patterns of

Figure 16.15 Comparing proteins. Below, partial amino acid sequences of mitochondrial cytochrome *b* from 19 species are aligned. This protein is a crucial component of mitochondrial electron transfer chains. The honeycreeper sequence is identical in ten species of honeycreeper; amino acids that differ in the other species are shown in red. Dashes reflect insertion or deletion mutations.

FIGURE IT OUT
Based on this comparison, which species is the most closely related to honeycreepers?

Answer: The song sparrow

```
...CRDVQFGWLIRNLHANGASFFFICIYLHIGRGIYYGSYLNK--ETWNIGVILLLTLMATAFVGYVLPWGQMSFWG...  honeycreepers (10)
...CRDVQFGWLIRNLHANGASFFFICIYLHIGRGIYYGSYLNK--ETWNVGIILLLALMATAFVGYVLPWGQMSFWG...  song sparrow
...CRDVQFGWLIRNIHANGASFFFICIYLHIGRGLYYGSYLYK--ETWNVGVILLLTLMATAFVGYVLPWGQMSFWG...  Gough Island finch
...CRDVNYGWLIRYMHANGASMFFICLFLHVGRGMYYGSYTFT--ETWNIGIVLLFAVMATAFMGYVLPWGQMSFWG...  deer mouse
...CRDVHYGWIIRYMHANGASMFFICLFMHVGRGLYYGSYLLS--ETWNIGIILLFTVMATAFMGYVLPWGQMSFWG...  Asiatic black bear
...CRDVNYGWLIRNLHANGASFFFICIYLHIGRGLYYGSYLYK--ETWNIGVVLLLLVMGTAFVGYVLPWGQMSFWG...  bogue (a fish)
...TRDVNYGWIIRYLHANGASMFFICLFLHIGRGLYYGSFLYS--ETWNIGIILLLATMATAFMGYVLPWGQMSFWG...  human
...MRDVEGGWLLRYMHANGASMFLIVVYLHIFRGLYHASYSSPREFVWCLGVVIFLLMIVTAFIGYVLPWGQMSFWG...  thale cress (a plant)
...ETDVMNGWMVRSIHANGASWFFIMLYSHIFRGLWVSSFTQP--LVWLSGVIILFLSMATAFLGYVLPWGQMSFWG...  baboon louse
...MRDVHNGYILRYLHANGASFFFMVMFMHMAKGLYYGSYRSPRVTLWNVGVIIFTLTIATAFLGYCCVYGQMSHWG...  baker's yeast
```

Figure 16.16 Comparing vertebrate embryos. Below, all vertebrates go through an embryonic stage in which they have four limb buds, a tail, and divisions called somites along their back. From left to right: human, mouse, lizard, bat, alligator, chicken.

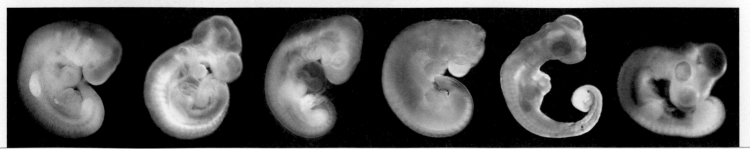

CREDITS: (16) From left: 1, © Lennart Nilsson/Bonnierforlagen AB; 2, Courtesy of Anna Bigas, IDIBELL-Institut de Recerca Oncologica, Spain; 3, Catherine May; 4, From "Embryonic staging system for the short-tailed fruit bat, Carollia perspicillata, a model organism for the mammalian order Chiroptera, based upon timed pregnancies in captive-bred animals" C.J. Cretekos et al., *Developmental Dynamics* Volume 233, Issue 3, July 2005, Pages: 721–738. Reprinted with permission of Wiley-Liss, Inc. a subsidiary of John Wiley & Sons, Inc.; 5, USGS; 6, Courtesy of Prof. Dr. G. Elisabeth Pollerberg, Institut für Zoologie, Universität Heidelberg, Germany.

Figure 16.17 Comparing master gene expression.
Expression of the *Hoxc6* gene is indicated by purple stain in two vertebrate embryos, chicken (left) and garter snake (right). Expression of this gene causes a vertebra to develop ribs as part of the back. Chickens have 7 vertebrae in their back and 14 to 17 vertebrae in their neck; snakes have upwards of 450 back vertebrae and essentially no neck.

embryonic development because the same master genes direct the process. Remember from Section 10.2 that embryonic development in animals is orchestrated by layer after layer of master gene expression. The failure of any single master gene to participate in this symphony of expression can result in a drastically altered body plan, an outcome that often has severe consequences on the organism's health. Because a mutation in a master gene can unravel development completely, these genes tend to be highly conserved. Even among lineages that diverged a very long time ago, many master genes retain similar sequences and functions.

If the same genes direct development in all vertebrate lineages, how do the adult forms end up so different? Part of the answer is that there are differences in the timing of early steps in development. These differences are brought about by variations in master gene expression patterns. The variation has arisen at least in part as a result of gene duplications followed by mutation, the same way that multiple globin genes evolved in primates (Section 14.4).

Consider homeotic genes called *Hox*, which, like other homeotic genes, help sculpt details of the body's form during embryonic development. Insects have ten *Hox* genes, and vertebrates have multiple versions of all of them. You learned about one insect *Hox* gene, *antennapedia*, in Section 10.2; this gene triggers leg formation in insect embryos, so it normally determines the identity of the thorax (the part with legs). Humans and other vertebrates have a version of *antennapedia* called *Hoxc6*. Expression of *Hoxc6* in a vertebrate embryo causes ribs to develop on a vertebra, so it normally determines the identity of the back. Vertebrae of the neck and tail develop with no *Hoxc6* expression, and no ribs (**Figure 16.17**).

K–Pg boundary sequence

Figure 16.18 The K–Pg boundary sequence, an unusual, worldwide sedimentary rock formation that formed 66 million years ago. This rock formation marks an abrupt transition in the fossil record that implies a mass extinction of the dinosaurs. It also contains materials consistent with a meteorite impact. The red pocketknife gives an idea of scale.

REFLECTIONS OF A DISTANT PAST

The foundation for scientific understanding of the history of life is this premise: Natural phenomena that occurred in the ancient past can be explained by the same physical, chemical, and biological processes that operate today. Consider what caused the dinosaurs to become extinct at the end of the Cretaceous period. Most scientists accept the hypothesis that dinosaurs perished in the aftermath of a catastrophic meteorite impact. No humans were around 66 million years ago to witness anything, so how could anyone know what happened? The event is marked by an unusual, worldwide formation of sedimentary rock called the K–Pg boundary sequence (it was formerly known as the K–T boundary). There are plenty of dinosaur fossils below this formation. Above it, in layers of rock that were deposited more recently, there are no dinosaur fossils, anywhere.

The K–Pg boundary sequence consists of an unusual clay (**Figure 16.18**) that is rich in iridium, an element rare on Earth's surface but common in meteorites. It also contains shocked quartz (left) and small glass spheres called tektites; both form when quartz or sand (respectively) undergoes a sudden, violent application of extreme pressure. The only processes on Earth known to produce these minerals are nuclear explosions and meteorite impacts.

Geologists hypothesized that the K–Pg boundary layer must have originated with extraterrestrial material, and began looking for evidence of a meteorite that hit Earth 66 million years ago—one big enough to cover the entire planet with its debris. Twenty years later, they found it: an impact crater the size of Ireland off the coast of the Yucatán Peninsula. To make a crater this big, a meteorite 20 km (12 miles) wide would have slammed into Earth with the force of 100 trillion tons of dynamite—enough to cause an ecological disaster of sufficient scale to wipe out almost all life on Earth.

Section 16.1 Expeditions by nineteenth-century explorers yielded increasingly detailed observations of nature that did not fit with prevailing beliefs. Cumulative findings from geology, **biogeography**, and **comparative morphology** of organisms and their **fossils** led naturalists to question traditional ways of interpreting the natural world.

Section 16.2 In the 19th century, attempts to reconcile traditional beliefs with physical evidence led to new ways of thinking about the natural world. Evidence found in the 1800s led to the idea that Earth and the species on it had changed over time. This idea set the stage for a theory proposed by Darwin and Wallace, stated here in modern terms: A population tends to grow until it exhausts resources in its environment. As that happens, competition for those resources intensifies among the population's members. Individuals with forms of shared, heritable traits that make them more competitive for the resources tend to produce more offspring. Thus, **adaptive traits (adaptations)** imparting greater **fitness** tend to become more common in a population over generations. The process in which environmental pressures result in the differential survival and reproduction of individuals of a population is called **natural selection**. It is one of the processes that drive **evolution**, which is change in a line of descent (a **lineage**).

Section 16.3 Fossils—a stone-hard record of life—are typically found in stacked layers of sedimentary rock. The most recent fossils usually occur in layers deposited most recently, on top of older fossils in older layers. The fossil record will never be complete because fossils are relatively rare.

Section 16.4 Each radioisotope has a characteristic **half-life**. With a technique called **radiometric dating**, radioisotope content and half-life are used in calculations that reveal the age of rocks and fossils.

Section 16.5 By the **plate tectonics theory**, Earth's crust is cracked into giant plates that carry landmasses to new positions as they move. The course of life's evolution has been profoundly influenced by this movement, for example during the formation and breakup of **Gondwana** and **Pangea**.

Section 16.6 Layers of rock in similar formations found worldwide reflect intervals of time in the **geologic time scale**. This chronology of Earth's history correlates time with geological and evolutionary events of the ancient past.

Section 16.7 Comparative morphology can reveal evidence of evolutionary connections among lineages. **Homologous structures** are derived from a common ancestor. By **morphological divergence**, these structures became modified differently in different lineages. **Analogous structures**, which arise by the process of **morphological convergence**, appear similar but evolved independently in different lineages.

Section 16.8 We can discover and clarify evolutionary relationships by comparing amino acid sequences of proteins, or nucleotide sequences of DNA, from different organisms. In general, these sequences are more similar among lineages that diverged more recently. Master genes that affect development tend to be highly conserved, so similarities in patterns of embryonic development reflect shared ancestry that can be evolutionarily ancient.

Application Events of the ancient past can be explained by the same physical, chemical, and biological processes that operate today. A mass extinction 66 million years ago may have been caused by a catastrophic meteorite impact that left traces in a worldwide sedimentary rock formation.

SELF-ASSESSMENT
ANSWERS IN APPENDIX VII

1. The number of species on an island usually depends on the size of the island and its distance from a mainland. This statement would most likely be made by _____ .
   a. an explorer   c. a geologist
   b. a biogeographer   d. a philosopher

2. The bones of a bird's wing are similar to the bones in a bat's wing. This observation is an example of _____ .
   a. uniformity   c. comparative morphology
   b. evolution   d. a lineage

3. Evolution _____ natural selection.
   a. is the same as   c. is the goal of
   b. can occur by   d. explains the origin of life by

4. A trait is adaptive if it _____ .
   a. arises by mutation   c. is passed to offspring
   b. increases fitness   d. occurs in fossils

5. In which type of rock are you more likely to find a fossil?
   a. basalt, a dark, fine-grained volcanic rock
   b. limestone, composed of calcium carbonate sediments
   c. slate, a volcanically melted and cooled shale
   d. granite, which forms by crystallization of molten rock below the surface of the Earth

6. True or false? Wrinkly textures in rock that formed from ancient biofilms living in marine sediments are considered trace fossils.

7. If the half-life of a radioisotope is 20,000 years, then a sample in which three-quarters of that radioisotope has decayed is _____ years old.
   a. 15,000   b. 26,667   c. 30,000   d. 40,000

8. Did Pangea or Gondwana form first?

DISCOVERY OF IRIDIUM IN THE K–PG BOUNDARY LAYER In the late 1970s, geologist Walter Alvarez was investigating the composition of the K–Pg boundary sequence in different parts of the world. He asked his father, Nobel Prize-winning physicist Luis Alvarez, to help him analyze the elemental composition of the layer (right, Luis and Walter Alvarez with a section of the boundary). The Alvarezes and their colleagues tested the K–Pg boundary sequence in Italy and Denmark. They discovered that it has a much higher iridium content than the surrounding rock layers. Some of their results are shown in **Figure 16.19**.

Sample Depth	Average Abundance of Iridium (ppb)
+ 2.7 m	< 0.3
+ 1.2 m	< 0.3
+ 0.7 m	0.36
boundary layer	41.6
− 0.5 m	0.25
− 5.4 m	0.30

Figure 16.19 Abundance of iridium in and near the K–Pg boundary sequence. Luis and Walter Alvarez tested the iridium content of many rock samples above, below, and at this boundary in different parts of the world. These are some of the Alvarezes' results from Stevns Klint, Denmark. Sample depths are given as meters above or below the boundary. ppb, parts per billion: An average Earth rock contains 0.4 ppb iridium. The average meteorite contains about 550 ppb iridium.

1. What was the iridium content of the K–Pg boundary layer?

2. How much higher was the iridium content of the boundary layer than the sample taken 0.7 meter above the sequence?

---

9. Forces that cause geologic change do not include _____ .
   - a. erosion
   - b. natural selection
   - c. volcanic activity
   - d. tectonic plate movement
   - e. wind
   - f. meteorite impacts

10. Through _____ , a body part of an ancestor is modified differently in different lines of descent.
    - a. homologous evolution
    - b. morphological convergence
    - c. adaptive divergence
    - d. morphological divergence

11. Homologous structures among major groups of organisms may differ in _____ .
    - a. function, but not size and shape
    - b. size, shape, and function

12. A mutation that alters the embryonic expression pattern of a(n) _____ may lead to major differences in the adult form.
    - a. derived trait
    - b. homeotic gene
    - c. homologous structure
    - d. analogous structure

13. The last dinosaurs died _____ million years ago.

14. All of the following data types can be used as evidence of shared ancestry except similarities in _____ .
    - a. amino acid sequence
    - b. DNA sequence
    - c. fossil morphology
    - d. embryonic development
    - e. form due to convergence
    - f. all are appropriate

15. Match the terms with the most suitable description.
    - ___ fitness
    - ___ fossils
    - ___ natural selection
    - ___ homeotic genes
    - ___ half-life
    - ___ analogous structures
    - ___ homologous structures
    - ___ sedimentary rock
    - ___ neutral mutation

    - a. does not affect fitness
    - b. best fossil content
    - c. survival of the fittest
    - d. characteristic of a radioisotope
    - e. similar across diverse taxa
    - f. evidence of life in distant past
    - g. insect wing and bird wing
    - h. human arm and bird wing
    - i. measured by reproductive success

CLASS ACTIVITY
CRITICAL THINKING

1. Radiometric dating does not measure the age of an atom. It is a measure of the age of a quantity of atoms—a statistic. As with any statistic, its values may deviate around an average (see sampling error, Section 1.7). Imagine that one sample of rock is dated ten different ways. Nine of the tests yield an age close to 225,000 years. One test yields an age of 3.2 million years. Do the nine consistent results imply that the one that deviates is incorrect, or does the one odd result invalidate the nine that are consistent?

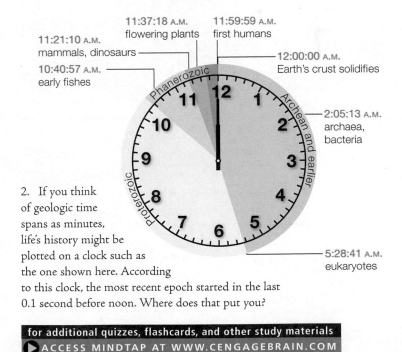

2. If you think of geologic time spans as minutes, life's history might be plotted on a clock such as the one shown here. According to this clock, the most recent epoch started in the last 0.1 second before noon. Where does that put you?

for additional quizzes, flashcards, and other study materials
▶ **ACCESS MINDTAP AT WWW.CENGAGEBRAIN.COM**

LEARNING OBJECTIVES

Explain the effects of directional selection on a population.

Use appropriate examples to explain how environmental factors can cause directional selection.

SELECTION FOR AN EXTREME FORM

With **directional selection**, forms of a trait at one end of a range of variation are adaptive (**Figure 17.4**). The following examples show how field observations provide evidence of directional selection.

The Peppered Moth A well-documented case of directional selection involves coloration changes in the peppered moth, which has two color morphs: light with black speckles, and black. These moths feed and mate at night, then rest on trees during the day. In preindustrial England, the vast majority of peppered moths were light, and a small number were dark. At the time, the air was clean, and light-gray lichens grew on the trunks and branches of most trees. Light moths that rested on lichen-covered trees were well camouflaged, but black moths were not (**Figure 17.5A**). By the 1850s, black moths had become much more common than light moths.

Scientists suspected that predation by birds was the selective pressure that shaped moth coloration in local populations. The industrial revolution had begun, and smoke emitted by coal-burning factories was killing lichens; black moths were better camouflaged from predatory birds on lichen-free, soot-darkened trees (**Figure 17.5B**). In the 1950s, H. B. Kettlewell set out to test this hypothesis. He bred both color morphs in captivity, marked them for easy identification, then released them in several areas. His team recaptured more of the black moths in the polluted areas, and more of the light moths in the less polluted areas. The researchers also observed predatory birds eating more light moths in soot-darkened forests, and more black moths in cleaner, lichen-rich forests. Black moths were clearly at a selective advantage in industrialized areas.

Pollution controls went into effect in 1952. As a result, tree trunks gradually became free of soot, and lichens made a comeback. Kettlewell observed that moth phenotypes shifted too: Wherever pollution decreased, the frequency of black moths in local populations decreased as well. Recent research has confirmed Kettlewell's results implicating birds as selective agents of peppered moth coloration, and also that this selection causes a shift in the frequency of alleles underlying the coloration. Peppered moth color is determined by a single gene; individuals with a dominant allele of this gene are black, and those homozygous for a recessive allele are light.

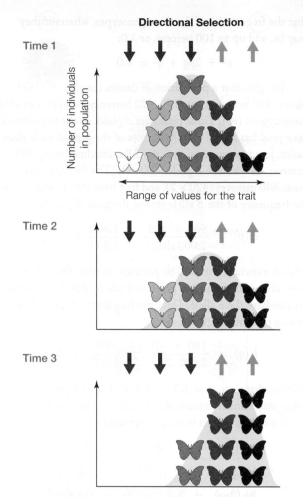

Figure 17.4 With directional selection, a form of a trait at one end of a range of variation is adaptive. Bell curves indicate continuous variation. Red arrows indicate which forms are being selected against; green, forms that are adaptive.

A Light moths on a nonsooty tree trunk (top) are hidden from predators; black ones (bottom) stand out.

B In places where soot darkens tree trunks, the black color (bottom) provides more camouflage than the light color (top).

Figure 17.5 Adaptive value of two color morphs of the peppered moth.

CREDITS: (5) J. A. Bishop, L. M. Cook.

Figure 17.6 Rats thrive wherever people do. Spreading poisons around buildings and soil does not usually exterminate rat populations, which recover quickly. Rather, the practice exerts directional selection favoring resistant rats.

Warfarin-Resistant Rats Human attempts to control the environment can result in directional selection. Consider that the average large city sustains about one rat for every ten people. Rats thrive in urban centers, where garbage is plentiful and natural predators are not (**Figure 17.6**). Part of their success stems from an ability to reproduce very quickly: Rat populations can expand within weeks to match the amount of garbage available for them to eat.

For decades, people have been fighting rats with poisons. Baits laced with warfarin, an organic compound that interferes with blood clotting, became popular in the 1950s. Rats that ate the poisoned baits died within days after bleeding internally or losing blood through cuts or scrapes. Warfarin was extremely effective, and its impact on harmless species was much lower than that of other rat poisons. It quickly became the rat poison of choice. By 1980, however, about 10 percent of rats in urban areas were resistant to warfarin. What happened?

Warfarin interferes with blood clotting because it inhibits the function of an enzyme called VKORC1. This enzyme regenerates vitamin K, which functions as a coenzyme (Section 5.5) in the production of blood clotting factors (Section 14.3). When vitamin K is not regenerated, the clot-

ting factors are not properly produced, and clotting cannot occur. Rats resistant to warfarin have a mutated version of the VKORC1 gene, and the enzyme encoded by this allele is insensitive to warfarin. "What happened" was evolution by natural selection: Rats with the normal allele died after eating warfarin; the lucky ones with a mutated allele survived and passed it to offspring. The rat populations recovered quickly, and a higher proportion of individuals in the next generation carried a mutation. With each onslaught of warfarin, the frequency of the mutation in rat populations increased. Exposure to warfarin had exerted directional selection.

Mutations that result in warfarin resistance also reduce the activity of the VKORC1 enzyme, so rats that have them require a lot of extra vitamin K. However, being deficient in vitamin K is not so bad when compared with being dead from rat poison. In the absence of warfarin, though, rats with the allele are at a serious disadvantage because they cannot easily obtain enough vitamin K from their diet to sustain normal blood clotting and bone formation. Thus, the frequency of a warfarin resistance allele in a rat population declines quickly after warfarin exposure ends—another example of directional selection.

Now, savvy exterminators in urban areas know that the best way to control a rat infestation is to exert another kind of selection pressure: Remove the rats' source of food, which is usually garbage. Then they will eat each other.

directional selection Mode of natural selection in which phenotypes at one end of a range of variation are favored.

LEARNING OBJECTIVES

Using appropriate examples, explain sexual selection and its outcomes.

Describe two ways that a balanced polymorphism can be maintained.

Explain why the harmful *HbS* allele persists at high frequency in areas where malaria is rampant.

SURVIVAL OF THE SEXIEST

Not all evolution is driven by selection for traits that enhance survival. Competition for mates is another selective pressure that can shape form and behavior. Consider how individuals of many sexually reproducing species have a distinct male or female phenotype. A trait that differs between males and females is called a **sexual dimorphism**. Individuals of one sex (often males) are more colorful, larger, or more aggressive than individuals of the other sex. These traits can seem puzzling because they take energy and time away from activities that enhance survival, and some actually hinder an individual's ability to survive. Why, then, do they persist?

The answer is **sexual selection**, in which the evolutionary winners outreproduce others of a population because they are better at securing mates. With this mode of natural selection, the most adaptive forms of a trait are those that help individuals defeat rivals for mates, or are most attractive to the opposite sex. For example, the females of some species cluster in defensible groups when they are sexually receptive, and males compete for sole access to the groups. Competition for the ready-made harems favors brawny, combative males (**Figure 17.11A**).

Males or females that are choosy about mates act as selective agents on their own species. The females of some species shop for a mate among males that display species-specific cues such as a highly specialized appearance or courtship behavior (**Figure 17.11B**). The cues often include flashy body parts or movements (the chapter opener photo shows an example)—traits that attract predators and in some cases are a physical hindrance. However, to a female member of the species, a flashy male's survival despite his obvious handicap may imply health and vigor, two traits that are likely to improve her chances of bearing healthy, vigorous offspring. Selected males pass alleles for their attractive traits to the next generation of males, and females pass alleles that influence mate preference to the next generation of females. Highly exaggerated traits can be an evolutionary outcome (**Figure 17.11C**).

MAINTAINING MULTIPLE ALLELES

Any mode of natural selection may maintain two or more alleles of a gene at relatively high frequency in a population's

A Male elephant seals engaged in combat. Males of this species typically compete for access to clusters of females.

B A male peacock engaged in a flashy courtship display has caught the eye (and, perhaps, the sexual interest) of a female. Females are choosy; a male mates with any female that accepts him.

C Mating stalk-eyed flies. Female stalk-eyed flies prefer to mate with males that have the longest eyestalks, a trait that provides no obvious selective advantage other than sexual attractiveness.

Figure 17.11 Sexual selection in action.

gene pool, a state called **balanced polymorphism**. For example, sexual selection maintains multiple alleles that govern eye color in populations of *Drosophila* fruit flies (**Figure 17.12**). Female flies prefer to mate with rare white-eyed males, until the white-eyed males become more common than red-eyed males, at which point the red-eyed flies are again preferred. This is an example of **frequency-dependent selection**, in which the adaptive value of a particular form of a trait depends on its frequency in a population.

Balanced polymorphism can also arise in environments that favor heterozygous individuals. Consider the gene that encodes the beta globin chain of hemoglobin (Sections 9.5 and 12.1). *HbA* is the normal allele; the codominant *HbS* allele carries a mutation that causes sickle-cell anemia. Without treatment, the vast majority of individuals homozygous for the *HbS* allele die in early childhood.

Despite being so harmful, the *HbS* allele persists at very high frequency among the human populations in tropical and subtropical regions of Asia, Africa, and the Middle East. Why? Populations with the highest frequency of the *HbS* allele also have the highest incidence of malaria. Mosquitoes transmit *Plasmodium*, the parasitic protist that causes malaria, to human hosts (more about this in Section 20.5). *Plasmodium* multiplies in the liver and then in red blood cells, which rupture and release new parasites during recurring bouts of severe illness. People who are *HbA/HbS* heterozygous make both normal and sickle hemoglobin. Red blood cells of these individuals can sickle under some circumstances, but not enough to cause severe symptoms. One of the circumstances under which sickling occurs is infection with *Plasmodium*. The abnormal shape brings the cells to the attention of the immune system, which destroys them along with the parasites they harbor. The action of the immune system can prevent the infection from spreading to other red blood cells. Thus, heterozygous individuals are more likely to survive malaria than individuals homozygous for the normal *HbA* allele. *Plasmodium*-infected red blood cells of people who make only normal hemoglobin do not sickle, so the parasite may remain hidden from the immune system.

In areas where malaria is common, the persistence of the *HbS* allele is a matter of relative evils. Malaria and sickle-cell anemia are both potentially deadly. Heterozygous individuals may not be completely healthy, but they do have a better chance of surviving malaria than people homozygous for the normal allele (*HbA/HbA*). With or without malaria, people who have both alleles (*HbA/HbS*) are more likely to live long enough to reproduce than individuals homozygous for the sickle allele (*HbS/HbS*). Thus, the *HbS* allele persists at high frequency in populations inhabiting the most malaria-ridden regions of the world (**Figure 17.13**).

Figure 17.12 Red and white eye color forms in fruit flies. Sexual selection maintains multiple alleles that govern this trait: Female flies prefer males with the least common eye color.

A Distribution of people who carry the sickle-cell allele in Gabon.

Yellow indicates regions where more than 10 percent of the population are carriers of the *HbS* allele. In the white regions, less than 10 percent of the population are carriers.

B Distribution of malaria cases in Gabon.

Yellow indicates regions where more than 30 percent of the population have malaria. In the white regions, less than 30 percent of the population have the disease.

Figure 17.13 Incidence in 2014 of sickle-cell trait and malaria in Gabon, Africa. Notice the correlation between the two maps.

---

balanced polymorphism Maintenance of two or more alleles of a gene at high frequency in a population.

frequency-dependent selection Natural selection in which a trait's adaptive value depends on its frequency in a population.

sexual dimorphism A trait that differs between males and females of a species.

sexual selection Mode of natural selection in which some individuals outreproduce others of a population because they are better at securing mates.

LEARNING OBJECTIVES

Describe speciation in terms of reproductive isolation.

Using suitable examples, explain the mechanisms of reproductive isolation.

When two populations of a species do not interbreed, the number of genetic differences between them increases because mutation, natural selection, and genetic drift occur independently in each one. Over time, the populations may become so different that we consider them to be different species. The emergence of a new species is an evolutionary process called **speciation**.

Evolution is a dynamic, extravagant, messy, and ongoing process that can be challenging for people who like neat categories. Speciation offers a perfect example, because it rarely occurs at a precise moment in time: Individuals often

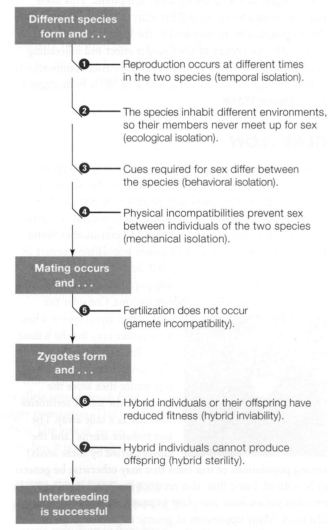

Different species form and . . .

① — Reproduction occurs at different times in the two species (temporal isolation).

② — The species inhabit different environments, so their members never meet up for sex (ecological isolation).

③ — Cues required for sex differ between the species (behavioral isolation).

④ — Physical incompatibilities prevent sex between individuals of the two species (mechanical isolation).

Mating occurs and . . .

⑤ — Fertilization does not occur (gamete incompatibility).

Zygotes form and . . .

⑥ — Hybrid individuals or their offspring have reduced fitness (hybrid inviability).

⑦ — Hybrid individuals cannot produce offspring (hybrid sterility).

Interbreeding is successful

Figure 17.16 How reproductive isolation prevents interbreeding between species.

continue to interbreed even as populations are diverging, and populations that have already diverged may come together and interbreed again.

Every time speciation happens, it happens in a unique way, which means that each species is a product of its own unique evolutionary history. However, there are recurring patterns. For example, **reproductive isolation**, the end of gene flow between populations, is always part of the process by which sexually reproducing species achieve and maintain separate identities. Several mechanisms of reproductive isolation prevent successful interbreeding, and thus reinforce differences between diverging populations (**Figure 17.16**).

① Some closely related species cannot interbreed because the timing of their reproduction differs, an effect called tempo-

ral isolation. Consider the periodical cicada (left). Larvae of these insects feed on roots as they mature underground, then the adults emerge to reproduce. Three cicada species reproduce every 17 years. Each has a sibling species: a related species with nearly identical form and behavior that emerges on a 13-year cycle instead of a 17-year

cycle. Sibling species have the potential to interbreed, but they can only get together once every 221 years!

② Adaptation to different environmental conditions may prevent closely related species from interbreeding, a mechanism called ecological isolation. For example, two species of manzanita, a plant native to the Sierra Nevada mountain range, rarely hybridize. One species that lives on dry, rocky hillsides is better adapted for conserving water. The other species, which requires more water, lives on lower slopes where water stress is not as intense. The physical separation makes interbreeding unlikely.

③ Behavioral isolation occurs when differences in behavior can prevent mating between related animal species. For example, males and females of many animal species engage in courtship displays before sex (the chapter opener photo shows one example). In a typical pattern, the female recognizes the sounds and movements of a male of her species as an overture to sex; females of different species do not.

④ Mechanical isolation occurs when the size or shape of an individual's reproductive parts prevent it from mating with members of related species. For example, closely related species of sage plants grow in the same areas, but their flowers are specialized for different pollinators so cross-pollination rarely occurs and hybrids rarely form. Small-flowered black sage is pollinated most effectively by small bees and other insects; large-flowered white sage, by larger insects (**Figure 17.17**).

⑤ Even if gametes of different species do meet up, they often have molecular incompatibilities that prevent a zygote from forming. For example, the molecular signals that trigger

A Black sage is pollinated mainly by small bees and other insects that touch the reproductive parts of the flowers as they sip nectar (top). Larger insects cannot perch on small sage flowers; they access nectar by piercing the petals (bottom), so they do not touch the flower's reproductive parts (stigma and pollen-bearing stamens).

B Flowers of white sage are pollinated mainly by insects heavy enough to activate a tripping mechanism. When a big insect lands on the flower to sip nectar, it pushes down the large top petal (top). This causes stamens and stigma to move inward and touch the pollinator's body. Insects that do not weigh enough to trigger this mechanism do not touch the flower's reproductive parts (bottom).

Figure 17.17 Mechanical isolation in sage.

pollen germination in flowering plants are species-specific; thus, pollen grains typically will not germinate if they land on a flower of a different species. (Section 27.2 returns to pollen germination and other aspects of flowering plant reproduction.) Gamete incompatibility may be the primary speciation route among animals that release eggs and free-swimming sperm into water.

❻ Genetic changes are the basis of divergences in form, function, and behavior. Even chromosomes of species that diverged relatively recently may be different enough that a hybrid zygote ends up with extra or missing genes, or genes with incompatible products. Such outcomes typically disrupt embryonic development. Hybrid individuals that do survive embryonic development often have reduced fitness. For example, hybrid offspring of lions and tigers have more health problems and a shorter life expectancy than individuals of either parent species. If hybrids live long enough to reproduce, their offspring often have lower and lower fitness with each successive generation. Incompatible nuclear and mitochondrial DNA may be the cause (mitochondrial DNA is inherited from the mother only).

❼ Some interspecies crosses produce robust but sterile offspring. For example, mating between a female horse (64 chromosomes) and a male donkey (62 chromosomes) produces a mule (63 chromosomes: 32 from the horse, and 31 from the donkey). Mules are healthy but their chromosomes cannot pair up properly during meiosis, so this animal makes few viable gametes.

reproductive isolation The end of gene flow between populations.
speciation (spee-see-A-shun) Emergence of a new species.

Use an example to explain how sympatric speciation occurs.

Describe parapatric speciation with an example.

SYMPATRIC SPECIATION

Sympatric speciation takes place in the absence of a physical barrier between populations (*sym-* means together). Sympatric speciation can occur in a single generation when the chromosome number multiplies. Polyploidy (having three or more sets of chromosomes, Section 14.5) often arises because meiosis fails to reduce the chromosome number in gametes. If two diploid gametes fuse, the resulting zygote is tetraploid ($4n$). While this outcome is invariably fatal in humans, it is not uncommon in some plants: About 95 percent of modern ferns and 70 percent of modern flowering plant species are polyploid (**Figure 17.20**). Polyploidy offers advantages for plants. Having an extra set of genes fuels diversification, for example: One set is free to mutate while the other carries out its original function (Section 14.4). Crossing over among multiple sets of chromosomes during meiosis also multiplies variation among offspring (Section 12.4).

Sympatric speciation can also occur without chromosome number multiplication. The mechanically isolated sage plants you learned about in Section 17.7 speciated like this.

As another example, more than 500 species of cichlid fishes arose by sympatric speciation in the shallow waters of Lake Victoria. This large freshwater lake sits isolated from river inflow on an elevated plain in Africa's Great Rift Valley. Lake Victoria has dried up completely several times during its history. DNA sequence comparisons indicate that almost all of the cichlid species in this lake arose since it was last dry, about 12,400 years ago. How could hundreds of species arise so quickly? The answer involves sexual selection. Consider how the color of ambient light differs in different parts of the lake. The light in the lake's shallower, clear water is mainly blue; light that penetrates the deeper, muddier water is mainly red. The cichlid species vary in color (**Figure 17.21**), and female cichlids prefer to mate with brightly colored males. Their preference has a genetic basis, in alleles of genes for light-sensitive pigments of the retina (part of the eye). Retinal pigments made by species that live mainly in shallow areas of the lake are more sensitive to blue light. The males of these species are also the bluest. Retinal pigments made by species that prefer deeper areas of the lake are more sensitive to red light. Males of these species are redder. In other words, the colors that a female cichlid sees best are the same colors displayed by males of her species. Thus, mutations that affect color perception are likely to affect a female's choice of mates. Mutations like these are probably the way sympatric speciation begins in these fishes.

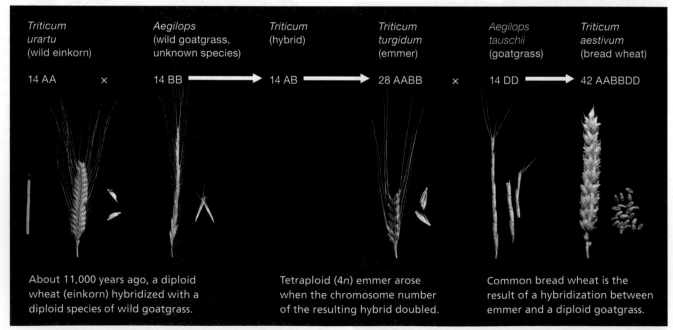

Triticum urartu (wild einkorn)		*Aegilops* (wild goatgrass, unknown species)		*Triticum* (hybrid)		*Triticum turgidum* (emmer)		*Aegilops tauschii* (goatgrass)		*Triticum aestivum* (bread wheat)
14 AA	×	14 BB	→	14 AB	→	28 AABB	×	14 DD	→	42 AABBDD

About 11,000 years ago, a diploid wheat (einkorn) hybridized with a diploid species of wild goatgrass.

Tetraploid ($4n$) emmer arose when the chromosome number of the resulting hybrid doubled.

Common bread wheat is the result of a hybridization between emmer and a diploid goatgrass.

Figure 17.20 Sympatric speciation in wheat. The wheat genome, which consists of seven chromosomes, occurs in slightly different forms called A, B, C, D, and so on. Many wheat species are polyploid, carrying more than two copies of the genome. For example, modern bread wheat (*Triticum aestivum*) is hexaploid, with six copies of the wheat genome: two each of genomes A, B, and D (or 42 AABBDD).

Figure 17.21 Red fish, blue fish: Males of four closely related species of cichlid native to Lake Victoria, Africa.

Hundreds of cichlid species arose by sympatric speciation in this lake. Mutations that affect female cichlids' perception of the color of ambient light in deeper or shallower regions of the lake also affect their choice of mates. Female cichlids prefer to mate with males that they perceive to be most brightly colored.

FIGURE IT OUT
What form of natural selection has been driving sympatric speciation in Lake Victoria cichlids?

Answer: Sexual selection

PARAPATRIC SPECIATION

With **parapatric speciation**, adjacent populations speciate despite being in contact across a common border. Divergences spurred by local selection pressures are reinforced because hybrids that form in the contact zone are less fit than individuals on either side of it.

Consider velvet walking worms, which resemble caterpillars but may be more related to spiders: They are predatory,

and shoot streams of glue from their head (left) to entangle insect prey. Two rare species of velvet walking worm are native to the island of Tasmania. The giant velvet walking worm and the blind velvet walking worm can interbreed, but they only do so in a tiny area where their habitats overlap. Hybrid offspring must be less fit than their parents, because the two species are maintaining separate identities in the absence of a physical barrier between their neighboring populations.

---

parapatric speciation Speciation pattern in which populations speciate while in contact along a common border.
sympatric speciation Speciation pattern in which speciation occurs in the absence of a physical barrier to gene flow.

Dr. Julia J. Day

Understanding processes that lead to speciation is fundamental to explaining the diversity of life. National Geographic Explorer Julia Day studies why some environments give rise to and maintain higher species richness than others. Her research focuses on comparing the staggering biodiversity of three East African great lakes: Tanganyika, Victoria, and Malawi. Cichlid fishes in these lakes evolved into "flocks" of many hundred species, most of which are close relatives. Each species flock displays astonishing levels of ecological, phenotypic, and behavioral diversity. In addition to being generally very colorful, these fish have a stunning variety of adaptations: unusual dietary habits, clever feeding behaviors, and sophisticated reproductive behaviors and parental care strategies.

Lake Tanganyika, despite being larger and older than Lakes Victoria and Malawi, harbors only about one-third the number of cichlid species in the other two lakes. Why the difference? Day and her team analyzed mitochondrial DNA samples from almost all of the Lake Tanganyika cichlids. Her results indicate that species in this lake diversified from several distinct lineages, rather than from a single ancestor (as occurred in Lakes Victoria and Malawi). Cichlid diversifications in Lake Tanganyika occurred several times, possibly coinciding with periods of changing water levels in the lake, or with successive invasions of new ancestral cichlid species. Speciation also occurred about six times more slowly than in the other lakes—a rate more in line with typical speciation rates of plants and animals in other environments.

Something about Lake Tanganyika was less conducive to cichlid speciation than the other lakes. Day thinks that diversification may have been relatively inhibited by the presence of older species already filling Lake Tanganyika's niches. Unlike Lakes Victoria and Malawi, Lake Tanganyika has not completely dried up in the past.

As these unique ecosystems are threatened by increasing populations and climate change, understanding and assessing the origins and maintenance of their biodiversity are of high priority.

CREDITS: (21) Kevin Bauman, www.african-cichlid.com; (in text) left, John Downer Productions/Minden Pictures; right, Courtesy of Julia J. Day.

CHAPTER 17 PROCESSES OF EVOLUTION **291**

LEARNING OBJECTIVES

Distinguish between microevolution and macroevolution.

Use examples to describe some patterns of macroevolution.

Microevolution is change in allele frequency within a single species or population. Evolution also occurs on a larger scale than microevolution, and we call large-scale evolutionary patterns **macroevolution**. Macroevolution includes trends such as land plants evolving from green algae, the dinosaurs disappearing in a mass extinction, a burst of divergences from a single species, and so on.

PATTERNS OF MACROEVOLUTION

Stasis Very little change may occur over a very long period of time, a macroevolutionary pattern called **stasis**. Consider coelacanths, an ancient order of lobe-finned fish that had been assumed extinct for at least 70 million years until a fisherman caught one in 1938. In its unique form and other traits, modern coelacanth species are similar to fossil specimens hundreds of millions of years old (**Figure 17.22**).

Exaptations Major evolutionary novelties often stem from the use of an existing trait for an entirely new purpose. A trait that has been evolutionarily repurposed is called an **exaptation**. For example, feathers that allow modern birds to fly are derived from feathers that first appeared in some dinosaurs. Dinosaur feathers could not have evolved for flight; they probably evolved for insulation. Thus, we say that bird feathers are an exaptation for flight.

Mass Extinctions By current estimates, more than 99 percent of all species that ever lived are now **extinct**, which means they no longer have living members. In addition to continuing extinctions of individual species, the fossil record indicates that there have been more than twenty mass extinctions, which are simultaneous losses of many lineages. These include five catastrophic events in which the majority of species on Earth disappeared (Section 16.6).

Adaptive Radiation With **adaptive radiation**, one lineage rapidly diversifies into several new species. Adaptive radiation can occur after a population colonizes a new environment that has a variety of different habitats and few competitors. Speciation occurs as adaptations to the different habitats evolve. The Lake Victoria cichlids that you learned about in the previous section arose by adaptive radiation, as did the Hawaiian honeycreepers.

A geologic or climatic event that eliminates some species from a habitat can spur adaptive radiation; species that survive the event then have access to resources from which they had previously been excluded. This is the way mammals were able to undergo an adaptive radiation after the dinosaurs disappeared 66 million years ago. Adaptive radiation may also occur after a key innovation evolves. A **key innovation** is an adaptive trait that allows its bearer to exploit a habitat more efficiently or in a novel way. The evolution of lungs offers an example, because lungs were a key innovation that opened the way for an adaptive radiation of vertebrates on land.

Coevolution Two species that have close ecological interactions may evolve jointly, a pattern called **coevolution**. One

Notochord
This tough, elastic tube, which is partially hollow and filled with fluid, is ancestral to the spinal cord.

Lobed fins
These fleshy fins retain a few of the ancestral bones that gave rise to legs and arms in other lineages.

Long gestation
Coelacanths give birth to litters of up to 26 fully developed "pups" after gestation of more than a year.

Rostral organ
A sensory organ that perceives electrical impulses in water, it probably helps the fish locate prey in dark ocean depths.

Figure 17.22 An example of stasis. Left, compare a 320-million-year-old coelacanth fossil found in Montana with a live coelacanth. Right, a few of the coelacanth's unusual ancestral features that have been lost in almost all other fish lineages over evolutionary time.

A A *Maculinea arion* caterpillar interacting with a *Myrmica sabuleti* ant. This beguiled ant is preparing to carry the honey-exuding, hunched-up caterpillar back to its nest, where the caterpillar will feed on ant larvae for the next 10 months until it becomes a pupa.

B *Maculinea arion* butterflies emerge from pupae to feed, mate, and lay eggs on wild thyme flowers. Larvae that emerge from the eggs will survive only if a colony of *Myrmica sabuleti* ants adopts them.

Figure 17.23 An example of coevolved species: *Maculinea arion* and *Myrmica sabuleti*.

species acts as an agent of selection on the other, and each adapts to changes in the other. Over evolutionary time, the two species may become so interdependent that they can no longer survive without one another.

Relationships between coevolved species can be incredibly intricate. Consider the large blue butterfly (*Maculinea arion*), a parasite of ants. After hatching, the butterfly larvae (caterpillars) feed on wild thyme flowers and then drop to the ground. An ant that finds a caterpillar strokes it, which makes the caterpillar exude honey. The ant eats the honey and continues to stroke the caterpillar, which secretes more honey. This interaction continues for hours, until the caterpillar suddenly hunches itself up into a shape that appears (to an ant) very much like an ant larva (**Figure 17.23A**). The deceived ant then picks up the caterpillar and carries it back to the ant nest, where, in most cases, other ants kill it—unless the ants are of the species *Myrmica sabuleti*. The caterpillar secretes the same chemicals

as *Myrmica sabuleti* larvae, and makes the same sounds as their queen—behaviors that trick the ants into adopting the caterpillar and treating it better than their own larvae. The adopted caterpillar feeds on ant larvae for about 10 months, then undergoes metamorphosis, changing into a butterfly that emerges from the ground to mate (**Figure 17.23B**). Eggs are deposited on wild thyme near another *M. sabuleti* nest, and the cycle starts anew. This relationship between ant and butterfly is typical of coevolved relationships in that it is extremely specific. Any increase in the ants' ability to identify a caterpillar in their nest selects for caterpillars that better deceive the ants, which in turn select for ants that can better identify the caterpillars. Each species exerts directional selection on the other.

EVOLUTIONARY THEORY

Biologists do not doubt that macroevolution occurs, but many disagree about *how* it occurs. However we choose to categorize evolutionary processes, the very same genetic change may be at the root of all evolution—fast or slow, large-scale or small-scale. Dramatic jumps in form, if they are not artifacts of gaps in the fossil record, may be the result of mutations in homeotic or other regulatory genes. Macroevolution may include more processes than microevolution, or it may not. It may be an accumulation of many microevolutionary events, or it may be an entirely different process. Evolutionary biologists may disagree about these and other hypotheses, but all of them are trying to explain the same thing: how all species are related by descent from common ancestors.

---

adaptive radiation A lineage undergoes a burst of genetic divergences that gives rise to many species.

coevolution The joint evolution of two closely interacting species; each species is a selective agent for traits of the other.

exaptation (eggs-app-TAY-shun) A trait that has been repurposed during evolution.

extinct Refers to a species that no longer has living members.

key innovation An evolutionary adaptation that gives its bearer the opportunity to exploit a particular environment much more efficiently or in a new way.

macroevolution Large-scale evolutionary patterns and trends.

stasis (STAY-sis) Evolutionary pattern in which a lineage persists with little or no change over evolutionary time.

17.11 Phylogeny

LEARNING OBJECTIVES

Explain how traditional (Linnaean) taxonomy differs from modern evolutionary biology.

Distinguish between a character and a derived trait.

Classifying life's tremendous diversity into a series of taxonomic ranks (Section 1.4) is a useful endeavor, in the same way that it is useful to organize a telephone book or contact list in alphabetical order: The result is convenient. Traditional (Linnaean) classification schemes have ranked species into successively higher taxa based on shared traits—birds have feathers, cacti have spines, and so on—but these rankings do not necessarily reflect evolutionary relationships.

Today's biologists work from the premise that every living thing is related if you just look back far enough in time. Grouping species according to evolutionary relationships is a way to fill in the details of this bigger picture of evolution. Thus, reconstructing **phylogeny**, the evolutionary history of a species or a group of species, is a priority. Phylogeny is a kind of genealogy that follows evolutionary relationships through time.

Humans were not around to witness the evolution of most species, but there is plenty of evidence to help us understand ancient events (Chapter 16). Consider how each species bears traces of its own unique evolutionary history in its characters. A **character** is a quantifiable, heritable trait such as the number of segments in a backbone, the nucleotide sequence of ribosomal RNA, or the presence of wings (**Table 17.2**). Evolutionary biologists group (rather than rank) species based on shared characters. They focus on what makes the species share the traits in the first place: a common ancestor. Common ancestry is determined by a **derived trait**—a character that is present in a group under consideration, but not in any of the group's ancestors.

A grouping whose members share one or more defining derived traits is called a **clade**. A clade is a **monophyletic group**, which is a grouping that consists of an ancestor (in

TABLE 17.2

Examples of Characters

	Bird	Bat	Lion
Warm-blooded	Y	Y	Y
Hair	N	Y	Y
Milk	N	Y	Y
Teeth	N	Y	Y
Wings	Y	Y	N
Feathers	Y	N	N

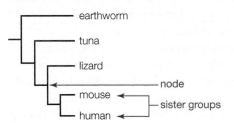

A Evolutionary connections among clades are represented as lines on a cladogram. Sister groups emerge from a node, which represents a common ancestor.

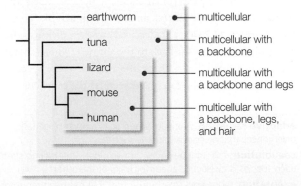

B A cladogram can be viewed as "sets within sets" of derived traits. Each set (an ancestor together with all of its descendants) is a clade.

Figure 17.24 An example of a cladogram.

FIGURE IT OUT

Which sister groups in this cladogram diverged most recently?

Answer: Human and mouse

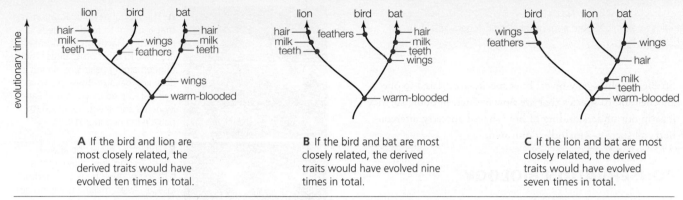

A If the bird and lion are most closely related, the derived traits would have evolved ten times in total.

B If the bird and bat are most closely related, the derived traits would have evolved nine times in total.

C If the lion and bat are most closely related, the derived traits would have evolved seven times in total.

Figure 17.25 An example of cladistics, using parsimony analysis with the characters in Table 17.2. **A**, **B**, and **C** show the three possible evolutionary pathways that could connect birds, bats, and lions; red indicates the evolution of a derived trait. The pathway most likely to be correct (**C**) is the simplest—the one in which the derived traits would have had to evolve the fewest number of times in total.

which a derived trait evolved) together with all of its descendants. It is the relative newness of a derived trait that defines a clade. Humans and bacteria use some of the same proteins to repair DNA, for example, but humans and bacteria are not close relatives. As another example, consider how alligators look a lot more like lizards than birds. In this case, the similarity in appearance does indicate shared ancestry, but it is a more distant relationship than alligators have with birds. A unique set of traits that include a gizzard and a four-chambered heart evolved in the lineage that gave rise to alligators and birds, but not in the lineage that gave rise to lizards.

Many traditional taxonomic rankings are equivalent to clades—flowering plants, for example, are both a phylum and a clade—but some are not. For example, the traditional Linnaean class Reptilia ("reptiles") includes crocodiles, alligators, tuataras, snakes, lizards, turtles, and tortoises. While it is convenient to classify these animals together, they would not constitute a clade unless birds are also included, as you will see in Chapter 24.

CLADISTICS

All species are interconnected in the big picture of evolution; an evolutionary biologist's job is to figure out where the connections are. Making hypotheses about evolutionary relationships among clades is called **cladistics**. Part of cladistics involves making **evolutionary trees**—diagrams of branching evolutionary connections. A **cladogram** is an evolutionary tree that visually summarizes a hypothesis about how a group of clades are related (**Figure 17.24**). Data from an outgroup (a species not closely related to any member of the group under study) may be included in order to "root" the tree. Each line is a lineage, which may branch into two lineages at a node. The node represents a common ancestor of two lineages, and the two lineages that emerge from a node

are called **sister groups**. Any complete branch that can be cut from a cladogram—including the smallest branch possible (a species)—is a clade.

One way of constructing cladograms involves the logical rule of simplicity: When there are several possible ways that a group of clades can be connected, the simplest evolutionary pathway is probably the correct one. By comparing all of the possible connections among the clades, we can identify the simplest—the one in which the defining derived traits evolved the fewest number of times (**Figure 17.25**). The process of finding the simplest pathway is called parsimony analysis.

Evolutionary history does not change because of events in the present: A species' ancestry remains the same no matter how it evolves. However, as with traditional taxonomic rankings, we can make mistakes when we group organisms based on incomplete information. Thus, a clade or cladogram may change when new discoveries are made. As with all hypotheses, the more data in support of an evolutionary grouping, the less likely it is to require revision.

character Quantifiable, heritable trait.

clade (CLAYD) A group whose members share one or more defining derived traits.

cladistics (cluh-DISS-ticks) Making hypotheses about evolutionary relationships among clades.

cladogram (CLAD-oh-gram) Evolutionary tree diagram that shows evolutionary connections among a group of clades.

derived trait A novel trait present in a clade but not in any of the clade's ancestors.

evolutionary tree Diagram showing evolutionary connections.

monophyletic group (ma-no-fill-EH-tick) An ancestor in which a derived trait evolved, together with all of its descendants.

phylogeny (fie-LA-juh-knee) Evolutionary history of a species or group of species.

sister groups The two lineages that emerge from a node on a cladogram.

Studies of phylogeny reveal how species are related to one another and to species that are now extinct. In doing so, they inform our understanding of how shared ancestry interconnects all species—including our own.

CONSERVATION BIOLOGY

The story of the Hawaiian honeycreepers is a dramatic illustration of how evolution works. It also shows how finding ancestral connections can help species that are still living.

The first Polynesians arrived on the Hawaiian Islands before 1000 A.D., and Europeans followed in 1778. Forests were cleared to grow imported crops, and plants that escaped cultivation began to crowd out native plants. Escaped livestock ate and trampled rain forest plants that had provided the honeycreepers with food and shelter. Mosquitoes accidentally introduced in 1826 spread diseases such as avian malaria from imported chickens to native bird species. Stowaway rats ate their way through populations of native birds and their eggs; mongooses deliberately imported to eat the rats preferred to eat birds and bird eggs.

The isolation that had allowed honeycreepers to arise by adaptive radiation also made them vulnerable to extinction. Divergence from the ancestral species had led to the loss of unnecessary traits such as defenses against mainland predators and diseases. Traits that had been adaptive—a long, curved beak matching the flower of a particular plant, for example—became hindrances when habitats suddenly changed or disappeared. Thus, at least 43 Hawaiian honeycreeper species that had thrived on the islands before humans arrived were extinct by 1778. Conservation efforts began in the 1960s, but another 43 species have since disappeared.

Today, the few remaining Hawaiian honeycreepers are still being heavily pressured by established populations of nonnative species of plants and animals (**Figure 17.26A**). Rising global temperatures are also allowing mosquitoes to invade high-altitude habitats that had previously been too cold for the insects, so honeycreeper species remaining in these habitats are now succumbing to mosquito-borne diseases (**Figure 17.26B**). As more and more honeycreeper species become extinct, the group's reservoir of genetic diversity dwindles. The lowered diversity means the group as a whole is less resilient to change, and more likely to suffer catastrophic losses. Deciphering their phylogeny can tell us which honeycreeper species are most different from the others—and those are the ones most valuable in terms of preserving the

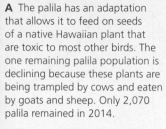

A The palila has an adaptation that allows it to feed on seeds of a native Hawaiian plant that are toxic to most other birds. The one remaining palila population is declining because these plants are being trampled by cows and eaten by goats and sheep. Only 2,070 palila remained in 2014.

B The lower bill of the akekee points to one side, allowing this bird to pry open buds that harbor insects. Avian malaria is wiping out the last population of this species. Between 2007 and 2012, the number of akekee dropped from 3,536 birds to 945.

C This poouli—rare, old, and missing an eye—died in 2004 from avian malaria. There were two other poouli alive at the time, but neither has been seen since then.

Figure 17.26 Three honeycreeper species: going, going, gone.

group's diversity. Such research allows us to concentrate our resources and conservation efforts on those species whose extinction would mean a greater loss to biodiversity. For example, we now know the poouli (**Figure 17.26C**) to be the most distant relative in the Hawaiian honeycreeper family. Unfortunately, the knowledge came too late; the poouli is probably extinct. Its extinction means the loss of a large part of evolutionary history of the group: One of the longest branches of the honeycreeper family tree is gone forever.

MEDICAL RESEARCH

Researchers often study the evolution of viruses and other infectious agents by grouping them into clades based on biochemical traits. Viruses are not alive, but they can mutate every time they infect a host, so their genetic material changes quickly. Consider the H5N1 strain of influenza (flu) virus, which infects birds and other animals. H5N1 has a very high mortality rate in humans, but human-to-human transmission has been rare to date. The virus replicates in pigs without causing symptoms. A phylogenetic analysis of H5N1 isolated from pigs showed that the virus "jumped" from birds to pigs at least three times since 2005, and that one of the isolates had acquired the potential to be transmitted among humans. Our increased understanding of the evolutionary history of this virus is helping us develop strategies to prevent it from spreading to humans again.

Figure 17.27 Agricultural operations such as this chicken farm are hot spots for the evolution of antibiotic-resistant bacteria.

DIRECTIONAL SELECTION: SUPERBUG FARMS

Scarlet fever, tuberculosis, and pneumonia once caused one-fourth of the annual deaths in the United States. Since the 1940s, we have been relying on antibiotics to fight these and other dangerous bacterial diseases. We have also been using them in other, less dire circumstances. For an unknown reason, antibiotics promote growth in cattle, pigs, poultry, and even fish. The agricultural industry uses a lot of antibiotics, mainly for this purpose: 13.7 million kilograms (about 30 million pounds) in 2011—four times the amount used for human medical purposes in the same year in the U.S. Despite recommendations to stop the practice, agricultural use of antibiotics is still rising.

A natural population of bacteria is diverse, and it can evolve astonishingly fast. Consider how each cell division is an opportunity for mutation. The common intestinal bacteria *E. coli* can divide every 17 minutes, so even if a population starts out as clones, its cells diversify quickly. Bacteria can swap genes even among distantly related species, and this adds even more genetic diversity to their respective gene pools. When a natural bacterial population is exposed to an antibiotic, some cells in it are likely to survive because they carry an allele that offers resistance. As susceptible cells die and the survivors reproduce, the frequency of antibiotic-resistance alleles increases in the population. A typical two-week course

of antibiotics can exert selection pressure on over a thousand generations of bacteria. The pressure drives genetic change in bacterial populations so they become composed mainly of antibiotic-resistant cells (an example of directional selection). Thus, using antibiotics on an ongoing basis effectively guarantees the production of antibiotic-resistant bacterial populations.

Farms where antibiotics are used to promote growth are hot spots for the evolution of antibiotic-resistant bacteria and their spread to humans (**Figure 17.27**). Veterinarians and other people who work with the animals on these farms tend to carry more antibiotic-resistant bacteria in their bodies. So do neighbors who live within a mile of the farms. The bacteria spread much farther, however. Bacteria on an animal's skin or in its digestive tract can easily contaminate its meat during slaughter, and contaminated meat ends up in restaurant and home kitchens. A 2013 investigation found "worrisome" amounts of bacteria in 97% of the chicken meat in stores across the United States. About half of the samples tested were contaminated with superbugs—bacteria that are resistant to multiple antibiotics—and one in ten contained multiple types of superbugs. An earlier study found antibiotic-resistant bacteria in more than half of supermarket ground beef and pork chops, and in over 80 percent of ground turkey. Bacteria can be killed by the heat of cooking, but it is

almost impossible to prevent them from spreading from contaminated meat to kitchen surfaces—and to people—during the process.

We have only a limited number of antibiotic drugs, and developing new ones is much slower than bacterial evolution. As resistant bacteria become more common, the number of antibiotics that can be used to effectively treat infections in humans dwindles. Using a particular antibiotic only in animals, or only in humans, is not a solution to this problem because there are just a few mechanisms by which these drugs kill bacteria; resistance to one antibiotic often confers resistance to others. For example, bacteria that become resistant to Flavomycin® (an antibiotic used only in animals) also resist vancomycin (an antibiotic used only in humans). Superbugs that are resistant to most currently available antibiotics are turning up at a very alarming rate.

All of this amounts to bad news. We are now paying the price for overuse of antibiotics: An infection with antibiotic-resistant bacteria tends to be longer, more severe, and more likely to be deadly than one more easily treatable with antibiotics. Superbugs cause more than 2 million cases of serious illness each year in the United States alone, and they outright kill 23,000 of these people. Many, many more die because the infection complicates another, preexisting illness.

Sections 17.1, 17.2 All alleles of all genes in a **population** constitute a **gene pool**. Mutations may be **neutral**, **lethal**, or adaptive. **Microevolution** is change in **allele frequency** of a population. Deviations from **genetic equilibrium** indicate that a population is evolving.

Sections 17.3–17.5 **Directional selection** is a mode of natural selection in which a form of a trait at one end of a range of variation is adaptive. An intermediate form of a trait is adaptive in **stabilizing selection**; extreme forms are adaptive in **disruptive selection**. **Sexual dimorphism** is one outcome of **sexual selection**. **Frequency-dependent selection** or any other type of natural selection can maintain a **balanced polymorphism**.

Section 17.6 **Genetic drift**, which is most pronounced in small or **inbreeding** populations, can cause alleles to become **fixed**. The **founder effect** may occur after an evolutionary **bottleneck**. **Gene flow** can counter the effects of mutation, natural selection, and genetic drift.

Sections 17.7–17.9 **Reproductive isolation** is always a part of **speciation** (**Table 17.3**). With **allopatric speciation**, a geographic barrier interrupts gene flow between populations. **Sympatric speciation** occurs by divergence within a population. With **parapatric speciation**, populations in contact along a common border speciate.

Section 17.10 **Macroevolution** refers to large-scale patterns of evolution. With **stasis**, very little change occurs over long spans of time. An **exaptation** is a trait that has been evolutionarily repurposed for a new use. A **key innovation** can result in an **adaptive radiation**. **Coevolution** occurs when two species act as agents of selection upon one another. A lineage with no more living members is **extinct**.

TABLE 17.3

Comparison of Speciation Patterns

	Allopatric	Parapatric	Sympatric
Original population(s)	●	●●	●
Initiating event:	physical barrier arises	selection pressures differ	genetic change
Reproductive isolation occurs			
New species arises:	in isolation	in contact along common border	within existing population

Section 17.11 Evolutionary biologists reconstruct evolutionary history (**phylogeny**) by comparing **characters** among species. A **clade** is a **monophyletic group** that consists of an ancestor in which one or more **derived traits** evolved, together with all of its descendants. Making hypotheses about the evolutionary history of a group of clades is called **cladistics**. These hypotheses are often represented as **evolutionary trees** called **cladograms**, with each line being a lineage, and each branch being a clade. A lineage branches into two **sister groups** at a node.

Section 17.12 Reconstructing phylogeny is part of our efforts to preserve endangered species. Phylogeny is also used for studying the spread of viruses and other agents of disease.

Application Our overuse of antibiotics has exerted directional selection favoring resistant bacterial populations, which are now common in the environment. We are running out of effective antibiotics to use as human drugs.

1. _____ is the original source of new alleles.
   - a. Mutation
   - b. Natural selection
   - c. Gene flow
   - d. Genetic drift

2. Which is required for evolution to occur in a population?
   - a. genetic diversity
   - b. selection pressure
   - c. gene flow
   - d. none of the above

3. Match the modes of natural selection with their best descriptions.
   - ____ stabilizing
   - ____ directional
   - ____ disruptive
   - a. eliminates extreme forms of a trait
   - b. eliminates midrange forms of a trait
   - c. shifts phenotypes in one direction

4. Sexual selection frequently influences aspects of body form and can lead to _____ .
   - a. sister groups
   - b. males and females
   - c. exaggerated traits
   - d. genetic equilibrium

5. The persistence of the sickle allele at high frequency in a population is an example of _____ .
   - a. bottlenecking
   - b. inbreeding
   - c. the founder effect
   - d. a balanced polymorphism

6. _____ among populations can keep them similar to one another.
   - a. Genetic drift
   - b. Gene flow
   - c. Mutation
   - d. Natural selection

7. In many bird species, sex is preceded by a courtship dance. If a male's dance is unrecognized by the female, she will not mate with him. This is an example of _____ .
   - a. reproductive isolation
   - b. behavioral isolation
   - c. sexual selection
   - d. all of the above

RESISTANCE TO RODENTICIDES IN WILD RAT POPULATIONS

Beginning in 1990, rat infestations in northwestern Germany started to intensify despite continuing use of rat poisons. In 2000, Michael Kohn and his colleagues tested wild rat populations around Münster. In five towns, they trapped and tested wild rats for resistance to warfarin and the more recently developed poison bromadiolone. The results are shown in **Figure 17.28**.

1. In which of the five towns were most of the rats susceptible to warfarin?

2. Which town had the highest percentage of poison-resistant wild rats?

3. What percentage of rats in Olfen were warfarin resistant?

4. In which town do you think the application of bromadiolone was most intensive?

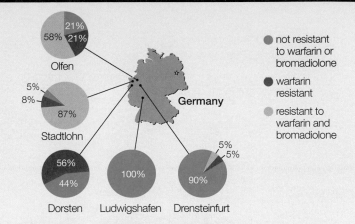

Figure 17.28 Poison resistance in wild rats in Germany, 2000.

8. Which of the following is not part of how we define a species?
   a. Its individuals appear different from other species.
   b. It is reproductively isolated from other species.
   c. Its populations can interbreed.
   d. Fertile offspring are produced.

9. Natural selection does not explain _____ .
   a. genetic drift
   b. the founder effect
   c. gene flow
   d. how mutations arise
   e. inheritance
   f. any of the above

10. After fire devastates all of the trees in a wide swath of forest, populations of a species of tree-dwelling frog on either side of the burned area diverge to become separate species. This is an example of _____ .

11. _____ is a way of reconstructing evolutionary history based on derived traits.
   a. Natural selection
   b. Coevolution
   c. Gene flow
   d. Cladistics

12. The only taxon that is always the same as a clade is _____ .

13. In evolutionary trees, each node represents a(n) _____ .
   a. single lineage
   b. extinction
   c. divergence
   d. adaptive radiation

14. Match the evolutionary concepts.
   ___ gene flow
   ___ sexual selection
   ___ extinct
   ___ genetic drift
   ___ phylogeny
   ___ adaptive radiation
   ___ derived trait
   ___ coevolution
   ___ natural selection
   ___ cladogram

   a. can lead to interdependent species
   b. changes in a population's allele frequencies due to chance alone
   c. alleles enter or leave a population
   d. evolutionary history
   e. operates on variation in shared traits
   f. adaptive traits make their bearers better at securing mates
   g. no more living members
   h. diagram of sets within sets
   i. burst of divergences from one lineage into many
   j. present in a group, but not in any of the group's ancestors

15. The evolution of wings helped the insect clade to be very successful. In this example, wings are a(n) _____ .
   a. derived trait
   b. adaptive trait
   c. key innovation
   d. all of the above

1. Species have traditionally been characterized as "primitive" and "advanced." For example, mosses were considered to be primitive, and flowering plants advanced; crocodiles were primitive and mammals were advanced. Why do most biologists of today think it is incorrect to refer to any modern species as primitive?

2. Rama the cama, a llama–camel hybrid, was born in 1998. The idea was to breed an animal that has the camel's strength and endurance, and the llama's gentle disposition. However, instead of being large, strong, and sweet, Rama is smaller than expected and has a camel's short temper. The breeders plan to mate him with Kamilah, a female cama. What potential problems with this mating should the breeders anticipate?

3. Two species of antelope, one from Africa, the other from Asia, are put into the same enclosure in a zoo. To the zookeeper's surprise, individuals of the different species begin to mate and produce healthy, hybrid baby antelopes. Explain why a biologist might not view these offspring as evidence that the two species of antelope are in fact one.

4. Some human traits may have arisen by sexual selection. Over thousands of years, women attracted to charming, witty men perhaps prompted the development of human intellect beyond what was necessary for mere survival. Men attracted to women with juvenile features may have shifted the species as a whole to be less hairy and softer featured than any of our simian relatives. Can you think of a way to test these hypotheses?

for additional quizzes, flashcards, and other study materials
▶ **ACCESS MINDTAP AT WWW.CENGAGEBRAIN.COM**

18 Life's Origin and Early Evolution

Links to Earlier Concepts

This chapter describes investigations into life's origin and what is known about life's early history. It draws on your knowledge of scientific method (Section 1.5), plate tectonics (16.5), and the geologic time scale (16.6). We discuss the origin of eukaryotic organelles (4.5) and return to the link between photosynthesis and aerobic respiration (7.1).

Core Concepts

The field of biology relies upon experimentation and the collection and analysis of scientific evidence.

Hypotheses supported by scientific evidence provide a natural explanation for the origin of life. Simulations show that chemical and physical processes could have produced the organic building blocks of life from inorganic precursors on early Earth and that more complex molecules assemble spontaneously from these simple subunits.

All organisms are linked by lines of descent from a shared ancestor.

The shared core processes and characteristics observed among modern and fossil organisms provide evidence for life's common origin. All living things consist of the same components. All cells have a plasma membrane that separates the cell from its environment, and a genome of DNA that provides the instructions for all cellular processes.

Natural selection leads to the evolution of structures and processes that increase fitness in a specific environment.

The first cells evolved in a low-oxygen or oxygen-free environment. The later evolution of oxygen-producing photosynthetic bacteria altered this environment. Most organisms that had previously adapted to the absence of oxygen could not tolerate the change and died out; those capable of living in the presence of oxygen repopulated Earth. Aerobic respiration evolved in some of the survivors.

◄ Stromatolites in Australia's Shark Bay consist of cells and sediments. Fossil stromatolites provide evidence of early life.

Photograph by Frans Lanting, National Geographic Creative.

LEARNING OBJECTIVES

Explain what the study of ancient rocks can reveal about conditions on Earth 4 billion years ago.

Describe an experiment that tested whether the building blocks of life could have formed by a nonbiological process on early Earth.

List three locations where scientists think the building blocks of Earth's first life could have formed.

CONDITIONS ON EARLY EARTH

To understand how life on Earth could have begun, it helps to know a bit about our planet's history. Our planet came into existence by about 4.6 billion years ago, through the aggregation of dust and rock bits that were in orbit around our sun. The interval from 4.6 billion to 4.0 billion years ago is known as the Hadean eon. *Hadean* means "hell-like," and early Earth lived up to this name (**Figure 18.1**). Planet formation did not remove all the material orbiting the sun: Interplanetary objects such as asteroids (large rocky bodies), meteoroids (smaller rocks), and comets (balls of ice and dust) constantly bombarded Earth. These extraterrestrial materials, together with gases and rocky material released by frequent volcanic eruptions, provided the components of Earth's land, oceans, and atmosphere.

No rocks that date to the time of Earth's origin persist, but geologists have discovered crystals of zircon (a type of mineral) that date back 4.3 billion years. As Section 16.4 explained, zircon crystals form when molten rock cools, and radiometric dating can determine the time since their formation. The composition of some of these 4.3-billion-year-old crystals implies they formed in water, suggesting that by this time Earth had cooled enough for water to pool on its surface. The presence of liquid water is important because metabolic reactions that sustain life involve molecules dissolved in water.

The exact composition of Earth's early atmosphere remains a matter of debate. We do know that there was little or no free oxygen (O_2), because the oldest iron-containing rocks show no evidence of iron oxidation (rusting). The apparent lack of free oxygen was fortunate. Had such oxygen been present, oxidation reactions would have broken apart small organic molecules as soon as they formed.

SOURCES OF ORGANIC SUBUNITS

All living things are made from the same organic subunits: amino acids, fatty acids, nucleotides, and simple sugars. Until the early 1800s, people thought that such molecules possessed a special "vital force" and that only living organisms could

Figure 18.1 Artist's depiction of the Hadean Earth. Frequent volcanic eruptions released heat, rock fragments, and gases such as carbon dioxide, hydrogen, and water vapor. Condensation of water vapor contributed to the first seas.

Figure 18.2 The Miller–Urey experiment. Stanley Miller and Harold Urey used this glass apparatus to test whether organic compounds could have formed by lightning-driven chemical reactions in Earth's early atmosphere. Water, hydrogen gas (H_2), methane (CH_4), and ammonia (NH_3) circulated through the apparatus as sparks from an electrode simulated lightning.

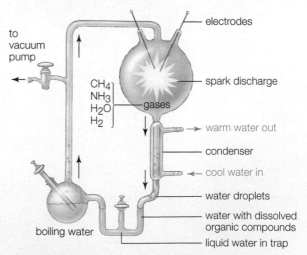

- electrodes
- to vacuum pump
- CH_4
- NH_3
- H_2O
- H_2
- gases
- spark discharge
- warm water out
- condenser
- cool water in
- water droplets
- water with dissolved organic compounds
- boiling water
- liquid water in trap

FIGURE IT OUT
Which gas in this mixture provided the nitrogen for the amino group in the amino acids?

Answer: Ammonia

Figure 18.3
A type of deep-sea hydrothermal vent called a black smoker.

Dr. Robert Ballard

National Geographic Explorer Bob Ballard is best known for finding wrecks of historic ships such as the RMS *Titanic*. However, he says his proudest accomplishment is discovering that life thrives in the deep, dark ocean depths near hydrothermal vents. In 1977, Ballard led a team that used a submersible vehicle to explore a region of the seafloor where tectonic plates are moving apart (as described in Section 16.5). As Ballard and others had predicted, the team found openings (vents) in the seafloor. Mineral-rich water heated by geothermal energy streamed from the vents into the frigid waters of the deep ocean. To Ballard's astonishment, the team also found a wealth of previously unknown life. He says, "We discovered that this whole life system was living not off the energy from the sun, but from the energy of the Earth." Dr. Ballard's discovery led other scientists to reconsider life's origins. Many scientists now think that Earth's earliest life arose deep in the sea near hydrothermal vents.

make them. Then, in 1828, a chemist synthesized urea, an organic molecule abundant in urine. Later, another chemist synthesized alanine, an amino acid. These synthesis reactions proved that under the right conditions, nonbiological processes can produce organic molecules from inorganic ones.

With this in mind, some scientists began to think about how and where the organic molecules that served as life's first building blocks formed. We consider three possible mechanisms for this process below. Keep in mind that they are not mutually exclusive. All three mechanisms may have operated simultaneously and contributed to an accumulation of simple organic compounds in Earth's early seas. Once formed, these compounds would have persisted longer than they would today. There was no oxygen to break them down, and organisms capable of metabolizing them had not yet evolved.

Lightning-Fueled Atmospheric Reactions
The first test of the hypothesis that organic compounds could have formed spontaneously on early Earth took place in 1953. Stanley Miller and Harold Urey used the apparatus illustrated in **Figure 18.2** to determine if lightning-fueled atmospheric reactions could have produced the building blocks of life. To simulate Earth's early atmosphere, the scientists filled a reaction chamber with methane, hydrogen gas, ammonia, and water. They then circulated the mixture and zapped it with sparks to simulate lightning. Within a week, amino acids and other organic molecules common to all modern organisms had formed.

We now know that the mix of gases in this initial experiment probably did not accurately represent those present

on early Earth. Later experimental variations that included hydrogen sulfide gas and mimicked atmospheric conditions around ancient volcanoes produced even more amino acids.

Reactions at Hydrothermal Vents
Deep sea **hydrothermal vents** are underwater hot springs—places where hot, mineral-rich water flows out through an opening in the seafloor (**Figure 18.3**). When this water mixes with cool seawater, the dissolved minerals come out of solution to form mineral-rich rock. Simulations designed to mimic conditions near hydrothermal vents have produced amino acids and other organic compounds.

Delivery From Outer Space
Modern-day meteorites (meteoroids that fell to Earth) sometimes contain amino acids, sugars, and nucleotide bases. These compounds (or precursors of them) have also been discovered in gas clouds surrounding nearby stars. Thus it is possible that some of the many extraterrestrial objects that fell on early Earth delivered organic monomers from outer space. Keep in mind that in Earth's early years, such materials fell to Earth thousands of times more frequently than they do today.

---

hydrothermal vent Underwater opening where hot, mineral-rich water streams out.

CREDITS: (3) From the IMAX film "Volcanoes of the Deep Sea" produced by The Stephen Low Company in association with Rutgers University; (inset) O. Louis Mazzatenta/National Geographic Creative.

PROPERTIES OF CELLS

In addition to sharing the same molecular components, all cells have a plasma membrane composed primarily of a lipid bilayer. As Chapter 9 explained, cells have a genome of DNA that enzymes transcribe into RNA, and ribosomes that translate RNA into proteins. All cells replicate and pass on copies of their genetic material to their descendants. The similarities in structure, metabolism, and replication processes among all known organisms are considered evidence that they all are descendants of the same cellular ancestor.

Time has erased all evidence of the earliest cells, but scientists can still investigate this first chapter in life's history. They use their knowledge of chemistry to design experiments that test whether a particular hypothesis about how life began is plausible. Results of such studies support the hypothesis that cells arose as a result of a stepwise process, beginning with inorganic materials (**Figure 18.4**). By this hypothesis, each step on the road to life can be explained by chemical and physical mechanisms that still operate today.

ORIGIN OF METABOLISM

Modern cells take up organic subunits and link them into organic polymers. Enzymes speed polymer synthesis, but it can also happen on its own; it just occurs more slowly. Remember from Section 5.3 that the closer substrates are to one another, the more likely they are to combine. Before there were cells, a nonbiological process that concentrated organic subunits in one place would have increased the chance that the subunits would combine on their own.

By one hypothesis, this process occurred on clay-rich tidal flats. Clay particles have a slight negative charge, so positively charged molecules in seawater stick to them. At low tide, evaporation would have dried the clay, concentrating the subunits even more, and energy from the sun could have helped them bond together as polymers. In simulated tidal flat conditions, amino acids bond together in short polypeptide chains, and nucleotides assemble as RNA-like chains.

By another hypothesis, metabolic reactions began in high-temperature, high-pressure seawater near hydrothermal vents. Rocks around some of these vents contain iron sulfide (pyrite) and are porous, with many tiny chambers about the

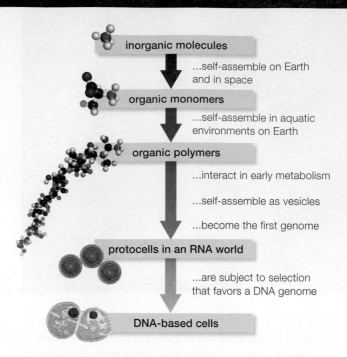

Figure 18.4 Proposed sequence for the evolution of cells. Scientists investigate this process by carrying out experiments and simulations that test hypotheses about feasibility of individual steps.

size of cells. Metabolism may have begun when iron sulfide in the rocks donated electrons to dissolved carbon monoxide (CO), a reaction that forms organic compounds. Under simulated vent conditions, organic compounds such as pyruvate do form and accumulate in the rocky chambers. In addition, all modern organisms have proteins that use iron–sulfur clusters as cofactors (Section 5.5). The universal requirement for these cofactors may be a legacy of life's rocky beginnings.

ORIGIN OF THE GENOME

All modern cells have a DNA genome. They pass copies of their DNA to descendant cells, which use instructions encoded in the DNA to build proteins. Some of these proteins are enzymes that synthesize new DNA, which is passed to descendant cells, and so on. Thus, protein synthesis depends on DNA, which is built by proteins. How did this cycle begin?

In the 1960s, Francis Crick and Leslie Orgel suggested there was time during which RNA both stored genetic information and functioned as an enzyme that catalyzed protein synthesis. This proposed interval is referred to as **RNA world**. Like DNA, RNA consists of a sequence of nucleotides, and can serve as a template to direct the synthesis of a complementary sequence. Some modern viruses use RNA as their genetic material. RNAs can also act as enzymes. **Ribozymes** are RNAs that have enzymatic activity, in living cells. For example, the rRNA component of ribosomes speeds formation of peptide bonds during protein synthesis (Section 9.3). Other ribozymes cut noncoding bits (introns) out of newly formed RNAs (Section 9.2).

PROTOCELL FORMATION

Molecules formed by early synthetic reactions would have floated away from one another unless something enclosed them. In modern cells, a plasma membrane serves this role. If the first reactions took place in tiny rock chambers, the rock would have served as a boundary. Over time, lipids produced by reactions inside a chamber could have accumulated and lined the chamber wall. Such lipid-enclosed collections of interacting molecules may have been the first protocells. A **protocell** is a membrane-enclosed collection of molecules that takes up material and replicates itself.

Membranes of early protocells probably consisted of a mix of fatty acids, rather than the phospholipids that make up the bulk of modern cell membranes (Section 4.3). Like phospholipids, fatty acids have water-loving (hydrophilic) and water-hating (hydrophobic) parts, so they form a lipid bilayer when mixed with water and can self-assemble as vesicles (membranous, fluid-filled chambers). However, fatty acids are structurally simpler than phospholipids and would have been more abundant on early Earth. A fatty acid membrane also provides an advantage in terms of permeability. Nucleotides and ions move through a fatty acid bilayer, but they cannot cross a phospholipid bilayer. Having a highly permeable bilayer would have been especially important because, unlike modern cells, protocells would not have had transport proteins embedded in their membrane.

Simulating formation of protocells provides insight into how cellular life could have originated. Biochemist Jack Szostak carries out such simulations to investigate how protocells form and replicate. **Figure 18.5A** is a computer model of the type of protocell that he investigates, and **Figure 18.5B** shows a protocell that formed in his laboratory. The laboratory-produced protocell consists of a fatty acid membrane enclosing a clay particle with RNA bound to its surface. The clay serves a dual purpose; it encourages the assembly of both nucleotides into RNA and fatty acids into vesicles.

---

protocell Membranous sac that contains interacting organic molecules; hypothesized to have formed prior to the first life forms.
ribozyme (RYE-boh-zime) RNA that has enzymatic activity.
RNA world Hypothetical early interval when RNA served as genetic material.

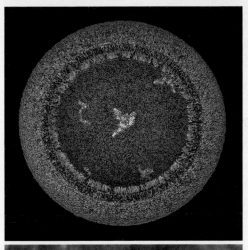

A Computer model of a protocell with a membrane of fatty acids around strands of RNA.

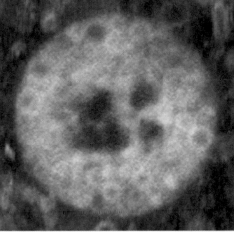

B Laboratory-formed protocell consisting of fatty acids (green) surrounding RNA-coated clay (red).

Figure 18.5 Protocells. Computer models and laboratory experiments help scientists determine how and where protocells could have formed, as well as the protocells' most likely components.

Laboratory-produced protocells "grow" by incorporating fatty acids into their membrane and nucleotides into their RNA. New protocells are produced when large ones break into smaller units as a result of mechanical force. However, the resulting "daughter cells" do not have the same RNA sequence as their parent. Scientists have not yet produced a synthetic protocell with an RNA genome that can replicate itself. Szostak and others continue to investigate the conditions that would make such lifelike replication of the genome possible.

If the earliest self-replicating protocells that formed naturally were RNA-based, why do all organisms now have a genome of DNA? The difference in the stability of the two molecules was probably a factor. Compared to a double-stranded DNA molecule, a single-stranded RNA breaks apart more easily and is more prone to replication errors. Thus, a switch from RNA to DNA would make larger, more stable genomes possible.

CREDITS: (5A) © Janet Iwasa; (5B) From Hanczyc, Fujikawa, and Szostak, "Experimental Models of Primitive Cellular Compartments: Encapsulation, Growth, and Division"; www.sciencemag.org, *Science* 24 October 2003; 302;529, Fig. 2, p. 619. Reprinted with permission of the authors and AAAS.

LEARNING OBJECTIVES

List the presumed traits of the first cells.

Describe the oldest known fossil prokaryotes and eukaryotes.

Explain how the evolution of cyanobacteria altered conditions on Earth.

TRAITS OF THE FIRST CELLS

The processes described in Section 18.2 may have produced cellular life more than once. However, the many genetic and metabolic similarities among living organisms suggest that all modern life descended from a common ancestor. Such a high degree of similarity would be unlikely if independent lineages arose and persisted. Given what scientists know about relationships among modern species, most assume that the common ancestor of all life was prokaryotic, meaning it did not have a nucleus. Oxygen was scarce or absent on early Earth, so the ancestral cell must also have been anaerobic (capable of living without oxygen). Other aspects of this cell's metabolism are less clear. It may have been a heterotroph that fermented organic compounds. Alternatively, it may have been an autotroph that stripped electrons from inorganic material and assembled its own food from carbon dioxide. It was most likely not photosynthetic. Photosynthesis is a relatively complicated process that requires the evolution of much specialized metabolic machinery.

SEARCHING FOR ANCIENT LIFE

Finding and identifying signs of early cells is a challenge. Cells are microscopic, and most have no hard parts to fossilize. In addition, few ancient rocks that might hold early fossils still exist. Tectonic plate movements have destroyed nearly all rocks older than about 4 billion years, and most of the slightly younger rocks have been subject to volcanic heating and other processes that destroy all traces of biological material. To add to the difficulty, structures formed by nonbiological processes sometimes resemble fossils. To avoid mistakenly accepting these structures as genuine fossils, scientists constantly reanalyze purported fossil finds and they often question one another's conclusions.

THE OLDEST FOSSIL CELLS

Life may have originated as early as the onset of the Archean eon, 4.0 billion years ago. (*Archean* is Greek for beginning.) However, the oldest material widely accepted as fossil cells comes from 3.4-billion-year-old Australian sandstone rocks. These rocks hold spherically shaped fossils nestled among sand grains from an ancient beach (**Figure 18.6**).

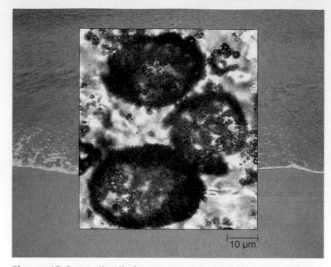

Figure 18.6 Fossil cells from an ancient beach. Microscopic 3.4-billion-year-old fossils nestle among sand grains in a sandstone. The cells may be relatives of modern bacteria.

The presence of the mineral pyrite (iron sulfide) in and around the fossil cells suggests they had a metabolism similar to that of modern sulfate-reducing bacteria. These anaerobic bacteria commonly live in mudflats, and they produce hydrogen sulfide gas as a by-product of their metabolism. When this gas reacts with iron, it forms pyrite.

At another Australian site, 3.2-billion-year-old rocks that formed near a deep-sea hydrothermal vent contain fossil filaments whose shape and arrangement are similar to that of cells found in the vicinity of modern vents.

Taken together, these two sets of fossils suggest that by 3.2 billion years ago, cellular life was widespread in the seas, from sandy shorelines to the ocean floor.

STROMATOLITES AND THE RISE IN OXYGEN

The most widespread evidence of early cellular life comes from fossil stromatolites. A **stromatolite** is a rocky conical or dome-shaped structure composed of layers of cells, cell remains, and sediment (**Figure 18.7A**). Some structures that may be stromatolites are 3.4 billion years old, but the oldest undisputed stromatolites date to 2.8 billion years ago. Stromatolites reached their peak distribution 1.25 billion years ago, when they were common in seas worldwide.

To learn how fossil stromatolites could have formed, scientists study modern stromatolites such as those in Australia's Shark Bay (shown in the chapter's opening photo). The Shark Bay stromatolites began growing an estimated 2,000 years ago and now stand up to 1.5 meters high. Photosynthetic bacteria that live on the stromatolites' upper, sunlit surface grow as a mat that traps sediments. As new sediment is added, bacterial cells grow up through it and trap more sediment. Archaea and nonphotosynthetic bacteria live in the stromatolite's deeper layers. Ancient stromatolites probably contained both bacteria and archaea as well. We know the two prokaryotic domains, Bacteria and Archaea, diverged early in life's history, although we cannot pinpoint the exact timing.

Most photosynthetic cells found in both modern and fossil stromatolites are cyanobacteria (**Figure 18.7B**). Photosynthesis evolved in many bacteria, but only cyanobacteria use the noncyclic pathway that produces oxygen. The evolution of oxygen-producing photosynthesis in this group had a dramatic effect on early life. About 2.5 billion years ago, the Proterozic eon had just begun. (The name of this eon means first life.) Oxygen released by cyanobacteria was accumulating in the atmosphere and seas, creating a new, global selection pressure. The oxygen was toxic to most species that had evolved in its absence. Inside cells, oxygen reacts with metal ions, forming free radicals that damage DNA and other essential cell components (Sections 2.2 and 5.5). Species unable to detoxify the free radicals either went extinct or became restricted to the low-oxygen environments that remained. By contrast, species with metabolic machinery capable of detoxifying the free radicals thrived. Some of these survivors began to use this machinery for aerobic respiration (Section 7.1). Later, multicelled eukaryotes would use this very efficient energy-releasing pathway to meet their high energy needs.

The increase in atmospheric oxygen would also lead, over time, to the formation of the **ozone layer**, a region of the upper atmosphere that has a high concentration of ozone gas (O_3). Before the ozone layer formed, life could exist only underwater, where the water shielded it from some incoming UV radiation. Only after the ozone layer formed and screened out much of the mutation-causing UV radiation could life begin to move onto land.

EARLY EUKARYOTES

Nuclei are seldom seen in fossils, but other traits can indicate a fossil cell was eukaryotic. Eukaryotic cells are generally larger than prokaryotic ones. A cell wall with complex patterns, spines, or spikes likely housed a eukaryote. Certain compounds are indicative of eukaryotes. For example, steroids in ancient rocks suggest eukaryotes were present by 2.7 billion years ago; prokaryotes do not make steroids. The oldest fossils that most scientists accept as eukaryotic date to about 1.8 billion years ago. They resemble resting stage cells (cysts) of some modern single-celled algae.

A simple sort of multicellularity, in which cells stay attached to one another, evolved early in the history of life.

A In cross section, a stromatolite reveals the layered remains of countless photosynthetic bacteria and sediment they captured.

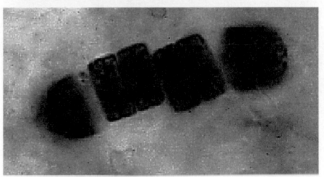

B Cyanobacteria such as those in this 850-million-year-old fossil are the main photosynthetic cells in stromatolites.

Figure 18.7 Fossil stromatolites.

Some modern cyanobacteria form a chain of cells that is held together by a secreted mucus sheath. The cells in the fossil in **Figure 18.7B** were probably held together in this way. Some cyanobacteria also show differentiation; cells become a few specialized types. However, multicelled eukaryotes have a variety of cell types, each specialized for a different task, with special cells (germ cells) set aside for reproduction.

Sexual reproduction occurs only in eukaryotes. The earliest evidence of a sexual reproduction is a fossil of a seaweed-like red alga that lived 1.2 billion years ago. Some of the alga's cells formed a holdfast to anchor it in place, others were part of a stalk, and still others produced spores that functioned in sexual reproduction.

---

ozone layer (OH-zone) Atmospheric region rich in ozone that screens out incoming ultraviolet (UV) radiation.
stromatolite (stroh-MAT-oh-lite) Rocky structure composed of layers of bacterial cells and sediments.

Describe the process by which the nuclear membrane and the endomembrane system could have formed.

Explain the ancestry of mitochondria and of chloroplasts.

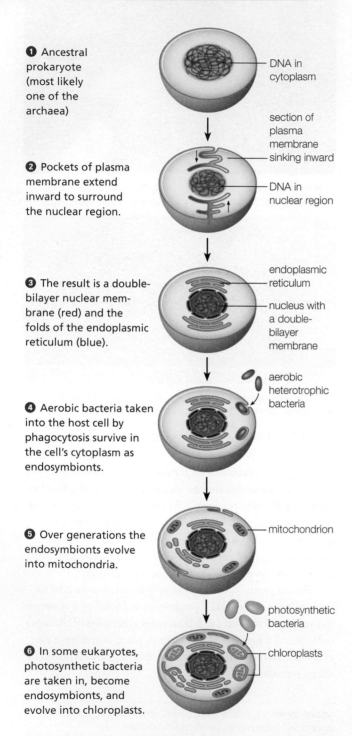

❶ Ancestral prokaryote (most likely one of the archaea)

DNA in cytoplasm

section of plasma membrane sinking inward

❷ Pockets of plasma membrane extend inward to surround the nuclear region.

DNA in nuclear region

❸ The result is a double-bilayer nuclear membrane (red) and the folds of the endoplasmic reticulum (blue).

endoplasmic reticulum

nucleus with a double-bilayer membrane

❹ Aerobic bacteria taken into the host cell by phagocytosis survive in the cell's cytoplasm as endosymbionts.

aerobic heterotrophic bacteria

❺ Over generations the endosymbionts evolve into mitochondria.

mitochondrion

❻ In some eukaryotes, photosynthetic bacteria are taken in, become endosymbionts, and evolve into chloroplasts.

photosynthetic bacteria

chloroplasts

Figure 18.8 Proposed evolutionary mechanisms for the origin of some eukaryotic organelles.

UNIQUE EUKARYOTIC TRAITS

Let's take a moment to review how eukaryotes differ from both bacteria and archaea. The genome of a typical bacterial or archaeal cell consists of a circle of DNA that, in almost all species, is in contact with the cytoplasm, rather than enclosed within a membrane. By contrast, the DNA of a eukaryotic cell is distributed among multiple linear chromosomes and is always inside a nucleus. The nucleus is associated with an endomembrane system unique to eukaryotes that includes the endoplasmic reticulum (ER) and Golgi bodies. All eukaryotes also have mitochondria or similar energy-releasing organelles derived from mitochondria, and many have chloroplasts. No bacteria or archaea have either mitochondria or chloroplasts.

A MIXED HERITAGE

The first eukaryote must have evolved from a prokaryotic ancestor, so we can ask whether that ancestor was bacterial or archaeal. The answer is complicated and still not fully understood. Eukaryotes are similar to archaea with regard to how they store their DNA (wrapped around histone proteins) and in some details of how they transcribe, replicate, and repair their DNA. On the other hand, many eukaryotic genes, especially those involved in energy metabolism, are more similar to bacterial genes than archaeal ones. In addition, mitochondria and chloroplasts resemble bacteria in their size and shape, and these organelles replicate independently of the eukaryotic cell that holds them. These organelles have their own DNA which, as in most prokaryotes, is in a single circular chromosome. They also have at least two outer membranes, with the innermost membrane structurally similar to a bacterial plasma membrane.

So, were the ancestors of eukaryotes archaea or bacteria? The mix of many bacterial and archaeal traits in eukaryotes indicates that this group probably has a composite heritage. Long ago, an archaeon or a member of a pre-eukaryotic lineage that had recently diverged from the archaea partnered with bacteria, bringing together the traits and genes we observe in modern eukaryotes.

EVOLUTION OF ORGANELLES

Nucleus and Endoplasmic Reticulum The nucleus and endoplasmic reticulum probably evolved from pockets of plasma membrane that had been drawn inward (**Figure 18.8 ❶, ❷**). The membrane that surrounded the DNA became the nuclear membrane, and membrane associated with it became the endoplasmic reticulum ❸.

Studies of the few types of prokaryotes that have internal membranes illustrates that such infolding has occurred

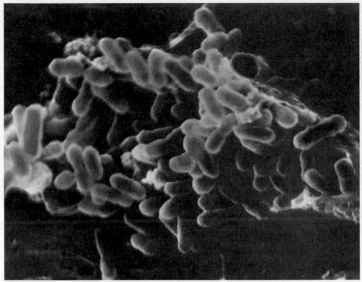

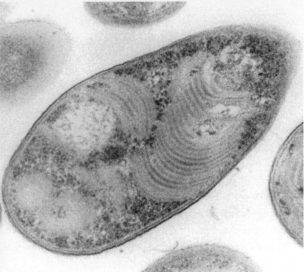

A *Rickettsia prowazekii*, the bacteria that cause typhus.

B *Methylobacter*, a methane-feeder with stacked internal membranes.

Figure 18.9 Modern bacteria that may be related to the bacterial ancestors of mitochondria.

and how it can be advantageous. For example, some modern marine bacteria have membrane infoldings that increase the surface area available to hold membrane-associated enzymes. Internal membranes also protect a genome from physical or biological threats such as viruses.

Mitochondria and Chloroplasts According to the **endosymbiont theory** for the origin of mitochondria and chloroplasts, these organelles descended from bacteria that entered a host cell and lived inside it. We refer to organisms that live inside another organism as **endosymbionts**; *endo–* means inside and *symbiosis* means living together. The term is usually reserved for cells that help their host, or at least do not harm it.

Consider how mitochondria are thought to have arisen. The process began when an anaerobic host cell engulfed heterotrophic bacteria capable of aerobic respiration ❹.

The endosymbionts lived inside their host, where they continued to carry out aerobic respiration and to reproduce. When the host cell divided, it passed on "guest" cells and the ability to carry out this efficient energy-releasing pathway.

As the two species lived together over many generations, the endosymbiont lost some genes that duplicated the function of host genes, and donated other genes to the chromosome of the host cell. At the same time, the host became dependent on the ATP produced by its aerobic partner. Eventually, the host and endosymbiont became incapable of living independently. The endosymbiont had evolved into the organelle we call mitochondria ❺.

The high degree of similarity among the genomes of all mitochondria indicates that these organelles descended from one species of bacteria. To learn what that species could have been like, researchers look for related bacteria still alive today. Metabolic and genetic similarities between mitochondria and specific modern bacterial groups are taken as evidence of shared ancestry.

One proposed mitochondrial relative is *Rickettsia prowazeki* (**Figure 18.9A**), which causes the deadly disease typhus in humans. Rickettsias are tiny bacteria that invade eukaryotic cells and replicate inside them. Alternatively, the closest living relatives of mitochondria may be some methane-feeding (methanotropic) bacteria. Similarities in the metabolic pathways of these groups suggest a link between them. In addition, some methanotrophic bacteria have stacked membranes in their cytoplasm (**Figure 18.9B**). These membranes resemble the stacked internal membranes of mitochondria.

Tracing the bacterial ancestry of chloroplasts is easier. Because chloroplasts carry out photosynthesis by the oxygen-producing pathway that evolved only in cyanobacteria, these bacteria are chloroplasts' most likely ancestors. Genetic similarities confirm this ancestor-descendant relationship. The first chloroplasts evolved after cyanobacteria were engulfed by an early eukaryote ❻.

endosymbiont (en-doh-SIM-bee-ont) Organism that lives inside another organism and does not harm it.
endosymbiont theory Theory that mitochondria and chloroplasts evolved from bacteria that entered and lived inside a host cell.

CREDITS: (9A) CNRI/Science Source; (9B) Julia Gebert, Alexander Gröngröft, Michael Schloter, Andreas Gattinger, Community structure in a methanotroph biofilter as revealed by phospholipid fatty acid analysis, FEMS Microbiology Letters, 2004, Volume 240, Issue 1, 61–68, by permission of Oxford University Press.

LEARNING OBJECTIVES

Describe the length of the Precambrian relative to the history of Earth.

Discuss the environmental changes that occurred during the Precambrian.

List the types of organisms that were living in the sea at the end of the Precambrian.

The evolutionary events discussed earlier in this chapter took place during the long interval of time commonly referred to as the Precambrian (**Figure 18.10**). The **Precambrian** encompasses the vast majority of Earth's history and three of the four eons on the geologic time scale. It begins with Earth's origin 4.6 billion years ago, and ends 541 million years ago at the onset of our present eon, the Phanerozoic. The Precambrian is so named because it precedes the Cambrian period, which is the first period in the Phanerozoic.

The Precambrian opened on a Hadean Earth that was too hot for life and devoid of oxygen gas. By 4.3 billion years ago, Earth had cooled enough for water to pool and form seas; those seas and the atmosphere held little or no oxygen. The period described as the RNA world presumably occurred in this environment around 4 billion years ago, as the Hadean eon gave way to the Archean eon.

The first prokaryotic cellular life appears in the fossil record during the Archean, about 3.4 billion years ago. Given how difficult it is to locate fossils of cells, scientists assume that these prokaryotes actually arose a bit earlier.

Stromatolites appear in the fossil record beginning at least 2.8 billion years ago, near the border between the Archean and Proterozoic eons. They increased in abundance as the Proterozoic continued, and by about 2.5 billion years ago, oxygen gas released by cyanobacteria living in stromatolites and elsewhere had begun to transform the atmosphere.

Some of the aerobic bacteria favored by the increase in oxygen would later evolve the ability to carry out aerobic respiration. One such species would undergo additional evolution, becoming mitochondria in the new eukaryotic domain. The first eukaryotes were single-celled members of the collection of lineages that we refer to collectively as protists. As you will learn in Chapter 20, branches from various protist lineages would later give rise to the more familiar eukaryotic groups: plants, fungi, and animals.

The oldest animal fossils date to about 650 million years ago, during the final period of the Proterozic (the Ediacaran period). Animal lineages that arose during the Ediacaran period and survived to the present include sponges, cnidarians (such as jellies and sea anemones), and annelids (such as sandworms and earthworms). However, no known fossils of snails, crabs, sea stars, or fish date to the Precambrian. The groups to which these animals belong first appear in the fossil record during an adaptive radiation in the Cambrian. We discuss animal evolution in detail in Chapters 23 and 24.

Fungi were also present in the seas of the late Proterozoic. However, they were not the sort of mushroom-making fungi with which you are probably familiar. The fungi that evolved in the late Proterozoic lived as aquatic, flagellated cells. As you will learn in Chapter 22, some fungi still produce flagellated spores, a legacy of the group's aquatic origins.

The green algal ancestors of plants were present in the Proterozoic, but plants themselves did not appear until about 400,000 years ago. We discuss plant evolution in Chapter 21.

Note that Precambrian history is the story of life in the seas. For most of this interval, the ozone layer remained so thin that the amount of incoming ultraviolet radiation made life on land impossible. Life survived only in the seas, where seawater screened out some of this damaging radiation.

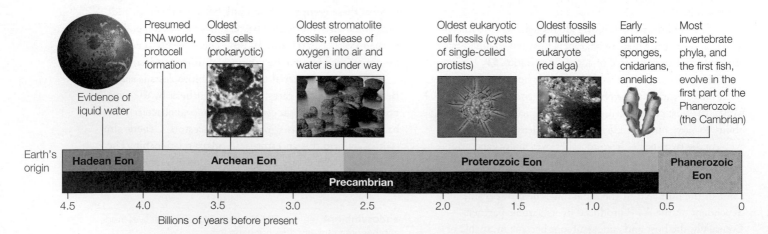

Figure 18.10 Some milestones in the Precambrian and its position in Earth's history.

Precambrian Period of Earth's history from 4.6 billion to 541 million years ago: the Hadean, Archean, and Proterozoic eons.

Figure 18.11 Exploring Mars. This composite selfie of NASA's Curiosity Rover was taken in 2015 on Mars, where the Rover is assessing whether conditions support or could have supported life. The inset photo shows the relative sizes of Earth and Mars.

LOOKING FOR LIFE

We live in a vast universe that we have only begun to explore. So far, we know of only one planet that has life—Earth. In addition, biochemical, genetic, and metabolic similarities among Earth's species imply that all evolved from a common ancestor that lived billions of years ago. What properties of ancient Earth allowed life to arise, survive, and diversify? Could similar processes occur on other planets? These questions are part of astrobiology, the study of life's origins and distribution in the universe.

Astrobiologists study Earth's extreme habitats to determine the range of conditions that living things can tolerate. One group discovered a community of bacteria living about 30 centimeters (1 foot) below the soil surface of Chile's Atacama Desert, a place said to be the driest on Earth. Another group drilled 3 kilometers (almost 2 miles) beneath the soil surface in Vir-

ginia and found bacteria thriving at high pressure and temperature. They named their find *Bacillus infernus*, or "bacterium from hell."

Knowledge gained from studies of life on Earth inform the search for extraterrestrial life. On Earth, all metabolic reactions involve interactions among molecules in aqueous solution (dissolved in water). We assume that the same physical and chemical laws operate throughout the universe, so liquid water is considered an essential requirement for life. Thus, scientists were excited when a robotic lander discovered ice in the soil of Mars, our closest planetary neighbor (**Figure 18.11**). If there is life on Mars, it is likely to be deep underground. Mars has no ozone layer, so ultraviolet radiation would fry organisms at the planet's surface. However, Martian life may exist in deep rock just as it does on Earth.

Any Martian life is also almost certainly anaerobic. Mars is only about half the size of Earth. As a result of its smaller size, Mars has less gravity than Earth and is less able to keep atmospheric gases from drifting off into space. The relatively small amount of atmosphere that remains consists mainly of carbon dioxide, some nitrogen, and only traces of oxygen.

Suppose scientists do find evidence of microbial life on Mars or another planet. Why would it matter? Such a discovery would support the hypothesis that life on Earth arose as a consequence of physical and chemical processes that occur throughout the universe. Discovery of extraterrestrial microbes would also make the possibility of nonhuman intelligent life in the universe more likely. The more places life exists, the more likely it is that complex, possibly intelligent life evolved on other planets.

CREDITS: (11) photo, NASA/JPL-Caltech/MSSS; Inset, NASA/JPL.

Section 18.1   Earth formed by about 4.6 billion years ago, and by 4.3 billion years ago water had begun to pool in shallow seas. Earth's early seas and atmosphere were free of oxygen gas, as demonstrated by the lack of rust in the oldest remaining rocks.

Laboratory simulations demonstrate that the simple organic building blocks of life could have formed by lightning-fueled reactions in the atmosphere or as a result of reactions in hot, mineral-rich water around **hydrothermal vents**. Such compounds also form in deep space and are carried to Earth by meteorites, which were more plentiful in Earth's early years.

Section 18.2   Researchers carry out experiments to determine whether their hypotheses about steps on the road to life are plausible. Such experiments show that metabolic reactions could have begun after proteins formed spontaneously on clay-rich tidal flats. Clay attracts amino acids that can bond under the heat of the sun. Metabolic reactions could also have begun in tiny cavities in rocks near deep-sea hydrothermal vents.

DNA now serves as the molecule of inheritance for all life. However, an **RNA world** may have preceded the current DNA-based system. RNA still is a part of ribosomes that carry out protein synthesis in all organisms. The existence of **ribozymes** (RNAs with enzymatic activity) lends support to the RNA world hypothesis. A later switch from RNA to DNA would have made the genome more stable.

Laboratory-produced vesicles with a fatty acid membrane serve as a model for the **protocells** that likely preceded cells.

Section 18.3   The many similarities among modern organisms indicate that all descended from the same ancestor. That ancestral cell lineage was anaerobic and prokaryotic, and it most likely was not photosynthetic.

It is difficult to find and identify evidence of very early life because cells are small and lack hard parts, and most ancient rocks have been altered or destroyed by geologic processes. Cells may have existed as early as 4 billion years ago, but the oldest fossils of prokaryotic cells date to about 3.5 billion years ago. Some of these fossils are of cells that lived on mudflats, others of cells from the deep sea near hydrothermal vents.

Stromatolites are the most abundant fossils of early life. These layered rocky structures form over thousands of years as colonies of photosynthetic bacteria trap sediment. In some locations, cyanobacteria-containing stromatolites are still forming.

Oxygen-releasing photosynthesis carried out by cyanobacteria added oxygen to Earth's air and seas, producing a new selective pressure that favored cells capable of detoxifying oxygen. Aerobic respiration evolved as a modification of such detoxification pathways. In addition, oxygen molecules reacted in Earth's upper atmosphere to form a protective **ozone layer**. The ozone layer provided protection from UV radiation, so it allowed life to move onto land.

Fossils of large cells and cells with elaborate cell walls are considered likely eukaryotic. The earliest well-accepted fossil eukaryotes date to about 1.8 billion years ago and appear to be resting forms of single-celled algae. Multicelled algae that reproduced sexually appeared by about 1.2 billion years ago.

Section 18.4   Eukaryotic organisms have some genes derived from archaea and others derived from bacteria. Genes involving nuclear functions (such as DNA replication) resemble archaeal genes and, like archaea, eukaryotes wrap their DNA around histone proteins. Eukaryotic genes governing metabolism resemble bacterial genes. This indicates that eukaryotes have both archaeal and bacterial ancestry.

Internal membranes typical of eukaryotic cells are thought to have evolved through infoldings of the plasma membrane in an ancestral archaeon.

The **endosymbiont theory** holds that mitochondria and chloroplasts evolved from bacteria. Bacteria entered a host cell and, over generations, host and **endosymbionts** came to depend on one another for essential metabolic processes. Mitochondria are descendents of aerobic bacteria that may be related to modern typhus-causing rickettsias or to marine methane-feeding bacteria. Chloroplasts evolved from cyanobacteria.

Section 18.5   The interval from 4.6 billion years ago to 541 million years ago is called the **Precambrian**. It encompasses three of the four geological eons (the Hadean, Archean, and Proterozoic). All three domains of life (Bacteria, Archaea, and Eukarya) arose during the Precambrian.

Application   Astrobiology is the study of life's origin and distribution in the universe. The discovery of cells in deserts and deep below Earth's surface suggests that life may exist in similar settings on other planets. If life exists on Mars, it must be anaerobic, because the thin Martian atmosphere contains little oxygen.

SELF-ASSESSMENT
ANSWERS IN APPENDIX VII

1. An abundance of _____ in Earth's early atmosphere would have prevented the spontaneous assembly of organic compounds on early Earth.
   a. hydrogen   b. methane   c. oxygen   d. nitrogen

2. The first interval of Earth's existence is called the _____ .
   a. Cambrian   c. Proterozoic eon
   b. Hadean eon   d. Archean eon

3. The prevalence of iron–sulfur cofactors in organisms supports the hypothesis that life arose _____ .
   a. in outer space   c. near deep-sea vents
   b. on tidal flats   d. in the upper atmosphere

A CHANGING EARTH Modern conditions on Earth are unlike those when life first evolved. **Figure 18.12** shows how the frequency of asteroid impacts and the composition of the atmosphere have changed over time. Use this figure and information in the chapter to answer the following questions.

1. Which occurred first: a decline in asteroid impacts, or a rise in the atmospheric level of oxygen?

2. How do modern levels of carbon dioxide and oxygen compare to those at the time when the first cells arose?

3. Which is now more abundant, oxygen or carbon dioxide?

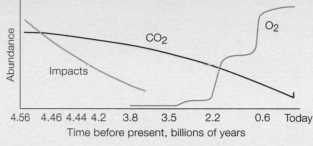

Figure 18.12 How asteroid impacts (green) and atmospheric concentrations of carbon dioxide (pink) and oxygen (blue) changed over geologic time.

4. Miller and Urey created a reaction chamber that simulated conditions in Earth's early atmosphere to test the hypothesis that _____ .
   a. lightning-fueled atmospheric reactions could have produced organic compounds
   b. meteorites contain organic compounds
   c. organic compounds form at hydrothermal vents
   d. oxygen prevents formation of organic compounds

5. RNA in ribosomes catalyzes formation of peptide bonds in all organisms. This supports the hypothesis that _____ .
   a. RNA can hold more information than DNA
   b. RNA is more stable than DNA
   c. an RNA world existed prior to the rise of DNA
   d. proteins evolved before RNA

6. By one hypothesis, clay _____ .
   a. facilitated assembly of early polypeptides
   b. was present at hydrothermal vents
   c. provided energy for early metabolism
   d. served as an early genome

7. Photosynthesis carried out by one group of _____ resulted in Earth's first increase in atmospheric oxygen levels.
   a. archaea   c. protists
   b. bacteria   d. algae

8. Mitochondria are most closely related to _____ .
   a. archaea   c. rickettsias
   b. cyanobacteria   d. algae

9. The presence of ozone in the upper atmosphere protects life from _____ .
   a. free radicals   c. viruses
   b. ultraviolet (UV) radiation   d. ionizing radiation

10. A ribozyme consists of _____ .
    a. clay   c. DNA
    b. lipids   d. RNA

11. A rise in oxygen in Earth's air and seas set the stage for the evolution of _____ .
    a. aerobic respiration   c. photosynthesis
    b. fermentation   d. sexual reproduction

12. Which of the following was not present by the end of the Precambrian?
    a. archaea   c. fungi
    b. bacteria   d. fish

13. The first chloroplasts evolved from _____ .
    a. archaea   c. cyanobacteria
    b. aerobic bacteria   d. early eukaryotes

14. During the Precambrian, _____ .
    a. atmospheric oxygen concentration declined
    b. bacteria, archaea, and early eukaryotes arose
    c. dinosaurs became extinct
    d. all of the above

15. Arrange these events in order of occurrence, with 1 being the earliest and 6 the most recent.
    ___ 1   a. Earth's first seas form
    ___ 2   b. origin of mitochondria
    ___ 3   c. first protocells form
    ___ 4   d. Precambrian ends
    ___ 5   e. origin of chloroplasts
    ___ 6   f. first animals appear

1. Researchers looking for fossils of the earliest life forms face many hurdles. For example, few sedimentary rocks date back more than 3 billion years. Review what you learned about plate tectonics (Section 16.5). Explain why so few remaining samples of these early rocks remain.

2. Rickettsia bacteria always live as parasites inside eukaryotic cells, and their genomes are much smaller than those of free-living bacteria. Organisms that live only as parasites often have reduced genomes compared to their free-living relatives. How could a parasitic lifestyle contribute to a reduction in genome size?

for additional quizzes, flashcards, and other study materials
▶ **ACCESS MINDTAP AT WWW.CENGAGEBRAIN.COM**

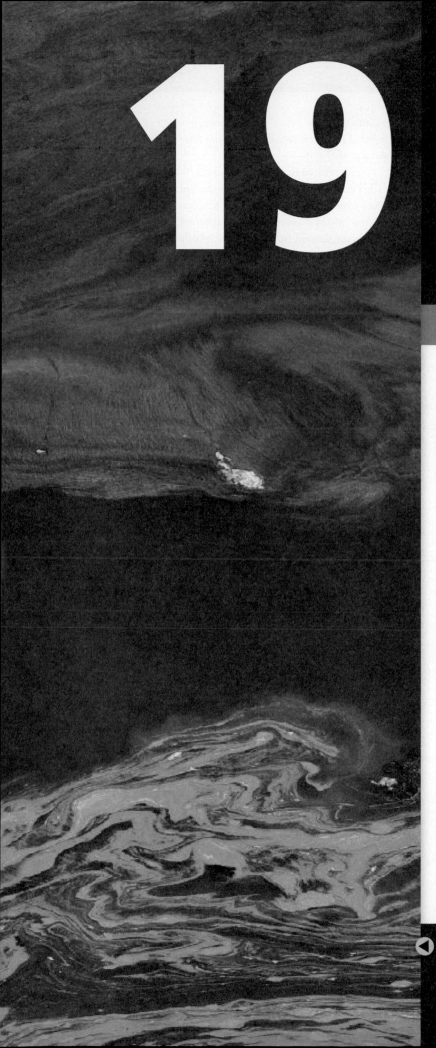

19

Viruses, Bacteria, and Archaea

Links to Earlier Concepts

This chapter's discussion of viruses touches on reverse transcription (Section 15.1) and cancer (11.5). The chapter also covers bacteria and archaea (4.4). You will learn more about comparing genomes (15.4), and defining species (1.4), domains (1.4), and clades (17.11).

Core Concepts

Organisms exchange matter and energy with the environment in order to grow, maintain themselves, and reproduce.

As a group, bacteria and archaea have more ways of meeting their energy and nutrient needs than eukaryotes, so they can exploit a more diverse range of environments. These prokaryotic organisms reproduce rapidly, using a mechanism of asexual reproduction unlike that of eukaryotes. Viruses, which are noncellular, have no metabolic machinery and cannot reproduce on their own.

Evolution underlies the unity and diversity of life.

Bacteria, archaea, and viruses have a capacity for genetic change through mutation. They also often exchange genes among existing individuals. In populations of these microbes, traits that increase fitness in a particular environment tend to become more common over time. Viruses in particular have replication strategies that foster rapid evolution.

Biological systems are affected by factors that disrupt their homeostasis.

Disruptions at the molecular and cellular levels can negatively affect the health of the organism. Viruses disrupt cell function in their hosts, causing the host to put aside essential activities in favor of virus production. Toxins made by disease-causing bacteria either harm cells directly or elict a host defensive response that causes disease symptoms.

◀ A population explosion of photosynthetic cyanobacteria colors water of California's Klamath River green.

Photograph by David McLain/National Geographic Creative.

19.1 The Viruses

List the components common to all free viral particles.

Explain the difference between the two bacteriophage replication pathways.

Describe how HIV replicates in a white blood cell.

A **virus** is a noncellular infectious particle that can replicate only inside a living cell. By many definitions, viruses are not alive. Unlike cells, viruses do not maintain homeostasis or engage in metabolic processes such as ATP production, and they cannot reproduce on their own. However, viruses are replicators with a genome of RNA or DNA that can mutate. Thus they do evolve by natural selection.

VIRUS STRUCTURE

A virus that is not inside a host cell is referred to as a **viral particle**. With rare exceptions, viral particles are so small that they can only be viewed with an electron microscope. A viral particle always includes a viral genome inside a protein coat. The genome consists of RNA or DNA that may be single-stranded or double-stranded. The protein coat protects the nucleic acid and facilitates its delivery into a host cell. In all viruses, some proteins of the viral coat must bind to proteins at the surface of a host cell during infection.

A viral coat consists of many protein subunits arranged in a repeating pattern. When viral proteins assemble as a helix around a strand of nucleic acid, the result is a rod-shaped or filamentous virus. Many plant viruses, including the tobacco mosaic virus, have this type of structure (**Figure 19.1A**).

Bacteriophages are viruses that infect bacteria, and most have a complex structure. Consider the T4 bacteriophage, a virus common in the human gut (**Figure 19.1B**). The viral genome (DNA) is enclosed within a polyhedral (many-sided) "head." Other protein components of the T4 viral particle

include a hollow helical sheath through which viral DNA is injected into a bacterial cell, and "tail fibers" that attach the virus to that cell.

The tobacco mosaic virus and the T4 bacteriophage are called nonenveloped ("naked") viruses because their outermost layer is the protein coat. By contrast, the coat of most animal viruses is covered by a **viral envelope**, a layer of cell membrane derived from the cell in which the virus formed. **HIV (human immunodeficiency virus)** is an enveloped RNA virus (**Figure 19.1C**). It causes AIDS (acquired immunodeficiency syndrome), which we cover in detail in Section 34.9.

VIRAL REPLICATION

A virus cannot move itself toward a host, so an infection begins with a chance encounter. Typically, viral coat proteins bind proteins of a host's cell membrane, then viral genes enter the cell. Viral genes hijack the cell's internal machinery, causing it to make viral components that self-assemble as viral particles. The particles are then released into the environment.

Bacteriophage Replication Replication of a bacteriophage such as T4 proceeds by one of two pathways. Both start when the virus attaches to a bacterium and injects DNA (**Figure 19.2**). In the **lytic pathway**, viral DNA is put to use immediately, causing the host cell to produce new viral DNA and proteins. These viral components assemble as virus particles, which escape when the cell breaks open (lyses) and dies.

In the **lysogenic pathway**, viral DNA becomes integrated into the host cell's genome. However, viral genes are not expressed, so the host cell remains healthy. When this cell reproduces, a copy of the viral DNA is passed on along with the host's genome. Like miniature time bombs, the viral DNA in the host's descendants awaits a cue, such as increased host density, that will trigger a shift to the lytic pathway.

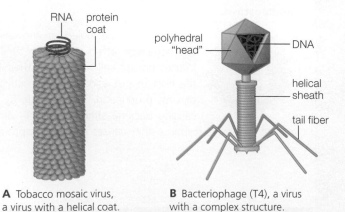

A Tobacco mosaic virus, a virus with a helical coat.

B Bacteriophage (T4), a virus with a complex structure.

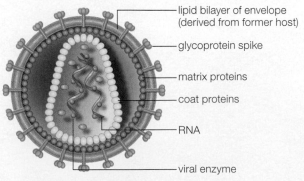

C HIV, an enveloped virus.

Figure 19.1 Examples of virus structure.

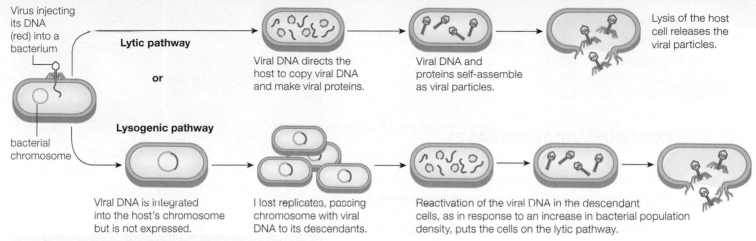

Virus injecting its DNA (red) into a bacterium

Lytic pathway

or

Lysogenic pathway

bacterial chromosome

Viral DNA directs the host to copy viral DNA and make viral proteins.

Viral DNA and proteins self-assemble as viral particles.

Lysis of the host cell releases the viral particles.

Viral DNA is integrated into the host's chromosome but is not expressed.

Host replicates, passing chromosome with viral DNA to its descendants.

Reactivation of the viral DNA in the descendant cells, as in response to an increase in bacterial population density, puts the cells on the lytic pathway.

Figure 19.2 Bacteriophage replication pathways.

HIV Replication

HIV replicates in human white blood cells (**Figure 19.3**). During infection, spikes of viral glycoprotein attach to proteins in the cell's plasma membrane ❶. The host cell's plasma membrane and the viral envelope fuse, allowing viral enzymes and RNA to enter the cell ❷. One of these enzymes, reverse transcriptase, converts viral RNA into double-stranded DNA that the cell can transcribe ❸. RNA viruses that use reverse transcriptase to produce viral DNA are called **retroviruses**. Once viral DNA has been produced, it is moved into the nucleus, along with another viral enzyme that integrates the DNA into the host's chromosome ❹. Viral DNA is transcribed into RNA along with the host's genes ❺. Some of the resulting RNA encodes viral proteins and is translated ❻. Other virus-encoded RNA becomes the genetic material of new HIV particles ❼ that self-assemble at the host cell's plasma membrane ❽. A portion of this membrane becomes a viral envelope when the new viruses bud from the cell ❾.

Drugs that fight HIV interfere with viral binding to the host, reverse transcription, integration of DNA, or processing of viral polypeptides to form viral proteins.

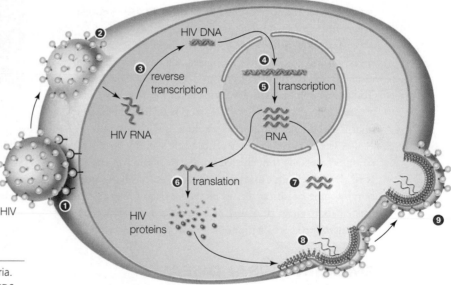

Figure 19.3 Replication of HIV, an enveloped retrovirus.
❶ Virus binds to white blood cell.
❷ Viral RNA and enzymes enter the cell.
❸ Viral reverse transcriptase uses viral RNA to make double-stranded viral DNA.
❹ Viral DNA enters the nucleus and is integrated into the host genome.
❺ Transcription produces viral RNA.
❻ Some viral RNA is translated into viral proteins.
❼ Other viral RNA forms the new viral genome.
❽ Viral proteins and viral RNA self-assemble at the host plasma membrane.
❾ New virus buds from the host cell, with an envelope of host plasma membrane.

bacteriophage (bac-TEER-ee-oh-fayj) Virus that infects bacteria.
HIV (human immunodeficiency virus) Virus that causes AIDS.
lysogenic pathway (lice-oh-JEN-ick) Bacteriophage replication path in which viral DNA becomes integrated into the host's chromosome and is passed to the host's descendants.
lytic pathway (LIH-tik) Bacteriophage replication pathway in which a virus immediately replicates in its host and kills it.
retrovirus RNA virus that uses the enzyme reverse transcriptase to produce viral DNA in a host cell.
viral envelope Of some viral particles, an outer membrane derived from the host cell in which the particle formed.
viral particle Virus that is not inside a host cell.
virus Noncellular, infectious particle of protein and nucleic acid; replicates only in a host cell.

LEARNING OBJECTIVES

List some common diseases caused by viruses.

Discuss the different ways in which viral diseases can spread.

Describe two emerging viral diseases.

Explain how new flu strains arise.

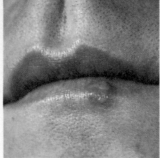 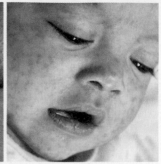

A Fluid rich in herpesvirus particles leaks from a sore. **B** A rash of flat, red bumps caused by measles virus.

Figure 19.4 Symptoms of viral infection.

COMMON VIRAL DISEASES

Many viruses live in our body without any ill effects, but others are **pathogens**, meaning they cause disease. Viral diseases usually produce mild symptoms and sicken us only briefly, as by causing common colds. Viruses that infect cells in the lining of our gastrointestinal tract often cause a brief bout of vomiting and diarrhea. Viruses also cause seasonal flus.

Infections by some viral pathogens can persist for long periods, or even for life. An initial infection by one of these viruses causes symptoms that subside quickly. However, the virus remains present in a latent (resting) state and can reawaken. For example, herpes simplex virus 1 (HSV-1) can remain dormant in nerve cells for years before becoming activated by stress. When this virus replicates, it causes painful "cold sores" at the edge of the lips (**Figure 19.4A**). Another herpesvirus causes similar sores on the genitals.

Measles (**Figure 19.4B**), mumps, rubella (German measles), and chicken pox are potentially deadly viral diseases of childhood that, until recently, were common worldwide. Today, most children in developed countries have been vaccinated against these illnesses. A vaccine primes the body to fight a pathogen. (We discuss vaccines in Chapter 34.)

A few viruses can cause cancer. Infection by certain strains of sexually transmitted human papillomaviruses (HPV) is the main cause of cervical cancer. Similarly, infection by some hepatitis viruses raises the risk of liver cancer. Epstein-Barr virus, the cause of infectious mononucleosis, also raises the risk of lymphoma (a blood cell cancer).

EMERGING VIRAL DISEASES

Viruses cause a number of emerging diseases. An **emerging disease** is a disease that has only recently been detected in humans, or has recently expanded its range. AIDS is an emerging viral disease that was first identified in humans in 1981. Since then, it has caused about 39 million deaths. AIDS is a communicable disease, meaning HIV spreads from one infected person to another. Currently, about 37 million people are infected with HIV.

By contrast, the virus that causes West Nile disease cannot spread directly from person to person. Mosquitoes carry this virus from host to host, so these insects are called

Figure 19.5 Fighting Ebola Dr. Tom Frieden, head of the U.S. Centers for Disease Control and Prevention, at an Ebola treatment center in Liberia during the 2014 outbreak. The micrograph shows the Ebola virus, which has a threadlike structure.

vectors. Birds are a "reservoir" for West Nile virus, which means the virus can replicate in birds, and that birds serve as a source of virus for new human infections. West Nile virus was unknown in North America until 1999, when it emerged in New York. It has since spread across the continent. From 1999 through 2014, there were about 42,000 reported cases of West Nile disease in the United States and 1,765 deaths.

The Zika virus is a relative of West Nile virus. It is transmitted by mosquitoes, and also by sexual contact. Zika has been known in Africa since 1947, but was recently introduced to South America. Most people have mild symptoms, but some develop a temporary paralysis. An infection during pregnancy raises the risk of miscarriage and of microcephaly (a unusually small brain and head) in the developing child.

Ebola is caused by an enveloped RNA virus first identified in Africa in 1976. Within three weeks of infection, a person develops flulike symptoms, followed by a rash, vomiting, diarrhea, and bleeding from all body openings. The virus

CREDITS: (4A) CDC/Dr. Hermann; (4B) CDC/Molly Kurnit, M.P.H.; (5 background) CDC/NIAID; (5 inset) CDC/Sally Ezra.

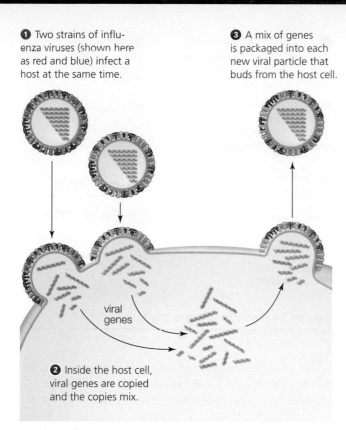

❶ Two strains of influenza viruses (shown here as red and blue) infect a host at the same time.

❸ A mix of genes is packaged into each new viral particle that buds from the host cell.

viral genes

❷ Inside the host cell, viral genes are copied and the copies mix.

Figure 19.6 Viral reassortment.

is transmitted by contact with body fluids, so caregivers must wear protective gear (**Figure 19.5**).

Until recently, Ebola outbreaks had occurred only in limited regions of Africa and had affected fewer than 500 people. However, an outbreak that began in Guinea in December 2013 had killed more than 11,000 people by the end of 2015. Nearly all infections and deaths occurred in west Africa, although a few people with the virus traveled or were evacuated to other continents.

NEW FLUS—VIRAL REASSORTMENT

Seasonal flus are caused by influenza viruses, which are enveloped RNA viruses. To keep up with changes in these viruses, scientists create a new flu vaccine every year. However, determining which flu strains will be around in the future is not an

---

emerging disease A disease that was previously unknown or has recently begun spreading to a new region.

pathogen Disease-causing agent.

vector Animal that carries a pathogen from one host to the next.

viral reassortment Multiple strains of virus infect a host simultaneously and swap genes.

ENGAGE: COLLABORATIVE SCIENCE

Dr. Nathan Wolfe

National Geographic Explorer Dr. Nathan Wolfe is working to create an early warning system that can forecast and contain new plagues before they kill millions. He says, "The way that pandemics (worldwide epidemics) come to us is through our interaction with animals. That's true whether it's swine flu (H1N1), bird flu (H5N1), Ebola, or HIV. These are all animal diseases that jumped to humans." Wolfe monitors people who have high levels of contact with wild animals. Wolfe recalls, "When I started this work in 1999, we were particularly interested in retroviruses because we knew they had the potential to cause devastating pandemics like HIV. But we didn't have a good idea of the frequency with which they were crossing over from animals to humans. Our results were shocking. We discovered that cross-species transmission wasn't rare; it was happening on a regular basis." Despite the threats, Wolfe remains optimistic. "It's really the dawn of a new scientific era. As we try to detect viruses that can do great harm, we could also discover the next generation of vaccines and cures. If we can provide even a few months of early warning for just one pandemic, the benefits will outweigh all the time and energy we're devoting. Imagine preventing health crises, not just responding to them."

exact science. Even after a flu shot, a person is susceptible to a virus that differs from the strains targeted by the vaccine.

New influenza strains arise through mutation, and also by **viral reassortment**: gene-swapping between related viruses that infect a host at the same time (**Figure 19.6**). Viral reassortment occurs all the time, and it can be dangerous. Consider what could happen if two current influenza strains reassorted. Influenza strains are defined by and named for the varying structures of two viral proteins—hemagglutinin (H) and neuraminidase (N)—that extend through the viral envelope. The H5N1 strain is a bird flu that has a high mortality rate in humans but is not easily transmitted. By contrast H1N1 (the swine flu) is rarely deadly, but it is transmitted easily. If both strains were to infect the same individual at the same time, the result could be a new strain that has a high mortality rate and is also highly transmissible.

19.3 Bacterial Structure and Function

LEARNING OBJECTIVES

Contrast the structure of a bacterial cell with that of a eukaryotic cell.

Describe the process by which bacteria reproduce.

Discuss how bacterial conjugation differs from reproduction.

Distinguish between the four modes of nutrition in bacteria.

Prokaryotes is an informal term for cellular organisms whose DNA is not contained within a nucleus (Section 4.4). There are two lineages of prokaryotes: domains Bacteria and Archaea. We begin our survey of prokaryotic life with the better known of these groups, the **bacteria**.

CELL SIZE AND STRUCTURE

All bacteria are single-celled and, with rare exceptions, too small to be observed without a microscope. Nearly all bacteria have a porous cell wall around their plasma membrane. The wall contains peptidoglycan, a polymer of amino acids and sugars that is unique to bacteria. The wall gives the cell its shape, which may be spherical, spiral, or rod-shaped (**Figure 19.7A**). A spherical cell is a coccus, a spiral-shaped one is a spirillum, and a rod-shaped one is a bacillus.

A secreted slime layer or a capsule covers the cell wall of many bacteria. A slime layer is a coating that washes off easily. A capsule is a coating that adheres strongly to the cell wall and cannot be washed off. Slime helps a cell stick to surfaces. A capsule helps some pathogenic bacteria evade the immune defenses of their vertebrate hosts.

In most bacterial cells, a single circular chromosome (a ring of double-stranded DNA) lies exposed in the cytoplasm (**Figure 19.7B**). A few types of bacteria have internal membranes, but none has a nuclear envelope like that of eukaryotes. Ribosomes, the structures on which proteins are assembled, are scattered in the cytoplasm.

Bacteria often have one or more flagella that rotate like a propeller. By contrast, a eukaryotic flagellum whips from side

❶ A bacterium has one circular chromosome that attaches to the inside of the plasma membrane.

❷ The cell duplicates its chromosome, attaches the copy beside the original, and adds membrane and wall material between them.

❸ When the cell has almost doubled in size, new membrane and wall are deposited across its midsection.

❹ Two genetically identical cells result.

Figure 19.8 Binary fission in bacteria.

to side. Many bacteria also have hairlike projections called pili that function in adhesion or locomotion. A type of retractable pilus can draw cells together for gene exchanges.

REPRODUCTION AND GENE EXCHANGE

Bacteria usually reproduce by **binary fission**, a type of asexual reproduction (**Figure 19.8**). The process begins when the cell replicates its single chromosome, which is attached to the inside of the plasma membrane **❶**. The DNA replica attaches to the plasma membrane adjacent to the parent molecule. Addition of new membrane and wall material elongates the cell and moves the two DNA molecules apart **❷**. Then, membrane and cell wall material are deposited across the cell's midsection **❸**. This material partitions the cell, yielding two identical descendant cells **❹**.

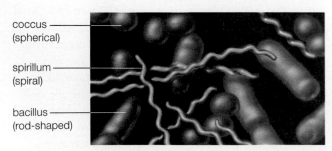

coccus (spherical)

spirillum (spiral)

bacillus (rod-shaped)

A Three cell shapes common among bacteria.

Figure 19.7 Bacterial shapes and structural features.

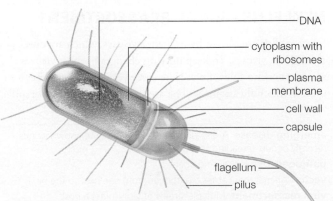

DNA

cytoplasm with ribosomes

plasma membrane

cell wall

capsule

flagellum

pilus

B Structural components typical of bacterial cells.

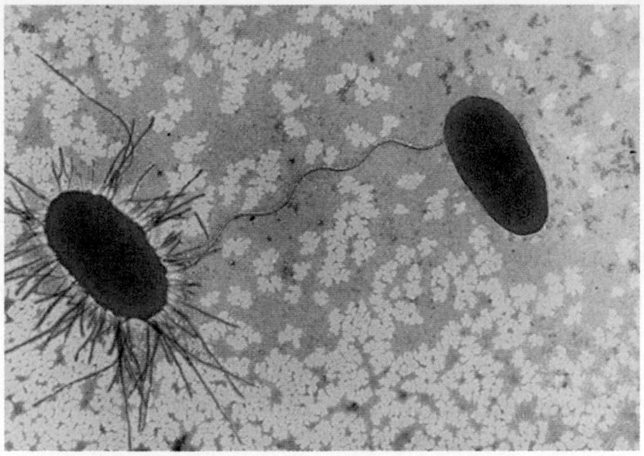

Figure 19.9 Prokaryotic conjugation. One cell extends a sex pilus out to another, draws it close, and gives it a copy of a plasmid.

CARBON SOURCE	ENERGY SOURCE: Light	ENERGY SOURCE: Chemicals
Inorganic source such as CO_2	**Photoautotrophs** bacteria, archaea, photosynthetic protists, plants	**Chemoautotrophs** bacteria, archaea
Organic source such as glucose	**Photoheterotrophs** bacteria, archaea	**Chemoheterotrophs** bacteria, archaea, fungi, animals, nonphotosynthetic protists

Figure 19.10 Modes of obtaining energy and carbon, and the groups that can use them.

FIGURE IT OUT
Does conjugation increase the number of cells?
Answer: No. It is not a mode of reproduction.

FIGURE IT OUT
Which group of organisms can build their own food from CO_2 in the dark?
Answer: Chemoautotrophs

Binary fission is the way bacteria inherit DNA "vertically" from a parent. Bacteria can also exchange genetic material by passing DNA among existing individuals. This process, which is called **horizontal gene transfer**, does not increase the number of cells; rather, it alters existing cells. **Conjugation** is a form of horizontal gene transfer that moves a plasmid from one individual to another. A plasmid is a small circle of DNA that is separate from the bacterial chromosome (Section 4.4). During conjugation, a cell with a plasmid extends a sex pilus to another cell (**Figure 19.9**), draws it close, then puts a copy of the plasmid into its partner.

Genes also move between bacteria by transduction and transformation. Transduction occurs when a virus picks up a bit of DNA from one bacterial host cell, then transfers the DNA to its next host. Transformation occurs when bacteria take up DNA from their environment and integrate it into their genome.

METABOLIC DIVERSITY

Metabolic diversity gives bacteria the ability to thrive in a variety of environments. Many bacteria are anaerobic: They can (or must) live where there is no oxygen. Others are aerobic, which means they require (or at least tolerate) oxygen.

There are four possible mechanisms of obtaining energy and nutrients (**Figure 19.10**) and, as a group, bacteria use them all. Autotrophs build their own food from an inorganic source of carbon such as carbon dioxide (CO_2). Many bacteria are **photoautotrophs**, which carry out photosynthesis. Like plants, they assemble their food using light as an energy source. **Chemoautotrophs** fuel the assembly of their food by oxidizing (removing electrons from) inorganic substances such as hydrogen sulfide. Chemoautotrophic bacteria and archaea are producers in dark places such as the seafloor.

Heterotrophs get carbon by breaking down organic molecules from their environment. **Photoheterotrophs** use light energy to fuel this breakdown. **Chemoheterotrophs** obtain both energy and carbon through the breakdown of organic compounds. All pathogenic bacteria are chemoheterotrophs that extract the organic compounds they need to live from a host. Other bacterial chemoheterotrophs are decomposers.

bacteria One of two lineages of prokaryotic cells; cell walls when present contain peptidoglycan.
binary fission Method of asexual reproduction in which one prokaryotic cell divides into two identical descendant cells.
chemoautotroph (keem-oh-AWE-toe-trof) Organism that uses carbon dioxide as its carbon source and obtains energy by oxidizing inorganic molecules.
chemoheterotroph (keem-oh-HET-ur-o-trof) Organism that obtains energy and carbon by breaking down organic compounds.
conjugation (con-juh-GAY-shun) Mechanism of horizontal gene transfer in which one prokaryote passes a plasmid to another.
horizontal gene transfer Transfer of genetic material between existing individuals.
photoautotroph (foe-toe-AWE-toe-trof) Organism that obtains carbon from carbon dioxide and energy from light.
photoheterotroph (foe-toe-HET-ur-o-trof) Organism that obtains its carbon from organic compounds and its energy from light.

List four ways in which bacteria benefit other organisms.

Describe how bacteria are used in biotechnology.

Explain the function of an endospore.

Differentiate between Gram-positive and Gram-negative bacteria.

There are many bacterial lineages, and new ones are constantly being discovered. Here we consider a few major groups to provide insight into bacterial diversity and ecological importance.

CYANOBACTERIA

Photosynthesis evolved in many bacterial lineages, but only **cyanobacteria** release oxygen as a by-product. If, as evidence suggests, chloroplasts evolved from ancient cyanobacteria, we have cyanobacteria and their chloroplast descendants to thank for nearly all the oxygen in Earth's atmosphere.

Some cyanobacteria also carry out **nitrogen fixation**, which means they incorporate nitrogen from the air into ammonia (NH_3). Nitrogen fixation is an important ecological service. Photosynthetic eukaryotes need nitrogen but they

cannot use the gaseous form ($N\equiv N$) because they do not have an enzyme that can break the molecule's triple bond. Plants and photosynthetic protists can, however, take up ammonia released by nitrogen-fixing bacteria.

Some cyanobacteria partner with fungi to form lichens (which we discuss in Section 22.3), and others grow on the surface of soil, but most are aquatic. The photo that opens this chapter shows a river tinted green by cyanobacteria. The group is named for the color cyan, which is a blue-green. Aquatic cyanobacteria grow as single cells or as long filaments of cells arranged end to end (**Figure 19.11**). When conditions become unfavorable for growth, some filamentous cyanobacteria produce thick-walled resting cells. These easily dispersed cells remain dormant until conditions improve.

GRAM-POSITIVE BACTERIA

Gram-positive bacteria are a bacterial lineage with cell walls that have a thick outer layer of peptidoglycan. This layer retains purple stain when a technique called Gram staining is used to prepare samples for microscopy. In other bacterial lineages, the cell wall has a thinner peptidoglycan layer that retains very little dye during Gram staining, so these lineages are described as Gram-negative.

Most Gram-positive bacteria are chemoheterotrophs, and many serve as decomposers. **Decomposers** break down complex organic molecules in the wastes and remains of other organisms, leaving inorganic leftovers that serve as nutrients for plants. Decomposers also break down pesticides and pollutants, thus improving the environment for other organisms.

Actinomycetes are Gram-positive decomposers that grow through soil as long, branching chains of cells. Their presence

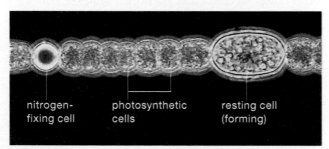

nitrogen-fixing cell   photosynthetic cells   resting cell (forming)

Figure 19.11 Chain of aquatic cyanobacterial cells.

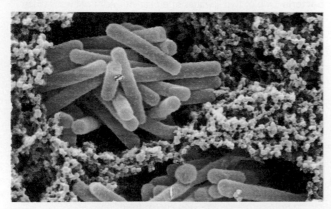

A Lactic acid bacteria amidst curdled milk proteins in yogurt.
Figure 19.12 Gram-positive bacteria.

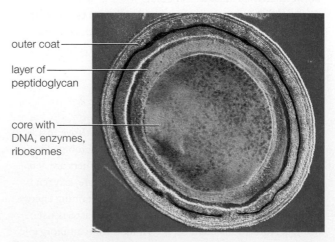

outer coat

layer of peptidoglycan

core with DNA, enzymes, ribosomes

B Endospore of *Clostridium tetani* in cross-section.

CREDITS: (11) © P. W. Johnson and J. MeN. Sieburth, Univ. Rhode Island/BPS; (12A) SciMAT/Science Source; (12B) © PTP/Phototake.

gives freshly exposed soil its distinctive "earthy" smell. Many antibiotics, including streptomycin and vancomycin, were first isolated from actinomycetes.

Lactic acid bacteria are another Gram-positive subgroup. These cells ferment sugars in milk and produce lactic acid as a by-product. *Lactobacillus* species can spoil milk, but they are also used to produce cheese and yogurt (**Figure 19.12A**). Lactic acid released by these bacteria produces a sour taste and denatures milk proteins, causing the milk to thicken.

Soil bacteria in the genera *Clostridium* and *Bacillus* are Gram-positive cells that form endospores when conditions are unfavorable. An **endospore** contains the cell's DNA and a bit of cytoplasm in a protective coat (**Figure 19.12B**). It is functionally similar to a cyanobacterial resting cell, but much tougher. Unlike active cells, endospores withstand heating, freezing, drying out, and exposure to ultraviolet radiation.

Many human pathogens are Gram-positive bacteria. We consider their effects later in the chapter.

PROTEOBACTERIA

Proteobacteria, the largest bacterial group, are named for Proteus, a Greek ocean god who could take many forms. This diverse group of Gram-negative bacteria is defined on the basis of ribosomal RNA sequences rather than any structural or metabolic feature. Most proteobacteria are aerobic. Some are photosynthetic, but they do not produce oxygen. Other proteobacteria are chemoautotrophs or chemoheterotrophs. The chemoautotrophic species *Thiomargarita namibiensis* is among the largest prokaryotes. It can be up to 0.75 mm wide (**Figure 19.13A**).

Many chemoheterotrophic proteobacteria live inside other organisms. Members of the genus *Rhizobium* live in the roots of legumes such as peas and beans. The bacteria receive sugars from the plant, and in turn provide the plant with ammonia they produced by fixing nitrogen.

The most well-studied prokaryote is *Escherichia coli* (**Figure 19.13B**), a bacillus that lives in the mammalian intes-

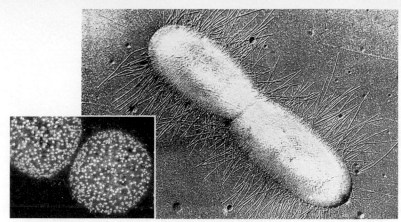

A The largest prokaryote, *Thiomargarita namibiensis*. Cells can be seen with the naked eye.

B The best-studied prokaryote, *E. coli*, lives in the mammalian gut and can divide by binary fission every 20 minutes.

Figure 19.13 Proteobacteria.

tines and is easily grown in the laboratory. Researchers often use it when investigating genetic and metabolic processes. *E. coli* is also widely used in industrial biotechnology: Genetically modified *E. coli* now produce synthetic insulin and many other proteins for medical use.

When biotechnologists want to alter a plant's genome, they may turn to *Agrobacterium*. Soil proteobacteria of this genus have a plasmid that gives them the capacity to infect plants and cause a tumor. As Section 15.5 explained, scientists use these bacteria to genetically engineer plants. They replace the Ti plasmid's tumor-inducing genes with desirable foreign genes. Bacteria with the resulting recombinant plasmid are then allowed to infect a plant and transfer the desirable foreign genes into the plant.

Rickettsias are very tiny proteobacteria that live inside cells. Of all bacteria, the genomes of mitochondria are most similar to those of proteobacteria such as rickettsias. This indicates that the bacteria ancestral to mitochondria was a member of the proteobacterial lineage or ancestral to it.

SPIROCHETES

Spirochetes are tiny Gram-negative bacteria shaped like a stretched-out spring (**Figure 19.14**). All are chemoheterotrophs. Some infect humans and cause diseases such as Lyme disease and syphilis. Others live in the stomach of cattle and benefit their host by breaking down cellulose.

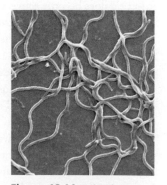

Figure 19.14 Spirochetes.

cyanobacteria (sigh-an-o-bac-TEER-ee-uh) Photosynthetic, oxygen-producing bacteria.
decomposer Organism that breaks down organic compounds in wastes and remains into their inorganic subunits.
endospore Resistant resting stage of some soil bacteria.
Gram-positive bacteria Bacteria with thick cell walls that are colored purple when prepared for microscopy by Gram staining.
nitrogen fixation Incorporation of nitrogen gas into ammonia.
proteobacteria (pro-tee-o-bac-TEER-ee-uh) Largest bacterial lineage; members show a wide range of metabolic diversity.
spirochetes (SPY-roh-keets) Group of small bacteria shaped like a stretched-out spring.

CREDITS: (13A) © Dr. Manfred Schloesser, Max Planck Institute for Marine Microbiology; (13B) Courtesy James Evarts; (14) CDC/Janice Haney Carr.

NORMAL MICROBIOTA

Bacteria and other microorganisms that normally live in or on our body are our **normal microbiota**. These cells outnumber our own cells by about ten to one, and they are our first defense against pathogens. For example, lactic acid produced by bacteria in the human mouth, gut, and vagina help keep acid-intolerant pathogenic bacteria from taking hold.

Some bacteria that live in our gut also benefit us by producing essential vitamins. Our intestines are home to lactic acid bacteria that synthesize some B vitamins that we cannot make, and also to *E. coli* that produce vitamin K.

TOXINS AND DISEASE

Bacteria cause many common diseases (**Table 19.1**). Most pathogenic bacteria harm us by way of toxins that disrupt our health. The toxin may be a substance that bacteria release into their environment (an exotoxin), or a molecule integral to the cell's wall (an endotoxin). Exotoxins directly harm cells. For example, botulinum toxin released by *Clostridium botulinum* disables nerve cells, which is why a dangerous paralyzing food poisoning occurs after ingesting the bacteria. Endotoxins do not directly harm cells, but rather elicit an immune response that produces symptoms such as fever and aches. Entry of bacteria with endotoxins into the bloodstream can elicit a runaway immune response called shock.

Whooping cough (pertussis) and tuberculosis are bacterial respiratory diseases spread by coughs or sneezes. *Mycobacterium tuberculosis* (**Figure 19.15**), the cause of TB, now infects about a third of the human population. In most people, the infection is inactive and does no harm. If the infection does become active, bacteria grow in the lungs and cause coughing, chest pain, fever, and weight loss. Each year, more than a million people die as a result of active TB.

Impetigo is a skin disease caused by *Streptococcus* and *Staphylococcus*, Gram-positive cocci that infect outer layers of the skin. *Streptococcus* also causes strep throat.

Gonorrhea, syphilis, and chlamydia are bacterial diseases transmitted by sexual contact.

Bacterial pathogens sometimes enter the body in tainted food or water. Most cases of bacterial food poisoning occur after bacteria-rich animal feces contaminate food. In regions where safe drinking water is not readily available,

TABLE 19.1

Examples of Bacterial Diseases

Disease	Description
Whooping cough	Childhood respiratory disease
Tuberculosis	Respiratory disease
Impetigo, boils	Blisters, sores on skin
Strep throat	Sore throat, can damage heart
Cholera	Diarrheal illness
Syphilis	Sexually transmitted disease
Gonorrhea	Sexually transmitted disease
Chlamydia	Sexually transmitted disease
Lyme disease	Rash, flulike symptoms, spread by ticks
Botulism, tetanus	Muscle paralysis by bacterial toxin

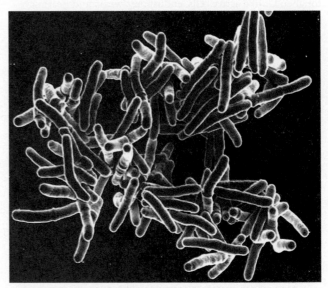

Figure 19.15 Bacteria that cause tuberculosis. *Mycobacterium tuberculosis* infects about a third of the human population.

cholera sickens millions of people each year and kills about 100,000. Cholera-causing bacteria (*Vibrio cholerae*) produce an exotoxin that causes intestinal cells to malfunction, and this results in a potentially fatal diarrhea. Tainted water also spreads *Helicobacter pylori*, which causes stomach ulcers. *H. pylori* is the only bacterial species known to cause cancer. A long-term infection increases the risk of stomach cancer.

Lyme disease is a vector-borne bacterial disease. Ticks convey spirochetes that cause the disease between vertebrate hosts. Infection may initially cause a bull's-eye-shaped rash at the site of the tick bite. Flulike symptoms follow. A long-term infection can harm joints and the nervous system.

normal microbiota Collection of normally harmless or beneficial microorganisms that typically live in or on a body.

List some traits common to bacteria and archaea, and some ways in which they differ.

Give examples of locations where archaea have been discovered.

Explain why archaea are not considered a threat to human health.

COMPARISONS WITH BACTERIA

Archaea, the most recently named lineage of prokaryotic organisms, are similar to bacteria in many respects. They are about the same size as bacteria, and typical archaeal cells are rod-shaped or spherical like typical bacteria. Also like bacteria, archaea have a single circular chromosome, reproduce by binary fission, and can exchange plasmids among existing cells by conjugation. Researchers are finding that archaea live alongside bacteria nearly everywhere.

As a result of their similarities to bacteria, archaea were not recognized as a distinct lineage until the 1970s. At that time, microbiologist Carl Woese discovered these distinctive prokaryotic species and proposed that they be put in a separate domain, which he called the Archaea. *Archaea* means "ancient ones" in Greek.

Since the discovery of archaea, scientists have documented many ways in which they differ from bacteria. Archaeal cell walls never contain peptidoglycan, and their plasma membrane contains lipids unknown in bacteria. Like eukaryotes, archaea organize their DNA around histone proteins, which bacteria do not have. The enzymes and signals that archaea use in transcription and translation are also similar to those of eukaryotes, and unlike those of bacteria.

ARCHAEAL DIVERSITY

Archaea can live in places where few or no other cells survive. For example, they are the most abundant cells in the ocean's deepest, darkest waters. Archaea were first discovered in hot springs, and many are **extreme thermophiles**, meaning they grow at a very high temperature (**Figure 19.16A**). Some archaea that live near deep-sea hydrothermal vents grow even at 110°C (230°F).

Other archaea are **extreme halophiles**, meaning they live in highly salty water. Salt-loving archaea live in the Dead Sea, the Great Salt Lake, and smaller brine-filled lakes (**Figure 19.16B**). Most are photoheterotrophs that capture light energy using a red pigment called bacteriorhodopsin.

Many archaea, including some extreme thermophiles and extreme halophiles, are methanogens. **Methanogens** are chemoautotrophs, and their ATP-producing reactions also produce methane (CH_4), an odorless gas. Methanogenic archaea are strict anaerobes, meaning they cannot live in the

A Thermally heated waters. Archaeal extreme thermophiles live in the waters of this hot spring in Nevada.

B Highly salty waters. Pigmented extreme halophiles give the brine in this California lake its color.

C The gut of many animals. Cattle must belch frequently to release the methane produced by archaea in their stomach.

Figure 19.16 Examples of archaeal habitats.

presence of oxygen. They abound in sewage, marsh sediments, and the animal gut (**Figure 19.16C**). Cattle have methanogens in their stomach and release methane gas by belching.

About a third of people have significant numbers of methanogens in their intestine, so their flatulence (farts) contains methane. Archaeal methanogens have also been discovered in the human mouth and vagina. So far, no archaea have been found to be human pathogens on their own. However, some methanogens that live in the mouth contribute to periodontitis (gum disease). The archaea seem to improve the environment for pathogenic bacteria that cause this disease.

archaea (ar-KEY-uh) Prokaryotic lineage that is most similar to eukaryotes.
extreme halophile (HAY-lo-file) Organism adapted to life in a highly salty environment.
extreme thermophile (THERM-o-file) Organism adapted to life in a very high-temperature environment.
methanogen (meth-ANN-o-gen) Organism that produces methane gas (CH_4) as a metabolic by-product.

Advances in the ability to compare genomes have allowed scientists to delve into evolutionary relationships among microbes. Here are two examples of what they've found.

IDENTIFYING SPECIES

In eukaryotes, a species is defined on the basis of the ability of its members to mate and produce fertile offspring (Section 1.4). However, this biological species concept is not easily applied to bacteria and archaea, which reproduce only asexually. In these groups, a species is defined as a group of individuals that share an ancestor and have a high degree of similarity in many independently inherited traits.

Historically, classification of bacteria was based on numerical taxonomy. By this process, an unidentified cell is compared against a known group on the basis of shape, metabolism, and properties of the cell wall. The more traits the cell shares with the known group, the closer is their inferred relatedness. This approach works best for cells that can be grown in the laboratory, where their responses to different conditions can be carefully observed. However, many bacteria and most archaea do not grow in the lab.

Genomics is the study of genomes (Section 15.4). The relatively new field of metagenomics assesses microbial diversity by sequence analysis of DNA in samples collected directly from an environment. Results of these studies show that the diversity of prokaryotic species is staggering. For example, researchers now estimate that humans collectively host about 10,000 different species of microbes.

RELATIONSHIPS AMONG DOMAINS

As previously noted, discovery of the archaea led to a major change in how biologists viewed the tree of life. On the basis of this discovery, they began to divide organisms into three domains: Bacteria, Archaea, and Eukarya. **Figure 19.17A** illustrates the pattern of evolutionary branching that underlies the three-domain hypothesis. The first branching occurs between bacteria and the common ancestor of eukaryotes and archaea. Then, a second branching puts archaea and the eukaryotes on separate evolutionary paths.

Recently, new genetic data have led some biologists to propose an alternative view of the relationship between archaea and eukaryotes. As more archaeal genomes have been sequenced, it has become clear that the archaea can be divided into two subgroups. To keep things simple, we will refer to them simply as Archaea 1 and Archaea 2. Results from recent gene studies also suggest that the eukaryotes are more similar to one of these subgroups than to the other. This result would not be expected if eukaryotes and archaea diverged before the two subgroups of archaea did. **Figure 19.17B** illustrates a pattern of evolutionary branching that would give rise to the observed data. As in the three-domain system, the first branching occurs between the bacteria and the common ancestor of eukaryotes and archaea. However, the next branching occurs *within* the Archaea. One branch gives rise to Archaea 1, and the other to Archaea 2, from which the eukaryotes are descended.

If this newly proposed branching pattern is correct, then archaea are not a valid biological group (a clade). Recall that a clade includes all descendants of a common ancestor (Section 17.11). Thus, some biologists have suggested that we should classify all organisms into two primary domains of life. The first is the Bacteria, and the second is the group composed of the Archaea and the Eukaryotes.

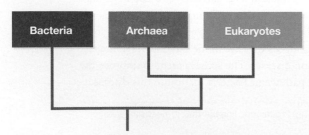

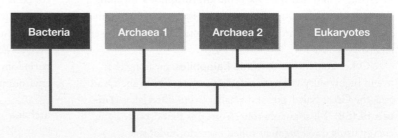

A **Three-domain hypothesis.** An early branching separated the archaea and the eukaryotes, then each evolved separately. By this hypothesis, similarities between eukaryotes and archaea are the result of both groups having evolved from a common, nonarchaeal ancestor.

B **Two-domain hypothesis.** An early branching separated one archaeal lineage (Archaea 1) from the common ancestor of other archaea (Archaea 2) and the eukaryotes. The higher degree of similarity between eukaryotes and one subgroup of archaea is the result of eukaryotes having an ancestor within that subgroup. In this case, the Archaea and Eukaryotes together could be considered as a single domain.

Figure 19.17 Two proposed trees of life.

Figure 19.18 Wild chimpanzees from a population infected by SIV, the chimpanzee equivalent of HIV.

SHARING VIRUSES

We share many traits with chimpanzees, including susceptibility to certain viruses. HIV is similar to simian immunodeficiency virus (SIV), which infects wild chimpanzees in Africa (**Figure 19.18**). SIV harms the chimpanzee immune system, just as HIV harms the human one. Researchers think HIV evolved after SIV entered and survived inside a human. The person may have become infected while butchering an infected animal. (Chimpanzees and other primates are hunted and butchered as "bushmeat" in many parts of Africa.) Alternatively, the person could have been bitten by an SIV-infected animal.

To find out when HIV first arose in humans, researchers looked for HIV in old tissue samples stored at hospitals. The earliest HIV samples known date to about 1960 and come from two people who lived in Africa's Democratic Republic of the Congo. Given what scientists know about the mutation rates for viruses, they estimate that HIV first appeared in humans during the early 1900s.

Comparing genes of HIV in stored and modern blood samples has allowed researchers to trace the virus's spread. In 1966, HIV was introduced from Africa to Haiti, where new mutations arose. In about 1969, someone infected by HIV with those new mutations introduced the virus to the United States. From there, it spread quietly for 12 years until AIDS was identified as a threat in 1981. Two years later, scientists showed that HIV causes AIDS.

Could other viruses also make their way from nonhuman primates into the human population? It seems that some are already trickling in. For example, people exposed to the blood or saliva of a wild Old World primate sometimes become infected by simian foamy virus (SFV). More than 100 cases of human SFV infection have been documented. As far as we know, SFV infection has no adverse effect on human health. However, with some viruses (including HIV), symptoms may not appear until years after the initial infection. There is also no evidence that SFV can spread from person to person. However, viruses do evolve, so that may change. In addition, opportunities for SFV and other currently unknown viruses to infect humans are likely to increase as human populations expand into previously remote areas of primate habitat.

20

The Protists

Links to Earlier Concepts

This chapter describes protists, a collection of eukaryotic lineages introduced in Section 1.3. We look at how various protist cells move (4.9) and how osmosis affects them (5.6). We look again at how organelles such as chloroplasts (4.8, 6.1) evolved by endosymbiosis (18.4) and consider the role of variations in photosynthetic pigments (6.2).

Core Concepts

Evolution underlies the unity and diversity of life.

All protists are eukaryotes (cells with a nucleus). Diverse structures and physiological processes adapt these organisms to a variety of environments. Some protists are single cells, others are colonial, and still others are multicellular. Some are anaerobic, but most need oxygen. Some are heterotrophs, but chloroplasts have evolved independently in many protist lineages.

All organisms alive today are linked by lines of descent from a shared ancestor.

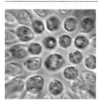

Phylogenetic trees and cladograms are constantly revised based on which biological data is used to make them, as well as new data and new ways of interpreting existing data. We now know that protists include lineages that are related to plants, to animals, and to fungi. We have also learned that protists are not themselves a clade.

Living things sense and respond appropriately to their internal and external environments.

Living things maintain themselves by responding dynamically to internal and external conditions. Some photosynthetic protists can sense and respond to light. Environmental challenges such as dwindling resources can cause protists to form resting structures such as spores. Single-celled freshwater protists maintain homeostasis by using special organelles to pump excess water out of their cytoplasm.

A stand of giant kelp, the largest protists, off the coast of California. Like trees in a forest, kelp shelter and feed a wide variety of organisms.

Photograph by Brian J. Skerry, National Geographic Creative.

20.7 Amoebozoans

20.7 Amoebozoans

LEARNING OBJECTIVES

Describe the structure and feeding of free-living amoeba.

Explain why slime molds are referred to as social amoebas.

Discuss the differences between cellular and plasmodial slime molds.

Amoebozoans (uh-me-buh-ZOE-ans) are members of the eukaryotic supergroup Amoebozoa. This supergroup is the sister group to the supergroup Opisthokonta, which we discuss in the next section. Amoebozoans are shape-shifting heterotrophic cells that typically lack a wall or pellicle. They move about and capture smaller cells by extending thick lobes of cytoplasm called pseudopods (Section 4.9). There are two groups of amoebozoan protists: amoebas and slime molds.

AMOEBAS

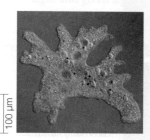

Figure 20.14 Freshwater amoeba.

Amoebas are single cells (**Figure 20.14**) that live in marine or freshwater sediments or in damp soil. They are important predators of bacteria. A few amoebas are human pathogens. *Entamoeba* infects the human gut, causing cramps, aches, and bloody diarrhea. Infections are most common in developing nations where amoebic cysts contaminate drinking water. The freshwater species *Naegleria fowleri* can enter the nose and cause a rare but deadly brain infection. Most people become infected while swimming in a warm pond or lake.

SLIME MOLDS

Slime molds are sometimes described as "social amoebas." They are common on the floor of temperate forests.

Cellular slime molds spend most of their existence as individual amoeba-like cells (**Figure 20.15**). Each cell ingests bacteria and reproduces by mitosis ❶. If food runs out, thousands of cells stream together ❷ to form a cohesive multicelled unit referred to as a "slug" ❸. The slug moves to a suitable area, where its component cells differentiate, forming a fruiting body with spores (thick-walled resting cells) atop a stalk ❹. After dispersal, each spore will release an amoeba-like cell ❺.

Plasmodial slime molds spend most of their life cycle as a plasmodium, a multinucleated mass that forms when a diploid cell undergoes mitosis many times without cytoplasmic division. A plasmodium streams along surfaces, feeding on microbes and organic matter (**Figure 20.16**). When food runs out, the plasmodium develops into many spore-bearing fruiting bodies.

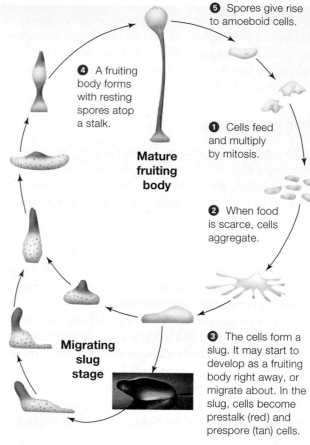

❺ Spores give rise to amoeboid cells.

❹ A fruiting body forms with resting spores atop a stalk.

Mature fruiting body

❶ Cells feed and multiply by mitosis.

❷ When food is scarce, cells aggregate.

❸ The cells form a slug. It may start to develop as a fruiting body right away, or migrate about. In the slug, cells become prestalk (red) and prespore (tan) cells.

Migrating slug stage

Figure 20.15 Life cycle of a cellular slime mold (*Dictyostelium discoideum*).

Figure 20.16 Plasmodial slime mold streaming across a log.

amoeba (uh-ME-buh) Single-celled protist that extends pseudopods to move and to capture prey.

cellular slime mold Amoeba-like protist that feeds as a single predatory cell; joins with others to form a multicellular spore-bearing structure under unfavorable conditions.

plasmodial slime mold (plaz-MOH-dee-ul) Protist that feeds as a multinucleated mass; forms a spore-bearing structure when environmental conditions become unfavorable.

20.8 Opisthokont Protists

LEARNING OBJECTIVES

Describe the protist group most closely related to animals.

Use adhesion proteins as an example to explain the process of exaptation in evolution of animals.

Opisthokonts (oh-PIS-thuh-konts) include fungi, animals, and the protist relatives of both groups. The closest living relatives of fungi are a group of amoeba-like cells (nucleariids) that live in soil and freshwater. We discuss the fungi in detail in Chapter 22.

The protist group most closely related to animals is the **choanoflagellates**. Each choanoflagellate cell has a flagellum surrounded by a "collar" of threadlike projections (**Figure 20.17**). Movement of the flagellum sets up a current that draws food-laden water through the collar. Some sponge cells have a similar structure and function.

Most choanoflagellate species live as single cells, but some are colonia. The colonies arise by mitosis, when descendant cells do not separate after division, but rather stick together with the help of adhesion proteins. The adhesion proteins of choanoflagellates are similar to those of animals, and researchers have discovered that even solitary choanoflagellates have these proteins. By one hypothesis, the common ancestor of animals and choanoflagellates was a single-celled protist with adhesion proteins that helped it capture prey. Later, these proteins were put to use in helping cells stick together to form multicelled colonies. Later still, the proteins allowed animal cells to adhere to one another in multicelled bodies. Such repurposing is an example of exaptation, the evolutionary process by which a trait that evolves with one function later takes on a different function (Section 17.10). Researchers continue to study choanoflagellates to learn more about what the earliest animals may have been like.

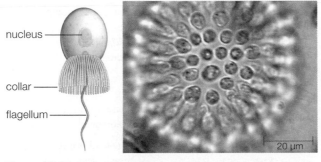

nucleus

collar

flagellum

20 µm

Figure 20.17 Choanoflagellates. The colony at the right is made up of about 230 cells of the type illustrated at the left.

choanoflagellates (ko-an-oh-FLAJ-el-ets) Flagellated heterotrophic protists closely related to animals.

CREDITS: (17) right, Courtesy of Damian Zanette; (18) © Peter M. Johnson/ www.flickr.com/photos/pmjohnso.

APPLICATION: SCIENCE & SOCIETY

Figure 20.18 A bloom of toxin-producing dinoflagellate cells colors seawater along Florida's Gulf coast red.

TOXIC ALGAL BLOOMS

Fertilizing plants enhances their growth. Similarly, adding nutrients to an aquatic habitat encourages photosynthetic protists to reproduce. We do not deliberately fertilize our waters, but fertilizers from agriculture, sewage from cities, and animal wastes from farms get into rivers and flow to the sea. The added nutrients cause an "algal bloom," a population explosion of aquatic protists.

If the protist involved secretes toxins, a bloom can poison other species. Toxic algal blooms affect every coastal region of the United States. Blooms of toxin-producing dinoflagellates are common in the Gulf of Mexico and along the Atlantic coast (**Figure 20.18**). Along the Pacific coast, population explosions of diatoms that secrete toxins are most common.

By a process called bioaccumulation, protist toxins build up in the bodies of plankton-eating animals over time. Plankton feeders such as shellfish, anchovies, and sardines often bioaccumulate high levels of diatom or dinoflagellate toxins.

Plankton feeders are unaffected by such toxins, but these chemicals interfere with function of mammalian nerve cells. Human ingestion of domoic acid, a diatom toxin, can result in dizziness, confusion, memory loss, seizures, and even death. Ingestion of the dinoflagellate toxin saxitoxin has a paralyzing effect and can lead to fatal respiratory failure. Domoic acid and saxitoxin have no color or odor, and both are unaffected by heating or freezing. Government agencies use laboratory tests to detect algal toxins in water samples and shellfish. When the toxins reach a threatening level, a shore is closed to shellfishing.

Sections 20.1, 20.2 **Protists** are a diverse collection that includes members of many eukaryotic groups. Most protists are single cells, but some are **colonial organisms** or **multicellular organisms**. Some single-celled protists have a **pellicle** that helps them maintain their shape. Others have a secreted cell wall or shell. Single-celled freshwater protists use a **contractile vacuole** to expel excess water. Some protists are heterotrophs and others have chloroplasts that evolved from cyanobacteria by **primary endosymbiosis**. Still others have chloroplasts that evolved from an alga by **secondary endosymbiosis**.

Section 20.3 Members of the eukaryotic supergroup Excavata are single-celled protists with flagella and no cell wall. Diplomonads and parabasalids are anaerobic. Both groups include species that infect humans. **Trypanosomes** are parasites with a single mitochondrion. Some cause deadly disease in humans. **Euglenoids** are free-living in lakes and ponds. Some have chloroplasts that evolved from a green alga.

Sections 20.4, 20.5 The SAR supergroup unites three lineages of protists. Stramenopiles include two photosynthetic groups. The silica-shelled **diatoms** are mostly single cells that are part of the **phytoplankton**, whereas **brown algae** range from microscopic strands to giant kelps. **Water molds**, a third stramenopile lineage, are decomposers and parasites that grow as a mesh of absorptive filaments.

The second SAR lineage, the alveolates, has tiny sacs (alveoli) beneath the plasma membrane. **Dinoflagellates** are whirling aquatic cells. They include autotrophs and heterotrophs. Most are free-living, but some live in corals. Some dinoflagellates can emit light (are bioluminescent). **Ciliates** use cilia to move and to feed. Most are aquatic predators, but some are parasites.

All **apicomplexans** are parasitic alveolates that live in animal cells. The disease malaria is caused by the apicomplexan parasite *Plasmodium* and transmitted by mosquitoes. The parasite lives in human liver and blood cells. Untreated malaria is deadly, but drugs can prevent and cure it.

Members of the third SAR lineage, the rhizaria, are single-celled heterotrophs with a perforated shell. Cytoplasmic projections extend through openings in the shell. **Foraminifera** have a calcium carbonate shell and most live on the seafloor. **Radiolaria** have a silica shell and float as part of the **zooplankton**.

Section 20.6 The Archaeplastida supergroup includes plants and their protist relatives: red algae and green algae. All have cell walls of cellulose and chloroplasts derived from cyanobacteria. **Red algae** are multicelled and marine. Accessory pigments called phycobilins allow them to carry out photosynthesis in deep waters. **Green algae** may be single cells, colonial, or multicelled. The charophyte green algae are the closest living relatives of land plants. Like all land plants, some algae have an **alternation of generations**, a life cycle that includes two kinds of multicelled bodies: a diploid, spore-producing **sporophyte** and a haploid, gamete-producing **gametophyte**.

Sections 20.7, 20.8 The supergroup Amoebozoa includes **amoebas** and slime molds. The **plasmodial slime molds** feed as a multinucleated mass. Cells of **cellular slime molds** aggregate when food is scarce. Both form resting spores when conditions do not favor growth.

A related supergroup, the Opisthokonta, includes fungi, animals, and the protist relatives of these groups. The closest relatives of animals are **choanoflagellates**, a groups of heterotrophic protists that can be solitary or colonial.

Application An algal bloom is a population explosion of aquatic protists that results from nutrient enrichment of an aquatic habitat. Blooms of toxin-secreting diatoms or dinoflagellates can adversely affect human health. Shellfish and small fish accumulate the secreted toxins in their tissues, and people who eat these animals can be poisoned.

SELF-ASSESSMENT
ANSWERS IN APPENDIX VII

1. All protists _____ .
   - a. lack mitochondria
   - b. are aquatic
   - c. live as single cells
   - d. have a nucleus

2. Deposits of shells from ancient _____ have been transformed to chalk and limestone.
   - a. dinoflagellates
   - b. diatoms
   - c. radiolaria
   - d. foraminifera

3. The presence of a contractile vacuole indicates that a single-celled protist _____ .
   - a. is marine
   - b. lives in fresh water
   - c. is photosynthetic
   - d. secretes a toxin

4. _____ is an anaerobic flagellated protist that causes a common sexually transmitted disease.
   - a. *Plasmodium*
   - b. *Chlorella*
   - c. *Trichomonas*
   - d. *Giardia*

5. An insect bite can transmit a malaria-causing _____ to a human host.
   - a. trypanosome
   - b. apicomplexan
   - c. ciliate
   - d. diplomonad

6. _____ are the closest relatives of the land plants.
   - a. Green algae
   - b. Red algae
   - c. Brown algae
   - d. Euglenoids

7. Accessory pigments of _____ allow them to carry out photosynthesis at greater depths than other algae.
   - a. euglenoids
   - b. green algae
   - c. brown algae
   - d. red algae

TRACKING CHANGES IN ALGAL BLOOMS Algal blooms are a natural phenomenon along Florida's southwest coast. However, their frequency has been increasing. University of Miami researchers suspected that increased nutrient runoff caused by human activity contributed to an increase in dinoflagellate populations. To find out if dinoflagellate numbers rose along with Florida's human population, the researchers looked at records for coastal waters that have been monitored for more than 50 years. **Figure 20.19** shows the average abundance and distribution of dinoflagellates during two time periods.

1. How did the number of dinoflagellates in waters less than 5 kilometers from shore change between the two time periods?

2. How did the number of dinoflagellates in waters more than 25 kilometers from shore change between the two time periods?

3. Do these data support the hypothesis that human activity increased the abundance of dinoflagellates by adding nutrients to coastal waters?

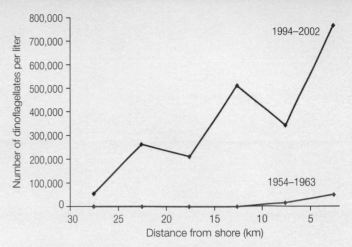

Figure 20.19 Average concentration of dinoflagellate cells detected at various distances from the shore during two time periods: 1954–1963 (blue line) and 1994–2002 (red line).

8. Some _____ are bioluminescent and other live inside the bodies of corals.
   a. red algae   b. diatoms   c. dinoflagellates   d. radiolaria

9. The _____ are important plant pathogens.
   a. dinoflagellates   c. water molds
   b. ciliates   d. slime molds

10. A high concentration of oil helps _____ stay afloat and makes them useful in biofuel production.
    a. foraminiferans   c. dinoflagellates
    b. radiolarians   d. diatoms

11. The sporophyte of a multicellular alga _____ .
    a. is haploid   c. produces spores
    b. is a single cell   d. produces gametes

12. Cellular slime molds most often live _____ .
    a. on the forest floor   c. in an animal gut
    b. in tropical seas   d. in mountain lakes

13. Some euglenoids have chloroplasts that evolved from _____ by secondary endosymbiosis.
    a. a green alga   c. a dinoflagellate
    b. a red alga   d. cyanobacteria

14. All green algae _____ .
    a. have a cell wall   c. are multicellular
    b. are marine   d. are descended from red algae

15. Match each item with its description.
    ___ choanoflagellate   a. chalky-shelled heterotroph
    ___ radiolarian   b. silica-shelled autotroph
    ___ foraminiferan   c. deep dweller with phycobilins
    ___ diatom   d. close relative of animals
    ___ brown alga   e. community of tiny heterotrophs
    ___ red alga   f. multicelled, with fucoxanthin
    ___ zooplankton   g. silica-shelled heterotroph

CLASS ACTIVITY
CRITICAL THINKING

1. Imagine you are in a developing country where sanitation is poor. Having read about parasitic protists in water and damp soil, what would you consider safe to drink? What foods might be best to avoid or which food preparation methods might make them safe to eat?

2. Which groups of protists are scientists most likely to find as fossils? Which groups are least likely to be fossilized? Explain your reasoning.

3. Water in abandoned swimming pools often turns green. If you examined a drop of this water with a microscope, how could you tell whether the water contains protists and, if so, which group they might belong to?

4. Melvin Calvin and Andrew Benson determined the steps in the light-independent reactions of photosynthesis by exposing the green alga *Chlorella* to CO_2 labeled with the radioisotope carbon 14. By looking at which compounds the C^{14} ended up in, they were able to identify all intermediates in this cyclic pathway. Why did they predict (correctly) that the same set of intermediates would be formed in the light reactions in land plants?

5. The protist that causes malaria evolved from a photosynthetic ancestor and has the remnant of a chloroplast. The organelle no longer functions in photosynthesis, but it remains essential to the protist. Why might targeting this organelle yield an antimalarial drug that produces minimal side effects in humans?

for additional quizzes, flashcards, and other study materials
▶ ACCESS MINDTAP AT WWW.CENGAGEBRAIN.COM

21

Plant Evolution

Links to Earlier Concepts

Section 20.6 introduced the algae that are the closest relative of plants. Section 12.2 introduced gamete formation in plants and here you will see specific examples. You will learn about evolution of cell walls strengthened with lignin (4.10) and a waxy cuticle perforated by stomata (6.5). We also discuss coevolution (17.10) in the context of pollinators.

Core Concepts

Natural selection leads to the evolution of structures and processes that increase fitness in a specific environment.

Plants evolved from an aquatic green alga. Over time, changes in structure, life cycle, and reproductive processes adapted plants to life in increasingly drier climates. The oldest plant lineages, such as mosses and ferns, are adapted to moist habitats, where their flagellated sperm can swim to eggs through films of water. Seed plants, which evolved later, are better suited to drier environments.

Adaptation of organisms to a variety of environments has resulted in diverse structures and physiological processes.

Plants range in size from low-growing mosses to towering seed-bearing trees that are supported by massive woody stems. Early plant lineages such as mosses and ferns dispersed by releasing spores, whereas the gymnosperms and angiosperms disperse by releasing seeds. Flowers and fruits are structures unique to angiosperms (flowering plants).

Organisms exchange matter and energy with the environment in order to grow, maintain themselves, and reproduce.

All organisms harvest atomic and molecular building blocks from their environment. Plants are autotrophs that capture energy from sunlight. The simplest plants absorb essential water and nutrients across their body surface. Most plants have roots that take water and nutrients up from the soil. Internal pipelines (vascular tissues) distribute water and nutrients through the body.

◄ Ancient bristlecone pines in California's White Mountains. These plants can live more than 4,500 years.

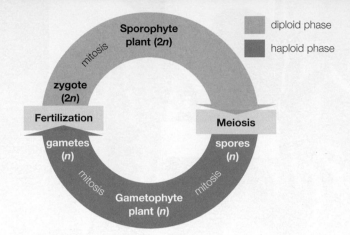

Figure 21.2 Generalized plant life cycle. Diploid phase is shown in orange, haploid in blue. This color coding is used in life-cycle diagrams throughout the chapter.

Plants (Kingdom Plantae) are multicelled, photosynthetic eukaryotes that typically live on land. The group evolved about 450 million years ago from a freshwater green alga (a charophyte alga) that grew at the water's edge. Like their green algal relatives, plants have cells with walls made of cellulose, they have chloroplasts that contain chlorophylls *a* and *b*, and they store sugars as starch. Unlike algae, plants shelter a multicelled embryo (a developing plant) on the parental body. Thus, the clade of plants is referred to as the embryophytes.

STRUCTURAL ADAPTATIONS

Moving onto land posed challenges to an aquatic lineage. A multicelled green alga absorbs all the water, dissolved nutrients, and gases it needs across its body surface. Water also buoys algal parts, helping the alga remain upright. By contrast, land plants face the threat of drying out, must take up water and nutrients from soil, and have to stand upright on their own.

A variety of structural features allow land plants to meet these challenges. Most plants secrete a waxy **cuticle** that reduces water loss (**Figure 21.1**). **Stomata** (singular, stoma) are adjustable pores that span the cuticle and epidermis (the

outermost tissue layer). Depending on conditions in the environment and inside the plant, stomata can open to allow gas exchange or close to conserve water.

Early plants were held in place by threadlike structures, but these structures did not deliver water or dissolved minerals to the rest of the plant body. True roots contain vascular tissue, which has internal pipelines that transport water and nutrients among plant parts. **Xylem** is the vascular tissue that distributes water and mineral ions taken up by roots. **Phloem** is the vascular tissue that distributes sugars made by photosynthetic cells. Of the 295,000 or so modern plant species, more than 90 percent are **vascular plants**, meaning plants that have xylem and phloem. Mosses and other ancient plant lineages are **nonvascular plants**, meaning they do not have these vascular tissues.

An organic compound called **lignin** stiffens cell walls in vascular tissue. Lignin-reinforced vascular tissue not only distributes materials, it also helps vascular land plants stand upright and supports their branches. Most vascular plants have branches with many leaves that increase the surface area for capturing light and for gas exchange.

LIFE-CYCLE CHANGES

Adapting to life in dry habitats also involved life-cycle changes. Like some algae, land plants have an alternation of generations (**Figure 21.2**). Meiosis of cells in a diploid, multicelled **sporophyte** produces spores, which in plants are walled haploid cells. A spore germinates, divides by mitosis, and develops into a haploid, multicelled **gametophyte** that produces gametes by mitosis. Gametes combine at fertilization to form a diploid zygote that grows and develops into a new sporophyte.

Figure 21.1 Diagram of a vascular plant leaf in cross-section. Some of the traits that contribute to the success of this group are shown.

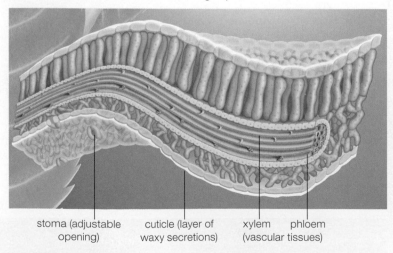

stoma (adjustable opening)   cuticle (layer of waxy secretions)   xylem (vascular tissues)   phloem

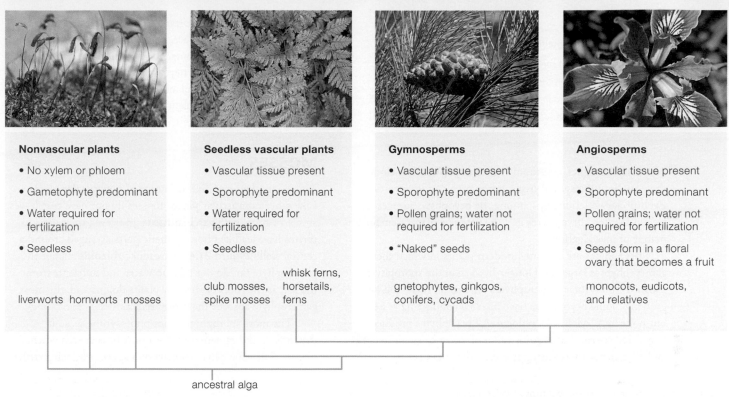

Nonvascular plants
- No xylem or phloem
- Gametophyte predominant
- Water required for fertilization
- Seedless

liverworts hornworts mosses

Seedless vascular plants
- Vascular tissue present
- Sporophyte predominant
- Water required for fertilization
- Seedless

club mosses, spike mosses

whisk ferns, horsetails, ferns

Gymnosperms
- Vascular tissue present
- Sporophyte predominant
- Pollen grains; water not required for fertilization
- "Naked" seeds

gnetophytes, ginkgos, conifers, cycads

Angiosperms
- Vascular tissue present
- Sporophyte predominant
- Pollen grains; water not required for fertilization
- Seeds form in a floral ovary that becomes a fruit

monocots, eudicots, and relatives

ancestral alga

Figure 21.3 Traits of the modern plant groups and relationships among them. Only the gymnosperms and angiosperms are clades.

In nonvascular plants, the gametophyte is larger and longer lived than the sporophyte. By contrast, the sporophyte dominates the life cycle of vascular plants. Living in dry conditions favors an increased emphasis on spore production, because spores withstand drying out better than gametes do.

POLLEN AND SEEDS

Evolution of new reproductive traits in seed plants gave this vascular plant lineage a competitive edge in dry habitats. Unlike bryophytes and seedless vascular plants, seed plants do not release spores. Instead, their spores give rise to gametophytes inside specialized structures on the sporophyte body.

A **pollen grain** is a walled, immature male gametophyte of a seed plant. Once released, it can be transported to another plant by wind or animals even in the driest of times. By contrast, plants that do not make pollen (nonvascular plants and seedless vascular plants) can only reproduce when a film of water allows their sperm to swim to eggs.

Fertilization of a seed plant takes place on the sporophyte body. It results in development of a **seed**—an embryo sporophyte packaged with a supply of nutritive tissue inside a protective seed coat. Seed plants disperse a new generation by releasing seeds.

There are two modern seed plant lineages. Gymnosperms include nonflowering seed producers such as pine trees. Angiosperms, which make flowers and package their seeds inside a fruit, are the most widely distributed and diverse plant group.

Figure 21.3 summarizes the relationships among the major plant groups and the traits of each group.

cuticle Secreted covering at a body surface.
gametophyte Multicelled, haploid, gamete-producing body.
lignin Material that stiffens cell walls of vascular plants.
nonvascular plant Plant that does not have xylem and phloem; a bryophyte such as a moss.
phloem (FLOE-em) Plant vascular tissue that distributes sugars.
plant Multicelled photosynthetic eukaryote that forms and nurtures embryos on the parental body.
pollen grain Walled, immature male gametophyte of a seed plant.
seed Embryo sporophyte of a seed plant packaged with nutritive tissue inside a protective coat.
sporophyte Multicelled, diploid, spore-producing body.
stoma Opening across a plant's cuticle and epidermis; can be opened for gas exchange or closed to prevent water loss.
vascular plant Plant with xylem and phloem.
xylem (ZYE-lum) Plant vascular tissue that distributes water and dissolved mineral ions.

21.2 Nonvascular Plants—Bryophytes

LEARNING OBJECTIVES

Compare the structure of a moss gametophyte and a moss sporophyte, and explain the role of each in the moss life cycle.

Discuss the traits common to all three bryophyte lineages.

Give two examples of asexual reproduction in bryophytes.

Discuss the ecological and commercial importance of bryophytes.

Nonvascular plants, commonly called **bryophytes**, include three modern lineages: mosses, hornworts, and liverworts. Evolutionary relationships among the bryophyte lineages, as well as between bryophytes and the vascular plants, remain a matter of investigation.

Bryophytes are the only modern plants in which the gametophyte is larger and longer-lived than the sporophyte. However, even their gametophytes tend to be small and low-growing. Some bryophytes have internal conducting vessels, but none have xylem and phloem. All bryophytes produce flagellated sperm that require a film of water to swim to eggs, and all disperse by releasing spores, rather than seeds.

Bryophytes play important ecological roles. They absorb nutrients across their surface, rather than withdrawing them from soil, so they can colonize rocky sites where vascular plants cannot take root. They also withstand drought and cold better than vascular plants. For these reasons, nonvascular plants are the only plant life in some parts of the Arctic and Antarctic.

MOSSES

Mosses are the most diverse lineage of nonvascular plants. Like other nonvascular plants, mosses do not have true leaves, stems, or roots with vascular tissue. However, their gametophytes have leaflike photosynthetic parts arranged around a central stalk (**Figure 21.4**). Threadlike **rhizoids** anchor the gametophyte but do not take up water and nutrients from the soil as the roots of vascular plants do. Instead, the moss gametophyte absorbs these materials across its entire body.

The moss sporophyte is not photosynthetic, so it depends on the gametophyte for nourishment even when mature. The sporophyte consists of a spore-producing struc-

Figure 21.4 Life cycle of a common moss (*Polytrichum*).

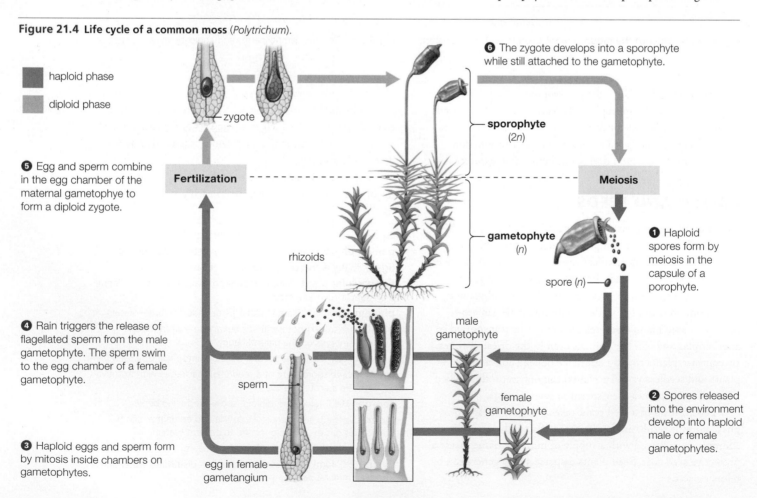

haploid phase

diploid phase

zygote

❻ The zygote develops into a sporophyte while still attached to the gametophyte.

sporophyte
(2n)

❺ Egg and sperm combine in the egg chamber of the maternal gametophye to form a diploid zygote.

Fertilization

Meiosis

gametophyte
(n)

❶ Haploid spores form by meiosis in the capsule of a porophyte.

spore (n)

rhizoids

❹ Rain triggers the release of flagellated sperm from the male gametophyte. The sperm swim to the egg chamber of a female gametophyte.

male gametophyte

sperm

female gametophyte

❷ Spores released into the environment develop into haploid male or female gametophytes.

❸ Haploid eggs and sperm form by mitosis inside chambers on gametophytes.

egg in female gametangium

ture (a sporangium) atop a stalk. Meiosis of cells inside the sporangium produces haploid spores ❶. After dispersal by the wind, a spore germinates and grows into a filament of cells from which one or more gametophytes develop ❷.

Multicellular gamete-producing structures (gametangia) develop in or on each gametophyte. (The moss we are using as our example has separate sexes, but in some species, each gametophyte produces both.) Eggs and sperm form by mitosis ❸. Rain stimulates the release of flagellated sperm that swim through a film of water to eggs ❹. Fertilization occurs inside the maternal egg chamber and produces a zygote ❺. The zygote grows and develops into a new sporophyte ❻.

Mosses also reproduce asexually by fragmentation when a bit of gametophyte breaks off and develops into a new plant.

Peat mosses (*Sphagnum*) are the most economically important nonvascular plants. They grow in peat bogs that cover hundreds of millions of acres in high-latitude regions of Europe, Asia, and North America. Many peat bogs have persisted for thousands of years, and layer upon layer of plant remains have become compressed as carbon-rich material called peat. Blocks of peat are cut, dried, and burned as fuel, especially in Ireland (**Figure 21.5**). Freshly harvested peat moss is also an important commercial product. The moss is dried and added to planting mixes to help soil retain moisture.

LIVERWORTS AND HORNWORTS

The name "liverworts" refers to the flat, multilobed body of some liverwort gametophytes, which were thought to resemble lobes of a liver (**Figure 21.6A**). These liverworts reproduce asexually by producing gemmae (little clusters of cells) in cups on the gametophyte surface. Raindrops disperse the gemmae, which grow into new gametophytes. Liverworts also reproduce sexually. In some species, the gamete-producing structures look like tiny umbrellas elevated above the flattened gametophyte body (**Figure 21.6B**).

Hornworts are named for a hornlike (pointy) sporophyte that develops on a flat, ribbonlike, or rosette-shaped gametophyte (**Figure 21.7**). Spores form in an upright capsule at the sporophyte's tip. When they mature, the tip of the capsule splits, releasing them into the environment. Unlike the sporophyte of a moss or liverwort, a hornwort sporophyte grows continually from its base, contains chloroplasts and, in some species, can even survive even after the gametophyte dies.

bryophyte (BRY-oh-fight) Nonvascular plant; a moss, liverwort, or hornwort.

rhizoid (RYE-zoyd) Threadlike structure that holds a nonvascular plant in place.

Figure 21.5 Harvesting peat. Blocks of peat (compressed remains of *Sphagnum* moss) being cut and dried for use as fuel.

A Multilobed gametophyte with cups of cell clusters that function in asexual reproduction.

B Umbrella-like structures on which female gametes formed. Sporophytes with yellow spores remain attached.

Figure 21.6 Liverwort (*Marchantia*).

Figure 21.7 Hornwort. Photosynthetic hornlike sporophytes grow from a flattened gametophyte body.

CREDITS: (5) gabriel12/Shutterstock.com; (6A) Henrik Larsson/Shutterstock.com; (6B) Dr. Annkatrin Rose, Appalachian State University; (7) Daniel Vega/Getty Images.

21.3 Seedless Vascular Plants

Compare the life cycle of a seedless vascular plant with that of a bryophyte.

Explain how evolution of an ability to produce lignin allowed innovations in the body form of vascular plants.

Compare the structure of a fern sporophyte and a fern gametophyte.

Name two nonfern lineages of seedless vascular plants and describe the structure of their sporophytes.

Only vascular plants have lignin-strengthened vascular tissue. This innovation allowed evolution of larger, branching sporophytes with roots, stems, and leaves. Such sporophytes are the predominant generation in all vascular plants. Like nonvascular plants, **seedless vascular plants** disperse by releasing spores directly into the environment. The spores develop into tiny, short-lived gametophytes. As in nonvascular plants, sperm are flagellated and must swim to the eggs.

Two lineages of seedless vascular plants survived to the present. One (monilophytes) includes ferns and related groups; the other (lycophytes) includes club mosses and their relatives. These lineages diverged from a common ancestor before leaves and roots evolved, so each developed these features in a different way.

FERNS AND CLOSE RELATIVES

With 12,000 or so species, ferns are the most diverse and familiar seedless vascular plants. Most live in the tropics. A typical fern sporophyte has fronds (leaves) and roots that grow out from a **rhizome**, a horizontal underground stem (**Figure 21.8**). Fiddleheads, the young, tightly coiled fronds of some ferns, are harvested from the wild as food.

Fern spores form by meiosis in a **sorus** (plural, sori), a cluster of capsules that develop on the lower surface of fronds ❶, ❷. The capsules are sporangia, and they pop open to release spores that disperse in the wind. After a spore germinates, it grows into a photosynthetic, heart-shaped gametophyte just a few centimeters wide ❸. Eggs and sperm form in chambers on the underside of the tiny gametophyte ❹. The chambers are gametangia that release sperm when rainwater wets them. The sperm then swim through a film of water to reach and fertilize an egg ❺. The resulting zygote develops into a new sporophyte ❻, and its parental gametophyte dies.

Fern sporophytes vary in size and form. Some floating ferns have fronds only 1 millimeter long. Tree ferns can be 25 meters (80 feet) high (**Figure 21.9A**). Many tropical ferns are epiphytes, meaning they attach to and grow on a trunk or branch of another plant without harming that plant.

Figure 21.8 Life cycle of a fern.

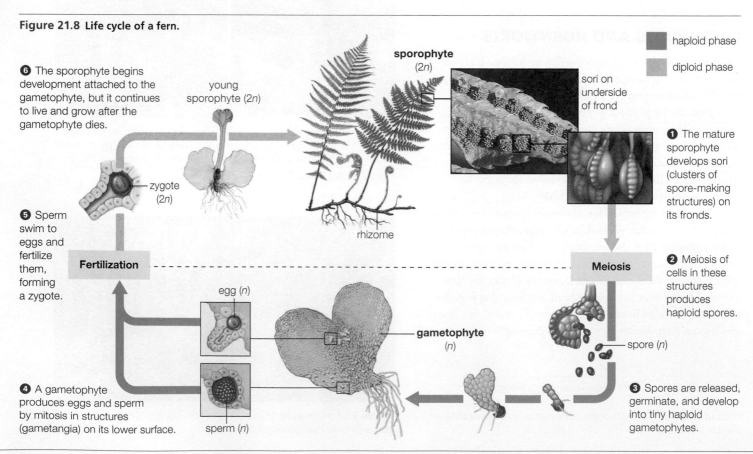

- haploid phase
- diploid phase

sporophyte (2n)

young sporophyte (2n)

sori on underside of frond

❻ The sporophyte begins development attached to the gametophyte, but it continues to live and grow after the gametophyte dies.

❶ The mature sporophyte develops sori (clusters of spore-making structures) on its fronds.

zygote (2n)

rhizome

Meiosis

❷ Meiosis of cells in these structures produces haploid spores.

❺ Sperm swim to eggs and fertilize them, forming a zygote.

Fertilization

egg (n)

gametophyte (n)

spore (n)

❹ A gametophyte produces eggs and sperm by mitosis in structures (gametangia) on its lower surface.

sperm (n)

❸ Spores are released, germinate, and develop into tiny haploid gametophytes.

A Tree ferns in Australia.   B Whisk fern (*Psilotum*).   C Horsetail (*Equisetum*).   D Club moss (*Lycopodium*).

Figure 21.9 Seedless vascular plant sporophytes. These plants disperse by releasing spores.

Whisk ferns and horsetails are close relatives of ferns. Most whisk ferns are in the genus *Psilotum*. Their sporophyte has leafless photosynthetic stems that grow from rhizomes (**Figure 21.9B**). Spores form in fused sporangia at the tips of short lateral branches. Florists use these whisk ferns to add visual interest to mixed bouquets.

The sporophyte of a horsetail (*Equisetum*) has hollow stems with tiny nonphotosynthetic leaves at the joints. Photosynthesis occurs in stems and leaflike branches (**Figure 21.9C**). Deposits of silica in the stem support the plant and give stems a sandpaper-like texture that helps fend off her-

bivores. Before scouring powders and pads became widely available, people used the stems of horsetails as pot scrubbers. Horsetail spores form on a **strobilus**, a soft cone-shaped structure composed of modified leaves or branches. Lycophytes and some gymnosperms also produce strobili.

LYCOPHYTES

Lycophytes include plants commonly known as quillworts, spikemosses, and club mosses. Club mosses of the genus *Lycopodium* are common in temperate forests (**Figure 21.9D**). Their sporophyte has a rhizome from which roots and upright stems with tiny leaves grow. When a *Lycopodium* plant is several years old, it begins to make spore-producing strobili seasonally. *Lycopodium* is gathered from the wild for a variety of uses. The stems are used in wreaths and bouquets. Spores, which have a waxy, flammable coating, are sold as "flash powder" for special effects. When the spores are sprayed out as a fine mist and ignited, they produce a bright flame.

---

rhizome (RYE-zohm) Stem that grows horizontally along or under the ground.

seedless vascular plant Plant that disperses by releasing spores and has xylem and phloem. Ferns and club mosses are examples.

sorus (SORE-us) Cluster of spore-producing capsules on a fern leaf.

strobilus (STROH-bill-us) Of some plants, a spore-forming, cone-shaped structure composed of modified leaves or branches.

LEARNING OBJECTIVES

Describe the types of plants that lived in a Carboniferous coal forest and explain how coal forms.

Explain how the reproductive traits of seed plants give them an advantage over seedless plants in dry habitats.

Compare the function of megaspores and microspores in the seed plant life cycle and explain where each type of spore forms.

Discuss how the evolution of wood provided an advantage to seed plants.

TINY BRANCHERS TO COAL FORESTS

We know from fossils that vascular plants evolved by about 450 million years ago. Early vascular plants were only a few centimeters tall and had a simple branching pattern, with no leaves or roots. Spores formed at branch tips. By the early Devonian, taller species with a more complex branching pattern were common. As the Devonian continued, larger seedless vascular plants arose. The oldest forest known existed about 385 million years ago in what is now upstate New York. Some plants of this forest stood about 8 meters (26 feet) high and resembled modern tree ferns in their structure.

In the Carboniferous period (359–299 million years ago), some ancient relatives of horsetails and lycophytes grew even taller (**Figure 21.10A**). Some stood 40 meters (more than 130 feet) high. After these forests first formed, climates changed, and the sea level repeatedly rose and fell. Over time, the resulting pressure and heat transformed the compacted organic remains of Carboniferous forests into **coal**.

RISE OF THE SEED PLANTS

Seed plants evolved late in the Devonian (about 365 million years ago), so gymnosperms lived beside seedless plants in Carboniferous forests. As Earth became cooler and drier during the Permian (299–252 million years ago), gymnosperms replaced seedless plants in many habitats. Angiosperms (flowering plants) first appear in the fossil record about 125 million years ago, during the Cretaceous period (**Figure 21.10B**).

Reproductive traits of seed-bearing plants provide an advantage in dry habitats. Gametophytes of seedless vascular plants develop from spores that are released into the environment. By contrast, gametophytes of seed plants form inside and are protected within the reproductive parts on the sporophyte body (**Figure 21.11**). Pollen grains, which give rise to male gametophytes, develop inside **pollen sacs** ❶ from haploid spores called **microspores** ❷. Egg-producing gametophytes develop in a protective chamber called an **ovule** ❸, from haploid spores called **megaspores** ❹.

A seed-bearing plant releases pollen grains, but holds onto its eggs. Wind or animals deliver pollen from one seed plant to another, a process called **pollination** ❺. Sperm of seed plants do not need to swim through a film of water to reach eggs, so these plants can reproduce when the weather is dry. After pollination, a pollen tube grows through the ovule ❻ and delivers the sperm to the egg. Fertilization occurs, then an ovule develops into a seed ❼. A seed, remember, contains food to nourish the embryo sporophyte. By contrast, seedless plants release single-celled spores without any stored food.

A Carboniferous coal forest. An understory of ferns is shaded by tree-sized relatives of modern horsetails and lycophytes.

Figure 21.10 Ancient vascular plants.

B Early angiosperms in the Cretaceous. Some of the oldest angiosperm lineages, including magnoliids (foreground), evolved while dinosaurs were the dominant animals on land.

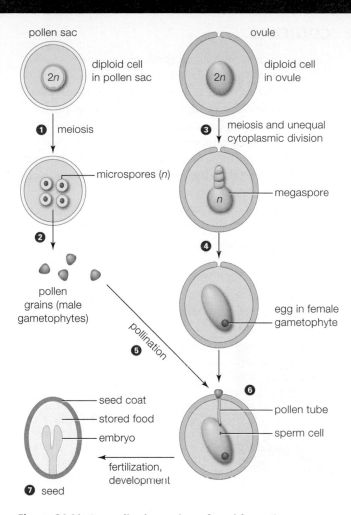

Figure 21.11 **Generalized overview of seed formation.**

Seed plants are the only plants that produce true wood. As a woody plant grows, addition of lignin-stiffened xylem cells widens its parts. Wood may have first evolved as a means of improving water delivery through the plant, but it also provided structural advantages. Wood makes a plant sturdier and more resistant to mechanical stress, so it can grow taller and overshadow its competitors.

coal Fossil fuel formed over millions of years by compaction and heating of plant remains.
megaspore Haploid spore formed in ovules of seed plants; gives rise to the female gametophyte.
microspore Haploid spore formed in pollen sacs of seed plants; gives rise to the male gametophyte.
ovule Of seed plants, reproductive structure in which the female gametophyte develops; after fertilization, it matures into a seed.
pollen sac Of seed plants, reproductive structure in which pollen grains develop.
pollination Delivery of a pollen grain to the pollen-receiving part of a seed plant.

Jeff Benca

Like many children, National Geographic Explorer Jeff Benca was fascinated by dinosaurs. As a teen he transferred his interest to another ancient group—the lycophytes. By the time he graduated from college, he had one of the world's largest lycophyte collections. Now he investigates what these early-diverging vascular plants can tell us about conditions in Earth's deep past.

Paleobotanists have long used the leaf shapes of fossilized woody flowering plants to understand ancient climatic conditions. For example, botanists know that woody angiosperms with more serrated leaf margins (like maples) are now prominent in temperate regions with cool, wet climates. By contrast, woody angiosperms with smooth leaf margins (such as magnolias) now predominate in tropical regions. If ferns and lycophytes also show climate-influenced changes in leaf shape, their fossils could yield a wealth of information about Earth's distant past. The fossil record of angiosperms extends back 125 million years, but ferns and lycophytes have fossil records exceeding 360 million years.

To investigate how climate affects leaves of ferns and lycophytes, Benca grows members of these groups under a variety of conditions in climate-controlled growth chambers, and then compares their leaf shapes. He is also looking at how the shape of fern and lycophyte leaves varies with elevation (and climate) on a tropical mountain.

Aside from his research, Benca advocates conservation of lycophytes. Although this ancient lineage survived several mass extinctions, many species are now declining swiftly due to habitat loss. As a result, lycophytes are in need of increased research attention and protection.

LEARNING OBJECTIVES

Explain the origin of the name "gymnosperm."

List the distinguishing traits of conifers and describe their life cycle.

Using appropriate examples, discuss the diversity of forms among gymnosperm lineages.

Gymnosperms are vascular seed plants whose seeds are "naked," meaning that unlike the seeds of angiosperms they are not encased in a fruit. (*Gymnos* means naked and *sperma* is taken to mean seed.) Many gymnosperms enclose their seeds in a fleshy or papery covering, but these are not fruits.

CONIFERS

Conifers, the most diverse and familiar gymnosperms, include 600 or so species of trees and shrubs that have needlelike or scalelike leaves, and produce seeds on woody cones (seed cones). Conifers are typically more resistant to drought and cold than flowering plants, and they abound in cool forests of the Northern Hemisphere. Conifers include the tallest trees in the Northern Hemisphere (redwoods), and the most abundant (pines). They also include the long-lived bristlecone pines shown in the chapter opening photo.

Conifers are of great commercial importance. We use fir bark to mulch gardens, and oils from cedar in cleaning prod-

Figure 21.12 Conifer (pine) life cycle.

■ haploid phase   ■ diploid phase

FIGURE IT OUT
Does the pollen tube grow through tissue of a male cone or a female cone?
Answer: A female cone. It grows after a pollen grain alights on a female cone.

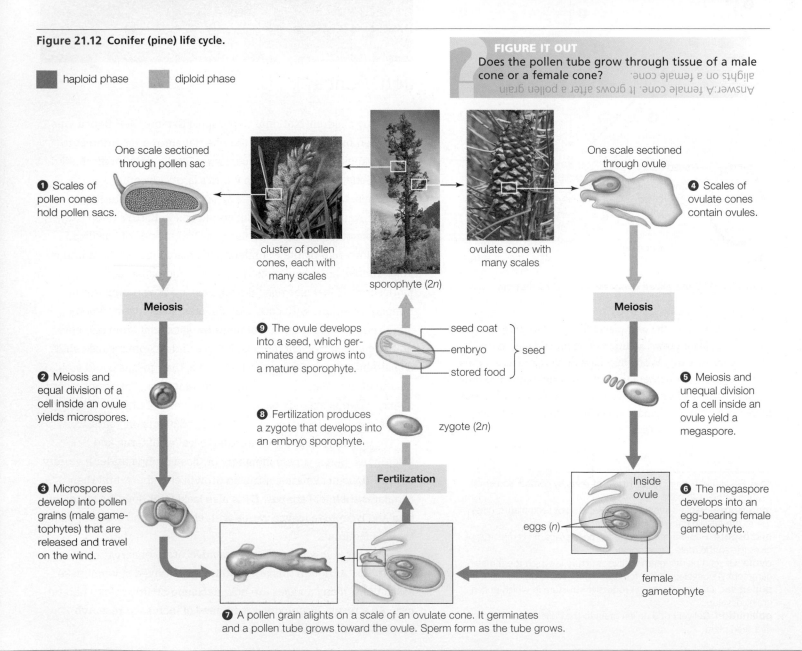

One scale sectioned through pollen sac

cluster of pollen cones, each with many scales

sporophyte (2n)

One scale sectioned through ovule

ovulate cone with many scales

❶ Scales of pollen cones hold pollen sacs.

❷ Meiosis and equal division of a cell inside an ovule yields microspores.

❸ Microspores develop into pollen grains (male gametophytes) that are released and travel on the wind.

Meiosis

❹ Scales of ovulate cones contain ovules.

Meiosis

❺ Meiosis and unequal division of a cell inside an ovule yield a megaspore.

❻ The megaspore develops into an egg-bearing female gametophyte.

Inside ovule

eggs (n)

female gametophyte

❼ A pollen grain alights on a scale of an ovulate cone. It germinates and a pollen tube grows toward the ovule. Sperm form as the tube grows.

Fertilization

❽ Fertilization produces a zygote that develops into an embryo sporophyte.

zygote (2n)

❾ The ovule develops into a seed, which germinates and grows into a mature sporophyte.

seed coat — embryo — stored food } seed

ucts. We also eat seeds, or "pine nuts," of some pines. Pines provide lumber for building homes. Some pines make a sticky resin that deters insects from tunneling into them. We use this resin to make turpentine, a paint solvent.

Pines have a life cycle typical of conifers (**Figure 21.12**). Like most gymnosperms they are wind-pollinated. The pine tree is a sporophyte that produces spores on specialized strobili called cones. In most pine species, each tree makes one type of cone: either small, soft, pollen cones or large, woody, ovulate cones (seed cones). Both types of cones have scales arranged around a central axis. Each scale of a pollen cone contains pollen sacs ❶. Microspores that form by meiosis in these sacs ❷ develop into pollen grains ❸. Pollen cones release these tiny grains to drift with the winds. Each scale of an ovulate cone encloses a pair of ovules ❹. Meiosis and unequal cytoplasmic division of diploid cells in an ovule yields megaspores ❺ that develop into egg-producing gametophytes ❻.

Before pollination, an ovule exudes a sugary fluid that can capture windborne pollen. The pollen grain then germinates, and a pollen tube grows through the ovule tissue to deliver sperm to the egg ❼. Fertilization produces a zygote ❽. The zygote develops into an embryo sporophyte that, along with tissues of the ovule, becomes a seed. The seed is released, germinates, then develops into a new sporophyte ❾.

LESSER-KNOWN LINEAGES

The species diversity of cycads and ginkgos peaked about 200 million years ago. Today, they are the only modern seed plants with flagellated sperm. Sperm emerge from pollen grains, then swim in fluid produced by the plant's ovule. Both cyads and ginkgos produce seeds with a fleshy covering.

About 130 modern cycad species are native to the dry tropics and subtropics (**Figure 21.13A**). Some resemble palm plants, but the two groups are not close relatives. The "sago palms" commonly used in landscaping are actually cycads.

The only living ginkgo species is *Ginkgo biloba*, the maidenhair tree (**Figure 21.13B**). It is a deciduous native of China. Deciduous plants drop all their leaves at once seasonally. The ginkgo's pretty fan-shaped leaves and resistance to insects, disease, and air pollution make it a popular tree along city streets. Female trees produce fleshy plum-sized seeds with a strong unpleasant odor, so male trees are preferred for urban landscaping. Extracts of *G. biloba* are touted as a memory aid and a treatment for Alzheimer's disease, but studies of these claims have not shown a consistent benefit.

Gnetophytes (NEAT-uh-fights) are a group defined by their vascular tissue, which differs somewhat in structure from that of other gymnosperms. They include woody vines, tropical trees, and shrubs. Members of the genus *Ephedra*

B Fan-shaped leaves and fleshy seeds of *Ginkgo biloba*.

A Cycad with fleshy seeds.

C Pollen cones of *Ephedra*, a gnetophye.

D *Welwitschia*, a gnetophyte native to Africa's Namib desert, with seed cones and two long, wide leaves that split repeatedly.

Figure 21.13 Gymnosperm diversity.

are evergreen desert shrubs with broomlike green stems and inconspicuous leaves (**Figure 21.13C**). Native Americans brewed tea using one species common in semiarid regions of the American West. Some Eurasian *Ephedra* species produce ephedrine, a stimulant similar to amphetamines.

Another gnetophyte, *Welwitschia*, grows only in Africa's Namib desert (**Figure 21.13D**). It has a taproot that grows deep into the soil, a short woody stem, and two long, straplike leaves. These leaves split lengthwise repeatedly, giving the plant a strange shaggy appearance. *Welwitschia* is very long-lived; some plants are more than a thousand years old.

gymnosperm (JIM-no-sperm) Seed plant whose seeds are not enclosed within a fruit; a conifer, cycad, ginkgo, or gnetophyte.

Describe the components of a flower and their functions.

Explain how endosperm forms and its function.

Discuss how components of a flower give rise to fruits and seeds.

Give examples of the two major lineages of flowering plants and explain the trait that distinguishes them.

FLOWERS AND FRUITS

Angiosperms, the most diverse seed plant lineage, are the only plants that make flowers and fruits. A **flower** is a specialized reproductive shoot. Floral structure varies, but most flowers include the parts shown in **Figure 21.14**. Sepals, which usually have a green leaflike appearance, ring the base

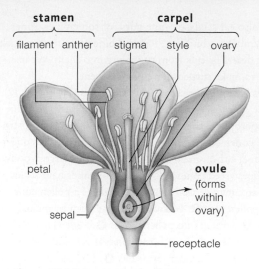

Figure 21.14 Anatomy of a flower.

Figure 21.15 Angiosperm life cycle.

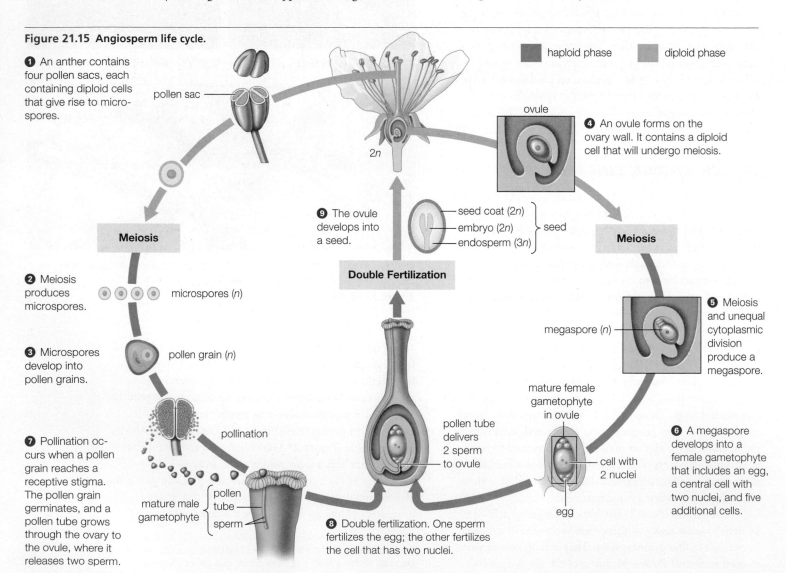

❶ An anther contains four pollen sacs, each containing diploid cells that give rise to microspores.

❷ Meiosis produces microspores.

❸ Microspores develop into pollen grains.

❼ Pollination occurs when a pollen grain reaches a receptive stigma. The pollen grain germinates, and a pollen tube grows through the ovary to the ovule, where it releases two sperm.

❽ Double fertilization. One sperm fertilizes the egg; the other fertilizes the cell that has two nuclei.

❾ The ovule develops into a seed.

❹ An ovule forms on the ovary wall. It contains a diploid cell that will undergo meiosis.

❺ Meiosis and unequal cytoplasmic division produce a megaspore.

❻ A megaspore develops into a female gametophyte that includes an egg, a central cell with two nuclei, and five additional cells.

pollen tube delivers 2 sperm to ovule

mature female gametophyte in ovule

cell with 2 nuclei

egg

mature male gametophyte

pollen tube

sperm

pollination

pollen grain (n)

microspores (n)

Meiosis

Meiosis

Double Fertilization

seed coat (2n)
embryo (2n) } seed
endosperm (3n)

megaspore (n)

ovule

pollen sac

2n

haploid phase   diploid phase

of a flower and enclose it until it opens. Inside the sepals are petals, which are often colored and scented. The petals surround the stamens. **Stamens** are organs that produce pollen. A typical stamen consists of a tall stalk (the filament), topped by an anther that holds four pollen sacs. The innermost part of the flower is the **carpel**, the organ that captures pollen and produces eggs. The carpel consists of a stigma, style, and ovary. The sticky or hairy stigma, which is specialized for receiving pollen, tops a stalk called the style. At the base of the style is an **ovary**, a chamber that contains one or more ovules. The name angiosperm refers to the fact that seeds form within an ovary. (*Angio–* means enclosed chamber, and *sperma*, seed.) After fertilization, an ovule matures into a seed and the ovary becomes a **fruit**. Chapter 27 discusses the functions of flowers and fruit and their many structural variations in more detail.

Figure 21.15 shows a flowering plant life cycle. Pollen forms inside an anther's pollen sacs. Inside those sacs, diploid cells ❶ give rise to microspores by meiosis ❷. The microspores then develop into pollen grains ❸. A flowering plant ovule forms on the wall of an ovary at the base of a carpel ❹. Meiosis and unequal cytoplasmic division of a diploid cell in an ovule yields one haploid megaspore ❺. The megaspore develops into a female gametophyte, which consists of a haploid egg, a cell with two nuclei, and a few additional cells ❻.

Pollination occurs when a pollen grain arrives on a receptive stigma ❼. The pollen grain germinates, and a pollen tube grows through the style to the ovary at the base of the carpel. Two nonmotile sperm form inside the tube as it grows. Double fertilization occurs when a pollen tube delivers the two sperm into the ovule ❽. One sperm fertilizes the egg to create a zygote. The other fuses with the cell that has two nuclei, forming a triploid (3*n*) cell. After double fertilization, the ovule matures into a seed ❾. The zygote develops into an embryo sporophyte and the triploid cell develops into a triploid tissue called **endosperm** that later serves as a source of nutrients for the developing embryo.

ANGIOSPERM LINEAGES

Water lilies (**Figure 21.16A**) are one of the oldest angiosperm lineages. Most flowering plants belong to one of three later-evolving lineages: magnoliids, eudicots (true dicots), and monocots (**Figure 21.16B–D**). The 9,200 species of magnoliids include avocados as well as magnolias.

The monocots and eudicots derive their group names from the number of seed leaves, or **cotyledons**, in the embryo. **Monocots** are flowering plants with one cotyledon. The 80,000 or so monocot species include palms, lilies, orchids, irises, and grasses such as the many grains we

A Water lily (basal angiosperm).   **B** Magnolia (magnoliid).

C Iris (monocot).   **D** Chickweed (eudicot).

Figure 21.16 Representatives of four angiosperm lineages. The vast majority of angiosperms are monocots or eudicots.

eat. **Eudicots** are flowering plants with two cotyledons. The 170,000 species of eudicots include most familiar broadleaf plants such as tomatoes, cabbages, poppies, and roses, as well as the cacti and all fruit trees. Some eudicots are woody, but no monocots produce true wood. Monocots and dicots also differ in the arrangement of their vascular tissues, number of flower petals, and other traits. We discuss monocot and dicot traits in more detail in Section 25.1.

---

angiosperms (AN-jee-oh-sperms) Highly diverse seed plant lineage; only plants that make flowers and fruits.

carpel (CAR-pull) Floral reproductive organ that produces female gametophytes; a stigma elevated on a style above an ovary.

cotyledon (cot-uh-LEE-dun) Seed leaf of a flowering plant embryo.

endosperm Nutritive tissue in the seeds of flowering plants.

eudicot (YOU-dih-cot) Flowering plant in which the embryo has two cotyledons.

flower Specialized reproductive shoot of a flowering plant.

fruit Mature ovary of a flowering plant; encloses a seed or seeds.

monocot (MAH-no-cot) Flowering plant in which the embryo has one cotyledon.

ovary In flowering plants, the enlarged base of a carpel, inside which one or more ovules form and eggs are fertilized.

stamen (STAY-men) Floral reproductive organ that produces male gametophytes; an anther atop a filament.

Nearly all human crop plants are angiosperms, and it would be nearly impossible to overestimate the ecological importance of this group. As the most abundant plants in most land habitats, angiosperms provide food and shelter for a variety of animals. Their ecological significance is a reflection of their abundance and diversity, which arise from multiple factors.

ACCELERATED LIFE CYCLE

Compared to gymnosperms, most angiosperms have a shorter life cycle. A dandelion or grass can grow from a seed, mature, and produce seeds of its own within a month or so. In contrast, gymnosperms such as pine trees take years to mature. Producing and dispersing seeds fast allows an angiosperm to expand its range faster than a gymnosperm.

PARTNERSHIPS WITH POLLINATORS

Flowers give angiosperms an edge by facilitating animal-assisted pollination. As seed plants evolved, some animals began feeding on their protein-rich pollen. The plants lost a bit of pollen but benefited when the insects inadvertently transferred pollen from one plant to another of the same species. An animal that facilitates pollination by transferring pollen between plants is a **pollinator**. Most pollinators are insects (**Figure 21.17A**), but birds, bats, and other animals fulfill this role for some plants.

Over time, many flowering plants coevolved with their pollinators. (As Section 17.10 explained, coevolution refers to the joint evolution of two or more species as a result of their close ecological interactions.) Producing sugary nectar encourages more pollinator visits, which improves pollination rates and enhances seed production. Conspicuous petals and distinctive scents that attract pollinator attention also provide a selective advantage. So do mutations that alter floral shapes and increase the likelihood that pollen will be transferred onto a pollinator or from a pollinator to a receptive stigma.

ANIMAL-DISPERSED FRUITS

Many angiosperms also benefit by having animals disperse their seeds (**Figure 21.17B**). Some plants make seeds with hooks or spines that can stick to animal fur. Others make brightly colored, fleshy fruits that attract fruit-eating birds or mammals. Such plants benefit when the fruit eater later regurgitates their seeds or passes them unharmed in its feces. Some fruits have features that fend off animals that tend to destroy seeds in the fruits they eat. For example, rodents tend to chew up seeds, so chili pepper fruits contain a compound (capsaicin) that tastes "hot" to rodents and other mammals. Birds, which do not chew seeds and cannot taste the compound, are the main dispersers of chili seeds.

pollinator Animal that moves pollen, thus facilitating pollination.

Figure 21.17 Animal assistants of flowering plants.

A Bees commonly serve as pollinators.

B Fruit-eating birds often disperse seeds.

Figure 21.18 The Svalbard Global Seed Vault. To learn more about the vault and the seeds stored inside it, visit www.croptrust.org.

SAVING SEEDS

Plant diversity is declining. We know of threats to about 12,000 wild plant species, and the actual number of threatened plants is no doubt much higher. As a result, many as yet unrecognized sources of food, medicine, and other products could disappear before we understand their potential value.

A decline in the diversity of cultivated plants raises additional concerns. In the past, food crops were locale-specific. People developed and grew plant varieties that did well in their region, and they saved seeds from their crops to plant the following year. Now, many traditional varieties of crop plants are disappearing as farmers worldwide increasingly rely on the same few large companies for the same strains of seeds. Different varieties of plants are resistant to different diseases,

so widespread planting of one variety increases the risk that a disease could decimate the global supply of a crop. The more varieties of a crop we plant, the more likely it is that some will resist a particular disease.

Sustaining the wild relatives of crop plants provides a form of insurance because it maintains a reservoir of genetic diversity. Plant breeders can draw upon that diversity to meet future challenges.

Storing seeds in seed banks is one way to ensure the survival of potentially useful plants. Today, there are more than 1,500 seed banks around the world. The most ambitious, the Svalbard Global Seed Vault, was built in 2008 on a Norwegian island about 700 miles from the North Pole. This so-called doomsday vault serves as a backup for other seed banks (**Figure**

21.18). The seeds are stored deep inside a mountain, in a permanently chilled, earthquake-free zone 400 feet above sea level. The location was chosen to keep the seeds high and dry even if global climate change causes the polar ice caps to melt. The vault also has an advanced security system and has been engineered to withstand any nearby explosions.

The Svalbard vault now holds the world's most diverse collection of seeds, with more than 750,000 samples and material from nearly every nation. The United States has stored seeds of its native crops such as chili peppers from New Mexico, as well as seeds that American scientists collected elsewhere in the world. Under the deep-freeze conditions in Svalbard, the seeds are expected to survive for about 100 years.

Section 21.1 **Plants** evolved from green algae. They are embryophytes, which means they form a multicelled embryo on the parental body. Key innovations that allowed plants to move into dry habitats include a waterproof **cuticle** perforated by **stomata**, and internal pipelines of vascular tissue (**xylem** and **phloem**) reinforced by **lignin**.

Plant life cycles involve an alternation of generations. Two types of multicelled bodies form: a haploid **gametophyte** and a diploid **sporophyte**. The gametophyte predominates in the life cycle of **nonvascular plants**, but the sporophyte is larger and longer-lived in **vascular plants**. **Seeds** and **pollen grains** that can be dispersed without water are adaptations that contribute to the success of seed plants.

Section 21.2 **Bryophytes** are short, nonvascular plants that disperse by releasing spores. They include three modern lineages: mosses, liverworts, and hornworts. All have a gametophyte-dominated life cycle and produce flagellated sperm that swim through a film of water to eggs. Mosses are the most diverse group. **Rhizoids** attach them to soil or a surface. Remains of some mosses form peat, which is dried and burned as fuel.

Section 21.3 In **seedless vascular plants**, sporophytes have vascular tissues, and they are the larger, longer-lived phase of the life cycle. Typically the sporophyte's roots and shoots grow from a horizontal stem, or **rhizome**. Tiny free-living gametophytes make flagellated sperm that require water for fertilization. Ferns, the most diverse group of seedless vascular plants, produce spores in **sori**. Many ferns grow as epiphytes. Club mosses and horsetails produce spores in conelike structures called **strobili**.

Section 21.4 Forests of giant seedless vascular plants thrived during the Carboniferous period. Later, heat and pressure transformed remains of these forests to **coal**. Seed plants rose to dominance during the Permian, as the climate became cooler and drier. Seed plant sporophytes have **pollen sacs**, where **microspores** form and develop into immature male gametophytes (pollen grains). They also have **ovules**, in which **megaspores** form and develop into female gametophytes. Delivery of pollen to the female parts of a seed plant is **pollination**.

Section 21.5 **Gymnosperms** are seed plants that do not enclose their seeds within a fruit. Conifers, which have woody seed cones, include commercially important groups such as the pines. Their pollen is dispersed by wind, and they are more resistant to drought and to cold than flowering plants. Cycads and ginkgos are gymnosperms that have flagellated sperm and fleshy seeds. Gnetophytes include vines and desert shrubs.

Section 21.6 **Angiosperms** are seed plants that make **flowers** and **fruits**. The **stamens** of a flower produce pollen inside pollen sacs. The **carpel** has a stigma specialized for receiving pollen, atop a stalk called the style. An **ovary** at the base of the carpel holds one or more ovules. After pollination, double fertilization occurs. One sperm delivered by the pollen tube fertilizes the egg and the other fertilizes a cell that has two nuclei. After fertilization, the flower's ovary develops into a fruit that contains one or more seeds. A flowering plant seed includes an embryo sporophyte and **endosperm**, a nutritious tissue. The two main angiosperm lineages, **eudicots** and **monocots**, differ in their number of **cotyledons** and other traits.

Section 21.7 Angiosperms are the most widely distributed and diverse plant group. Accelerated life cycles and partnerships with animal **pollinators** and seed dispersers contributed to their success.

Application Sustaining many varieties of crop plants and their wild relatives ensures that plant breeders will have a reservoir of genetic diversity to tap into if widely planted varieties fail. Seed banks can help us maintain a wide variety of potentially valuable plant species.

SELF-ASSESSMENT
ANSWERS IN APPENDIX VII

1. The first plants were _____ .
   a. ferns   c. bryophytes
   b. flowering plants   d. conifers

2. Which of the following statements is false?
   a. Ferns produce seeds inside sori.
   b. Conifers do not produce fruits.
   c. Both gymnosperms and angiosperms produce pollen.
   d. Only angiosperms produce flowers.

3. In bryophytes, eggs are fertilized in a chamber on the _____ and a zygote develops into a _____ .
   a. sporophyte; gametophyte
   b. gametophyte; sporophyte
   c. sorus; cone

4. Horsetails and ferns are _____ plants.
   a. multicelled aquatic   c. seedless vascular
   b. nonvascular seed   d. seed-bearing vascular

5. Coal consists primarily of compressed remains of the _____ that dominated ancient swamp forests.
   a. bryophytes   c. flowering plants
   b. conifers   d. seedless vascular plants

6. The _____ produce flagellated sperm.
   a. mosses   c. monocots
   b. conifers   d. eudicots

INSECT-ASSISTED FERTILIZATION IN MOSS Plant ecologist Nils Cronberg suspected that crawling insects facilitate fertilization of mosses. To test his hypothesis, he carried out an experiment. He placed patches of male and female moss gametophytes in dishes, either next to one another or with water-absorbing plaster between them so sperm could not swim between plants. He then looked at how the presence or absence of insects affected the number of sporophytes formed. **Figure 21.19** shows his results.

1. Why is sporophyte formation a good way to determine if fertilization occurred?

2. How close did the male and female patches have to be for sporophytes to form in the absence of insects?

3. Does this study support the hypothesis that insects aid moss fertilization?

4. How might a crawling insect aid moss fertilization?

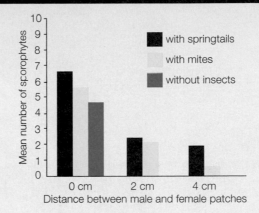

Figure 21.19 Effect of insects on moss fertilization. Two types of crawling insects were used, springtails and mites. No sporophytes formed in the insect-free dishes when moss patches were 2 or 4 centimeters apart.

7. A seed is a _____ .
   a. female gametophyte   c. mature pollen grain
   b. mature ovule   d. modified microspore

8. Spores of a fern or horsetail are _____ .
   a. haploid   b. diploid   c. triploid

9. Only angiosperms produce _____ .
   a. pollen   c. fruits
   b. seeds   d. all of the above

10. The _____ do not have xylem or phloem.
    a. mosses   b. ferns   c. monocots   d. conifers

11. A waxy cuticle helps land plants _____ .
    a. conserve water   c. reproduce
    b. take up oxygen   d. stand upright

12. Pollinators aid many _____ .
    a. conifers   c. angiosperms
    b. mosses   d. ferns

13. _____ produce seeds on woody cones.
    a. Cycads   c. Ginkgos
    b. Conifers   d. Hornworts

14. Match the terms appropriately.
    ___ bryophyte   a. seeds, but no fruit
    ___ seedless   b. flowers and fruits
        vascular plant   c. no xylem or phloem
    ___ gymnosperm   d. xylem and phloem,
    ___ angiosperm   but no ovule

15. Match the terms appropriately.
    ___ anther   a. a type of angiosperm
    ___ monocot   b. horizontal stem
    ___ rhizoid   c. where pollen forms
    ___ sorus   d. attaches moss to soil
    ___ fruit   e. mature ovary
    ___ endosperm   f. nutritive tissue in seed
    ___ rhizome   g. where fern spores form

1. Early botanists admired ferns but found their life cycle perplexing. In the 1700s, they learned to propagate ferns by sowing what appeared to be tiny dustlike "seeds" that they collected from the undersides of fronds. Despite many attempts, the botanists could not locate the pollen source, which they assumed must stimulate these "seeds" to develop. Imagine you could write to these botanists. Compose a note that explains the fern life cycle and clears up their confusion.

2. In animals, meiosis of germ cells produces gametes. Some plant cells also undergo meiosis, but the resulting cells are not gametes. Explain what the products of plant meiosis are, what they develop into, and how plants produce gametes.

3. In most plants the largest, longest-lived body is a diploid sporophyte. By one hypothesis, diploid dominance was favored because it allowed a greater level of genetic diversity. Suppose that a recessive mutation arises. It is mildly disadvantageous now, but it will be useful in some future environment. Explain why such a mutation would be more likely to persist in a fern than in a moss.

4. Consider a cherry pit, which is a seed with a cherry embryo inside it. Trace the paternal ancestry of that embryo. Explain how the sperm that fathered the embryo came to unite with the egg, and the process by which that sperm originated.

5. Suppose you wanted to make a science fiction film in which human time travelers went back to the Carboniferous. What types of modern plants would you use and which would you avoid if you wanted to depict the Carboniferous setting as accurately as possible?

22

Fungi

Links to Earlier Concepts

Before starting, review Figure 20.1 to get a sense of where fungi fit in the eukaryotic family tree. In this chapter you will learn about fungal interactions with cyanobacteria (19.4) and green algae (20.6). You will also draw on your knowledge of the organic molecules chitin (3.2) and lignin (21.1), and your understanding of the processes of nutrient cycling (1.2) and fermentation (7.5).

Core Concepts

The process of evolution is responsible for the diversity and unity of life.

Fungi are eukaryotes that share a protist ancestor with animals. Fungi range in size from single cells to multicelled networks that extend through the soil for miles. They produce spores asexually and sexually. Fungal groups are defined by their mechanism of sexual reproduction. Elaborate spore-producing structures evolved in some groups.

Organisms exchange matter and energy with the environment in order to grow, maintain themselves, and reproduce.

Fungi are heterotrophs and most are decomposers. All fungi meet their nutritional needs by secreting digestive enzymes onto organic matter and absorbing the resulting breakdown products. The porous structure of fungal cell walls ensures that food and water taken up by one part of a fungus can be shared with cells in other regions of the fungal body.

Interactions among the components of each biological system give rise to complex properties.

Populations of different organisms in a biological community interact in complex ways. Nutrients released by fungal decomposers can be taken up by other organisms in the environment. Some fungi partner with algae to form lichens, and others partner with plant roots. Still other fungi are parasites or pathogens that affect the health of their plant or animal hosts.

Fruiting body (spore-producing structure) of a stinkhorn. Flies attracted by the stinkhorn's scent will spread its spores.

CHARACTERISTICS OF FUNGI

A **fungus** (plural, fungi) is a eukaryotic heterotroph that feeds by secreting digestive enzymes onto organic material, then absorbing the resulting breakdown products. **Table 22.1** compares traits of plants, fungi, and animals. Like plants, fungi have cell walls and a life cycle that involves producing spores. Chitin, a nitrogen-containing polysaccharide (Section 3.2) is the main component of fungal cell walls. Like animals, to whom they are more closely related (Sections 20.1 and 20.8), fungi are heterotrophs. Unlike either plants or animals, fungi include both single-celled and multicellular species.

Organization and Structure Single-celled fungi are commonly referred to as **yeasts** (**Figure 22.1A**). We use some yeasts to bake bread and make beer. Others cause human yeast infections. Yeasts most often reproduce asexually by budding, a process in which mitosis is followed by unequal cytoplasmic division. During budding, a descendant cell with a small amount of cytoplasm pinches off from its parent.

Most fungi are multicelled. Powdery mildews (**Figure 22.1B**), shelf fungi (**Figure 22.1C**), and mushrooms are examples. A multicelled fungus consists of a mesh of microscopic threadlike filaments collectively referred to as a **mycelium** (plural, mycelia). Each filament in the mycelium is a **hypha** (plural, hyphae) (**Figure 22.1D**). Depending on the fungal group, there may or may not be cross-walls at regular intervals within a hypha.

Metabolism and Nutrition Most fungi require oxygen, but some species are full-time or part-time anaerobes. For example, yeasts that are used in baking bread and making wine can switch between aerobic respiration and fermentation depending on the availability of oxygen in their environment.

Most fungi are decomposers. **Decomposers** are organisms that feed on organic remains by breaking down molecular components into smaller, absorbable subunits. A lesser number of fungi live on or inside living organisms and draw nourishment from their host.

TABLE 22.1

Comparison of Plants, Fungi, and Animals

Kingdom	Cell Wall	Level of Organization	Mode of Nutrition	Spore Formation	Eukaryotic Supergroup
Plants	Present (cellulose)	All multicelled	Autotrophs (photosynthetic)	Produce spores by meiosis	Archaeplastids
Fungi	Present (chitin)	Most multicelled, some single-celled	Heterotrophs (external digestion and absorption)	Produce spores by meiosis and mitosis	Opisthokonts
Animals	Absent	All multicelled	Heterotrophs (ingest and digest)	Do not produce spores	Opisthokonts

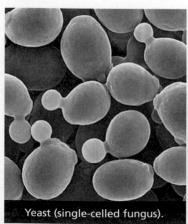
Yeast (single-celled fungus).

Powdery mildew on leaves.

Shelf fungus on a tree trunk.

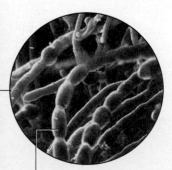
one cell of a hypha in the mycelium

D Mycelium of a club fungus.

Figure 22.1 A sampling of fungal forms.

CREDITS: (1A) © Dr. Dennis Kunkel/Visuals Unlimited; (1B) Nigel Cattlin/Science Source; (1C) © Robert C. Simpson/Nature Stock; (1D) © Garry T. Cole, University of Texas, Austin/BPS.

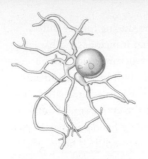

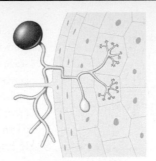

A Chytrids. Mostly aquatic, only group with flagellated spores.

B Zygote fungi. Mostly molds, some parasites.

C Glomeromycetes. Soil fungi with hyphae that branch inside root cells.

D Sac fungi. Most diverse group. Yeasts, molds, parasites, species that produce fruiting bodies.

E Club fungi. Mushrooms, other species that produce fruiting bodies, some plant pathogens.

Figure 22.2 Major fungal groups. Graphics show examples of spore-bearing structures.

Spore Production Fungi disperse by producing spores. Spores may be produced asexually by mitosis or sexually by meiosis. Fungal spores can be single-celled or multicellular, but all are microscopic. With the exception of one group, spores are nonmotile.

MAJOR GROUPS

The five major subgroups of fungi are defined largely by the structure of their spores and how those spores form.

Chytrids include the oldest fungal lineages, and are the only modern fungi that make flagellated spores (**Figure 22.2A**). Most chytrids are aquatic decomposers, but some live inside plants or animals. Chytrids usually reproduce asexually, producing spores at the tips of hyphae.

Zygote fungi (zygomycetes) produce thick-walled resting spores during sexual reproduction, which we describe in detail in Section 22.2. However, zygote fungi most often exist as a **mold**, a mass of hyphae that grows rapidly over organic material and produces spores asexually at the tips of specialized hyphae (**Figure 22.2B**).

Glomeromycetes are soil fungi that form a mutually beneficial partnership with plant roots. Their hyphae extend inside the cell wall of plant root cells. Glomeromycetes produce spores asexually at the tips of hyphae that extend out of the host's root (**Figure 22.2C**).

Hyphae of chytrids, zygote fungi, and glomeromycetes do not have regular cross-walls. In these groups, a hypha is a chitin-reinforced tube filled with cytoplasm and many nuclei. By contrast, hyphae of sac fungi and club fungi, which we will discuss next, have cross-walls between cells. The cross-walls are porous, so nutrients and water taken up in one part of a mycelium can be shared with cells elsewhere in the mycelium.

The presence of cross-walls makes hyphae stronger and less vulnerable to drying out. In some sac fungi and club fungi, a fruiting body composed of many intertwined hyphae forms during sexual reproduction. Specialized cells on the fruiting body produce spores by meiosis.

Sac fungi (ascomycetes), the most diverse fungal group, reproduce sexually by producing spores in sac-shaped cells. In some species such as morels, these cells form on a fruiting body (**Figure 22.2D**). Sac fungi include yeasts, molds, parasitic species, and species that partner with photosynthetic cells to form lichens.

Club fungi (basidiomycetes) reproduce sexually by producing spores on club-shaped cells. Mushrooms are the fruiting bodies of club fungi (**Figure 22.2E**). Shelf fungi (**Figure 22.1C**), puffballs, and stinkhorns such as the one shown in the chapter opening photo are club fungi too. The group also includes plant pathogens such as smuts and rusts.

---

chytrids (KIH-trid) Fungi with flagellated spores.

club fungi Fungi that produce spores on club-shaped structures during sexual reproduction.

decomposer Heterotroph that feeds on organic remains by breaking them down into smaller, absorbable subunits.

fungus Single-celled or multicellular eukaryotic heterotroph that digests food outside the body, then absorbs the resulting breakdown products. Has chitin-containing cell walls.

glomeromycetes (gloh-mere-oh-MY-seats) Fungi that partner with plant roots; their hyphae grow inside cell walls of root cells.

hypha (HI-fah) One filament in a fungal mycelium.

mold A mass of fungal hyphae that grows rapidly over organic material and produces spores asexually.

mycelium (my-SEAL-ee-um) Mass of threadlike filaments (hyphae) that make up the body of a multicelled fungus.

sac fungi Fungi that form spores in a sac-shaped structure during sexual reproduction. Most diverse fungal group.

yeast Single-celled fungus.

zygote fungi (ZY-goat) Fungi that form a thick-walled zygospore during sexual reproduction.

Explain how sexual reproduction of a multicelled fungus differs from that of a plant or animal.

Describe the sexual and asexual reproduction of bread mold.

Explain the role of mushrooms in the life cycle of club fungi.

Draw haploid, diploid, and dikaryotic fungal cells showing the nuclei.

Many fungi have a life cycle that involves both asexual and sexual reproduction. During asexual reproduction, multicelled fungi form spores by mitosis at the tips of specialized hyphae.

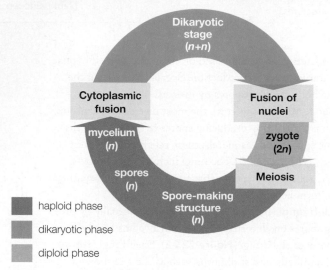

Figure 22.3 Sexual reproduction cycle for a multicellular fungus.

The life cycle of a sexually reproducing fungus differs from that of a plant or animal in that it includes a dikaryotic stage (**Figure 22.3**). A **dikaryotic** structure contains two genetically distinct types of nuclei ($n+n$). In plants and animals, the nuclei of egg and sperm fuse immediately after their cytoplasms fuse. In fungi, fusion of the cytoplasm of two partners is not immediately followed by fusion of their nuclei. Instead, the nuclei remain separate for a while. The length of this dikaryotic stage differs among fungal groups, as you will see below. When the nuclei do fuse, the result is a diploid zygote that undergoes meiosis to produce haploid spores. Each spore can germinate and grow into a new mycelium.

ZYGOTE FUNGI

Black bread mold (*Rhizopus stolonifer*) is a zygote fungus. When food is plentiful, it grows as a haploid mycelium and produces spores by mitosis (**Figure 22.4 ❶**). If the food supply dwindles and a compatible partner is nearby, both partners extend special hyphae toward one another. As each of these hyphae grows, a wall forms behind its tip, isolating multiple haploid nuclei in a structure called a gametangium ❷. When two gametangia come into contact, their cytoplasm fuses. The result is an immature, dikaryotic zygospore (a zygospore with multiple nuclei from each parent) ❸. Nuclei within the zygospore then pair up and fuse. The result is a mature diploid zygospore that develops a thick protective coat ❹. When the zygospore later germinates, it undergoes meiosis and the hypha that emerges produces haploid spores at its tip ❺.

Figure 22.4 Life cycle of black bread mold, a zygote fungus.

❶ A haploid mycelium grows in size and produces spores by mitosis.

❷ When nutrients are limited, two individuals produce special hyphae that grow toward one another. As these hyphae develop, a wall forms behind their tips, isolating multiple nuclei inside each of the resulting gametangia.

❸ Cytoplasmic fusion of the gametangia produces a zygospore that contains many haploid nuclei from each parent.

❹ Nuclei within the zygospore pair up and fuse to produce a mature zygospore with many diploid nuclei. A thick protective wall develops around the zygospore.

❺ The zygospore undergoes meiosis and germinates. A hypha emerges and produces spores, each capable of giving rise to a new haploid mycelium.

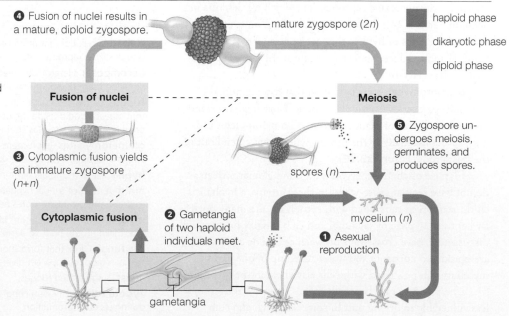

❹ Fusion of nuclei results in a mature, diploid zygospore. — mature zygospore ($2n$)

Fusion of nuclei

❸ Cytoplasmic fusion yields an immature zygospore ($n+n$)

Cytoplasmic fusion

Meiosis

❺ Zygospore undergoes meiosis, germinates, and produces spores.

spores (n)

❷ Gametangia of two haploid individuals meet.

mycelium (n)

❶ Asexual reproduction

gametangia

haploid phase

dikaryotic phase

diploid phase

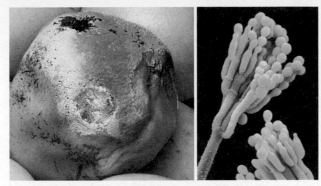  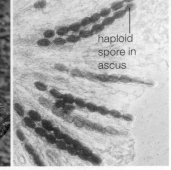

A *Penicillium* growing on grapefruit (left) reproduces asexually by producing mitotic spores atop specialized hyphae (right).

B Cup fungi that grow on logs reproduce sexually by producing an ascocarp (left). Spores (right) form by meiosis of cells on the cup's concave surface.

Figure 22.5 Reproduction in sac fungi.

SAC FUNGI

Sac fungi that grow as molds reproduce mainly asexually; they produce haploid spores by mitosis at the tips of specialized hyphae (**Figure 22.5A**). Other sac fungi such as morels and cup fungi reproduce sexually by means of a fruiting body called an ascocarp (**Figure 22.5B**). Spores form in a saclike cell (an ascus) on the fruiting body. This cell, which is initially dikaryotic, undergoes fusion of nuclei, followed by meiosis, then mitosis. The result is eight haploid spores.

CLUB FUNGI

The most prolonged dikaryotic phase occurs in club fungi (**Figure 22.6**). When a spore of a mushroom-forming club fungus germinates, it grows into a haploid mycelium ❶. If hyphae of two compatible individuals meet, their cytoplasms will fuse, giving rise to a dikaryotic mycelium ❷. This mycelium grows through the soil unseen, sometimes for years, and may reach great size. Mycelia of some mushroom-forming fungi extend for miles. Embryonic mushrooms form on the mycelium and, when it rains, these tiny mushrooms expand and break through the soil surface ❸. Producing fruiting bodies after a rain ensures that spores will disperse when conditions favor their survival.

The underside of a mushroom's cap has thin tissue sheets (gills) fringed with club-shaped dikaryotic cells. Fusion of the nuclei in these cells forms a diploid zygote ❹. The zygote undergoes meiosis, forming four haploid spores ❺. After these spores are released from the fruiting body, they will germinate and begin the life cycle again.

dikaryotic (die-kary-OTT-ik) Having two genetically distinct nuclei (*n+n*).

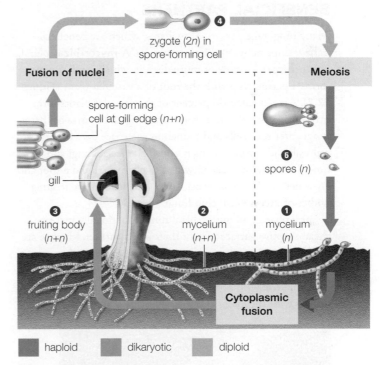

Figure 22.6 Life cycle of a button mushroom, a club fungus.

❶ A haploid spore gives rise to a haploid mycelium. (Red dots and blue dots represent genetically distinct nuclei.)

❷ Cytoplasmic fusion of two haploid mycelia produces a dikaryotic cell that gives rise to a dikaryotic mycelium.

❸ A fruiting body (mushroom) composed of dikaryotic hyphae develops.

❹ Fusion of nuclei in cells at edge of a mushroom gill produces zygotes.

❺ Meiosis of a zygote produces haploid spores.

FIGURE IT OUT
What process produces a dikaryotic cell from two haploid cells?

Answer: Cytoplasmic fusion

LEARNING OBJECTIVES

Using appropriate examples, describe the ways that fungi can benefit and harm plants and animals.

Describe the effects of two types of fungal infections common in humans.

NATURE'S RECYCLERS

Fungi provide an important ecological service by breaking down complex compounds in organic wastes and remains. Some soluble nutrients released by this process enter soil or water. Plants and other producers can then take up the nutrients to meet their own needs. Bacteria also serve as decomposers, but they tend to grow mainly on surfaces. By contrast, fungal hyphae can extend deep into a decaying log or other bulky food source.

BENEFICIAL PARTNERS

Many fungi take part in mutualisms, which are beneficial partnerships between different species. A **mycorrhiza** (plural, mycorrhizae) is a type of mutualism in which a soil fungus forms a partnership with the root of a vascular plant (**Figure 22.7**). An estimated 80 percent of the vascular plants form mycorrhizae with a glomeromycete fungus. Hyphae of these fungi enter root cells and branch in the space between the cell wall and the plasma membrane. Some sac fungi and club fungi form mycorrhizae in which hyphae surround a root and grow between its cells. Most forest mushrooms are fruiting bodies of mycorrhizal club fungi.

Hyphae of all mycorrhizal fungi functionally increase the absorptive surface area of their plant partner. Hyphae are thinner than even the smallest roots and can grow between soil particles. The fungus shares water and nutrients taken up by its hyphae with root cells. In return, the plant supplies the fungus with sugars.

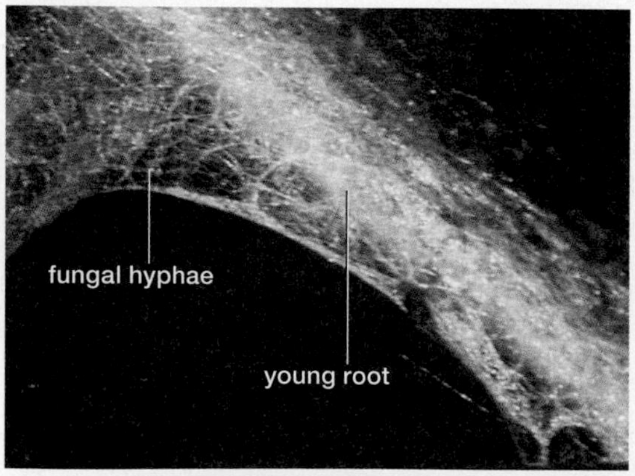

A **lichen** is a composite organism that consists of a sac fungus and either cyanobacteria or green algae (**Figure 22.8A**). The fungus makes up the bulk of a lichen's mass, with its hyphae providing structural support to the photosynthetic cells. These cells provide the fungus with sugars and, if the cells are cyanobacteria, with fixed nitrogen (Section 19.4). Lichens play an important ecological role by colonizing places too hostile for most organisms, such as exposed rocks (**Figure 22.8B**). They break down the rock and produce soil by releasing acids and by holding water that freezes and thaws. When soil forms, plants can move in and take root. Long ago, lichens may have preceded plants onto land.

Fungal partners also enhance the nutrition of some animals. Chytrid fungi that live in the stomachs of grazing hoofed mammals such as cattle, deer, and moose aid their hosts by breaking down otherwise indigestible cellulose. Similarly, fungal partners of some ants and termites serve as an external digestive system. Leaf-cutter ants gather bits of leaves to feed a fungus that lives in their nest. The ants cannot digest leaves, but they do eat some of the fungus.

PARASITES AND PATHOGENS

Many sac fungi and club fungi are plant parasites. Powdery mildews (sac fungi) and rusts and smuts (club fungi) are parasites that grow in living plants. Their hyphae extend into cells of stems and leaves, where they suck up photosynthetically produced sugars. The resulting loss of nutrients stunts the plant, minimizes seed production, and may eventually kill it. However, the plant usually does not die before the fungus has produced spores on the surface of its infected parts.

Other fungi are pathogens that kill plant tissue with their toxins, then feed on the remains. Consider the club fungus *Armillaria*, which causes root rot by infecting trees and woody shrubs in forests worldwide. Once an infected

Figure 22.7 Mycorrhiza. This photo shows a fungus partnered with the root of a hemlock tree.

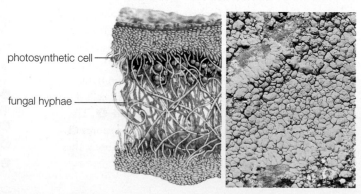

photosynthetic cell

fungal hyphae

A Structure of a leafy lichen.

B Lichen growing on a rock.

Figure 22.8 Lichens. A lichen is a composite organism that consists of a fungus and either green algal cells or cyanobacteria.

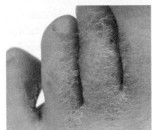
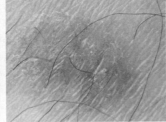

A Remains of a "zombie fly." A fly infected by the zygote fungus *Entomophthora muscae* ceases normal activities, climbs to the top of a stem, and clings there with outstretched wings. It soon dies, and spore-bearing hyphae erupt through its abdominal wall to shed spores that can cause new infections.

B Athlete's foot.   **C** Ringworm.

Figure 22.9 Fungi as parasites.

tree dies, the fungus decomposes the stumps and logs left behind. In one Oregon forest, the mycelium of a single honey mushroom (*A. solidipes*) extends across nearly 4 square miles (10 km²). It has been growing for an estimated 2,400 years.

Many more fungi infect plants than animals. Among animals, those that do not maintain a high body temperature are most vulnerable to fungal infections. Hundreds of fungal species infect insects, and some turn their hosts into seeming zombies, the better to disperse their spores (**Figure 22.9A**).

Most human fungal infections involve sac fungi on body surfaces. Infected areas are raised, red, and itchy. For example, several species of sac fungi infect skin of the feet, causing "athlete's foot" (**Figure 22.9B**). Fungal vaginitis (a vaginal yeast infection) occurs when a sac fungal yeast (*Candida*) that normally lives in the vagina in low numbers undergoes a population explosion. Sac fungi also cause skin infections misleadingly known as "ringworm." No worm is involved. A ring-shaped lesion forms as fungal hyphae grow out from the

Dr. DeeAnn Reeder

DeeAnn Reeder is an expert on bats, with a particular interest in species that hibernate. As a National Geographic Explorer, she has documented bat diversity in the new country of South Sudan. She is also part of a team investigating white nose syndrome, a bat disease first described in New York in 2006. By early 2016, it had been discovered in 26 states and five Canadian provinces, and it had killed millions of North America's bats.

Bats with white nose syndrome fly when they should be hibernating, lose weight, and have fuzzy white filaments of the fungus *Pseudogymnoascus destructans* on their wings, ears, and muzzle. Most fungal infections do not kill mammals, and fungi often infect animals weakened by other pathogens. Thus, Reeder and other biologists initially assumed the fungus was a symptom of a disease, rather than its cause. However, when they experimentally infected healthy bats with *P. destructans*, they found the fungus alone is sufficient to cause the disease. Reeder and others are now working to understand why the fungus has such deadly effects, which types of bat colonies are most susceptible to infection, and what can be done to halt the spread of the fungus.

initial infection site (**Figure 22.9C**). Life-threatening fungal infections are rare and most often occur in people whose immune response is weak because of other factors.

lichen (LIE-kin) Composite organism consisting of a fungus and green algae or cyanobacteria.
mycorrhiza (my-kuh-RYE-zah) Mutually beneficial partnership between a fungus and a plant root.

22.4 Human Uses of Fungi

Many fungal fruiting bodies serve as human food. Button mushrooms, shiitake mushrooms, and oyster mushrooms decompose organic matter, so these species are easily cultivated. By contrast, edible mycorrhizal fungi such as chanterelles, porcini mushrooms, morels, and truffles need a living plant partner, so they are typically gathered from the wild. Picking wild mushrooms is best done with the aid of someone experienced in identifying species. Each year thousands of people become ill after eating poisonous mushrooms they mistook for edible ones.

An interesting dispersal strategy has evolved in truffle-producing fungi, which are highly prized by gourmets. Truffles are the fruiting bodies of some mycorrhizal fungi that partner with tree roots. Truffles form underground near their host trees and, when mature, produce an odor similar to that of an amorous male wild pig. Female wild pigs detect the scent and root through the soil in search of their seemingly subterranean suitor. When they unearth the truffles, they eat them, then disperse fungal spores in their feces. Human truffle hunters usually rely on trained dogs to sniff out the fungi.

Fungal fermentation reactions helps us make a variety of products. Fermentation by one mold (*Aspergillus*) helps make soy sauce. Another mold (*Penicillium*) produces the tangy blue veins in cheeses such as Roquefort (**Figure 22.10**). A packet of baker's yeast holds spores of the sac fungus *Saccharomyces cerevisiae*. When these spores germinate in bread dough, they ferment sugar and produce carbon dioxide that causes the dough to rise. Other strains of *S. cerevisiae* are used to make wine and beer, and also to produce ethanol biofuel.

Geneticists and biotechnologists also make use of yeasts. Like *E. coli* bacteria, yeasts grow readily in laboratories and they have the added advantage of being eukaryotes, like us. Checkpoint genes that regulate the eukaryotic cell cycle (Section 11.1) were first discovered in the yeast *S. cerevisiae*. This discovery was the first step toward our current understanding of how mutations in these genes cause human cancers. Genetically engineered *S. cerevisiae* and other yeasts are now used to produce proteins that serve as vaccines or medicines.

Some naturally occurring fungal compounds have medicinal or psychoactive properties. The initial source of the antibiotic penicillin was a soil fungus (*Penicillium chrysogenum*). Another soil fungus gave us cyclosporin, an immune suppressant used to prevent rejection of transplanted organs. Ergotamine, a compound used to relieve migraines, was first isolated from ergot (*Claviceps purpurea*), a club fungus that infects rye plants. Ergotamine is also used in synthesis of the hallucinogen LSD.

Another hallucinogen, psilocybin, is the active ingredient in so-called "magic mushrooms." Psilocybin affects perception by increasing communication among different regions of the brain. Users often describe the resulting experience as mystical; they report altered sensations accompanied by feelings of joy, openness, and connectedness. A minority of users have increased fear and anxiety. Researchers recently began investigating whether medically supervised use of psilocybin can help treat some psychological disorders and conditions. Preliminary studies indicate that doctor-supervised use of psilocybin can lessen symptoms of depression and post-traumatic stress disorder (PTSD), relieve the anxiety of people who have been diagnosed with fatal cancers, and assist in overcoming an addiction to tobacco or alcohol. Possession of psilocybin-containing mushrooms outside of a medical setting remains illegal in the United States.

Figure 22.10 Fungi as food. Sac fungi help us produce breads, wine, and blue cheeses.

Figure 22.11 Frogs killed by the chytrid fungus Bd at a high-elevation lake in California's Sierra Nevada.

SPREAD OF FUNGAL PATHOGENS

The dispersal of fungal pathogens by global trade and travel can have devastating effects on ecosystems. In the early twentieth century, a plant-infecting sac fungus native to China was introduced to North America in imported plants. The fungus caused a chestnut blight that eliminated all mature American chestnut trees. Some American chestnuts persist as root systems, but the blight always kills them before they mature and reproduce.

Today, human-facilitated spread of a fungal pathogen is among the foremost causes of an amphibian extinction crisis. The fungus, *Batrachochytrium dendroba-* *tidis*, is a chytrid known as Bd. It is native to Africa, where it infects African clawed frogs without sickening them.

International trade in African clawed frogs began in 1934, when scientists discovered the frogs could be used in pregnancy tests. If a female frog is injected with blood from a pregnant woman, hormones in her blood cause the frog to lay eggs. Until the 1950s, this frog-based test was the main method of determining pregnancy, so millions of African clawed frogs were exported for this purpose. Unfortunately, the Bd chytrid traveled with them. It became established in new regions when people released infected frogs or dumped water with Bd spores into the environment. Bd now occurs on all continents except Antarctica, and trade in amphibians continues to introduce it into previously uninfected habitats.

The effect of Bd varies among amphibian species. Many frogs outside of Africa have little or no resistance to Bd, so their death rate from infection is extremely high (**Figure 22.11**). Growth of the fungus on an amphibian's skin causes the skin to thicken. Thickened skin prevents the frog from absorbing water properly, so it eventually dies of dehydration.

Section 22.1 **Fungi** are heterotrophs, mainly **decomposers**, that secrete digestive enzymes onto organic matter and absorb released nutrients. Their cell walls include the compound chitin and they disperse by releasing spores. In a multicelled species, spores germinate and give rise to filaments called **hyphae**. The filaments typically grow as an extensive meshlike **mycelium**. Water and nutrients move freely between cells of a hypha.

The oldest fungal lineage, the **chytrids**, are a mostly aquatic group and the only fungi with flagellated spores. **Zygote fungi** include many familiar **molds** that grow on fruits, breads, and other foods. **Glomeromycetes** live in soil and extend their hyphae into plant roots. **Sac fungi** are the most diverse fungal group. They include single-celled **yeasts**, molds, and species that produce multicelled fruiting bodies. Most familiar mushrooms are fruiting bodies of **club fungi**. Multicelled sac fungi and club fungi can form large fruiting bodies because their hyphae, unlike those of other fungi, have cross-walls between the cells. These walls reinforce the structure of a hypha.

Section 22.2 Depending on the group, the cells of a hypha may be haploid (*n*) or **dikaryotic** (*n*+*n*). Mechanisms of spore formation differ among fungal groups. Zygote fungi such as bread molds usually reproduce asexually by producing spores atop specialized hyphae until food runs out. Then hyphae from two individuals fuse to produce a diploid zygospore that undergoes meiosis and releases haploid spores.

During sexual reproduction, a sac fungus produces spores in a sac-shaped cell (an ascus). In some sac fungi, including the cup fungi, the asci form on a multicelled fruiting body (an ascocarp).

Club fungi produce spores on club-shaped structures on their fruiting bodies. Typically, a dikaryotic mycelium grows by mitosis. When conditions favor reproduction, a fruiting body made of dikaryotic hyphae develops. A mushroom is an example. Fusion of nuclei in dikaryotic club-shaped cells at the edges of gills produces diploid cells. Meiosis in these cells produces haploid spores.

Section 22.3 Fungi have many important roles in ecosystems. Most are decomposers. By breaking down organic wastes and remains, they free up nutrients that had been tied up in these materials. Producers use these nutrients.

A **lichen** is a composite organism composed of a fungus and photosynthetic cells of a green alga or cyanobacterium. The fungus, which makes up the bulk of the lichen body, obtains a supply of nutrients from its photosynthetic partner. Lichens are important pioneers in new habitats because they facilitate the breakdown of rock to form soil.

A **mycorrhiza** is an interaction between a fungus and a plant root. Fungal hyphae penetrate roots and supplement their absorptive surface area. The fungus shares absorbed nutrients with the plant and obtains photosynthetically produced sugars in return.

Fungi that parasitize plants either insert their hyphae into plant cells to steal sugars, or kill plant cells and absorb the released nutrients. Fungi parasitize animals less frequently than they do plants. Animals with a low body temperature, such as insects, are the most likely to be targeted.

In humans, fungal infections tend to occur on body surfaces, as when they cause athlete's foot or a vaginal yeast infection. Typically, fungi cause life-threatening infections only in people whose immune response is impaired.

Section 22.4 Many fungal fruiting bodies are edible, although some produce dangerous toxins. Fungi that carry out fermentation reactions help us produce food products, alcoholic beverages, and ethanol for use as a biofuel.

Fungi also are used in scientific studies and to produce medicines. Study of yeasts can provide insights into eukaryotic genetics. Recombinant yeasts produce vaccines and other desired proteins. Some compounds extracted from fungi are useful as medicines or psychoactive drugs.

Application Human activities have introduced some fungal pathogens to new environments where they have negative effects on the health of other organisms. One sac fungus introduced to North America wiped out American chestnuts. Today, a chytrid fungus native to Africa threatens amphibian species worldwide.

SELF-ASSESSMENT
ANSWERS IN APPENDIX VII

1. All fungi _____ .
   a. are multicelled   c. are heterotrophs
   b. form flagellated spores   d. all of the above

2. Most fungi obtain nutrients from _____ .
   a. wastes and remains   c. living animals
   b. living plants   d. photosynthesis

3. A fungus that usually lives as a mass of haploid hyphae and reproduces asexually is called a _____ .
   a. lichen   c. yeast
   b. mold   d. mushroom

4. A _____ steals sugars from a living plant cell.
   a. rust or smut   c. lichen
   b. mushroom   d. mold

5. In many _____ , a dikaryotic mycelium is the largest and longest-lived component of the life cycle.
   a. chytrids   c. sac fungi
   b. zygote fungi d. club fungi

6. A _____ produces spores by meiosis in an ascus.
   a. chytrid   c. sac fungus
   b. zygote fungus d. club fungus

FIGHTING A FOREST FUNGUS The club fungus *Armillaria solidipes*, commonly known as the honey mushroom, infects living trees and withdraws nutrients from them. After an infected tree dies, the fungus continues to feed on its remains. Fungal hyphae grow out from the roots of infected trees as well as the roots of dead stumps. If these hyphae contact roots of a healthy tree, the hyphae can invade these roots and cause a new infection.

Canadian forest pathologists hypothesized that removing fungus-infected stumps after infected trees were logged could help prevent additional tree deaths. To test this hypothesis, they carried out an experiment. In one region of a forest they removed stumps after logging. In another, they left stumps behind as a control. For more than 20 years, they recorded tree deaths and whether the deaths resulted from infection by *A. solidipes* or some other cause. **Figure 22.12** shows the results.

1. Which tree species was most often killed by *A. ostoyae* in control forests? Which was least affected by the fungus?

2. For the most-affected species, what percentage of deaths did *A. solidipes* cause in control and in experimental regions?

3. Looking at the overall results, do the data support the hypothesis? Does stump removal reduce tree mortality from *A. solidipes*?

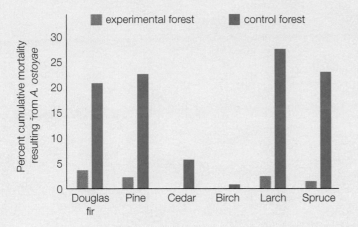

Figure 22.12 Effect of logging practices on tree deaths caused by the fungus *A. solidipes*. In the experimental forest, whole trees—including stumps—were removed (brown bars). The control half of the forest was logged conventionally, with the stumps left behind (blue bars).

---

7. A mushroom is _____ .
   a. the digestive organ of a club fungus
   b. the only part of the fungal body made of hyphae
   c. a reproductive structure that releases sexual spores
   d. the only diploid phase in the club fungus life cycle

8. Spores released from a mushroom's gills are _____ .
   a. dikaryotic   c. flagellated
   b. diploid   d. produced by meiosis

9. _____ by yeasts helps us produce bread, soy sauce, and ethanol as a biofuel.
   a. Photosynthesis   c. Decomposition
   b. Fermentation   d. Budding

10. A _____ helps to break down rocks and form soil.
    a. chytrid   c. mycorrhiza
    b. glomeromycete   d. lichen

11. _____ are mycorrhizal fungi with hyphae that grow into a root cell and branch inside it.
    a. Glomeromycetes   c. Zygote fungi
    b. Chytrids   d. Club fungi

12. A truffle is an example of a _____ .
    a. spore   c. mycorrhizal fungus
    b. lichen   d. plant pathogen

13. Frogs worldwide are now threatened by a pathogenic _____ native to Africa.
    a. glomeromycete   c. zygote fungus
    b. chytrid   d. club fungus

14. Human fungal infections usually _____ .
    a. alter brain function   c. are deadly
    b. involve the skin   d. are caused by club fungi

15. Match the terms appropriately.
    ___ hypha   a. produces flagellated spores
    ___ chitin   b. component of fungal cell walls
    ___ chytrid   c. filament of a mycelium
    ___ zygote fungus d. fungus–root partnership
    ___ club fungus   e. bread mold is an example
    ___ lichen   f. many form mushrooms
    ___ mycorrhiza   g. single-celled fungus
    ___ yeast   h. partnership between a fungus
                          and photosynthetic cells

CLASS ACTIVITY
CRITICAL THINKING

1. Developing antifungal drugs is more difficult than developing antibacterial drugs. Compounds that harm fungi more frequently cause side effects in humans than compounds that harm bacteria. Explain why this is the case, given the evolutionary relationships among bacteria, fungi, and humans.

2. Bakers who want to be sure their yeast is alive "proof" it. They test the yeast's viability by putting a bit of it in warm water with some sugar. If the yeast is alive, bubbles will appear in the mix. What gas in the bubbles and how is it formed?

3. Molds usually reproduce sexually when food is running low. Why might sexual reproduction be more advantageous at this time than when food remains plentiful?

for additional quizzes, flashcards, and other study materials
▶ **ACCESS MINDTAP AT WWW.CENGAGEBRAIN.COM**

23 Animals I: Major Invertebrate Groups

Links to Earlier Concepts

This chapter draws on your knowledge of animal tissues and organs (Section 1.1) and of homeotic genes (10.2), fossils (16.3, 16.4), analogous structures (16.7), and speciation (17.8). You will learn about another use of chitin (3.2) and more effects of osmosis (5.6). You will learn how animals interact with dinoflagellates (20.4), protists that cause malaria (20.5), and flowering plants (21.7).

Core Concepts

All organisms alive today are linked by lines of descent from shared ancestors.

Shared core processes and features provide evidence that all living things are descended from a common ancestor. All animals have unwalled cells connected by cell junctions and a collagen-containing extracellular matrix. Animals evolved from a colonial protist ancestor by about 600 million years ago.

Organisms exchange matter and energy with the environment in order to grow, maintain themselves, and reproduce.

Animals are heterotrophs that ingest and digest their food. In the simplest animals, the gut is saclike and nutrients diffuse through tissues. Animals with more complex body plans have a tubular gut and some have a circulatory system. All animals undergo development and move from place to place during part or all of the life cycle.

Living things sense and respond appropriately to their internal and external environments.

Homeostatic mechanisms that involve feedback control allow living things to maintain themselves by responding dynamically to internal and external conditions. Many animals have organ systems that allow them to control the solute level of their body fluids. Most have sensory organs that alert them to changes in their environment. All move from place to place during part or all of the life cycle.

◀ Sea stars, sea anemones, and pink encrusting sponges in a Pacific Northwest tide pool. All are invertebrates.

Photograph by George Grall, National Geographic Creative.

23.1 Animal Traits and Trends

LEARNING OBJECTIVES

List some traits shared by all animals.

Using appropriate examples, explain the difference between a body plan with radial symmetry and one with bilateral symmetry.

Explain the trait that defines the two subgroups of bilateral animals and gives them their names.

Draw cross-sections of animals with and without a coelom.

SHARED TRAITS

Animals (kingdom Animalia or clade Metazoa) are multi-celled heterotrophs that ingest food (take it into their body). All consist of multiple types of unwalled cells held together by tight junctions and adhering junctions, and by a secreted extracellular matrix that contains the protein collagen. (Cell junctions and extracelluar matrix are described in Section 4.10.) Most animals are diploid. Gametes (immobile eggs and flagellated sperm) are the only haploid components in their life cycle. Cells become specialized as an animal develops from an embryo (an individual in the earliest stage of development) to an adult. All animals are motile (move from place to place) during part or all of their life.

BODY PLAN VARIATIONS

This chapter describes major groups of **invertebrates**, animals that do not have a backbone. The next chapter focuses on vertebrates (animals with a backbone) and their closest

invertebrate relatives. Although invertebrates are typically smaller and structurally simpler than vertebrates, they are just as well adapted to their environments. Some invertebrate lineages have survived for more than 500 million years, and about 95 percent of animals are invertebrates.

Figure 23.1 shows an evolutionary tree for the major animal groups covered in this book, and we will use it as a guide to discuss evolutionary trends. All animals are multicellular ❶ and constitute the clade Metazoa. The earliest animals were probably aggregations of cells, and this level of organization persists in sponges. However, most other modern animals have cells organized as tissues ❷.

Tissue organization begins in animal embryos. Embryos of jellies and other cnidarians have two tissue layers: an outer **ectoderm** and an inner **endoderm**. In other modern animals, embryonic cells typically rearrange themselves to form a middle tissue layer called **mesoderm**:

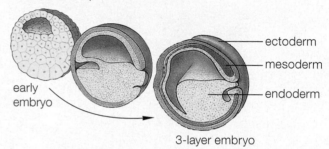

early embryo

3-layer embryo

— ectoderm
— mesoderm
— endoderm

Evolution of a three-layer embryo allowed an important increase in structural complexity. Most internal organs in animals develop from embryonic mesoderm.

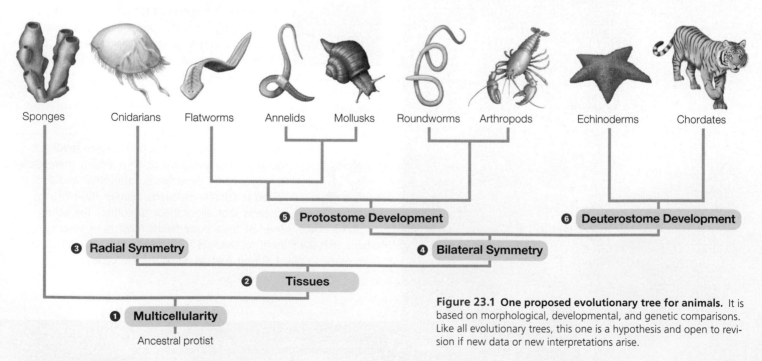

Sponges　Cnidarians　Flatworms　Annelids　Mollusks　Roundworms　Arthropods　Echinoderms　Chordates

❺ **Protostome Development**　❻ **Deuterostome Development**

❸ **Radial Symmetry**　❹ **Bilateral Symmetry**

❷ **Tissues**

❶ **Multicellularity**

Ancestral protist

Figure 23.1 One proposed evolutionary tree for animals. It is based on morphological, developmental, and genetic comparisons. Like all evolutionary trees, this one is a hypothesis and open to revision if new data or new interpretations arise.

Animals with the simplest structural organization are asymmetrical, meaning their body cannot be divided into halves that are mirror images. Jellies, sea anemones, and other cnidarians have **radial symmetry**: Their body parts are repeated around a central axis, like spokes of a wheel ❸. Radial animals attach to an underwater surface or drift along. A radial body plan allows them to capture food that can arrive from any direction. Animals with a three-layer body plan typically have **bilateral symmetry**: The body's left and right halves are mirror images ❹. Bilateral animals such as worms,

snails, and lobsters have a distinctive head end with a concentration of nerve cells, and typically move headfirst through their environment.

The two lineages of bilateral animals are defined in part by developmental differences. In **protostomes**, the first opening that appears on the embryo becomes the mouth ❺. *Proto–* means first and *stoma* means opening. In **deuterostomes**, the mouth develops from the second embryonic opening ❻. *Deutero–* mean second.

Some animals digest food in a saclike cavity with a single opening, but most have a tubular gut, with a mouth at one end and an anus at the other. Parts of the tube are typically specialized for taking in food, digesting it, absorbing nutrients, or compacting wastes. A tubular gut can carry out these tasks simultaneously, whereas a saclike cavity cannot.

The gut of a flatworm is surrounded by a mass of tissues and organs (**Figure 23.2A**). Most other animals have a "tube within a tube" body plan. Their gut runs through a fluid-filled body cavity. In roundworms, this cavity is partially lined with tissue derived from mesoderm and is called a **pseudocoelom** (**Figure 23.2B**). By contrast, most bilateral animals have a **coelom**, a body cavity that is fully lined with tissue derived from mesoderm (**Figure 23.2C**). Sheets of tissue called mesentery suspend the gut in the center of a coelom. Coelomic fluid cushions the gut and keeps it from being distorted by body movements. The fluid also helps distribute material through the animal body, and in some animals it plays a role in locomotion.

Figure 23.2 Animal body cavities. Most invertebrates and all vertebrates have a coelomate body plan, as in **C**.

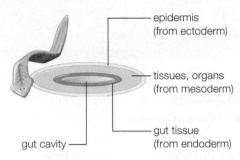

epidermis
(from ectoderm)

tissues, organs
(from mesoderm)

gut tissue
(from endoderm)

gut cavity

A Flatworms have no body cavity except their gut.

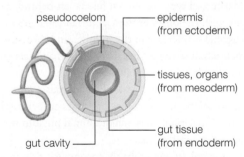

pseudocoelom

epidermis
(from ectoderm)

tissues, organs
(from mesoderm)

gut tissue
(from endoderm)

gut cavity

B Roundworms have a fluid-filled pseudocoelom.

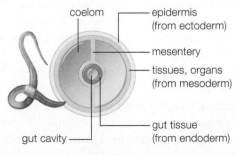

coelom

epidermis
(from ectoderm)

mesentery

tissues, organs
(from mesoderm)

gut tissue
(from endoderm)

gut cavity

C Annelids have a fluid-filled coelom with sheets of tissue (mesentery) that hold the gut in place.

FIGURE IT OUT

Which of the animals shown above develops from a three-layered embryo?

Answer: All of them.

animal Multicelled heterotroph with unwalled cells. Most ingest food and are motile during at least part of the life cycle.

bilateral symmetry Having paired structures so the right and left halves are mirror images.

coelom (SEE-lum) A fluid-filled cavity between the gut and body wall that is lined by tissue derived from mesoderm; the main body cavity of most animals.

deuterostomes (DUE-ter-oh-stomes) Lineage of bilateral animals in which the second opening on the embryo surface develops into a mouth.

ectoderm (EK-toe-derm) Outermost tissue layer of an animal embryo.

endoderm (EN-doe-derm) Innermost tissue layer of an animal embryo.

invertebrate Animal that does not have a backbone.

mesoderm (MEEZ-o-derm or MEZ-o-derm) Middle tissue layer of a three-layered animal embryo.

protostomes (PRO-toe-stomes) Lineage of bilateral animals in which the first opening on the embryo surface becomes a mouth.

pseudocoelom (SUE-doe-see-lum) Fluid-filled cavity between the gut and body wall; partially lined by tissue derived from mesoderm.

radial symmetry Having parts arranged around a central axis, like the spokes of a wheel.

23.2 Animal Origins and Diversification

LEARNING OBJECTIVES

Explain how a colonial protist might have given rise to the first animal.

Discuss what the fossil record reveals about early animal evolution.

List some environmental and biological factors that may have encouraged the adaptive radiation that occurred during the Cambrian period.

A Reconstruction of an Ediacaran sea based on body fossils.

B Fossil of *Spriggina*, about 3 centimeters (1 inch) long. It may have been a soft-bodied ancestor of arthropods, a group that includes modern crabs and insects.

Figure 23.3 Ediacaran animals.

COLONIAL ORIGINS

According to the **colonial theory of animal origins**, animals evolved from a colonial protist. At first, all cells in the colony were similar. Each could reproduce and carry out all other essential tasks. Later, mutations produced cells that specialized in some tasks and did not carry out others. Perhaps these cells captured food more efficiently but did not make gametes, whereas others made gametes but did not catch food. The division of labor among interdependent cells made the colony as a whole more efficient, allowing the colonies with mutations to obtain more food and produce more offspring. Over time, additional specialized cell types evolved.

What was the protist ancestral to animals like? Choano-flagellates, the modern protists most closely related to animals (Section 20.8), may provide some clues. Choanoflagellates are flagellated cells that live either as single cells or as a colony of genetically identical cells. Choanoflagellate cells closely resemble the flagellated cells that line the interior cavities of modern sponges. In addition, choanoflagellates and animals are the only living organisms that make both cadherins (a type of adhesion molecule) and collagen (a structural protein that is the main component of their extracellular matrices).

EVIDENCE OF EARLY ANIMALS

The oldest widely accepted evidence of animals dates to the final period of the Precambrian: the Ediacaran (635–541 million years ago). There are many Ediacaran trace fossils—rocks with trails or burrows created when unidentified early animals moved across or tunneled through sediments on the ancient seafloor. There are also a variety of body fossils (**Figure 23.3A**). All known Ediacaran animals were soft-bodied. Some were unlike any modern animal. Others resemble and may be related to some modern invertebrates (**Figure 23.3B**). Various late Ediacaran fossils are considered possible relatives of sponges, cnidarians, annelids, mollusks, or arthropods. Thus, the origin of bilateral body plans and the divergence between protostomes and deuterostomes are thought to have occurred during the Ediacaran.

---

colonial theory of animal origins Hypothesis that the first animals evolved from a colonial protist.

AN EXPLOSION OF DIVERSITY

The abundance and types of animal fossils left behind rose dramatically with the onset of the Cambrian (541–485 million years ago). The increase in the number and diversity of fossil animals resulted in part because the first animals with hard parts evolved: Remains of these animals were more likely to become fossils than soft-bodied animals. Other evolutionary innovations also took place during this period. By the end of the Cambrian, every major animal phylum now known was present in the seas.

Environmental factors probably encouraged this adaptive radiation (Section 17.10). During the Cambrian, Earth warmed and the oxygen concentration rose in the seas. Both factors made the environment more hospitable to animal life. Also during the Cambrian, the supercontinent Gondwana (Section 16.5) underwent a dramatic rotation. Movement of this landmass would have isolated populations, increasing the likelihood of allopatric speciation events (Section 17.8).

Biological factors encouraged diversification as well. After the first predatory animals arose, evolution of novel prey defenses such as protective hard parts would have been favored. Duplications and divergence of homeotic genes (Section 10.2) would have facilitated diversification. Changes in these genes have dramatic effects on body plans. Some mutations in these genes could have produced adaptive traits that allowed animals to more easily escape predators or to survive in novel habitats.

23.3 Sponges

LEARNING OBJECTIVES

Describe the types of cells in a sponge body and explain how they work together to keep the animal alive.

Explain how sponges reproduce sexually.

Describe how sponges can reproduce asexually and their capacity to regenerate.

Sponges (phylum Porifera) are aquatic animals with a porous body that does not have tissues or organs. The vast majority of sponges live in tropical seas, and nearly all adults are sessile, meaning they live fixed in place. Sponge bodies vary in size and shape. Some fit on a fingertip, whereas others stand meters tall. Asymmetrical vaselike or columnar forms are most common, but some sponges grow as a thin crust.

Flat, nonflagellated cells cover a sponge's outer surface, flagellated collar cells line the inner surface, and a jellylike extracellular matrix rich in collagen lies in between (**Figure 23.4**). In many sponges, cells in this matrix secrete fibrous proteins, glassy silica spikes, or both. These materials structurally support the body and help fend off predators. Some protein-rich sponges are harvested from the sea, dried, cleaned, and bleached. Their rubbery protein remains (left) are sold for bathing and cleaning.

A typical sponge is a **suspension feeder**, meaning it filters food from the surrounding water. Whiplike motion of flagella on collar cells draws food-laden water through pores in a sponge's body wall. The collar cells engulf food by phagocytosis, then digest it intracellularly. Amoeba-like cells in the matrix receive breakdown products of digestion from collar cells and distribute them to other cells in the sponge body.

Most sponges are **hermaphrodites**, meaning each individual makes both eggs and sperm. Typically, a sponge releases its motile sperm into the water (**Figure 23.5**) but holds onto its eggs. Fertilization produces a zygote that develops into a ciliated larva. A **larva** (plural, larvae) is a sexually immature individual, with a body plan unlike that of an adult. Larvae form in the life cycle of many invertebrates. Sponge larvae swim briefly, then settle and become sessile adults.

Of all animals, sponges have the greatest ability to regenerate. Many sponges reproduce asexually when small buds or fragments break away and grow into new sponges. Some can even recover after being broken into individual cells. When thus isolated, amoeba-like sponge cells reaggregate. Some then differentiate to become missing cell types. If cells of two different sponge species are mixed, cells of each species will seek out and reaggregate with their own kind alone.

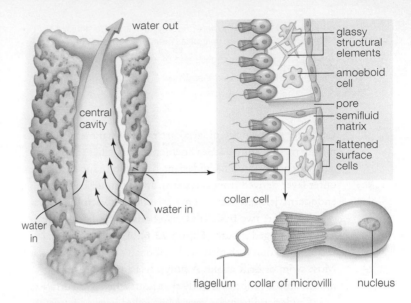

Figure 23.4 Body plan of a simple sponge.

Figure 23.5 Barrel sponge releasing sperm.

hermaphrodite (her-MAFF-roe-dite) Individual animal that makes eggs and sperm.
larva In some animal life cycles, a sexually immature form that differs from the adult in its body plan.
sponge Aquatic invertebrate that has no tissues or organs and filters food from the water.
suspension feeder Animal that filters food from water around it.

CREDITS: (in text) © ultimathule/Shutterstock; (5) Marty Snyderman/Planet Earth Pictures.

LEARNING OBJECTIVES

Compare the two cnidarian body plans, noting how they are similar and how they differ. Give an example of a cnidarian with each body plan.

Discuss how cnidarians capture, ingest, and digest their food.

Describe two types of colonial cnidarians.

Cnidarians (phylum Cnidaria) are radially symmetrical, mostly marine animals. Sea anemones and jellies (also called jellyfish) are examples. Cnidarians have two tissue layers: an outer layer derived from ectoderm, and an inner layer from endoderm. Jellylike secreted material lies between the layers.

There are two basic cnidarian body plans, both with a tentacle-ringed mouth (**Figure 23.6**). A **medusa** (plural, medusae) is dome-shaped, with a mouth on the lower surface. Most swim or drift about. A **polyp** has an upward-facing mouth atop a cylindrical body that is attached to a surface.

Cnidarian life cycles vary. Most jellies have a life cycle that alternates between polyps that reproduce asexually by budding, and medusae that produce gametes. Sea anemones, corals, and the hydras (a freshwater group) exist only as polyps that can both bud and produce gametes. In all cnidarians, the zygote produced by sexual reproduction develops into a bilaterally symmetrical ciliated larva called a planula.

The name Cnidaria is from *cnidos*, the Greek word for the stinging nettle plant, and refers to the animals' mechanism

Figure 23.6 Cnidarian body plans and representative examples.

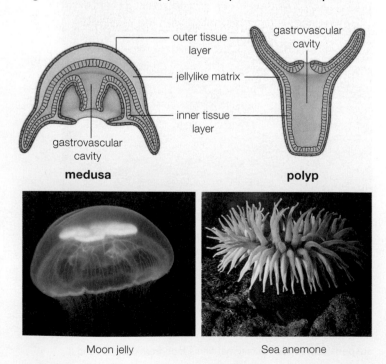

outer tissue layer

gastrovascular cavity

jellylike matrix

inner tissue layer

gastrovascular cavity

medusa　　　　**polyp**

Moon jelly　　　　Sea anemone

In response to a touch of the trigger, the capsule around the nematocyst becomes more permeable. As water diffuses in, pressure builds up, and the thread is forced to turn inside out. The barbed tip pierces prey.

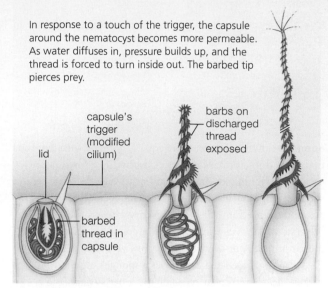

capsule's trigger (modified cilium)

lid

barbs on discharged thread exposed

barbed thread in capsule

Figure 23.7 Nematocyst action.

of feeding. Cnidarians are predators. Their tentacles have specialized stinging cells (**cnidocytes**) that help them capture prey. A cnidocyte contains a nematocyst, an organelle that functions like a jack-in-the-box (**Figure 23.7**). When prey brushes a nematocyst's trigger, a barbed thread pops out and delivers a dose of venom. People that brush against a jellyfish can trigger this same response and end up with a painful sting. Some box jellies that live near Australia produce a venom so powerful that their sting occasionally kills people.

Tentacles move captured food to the mouth, which opens to a **gastrovascular cavity**: a saclike region that functions in digestion and gas exchange. Cnidarians digest food extracellularly. Enzymes secreted into the gastrovascular cavity break down food, then the released nutrients are absorbed and distributed through the body by diffusion. Digestive waste exits the gastrovascular cavity through the mouth.

Cnidarians are brainless, but interconnecting nerve cells extend through their tissues as a **nerve net**. Body parts move when these nerve cells signal contractile cells to shorten. Contraction alters the body shape by redistributing fluid trapped in the gastrovascular cavity. By analogy, think of what happens if you squeeze a water-filled balloon. A fluid-filled cavity that contractile cells act against is a **hydrostatic skeleton**. Many soft-bodied invertebrates have this type of skeleton.

Most cnidarians are solitary, but colonial groups exist. Coral reefs are built by colonies of polyps enclosed in a skeleton of secreted calcium carbonate. In a mutually beneficial relationship, photosynthetic dinoflagellates (Section 20.4) live inside each polyp's tissues (**Figure 23.8A**). The protists receive shelter and carbon dioxide from the coral, and give it

A (Above) Coral reef. Corals that build reefs are colonies of polyps with dinoflagellates living inside them. The hard part of the reef is calcium-rich material secreted by the polyps.

B (Left) Siphonophore. The Portuguese man-of-war (*Physalia*) consists of a gas-filled polyp (the float), stinging polyps, and polyps that specialize in digestion.

Figure 23.8 Colonial cnidarians.

sugars and oxygen in return. If a reef-building coral loses its protist partners, an event called "coral bleaching," it may die. Siphonophores such as the Portuguese man-of-war (*Physalia*) look like a single animal (**Figure 23.8B**) but are actually colonies. As the colony drifts along, its stinging polyps capture and kill fish. Other specialized polyps digest the prey.

---

cnidarian (nigh-DARE-ee-en) Radially symmetrical invertebrate with two tissue layers; uses tentacles with stinging cells to capture food.

cnidocyte (NIGH-duh-site) Stinging cell unique to cnidarians.

gastrovascular cavity A saclike gut that also functions in gas exchange.

hydrostatic skeleton Of soft-bodied invertebrates, a fluid-filled chamber that contractile cells exert force on.

medusa (meh-DUE-suh) Dome-shaped cnidarian body plan with the mouth on the lower surface. Jellies are examples.

nerve net Decentralized mesh of nerve cells that allows movement in cnidarians.

polyp (PAH-lup) Cylindrical cnidarian body plan with the mouth on the upper surface. Sea anemones are examples.

Dr. David Gruber

National Geographic Explorer Dr. David Gruber studies biofluorescence—the process by which organisms absorb light of one wavelength and emit light of another. **Figure 23.9** is a photo Gruber took of a coral polyp illuminated by blue light. A protein in this coral absorbs violet wavelengths, then emits green, orange, and red wavelengths.

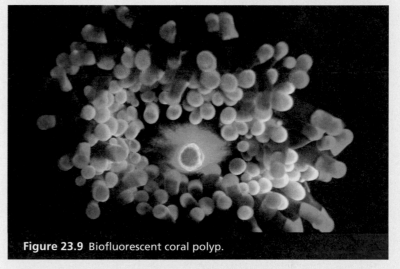

Figure 23.9 Biofluorescent coral polyp.

Most of the known biofluorescent animals are cnidarians, so Gruber searches the world's reefs for species that glow. He and his research team have discovered over 30 new fluorescent proteins from corals, including the brightest one found to date. Recently, he has also been finding biofluorescence in several species of sharks, rays, and bony fishes. Fluorescent proteins are useful in science and medicine. They allow researchers to visualize specific cells and structures using fluorescence microscopy (Section 4.2). Among other applications, this technique can be used in early detection of cancers and to study how signals travel through the brain. We do not know how having biofluorescent proteins benefits cnidarians, but these proteins are linked to coral health. Gruber also suspects that other reef-dwelling organisms recognize the fluorescence as a signal of some sort.

To see additional examples of biofluoresence and learn more about the research of Gruber and his colleagues, visit their website Luminescent Labs at luminescentlabs.org.

CREDITS: (8A) George Grall, National Geographic Creative; (8B) Peter Scoones/Science Source; (in text, 9) David Gruber/National Geographic.

Compare the structural organization and symmetry of a flatworm with that of a cnidarian.

Explain how nutrients and gases are distributed through a flatworm's body.

Describe the life cycle of a tapeworm.

Figure 23.10 Marine flatworm. *Bifurcus*, a scavenger common on coral reefs.

With this section, we begin our survey of protostomes, one of the two lineages of bilaterally symmetrical animals. All of these animals develop from a three-layered embryo.

Flatworms (phylum Platyhelminthes) are the simplest protostomes. They have a flattened body with an array of organ systems, but no body cavity other than a gastrovascular cavity. Like cnidarians, they rely entirely on diffusion to move nutrients and gases through their body. There is no heart or blood vessels. Some flatworms are free-living and others are parasites. Nearly all are hermaphrodites. Free-living flatworms typically glide along, propelled by the action of cilia that cover the body surface. Some marine species can also swim with an undulating motion.

FREE-LIVING FLATWORMS

Most free-living flatworms live in tropical seas, and many of these are brightly colored (**Figure 23.10**). Most colorful flatworms contain toxins that make them distasteful to fish. Having a bright, distinctive color pattern benefits distasteful flatworms by making it easier for fish to recognize and avoid them after naively taking an initial bite. A small number of free-living flatworms live in damp places on land. They are predators that feed on other invertebrates.

Planarians are free-living flatworms about a centimeter long. They are common in ponds and streams. Some are scavengers and others are predators. All have a highly branched gastrovascular cavity (**Figure 23.11A**), and nutrients diffuse from the fine branches to all body cells. There is no anus;

food enters and wastes leave through the mouth. The mouth is at the tip of a muscular tube (the pharynx) that extends from the animal's lower surface.

A planarian's head has chemical receptors and eyespots that detect light. These sensory structures send messages to a simple brain that consists of paired groupings of nerve cell bodies (ganglia). A pair of nerve cords extend from the brain and run the length of the body (**Figure 23.11B**).

A planarian's body fluid has a higher solute concentration than its freshwater environment, so water tends to move into the body by osmosis. A system of tubes (**Figure 23.11C**) regulates internal water and solute levels by driving excess fluids out through a pore at the body surface.

A planarian has both female and male sex organs (**Figure 23.11D**) but cannot fertilize its own eggs. Freshwater planarians typically swap sperm. By contrast, some marine flatworms battle over who will assume the male role. In a behavior described as "penis fencing," each flatworm attempts to stab its penis into a partner's body and squirt in some sperm, while fending off its partner's attempts to do the same.

Planarians also reproduce asexually. During asexual reproduction, the body splits in two near the middle, then each piece regrows missing parts. This capacity for regrowth also allows regeneration after a planarian is injured, as by a

Figure 23.11 Organ systems of a planarian, a free-living, freshwater flatworm.

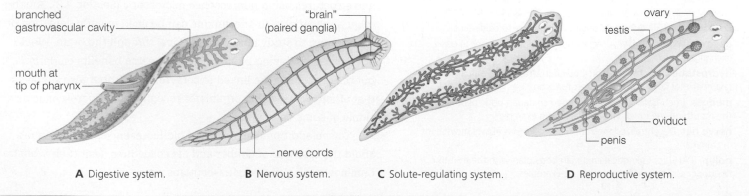

branched gastrovascular cavity

mouth at tip of pharynx

A Digestive system.

"brain" (paired ganglia)

nerve cords

B Nervous system.

C Solute-regulating system.

ovary

testis

oviduct

penis

D Reproductive system.

predator. A planarian can regenerate itself even when a large portion of its body is lost.

PARASITIC FLATWORMS

Flukes and tapeworms are parasitic flatworms whose life cycle often involves multiple hosts. Typically, larvae reproduce asexually in one or more intermediate hosts before developing into adults that reproduce sexually in a final or definitive host.

The tropical disease schistosomiasis is caused by a fluke (*Schistosoma*) that reproduces asexually in freshwater snails and sexually in mammals. People become infected when they swim or stand in fresh water where infected snails live. Free-swimming fluke larvae (**Figure 23.12A**) that developed inside the snails can cross human skin. Once inside a human host, the larvae enter the bloodstream and travel to veins where they mature into sexually reproducing adults (**Figure 23.12B**). The presence of actively reproducing adult flukes in the human body results in weakness, fever, and damage to internal organs. More than 200 million people currently suffer from schistosomiasis, most of them in Southeast Asia and northern Africa.

Tapeworms are parasitic flatworms that live and reproduce in the vertebrate gut. The head has a scolex, a structure with hooks or suckers that allow the worm to attach to the gut wall. Behind the head are body units called proglottids. Unlike planarians and flukes, tapeworms do not have a gastrovascular cavity. Instead, the worm absorbs nutrients released when the host digests the food in its gut. **Figure 23.13** shows the life cycle of a beef tapeworm.

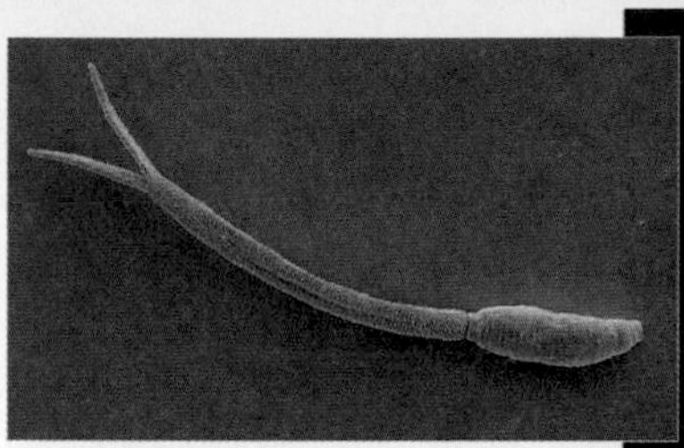

A Free-swimming larva that develops in freshwater snails, then infects humans (colorized SEM). The larva is about 300 micrometers long.

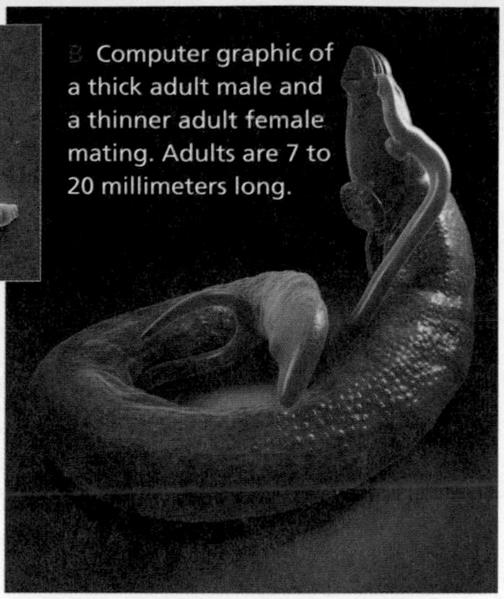

B Computer graphic of a thick adult male and a thinner adult female mating. Adults are 7 to 20 millimeters long.

Figure 23.12 Blood fluke. Two stages in the life cycle of a blood fluke (*Schistosoma*).

flatworm Bilaterally symmetrical invertebrate with organs but no body cavity; for example, a planarian or tapeworm.

Figure 23.13 Life cycle of a beef tapeworm.

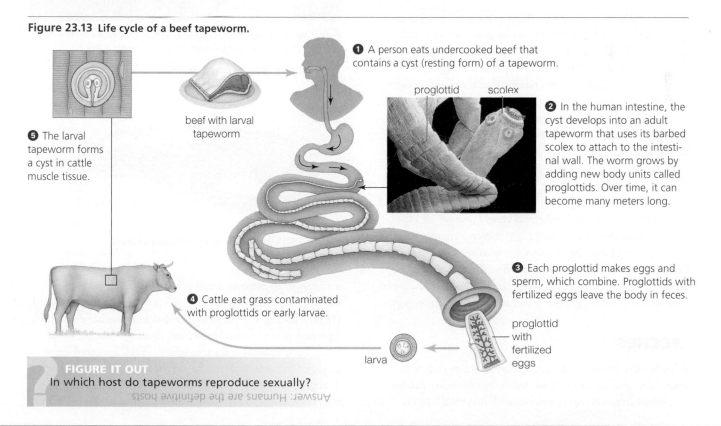

❶ A person eats undercooked beef that contains a cyst (resting form) of a tapeworm.
beef with larval tapeworm
proglottid
scolex
❷ In the human intestine, the cyst develops into an adult tapeworm that uses its barbed scolex to attach to the intestinal wall. The worm grows by adding new body units called proglottids. Over time, it can become many meters long.
❸ Each proglottid makes eggs and sperm, which combine. Proglottids with fertilized eggs leave the body in feces.
proglottid with fertilized eggs
larva
❹ Cattle eat grass contaminated with proglottids or early larvae.
❺ The larval tapeworm forms a cyst in cattle muscle tissue.

FIGURE IT OUT
In which host do tapeworms reproduce sexually?
Answer: Humans are the definitive hosts.

CREDITS: (12A) David Scharf/Science Source; (12B) Science Picture Co/Science Source; (13) right, Andrew Syred/Science Source.

23.6 Annelids

Annelids (phylum Annelida) are bilateral worms with a coelom and a segmented body plan; their body consists of multiple similar units arranged one after the other. A tubular digestive tract runs the length of the body inside the coelom.

Gases and nutrients are distributed through the body by a **closed circulatory system**, a system in which the blood pumped out of a heart travels back to the heart by way of a continuous series of blood vessels. In a closed circulatory system, all exchanges between blood and the tissues take place across the walls of blood vessels.

There are three annelid subgroups. Two are named for the chitin-reinforced bristles called chaetae on their segments. Polychaetes have many bristles and oligochaetes have few. (*Poly–* means many; *oligo–* means few.) The third subgroup, the leeches, does not have any bristles.

POLYCHAETES

Most polychaetes are marine. Those commonly known as bristleworms or sandworms are active predators that use their chitin-hardened jaws to capture other soft-bodied invertebrates (**Figure 23.14A**). Each body segment has a pair of bristle-tipped paddlelike appendages that the worm uses to burrow in sediments and pursue prey.

Other polychaetes are suspension feeders. Fan worms and feather duster worms live in a tube that they make from sand grains and mucus. The worm's head, which protrudes from the tube, has elaborate tentacles that capture food as it drifts by (**Figure 23.14B**).

Still other tube-dwelling polychaete worms have adapted to life near deep-sea hydrothermal vents (Section 18.1). These worms do not have tentacles or even a mouth. Rather than ingesting food, they rely on chemoautotrophic bacteria living inside them to provide nutrients. The bacteria take up dissolved materials from the water and use them to assemble organic compounds that the worm breaks down for energy.

LEECHES

A leech lacks bristles and has a sucker at either end of its body. Most leeches live in fresh water, but some are marine and others live in damp places on land. A typical leech is

A The bristleworm uses its many appendages to burrow into sediment on marine mudflats. It is an active predator with hard jaws.

B Feather duster worm in its tube. Feathery extensions on the head capture food, which cilia then move to the mouth.

Figure 23.14 Marine polychaetes.

Figure 23.15 Leech feeding on human blood.

a scavenger or preys on small invertebrates. An infamous minority attach to a vertebrate, pierce its skin, and suck blood (**Figure 23.15**). The saliva of bloodsucking leeches contains a protein that prevents blood from clotting while the leech feeds. For this reason, doctors who reattach a severed finger or ear sometimes apply leeches to the reattached body part. The presence of the feeding leeches prevents clots from forming inside the newly reconnected blood vessels.

CREDITS: (14A) Darlyne A. Murawski, National Geographic Creative; (14B) © Jon Kenfield/ Bruce Coleman Ltd.; (15) sydeen/Shutterstock.com.

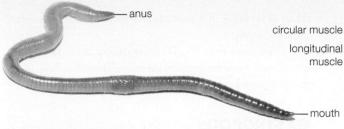

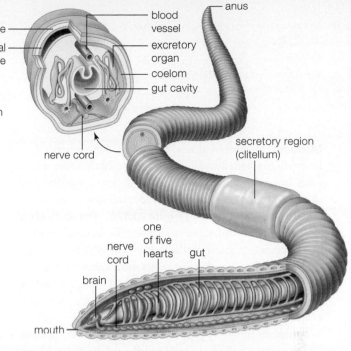

Figure 23.16 Earthworm. Internal anatomy is shown to the right.

OLIGOCHAETES

Oligochaetes include marine and freshwater species, but the land-dwelling earthworms are most familiar.

Earthworm Body Plan The anatomy of an earthworm is illustrated in **Figure 23.16**. The earthworm's skin is protected by a cuticle of secreted proteins, and a layer of mucus. Visible grooves at the body surface correspond to internal partitions between coelomic chambers.

Gas exchange occurs across the body surface. A closed circulatory system speeds the distribution of gases. Five paired, heavily muscled blood vessels in the front of the worm function like hearts. Their contractions provide the pumping power that moves the worm's blood. Large blood vessels run the length of the worm along its the upper and lower surfaces.

A simple brain at the worm's head end connects to a nerve cord that extends along the lower length of the body. The brain coordinates locomotion and receives sensory information. Earthworms can sense light (which they avoid), detect touch and vibration, and distinguish specific odors emitted by their food.

A tubular gut extends the length of the body inside the coelom. Earthworms eat their way through soil or decaying material. They digest both organic debris and any microorganisms that cling to it.

Also inside the coelom, most body segments have a pair of excretory organs that regulate the solute composition and volume of coelomic fluid. The organs take in coelomic fluid, adjust its composition, and expel wastes and excess solutes through a pore in the body wall.

Two sets of muscles in the body wall function in locomotion. Longitudinal muscles parallel the body's long axis, and circular muscles ring the body. The worm's coelomic fluid is a hydrostatic skeleton. Contraction of muscles puts pressure on coelomic fluid trapped inside body segments, causing the segments to change shape. When a segment's longitudinal muscles contract, the segment gets shorter and fatter. When circular muscles contract, a segment gets longer and thinner. Coordinated waves of contraction that run along the body propel the worm through soil.

FIGURE IT OUT
Which muscles contract to lengthen a segment?
Answer: Circular muscles.

Earthworms are hermaphrodites, but they cannot fertilize themselves. During mating, a secretory organ (the clitellum) produces mucus that glues two worms together while they exchange sperm. Later, the same organ secretes a silky case that protects the fertilized eggs outside the body.

Earthworm Ecology Earthworms play an important role in soil ecosystems. They improve soil by loosening its particles so that it drains more easily, and by excreting tiny bits of organic matter that decomposers can easily break down. Excreted earthworm "castings" are sold as a natural nitrogen-rich fertilizer, and many studies have demonstrated that the presence of earthworms increases crop yields.

In the United States, most earthworms observed in backyards and compost heaps are introduced species native to Europe. However, some undisturbed habitats retain native earthworms, including a species in Oregon that can reach more than a meter in length.

annelid (ANN-uh-lid) Segmented worm with a coelom, tubular digestive tract, and closed circulatory system; a polychaete, oligochaete, or leech.
closed circulatory system Circulatory system in which blood flows through a continuous network of vessels; all materials are exchanged across the walls of those vessels.

Explain the structure and functions of the mollusk mantle.

Using appropriate examples, name the three major mollusk groups and discuss the traits that define them.

Compare an open circulatory system with a closed circulatory system and note which groups of mollusks have each.

Mollusks (phylum Mollusca) are bilaterally symmetrical invertebrates with a small coelom. Most dwell in the sea, but some live in fresh water or on land. All have a **mantle**; this skirtlike extension of the upper body wall drapes over an organ-filled mantle cavity (**Figure 23.17A**). The shell, when present, consists of calcium-rich, bonelike material secreted by the mantle. Aquatic mollusks have one or more gills in their fluid-filled mantle cavity. **Gills** are respiratory organs that exchange gases with the fluid around them. All mollusks have a tubular gut. In most mollusks, the mouth contains a radula, a tonguelike organ hardened with chitin.

Aquatic mollusks typically have separate sexes and produce a ciliated swimming larva called a trochophore. Marine polychaetes have the same sort of larva, and this similarity is taken as evidence that annelids and mollusks shared a common ancestor.

With more than 85,000 living species, mollusks are second only to arthropods in diversity. There are three main groups: gastropods, bivalves, and cephalopods.

GASTROPODS

Gastropods are the most diverse mollusk lineage. Their name means "belly foot," and most species glide about on the broad muscular foot that makes up most of the lower body mass. A gastropod shell, when present, is one piece and often coiled.

Gastropods have a distinct head that, in most species, has eyes and sensory tentacles. In many aquatic species, a part of the mantle forms a siphon, a tube that the animal uses to draw water into its mantle cavity. Gastropods and bivalves have an **open circulatory system**, in which vessels do not form a continuous loop. Instead, the circulatory fluid (called hemolymph) leaves vessels and seeps among tissues before returning to the heart. Cells exchange substances with hemolymph while it is outside of vessels.

Gastropods include the only terrestrial mollusks. In snails and slugs that live on land, a lung replaces the gill. A **lung** is a respiratory organ that exchanges gases with the air. Mucus secreted by glands on the foot protects the animal as it moves across dry or abrasive surfaces. Most mollusks have separate sexes, but most land snails and slugs are hermaphrodites. Unlike other mollusks, which produce a swimming larva, embryos of these groups develop directly into adults.

Nudibranchs, commonly called sea slugs, are marine gastropods with a greatly reduced shell. Nudibranch means "naked gills." Many nudibranchs defend themselves from predators using defensive weapons obtained from their prey. Sponge-eating nudibranchs store sponge toxins and secrete them when disturbed. Cnidarian-eating nudibranchs store the cnidarians' stinging cells in outpouchings on their body (**Figure 23.17B**). A predator that bites such a nudibranch is stung when the cells discharge. Most well-defended nudibranchs have bright coloration that predators learn to avoid.

BIVALVES

Mussels, oysters, scallops, clams, and most other mollusks that commonly end up on dinner plates are bivalves. **Bivalves** have a hinged, two-part shell (**Figure 23.18**). A band of elastic protein forms the hinge, and powerful adductor muscles inside the shell connect the shell's two halves. When these muscles contract, they pull the shell shut and stretch the elastic hinge. When the muscles relax, the elastic hinge springs back to its unstretched length and the shell opens.

Figure 23.17 Gastropod mollusks (belly-footed mollusks).

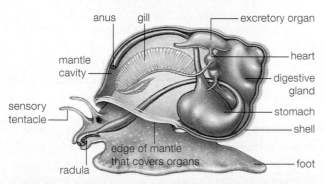

A Body plan of an aquatic snail. Mantle is shown as transparent.

anus — gill — excretory organ — mantle cavity — heart — digestive gland — sensory tentacle — stomach — shell — radula — edge of mantle that covers organs — foot

Orange projections on the nudibranch's back hold stinging cells derived from its cnidarian prey.

B Nudibranch (sea slug).

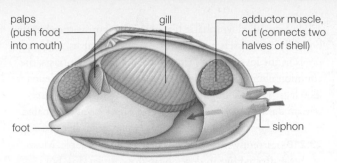

Figure 23.18 **Bivalve mollusk.** A clam, shown with one part of its two-part shell removed. Blue arrows show water flow.

A bivalve has no obvious head, but many have simple eyes around the edge of the mantle. Most bivalves use their large foot to burrow into sediments. They feed by drawing water into their mantle cavity through a siphon so that food particles become trapped in mucus on the gills. Cilia direct food-laden mucus to a pair of fleshy extensions (palps) that sweep this material into the mouth.

CEPHALOPODS

Cephalopods are mollusks that move by jet propulsion. They draw water into their mantle cavity, then force it out through a funnel-shaped siphon. All cephalopods are predators and most have beaklike, biting mouthparts in addition to a radula. The name cephalopod, which means "head-footed," refers to arms and/or tentacles extending from the head. These structures are evolutionarily derived from the ancestral foot.

During the late Cambrian, large cephalopods with a long, conelike shell were the top predators in the seas. Nautiluses (**Figure 23.19A**) are the only modern cephalopods with an external shell, and theirs is tightly coiled. It contains many gas-filled compartments that make the animal buoyant.

Evolutionary reduction or loss of the shell in other cephalopod groups may have been driven by competition for prey with jawed fishes, which evolved about 400 million years ago. Cephalopods with smaller shells would have been faster and more agile. Other changes also supported a speedier lifestyle. Of all mollusks, only cephalopods have a closed circulatory system. They also have surprisingly good vision. Like vertebrates, cephalopods have eyes with a lens that focuses light. This type of eye presumably evolved independently in the two groups; mollusks and vertebrates are distant relatives.

Squids (**Figure 23.19B**) and octopuses (**Figure 23.19C**) have eight arms. Squids have a streamlined body reinforced by a stiff internal rod. Most prey on fish in the ocean's upper waters, although some, such as the giant squid, hunt in the ocean's depths. The giant squid is the largest invertebrate known; individuals can weigh nearly a ton.

A Nautilus

B Squids

C Octopus

Figure 23.19 **Examples of cephalopod mollusks.**

Most octopuses hunt prey on the seafloor. They can slip into tiny spaces between rocks because mouthparts are their only hard parts. A 600-pound octopus can wriggle through a hole the width of a quarter. Octopuses are also the smartest invertebrates. Relative to body size, an octopus brain is larger than that of most fish and amphibians.

bivalve (BYE-valv) Mollusk with a hinged, two-part shell.
cephalopod (CEFF-uh-luh-pod) Predatory mollusk with a closed circulatory system; moves by jet propulsion.
gastropod Mollusk in which the lower body is a broad "foot."
gill Respiratory organ that exchanges gases with fluid around it.
lung Respiratory organ that exchanges gases with the air.
mantle Of mollusks, a skirtlike extension of the body wall.
mollusk Invertebrate with a reduced coelom and a mantle.
open circulatory system System in which hemolymph leaves vessels and seeps among tissues before returning to the heart.

CREDITS: (19A) Vudhikrai/Shutterstock.com; (19B) George Grall/National Geographic Creative; (19C) NURC/UNCW and NOAA/FGBNMS.

Describe some human diseases caused by roundworms.

Explain why roundworms are useful as models in scientific studies.

Roundworms, or nematodes (phylum Nematoda), are cylindrical, unsegmented worms with a pseudocoelom (**Figure 23.20**). They have a tubular digestive tract, excretory organs, and a nervous system, but no circulatory or respiratory organs. Most feed on organic debris in soil or water and are less than a millimeter long. Parasitic species tend to be larger. Like arthropods, roundworms have a cuticle of secreted proteins that they periodically **molt** (shed and replace) as they grow. Arthropods and roundworms are grouped together in the clade ecdysozoans, which means animals that molt.

The soil roundworm *Caenorhabditis elegans* is commonly used in research. It is a good model for studies of development and aging because it has the same kinds of tissues as more complex organisms, and it is transparent, has fewer than 1,000 body cells, and reproduces fast. *C. elegans* also has a small genome—about 1/30 the size of the human genome. It was the first multicelled organism to have its genome sequenced. Studies involving *C. elegans* showed how death of specific cells shapes organs during development, and how mutations in some genes can extend life span.

Several kinds of parasitic roundworms infect humans. Mosquito-transmitted roundworms cause a disfiguring tropical disease called lymphatic filariasis. The worms travel in the body's lymph vessels and destroy the vessels' valves, so lymph pools in the lower limbs (**Figure 23.21A**). Elephantiasis, the common name for this disease, refers to fluid-filled, "elephant-like" legs. The swelling is permanent because lymph vessels remain damaged even after the worms have been eliminated.

The intestinal parasite *Ascaris lumbricoides* (**Figure 23.21B**) currently infects more than 1 billion people. Most of those affected live in developing tropical nations, but occasional infections occur in rural parts of the American Southeast. Typically an infected person has no symptoms, but a heavy infection can block the intestine.

Pinworms (*Enterobius vermicularis*) commonly infect children worldwide. The small worms, about the size of a staple, live in the rectum. At night, females crawl out and lay eggs on the skin near the anus. Their movement produces an itching sensation. Scratching to relieve the itch puts eggs onto fingertips, from which they can be spread through the environment. Swallowing eggs starts a new infection.

Parasitic roundworms also infect our livestock, pets, and crop plants. A roundworm that lives in pigs can also infect humans who eat undercooked pork, causing the disease trichinosis. Dogs are susceptible to heartworms transmitted by mosquitoes. Cats usually become infected by eating an infected rodent that serves as an intermediate host. Many crop plants are susceptible to nematode parasites. In some cases, the worms suck on plant roots; in others, they enter the plant (**Figure 23.21C**). Either way, the infection stunts the plant's growth and lowers crop yields.

mouth   pseudocoelom   eggs in uterus   intestine   anus

Figure 23.20 Body plan of a free-living roundworm.

molt To shed and replace an outer body layer or part.
roundworm Cylindrical worm with a pseudocoelom.

A Lymphatic filariasis (elephantiasis).

B Intestinal parasite (*Ascaris*) passed by a child. These worms can be up to a meter long.

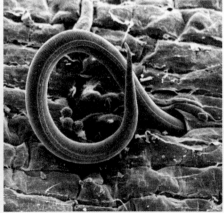

C Plant-infecting roundworm (about 0.5 mm long) entering a root.

Figure 23.21 Roundworms as parasites.

CREDITS: (21A) Courtesy of © Emily Howard Staub and The Carter Center; (21B) CDC/Henry Bishop; (21C) William Wergin and Richard Sayre. Colorized by Stephen Ausmus.

Arthropods (phylum Arthropoda) are bilaterally symmetrical invertebrates with a tubular gut, an open circulatory system, and a small coelom. Modern representatives include spiders, lobsters, barnacles, centipedes, and insects. Trilobites, a now-extinct lineage, were the most abundant and diverse arthropods in Cambrian seas, and they left many fossils (**Figure 23.22**). The last trilobites disappeared about 250 million years ago. Despite this loss, arthropods remain the most diverse animal phylum. In this section, we consider traits that contribute to their success. In the next two sections we take a closer look at specific arthropod groups.

A JOINTED EXOSKELETON

A tough cuticle stiffened by chitin (Section 3.2) protects and supports the arthropod body. An arthropod cuticle is an **exoskeleton**, an external skeleton to which muscles attach. The exoskeleton does not inhibit movement because it is thin at the joints where two body parts connect. In fact, *arthropod* means jointed leg. The cuticle consists of secreted materials and does not grow, so, like roundworms, arthropods must periodically molt their cuticle and replace it with a larger one.

The exoskeleton has additional functions among arthropods that live on land. It supports their body against the force of gravity and helps them conserve water. Thus, arthropods thrive even in dry environments.

SPECIALIZED SEGMENTS

Arthropods have a segmented body. In most groups, some segments have fused to form distinct body regions. In some groups, including lobsters, segments of the head and thorax (midbody) form a unit called the cephalothorax (**Figure 23.23**). Segments and appendages became modified in different ways in different arthropod groups. In some species, including American lobsters, the appendages on one segment evolved into large, muscular claws that can crush food. In flying insects, one or two body segments have wings that evolved from extensions of the body wall.

Well-developed sensory structures on some segments allow arthropods to monitor their environment. All arthropod lineages have one or more pairs of eyes. Some groups have **compound eyes**, which consist of many image-forming

Figure 23.22 Fossil trilobite. Over 20,000 species of this now-extinct group have been identified from fossils.

units that are highly sensitive to motion. Most arthropods also have one or more pairs of **antennae** (singular, antenna), sensory structures that can detect chemicals in water or air.

METAMORPHOSIS

The body plan of many arthropods changes during the life cycle. Individuals undergo **metamorphosis**: A dramatic remodeling of the body plan occurs as larvae develop into adults. For example, crab larvae float as part of plankton, but adults are bottom-feeders. The body form in each stage is specialized for a different lifestyle, so adults and juveniles do not compete for resources.

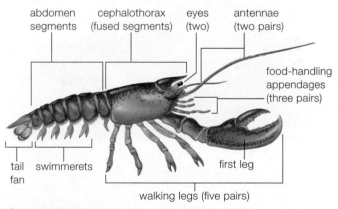

Figure 23.23 Body plan of an American lobster.

antenna Of some arthropods, sensory structure on the head that detects touch and odors.

arthropod Invertebrate with jointed legs and a hard exoskeleton that is periodically molted.

compound eye Of some arthropods, a motion-sensitive eye made up of many image-forming units.

exoskeleton Hard external parts that muscles attach to and move.

metamorphosis Dramatic remodeling of body form during the transition from larva to adult.

CHELICERATES

Chelicerates are arthropods that have a pair of special feeding appendages (chelicerae) in front of the mouth. Their head has one or more pairs of eyes, but no antennae. The body is usually divided into a cephalothorax and an abdomen.

Four species of horseshoe crabs are the only marine chelicerates and members of the oldest surviving arthropod lineage. Horseshoe crabs are bottom-feeders that eat clams and worms. A horseshoe-shaped shield covers their cephalothorax, and their last abdominal segment has evolved into a long spine that serves as a rudder when swimming. Each spring, large numbers of Atlantic horseshoe crabs come ashore on sandy beaches to lay their eggs (**Figure 23.24**). The eggs feed millions of migratory shorebirds.

Figure 23.24 Horseshoe crabs. These marine chelicerates come ashore once a year to mate and lay their eggs.

Arachnids are chelicerates that have four pairs of walking legs attached to their cephalothorax (**Figure 23.25**). They include spiders, scorpions, ticks, and mites, nearly all of which live on land.

Spiders and scorpions are venomous predators. Spiders dispense venom through their fangs, which are modified chelicerae. Of more than 45,000 known spider species, about 30 produce venom harmful to humans. Most spiders benefit us by eating insect pests. Some catch prey in a web made of silk ejected by glands on their abdomen. Others such as tarantulas actively hunt prey (**Figure 23.25A**). Scorpions are active hunters too. They capture prey with pincers and dispense venom through a stinger on their abdomen (**Figure 23.25B**).

Ticks are parasites of vertebrates. They pierce a host's skin with their chelicerae, then suck blood. An unfed tick has a flattened body that can expand dramatically when its gut fills with blood (**Figure 23.25C**). Some ticks transmit human pathogens, such as the spirochete bacteria that cause Lyme disease (Section 19.5).

Mites include predators, scavengers, and parasites (**Figure 23.25D**). Most are less than a millimeter long. Dust mites are scavengers whose feces cause allergies in some people. Chiggers are larval mites that attach to human skin at hair follicles and sip our tissue fluids. Mites that burrow into the skin cause scabies in humans and mange in dogs. Affected individuals have red, itchy skin.

MILLIPEDES AND CENTIPEDES

Millipedes and centipedes have a head with one pair of antennae attached to an elongated body with many similar segments. Centipedes have a low-slung, flattened body with one pair of legs per segment, for a total of 30 to 50 legs (**Figure 23.26A**). These fast-moving predators use fangs to inject paralyzing venom. Most centipedes prey on insects, but some big tropical species eat small vertebrates. Millipedes

Figure 23.25 Arachnids. All have eight walking legs and no antennae.

abdomen   cephalothorax

A Spider, a predator.

B Scorpion, a predator.

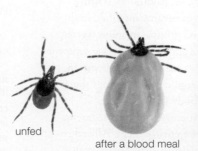

unfed

after a blood meal

C Ticks, parasites.

D Mite, a parasite of citrus. (colorized SEM)

A Centipede, a speedy predator.

B Millipede, a scavenger of decaying plant material.

Figure 23.26 Millipede and centipede.

A Antarctic krill.

B Copepod.   **C** Barnacle.

Figure 23.27 Crustaceans. See also Figure 23.23.

are slower-moving animals that feed on decaying vegetation. Their cylindrical body has a cuticle hardened by calcium carbonate. There are two pairs of legs per segment, for a total of a few hundred (**Figure 23.26B**).

CRUSTACEANS

Most marine arthropods are **crustaceans**. The great diversity and abundance of crustaceans is reflected in their nickname, "insects of the seas." Crustaceans have two pairs of antennae and a distinct cephalothorax and abdomen. In many, the exoskeleton is hardened by calcium.

You are probably familiar with some of the decapod crustaceans. This group of bottom-feeding scavengers includes lobsters, crayfish, crabs, and shrimps. Decapod means ten-legged; members of this group typically have five pairs of walking legs on their cephalothorax.

Krill are crustaceans with a shrimplike body a few centimeters long (**Figure 23.27A**). They swim through cool marine waters worldwide. Krill have historically been so abundant that a 30-ton blue whale could sustain itself largely on the krill that it filtered from the sea. Recent overharvesting of krill in some regions has caused steep population declines and may endanger animals that depend on this food source. (People harvest krill to feed farm-raised salmon and as a source of omega-3 fatty acids.)

Tiny, free-swimming copepods (**Figure 23.27B**) about 1 millimeter long are the most numerous of all crustaceans. They live anywhere there is water and, like krill, are an impor-

tant source of food for larger aquatic species. Some copepod species affect human health by serving as intermediate hosts for pathogens such as cholera-causing bacteria.

Larval barnacles swim, but adults glue themselves in place by their antennae and secrete a thick calcified shell. They filter food from the water with feathery legs (**Figure 23.27C**). Some barnacles are notable for the length of their penis, which can be eight times that of their body.

Crustaceans that have adapted to life on land include some crabs and some isopods. The isopods commonly called wood lice or pill bugs (right) live in damp places where they feed on organic debris or on soft, young plant parts.

INSECTS

Insects are the most diverse arthropods. Historically, they were thought to be close relatives of centipedes and millipedes because of structural similarities. Then, gene comparisons made scientists rethink the connections. Genetic data indicate that insects descended from freshwater crustaceans. We consider insect diversity and ecology in the next section.

arachnids (uh-RAK-nidz) Land-dwelling arthropods with no antennae and four pairs of walking legs; spiders, scorpions, mites, and ticks.
chelicerates (kell-ISS-er-ates) Arthropod group with specialized feeding structures (chelicerae) and no antennae; arachnids and horseshoe crabs.
crustaceans (krus-TAY-shuns) Mostly marine arthropods with a calcium-hardened cuticle and two pairs of antennae; for example lobsters, crabs, krill, and barnacles.

LEARNING OBJECTIVES

Describe the traits that allow insects to thrive on land.

List some ways that insects benefit other animals and some ways that they benefit humans.

Using appropriate examples, explain how some insects harm human health or have a negative economic impact.

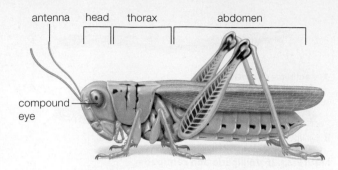

Figure 23.28 Body plan of an insect, a grasshopper.

With more than a million named species, **insects** are the most diverse class of animals. They are also breathtakingly abundant. By some estimates, ants alone make up about 10 percent of the total weight of all living land animals. Insects live on every continent, including Antarctica. Here we consider the traits that contribute to insect diversity and the ways that they interact with us and with other organisms.

CHARACTERISTICS OF INSECTS

Insects have a three-part body plan, with a distinct head, thorax, and abdomen (**Figure 23.28**). The head has one pair of antennae and two compound eyes. Three pairs of legs attach to the thorax. In winged insects, the thorax also has one or two pairs of wings. Digestive organs and sex organs are in the abdomen. Like other arthropods, insects have a tubular gut and an open circulatory system. The insect respiratory system is one adaptation to life on land. It consists of tracheal tubes that convey air from the insect's surface to deep in the body where gas exchange occurs. Water-conserving excretory organs (Malpighian tubules) that serve the same function as vertebrate kidneys are another adaptation to life on land.

The earliest insects were wingless, ground-dwelling scavengers that did not undergo metamorphosis. A few modern insects such as bristletails and silverfish retain this type of body form and development. When they hatch from an egg, they look like a tiny adult, and they simply grow bigger with each molt. Most other modern insects have wings and

undergo metamorphosis. Insects are the only winged invertebrates, and their ability to fly gives them dispersal abilities unrivaled among other land invertebrates.

Cockroaches, grasshoppers, and dragonflies are winged insects that undergo incomplete metamorphosis: An egg hatches into a nymph that takes on the adult form over the course of several molts. A dragonfly, for example, lives as an aquatic nymph for one to three years before emerging from the water and undergoing a final molt to the winged adult form.

With complete metamorphosis, a larva grows and molts without altering its form, then undergoes pupation. A pupa is a nonfeeding body in which larval tissues are remodeled into the adult form (**Figure 23.29**). Members of the four most diverse insect orders have wings and undergo complete metamorphosis. There are about 150,000 species of flies (Diptera), and at least as many beetles (Coleoptera). The order Hymenoptera includes 130,000 species of wasps, ants, and bees. Moths and butterflies are Lepidoptera, a group of about 180,000 species. As a comparison, consider that there are about 5,500 species of mammals.

INSECT ECOLOGY

Insects play essential roles in just about every land ecosystem. The vast majority of flowering plants are pollinated by members of the four diverse insect orders discussed above. Other insect orders contain few or no pollinators. Close interactions between pollinator and flowering plant lineages likely contributed to an accelerated rate of speciation that increased the diversity of both.

Insects serve as food for a variety of wildlife. Most songbirds nourish their nestlings on an insect diet. Those that migrate often travel long distances in order to nest and raise their young in areas where insect abundance is seasonally high. Aquatic larvae of insects such as dragonflies and mayflies serve as food for trout and other freshwater fish. Most amphibians and reptiles feed mainly on insects. Even humans

Figure 23.29 Complete metamorphosis in a butterfly.

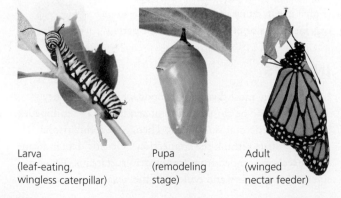

Larva
(leaf-eating, wingless caterpillar)

Pupa
(remodeling stage)

Adult
(winged nectar feeder)

CREDITS: (28) From Russell/Wolfe/Hertz/Starr, *Biology*, 1e. © 2008 Brooks Cole, a part of Cengage Learning, Inc.; (29) left and middle, © Jacob Hamblin/Shutterstock.com; right, © Laurie Barr/Shutterstock.com.

Dung beetles feed on feces. The beetles sniff out their stinky food, then form it into balls that they roll away and bury. The adult beetles may later eat the dung themselves, or the female may lay her eggs in it, in which case it feeds her larvae.

Figure 23.30 Dung beetles, part of nature's clean-up crew.

eat insects; in many cultures, they are considered a tasty source of protein.

Insects dispose of wastes and remains. Flies and beetles quickly discover an animal corpse or a pile of feces (**Figure 23.30**). They lay their eggs in or on this organic material, which serves as food for their larvae. By making use of this material, these insects keep wastes and remains from piling up, and help distribute nutrients through ecosystems.

HEALTH AND ECONOMIC EFFECTS

Plant-eating insects are our main competitors for plant products. Each year, they devour as much as one-third of all crops grown in the United States. Pine beetles and other insects bore into living trees and reduce lumber harvests. Other beetles can damage wooden buildings, as do some termites and ants.

Parasitic insects can pose a threat to human health and well-being. Mosquitoes cause more than a million deaths per year by transmitting malaria, West Nile fever, and other

deadly diseases. Fleas transmit bubonic plague, and body lice spread typhus. Bedbugs (left) do not cause disease, but infestations are stressful and eliminating them can be costly. On the other hand, some insects help us keep pest species in check. For example, larvae of ladybird beetles devour aphids that would otherwise suck juices from valuable crops.

Economically valuable products produced by insects include honey, silk, and shellac. Honey is floral nectar concentrated and stored by honeybees. Larval silk moths produce silk as they spin a cocoon in which to pupate. Asian bark-feeding bugs excrete shellac, a resin we use to give a shiny coating to wood products and to candies such as jelly beans.

insects Most diverse arthropod group; members have six legs, two antennae, and, in most subgroups, wings.

LEARNING OBJECTIVES

Explain the structure and function of the echinoderm water–vascular system.
Describe how a sea star feeds.

With this section we begin our survey of the deuterostome lineages. Echinoderms are the most diverse group of invertebrate deuterostomes. We discuss other invertebrate deuterostomes and their vertebrate relatives in the next chapter.

Figure 23.31 Echinoderms.

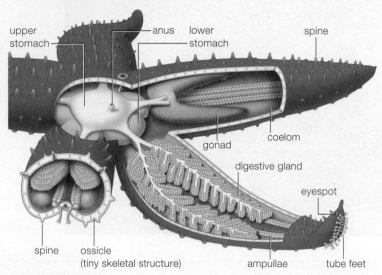

A Sea star body plan.

B Brittle star.

C Sea urchin.

D Sea cucumber.

The name **echinoderms** (phylum Echinodermata) means "spiny-skinned" and refers to interlocking spines and plates of calcium carbonate embedded in the skin. Adults are coelomate animals and most have a radial body, with five parts around a central axis. Larvae are bilateral, and echinoderms are thought to have evolved from a bilateral ancestor.

Sea stars (also called starfish) are the most familiar echinoderms (**Figure 23.31A**). Sea stars lack a brain but do have a nerve net. A nerve ring surrounds the mouth, and nerves that branch from this ring extend into the arms. Eyespots at the tips of arms detect light and movement.

Most sea stars are active predators that move about on fluid-filled tube feet. Tube feet are part of a **water–vascular system** unique to echinoderms. The system includes a central ring and fluid-filled canals that extend into each arm. Side canals deliver coelomic fluid into tiny bulbs that function like the bulb on a medicine dropper. Contraction of an ampulla forces fluid into an attached tube foot, extending the foot. A sea star glides along as coordinated contraction and relaxation of the bulbs redistribute fluid among hundreds of tube feet. Gas exchange occurs by diffusion across the tube feet and tiny skin projections at the body surface.

Sea stars feed on bivalve mollusks. A sea star's mouth is on its lower surface. To feed, the animal slides its stomach out through its mouth and into a bivalve's shell. The stomach secretes acid and enzymes that kill the mollusk and begin to digest it. Partially digested food is taken into the stomach, then digestion is completed with the aid of digestive glands in the arms. Arms also hold gonads (reproductive organs). Each sea star is either male or female.

In addition to sea stars, echinoderms include brittle stars, sea urchins, and sea cucumbers. Brittle stars are the most diverse and abundant echinoderms (**Figure 23.31B**). These scavengers generally live in deep water. They have a central disk and thin, flexible arms that move in a snakelike way. Sea urchins have a stiff, rounded cover of calcium carbonate plates through which movable spines protrude (**Figure 23.31C**). Most graze on algae. Sea cucumbers are, as their name implies, cucumber-shaped, with microscopic plates embedded in a soft body (**Figure 23.31D**). Most feed by burrowing through sediments and ingesting them. If attacked, a sea cucumber expels organs through its anus to distract the predator. If the sea cucumber escapes, its missing parts grow back. Sea stars also have great powers of regeneration; an animal can grow from one arm and a bit of central disk.

echinoderms (eh-KYE-nuh-derms) Invertebrates with a water–vascular system, and hardened plates and spines in the skin or body.
water–vascular system Of echinoderms, a system of fluid-filled tubes and tube feet that function in locomotion.

CREDITS: (31B) Stubblefield Photography/Shutterstock.com; (31C) © Derek Holzapfel/photos.com; (31D) Andrew David, NOAA/NMFS/SEFSC Panama City, Lance Horn, UNCW/ NURC-Phantom II ROV operator.

Figure 23.32 A cone snail eating a fish that it has injected with venom.

MEDICINES FROM THE SEA

There are over 800 species of cone snails. Each species of these predatory snails produces a venom that contains around a hundred different peptides. Cone snail peptides typically act on a prey animal's nervous system. The snail catches prey such as a small fish by harpooning it with a modified radula, injecting it with venom, then engulfing it (**Figure 23.32**). The venom sedates the fish and dulls its pain, so it does not struggle and possibly damage the snail's harpoon.

The human nervous system uses the same chemical signals a fish nervous system does, so compounds used by a snail to sedate or anesthetize fish can be useful as human medicines. Consider the injectable painkiller ziconitide (Prialt). It is a synthetic version of a cone snail peptide that is more powerful than morphine, yet not addictive. Another synthetic version of a snail peptide is being tested as a treatment for epilepsy. Researchers have only begun to explore the potential of cone snail venom. Given the large number of cone snails and the complexity of their venoms, there is plenty to do.

Nonvenomous invertebrates can also be a source of new medicines. Many invertebrates produce chemicals that have antibacterial, antiprotozoal, or antifungal activity. To develop new drugs with these properties, researchers extract compounds from invertebrates, then test the ability of the compounds to kill pathogens cultured in the laboratory. Researchers also test the effect of each compound on cultured human cells. An ideal candidate drug is one that kills pathogens and does no harm to human cells. For example, some compounds made by marine sponges kill disease-causing protists (such as trypanosomes and apicomplexans) and have little effect on human cells growing in culture.

A new appreciation of the potential medicinal value of compounds produced by marine invertebrates has many researchers worried about the declining state of our oceans. Pollution, destructive harvesting methods, overharvesting, and climate change are likely to drive many potentially valuable species to extinction before we have time to discover the beneficial compounds they make.

Section 23.1 **Animals** are multicelled heterotrophs that digest food in their body. Most have an embryo with three layers: **ectoderm**, **endoderm**, and **mesoderm**. Most animals are **invertebrates**. Early invertebrates had no body symmetry. Cnidarians have **radial symmetry**, but most animals have **bilateral symmetry**. Most bilateral animals have a body cavity (**pseudocoelom** or a **coelom**). There are two bilateral animal lineages: **protostomes** and **deuterostomes**. **Table 23.1** summarizes the traits of major animal lineages discussed in this chapter.

Section 23.2 According to the **colonial theory of animal origins**, animals evolved from a colonial protist. Trace fossils and body fossils provide evidence that animals arose more than 600 million years ago. A great adaptive radiation during the Cambrian gave rise to most modern animal lineages.

Section 23.3 **Sponges** have a porous body with no tissues or organs. These **suspension feeders** filter food from water. Each is a **hermaphrodite**, producing both eggs and sperm. The ciliated **larva** is the only motile stage.

Section 23.4 **Cnidarians** are **medusae** or **polyps** that have stinging cells called **cnidocytes**. A **gastrovascular cavity** functions in both respiration and digestion. A **nerve net** commands contractile cells that redistribute fluid of the **hydrostatic skeleton** to alter the body's shape.

Section 23.5 **Flatworms**, the simplest animals with organ systems, include marine species and the freshwater planarians, as well as parasitic tapeworms and flukes. Some tapeworms and flukes infect humans.

Section 23.6 **Annelids** are segmented worms. They have a **closed circulatory system** and digestive, solute-regulating, and nervous systems. Annelids move when muscles exert force on coelomic fluid inside their segments, altering the shape of segments in a coordinated manner. Oligochaetes include aquatic species and the familiar earthworms. Polychaetes are predatory or filter-feeding marine worms. Leeches are scavengers, predators, or bloodsucking parasites.

Section 23.7 **Mollusks** have a small coelom and a sheetlike **mantle**. They include **gastropods** (such as snails), **bivalves** (such as clams), and **cephalopods** (such as squids and octopuses). Aquatic mollusks have **gills**; land-dwelling gastropods have **lungs**. Gastropods and bivalves have an **open circulatory system**; cephalopods have a closed circulatory system.

Section 23.8 **Roundworms** (nematodes) have an unsegmented, cylindrical body covered by a cuticle that they **molt** periodically as they grow. Most roundworms are decomposers in soil, but some are parasites of plants or animals, including humans. One free-living soil roundworm is commonly used in scientific studies of development and aging.

Sections 23.9–23.11 There are more than 1 million **arthropod** species. Their diversity is attributed to structural traits such as a hardened **exoskeleton**, jointed appendages, specialized segments, and sensory structures such as **antennae** and **compound eyes**. Dramatic body changes during **metamorphosis** allow some arthropods to exploit multiple resources during their lifetime.

Chelicerates include marine horseshoe crabs and the mostly terrestrial **arachnids** (spiders, scorpions, ticks, and mites). The most abundant arthropods in aquatic habitats are **crustaceans**. They include bottom-feeding crabs, free-swimming krill and copepods, and the sessile, shelled barnacles. Centipedes and millipedes have segmented bodies with many legs. **Insects** include the only winged invertebrates. Some insects are decomposers and pollinators; others harm crops and transmit diseases.

Section 23.12 **Echinoderms** such as sea stars, brittle stars, sea urchins, and sea cucumbers belong to the deuterostome lineage. Spines and other hard parts embedded in their skin support the body. There is no central nervous system. A **water–vascular system** with tube feet functions in locomotion. Adults are radial, but larvae are bilateral, and the group is thought to have bilateral ancestry.

Application Invertebrate species are a largely untapped source of medicines. Compounds that predatory snails use to subdue fish prey can sometimes be used as drugs because all vertebrates have similar nervous systems.

TABLE 23.1

Comparative Summary of Animal Body Plans

Group	Adult Symmetry	Embryonic Layers	Digestive Cavity	Circulatory System
Sponges	None	None	None	None
Cnidarians	Radial	2	Gastrovascular cavity	None
Protostomes				
Flatworms	Bilateral	3	Gastrovascular	None
Annelids	Bilateral	3	Tubular gut	Closed
Mollusks	Bilateral	3	Tubular gut	Most open; cephalopods closed
Roundworms	Bilateral	3	Tubular gut	None
Arthropods	Bilateral	3	Tubular gut	Open
Deuterostomes				
Echinoderms	Radial	3	Tubular gut	Open
Vertebrates	Bilateral	3	Tubular gut	Closed

DATA ANALYSIS

CLASS ACTIVITY

USE OF HORSESHOE CRAB BLOOD Horseshoe crab blood clots immediately upon exposure to bacterial endotoxins (Section 19.5), so it can be used to test injectable drugs for the presence of dangerous bacteria. To keep horseshoe crab populations stable, blood is extracted from captured animals, which are then returned to the wild. Concerns about the survival of animals after bleeding led researchers to do an experiment. They compared survival of animals captured and maintained in a tank with that of animals captured, bled, and kept in a similar tank. **Figure 23.33** shows the results.

1. In which trial did the most control crabs die? In which did the most bled crabs die?

2. Looking at the overall results, how did the mortality of the two groups differ?

3. Based on these results, would you conclude that bleeding harms horseshoe crabs more than capture alone does?

	Control Animals		Bled Animals	
Trial	Number of crabs	Number that died	Number of crabs	Number that died
1	10	0	10	0
2	10	0	10	3
3	30	0	30	0
4	30	0	30	0
5	30	1	30	6
6	30	0	30	0
7	30	0	30	2
8	30	0	30	5
Total	200	1	200	16

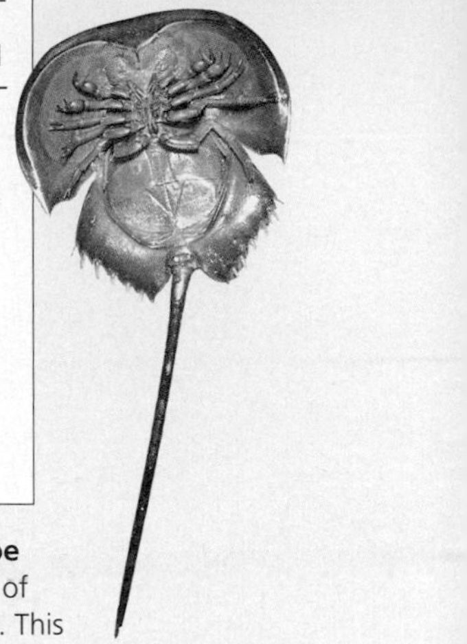

Figure 23.33 Effect of blood extraction on horseshoe crabs. Blood was taken from half the animals on the day of their capture. Control animals were handled, but not bled. This procedure was repeated eight times with different sets of horseshoe crabs. Animals were monitored for two weeks.

SELF-ASSESSMENT

ANSWERS IN APPENDIX VII

1. A(n) _______ develops from a two-layer embryo.
   a. sea star
   b. sea anemone
   c. butterfly
   d. earthworm

2. A great adaptive radiation of animals that occurred during the _______ gave rise to modern animal lineages.
   a. Precambrian
   b. Ediacaran
   c. Cambrian
   d. Jurassic

3. A coelom is a _______ .
   a. type of bristle
   b. resting stage
   c. sensory organ
   d. lined body cavity

4. Cnidarians alone have _______ .
   a. cnidocytes
   b. a mantle
   c. a hydrostatic skeleton
   d. a radula

5. Flukes are most closely related to _______ .
   a. tapeworms
   b. roundworms
   c. leeches
   d. echinoderms

6. Which group has six legs and two antennae?
   a. oligochaetes
   b. insects
   c. spiders
   d. horseshoe crabs

7. The _______ are mollusks with a hinged shell.
   a. bivalves
   b. barnacles
   c. gastropods
   d. cephalopods

8. Which of these groups includes the most species?
   a. protostomes
   b. roundworms
   c. arthropods
   d. mollusks

9. The _______ include the only winged invertebrates.
   a. cnidarians
   b. echinoderms
   c. arthropods
   d. mollusks

10. The _______ move about on tube feet.
    a. cnidarians
    b. echinoderms
    c. annelids
    d. flatworms

11. Annelids and cephalopods have a(n) _______ circulatory system.
    a. open
    b. closed

12. A pupa forms during the life cycle of some _______ .
    a. crustaceans
    b. echinoderms
    c. annelids
    d. insects

13. Match the organisms with their descriptions.

___ echinoderms	a. tubular gut, pseudocoelom
___ mollusks	b. spiny skin
___ sponges	c. simplest organ systems
___ cnidarians	d. lots of pores, no symmetry
___ flatworms	e. jointed exoskeleton
___ roundworms	f. mantle over body mass
___ annelids	g. segmented worms
___ arthropods	h. tentacles with stinging cells

CLASS ACTIVITY

CRITICAL THINKING

1. Most hermaphrodites cannot fertilize their own eggs, but the tapeworms can. Explain the advantages and disadvantages of self-fertilization in these flatworms.

2. Increased acidity makes it more difficult for invertebrates to build structures that contain calcium carbonate. List the types of invertebrates that will be harmed by ongoing ocean acidification.

3. Some insecticides that act by preventing molting also can harm crustaceans accidentally exposed to them. Explain why.

for additional quizzes, flashcards, and other study materials
ACCESS MINDTAP AT WWW.CENGAGEBRAIN.COM

CREDITS: (33) left, Data: Walls, E., Berkson, J., *Fish. Bull.* 101:457–459 (2003); right, Jane Burton/ Bruce Coleman, Ltd.

24

Animals II: The Chordates

Links to Earlier Concepts

As we continue to survey animals, reviewing the geologic time scale (16.6) may help put events in perspective. We discuss a variety of fossils (16.3) and consider homologous structures (16.7). You will draw on your understanding of genome comparisons (15.4) speciation (17.7), adaptive radiation and extinction (17.10), and cladistics (17.11).

Core Concepts

All organisms alive today are linked by lines of descent from shared ancestors.

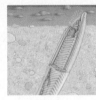

Shared core processes and features provide evidence that organisms are descended from a common ancestor. The four shared traits that link all chordates (invertebrate and vertebrate) can be observed in their embryos: a supporting rod (notochord), a dorsal nerve cord, a pharynx with gill slits, and a tail that extends past the anus.

Adaptation of organisms to a variety of environments has resulted in diverse structures and physiological processes.

Fishes are adapted to a life spent in water by gills and fins. Lungs, a more efficient heart, and limbs with supporting bones allow amphibians to spend time on land. Waterproof skin and eggs allow amniotes (reptiles, birds, and mammals) to live entirely on land. Most primates are adapted to life in trees, but humans and their extinct relatives have traits that facilitate upright walking.

Biological systems are affected by factors that disrupt their homeostasis.

Many species that once existed have become extinct, and extinction is an ongoing process. With the exception of birds, the dinosaur lineage is extinct. Many human relatives such as Neanderthals are known only from the fossil record. Amphibians are currently in the midst of an extinction crisis. Rising human populations are destroying amphibian habitats, introducing pollutants, and distributing pathogens on a global basis.

◄ Excavation of fossil dinosaur bones in northern Wyoming. Much of our information about vertebrate evolution comes from studies of such fossil skeletons.

Photograph by Lynn Johnson, National Geographic Creative.

LEARNING OBJECTIVES

List the four traits that define chordates.

Describe the two groups of invertebrate chordates and draw a tree showing how they relate to one another and to vertebrates.

Explain how an endoskeleton differs from an exoskeleton.

CHORDATE CHARACTERISTICS

The previous chapter ended with a discussion of echino-derms, one of the deuterostome lineages. The other major deuterostome lineage is the **chordates** (phylum Chordata). These bilaterally symmetrical, coelomate animals typically have a tubular digestive tract and a closed circulatory system. Four traits of chordate embryos define the lineage:

1. A **notochord**, a rod of stiff but flexible connective tis-sue, extends the length of the body and supports it.
2. A dorsal, hollow nerve cord parallels the notochord. In other animals, any nerve cord or cords runs along the body's ventral (lower) surface.
3. Narrow gill slits open across the wall of the pharynx (the throat region).
4. A muscular tail extends beyond the anus.

Depending on the subgroup, some, none, or all of these embryonic traits persist in the adult.

Most chordate species are **vertebrates** (subphylum Ver-tebrata), animals that have a backbone. However, the chor-dates also include two lineages of marine invertebrates.

INVERTEBRATE CHORDATES

Lancelets (Cephalochordata) are invertebrates with an elon-gated body that retains all characteristic chordate traits into adulthood (**Figure 24.1A**). A lancelet is about 5 centimeters (2 inches) long. Its dorsal nerve cord connects to a simple brain. An eyespot at the end of the nerve cord detects light, but there are no paired sensory organs like those of fishes. Lancelets usually bury themselves tail down in sand, then draw water in through their mouth. Food particles are filtered out as the water exits through the gill slits. Cilia on the gills move food particles to the gut. Lancelets have separate sexes: Each individual is either male or female.

Tunicates (Urochordata) have larvae that retain all four characteristic chordate features (**Figure 24.1B**). Larvae swim about briefly, then undergo metamorphosis. The tail breaks down and other parts are remodeled. Adults retain only the pharynx with gill slits (**Figure 24.1C**). They secrete a carbohydrate-rich covering, or "tunic," that gives the group its common name. An adult feeds by drawing water in through an oral opening, capturing food on its gill slits, then expelling water through a second opening. Most tunicates are sessile,

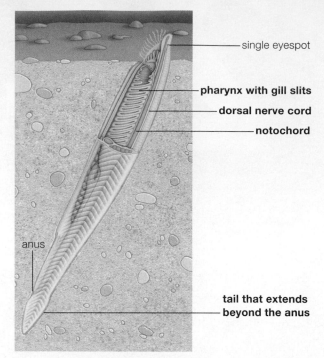

A Lancelet. Both adults and larvae retain all four of the defining traits of chordate embryos.

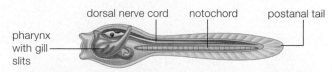

B Tunicate larva. The free-swiming larvae retain all four of the defining traits of chordate embryos.

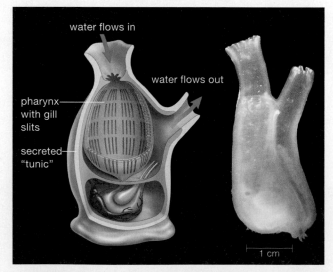

C Adult tunicate. The only defining chordate trait retained by the adult is the pharynx with gill slits. The species shown at right attaches to a surface.

Figure 24.1 Invertebrate chordates.

CREDITS: (1B, 1C left) From Russell/Wolfe/Hertz/Starr. Biology, 1E. © Cengage Learning Inc.; (1C right) © California Academy of Sciences.

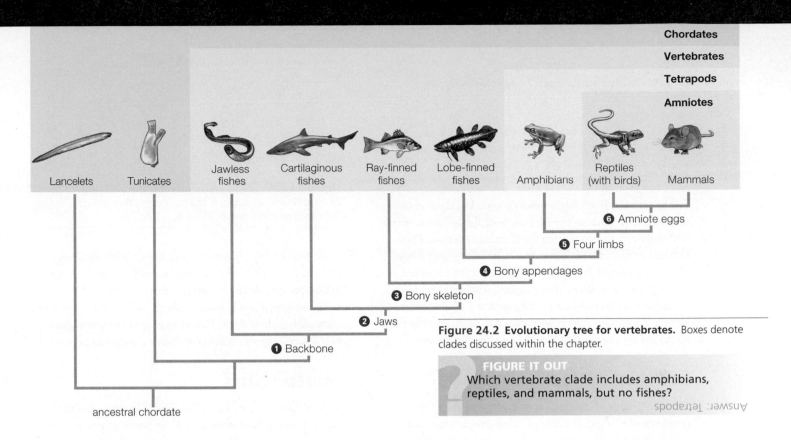

Figure 24.2 Evolutionary tree for vertebrates. Boxes denote clades discussed within the chapter.

FIGURE IT OUT
Which vertebrate clade includes amphibians, reptiles, and mammals, but no fishes?

Answer: Tetrapods

but those called salps drift or swim in the sea. Tunicates are hermaphrodites: An individual releases both eggs and sperm.

OVERVIEW OF CHORDATE EVOLUTION

Until recently, lancelets were considered the closest invertebrate relatives of vertebrates. An adult lancelet looks more like a fish than an adult tunicate does, but such apparent similarities can be deceiving. Comparisons of developmental processes and gene sequences have revealed that tunicates are closer to the vertebrates, as illustrated in **Figure 24.2**.

Most chordates are vertebrates ❶. Their backbone and other skeletal elements are components of the vertebrate **endoskeleton**, or internal skeleton. The vertebrate endoskeleton consists of living cells and grows with an animal. Thus, it does not have to be molted as an exoskeleton does.

The first vertebrates were jawless fishes that sucked up or scraped up food. Later, hinged skeletal elements called jaws evolved ❷. Jawless fishes had a skeleton of cartilage, and one group of jawed fishes (cartilaginous fishes) retains this trait. In all other jawed animals, bone replaces cartilage as the major component of the skeleton ❸.

Thick, lobe-like fins with muscle and bone inside them define one group of bony fishes (lobe-finned fishes) ❹. In one lineage that descended from these fishes, these fins became modified as bony limbs.

Tetrapods are vertebrates that have four bony limbs or are descended from a four-limbed ancestor ❺. The earliest tetrapods were fully aquatic. Amphibians are tetrapods that can spend time on land, but must return to water to breed. They gave rise to **amniotes,** whose eggs have a series of waterproof membranes that enclose the embryo in fluid ❻. These specialized eggs and other adaptations to a dry habitat allowed amniotes to disperse widely on land. Modern amniotes include the animals traditionally known as reptiles (lizards, snakes, and turtles), as well as birds and mammals.

amniote (AM-nee-oat) Vertebrate whose egg has waterproof membranes that allow it to develop away from water; a reptile, bird, or mammal.

chordate (CORE-date) Animal with an embryo that has a notochord, dorsal nerve cord, pharyngeal gill slits, and a tail that extends beyond the anus. For example, a lancelet or a vertebrate.

endoskeleton Internal skeleton made up of hardened components such as bones.

lancelet (LANCE-uh-lit) Invertebrate chordate that has a fishlike shape and retains the defining chordate traits into adulthood.

notochord (NO-toe-cord) Stiff rod of connective tissue that runs the length of the body in chordate larvae or embryos.

tetrapod (TEH-trah-pod) A vertebrate with four bony limbs, or a descendant of a four-legged ancestor.

tunicate (TOON-ih-kit) Invertebrate chordate that loses most of its defining chordate traits during the transition to adulthood.

vertebrate Animal with a backbone.

Describe the two modern groups of jawless fishes.

List some anatomical differences between cartilaginous fishes and ray-finned fishes.

Explain the function of a cloaca.

Using appropriate examples, describe the two lineages of bony fishes and the anatomical traits that distinguish them.

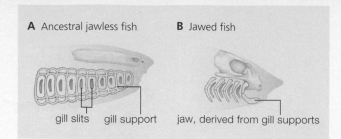

A Ancestral jawless fish   **B** Jawed fish

gill slits   gill support   jaw, derived from gill supports

Figure 24.4 Proposed mechanism for the evolution of jaws. Modification of gill supports of a jawless fish **A** produced a hinged, two-part jaw in early jawed fishes **B**.

Fishes were the first vertebrates to evolve, and they remain the most fully aquatic. The earliest fossil fishes date to about 530 million years ago, during the Cambrian period. These fish had a tapered body a few centimeters long, and a head with a pair of eyes, but no jaws. Their skeleton consisted of cartilage, the same tissue that supports your ears and nose. All fishes have a closed circulatory system with one heart. All also have a urinary system with paired kidneys that filter and adjust the composition of blood, and eliminate wastes.

JAWLESS FISHES

The two groups of modern **jawless fishes**—hagfishes and lampreys—belong to the clade Cyclostomata, which means round-mouthed. Both groups have a cartilage skeleton and a cylindrical body about a meter long. They do not have fins and, like lancelets, they move by wiggling. Rather than jaws, their mouth has hard parts made of keratin, the same protein that makes up your nails. Multiple gill slits are visible on either side of the body. On the ventral (lower) surface there is a **cloaca**, a multipurpose opening that serves as the exit for digestive and urinary waste, and also functions in reproduction.

Hagfishes (**Figure 24.3A**) are all marine bottom feeders. They eat soft-bodied invertebrates and scavenge dead or dying fish. When threatened, a hagfish secretes proteins that mix with water to form a slimy mucus. The cloud of slime deters most predators, but humans still harvest hagfish. Most products sold as "eelskin" are actually hagfish skin.

Lampreys breed in fresh water, although some spend time in the ocean. Many lamprey species parasitize other fish. A parasitic lamprey (**Figure 24.3B**) attaches to a host fish, then uses its hard mouthparts to scrape off bits of flesh. In the early 1900s, parasitic Atlantic lampreys that entered the Great Lakes by way of newly built canals decimated populations of native fish such as trout. Fishery managers in these regions now lower lamprey numbers with dams, nets, and poisons.

JAWED FISHES

Jawless fishes have skeletal elements called gill arches that support their gills (**Figure 24.4A**). In jawed fishes, some of these skeletal elements have been modified and now serve as components of the jaws. Jawed fishes were also the first animals with scales and paired fins. Scales are hard, flattened structures that grow from and often cover the skin. Fins are flattened appendages that propel and steer a body while swimming. Most jawed fishes have paired fins.

Cartilaginous Fishes The **cartilaginous fishes** (Chondrichthyes) are jawed fishes with a skeleton made primarily of cartilage. They have paired immobile fins and multiple rows of teeth that are continually shed and replaced. As in jawless fishes, gill slits are visible at the body surface. Most cartilaginous fishes have a cloaca. Sharks, rays, and skates are examples of cartilaginous fishes.

The most well-known sharks are speedy predators that chase down and tear apart prey (**Figure 24.5A**). Other sharks are bottom-feeders that suck up invertebrates and act as scavengers. Still others strain plankton from the seawater. The largest living fish, the whale shark, is a plankton feeder. It can weigh several tons.

Rays have a flattened body with large pectoral fins and a long, thin tail. For example, manta rays glide through seawater and filter out plankton (**Figure 24.5B**). Stingrays are bottom-feeders. A sharp, spearlike point at the tip of their tail defends them against predators. In some stingrays, the tail can deliver a dose of venom. Skates are bottom-feeders that resemble rays but have a shorter, thicker tail.

Figure 24.3 Modern jawless fishes.

A Hagfish. It feeds on worms and scavenges on the seafloor.

B Parasitic lamprey. It attaches to another fish with its oral disk and scrapes off bits of flesh.

A Galápagos shark, a fast-swimming predator.

B Manta ray, a slow-moving plankton feeder.

Figure 24.5 Cartilaginous fishes. Note the gill slits visible at the body surface of both fishes.

Bony Fishes Bony fishes are not a clade, they are an informal grouping of fishes in which the skeleton consists largely of bone. There are two modern lineages of bony fishes: ray-finned fishes and lobe-finned fishes. Most bony fishes have movable pelvic and pectoral fins and a movable cover over their gill slits. Bony fishes also have a gas-filled organ or organs derived from outpouchings of their pharynx (throat). In most bony fishes, urinary waste, digestive waste, and gametes exit through three separate body openings.

Ray-finned fishes have thin, weblike fins with flexible supports derived from skin (**Figure 24.6A**). With more than 21,000 freshwater and marine species, they are the most diverse vertebrate group. Most ray-finned fishes have a **swim bladder**, a gas-filled organ that allows them to adjust their buoyancy (**Figure 24.6B**).

Lobe-finned fishes have pelvic and pectoral fins supported by internal bones. This lineage includes coelacanths (Section 17.10) and lungfishes (**Figure 24.6C**). Typical lungfishes have gills and one or two lungs. **Lungs** are respiratory organs—air-filled sacs that exchange gases with an associated network of tiny blood vessels. A lungfish gulps air into its lungs, then oxygen diffuses from its lungs into the blood. As the next section explains, bony fins and simple lungs facilitated the evolutionary move to land.

---

cartilaginous fish (car-tih-LAJ-ih-nuss) Jawed fish with a skeleton of cartilage; a shark, ray, or skate.

cloaca (klo-AY-kuh) Body opening that serves as the exit for digestive waste and urine; also functions in reproduction.

jawless fish Fish with a skeleton of cartilage, no fins or jaws; a lamprey or hagfish.

lobe-finned fish Jawed fish with fins supported by internal bones; a coelacanth or lungfish.

lung Respiratory organ; an air-filled sac that exchanges gases with an associated network of blood vessels.

ray-finned fish Jawed fish with fins supported by thin rays derived from skin; member of most diverse lineage of fishes.

swim bladder Of many ray-finned fishes, a gas-filled organ that adjusts buoyancy.

A Ray-finned fish (a carp).

gill cover
pectoral fin
pelvic fin

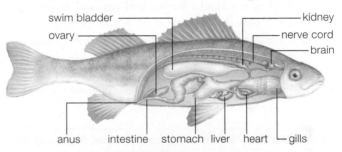

swim bladder
ovary
kidney
nerve cord
brain
anus
intestine
stomach
liver
heart
gills

B Ray-finned fish body plan (a perch).

pelvic fin
pectoral fin

C Lobe-finned fish (lungfish).

Figure 24.6 Bony fishes.

FIGURE IT OUT
Which of these fishes has an air-filled organ that functions in respiration?

Answer: The lungfish

CREDITS: (5A) Jonathan Bird/Oceanic Research Group, Inc.; (5B) © Gido Braase/Deep Blue Productions; (6A) Ultrachock/Shutterstock.com; (6B) From Solomon, L. Berg, and D. W. Martin, *Biology*, Seventh Edition, Cengage Learning Inc.; (6C) © Wernher Krutein/photovault.com.

Describe the anatomical and physiological adaptations that arose during the transition to life on land.

Explain how the body form and development of salamanders differ from those of frogs and toads.

List some causes of the ongoing decline in amphibian diversity.

Figure 24.8 Salamander, with equal-sized forelimbs and hindlimbs.

THE MOVE TO LAND

Amphibians are scaleless, land-dwelling vertebrates that typically breed in water. All are carnivores. Amphibians were the first tetrapods. Recently discovered fossil footprints from Poland show that amphibians were walking on land by about 395 million years ago, in the Devonian. The animal that left the footprints was about 2.5 meters (8 feet) long.

Amphibians evolved from a lobe-finned fish. Fossils demonstrate how fishes adapted to swimming evolved into four-legged walkers (**Figure 24.7**). Bones of a lobe-finned fish's pectoral fins and pelvic fins are homologous with those of an amphibian's front and hind limbs. During the transition to land, these bones became larger and better able to bear weight. Ribs enlarged and a distinct neck emerged, allowing the head to move independently of the rest of the body.

The transition to land was not simply a matter of skeletal changes. Lungs, which had previously served an accessory purpose, became larger and more complex. Division of the previously two-chambered heart into three chambers allowed blood to flow in two circuits, one to the body and back, and one to and from the increasingly important lungs. (We discuss evolution of vertebrate hearts further in Section 33.1.) Changes to the inner ear improved detection of airborne sounds. Eyes became protected from drying out by eyelids.

What drove the move onto land? Evolution of an ability to spend time out of water would have been favored in seasonally dry places. In addition, it would have allowed individuals to escape aquatic predators and to access a new source of food—insects—which also arose during the Devonian.

MODERN AMPHIBIANS

Salamanders and newts have a body form similar to early tetrapods, with a long tail, and forelimbs about the same size as their back limbs (**Figure 24.8**). When a salamander walks, its body movements resemble the swimming movements made

---

Figure 24.7 Fossil fish from the late Devonian. Fossils illustrate the modification.

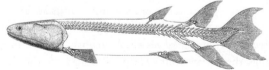

❶ Fish (*Eusthenopteron*) with bony fins.

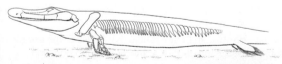

❷ Fish (*Tiktaalik*) with sturdier, weight-bearing pectoral fins, wristlike bones, and enlarged ribs.

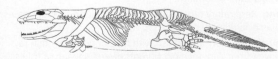

❸ Early amphibian (*Icthyostega*) with well-developed ribs, thick limbs, and distinct digits (toes).

by a fish. Early tetrapods initially walked in the water before one lineage ventured onto land.

Larval salamanders look like small versions of adults, except for the presence of gills. In a few species (axolotls), gills are retained into adulthood. More typically, gills disappear and lungs develop as the animal matures.

Frogs and toads belong to the most diverse amphibian lineage (Anura), with more than 5,000 species. The long, muscular hindlimbs of an adult frog allow it to swim, hop, and make spectacular leaps (**Figure 24.9**). The much smaller forelimbs help absorb the impact of landings. Toads can hop, but they usually walk, and have somewhat shorter hind legs than frogs. Toads are also better adapted to dry conditions. Both frogs and toads undergo metamorphosis, during which the gilled, tailed larva transforms itself into an adult with lungs and no tail.

DECLINING DIVERSITY

Amphibians are declining and disappearing throughout the world (**Figure 24.10**). Shrinking or deteriorating habitats threaten the remaining populations. Consider that nearly all amphibians deposit their eggs and sperm in water, and their larvae must develop in water. People destroy amphibian breeding spots when they fill in or develop areas where seasonal pools form. They also degrade amphibian habitats by allowing chemicals such as weed killers, fertilizers, and insecticides to pollute freshwater environments. Some pollutants harm amphibians directly, and others increase their vulnerability to pathogens. For example, some agricultural chemicals make tadpoles more easily infected by flukes that can deform their developing limbs.

The spread of introduced pathogens and parasites is another pressing threat. The Chapter 22 Application described the devastating effects of *Batrachochytrium dendrobatidis*

Figure 24.9 Frogs.
Adult frogs have longer hindlimbs than forelimbs and are capable of hopping or jumping. Their aquatic larvae (inset photo) have gills, a long tail, and no limbs.

(Bd), a pathogenic chytrid fungus. Bd is native to Africa, but humans have spread it worldwide. A related chytrid fungus from Asia is now killing salamanders in Europe. As of 2016, this fungus had not been detected in North America, where salamander diversity is greatest. However, scientists worry that it could be spread worldwide by the largely unregulated trade in exotic pets.

The decline of amphibians is one aspect of an ongoing vertebrate extinction crisis that has accompanied the rise in human population. Since the 1500s, vertebrate species have been going extinct at a rate 20 to 80 times faster than they did during the period in which the dinosaurs perished.

amphibian (am-FIB-ee-un) Tetrapod with scaleless skin; it typically develops in water, then lives on land as a carnivore with lungs. For example, a frog or salamander.

Figure 24.10 Amphibian diversity and decline by region.*

*Adapted from amphibiaweb.org.

Region	Named Species	Threatened	Extinct or Nearly So
Afrotropical	760	162	26
Australian	229	35	19
Madagascan	245	60	7
Nearctic	440	107	41
Neotropical	2231	406	127
Oceania	411	20	1
Oriental	1021	266	41
Palearctic	198	233	6
Panamanian	1098	350	231
Saharo-Arabian	73	5	4
Sino-Japanese	356	66	10

24.4 Amniotes

LEARNING OBJECTIVES

Describe amniote adaptations to life on land.

Explain which groups biologists include in the clade Reptilia.

Compare the physiology of endotherms and ectotherms and provide examples of each.

Amniotes branched off from an amphibian ancestor about 300 million years ago, during the Carboniferous. A variety of adaptations allow them to live in dry places. Amniote eggs enclose the embryo in fluid, so amniote development can occur entirely on land. Inside the eggs, a series of membranes function in gas exchange, nutrition, and waste removal. Amniotes also have lungs, rather than gills, throughout their life. Their skin is rich in keratin, a protein that makes it waterproof. A pair of well-developed kidneys help conserve water, and fertilization usually takes place inside the female's body.

An early branching of the amniote lineage separated the ancestor of mammals from the ancestor of the birds and **reptiles** (an informal group that includes crocodilians, turtles, lizards, snakes). You probably do not think of birds as reptiles, but biologists place them in the clade Reptilia:

Mammals | Lizards, snakes | Turtles | Crocodilians | Birds

Clade Reptilia

Dinosaurs are extinct members of the Reptilia. Biologists define them by skeletal features such as the shape of their pelvis and hips. Like other reptiles, dinosaurs produced amniote eggs. One dinosaur group, the theropods, includes many feathered species (**Figure 24.11**). Most feathered theropods were too large and heavy to fly, so the feathers presumably benefited them by providing insulation. Like modern birds and mammals, feathered dinosaurs may have been **endotherms**, animals that maintain their body temperature by adjusting their production of metabolic heat. Endotherm means "heated from within." By contrast, **ectotherms** are animals whose internal temperature varies with that of their environment. Ectotherms regulate their body temperature by their behavior, as by basking on a sunny rock to warm up.

Birds branched from a theropod dinosaur lineage during the Jurassic. They are the only surviving descendants of dinosaurs. All dinosaurs became extinct by the end of the Cretaceous, as a result of an asteroid impact.

Mammals also arose during the Jurassic, so some lived alongside dinosaurs. However, the group to which most modern mammals belong—placental mammals—probably did not evolve until after the dinosaurs' demise.

dinosaur Group of reptiles that include the ancestors of birds; became extinct at the end of the Cretaceous.

ectotherm Animal whose body temperature varies with that of its environment.

endotherm Animal that maintains its temperature by adjusting its production of metabolic heat; for example, a bird or mammal.

reptiles Informal amniote subgroup that includes lizards, snakes, turtles, crocodilians.

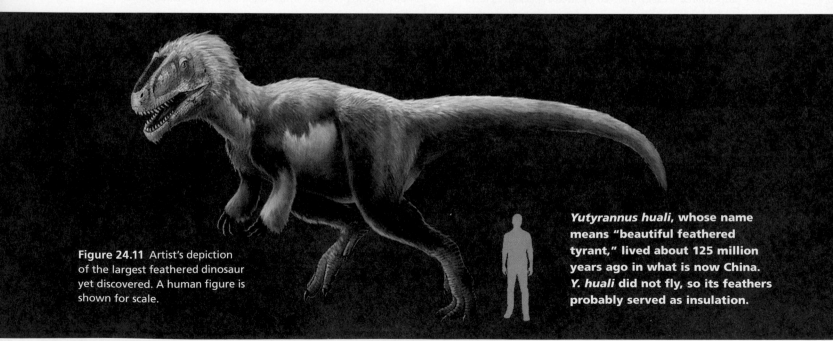

Figure 24.11 Artist's depiction of the largest feathered dinosaur yet discovered. A human figure is shown for scale.

Yutyrannus huali, whose name means "beautiful feathered tyrant," lived about 125 million years ago in what is now China. Y. huali did not fly, so its feathers probably served as insulation.

CREDIT: (11) Xing Lida, National Geographic Creative.

In this section we consider the animals traditionally called "reptiles." All are ectotherms that have a cloaca and a body covered with scales.

LIZARDS AND SNAKES

Lizards and snakes constitute the most diverse group of modern reptiles. Their body is covered with overlapping scales. Most lay eggs (**Figure 24.12A**), but females of some species brood eggs in their body and give birth to well-developed young. The eggs are self-contained; the mother does not provide any nourishment to offspring inside her body. Most lizards are predators, although iguanas are herbivores.

Snakes evolved from short-legged, long-bodied lizards during the Cretaceous. Some modern snakes retain bony remnants of hindlimbs, but most lack limb bones. All are predators. Many have flexible jaws and can swallow prey whole. All snakes have teeth, but not all have fangs. Rattlesnakes and other fanged snakes bite and subdue prey with venom they produce in modified salivary (saliva-producing) glands.

TURTLES

Turtles have a bony, keratin-covered shell attached to their skeleton. A 200-million-year-old fossil turtle has a protective plate derived from expanded ribs on its belly side, but no shell on its back. The fossil turtle also has teeth. By contrast, modern turtles are toothless. As in birds, a thick layer of keratin covers their jaws and forms a horny beak (**Figure 24.12B**). Most turtles that live in the sea feed on invertebrates such as sponges or jellies; others feed mainly on sea grass. Freshwater turtles prey on fish and invertebrates. Land-dwelling turtles, commonly called tortoises, feed on plants.

CROCODILIANS

Crocodilians, which include alligators and crocodiles, are stealthy predators of the tropics that spend much of their time in water. They have a long, powerful tail, a prominent snout, and many sharp, peglike teeth (**Figure 24.12C**). Most alligators live in the Americas; crocodiles live in the Americas, Africa, Asia, and Australia.

Crocodilians are the closest living relatives of birds. Like most birds, crocodilians are highly vocal and engage in com-

A Hognose snakes emerging from leathery amniote eggs.

B Turtle in a defensive posture. Note the shell and the horny beak.

C Crocodile with its fish prey. All crocodilians are predators with peglike teeth. They spend much of their time in water.

Figure 24.12 Examples of nonbird reptiles.

plex parental behavior. During courtship, males and females grunt and bellow. After a female mates, she digs a nest, lays eggs, then buries and guards them. Young that are ready to hatch call their mother, who helps them dig their way out.

Crocodilians and birds also share another trait: a four-chambered heart. In lizards, snakes, and turtles (as in amphibians) the heart has three chambers and allows some mixing of oxygen-rich and oxygen-poor blood. In a four-chambered heart, oxygen-poor blood from body tissues never mixes with oxygen-rich blood from the lungs.

24.6 Birds—Adapted to Flight

LEARNING OBJECTIVES

Give examples of structural and physiological traits that adapt a bird to flight.

List some functions of bird feathers.

Describe the components of a fertilized bird egg.

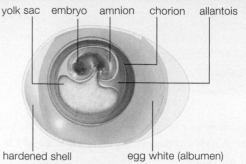

Figure 24.14 **Fertilized bird's egg.** The amnion, chorion, and allantois are membranes characteristic of amniote eggs.

Birds are the only living animals with feathers, which are filamentous keratin structures derived from scales. Birds are endotherms, and their downy feathers help them maintain their temperature. Feathers slow the loss of metabolic heat in cool environments, and prevent heat gain in hot ones. Feathers also shed water, so they help to keep a bird's skin dry.

In many birds, colorful feathers play a role in courtship.

Flight feathers are one of a bird's many adaptations for flight. Like humans, birds stand upright on their hindlimbs, and their wings are homologous with our arms. Long flight feathers that extend outward from each wing increase its surface area (above).

The bird's sternum (breastbone) has a large extension called a keel. Flight muscles run from the keel of the sternum to bones of the upper limb (**Figure 24.13**). Contraction of one set of flight muscles produces a powerful downstroke that lifts the bird. A less powerful set of muscles contracts to raise the wing.

Other adaptations for flight minimize a bird's weight. Bird bones are stronger and stiffer than mammal bones, but internal air cavities make bird bones lighter than their mammalian equivalents. A beak that consists of keratin weighs much less than bony teeth. Birds also have no bladder (an organ that stores urinary waste in many other vertebrates).

Birds use a lot of energy, both in flight and to maintain their body temperature. Several adaptations support their high metabolism. A unique system of air sacs keeps air flowing continually through the lungs, ensuring an adequate oxygen supply. (Section 35.3 discusses bird respiration in detail.) Birds also have a four-chambered heart that is large for their size, and a rapid heartbeat speeds the flow of blood.

Flying requires good eyesight and a great deal of coordination. Compared to a lizard of a similar body weight, a bird has much larger eyes and a bigger brain. Bird eyes are largely immobile, so a bird has a highly flexible neck that allows it to rotate its head in all directions.

As in other reptiles, fertilization is internal. All birds have a cloaca and most male birds do not have a penis. To inseminate a female, a male must press his cloaca against hers, a maneuver poetically described as a cloacal kiss. After fertilization occurs in the female's body, she lays an egg with the characteristic amniote membranes (**Figure 24.14**). Nutrients from the egg's yolk and water from its albumin sustain the developing embryo. Birds encase their eggs in a hard calcium carbonate shell. The egg must be kept warm in order for the embryo it contains to develop. In nearly all bird species, one or both parents incubate the eggs until they hatch.

More than half of all bird species belong to a subgroup called perching birds. Sparrows, finches, jays, and other birds commonly seen at backyard feeders are in this group. The next most diverse group, the hummingbirds, includes the most agile fliers and the only birds capable of flying backward. At the other extreme, penguins and the ratite birds such as ostriches can no longer fly.

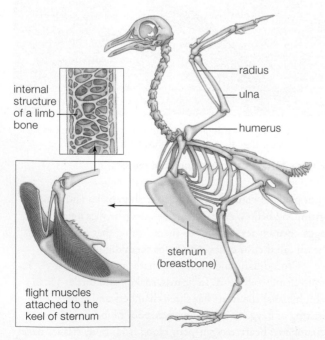

Figure 24.13 **Bird skeletal anatomy.**

bird Feathered amniotes; most have a body adapted for flight.

24.7 Mammals

Describe the traits unique to mammals.

Explain the differences among the three mammalian subgroups and provide examples of each.

FURRY OR HAIRY MILK MAKERS

Mammals are vertebrates in which females nourish their young with milk secreted by mammary glands. The group name is derived from the Latin *mamma*, meaning breast. Mammals are endotherms, and a coat of fur or head of hair helps them maintain their body temperature. Mammals also include the only animals that sweat, although not all mammals do so. The mammalian heart is four-chambered, and gas exchange occurs in a pair of well-developed lungs.

Mammals have distinctive skeletal and dental traits. Their lower jaw consists of a single bone, whereas other jawed vertebrates have multiple lower jaw bones. Mammals are also the only vertebrates that have three bones in their middle ear. Mammalian teeth are another distinguishing trait; there are four different types, each with a different shape and function. Incisors are used to gnaw, canines tear and rip flesh, and premolars and molars grind and crush hard foods. Not all mammals have teeth of all four types, but most have some combination. In other vertebrates, an individual's teeth may differ in size, but they are all the same shape.

MODERN SUBGROUPS

Monotremes, the oldest mammalian lineage, lay eggs with a leathery shell. Only three monotreme species survive: the duck-billed platypus (**Figure 24.15A**) and two kinds of spiny anteater (echidna). Newly hatched monotremes are tiny, hairless, and blind; they cling to the mother or are held in a skin fold on her belly. A monotreme has mammary glands but no nipples. Young lap up milk that oozes from openings on their mother's skin.

Marsupials are pouched mammals. Young marsupials develop briefly in their mother's body, nourished by egg yolk and by nutrients that diffuse from the mother's tissues. They are born at an early developmental stage, then crawl to a permanent pouch on their mother's belly. A nipple in the pouch supplies milk to sustain development. About 230 marsupial species, including koalas and kangaroos, live in Australia and on nearby islands. Another 100 or so species, mainly opposums (**Figure 24.15B**), live in South or Central America. Opposums are the only marsupials native to North America.

In **placental mammals** (eutherians), maternal and embryonic tissues combine as an organ called the placenta. A placenta allows maternal and embryonic bloodstreams to

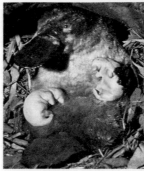

A Monotremes (platypus).   **B** Marsupials (opposums).

C Placental mammals (grizzly bears).

Figure 24.15 Mammalian mothers and their young. All female mammals produce milk to feed offspring.

transfer materials without mixing. Placental embryos grow faster than those of other mammals, and the offspring are born more fully developed. After birth, young suck milk from nipples on their mother's surface (**Figure 24.15C**). Monotremes and marsupials have a cloaca, but placental mammals have separate openings that service the urinary, reproductive, and excretory systems.

Placental mammals tend to outcompete other mammals and are now dominant on all continents except Australia. Of the approximately 5,000 species of placental mammals, nearly half are rodents. The next most diverse group is bats, with 1,200 species. Primates, the mammalian group to which humans belong, are the topic of the next section.

---

mammal Animal with hair or fur; females secrete milk from mammary glands.

marsupial (mar-SUE-pee-uhl) Mammal in which young are born at an early stage and complete development in a pouch on the mother's surface.

monotreme (MON-oh-treem) Egg-laying mammal.

placental mammal Mammal in which maternal and embryonic bloodstreams exchange materials by means of a placenta.

Section 24.1 Four features define **chordate** embryos: a **notochord**, a dorsal hollow nerve cord, a pharynx with gill slits, and a tail extending past the anus. Some or all of these features persist in adults. Chordates include two groups of marine invertebrates, the **tunicates** and **lancelets**, but most are **vertebrates**, which have an **endoskeleton** that includes a backbone. The first vertebrates were jawless fishes. **Tetrapods** are vertebrates with four limbs. **Amniotes** are tetrapods that produce eggs that allow embryos to develop on land.

Section 24.2 The first fishes had no jaws, fins, or scales. Modern **jawless fishes** (hagfishes and lampreys) also lack these traits. Jaws evolved by a modification of structures that support gill slits. **Cartilaginous fishes** have a skeleton of cartilage. Like jawless fishes, they have a **cloaca** that functions in excretion and reproduction. There are two lineages of bony fishes. A **swim bladder** allows **ray-finned fishes** (the most diverse vertebrate group) to adjust their buoyancy. **Lobe-finned fishes** have bones in their fins, and some have a **lung** or lungs.

Section 24.3 **Amphibians**, the first lineage of tetrapods, evolved from a lobe-finned fish ancestor. A variety of modifications, including a three-chambered heart, adapt amphibians to life on land. All modern amphibians, such as frogs, toads, and salamanders, are carnivores. Most spend part of their life cycle on land, but return to the water to reproduce. Many amphibian species face the threat of extinction.

Section 24.4 A variety of traits adapt amniotes to life away from water. Their kidneys and keratin-rich skin help minimize water loss. Fertilization usually takes place inside the female's body. Amniote eggs encase a developing embryo in fluid, so amniotes can develop on dry land.

Reptiles and birds belong to one amniote clade (Reptilia), and mammals to another. Birds are descended from **dinosaurs**. Some dinosaurs may have been **endotherms**, like modern birds and mammals. Modern reptiles are **ectotherms** that rely on environmental sources of heat.

Section 24.5 Lizards and snakes belong to the most diverse reptile lineage. All have scales, and most are carnivores. Turtles have a bony, keratin-covered shell and a horny beak rather than teeth. Crocodilians are aquatic carnivores and the closest relatives of birds. Like birds, they have a four-chambered heart and assist their young.

Section 24.6 **Birds** are the only living animals with feathers. Most have a body adapted to flight, with lightweight bones, a four-chambered heart, a beak rather than teeth, and a highly efficient respiratory system.

Section 24.7 **Mammals** nourish young with milk secreted by mammary glands, have fur or hair, and have more than one kind of tooth. Existing lineages are egg-laying mammals (**monotremes**), pouched mammals (**marsupials**), and **placental mammals**, the most diverse group. Placental embryos develop faster than those of other mammals and are born at a later stage of development. They have become the primary mammal group in most regions.

Sections 24.8–24.10 **Primates** are adapted to climbing. They have flexible shoulder joints, grasping hands tipped by nails, and good depth perception. Old World monkeys, New World monkeys, **apes**, and humans share a common ancestor and are **anthropoid primates**. Apes and humans do not have a tail. The lineage leading to humans and the chimpanzee/bonobo lineage diverged from a common ancestor an estimated 4 to 13 million years ago. Humans and our closest extinct ancestors are **hominins**, a group defined by **bipedalism**. We also have larger brains, finer hair, and more flexible hands than other primates. These traits evolved at different rates.

The earliest fossils that may be hominins date to 7 million years ago (**Figure 24.21**). *Ardipithecus ramidus* lived about 5 million years ago and walked upright on the ground, but also climbed among tree branches. The australopiths are a genus of hominins that likely include some human ancestors. They walked upright and had small brains.

Sections 24.11 and 24.12 The human genus (*Homo*) arose by 2 million years ago in Africa. The oldest named species, **H. habilis**, resembled australopiths but had a slightly larger brain and may have made tools. **H. erectus** had a much larger brain and a body form like that of modern humans. Some members of this species left Africa and became established in Europe and Asia. *H. erectus* is thought to be the ancestor of both Neanderthals (**Homo neanderthalensis**) and modern humans (*H. sapiens*). Neanderthals, our closest extinct relatives, had a big brain and a stocky body. They lived in Africa, Europe, and Asia, then disappeared about 28,000 years ago. *H. floresiensis* and *H. naledi* are recently proposed additions to the genus.

Our species, **H. sapiens**, arose by about 195,000 years ago in East Africa. Geneticists can determine the paths taken by early humans by looking at differences in the frequency of genetic markers among modern ethnic groups. Study of modern human genomes and of fossils has revealed evidence of long-ago matings between humans and Neanderthals.

Application Skeletal modifications that adapt the human body to bipedalism also put us at risk for knee and back problems. Coupled with our large skull size, they make childbirth more difficult in humans than other primates.

NEANDERTHAL HAIR COLOR The *MC1R* gene regulates pigmentation in humans (Sections 14.1 and 15.1 revisited), so loss-of-function mutations in this gene affect hair and skin color. A person with two mutated alleles for this gene makes more of a reddish pigment than a brownish one, resulting in red hair and pale skin. DNA extracted from two Neanderthal fossils has an *MC1R* mutation that has not yet been found in humans. To see how this mutation affects activity of the *MC1R* gene, Carles Lalueza-Fox and his team introduced the mutant allele into cultured monkey cells (**Figure 24.26**).

1. How did *MC1R* activity in monkey cells with the mutant allele differ from that in cells with the normal allele?

2. What does this imply about the mutation's effect on Neanderthal hair color?

3. What purpose do the cells with the jellyfish gene for green fluorescent protein serve in this experiment?

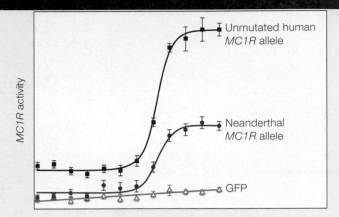

Figure 24.26 *MC1R* activity in monkey cells transgenic for an unmutated *MC1R* gene, the Neanderthal *MC1R* allele, or the gene for green fluorescent protein (GFP). GFP is not related to *MC1R*.

1. All chordates have (a) _____ as embryos.
   a. backbone   b. jaws   c. notochord   d. kidneys

2. The _____ are invertebrate chordates.
   a. echinoderms   c. lancelets
   b. hagfishes   d. lampreys

3. Vertebrate jaws evolved from _____ .
   a. gill supports   b. ribs   c. scales   d. teeth

4. Sharks and rays are _____ .
   a. ray-finned fishes   c. jawless fishes
   b. cartilaginous fishes d. lobe-finned fishes

5. A divergence from _____ gave rise to tetrapods.
   a. ray-finned fishes   c. cartilaginous fishes
   b. lizards   d. lobe-finned fishes

6. Reptiles, including birds, belong to one major lineage of amniotes, and _____ belong to another.
   a. sharks   c. mammals
   b. frogs and toads   d. salamanders

7. Reptiles are adapted to life on land by _____ .
   a. tough skin   d. amniote eggs
   b. internal fertilization e. both a and c
   c. good kidneys   f. all of the above

8. The closest modern relatives of birds are _____ .
   a. crocodilians   b. mammals   c. turtles   d. lizards

9. Among living animals, only birds have _____ .
   a. a cloaca   c. feathers
   b. a four-chambered heart d. amniote eggs

10. The defining trait of hominins is _____ .
    a. tool use   b. bipedalism   c. a big brain   d. endothermy

11. On what continent did *Homo sapiens* arise?

12. Match the organisms with the appropriate description.
    ___ tunicates   a. pouched mammals
    ___ fishes   b. invertebrate chordates
    ___ amphibians   c. feathered amniotes
    ___ primates   d. egg-laying mammals
    ___ birds   e. oldest vertebrate lineage
    ___ monotremes   f. have grasping hands with nails
    ___ marsupials   g. first land tetrapods
    ___ placental   h. most diverse mammal
        mammals   lineage

13. Match each term with the appropriate description.
    ___ keratin   a. homologous to human arm
    ___ cloaca   b. waterproofs amniote skin
    ___ endotherm   c. behaviorally alters temperature
    ___ ectotherm   d. metabolically alters temperature
    ___ mammary gland   e. multipurpose opening
    ___ bird wing   f. secretes milk

1. Male aggression is rare in bonobo society and common in chimpanzee society. Various authors have argued that either one species or the other should be considered a model for "natural" human behavior. Explain why, from the standpoint of relatedness, there is no reason to think that one of these species is a better model for human behavior than the other.

2. Consider the human dispersal pattern described and illustrated in **Figure 24.23**. Given what you know about the founder effect (Section 17.6), would you expect populations native to South America to be more or less genetically diverse than those native to Asia? Explain your reasoning.

for additional quizzes, flashcards, and other study materials
▶ **ACCESS MINDTAP AT WWW.CENGAGEBRAIN.COM**

25 Plant Tissues

Links to Earlier Concepts

This chapter builds on the discussion of plant structure and life cycles introduced in Section 21.1. It examines the anatomy of flowering plants (21.6) in terms of life's organization (1.1), and revisits carbohydrates (3.2), plant cell specializations (4.8, 4.10), photosynthesis (6.1, 6.5), and differentiation (10.1).

Core Concepts

Organisms must exchange matter with the environment in order to grow, maintain themselves, and reproduce.

Living things have various strategies to eliminate wastes and to obtain nutrients and energy for use in biological processes. Plant leaves are specialized to intercept sunlight and to exchange gases for photosynthesis and respiration. Roots are specialized to take up water and nutrients from soil.

Evolution underlies the unity and diversity of life.

All plant parts consist of the same tissues, but the arrangement of the tissues differs between monocots and eudicots. Plants specialized for different environments have different adaptations for acquiring and retaining water and nutrients, supporting growth and reproduction, and optimizing photosynthetic performance.

Interactions among the components of each biological system give rise to complex properties.

Stacks of cells in vascular tissue form conducting tubes that deliver water and nutrients to all of a plant's living cells. All plant parts arise from continually dividing undifferentiated cells in the tips of roots and shoots. The activity of these cells lengthens plant parts. Thickening arises from cylinders of undifferentiated cells that run lengthwise through older structures.

◄ Trichomes are outgrowths of leaf epidermal cells. Glandular types such as the ones on the surface of this marijuana leaf secrete substances that deter plant-eating animals. Marijuana trichomes secrete a chemical (tetrahydrocannabinol, or THC) that has a psychoactive effect in humans.

With more than 260,000 species, angiosperms—flowering plants—dominate the plant kingdom. Magnoliids, eudicots (true dicots), and monocots (Section 21.6) are the major angiosperm groups. In this chapter, we focus mainly on the structure of eudicots and monocots. Eudicots include flowering shrubs and trees, vines, and many nonwoody plants such as tomatoes and dandelions. Lilies, orchids, grasses, and palms are examples of monocots.

Like cells of other complex, multicelled organisms, those in plants are organized as tissues, organs, and organ systems. A vascular plant's body has two organ systems: shoots and roots (**Figure 25.1**). Stems, leaves, and reproductive organs such as flowers are components of the shoot system. Roots absorb water and minerals dissolved in it, and they often serve to anchor the plant in soil. Most shoots grow above the ground and most roots grow below it, but there are many exceptions to this general rule.

Roots and shoots are composed of three tissue systems: dermal, vascular, and ground. The **dermal tissue system** consists of tissues that cover and protect the plant's exposed surfaces. The **vascular tissue system** includes the tissues that carry water and nutrients from one part of the plant body to another. The **ground tissue system** is essentially everything that is not part of the dermal or vascular tissue systems: Cells that carry out the most basic processes necessary for survival—growth, photosynthesis, and storage—are part of the ground tissue system.

Monocots and eudicots have the same types of cells and tissues, but the two lineages differ in many aspects of their organization and hence in the details of their structure (**Figure 25.2**). The names of the groups refer to one difference, the number of **cotyledons** (seed leaves) in their seeds. Monocot seeds have one cotyledon; eudicot seeds have two (Chapter 27 returns to embryonic development in plants). As you will see, the tissue organization inside shoots and roots also differs between eudicots and monocots.

---

cotyledon (cot-uh-LEE-dun) Seed leaf of a flowering plant.
dermal tissue system Tissues that cover and protect the plant.
ground tissue system All tissues that are not dermal or vascular tissue; makes up the bulk of the plant body. Carries out basic processes of survival such as photosynthesis, storage, and growth.
vascular tissue system All tissues that carry water and nutrients through a plant body.

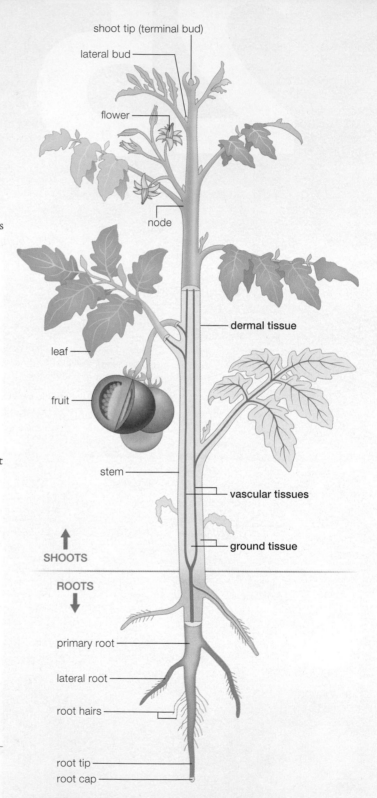

Figure 25.1 Body plan of a tomato plant.
Vascular tissues (purple) conduct water and solutes. They thread through ground tissues that make up most of the plant body. Dermal tissue covers the plant's surfaces.

Eudicots	**Monocots**

| In seeds, two cotyledons (seed leaves) | In seeds, one cotyledon (seed leaf) |

| Flower parts in fours or fives (or multiples of four or five) | Flower parts in threes (or multiples of three) |

| Leaf veins usually forming a netlike array | Leaf veins usually running parallel with one another |

 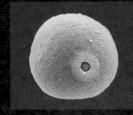

| Pollen grains with three pores or furrows | Pollen grains with one pore or furrow |

| Vascular bundles in a ring in ground tissue of stem | Vascular bundles all through ground tissue of stem |

Figure 25.2 Some structural differences between eudicots and monocots.

Dr. Mark Olson

The ghosts of lost plants haunt National Geographic Explorer Mark Olson. "When you know a plant from just one collection, all you know about it is that it exists . . . and where it was found, but we can't say much about it," Olson said. "We don't know if it's very rare and very restricted or if people have just been overlooking it."

All across the globe, plants that have been found, filed away, and forgotten are his passion. These plants are what drive the plant biologist to explore some of the most remote and dangerous spots on the planet. "The search for and rediscovery of these species include some of the highlights of my fieldwork—and some of the most heartbreaking—when it becomes clear that a species has probably become extinct," he said. Olson conducts field research from a powered paraglider, a flying machine that he describes as "like a moped in the air."

Olson knew early on that he wanted to be a biologist. "Ever since I was a kid, I was obsessed with plants and animals," he says. Studying the newts, insects, and the varied oak and madrona trees that flourished outside his back door made him confident that one day he'd be a scientist. After high school, Olson discovered a love for fieldwork during a turtle-tagging expedition to Costa Rica. Less appealing was one collection method for zoological work. "We put tags on the sea turtles," Olson recalled, "[and the tags] crunched through their flippers. They gasp and wince and bleed all over the place. Obviously it hurts them." Plant collecting—a branch snipped here, a leaf pressed there—seemed less harmful to him, with lower impacts on sensitive populations. Besides, he said, "I thought plants were more interesting."

Distinguish between simple and complex plant tissues.

Describe the three types of simple tissues in plants.

Explain the function of plant epidermal tissue.

Describe plant vascular tissues and some of their functions.

TABLE 25.1

Simple and Complex Plant Tissues

Tissue Type	Main Components	Main Functions
Simple Tissues		
Parenchyma	Parenchyma cells	Photosynthesis, storage, secretion, tissue repair
Collenchyma	Collenchyma cells	Pliable structural support
Sclerenchyma	Sclerenchyma cells (fibers or sclereids)	Structural support
Complex Tissues		
Dermal		
Epidermis	Epidermal cells (including specialized types) and their secretions	Secretion of cuticle; protection; control of gas exchange and water loss
Periderm	Cork cambium, cork	Forms protective cover on older stems and roots
Vascular		
Xylem	Tracheids; vessel elements; parenchyma cells; sclerenchyma cells	Water-conducting tubes; structural support
Phloem	Sieve elements, parenchyma cells; sclerenchyma cells	Sugar-conducting tubes and their supporting cells

Plant tissues may be simple or complex. Simple tissues, with one cell type, are named after the cell type that makes them up: parenchyma, collenchyma, and sclerenchyma (**Table 25.1**). Dermal and vascular tissues are complex tissues, which means they consist of two or more cell types. Some plant tissues include the remains of dead cells, the rigid walls of which lend support to the mature structure. **Figure 25.3** shows some locations of simple and complex tissues in a stem.

SIMPLE TISSUES

Parenchyma **Parenchyma** is a simple tissue that consists of parenchyma cells. The shape of these cells varies, but all typically have a thin, flexible cell wall. Parenchyma cells remain alive at maturity and retain the ability to divide; as you will see in Section 25.6, all plant growth arises from parenchyma. Other specialized functions include structural support, nectar secretion, storage, and photosynthesis. The photo above left shows parenchyma cells in the ground tissue of a buttercup root; starch-packed amyloplasts are visible inside these cells. The photosynthetic tissue called **mesophyll** consists of parenchyma cells that are filled with chloroplasts (below left).

Collenchyma **Collenchyma** is a simple tissue that provides structural support to rapidly growing plant parts such as young stems. Collenchyma cells remain alive in mature tissue. A complex polysaccharide called pectin imparts flexibility to their primary wall, which is thickened unevenly where three or more of the cells abut (left).

Sclerenchyma Variably shaped cells of **sclerenchyma** die after they mature. Their thick cell walls, which contain a high proportion of durable polymers such as cellulose and lignin, lend sturdiness to plant parts and help them resist stretching and compression. Fibers are long, tapered sclerenchyma cells; they occur in bundles that support and protect vascular tissues in stems and leaves (the photo on the left shows a bundle of fibers in vascular tissue

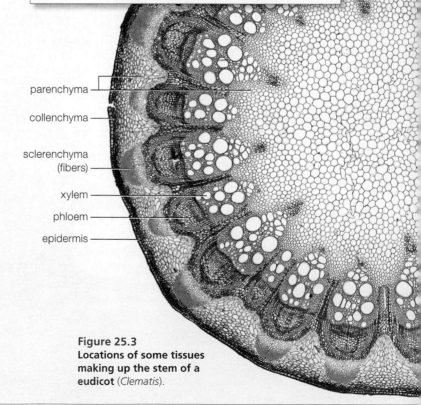

parenchyma

collenchyma

sclerenchyma (fibers)

xylem

phloem

epidermis

Figure 25.3
Locations of some tissues making up the stem of a eudicot (*Clematis*).

CREDITS: (in text) from top, Scientifica/Visuals Unlimited, Inc.; © ISM/Phototake; Dr. Keith Wheeler/Science Source; © Ross E. Koning, http://plantphys.info; (3) Dr. Keith Wheeler/Science Source.

of a sunflower stem). Fibers of some plants are used to make cloth, rope, paper, and other commercial products. Sclereids (left) are sclerenchyma cells that strengthen hard seed coats, and they also make pear flesh gritty.

COMPLEX TISSUES

Dermal Tissue The first dermal tissue to form on a

plant is **epidermis**, which in most species consists of a single layer of epidermal cells on the plant's outer surface. These cells deposit a waterproof, protective cuticle onto their outward-facing cell walls. Hairs and other outgrowths of epidermal cells are common (the photo above left shows some of these structures on the surface of a coleus leaf). Epidermis of leaves and young stems includes specialized cells such as those that form stomata. Plants close these

tiny gaps to limit water loss from internal tissues. In older woody stems and roots, a dermal tissue called periderm replaces epidermis. The photo below left shows multiple cell types in periderm of an elder stem.

Vascular Tissue Vascular tissues, which distribute water and nutrients through the plant body, are complex tissues. There are two types of vascular tissues, both composed of elongated conducting tubes bundled with sclerenchyma fibers and parenchyma.

Xylem is the vascular tissue that conducts water and minerals dissolved in it. Xylem pipelines consist of two types of cells that are dead in mature tissue: vessel elements (**Figure 25.4A**) and tracheids (**Figure 25.4B**). The stiff,

waterproof walls of these dead cells form tubes that thread through the plant (left). In addition to conducting water, xylem tubes lend structural support to plant parts. Interconnecting perforations and pits in the walls of adjacent cells allow water to move laterally between the tubes as well as vertically through them.

Phloem is the vascular tissue that conducts sugars and other organic solutes through the plant. Phloem pipelines

are called sieve tubes (**Figure 25.4C**). A sieve tube is a stack of living cells—sieve elements—connected end to end at perforated sieve plates (left). Each sieve element has an associated parenchyma cell. This "companion cell" provides the sieve element with metabolic support, and also transfers sugars into it. The sugars move through the sieve tubes to all parts of the plant.

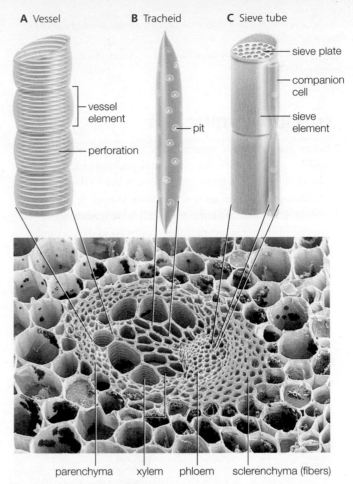

Figure 25.4 Cells that make up plant vascular tissues. Xylem and phloem are vascular tissues. Both consist of cells that make up conducting tubes bundled with fibers and parenchyma.

FIGURE IT OUT
What are the green structures in the cells surrounding the vascular tissues?
Answer: Chloroplasts

collenchyma (coal-EN-kuh-muh) In plants, simple tissue composed of living cells with unevenly thickened walls; provides flexible support.

epidermis (epp-ih-DUR-miss) Dermal tissue; outermost layer of a young plant.

mesophyll (MEZ-uh-fill) Photosynthetic parenchyma.

parenchyma (puh-REN-kuh-muh) In plants, simple tissue composed of living cells with functions that depend on location.

phloem (FLOW-um) Complex vascular tissue of plants; its living sieve elements compose sieve tubes that distribute sugars and other organic solutes.

sclerenchyma (sklair-EN-kuh-muh) In plants, simple tissue composed of cells that die when mature; their tough walls structurally support plant parts. Includes fibers, sclereids.

xylem (ZY-lum) Complex vascular tissue of plants; cell walls of its dead tracheids and vessel elements form tubes that distribute water and mineral ions.

CREDITS: (in text) left column from the top, © Kingsley R. Stern; Biodisc/Visuals Unlimited/Corbis; Dr. Keith Wheeler/Science Source; Forestry and Forest Products Research Institute, Japan; Spike Walker/Wellcome Images; (4) bottom, Andrew Syred/Science Source.

Describe vascular bundles and their arrangement in the stems of monocots and eudicots.

Using appropriate examples, list some types of modified stems and explain their defining features.

INTERNAL STRUCTURE

Stems form the basic framework of a flowering plant, providing support and keeping leaves positioned for photosynthesis. They can grow above or below the soil, and in many species they are specialized for storage or asexual reproduction. Stems characteristically have **nodes**, which are regions of the stem where leaves attach. New shoots (and roots) can form at nodes.

Xylem and phloem are organized in long, multistranded **vascular bundles** that run lengthwise through a stem. The main function of vascular bundles is to conduct water, ions, and nutrients between different parts of the plant. Some components of the bundles—fibers and the lignin-reinforced walls of tracheids—also play an important role in supporting upright stems of vascular plants.

Vascular bundles extend through the ground tissue of all stems and leaves, but the arrangement of the bundles differs between monocots and eudicots. In monocot stems, vascular bundles are typically distributed throughout the ground tissue (**Figure 25.5A**). By contrast, a typical eudicot stem has all of the vascular bundles arranged in a characteristic ring (**Figure 25.5B**). The ring divides the stem's ground tissue into distinct regions of pith (inside the ring) and cortex (outside the ring).

Figure 25.5 Comparing the structure of a young shoot from A a monocot and B a eudicot.
The photos show stem sections that have been stained with different dyes, so the colors of the cell types vary.

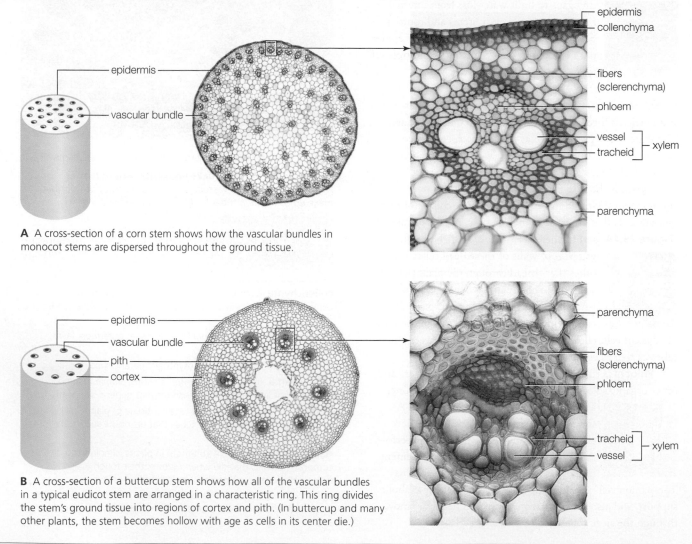

A A cross-section of a corn stem shows how the vascular bundles in monocot stems are dispersed throughout the ground tissue.

B A cross-section of a buttercup stem shows how all of the vascular bundles in a typical eudicot stem are arranged in a characteristic ring. This ring divides the stem's ground tissue into regions of cortex and pith. (In buttercup and many other plants, the stem becomes hollow with age as cells in its center die.)

VARIATIONS ON A STEM

Many plants have modified stem structures that function in storage and reproduction. Some examples are listed below.

stolon

Stolons Stolons are stems that branch from the main stem of the plant and grow horizontally on the ground or just under it. They are commonly called runners because in many plants they "run" along the surface of the soil. Stolons may look like roots, but they have nodes (roots do not have nodes). Roots and shoots that sprout from the nodes develop into new plants. The photo above shows new plants sprouting from nodes on strawberry plant stolons.

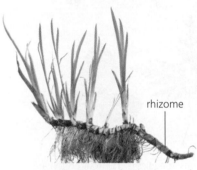

rhizome

Rhizomes Ginger, irises, and some grasses have rhizomes, which are fleshy stems that typically grow under the soil and parallel to its surface. In many plants, the main stem is a rhizome, and it often serves as the primary region for storing food. Shoots that sprout from nodes grow aboveground for photosynthesis and flowering. The photo above shows shoots and roots sprouting from the rhizome of a calamus plant.

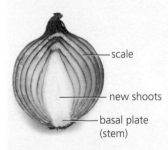

scale

new shoots

basal plate (stem)

Bulbs A bulb consists of a short, flattened stem (a basal plate) encased in overlapping layers of thickened modified leaves called scales. The photo on the left shows the fleshy scales of an onion, which is the bulb of an *Allium cepa* plant. The lower portion of a bulb's basal plate gives rise to roots; the upper portion, to scales and new shoots. Bulb scales contain starch and other substances that a plant holds

node A region of stem where leaves attach and new shoots form.
vascular bundle Multistranded bundle of xylem, phloem, and sclerenchyma fibers running through a stem or leaf.

in reserve during times when conditions in the environment are unfavorable for growth. When favorable conditions return, the plant uses these stored substances to sustain rapid growth. The dry, paperlike outer scale of many bulbs serves as a protective covering.

Corms A corm is a swollen base of a stem with a papery or netlike covering. Corms are like bulbs in that they store nutrients during times when conditions in the environment are unfavorable for growth, and they have a basal plate that gives rise to roots. Unlike a bulb, however, a corm is solid—it has no fleshy scales—and nodes form in rings on its outer surface. The photo above shows corms of crocus plants. New roots are developing from their basal plates, and new shoots are sprouting from their nodes.

Tubers Stem tubers are thick, fleshy storage structures that form on the stolons or rhizomes of some plant species. Most are underground and temporary. Unlike bulbs and corms, tubers have no basal plate; all new shoots and roots form at nodes on their surface. The photo on the left shows how potatoes, which are stem tubers, grow on stolons. The "eyes" of a potato are its nodes.

Cladodes Many types of cacti and other succulents have cladodes, which are flattened, photosynthetic stems specialized to store water. The cladodes of some plants appear leaflike, but most are unmistakably fleshy. Flowers, spikes, small leaves, or entire plants may form at nodes. The photo above shows a cladode of a prickly pear plant growing from a node on another cladode.

Draw the arrangement of tissues composing a typical eudicot leaf.

List some of the variations in eudicot leaf structure.

Describe the main structural differences between the leaves of typical monocots and eudicots.

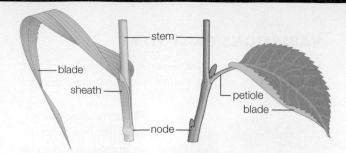

A Typical leaves of monocots (left) and eudicots (right).

Leaves are the main organs of photosynthesis in most flowering plant species. They also function in gas exchange, and they are the major site of evaporative water loss. Typical leaves are thin, with a high surface-to-volume ratio.

Leaves of most monocots are long and narrow, and the base of the leaf wraps around the stem to form a sheath around it (**Figure 25.6A**). In most eudicots, a short stalk called a petiole attaches the base of a leaf to a stem. Eudicot leaves vary widely in their structure. Simple leaves are undivided, although many are lobed; compound leaves are divided into leaflets (**Figure 25.6B**). Most cacti appear to be leafless, but they are not. In many cactus species, leaves form and fall off as the stem ages. They are often replaced by spines, which are modified leaves. Other eudicot leaves have different specializations (**Figure 25.6C**).

INTERNAL STRUCTURE

Figure 25.7 illustrates the internal structure of a typical eudicot leaf. The bulk of the leaf consists of photosynthetic parenchyma—mesophyll—that lies between an upper and a lower layer of epidermis ❶. Eudicot leaves are typically oriented perpendicular to the sun's rays, and have two layers of mesophyll. The uppermost layer is palisade mesophyll ❷. Cells in this layer are elongated, and have more chloroplasts than cells of the spongy mesophyll layer below them. Cells of spongy mesophyll ❸ are irregularly shaped, and have larger air spaces between them than cells of palisade mesophyll.

Vascular bundles in leaves are called **leaf veins**. Inside each vein, strands of xylem transport water and dissolved mineral ions to photosynthetic cells, while strands of phloem transport products of photosynthesis (sugars) away from them ❹. Layers of sclerenchyma fibers surrounding the xylem and phloem stiffen the vein, and also provide support for the leaf's softer tissues. In most eudicot leaves, large veins branch into a network of minor veins.

The upper and lower surfaces of a leaf are sheets of epidermis one cell thick. Either or both surfaces may be smooth, sticky, or slimy, and have epidermal cell outgrowths such as hairs, scales, spikes, hooks, and other structures (some are shown in the chapter opening photo). Epidermal cells secrete a translucent, waxy cuticle that slows water loss. A leaf's upper surface, which typically receives the most direct

B Examples of common leaf forms in eudicots. The four on top are simple leaves; the three on the bottom are compound leaves.

C Example of specialized leaf form. Carnivorous plants of the genus *Nepenthes* grow in places where nitrogen is scarce. They secrete acids and protein-digesting enzymes into fluid in a cup-shaped modified leaf. The enzymes release nitrogen from small animals (such as insects, frogs, and mice) that are attracted to odors from the fluid and then drown in it. The leaves then take up the released nitrogen.

Figure 25.6 Leaf structure.

Figure 25.7 Anatomy of a eudicot leaf.

The leaf surface is epidermis with a secreted layer of cuticle ❶. The bulk of the leaf is mesophyll, a type of photosynthetic parenchyma. Eudicot leaves often have two distinct layers of this tissue: palisade mesophyll with elongated cells ❷, and spongy mesophyll with irregularly shaped cells ❸.

❹ Inside leaf veins, vascular bundles of xylem (blue) and phloem (pink) transport materials to and from photosynthetic cells.

❺ Gas exchanges between air inside and outside of the leaf occur at stomata.

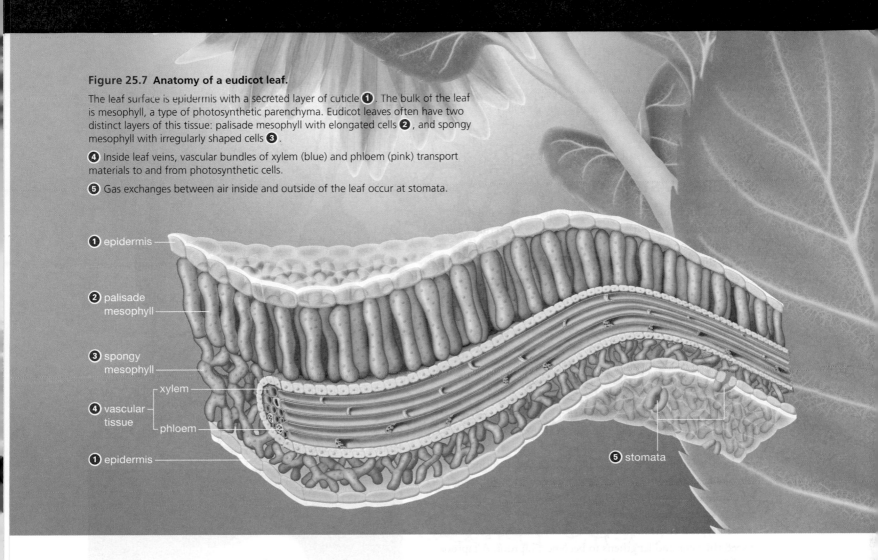

❶ epidermis
❷ palisade mesophyll
❸ spongy mesophyll
❹ vascular tissue — xylem — phloem
❶ epidermis
❺ stomata

sunlight, may have a thicker cuticle than the lower surface, which tends to be shaded. The lower surface usually has more stomata ❺. These openings can be closed to prevent water loss, or opened to allow gases to cross the epidermis. Carbon dioxide needed for photosynthesis enters a leaf through open stomata, then diffuses through air spaces to mesophyll cells. Oxygen released by photosynthesis diffuses in the opposite direction. Section 26.4 returns to stomata function.

The structure of leaves differs somewhat between eudicots and monocots (**Figure 25.8**). Blades of grass and other monocot leaves that grow vertically can intercept light from all directions. Unlike eudicot leaves, monocot leaves typically have a single layer of mesophyll. The arrangement of veins also differs. A eudicot leaf has branching veins; almost all of the veins in a monocot leaf are similar in length and run parallel with the leaf's long axis (an example is shown in Figure 25.2).

leaf vein A vascular bundle in a leaf.

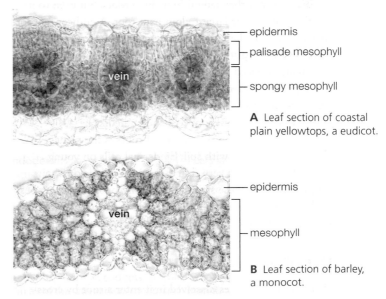

epidermis
palisade mesophyll
spongy mesophyll
vein

A Leaf section of coastal plain yellowtops, a eudicot.

epidermis
vein
mesophyll

B Leaf section of barley, a monocot.

Figure 25.8 Comparing the arrangement of cells and tissues in eudicot and monocot leaves.

CREDIT: (8) Eri Maai, Shouu Shimada, Masahiro Yamada, Tatsuo Sugiyama, Hiroshi Miyake, Mitsutaka Taniguchi; The avoidance and aggregative movements of mesophyll chloroplasts in C4 monocots in response to blue light and abscisic acid; Journal of Experimental Botany, May 2011, Volume 62, issue number 9, 3213–3221, by permission of Oxford University Press.

Section 25.1 Most flowering plants have belowground roots and aboveground shoots (stems, leaves, and flowers). **Ground tissues** make up most of the soft internal tissues of a plant, and **dermal tissues** protect its surfaces. **Vascular tissues** conduct water and nutrients to all parts of the plant. Monocots and eudicots have the same tissues organized in different ways. For example, monocot seeds have one **cotyledon**; eudicot seeds have two.

Section 25.2 **Parenchyma, collenchyma,** and **sclerenchyma** are simple tissues (each consists of only one type of cell). Parenchyma stays alive at maturity; **mesophyll** is photosynthetic parenchyma. Living cells in collenchyma have flexible walls that support fast-growing plant parts. Fibers and sclereids that support plant tissues are sclerenchyma cells, which die at maturity. Dermal and vascular tissues are complex (they consist of multiple cell types). Stomata open across **epidermis**, a dermal tissue that covers soft plant parts. In vascular tissue, water and dissolved minerals flow through vessels of **xylem**, and sugars travel through vessels of **phloem.**

Section 25.3 **Vascular bundles** extending through stems conduct water and nutrients between different parts of the plant, and also help structurally support the plant. In most eudicot stems, vascular bundles form a ring that divides the ground tissue into cortex and pith. In monocot stems, the vascular bundles are distributed throughout the ground tissue. New shoots and roots form at **nodes** on stems. Stem specializations such as rhizomes, corms, stem tubers, bulbs, cladodes, and stolons are adaptations for storage or reproduction in many types of plants.

Section 25.4 Leaves, which are specialized for photosynthesis, contain mesophyll and vascular bundles (**leaf veins**) between upper and lower epidermis. Eudicots typically have two layers of mesophyll; monocots have one. Water vapor and gases cross the cuticle-covered epidermis at stomata.

Section 25.5 Roots are specialized for absorbing water and nutrients from soil. Inside each is a **vascular cylinder** enclosed by a layer of **endodermis. Root hairs** increase the surface area of roots. Monocots have a **fibrous root system** of similar-sized adventitious roots. Most eudicots have a **taproot system**, which is an enlarged primary root with its lateral root branchings. Lateral roots arise from divisions of **pericycle** cells inside the root vascular cylinder.

Section 25.6 All plant tissues originate at **meristems**, which are regions of undifferentiated cells that retain their ability to divide. **Primary growth** (lengthening) arises at **apical meristems** in shoot and root tips. **Secondary growth**

(thickening) arises at **lateral meristems** (**vascular cambium** and **cork cambium**) in older stems and roots. Vascular cambium produces secondary xylem (**wood**) on its inner surface, and secondary phloem on its outer surface. Cork cambium gives rise to **cork**, which is part of **periderm. Bark** is all tissue outside of the vascular cambium of a woody plant.

Section 25.7 In many trees, one growth ring (tree ring) forms each year. Tree rings hold information about environmental conditions that prevailed while the rings were forming.

Application By the process of photosynthesis, plants naturally remove carbon from the atmosphere and incorporate it into their tissues. Carbon that is locked in molecules of wood and other durable plant tissues can stay out of the atmosphere for thousands of years.

SELF-ASSESSMENT
ANSWERS IN APPENDIX VII

1. In plants, fibers are a type of _____ cell.
   a. parenchyma
   b. sclerenchyma
   c. collenchyma
   d. mesophyll

2. All plant growth arises from _____ .
   a. parenchyma
   b. sclerenchyma
   c. collenchyma
   d. mesophyll

3. Which of the following cell types remain alive in mature plant tissue?
   a. sclerenchyma
   b. sieve elements
   c. tracheids
   d. vessel elements

4. All of the vascular bundles inside a typical _____ are arranged in a ring.
   a. monocot stem
   b. eudicot stem
   c. monocot root
   d. eudicot root

5. True or false? Lateral roots form at nodes on roots.

6. Epidermis and periderm are _____ tissues.
   a. ground   b. vascular   c. dermal

7. A vascular bundle in a leaf is called _____ .
   a. a vascular cylinder
   b. mesophyll
   c. a vein
   d. vascular cambium

8. Typically, vascular tissue is organized as _____ in stems and as _____ in roots.
   a. multiple vascular bundles; one vascular cylinder
   b. one vascular bundle; multiple vascular cylinders
   c. one vascular cylinder; multiple vascular bundles
   d. multiple vascular cylinders; one vascular bundle

9. Is an onion a root or a stem?

10. In a(n) _____ , the primary root is typically the largest.
    a. lateral meristem
    b. adventitious root system
    c. fibrous root system
    d. taproot system

TREE RINGS AND DROUGHTS El Malpais National Monument, in west central New Mexico, has pockets of vegetation that have been surrounded by lava fields for about 3,000 years, so they have escaped wildfires, grazing animals, agricultural activity, and logging. Henri Grissino-Mayer generated a 2,129-year annual precipitation record using tree ring data from living and dead trees in this park (**Figure 25.17**).

1. Around 770 A.D., the Mayan civilization began to suffer a massive population loss, particularly in the southern lowlands of Mesoamerica. The El Malpais tree ring data show a drought during that time. Was it more or less severe than the Dust Bowl drought in 1933–1939?

2. One of the worst population catastrophes ever recorded occurred in Mesoamerica between 1519 and 1600 A.D., when around 22 million people native to the region died. Which period between 137 B.C. and 1992 had the most severe drought? How long did that drought last?

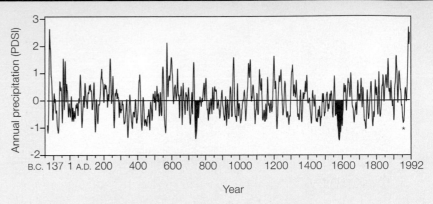

Figure 25.17 Annual precipitation record for 2,129 years, inferred from compiled tree ring data in El Malpais National Monument, New Mexico. Data were averaged over 10-year intervals; graph correlates with other indicators of rainfall collected in all parts of North America. PDSI, Palmer Drought Severity Index: 0, normal rainfall; increasing numbers mean increasing excess of rainfall; decreasing numbers mean increasing severity of drought.

*A severe drought contributed to a series of catastrophic dust storms that turned the midwestern United States into a "dust bowl" between 1933 and 1939.

---

11. Root hairs _____ .
    a. conduct water from cortex to aboveground shoots
    b. increase the root's surface area for absorption
    c. anchor the plant in soil

12. Roots and shoots lengthen through activity at _____ .
    a. apical meristems   c. the vascular cambium
    b. lateral meristems   d. the cork cambium

13. The activity of lateral meristems _____ older roots and stems.
    a. lengthens   b. thickens   c. ages

14. Tree rings occur _____ .
    a. when there are droughts during the time the rings form
    b. where environmental conditions influence xylem cell size
    c. if heartwood alternates with sapwood
    d. as epidermis replaces periderm

15. Match the plant parts with the best description.
    ____ vascular cambium   a. ground tissue
    ____ mesophyll   b. modified stem structure
    ____ wood   c. a lateral meristem
    ____ cortex   d. photosynthetic parenchyma
    ____ potato   e. mass of secondary xylem
    ____ cotyledons   f. in epidermis
    ____ stomata   g. only one in a monocot seed

1. Aboveground plant surfaces are covered with a waxy cuticle. Why do roots lack this protective coating?

2. Why do eudicot trees tend to be wider at the base than at the top?

3. Is the plant with the yellow flower (above) a eudicot or a monocot? What about the plant with the purple flower?

4. Oscar and Lucinda meet in a tropical rain forest and fall in love, and he carves their initials into the bark of a tree. They never do get together, though. Ten years later, still heartbroken, Oscar searches for the tree. Given what you know about primary and secondary growth, will he find the carved initials higher relative to ground level? If he goes berserk and chops down the tree, what kinds of growth rings will he see?

5. Was the micrograph shown at right taken from a section of stem or root? Monocot or eudicot? How do you know?

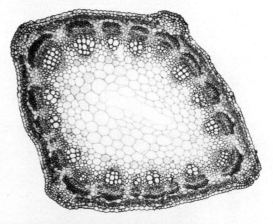

for additional quizzes, flashcards, and other study materials
▶ ACCESS MINDTAP AT WWW.CENGAGEBRAIN.COM

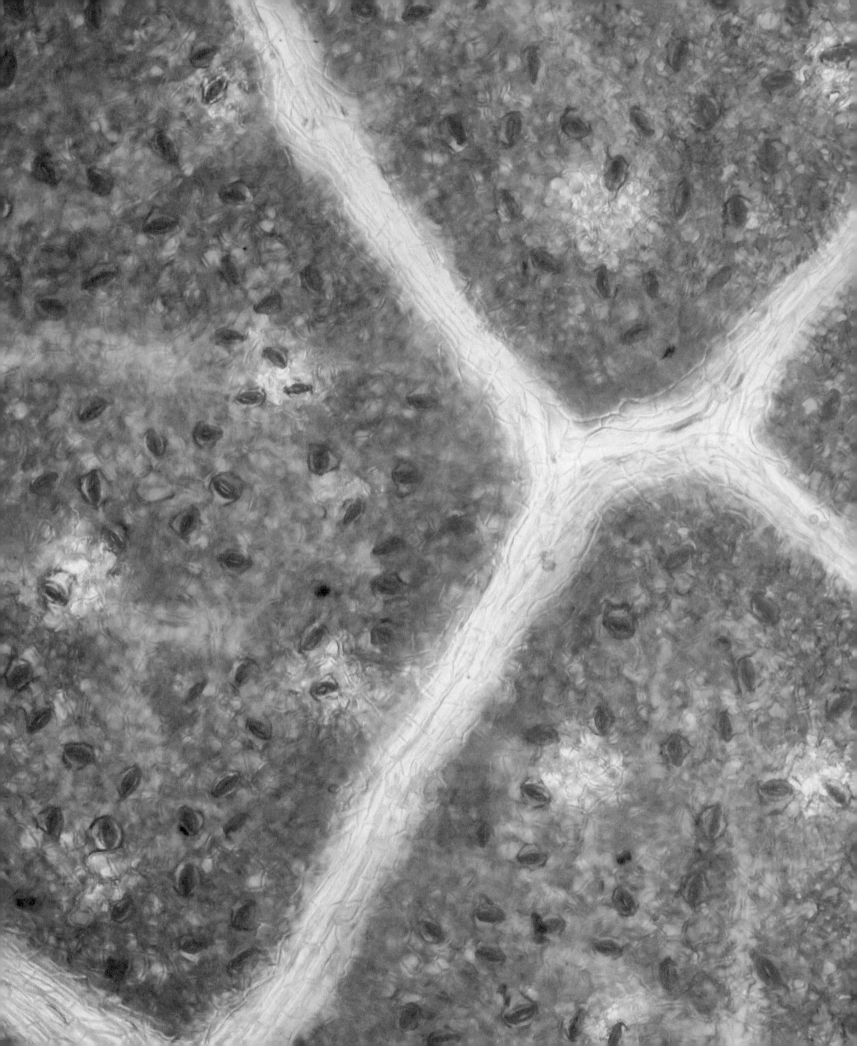

26

Plant Nutrition and Transport

Links to Earlier Concepts

Before beginning this chapter, be sure you understand plant anatomy (Sections 25.1–25.5) and adaptations to life on land (4.10, 21.1). Water movement through a plant depends on hydrogen bonding and cohesion (2.4). We revisit nutrients (1.2), ions (2.2), carbohydrates (3.2), osmosis (5.6), membrane transport (5.7), photosynthesis (6.3–6.5), aerobic respiration (7.1, 7.2), nitrogen fixation (19.4), and mycorrhizae (22.3).

Core Concepts

Organisms must exchange matter with the environment in order to grow, maintain themselves, and reproduce.

Organisms have various strategies to eliminate wastes, and to obtain nutrients and energy for use in biological processes. Nutrients required for plant growth are available in water, air, and soil. Open stomata allow plants to exchange gases with air, but they also allow the loss of water by evaporation. Water must be continually taken up from soil to replace this loss.

Interactions among the components of each biological system give rise to complex properties.

Passive and active mechanisms drive the movement of fluids through conducting tubes of vascular tissue in plants. Water's special properties allow it to be drawn upward through a plant, from roots to shoots, through narrow conduits formed by dead xylem cells. Osmosis and active transport drive the movement of sugars through stacks of living phloem cells, from regions of the plant where sugars are being produced to regions where they are being used.

Evolution underlies the unity and diversity of life.

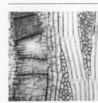

Evolutionary adaptations of plants include external specializations that facilitate root uptake of water and nutrients from soil. Internal specializations allow selectivity over substances that enter vascular tissue for distribution to the rest of the plant. The special structure of xylem tubes reflects the properties of water that must be conducted through them from soil to the tips of shoots.

◄ Highways of vascular tissue, shown here crisscrossing a leaf, carry substances to and from all living cells in a plant body. Xylem carries water and dissolved minerals taken up by roots; phloem carries sugars.

26.1 Plant Nutrients and Availability in Soil

LEARNING OBJECTIVES

Give some examples of plant macronutrients and their sources.

State the difference between plant macronutrients and micronutrients.

Describe the components of soil and their importance for plant growth.

Explain two ways that soil can lose nutrients.

Sixteen elements are required for growth and maintenance of a plant body. Nine are macronutrients, meaning they are required in amounts above 0.5 percent of a plant's dry weight. Carbon, oxygen, and hydrogen are macronutrients, as are nitrogen, phosphorus, and sulfur (components of proteins and nucleic acids), potassium and calcium (which affect processes such as cell signaling), and magnesium (a component of chlorophyll). Chlorine, iron, boron, manganese, zinc, copper, and molybdenum are micronutrients, meaning they make up traces of the plant's dry weight. Some of these micronutrients serve as enzyme cofactors (Section 5.5).

PROPERTIES OF SOIL

Three nutrients required by plants are abundantly available in air and water: Carbon dioxide (CO_2) in air provides carbon and oxygen, and water (H_2O) provides hydrogen and more oxygen. The remaining elements occur in soil. Plants take them up in the form of mineral ions dissolved in soil water.

Soils vary in their nutrient content and other properties, so some support plant growth better than others. Soil consists mainly of mineral particles—sand, silt, and clay—that form by the weathering of rocks. Sand grains are about one millimeter in diameter. Silt particles are hundreds or thousands of times smaller than sand grains, and clay particles are even smaller. Soils rich in clay can retain dissolved nutrients that might otherwise trickle past roots too quickly to be absorbed. This is because tiny clay particles carry a negative charge that attracts and traps positively charged mineral ions in soil water. However, clay particles can pack so tightly that they exclude air—and the oxygen in it. Cells in a plant's roots, like cells in

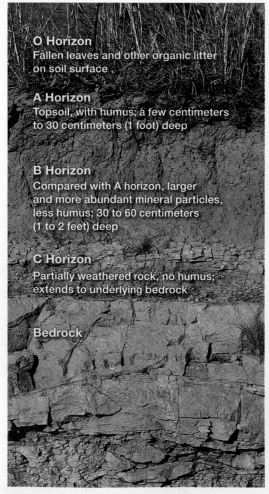

O Horizon
Fallen leaves and other organic litter on soil surface

A Horizon
Topsoil, with humus; a few centimeters to 30 centimeters (1 foot) deep

B Horizon
Compared with A horizon, larger and more abundant mineral particles, less humus; 30 to 60 centimeters (1 to 2 feet) deep

C Horizon
Partially weathered rock, no humus; extends to underlying bedrock

Bedrock

Figure 26.1 An example of soil horizons.

Figure 26.2 Runaway erosion in Providence Canyon, Georgia.

its shoots, require oxygen for aerobic respiration. Sand and silt intervene between clay particles, so they loosen soil. **Loam** is soil with roughly equal proportions of sand, silt, and clay. Most plants do best in loam that contains between 10 and 20 percent **humus**, which is partially decomposed organic material—fallen leaves, feces, and so on. Humus supports plant growth because it releases nutrients as it breaks down, and its negatively charged organic acids can trap positively charged mineral ions in soil water. Humus also swells and shrinks as it absorbs and releases water, and these changes in size aerate soil by opening spaces for air to penetrate.

Humus tends to accumulate in swamps, bogs, and other perpetually wet areas because waterlogged soil holds little air. Humus can form in anaerobic conditions, but its breakdown is especially slow. Soils in these areas often contain more than 90 percent humus. Very few plants can grow in these areas.

HOW SOILS CHANGE

Soils develop over thousands of years. They exist in different stages of development in different regions. Most form in layers called horizons that are distinct in color and other properties (**Figure 26.1**). Identifying the layers helps us compare soils in different places. For instance, the A horizon (the uppermost) is **topsoil**, and it contains the greatest amount of organic matter. The roots of most plants grow most densely in this layer. Grasslands typically have a deep layer of topsoil; tropical forests do not.

Minerals, salts, and other molecules dissolve in water as it filters through soil. **Leaching** is the process by which water removes nutrients from soil and carries them away. Leaching is fastest in sandy soils, which do not trap nutrients as well as clay soils. Wind and water can remove soil from an area, a process called **soil erosion**. Poor farming practices result in erosion. In 1800, for example, European settlers who arrived in what is now Providence Canyon, Georgia, plowed the land straight up and down the hills. The furrows made excellent conduits for rainwater, which carved deep crevices that made even better rainwater conduits. The area became useless for farming by 1850. It now consists of about 445 hectares (1,100 acres) of deep canyons that continue to expand at the rate of about 2 meters (6 feet) per year (**Figure 26.2**).

---

humus (HUE-muss) Partially decayed organic material in soil.
leaching Process by which water moving through soil removes nutrients from it.
loam Soil with roughly equal amounts of sand, silt, and clay.
soil erosion Loss of soil under the force of wind and water.
topsoil Uppermost soil layer; contains the most organic matter and nutrients for plant growth.

Dr. Jerry Glover

What if a new grain of wheat holds the key to saving biodiversity, polluted ecosystems, and starving people? National Geographic Explorer Jerry Glover is working on it, and he claims the future starts with the past. "Before agriculture, natural plant communities ruled the earth and kept ecosystems in perfect balance. How? Those plants were perennials, alive year-round and incredibly efficient at regulating processes like nutrient cycling and water management that protect ecosystem health. Their roots stretch deep below ground [right], controlling erosion and improving soil quality." Then everything changed. Humans harnessed vast swaths of the planet, replacing natural grasses with billions of acres of annual crops—plants that take much more than they give to crucial natural systems. Agriculture is now the number one man-made threat to global biodiversity and ecosystem function.

"It's time to put natural plant communities back in charge of the landscape," says Glover. "Annual crops must be planted from seed every year, literally starting over from scratch. That requires tremendous amounts of time, effort, and expense because annual crop species are very inefficient. They allow half of the nitrogen fertilizer farmers spread over fields to escape below the root zone or run off the soil surface. Growing annual crops requires extra fertilizer, heavy machinery, and disturbance of the land; and every year it's required again." Glover is part of a research team developing perennial crops—mainly grains—that could revolutionize agriculture and solve problems far beyond farm fields. Grains are cultivated on almost 70 percent of the world's agricultural land.

Why does this world beneath our feet matter so much? Many nutrients essential to human health come from the soil. Plants are the delivery system. "When we lose the health of our soil through erosion or degradation," says Glover, "crucial nutrients are no longer carried up to plants and passed on to humans. Studying this helps us see how perennial crops of the future could be farmed with less effort, more nutritional value, and at the high yields we'll need to feed a planet of seven to nine billion hungry people."

LEARNING OBJECTIVES

Describe the paths that water and minerals travel as they move from soil into a root's vascular cylinder.

Explain the formation and benefits of mycorrhizae and root nodules.

THE FUNCTION OF ENDODERMIS

Water moves from soil, through a root's epidermis and cortex, to the vascular cylinder (**Figure 26.3**). Osmosis (Section 5.6) drives this movement, because fluid in the plant typically contains more solutes than soil water. After soil water enters a vascular cylinder, xylem distributes it to the rest of the plant.

Most of the soil water that enters a root moves through cell walls (Section 4.10) of adjacent plant cells ❶. Cell walls form a more or less continuous conduit through the cortex of a root, so water can seep from epidermis to vascular cylinder without ever entering a cell.

Although soil water can reach the vascular cylinder by diffusing through cell walls, it cannot enter the cylinder the same way. Why not? A vascular cylinder is surrounded by endodermis, a tissue that consists of a single layer of tightly packed endodermal cells (Section 25.5). These cells secrete a waxy substance into their walls wherever they abut. The substance forms a **Casparian strip**, which is a waterproof

band between the plasma membranes of endodermal cells (**Figure 26.4**). A Casparian strip prevents water from diffusing across endodermis through cell walls ❷. Thus, soil water can enter a vascular cylinder only by passing through the cytoplasm of an endodermal cell.

Water enters an endodermal cell by diffusing across its plasma membrane from the cell wall. It can also diffuse into the cell from adjacent cells in root cortex via plasmodesmata (Section 4.10). Once in endodermal cell cytoplasm, the water moves through plasmodesmata into a pericycle cell, then into a conducting tube of xylem.

Unlike water, ions cannot diffuse directly across lipid bilayers (Sections 5.6 and 5.7). Mineral ions in soil water enter or exit a cell only through transport proteins in its plasma membrane. Thus, transport proteins in root cell membranes control the types and amounts of ions that move from soil water into the plant body. This mechanism offers protection from some toxic substances that may be in soil water. In addition, most mineral ions important for plant nutrition are more concentrated in the root than in soil, so they will not spontaneously diffuse into the root. These nutrients must be actively transported into root cells. In young roots, much of this transport occurs at the membrane of root hairs ❸. After entering the cytoplasm of a root cell, mineral ions diffuse

---

Figure 26.3 Uptake of soil water by a root. In most flowering plants, plasma membrane transport proteins control the uptake of substances dissolved in soil water.

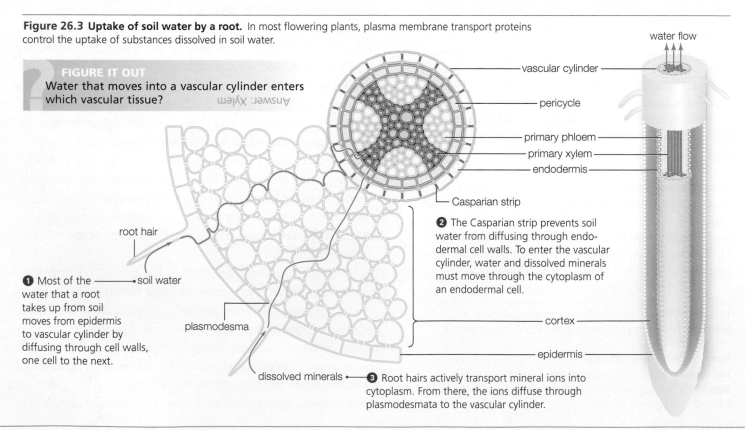

FIGURE IT OUT Water that moves into a vascular cylinder enters which vascular tissue?
Answer: Xylem

water flow

vascular cylinder

pericycle

primary phloem

primary xylem

endodermis

Casparian strip

❷ The Casparian strip prevents soil water from diffusing through endodermal cell walls. To enter the vascular cylinder, water and dissolved minerals must move through the cytoplasm of an endodermal cell.

root hair

❶ Most of the water that a root takes up from soil moves from epidermis to vascular cylinder by diffusing through cell walls, one cell to the next.

soil water

plasmodesma

cortex

epidermis

dissolved minerals ❸ Root hairs actively transport mineral ions into cytoplasm. From there, the ions diffuse through plasmodesmata to the vascular cylinder.

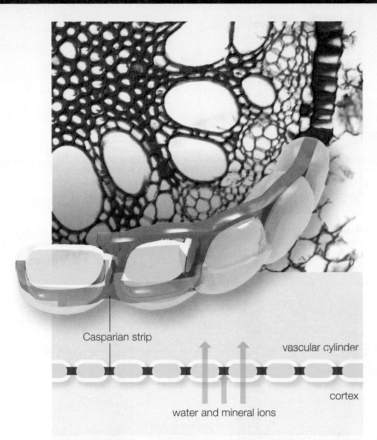

Casparian strip

vascular cylinder

cortex

water and mineral ions

Figure 26.4 The Casparian strip.
The micrograph shows a vascular cylinder in the root of an iris plant. Parenchyma cells that make up endodermis are specialized to secrete a waxy substance into their walls wherever they touch. The secretions, which form a Casparian strip, prevent soil water from diffusing through endodermal cell walls to enter the vascular cylinder.

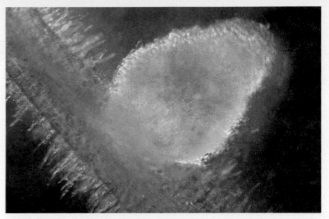

A Root nodule. The red color is from a molecule similar to hemoglobin that delivers oxygen for aerobic respiration. Bound to this molecule, oxygen cannot interfere with the oxygen-sensitive bacterial enzyme that fixes nitrogen.

B Soybean plants in nitrogen-poor soil show how root nodules affect growth. Only the darker green plants growing in the rows on the right were infected with nitrogen-fixing *Rhizobium* bacteria.

Figure 26.5 Root nodules.

from cell to cell through plasmodesmata until they enter pericycle cells inside the vascular cylinder. From there, transport proteins in pericycle cell plasma membranes selectively load the ions into conducting tubes of xylem.

MUTUALISMS

Most flowering plants maximize their ability to take up nutrients by taking part in mutualisms. As Section 22.3 explained, a mycorrhiza is a mutually beneficial interaction between a root and a fungus that grows on or in it. Filaments of the fungus (hyphae) form a velvety cloak around roots or penetrate their cells. Collectively, the hyphae have a large surface area,

Casparian strip Waxy, waterproof band that seals abutting cell walls of root endodermal cells.
root nodules Of some plant roots, swellings that contain mutualistic nitrogen-fixing bacteria.

so they absorb mineral ions from a larger volume of soil than roots alone. The fungus absorbs some sugars and nitrogen-rich compounds from root cells. In return, the root cells get some scarce minerals that the fungus is better able to absorb.

Legumes and some other plants can form mutualisms with nitrogen-fixing *Rhizobium* bacteria (Section 19.4). The roots of these plants release certain compounds into the soil that are recognized by compatible bacteria. The bacteria respond by releasing signaling molecules that, in turn, trigger the roots to grow around the bacteria and encapsulate them inside swellings called **root nodules** (**Figure 26.5**). The association is beneficial to both parties. The plants require a lot of nitrogen, but cannot use nitrogen gas that is abundant in air. The bacteria in root nodules convert this gas to ammonia, which is a form of nitrogen that the plant can use. In return for this valuable nutrient, the plant provides a low-oxygen environment for the anaerobic bacteria, and shares its photosynthetically produced sugars with them.

CREDITS: (4) photo, Dr Keith Wheeler/Science Photo Library; (5A) Ninjatacoshell; (5B) © NifTAL Project, Univ. of Hawaii, Maui.

26.3 Water Movement Inside Plants

LEARNING OBJECTIVES

Describe the movement of water through xylem tubes.

Explain how water can move from the roots of a vascular plant to leaves that may be hundreds of feet above them.

Water that enters a root travels to the rest of the plant inside conducting tubes of xylem. Remember from Section 25.2 that these tubes consist of the stacked, interconnected walls of dead cells. Rings or spirals of sturdy lignin in the cell walls impart strength to the tubes, which in turn impart strength to the stem or other organ where they occur. Water flows lengthwise through xylem tubes, and also laterally (from one tube to another) through openings in their walls.

As xylem tissue is maturing, its still-living cells deposit secondary wall material on the inner surface of their primary wall (Section 4.10). The material is deposited unevenly, leaving the primary wall uncovered in patterns such as rows of spots or bands that match up between adjacent cells. When the cells die, the uncovered areas form cavities called pits. The primary wall exposed at the pits has tiny pores where plasmodesmata once connected the adjacent cells. In the mature tissue, water moves through these pores laterally from one xylem tube to another (**Figure 26.6A**). Pits are often bordered, which means that a circular flap of secondary wall extends over the edge of the exposed primary wall. The types and concentrations of ions in water moving through a bordered pit determines the rate of flow through it.

Angiosperms have two types of xylem tubes, each composed of one cell type: vessel elements or tracheids. Vessels, the main water-conducting tubes in angiosperms, consist of **vessel elements** stacked end to end (**Figure 26.6B**). In maturing xylem tissue, enzymes digest very large holes in the end walls of stacked vessel elements (where they meet) just before they die. The resulting perforation plates allow water to flow freely through vessel elements in a stack. Some perforation plates have ladderlike bars that break up air bubbles in water flowing through larger tubes.

Tracheids, like vessel elements, die in stacks to form water-conducting tubes in xylem. Tracheids are narrower than vessel elements, so they are more resistant to vertical compression. Thus, in addition to conducting water, tracheids have a substantial role in structural support. Unlike vessel elements, tracheids have no perforation plates; their ends are typically pointier and closed (**Figure 26.6C**). Pits in the narrowed ends of stacked tracheids match up. Water flows vertically through the tube by moving through these matched pits. Vessel elements evolved more recently than tracheids, and may have contributed to the evolutionary success of angiosperms (most gymnosperms have only tracheids).

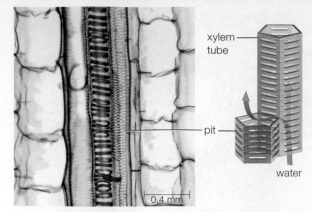

A Cells in tubes of xylem meet side to side at rows of holes (pits) in their walls. The pits match up, so water can flow through them into an adjacent cell. Pits in a xylem vessel are visible in the micrograph (left) of a lengthwise section through a corn stem. Rings of lignin deposited in the walls of an older vessel are also visible.

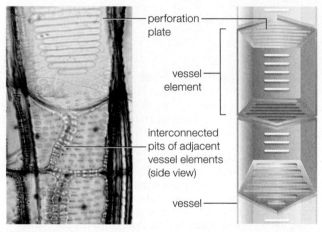

B Xylem vessels consist of stacked vessel elements connected end to end at perforation plates. Water flows vertically between cells through these plates. The perforation plate in this micrograph of a section of Japanese holly (left) has ladderlike bars. The bars break up large air bubbles in fluid moving lengthwise through a vessel.

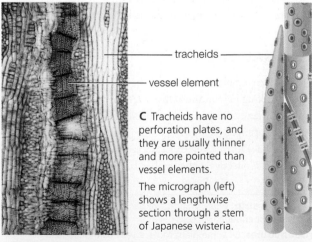

C Tracheids have no perforation plates, and they are usually thinner and more pointed than vessel elements.

The micrograph (left) shows a lengthwise section through a stem of Japanese wisteria.

Figure 26.6 Pipelines of xylem.

CREDITS: (6A) Garry DeLong/Science Source; (6B) Elisabeth Wheeler, North Carolina State University; (6C) Forestry and Forest Products Research Institute, Japan.

COHESION–TENSION THEORY

How does water in xylem move all the way from roots to leaves? Tracheids and vessel elements that compose xylem tubes are dead, so these cells cannot be expending any energy to pump water upward against gravity. Rather, the upward movement of water in vascular plants occurs as a consequence of evaporation and cohesion (Section 2.4). By the **cohesion–tension theory**, water in xylem is pulled upward by air's drying power, which creates a continuous negative pressure called tension. **Figure 26.7** illustrates this mechanism, which begins with water evaporating from leaves and stems ❶. The evaporation of water from a plant's aboveground parts is called **transpiration**. Transpiration's effect on water inside a plant is a bit like what happens when you suck a drink through a straw. Transpiration exerts negative pressure (it pulls) on water. Because the water molecules are connected by hydrogen bonds, a pull on one tugs all of them (cohesion). Thus, the negative pressure (tension) created by transpiration pulls on entire columns of water that fill xylem tubes ❷. The tension extends all the way from leaves that may be hundreds of feet in the air, down through stems, into young roots where water is being taken up from soil ❸.

Water is pulled upward in continuous columns because xylem tubes are narrow, and water moving through a narrow conduit (such as a straw or a xylem tube) resists breaking into droplets. This phenomenon is partly an effect of water's cohesion, and partly because water molecules are attracted to hydrophilic materials (such as cellulose in the walls of xylem) making up the tube.

Transpiration drives almost all of the upward movement of water through a vascular plant, but many metabolic pathways also contribute to the negative pressure that sucks water through xylem. For example, the noncyclic reactions of photosynthesis (Section 6.3) split water molecules into oxygen and hydrogen ions. Thus, cells carrying out photosynthesis must receive a constant supply of water molecules, and these are delivered by xylem.

---

cohesion–tension theory Explanation of how transpiration creates a tension that pulls a cohesive column of water upward through xylem.

tracheids (TRAY-key-idz) Tapered cells of xylem that die when mature; their interconnected, pitted walls remain and form water-conducting tubes.

transpiration (trans-purr-A-shun) Evaporation of water from aboveground plant parts.

vessel elements Of xylem, cells that form in stacks and die when mature; their pitted walls remain to form water-conducting tubes. Each tube consists of a stack of vessel elements that meet end to end at perforation plates.

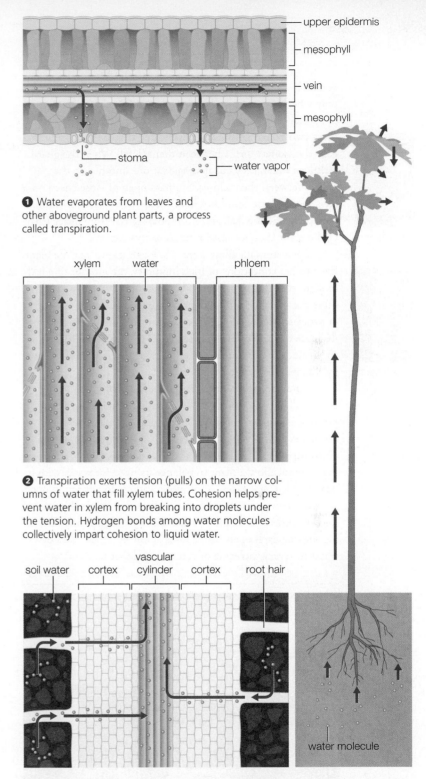

❶ Water evaporates from leaves and other aboveground plant parts, a process called transpiration.

❷ Transpiration exerts tension (pulls) on the narrow columns of water that fill xylem tubes. Cohesion helps prevent water in xylem from breaking into droplets under the tension. Hydrogen bonds among water molecules collectively impart cohesion to liquid water.

❸ The tension inside xylem tubes extends from leaves to roots, where osmosis is driving the uptake of water from soil.

Figure 26.7 Cohesion–tension theory of water transport in vascular plants.

26.4 Conserving Water

Describe the function of stomata in plants.

guard cells

open stoma

closed stoma

A cuticle stops most water loss from aboveground parts, but only when stomata are closed (Sections 6.4 and 21.1). A pair of specialized epidermal cells borders each stoma. When these two **guard cells** swell with water, they bend slightly so a gap (the stoma) forms between them (left). When the guard cells lose water, they collapse against one another, so the stoma between them closes. Stomata open or close based on cues that include humidity, light intensity, the level of carbon dioxide inside the leaf, and hormonal signals from other parts of the plant (Section 27.8 returns to this topic).

Open stomata allow water—a lot of it—to exit the plant. Even under conditions of high humidity, the interior of a leaf or stem contains more water than air. Thus, water vapor diffuses out of a stoma whenever it is open. Indentations, ledges, or other structures that reduce air movement next to a stoma decrease the rate of evaporation from it (**Figure 26.8**). Even so, around 95 percent of the water taken up from soil is lost by transpiration from open stomata.

Water is important for land plants, so why do they let almost all of it evaporate away? A cuticle is not only waterproof, it is also gasproof. When a plant's stomata are closed, it cannot exchange enough carbon dioxide and oxygen with the air to support critical metabolic processes. Opening and closing stomata allow a plant to balance its need for water with its need to exchange gases. A lot more water is lost through open stomata than gases are gained, however, and this is the reason for the fantastic amount of water loss by transpiration. During the day, each open stoma loses about 400 molecules of water for every molecule of carbon dioxide it takes in.

guard cell One of a pair of cells that define a stoma.

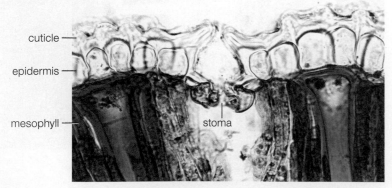

cuticle

epidermis

mesophyll   stoma

Figure 26.8 Water-conserving structures of a pincushion leaf. In this plant, leaf stomata are recessed beneath a ledge of cuticle. The ledge creates a small cup that traps moist air above the stoma.

26.5 Phloem Function

Describe the structure of sieve tubes and the role of companion cells in their functioning.

Explain the steps involved in translocation of sucrose.

Phloem is the vascular tissue that conducts sugars and other organic solutes through the plant (Section 25.2). Conducting tubes of phloem are called **sieve tubes**, and each is composed of a stack of living cells called **sieve elements** (**Figure 26.9**). Unlike the cells that make up xylem tubes, sieve elements have no pits in their sides; fluid flows through their ends only.

As phloem tissue matures, plasmodesmata connecting the sieve elements in a stack enlarge up to 100 times their original size, leaving very large holes in the cell walls. The resulting perforated end walls, which are called sieve plates after their appearance, separate stacked sieve elements. The stacked cells' plasma membranes then merge through matching holes in their sieve plates, so their cytoplasm becomes continuous throughout the sieve tube. The tube becomes an open conduit for fluid when the sieve elements lose most of their organelles (including the nucleus).

How does a sieve element stay alive without organelles? Each sieve element has an associated **companion cell** that arises by division of the same parent cell. The companion cell retains its nucleus and other components. Many plasmodesmata connect the cytoplasm of the two cells, so the companion cell can provide all metabolic functions necessary to sustain its paired sieve element.

PRESSURE FLOW THEORY

A plant produces sugar molecules during photosynthesis. Some of these molecules are used by or stored in the cells that make them; the rest are conducted through sieve tubes to other parts of the plant. The movement of sugars and other organic molecules through phloem is called **translocation**.

Inside sieve tubes, fluid rich in sugars—mainly sucrose—flows from a **source** (a region of plant tissue where sugars are being produced or released from storage) to a **sink** (a region where sugars are being broken down or put into storage for later use). Photosynthetic tissues are typical source regions; tissues of developing roots and fruits are typical sink regions.

Why does sugar-rich fluid flow from a source to a sink? A pressure gradient between the two regions drives the movement (**Figure 26.10**). Consider mesophyll cells in a leaf, which produce far more sugar than they use for their own metabolism. Photosynthesis in these cells sets up the leaf as a source region. Excess sugar molecules produced by mesophyll cells diffuse into adjacent companion cells through plasmo-

CREDITS: (in text) Courtesy of E. Raveh; (8) Dr. Keith Wheeler/Science Source.

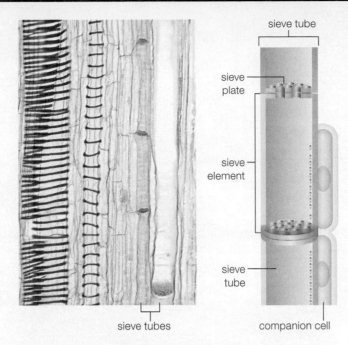

Figure 26.9 Sieve tubes of phloem. Each sieve tube consists of a stack of sieve elements that meet at sieve plates. Plasmodesmata connect each sieve element with a companion cell.

desmata. Depending on the species and the source region, sugar molecules may also be pumped into a companion cell through active transport proteins (Section 5.7).

Once inside a companion cell, sugar molecules diffuse through plasmodesmata into associated sieve elements **❶**. The sugar loaded into sieve tubes at a source region increases the solute concentration of a sieve tube's cytoplasm, so it becomes hypertonic (Section 5.6) with respect to the surrounding cells. Water follows the sugars by osmosis **❷**, moving into the sieve tube from cells surrounding it.

The rigid cell walls of mature sieve elements cannot expand very much, so the influx of water raises pressure in the tube. Pressure inside a sieve tube can be very high—up to five times higher than that of an automobile tire. Pressure that a fluid exerts against a structure that contains it is called turgor (Section 5.6). High fluid pressure (turgor) inside a sieve tube pushes sugar-rich cytoplasm from one sieve element to the next, toward a sink region where the turgor is lower **❸**. Pressure inside a sieve tube decreases at a sink because sugars leave the tube in this region. Sugar molecules diffuse into companion cells, and then move into adjacent sink cells (either through plasmodesmata or by active transport). Water follows, again by osmosis **❹**, so turgor inside sieve elements decreases in sink regions. This explanation of how turgor pushes sugar-rich fluid inside a sieve tube from source to sink is called the **pressure flow theory.**

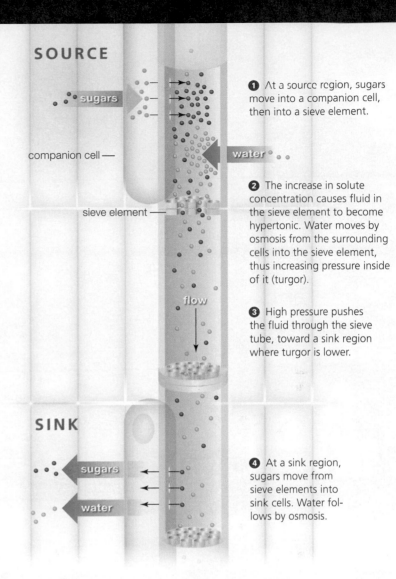

❶ At a source region, sugars move into a companion cell, then into a sieve element.

❷ The increase in solute concentration causes fluid in the sieve element to become hypertonic. Water moves by osmosis from the surrounding cells into the sieve element, thus increasing pressure inside of it (turgor).

❸ High pressure pushes the fluid through the sieve tube, toward a sink region where turgor is lower.

❹ At a sink region, sugars move from sieve elements into sink cells. Water follows by osmosis.

Figure 26.10 Translocation of organic compounds in phloem from source to sink. Water molecules are represented by blue balls; sugar molecules, by red balls. Each sieve tube is a stack of sieve elements that meet end to end at sieve plates. Sugars move into and out of a sieve element via its associated companion cell.

companion cell In phloem, cell that provides metabolic support to its partnered sieve element.
pressure flow theory Explanation of how a difference in turgor between sieve elements in source and sink regions pushes sugar-rich fluid through a sieve tube.
sieve elements Living cells that compose sugar-conducting sieve tubes of phloem. Each sieve tube consists of a stack of sieve elements that meet end to end at sieve plates.
sieve tube Sugar-conducting tube of phloem; consists of stacked sieve elements.
sink Region of plant tissue where sugars are being used or put into storage.
source Region of a plant tissue where sugars are being produced or released from storage.
translocation Movement of organic compounds through phloem.

Figure 26.11 Today, poplar trees are removing toxic waste in the soil of of J-Field, a legacy of 60 years of weapons testing and disposal.

LEAFY CLEANUP

From 1940 until the 1970s, the United States Army tested and disposed of weapons at J-Field, Aberdeen Proving Ground in Maryland (left). Obsolete chemical weapons and explosives were burned in open pits, together with plastics, solvents, and other wastes. Lead, arsenic, mercury, and other metals heavily contaminated the soil and groundwater. So did highly toxic organic compounds, including one called TCE (trichloroethylene). TCE damages the nervous system, lungs, and liver, and exposure to large amounts of it can be fatal.

To clean up the soil at J-Field and protect nearby Chesapeake Bay, the Army and the Environmental Protection Agency turned to phytoremediation: the use of plants to take up and concentrate or degrade environmental contaminants. In the case of organic toxins such as

CREDITS: (11) © Billy Wrobel, 2004; (in text) © OPSEC Control Number #4 077-A-4.

TCE, the best phytoremediation strategies use plants that break down the compounds to less toxic molecules. So, in 1996, the Army planted poplar trees that cleanse groundwater by taking up TCE and other organic pollutants in it (**Figure 26.11**). Like other vascular plants, the trees take up soil water through their roots. Along with the water come nutrients and chemical contaminants, including TCE. Though TCE is toxic to animals, it does not harm the trees. The poplars break down some of the toxin, but release most of it into the atmosphere. Airborne TCE is the lesser of two evils: It breaks down more quickly in air than it does in groundwater.

Analyses at J-Field since the poplars were planted show that the phytoremediation strategy is working splendidly. By 2001, the trees had removed 60 pounds of TCE from the soil and groundwater. In addition, they had helped foster communities of microorganisms that were also breaking down toxins. Researchers estimate that the trees will have removed the majority of contaminants from J-Field by 2030.

Compared to other methods of cleaning up toxic waste sites, phytoremediation is usually less expensive—and it is more appealing to neighbors. Consider Ford Motor Company's Rouge Center in Dearborn, Michigan, where decades of steelmaking left soil contaminated with highly carcinogenic compounds known as polycyclic aromatic hydrocarbons, or PAHs. With funding from the automaker, researchers developed a phytoremediation system based on native plants, which also boosted a sitewide initiative to restore the facility's native wildlife habitat (**Figure 26.12**). The site's plantings are attracting insects, birds, and other wildlife while aggressively accelerating the natural degradation of toxins. "If left undisturbed, it would take decades or centuries for these contaminants to naturally decompose," said researcher Clayton Rugh. "What

Figure 26.12 Phytoremediation at Ford's Rouge Center, Dearborn, Michigan. Top, the Rouge Center in 1927. Bottom, the Rouge Center today. This 10-acre green roof over a truck manufacturing plant collects and filters rain, improving the water quality of runoff flowing into nearby Rouge River.

our research is indicating is that we can achieve at least 50 percent degradation in three to five years—and that's at the very least. Some [species] seem to be approaching 70 percent in just three growing seasons."

With metal pollutants, the best phytoremediation strategies use plants that take up toxins and store them in aboveground tissues. The toxin-laden plant parts can then be harvested for safe disposal. This strategy is being tested in Ukraine, where in 1986 an accident at the Chernobyl nuclear power plant released huge amounts of radioactive material that still heavily contaminates the region. Transgenic sunflowers planted on rafts in heavily contaminated ponds quickly take up radioactive cesium (a metal) from the water.

CREDITS: (12) top, Library of Congress Prints and Photographs Division Washington, D.C. [LC-D414-K3461]; bottom, © Ford Photomedia.

Section 26.1 Plant growth requires elemental nutrients. Oxygen, carbon, and hydrogen atoms are abundant in air and water; nitrogen, phosphorus, sulfur, and other elements occur in soil minerals.

The availability of water and mineral ions in a particular soil depends on the proportions of sand, silt, and clay, and also on **humus** content. **Loams** have roughly equal proportions of sand, silt, and clay. **Leaching** and **soil erosion** deplete nutrients from soil, particularly **topsoils**.

Section 26.2 Endodermal cells surrounding a root's vascular cylinder deposit a **Casparian strip** into their abutting walls. This waxy, waterproof band prevents water from diffusing around the cells into vascular tissue. To enter xylem, substances dissolved in soil water must first pass through the cytoplasm of an endodermal cell. Thus, transport proteins in root cell plasma membranes control the plant's uptake of minerals dissolved in soil water.

Many plants form mutualisms with microorganisms. Fungi associate with young roots in mycorrhizae, which enhance a

plant's ability to absorb mineral ions from soil. Nitrogen fixation by bacteria in **root nodules** gives a plant extra nitrogen. In both cases, the microorganisms receive some sugars made by the plant.

Section 26.3 Water flows through xylem tubes from roots to shoot tips. These tubes consist of the walls of **tracheids** and **vessel elements** that formed in stacks and then died. Water moves through the pitted and lignin-reinforced secondary walls of these cells.

The **cohesion–tension theory** explains how water moves upward through xylem: **Transpiration** (evaporation from above-ground plant parts, mainly at stomata) pulls water upward inside xylem tubes. Cohesion helps water in these tubes to resist breaking into droplets under the tension exerted by the pull.

Section 26.4 A cuticle helps a plant conserve water; stomata help it balance water conservation with gas exchange required for metabolism. A stoma, which is a gap between two **guard cells**, may be surrounded by an indentation, protrusions, or other specializations that decrease evaporation from it. A stoma closes when the guard cells lose water and collapse against one another. It opens when the guard cells swell with water.

Section 26.5 Sugars move through a plant by **translocation** in phloem's sieve tubes, which consist of stacked **sieve elements** separated by perforated sieve plates. By the **pressure flow theory**, the movement of sugar-rich fluid through a **sieve tube** is driven by a pressure gradient between

source and **sink** regions. **Companion cells** load sugars into sieve elements at sources, and unload them at sinks.

Application The ability of plants to take up substances from soil water is the basis for phytoremediation. This environmental cleanup method uses plants specifically to remove pollutants from a contaminated area.

SELF-ASSESSMENT
ANSWERS IN APPENDIX VII

1. The main source(s) of hydrogen and oxygen for plants is (are) _____ .
   a. soil and air   c. water and fertilizer
   b. water and soil   d. water and air

2. Decomposing matter in soil is called _____ .
   a. clay   c. silt
   b. humus   d. sand

3. Most water moves from soil to vascular cylinder _____ .
   a. through root hairs   c. through root cell walls
   b. between root cells   d. in xylem

4. A _____ strip between abutting endodermal cell walls forces water and solutes to move through these cells rather than around them.
   a. cutin   b. Casparian   c. cohesion   d. cellulose

5. The nutrition of some plants is enhanced by a mutually beneficial association between a root and a fungus. The association is known as a _____ .
   a. root nodule   c. root hair
   b. mycorrhiza   d. hypha

6. Water evaporation from plant parts is called _____ .
   a. translocation   c. transpiration
   b. respiration   d. tension

7. Water transport from roots to leaves occurs by _____ .
   a. a pressure gradient inside sieve tubes
   b. different solutes at source and sink regions
   c. the pumping force of xylem vessels
   d. transpiration, tension, and cohesion of water

8. Tracheids are part of _____ .
   a. cortex   c. phloem
   b. mesophyll   d. xylem

9. Sieve tubes are part of _____ .
   a. cortex   c. phloem
   b. mesophyll   d. xylem

10. When stomata are open, a waterproof cuticle _____ .
    a. minimizes water loss through plant surfaces
    b. inhibits gas exchange between the plant and the air
    c. allows translocation to occur

11. When guard cells swell, _____ .
    a. transpiration ceases   c. stomata open
    b. sugars enter phloem   d. root cells die

TCE UPTAKE BY TRANSGENIC PLANTS Plants used for phytoremediation take up organic pollutants, then transport the chemicals to tissues where they are stored or broken down. Researchers are now designing transgenic plants with enhanced ability to take up or break down toxins.

In 2007, Sharon Doty and her colleagues published the results of their efforts to design plants for phytoremediation of soil and air contaminated with organic solvents. The researchers used *Agrobacterium tumefaciens* (Section 15.5) to deliver a mammalian gene into poplar plants. The gene encodes cytochrome P450, an enzyme involved in the breakdown of a range of organic molecules, including solvents such as TCE. **Figure 26.13** shows data from one test on the resulting transgenic plants.

1. How many transgenic plants did the researchers test?

2. In which group did the researchers see the slowest rate of TCE uptake? The fastest?

3. On day 6, what was the difference between the TCE content of air around planted transgenic plants and that around vector control plants?

4. Assuming no other experiments were done, what two explanations are there for the results of this experiment? What other control might the researchers have used?

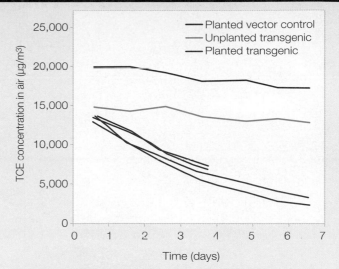

Figure 26.13 TCE uptake from air by transgenic poplar plants. Individual potted plants were kept in separate sealed containers with an initial level of TCE (trichloroethylene) around 15,000 micrograms per cubic meter of air. Samples of the air in the containers were taken daily and measured for TCE content.

Controls included a tree transgenic for a Ti plasmid with no cytochrome P450 in it (vector control), and a bare-root transgenic tree (one that was not planted in soil).

---

12. In phloem, organic compounds flow through _____ .
    a. collenchyma cells   c. vessels
    b. sieve elements   d. tracheids

13. Sugar transport from leaves to roots occurs by _____ .
    a. a pressure gradient inside sieve tubes
    b. different solutes at source and sink regions
    c. the pumping force of xylem vessels
    d. transpiration, tension, and cohesion of water

14. Match the concepts of plant nutrition and transport.
    ___ stomata   a. end of a vessel element
    ___ nutrient   b. takes up soil water and nutrients
    ___ sink   c. required element
    ___ root system   d. cohesion in xylem tubes
    ___ sieve plate   e. sugars unloaded from sieve tubes
    ___ xylem   f. distributes sugars
    ___ phloem   g. separates cells in phloem tubes
    ___ perforation plate h. distributes water
    ___ hydrogen bonds   i. balance water loss with gas
                            exchange

CLASS ACTIVITY
CRITICAL THINKING

1. Nitrogen deficiency stunts plant growth; leaves yellow and then die. Why does nitrogen deficiency causes these symptoms? *Hint*: Which biological molecules incorporate nitrogen atoms?

2. If a plant's stomata are made to stay open at all times, or closed at all times, it will die. Why?

3. You just returned home from a three-day vacation. Your severely wilted plants tell you they were not watered before you left. Use the cohesion–tension theory of water transport to explain what happened to them.

4. When you dig up a plant to move it from one spot to another, the plant is more likely to survive if some of the soil around the roots is transferred along with the plant. Formulate a hypothesis that explains this observation.

5. What are the three structures shown in the micrographs in **Figure 26.14**? From which tissue(s) do the structures originate?

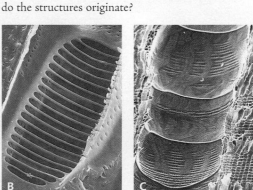

Figure 26.14 Name the mystery structures.

for additional quizzes, flashcards, and other study materials
▶ **ACCESS MINDTAP AT WWW.CENGAGEBRAIN.COM**

CREDITS: (14A) Alison W. Roberts, University of Rhode Island; (14 B–C) H. A. Core, W. A. Cote, and A. C. Day, Wood Structure and Identification, 2nd Ed., Syracuse University Press, 1979.

27 Reproduction and Development of Flowering Plants

Links to Earlier Concepts

This chapter revisits plant structure (Sections 4.8, 4.10, 6.4, 6.5, 25.1–25.6, 26.3, 26.5), life cycles (12.2, 21.1, 21.6, 21.7), and evolution (21.4); carbohydrates (3.1, 3,2); membrane transport (5.7); gene expression (9.1, 10.1–10.3); reproductive modes (12.1,12.3); polyploidy (14.5, 17.9); and coevolution (17.10).

Core Concepts

Evolution underlies the unity and diversity of life.

Sexual reproduction in flowering plants involves the transfer of pollen among individuals. Specializations in flower form are adaptations for particular pollination vectors, and many plant species have coevolved animal pollinators. Diversity in fruit form reflects adaptations for particular seed-dispersal vectors. Secondary metabolites produced by plants also affect other organisms, including humans.

Living things sense and respond appropriately to their internal and external environments.

Plants produce and respond to hormones. As in animals, hormones orchestrate development. Unlike animals, plants continue to develop after maturity, and in response to environmental cues. Feedback mechanisms involving plant hormones are part of development and homeostatic control. Specialized physiological defenses help plants survive environmental stresses and attacks by herbivores and pathogens.

Living systems store, retrieve, transmit, and respond to information essential for life.

The timing and coordination of specific molecular and cellular events are regulated by mechanisms that govern gene expression. Many types of environmental cues can trigger internal molecular signals in plants. In turn, these signals result in adjustments to a complex interplay of hormones and other molecules that affect gene expression.

◄ Flowers of raspberry and other *Rubus* species can fertilize themselves, but the fruit that forms from a self-fertilized flower is smaller and of lower quality than that of a flower cross-fertilized by a bee.

Photograph by © Alan McConnaughey. www.flickr.com/photos/engripiman.

LEARNING OBJECTIVES

Describe the components of a complete flower and their function.

Using suitable examples, describe the adaptations of flowers to particular pollination vectors.

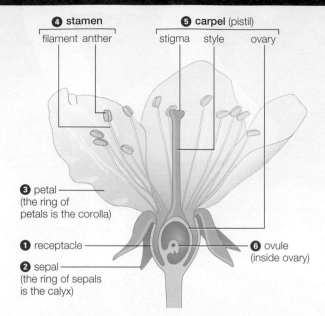

Figure 27.1 Anatomy of a typical flower. Male gametophytes are produced by stamens; female gametophytes, by carpels.

FLORAL STRUCTURE

A **flower** is a reproductive structure of an angiosperm (Section 21.6). The parts of a flower are modified leaves that develop from a thickened region of stem called a receptacle (**Figure 27.1 ❶**). A typical flower consists of four rings (whorls) of modified leaves. The outermost whorl is a calyx, a ring of leaflike sepals ❷. Sepals are photosynthetic and inconspicuous; they enclose and protect internal tissues before the flower opens. Just inside the calyx is the corolla (from the Latin *corona*, or crown), a ring of petals ❸. **Petals** are often the largest and most brightly colored parts of a flower. Inside the corolla is a whorl of **stamens**, the reproductive organs that produce the plant's male gametophytes (gamete-making structures). A typical stamen is a thin filament with an **anther** at the tip ❹. A typical anther consists of four pouches called pollen sacs. **Pollen grains**, which are immature male gametophytes of seed plants, form in pollen sacs. The modified leaves making up the flower's innermost whorl fold and fuse into one or more **carpels**: reproductive organs that produce female gametophytes ❺. The flowers of some species have one carpel; those of other species have several (a carpel—or a compound structure that consists of multiple fused carpels—is commonly called a pistil). The upper region of a carpel is a sticky or hairy **stigma** that is specialized to receive pollen grains. Typically, the stigma sits on top of a slender stalk called a style. The lower, swollen region of a carpel is the **ovary**, which contains one or more ovules. In seed plants, an **ovule** ❻ is

the part of the carpel where the female gametophyte forms. As you will see in Section 27.3, seeds are mature ovules.

Expression of floral identity genes (Section 10.3) determines the fate of a flower's four whorls. A variety of floral forms arise from variations in these and other genes. Some flowers are single blossoms (**Figure 27.2A**); others occur in clusters called inflorescences (**Figure 27.2B**). A composite flower is an inflorescence of many flowers that cluster in a single head (**Figure 27.2C**). A "regular" flower is radially symmetric: If it were cut like a pie, the pieces would be roughly identical. "Irregular" flowers are not radially symmetric. A flower with all four sets of modified leaves (sepals, petals, stamens, and carpels) is a "complete" flower. An "incomplete" flower lacks one or more of these structures (**Figure 27.2D**).

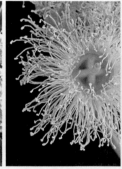

A The solitary flower of the lady's slipper orchid is irregular.

B A hyacinth flower is an elongated inflorescence.

C A daisy is a composite flower with many individual florets.

D The incomplete flower of eucalyptus has no petals.

E Some begonias have imperfect flowers: Female blossoms (with no stamens, left) form on the same plants as male blossoms (with no carpels, right).

Figure 27.2 Examples of structural variation in flowers.

POLLINATION

Pollination is the arrival of pollen on a pollen-receiving reproductive part of a seed plant. "Perfect" flowers have both stamens and carpels, so they can potentially pollinate themselves. "Imperfect" flowers, which lack stamens or carpels (**Figure 27.2E**), cannot. Self-pollination can be adaptive because it does not require a partner, but it lowers a population's genetic diversity over time (Section 17.6) and often reduces the quality of fruit and offspring (**Figure 27.3**). Accordingly, many plants have adaptations that encourage or even require cross-pollination by another individual. For example, a perfect flower may release pollen only after its stigma is no longer receptive to being fertilized. In some species, stamen-bearing and carpel-bearing flowers form on different plants.

The diversity of flower form reflects a dependence on **pollination vectors**: animals or environmental agents that transfer pollen from anther to stigma. Animal pollination vectors are called **pollinators**. Fragrance and other flower traits not directly related to reproduction are evolutionary adaptations that attract coevolved pollinators (Section 21.7). An animal attracted to a particular flower often picks up pollen on a visit, then inadvertently transfers it to another flower on a later visit. The more specific the attraction, the more efficient the transfer of pollen among plants of the same species. Thus, a flower's traits typically reflect the sensory abilities and preferences of its coevolved pollinator. In an example of morphological convergence (Section 16.7), the flowers of angiosperm species that depend on the same pollinator (bees, for example, or birds) often share a certain set of traits. Consider that bees have a keen sense of smell, and they can see ultraviolet (UV) light; bee-pollinated flowers tend to be fragrant, with a bull's-

Figure 27.3 Self-pollination versus cross-pollination. The two raspberries on the left formed from self-pollinated flowers; the one on the right, from an insect-pollinated flower. Larger fruit with more seeds formed by cross-pollination.

eye pattern of UV-reflecting pigments in their petals. Bee-pollinated flowers also attract and communicate with bees via electric fields. Birds have excellent color vision and a relatively poor sense of smell, so many bird-pollinated flowers are bright red and unscented. Flowers pollinated by night-flying bats, which have a good sense of smell but poor vision, tend to be large and light-colored, with a strong nighttime fragrance. Not all flowers smell sweet; odors like dung or rotting flesh beckon beetles and flies.

Rewards offered by a flower reinforce a pollinator's memory of a visit, and encourage the animal to seek out other flowers of the same species. Many flowers exude a sweet fluid called **nectar** to attract pollinators. Nectar is the only food for most adult butterflies, and it is a favorite food of hummingbirds. Honeybees convert nectar to honey, which helps feed the bees and their larvae through the winter. Bees also collect protein-rich pollen for food. Many beetles feed primarily on pollen. Pollen is the only reward for a pollinator of roses, poppies, and other flowers that produce no nectar.

Specializations of some flowers prevent pollination by all but one type of pollinator. For example, nectar or pollen at the bottom of a long floral tube may be accessible only to an insect with a matching feeding device. In some plants, only a specific pollinator can trip a floral mechanism that brings stamens on stigma into contact with the animal's body (the white sage plants you learned about in Section 17.7 offer an example). Such relationships are to both species' mutual advantage: A flower that captivates the attention of an animal has a pollinator that spends its time seeking (and pollinating) only those flowers. Pollinators also benefit when they receive an exclusive supply of the flower's reward.

Grasses and other plants pollinated by wind do not benefit by attracting animals, so they do not waste resources making scented, brilliantly colored, or nectar-laden flowers. Wind-pollinated flowers are typically small, nonfragrant, and green, with insignificant petals and large stamens that produce a lot of pollen.

---

anther (AN-thur) Part of the stamen that produces pollen grains.

carpel (CAR-pull) Floral reproductive organ that produces the female gametophyte; consists of an ovary, stigma, and usually a style.

flower Specialized reproductive structure of an angiosperm.

nectar Sweet fluid exuded by some flowers; rewards pollinators.

ovary In flowering plants, the enlarged base of a carpel, inside which one or more ovules form.

ovule (AH-vule) Of a seed plant, the part of the carpel where the female gametophyte forms.

petal Unit of a flower's corolla; often showy and conspicuous.

pollen grain Immature male gametophyte of a seed plant.

pollination The arrival of pollen on a pollen-receiving reproductive part of a seed plant.

pollination vector Animal or environmental agent that moves pollen grains from one plant to another.

pollinator An animal pollination vector.

stamen (STAY-men) Floral reproductive organ that consists of an anther and, in most species, a filament.

stigma Upper part of a carpel; adapted to receive pollen.

LEARNING OBJECTIVE

Draw the life cycle of a typical flowering plant and explain the roles of meiosis and mitosis in it.

Flowers are made by the angiosperm sporophyte, a diploid, spore-producing plant body that grows by mitosis from a fertilized egg (Section 21.6). Spores that form by meiosis in flowers develop into haploid gametophytes, which produce gametes (Section 12.2). **Figure 27.4** and **Figure 27.5** show the life cycle of a typical angiosperm.

FEMALE GAMETE PRODUCTION

The production of female gametes begins when an ovule starts growing on the inner wall of an ovary ❶. In a common pattern, a cell in the ovule enlarges, then undergoes meiosis and cytoplasmic division to form four haploid **megaspores** ❷. Three of the four megaspores disintegrate, and the remaining one undergoes three rounds of mitosis without cytoplasmic division. One cell results from these divisions, and it has eight haploid nuclei ❸. The cytoplasm of this cell divides unevenly, forming a seven-celled embryo sac—the female gametophyte ❹. One of the cells is much larger than the others, and it has two nuclei ($n + n$); this cell is called the central cell. Another cell is the egg. In most species, the female gametophyte is surrounded by one or two protective outer layers called integuments.

MALE GAMETE PRODUCTION

The production of male gametes begins as masses of diploid, spore-producing cells form by mitosis inside anthers. Walls typically develop around the masses, so four pollen sacs form ❺. Each cell inside the pollen sacs undergoes meiosis and cytoplasmic division to form four haploid **microspores** ❻. Mitosis and differentiation of a microspore produce a pollen grain ❼. A pollen grain consists of two cells, one inside the cytoplasm of the other, enclosed by a durable coat that will protect the cells on their journey to meet an egg. After the pollen grains form, they enter a period of suspended metabolism—**dormancy**—before being released from the anther ❽ when the pollen sacs split open.

DOUBLE FERTILIZATION

A pollen grain that lands on a receptive stigma **germinates**, which means it resumes metabolic activity after dormancy. When a pollen grain germinates, its outer cell develops into a tubular outgrowth called a pollen tube ❾. The inner cell

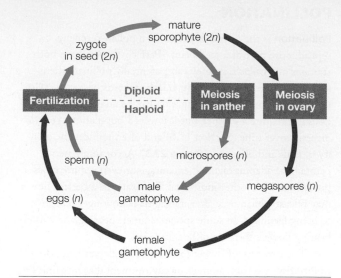

Figure 27.4 Overview of the life cycle of a typical flowering plant (compare Figure 27.5).

undergoes mitosis to produce two sperm cells (the male gametes) within the pollen tube. The mature male gametophyte is the pollen tube together with its contents of male gametes ❿.

The pollen tube elongates at its tip, burrowing into the tissues of the style and carrying its contents (the two sperm cells) toward the ovary. Chemical signals secreted by cells in the female gametophyte guide the pollen tube's growth to the ovule. Many pollen tubes may grow down into a carpel, but usually only one penetrates the female gametophyte.

A pollen tube that reaches and penetrates the ovule releases the two sperm cells ⓫. Flowering plants undergo **double fertilization**, in which one of the sperm cells delivered by the pollen tube fuses with (fertilizes) the egg and forms a diploid zygote. The other sperm cell fuses with the central cell to form a triploid ($3n$) cell. This cell gives rise to triploid **endosperm**, a nutritious tissue that forms only in seeds of flowering plants. When the seed sprouts, endosperm will sustain rapid growth of the seedling until its new leaves begin photosynthesis.

dormancy Period of temporarily suspended metabolism.
double fertilization Fertilization as it occurs in flowering plants. Two cells are fertilized: One sperm cell fertilizes the egg to form the zygote; the second sperm cell fuses with the central cell to form a triploid ($3n$) cell that gives rise to endosperm.
endosperm Nutritive tissue in the seeds of flowering plants.
germinate To resume metabolic activity after dormancy.
megaspore Of seed plants, haploid spore that forms in an ovule and gives rise to an egg-producing gametophyte.
microspore Of seed plants, haploid spore that gives rise to a pollen grain.

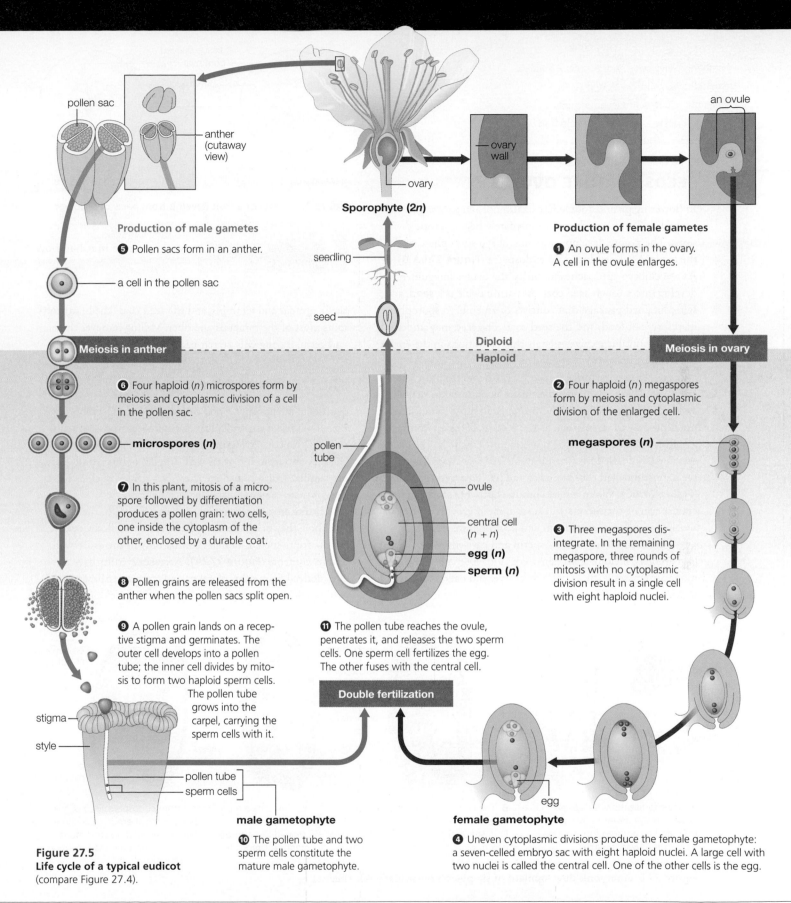

Production of male gametes

❺ Pollen sacs form in an anther.

a cell in the pollen sac

Meiosis in anther

❻ Four haploid (*n*) microspores form by meiosis and cytoplasmic division of a cell in the pollen sac.

microspores (*n*)

❼ In this plant, mitosis of a microspore followed by differentiation produces a pollen grain: two cells, one inside the cytoplasm of the other, enclosed by a durable coat.

❽ Pollen grains are released from the anther when the pollen sacs split open.

❾ A pollen grain lands on a receptive stigma and germinates. The outer cell develops into a pollen tube; the inner cell divides by mitosis to form two haploid sperm cells. The pollen tube grows into the carpel, carrying the sperm cells with it.

stigma

style

pollen tube
sperm cells

male gametophyte

❿ The pollen tube and two sperm cells constitute the mature male gametophyte.

pollen sac

anther (cutaway view)

ovary

Sporophyte (2*n*)

seedling

seed

Diploid
Haploid

pollen tube

ovule

central cell (*n* + *n*)

egg (*n*)
sperm (*n*)

⓫ The pollen tube reaches the ovule, penetrates it, and releases the two sperm cells. One sperm cell fertilizes the egg. The other fuses with the central cell.

Double fertilization

ovary wall

an ovule

Production of female gametes

❶ An ovule forms in the ovary. A cell in the ovule enlarges.

Meiosis in ovary

❷ Four haploid (*n*) megaspores form by meiosis and cytoplasmic division of the enlarged cell.

megaspores (*n*)

❸ Three megaspores disintegrate. In the remaining megaspore, three rounds of mitosis with no cytoplasmic division result in a single cell with eight haploid nuclei.

egg

female gametophyte

❹ Uneven cytoplasmic divisions produce the female gametophyte: a seven-celled embryo sac with eight haploid nuclei. A large cell with two nuclei is called the central cell. One of the other cells is the egg.

Figure 27.5
Life cycle of a typical eudicot
(compare Figure 27.4).

SEEDS: MATURE OVULES

In flowering plants, double fertilization produces a zygote and a triploid (3*n*) cell, and both immediately begin mitotic divisions. The zygote develops into an embryo sporophyte, and the triploid cell develops into endosperm (**Figure 27.6A,B**). As the embryo approaches maturity, the ovule's integuments develop into a tough seed coat. A mature ovule is a **seed**, a self-contained package that consists of an embryo sporophyte, its reserves of food, and the seed coat. The seed may undergo dormancy until it receives signals that conditions in the environment are appropriate for germination.

As a seed is forming, it receives nutrients from the parent plant. These nutrients accumulate in endosperm mainly as starch with some lipids and proteins. In most monocots, nutrients stay in endosperm as the seed matures. By contrast, a maturing eudicot embryo transfers nutrients from the endosperm to its cotyledons (seed leaves). In a mature eudicot seed, most nutrients are stored in the two enlarged cotyledons (**Figure 27.6C**). When a seed sprouts, nutrients stored in its endosperm or cotyledons will sustain rapid growth of the seedling until new leaves form and begin photosynthesis.

Nutrients stored in endosperm and cotyledons also sustain animals, including humans. We cultivate many cereals—monocot grasses such as rice, wheat, rye, and oats—for their nutritious seeds. The embryo (the germ) contains most of the

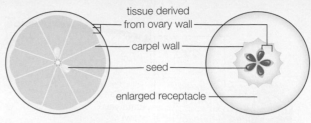

A The tissues of an orange develop from the ovary wall.

B The flesh of an apple is an enlarged receptacle.

Figure 27.7 Parts of a fruit develop from parts of a flower.

FIGURE IT OUT
The flower that gave rise to the orange in **A** had how many carpels?

Answer: Eight

seed's protein and vitamins, and the seed coat (the bran) contains most of the minerals and fiber. Milling removes the bran and germ, leaving only starch-packed endosperm.

FRUITS: MATURE OVARIES

Nutrients in endosperm and cotyledons nourish embryonic plants. Nutrients in fruits have a different purpose. A **fruit** is a seed-containing mature ovary, often with fleshy tissues that develop from the ovary wall. Apples, oranges, and grapes are familiar fruits, but so are many "vegetables" such as beans, peas, tomatoes, grains, eggplant, and squash.

Fruits are categorized by the composition of their tissues, how they originate, and whether they are fleshy or dry. "True" fruits such as oranges consist only of the ovary wall and its contents (**Figure 27.7A**). "Accessory" fruits have tissues derived from other floral parts—petals, sepals, stamens, or receptacle—that expand along with the developing ovary.

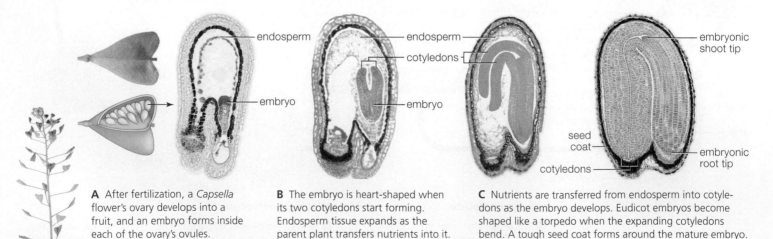

A After fertilization, a *Capsella* flower's ovary develops into a fruit, and an embryo forms inside each of the ovary's ovules.

B The embryo is heart-shaped when its two cotyledons start forming. Endosperm tissue expands as the parent plant transfers nutrients into it.

C Nutrients are transferred from endosperm into cotyledons as the embryo develops. Eudicot embryos become shaped like a torpedo when the expanding cotyledons bend. A tough seed coat forms around the mature embryo.

Figure 27.6 Embryonic development of shepherd's purse (*Capsella*), a eudicot.

A cherry is a true fruit, a simple fruit, and a fleshy fruit.

A blackberry is an aggregate of individual fleshy fruits.

A pineapple is an accessory fruit, and a multiple fruit formed from fused ovaries of an entire inflorescence.

A pea pod is a true fruit, a simple fruit, and a dry fruit.

A strawberry is an accessory fruit, and an aggregate of individual dry fruits.

An apple is an accessory fruit; most of its flesh is an enlarged receptacle (**Figure 27.7B**). "Simple" fruits are derived from one ovary or a few fused ovaries of one flower. Cherries (left), beans, acorns, tomatoes, and shepherd's purse are examples. By contrast, "aggregate" fruits such as blackberries are derived from multiple unfused ovaries of one flower that mature as a cluster. A "multiple" fruit such as a pineapple or a fig develops as a unit from several individually pollinated flowers.

Cherries, almonds, and olives are fleshy fruits, as are individual fruits of blackberries and other related species. Acorns, pea pods, and grains are dry fruits, as are the fruits of sunflowers, maples, and strawberries. Strawberries are not berries and their fruits are not fleshy. The red, juicy part of a strawberry is an enlarged receptacle with individual dry fruits on its surface. To a botanist, a "berry" is a fleshy fruit produced from one ovary. Grapes, tomatoes, citrus fruits, pumpkins, watermelons, and cucumbers are berries. Apples and pears are "pomes," fruits in which fleshy tissues derived from the receptacle enclose a core derived from the ovary wall.

The function of a fruit is to protect and disperse seeds. Dispersal increases reproductive success by minimizing competition for resources among parent and offspring. Just as flower structure is adapted to certain pollination vectors, so is fruit structure adapted to certain dispersal vectors: environmental agents of movement such as wind or water, or mobile organisms such as birds or insects. For example, fruits dispersed by wind are lightweight with breeze-catching specializations (**Figure 27.8A,B**). Fruits dispersed by water have water-repellent outer layers, and they float (**Figure 27.8C,D**). Many fruits have specializations that facilitate dispersal by animals (**Figure 27.8E,F**). Some have hooks or spines that stick to the feathers, feet, fur, or clothing of more mobile species. Colorful, fleshy, or fragrant fruits attract birds and mammals that disperse seeds. The animal may eat the fruit and discard the seeds, or eat the seeds along with the fruit. Seeds of apricots, plums, and many other fruits contain toxins that are released if the seed is crushed, thus discouraging animals from chewing them up completely and killing the embryo. However, abrasion of the seed coat by teeth or by digestive enzymes in an animal's gut can help the seed germinate after it departs in feces.

A Wind that lifts the hairy modified sepals of a dandelion fruit may carry it miles away from the parent plant.

B Dry outgrowths of the ovary wall of a maple fruit form "wings" that catch the wind and spin the seeds away from the parent tree.

C Fruits of sedges native to American marshlands have seeds encased in a bladderlike envelope that floats.

D Buoyant fruits of the coconut palm have tough, waterproof husks. They can float for thousands of miles in seawater.

E Curved spines attach cocklebur fruits to the fur of animals (and the clothing of humans) that brush past the plant.

F The red, fleshy fruits of hawthorn plants are an important food source for cedar waxwings, which disperse the fruits' seeds in feces.

Figure 27.8 Some adaptations that aid fruit dispersal.

fruit Mature ovary of a flowering plant, often with accessory parts; encloses a seed or seeds.

seed Mature ovule of a seed plant: embryo sporophyte packaged with nutritive tissue inside a protective coat.

LEARNING OBJECTIVES

Describe vegetative reproduction and give some examples.

List some of the ways that humans make use of vegetative reproduction.

Explain why some fruits are seedless.

Many flowering plants can reproduce asexually by a natural process called **vegetative reproduction**, in which new roots and shoots grow from extensions or pieces of a parent plant. Each new plant is a clone, a genetic replica of its parent. For example, new roots and shoots can sprout from nodes on stems (**Figure 27.9A**). Entire forests of quaking aspen trees (*Populus tremuloides*) are actually stands of clones that grew from root suckers—shoots that sprout from shallow lateral roots (**Figure 27.9B**). One quaking aspen forest in Utah consists of about 47,000 shoots and stretches for 107 acres. This stand of clones, which has been named Pando, is estimated to be around 80,000 years old! Such clones are as close to immortal as any organism can be.

AGRICULTURAL APPLICATIONS

For thousands of years, we humans have been taking advantage of the natural capacity of many plants to reproduce asexually. For example, almost all houseplants, woody ornamentals, and orchard trees are clones grown from stem fragments (cuttings) of a parent plant. Propagating plants from cuttings may be as simple as jamming a broken stem into soil. This method harnesses the root- and shoot-forming ability of nodes on a stem. Plants grown from cuttings may mature more quickly when they are grafted (induced to fuse with and become supported by another plant).

Propagating a plant from cuttings ensures that its offspring will have the same desirable traits as the parent. Consider familiar orchard apples, which are descendants of a wild species native to central Asia. The chromosome number of wild apples doubled between 50 and 60 million years ago, and subsequent chromosomal recombinations and rearrangements made a genetic mashup of duplicated genes. Today, each apple tree is highly heterozygous, and this is the reason why apples do not breed true for fruit traits (**Figure 27.10**). Every seed—even ones from the same apple—will produce a tree that bears unique fruit.

In fact, most apple trees grown from seed produce fruit that is uniquely unpalatable. In the early 1800s, John Chapman (known as Johnny Appleseed) grew millions of apple trees from seed in the midwestern United States. He sold the trees to homesteading settlers, who made hard cider from the otherwise inedible apples. About one of every hundred trees produced tasty fruit. Its lucky owner would patent the

A Potatoes are modified stems; new roots and shoots sprout from their nodes, which we call "eyes" (Section 25.3). The new plants are clones, genetically identical to the parent.

B In some plants, new individuals can form from pericycle in a root's vascular cylinder. The resulting clones are called root suckers. This photo shows suckers forming on an aspen root.

Figure 27.9 Examples of asexual reproduction in plants.

tree, then propagate it from cuttings grafted onto other, less desirable apple trees. An estimated 16,000 varieties of edible apples were discovered during this era. Few of them remain; only 15 varieties now account for 90 percent of the apples sold in U.S. grocery stores. All are clones of Chapman's original trees, and all are still grafted.

Grafting is also used to increase disease resistance of a desirable plant. In 1862, the plant louse *Phylloxera* was accidentally introduced into France via imported American grapevines. European grapevines had little resistance to this tiny insect, which attacks and kills the root systems of the vines. By 1900, *Phylloxera* had destroyed two-thirds of the vineyards

tissue culture propagation Laboratory method in which individual plant cells (typically from meristem) are induced to form embryos.

vegetative reproduction Growth of new roots and shoots from extensions or fragments of a parent plant; form of asexual reproduction in plants.

Figure 27.10 Wild apples. Color, flavor, size, sweetness, and texture vary in the fruits of wild apple trees (21 are shown in this photo). Apple trees are grafted because they do not breed true for these valuable traits.

in Europe, thus devastating the wine-making industry for decades. Today, French vintners routinely graft prized grapevines onto the roots of *Phylloxera*-resistant American vines.

Grafting allows us to propagate plants that produce fruit that is "seedless," which means it has underdeveloped seeds or none at all. Fruit can form on some plants even in the absence of fertilization, and fruits that develop from unfertilized flowers are seedless. Plants with mutations that cause ovules or embryos to abort during development also make seedless fruit; seedless grapevines and navel orange trees are like this. All commercially produced bananas are seedless because the plants that bear them are triploid (3*n*). During meiosis, the three chromosomes cannot be divided equally between the two spindle poles, so viable gametes do not form and neither do seeds. Grafting bananas and other monocots is notoriously difficult, so seedless banana plants are propagated from shoots that sprout from nodes on their corms (Section 25.3), or by tissue culture. With **tissue culture propagation**, individual cells (typically from meristem) are coaxed to divide and form embryos in a laboratory. This technique can yield millions of genetically identical offspring from a single parent plant. It is frequently used in research aimed at improving food crops, and also to propagate rare or hybrid ornamental plants.

Plant breeders can artificially increase the frequency of polyploidy by treating plants with colchicine, a microtubule poison that disrupts spindle function during meiosis. Tetraploid (4*n*) plants produced by colchicine treatment are then crossed with diploid (2*n*) individuals. The resulting offspring are triploid and sterile: They make seedless fruit on their own or after pollination (but not fertilization) by a diploid plant. Seedless watermelons are produced this way.

LEARNING OBJECTIVE

Discuss the roles of hormones in plants.

Most animal development occurs before adulthood. By contrast, development in plants continues throughout the individual's lifetime. A plant adjusts to unfavorable conditions by altering growth patterns. Changes in form and physiology are triggered by environmental cues such as temperature, gravity, night length, the availability of water and nutrients, and the presence of pathogens or herbivores. This developmental flexibility depends on extensive coordination among a plant's cells. Cells in different tissues or even different parts of a plant coordinate activities by communicating with one another. As an example, a leaf being chewed by a caterpillar signals other parts of the plant to produce appropriate caterpillar-deterring chemicals. Cell-to-cell communication in plants involves **plant hormones**: extracellular signaling molecules that exert effects at very low concentrations.

All cells in a plant have the ability to make and release hormones. Plant hormones can be released far from the tissue they affect. For example, cells in an actively growing root release a hormone that keeps shoot tips actively growing too. However, plant hormones also act locally, as when a cell in a ripening fruit releases a hormone that affects itself as well as its neighbors.

Hormones work by binding to receptor proteins on target cells. When a receptor binds to a hormone, it triggers a change in cellular activities (Section 4.3). Differences in a hormone's effect arise from differences in how these receptors work. Consider how some plant hormones act as hormones in animals, and vice versa. Both plants and mammals make estrogen, progesterone, testosterone, and other similar steroid hormones, for example, but the effects of these molecules differ between the two groups. In mammals, they are sex hormones; in plants, they function in stress responses.

As you will see in the following sections, a plant cell's response to a hormone often varies with the hormone's concentration. One plant hormone can enhance or oppose another's effects on the same cell. Many plant hormones inhibit their own expression, a mechanism of negative feedback that maintains ongoing activities and circadian rhythms (Section 10.4). Positive feedback loops in which a plant hormone promotes its own transcription are part of intermittent processes.

---

plant hormone Extracellular signaling molecule of plants; exerts an effect on target cells at very low concentration.

CREDITS: (10) Photo by Peggy Greb, USDA, ARS; (in text) siamionau pavel/Shutterstock.com.

CHAPTER 27 REPRODUCTION AND DEVELOPMENT OF FLOWERING PLANTS   **463**

27.6 Auxin and Cytokinin

LEARNING OBJECTIVES

Explain auxin's effect on growth and how it causes plant cells to enlarge.

Summarize the interaction of auxin and cytokinin in apical dominance.

Auxin is a plant hormone that was first discovered for its growth-promoting effect. It was later shown to have a critical role in all aspects of plant development (**Table 27.1**). Auxin is involved in polarity and tissue patterning in the embryo, formation of plant parts, differentiation of vascular tissues, formation of lateral roots, and, as you will see in later sections, shaping the plant body in response to environmental stimuli.

Researchers are still working out the mechanisms of many of auxin's effects, but they do understand one way in which auxin promotes growth directly: It causes young plant cells to expand by softening their cell wall. Turgor, the pressure exerted by fluid inside the softened wall (Section 5.6), enlarges the cell irreversibly.

Auxin is made mainly in shoot apical meristems (Section 25.6) and in young leaves, then transported elsewhere. Some of the auxin produced in shoot tips is loaded into phloem, travels to roots, and then is unloaded into root cells. A unique cellular mechanism also distributes auxin directionally through adjacent cells, thus establishing localized auxin concentration gradients. These gradients have a central role in coordinating the release and effects of other hormones.

The plant hormone **cytokinin** affects division and differentiation of meristem cells. It also stimulates development of

TABLE 27.1

Some Effects of Auxin and Cytokinin

Auxin

Coordinates effects of other plant hormones during development of shoots and roots
Promotes apical dominance
Promotes growth by cell enlargement
Stimulates division of meristem cells

Cytokinin

Stimulates differentiation of cells in root apical meristem
Stimulates cell division in shoot apical meristem
Inhibits formation of lateral roots

lateral buds, and inhibits development of lateral roots. Cytokinin and auxin influence one another's expression, a homeostatic mechanism that dynamically regulates their relative concentrations. They also influence the same cells. In shoots, the two hormones act on shoot apical meristem cells to support cell division and to prevent differentiation. In most other contexts, cytokinin opposes auxin's effects. For example, the two hormones work in opposition to maintain the balance of differentiating and undifferentiated cells in root apical meristem: Auxin supports division of undifferentiated meristem cells, and cytokinin signals the cells to differentiate.

Consider how auxin and cytokinin help balance the growth of apical and lateral shoots. When a shoot is lengthening, its lateral buds are typically dormant, an effect called **apical dominance** (**Figure 27.11**). If the shoot's tip breaks off, its lateral buds begin to grow, an effect exploited by gardeners who pinch off shoot tips to make a plant bushier. Auxin is produced by a shoot's apical meristem and actively transported from cell to cell down through the stem ❶. The flow of auxin through these cells causes the stem to lengthen, and it also keeps the cytokinin level low. When the shoot's tip is removed, its source of auxin disappears. The auxin level declines in the stem, allowing the cytokinin level to rise ❷. The cytokinin moves into lateral buds and stimulates cell divisions of apical meristem inside them ❸. The newly active meristem cells produce auxin, which is then transported away from the bud tips and down the stem. The auxin flow causes the lateral buds to lengthen, so new branches develop ❹.

Figure 27.11 Interaction of auxin and cytokinin in the release of apical dominance.

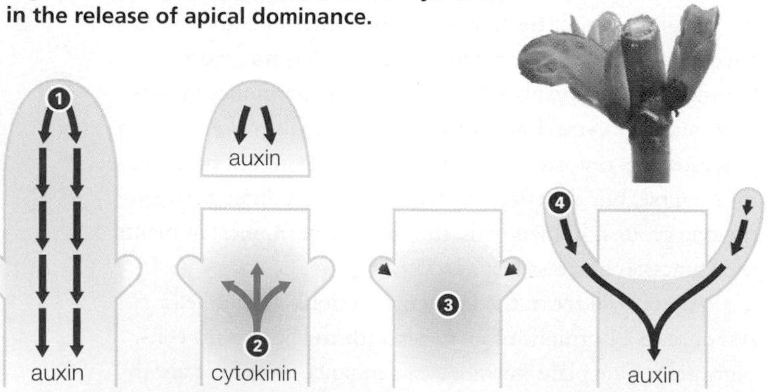

❶ Auxin being transported away from shoot apical meristem supports stem lengthening. Its presence in the stem keeps the level of cytokinin low.

❷ Removing the tip also removes the source of auxin, so auxin flow ends in the stem. As the auxin level declines, the cytokinin level rises.

❸ The rise in the cytokinin level in the stem stimulates cell division in apical meristem of lateral buds. The cells begin to produce auxin.

❹ Auxin is transported away from the lateral buds and down through the stem. This movement causes the lateral buds to lengthen.

apical dominance Effect in which a lengthening shoot tip inhibits the growth of lateral buds.

auxin (OX-in) Plant hormone that causes cell enlargement; also has a central role in growth and development by coordinating the effects of other hormones.

cytokinin (site-oh-KINE-in) Plant hormone that promotes cell division in shoot apical meristem and cell differentiation in root apical meristem. Often opposes auxin's effects.

27.7 Gibberellin

Describe the way gibberellin causes plant growth.

Explain gibberellin's role in seed germination.

Figure 27.12
Stem-lengthening effect of gibberellin. The three tall cabbage plants to the right of the ladder were treated with gibberellin. The two short plants were not treated.

In 1926, researcher Eiichi Kurosawa was studying what Japanese call *bakane*, the "foolish seedling" effect. The stems of rice seedlings infected with a fungus, *Gibberella fujikuroi*, grew twice the length of uninfected seedlings. Kurosawa discovered that he could make this lengthening happen experimentally by applying extracts of the fungus to healthy seedlings. Other researchers later purified the substance in the fungal extracts that brought about stem lengthening (**Figure 27.12**). They named it gibberellin, after the fungus, before discovering that plants make the same compound.

As we now know, **gibberellin** is a plant hormone that promotes growth of flowering plants, among other functions (**Table 27.2**). Gibberellin causes a stem to lengthen between the nodes by inducing cell division and elongation. Along with auxin, gibberellin stimulates expansion along the long axis of a plant organ. Thus, it increases the length of the organ, and of the plant itself. Gibberellin is also involved in slowing the aging of leaves and fruits, breaking dormancy in seeds, and, in some species, flowering.

Gibberellin made by cells in young leaves and root tips is transported through phloem to the rest of the plant. Seeds also make this hormone. Seedless grapes tend to be smaller than seeded varieties because they have tiny, underdeveloped seeds that do not produce normal amounts of gibberellin. Farmers spray their seedless grape plants with synthetic gibberellin to increase fruit size.

Gibberellin works by inhibiting inhibitors, thus removing the brakes on some cellular processes. Consider how the hormone works during germination of a barley seed. Water absorbed by the seed causes cells of the embryo to release

TABLE 27.2

Some Effects of Gibberellin

Stimulates cell division, elongation in stems

Mobilizes food reserves in germinating seeds

Stimulates flowering in some plants

Delays senescence (Section 11.4)

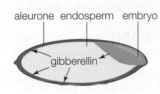

A Absorbed water causes cells of a barley embryo to release gibberellin, which diffuses through the seed into the aleurone layer of the endosperm.

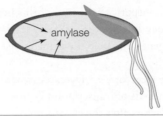

B Gibberellin causes cells of the aleurone layer to produce amylase. This enzyme diffuses into the starch-packed endosperm.

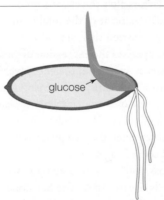

C The amylase breaks down starch into glucose. The glucose diffuses into the embryo and is used for aerobic respiration. Energy released by the reactions of aerobic respiration fuels meristem cell divisions in the embryo.

Figure 27.13 Gibberellin and germination, as illustrated in a seed of barley (a monocot).

gibberellin, which then diffuses into the aleurone, a single layer of living cells that surrounds endosperm (**Figure 27.13**). Gibberellin causes these cells to destroy a transcription factor (Section 10.1) that suppresses transcription of the gene for amylase (the enzyme that breaks the bonds between glucose monomers in starch, Section 3.2). The cells start producing amylase and releasing it into the endosperm's interior, where the enzyme breaks down stored starch molecules. The embryo then uses the released glucose for aerobic respiration, which fuels rapid growth of the embryonic shoot and root.

gibberellin (jib-er-ELL-in) Plant hormone that induces stem elongation; also helps seeds break dormancy.

27.8 Abscisic Acid and Ethylene

LEARNING OBJECTIVES

Describe in general terms the mechanism by which abscisic acid helps minimize water loss.

Explain the feedback loops involved in ethylene synthesis.

Describe the role of ethylene in fruit ripening.

TABLE 27.3

Some Effects of Abscisic Acid (ABA) and Ethylene

Abscisic acid (ABA)	Ethylene
Mediates stress responses	Stimulates fruit ripening
Closes stomata	Stimulates abscission
Inhibits shoot growth	Involved in breaking dormancy
Inhibits seed germination	Involved in stress responses

ABSCISIC ACID

The plant hormone **abscisic acid** (**ABA**) was named because its discoverers thought it mainly mediated **abscission**, the process by which a plant sheds leaves or other parts. It was later discovered to have a much greater role in plant stress responses (**Table 27.3**). Temperature extremes, water shortages, and other environmental stresses trigger an increase in ABA synthesis, release, and transport, so the level of this hormone rises in the plant's tissues. In turn, the increased level of ABA induces expression of genes that help the plant survive the adverse conditions.

When ABA binds to its receptors, it triggers a cascade of reactions that activate hundreds of transcription factors. In turn, these transcription factors alter the expression of thousands of genes—around 10 percent of the total number in plants and a much larger proportion than any other plant hormone. In general, ABA suppresses the expression of genes involved in growth, and enhances expression of genes involved in metabolism, stress responses, and embryonic development.

Consider a mechanism that closes stomata when a plant is stressed by drought (a lack of water). Remember from Section 26.4 that a stoma closes when the two guard cells that define it lose water and collapse against each other. This response minimizes water loss, and it is mediated by ABA. When soil dries out, root cells release ABA. The hormone travels through the plant's vascular system to leaves and stems, where it binds to receptors on guard cells (**Figure 27.14**). ABA binding triggers the guard cells to make and release a burst of nitric oxide (NO), a gas that functions as a hormone in plants (and also in animals). The burst of NO activates transport proteins (Section 5.7) that allow positively charged calcium ions (Ca^{2+}) to flow into guard cell cytoplasm ❶. An increase in calcium ion concentration in guard cell cytoplasm has two effects. First, it activates another set of transport proteins that pump chloride (Cl^-) and other negatively charged ions out of the cells ❷. Second, the inflow of positively charged calcium ions (and the exit of negatively charged ions) alters the voltage (distribution of charge) across the guard cell plasma membrane. The change affects voltage-gated ion channels: transport proteins with a "gate" that opens or closes in response to a change in voltage (as you will learn in Chapter 29, voltage-gated ion channels are integral to nervous signaling in animals). Guard cell plasma membranes have voltage-gated potassium ion (K^+) channels that open in response to a change in voltage. When the gates open, potassium ions follow the charge gradient across the membrane and exit the cells ❸. Water follows the solutes by osmosis, and stomata close as the guard cells lose turgor and collapse against one another ❹. Note that the release of nitric oxide also triggers production of an enzyme that breaks down ABA—a feedback loop that dynamically controls ABA signaling.

Figure 27.14 **ABA causes stomata to close** by triggering the decrease of turgor in guard cells.

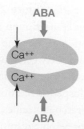

❶ ABA binds to its receptor on guard cell membranes. The binding triggers a series of events that activate calcium transport proteins, so these ions enter cytoplasm.

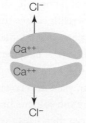

❷ The calcium ions activate transport proteins that pump negatively charged ions out of the cells.

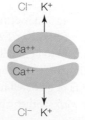

❸ The shift in charge across the guard cell plasma membranes opens gated transport proteins that allow potassium ions to exit the cells.

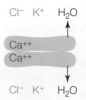

❹ Water follows the solutes by osmosis. The guard cells lose turgor and collapse against one another, so the stoma closes.

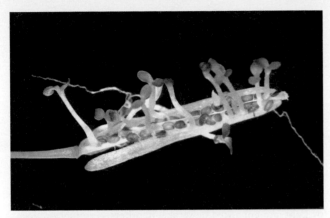

Figure 27.15 ABA prevents seed germination.
Part of the ABA signaling pathway in this *Arabidopsis* plant has been knocked out. Its seeds germinated on the parent plant, before they had a chance to disperse.

In addition to mediating stress responses, ABA has an important role in embryo maturation, fruit ripening, and germination. ABA accumulates in a seed as it forms, matures, and dries out. The hormone prevents premature germination (**Figure 27.15**), in part by inhibiting expression of genes involved in cell wall loosening and expansion—both critical processes for growth of an embryonic plant. ABA also inhibits expression of genes involved in gibberellin synthesis.

ETHYLENE

The plant hormone **ethylene** is a gas, and it is part of regulatory pathways that govern a wide range of metabolic and developmental processes in plants. Two enzymes are involved in ethylene synthesis, and each occurs in multiple versions encoded by slightly different genes. Expression of some of these genes is inhibited by ethylene (a negative feedback loop); expression of the others is enhanced by ethylene (a positive feedback loop). Negative feedback loops maintain a low level of ethylene that helps fine-tune ongoing metabolic and developmental processes such as growth and cell expansion. Positive feedback loops produce surges of ethylene required for intermittent processes such as germination, defense responses, abscission, and ripening.

Consider how ABA accumulates in a fruit as it forms and enlarges, eventually triggering production of ethylene in a positive feedback loop. The resulting ethylene surge enhances transcription of genes whose products have several effects associated with ripening (**Figure 27.16**). The fruit changes color as chloroplasts are converted to chromoplasts (Section 4.8); inside the organelles, red, orange, yellow, or purple accessory pigments are produced as green chlorophylls break down. Firm cell walls weaken. Starch and acids are converted to sugars, and aromatic molecules are produced. The resulting color change, softening, and increased palatability attract animals that can disperse the fruit's seeds.

Because ethylene is a gas, it can diffuse from one fruit to initiate ripening in another. Humans use this property to artificially ripen several types of fruit. Hard, unripe fruit can be transported long distances with less damage than soft, ripe fruit. Upon arrival at its final destination, the unripe fruit is exposed to synthetic ethylene, which jump-starts the positive feedback ripening loop. If you purchase unripe fruit, you can encourage ripening by storing it in a closed container with ripe apples or bananas. These ripe fruits emit ethylene.

---

abscisic acid (ABA) (ab-SIH-sick) Plant hormone that has a major role in stress responses; also inhibits germination.
abscission (ab-SIH-zhun) Process by which plant parts are shed.
ethylene (ETH-ill-een) Gaseous plant hormone that participates in germination, abscission, ripening, and stress responses.

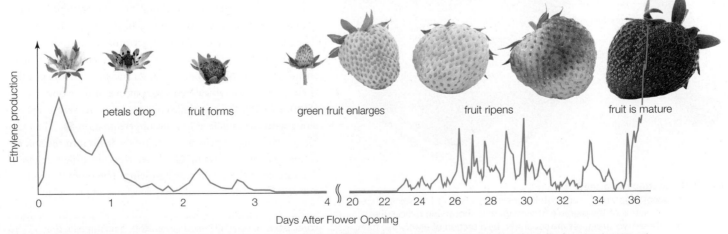

Figure 27.16 Ethylene production during strawberry formation and ripening. Oscillations are normal daily cycles (see Section 27.11).

CREDITS: Image courtesy Kazuo NAKASHIMA, Ph.D. Taishi Umezawa, Kazuo Nakashima, Takuya Miyakawa, Takashi Kuromori, Masaru Tanokura, Kazuo Shinozaki, and Kazuko Yamaguchi-Shinozaki. Molecular Basis of the Core Regulatory Network in ABA Responses: Sensing, Signaling and Transport. Plant Cell Physiol (2010) 51(11): 18211839 first published online October 26, 2010. doi:10.1093/pcp/pcq156; (16) photos from left, (1) © Madlen/Shutterstock; (2) © Westend61/SuperStock; (3–4) © Anest/Shutterstock.com; (5–8) © Alena Brozova/ Shutterstock; Graph data courtesy of Dr. Frans J.M. Harren and Dr. Simona M. Cristescu, Radboud University Nijmegen, The Netherlands.

LEARNING OBJECTIVES

Explain why seeds undergo dormancy, and list some triggers for germination.

Describe the process of germination of a seed.

Compare the pattern of early growth that occurs in eudicots and monocots after germination.

An embryonic plant complete with shoot and root apical meristems forms as part of a seed (**Figure 27.17**). A tiny section of embryonic stem separates the embryonic shoot (the plumule) from the embryonic root (radicle). As the seed matures, the embryo may dry out and enter a period of dormancy, which, as you know, is induced and maintained by a high level of ABA and a low level of gibberellin. A dormant embryo can rest in its protective seed coat for many years before it resumes metabolic activity and germinates.

Seed dormancy is a climate-specific adaptation that allows germination to occur when conditions in the environment are most likely to support the growth of a seedling. For example, seeds of many annual plants native to cold winter regions are dispersed in autumn. If the seeds germinated immediately, the tender seedlings would not survive the upcoming winter. Instead, the seeds remain dormant until spring, when milder temperatures and longer day length favor growth. By contrast, the weather in regions near the equator does not vary by season. Seeds of most plants native to such regions do not enter dormancy; these seeds germinate as soon as they mature.

BREAKING DORMANCY

Other than the presence of water, the triggers for germination differ by species. Some seed coats are so dense that they must be abraded or broken (by being chewed, for example) before germination can occur. Seeds of many cool-climate plants require exposure to freezing temperatures; those of some lettuce species, to bright light. In seeds of some species native to regions that have periodic wildfires, germination is inhibited by light and enhanced by smoke; in others, germination does not occur unless the seeds have been previously burned. Such requirements are evolutionary adaptations to life in a particular environment.

A seed can germinate only after its ABA level declines (exactly how this decline is triggered is unknown). Germination typically begins with water seeping into the seed. The water triggers cells of the embryo to release gibberellin, which results in the breakdown of stored starches into glucose. Absorbed water also swells the seed's internal tissues, causing the seed coat to rupture. Air now diffuses into the embryo's cells. Together with the released glucose, oxygen in the air allows meristem cells in the embryo to begin aerobic respiration, and these cells begin to divide rapidly. Meanwhile, the decline in ABA level has triggered production of ethylene and other molecules that softened the cells' walls, allowing them to expand. The embryonic plant begins to grow as its cells divide and enlarge. Germination ends when the radicle emerges from the seed coat to become a primary (first) root.

AFTER GERMINATION

Remember that monocots and eudicots differ in some details of their structure (Section 25.1). For example, monocot embryos have one cotyledon; eudicot embryos have two. The pattern of early growth after germination also differs between the two groups. In corn and other monocots, a rigid sheath called a **coleoptile** surrounds and protects the primary (first) shoot. In a typical monocot pattern of development (**Figure 27.18**), the coleoptile emerges from the seed coat along with the radicle at the end of germination ❶. The radicle, which is now a primary root, grows down into the soil ❷ as the coleoptile grows up ❸. When the coleoptile reaches the surface of the soil, it stops growing. The plumule then emerges from the coleoptile as it develops into the plant's

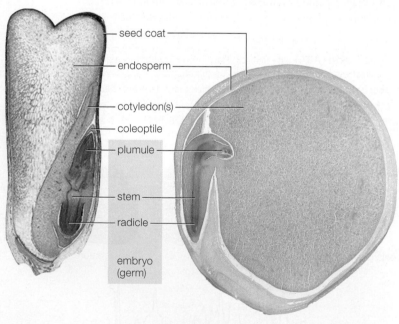

Figure 27.17 Anatomy of a seed.
Left, corn, a monocot; right, pea, a eudicot. As dormancy ends, cell divisions resume mainly at apical meristems of the plumule (the embryonic shoot) and radicle (the embryonic root). These two structures are separated by a section of embryonic stem. In monocot grasses such as corn, the plumule is protected by a sheathlike coleoptile.

seed coat
endosperm
cotyledon(s)
coleoptile
plumule
stem
radicle
embryo (germ)

coleoptile (coal-ee-OPP-till) In monocots, a rigid sheath that protects the plumule (embryonic shoot).

Figure 27.18 Early growth of corn, a monocot.

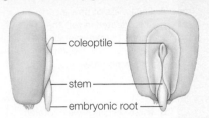

coleoptile

stem

embryonic root

primary shoot

coleoptile

❶ At the end of germination of a corn grain (seed), the embryonic root (radicle) emerges from the seed coat. The root is now a primary (first) root.

❷ The primary root grows down into the soil.

❸ The coleoptile grows up and opens a channel through the soil to the surface, where it stops growing. Note the adventitious roots; these will replace the primary root in a fibrous root system typical of monocots (Section 25.5).

❹ The embryonic shoot develops into a primary shoot as it emerges from the coleoptile.

❺ The single cotyledon remains under the soil, transferring nutrients to the seedling until its new leaves can produce enough sugars by photosynthesis to sustain growth.

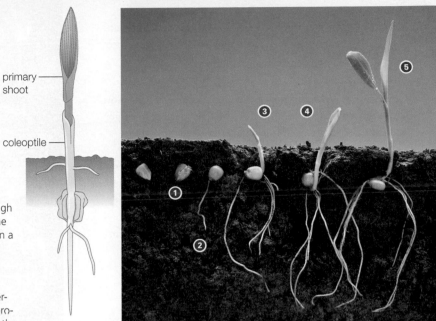

Figure 27.19 Early growth of the common bean, a eudicot.

❶ A bean's seed coat splits and the embryonic root tip emerges.

❷ The embryonic stem emerges from the seed and bends in the shape of a hook.

❸ As the stem lengthens, it drags the two cotyledons upward toward the surface of the soil.

❹ Exposure to light causes the stem to straighten.

❺ Primary leaves emerge from between the cotyledons and begin photosynthesis.

❻ Cotyledons undergo a period of photosynthesis before withering and falling off the stem.

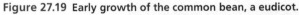

primary (first) shoot ❹. The single cotyledon, which typically stays beneath the soil, functions mainly to transfer endosperm breakdown products to the seedling until leaves form on the shoot and begin photosynthesis ❺.

In eudicot seedlings, the primary shoot is not protected by a coleoptile. In a common eudicot pattern of development (**Figure 27.19**), the embryonic stem emerges from the seed along with the radicle ❶, and then it bends into the shape of a hook ❷. The bent stem lengthens and pulls the cotyledons upward as the primary root grows downward ❸. When the stem reaches the surface, exposure to light makes it straighten ❹. Primary leaves emerge from between the

cotyledons and begin photosynthesis ❺. The cotyledons typically undergo a period of photosynthesis before shriveling ❻. Eventually, the cotyledons fall off the lengthening stem, and the young plant's new leaves produce all of its food.

Germination is the first step in plant development. As a sporophyte grows and matures, its tissues and organs develop in characteristic patterns. Leaves form in predictable shapes and sizes, stems and roots lengthen and thicken, flowering occurs at a certain time of year, and so on. Such patterns arise from hormonal control of gene expression in response to environmental cues. The remainder of this chapter explores some of the ways that plants respond to environmental cues.

Plants cannot move themselves from place to place, but movement of plant parts is common. Reversible changes in turgor can move a structure, as can a change in a growth pattern (such as when cells on one side of a stem start to elongate more than cells on the other side). Plant movement occurs in response to environmental stimuli such as light or contact, but only some of that movement occurs in a direction that depends on the stimulus (as when a stem bends toward a light source). A directional movement toward or away from a stimulus that triggers it is called a **tropism**.

ENVIRONMENTAL TRIGGERS

Following are examples of environmental cues that can provoke movements in plants.

Gravity Even if a seedling is turned upside down just after germination, its primary root and shoot will curve so the root grows down and the shoot grows up (**Figure 27.20**). A directional response to gravity is called **gravitropism**. A root or shoot bends because of shifts in the direction of local auxin transport. In shoots, auxin enhances cell elongation, so when auxin flow is directed toward one side of a shoot, the shoot bends away from that side. Auxin has the opposite effect in roots: When it is directed toward one side of a growing root, the root bends toward that side.

In plants and many other organisms, the ability to sense gravity is based on organelles called statoliths. Plant statoliths are amyloplasts (plastids fills with dense grains of starch, Section 4.8). Amyloplasts occur in endodermal cells (of stems, Sections 25.5 and 26.2) and in root cap cells (Section 25.6). Starch grains are heavier than cytoplasm, so statoliths tend to sink to the lowest region of the cell, wherever that is (**Figure 27.21**). When the position of statoliths changes, the direction of auxin flow through the cells changes too, so auxin is always directed toward the down-facing side of the structure. A shift in the direction of auxin flow results in a directional response of the structure.

Light A **phototropism** is a directional response to light. Many plants can change the orientation of leaves or other parts in a direction that depends on a light source. For example, an elongating stem curves toward a light source, thus optimizing the amount of light captured by leaves for photo-

A Regardless of how a corn seed is oriented in soil, the seedling's primary root always grows down, and its primary shoot always grows up.

B These seedlings were rotated 90° counterclockwise after they germinated. The plants adjusted to the change by redistributing auxin, and the direction of growth shifted as a result.

C Auxin transport was inhibited in these seedlings. They were also rotated 90° counterclockwise after germination, but the direction of growth did not shift.

Figure 27.20 Gravitropism.

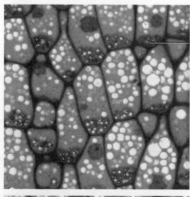

statoliths

A This micrograph shows heavy, starch-packed statoliths settled on the bottom of gravity-sensing cells in a corn root cap.

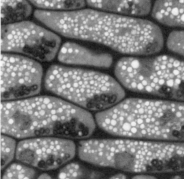

B This micrograph was taken ten minutes after the root in **A** was rotated 90°. The statoliths are already settling to the new "bottom" of the cells.

Figure 27.21 Gravity and statoliths.

FIGURE IT OUT
In which direction was this root rotated?

Answer: Counterclockwise

CREDITS: (20A) Michael Clayton, University of Wisconsin; (20B,C) © Muday, G.K. and P. Haworth (1994), "Tomato root growth, gravitropism, and lateral development: Correlations with auxin transport." *Plant Physiology and Biochemistry* 32, 193–203, with permission from Elsevier Science; (21) Micrographs courtesy of Randy Moore from "How Roots Respond to Gravity," M. L. Evans, R. Moore, and K. Hasenstein, *Scientific American*, December 1986.

synthesis in low-light environments. In shoots, phototropism is mediated by nonphotosynthetic pigments called phototropins, which absorb blue light. By an unknown mechanism, light-energized phototropins cause auxin to be directed away from the illuminated side of a shoot tip or coleoptile. Cells on the shaded side elongate more than cells on the illuminated side (**Figure 27.22A**), and the difference causes the entire structure to bend toward the light as it lengthens (**Figure 27.22B**).

Leaves or flowers of some plants change position in response to the changing angle of the sun throughout the day, a phototropic response called heliotropism (from Greek *helios*, sun). The mechanism that drives heliotropism is not understood, but it may be similar to phototropism because it has been observed to coincide with differential elongation of cells in stems, and it also occurs in response to blue light.

Contact A directional response to contact with an object is called **thigmotropism** (*thigma* means touch). We see the effects of thigmotropism in plants when a vine's tendrils coil around an object (left). The mechanism that gives rise to the response is not well understood, but it probably involves the immediate increase in cytoplasmic calcium concentration that typically accompanies contact. Plants have plasma membrane transport proteins that flood cytoplasm with calcium ions upon mechanical disturbance of the membrane. Several gene products that can sense calcium ions are involved in the response to the flood of ions. These molecules trigger unequal growth rates of cells on opposite sides of the shoot tip, and this causes a growing shoot to coil around an object as it lengthens. A similar mechanism causes roots to grow away from contact, so they "feel" their way around rocks and other impassable objects in soil.

The fastest movements of plant parts occur in response to contact. For example, modified leaves of Venus flytraps and some other carnivorous plants spring shut in a fraction of a second upon being triggered by insect prey (**Figure 27.23**). The mechanism involves electrical signaling and the redistribution of auxin, but it is not well understood.

---

gravitropism (grah-vih-TRO-pizm) Directional response to gravity.
phototropism (foe-toe-TRO-pizm) Directional response to light.
thigmotropism (thig-MA-truh-pizm) In plants, growth in a direction influenced by contact.
tropism (TROPE-izm) A directional movement toward or away from a stimulus that triggers it.

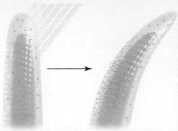 

A Sunlight strikes only one side of a coleoptile. Auxin flow is directed toward the shaded side, so cells on that side lengthen more.

B This shamrock plant is adjusting the direction of its growth in response to a directional light source.

Figure 27.22 Phototropism. Auxin-mediated differences in cell elongation between two sides of a shoot induce bending toward light. Red dots in **A** signify auxin molecules.

A The "trap" of a Venus flytrap plant is a highly modified leaf that is creased in the middle. Three sensitive hairs protrude from the upper surface of each half of the leaf. An insect that brushes a hair two times in succession triggers a mechanism that causes the leaf to snap shut very quickly—even faster than a fly.

B Secreted enzymes slowly digest trapped prey, releasing nitrogen that the plant requires for growth. The trap will not spring shut unless a hair is brushed twice in quick succession, so raindrops or debris do not trigger it.

Figure 27.23 Movement of the Venus flytrap.

DAILY CHANGE

Plants respond to cycles of environmental change with cycles of activity. A **circadian rhythm** is a cycle of biological activity that repeats every twenty-four hours or so (Section 10.4). For example, a bean plant holds its leaves horizontally during the day but folds them close to its stem at night. Bean plants exposed to constant light or constant darkness for a few days will continue to move their leaves in and out of the "sleep" position at the normal time of sunset and sunrise (**Figure 27.24**). Similar mechanisms cause flowers of some plants to open only at certain times of day. For example, the flowers of plants pollinated mainly by night-flying bats open, secrete nectar, and release fragrance only at night. Periodically closing flowers protects the delicate reproductive parts at times when the likelihood of pollination is lowest.

Nonphotosynthetic pigments called phytochromes and cryptochromes provide light input into internal circadian "clocks." **Phytochromes** absorb red light (660 nanometers) and far-red light (730 nanometers). Absorbing red light causes their structure to change from an inactive to an active form. Absorbing far-red light, which predominates in shade, changes the molecule's structure back to the inactive form. Blue light activates cryptochromes, which occur widely in organisms of all kingdoms. Plants use cryptochromes to control a variety of activities, including circadian ones. Human circadian clocks also use them. Activated cryptochromes associate with other molecules to become magnetic, and are part of a light-dependent magnetic compass in birds (Section 30.1 returns to this topic).

SEASONAL CHANGE

Except at the equator, the length of day varies with the season. Days are longer in summer than in winter, and the difference increases with latitude. These seasonal changes in light

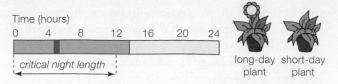

Time (hours)

A A flash of red light at 660 nm interrupting a long night causes plants to respond as if the night were short. Long-day plants flower; short-day plants do not.

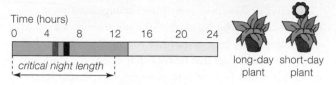

Time (hours)

B A flash of far-red light at 730 nm cancels the effect of a red light flash at night. Short-day plants flower; long-day plants do not.

Figure 27.25 Experiments showing that long- or short-day plants flower in response to night length. Each horizontal bar represents 24 hours. Blue indicates night length; yellow, day length.

availability trigger seasonally appropriate responses in plants. **Photoperiodism** refers to an organism's response to changes in the length of day relative to night.

Flowering is a photoperiodic response in many species. Such species are termed long-day or short-day plants depending on the season in which they flower. These names are somewhat misleading, as the main trigger for flowering is the length of night, not the length of day (**Figure 27.25**).

Photoperiodic flowering involves phytochromes and cryptochromes, as well as circadian cycles of regulatory gene expression. Inputs from both converge on a gene called *CO*, which encodes a transcription factor. All plants have this gene. In long-day plants, the transcription factor induces expression of the *FT* gene, which in turn activates expression of floral identity genes (Section 10.3). Irises and other long-day

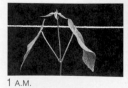    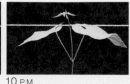 

Figure 27.24 A circadian rhythm: leaf movements by a young bean plant. Physiologist Frank Salisbury kept this plant in darkness for twenty-four hours. Despite the lack of light cues, the leaves kept on folding and unfolding at sunrise (6 A.M.) and sunset (6 P.M.).

1 A.M.   6 A.M.   Noon   3 P.M.   10 P.M.   Midnight

A Ripening horse chestnuts emit ethylene that causes abscission of the fruit stalks.

B Horse chestnut trees take their name from abscission zones that leave horseshoe-shaped scars in this species.

Figure 27.26 Abscission as part of the normal life cycle of deciduous plants such as the horse chestnut.

plants flower only when the hours of darkness fall below a critical value. These plants flower only in summer because the expression of their *CO* gene and the activity of its product peak 8 to 10 hours after dawn—which is late afternoon in summer, but after dusk in other seasons. The transcription factor is broken down at night, so it does not accumulate to a high enough level to promote flowering except in summer.

Cabbage, spinach, and other edible long-day plants grown for their leaves are said to "bolt" when they switch from producing leaves (desirable) to producing flowers (undesirable). Farmers minimize bolting in these plants by growing them in short-day seasons.

Chrysanthemums and other short-day plants flower only when the hours of darkness are greater than a critical value. In these plants, the same *CO* gene product inhibits *FT* gene expression, and flowering is stimulated in a separate pathway involving gibberellin.

Some plants native to temperate regions flower only after exposure to a long period of cold in the preceding winter, a response called **vernalization**. In these plants, expression of a gene called *VRN* increases with the duration of cold exposure. The gene's transcription factor product promotes rapid flowering when days start lengthening in spring. In other plants, exposure to cold does not affect *VRN* expression.

Yearly cycles of abscission and dormancy are photoperiodic responses too. Plants that drop their leaves before dormancy are typically native to regions too dry or too cold for optimal growth during part of the year. For example, deciduous trees of the northeastern United States lose their leaves around September. The trees remain dormant during the months of harsh cold weather that would otherwise damage tender leaves and buds. Growth resumes when milder conditions return, around March. On the opposite side of the world, many tree species native to tropical monsoon forests of south Asia lose their leaves during the dry season (between November and May). The region receives a lot of rain annually, but almost none of it falls during this period of the year. Dormancy offers the trees a way to survive the seasonal droughtlike conditions. New growth appears at the beginning of June, just in time to be supported by the ample water of monsoon rains.

Seasonal abscission of leaves and fruits is mediated by ethylene. Let's use a horse chestnut tree as an example. In this temperate zone species, most root and shoot growth occurs between March and June. By July, the tree is producing fruits and seeds. In August, the growing season is coming to a close, and nutrients are being routed to stems and roots for storage during the forthcoming period of dormancy. Ripe fruits and seeds (**Figure 27.26A**) release ethylene that diffuses into nearby twigs, petioles, and fruit stalks. The ethylene triggers formation of a dense layer of small cells. Cells above this layer begin to produce enzymes that digest their own walls. The cells bulge as their walls soften, and separate from one another as the extracellular matrix that cements them together dissolves. Tissue in this abscission zone weakens, and the structure above it drops. A scar often remains where the structure had been attached (**Figure 27.26B**).

circadian rhythm (sir-KAY-dee-un) A biological activity that is repeated about every 24 hours.

photoperiodism (foe-toe-PEER-ee-ud-is-um) Biological response to seasonal changes in the relative lengths of day and night.

phytochrome (FIGHT-uh-krome) A light-sensitive pigment that helps set plant circadian rhythms based on length of night.

vernalization (vurr-null-iz-A-shun) Stimulation of flowering in spring by long exposure to low temperature in winter.

Explain what happens when a pathogen enters a plant's tissues.

Distinguish between a hypersensitive response and systemic acquired resistance in plants.

Describe one way that plants defend themselves from herbivory.

A plant cannot run away from inhospitable conditions: It either adjusts to the conditions or dies. The adjustments involve defensive responses to abiotic stressors (caused by nonliving environmental conditions) and biotic stressors (imposed by pathogens and herbivores).

DEFENDING AGAINST DISEASE

Abiotic stressors such as temperature extremes or lack of water trigger ABA synthesis. You learned in Section 27.8 how ABA and nitric oxide are part of a response that causes a plant's stomata to close during a water shortage, thus minimizing water loss. ABA and nitric oxide also participate in plant responses to pathogens. Consider that receptors on plant cells recognize molecules specific to microbial pathogens. Some of the receptors recognize flagellin, a protein that makes up bacterial flagella (Section 4.4). When flagellin binds to these receptors, it triggers a burst of ethylene synthesis. In turn, the ethylene burst lifts the brakes on production of molecules used in defense. The cell also starts making more flagellin receptors—a positive feedback loop that makes the plant more sensitive to the presence of bacteria. If any additional flagellin binds to the accumulated receptors, an ABA-mediated nitric oxide burst immediately closes stomata. Bacteria cannot penetrate plant epidermis; they can enter plant tissues only through wounds or open stomata. Thus, closing stomata defends the plant from bacterial invasion.

Mutualistic bacteria and fungi avoid triggering a plant's defense responses by exchanging chemical signals with it. Remember that *Rhizobium* bacteria in soil are attracted to compounds released from root cells (Section 26.2). These bacteria respond to the plant compounds by producing molecules required for living in symbiosis with the plant. One of these molecules is secreted into the soil, and it is recognized by root cells as a signal to form a nodule.

When a pathogen does enter plant tissues, it triggers a large surge of hydrogen peroxide and nitric oxide that causes cells in the infected region to commit suicide. The details of this "hypersensitive" response are unknown, but its effects are clear: It can prevent a pathogen from spreading to other parts of the plant because it often kills the pathogen along with the infected tissue (**Figure 27.27**). Grapevines native to America are more resistant than European cultivated grape-

Figure 27.27 Plant hypersensitive response to a pathogen. Brown spots are dead tissue where germinating fungi penetrated the leaf's epidermis, triggering a release of hydrogen peroxide and nitric oxide that resulted in plant cell suicide. This hypersensitive response can prevent the spread of a pathogen into healthy tissue.

vines to *Phylloxera* (Section 27.4) because their hypersensitive response to this insect is much stronger. American grapevines make more resveratrol, a phytoestrogen hormone that mediates the hypersensitive response in these plants.

A hypersensitive response is often ineffective against pathogens that gain nutrients by killing cells outright. Some sac fungi, for example, use toxins to kill their plant hosts, then absorb nutrients released from their decomposing tissues. A different defense response, **systemic acquired resistance**, increases resistance of the whole plant body to attack by a wide range of fungi and other pathogens, and it enhances tolerance to environmental stresses. The increased resistance and hardiness persists for the plant's lifetime and for a few generations thereafter (epigenetic inheritance, Section 10.5).

Systemic acquired resistance begins when an infected tissue releases small molecules that travel to other parts of the plant, where they trigger cells to produce a plant hormone called salicylic acid. Salicylic acid increases transcription of hundreds of genes whose products confer general hardiness to the plant: lignin, for example, that strengthens plant cell walls; enzymes that break down chitin in fungal cell walls; phytoestrogens that inhibit fungal metabolic enzymes and attract symbiotic soil bacteria; antioxidants that detoxify molecules produced in a hypersensitive response; and so on.

PLANTS DO NOT WANT TO BE EATEN

Plants have the ability to release into the air thousands of volatile chemicals that can be used for communication with other plants and other organisms. The chemicals released depend on a plant's internal conditions and environmental

A A tobacco plant emits a particular combination of 11 volatile chemicals in response to being chewed by tobacco budworms.

B Red-tailed wasps, which parasitize tobacco budworms, recognize the unique chemical signature emitted by the plant. They follow the trail of chemicals back to the source. A wasp that finds a budworm deposits an egg in it; when the egg hatches, a larva emerges and begins to eat the budworm, which eventually dies.

Figure 27.28 A plant defense against herbivory.
Plant cells release a particular combination of volatile chemicals in response to being chewed by a particular insect species. The chemical signature attracts wasps that parasitize the insect.

cues. For example, wounding of a leaf, such as occurs when an insect chews on it, triggers the release of a particular mixture of volatiles. The chemical signature attracts wasps that parasitize insect herbivores (**Figure 27.28**). Volatile chemicals released in response to herbivory are also detected by neighboring plants, which respond by increasing production of chemicals that make them less palatable.

---

systemic acquired resistance In plants, inducible whole-body resistance to a wide range of pathogens and environmental stresses.

Dr. Grace Gobbo

When National Geographic Explorer Grace Gobbo walks through a Tanzanian rain forest, she doesn't see only trees, flowers, and vines. She sees cures. For centuries, medicinal plants used by traditional healers have been at the heart of effective health care in this lush African nation. With expensive imported pharmaceuticals unaffordable for most of the population, traditional cures often provide the only relief for illness. But today, both the lush landscape and indigenous medical knowledge are disappearing. Gobbo hopes her efforts to preserve natural remedies and native habitat will help reverse the trend.

Growing up in a Christian family with a doctor for a father, Gobbo dismissed traditional healing as witchcraft. Then, coursework in botany exposed her to evidence she could not ignore. "We studied cases where a particular plant successfully treated coughs," she recalls. "Laboratory results proved it stopped bacterial infection." Intrigued, Gobbo began interviewing traditional healers in her own area who described their success using plants to treat a wide range of ailments, including skin and chest infections, stomach ulcers, diabetes, heart disease, mental illness, and even cancer. To date, she has recorded information shared by more than 80 healers, entering notes and photographs about plants and their uses into a computer database. "Before now, these facts existed only as an oral tradition," she explains. "Nothing was written down. The knowledge is literally dying out with the elders since today's young generation considers natural remedies old-fashioned." Gobbo wants to capture the irreplaceable facts before they are lost, and convince young people to appreciate their value. "In this part of the world, modern medicine is not enough; most people can't afford it. Traditional healing has so much potential. Local communities are comfortable with this approach; it just needs to be more accessible. But the longer we wait, the more knowledge disappears. We need to act now, and act fast."

28

Animal Tissues and Organ Systems

Links to Earlier Concepts

This chapter applies what you learned about levels of organization (Section 1.1) to animal bodies, and it expands on the nature of multicelled body plans (23.1). You may wish to review cell junctions (4.10), the surface-to-volume ratio (4.1), diffusion and transport proteins (5.6, 5.7), aerobic respiration (7.1), and energy conversion pathways (7.6).

Core Concepts

Interactions among the components of a biological system give rise to complex properties.

Each animal tissue consists of specific cell types that collectively carry out a task or tasks. Tissues interact in organs, which in turn interact as components of organ systems. At all levels of organization, animal structure is shaped by natural selection, and constrained by physical and developmental factors.

All organisms alive today are linked by lines of descent from shared ancestors.

Shared core processes and features provide evidence that living things are related. Four types of tissues occur in all vertebrates. Epithelial tissues cover the body's surfaces and line its cavities. Connective tissues provide support. Muscle tissues bring about movement. Nervous tissue receives, integrates, and communicates information.

Living things sense and respond appropriately to their internal and external environments.

Homeostatic mechanisms that involve negative feedback allow animals to maintain themselves by responding dynamically to internal and external conditions. Organisms keep their internal environment stable by returning a changed condition back to its target set point.

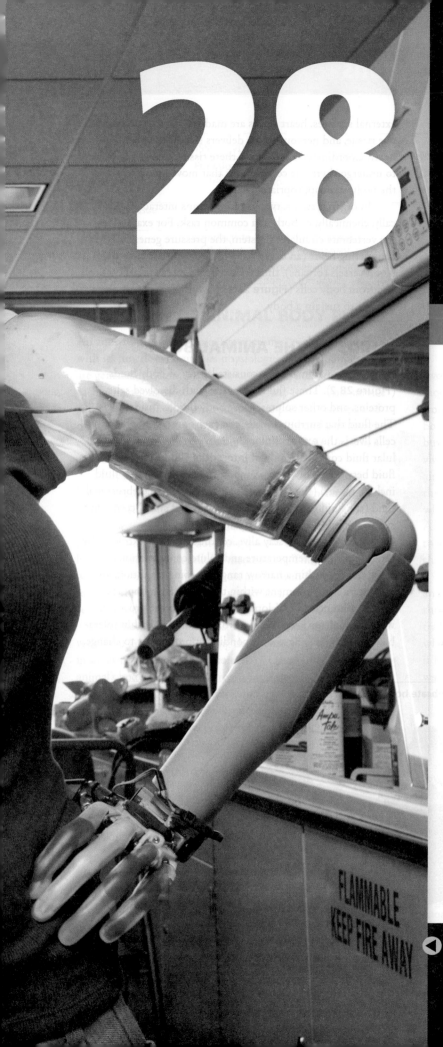

Amputee Amanda Kitts tests a bionic arm developed at Johns Hopkins. Like an arm made of skin, muscle, and bone, it responds to signals from her nervous system.

481

Photograph by Mark Thiessen, National Geographic Creative.

Using appropriate examples, list the different types of connective tissues and describe their functions.

Explain why a deficiency in vitamin C harms connective tissue.

GENERAL CHARACTERISTICS

Connective tissues consist of cells scattered in an extensive extracellular matrix (Section 4.10) of their own secretions. Most connective tissue contains fibroblasts, which are cells that secrete an extracellular matrix containing polysaccharides and the proteins collagen and elastin. Collagen and elastin are fibrous proteins—proteins in which molecules aggregate as long, insoluble strands. Collagen is the most abundant protein in an animal body. Elastin is springier than collagen, and it abounds in tissues that routinely stretch out and spring back.

Synthesizing collagen requires vitamin C, so a deficiency in this vitamin causes the nutritional disorder scurvy. Early in this disorder, connective tissue surrounding tiny blood vessels breaks down, so gums bleed and small bruiselike spots appear. If scurvy continues untreated, bones weaken. Collagen is the main component of scar tissue, so a person with scurvy heals poorly. Eventually, this disorder can be fatal.

TYPES OF CONNECTIVE TISSUE

Loose connective tissue is the most common connective tissue in a vertebrate body. It consists mainly of fibroblasts, collagen fibers, and elastin fibers widely scattered in a gel-like matrix (**Figure 28.5A**). Loose connective tissue underlies most epithelia, and it surrounds nerves and blood vessels.

In **dense, irregular connective tissue**, fibroblasts are interspersed among a tight mesh of randomly arranged collagen fibers (**Figure 28.5B**). The random arrangement of the fibers allows the tissue to withstand stretching in any direction. Dense, irregular connective tissue makes up deep skin layers, underlies the lining of the gut, and forms a protective capsule around the kidneys and the testes.

Dense, regular connective tissue consists of fibroblasts in orderly rows between parallel, tightly packed bundles of collagen fibers (**Figure 28.5C**). This parallel organization maximizes the strength of a tissue that is typically pulled in a single direction, parallel to the length of the fibers. Dense, regular connective tissue is the main tissue in tendons and ligaments. Tendons connect skeletal muscle to bones. Ligaments attach one bone to another.

Adipose tissue is a connective tissue consisting of special lipid-storing cells with little matrix between them. It is the body's main energy reservoir. It also functions as insulation and as protective padding. Many small blood vessels run through adipose tissue. In most adipose tissue, white fat cells are the most abundant cells. A vacuole with triglycerides fills their interior, restricting the nucleus and cytoplasm to a thin layer around the edge (**Figure 28.5D**). Some mammals, including humans, also have adipose tissue rich in brown fat cells. Brown fat cells have a centrally located nucleus, lots of lipid droplets in the cytoplasm, and an abundance of mitochondria that specialize in heat production. We discuss the function of brown fat in more detail in Section 37.6.

Figure 28.5 Vertebrate connective tissues.

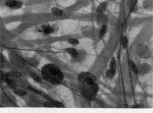
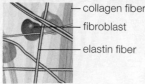

A Loose connective tissue
- Underlies most epithelia
- Provides elastic support; stores fluid

collagen fiber
fibroblast
elastin fiber

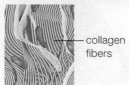

B Dense, irregular connective tissue
- Deep layer of skin (dermis) and wall of gut, in kidney capsule
- Binds parts together; stretches in all directions

collagen fibers

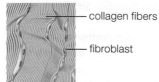

C Dense, regular connective tissue
- Tendons and ligaments
- Fibrous attachment between parts; allows stretching, but in one direction only

collagen fibers
fibroblast

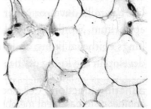
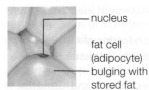

D Adipose tissue
- Under skin and around the heart and kidneys
- Stores energy-rich lipids; insulates the body and cushions its parts

nucleus
fat cell (adipocyte) bulging with stored fat

Walruses and other marine mammals have a special type of adipose tissue called blubber (**Figure 28.6**). Fat is less dense than water, so being encased in blubber helps an animal float. Unlike most adipose tissue, blubber includes collagen and elastin fibers, so it spring backs when compressed. This springiness reduces the amount of energy required to swim. A coat of blubber also insulates a walrus. Even when the outside temperature dips far below freezing, the core body temperature of a walrus remains about the same as yours.

Cartilage (**Figure 28.5E**) is a connective tissue in which cells secrete a rubbery matrix containing collagen fibers and proteoglycans (proteins with bound polysaccharides). Some fish such as lampreys and sharks have a cartilage skeleton. In humans, as in all bony fishes and tetrapods, cartilage forms a model for the developing skeleton, then bone replaces most of it. After birth, cartilage still supports the outer ears, nose, and large airways. It covers ends of bones and is a shock absorber between segments of the backbone. Blood vessels do not extend through cartilage, so cartilage injuries are slow to heal.

Bone tissue is a connective tissue in which bone cells secrete and are surrounded by a collagen-rich matrix hardened by calcium and phosphorus (**Figure 28.5F**). Bone tissue is the main tissue of the skeleton that structurally supports the vertebrate body.

Blood (**Figure 28.5G**) is a fluid connective tissue. It contains cellular components that drift along in plasma, which is a fluid extracellular matrix. Bone marrow produces the cellular components of blood: red blood cells (which transport oxygen), white blood cells (which defend the body against pathogens), and platelets (which help blood clots form).

Figure 28.6 Blubber-encased walrus resting on ice. Blubber is a special type of adipose tissue that is especially springy because it contains collagen and elastin fibers. It makes a walrus buoyant, helps it swim, and keeps it warm.

adipose tissue Connective tissue specializing in fat storage.

blood Fluid connective tissue consisting of plasma and cellular components (red cells, white cells, platelets) that form in bones.

bone tissue Connective tissue that consists of cells surrounded by a mineral-hardened matrix.

cartilage Connective tissue that consists of cells surrounded by a rubbery matrix.

connective tissue Animal tissue with an extensive extracellular matrix; provides structural and functional support.

dense, irregular connective tissue Connective tissue that consists of randomly arranged fibers and scattered fibroblasts.

dense, regular connective tissue Connective tissue that consists of fibroblasts arrayed between parallel arrangements of fibers.

loose connective tissue Connective tissue that consists of fibroblasts and fibers scattered in a gel-like matrix.

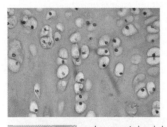
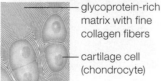

glycoprotein-rich matrix with fine collagen fibers

cartilage cell (chondrocyte)

E Cartilage
- Internal framework of ears, nose, airways; covers ends of bones
- Supports soft tissues; cushions, reduces friction at joints

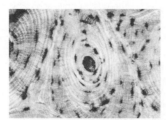
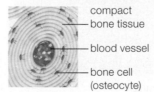

compact bone tissue

blood vessel

bone cell (osteocyte)

F Bone tissue
- Bulk of most vertebrate skeletons
- Protects soft tissues; functions in movement; stores minerals; produces blood cells

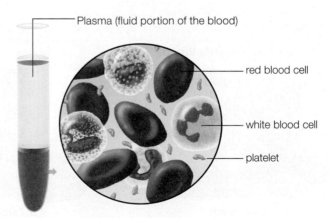

Plasma (fluid portion of the blood)

red blood cell

white blood cell

platelet

G Blood
- Flows through blood vessels, and the heart
- Distributes essential gases, nutrients to cells; removes wastes from them

In **muscle tissue**, cells contract (shorten) in response to stimulation, then they relax and passively lengthen. Coordinated contractions of muscles move body parts or propel material through the body. We discuss the molecular components of muscle and the mechanism of muscle contraction in detail in Chapter 32.

SKELETAL MUSCLE

Skeletal muscle tissue, the functional partner of bone (or cartilage), helps move and maintain the positions of the body and its parts. Skeletal muscle tissue has parallel arrays of long, cylindrical muscle fibers (**Figure 28.7A**). These fibers are multinucleated cells that formed when groups of cells fused together during embryonic development. The interior of each skeletal muscle fiber is filled by many long threadlike structures (myofibrils) that consist of row after row of contractile units (sarcomeres). Because these rows are aligned the same way in the many myofibrils, skeletal muscle has a striated, or striped, appearance under the microscope.

Skeletal muscle tissue makes up 40 percent or so of the weight of an average human. Some skeletal muscles contract reflexively, but most contract when we want them to. Thus, skeletal muscles are commonly called "voluntary" muscles. Along with the liver, skeletal muscle is a major site for glycogen storage. Metabolic activity in skeletal muscles is the major source of body heat.

CARDIAC MUSCLE

Cardiac muscle tissue is found only in the heart wall (**Figure 28.7B**). Like skeletal muscle tissue, it appears striated. Unlike skeletal muscle, it has branching cells, each with a single nucleus. Cardiac muscle cells abut at their ends, where there are two types of junctions. Adhering junctions prevent cells from being ripped apart during forceful contractions. Gap junctions form a channel that connects the cytoplasm of the abutting cells, allowing ions that trigger contraction to pass swiftly from cell to cell. The presence of gap junctions between cells allows cardiac muscle cells to contract as a unit.

SMOOTH MUSCLE

We find **smooth muscle tissue** in the wall of some blood vessels and the wall of soft internal organs, such as the stomach, uterus, and bladder. This tissue's unbranched cells have one nucleus at their center and are tapered at both ends (**Figure 28.7C**). Contractile units are not arranged in an orderly repeating fashion, so smooth muscle tissue does not appear striated. Smooth muscle contracts more slowly than skeletal muscle, but contractions can be sustained longer. These contractions propel material through the gut, alter the diameter of blood vessels, and adjust how much light enters the eye.

Smooth muscle and cardiac muscle are called "involuntary" muscle because they cannot be contracted at will.

---

cardiac muscle tissue Muscle of the heart wall.
muscle tissue Tissue that consists mainly of contractile cells.
skeletal muscle tissue Muscle that interacts with skeletal components to move body parts; under voluntary control.
smooth muscle tissue Muscle that lines blood vessels and forms the wall of hollow organs; not under voluntary control.

Figure 28.7 Muscle tissues.

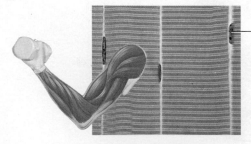

A Skeletal muscle
- Long, multinucleated, cylindrical cells with conspicuous striping (striations)
- Pulls on bones to bring about movement, maintain posture
- Under voluntary control, but some activated in reflexes

B Cardiac muscle
- Striated, branching cells (each with a single nucleus) attached end to end
- Found only in the heart wall
- Contraction is not voluntary

C Smooth muscle
- Cells with a single nucleus, tapered ends, and no striations
- Found in the walls of arteries, the digestive tract, the reproductive tract, the bladder, and other organs
- Contraction is not voluntary

28.5 Nervous Tissues

Nervous tissue allows an animal to collect information about its internal and external environment, to integrate that information, and to control glands and muscles. It is the main tissue of the vertebrate brain and spinal cord, and of the peripheral nerves that extend throughout the body.

Neurons are specialized signaling cells characteristic of nervous tissue (**Figure 28.8**). A neuron has a central cell body that contains a nucleus and other organelles. Cytoplasmic extensions that project from the cell body receive and send electrochemical signals. Extensions called dendrites receive signals, and an extension called the axon sends signals. When a neuron receives sufficient stimulation, an electrical signal travels along its plasma membrane to the end of its axon. Here, the arrival of an electrical signal causes the neuron to release chemical signaling molecules. These molecules diffuse across a small gap to an adjacent neuron, muscle fiber, or gland cell, and alter that cell's behavior.

Your nervous system has more than 100 billion neurons. There are three types of neurons. Sensory neurons are excited by specific stimuli, such as light or pressure. Interneurons receive and integrate sensory information. They coordinate responses to stimuli and function in memory. Vertebrates have interneurons in their brain and spinal cord. Motor neurons relay commands from the brain and spinal cord to glands and muscle cells.

Nervous tissue also includes a variety of cells collectively referred to as neuroglial cells, or neuroglia. These cells constitute a significant portion of nervous tissue—at least half of the cells in your brain are neuroglia. Neuroglia keep neurons positioned where they should be and provide them with metabolic support. Some also help neurons send signals: Neuroglial cells that wrap around the signal-sending axons of most neurons speed the rate at which electrical signals travel along them.

The brain and spinal cord include two visually distinct types of nervous tissue: white matter and gray matter (**Figure 28.9**). White matter consists primarily of the signal-sending axons of neurons, together with the neuroglial cells wrapped around them. Gray matter includes other types of neuroglia, together with most neuron cell bodies and their signal-receiving dendrites. Gray matter is found at the brain's outer surface, and white matter occurs in its interior.

We discuss the function of neurons and the anatomy of the nervous system in detail in the next two chapters.

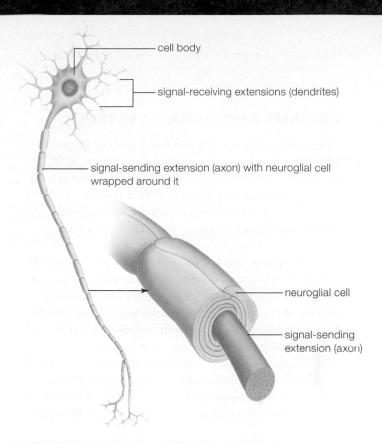

Figure 28.8 Neuron and associated neuroglial cell.

FIGURE IT OUT
What region of the neuron contains most of its organelles?

Answer: The cell body

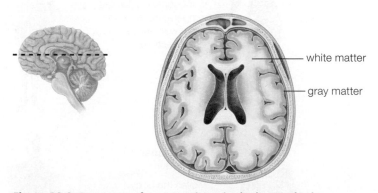

Figure 28.9 Two types of nervous tissue in the human brain.
A section through the brain shows the outer layer of gray matter and the inner layer of white matter.

nervous tissue Animal tissue composed of neurons and neuroglia; detects stimuli and controls responses to them.
neuron (NUHR-on) Specialized signaling cell in nervous tissue; transmits electrical signals along its plasma membrane and sends chemical messages to other cells.

LEARNING OBJECTIVE

List the organ systems in a human body and provide a brief description of the components and function of each.

ORGANS AND ORGAN SYSTEMS

Tissues interact structurally and functionally in organs. Organs, in turn, interact in organ systems. **Figure 28.10** shows the organ systems of a human body, but keep in mind that all vertebrates have the same array of systems.

The integumentary system includes skin and structures derived from it such as hair and nails ❶. We discuss skin's functions in detail in the next section.

The nervous system is the body's main control center ❷. The brain, spinal cord, and nerves are part of this system, as are sensory organs such as the eyes. The endocrine system consists of hormone-secreting endocrine glands and cells ❸. The endocrine system and nervous system work together and both control the activity of other organ systems.

The muscular system consists of individual muscles that move the body and its parts ❹. This system also plays an important role in regulating body temperature by generating heat. The skeletal system is the body's framework ❺. It pro-

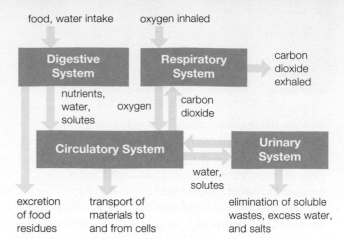

food, water intake   oxygen inhaled

excretion of food residues   transport of materials to and from cells   elimination of soluble wastes, excess water, and salts

Figure 28.11 Organ system interactions that keep the body supplied with essential substances and eliminate unwanted wastes.

tects internal organs and serves as a point of attachment for skeletal muscles. It also stores minerals and produces blood cells. Bones are organs of the skeletal system.

The circulatory system consists of the heart, blood, and blood vessels ❻. It cooperates with the respiratory, digestive, and urinary systems in delivering oxygen and nutrients to cells throughout the body, and clearing away their wastes

Figure 28.10 Human organ systems and their functions.

❶ Integumentary System

Protects body from injury, dehydration, pathogens; moderates temperature; excretes some wastes; detects external stimuli.

❷ Nervous System

Detects external and internal stimuli; coordinates responses to stimuli; integrates organ system activities.

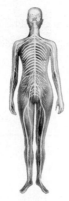

❸ Endocrine System

Secretes hormones that control activity of other organ systems. (Male testes added.)

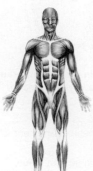

❹ Muscular System

Moves the body and its parts; maintains posture; produces heat to maintain body temperature.

❺ Skeletal System

Supports and protects body parts; site of muscle attachment; produces blood cells; stores minerals.

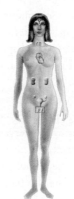

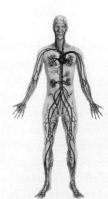

❻ Circulatory System

Distributes materials and heat throughout the body.

(**Figure 28.11**). It also helps regulate body temperature by conveying heat from the body's warm core to the skin where heat can be lost to the environment. The lymphatic system consists of vessels that move fluid (lymph) from tissues to the blood, and of lymphatic organs (lymph nodes, tonsils, and the spleen) that help protect the body against pathogens ❼.

The respiratory system includes lungs and the airways that lead to them ❽. It delivers oxygen from the air to the blood, and expels carbon dioxide from blood into the air. Oxygen is needed for cells to carry out aerobic respiration.

The digestive system takes in food, breaks it down, delivers nutrients to the blood, and eliminates undigested wastes ❾. It includes organs of the gut (esophagus, stomach, intestine), as well as glandular organs (such as the liver and pancreas) that supply substances that function in digestion.

The urinary system consists of kidneys (organs that filter blood and make urine), a bladder, and ducts that deliver urine

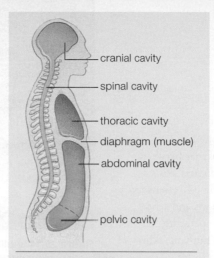

Figure 28.12 **Body cavities that hold human organs.**

cranial cavity
spinal cavity
thoracic cavity
diaphragm (muscle)
abdominal cavity
pelvic cavity

out of the body ❿. It removes wastes from blood, and adjusts blood volume and solute composition.

The reproductive system of both sexes includes gamete-making organs (ovaries or testes) and the ducts through which gametes travel ⓫. In females, it also includes a uterus, the organ in which offspring develop.

Some organs are components of more than one organ system. For example, the pancreas is both an endocrine gland and a digestive organ.

Figure 28.12 shows body cavities that hold organs. A cranial cavity holds the brain, and a spinal cavity holds the spinal cord. Like other vertebrates, humans have a lined body cavity, or coelom (Section 23.1). A skeletal muscle called the diaphragm divides the coelom into two cavities. The upper thoracic cavity holds the heart and lungs. The lower abdominopelvic cavity has two regions. The upper region, called the abdominal cavity, holds digestive organs such as the stomach and intestines. The lower region, called the pelvic cavity, holds the bladder, rectum, and reproductive organs.

❼ **Lymphatic System**

Collects and returns excess tissue fluid to the blood; defends the body against infection and cancers.

❽ **Respiratory System**

Takes in oxygen needed for aerobic respiration; expels carbon dioxide; helps maintain pH.

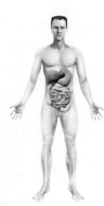

❾ **Digestive System**

Takes in food and water; breaks food down and absorbs nutrients, then eliminates food residues.

❿ **Urinary System**

Maintains volume and composition of blood; excretes excess fluid and wastes; helps maintain pH.

⓫ **Reproductive System**

Female: Produces eggs; nourishes and protects developing offspring.

Male: Produces and transfers sperm to a female.

LEARNING OBJECTIVES

List the functions of human skin.

Describe the structure of the epidermis and dermis.

Explain how a hair forms.

Explain how variations in skin color adapt people to life at different latitudes.

Figure 28.14 Vitiligo. This noncontagious disorder has caused death of melanocytes on the hands, arms, and face of television reporter Lee Thomas.

Skin, a vertebrate's largest organ, serves as a barrier between an animal and its external environment. We discuss skin's role in excluding pathogens in detail in Section 34.2. Sensory receptors in the skin monitor external conditions and send signals to the brain. Skin also functions in temperature control and helps prevent excessive water loss. In humans, skin produces vitamin D.

COMPONENTS OF HUMAN SKIN

Skin consists of two layers, a thin upper epidermis and the dermis beneath it (**Figure 28.13**).

Epidermis is a stratified squamous epithelium with an abundance of adhering junctions and no extracellular matrix. Human epidermis consists mainly of keratinocytes, cells that make the waterproof protein keratin.

Mitotic cell divisions in the deepest epidermal layer continually produce new keratinocytes that displace older cells upward toward the skin's surface. As cells move upward, they travel farther and farther away from underlying connective tissues where blood vessels deliver nutrients and gases. Eventually the cells, deprived of these essential substances, die and

Figure 28.13 Skin structure. Skin components, such as glands, differ from one body region to the next.

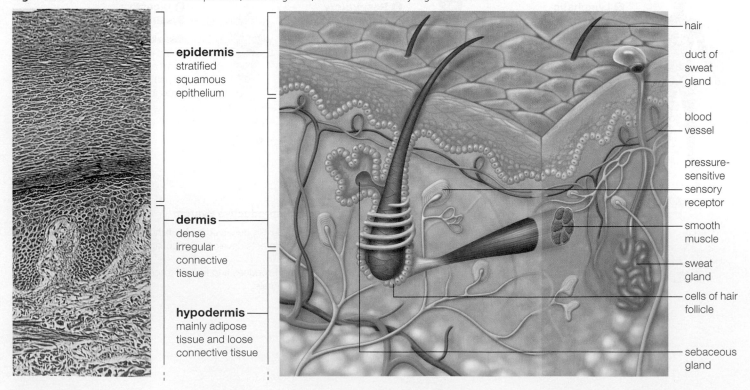

epidermis
stratified
squamous
epithelium

dermis
dense
irregular
connective
tissue

hypodermis
mainly adipose
tissue and loose
connective tissue

hair

duct of
sweat
gland

blood
vessel

pressure-
sensitive
sensory
receptor

smooth
muscle

sweat
gland

cells of hair
follicle

sebaceous
gland

become flattened. The upper layers of the epidermis consist of the keratin-rich remains of keratinocytes. This material forms an abrasion-resistant layer that helps prevent water loss.

The deepest epidermal layer also contains melanocytes, which make pigments called melanins and donate them to keratinocytes. Variations in human skin color arise from differences in the distribution and activity of melanocytes, and in the type of melanin they produce. One melanin is brown to black. Another is red to yellow. The effect of melanocytes can be seen with vitiligo, a skin disorder in which destruction of these cells results in light patches of skin (**Figure 28.14**).

Dermis consists primarily of dense, irregular connective tissue with springy elastin fibers and supportive collagen fibers. Blood vessels, lymph vessels, and sensory receptors weave through the dermis. The dermis is thicker than the epidermis, and more resistant to tearing. Leather is animal dermis that has been treated with chemicals.

The dermis connects to the hypodermis, a layer of loose connective tissue and adipose tissue that underlies the skin. The thickness of hypodermis varies among body regions. The hypodermis beneath the skin of our eyelids is thin, with few adipose cells. By contrast, the hypodermis of our buttocks contains many adipose cells.

Sweat glands, sebaceous glands, and hair follicles are pockets of epidermal cells that migrated into the dermis during early development. Sweat is mostly water. As it evaporates, it cools the skin surface and helps keep the body from overheating. Sebaceous glands produce sebum, an oily mix of triglycerides, fatty acids, and other lipids. Sebum helps keep skin and hair soft. It also has antimicrobial properties.

The portion of a hair that you can see consists of the remains of cells that began life deep in the dermis, at the base of the hair follicle. Keratinocytes at a follicle's base divide every 24 to 72 hours, making them among the fastest-dividing cells in the body. As a result of these divisions, newly formed keratinocytes are continually forced away from the follicle's base. As keratinocytes are pushed along through a sheath that extends to the skin's surface, they take up pigment from melanocytes and they produce and store keratin. Once filled with keratin, they die—before ever reaching the skin's surface. The part of the threadlike stack of dead, keratin-rich cells that protrudes above the skin surface is referred to as a shaft of hair.

Our hair can "stand up" because a smooth muscle (the arrector pili muscle) attaches to a sheath that surrounds the base of each hair shaft. When this muscle reflexively contracts in response to cold or fright, the hair is pulled upright. This reflex is of little use to modern humans, but in our hairier ancestors it helped retain body heat (in the cold) and made the body appear bigger (in response to a fright).

SUN AND THE SKIN

Melanin produced by the epidermis functions as a sunscreen, absorbing ultraviolet (UV) radiation that could otherwise damage the DNA of underlying skin layers. When a patch of skin is exposed to sunlight, melanocytes in that area make more of the brownish-black melanin, resulting in a "tan."

A bit of UV exposure is a good thing; it stimulates skin to produce a molecule that the body converts to vitamin D. We need vitamin D to absorb calcium ions from food. However, UV exposure also causes the breakdown of folate, which is one of the B vitamins. Among other problems, a folate deficiency during pregnancy damages an offspring's developing nervous system.

Variations in skin color among human populations evolved as adaptations to differences in sunlight exposure. Humans arose in equatorial Africa, where the sun's rays are intense and day length does not decline dramatically in the winter. In this environment, melanin-rich skin that protected folate still made plenty of vitamin D. Later, some humans moved to more northerly regions, where sunlight is less intense, winter days are short, and more time is spent indoors or bundled in clothing. Under these circumstances, skin with fewer melanocytes is advantageous. During cold, dark winters, having lighter skin makes it easier to get enough sunlight to maintain an adequate level of vitamin D.

Dark skin provides more UV protection than light skin, but in anyone, prolonged or repeated UV exposure damages collagen and causes elastin fibers to clump. Chronically tanned skin becomes less resilient and begins to look leathery. UV harms DNA, and the damage increases the risk of skin cancer. Melanoma, the most dangerous skin cancer, arises when melanocytes divide uncontrollably.

As people age, their epidermal cells divide less frequently. Glandular secretions that kept the skin soft and supple dwindle. Thickness and elasticity of the dermis decline as collagen and elastin fibers become sparse. Permanent wrinkles appear in places where making certain facial expressions once produced temporary creases.

Excessive tanning accelerates this aging process, as does smoking. Smoking lessens the flow of blood to the skin, depriving it of oxygen and nutrients. Damaging effects of sun are localized to sun-exposed regions, but smoking harms skin throughout the body.

---

dermis Deep layer of skin that consists of connective tissue with nerves and blood vessels running through it.

epidermis (ep-eh-DERM-iss) Outermost tissue layer; in animals, the epithelial layer of skin.

28.8 Negative Feedback in Homeostasis

Using an appropriate biological example, explain how negative feedback mechanisms contribute to homeostasis.

Homeostasis, again, is the process of maintaining the body's internal environment within a tolerable range. In vertebrates, homeostasis involves interactions among sensory receptors, the brain, and muscles and glands. A **sensory receptor** is a cell or cell component that responds to a change in a specific stimulus such as temperature or blood pressure. Sensory receptors involved in homeostasis function like internal watchmen; they monitor the body for changes. Information from sensory receptors throughout the body flows to an integrator. In vertebrates, the brain or spinal cord most often serves as the integrator. It processes the incoming information, then signals effectors—muscles and glands—to take any necessary actions to keep the body functioning properly.

Homeostasis typically involves a **negative feedback mechanism**, in which a change brings about a response that reverses the change. An air conditioner with a thermostat provides a familiar nonbiological example of a negative feedback mechanism. A person sets the air conditioner to a desired temperature. When the temperature rises above this preset point, a sensor in the air conditioner detects the change and turns the unit on. When the temperature declines to the desired level, the thermostat detects the change and turns off the air conditioner. (Note that the use of "negative" in this context does not mean the response is bad or undesirable, but rather that it negates the stimulus that elicited it.)

A similar negative feedback mechanism operates when you exercise on a hot day. Muscle activity generates heat, and your body's internal temperature rises (**Figure 28.15**). Sensory receptors in the skin detect the increase and signal your brain (the integrator). In response, the brain signals effectors. These signals increase the flow of blood from the body's hot interior to the skin. The shift maximizes the amount of heat given off to the surrounding air. At the same time, sweat glands in the skin increase their output. Evaporation of sweat helps cool the body surface. You breathe faster and deeper, speeding the transfer of heat from the blood in your lungs to the air. Your rate of heat loss increases, and your body's internal temperature declines back to normal.

Receptors in the skin also notify your brain when the external environment becomes chilly. The brain then sends out signals that decrease blood flow to your skin, thus reducing the amount of heat sent to the body surface, where it would be lost to the surrounding air. With prolonged cold, the brain commands skeletal muscles to contract ten to twenty times a second. This shivering response dramatically increases heat production by muscles.

---

negative feedback mechanism A change causes a response that reverses the change; important mechanism of homeostasis.
sensory receptor Cell or cell component that responds to a specific stimulus, such as temperature or light.

Figure 28.15 Negative feedback mechanism that maintains human body temperature under hot conditions.

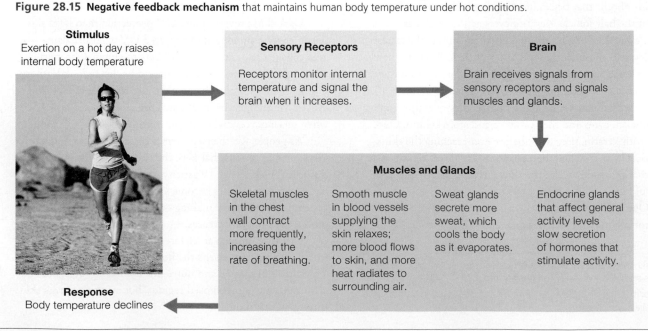

Stimulus
Exertion on a hot day raises internal body temperature

Sensory Receptors
Receptors monitor internal temperature and signal the brain when it increases.

Brain
Brain receives signals from sensory receptors and signals muscles and glands.

Muscles and Glands

Skeletal muscles in the chest wall contract more frequently, increasing the rate of breathing.

Smooth muscle in blood vessels supplying the skin relaxes; more blood flows to skin, and more heat radiates to surrounding air.

Sweat glands secrete more sweat, which cools the body as it evaporates.

Endocrine glands that affect general activity levels slow secretion of hormones that stimulate activity.

Response
Body temperature declines

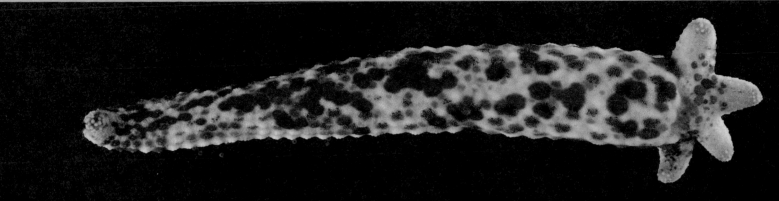

Figure 28.16 A sea star in the process of regenerating four arms. Stem cells in the remaining bit of the central disk give rise to replacement parts.

GROWING REPLACEMENT TISSUES

All animals replace tissues lost to injury, but invertebrates have the greatest capacity for regeneration. Some sea stars, for example, can regrow an entire body from a single arm and a bit of the central disk (**Figure 28.16**). No vertebrate can grow a body from a limb, but many salamanders and lizards can replace a lost tail. Mammals do not replace limbs or tails, although like other animals they replace skin and blood cells.

Stem cells are key to producing new or replacement tissues. A **stem cell** is a cell that can divide and produce more stem cells, or differentiate into one of the specialized cells that characterize specific tissues (**Figure 28.17**). In short, all cells in an animal body "stem" from stem cells.

Human embryonic stem cells form soon after fertilization, when mitotic divisions produce a cluster of cells collectively smaller than the head of a pin. Each cell in the cluster is "pluripotent," meaning it can develop into any of the cell types in a human body. Stem cells become more specialized as development continues. In an adult, for example, stem cells in bone can become blood cells, but not muscle cells or nerve cells.

Skin cells, blood cells, and other cell types that your body replaces on an ongoing basis arise continually from adult stem cells. This is why you can replace skin or blood lost to injury. However, human adults have few stem cells capable of becoming cardiac muscle or neurons. Thus, some tissues are not replaced if they are damaged or lost. This is why death of cardiac muscle cells during a heart attack permanently weakens the heart, and why spinal cord injuries result in permanent paralysis.

Stem cell researchers are investigating ways to replace dead or malfunctioning cells that the body cannot replace on its own. One way to do this is to produce the desired cell types from embryonic stem cells (these cells are derived from human embryos that have been donated or nonviable embryos produced specifically for this purpose).

An alternative is to alter adult cells such as fibroblasts through biotechnology so we can use them to grow replacement tissues. Adult cells that have been persuaded to behave like embryonic stem cells are called induced pluripotent stem cells (IPSCs).

To produce IPSCs, adult cells must be treated in a way that causes them to dedifferentiate (reverse the process of differentiation that gave them their specific character). For example, fibroblasts from skin can be turned into IPSCs by introducing synthetically modified RNAs.

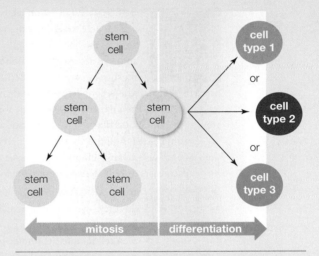

Figure 28.17 Stem cells. Each stem cell can divide to form two new stem cells, or differentiate into a specialized cell.

The RNAs are taken up by fibroblasts (by endocytosis), then translated into proteins that bring about dedifferentiation.

Stem cell research is in its early stages, but results are promising. A stem cell–based treatment for a degenerative form of blindness has improved vision in a small number of patients. Stem cell–based treatments for a variety of conditions have also proven effective in animal models. For example, researchers have used embryonic stem cells to cure diabetes in mice.

stem cell Cell that can divide to produce more stem cells or differentiate into a specialized cell type.

Section 28.1 Most animals have four types of **tissues** organized as **organs** and organ systems. The body consists largely of fluid. Most fluid is inside cells. **Extracellular fluid** serves as the cell's environment. In humans, extracellular fluid consists mainly of **interstitial fluid** and plasma. Physical constraints and evolutionary history influence animal body plans.

Section 28.2 **Epithelial tissue** covers and lines body surfaces and cavities. Its cells have little matrix between them. A secreted **basement membrane** attaches epithelium to an underlying tissue. **Tight junctions** form a waterproof barrier between cells and **adhering junctions** hold cells together. Some epithelial cells have cilia or **microvilli**. Gland cells are special epithelial cells. **Endocrine glands** are ductless and secrete hormones. **Exocrine glands** secrete substances such as milk through ducts.

Section 28.3 **Connective tissues** consist of cells in an extracellular matrix of their own secretions. They structurally and functionally connect tissues. **Loose connective tissue**, the most common type, underlies most epithelia. Stretchy **dense, regular connective tissue** makes up ligaments and tendons. **Dense, irregular connective tissue** forms capsules, as around the kidney. **Cartilage** and **bone tissue** are components of the skeletal system. **Adipose tissue** stores lipids, serves as insulation, and acts as padding. **Blood** is considered a fluid connective tissue; it consists of cells derived from bone in a fluid plasma.

Section 28.4 **Muscle tissues** contract (shorten) when stimulated. They help move the body and its component parts. The three types are **skeletal muscle**, **cardiac muscle**, and **smooth muscle tissue**. Skeletal muscle, which interacts with bone, and cardiac muscle, which is the muscle of the heart, appear striated. Smooth muscle does not. Only skeletal muscle is under voluntary control.

Section 28.5 **Nervous tissue** coordinates activities within a body. It contains **neurons**, which relay electrical signals along their length and send and receive chemical messages. Nervous tissue also contains neuroglial cells that protect and support neurons. The white matter and gray matter of the brain and spinal cord are nervous tissues, as are the peripheral nerves that extend throughout the body.

Section 28.6 An organ system consists of two or more organs that interact chemically, physically, or both in tasks that help keep individual cells as well as the whole body functioning. All vertebrates have the same set of organ systems. Many internal organs reside inside body cavities. The human thoracic, abdominal, and pelvic cavities are regions of our coelom. Organ systems interact to provide cells with the materials that they need, and to rid the body of wastes.

Section 28.7 Vertebrate skin is a large organ. It provides protection, helps control temperature, monitors external conditions, and aids in production of vitamin D. Skin consists of an outer **epidermis** and a deeper **dermis**. Keratin makes skin waterproof and melanin gives its color. Variations in melanin affect skin color and adapt people to different latitudes.

Section 28.8 Homeostasis requires **sensory receptors** that detect changes, an integrating center (the brain or spinal cord), and effectors (muscles and glands) that bring about responses. Homeostasis often involves a **negative feedback mechanism**: A change causes a response that reverses the change.

Application A **stem cell** can divide to form more stem cells or differentiate to form specialized cells. Embryonic stem cells can produce any cell type in the body. Cells of adults are less versatile, but scientists are developing methods to make them behave like embryonic stem cells. Use of stem cells may provide a way to replace cells that the body normally cannot replace, such as cardiac muscle cells.

SELF-ASSESSMENT
ANSWERS IN APPENDIX VII

1. _____ tissues are sheetlike with one free surface.
   a. Epithelial   b. Connective   c. Nervous   d. Muscle

2. _____ allow cardiac muscle cells to contract in unison.
   a. Tight junctions   c. Gap junctions
   b. Adhering junctions   d. all of the above

3. Glands are derived from _____ tissue.
   a. epithelial   b. connective   c. muscle   d. nervous

4. Most _____ have many collagen and elastin fibers.
   a. epithelial tissues   c. muscle tissues
   b. connective tissues   d. nervous tissues

5. _____ is mostly plasma.
   a. Adipose tissue   c. Cartilage
   b. Blood   d. Bone

6. Your body converts excess carbohydrates to fats. _____ specializes in storing the fats.
   a. Epithelial tissue   c. Adipose tissue
   b. Dense connective tissue   d. Bone tissue

7. Cells of _____ can shorten (contract).
   a. epithelial tissue   c. muscle tissue
   b. connective tissue   d. nervous tissue

8. _____ muscle tissue has a striated appearance and is under voluntary control.
   a. Skeletal   b. Smooth   c. Cardiac   d. a and c

9. _____ detects and integrates information about changes and controls responses to those changes.
   a. Epithelial tissue   c. Muscle tissue
   b. Connective tissue   d. Nervous tissue

GROWING SKIN TO HEAL WOUNDS Maintaining the level of sugar in the blood is an important aspect of homeostasis. Diabetes is a disorder in which the blood sugar level is not properly controlled. Among other complications, diabetes reduces blood flow to the lower legs and feet. Impaired blood flow slows healing.

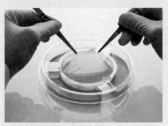

Apligraf®, a bioengineered, bilayered living cell technology that is FDA approved to treat chronic diabetic foot ulcers and venous leg ulcers.

As a result, about 3 million diabetes patients have ulcers (open wounds that do not heal) on their feet.

Fibroblasts and keratinocytes can be grown to produce a cultured skin product (left) that is placed over wounds to help them heal. **Figure 28.18** shows the results of a clinical experiment that tested the effect of the cultured skin product versus standard treatment for diabetic foot wounds. Patients were randomly assigned either to the experimental treatment group or to the control group. Their progress was monitored for 12 weeks.

1. What percentage of wounds had healed at 8 weeks with standard treatment? With cultured skin treatment?

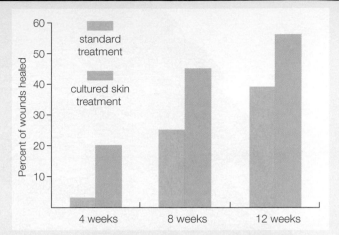

Figure 28.18 Results of a study comparing standard treatment for diabetic foot ulcers to use of a cultured cell product. Bars show the percentage of foot ulcers that had completely healed.

2. What percentage of wounds had healed at 12 weeks with standard treatment? With cultured skin treatment?

3. How early was the healing difference between the control and treatment groups apparent?

10. Skin darkens when exposed to sunlight because _____ produce more pigment.
    - a. melanocytes
    - b. keratinocytes
    - c. neuroglial cells
    - d. fibroblasts

11. The heart and lungs are in the _____ cavity.
    - a. thoracic
    - b. pelvic
    - c. cranial
    - d. abdominal

12. With negative feedback, a change induces a response that _____ that change.
    - a. increases
    - b. reverses

13. Lack of vitamin C impairs production of _____ .
    - a. keratin
    - b. folate
    - c. vitamin D
    - d. collagen

14. A hair is a keratin-rich structure formed by _____ tissue.
    - a. muscle
    - b. nervous
    - c. epithelial
    - d. connective

15. Match the terms with the most suitable description.
    - ___ exocrine gland
    - ___ endocrine gland
    - ___ cartilage
    - ___ dermis
    - ___ smooth muscle
    - ___ bone
    - ___ melanin
    - ___ blood
    - ___ neuroglia
    - ___ brown fat
    - ___ simple squamous epithelium

    - a. strong, pliable; like rubber
    - b. secretion through duct
    - c. deep skin layer
    - d. contracts, not striated
    - e. assist and support neurons
    - f. makes skin dark
    - g. lines lungs
    - h. cells in a hardened matrix
    - i. fluid connective tissue
    - j. ductless secretion
    - k. many mitochondria produce heat

1. Many people oppose the use of animals for testing the safety of cosmetics. They argue that alternative test methods are available, such as use of lab-grown human tissues. Given what you learned in this chapter, speculate on the advantages and disadvantages of tests that use lab-grown tissues as opposed to living animals.

2. In winter, dark-skinned immigrants to Sweden are more likely to have a vitamin D deficiency than light-skinned people native to this region. Why?

3. Each level of biological organization has emergent properties that arise from interactions at a lower level. For example, cells have a capacity for inheritance that the molecules making up the cell do not. Can you think of an emergent property of a tissue? Of an organ that contains that tissue?

4. The micrograph to the left shows cells from the lining of an airway leading to the lungs. The gold cells are ciliated and the darker brown ones secrete mucus. What type of tissue is this? How can you tell?

for additional quizzes, flashcards, and other study materials
▶ ACCESS MINDTAP AT WWW.CENGAGEBRAIN.COM

CREDITS: (in text, left) Courtesy of © Organogenesis, Inc., www.organogenesis.com; (in text, right) CNRI/Science Source.

29

Neural Control

Links to Earlier Concepts

In this chapter, you will draw on your knowledge of potential energy (Section 5.1), diffusion and concentration gradients (5.6), passive and active transport (5.7), and exocytosis (5.8). You will revisit body plans of a few invertebrates (23.4, 23.5, 23.11) and typical chordate traits (24.1). You will increase your knowledge of nervous tissue (28.5) and of how alcohol exerts its effects (Chapter 5 Application).

Core Concepts

Living things sense and respond appropriately to their internal and external environments.

In animals, the nervous system detects changes in the internal and external environment and issues the signals that bring about appropriate responses to those changes. The structure of animal nervous systems varies among groups; there has been an evolutionary trend toward more complex, centralized systems.

Cellular homeostasis involves maintenance of an internal environment that differs from the external environment.

Transport proteins embedded in plasma membranes allow cells to establish electrical and concentration gradients across their plasma membrane. In neurons, controlled opening and closing of specialized tranport proteins allows transmission of electrical signals within a cell. Release of chemical signals allows communication between cells.

Interactions among the components of a biological system give rise to complex properties.

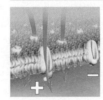

The neurons and supporting cells of the vertebrate nervous system are organized into a brain, a spinal cord, and the many peripheral nerves that extend throughout the body. In humans, interactions among these components not only maintain homeostasis but also give rise to emotion, memory, a capacity for language, and consciousness.

A volunteer wired for a study of brain activity. The electrodes will detect traces of electrical signals produced by neurons inside his skull.

Photograph by Maggie Steber, National Geographic Creative.

LEARNING OBJECTIVES

List the three steps involved in intercellular communication.

Explain how a sea anemone's nerve net differs from the nervous system of a flatworm in its structure and function.

Name the two functional divisions of the vertebrate nervous system and describe how they interact.

For an animal body to work as a whole, cells must constantly signal one another. In some tissues, gap junctions connect adjacent cells and allow direct transfer of signals between their cytoplasms. In other cases, cell–cell communication involves secretion of signaling molecules into interstitial fluid (the fluid between cells). These molecules exert effects only when they bind to a receptor on or inside another cell. Any cell with receptors that bind and respond to a specific signaling molecule is a "target" of that molecule.

Secreted signaling molecules exert their effect by a three-step process. The molecules bind to a receptor protein on or in the target cell, the receptor transduces the signal (changes it into a form that affects target cell behavior), and the target cell responds:

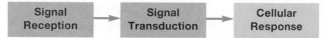

| Signal Reception | → | Signal Transduction | → | Cellular Response |

Two organ systems—the nervous system and the endocrine system—allow long-distance communication in an animal body. The endocrine system coordinates cell activities by way of signaling molecules called hormones that travel in the blood. We discuss this system in Chapter 31. Here, our focus is the **nervous system**, the organ system that gathers information about the internal and external environment, integrates this information, and issues commands to muscles and glands. **Neurons** make up the communication lines of a nervous system. We discuss their function in later sections. Here we survey the different types of nervous systems.

INVERTEBRATE NERVOUS SYSTEMS

Nerve Nets Animals with radial symmetry have a mesh of interconnected neurons called a **nerve net**. Information flows in all directions among cells of the nerve net, and there is no centralized, controlling organ. This type of system is adaptive when food or threats can arrive from any direction. Sea anemones and other cnidarians (Section 23.4) are the simplest animals with a nerve net (**Figure 29.1A**). By commanding cells in the body wall to contract, a cnidarian nerve net can alter the size of the mouth, change the body's shape, or move the tentacles.

Echinoderms (Section 23.12) are also radial, and they too have a nerve net. However, the echinoderm system is a bit more complex because it includes nerves. A **nerve** consists of a bundle of neuron axons (cytoplasmic extensions) wrapped in connective tissue. In sea stars, a ring of nerves surrounds the mouth, and nerves from this ring extend into each arm.

Bilateral Cephalized Systems Most animals have bilateral symmetry. Evolution of bilateral body plans was accompanied by **cephalization**: an evolutionary trend whereby many neurons that detect and process information become concentrated at the body's anterior (front) end.

Planarian flatworms are bilateral animals with a ladder-like, cephalized nervous system. A pair of ganglia in the head serves as a center that integrates information (**Figure 29.1B**). Each **ganglion** (plural, ganglia) consists of a cluster of neuron cell bodies. (The cell body is the part of the neuron that holds the cell's nucleus and other organelles). A planarian's ganglia receive signals from eyespots and chemical-detecting cells on the head. They also connect to a pair of nerve cords that extend the length the animal's ventral (lower) surface. A **nerve cord** is a bundle of many nerves that runs the length of the body. Nerves branch from the nerve cords and cross the body in a ladderlike series of connections.

increasing cephalization

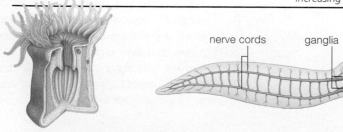

nerve cords   ganglia   brain   nerve cords with ganglia

A Nerve net (purple) of sea anemone. No central organ integrates signals.

B Planarian nervous system. Two ganglia in the head integrate information. Nerve cords extend the length of the body along the ventral (lower) surface.

C Insect nervous system. A brain with hundreds of thousands of neurons connects to a ventral nerve cord that has a ganglion in each segment. The ganglia serve as local control centers. The head has a variety of sensory receptors.

Figure 29.1 Examples of invertebrate nervous systems.

Compared to planarians, arthropods are more highly cephalized, having a head with a brain and a variety of sensory structures, including eyes and antennae. Paired nerve cords that run along the ventral surface connect to a brain (**Figure 29.1C**). A **brain** is a central control organ of a nervous system. It receives and integrates sensory information, regulates internal processes, and sends out signals that bring about movements.

Mollusks vary in their degree of cephalization. Clams and other bivalves are headless and brainless, having instead widely separated ganglia that serve as local control centers. Octopuses are highly cephalized, having a brain that is proportionately larger than that of many fish or reptiles.

THE VERTEBRATE NERVOUS SYSTEM

A dorsal nerve cord is one of the defining features of chordate embryos (Section 24.1). In vertebrates, the dorsal nerve cord evolved into a brain and spinal cord, which together constitute the **central nervous system**. The central nervous system is one of the two divisions of the vertebrate nervous system (**Figure 29.2**). The second, the **peripheral nervous system**, consists of nerves that extend from the central nervous system out through the rest of the body (**Figure 29.3**). The human peripheral nervous system consists of 12 cranial nerves (nerves that connect to the brain) and 31 spinal nerves (nerves that connect to the spinal cord). Most peripheral nerves carry signals both to and from the central nervous system. For example, the sciatic nerve is a spinal nerve that relays signals from the skin of your leg toward your spinal cord, and from your spinal cord to muscles in your leg.

---

brain Central control organ of a nervous system; receives and integrates sensory information, regulates internal conditions, and sends out signals that result in movements.

central nervous system Of vertebrates, a brain and spinal cord.

cephalization Evolutionary trend whereby neurons that detect and process information become concentrated in the head of a bilateral animal.

ganglion (GANG-lee-on) Cluster of nerve cell bodies.

nerve Neuron axons bundled inside a sheath of connective tissue.

nerve cord Bundle of nerves that runs the length of a body.

nerve net Of cnidarians and echinoderms, a mesh of interacting neurons with no central control organ.

nervous system Organ system that gathers information about the internal and external environment, integrates this information, and issues commands to muscles and glands, all by way of neurons.

neuron One of the cells that make up communication lines of nervous systems; transmits electrical signals along its plasma membrane and sends chemical messages to other cells.

peripheral nervous system Of vertebrates, nerves that carry signals between the central nervous system and the rest of the body.

CENTRAL NERVOUS SYSTEM Brain, Spinal Cord		
⬇⬆		
PERIPHERAL NERVOUS SYSTEM Cranial and Spinal Nerves		
Autonomic Nerves Nerves that carry signals to smooth muscle, cardiac muscle, and glands		**Somatic Nerves** Nerves that carry signals to and from skeletal muscle, tendons, and the skin
Sympathetic Division	Parasympathetic Division	
Two sets of nerves that often signal the same effectors; have opposing effects		

Figure 29.2 Functional divisions of vertebrate nervous systems. The spinal cord and brain constitute the central nervous system. The peripheral nervous system includes cranial nerves and spinal nerves that extend throughout the body. Peripheral nerves carry signals to and from the central nervous system.

Sections 29.6 to 29.12 detail the structure and function of the two divisions of the human nervous system.

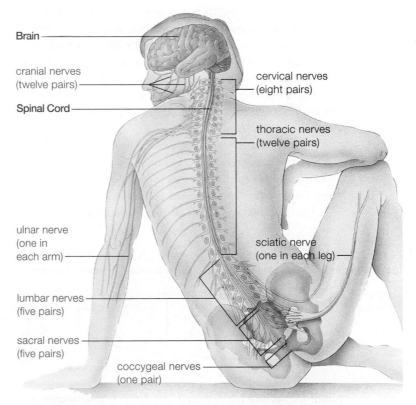

Brain

cranial nerves (twelve pairs)

Spinal Cord

cervical nerves (eight pairs)

thoracic nerves (twelve pairs)

ulnar nerve (one in each arm)

sciatic nerve (one in each leg)

lumbar nerves (five pairs)

sacral nerves (five pairs)

coccygeal nerves (one pair)

Figure 29.3 Some components of the human nervous system.

List the three types of neurons and explain how their structure reflects their functions.

Describe some of the functions of neuroglial cells.

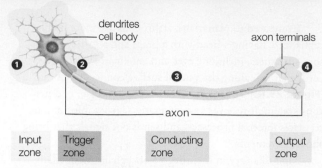

Figure 29.5 **Functional zones of a motor neuron.**

THREE TYPES OF NEURONS

Neurons are the signaling cells in nervous tissue. All neurons have a cell body that contains the nucleus and most other organelles. Cytoplasmic extensions from the cell body receive and send out signals. All neurons have one **axon**, a cytoplasmic extension that transmits electrical signals along its length and releases chemical signals at the axon terminals (endings). Some neurons also have **dendrites** that receive chemical signals from other neurons.

In vertebrate nervous systems, information typically flows from sensory neurons, to interneurons, to motor neurons (**Figure 29.4**). **Sensory neurons** are activated by a specific stimulus such as light, pressure, or the presence of a particular chemical. They do not receive signals from other neurons, so the axon is their only cytoplasmic extension. One end of the axon has receptor endings that detect the stimulus, and the other has axon terminals that signal another neuron, usually an interneuron (**Figure 29.4A**). **Interneurons** integrate information and issue commands. They receive signals from and send signals to other neurons, so they have dendrites as well as a short axon (**Figure 29.4B**). In vertebrates, interneurons are located in the brain or spinal cord. A **motor neuron** controls a muscle or a gland. It receives signals from interneurons, and sends signals to the muscle or gland that it controls. Like interneurons, motor neurons have dendrites as well as an axon (**Figure 29.4C**).

Let's take a closer look at a motor neuron (**Figure 29.5**). A neuron's dendrites and the cell body are input zones, where the cell receives chemical signals and converts them to electrical signals ❶. The part of the axon nearest the cell body is a trigger zone ❷. Excitation of this region triggers electrical

signals that flow along the conducting zone of the axon ❸ to axon terminals. Axon terminals are the output zone: They release signaling molecules that affect other cells ❹.

NEUROGLIA

Neuroglia (also called glia) are cells that support and assist neurons. They form a framework that holds neurons in place (glia means "glue" in Latin). Some neuroglia wrap around axons and speed the transmission of nerve impulses (more about this in Section 29.4). Ciliated neuroglia that line the fluid-filled cavities inside the brain and spinal cord keep nutrient-rich fluid moving around these organs. Other glia wrap around blood vessels of the brain and help to prevent bloodborne toxins from reaching this organ. Still others engulf and dispose of dead or damaged neurons.

---

axon Of a neuron, a cytoplasmic extension that transmits electrical signals along its length and secretes chemical signals at its endings.

dendrite Of a neuron, a cytoplasmic extension that receives chemical signals sent by other neurons.

interneuron Neuron that integrates information; both receives signals from and sends signals to other neurons.

motor neuron Neuron that controls a muscle or gland.

neuroglia (nur-oh-GLEE-uh) Nervous system cells that assist neurons and facilitate their function in a variety of ways.

sensory neuron Neuron that is activated when its receptor endings detect a specific stimulus, such as light or pressure.

Figure 29.4 **Information flow among the three functional types of neurons.** Blue arrows indicate the direction of information flow.

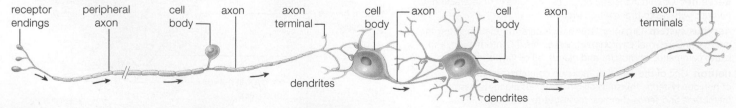

A Sensory neuron. Its long axon has receptor endings at one end and axon terminals at the other.

B Interneuron. It has many dendrites and a short axon.

C Motor neuron. It has many dendrites and a long axon that extends out to a muscle or gland.

29.3 Membrane Potentials

Describe the ion concentration gradients across the plasma membrane of a neuron at resting potential.

Explain how sodium–potassium pumps contribute to resting potential.

With this section, we turn from the structure of neurons and the nervous system to neuron function. To understand how signals are transmitted along a neuron, you need to know a bit about the electrical properties of cells. All living cells have an electrical gradient across their plasma membrane. As in a battery, the separation of charge constitutes potential energy (Section 5.1) that we measure as voltage. The voltage across a cell's membrane is called **membrane potential**.

RESTING POTENTIAL

The membrane potential of a neuron that is not excited is the neuron's **resting potential**. This potential is usually about −70 millivolts. (A millivolt [mV] is one-thousandth of a volt.) The negative sign indicates that the inside of the neuron is more negative than the outside.

Resting potential arises from the distribution of charged proteins and small ions. A neuron's cytoplasm contains more negatively charged proteins than the interstitial fluid. Being large and charged, these proteins cannot diffuse across the plasma membrane. Positively charged potassium ions (K⁺) and positively charged sodium ions (Na⁺) also contribute to the resting potential. These ions can move in and out of the neuron only with the assistance of transport proteins. Active transport proteins called sodium–potassium pumps (Section 5.7) move two potassium ions into the cell for every three sodium ions they move out (**Figure 29.6A**). By moving more positively charged ions out of the cell than into it, these pumps increase the charge gradient across the membrane. They also contribute to concentration gradients for sodium and potassium ions across the neuron's plasma membrane.

Sodium ions do not enter or exit a resting neuron, but some potassium ions leave the cell through passive transport proteins that are always open (**Figure 29.6B**). Movement of potassium ions out of the cell increases the charge difference across its membrane.

In summary, a resting neuron's cytoplasm has more negatively charged proteins, fewer sodium ions (Na⁺), and more potassium ions (K⁺) than the interstitial fluid. We illustrate

---

action potential Abrupt reversal of the charge difference across a neuron membrane.
membrane potential Voltage difference across a cell membrane.
resting potential Membrane potential of a neuron at rest.

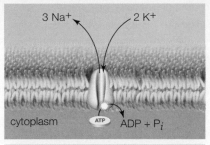

A Sodium–potassium pumps actively transport 3 sodium ions (Na⁺) out of a neuron for every 2 potassium ions (K⁺) they pump into it.

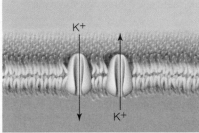

B Passive transport proteins allow K⁺ ions to diffuse across the plasma membrane, down their concentration gradient.

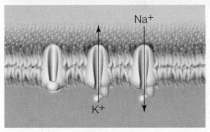

C Voltage-gated channels for K⁺ or Na⁺ ions are shut when a neuron is at rest. They open during an action potential.

Figure 29.6 Transport proteins in a neuron's plasma membrane.

these differences below (larger text represents higher concentrations; the green ball is negatively charged proteins):

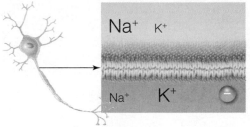

interstitial fluid

plasma membrane

neuron's cytoplasm

ACTION POTENTIAL

Neurons and muscle cells are said to be "excitable" because they can undergo an **action potential**—an abrupt reversal in the electric gradient across the plasma membrane. The reversal occurs when sodium and potassium ions flow through voltage-gated ion channels. The channels are passive transport proteins that open only during an action potential (**Figure 29.6C**). Neurons have voltage-gated channels in the membrane of their trigger zone and conducting zone. You will see how voltage-gated ion channels operate in the next section, where we continue our discussion of action potentials.

Describe the flow of ions during an action potential, and explain how voltage-gated ion channels regulate that flow.

Explain how an action potential moves along an axon.

Explain the importance of myelin in nerve conduction.

Action potentials are commonly known as nerve impulses. They are electrical signals that travel along neuron axons.

REACHING THRESHOLD

An action potential begins in a neuron's trigger zone, the region of the axon adjacent to the cell body. Consider what happens in a motor neuron. When the neuron is at rest, all of its voltage-gated ion channels are closed, and the membrane potential is negative (**Figure 29.7 ❶**).

A signal from another neuron shifts the membrane potential in the neuron's trigger zone. If the stimulus is large enough, the membrane potential reaches **threshold potential**, the potential at which voltage-gated sodium channels in the trigger zone open and an action potential begins.

AN ALL-OR-NOTHING SIGNAL

Opening of voltage-gated sodium channels in the trigger zone allows sodium ions to follow their concentration gradient and diffuse from interstitial fluid into the neuron's cytoplasm ❷. As these positively charged ions flow into the cell, they reduce the charge difference across the membrane. The change in membrane potential causes voltage-gated sodium channels in nearby regions of membrane to open, allowing more sodium ions to flow into the cytoplasm, and so on. This increasing flow of sodium ions inward is an example of a **positive feedback mechanism**, in which an activity intensifies because of its own occurrence. Positive feedback ensures that an action potential is an all-or-nothing event: Once the membrane potential in the trigger zone reaches threshold potential, an action potential always occurs.

All action potentials are the same size. When an action potential is triggered, the membrane potential quickly shoots up to about +30 mV. The positive membrane potential means that the cytoplasm has more positively charged ions than the interstitial fluid. Membrane potential cannot exceed +30 mV, because when it reaches this level, voltage-gated sodium channels close, and voltage-gated potassium channels open ❸. As positively charged potassium ions (K^+) diffuse out of the cell through the now-open voltage-gated potassium channels, the axon's cytoplasm once again becomes more negatively charged than interstitial fluid. Once that happens, voltage-gated potassium channels close.

In summary, during an action potential, membrane potential rises from the resting potential to a peak value, then declines once again to resting potential. The whole process takes only a few milliseconds.

Figure 29.7 Action potential.
Right, plot of a neuron's membrane potential over time. The numbers correlate with those in the graphics below, which illustrate the events at one region of the axon membrane.

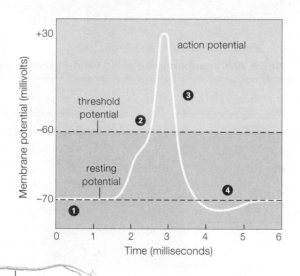

❶ At resting potential, all voltage-gated channels at a trigger zone are closed. White pluses and minuses indicate that the cytoplasm's charge is negative relative to that of the interstitial fluid.

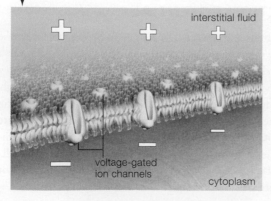

❷ At threshold potential, voltage-gated sodium (Na^+) channels in the trigger zone open, and Na^+ flows inward (blue arrows). This inward flow of positively charged ions causes a local reversal of the charge across the neuron membrane.

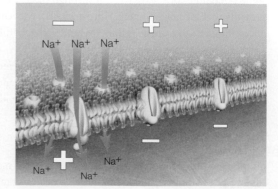

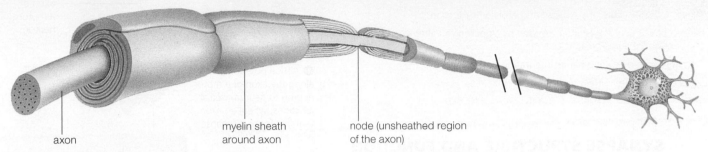

axon

myelin sheath
around axon

node (unsheathed region
of the axon)

Figure 29.8 Myelinated axon. Neuroglial cells form a discontinuous wrapping of myelin around the axon.

PROPAGATION ALONG AN AXON

An action potential is self-propagating. It travels along an axon, affecting each patch of membrane only briefly. When voltage-gated sodium channels open, some of the sodium ions that rush inward diffuse into adjacent regions of cytoplasm. The arrival of these positive ions raises the membrane potential in these regions to threshold level, setting in motion an action potential in the adjacent patches of membrane ❹. Voltage-gated sodium channels open in one region of membrane after another, and the action potential moves along the axon from trigger zone to terminals. The action potential moves only in one direction because it cannot move backward. After a voltage-gated sodium channel closes, it cannot open for a brief period. However, voltage-gated sodium channels in regions farther along the axon's membrane can and do swing open as these regions reach threshold.

---

myelin (MY-uh-lin) Fatty material produced by neuroglial cells; insulates axons and thus speeds conduction of action potentials.

positive feedback mechanism A response intensifies the conditions that caused its occurrence.

threshold potential Neuron membrane potential at which gated sodium channels open, causing an action potential to occur.

In most vertebrate nerves, axons associate with neuroglial cells that wrap the axon in **myelin**, a fatty material that functions like insulation around a wire (**Figure 29.8**). A myelinated axon has voltage-gated channels for sodium only at gaps (called nodes) in the myelin wrapping. Thus action potentials can occur only at nodes. When an action potential occurs at one node, sodium ions flow into the neuron, then diffuse through the cytoplasm until they reach the next node, where their arrival triggers an action potential. By "jumping" from node to node, an action potential can travel along a myelinated axon as quickly as 150 meters per second. In unmyelinated axons, the maximum speed is about 10 meters per second.

The importance of myelin is illustrated by the effects of multiple sclerosis (MS). This nervous system disorder arises when a person's immune system mistakenly attacks and destroys myelin in the brain and spinal cord. As myelin is replaced by scar tissue, the speed at which action potentials travel declines. The result is weakness, fatigue, impaired coordination, numbness, and vision problems. Women are affected twice as often as men. The trigger for MS is unknown, although genetics certainly plays a role. In the general population, the risk of developing MS is about 1 in 1,000, but a person whose identical twin has MS has a risk of 1 in 4.

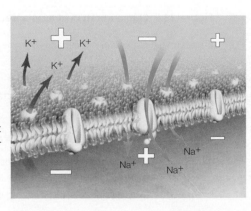

❸ When membrane potential reaches +30 mV, the voltage-gated Na+ channels close, and voltage-gated potassium (K+) channels open. K+ flows out of the axon (red arrows). Na+ that diffuses through cytoplasm drives adjacent membrane to threshold potential, so Na+ channels open there.

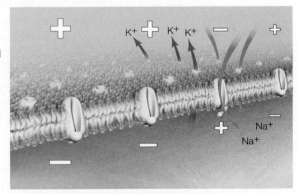

❹ Voltage-gated K+ channels in the trigger zone close. The action potential propagates along the axon as Na+ diffusing through cytoplasm triggers voltage-gated Na+ channels to open farther and farther away from the trigger zone.

29.5 The Synapse

SYNAPSE STRUCTURE AND FUNCTION

Action potentials cannot pass directly from a neuron to another cell. Instead, signaling molecules relay a message between neurons, or from a neuron to a muscle or gland. The region where a neuron transmits chemical signals to another cell is a **synapse**. The signal-sending neuron at a synapse is called a presynaptic cell. A fluid-filled synaptic cleft about 20 nanometers wide separates its axon terminal from the input zone of a postsynaptic cell (the cell that receives the signal). For comparison, a hair is 100,000 nanometers wide.

Figure 29.9 illustrates how signals are transmitted at a **neuromuscular junction**, a synapse between a motor neuron and a skeletal muscle fiber. Action potentials travel along the neuron's axon to the axon terminals ❶. The terminals have vesicles filled with **neurotransmitter**, a type of signaling molecule that relays messages between cells at a synapse ❷. Neurotransmitter is made in a neuron's cell body, then moved to axon terminals where it is stored until an action potential arrives. Arrival of action potential at an axon terminal triggers exocytosis: The neurotransmitter-filled vesicles move to the plasma membrane and fuse with it, releasing the neurotransmitter into the synaptic cleft ❸.

The plasma membrane of a postsynaptic cell has receptor proteins that reversibly bind neurotransmitter ❹. Motor neurons release a neurotransmitter called acetylcholine (ACh), and cells that make up skeletal muscle have receptors for this molecule. ACh receptors are passive transport proteins. When they bind to ACh, they change shape so a channel opens up inside them. Sodium ions travel through these channels, from interstitial fluid into the muscle cell ❺. Thus, the chemical signal (ACh) is transduced into an electrical signal (a change in the ion concentration across the membrane).

The response to this signal is an action potential. The influx of sodium caused by the binding of ACh drives the muscle cell's membrane toward threshold potential. Once this threshold is reached, action potentials stimulate muscle contraction. (Section 32.7 describes in detail how nervous signals bring about muscle contraction.)

After neurotransmitter molecules do their work, they must be removed from the synaptic cleft to make way for new signals. Reuptake of neurotransmitter by presynaptic cells is

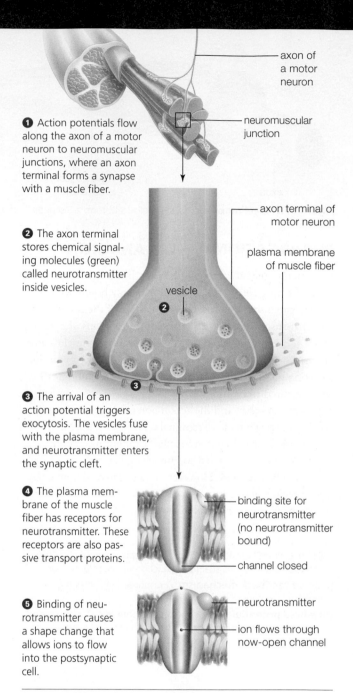

❶ Action potentials flow along the axon of a motor neuron to neuromuscular junctions, where an axon terminal forms a synapse with a muscle fiber.

axon of a motor neuron

neuromuscular junction

❷ The axon terminal stores chemical signaling molecules (green) called neurotransmitter inside vesicles.

axon terminal of motor neuron

plasma membrane of muscle fiber

vesicle

❸ The arrival of an action potential triggers exocytosis. The vesicles fuse with the plasma membrane, and neurotransmitter enters the synaptic cleft.

❹ The plasma membrane of the muscle fiber has receptors for neurotransmitter. These receptors are also passive transport proteins.

binding site for neurotransmitter (no neurotransmitter bound)

channel closed

❺ Binding of neurotransmitter causes a shape change that allows ions to flow into the postsynaptic cell.

neurotransmitter

ion flows through now-open channel

Figure 29.9 Communication at a synapse. This example depicts what occurs at a neuromuscular junction, a synapse between an axon terminal of a motor neuron and a muscle cell.

part of this process. In addition, enzymes in the membrane of postsynaptic cells can break down neurotransmitter.

SIGNAL AND RECEPTOR VARIETY

The effect a neurotransmitter has on a postsynaptic cell depends on the type of neurotransmitter and the type of receptor it binds to. For example, binding of ACh to receptors on skeletal muscle encourages muscle contraction, whereas binding of ACh to receptors on cardiac muscle inhibits contraction. The two types of muscle respond differently to ACh because they have different ACh receptors. ACh also acts in the brain, where it affects alertness and memory.

ACh is one of many vertebrate neurotransmitters. Norepinephrine and epinephrine (collectively known as adrenaline) are neurotransmitters that prepare the body to respond to stress or excitement. Serotonin influences mood and memory. Glutamate is the main excitatory signal in the central nervous system. Endorphins are the body's natural pain relievers. Dopamine functions in fine motor control and in reward-based learning. The brain releases a surge of dopamine in response to behaviors such as eating, which enhance survival, or engaging in sex, which enhances reproduction.

SYNAPTIC INTEGRATION

A postsynaptic cell typically receives messages from many neurons. An interneuron in the brain can be on the receiving end of hundreds to thousands of synapses. At each synapse, the incoming signal may be excitatory and boost the membrane potential closer to threshold. Or the signal may be inhibitory and move the membrane potential away from threshold. By the process of **synaptic integration**, a neuron sums all inhibitory and excitatory signals arriving at its input zone during a short time. When excitatory signals cumulatively outweigh inhibitory ones, an action potential occurs.

DRUGS THAT ACT AT SYNAPSES

Psychoactive drugs are chemicals that enter the brain and alter mood or perception by acting at synapses (**Figure 29.10**). Stimulants make users feel alert but also anxious, and they can interfere with fine motor control. Nicotine is a stimulant that, among other effects, mimics ACh's excitatory effects in the brain. Caffeine stimulates users by blocking receptors for neurotransmitters that slow brain activity. Cocaine and amphetamines produce their stimulating effects by blocking the reuptake of excitatory neurotransmitters from the synaptic cleft. Ecstasy is a type of amphetamine, as is methamphetamine.

Analgesics relieve pain. Opiate analgesics, such as morphine, codeine, heroin, and oxycodone, mimic endorphins, the body's natural painkillers. They dull pain and, in larger doses, produce a rush of euphoria.

Depressants such as alcohol and barbiturates have a calming effect and slow motor responses. Alcohol also stimulates the release of endorphins, so users experience a brief euphoria followed by depression.

Figure 29.10 Commonly used psychoactive drugs. Alcohol, coffee, and nicotine alter the function of synapses in the brain.

Hallucinogens such as LSD (lysergic acid diethylamide) distort sensory perception and bring on a dreamlike state. Two related drugs, mescaline and psilocybin (Section 22.4), have similar but weaker effects.

Marijuana, which is dried parts of a *Cannabis* plant, is classified as a hallucinogen because large doses can cause hallucinations. At more typical doses, users become relaxed, sleepy, uncoordinated, and inattentive. The main psychoactive ingredient in marijuana, THC (delta-9-tetrahydrocannabinol), stimulates receptors for the neurotransmitter anandamide. Anandamide increases appetite, decreases pain perception and anxiety, and has a role in deleting memories.

Many psychoactive drugs can lead to addiction. Addiction is a state in which a person feels physically and/or psychologically compelled to use a particular drug, regardless of the consequences. Dopamine plays an important role in creating addiction. Addictive drugs trigger dopamine release or prevent its reuptake, thus causing pleasure sensations that tap into the mechanism for reward-based learning. Thus, drug users inadvertently teach themselves that the drug is essential to their well-being.

Continual use of a drug can result in tolerance, meaning the effect of a given dose of the drug decreases over time. Although two beers can make a nondrinker feel drunk, they have little effect on an alcoholic. Tolerance arises from drug-induced changes to the user's body.

---

neuromuscular junction Synapse between a neuron and a muscle cell.
neurotransmitter Chemical signal released by axon terminals.
synapse (SIN-apps) Region where a neuron's axon terminals transmit signaling molecules (neurotransmitter) to another cell.
synaptic integration The summing of excitatory and inhibitory signals by a postsynaptic cell.

Describe the structure and location of peripheral nerves.

Compare the organs controlled by somatic nerves and autonomic nerves.

Using several organs as examples, compare the effects of sympathetic and parasympathetic stimulation.

The vertebrate peripheral nervous system relays messages to and from the central nervous system. In humans, it consists of 12 cranial nerves that connect to the brain, and 31 spinal nerves that connect to the spinal cord. Each of these nerves consists of an outer layer of connective tissue surrounding bundles of axons (**Figure 29.11**). Most cranial nerves, and all spinal nerves, carry signals both to and from the central nervous system.

There are two functional divisions of the peripheral nervous system: the somatic nervous system and the autonomic nervous system (**Table 29.1**).

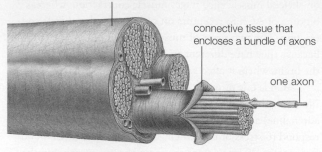

outer connective tissue sheath around the nerve

connective tissue that encloses a bundle of axons

one axon

Figure 29.11 Structure of a peripheral nerve.

SOMATIC NERVOUS SYSTEM

The **somatic nervous system** controls skeletal muscles and informs the central nervous system (CNS) about external conditions and the body's position. This system allows you to make voluntary movements and feel conscious sensations, as when someone touches you.

The somatic system also plays a role in some reflexes. A **reflex** is an automatic response to a stimulus, a movement or other action that does not require conscious thought. The stretch reflex is an example (**Figure 29.12**). Suppose you hold a bowl as someone drops fruit into it ❶. The added weight stretches the biceps muscle in your upper arm. Receptor endings of a somatic sensory neuron wrap around some cells of the biceps, and lengthening of the muscle excites this neuron ❷. As a result, an action potential travels along the sensory neuron's axon to the spinal cord ❸. In the spinal cord, the axon of the sensory neuron synapses with one of the somatic motor neurons that controls the biceps ❹. Excited by neurotransmitter from the sensory neuron, the motor neuron undergoes an action potential that travels along its axon to the biceps ❺. At a synapse with the biceps, the motor neuron's axon terminals release ACh. The biceps responds to the ACh by contracting, so the arm steadies against the additional weight ❻.

As this example illustrates, the somatic nervous system requires only one neuron to carry a signal to the CNS and one to carry a signal from it. In this system, there are no synapses outside the CNS.

Figure 29.12 Stretch reflex. This reflex involves sensory and motor neurons of the somatic division of the peripheral nervous system.

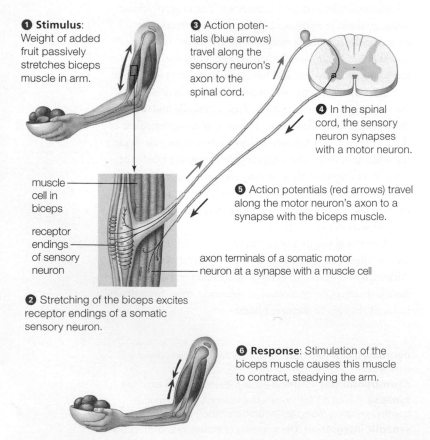

❶ **Stimulus**: Weight of added fruit passively stretches biceps muscle in arm.

❸ Action potentials (blue arrows) travel along the sensory neuron's axon to the spinal cord.

❹ In the spinal cord, the sensory neuron synapses with a motor neuron.

muscle cell in biceps

receptor endings of sensory neuron

❺ Action potentials (red arrows) travel along the motor neuron's axon to a synapse with the biceps muscle.

axon terminals of a somatic motor neuron at a synapse with a muscle cell

❷ Stretching of the biceps excites receptor endings of a somatic sensory neuron.

❻ **Response**: Stimulation of the biceps muscle causes this muscle to contract, steadying the arm.

autonomic nervous system Division of the peripheral nervous system that relays signals to and from internal organs and glands.

parasympathetic neurons Neurons of the autonomic system that encourage digestion and tasks that maintain the body.

reflex Automatic response that occurs without conscious thought.

somatic nervous system Division of the peripheral nervous system that controls skeletal muscles and relays sensory signals about movements and external conditions.

sympathetic neurons Neurons of the autonomic system that are activated during stress and danger.

AUTONOMIC NERVOUS SYSTEM

The **autonomic nervous system** controls glands and involuntary muscle (smooth muscle and cardiac muscle), and it informs the CNS about conditions inside the body. The autonomic nervous system both informs your brain when your blood pressure drops too low and adjusts your heart rate to restore the correct blood pressure.

In contrast to the somatic nervous system, the autonomic system requires two neurons to carry a signal to the CNS or from the CNS. Consider how a signal to increase the heart rate reaches the heart. The signal travels first along the axon of an autonomic neuron. The cell body of an autonomic neuron is in the spinal cord, and its axon extends out of the spinal cord to a ganglion (a collection of neuron cell bodies) some distance away. In the ganglion, the axon synapses with a second neuron. The axon of the second neuron extends to the heart, where it synapses with a cardiac muscle cell.

There are two subdivisions of the autonomic system: sympathetic and parasympathetic. They operate antagonistically, meaning signals of one type oppose signals from the other (**Figure 29.13**). Keep in mind that most organs continually receive opposing signals from the sympathetic and parasympathetic subdivisions. Synaptic integration determines the outcome of these conflicting commands.

Resting and Digesting
Under most circumstances, signals from **parasympathetic neurons** dominate. These neurons promote housekeeping tasks, such as digestion and urine formation, bringing about what is called a rest and digest response. Parasympathetic neurons release ACh at their synapses with target organs.

Fight or Flight
Signals from **sympathetic neurons** increase in times of stress, excitement, and danger. When you are startled, frightened, or angry, signals from these neurons raise your heart rate and blood pressure, and make you sweat

Organ(s)	Parasympathetic Effects	Sympathetic Effects
Eyes	Constricts pupils	Widens pupils
Heart	Slows heart rate	Speeds heart rate
Airways	Constricts airways	Widens airways
Stomach, intestines	Increases secretions, muscle contractions	Slows secretions, muscle contractions
Bladder	Stimulates urination	Inhibits urination

Figure 29.13 Examples of effects of parasympathetic and sympathetic simulation.

FIGURE IT OUT
Which of type of stimulation reflects a fight–flight response?

Answer: Sympathetic stimulation

more and breathe faster. In what is called the fight–flight response, sympathetic signals put an individual in a state of arousal, ready to fight or make a fast getaway.

Sympathetic neurons release norepinephrine at synapses with target organs. Thus, drugs that mimic norepinephrine produce a effect similar to that of sympathetic stimulation. This is why amphetamines dilate the pupils (the dark area in the eye) and increase heart rate. On the other hand, drugs that block the effect of sympathetic neurons are used to treat some disorders. For example, drugs called beta blockers lower heart rate and blood pressure by binding to and blocking receptors for norepinephrine in blood vessels and the heart.

TABLE 29.1

Comparison of the Components of the Peripheral Nervous System

	Somatic Nervous System	Autonomic Nervous System	
Motor functions	Carries signals for voluntary or reflexive contraction from the CNS to skeletal muscle	Carries signals for involuntary contraction from the CNS to cardiac muscle and smooth muscle, and carries CNS signals for gland activity to glands	
Sensory functions	Monitors the external environment and the position of the body; relays signals from receptors in skin and near joints to the CNS	Monitors the internal environment; relays signals from receptors in the walls of internal organs to the CNS	
Signaling pathway	Single-neuron pathway	Two-neuron pathway, with a synapse between neurons	
Subdivisions	None	**Parasympathetic division** Promotes maintenance tasks (resting and digesting)	**Sympathetic division** Prepares body for intensive activity (fight–flight)

Compare the components of white matter and gray matter.

Describe how cerebrospinal fluid is produced and its functions.

Explain the importance of the blood–brain barrier.

We turn now to the vertebrate central nervous system (CNS), which consists of the spinal cord and brain. This section provides an overview of the tissues that compose and protect these organs, and the special type of fluid that bathes and fills them. Later sections detail the structure and function of the spinal cord and brain.

WHITE MATTER AND GRAY MATTER

Two visibly different types of tissue make up the bulk of both the brain and spinal cord. **White matter** consists mainly of bundles of myelin-wrapped axons (myelin is white). In the CNS, these bundles are called **tracts**, rather than nerves.

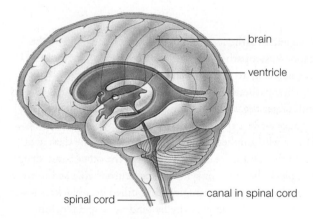

brain

ventricle

spinal cord

canal in spinal cord

Figure 29.14 Location of cerebrospinal fluid (shown in blue) in and around the brain and spinal cord.

---

blood–brain barrier Protective mechanism that prevents unwanted substances from entering cerebrospinal fluid.

cerebrospinal fluid Clear fluid that surrounds the brain and spinal cord and fills ventricles within the brain.

gray matter Central nervous system tissue that consists of neuron axon terminals, cell bodies, and dendrites, and some neuroglia.

meninges (meh-NIN-jeez) Layers of connective tissue that enclose the brain and spinal cord.

tract Bundle of axons in the central nervous system.

white matter Central nervous system tissue consisting primarily of axons with a wrapping of myelin.

Tracts relay information between different regions of the central nervous system. **Gray matter** consists of cell bodies and dendrites, along with supporting neuroglia.

CEREBROSPINAL FLUID

The brain and spinal cord are enclosed within protective bony cases (the skull and the backbone, respectively). Beneath the bone, the nervous tissue is surrounded by multiple layers of connective tissue collectively known as **meninges**.

The brain and spinal cord float in a clear liquid called **cerebrospinal fluid**. Cerebrospinal fluid also fills brain ventricles, which are large hollow spaces inside the brain (**Figure 29.14**). This fluid serves as a liquid cushion for the delicate tissues of the brain and spinal cord. Cerebrospinal fluid consists of water with ions and nutrient molecules that have been selectively transported out of blood. It is rich in oxygen and in glucose, which neurons use for aerobic respiration. Glial cells in the lining of brain ventricles are covered with cilia that keep cerebrospinal fluid circulating continuously.

A mechanism known as the **blood–brain barrier** maintains tight control over the composition and concentration of cerebrospinal fluid. The blood brain–barrier functions like a chemical filter. In the walls of brain capillaries (tiny blood vessels) epithelial cells are connected by an abundance of tight junctions. These cell junctions prevent fluid from seeping between the cells and leaking out of the capillaries. Glial cells (called astrocytes) positioned next to capillaries assist in the formation of the tight junctions and provide another layer of protection against leakage. The only way for ions or molecules in blood to enter cerebrospinal fluid is by passing through epithelial cells. Water and oxygen diffuse through these cells. Active transport proteins in the plasma membrane of these cells facilitate the necessary movement of glucose and other essential nutrients and ions.

The blood–brain barrier prevents most drugs from entering the brain. Psychoactive drugs such as alcohol (ethanol) and cocaine are notable exceptions. These substances are small and lipid soluble, so they diffuse easily across the lipid bilayers of both epithelial and glial plasma membranes. Once in the cerebrospinal fluid, they disrupt brain function by affecting synapse function, as described in Section 29.5.

The diagnostic procedure known as a lumbar puncture, or spinal tap, involves sampling and analyzing cerebrospinal fluid. A hollow needle is inserted between two segments of the backbone in the lower back so a small amount of fluid can be withdrawn from around the spinal canal. The presence of microorganisms in this fluid indicates an infection. Red blood cells indicate a torn blood vessel. The presence of breakdown products of myelin can be an indicator of multiple sclerosis.

Describe the structure of the spinal cord.

Explain why a spinal cord injury can cause permanent paralysis.

STRUCTURE AND FUNCTION

The **spinal cord** is the portion of the central nervous system that runs through the backbone and conveys signals between peripheral nerves and the brain. It also plays a role in reflexes that do not involve the brain. The stretch reflex, described in Section 29.6, is an example.

The human spinal cord is a cylinder about as thick as a thumb (**Figure 29.15**). Its outermost portion is white matter, and its interior consists of gray matter. All synapses within the spinal cord are in this gray matter. The backbone that encloses the spinal cord is a segmented structure consisting of individual bones called vertebrae. Axons of peripheral nerves that connect to the spinal cord reach the cord by way of small openings between the vertebrae. The regions where axons of peripheral nerves enter or exit the spinal cord are called nerve roots. Dorsal roots (roots toward the rear of the backbone) contain axons of sensory neurons. Ventral roots (roots near the front of the backbone) contain axons of motor neurons. Within the cord itself, the axons of motor and sensory neurons are bundled into separate tracts that run through spatially distinct areas of the cord.

ANESTHESIA AND INJURY

Physicians sometimes deliberately halt transmission of signals through the spinal cord to provide pain relief to a specific region of the body. For example, the most common way of lessening pain during childbirth is through epidural anesthesia, commonly called "an epidural." During this procedure, a doctor injects an anesthetic (painkiller) into the space between the meninges and spinal cord. The drug is administered where it will partially numb the body from the waist down, without otherwise impairing perception or interfering with motor function.

An injury to the spinal cord can result in the loss of sensation, or motor function (paralysis). Severed peripheral nerves can regrow a bit, but tracts within the spinal cord do not. As a result, spinal cord injuries usually cause permanent disability. Symptoms depend on where the cord is damaged. Tracts carrying signals to and from the upper body are higher in the cord than those that govern the lower body. An injury

spinal cord Central nervous system organ that extends through the backbone and connects peripheral nerves with the brain.

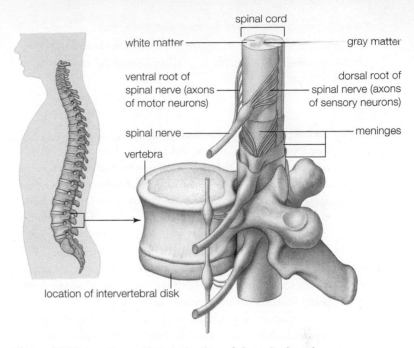

Figure 29.15 **Location and organization of the spinal cord.**

spinal cord

white matter — gray matter

ventral root of spinal nerve (axons of motor neurons) — dorsal root of spinal nerve (axons of sensory neurons)

spinal nerve — meninges

vertebra

location of intervertebral disk

to the lumbar (lower back) region of the cord often paralyzes the legs. An injury to higher regions can paralyze all limbs, as well as muscles used in breathing.

Robotic exoskeletons are a new technology that allows paralyzed individuals to stand and walk (**Figure 29.16**). These devices consist of a system of battery-powered, computer-assisted, motorized braces that move a user's hips and legs. In addition to enhancing mobility, walking with the assistance of a robotic exoskeleton provides exercise that maintains muscle mass, improves cardiovascular fitness, and lessens the risk of pressure sores (sores that often develop as a result of prolonged sitting).

Researchers also hope to restore spinal cord function through the use of stem cells. Several companies are now testing transplants of cells derived from human embryonic stem cells. So far, only a few people have received these treatments, so it is too early to say whether the treatments are safe and how much they improve function, if at all.

Figure 29.16 **Walking with aid of a robotic exoskeleton.**
Marine Captain Derek Herrera was paralyzed from the chest down after a sniper's bullet struck his spine while he was serving in Afghanistan. He now uses a robotic exoskeleton that he controls by punching buttons on his wrist. With the use of this device, Herrera can stand up, walk, and sit down.

CREDIT: (16) Joanne Rathe/The Boston Globe via Getty Images.

CHAPTER 29 NEURAL CONTROL   **511**

29.9 The Vertebrate Brain

LEARNING OBJECTIVES

Explain the embryonic origin of the spinal cord and brain.

Describe the location, function, and components of the hindbrain.

List the components of the forebrain and describe their functions.

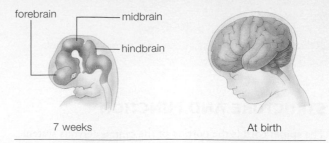

7 weeks   At birth

Figure 29.17 Development of the human brain.

As a vertebrate embryo develops, its embryonic neural tube develops into a spinal cord and a brain. The brain becomes organized as three functional regions: the forebrain, midbrain, and hindbrain (**Figure 29.17**). The hindbrain connects to the spinal cord and functions in reflexes and coordination. It is the brain region that has changed least over evolutionary time. By contrast, there has been an evolutionary trend toward a decrease in the size of the midbrain and an increase in the size of the forebrain. Relative to reptiles, birds and mammals have a smaller midbrain. Their larger forebrain has taken over many functions of the reptilian midbrain.

An adult human brain weighs about 1,400 grams, or 3 pounds. It contains about 100 billion interneurons, and neuroglia make up more than half of its volume. **Figure 29.18** shows the structure of a human brain. Other vertebrate brains have the same functional areas, although the relative size of the areas varies among species.

The hindbrain has three regions. Connected to the spinal cord is the **medulla oblongata**, which influences heartbeat and breathing. It also controls reflexes such as swallowing, coughing, vomiting, and sneezing. Just above the medulla oblongata is the **pons**, which also affects breathing. Pons means "bridge," a reference to tracts that extend through the pons to the midbrain. The peach-sized **cerebellum** at the rear of the skull controls posture, coordinates voluntary movements, and plays a role in learning new motor skills. It is densely packed with neurons, having as many as all other brain regions combined. A drunk person becomes uncoordinated because alcohol impairs cerebellar function. Police officers evaluate the extent of this impairment by asking a person

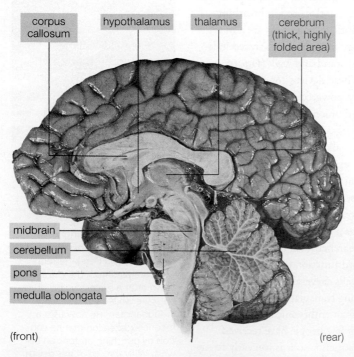

(front)   (rear)

Figure 29.18 Location and function of some major components of the human brain. Right half of the brain is shown.

Forebrain	
Cerebrum	Localizes, processes sensory inputs; initiates, controls skeletal muscle activity; governs memory, emotions, and, in humans, abstract thought
Corpus callosum	Connects the two cerebral hemispheres
Thalamus	Relays sensory signals to and from cerebral cortex; has a role in memory
Hypothalamus	With pituitary gland, functions in homeostatic control. Regulates volume, composition, temperature of internal environment; governs feelings related to homeostasis (e.g., thirst, hunger)

Midbrain	Relays signals, has role in voluntary movements

Hindbrain	
Pons	Bridges cerebrum and cerebellum; also connects spinal cord with midbrain. With the medulla oblongata, controls rate and depth of respiration
Cerebellum	Coordinates motor activity for moving limbs and maintaining posture, and for spatial orientation
Medulla oblongata	Relays signals between spinal cord and pons; functions in reflexes that affect heart rate, blood vessel diameter, and respiratory rate. Also involved in vomiting, coughing, other reflexive functions

CREDIT: (18) left, © C. Yokochi and J. Rohen, *Photographic Anatomy of the Human Body*, 2nd Ed., Igaku-Shoin, Ltd., 1979.

suspected of driving while drunk to walk a straight line. Over the long term, excessive alcohol consumption kills cells in the cerebellum, so alcoholics often walk with a shuffling gait. On the other hand, routinely practicing an athletic skill improves function of the cerebellum.

The pons, medulla, and midbrain are collectively referred to as the brain stem. In humans, the midbrain is the smallest of the three brain regions. It relays sensory information to the forebrain, acts in reward-based learning, and has a role in voluntary movement. Death of dopamine-producing neurons in part of the midbrain results in the symptoms of Parkinson's disease. Affected people first develop tremors (unintentional shaky movements). Later, all voluntary movement, including speech, may become difficult.

The forebrain is the largest part of the brain. The **cerebrum**, which makes up the bulk of the forebrain, receives sensory signals, integrates information, and initiates voluntary actions. A fissure divides the cerebrum into right and left hemispheres connected by the corpus callosum, a thick band of tissue that relays signals between them. Each hemisphere has an outer layer of gray matter called the cerebral cortex. Our large cerebral cortex is the part of the brain responsible for human capacities such as language and abstract thought. The next section discusses its functions in more detail.

The **thalamus** is a two-lobed forebrain structure that sorts sensory signals and sends them to the proper region of the cerebral cortex. It also influences sleep and wakefulness. In the rare genetic disorder called fatal familial insomnia, deterioration of the thalamus causes an inability to sleep, followed eventually by a coma, then death.

The **hypothalamus** ("under the thalamus") is the center for homeostatic control of the internal environment. It receives information about the state of the body, and regulates thirst, appetite, sex drive, and body temperature. It also interacts with the adjacent pituitary gland as a central control center for the endocrine system. We discuss endocrine functions of the hypothalamus in Chapter 31.

---

cerebellum (cer-uh-BELL-um) Hindbrain region that coordinates voluntary movements.

cerebrum (suh-REE-brum) Forebrain region that controls higher functions.

hypothalamus (hi-poe-THAL-uh-muss) Forebrain control center for homeostasis-related and endocrine functions.

medulla oblongata (meh-DOOL-uh ob-long-GAH-tah) Hindbrain region that influences breathing and controls reflexes such as coughing and vomiting.

pons (PONZ) Hindbrain region that influences breathing and serves as a bridge to the adjacent midbrain.

thalamus (THAL-ah-mus) Forebrain region that relays signals to the cerebrum; affects sleep–wake cycles.

Dr. Diana Reiss

National Geographic Explorer Diana Reiss has worked with dolphins in aquariums to learn about the nature of the intelligence of these socially complex mammals. She says, "I became fascinated with dolphins because they have such large and complex brains and are highly gregarious, yet so little was known about the nature of their communication and type of intelligence. These remarkable mammals share with us many abilities we once thought were uniquely human or that we shared only with our closest relatives, the great apes." The ratio of brain mass to body mass has been suggested to be a general estimate of the intelligence of an animal. When brain size is adjusted for body size, dolphins have the second largest brain after humans.

To investigate dolphin intelligence, Reiss tested dolphins' capacity for self-recognition. She marked animals with a dark spot and provided them with a mirror. The dolphins oriented themselves so that they could examine the mark on their body in the reflection, indicating that they understood the mirror showed their reflection, rather than another dolphin. Most children do not recognize themselves in a mirror until they are 18 to 24 months old. The only other animals that have demonstrated this capacity for self-recognition are great apes, magpies, and some elephants.

Although dolphin brains and human brains do not look much alike, the two groups show what Reiss calls cognitive convergence. Both live in social groups and have impressive mental abilities. They recognize themselves, learn by imitation, and can be taught to use a simple symbol-based keyboard to communicate with humans. Reiss notes, "Much of dolphin vocal learning is similar to what we see with young children."

Reiss's work with dolphins has made her a strong advocate for their conservation. She continues to speak out against the Japanese harvest of dolphins for use as human food and the capture of wild dolphins for aquariums and tourist attractions.

LEARNING OBJECTIVES

Describe the location of the cerebral cortex and the tissue that comprises it.

Describe some regions of the cerebral cortex that are known to specialize in specific tasks.

The human **cerebral cortex**, the outermost portion of the cerebrum, is a 2-millimeter-thick layer of gray matter. Over the course of human evolution, this layer has become increasingly folded. Folds allowed the evolutionary addition of gray matter while minimizing the increase in brain volume; a larger brain requires a larger skull, which can pose problems during childbirth. Prominent folds in the cortex are used as landmarks to define the lobes of each cerebral hemisphere (**Figure 29.19**).

The cortex of the frontal lobe allows us to make reasoned choices, concentrate on tasks, plan for the future, and behave appropriately in social situations. During the 1950s, more than 20,000 people had their frontal lobes damaged by frontal lobotomy, the surgical destruction of frontal lobe tissue. Lobotomy was used to treat mental illness, personality disorders, and even headaches. It made patients calmer, but blunted their emotions and impaired their ability to plan, concentrate, and behave appropriately.

Figure 29.20 The primary motor cortex. The graphic of the brain (below right) shows the location of the left hemisphere's primary motor cortex in pink. The graphic beside it shows a cross section of this region, the location of neurons that control various body regions (in teal), and the degree of control of those regions. Body parts that appear disproportionately large, such as hands, are the most finely controlled.

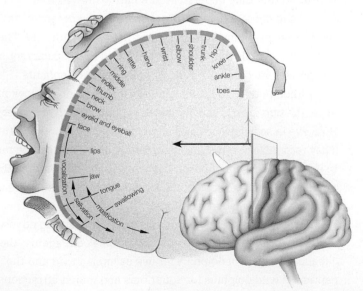

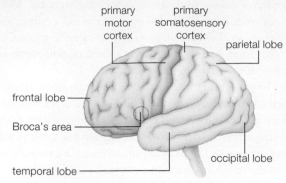

Figure 29.19 Lobes of the brain (indicated by pink, yellow, green, and blue) and some named functional regions.

The two cerebral hemispheres differ somewhat in their function. In most people who are right-handed, the left hemisphere plays the major role in movement and in language, whereas mathematical tasks, such as adding two numbers, typically activate areas of the right hemisphere. Broca's area, an area that translates thoughts into speech, is usually in the left frontal lobe. Damage to Broca's area often prevents a person from speaking, but it does not affect understanding of language. Note that, despite their differences, both hemispheres are flexible. If a stroke or injury damages one side of the brain, the other hemisphere can take on new tasks. People can even function with a single hemisphere.

Although most functions involve many brain regions, we can pinpoint regions of the cortex that have primary responsibility for certain tasks. The main region involved in voluntary movement is the **primary motor cortex** at the rear of each frontal lobe. Neurons of this region are laid out like a map of the body, with body parts capable of fine movements taking up the greatest area (**Figure 29.20**). The primary somatosensory cortex at the front of the parietal lobe receives sensory input from the skin and joints. A primary visual cortex at the rear of each occipital lobe receives signals from eyes. A primary auditory cortex in the temporal lobe receives signals regarding sounds from the ears.

The brain is "cross wired," meaning each cerebral hemisphere receives input from and sends commands to the opposite side of the body. Commands to move parts on the body's left side originate in the motor cortex of the right hemisphere, and sensory signals from the left ear end up in the right hemisphere's temporal lobe.

cerebral cortex Outer gray matter layer of the cerebrum; region responsible for most complex behavior.

primary motor cortex Region of frontal lobe that controls voluntary movement.

CREDIT: (20) left, After Penfield and Rasmussen, *The Cerebral Cortex of Man*, © 1950 Macmillan Library Reference. Renewed 1978 by Theodore Rasmussen.

29.11 Emotion and Memory

LEARNING OBJECTIVES

Give examples of how components of the limbic system affect emotions.

Explain the difference between skill memories and declarative memories.

Describe some of the evidence that the hippocampus plays an important role in formation of long-term memories.

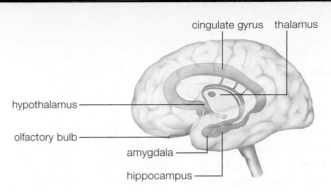

Figure 29.21 Components of the limbic system.

Most of the brain's gray matter is in the cerebral cortex. A few structures deep in the brain are also composed of gray matter. Among other functions, these structures play a role in emotion and in memory.

THE LIMBIC SYSTEM

Emotions such as sadness, fear, and fury arise in part from the forebrain structures collectively called the **limbic system** (**Figure 29.21**). Exactly how the limbic system gives rise to different emotions is poorly understood, but we do know a bit about the emotion-related functions of some of its components. For example, the hypothalamus, which sits just above the brain stem, summons up the physiological changes that accompany emotions. Signals from the hypothalamus make our heart pound and our palms sweat when we are fearful. The cingulate gyrus is an arch-shaped region of tissue above the corpus callosum (the tissue that bridges the two hemispheres). The cingulate gyrus helps us to control our emotions and to learn from negative experiences. Defects in or damage to this region can cause symptoms of obsessive–compulsive disorder. A person with this disorder has recurrent unwanted thoughts, and feels driven by fear or anxiety to repeat certain behaviors. The amygdala is a small region in the temporal lobe that becomes active when we hear or see something that could be a threat. Its activity gives rise to our emotion of fear. The amygdala is often overactive in people who suffer from panic disorders.

MAKING MEMORIES

At a cellular level, memory formation involves altering the number and placement of synaptic connections among existing neurons. The cerebral cortex receives information continually, but only a fraction of it is stored as memory.

Memory forms in stages. Short-term memory lasts seconds to hours. This type of memory holds a few bits of information: a set of numbers, words of a sentence, and so forth. In long-term memory, larger chunks of information can be stored more or less permanently. Different types of memories are stored and brought to mind by different mechanisms.

Repetition of motor tasks creates skill memories that can be highly persistent. Learn to ride a bicycle, dribble a basketball, or play the piano, and you are unlikely to lose that skill. Forming skill memories involves neurons in the cerebellum, which controls motor activity. Declarative memory stores facts and impressions. It helps you remember how a lemon smells, that a quarter is worth more than a dime, and the address of your home.

The **hippocampus** (plural, hippocampi), a structure adjacent to the amygdala, plays an essential role in the formation of declarative memories. This role was discovered in the 1950s after a patient (called HM in scientific papers) had both his hippocampi surgically removed to treat his seizures. The surgery alleviated HM's seizures, but it also destroyed his ability to form new memories. Five minutes after meeting someone, HM was unable to remember that they had ever met. He still retained memories of pre-surgery events, indicating that the hippocampus is not the site for long-term memory storage.

Additional evidence that the hippocampus plays a role in memory comes from studies of people with Alzheimer's disease. The hippocampus is one of the first brain regions destroyed by this disorder. Like HM, people in early-stage Alzheimer's usually have impaired short-term memory but retain memories of long-ago events. Alzheimer's disease is a progressive disorder, and it eventually destroys all memory, as well as the ability to think and to reason.

The hippocampus is one of the very few brain regions in which new neurons continue to arise in adulthood. It is not yet clear how the new neurons assist in memory formation. By one hypothesis, they help make room for new memories by disrupting connections among existing neurons.

hippocampus Brain region essential to formation of declarative memories.
limbic system Group of structures deep in the brain that function in expression of emotion.

LEARNING OBJECTIVES

Describe how EEGs, PET scans, and fMRIs are used to study the brain and the types of information each method provides.

Explain the function of brain banks.

Scientists can learn about brain function by studying the behavior of individuals who have suffered brain damage in a specific region as a result of stroke, accidents, or surgery. Additional information comes from techniques that allow observations of brain activity in living people, and from direct examination of brain tissue in the deceased.

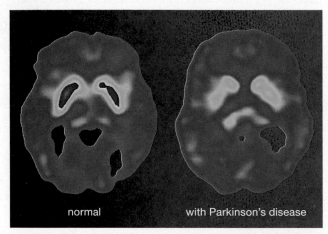

Figure 29.22 PET scan images from a normal brain and the brain from a person affected by Parkinson's disease. Red and yellow indicate areas of high glucose uptake by dopamine-secreting neurons. Death of these neurons in structures called basal nuclei gives rise to the symptoms of Parkinson's disease.

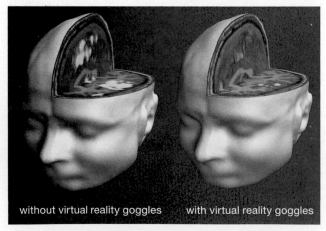

Figure 29.23 Pain-related brain activity with or without virtual reality goggles as revealed by fMRI. The goggles allow patients to enter a virtual world, distracting them as they undergo a painful procedure. Note the reduced extent of high-activity areas (red, orange, and yellow) with the goggles.

OBSERVING ELECTRICAL ACTIVITY

A technique called electroencephalography uses electrodes attached to the scalp (as shown in this chapter's opening photo) to detect voltage fluctuations in neurons of the cerebral cortex. It records the fluctuations as an electroencephalogram (EEG): a graph of the brain's electrical activity over time. EEGs provide information about a person's level of wakefulness, so they are frequently used to monitor people during sleep studies. (There are different stages of sleep, each characterized by a specific EEG pattern.) EEGs provide information about the level of neural activity across the entire cerebrum over intervals as short as a millisecond. However, they do not provide much information about the exact location of the active cells.

MONITORING METABOLISM

Other methods allow scientists to study events in specific brain regions. Positron-emission tomography (PET) reveals areas of activity in the brain by tracing uptake of radioactively labeled glucose. Glucose is the main energy source for brain cells, and active cells take up more than resting cells. **Figure 29.22** shows results of a PET study of Parkinson's disease.

Functional magnetic resonance imaging (fMRI) reveals details of brain activity by detecting changes in blood flow. Active brain cells use more oxygen than resting cells, and increased blood flow provides the necessary oxygen. **Figure 29.23** shows results of an fMRI study in which virtual reality goggles were used to reduce the patient's perception of pain.

EXAMINING BRAIN TISSUE

To study the structure, genetics, and biochemistry of human brain cells, scientists need brain tissue. "Brain banks" are repositories of such tissue. The largest brain bank, the Harvard Brain Tissue Resource Center, has thousands of donated brains, most from people who had degenerative brain disorders (such as Parkinson's disease) or psychiatric illnesses (such as schizophrenia). This federally funded center supplies tissue samples to researchers who investigate the underlying causes of these conditions. Recently, microscopic analysis of brain samples provided by the center allowed researchers to determine why schizophrenics tend to have an unusually small prefrontal cortex. The study showed that schizophrenics have the normal number of brain cells in their prefrontal cortex, but these cells are unusually tightly packed, with fewer synapses among them. Examination of donated brain tissue is also integral to the ongoing study of how physical trauma affects the brain, the topic we consider next.

CREDITS: (22) © From Neuro Via Clinical Research Program, Minneapolis VA Medical Center; (23) Image by Todd Richards and Aric Bills, U.W., copyright Hunter Hoffman, U.W.

Figure 29.24 Despite wearing helmets, football players are at a high risk for head injuries.

IMPACTS OF CONCUSSIONS

The brain is a surprisingly delicate organ that is only about as firm as Jell-O or warm butter. Although it fits quite tightly into the skull, there is a small cerebrospinal fluid–filled gap between the two. The gap allows the brain to move a bit within its bony enclosure. An impact or sudden stop can cause enough movement to tear, bruise, or stretch brain tissue. When this happens, axons twist and become torn, impairing their ability to signal. Neurotransmitters pour into the cerebrospinal fluid, causing the brain's neurons to undergo spontaneous action potentials. Tiny blood vessels tear also, compromising the blood–brain barrier and reducing blood flow to the brain.

The result is a traumatic brain injury called a concussion. Symptoms of a concussion can include confusion, dizziness, blurred vision, headache, difficulty concentrating, altered sleep patterns, and, in some severe cases, loss of consciousness.

There is no treatment for concussion, other than physical and mental rest. In most cases, symptoms disappear within about 10 days of the injury. However, repeated head injuries can cause chronic traumatic encephalopathy (CTE), a neurodegenerative disorder characterized by memory loss, depression and suicidal impulses, and eventually dementia.

Definitive diagnosis of CTE requires a brain autopsy. Upon examination, the brain of a person with CTE resembles that of someone with Alzheimer's disease. In both cases, the brain is smaller than normal, has enlarged ventricles, and accumulates deposits of a misfolded protein.

Concussions are an occupational hazard for professional football players (**Figure 29.24**), and some suffer the consequences in terms of brain health. In 2011, Dave Duerson, a 50-year-old former player for the Chicago Bears football team, shot himself in the chest. He left behind a request

that his brain be donated to a brain bank devoted to the study of sport-related brain injuries. Examination of his brain revealed the extensive degenerative brain damage characteristic of CTE. In 2015, researchers investigating the possible connection between football and CTE reported that they had examined the brains of 91 deceased football players and found signs of CTE in 87 of them.

Concerns about the long-term effects of concussions have prompted rule changes by several sports organizations. The National Hockey League, for example, has tightened its rules on head shots, and some youth soccer leagues have banned heading the ball. Sports organizations are also increasing efforts to educate coaches and athletes about symptoms of concussion. The goal is to maintain the positive aspects of participation in team sports, while minimizing the risk that participants will sustain brain-altering injuries.

Section 29.1 Most animals have a **nervous system** with **neurons** and supporting neuroglial cells. Cnidarians and echinoderms have a **nerve net**. Most bilateral lineages have undergone **cephalization**, which means they have paired **ganglia** or a **brain** at the head end. In bilateral invertebrates, ventral **nerve cords** run the length of the body and **nerves** branch from them. Vertebrates have a **central nervous system** (brain and spinal cord) and a **peripheral nervous system**.

Sections 29.2, 29.3 **Sensory neurons** detect stimuli. **Interneurons** relay signals between neurons. **Motor neurons** signal muscles and glands. A neuron's **dendrites** receive signals and its **axon** transmits signals. **Neuroglia** support and assist neurons. The voltage difference across a neuron plasma membrane is a **membrane potential**. At **resting potential** the interior of a neuron is negative relative to the interstitial fluid. An **action potential** is a brief reversal of this charge difference.

Section 29.4 An action potential occurs when a region of membrane reaches **threshold potential**, causing voltage-gated sodium channels to open. Inward flow of sodium causes more of these gated channels to open—an example of **positive feedback**. The influx of sodium makes the axon's cytoplasm more positive than the extracellular fluid. As a result, gated potassium channels open and potassium ions rush out, making the neuron cytoplasm once again more negative than the interstitial fluid. All action potentials are the same size and travel away from the cell body and toward the axon terminals. A covering of **myelin** speeds transmission along most axons.

Section 29.5 A neuron signals another cell at a **synapse**. An action potential triggers the release of **neurotransmitter** from a presynaptic cell's axon terminals. Neurotransmitter molecules diffuse across the synaptic cleft and bind to receptors on the postsynaptic cell. At **neuromuscular junctions**, a motor neuron releases acetylcholine that binds to receptors on a muscle fiber.

Different neurotransmitters have inhibitory or excitatory effects on a postsynaptic cell. The postsynaptic cell's response is determined by **synaptic integration** of signals.

Psychoactive drugs mimic neurotransmitters, or disrupt their release or uptake. Drug addiction taps into reward-based learning pathways that involve dopamine release.

Section 29.6 Nerves are bundles of axons that carry signals through the body. The peripheral nervous system has two functional divisions. The first, the **somatic nervous system**, controls voluntary movement of skeletal muscles and takes part in some **reflexes** that involve these muscles. The second, the **autonomic nervous system**, controls internal organs and glands. The autonomic system in turn has two types of func-

tionally different neurons. **Parasympathetic neurons** encourage "resting and digesting." **Sympathetic neurons** prepare the body for "fight or flight" in times of stress or danger.

Sections 29.7, 29.8 The brain and **spinal cord** consist of **white matter**, which contains myelinated axons in **tracts**, and **gray matter**, which contains cell bodies, dendrites, and axon terminals. The spinal cord and brain are enclosed by **meninges** and cushioned by **cerebrospinal fluid**. A mechanism known as the **blood–brain barrier** controls the composition of the cerebrospinal fluid.

Sections 29.9–29.12 The neural tube of a vertebrate embryo develops into the spinal cord and brain. The hindbrain connects to the spinal cord. It includes the **pons** and **medulla oblongata**, which regulate breathing and other vital tasks, and the **cerebellum**, which coordinates motor activity. The midbrain functions in movement and reward-based learning. The forebrain includes the **thalamus**, which affects wakefulness; the **hypothalamus**, a control center for homeostasis; and the **cerebrum**. The outer portion of the cerebrum, the **cerebral cortex**, governs complex functions. Its **primary motor cortex** controls voluntary movement. The cerebral cortex interacts with the **limbic system**, including the **hippocampus**, in emotions and memory.

EEGs record the electrical activity of neurons in the brain. PET scans and fMRI studies pinpoint areas of high neuron activity. Brain banks are repositories of donated tissue used for studies of brain anatomy and function.

Application Sudden jolts or blows to the head, as often occur in football, can cause brain trauma called concussion. Repeated brain trauma damages the brain, causing a condition with symptoms similar to Alzheimer's disease.

SELF-ASSESSMENT
ANSWERS IN APPENDIX VII

1. _____ relay messages from the brain and spinal cord to muscles and glands.
   - a. Motor neurons
   - c. Interneurons
   - b. Sensory neurons
   - d. Neuroglia

2. When a neuron is at rest, _____ .
   - a. it is at threshold potential
   - b. gated sodium channels are open
   - c. it has fewer sodium ions than the interstitial fluid
   - d. sodium ions cross its membrane freely

3. Action potentials occur when _____ .
   - a. potassium gates close
   - b. a stimulus pushes membrane potential to threshold
   - c. sodium–potassium pumps become active
   - d. neurotransmitter is reabsorbed

PRENATAL EFFECTS OF ECSTASY Animal studies are often used to assess effects of prenatal exposure to illicit drugs. Jack Lipton used rats to study the effect of prenatal exposure to MDMA, the active ingredient in the drug Ecstasy. He injected pregnant female rats with either MDMA or saline solution (the control) during the period of prenatal brain formation. Twenty-one days after the pups were born, Lipton tested their response to a new environment. He placed each young rat in a new cage and used a photobeam system to record how much each rat moved around before settling down. **Figure 29.25** shows his results.

1. Which rats moved most (caused the most photobeam breaks) during the first 5 minutes in a new cage: those prenatally exposed to MDMA or the controls?

2. How many photobeam breaks did MDMA-exposed rats make during their second 5 minutes in the new cage?

3. Which rats moved most in the last 5 minutes?

4. Does this study support the hypothesis that exposure to MDMA affects a developing rat's brain?

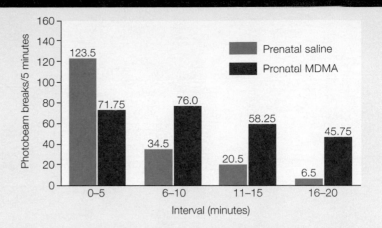

Figure 29.25 Effect of prenatal exposure to MDMA on activity levels of 21-day-old rats placed in a new cage. Movements were detected when the rat interrupted a photobeam. Individual rats were monitored for 20 minutes, and the results compiled in 5-minute intervals. Blue bars are average numbers of photobeam breaks by rats whose mothers received saline; red bars are photobeam breaks for rats whose mothers received MDMA.

4. Neurotransmitters are released by _____ .
   a. axon terminals   c. dendrites
   b. the cell body   d. the myelin sheath

5. Myelin is produced by _____ .
   a. neuroglial cells   c. sensory neurons
   b. motor neurons   d. interneurons

6. Skeletal muscles are controlled by _____ .
   a. sympathetic neurons   c. somatic motor neurons
   b. parasympathetic neurons d. somatic sensory neurons

7. When something frightens you, _____ neurons increase their output.
   a. sympathetic   b. parasympathetic

8. Damage to the _____ can disrupt short-term memory.
   a. hypothalamus   c. peripheral nerves
   b. hippocampus   d. spinal cord

9. EEGs detect _____ .
   a. sodium concentration   c. electrical activity
   b. changes in blood flow d. glucose uptake

10. _____ have a pair of ventral nerve cords.
    a. sea stars   b. vertebrates   c. insects   d. sea anemones

11. Alcohol affects coordination by its effect on the _____ .
    a. pons   c. spinal cord
    b. cerebellum d. hypothalamus

12. Myelinated axons make up the brain's _____ .
    a. white matter   c. gray matter
    b. ventricles   d. meninges

13. Match the terms with their descriptions.
    ___ thalamus   a. coordinates motor activity
    ___ dopamine   b. connects the hemispheres
    ___ limbic system   c. protects brain and spinal
    ___ corpus callosum   cord from some toxins
    ___ cerebral cortex   d. one type of neurotransmitter
    ___ cerebellum   e. support team for neurons
    ___ neuroglia   f. wrap brain and spinal cord
    ___ ganglion   g. roles in emotion, memory
    ___ blood–brain   h. most complex integration
        barrier   i. cluster of neuron cell bodies
    ___ meninges   j. regulates sleep–wake cycle

CLASS ACTIVITY
CRITICAL THINKING

1. Most tumors that originate in the brain are not cancer, but growth of any tumor within the confined space of the skull can put pressure on surrounding nervous tissue and thus harm neurons. Treating brain tumors with drugs is difficult because of the blood–brain barrier. Lipid-soluble drugs can penetrate this barrier, but they get into and affect all other cells too. Explain why.

2. Injections of botulinum toxin (Botox) into skeletal muscles of the face can prevent movements that cause facial skin to wrinkle. Botox acts by interfering with the release of a neurotransmitter. Which one?

3. Some survivors of disastrous events develop post-traumatic stress disorder (PTSD). Brain-imaging studies revealed that people with PTSD have a shrunken hippocampus and an unusually active amygdala. Given the functions of these regions, what sort of symptoms would you expect in people with PTSD?

30

Sensory Perception

Links to Earlier Concepts

This chapter draws on your understanding of sensory neurons (Section 29.2), action potentials (29.4), and reflexes (29.6). The discussion of vision touches on properties of light (6.2). There are examples of X-linked inheritance (14.3) and morphological convergence (16.7). You will also reconsider sensory changes involved in adapting to life on land (24.3).

Core Concepts

Living things sense and respond appropriately to their internal and external environments.

An animal's sensory system provides it with information about the state of its internal and external environment. Stimulation of sensory receptors causes action potentials to travel along nerves to the brain, where integration of multiple signals produces perceptions.

Adaptation of organisms to a variety of environments has resulted in diverse structures and physiological processes.

The types of sensory receptors an animal has, the range of stimuli that those receptors respond to, and the location of those receptors vary among animal groups and are adaptations to specific environments. Most animals can sense temperature, touch, tissue injury, light, and the presence of some chemicals. Some also have a capacity to sense other stimuli such as magnetic or electric fields.

Interactions among biological systems give rise to complex properties.

Sensory organs such as eyes and ears are complex structures. In addition to sensory receptors, they contain a variety of tissues that collectively ensure that the proper stimulus reaches these receptors, and that nerve signals from these receptors are sent on their way toward the brain. Input from sense organs throughout our body contributes to our perception of our world.

Wolves investigating a scent. Animals differ in their sensory abilities, and a wolf's sense of smell is more than one hundred times keener than a human's.

Photograph by Jim and Jamie Dutcher, National Geographic Creative.

LEARNING OBJECTIVES

Describe the vertebrate organs of equilibrium.

Explain the difference between dynamic and static equilibrium.

Organs of equilibrium are sensory organs that monitor the body's position and motions. They are the basis for what we commonly call our sense of balance. Many invertebrates have gravity-sensing organs called statocysts. A statocyst is a hollow, fluid-filled sphere containing one or more dense particles that sink to the lowest point in the sphere. Mechanoreceptors lining the walls of the sphere become excited when the particle or particles press on them.

Vertebrate organs of equilibrium are in the **vestibular apparatus**, a system of fluid-filled sacs and canals in the inner ear (**Figure 30.14A**). The organs of equilibrium contain mechanoreceptive hair cells somewhat similar to those that function in hearing. Changes in fluid pressure inside the various sacs and canals make the cilia of these cells bend, and the bending triggers action potentials. These action potentials travel along the vestibular nerve to the brain.

The three **semicircular canals** of the vestibular apparatus function in *dynamic* equilibrium, the special sense related to angular movement and rotation of the head. Among other things, you can use this sense to keep your eyes locked on an object when you turn your head or nod. The semicircular canals are oriented at right angles to one another, so moving the head in any direction—front/back, up/down, or left/right—moves the fluid inside at least one of them. At the bulging base of each semicircular canal is a structure called an ampulla. Inside the ampulla are hair cells whose cilia are embedded in a jellylike mass (**Figure 30.14B**). Movement of

fluid inside the canal pushes this mass, and the pressure triggers action potentials in the hair cells. The brain integrates signals received from semicircular canals in both ears.

The vestibular apparatus also contains sacs called the saccule and utricle that function in *static* equilibrium. They help the brain monitor the head's position in space when you move in a straight line. They also help keep the head upright and maintain posture. In the saccule and utricle, a jellylike mass weighted with calcium carbonate crystals known as "ear stones" lies just above hair cells (**Figure 30.14C**). Tilt your head, or start or stop moving, and this mass shifts. As it shifts, hair cells bend and alter their rate of action potentials.

A stroke, an inner ear infection, or loose particles in the semicircular canals can cause vertigo—a sensation that the world is spinning. Vertigo can also arise from conflicting sensory inputs, as when you stand at a great height and look down. The vestibular apparatus reports that you are motionless, but your eyes tell your brain that your body is floating in space. Mismatched signals also cause motion sickness. On a curvy road, passengers in a car feel changes in acceleration and direction that scream "motion" to their vestibular apparatus. At the same time, their eyes, which view the interior of the car, tell their brain that the body is at rest. Driving minimizes motion sickness because the driver focuses on sights outside the car, making visual signals consistent with vestibular ones.

---

organs of equilibrium Sensory organs that respond to body position and motion; function in sense of balance.

semicircular canals Organs of equilibrium that respond to rotation and angular movement of the head; part of the vestibular apparatus.

vestibular apparatus System of fluid-filled sacs and canals in the inner ear; contains organs of equilibrium.

Figure 30.14 Organs of equilibrium.

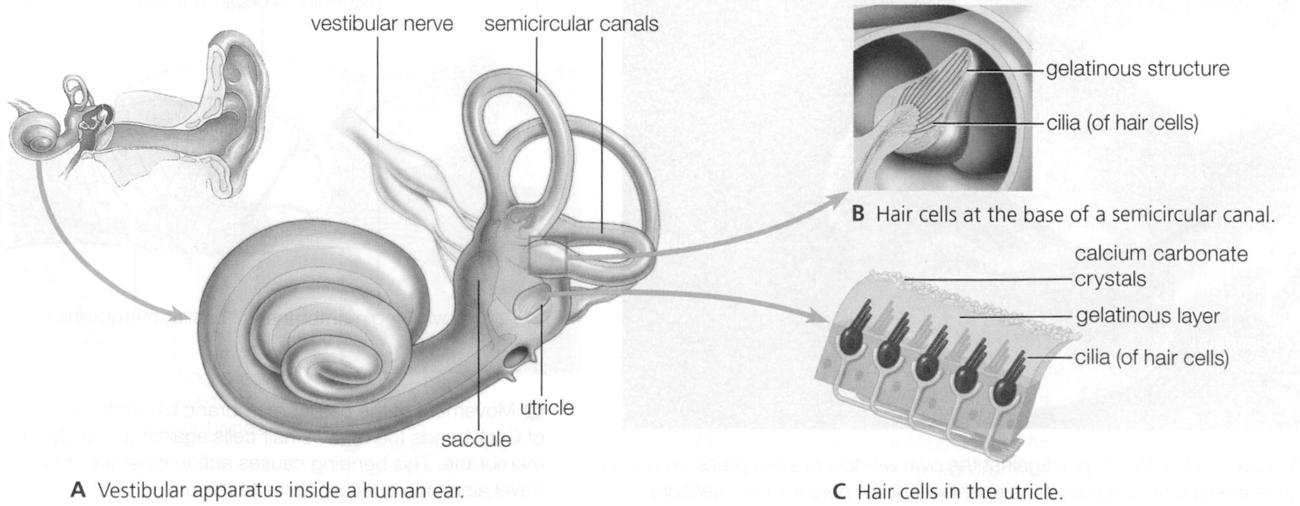

vestibular nerve   semicircular canals

gelatinous structure

cilia (of hair cells)

B Hair cells at the base of a semicircular canal.

calcium carbonate crystals

gelatinous layer

cilia (of hair cells)

utricle

saccule

A Vestibular apparatus inside a human ear.

C Hair cells in the utricle.

Figure 30.15 Retinal implant recipient Larry Hester sightseeing in New York city with his wife Jerry.

NEUROPROSTHESES

All sensory input reaches the brain in the form of action potentials, which are electrical signals. As a result, it is possible to produce sensory sensations using devices that electrically stimulate sensory nerves. Devices that assist users by manipulating their nervous system are known as neuroprostheses. (A prosthesis is a medical device that replaces or assists a missing or damaged body part.)

The most widely used neuroprostheses are cochlear implants, devices that replace the function of the inner ear in people who are deaf. Cochlear implants first became available about 30 years ago. Today, more than 350,000 people around the world benefit from these devices.

A person with a cochlear implant has a microphone outside each ear. The microphone picks up sounds and sends them to a processor that converts the sounds into electrical signals. Electrodes in the cochlear implant pick up these signals and generate action potentials in the appropriate cochlear regions. These action potentials then travel along the auditory nerves to the brain.

Retinal implants are a more recently developed neuroprosthesis. The first such device to come onto the market, the Argus II retinal prosthesis system, became available for use in the United States in 2013.

The Argus II system consists of a pair of glasses with a video camera that captures a scene and sends information to a processing unit mounted on the glasses. The processor converts the visual information to instructions that are sent wirelessly to an electrode-bearing implant on the surface of the retina. In response to the wireless signal, the electrodes in the implant electrically excite the optic nerve, triggering action potentials that travel to the brain.

Larry Hester (**Figure 30.15**) received an Argus II retinal implant in 2014. Hester had lost his sight in his early 30s as the result of an inherited retinal disorder called retinosis pigmentosa. Before Hester received the retinal implant, he had no vision at all; he perceived only blackness even in the brightest light. Hester's retinal implant cannot restore normal vision, but it allows him to perceive light and dark, so he can see the outlines of objects and detect the direction of motion. During a 2015 visit to the city of New York, the implant allowed him to marvel at the height of skyscrapers and to enjoy the display of flashing lights in Times Square.

Section 30.1 The sensory world in which an animal lives depends on the type of sensory receptors it has. **Chemoreceptors** respond to chemicals, **thermoreceptors** respond to temperature, **pain receptors** respond to tissue injury, **photoreceptors** respond to light, and **mechanoreceptors** respond to mechanical stimulation. Some animals have sensory abilities that we do not, such as an ability to detect magnetic fields.

The brain evaluates action potentials from sensory receptors based on which nerves deliver them, their frequency, and the number of axons firing. Continued stimulation of a receptor can lead to **sensory adaptation**. **Sensation** is detection of a stimulus, whereas **perception** involves assigning meaning to a sensation.

Section 30.2 The general senses arise from receptors throughout the body. **Somatic sensations** arise from receptors in the skin and in skeletal muscles. Signals from these receptors flow to the somatosensory cortex, where neurons are organized like maps of individual parts of the body surface. **Visceral sensations** originate from receptors in the walls of soft organs and tend to be diffuse. **Pain** is the perception of tissue damage. **Endorphins** are natural painkillers, and **substance P** enhances perception of pain.

Section 30.3 The special senses of taste and smell involve chemoreceptors. In humans, **taste receptors** are concentrated in taste buds of the tongue and mouth. **Olfactory receptors** line the nasal passages. Many vertebrates have a **vomeronasal organ** that responds to **pheromones**, which are chemical signals that are released and bring about changes in other members of the same species.

Sections 30.4–30.6 An eye is a sensory organ that functions in **vision**. Insects have a **compound eye**, with many individual units. Each unit has a **lens**, a structure that bends light rays so they fall on photoreceptors. Humans have **camera eyes**. The transparent **cornea** that covers the front of the human eye helps bend light rays. At the center of the colored **iris** is the **pupil**, the adjustable opening through which light enters the eye. **Visual accommodation** is the alteration of **lens** shape so light from objects at different distances falls on the **retina**. The retina's **rod cells** function in night vision. **Cone cells** provide color vision and are most abundant in the **fovea**. The optic nerve exits the retina through the **optic disk**. Common vision disorders result from defective or degenerating photoreceptors, misshapen eyes, a clouded lens, or excess aqueous humor.

Sections 30.7–30.8 Sound is a form of mechanical energy—pressure waves that vary in amplitude and frequency. Humans have a pair of ears with three functional regions. The **outer ear** collects sound waves. The eardrum and tiny bones of the **middle ear** amplify sound waves and transmit them to the **inner ear**. The inner ear includes the **vestibular apparatus** and the **cochlea**: a coiled, fluid-filled structure with three ducts. Pressure waves traveling through the fluid inside the cochlea bend hair cells embedded in one of the cochlear membranes. Sounds get sorted out according to their frequency. Mechanical energy of pressure waves is converted to action potentials that are relayed along auditory nerves to the brain.

Organs of equilibrium detect forces related to a body's position and motion. In many invertebrates, statocysts serve this purpose. Vertebrates have a **vestibular apparatus**, a system of fluid-filled sacs and canals in the inner ear. The **semicircular canals** of the vestibular apparatus respond to rotational and angular movements of the head. The saccule and utricle respond to motion in a straight line. Signals from the vestibular apparatus flow to the brain along the vestibular nerve.

Application Neuroprostheses produce sensory sensations by electrically stimulating nerves. Cochlear implants provide information about sound to the deaf, and retinal implants provide visual information to the blind.

SELF-ASSESSMENT
ANSWERS IN APPENDIX VII

1. The pain of a stomachache is an example of a _____ .
   a. somatic sensation   c. sensory adaptation
   b. visceral sensation   d. spinal reflex

2. _____ is a decrease in the response to an ongoing stimulus.
   a. Perception   c. Sensory adaptation
   b. Visual accommodation   d. Somatic sensation

3. Which is a somatic sensation?
   a. touch   b. vision   c. smell   d. hearing

4. Chemoreceptors play a role in the sense of _____ .
   a. touch   b. vision   c. smell   d. hearing

5. In the _____ , neurons are arranged like maps that correspond to different parts of the body surface.
   a. visual cortex   c. organ of Corti
   b. retina   d. somatosensory cortex

6. Mechanoreceptors in the _____ send signals to the brain about the body's position and motion.
   a. eye   b. ear   c. tongue   d. nose

7. The middle ear functions in _____ .
   a. detecting shifts in body position
   b. amplifying and transmitting sound waves
   c. sorting sound waves out by frequency

8. Substance P _____ .
   a. increases pain-related signals
   b. prevents perception of pain
   c. is the active ingredient in aspirin
   d. is a type of pain receptor

DATA ANALYSIS

CLASS ACTIVITY

OCCUPATIONAL NOISE AND HEARING LOSS Frequent exposure to noise of a particular pitch can cause death of hair cells in the part of the cochlea's coil that responds to that pitch. Many workers are at risk for such frequency-specific hearing loss because they work with or around noisy power tools. Taking precautions such as using ear plugs to reduce sound exposure is important. Noise-induced hearing loss can be prevented, but once it occurs it is irreversible. Dead hair cells are not replaced.

Figure 30.16 shows the threshold decibel levels at which sounds of different frequencies can be detected by an average 25-year-old carpenter, a 50-year-old carpenter, and a 50-year-old who has not been exposed to on-the-job noise. Sound frequencies are given in hertz (cycles per second). The more cycles per second, the higher the pitch.

1. Which sound frequency was most easily detected by all three people?

2. How loud did a 1,000-hertz sound have to be for the 50-year-old carpenter to detect it?

3. Which of the three people had the best hearing in the range of 4,000 to 6,000 hertz? Which had the worst?

4. Based on these data, would you conclude that the hearing decline in the 50-year-old carpenter was caused by age or by job-related noise exposure?

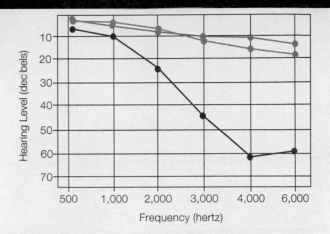

Figure 30.16 Effects of age and occupational noise exposure. The graph shows the threshold hearing capacities (in decibels) for sounds of different frequencies (given in hertz, or cycles per second) in a 25-year-old carpenter (●), a 50-year-old carpenter (●), and a 50-year-old who did not have any on-the-job noise exposure (●).

9. The organ of Corti contains receptors that signal in response to _____ .
   a. heat   b. sound   c. light   d. pheromones

10. Color vision begins with stimulation of _____ .
    a. hair cells   b. rod cells   c. cone cells   d. neuroglia

11. Visual accommodation involves adjustment to the shape of the _____ .
    a. conjunctiva   b. retina   c. optic nerve   d. lens

12. The fovea _____ .
    a. is a blind spot
    b. adjusts the amount of light entering the eye
    c. integrates signals from many rod cells
    d. has a dense concentration of cone cells

13. _____ in the vestibular apparatus function in balance.
    a. Hair cells   b. Rod cells   c. Cone cells   d. Neuroglia

14. _____ are chemical signals released by one individual that affect the behavior of another individual.
    a. Hormones   c. Pheromones
    b. Neurotransmitters   d. Endorphins

15. Match each structure with its description.
    ___ rod cell   a. protects eyeball
    ___ cochlea   b. detect head movements
    ___ lens   c. detects pheromones
    ___ sclera   d. detects dim light
    ___ cone cell   e. contains chemoreceptors
    ___ taste bud   f. focuses rays of light
    ___ semicircular   g. sorts out sound waves
        canals   h. detects color
    ___ pinna   i. collects sound waves
    ___ vomeronasal organ

CLASS ACTIVITY
CRITICAL THINKING

1. If you injure your leg, pain prevents you from putting too much weight on the affected leg. Shielding the injury gives it time to heal. An injured insect shows no such shielding response when its leg is injured. Some have cited the lack of such a response as evidence that insects do not feel pain. On the other hand, insects do produce substances similar to one of our natural painkillers. Is the presence of these compounds in insects sufficient evidence to conclude that they do feel pain?

2. Vertebrates that are nocturnal or live in the deep sea usually have a reflective layer at the back of the retina. The presence of this layer results in eyeshine; eyes appear to glow when light shines on them at night. Explain why having a reflective layer beneath the retina makes the eye more sensitive to light, but also reduces the ability to distinguish fine details.

3. Take a flash photo of a person in low light and the photo may show "red eye," in which blood vessels of the retina are visible through the pupil. A flash of light just prior to the flash used to take the picture can prevent red eye. Explain why.

4. A compound extracted from the leaves of the plant *Stevia rebaudiana* is used as a natural sugar substitute. The compound is 300 times sweeter than sugar, but it produces a slight bitter aftertaste. Given what you've learned about taste receptors, explain how a compound can be perceived as both sweet and bitter.

for additional quizzes, flashcards, and other study materials
▶ ACCESS MINDTAP AT WWW.CENGAGEBRAIN.COM

31 Endocrine Control

Links to Earlier Concepts

This chapter discusses steroids (Section 3.3) and proteins (3.4) that act as hormones. It draws on knowledge of lipid bilayers (5.6), sex determination (10.3), sympathetic nerve function (29.6), and brain anatomy (29.9). There are many examples of negative feedback mechanisms (28.8).

Core Concepts

Cells communicate by producing, transmitting, and receiving chemical signals.

Cells communicate with each other by direct contact, over short distances, and over long distances. Chemical signaling molecules of the endocrine system are used for long-distance communication. In multicelled organisms, cell-to-cell communication coordinates the activities within individual cells to support the function of the whole body.

Interactions among the components of each biological system give rise to complex properties.

Endocrine cells make and release chemicals that function as hormones. Binding of a hormone to a target cell triggers a cellular response. Signal transduction is a result of interactions between molecules that affect their structure. The nervous system interacts with the endocrine system to coordinate responses to chemical signals.

All organisms alive today are linked by lines of descent from shared ancestors.

Shared core processes and features provide evidence of shared ancestry. Common mechanisms of cell-to-cell communication offer examples. All vertebrates have the same types of endocrine glands. Some vertebrate hormones also have counterparts among the invertebrates, indicating that hormonal signaling evolved early in animal evolution.

An arctic hare sheds the white winter coat that hid it against snow to reveal a dark summer coat. The hormone melatonin governs such seasonal changes.

SIGNALS THAT TRAVEL IN BLOOD

Two organ systems—the nervous system and the endocrine system—facilitate long-distance communication among cells of an animal body. This chapter focuses on the **endocrine system**, which consists of all of the body's hormone-secreting glands and cells. Unlike a neurotransmitter (a chemical signal of the nervous system), an **animal hormone** is secreted by an endocrine cell and enters the blood. Compared to neurotransmitters, hormones travel farther, exert their effects on more cells, and persist longer in the body.

Like neurotransmitters, hormones are chemical signals that affect their target cells by a three-step process. The molecules bind to a target cell's receptor, the receptor transduces the signal (changes it into a form that affects target cell behavior), and the target cell responds:

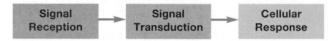

| Signal Reception | → | Signal Transduction | → | Cellular Response |

TYPES OF HORMONES

Most animal hormones are derived from amino acids. The **amino acid–derived hormones** include amine hormones (modified amino acids), peptide hormones (short chains of amino acids), and protein hormones (longer chains of amino acids). These hormones are typically polar, so they dissolve easily in blood, which is mostly water. Like other polar molecules, protein and peptide hormones cannot diffuse across a lipid bilayer. Thus, they can exert their effects only by binding to receptors at the plasma membrane of a target cell.

Binding of a hormone to its receptor in the plasma membrane results in the formation of a second messenger. The hormone is the first messenger. A **second messenger** is a molecule that forms in a target cell in response to binding of the first messenger, and in turn triggers a change in cellular activities (**Figure 31.1A**). In many cases, the second messenger sets in motion a cascade of reactions that ends in activation of an enzyme that is present in an inactive form. Consider the protein hormone glucagon. It binds to liver cells and triggers formation of cyclic AMP in their cytoplasm. The presence of cyclic AMP in turn triggers reactions that ultimately activate

an enzyme that breaks down glycogen. When activated, this enzyme breaks down stored glycogen into its component glucose subunits. As you will learn in Section 31.8, glucagon is part of a mechanism that regulates the concentration of glucose in your blood.

Some amino acid–derived hormones target endocrine cells. When receptors on these cells bind an amino acid hormone, the second messenger that forms influences the release of other hormones. A hormone that encourages secretion of a hormone by its target cell is called a **releasing hormone**. By contrast, **inhibiting hormones** discourage hormone secretion by their target endocrine cells. In still other cases, effects of second messenger formation extend into the nucleus and result in altered gene expression. Protein and peptide hormones cannot enter the nucleus to affect genes directly.

Steroid hormones are synthesized from cholesterol (Section 3.3). In humans, the gonads (testes and ovaries) and adrenal glands (glands above the kidney) are the only endocrine glands that produce steroid hormones. Being nonpolar, steroid hormones do not dissolve in water. They can travel in blood only by attaching to water-soluble plasma proteins. However, steroid hormones do dissolve in lipids, so they can diffuse directly across cell membranes. Once inside a target cell, a steroid hormone binds to a receptor protein, forming a hormone–receptor complex (**Figure 31.1B**). The receptor may be in the cytoplasm or in the nucleus. In either case, the hormone–receptor complex functions in the nucleus, where it binds to a promoter in the target cell's DNA (a promoter is a region where RNA polymerase binds, Section 9.2). Binding of a hormone–receptor complex to the promoter alters the rate of transcription of a nearby gene or set of genes. Thus, the hormone–receptor complex functions as a transcription factor (Section 10.1).

Although most effects of steroid hormones result from intracellular binding of the hormone, some steroid hormones can also bind receptors at the cell surface. As with other hor-

---

amino acid–derived hormone An amine (modified amino acid), peptide, or protein that functions as a hormone.

animal hormone Intercellular signaling molecule that is secreted by an endocrine gland or cell and travels in the blood.

endocrine system Organ system that coordinates responses among body cells by means of animal hormones.

inhibiting hormone Animal hormone that discourages secretion of another hormone by its target cells.

releasing hormone Animal hormone that encourages secretion of another hormone by its target cells.

second messenger Molecule that forms inside a cell when an animal hormone binds to a receptor in the plasma membrane; sets in motion reactions that alter the cell's activity.

steroid hormone Lipid-soluble hormone derived from cholesterol.

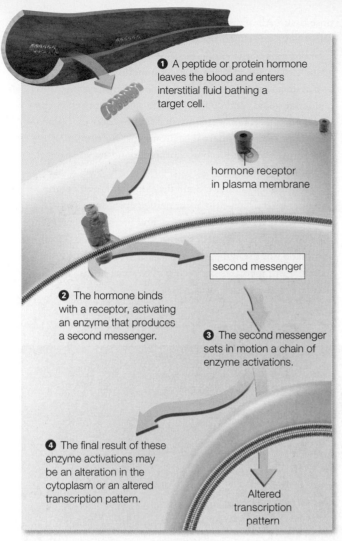

A Protein and peptide hormones act by way of a second messenger.

Figure 31.1 Mechanisms of hormone action.

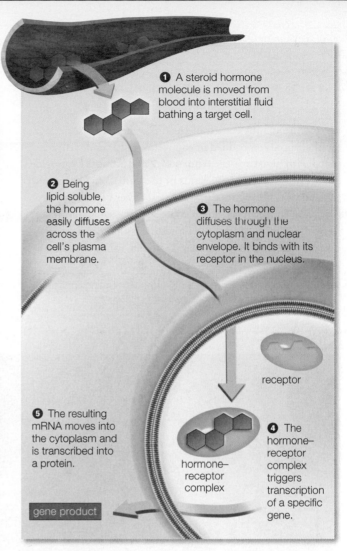

B Most steroid hormones act by entering a target cell.

mones, binding of a steroid hormone to a plasma membrane receptor results in formation of a second messenger.

The body continually secretes and breaks down hormones. Most hormones are taken up and destroyed by the kidney or the liver. Enzymes in the plasma (the fluid portion of the blood) also assist in hormone breakdown. Amino acid–derived hormones typically disappear from the blood more quickly than steroid hormones.

HORMONE RECEPTOR VARIATION

Many hormones target more than one type of cell, and they elicit a different response in each cell type. For example, ADH (antidiuretic hormone) affects urine formation when it binds to receptors on vertebrate kidney cells. It also triggers contraction when it binds to receptors on smooth muscle cells in blood vessel walls. These differing responses to ADH result from differences in the structure of ADH receptors. In each kind of cell, a different kind of receptor summons up a different response.

Mutations that alter receptor proteins can affect hormone function. For example, people with total androgen insensitivity syndrome are genetically male (XY) and make the male sex hormone testosterone, but lack functional receptors for it. In affected individuals, embryonic testes form, but do not descend into the scrotum. Genitals appear female, so these individuals are typically identified as girls at birth. Often their condition is not discovered until puberty. Being genetically male, they do not have ovaries, so they never develop breasts, or menstruate.

31.2 The Human Endocrine System

LEARNING OBJECTIVES

Give some examples of tissues or organs that include hormone-secreting cells but are not major endocrine glands.

Describe some interactions between the nervous and endocrine systems.

Endocrine glands consist of clusters of hormone-secreting epithelial cells embedded in connective tissue and well supplied with blood vessels. **Figure 31.2** and **Table 31.1** provide an overview of the main glands of the human endocrine system and the hormones they produce. Other vertebrates have the same endocrine glands and produce the same hormones. Some cells in the heart, small intestine, and other internal organs also produce hormones, as do fat-storing cells of adi-

pose tissue. All of an animal's hormone-secreting glands and cells collectively constitute the animal's endocrine system.

Portions of the endocrine system and nervous system are so closely linked that scientists sometimes refer to the two systems collectively as the neuroendocrine system. Both endocrine glands and neurons receive signals from the hypothalamus, a forebrain structure that is the center for homeostatic control of the internal environment (Section 29.9). Most organs respond to both hormones and signals from the nervous system.

Hormones influence brain development, both before and after birth. They also affect nervous processes such as sleep–wake cycles, emotion, mood, and memory. Conversely, the nervous system regulates hormone secretion. For example, stimulation of the sympathetic nervous system (as by a stressful situation) increases the secretion of some hormones and decreases the secretion of others.

---

endocrine gland Aggregation of hormone-secreting epithelial cells with associated connective tissue and blood vessels.

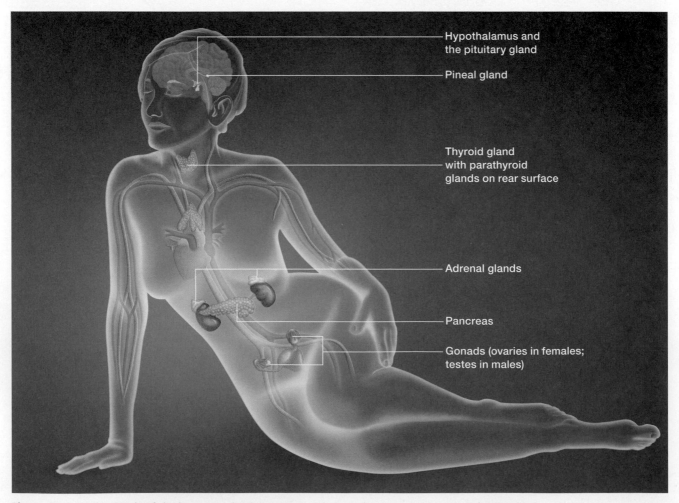

Figure 31.2 Major glands of the human endocrine system. Other vertebrates have the same endocrine glands.

Hypothalamus and the pituitary gland

Pineal gland

Thyroid gland with parathyroid glands on rear surface

Adrenal glands

Pancreas

Gonads (ovaries in females; testes in males)

TABLE 31.1

Examples of Human Endocrine Glands, Their Hormones, and Hormone Actions

Gland	Hormone(s)	Main Target(s)	Primary Actions
Hypothalamus	Releasing and inhibiting hormones	Anterior pituitary	Encourage or discourage release of other hormones
Posterior pituitary gland	Antidiuretic hormone (ADH; vasopressin)	Kidneys	Induces water conservation as required to control extracellular fluid volume and solute concentrations
	Oxytocin	Mammary glands	Induces milk movement into secretory ducts
		Reproductive tract	Induces muscle contractions during childbirth and orgasm
Anterior pituitary gland	Adrenocorticotropic hormone (ACTH)	Adrenal cortex	Stimulates release of cortisol
	Thyroid-stimulating hormone (TSH)	Thyroid gland	Stimulates release of thyroid hormone
	Follicle-stimulating hormone (FSH)	Ovaries, testes	In females, stimulates estrogen secretion, egg maturation; in males, helps stimulate sperm formation
	Luteinizing hormone (LH)	Ovaries, testes	In females, stimulates progesterone secretion, ovulation, corpus luteum formation; in males, stimulates testosterone secretion, sperm release
	Prolactin (PRL)	Mammary glands	Stimulates and sustains milk production
	Growth hormone (GH)	Most cells	Promotes growth in young; induces protein synthesis, cell division in adults
Thyroid gland	Thyroid hormone	Most cells	Regulates metabolism; has roles in growth, development
	Calcitonin	Bone	Lowers calcium level in blood
Parathyroid glands	Parathyroid hormone	Bone, kidney	Elevates calcium level in blood
Pineal gland	Melatonin	Brain	Affects sleep–wake cycles, seasonal changes
Adrenal glands			
Adrenal cortex	Cortisol*	Most cells	Promotes breakdown of glycogen, fats, and proteins as energy sources; raises blood level of glucose
	Aldosterone*	Kidney	Promotes sodium reabsorption (sodium conservation); helps control the body's salt–water balance
Adrenal medulla	Epinephrine (adrenaline) and norepinephrine	Most cells	Promote fight–flight response
Pancreas	Insulin	Liver, muscle, adipose tissue	Promotes cell uptake of glucose; lowers glucose level in blood
	Glucagon	Liver	Promotes glycogen breakdown; raises blood glucose
Gonads			
Testes (in males)	Testosterone*	Sex organs	Required in sperm formation, development of genitals, maintenance of sexual traits, growth, and development
Ovaries (in females)	Estrogens*	Uterus, breasts	Required for egg maturation and release; preparation of uterine lining for pregnancy and its maintenance in pregnancy; genital development
	Progesterone*	Uterus	Prepares, maintains uterine lining for pregnancy; stimulates development of breast tissues

* Denotes steroid hormones; all other hormones are derived from amino acids.

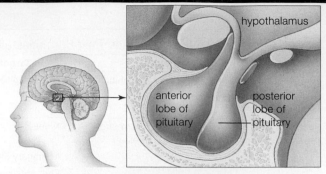

Figure 31.3 The two lobes of the pituitary gland.

The **hypothalamus** functions as the main center for control of the internal environment. It lies deep inside the forebrain and connects, structurally and functionally, with the **pituitary gland**. In humans, the pea-sized pituitary has two lobes (**Figure 31.3**). The posterior lobe releases hormones produced by neurosecretory cells in the hypothalamus. A **neurosecretory cell** is a special type of neuron that responds to an action potential by releasing a hormone into the blood. The anterior lobe of the pituitary makes its own hormones, and releases them in response to hormones produced by the hypothalamus. Together, the hypothalamus and pituitary regulate many other endocrine glands.

POSTERIOR PITUITARY FUNCTION

Figure 31.4A illustrates how the hypothalamus and the posterior lobe of the pituitary gland interact to produce and secrete a hormone. Cell bodies of neurosecretory cells in the hypothalamus synthesize peptide hormones. The hormones are then transported along the cells' axons to axon terminals in the posterior pituitary. An action potential causes these axon terminals to release hormones that then enter the blood.

Figure 31.4 Pituitary function.

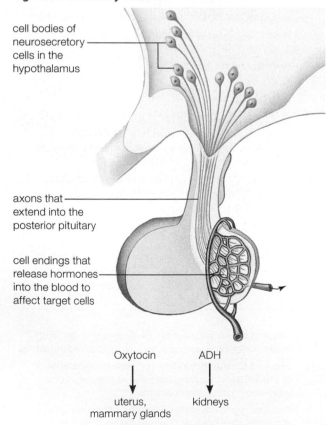

A Posterior pituitary function. Axon endings in the posterior pituitary secrete two hormones made by cell bodies in the hypothalamus.

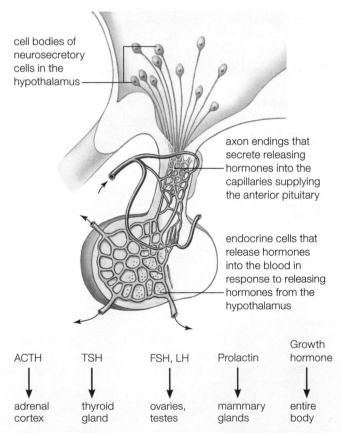

B Anterior pituitary function. Endocrine cells in the anterior pituitary produce six hormones and secrete them in response to hypothalamic releasing hormones.

The posterior pituitary releases antidiuretic hormone (ADH) and oxytocin, two peptide hormone that are similar in structure. **Antidiuretic hormone** (also known as vasopressin) causes kidneys to retain more water in the body and raises blood pressure by constricting blood vessels. (Section 37.3 details ADH function.) **Oxytocin** targets cells of smooth muscle in the reproductive tract and mammary glands (milk-secreting exocrine glands). It causes rhythmic contractions of muscles in the reproductive tract during orgasm (sexual climax), contractions of the uterus (womb) during childbirth, and contractions that move milk into milk ducts when a woman nurses a child. Chapter 38 discusses the functions of oxytocin related to reproduction. Both antidiuretic hormone and oxytocin also act in the brain, where they affect social bonds, a topic we consider in Chapter 39.

ANTERIOR PITUITARY FUNCTION

The anterior pituitary makes peptide hormones and secretes them in response to hormones from the hypothalamus (**Figure 31.4B**). Most hypothalamic hormones that act on the anterior pituitary are releasing hormones. However, the hypothalamus also produces inhibiting hormones.

Some anterior pituitary hormones act on endocrine glands. Section 31.4 discusses how **thyroid-stimulating hormone (TSH)** causes the thyroid gland to secrete thyroid hormone. Section 31.6 details how **adrenocorticotropic hormone (ACTH)** stimulates the release of cortisol by adrenal glands. **Follicle-stimulating hormone (FSH)** and **luteinizing hormone (LH)** affect sex hormone secretion and gamete production by gonads. We return to the function of FSH and LH in Section 31.7 and in Chapter 38.

The anterior pituitary also produces two hormones that do not act on endocrine glands. **Prolactin** acts on mammary glands and stimulates milk production after a woman gives birth. **Growth hormone (GH)** has target cells throughout the body. It affects metabolism, causing a decrease in fat storage and an increase in muscle protein synthesis. It also encourages production of new bone and cartilage.

HORMONAL GROWTH DISORDERS

GH production normally surges during teenage years (causing a growth spurt), then declines with age. Oversecretion of growth hormone during childhood leads to pituitary gigantism. A person affected by this disorder has a normal body form but is unusually tall. When excessive growth hormone secretion continues into or begins during adulthood, the result is acromegaly. With this disorder, continued deposition of new bone and cartilage enlarges and eventually deforms

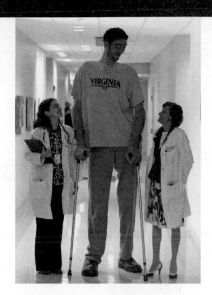

Figure 31.5 Sultan Kosen, the tallest living man, with members of his medical team at the University of Virginia. Kosen had a pituitary tumor that caused excessive growth hormone secretion. Treatment successfully halted his growth at 8 feet, 3 inches (2.5 meters).

the hands, feet, and face. The skin thickens, and the lips and tongue increase in size. Internal organs are also affected; the heart may become enlarged. Untreated, acromegaly can cause serious health problems. The most common cause of gigantism and acromegaly is a pituitary tumor (**Figure 31.5**).

Too little growth hormone during childhood can cause pituitary dwarfism. Adults affected by this disorder are normally shaped, but small. Injections of recombinant human growth hormone (rhGH) increase the growth rate of children who have a naturally low level of this hormone. Recombinant growth hormone is also used by athletes who want to encourage muscle growth. Administering hormones to enhance athletic ability is prohibited in all sports, but it continues.

adrenocorticotropic hormone (ACTH) Anterior pituitary hormone that stimulates cortisol secretion by the adrenal glands.
antidiuretic hormone (an-tie-dye-er-ET-ick) Posterior pituitary hormone that affects urine formation and blood pressure.
follicle-stimulating hormone (FSH) Anterior pituitary hormone that acts on the gonads to stimulate secretion of sex hormones.
growth hormone (GH) Anterior pituitary hormone that regulates growth and metabolism.
hypothalamus Forebrain region that controls processes related to homeostasis; produces and controls release of some hormones.
luteinizing hormone (LH) Anterior pituitary hormone that acts on gonads to stimulate secretion of sex hormones.
neurosecretory cell Specialized neuron that secretes a hormone into the blood in response to an action potential.
oxytocin (ox-ee-TOE-sin) Posterior pituitary hormone that acts on smooth muscle of the reproductive tract and breasts.
pituitary gland (pit-OO-it-airy) Pea-sized endocrine gland in the forebrain; interacts closely with the adjacent hypothalamus.
prolactin (pro-LAC-tin) Anterior pituitary hormone that stimulates milk production by mammary glands.
thyroid-stimulating hormone (TSH) Anterior pituitary hormone; stimulates secretion of thyroid hormone by the thyroid.

Describe the location and functions of the human thyroid gland.

Explain the role of the human parathyroid glands in maintaining the calcium ion concentration in the blood.

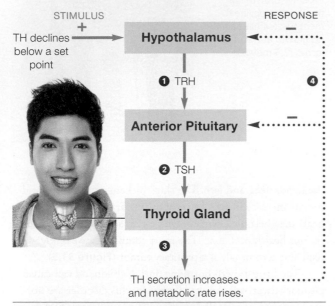

Figure 31.6 Control of thyroid hormone (TH) secretion.

FIGURE IT OUT

What effect does a high level of TH have on TSH secretion?

Answer: It lowers TSH secretion

METABOLIC AND DEVELOPMENTAL EFFECTS OF THYROID HORMONE

The **thyroid gland** at the base of the neck secretes two iodine-containing amines (triiodothyronine and thyroxine) collectively called **thyroid hormone (TH)**. This hormone increases metabolic activity of cells throughout the body.

The hypothalamus and anterior pituitary gland regulate thyroid hormone secretion by a negative feedback loop (**Figure 31.6**). A low blood concentration of thyroid hormone stimulates the hypothalamus to secrete thyroid-releasing hormone (TRH) ❶. This releasing hormone acts on cells of the anterior pituitary, causing them to secrete thyroid-stimulating hormone (TSH) ❷. TSH in turn acts on cells of the thyroid gland, causing them to secrete thyroid hormone ❸. When the blood level of thyroid hormone rises, the hypothalamus stops secreting TRH. As the TRH level declines, the secretion of TSH by the anterior pituitary slows ❹.

A deficiency in thyroid hormone, a condition called hypothyroidism, results in a reduced metabolic rate. Symptoms of hypothyroidism include fatigue, depression, increased sensitivity to cold, and weight gain. Hypothyroidism can arise from a TSH deficiency, a dietary deficiency in the iodine required to produce thyroid hormone, or an immune disorder in which the body mistakenly attacks the thyroid. By contrast, an excess of thyroid hormone (hyperthyroidism) causes nervousness and irritability, chronic fever, and weight loss. In some people, it also induces tissues behind the eyeball to swell, so the eyes bulge. Hyperthyroidism usually arises as a

Figure 31.7 Frog tadpole undergoing metamorphosis.

Thyroid hormone is necessary for amphibian metamorphosis, so frog larvae (tadpoles) can be used to test chemicals suspected of affecting thyroid function. Tadpoles are exposed to a chemical, then observed for any abnormalities as they develop. Such tests have shown that some widely used chemicals interfere with normal thyroid function.

result of an immune disorder or a thyroid tumor. Both hyperthyroidism and hypothyroidism can cause goiter, which is an abnormal enlargement of the thyroid gland.

Thyroid hormone has important developmental effects. In humans, maternal hypothyroidism during pregnancy can cause cretinism, a syndrome of stunted growth and impaired mental capacity. Cretinism can also be an outcome of hypothyroidism during infancy or early childhood. In frogs, thyroid hormone is essential for metamorphosis of the larva (a tadpole) to an adult (**Figure 31.7**). Tadpoles exposed to thyroid-disrupting chemicals develop abnormally, so frogs can be used to test substances suspected of interfering with thyroid function. For example, such tests showed that thyroid function is disrupted by perfluorooctane sulfonate, a chemical used as a stain repellent for fabrics and carpets.

HORMONAL REGULATION OF BLOOD CALCIUM LEVEL

Humans have four **parathyroid glands**, each about the size of a grain of rice, on the thyroid's posterior surface. These glands produce **parathyroid hormone (PTH)**, which is the main regulator of calcium ion concentration in the blood. When the blood's calcium ion concentration declines, PTH secretion increases. PTH increases the blood calcium level by causing the release of calcium ions from bone, increasing calcium ion reabsorption by kidneys, and increasing activation of vitamin D, which helps the intestine absorb calcium from food. Because parathyroid hormone encourages release of calcium from bone, a tumor or other disorder that increases parathyroid secretion can cause osteoporosis. In this disorder, bones lose calcium and become weak and easily broken.

Calcitonin, a hormone produced by the thyroid, opposes the effect of parathyroid hormone by encouraging bones to take up and incorporate calcium. In many animals, calcitonin plays an important role in calcium homeostasis. In humans, however, calcitonin secretion occurs mainly during childhood, and adult blood calcium is regulated primarily by the parathyroids. Calcitonin supplements are sometimes used as a treatment for osteoporosis.

---

calcitonin (cal-sih-TOE-nin) Thyroid hormone that encourages bones to take up and incorporate calcium.

parathyroid glands Four small endocrine glands on the rear of the thyroid that secrete parathyroid hormone.

parathyroid hormone (PTH) Hormone that regulates the concentration of calcium ions in the blood.

thyroid gland Endocrine gland in the base of the neck that secretes thyroid hormone and calcitonin.

thyroid hormone (TH) Iodine-containing hormones that collectively increase metabolic rate and play a role in development.

31.5 The Pineal Gland

LEARNING OBJECTIVES

Describe the location of the human pineal gland and name the hormone it secretes.

Explain the effect of light on pineal gland function.

Like the hypothalamus and pituitary, the pea-sized **pineal gland** lies deep inside the brain (**Figure 31.8**). This pinecone-shaped gland secretes the hormone **melatonin**, but only under low-light or dark conditions. Melatonin secretion slows when the retina detects light and action potentials flow along the optic nerve to the brain. The amount of light varies with time of day, and day length varies seasonally, so melatonin secretion rises and falls in daily and seasonal cycles.

Melatonin helps regulate human sleep–wake cycles. Under natural lighting, melatonin secretion increases each day after darkness falls. The increase triggers drowsiness and a drop in body temperature. Exposure to bright light at night can cause insomnia by altering melatonin secretion patterns. Melatonin supplements are sometimes taken as a sleep aid or to help minimize jet lag. This condition arises when the internal clock a person of a person traveling across multiple time zones does not match up with the new local time.

In many animals, changes in melatonin secretion patterns trigger seasonal adjustments in behavior or appearance. For example, changes in the amount of melatonin secreted determine when an arctic hare switches between its white winter coat and its darker summer coat, as shown in the photo that opens this chapter.

Figure 31.8 Location of the human pineal gland. This gland releases melatonin when the retina is not being stimulated by light. Typically, melatonin secretion peaks at about 2 A.M., during sleep.

pineal gland

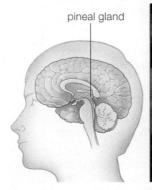

---

melatonin (mell-uh-TOE-nin) Pineal gland hormone that regulates sleep–wake cycles and seasonal changes.

pineal gland (pie-KNEE-ul) Endocrine gland in the brain that secretes melatonin under low-light or dark conditions.

31.6 The Adrenal Glands

There are two **adrenal glands**, one above each kidney (the Latin *ad–* means near, and *renal* refers to the kidney). Each is about the size of walnut, and has two functional regions: an outer cortex and an inner medulla.

The **adrenal cortex** releases steroid hormones. One of these hormones, aldosterone, acts in the kidneys and makes urine more concentrated. Section 37.3 explains this process in detail. The adrenal cortex also makes and secretes small amounts of the sex hormones, which we discuss in Section 31.7. Here we will focus on **cortisol**, an adrenal cortex hormone that affects metabolism and immune responses.

A negative feedback mechanism controls cortisol secretion (**Figure 31.9**). A decrease in cortisol triggers secretion of CRH (corticotropin-releasing hormone) by the hypothalamus ❶. CRH then stimulates secretion of ACTH (adrenocorticotropic hormone) by the anterior pituitary ❷. ACTH triggers cortisol release from the adrenal cortex ❸. The resulting rise in the level of cortisol results in a slowdown in the secretion of CRH and ACTH ❹.

Cortisol helps maintain the concentration of glucose in blood by inducing liver cells to break down their store of glycogen, and also by suppressing the uptake of glucose by most cells. Cortisol also induces adipose cells to break down fats, and skeletal muscles to break down proteins. Molecules produced by the breakdown of fats and proteins can serve as fuel for aerobic respiration (Section 7.6).

The **adrenal medulla** contains neurosecretory cells that respond to stimulation by releasing norepinephrine and epinephrine into the blood. The release has the same effect on target organs as stimulation by sympathetic nerves: It brings on a fight–flight response (Section 29.6).

HORMONES, STRESS, AND HEALTH

With injury, illness, or anxiety, the nervous system overrides the feedback loop governing cortisol, and the level of cortisol in the blood rises. This stress response helps the body deal with an immediate threat by diverting resources from maintenance tasks to support a state of arousal. A stress response is adaptive for short periods of time, as when an animal is fleeing from a predator. However, long-term elevation of cortisol (as by chronic stress) is unhealthy because it interferes with immunity, memory, and sexual function. It also raises the risk of cardiovascular problems. Similar effects occur with Cushing's syndrome, in which a chronically high level of cortisol results from an adrenal gland tumor, pituitary tumor that causes excess secretion of ACTH, or the ongoing use of the drug cortisone (cortisone is often prescribed to relieve chronic inflammation; the body converts it to cortisol).

An abnormally low level of cortisol, as occurs with Addison's disease, can cause health problems too. Symptoms include fatigue, depression, weight loss, and darkening of the skin. If the cortisol level falls too low, blood sugar and blood pressure can decline to life-threatening levels. Addison's disease can be treated with synthetic cortisone.

Figure 31.9 Control of cortisol secretion.

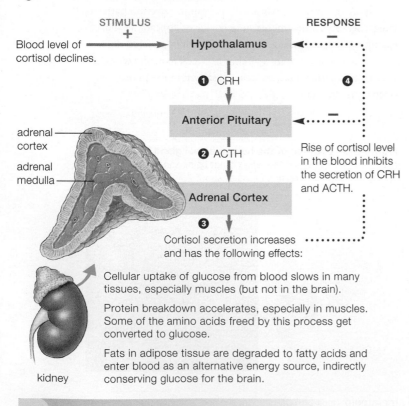

STIMULUS +

RESPONSE −

Blood level of cortisol declines. → **Hypothalamus**

❶ CRH

❹

Anterior Pituitary

❷ ACTH

adrenal cortex

adrenal medulla

Rise of cortisol level in the blood inhibits the secretion of CRH and ACTH.

Adrenal Cortex

❸

Cortisol secretion increases and has the following effects:

Cellular uptake of glucose from blood slows in many tissues, especially muscles (but not in the brain).

Protein breakdown accelerates, especially in muscles. Some of the amino acids freed by this process get converted to glucose.

Fats in adipose tissue are degraded to fatty acids and enter blood as an alternative energy source, indirectly conserving glucose for the brain.

kidney

FIGURE IT OUT

What effect does cortisol have on proteins in muscle?

Answer: It increases the breakdown of proteins.

adrenal cortex (uh-DREE-null) Outer portion of adrenal gland; secretes aldosterone and cortisol.

adrenal gland Endocrine gland that is located atop the kidney.

adrenal medulla Inner portion of adrenal gland; secretes epinephrine and norepinephrine.

cortisol (CORT-is-all) Adrenal cortex hormone that influences metabolism and immunity; secretions rise with stress.

31.7 The Gonads

List the main sex hormones made by males and by females and the organs that produce those hormones.

Give examples of secondary sexual characteristics.

Vertebrate **gonads** are primary reproductive organs, meaning they produce gametes (eggs or sperm). They also produce **sex hormones**, steroid hormones that control sexual development and reproduction. Sex hormones also affect **secondary sexual traits**—traits that differ between the sexes and arise as an individual becomes sexually mature. In lions, for example, sexually mature males are larger than females and have a mane that females lack (**Figure 31.10**).

Vertebrate gonads are the testes and the ovaries (**Figure 31.11**). Ovaries and testes produce the same sex hormones, but in very different proportions. A male's gonads—his testes (singular, testis)—secrete mainly testosterone. During embryonic development, **testosterone** causes development of male genitals. During sexual maturation, testosterone triggers sperm formation and development of secondary sexual characteristics. In humans, these traits include facial hair and an enlarged larynx (voice box) that lowers the voice.

Figure 31.10 Secondary sexual traits. Testosterone secreted by testes stimulates growth of a male lion's mane (right). A lioness (left) makes little testosterone, and does not have a mane.

Figure 31.11 Human gonads.

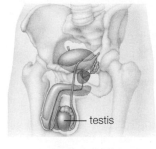
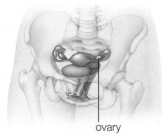

testis

ovary

Figure 31.12 Control of sex hormone secretion.

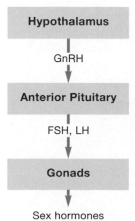

Hypothalamus

GnRH

The hypothalamus produces gonadotropin-releasing hormone that acts on the anterior pituitary.

Anterior Pituitary

FSH, LH

The anterior pituitary produces follicle-stimulating hormone and luteinizing hormone. Both hormones target cells in a male's testes and a female's ovaries.

Gonads

Sex hormones

The testes or ovaries produce sex hormones.

Ovaries (female gonads) produce mainly estrogens and progesterone. **Estrogens**, the primary female sex hormones, are responsible for maturation and maintenance of sex organs and for female sexual secondary traits. In humans, these traits include enlarged breasts and deposition of fat around the hips. **Progesterone** prepares the body for pregnancy and maintains the uterus if a pregnancy does occur.

The hypothalamus and anterior pituitary control sex hormone secretion (**Figure 31.12**). In both sexes, the hypothalamus produces gonadotropin-releasing hormone (GnRH). This releasing hormone causes the anterior pituitary to secrete follicle-stimulating hormone (FSH) and luteinizing hormone (LH), which target the gonads and cause them to produce and secrete sex hormones. FSH and LH are sometimes referred to as "gonadotropins." The suffix *–tropin* refers to a substance that has a stimulating effect. In this case, the effect is on the gonads.

We return to the topics of sex organs and sex hormones again when we discuss reproduction in Chapter 38.

estrogens Sex hormones that function in reproduction and cause development of female secondary sexual characteristics.

gonads Primary sex organ; produce gametes and sex hormones.

progesterone Sex hormone that prepares a female body for pregnancy and helps maintain a pregnancy.

secondary sexual traits Characteristics that differ between the sexes but do not play a direct role in reproduction.

sex hormone Steroid hormone produced by gonads; functions in reproduction and may affect secondary sexual characteristics.

testosterone Sex hormone responsible for development of male sex organs and secondary sexual characteristics.

LEARNING OBJECTIVES

Describe how pancreatic hormones regulate the concentration of sugar in the blood.

Compare the two types of diabetes.

The **pancreas**, located just behind the stomach (**Figure 31.13**), has exocrine and endocrine functions. Its exocrine cells secrete digestive enzymes into the small intestine. Its endocrine cells are grouped in clusters called pancreatic islets.

REGULATION OF BLOOD SUGAR

Beta cells, the most abundant cells in the pancreatic islets, secrete **insulin**—a hormone that causes its target cells to take up and store glucose. After a meal, a rise in the concentration of glucose in the blood stimulates beta cells to release insulin. Insulin's main targets are liver, fat, and skeletal muscle cells. Insulin triggers glucose uptake. In all target cells, it encour-

ages synthesis of fats and proteins and inhibits their breakdown. In liver and muscle it stimulates synthesis of glycogen. These cellular responses to insulin lower the level of glucose in the blood (**Figure 31.13 ❶ – ❺**).

Pancreatic islets also contain alpha cells. These endocrine cells secrete the peptide hormone **glucagon** when the concentration of glucose falls below a set point. The released glucagon binds to its receptors on liver cells, which respond by activating enzymes that break glycogen into glucose subunits. This response to glucagon raises the level of glucose in blood (**Figure 31.13 ❻ – ❿**). By working in opposition, glucagon and insulin secreted by the pancreas maintain the blood glucose level within a range that keeps cells functioning properly.

---

glucagon (GLUE-kah-gon) Pancreatic hormone that encourages glycogen breakdown; increases blood glucose level.

insulin Pancreatic hormone that encourages uptake of glucose by target cells; lowers level of glucose in the blood.

pancreas (PAN-kree-us) Organ that secretes digestive enzymes into the small intestine, and insulin and glucagon into the blood.

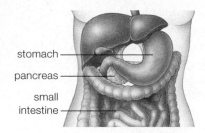

Figure 31.13 Pancreas function.
Above, the location of the pancreas.
Right, how cells that secrete insulin and glucagon work antagonistically to adjust the level of glucose in the blood.

❶ After a meal, glucose enters blood faster than cells can take it up, so blood glucose rises.

❷ The increase slows pancreatic secretion of glucagon and ❸ stimulates pancreatic secretion of insulin.

❹ In response to insulin, body cells increase their uptake of glucose; cells in the liver and muscle store more glucose as glycogen.

❺ The increased uptake of glucose reduces the blood glucose level to its normal level.

❻ Between meals, the blood glucose level falls as cells take it up and use it for metabolism.

❼ A drop in blood glucose encourages glucagon secretion and ❽ slows insulin secretion.

❾ In the liver, glucagon causes breakdown of glycogen into glucose, which enters the blood.

❿ Blood glucose increases to the normal level.

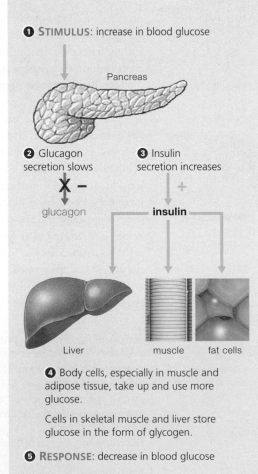

❶ **STIMULUS:** increase in blood glucose

Pancreas

❷ Glucagon secretion slows

❸ Insulin secretion increases

glucagon

insulin

Liver   muscle   fat cells

❹ Body cells, especially in muscle and adipose tissue, take up and use more glucose.

Cells in skeletal muscle and liver store glucose in the form of glycogen.

❺ **RESPONSE:** decrease in blood glucose

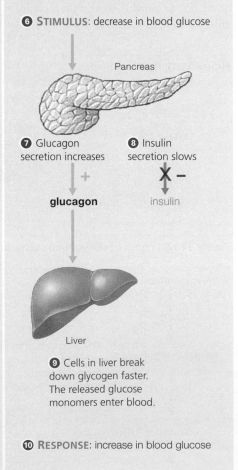

❻ **STIMULUS:** decrease in blood glucose

Pancreas

❼ Glucagon secretion increases

❽ Insulin secretion slows

glucagon

insulin

Liver

❾ Cells in liver break down glycogen faster. The released glucose monomers enter blood.

❿ **RESPONSE:** increase in blood glucose

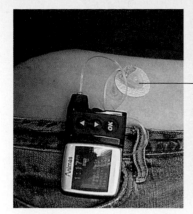
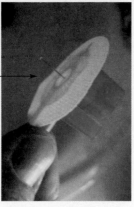

Figure 31.14 Insulin pump. This device delivers insulin into a diabetic's body to maintain blood glucose within normal levels.

DIABETES

Regulation of blood glucose is disrupted in diabetes mellitus, a common metabolic disorder. Its name, loosely translated as "passing honey-sweet water," is a reference to the sweet urine affected individuals produce. Their urine is sweet because cells do not take up and store glucose as they should. The resulting high blood sugar (hyperglycemia) disrupts normal metabolism throughout the body. Vision and hearing loss, skin infections, poor circulation, and nerve damage are among the many effects of diabetes.

There are two main types of diabetes mellitus, type 1 and type 2. About 10 percent of people with diabetes have the type 1 form, in which the body mistakenly mounts an immune response to its own insulin-secreting beta cells. The resulting destruction of beta cells results in symptoms that usually appear in childhood, so this disorder is known as juvenile-onset diabetes. All affected individuals require injected insulin. Devices called insulin pumps help smooth out variations in blood sugar by releasing the hormone slowly beneath the skin (**Figure 31.14**).

With type 2 diabetes, insulin levels are normal or even high. However, target cells do not respond properly to the hormone, so blood sugar remains elevated. Symptoms usually develop in middle age, and obesity increases one's risk. Most cases of type 2 diabetes can be treated with diet, exercise, and oral medications that encourage insulin production. However, if glucose levels are not lowered, pancreatic beta cells receive continual stimulation and eventually falter. When that happens, a type 2 diabetic may require insulin injections.

Rates of type 2 diabetes are soaring worldwide, due in part to the spread of Western diets and sedentary lifestyles. The prevention of diabetes and its complications is acknowledged to be among the most pressing public health priorities around the world.

31.9 Invertebrate Hormones

LEARNING OBJECTIVE

Explain the evolutionary significance of hormones and effects shared by vertebrates and invertebrates.

Until recently, the hormones and receptors of the vertebrate endocrine system were thought to be a derived trait (a trait that arose in and is unique to vertebrates). However, researchers have discovered that invertebrates such as sea stars, roundworms, annelids, and mollusks have steroid hormones and hormone receptors homologous to those of vertebrates. This homology suggests that steroid-based intercellular signaling evolved long ago, in the common ancestor of all bilateral animals. Invertebrates do not have the same glands as vertebrates do, but they produce homologous hormones in other glands. For example, female octopuses produce estrogens in a gland near their eye. As in vertebrates, estrogens and testosterone are sex hormones that function in reproduction.

On the other hand, some hormones are unique to invertebrate groups. For example, arthropods have special hormones that control molting (**Figure 31.15**). Arthropods have a hardened external cuticle that must be periodically shed as they grow (Section 23.9). A soft new cuticle forms beneath the old one before the animal molts. Although details vary among arthropod groups, molting is generally under the control of **ecdysone**, a steroid hormone. The arthropod molting gland produces and stores ecdysone, then releases it for distribution throughout the body when conditions favor molting. Chemicals that mimic ecdysone or interfere with its function are sometimes used as insecticides.

---

ecdysone (eck-DYE-sohn) Steroid hormone that controls molting in arthropods.

Figure 31.15 Effect of ecdysone. A "soft-shelled" blue crab (right) with the old cuticle that ecdysone stimulated it to molt.

APPLICATION: SCIENCE & SOCIETY

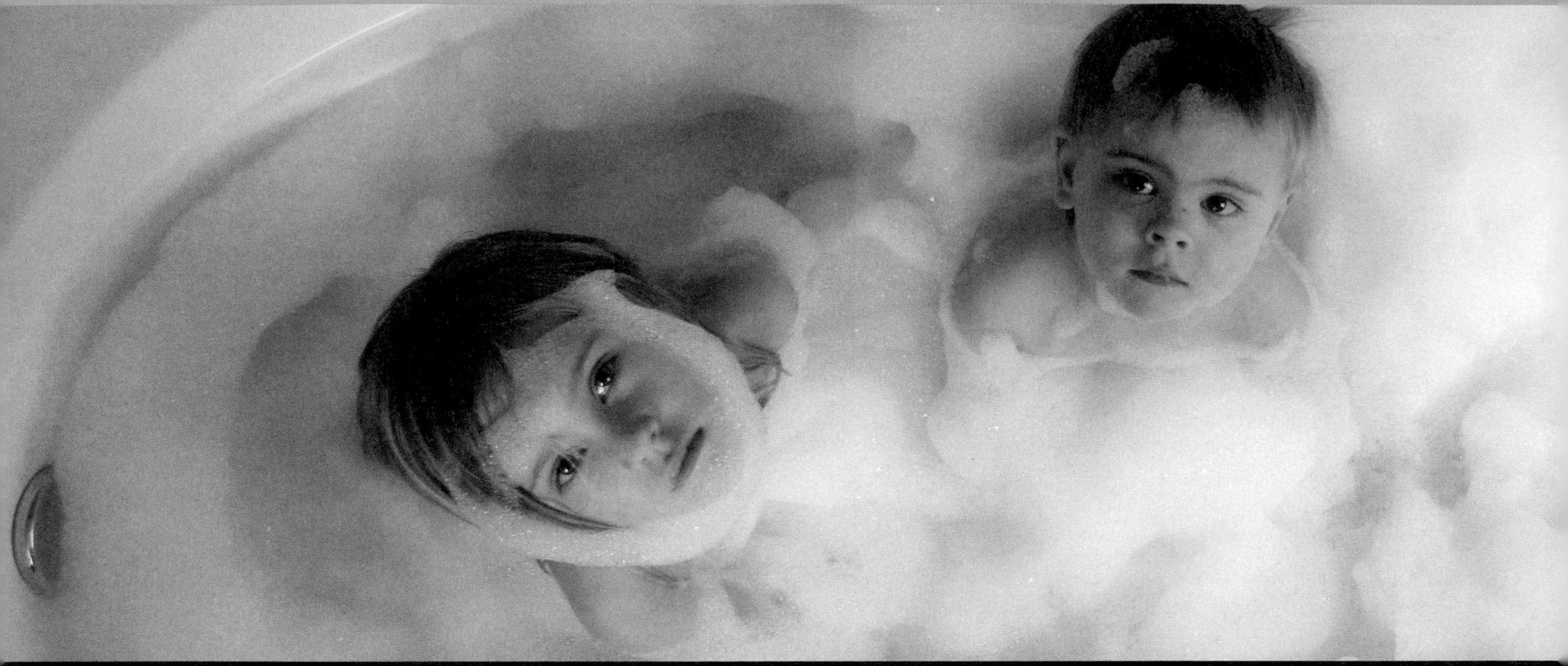

Figure 31.16 Potential source of endocrine disruption. Phthalates, common in scented products, can alter the levels of some hormones.

ENDOCRINE DISRUPTORS

We are awash in man-made chemicals. We drink from plastic bottles, wear synthetic fabrics that we wash in synthetic soaps, slather ourselves with synthetic skin products, and douse our food with synthetic pesticides. Synthetic chemicals enter our bodies when we ingest them, inhale them, or absorb them across our skin.

We know from experience that some synthetic chemicals are harmful. For example, DDT (a pesticide) and PCBs (used to make electronic products, caulking, and solvents) are endocrine disruptors. These substances interfere with hormone synthesis, metabolism, or function. In the United States, DDT was banned in 1972 and PCBs in 1979, but the chemicals persist in the environment because they are highly stable and were in widespread use for many years.

Phthalates (pronounced THAL-ates) are potential endocrine disruptors that remain in widespread use. These chemicals are used to make plastics more flexible and to stabilize fragrances in scented products. Laboratory studies have demonstrated that phthalates inhibit testosterone synthesis in rats, and exposure of male rats to phthalates during prenatal development interferes with development of their sex organs. Epidemiological studies suggest phthalates may have similar effects in humans. In adult men, a high concentration of phthalates is correlated with a low testosterone level and decreased sperm quality. In women, a high phthalate level during pregnancy is correlated with an increased likelihood of giving birth to a son with an undescended testicle.

Phthalates also disrupt the function of other hormones. Phthalates are not approved for use in human foods, but in 2011 a Taiwanese company illegally added them to a variety of foods and drinks. Children who had ingested these items had a lowered level of TSH (the pituitary hormone that stimulates thyroid hormone production). The reduction in TSH was dose dependent; the more contaminated food and drink a child had taken in, the greater his or her reduction in TSH level. Results of this horrible unplanned experiment are consistent with epidemiological studies that implicate phthalates in disrupted thyroid function. Other studies have found a positive correlation between phthalate exposure and diabetes.

In light of the increasing evidence that exposure to phthalates has negative health effects, some efforts have been made to reduce the use of these chemicals. Phthalates were widely used in pacifiers and teething toys until 2007, when a U.S. law went into effect limiting the phthalate content of products intended for children under age twelve. Phthalates have also been largely eliminated from plastic wraps for food and from other American-made plastics designed to come into contact with food.

However, opportunities for phthalate exposure remain. Vinyl, which

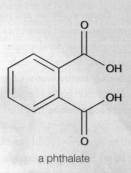

a phthalate

symbol for polyvinyl
chloride (PVC)

releases phthalates into the air, is common in
furniture, flooring, mattress covers, and shower
curtains. Phthalates also remain in artificially
scented products such as air fresheners, laundry
detergents, fabric softeneners, dryer sheets, and
personal care products such as shampoos, cosmetics, and bubble baths (**Figure 31.16**).

Pediatrician Sheela Sathyanarayana found
that the more scented products a mother used
on her infant, the higher the level of phthalates
in the child's urine. Sathyanarayana's study did
not demonstrate any ill effects related to children's elevated phthalate level. Nevertheless,
the American Academy of Pediatrics recommends that parents use only unscented products
on infants and young children. Similarly, many
obstetricians recommend that women who are
pregnant, are trying to become pregnant, or are
currently breast-feeding use personal care products specifically labeled as phthalate-free.

If you wish to avoid plastic products that
contain phthalates, check the recycling symbol
on the product. If an item has a "3" inside the
triangle of arrows (shown above), it is polyvinyl
chloride (PVC) and may contain phthalates.

Dr. Tyrone Hayes

Frogs have fascinated Dr. Tyrone Hayes since he began exploring the
swamps of South Carolina as a child. Today this National Geographic
Explorer studies the effects of synthetic pesticides on amphibians.
Hayes is best known for discovering the endocrine-disrupting effects of
atrazine, a widely used synthetic herbicide. Hayes found that exposure to
atrazine feminizes male tadpoles, and also causes some genetically male
frogs to develop ovaries in addition to or instead of testes.

Hayes's findings inspired a variety of other researchers to carry
out their own investigations of atrazine's hormonal effects. These
studies confirmed that atrazine has detrimental effects on frogs, and
also demonstrated endocrine-disrupting effects in fish, birds, and
rodents. They also uncovered mechanisms by which atrazine may exert
these effects. We now know that atrazine encourages conversion of
testosterone (the main sex hormone in males) to estrogen (the main sex
hormone in females). Atrazine also binds to a plasma membrane estrogen
receptor, so it mimics the effects of estrogen.

Atrazine has been banned in Europe but remains in wide use in the
United States, where it is a common water pollutant in agricultural regions.
In 2013, the company that produces and sells atrazine began paying a $105
million settlement to community water suppliers across the United States.
The payments are meant to reimburse communities for money spent to
filter atrazine out of their drinking water supply. Runoff of atrazine from
agricultural fields in these communities has resulted in water with an
atrazine level too high to be considered safe by regulatory agencies.

32 Structural Support and Movement

Links to Earlier Concepts

This chapter elaborates on animal traits and evolutionary trends introduced in Chapters 23 and 24. You will build on your knowledge of connective tissue (28.3), muscle tissue (28.4), neuromuscular junctions (29.5), and hormonal effects on bone (31.4). There is an example of active transport (5.7), and the discussion of muscle contraction involves a detailed look at the role of a motor protein (4.9).

Core Concepts

Organisms must exchange matter with the environment in order to grow, maintain themselves, and reproduce.

All animals can move their body parts, and most move about in search of food and mates. Moving from place to place requires overcoming the forces of friction and gravity. ATP-fueled muscle contractions exert force against a skeleton to bring about movements.

All organisms alive today are linked by lines of descent from shared ancestors.

All vertebrates have an internal skeleton. The structural and functional similarities among the components of this skeleton reflect a common ancestry. Natural selection leads to the evolution of structures that increase fitness in a specific environment, and evolved variations in body form allow efficent locomotion in different environments.

Interactions among the components of each biological system give rise to complex properties.

Vertebrate skeletal muscles move body parts by pulling on bones. These muscles often work in pairs, with the action of one opposing the action of the other. Signals from the nervous system initiate muscle contraction. Interactions among organized protein filaments in muscle cells bring about the contraction.

A cheetah, the fastest animal on land. Long legs and a flexible backbone help it achieve remarkable speeds.

Photograph by Frans Lanting, National Geographic Creative.

32.1 Animal Locomotion

LEARNING OBJECTIVES

Describe some adaptations that reduce the energy animals must expend to overcome friction and gravity when they move.

Explain how squid locomotion differs from fish locomotion.

List some structural adaptations that contribute to a cheetah's speed.

Figure 32.3 Dog walking. Elevating the body above the ground and having only some feet contact the ground at any given time minimizes the effects of friction.

All animals move. During part or all of their life, they are capable of **locomotion**—self-propelled movement from place to place. As adults, even sessile animals (those that live fixed in place) usually move some body parts. Consider barnacles, which begin life as free-swimming larvae, then settle and secrete a protective shell around their body. Although adults remain in one place, they still wave their feathery legs to capture food from the water around them. Other animals move about on a daily basis to find food and escape from predators.

Animals use diverse mechanisms of locomotion. They swim through water; fly through air; run, walk, or crawl along surfaces; or burrow through soil and sediments. In all cases, the same physical law governs all movement: For every action, there is an equal and opposite reaction. For an animal to move in one direction, it must exert a force in the opposite direction. Jet propulsion by squids provides a simple example of this effect (**Figure 32.1**). To move, a squid fills an internal cavity (the mantle cavity) with water, then shoots that water out through a small opening. As the water streams out in one direction, the squid moves in the opposite direction. By analogy, consider what happens if you inflate a balloon, then release

Figure 32.1 Jet-propelled squid.

squid moves this way

water shoots this way

it without tying it shut. As the rubber of the balloon recoils to its original shape, force exerted by air rushing out in one direction makes the balloon shoot the opposite direction.

Most other animals that swim do so by pushing at water through body movements and movements of flippers or fins. Animals also move along surfaces, both underwater and on land. When doing so, they propel themselves by pushing against those surfaces.

The effects of friction and gravity increase the amount of energy an animal must expend to move. Water is denser than air, so moving through water produces more friction. Think of how much harder it is to walk through waist-deep water than to walk on land. Aquatic animals that benefit by swimming fast typically have a streamlined body to minimize friction with water (**Figure 32.2**). Aquatic animals are generally somewhat buoyant, so gravity is less of a constraint for them than it is for land animals.

Snails, snakes, and some other animals expend energy mainly to overcome friction between their body and the surface they creep or slither across. Snails reduce friction by secreting a lubricating mucus from their large foot. A snake's scales can have a similar effect: Friction is minimal in body areas where scales are held flat against the skin. By adjusting the angle of the scales in certain parts of its body, a snake can increase friction only in the areas where traction is required for movement.

Animals that walk or run minimize friction by reducing the surface area in contact with the ground. Limbs partially or fully elevate their body and only some feet touch the ground at any given time (**Figure 32.3**).

The cheetah (shown in the chapter opening photo) is the fastest land animal. It can sprint as fast as 65 miles per hour (about 100 kilometers per hour), although it cannot sustain that speed. A variety of traits contribute to the cheetah's

Figure 32.2 Tuna, a streamlined swimmer.

speed. Compared to other big cats, a cheetah has longer legs relative to its body size, a wider range of motion at its hips and shoulders, and a longer, more flexible backbone. Together, these features maximize the cheetah's stride length.

The cheetah also has a highly elastic spine: When its back legs land, its backbone becomes bent into a tight C shape. When the animal pushes off again, its spine recoils, forcing the front of the body forward. Elastic tissue plays a locomotive role in many other animals too. Consider kangaroos, which usually move by hopping. Each time a kangaroo lands, elastic tissues in its hind legs compress. The compression is a form of stored energy that is released upon expansion; released energy helps propel the animal in the next hop.

A capacity for powered flight evolved independently in insects, birds, bats, and an extinct group of reptiles called pterosaurs. To fly, an animal must overcome gravity to become and stay airborne, and must generate a force to propel itself forward. Wing shape and movements facilitate flight. Consider what happens when the bird is gliding (flying without flapping its wings). A bird wing has the same shape as an airplane wing. In cross-section, the wing's upper surface is convex (bulges outward), and the lower surface is flat or slightly concave (sunken). As the bird moves forward, air flowing over the upper surface of its wings moves faster than air flowing over the wings' lower surface. This difference in air velocity across the two surfaces creates lift, an upward force that opposes gravity and keeps the bird aloft (**Figure 32.4**). To get off the ground in the first place, the bird has to flap its wings. It spreads its feathers to maximize the size of the wings, then contracts its chest muscles to bring its wings downward. As the air beneath the wings is forced downward, the bird moves in the opposite direction. The bird then folds its wings to minimize their size as it pulls them upward in preparation for another lift-producing downstroke.

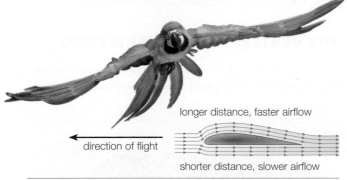

longer distance, faster airflow

← direction of flight

shorter distance, slower airflow

Figure 32.4 Parrot in flight. Air flows faster over the rounded upper surface of its wings than the flatter lower surface. This difference in air velocity creates an upward force called lift.

locomotion Self-propelled movement from place to place.

Dr. Kakani Katija Young

When aquatic animals move, they set the water around them in motion. National Geographic Explorer Dr. Kakani Katija Young is a bioengineer who studies these movements and their effects on aquatic habitats. Together with her colleagues, she devised a system that uses lasers and a high-speed camera to visualize and analyze currents created by swimming animals.

Young began her studies of aquatic locomotion with jellies because they have a simple symmetrical body (Section 23.4). She discovered that when a jelly propels itself forward by rhythmic contractions of its bell, its motion creates doughnut-shaped rings of water that quickly dissipate into the surrounding ocean. The rings and the water that the jellies transport with them as they move create a rotational flow that mixes fluid in the ocean.

The effect of individual jellies moving might seem insignificant, but Young explains that enormous numbers of jellies, small fish, and free-swimming arthropods such as shrimp, krill, and copepods churn the seas on a daily basis. She thinks the collective movement of these creatures in our seas could be as important to ocean circulation and global climate as the winds and tides. Her research could provide important new reasons to protect endangered sea life because, while winds and tides operate constantly, animals can disappear. "The decline and collapse of fisheries may have already severely changed the amount of biogenic energy animals can contribute to mixing our oceans," she warns. "Our data could underscore the need for drastic conservation measures."

32.2 Types of Skeletons

Using appropriate examples, describe the three types of skeletons.

Describe how interactions between muscles and a skeleton allow a fly to flap its wings and an earthworm to burrow through soil.

Explain the function of the vertebrate pelvic girdle and pectoral girdle.

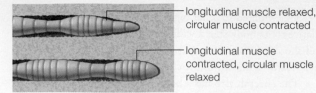

longitudinal muscle relaxed, circular muscle contracted

longitudinal muscle contracted, circular muscle relaxed

Figure 32.5 Hydrostatic skeleton of an earthworm. A worm moves when two sets of muscles put pressure on coelomic fluid in individual body segments, causing the segments to change shape.

Muscles bring about movement by interacting with a skeleton. Your bony internal framework is one type of skeleton. In many other animals, muscles exert force against fluid-filled chambers or external hard parts.

INVERTEBRATE SKELETONS

A system of fluid-filled internal chambers makes up the **hydrostatic skeleton** of soft-bodied invertebrates such as sea anemones and worms. For example, an earthworm's coelom is divided into many fluid-filled chambers, one per segment (Section 23.6). Movement occurs when muscles exert force against these chambers (**Figure 32.5**). Two sets of muscles

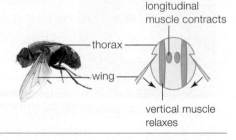

A The fly's wings pivot down when the vertical muscles relax and the longitudinal muscles contract, pulling the sides of the thorax inward.

thorax

wing

longitudinal muscle contracts

vertical muscle relaxes

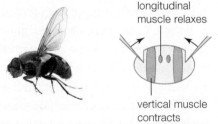

B The wings pivot up when the longitudinal muscles relax and the vertical muscles contract, flattening the thorax.

longitudinal muscle relaxes

vertical muscle contracts

Figure 32.6 Exoskeleton of a fly. Fly flight muscles attach to the endoskeleton of the thorax. By altering the shape of the thorax, these muscles cause the attached wings to pivot up or down.

squeeze the chambers to alter the shape of body segments, much as squeezing a water-filled balloon changes the balloon's shape. Circular muscle rings each earthworm segment; when this muscle contracts, the segment gets longer and narrower. Longitudinal muscle runs the length of each segment; when this muscle contracts, the segment gets shorter and wider. Coordinated changes in the shape of the earthworm's many body segments move it through the soil.

An **exoskeleton** is a cuticle, shell, or other hard external body part that receives the force of a muscle contraction. For example, muscles attached to the cuticle of a fly's thorax alter the shape of the thorax, causing wings attached to the thorax to flap up and down (**Figure 32.6**). The arthropod exoskeleton has the advantage of also protecting the soft body tissues inside it. However, external skeletons have some drawbacks. An exoskeleton consists of noncellular secreted material, so it cannot grow with the animal. An arthropod must periodically molt (shed its old skeleton and replace it with a larger one). Repeatedly producing new exoskeleton uses resources that could otherwise be put to use in growth or reproduction. In addition, after an arthropod molts, it remains highly vulnerable to predators until its soft new exoskeleton hardens.

An **endoskeleton** is an internal framework of hard parts. For example, the endoskeleton of echinoderms such as sea stars consists of hardened calcium-rich plates.

THE VERTEBRATE ENDOSKELETON

All vertebrates have an endoskeleton. In sharks and other cartilaginous fishes, it consists of cartilage, a rubbery connective

Figure 32.7 Skeletal elements typical of early reptiles.

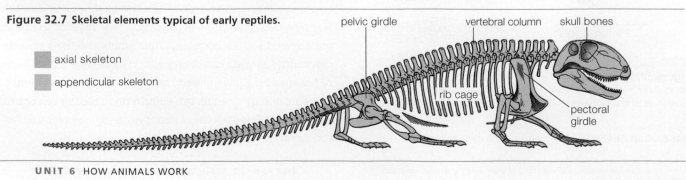

axial skeleton

appendicular skeleton

pelvic girdle

vertebral column

skull bones

rib cage

pectoral girdle

tissue. Other vertebrate skeletons include some cartilage, but consist mostly of bone tissue (Section 28.3).

The term "vertebrate" refers to the **vertebral column**, or backbone, a feature common to all members of this group (**Figures 32.7** and **32.8**). The backbone supports the body, serves as an attachment point for muscles, and protects the spinal cord that runs through a canal inside it. Bony segments called **vertebrae** (singular, vertebra) make up the backbone. **Intervertebral disks** of cartilage between vertebrae act as shock absorbers.

The vertebral column and bones of the head and rib cage constitute the **axial skeleton**, the bones that make up the main axis of the body. The **appendicular skeleton** consists of bones of the appendages (limbs or bony fins), and those that connect appendages to the axial skeleton (the pectoral girdle at the shoulders and the pelvic girdle at the hip).

You learned earlier how vertebrate skeletons have evolved over time. For example, jaws are derived from the gill supports of ancient jawless fishes (Section 24.2). As another example, bones in the limbs of land vertebrates are homologous to those inside the fins of lobe-finned fishes (Section 24.3).

For a closer look at vertebrate skeletal features, consider a human skeleton (**Figure 32.8**). The skull's flattened cranial bones fit together as a braincase, or cranium. Facial bones include cheekbones, bones around the eyes, the bone that forms the bridge of the nose, and the bones of the jaw.

Ribs and the breastbone, or sternum, form a protective cage around the heart and lungs. Both males and females have twelve pairs of ribs.

The vertebral column extends from the base of the skull to the pelvic girdle. Viewed from the side, our backbone has an S shape that keeps our head and torso centered over our feet when we stand upright.

The scapula (shoulder blade) and clavicle (collarbone) are bones of the human pectoral girdle. The thin clavicle transfers force from the arms to the axial skeleton. The upper arm has a single bone called the humerus. The forearm has two bones, the radius and ulna. Carpals are bones of the wrist, metacarpals are bones of the palm, and phalanges (singular, phalanx) are finger bones.

The pelvic girdle consists of two hip bones that connect to the sacrum to form a basin-shaped ring. The pelvic girdle protects organs inside the pelvic cavity and supports the weight of the upper body when you stand upright. Bones of the leg include the femur (thighbone), patella (kneecap), and the tibia and fibula (bones of the lower leg). The fibula serves as a point of muscle attachment, but does not bear weight. Tarsals are ankle bones, and metatarsals are bones of the sole of the foot. Like the bones of the fingers, those of the toes are called phalanges.

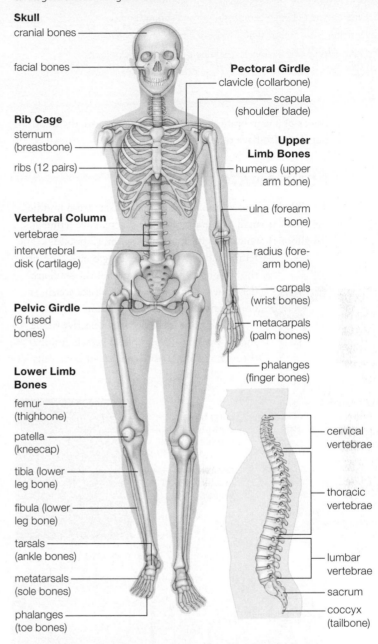

Figure 32.8 Human skeleton. Major bones are shown in tan and cartilage elements in light blue.

Skull
cranial bones
facial bones

Rib Cage
sternum (breastbone)
ribs (12 pairs)

Vertebral Column
vertebrae
intervertebral disk (cartilage)

Pelvic Girdle
(6 fused bones)

Lower Limb Bones
femur (thighbone)
patella (kneecap)
tibia (lower leg bone)
fibula (lower leg bone)
tarsals (ankle bones)
metatarsals (sole bones)
phalanges (toe bones)

Pectoral Girdle
clavicle (collarbone)
scapula (shoulder blade)

Upper Limb Bones
humerus (upper arm bone)
ulna (forearm bone)
radius (fore-arm bone)
carpals (wrist bones)
metacarpals (palm bones)
phalanges (finger bones)

cervical vertebrae
thoracic vertebrae
lumbar vertebrae
sacrum
coccyx (tailbone)

---

appendicular skeleton Bones of the limbs or fins and the bones that connect them to the axial skeleton.

axial skeleton Bones of the main body axis; skull, backbone.

endoskeleton Internal skeleton consisting of hard parts.

exoskeleton External skeleton of some invertebrates; hard external parts that muscles attach to and move.

hydrostatic skeleton In soft-bodied invertebrates, a fluid-filled chamber that muscles exert force against to move the body.

intervertebral disk Cartilage disk between two vertebrae.

vertebrae Bones of the backbone, or vertebral column.

vertebral column Backbone.

LEARNING OBJECTIVES

List the functions of bone.

Differentiate between compact bone and spongy bone.

Describe some factors that affect bone mass and bone density.

Using appropriate examples, describe the three types of joints.

TABLE 32.1

Functions of Bone

1. Movement. Bones interact with skeletal muscle to change or maintain positions of the body and its parts.

2. Support. Bones support and anchor muscles.

3. Protection. Many bones form hardened chambers or canals that enclose and protect soft internal organs.

4. Mineral storage. Bones are a reservoir of calcium and phosphorus ions. Deposits and withdrawals of these ions help maintain their consistent concentration in body fluids.

5. Blood cell formation. Blood cells form in the red marrow that fills the interior of some bones.

Bones of a vertebrate skeleton support and move the body, but they also have other functions (**Table 32.1**).

BONE ANATOMY

The 206 bones of a human adult range in size from middle ear bones as small as a grain of rice to the massive femur (thighbone), which weighs about a kilogram (2 pounds). The femur and limb bones are long bones. The ribs, the sternum, and most bones of the skull are flat bones. The carpals in the wrists are short bones, and are roughly squarish in shape.

Each bone has a dense connective tissue sheath with nerves and blood vessels running through it. Bone tissue consists of bone cells

in a secreted extracellular matrix (Section 4.10). The matrix consists mainly of the protein collagen, and it is hardened by deposits of calcium and phosphorus.

A long bone such as the femur includes two types of bone tissue, compact bone and spongy bone (**Figure 32.9**). **Compact bone** is dense bone that makes up the shaft of long bones such as the femur. It consists of many concentric rings of mineralized bone tissue, with living bone cells in spaces between the rings. Blood vessels and nerves run through the central canal in the middle of each ring. The knobby ends of long bones are filled with **spongy bone**. Spongy bone is strong yet lightweight; its matrix has many cavities.

Bone marrow is the soft tissue that fills the cavities inside a bone. The **red marrow** that fills spaces in spongy bone is the major site of blood cell formation. The **yellow marrow** that fills the central cavity of an adult femur and most other long bones consists mainly of fat. In cases of extreme blood loss, yellow marrow can convert to red marrow.

BONE FORMATION AND TURNOVER

A cartilage skeleton forms in all vertebrate embryos. In sharks and other cartilaginous fishes, the cartilage skeleton persists into adulthood. In other vertebrates, embryonic cartilage serves as a model for most of a bony skeleton. Before birth, mineralization of this model transforms most of it to bone.

Until a person is about twenty-four years old, bone cells secrete more matrix than they break down, so bone mass increases. Later in life, bone-producing cells become less active and bone mass gradually declines. However, bone remodeling occurs throughout life. The body must constantly repair the microscopic fractures that result from normal movements. In addition, bone is built and broken down in response to hormonal signals. Bones store most of the body's calcium. Parathyroid hormone (Section 31.4) raises the concentration of calcium ions in the blood by encouraging the

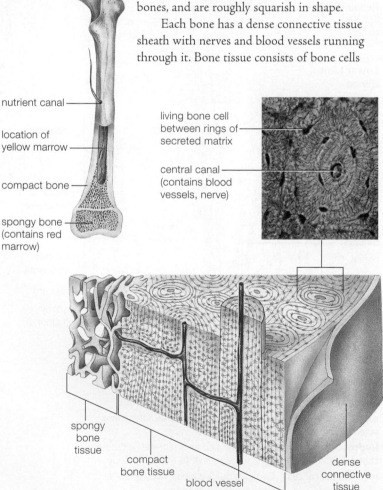

nutrient canal

location of yellow marrow

compact bone

spongy bone (contains red marrow)

living bone cell between rings of secreted matrix

central canal (contains blood vessels, nerve)

spongy bone tissue

compact bone tissue

blood vessel

dense connective tissue

Figure 32.9 Structure of a long bone (human femur).

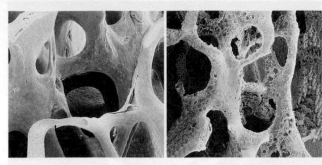

A Normal bone.   **B** Bone weakened by osteoporosis.

Figure 32.10 Effect of osteoporosis on spongy bone.

breakdown of calcium-rich bone matrix. Other hormones also affect bone turnover. The sex hormones estrogen and testosterone encourage deposition of bone. Cortisol, the stress hormone, slows it.

Osteoporosis is a disorder in which bone loss outpaces bone formation. As a result, the bones become weaker and more likely to break (**Figure 32.10**). Osteoporosis is most common in postmenopausal women because they produce little of the sex hormones that encourage bone deposition. However, osteoporosis can also occur in men.

To reduce your risk of osteoporosis, ensure that your diet provides adequate levels of calcium and of vitamin D, which facilitates calcium absorption from the gut. Avoid smoking and excessive alcohol intake (both slow bone deposition), and exercise regularly to encourage bone renewal.

JOINTS: WHERE BONES MEET

A **joint** is an area where bones come together. There are three types: fibrous joints, cartilaginous joints, and synovial joints.

At a fibrous joint, dense, fibrous connective tissue holds bones firmly in place. Fibrous joints connect bones of the skull and hold teeth in their sockets in the jaw.

Pads or disks of cartilage connect bones at cartilaginous joints. The flexible connection allows just a bit of movement. Cartilaginous joints connect vertebrae to one another and connect some ribs to the sternum.

At the most common type of joint, a **synovial joint**, the cartilage-covered ends of bones are separated by a small gap filled with lubricating fluid. Cords of dense, regular connective tissue called **ligaments** hold bones of a synovial joint in place, and dense, irregular connective tissue forms a capsule around the joint. The capsule's inner lining secretes the fluid that lubricates the joint.

Different synovial joints allow different movements. Ball-and-socket joints at the shoulders and hips allow rotational motion and movement in all planes. At other joints, including

some in the wrists and ankles, bones glide over one another. Joints at elbows and knees function like a hinged door, allowing bones to move back and forth in one plane.

Sprains, the most common joint injury, occur when ligaments at a synovial joint become overstretched or torn. A tear of cruciate ligaments in the knee joint often requires surgery. "Cruciate" means cross, and the two cruciate ligaments cross one another and stabilize the knee (**Figure 32.11**). When they tear, bones may shift so the knee gives out when a person tries to stand. A torn meniscus is another common knee injury. A meniscus is a crescent-shaped wedge of cartilage that cushions bones at a joint and reduces friction between them.

Arthritis is chronic inflammation of a joint. The most common type of arthritis, osteoarthritis, usually appears in old age, after cartilage on bones at a specific synovial joint or joints wears down. Rheumatoid arthritis is a disorder in which the immune system mistakenly attacks all of the body's synovial joints. It can occur at any age, and women are two to three times more likely than men to be affected.

Figure 32.11 The knee, a hinge-type synovial joint. Ligaments hold bones in place. Wedges of cartilage called menisci (singular, meniscus) provide additional stability.

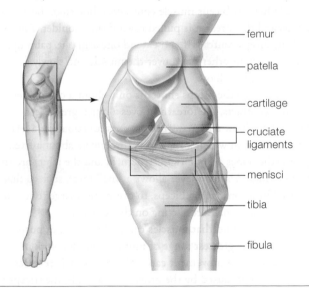

femur

patella

cartilage

cruciate ligaments

menisci

tibia

fibula

compact bone Dense bone; makes up the shaft of long bones.

joint Region where bones meet.

ligament (LIG-uh-ment) Strap of dense connective tissue that holds bones together at a joint.

red marrow Bone marrow that makes blood cells.

spongy bone Lightweight bone with many internal spaces; contains red marrow.

synovial joint (sin-OH-vee-ul) Capsule-enclosed joint at which cartilage-covered ends of bones are separated by a fluid-filled space.

yellow marrow Bone marrow that is mostly fat; fills the interior of most adult long bones.

32.6 The Sliding-Filament Model

LEARNING OBJECTIVES

Explain how interactions among filaments cause a sarcomere to shorten.

Draw a sarcomere in its relaxed and its contracted states.

Explain why ATP is necessary for muscle contraction.

The **sliding-filament model** explains how interactions between thick and thin filaments bring about muscle contraction. Neither actin nor myosin filaments change length, and the myosin filaments do not change position. Instead, myosin heads bind to actin filaments and pull them toward the center of a sarcomere. As the actin filaments are pulled, they slide inward, the ends of the sarcomere are drawn closer together, and the sarcomere shortens:

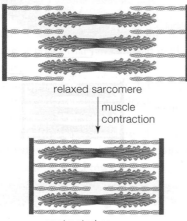

relaxed sarcomere

↓ muscle contraction

contracted sarcomere

Figure 32.14 provides a step-by-step look at events during muscle contraction. We begin by looking at a sarcomere in a resting muscle fiber. In a relaxed sarcomere, sites where myosin could bind to actin are blocked by other protein components of the thin filament. The myosin heads of thick filaments have bound ATP molecules and are in a low-energy conformation ❶. Removing a phosphate ion (P_i) from a bound ATP energizes a myosin head in a manner analogous to stretching a spring ❷. The myosin head remains in a high-energy state, with ADP and phosphate (P_i) still bound, until a signal from the nervous system excites the muscle. (The process by which nerves signal muscle is covered in detail in the next section.)

When the nervous signal arrives, myosin-binding sites on the thin filament are unblocked and myosin heads attach to actin. The attachment constitutes a cross-bridge between the thin and thick filaments ❸. Attachment releases a phosphate group from myosin and triggers a power stroke ❹. Like a stretched spring returning to its original shape, a myosin head snaps back toward the sarcomere center. As it does, it pulls the attached thin filament with it.

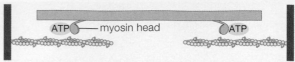

❶ In a relaxed sarcomere, myosin-binding sites on actin are blocked. The myosin heads have bound ATP, and are in their low-energy conformation.

❷ The myosin heads remove a phosphate group from the ATP. Absorbing energy released by breaking the phosphate bond boosts the myosin heads to a high-energy conformation. ADP and phosphate remain bound to each head.

❸ A signal from the nervous system opens up the myosin-binding sites on actin. Each myosin head attaches to actin and releases its bound phosphate group, thus forming a cross-bridge between thick and thin filaments.

❹ The release of the phosphate group triggers a power stroke, in which the myosin heads snap inward and release ADP. As the heads contract, they pull the attached thin filaments inward.

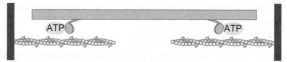

❺ Another ATP binds to each myosin head, causing it to release actin and return to its low-energy conformation.

Figure 32.14 Sliding-filament model of muscle contraction.

After a myosin head releases bound ADP during a power stroke, it can bind to a new molecule of ATP and return to its low-energy conformation ❺. In the process, the myosin head releases actin, and the cross-bridge is broken. Loss of one cross-bridge does not allow a thin filament to slip backward because other cross-bridges hold it in place.

A muscle as a whole contracts when many myosin heads of many sarcomeres in many muscle fibers repeatedly bind, move, and release the adjacent thin filaments.

sliding-filament model Explanation of how interactions among actin and myosin filaments shorten a sarcomere and bring about muscle contraction.

32.7 Control of Muscle Contraction

LEARNING OBJECTIVES

Trace the pathway a signal for voluntary skeletal muscle contraction takes, from the brain to a neuromuscular junction.

Describe how arrival of an action potential at a neuromuscular junction leads to muscle contraction.

Using appropriate examples, explain how variation in the number of fibers controlled by a motor unit affects muscle tension.

THE MOTOR SIGNAL PATHWAY

Commands for voluntary movement originate in the brain's frontal lobe, in a region called the primary motor cortex (Section 29.10). Signals from this region travel through the brain stem and into the spinal cord, where they excite a motor neuron (**Figure 32.15 ❶**). The axon of a motor neuron extends to a skeletal muscle and synapses with it at a neuromuscular junction ❷. Arrival of an action potential at this synapse triggers release of the neurotransmitter acetylcholine (ACh), as described in Section 29.5.

Like a neuron, a muscle fiber can undergo an action potential. A muscle fiber has receptors for ACh in its plasma membrane. Binding of ACh by these receptors triggers an action potential that travels along the plasma membrane, then down membrane extensions called transverse tubules or T tubules ❸. Action potentials travel along T tubules to the **sarcoplasmic reticulum**, a specialized endoplasmic reticulum that wraps around myofibrils and stores calcium ions. The sarcoplasmic reticulum has voltage-gated calcium channels that open when an action potential arrives. Calcium ions then follow their gradient and diffuse out of the organelle, thus raising the calcium concentration around the myofibrils.

Calcium ion concentration affects the ability of myosin to bind actin. When calcium ions are not present, non-actin proteins of the thin filament block the sites where myosin binds actin. Arrival of calcium ions changes the shape of these proteins, allowing the actin-myosin interactions that cause muscle contraction. After contraction ends, calcium pumps located in the membrane of the sarcoplasmic reticulum's actively transport calcium ions back into the organelle. Calcium ion concentration in the cytoplasm drops, myosin-binding sites become blocked, and the muscle fiber relaxes.

MOTOR UNITS

A motor neuron has many terminal endings, and each synapses on a different fiber inside a muscle. A motor neuron and all of the muscle fibers it synapses with constitute one **motor unit**. Stimulate a motor neuron, and all the muscle fibers in its motor unit contract. The nervous system cannot make only some fibers in a motor unit contract.

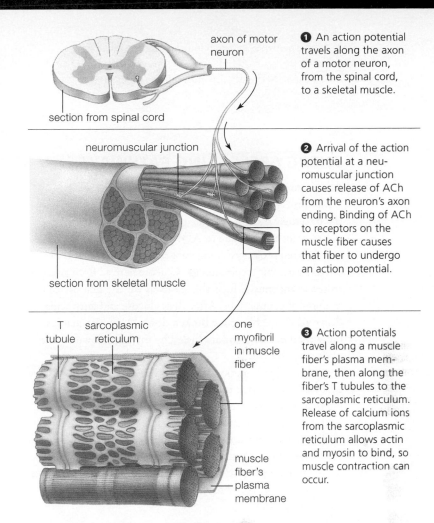

❶ An action potential travels along the axon of a motor neuron, from the spinal cord, to a skeletal muscle.

❷ Arrival of the action potential at a neuromuscular junction causes release of ACh from the neuron's axon ending. Binding of ACh to receptors on the muscle fiber causes that fiber to undergo an action potential.

❸ Action potentials travel along a muscle fiber's plasma membrane, then along the fiber's T tubules to the sarcoplasmic reticulum. Release of calcium ions from the sarcoplasmic reticulum allows actin and myosin to bind, so muscle contraction can occur.

Figure 32.15 Nervous control of skeletal muscle contraction.

The mechanical force generated by a contracting muscle—the **muscle tension**—depends on the number of muscle fibers contracting. Some tasks require more muscle tension than others, so the number of muscle fibers controlled by a single motor neuron varies. In motor units that bring about small, fine movements, one motor neuron synapses with only 5 or so muscle fibers. Motor units controlling eye muscles are like this. By contrast, muscles that must frequently exert a large amount of force have many fibers per motor unit. For example, the biceps of the arm has about 700 muscle fibers per motor unit. Having many fibers all pulling the same way at once increases the force a motor unit can generate.

motor unit One motor neuron and the muscle fibers it controls.
muscle tension Force exerted by a contracting muscle.
sarcoplasmic reticulum (sar-co-PLAZ-mic reh-TIC-you-lum) Specialized endoplasmic reticulum in muscle cells; stores and releases calcium ions.

LEARNING OBJECTIVES

Describe the three pathways by which muscles produce the ATP they need to contract.

Explain the traits used to characterize muscle fibers as fast or slow and red or white.

ENERGY-RELEASING PATHWAYS

Muscle contraction requires ATP, and muscles have three ways of meeting their ATP needs (**Figure 32.16**). First, they tap into their store of creatine phosphate, a compound that can transfer a phosphate to ADP to form ATP ❶.

During prolonged, moderate activity most ATP is produced by aerobic respiration ❷. Glucose derived from glycogen stored in the muscle fuels this pathway for the first five to ten minutes of contractions. After that, glucose and fatty acids delivered by the blood are broken down. Activities that last more than half an hour are fueled by the breakdown of fatty acids (Section 7.6).

Lactate fermentation is the third source of ATP ❸. Some pyruvate is converted to lactate by fermentation even in resting muscle, but lactate fermentation increases during exercise. Lactic fermentation produces ATP less efficiently than aerobic respiration, but it yields ATP more quickly and it operates even when the oxygen level in a muscle is low. It predominates during intense exercise, when a muscle's energy needs outpace the ability of the blood to supply the oxygen required for aerobic respiration.

TYPES OF MUSCLE FIBERS

There are two types of skeletal muscle fiber, and they differ both in color and metabolism. Red fibers have an abundance

Figure 32.16 Sources of ATP for muscle contraction.

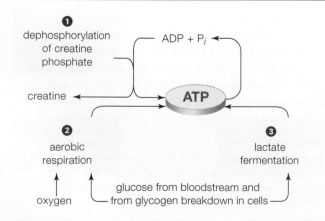

Figure 32.17 Loris, a slow-moving primate, whose limbs have a large proportion of slow red fibers. Creeping along branches very slowly helps a loris escape the attention of predators.

of mitochondria and produce ATP mainly by aerobic respiration. They are colored bright red by **myoglobin**, a protein that, like hemoglobin, reversibly binds oxygen. When the oxygen level in blood is high, oxygen diffuses into red muscle fibers and binds to myoglobin. During periods of muscle activity, the myoglobin releases bound oxygen. By contrast, white skeletal muscle fibers have no myoglobin and few mitochondria, so they make ATP mainly by lactate fermentation.

Muscle fibers can also be categorized as fast fibers or slow fibers depending on the speed with which their myosin converts ATP to ADP. All white fibers are fast fibers. They predominate in muscles that contract rapidly but do not sustain contractions. Muscles that move your eye consist mainly of white fibers. Some red fibers are fast and others are slow. Fast red fibers predominate in the human triceps muscle, which must often react quickly. Muscles that play a role in maintaining an upright posture, such as some muscles of the back, consist mainly of slow red fibers.

For any given muscle, the relative proportions of fiber types vary among species and reflect the pattern of muscle usage. Limb muscles of cheetahs, animals renowned for their sprinting, contain a large proportion of white fibers. By contrast, limb muscles of slow-moving lorises (**Figure 32.17**) have mostly slow red fibers. Similarly, among human athletes, leg muscles of successful sprinters tend to have a high percentage of fast white fibers; leg muscles of marathoners tend to be rich in slow red fibers.

myoglobin (MY-oh-glow-bin) Muscle protein that reversibly binds oxygen.

Figure 32.18 Prolonged periods of sitting in front of a television or computer alter muscle metabolism and raise the risk of chronic disease.

EXERCISE AND INACTIVITY

Your body is like a machine that improves with use. Aerobic exercise—low intensity, but long duration—makes your skeletal muscles more resistant to fatigue. It increases a muscle's blood supply by spurring growth of new capillaries, increases the number of mitochondria and amount of myoglobin in existing red muscle fibers, and encourages conversion of white fibers to red ones. Engaging in resistance exercise, such as weight lifting, encourages synthesis of additional actin and myosin filaments. The resulting increase in muscle mass allows for stronger contractions.

On the other hand, prolonged sitting can be hazardous to your health (**Figure 32.18**). When you are seated, your leg muscles decrease their production of lipoprotein lipase (LPL), an enzyme that facilitates uptake of fatty acids and triglycerides from the blood. A decrease in LPL activity leads to an increase in the concentration of lipids in the blood, which in turn increases the risk of cardiovascular disease and diabetes.

Surprisingly, the health risks associated with prolonged muscle inactivity persist even if a person also gets regular

exercise. In other words, exercising for an hour each morning, although it improves your health in many respects, does not cancel out the negative metabolic effects of sitting for hours later in the day.

The best way to prevent the health problems associated with sitting is to avoid remaining seated for long, uninterrupted intervals. When you must sit to carry out some task, take periodic short breaks to stand or walk around. Any activity that requires your leg muscles to support your body weight leads to increased production of LPL.

Section 32.1 All animals are capable of **locomotion** during part or all of their life cycle. Friction and gravity oppose efforts to move. In water, buoyancy reduces the effects of gravity, but friction is greater than that on land. Streamlined bodies help swimming animals minimize drag in water. Many animals that creep along the ground reduce friction by having smooth body parts or an ability to secrete mucus. Air is less dense than water, so most land animals that walk, run, hop, or fly expend energy mainly to overcome gravity. The wing shape of flying animals helps lift them off the ground.

Section 32.2 Animals move when their muscles exert pressure on skeletal elements. Soft-bodied invertebrates have a **hydrostatic skeleton**, chambers full of fluid that muscles exert force against. Arthropods such as flies have an **exoskeleton** of secreted hard parts at the body surface. An **endoskeleton** consists of hardened parts inside the body. Echinoderms and vertebrates have an endoskeleton.

All vertebrates have similar skeletal components. The vertebrate skull, vertebral column, and ribs constitute the **axial skeleton**. The **vertebral column** consists of **vertebrae** with **intervertebral disks** between them. Bony fins or limbs and the pectoral and pelvic girdles that attach them to the backbone constitute the **appendicular skeleton**.

Section 32.3 Bones consist of living cells in a secreted matrix of collagen hardened with calcium and phosphorus. In addition to having a role in movement, bones store minerals and protect organs. The shaft of a long bone such as a femur consists of **compact bone** that contains **yellow marrow**. Lightweight **spongy bone**, as in the ends of long bones, contains **red marrow** that makes blood cells. In a human embryo, many bones develop from a cartilage model. Even in adults, bones are continually remodeled. A **joint** is an area of close contact between bones. At **synovial joints**, cartilage-covered ends of bones are separated by a tiny fluid-filled gap and held in place by **ligaments**.

Section 32.4 Skeletal muscles bring about voluntary movements of body parts, take part in reflexes, function in breathing, and generate body heat. A sheath of connective tissue surrounds each skeletal muscle and extends beyond it as a **tendon**. Most often, the tendon connects the muscle to a bone. A muscle can only exert force in one direction; it can pull but not push. Some skeletal muscles work in pairs to oppose one another's actions. The biceps and triceps of the arm are an example. Skeletal muscles also function as **sphincters**.

Sections 32.5, 32.6 The internal organization of a skeletal muscle promotes a strong, directional contraction. A **skeletal muscle fiber** contains many **myofibrils**. Each myofibril consists of **sarcomeres**, basic units of muscle contraction, lined up end to end. A sarcomere has parallel arrays of **actin** and **myosin** filaments. The **sliding-filament model** describes how ATP-driven sliding of actin filaments past myosin filaments shortens a sarcomere. Shortening of all sarcomeres in all myofibrils of a stimulated muscle fiber causes the fiber to contract.

Sections 32.7, 32.8 Motor neurons control skeletal muscle. Each motor neuron and the skeletal muscle fibers it synapses on constitute a **motor unit**. Acetylcholine released at a neuromuscular junction triggers an action potential in a muscle fiber. The action potential travels along the plasma membrane of the fiber, and along T tubules to the **sarcoplasmic reticulum**. Calcium ions released by this organelle allow actin and myosin heads to interact so muscle contraction can occur. **Muscle tension** is the force generated by muscle contraction.

Aerobic respiration, lactate fermentation, and phosphate transfers from creatine phosphate provide muscle fibers with the ATP needed for contraction. Red fibers have oxygen-storing **myoglobin** and many mitochondria. They make ATP mainly by aerobic respiration. White fibers have no myoglobin and they make ATP mainly by fermentation. Most muscles include a mix of red and white fibers.

Application Exercise increases muscle strength and makes a muscle less easily fatigued. Prolonged muscle inactivity, as by sitting for long periods, causes changes in muscle metabolism that increase the risk of health problems.

SELF-ASSESSMENT
ANSWERS IN APPENDIX VII

1. A hydrostatic skeleton consists of _____ .
   a. a fluid-filled chamber or chambers
   b. hardened plates at the surface of a body
   c. internal hard parts

2. Release of _____ from the sarcoplasmic reticulum opens binding sites and allows actin and myosin to interact.
   a. ACh   c. calcium ions
   b. potassium ions d. oxygen

3. Bones move when _____ muscles contract.
   a. cardiac   c. smooth
   b. skeletal   d. any

4. A ligament connects _____ .
   a. bones at a joint   c. a muscle to a tendon
   b. a muscle to a bone   d. a tendon to bone

5. Skeletal muscle can only _____ bones.
   a. pull on   b. push against

6. Movement brought about by contraction of the arm's biceps muscle is _____ by contraction of the triceps muscle.
   a. prevented   c. reversed
   b. accelerated   d. unaffected

BUILDING STRONGER BONES Tiffany (right) was born with multiple fractures in her limbs. By age six, she had undergone surgery to correct more than 200 bone fractures. Her fragile bones are symptoms of osteogenesis imperfecta (OI), a genetic disorder caused by a mutation in a gene for collagen. As bone is laid down, collagen forms a scaffold for deposition of mineralized bone tissue. The scaffold forms improperly in children with OI. **Figure 32.19** shows the results of a test of a drug treatment for OI. Bone growth and health of treated children, all less than two years old, were compared to that of a control group of similarly affected, same-aged children who did not receive the drug.

1. How many treated children had bone growth (an increase in vertebral area)? How many untreated children?

2. How did the rate of fractures in the two groups compare?

3. Do the results support the hypothesis that this drug increases bone growth and reduces fractures in children with OI?

Treated child	Vertebral area in cm^2 (Initial)	(Final)	Fractures per year
1	14.7	16.7	1
2	15.5	16.9	1
3	6.7	16.5	6
4	7.3	11.8	0
5	13.6	14.6	6
6	9.3	15.6	1
7	15.3	15.9	0
8	9.9	13.0	4
9	10.5	13.4	4
Mean	11.4	14.9	2.6

Control child	Vertebral area in cm^2 (Initial)	(Final)	Fractures per year
1	18.2	13.7	4
2	16.5	12.9	7
3	16.4	11.3	8
4	13.5	7.7	5
5	16.2	16.1	8
6	18.9	17.0	6
Mean	16.6	13.1	6.3

Figure 32.19 Results of a clinical trial of a drug treatment for osteogenesis imperfecta (OI). This disorder is also known as brittle bone disease.

Nine children with OI received the drug. Six others were untreated controls. The surface area of specific vertebrae was measured before and after treatment. An increase in vertebral area during the 12-month period of the study was used as an indicator of bone growth. Researchers also recorded the number of fractures that occurred during the 12-month trial.

7. Calcium release from bone is stimulated by _____ .
   a. parathyroid hormone   c. myoglobin
   b. estrogen   d. creatine

8. _____ is a motor protein that has binding sites for ATP and actin.
   a. Myoglobin   b. Myosin   c. Lactate   d. Creatine

9. A sarcomere shortens when _____ .
   a. thick filaments shorten
   b. thin filaments shorten
   c. both thick and thin filaments shorten
   d. none of the above

10. ATP for muscle contraction can be provided by _____ .
    a. aerobic respiration
    b. lactate fermentation
    c. transfer of a phosphate from creatine phosphate
    d. all of the above

11. Red muscle fibers get their color from _____ .
    a. ATP   b. myosin   c. myoglobin   d. collagen

12. A motor unit is _____ .
    a. a muscle and the bone it moves
    b. two muscles that work in opposition
    c. the amount a muscle shortens during contraction
    d. a motor neuron and the muscle fibers it controls

13. Cheetahs and human sprinters tend to have a high proportion of _____ fibers in their leg muscles.
    a. white   b. fast red   c. slow red

14. The _____ helps reduce friction during locomotion.
    a. external skeleton of a fly
    b. elongated backbone of a cheetah
    c. streamlined body of a predatory fish
    d. hydrostatic skeleton of an earthworm

15. Match each item with its description.
    ___ tendon
    ___ Z line
    ___ myoglobin
    ___ joint
    ___ actin
    ___ red marrow
    ___ metacarpals
    ___ muscle fiber
    ___ sarcoplasmic reticulum

    a. stores and releases calcium
    b. all in the hands
    c. produces blood cells
    d. multinucleated contractile cell
    e. area of contact between bones
    f. at sarcomere ends
    g. reversibly binds oxygen
    h. connects muscle to bone
    i. main component of thin filaments in sarcomere

1. People with some muscle-weakening disorders such as muscular dystrophy sometimes benefit from creatine supplements. How might these supplements be beneficial?

2. After death, a person no longer makes ATP, so calcium stored in the sarcoplasmic reticulum diffuses down its concentration gradient into the muscle cytoplasm. The result is rigor mortis—an unbreakable state of muscle contraction that stiffens the body for a few days until muscles begin to decay. Explain why this contraction occurs.

3. Continued strenuous activity can cause lactate to accumulate in muscles. After the activity stops, the lactate is converted into pyruvate and used as an energy source. Explain how pyruvate can be used to produce ATP.

for additional quizzes, flashcards, and other study materials
► ACCESS MINDTAP AT WWW.CENGAGEBRAIN.COM

33

Circulation

Links to Earlier Concepts

This chapter expands on earlier introductions to animal circulatory systems (Sections 23.6, 23.7, 24.3, 24.5). It draws on your knowledge of hemoglobin (3.4, 9.5), cell junctions (4.10), and diffusion and osmosis (5.6). We consider the function of epithelium (28.2), cardiac muscle (28.4), and autonomic nerves (29.6), discuss the role of cholesterol (3.3), and see more examples of morphological convergence (16.7).

Core Concepts

Adaptation of organisms to a variety of environments has resulted in diverse structures and physiological processes.

 In the simplest animals, materials move into, out of, and through a body by diffusion alone. However, most animals have a circulatory system that facilitates distribution of materials through the body. Differences among those systems reflect adaptations to different environments.

All organisms alive today are linked by lines of descent from shared ancestors.

 Shared core processes and features provide evidence of shared ancestry. All vertebrates have a closed circulatory system that includes a single heart. In fishes, that heart has two chambers and pumps blood through a single circuit. Evolutionary modification of these chambers produced hearts that pump blood through two circuits.

Interactions among the components of a biological system give rise to complex properties.

 The cellular components of vertebrate blood arise in bone and are suspended in a liquid plasma. Circulation occurs when coordinated contractions of cardiac muscle cells force blood through a system of blood vessels. Adjustment to the width of some these vessels alters the distribution of blood based on the body's metabolic requirements.

◀ Hummingbirds have the fastest heart rate of any vertebrate. During flight, their heart beats up to 1,200 times per minute to keep flight muscles well supplied with oxygen.

33.1 Circulatory Systems

LEARNING OBJECTIVES

Using appropriate examples, explain how some animals survive without a circulatory system.

Describe an open circulatory system and a closed circulatory system.

Compare the path of blood flow in a fish and a mammal.

Discuss the advantages of a four-chambered heart.

All animals must keep their cells supplied with nutrients and oxygen, and all must rid themselves of cellular wastes. Some invertebrates, including cnidarians and flatworms, rely on diffusion alone to accomplish these tasks. In these animals, nutrients and gases reach cells by diffusing across a body surface and then diffusing through interstitial fluid (the fluid between cells). Wastes diffuse in the opposite direction. Diffusion occurs too slowly to move materials quickly over a long distance, so animals that rely entirely on diffusion to distribute materials have a body plan in which all cells are close to a body surface. The evolution of circulatory systems allowed larger bodies and more complex body plans. A **circulatory system** is an organ system that speeds distribution of materials within an animal's body.

OPEN AND CLOSED SYSTEMS

A circulatory system typically includes one or more **hearts** (muscular pumps) that propel a circulatory fluid through vessels that extend through the body.

Different types of circulatory systems evolved in different animal lineages. Arthropods and most mollusks have an **open circulatory system**, in which a heart or hearts pump circulatory fluid into open-ended vessels. The fluid, which is called **hemolymph**, flows out of vessels and comes in direct contact with body tissues before being drawn back into the heart through openings in the heart wall (**Figure 33.1A**).

By contrast, annelids (**Figure 33.1B**), cephalopod mollusks, and all vertebrates have a **closed circulatory system**, in which vessels form a continuous loop. The circulatory fluid of a closed system is called **blood.** The system is "closed" in that blood does not leave blood vessels to bathe tissues. Instead, most exchanges between blood and the cells of other tissues take place across the walls of the smallest-diameter blood vessels, which are called **capillaries**.

A closed circulatory system distributes substances faster than an open one. It also provides a greater degree of control over distribution of materials to specific body regions. In a closed system, blood vessels that supply a specific organ or structure can be widened or narrowed as needed.

VERTEBRATE CIRCULATORY SYSTEMS

All vertebrates have a closed circulatory system with a single heart. However, the structure of the heart and the circuits through which blood flows vary among vertebrate groups. In most fishes, the heart has two chambers, and blood flows in a single circuit (**Figure 33.2A**). One chamber of the heart, an **atrium** (plural, atria), receives blood. From there, the blood

Figure 33.1 Comparison of open and closed invertebrate circulatory systems.

A Open circulatory system. The long, tubular heart in a grasshopper's abdomen pumps yellowish hemolymph through a large vessel and out into tissue spaces. Hemolymph exchanges materials with cells, then reenters the heart through openings in the heart wall.

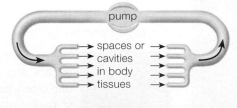

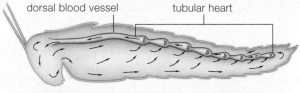

B Closed circulatory system. An earthworm's hearts pump blood through a continuous system of vessels that extend through the body. Exchanges between blood and the tissues take place across the walls of the smallest vessels (the capillaries).

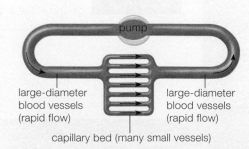

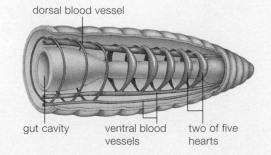

CREDITS: (1) right, After M. Labarbera and S. Vogel, *American Scientist*, 1982, 70:54–60.

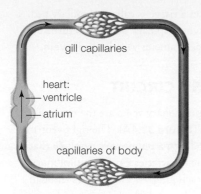 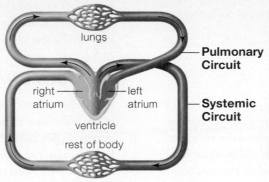 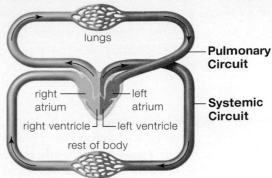

A The fish heart has one atrium and one ventricle. Contraction of the single ventricle propels blood through a single circuit.

B In amphibians and most reptiles, the heart has three chambers: two atria and one ventricle. Blood flows in two partially separated circuits. Oxygenated blood and oxygen-poor blood mix a bit in the ventricle.

C In birds and mammals, the heart has four chambers: two atria and two ventricles. Oxygenated blood and oxygen-poor blood do not mix.

Figure 33.2 Vertebrate circulatory systems. In this and other diagrams, red represents vessels carrying oxygenated blood and blue represents vessels carrying oxygen-poor blood.

enters a **ventricle**, a chamber that pumps blood out of the heart. Pressure exerted by contractions of the ventricle drives blood through a series of vessels, into capillaries inside each gill, through capillaries in body tissues and organs, and back to the heart. Pressure imparted to the blood by the ventricle's contraction dissipates as blood travels through capillaries, so the blood is not under much pressure when it leaves the gill capillaries, and even less as it travels back to the heart.

Adapting to life on land involved coordinated modifications of respiratory and circulatory systems. Amphibians and most reptiles have a three-chambered heart, with two atria emptying into one ventricle (**Figure 33.2B**). Evolution of a three-chambered heart allowed blood to move through two partially separated circuits and to move more quickly. The force of one contraction propels blood through the **pulmonary circuit**, to the lungs and then back to the heart. (The Latin word *pulmo* means lung.) Pressure from a second contraction speeds the oxygenated blood through the longer **systemic circuit**. This circuit encompasses the capillaries in body tissues and returns to the heart.

The ventricle became fully separated into two chambers in birds and mammals. Their four-chambered heart has two atria and two ventricles (**Figure 33.2C**). With two fully separate circuits, blood flow is rapid and oxygen-rich blood and oxygen-poor blood never mix. Such a system has the added advantage of allowing blood pressure to be regulated independently in each circuit. Strong contraction of the heart's left ventricle moves blood quickly through the lengthy systemic circuit. The right ventricle contracts more gently, protecting the delicate lung capillaries that would be blown apart by higher pressure.

The four-chambered heart of mammals and birds is an example of morphological convergence (Section 16.7). The two groups do not share an ancestor that had a four-chambered heart. Rather, this trait evolved independently in each group. Enhanced oxygen delivery provided by a four-chambered heart supports the high metabolism of these endothermic (heated from within) animals (Section 24.4). Endotherms have greater energy needs than comparably sized ectotherms because more of the energy they release by aerobic respiration is lost to the environment as heat.

atrium (AY-tree-um) Heart chamber that receives blood and pumps it into a ventricle.

blood Circulatory fluid of a closed circulatory system.

capillary Smallest diameter blood vessel; exchanges with interstitial fluid occur across its wall.

circulatory system Organ system consisting of a heart or hearts and vessels that distribute circulatory fluid through a body.

closed circulatory system Circulatory system in which blood flows through a continuous system of vessels, and substances are exchanged across the walls of the smallest vessels.

heart Muscular organ that pumps blood through a body.

hemolymph (HEEM-oh-limf) Fluid that circulates in an open circulatory system.

open circulatory system Circulatory system in which hemolymph leaves vessels and flows among tissues before returning to a heart.

pulmonary circuit (PUL-mon-erry) Circuit through which blood flows from the heart to the lungs and back.

systemic circuit (sis-TEM-ick) Circuit through which blood flows from the heart to the body tissues and back.

ventricle (VEN-trick-uhl) Heart chamber that receives blood from an atrium and pumps it out of the heart.

LEARNING OBJECTIVES

Describe the path of blood flow through the systemic and pulmonary circuits.

Explain the difference between arteries and veins.

Describe the function of the hepatic portal vein.

A network of capillaries in a tissue is called a capillary bed. After blood passes through a capillary bed, it is returned to the heart by way of a large-diameter vessel called a **vein**.

THE PULMONARY CIRCUIT

Pulmonary arteries and pulmonary veins are the main vessels of the pulmonary circuit (**Figure 33.4A**). The right ventricle pumps oxygen-poor blood into a pulmonary trunk ❶ that branches into two pulmonary arteries ❷. One **pulmonary artery** delivers blood to each lung. As blood flows through pulmonary capillaries (capillaries in the lungs), it picks up oxygen and gives up carbon dioxide. Oxygen-enriched blood then returns to the heart in **pulmonary veins** ❸, which empty into the left atrium.

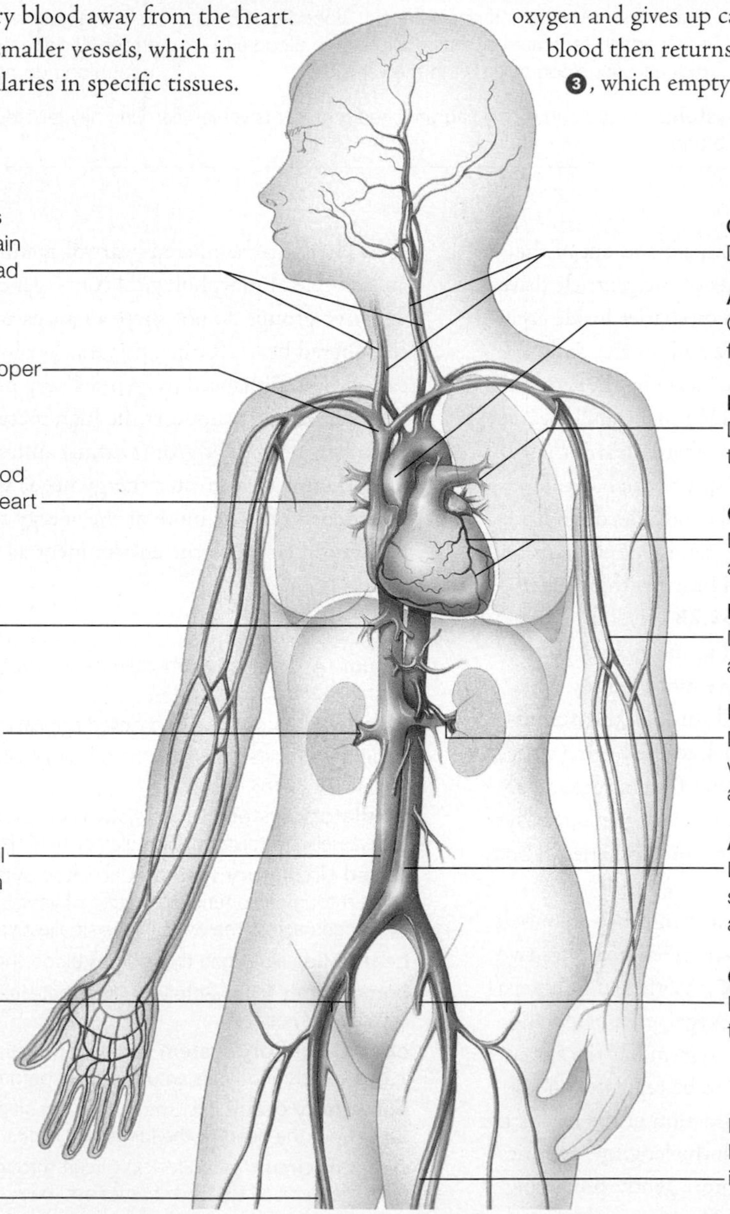

Like other mammals, humans have a circulatory system in which a four-chambered heart pumps blood through two circuits. **Figure 33.3** shows the location and function of our major blood vessels. In each circuit, the heart pumps blood out of a ventricle and into **arteries**, which are large-diameter blood vessels that carry blood away from the heart. Arteries branch into smaller vessels, which in turn branch into capillaries in specific tissues.

Internal Jugular Veins
Receive blood from brain and from tissues of head

Superior Vena Cava
Receives blood from upper limbs, neck, and head

Pulmonary Veins
Deliver oxygenated blood from the lungs to the heart

Hepatic Veins
Carry blood that has passed through small intestine and then liver

Renal Veins
Carry blood away from the kidneys

Inferior Vena Cava
Receives blood from all veins below diaphragm

Common Iliac Veins
Carry blood from the pelvis and the lower limbs

Femoral Veins
Carry blood from the lower limbs

Common Carotid Arteries
Deliver blood to neck, head, brain

Ascending Aorta
Carries oxygenated blood away from the heart

Pulmonary Arteries
Deliver oxygen-poor blood from the heart to the lungs

Coronary Arteries
Deliver blood to the constantly active muscle of the heart

Brachial Arteries
Deliver blood to the upper arm and forearm

Renal Arteries
Deliver blood to the kidneys, where its volume and composition are adjusted

Abdominal Aorta
Delivers blood to the digestive system, kidneys, pelvic organs, and the lower limbs

Common Iliac Arteries
Deliver blood to the pelvis and the lower limbs

Femoral Arteries
Deliver blood to the thigh and inner knee

Figure 33.3 Major blood vessels of a human. Red indicates oxygen-rich blood; blue indicates oxygen-poor blood.

A Pulmonary Circuit

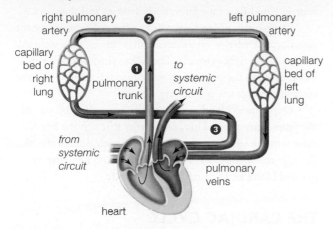

B Systemic Circuit

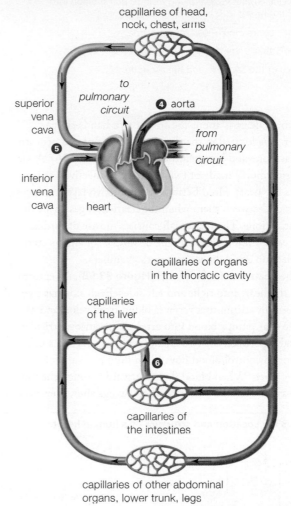

Figure 33.4 **Circuits of the human cardiovascular system.**

THE SYSTEMIC CIRCUIT

Oxygenated blood travels from the heart, to body tissues, and back by way of the systemic circuit (**Figure 33.4B**). The heart's left ventricle pumps blood into the body's largest artery, the **aorta** ❹. Branches from the aorta convey blood to regions throughout the body. The initial portion of the aorta (the ascending aorta) carries blood toward the head. The common carotid arteries, which service the brain, and coronary arteries, which service heart muscle, branch off from the aortic arch. The aorta then turns and descends through the thorax, continuing into the abdomen. Branches from descending portions of the aorta service most internal organs and the lower limbs. For example, renal arteries deliver blood to kidneys, and femoral arteries carry blood to each leg.

Blood traveling in the systemic circuit gives up oxygen and picks up carbon dioxide as it flows through capillaries in organs and tissues. Blood returns to the heart by way of two large veins ❺. The **superior vena cava** returns blood from the head, neck, upper trunk, and arms. Smaller veins carrying blood from the head and upper limbs drain into the superior vena cava. The longer **inferior vena cava** returns blood from the lower trunk and legs. Renal veins and femoral veins are among the vessels that drain into the inferior vena cava.

In most cases, blood flows through only one capillary bed before returning to the heart. However, blood that passes through the capillaries in the small intestine enters a vein (the hepatic portal vein) that delivers it to a capillary bed in the liver ❻. Journeying through these two capillary beds allows blood to pick up glucose and other substances absorbed from the gut, and deliver them to the liver. The liver stores some of the absorbed glucose as glycogen (Section 3.2). It also filters out and breaks down some ingested toxins, including alcohol.

---

aorta (ay-OR-tuh) Large artery that receives oxygenated blood pumped out of the heart's left ventricle.

artery Large-diameter vessel; carries blood away from the heart.

inferior vena cava (VEE-nah CAY-vuh) Vein that delivers oxygen-poor blood from the lower body to the heart.

pulmonary artery Artery that carries oxygen-poor blood from the heart to a lung.

pulmonary vein Vein that carries oxygen-rich blood from a lung to the heart.

superior vena cava (VEE-nah CAY-vuh) Vein that delivers oxygen-poor blood from the upper body to the heart.

vein Large-diameter vessel that returns blood to the heart.

State the major vessel(s) that receive blood from or pump blood into each chamber of the heart.

Explain why ventricles are more muscular than atria.

Describe the events that occur during one cardiac cycle.

Compare the roles of the SA node and the AV node.

The human heart is about the size of two hands clasped together. It is located in the thoracic cavity, centered beneath the breastbone and between the lungs (**Figure 33.5A**). A sac (the pericardium) made of two layers of connective tissue encloses the heart. Fluid between the sac's two layers reduces the friction between them when the heart changes shape during contractions. The heart wall consists mostly of cardiac muscle cells. Endothelium, a type of simple squamous epithelium (Section 28.2), lines the heart's chambers.

The heart is a double pump (**Figure 33.5B**). A septum divides the heart into right and left sides. Each side has two chambers: an atrium that receives blood from veins, and a ventricle that pumps blood into arteries. An atrioventricular (AV) valve between the two chambers functions like a one-way door to control blood flow. High fluid pressure forces the valve open. When blood flows outward through the valve, this pressure declines. The valve then swings shut, preventing

blood from moving backward. Other pressure-sensitive valves control blood flow into pulmonary arteries and the aorta.

Oxygen-poor blood delivered to the right atrium by the superior and inferior venae cavae flows through the right AV valve into the right ventricle. The right ventricle pumps this blood through the pulmonary valve into the pulmonary arteries, and through the pulmonary circuit.

Oxygenated blood returns to the left atrium by way of the pulmonary veins. This blood flows through the left AV valve into the left ventricle. The left ventricle pumps the blood through the aortic valve into the aorta. From here, the oxygenated blood flows to tissues of the body.

THE CARDIAC CYCLE

The series of events that occur from the onset of one heartbeat to the next are collectively called the **cardiac cycle** (Figure 33.6). During this cycle, the heart's chambers alternate through **diastole** (relaxation) and **systole** (contraction). First, the relaxed atria expand with blood ❶. Fluid pressure from the blood forces open the AV valves, so blood flows into the relaxed ventricles. Next, the atria contract ❷, forcing more blood into the ventricles. Once filled, the ventricles contract. Contraction raises the fluid pressure inside the ventricles, forcing the AV valves closed. The aortic and pulmonary valves

Figure 33.5 Location and structure of a human heart.

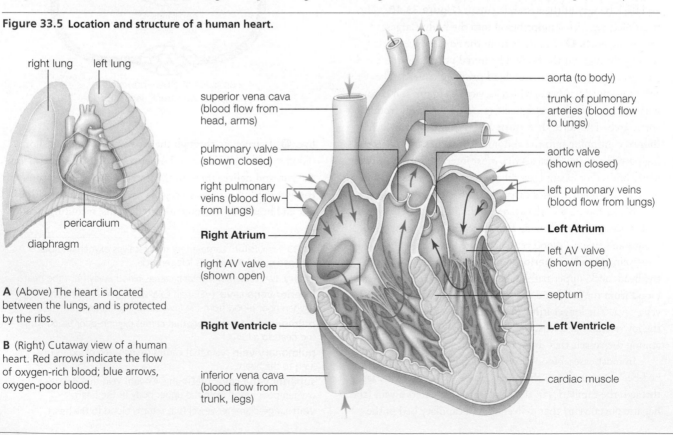

right lung   left lung

pericardium

diaphragm

A (Above) The heart is located between the lungs, and is protected by the ribs.

B (Right) Cutaway view of a human heart. Red arrows indicate the flow of oxygen-rich blood; blue arrows, oxygen-poor blood.

superior vena cava (blood flow from head, arms)

pulmonary valve (shown closed)

right pulmonary veins (blood flow from lungs)

Right Atrium

right AV valve (shown open)

Right Ventricle

inferior vena cava (blood flow from trunk, legs)

aorta (to body)

trunk of pulmonary arteries (blood flow to lungs)

aortic valve (shown closed)

left pulmonary veins (blood flow from lungs)

Left Atrium

left AV valve (shown open)

septum

Left Ventricle

cardiac muscle

1 Relaxed atria fill. Fluid pressure opens AV valves, so blood flows into the relaxed ventricles.

2 The atria contract, squeezing more blood into the relaxed ventricles.

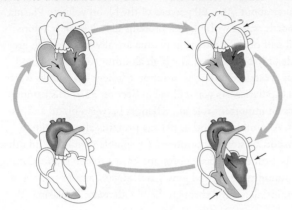

4 Blood flows into arteries, so pressure inside the ventricles declines. The decline causes aortic and pulmonary valves to close.

3 Ventricles start to contract. Rising pressure pushes AV valves shut. As the pressure continues to rise, aortic and pulmonary valves open.

Figure 33.6 The cardiac cycle.

FIGURE IT OUT
Which numbered graphic shows the onset of ventricular systole?

3 :rewsnA

are then forced open, allowing blood to flow through these valves and out of the ventricles **3**. Now emptied, the ventricles relax while the atria fill once again **4**.

During the cardiac cycle a "lub-dup" sound can be heard through the chest wall. Each "lub" is the sound of the heart's AV valves closing. Each "dup" is the sound of the aortic and pulmonary valves closing. If a valve does not close properly, some blood is forced backward through the defective valve. The result is a whooshing sound known as a heart murmur. Most heart murmurs do not threaten health, but a defective valve can be surgically repaired or replaced if necessary.

Contraction of ventricles provides the driving force for circulation; contraction of atria only helps fill ventricles. The structure of the cardiac chambers reflects their different functions. Atria need only generate enough force to squeeze blood

atrioventricular (AV) node (AY-tree-oh-ven-TRICK-you-lur) Clump of cells that conveys excitatory signals between the atria and ventricles.
cardiac cycle Sequence of contraction and relaxation of heart chambers that occurs with each heartbeat.
diastole (die-ASS-toll-ee) Relaxation phase of the cardiac cycle.
sinoatrial (SA) node (sigh-no-AY-tree-uhl) Cardiac pacemaker; group of cells that spontaneously emits rhythmic action potentials that result in contraction of cardiac muscle.
systole (SIS-toll-ee) Contractile phase of the cardiac cycle.

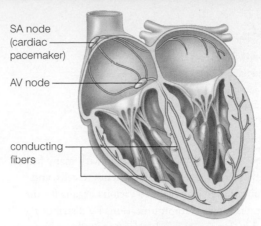

SA node (cardiac pacemaker)

AV node

conducting fibers

Figure 33.7 Cardiac conduction system.

into ventricles, so they have relatively thin walls. Ventricle walls are more thickly muscled because their contraction must generate enough pressure to propel blood through an entire circuit. The left ventricle, which pumps blood through the long systemic circuit, has thicker walls than the right ventricle, which pumps blood to the lungs and back to the heart.

SETTING THE PACE

Like skeletal muscle, cardiac muscle has orderly arrays of sarcomeres that contract by a sliding-filament mechanism (Section 32.6). Unlike skeletal muscle cells, cardiac muscle cells have gap junctions between adjacent cells. A gap junction is a tunnel-like cell junction that connects the cytoplasm of two adjacent cells (Section 4.10). The presence of gap junctions allows ions, and thus action potentials, to spread swiftly between cardiac muscle cells.

The **sinoatrial (SA) node**, a clump of specialized cells in the wall of the right atrium (**Figure 33.7**), is known as the cardiac pacemaker because it sets the heart rate. About seventy times a minute, the SA node generates an action potential that spreads across the atria, causing them to contract. Simultaneously, the action potential travels through specialized noncontractile muscle fibers to a clump of cells called the atrioventricular node. The **atrioventricular (AV) node** is the only place where action potentials can cross to ventricles. The time it takes for an action potential to cross the AV node allows blood from atria to fill ventricles before they contract.

From the AV node, the signal travels along conducting fibers in the septum between the heart's left and right halves. The fibers extend to the heart's lowest point and up the ventricle walls. As action potentials spread from the base of the ventricles upward via gap junctions, both ventricles contract from the bottom up, with a wringing motion.

PLASMA

The fluid portion of the blood, which is called **plasma**, constitutes about 50 to 60 percent of the blood volume. Plasma is mostly water with plasma proteins dissolved in it. More than half of the proteins in plasma are albumins, which are made only by the liver. The high albumin content of blood helps create a solute concentration gradient that draws water into capillaries. As you will see in Section 33.7, this osmosis plays an important role in exchanges between blood and tissues. Albumins and other plasma proteins also function in transport of steroid hormones, fat-soluble vitamins, and other lipids. Still other plasma proteins have a role in blood clotting or immunity. Dissolved ions, gases, sugars, amino acids, and vitamins also travel through the bloodstream in plasma.

CELLULAR COMPONENTS

The cellular portion of blood consists of various blood cells and platelets. All arise from stem cells in bone marrow.

Red Blood Cells Erythrocytes, or **red blood cells**, transport oxygen from lungs to aerobically respiring cells, and facilitate movement of carbon dioxide to the lungs. In all mammals, red blood cells lose their nucleus and other organelles as they develop. Mature red blood cells are flexible disks with a depression at their center. Their flexibility allows them to slip easily through narrow blood vessels, and their flattened shape facilitates gas exchange.

A mature red blood cell contains a large amount of hemoglobin, a respiratory protein you learned about in Section 3.4. Most oxygen that enters the blood travels to tissues while bound to the heme group of hemoglobin. In addition to hemoglobin, a mature red blood cell contains sugars, RNAs, and other molecules that sustain it for about 120 days. Ongoing replacements keep a healthy person's red blood cell count at a fairly stable level.

Red blood cells are the most abundant type of blood cell. Healthy human adults have between 4 and 6 million of these cells per microliter of blood. (A microliter is one-millionth of a liter.) During reproductive years, women typically have a lower red blood cell count than men, because women lose blood during menstruation.

Anemia is a disorder in which the red blood cell count declines or hemoglobin is defective. As a result, oxygen delivery slows and metabolism falters. Anemia can arise as a result of a dietary iron deficiency (iron is needed to make hemoglobin), destruction of red blood cells by pathogens (as occurs in malaria), excessive blood loss (as from unusually heavy menstrual bleeding), and genetic disorders. Sickle-cell anemia arises from a mutation that causes hemoglobin to form large

Vertebrate blood is a fluid connective tissue with many functions. It carries essential oxygen and nutrients to cells, and carries metabolic wastes from cells to various organs for disposal. It facilitates internal communications by distributing hormones, and it also serves as a highway for cells and proteins that protect and repair tissues. In birds and mammals, blood helps maintain a stable internal temperature by distributing heat generated by muscle activity to the skin, where it can be lost to the surroundings.

The 5 liters of blood (a bit more than 10 pints) in an average-sized human adult account for about 6 to 8 percent of the individual's body weight. Blood is—as the saying goes—thicker than water. Dissolved substances and suspended cells contribute to its greater viscosity. **Figure 33.8** shows the components of vertebrate blood.

plasma

cells and platelets

red blood cell · white blood cell · platelet

Figure 33.8 Components of vertebrate blood. The micrograph shows the cellular components.

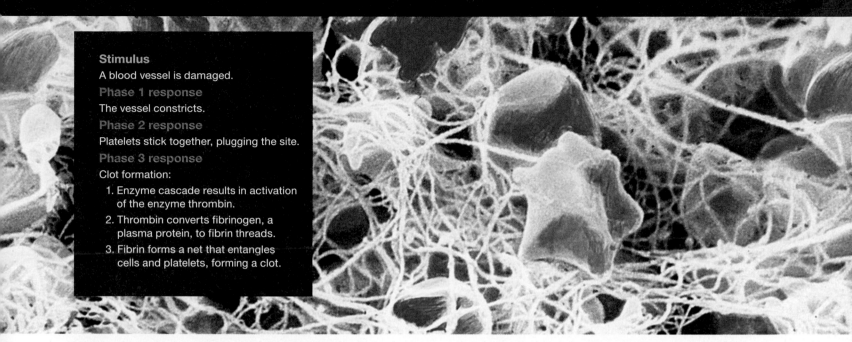

Stimulus
A blood vessel is damaged.
Phase 1 response
The vessel constricts.
Phase 2 response
Platelets stick together, plugging the site.
Phase 3 response
Clot formation:
1. Enzyme cascade results in activation of the enzyme thrombin.
2. Thrombin converts fibrinogen, a plasma protein, to fibrin threads.
3. Fibrin forms a net that entangles cells and platelets, forming a clot.

Figure 33.9 Hemostasis. The micrograph shows the final clotting phase—blood cells and platelets in a fibrin net.

clumps at low oxygen levels. The clumps distort red blood cells so they get stuck in small blood vessels (Section 9.5).

White Blood Cells Leukocytes, or **white blood cells**, dispose of cellular debris and defend the body against pathogens. We discuss the role of white blood cells in detail in the next chapter, but here is a brief preview. Neutrophils, the most abundant type of white blood cell, engulf bacteria and cellular debris. Eosinophils attack larger parasites, such as worms. Basophils and mast cells secrete chemicals that have a role in inflammation. Monocytes are white cells that circulate in the blood for a few days, then move into the tissues, where they develop into motile cells known as macrophages. Macrophages interact with lymphocytes to bring about immune responses. B cells and T cells are two types of lymphocytes. T cells mature in the thymus. Both protect the body against specific threats.

Leukemias are cancers that originate in stem cells of bone marrow. They are characterized by overproduction of abnormal white blood cells. Lymphomas are cancers that arise from B or T lymphocytes. Division of these cancerous lymphocytes produces tumors in lymph nodes and other parts of the lymphatic system.

Platelets A **platelet** is a membrane-wrapped fragment of cytoplasm that arises when a large cell (a megakaryocyte) breaks up. Once formed, a platelet is functional for up to nine days. Hundreds of thousands of platelets circulate in the blood, ready to take part in **hemostasis**, the process

that stops blood loss from a torn blood vessel and provides a framework for repairs. Hemostasis starts when an injured vessel constricts (narrows) so less blood leaks into the surrounding tissues (**Figure 33.9**). Platelets adhere to the injured site and release substances that attract more platelets. Plasma proteins convert blood plasma to a gel and form a clot. During clot formation, fibrinogen, a soluble plasma protein, is converted to insoluble threads of fibrin. **Fibrin** forms a mesh that traps cells and platelets.

Clot formation involves a cascade of enzyme reactions. Fibrinogen is converted to fibrin by the enzyme thrombin, which circulates in blood as the inactive precursor prothrombin. Prothrombin is activated by an enzyme that is activated by another enzyme, and so on. If a mutation affects any one of the enzymes that act in the cascade of clotting reactions, blood may not clot properly. Such mutations cause the genetic disorder hemophilia (Section 14.3). A vitamin K deficiency can also impair clotting, because this vitamin plays a role in the cascade of enzyme reactions (Section 17.3).

fibrin Threadlike protein that is formed from the soluble plasma protein fibrinogen during blood clotting.
hemostasis Process by which blood clots in response to injury.
plasma Fluid portion of blood.
platelet Cell fragment that helps blood clot.
red blood cell Hemoglobin-filled blood cell that carries oxygen (erythrocyte).
white blood cell Blood cell with a role in housekeeping and defense (leukocyte).

LEARNING OBJECTIVES

Describe how an artery's structure relates to its function.

Explain how arterioles adjust blood flow to body regions and describe some factors that trigger these adjustments.

Explain what you feel when you check a person's pulse.

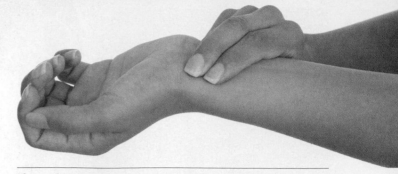

Figure 33.11 Checking the pulse in the radial artery. This artery delivers blood to the hand.

RAPID TRANSPORT IN ARTERIES

Blood pumped out of the left ventricle enters arteries. These large-diameter vessels are ringed by smooth muscle and have an outer covering of highly elastic connective tissue (**Figure 33.10A**). Like all other blood vessels and the heart, they are lined by endothelium.

Elastic properties of an artery help keep blood flowing, even when the ventricles relax. With each ventricular contraction, the pressure exerted by blood forced into an artery makes the artery wall bulge outward. Then, as the ventricle relaxes, the artery wall springs back like a rubber band that has been stretched. As the artery wall recoils, it pushes blood in the artery a bit farther away from the heart.

The bulging of an artery with each ventricular contraction is the **pulse**. You can feel a person's pulse by placing your finger on a pulse point, a body region where an artery runs close to the body surface. For example, to feel the pulse in

your radial artery, put your fingers on your inner wrist near the base of your thumb (**Figure 33.11**).

ADJUSTING FLOW AT ARTERIOLES

All blood pumped out of the right ventricle always flows through pulmonary arteries to your lungs. In the systemic circuit, about 20 percent of the blood always flows through the carotid arteries to the brain. The remainder of the blood is distributed across the rest of the systemic circulation according to the body's needs. Blood flow is adjusted primarily by altering the diameter of **arterioles**, blood vessels that branch from an artery and deliver blood to capillaries. Each arteriole (**Figure 33.10B**) is ringed by smooth muscle. When this muscle relaxes, an arteriole widens. When it contracts, the arteriole narrows.

Nerve signals influence the width of arterioles, as when sympathetic signals trigger a fight–flight response (Section 29.6). Sympathetic stimulation causes **vasodilation** (widening) of arterioles that supply skeletal muscles of the limbs, increasing flow to these muscles. At the same time, **vasoconstriction** (narrowing) of arterioles that deliver blood to the gut decreases their share of the blood supply. Activation of the parasympathetic nervous system has the opposite effect.

Arterioles also adjust blood flow in response to metabolic activity in nearby tissue. During exercise, skeletal muscle uses up oxygen and releases carbon dioxide and lactic acid. Arterioles delivering blood to the muscle widen in response to these changes.

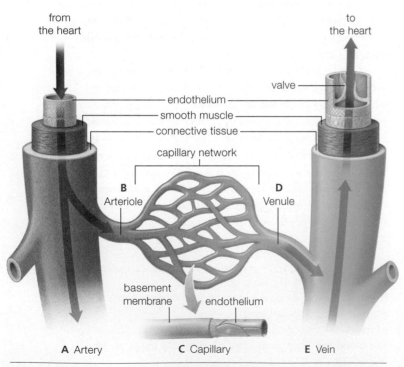

A Artery   **C** Capillary   **E** Vein

Figure 33.10 Structural comparison of blood vessels. The innermost layer of all vessels and the heart is endothelium, a type of simple squamous epithelium.

arteriole Blood vessel that conveys blood from an artery to capillaries.

pulse Brief expansion of artery walls that occurs when ventricles contract.

vasoconstriction Narrowing of a blood vessel when smooth muscle that rings it contracts.

vasodilation Widening of a blood vessel when smooth muscle that rings it relaxes.

CREDITS: (10) From Russell/Wolfe/Hertz/Starr, *Biology*, 1e. © 2008 Cengage Learning, Inc.; (11) caimacanul/Shutterstock.com.

33.6 Blood Pressure

LEARNING OBJECTIVES

Describe how blood pressure varies across the systemic circuit.

Explain how blood pressure is measured and the difference between systolic and diastolic pressure.

Blood pressure is the pressure blood exerts against the wall of the vessel that encloses it. The right ventricle contracts less forcefully than the left ventricle, so blood entering the pulmonary circuit is under less pressure than blood entering the systemic circuit. In both circuits, blood pressure is highest in arteries, and declines as blood flows through the circuit, being lowest in veins (**Figure 33.12**).

Systemic blood pressure is usually measured in the upper arm's brachial artery (**Figure 33.13**). Two pressures are recorded. **Systolic pressure**, the highest pressure of the cardiac cycle, occurs as contracting ventricles force blood into the arteries. **Diastolic pressure**, the lowest blood pressure of the cardiac cycle, occurs when the left ventricle is most relaxed. Blood pressure is measured in millimeters of mercury (mm Hg), a standard unit of pressure. It is written as systolic value/diastolic value. Normal blood pressure is less than 120/80 mm Hg, or "120 over 80."

Systemic blood pressure depends on the total blood volume, how much blood the left ventricle pumps out (cardiac output), and the degree of arteriole dilation. When blood pressure rises or falls, sensory receptors in the aorta and in carotid arteries of the neck signal the medulla oblongata, a region of the hindbrain (Section 29.9). In a reflexive response, the medulla oblongata sends signals that result in appropriate changes in cardiac output and arteriole diameter. Vasodilation of arterioles lowers blood pressure; vasoconstriction raises it. This reflex response adjusts blood pressure over the short term. Over the longer term, kidneys adjust blood pressure by regulating the amount of fluid lost as urine and thus determining the total blood volume.

Inability to regulate blood pressure results in hypertension, a disorder in which resting blood pressure exceeds 140/90. Often, the cause of hypertension is unknown. Heredity is a factor, and African Americans have an elevated risk. Diet plays a role; in some people, high salt intake causes water retention that raises blood pressure. High blood pressure is dangerous; it makes the heart and kidneys work harder, thus increasing risk of heart disease or kidney failure.

---

blood pressure Pressure exerted by blood against a vessel wall.
diastolic pressure (die-ah-STAHL-ic) Blood pressure when the left ventricle is most relaxed; lowest pressure of the cardiac cycle.
systolic pressure (sis-STAHL-ic) Blood pressure when the left ventricle is most contracted; highest pressure of the cardiac cycle.

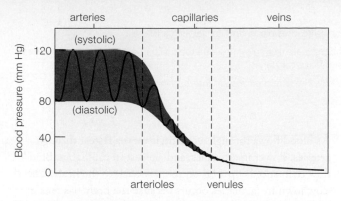

Figure 33.12 Plot of fluid pressure changes as a volume of blood flows through the systemic circuit. Systolic pressure occurs when ventricles contract; diastolic, when ventricles are relaxed.

Figure 33.13 Measuring blood pressure. Typically, a hollow inflatable cuff attached to a pressure gauge is wrapped around the upper arm. The cuff is inflated with air, putting pressure on blood vessels of the arm. Eventually, this pressure becomes so high that it cuts off blood flow through the brachial artery, the main artery of the upper arm.

As air in the cuff is slowly released, blood begins to spurt through the artery when ventricles contract and blood pressure is highest. Spurts of blood flow can be heard through a stethoscope as soft tapping sounds. The blood pressure at this point is the systolic pressure. It is typically about 120 mm Hg. Millimeters of mercury (mm Hg) is a standardized unit of pressure.

More air is released from the cuff. Eventually the sounds stop because blood is flowing continuously, even when ventricles are the most relaxed. The pressure when the sounds stop is the diastolic pressure, the lowest pressure during a cardiac cycle—usually about 80 mm Hg.

Right, compact monitors are now available that automatically test and record systolic/diastolic blood pressure.

33.7 Capillary Exchange

LEARNING OBJECTIVES

Explain what causes blood flow to slow in capillaries.

Using appropriate examples, describe the mechanisms by which substances enter and leave capillaries.

Explain how the lymphatic system interacts with blood capillaries.

As blood flows through a circuit, it moves fastest through arteries, slower in arterioles, and slowest in capillaries. Blood flow then speeds up a bit on the way back to the heart. The slowdown in capillaries occurs because the body has tens of billions of capillaries. Their collective cross-sectional area is far greater than that of the arterioles delivering blood to them, or the veins carrying blood away from them. By analogy, think about what happens if a narrow river (representing few larger vessels) delivers water to a wide lake (representing the many capillaries):

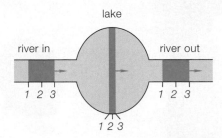

The flow rate is constant, with an identical volume moving from points 1 to 3 in each interval. However, flow velocity decreases in the lake. Here, the volume of water spreads out through a larger cross-sectional area, so it flows forward a shorter distance during any specified interval.

Slow flow through the small capillaries enhances the exchange of substances between the blood and interstitial fluid. The more time blood spends in a capillary, the more time there is for those exchanges to take place.

Materials move between capillaries and body cells in several ways. The capillary wall consists of a single layer of endothelial cells; there is no muscle layer. At the end of the capillary nearest the arteriole, there are spaces between the endothelial cells so the capillary wall is a bit leaky. Pressure exerted by the beating heart forces plasma fluid out through the spaces between cells and into the surrounding interstitial fluid (**Figure 33.14 ❶**). Thus, although we say that vertebrates have a "closed" circulatory system, a small bit of fluid does escape from this system. Plasma proteins, including albumin, are too large to slip through the spaces between cells, so they remain in the vessel.

Along the length of the capillary, oxygen (O_2) released by red blood cells diffuses from blood into the interstitial fluid, while nutrients such as glucose are transported in the same direction by membrane proteins ❷. Carbon dioxide (CO_2) diffuses from interstitial fluid into the capillary, and other metabolic wastes are transported into it ❸.

At end of the capillary nearest the vein, blood pressure is lower. Here, the main factor influencing the movement of fluid is the high protein content of the plasma. Plasma is hypertonic relative to interstitial fluid, so water moves by osmosis from the interstitial fluid into the capillary ❹.

As a result of all the leaking and osmotic movement of fluid, there is a small net outward flow of fluid from a capillary bed into interstitial fluid. Fluid that escapes from the bloodstream by this process is later returned by the lymphatic system. Some interstitial fluid enters lymph capillaries ❺, which drain into lymphatic ducts that return fluid to veins near the heart. We consider the structure and function of the lymph vascular system in more detail in Section 33.10.

Figure 33.14 Capillary exchange and connections with the lymphatic system.

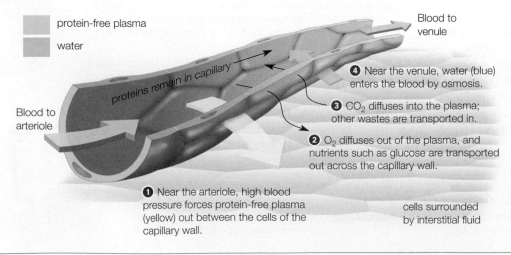

protein-free plasma

water

proteins remain in capillary

Blood to venule

Blood to arteriole

❹ Near the venule, water (blue) enters the blood by osmosis.

❸ CO_2 diffuses into the plasma; other wastes are transported in.

❷ O_2 diffuses out of the plasma, and nutrients such as glucose are transported out across the capillary wall.

❶ Near the arteriole, high blood pressure forces protein-free plasma (yellow) out between the cells of the capillary wall.

cells surrounded by interstitial fluid

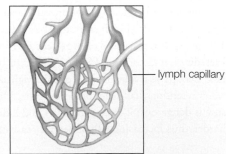

lymph capillary

❺ Some fluid leaked by blood capillaries enters neighboring lymph capillaries. This fluid, now called lymph, returns to the bloodstream when large lymph vessels drain into veins at the base of the neck near the heart.

33.8 Returning Blood to the Heart

LEARNING OBJECTIVES

Describe the functions of venules and of veins.

Explain the mechanisms that keep blood in the veins moving toward the heart.

VENULES AND VEINS

Blood from multiple capillaries flows into a **venule**, which is a small vessel that carries blood to veins. The tiniest venules, which connect directly to capillaries, have no smooth muscle in their walls. White blood cells that leave the circulation to act in other tissues usually exit the blood by squeezing out between cells in the wall of one of these venules. Larger venules and the veins have some elastic tissue and smooth muscle in their wall.

Veins are the body's main blood reservoir. When you are at rest, your veins hold about 60 percent of your total blood volume. Several mechanisms help blood, which is at its lowest pressure in the veins, move back toward the heart. First, veins have one-way valves that prevent backflow. These valves automatically shut when blood in the vein starts to reverse direction. For example, valves in the large veins of your leg prevent blood from moving downward in response to gravity when you stand. All vertebrates typically have the same types of arteries and veins, but the number and location of valves in their veins varies (**Figure 33.15**).

Contraction of smooth muscle in a vein's wall facilitates blood flow too. When this muscle contracts, a vein stiffens so it cannot hold as much blood. As a result, pressure on blood in the vein rises and the blood is forced toward the heart.

Skeletal muscles used in limb movements also help move blood through veins. When these muscles contract, they bulge and press on neighboring veins, squeezing the blood inside these veins toward the heart (**Figure 33.16**).

Exercise-induced deep breathing increases the pressure inside veins too. When the lungs and thoracic cavity expand during inhalation, they force adjacent organs against veins. As with skeletal muscles, the resulting increase in pressure forces the blood inside a vein forward through a valve.

IMPAIRED VENOUS RETURN

If valves in a vein become damaged, blood will pool in that vein. Varicose veins are twisted, blood-engorged veins that can arise as a result of valve damage or an inherited weakness of the vein wall. Chronic high blood pressure or an occupation that requires prolonged standing can increase the risk of varicose veins in the legs.

When blood pools in a vein, it may clot. A clot that forms inside a blood vessel and remains in place is called

Figure 33.15 Moving blood against the force of gravity. When a giraffe lowers its head to drink, gravity pulls oxygen-poor blood in its veins toward its brain. Seven valves in each jugular vein prevent unwanted backflow through the neck. By contrast, humans, who seldom lower their head below their heart, have only one valve in their jugular vein. When a giraffe lifts its head, high blood pressure keeps oxygen-rich blood flowing uphill to the brain. Giraffes have the highest blood pressure of any vertebrate.

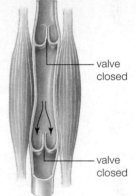

A When skeletal muscles contract, they bulge and press on adjacent veins, forcing blood through a one-way valve.

B When skeletal muscles relax, pressure in adjacent veins falls and valves shut, preventing the blood from moving backward.

Figure 33.16 How skeletal muscle activity encourages blood flow through veins.

a thrombus. A clot or part of a clot that breaks loose and travels through blood vessels to a new location is called an embolus. Either a thrombus or an embolus can endanger health by blocking blood flow to an essential organ. For example, a pulmonary embolus (an embolus in an artery in the lung) can damage the lung. A pulmonary embolus usually arises after a blood clot that formed in a vein of the thigh breaks off and travels through the heart into the lung.

venule (VEEN-yule) Blood vessel that conveys blood from capillaries to a vein.

33.9 Cardiovascular Disorders

LEARNING OBJECTIVES

Describe the different types of arrhythmias and explain which is most dangerous.

Explain the role of HDLs and LDLs in atherosclerosis.

Explain what occurs during a stroke and during a heart attack.

Cardiovascular disorders are conditions in which functions of the heart, the blood vessels, or both are impaired. In the United States, cardiovascular disorders kill about a million people every year. Tobacco smoking tops the list of risk factors for these disorders. Other factors include a family history of such disorders, hypertension, a high cholesterol level, diabetes mellitus, and obesity. Regular exercise helps lower the risk of these disorders even when the exercise is not particularly strenuous. Gender is another factor; until about age fifty, males are at greater risk than females.

ARRHYTHMIAS

Electrocardiograms, or ECGs, record the electrical activity of a beating heart (**Figure 33.17**). They can also reveal arrhythmias, which are abnormal heart rhythms. Bradycardia, an abnormally slow heart rate, is an arrhythmia that can arise from a thyroid hormone deficiency or an SA node malfunction. If the resulting slow blood flow impairs health, an artificial pacemaker is implanted to restore the normal heart rate.

Tachycardia is a faster-than-normal heart rate. Tachycardia is a normal response to exercise. However, many people experience palpitations, which are occasional episodes of tachycardia unrelated to exercise. Signals from sympathetic nerves increase heart rate (Section 29.6), so drugs that mimic or increase such signals can cause palpitations. Caffeine and amphetamines are examples of drugs that can have this effect. Palpitations can also be caused by an overactive thyroid or an SA node malfunction.

Fibrillation is an arrhythmia in which cardiac muscle undergoes irregular, uncoordinated contractions. With atrial fibrillation, the atria do not contract normally but instead quiver. This malfunction slows blood flow, raising the risk of clot formation. Ventricular fibrillation is an even more dangerous arrhythmia in which ventricles quiver rather than pumping normally. The result can be loss of consciousness and—if a normal rhythm is not restored—death.

A device called a defibrillator can sometimes restore a normal heart rhythm in a person with atrial or ventricular fibrillation. A defibrillator has paddles that deliver an electrical shock to the chest to reset the SA node. A defibrillator may be surgically implanted in a person who has had an episode of ventricular fibrillation. An implanted defibrillator automatically provides a shock if another episode occurs.

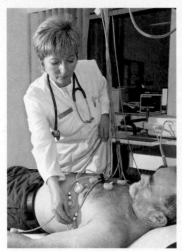
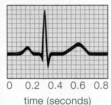

0   0.2 0.4 0.6 0.8
time (seconds)

An ECG (above) is a graph showing electrical changes that can be detected by electrodes attached to the skin (left). The changes arise as a result of the activity of the SA node and transmission of action potentials through the heart.

Figure 33.17 Electrocardiogram.

ATHEROSCLEROSIS

In **atherosclerosis**, the arterial wall thickens, narrowing the vessel's interior diameter. As you may know, cholesterol plays a role in this "hardening of the arteries." The human body requires cholesterol to make cell membranes, myelin sheaths, bile salts, and steroid hormones (Section 3.3). The liver makes enough cholesterol to meet these needs, but more is absorbed from food in the gut. Genetics affects how different people's bodies deal with an excess of dietary cholesterol.

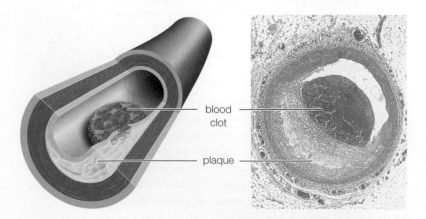

blood clot

plaque

Figure 33.18 Atherosclerosis. Plaque accumulates inside an artery making it more likely to rupture. If a rupture does occur, a clot forms at the rupture site and further narrows the vessel.

Most of the cholesterol dissolved in blood is bound to protein carriers (Section 3.4). The resulting complexes are known as low-density lipoproteins, or LDLs, and most cells can take them up. A lesser amount of cholesterol is bound up in high-density lipoproteins, or HDLs. Cells in the liver take up HDLs and use them to form bile, which the liver secretes into the gut. Eventually, bile leaves the body in feces. Thus HDLs help lower the body's cholesterol level.

Having a high LDL level, a low HDL level, or a combination of the two raises the risk of atherosclerosis. A buildup of lipids in an artery's endothelial lining attracts white blood cells, which invade the vessel wall. Smooth muscle cells divide repeatedly and, together with the white blood cells, form a fibrous cap. Eventually, a mass called an atherosclerotic plaque bulges into the vessel's interior, narrowing the vessel's interior diameter and slowing blood flow. A hardened plaque also makes an artery wall brittle and encourages clot formation (**Figure 33.18**). Atherosclerosis raises the risk that a blood vessel in the brain, heart, or elsewhere will become clogged.

HEART DISEASE AND STROKE

With heart disease, atherosclerosis affects one or more of the coronary arteries, which supply blood to heart muscle. A heart attack occurs when a coronary artery is completely blocked, most commonly by a clot. If the blockage is not removed, cardiac muscle cells die, permanently weakening the heart. Clot-dissolving drugs can restore blood flow, but only if they are given immediately after the onset of an attack.

Any suspected heart attack should receive prompt attention. The most common symptom of a heart attack in men is pain in the center of the chest or at its left side. Often there is also pain elsewhere in the upper body (especially in the back, shoulder, or arm), shortness of breath, and a feeling of fatigue.

A blocked coronary artery can be treated with coronary bypass surgery, in which doctors open a person's chest and use a blood vessel from elsewhere in the body to divert blood around the affected artery (**Figure 33.19A**). It can also be treated by angioplasty, a procedure in which the artery is mechanically widened. In balloon angioplasty, doctors inflate a small balloon in a blocked artery to flatten a plaque. Afterwards, a wire mesh tube called a stent may be inserted into the vessel to keep it open (**Figure 33.19B**).

A stroke is an interruption of blood flow inside the brain. Most strokes occur when a blood vessel in the brain becomes blocked. Atrial fibrillation raises the risk that a clot will form in the heart, travel to the brain, and cause a stroke. For this reason, people with atrial fibrillation are often treated with a drug that impairs clotting. Atherosclerosis raises the risk that a clot will form inside a vessel in the brain and block

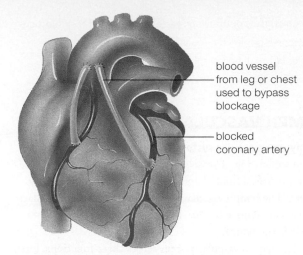

blood vessel from leg or chest used to bypass blockage

blocked coronary artery

A Coronary bypass surgery. Veins from another part of the body are used to divert blood past the blockages. This graphic shows a "double bypass," in which veins are placed to divert blood around two blocked coronary arteries.

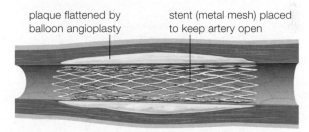

plaque flattened by balloon angioplasty

stent (metal mesh) placed to keep artery open

B Balloon angioplasty and placement of a stent. A balloon-like device is inflated inside an artery to open it and flatten the plaque, then a tube of metal (the stent) is left in place to keep the artery open.

Figure 33.19 Treatments for blocked coronary arteries.

it. A lesser number of strokes result from the bursting of a blood vessel in the brain. High blood pressure increases the risk of this type of stroke.

As with a heart attack, a suspected stroke requires immediate medical attention. The acronym FAST can help you remember how to evaluate and assist someone who may be having a stroke. *F* stands for face; a stroke causes one side of the face to droop, especially when the person smiles. *A* stands for arms; a stroke makes it difficult to hold one's arms out at the same level. *S* stands for speech; a stroke often causes slurred speech. *T* stands for time; if any signs of stroke are detected, minimizing the time to treatment increases the odds of survival and, in survivors, of avoiding brain damage.

atherosclerosis (ath-err-oh-scler-OH-sis) Cardiovascular disorder in which lipid-rich plaques form on the interior walls of blood vessels and narrow the vessels.

LEARNING OBJECTIVES
Describe the components of the lymph vascular system and their functions.
Explain how lymph is propelled through the lymph vessels.
Explain the roles of lymph nodes, spleen, and thymus.

LYMPH VASCULAR SYSTEM

The **lymph vascular system** is a portion of the lymphatic system consisting of vessels that collect water and solutes from interstitial fluid, then deliver them to the circulatory system. The lymph vascular system includes lymph capillaries and vessels (**Figure 33.20**). Fluid that moves through these vessels is the **lymph**.

The lymph vascular system serves three functions. First, its vessels serve as drainage channels for fluid that leaked out of capillaries and must be returned to the circulatory sys-

tem (Section 33.7). Second, it delivers to the blood the fats absorbed by the small intestine. Third, it transports cellular debris, pathogens, and foreign cells to the lymph nodes.

Fluid enters lymph capillaries through clefts between cells making up their walls. Lymph capillaries merge into larger-diameter lymph vessels. Two mechanisms move lymph through these vessels. First, slow, wavelike contractions of smooth muscle in the walls of large lymph vessels propel lymph forward. Second, as with veins, the bulging of adjacent skeletal muscles helps move fluid along. Also like veins, lymph vessels have one-way valves that prevent backflow.

The largest lymph vessels converge on collecting ducts that empty into veins near the heart. Each day, these ducts deliver about 3 liters of fluid to the blood.

LYMPHOID TISSUES

Lymphoid tissues are tissues that contain many lymphocytes (B cells and T cells) that help the body respond to injury and infection. The **thymus**, a hormone-producing organ beneath the sternum (breastbone), is central to immunity. T cells become capable of recognizing pathogens as they travel through the thymus. The thymus grows in size until puberty, then slowly dwindles with additional age.

Lymph nodes are small masses of tissue where many lymphocytes aggregate. Before entering blood, lymph trickles through at least one node. Lymphocytes that recognize a pathogen or toxin in lymph can divide rapidly to form cellular armies that eliminate the threat from the body.

The **spleen**, a fist-sized organ, is located in the upper left abdomen. In embryos, the spleen is a site of red blood cell formation. After birth, it functions in immunity and in quality control of the blood. Phagocytic white blood cells in the spleen engulf and digest any pathogens or worn-out red cells and platelets in the blood that passes through it.

Tonsils are patches of lymphoid tissue at or near the back of the throat. Tonsils help the body respond fast to inhaled pathogens. The lining of the intestine and the appendix also include lymphoid tissue.

The next chapter discusses the lymphatic system's role in immunity in more detail.

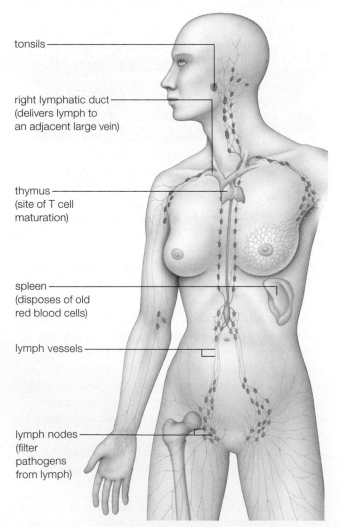

tonsils

right lymphatic duct
(delivers lymph to
an adjacent large vein)

thymus
(site of T cell
maturation)

spleen
(disposes of old
red blood cells)

lymph vessels

lymph nodes
(filter
pathogens
from lymph)

Figure 33.20 Components of the human lymphatic system.

---

lymph Fluid in the lymph vascular system.
lymph node Small mass of lymphoid tissue through which lymph filters; contains many lymphocytes.
lymph vascular system System of vessels that takes up interstitial fluid and carries it (as lymph) to the blood.
spleen Large lymphoid organ that functions in immunity and filters pathogens, old red blood cells, and platelets from the blood.
thymus Hormone-producing organ in which T cells mature.

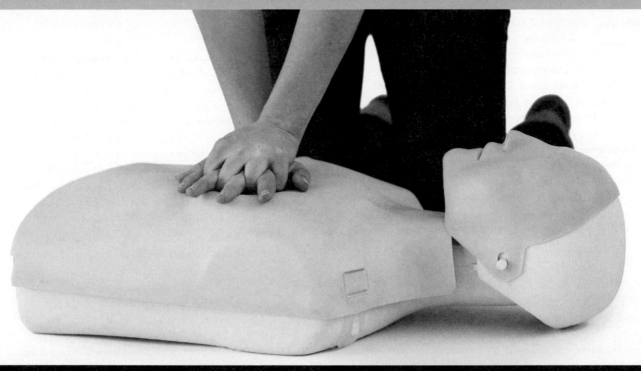

Figure 33.21 Chest compressions keep blood moving to brain cells of a person whose heart has stopped.

WHEN THE HEART STOPS

In sudden cardiac arrest, the heart stops, blood flow halts, and cells begin to die. Brain cells starved of oxygen are the first to be affected. An inborn pacemaker defect causes most cardiac arrests in people under age 35. In older people, heart disease usually makes the heart stop.

The likelihood of surviving a cardiac arrest increases if a person receives chest compressions immediately after the event (**Figure 33.21**). Chest compressions can keep oxygenated blood flowing to brain cells. With traditional cardiopulmonary resuscitation (CPR), a rescuer alternates between blowing into a person's mouth to inflate the lungs and compressing the chest to keep blood moving.

Cardiocerebral resuscitation (CCR) is a newer method that relies on chest compressions alone. It is also known as "hands-only CPR." This method involves strong, rapid chest compressions—about 100 a minute. Proponents of CCR argue that as long as a person's airway is clear, chest compressions will move enough air

into and out of a victim's lungs to oxygenate the blood flowing to the person's heart and brain. CCR has the added benefit of not requiring people to have mouth-to-mouth contact with a stranger.

Neither CPR nor CCR can restart a heart. Such a restart requires a defibrillator to deliver an electric shock to the chest and reset the natural pacemaker.

Automated external defibrillators (AEDs) that can accomplish this task are now available in most public places (**Figure 33.22**). An AED provides simple voice commands that direct the user to attach its electrodes to a person in distress. The AED then determines whether the person has a heartbeat and, if required, shocks the heart. Having AEDs widely available helps to ensure that a person who suffers a cardiac arrest does not have to wait for arrival of an ambulance to receive a shock that might restart the heart.

The presence of a bystander who is willing to carry out CPR or CCR or to use an AED often means the difference

between life and death. Yet only about 15 percent of sudden cardiac arrest victims receive such assistance before professional medical personnel arrive. Unfortunately, most people do not know how to administer chest compressions or use an AED. A half-day course given by the American Red Cross or another community health organization can teach you both skills. Learning life-saving skills is something we can all do to help one another.

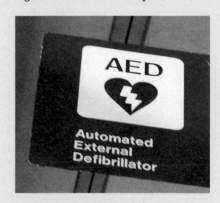

Figure 33.22 Sign indicating the location of an AED.

CREDITS: (21) Jan Wolak-Dyszynski/iStock by Getty Images; (22) Christine Evers.

Section 33.1 A **circulatory system** distributes substances through an animal's body. Some invertebrates have an **open circulatory system**; their **hemolymph** leaves vessels and seeps around tissues. In other invertebrates and all vertebrates, a **closed circulatory system** confines **blood** inside a **heart** and blood vessels. Exchanges of substances between blood and cells take place across the wall of **capillaries**.

Fish have a heart with two chambers: the **atrium** receives blood and the **ventricle** pumps bloods out and through the single circuit. Evolution of a two-circuit system accompanied the evolution of lungs. In this system, the **pulmonary circuit** moves blood to the lungs and back to the heart. The longer **systemic circuit** moves blood to other body tissues and back to the heart.

Section 33.2 Humans have a four-chamber heart. The two atria receive blood from **veins**, and the two ventricles pump blood into **arteries**. In the pulmonary circuit, **pulmonary arteries** carry blood to lungs and **pulmonary veins** return it to the heart. In the systemic circuit, blood is pumped out of the heart into the **aorta**, and returns to the heart via the **inferior vena cava** and **superior vena cava**. In most cases, blood flows through only one capillary system. Blood in intestinal capillaries will later flow through liver capillaries. The liver stores glucose delivered by the blood and neutralizes some bloodborne toxins.

Section 33.3 A human heart is partitioned into two sides, each with an atrium above a ventricle. Oxygen-poor blood from the body enters the right atrium and moves through a valve into the right ventricle, which pumps blood through a valve into the pulmonary circuit. Oxygen-rich blood enters the heart's left atrium and moves through a valve into the left ventricle. The left ventricle pumps the blood through a valve into the systemic circuit.

During one **cardiac cycle**, all heart chambers undergo rhythmic relaxation (**diastole**) and contraction (**systole**). Contraction of ventricles provides the force that powers movement of blood through blood vessels. Contraction of atria only fills the ventricles.

The **sinoatrial (SA) node** in the right atrium serves as the cardiac pacemaker. Action potentials that originate in the SA spread across atria by way of gap junctions and trigger atrial contraction. The action potentials then spread to ventricles by way of the **atrioventricular (AV) node**. The delay between atrial and ventricular contraction allows ventricles to fill fully before they contract.

Section 33.4 Blood consists of **plasma**, blood cells, and platelets. Plasma is mostly water. It carries dissolved nutrients and gases. The most abundant plasma protein, albumin, is made by the liver. Albumin helps transport lipids and encourages water to move into capillaries by osmosis. **Red blood cells** contain hemoglobin that transports oxygen. **White blood cells** provide defense against pathogens. **Platelets** and **fibrin**

act in **hemostasis** (clotting). Platelets and all blood cells arise from stem cells in bone.

Sections 33.5, 33.6 **Blood pressure** is the pressure blood exerts against vessel walls. It declines as blood proceeds through a circuit. Blood pressure is recorded as **systolic pressure** (pressure when ventricles contract) over **diastolic pressure** (pressure when ventricles relax). Thick elastic walls of arteries help move blood forward even as ventricles relax. A **pulse** is a brief expansion of an artery caused by ventricular contraction. **Arterioles** are the main site for regulation of blood flow in the systemic circuit. **Vasodilation** of arterioles supplies more blood to a region. **Vasoconstriction** decreases the blood supply.

Sections 33.7, 33.8 Blood flow slows more in capillaries than other vessels because of their collectively larger cross-sectional area. Capillaries are the site of exchanges between the blood and body cells. Blood pressure forces protein-free plasma out of a capillary at its arterial end, and water moves into the capillary by osmosis at its venous end.

Venules carry blood from capillaries to veins. Valves and the action of skeletal muscles help keep blood in veins moving.

Section 33.9 Abnormal heart rhythms can slow or halt blood flow. Ventricular fibrillation is a deadly arrhythmia that halts blood flow. With **atherosclerosis**, plaque narrows the interior of a blood vessel, increasing the risk of heart attack and stroke.

Section 33.10 The **lymph vascular system** takes up excess interstitial fluid, as well as fats absorbed from the small intestine, and delivers them to blood. It also delivers bloodborne pathogens to lymph nodes. **Lymph nodes** filter **lymph**, and white blood cells in the nodes attack any pathogens. The **spleen** filters the blood and removes old red blood cells. T lymphocytes mature in the **thymus**.

Application When the heart stops pumping, blood flow halts and cells begin to die. CPR and CCR can keep some oxygenated blood moving to cells, but restarting the heart requires a shock from a defibrillator.

SELF-ASSESSMENT
ANSWERS IN APPENDIX VII

1. All vertebrates have _____ .
   a. a closed circulatory system
   b. a two-chambered heart
   c. lungs
   d. hemolymph

RISKS OF HIGH BLOOD PRESSURE Eleni Rapsomaniki and her colleagues analyzed medical records of 1.25 million people in the United Kingdom to see how hypertension affects the risk of cardiovascular disorders. They looked at records of people 30 years old or older who initially had no history of heart disease or stroke, although some had high blood pressure. By looking at which of these people later had a heart attack or stroke, the researchers were able to estimate the lifetime risks of these disorders in people with and without high blood pressure. **Figure 33.23** shows the researchers' estimates of the lifetime risk for heart attack and for transient ischemic attack (TIA), a type of stroke.

1. How does age affect the overall risk of heart attack and TIA?

2. How much does having hypertension at age 30, affect the risk of having a heart attack by age 95? The risk of a TIA?

3. Is a person with normal blood pressure more likely to suffer a heart attack or a TIA by age 65?

4. Based on this data, how would preventing hypertension in people at age 30 affect the number of heart attacks and TIAs?

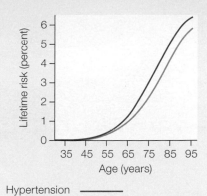

Figure 33.23 Lifetime risks of heart attack and TIA (a type of stroke), with and without hypertension at age 30. A transient ischaemic attack (TIA), sometimes called a mini-stroke, is caused by a temporary blockage of a blood vessel that supplies blood to the brain.

Source: © 2014 Rapsomaniki et al. Open Access article distributed under the terms of CC BY.

2. In _____ blood flows through a single circuit.
   a. birds   b. mammals   c. fishes   d. amphibians

3. The _____ circuit carries blood to and from lungs.

4. The most abundant protein in plasma is _____ .
   a. hemoglobin   c. albumin
   b. fibrin   d. collagen

5. Platelets function in _____ .
   a. oxygen transport   c. thermal regulation
   b. blood clotting   d. defense against worms

6. Most oxygen in blood is transported _____ .
   a. in red blood cells   c. in platelets
   b. in white blood cells d. dissolved in plasma

7. Blood flows directly from the left atrium to _____ .
   a. the aorta   c. the right atrium
   b. the left ventricle   d. the pulmonary arteries

8. Contraction of _____ drives the flow of blood through the aorta and pulmonary arteries.
   a. atria   b. veins   c. arterioles   d. ventricles

9. Blood pressure is highest in the _____ and lowest in the _____ .
   a. arteries; veins   c. veins; arteries
   b. arterioles; venules   d. capillaries; arterioles

10. At rest, the largest volume of blood is in _____ .
    a. arteries   b. veins   c. capillaries   d. arterioles

11. Which heart chamber has the thickest wall?
    a. right atrium   c. left atrium
    b. right ventricle   d. left ventricle

12. A pulmonary _____ carries oxygen-poor blood.
    a. artery   b. vein

13. Lymph nodes filter _____ .
    a. blood   c. plasma
    b. lymph   d. all of the above

14. Which is more dangerous?
    a. atrial fibrillation   b. ventricular fibrillation

15. Match the terms with their descriptions.
    ___ capillary   a. filters out pathogens
    ___ lymph node   b. cardiac pacemaker
    ___ atrium   c. blood vessel with valves
    ___ ventricle   d. largest artery
    ___ SA node   e. receives blood from veins
    ___ vein   f. exchange site
    ___ aorta   g. receives blood from atrium

CLASS ACTIVITY
CRITICAL THINKING

1. Long-distance flights can raise the risk of thrombus formation in the legs. Physicians suggest that air travelers drink plenty of fluids and periodically walk around the cabin. Explain how these precautions lower the risk of clot formation.

2. Explain why a blood clot that forms in the leg, then breaks free as an embolus, is more likely to become stuck in a lung than in the brain.

3. Chronic alcoholism harms the liver. Alcoholics often have a lower-than-normal level of albumin, and excess fluid tends to accumulate in their abdomen and legs. Explain how liver impairment contributes to these symptoms.

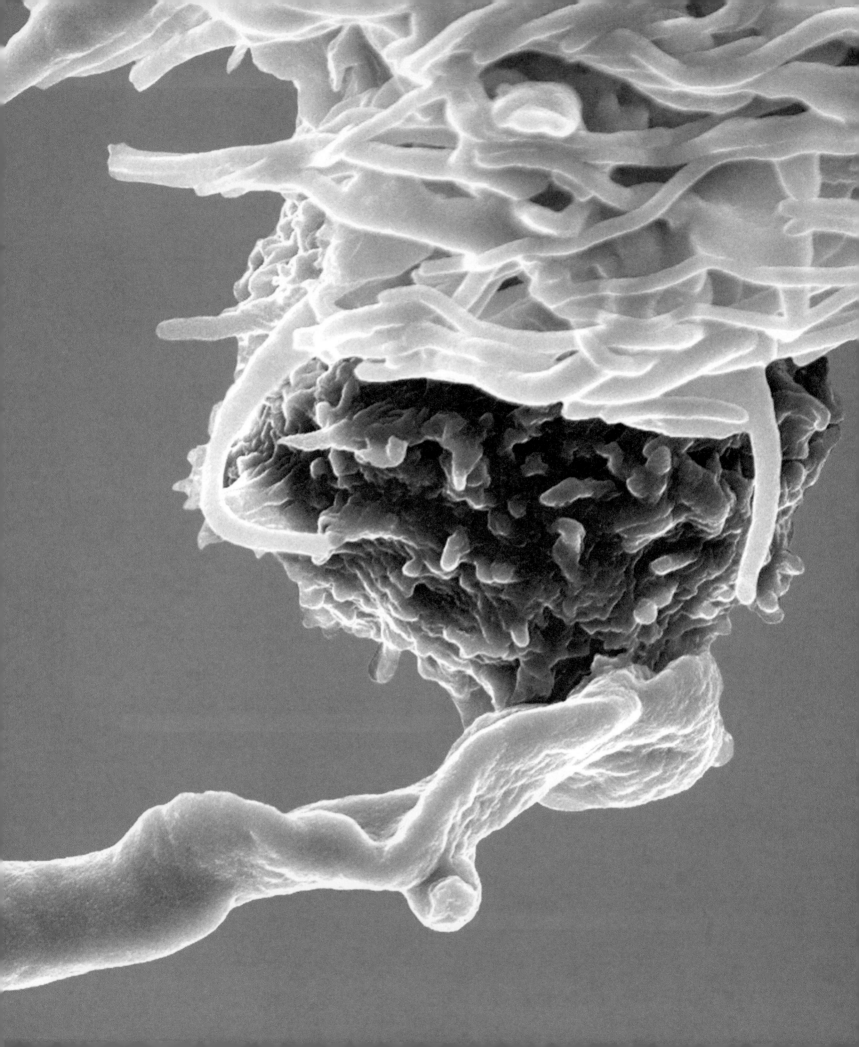

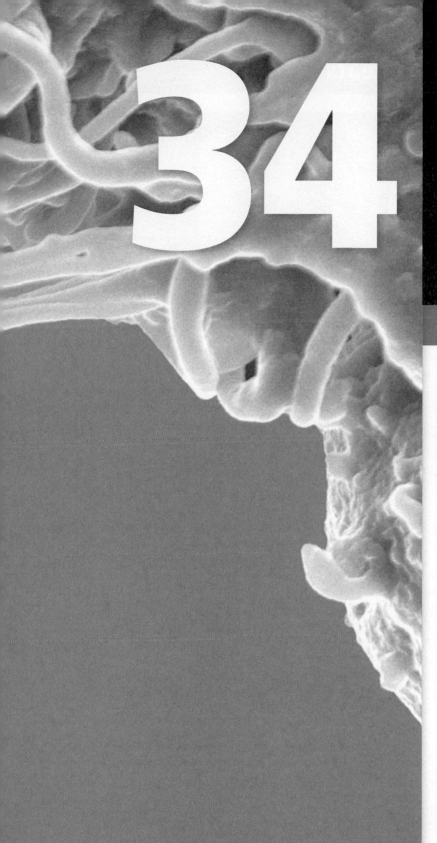

34

Immunity

Links to Earlier Concepts

This chapter explores bacteria and viruses (Sections 4.4, 19.1–19.5) in the context of immunity. You will also revisit proteins (3.4), cell structure and function (4.10, 5.8), blood type (13.4), the internal environment (28.1), plant immunity (27.12), epithelia and skin (28.2, 28.7), the nervous system (29.4, 29.9), the circulatory and lymphatic systems (33.4, 33.7, 33.10), and diseases (15.6, 33.9).

Core Concepts

All organisms alive today are linked by lines of descent from shared ancestors.

Shared core processes and features provide evidence that all living things are descended from a common ancestor. All plants and animals have a set of general mechanisms that defend the body from invasion by pathogens. Vertebrates have an extra level of protection in which defense responses are tailored to combat specific pathogens.

Interactions among the components of each biological system give rise to complex properties.

Vertebrate immunity comprises dynamic, multilevel responses to threats. Highly coordinated, complex interactions among a network of specialized cells occur via chemical signaling molecules. We can look at vertebrate immunity as a hierarchy of defenses, but immune responses occur simultaneously, in a nonlinear fashion.

Growth, reproduction, and homeostasis depend on proper timing and coordination of specific molecular events.

Disruptions to homeostasis affect biological systems. Infections, toxins, and allergens that disrupt the body at the cellular level can affect the overall health of the organism. Feedback mechanisms inherent in immune responses maintain the fine line between eliminating a threat and damaging the body.

A human T cell (pink) scans the surface of a dendritic cell (blue) for molecules that signal a threat such as a bacterial invader. If these molecules are detected, the T cell will initiate an immune response targeting the specific threat.

Photograph, Dr. Olivier Schwartz, Institut Pasteur/Science Source.

You continually cross paths with huge numbers of viruses, bacteria, fungi, parasitic worms, and other agents of disease, but humans coevolved with these pathogens so you have defenses against them. The evolution of **immunity**, an organism's capacity to resist and combat infection, began before multicelled eukaryotes evolved from free-living cells. Mutations introduced variations in membrane proteins of different cell lineages; as multicellularity evolved, so did mechanisms of identifying such molecules as self, or belonging to one's own body. Central to these mechanisms are receptor proteins—membrane proteins that trigger a change in cell activity in response to a specific stimulus (Section 4.3).

By about 1 billion years ago, the ability to recognize nonself molecules (that occur on other organisms) had also evolved. Cells of modern animals have a set of receptors that collectively can recognize around 1,000 nonself molecules called pathogen-associated molecular patterns, or **PAMPs**. (Plants and some algae have PAMP receptors too.) As their name suggests, PAMPs occur mainly on or in microorganisms that can be pathogenic. PAMPs include peptidoglycan of bacterial cell walls and flagellin of bacterial flagella (Section 4.4), chitin of fungal cell walls (Section 3.2), double-stranded RNA of some viruses, and so on. A PAMP is an example of **antigen**, any molecule or particle that is recognized by the body as nonself. Not all antigens are PAMPs: Molecules on body cells of other individuals can be antigens, for example.

THREE LINES OF DEFENSE

Anatomical and physiological barriers that prevent most microorganisms from entering the internal environment are the body's first line of defense against pathogens (**Figure 34.1**). A pathogen that breaches these barriers activates mechanisms of innate immunity, the second line of defense. **Innate immunity** is a set of immediate, general responses that help protect all multicelled organisms from infection. These responses are initiated by PAMP receptors that bind to an antigen-bearing particle such as a bacterium or virus.

In vertebrates, the activation of innate immunity triggers mechanisms of adaptive immunity, which is the third line of defense. **Adaptive immunity** tailors an immune response to a vast array of specific pathogens that an individual encounters during its lifetime. Two types of adaptive immune responses

Figure 34.1 A physical barrier to infection. In the lining of the airways leading to human lungs, goblet cells (gold) secrete mucus that traps bacteria and particles. Cilia (pink) on other cells sweep the mucus toward the throat for disposal.

TABLE 34.1

Comparing Vertebrate Innate and Adaptive Immunity

	Innate Immunity	Adaptive Immunity
Response time	Immediate	About a week
Antigen recognition	About 1,000 PAMPs	Billions of potential antigens
Specificity of response	General response	Response specific to pathogen that triggers it
Persistence	None	Long-term

adaptive immunity In vertebrates, a set of immune defenses that can be tailored to specific pathogens as an organism encounters them during its lifetime.

antigen (AN-tih-jenn) A molecule or particle that the immune system recognizes as nonself.

B cell B lymphocyte. Lymphocyte that can make antibodies.

basophil (BASE-uh-fill) Granular, circulating white blood cell.

cytokines (SITE-uh-kynes) Cell-to-cell signaling molecules. E.g., interleukins, interferons, tumor necrosis factors.

dendritic cell (den-DRIH-tick) Phagocytic white blood cell that specializes in antigen presentation.

eosinophil (ee-uh-SIN-uh-fill) Granular, circulating white blood cell that targets large, multicelled parasites.

immunity The body's ability to resist and fight infections.

innate immunity In all multicelled organisms, set of immediate, general responses that defend against infection.

macrophage (MACK-roe-faje) White blood cell that specializes in phagocytosis.

mast cell Granular white blood cell that stays anchored in tissues.

neutrophil (NEW-tra-fill) Circulating phagocytic white blood cell.

PAMPs Pathogen-associated molecular patterns. Molecules found on or in microorganisms that can be pathogenic.

T cell T lymphocyte. Lymphocyte central to adaptive immunity; some kinds target infected or cancerous body cells.

(antibody-mediated and cell-mediated) target different types of threats. Innate and and adaptive immunity have different features (**Table 34.1**), but they work together.

THE DEFENDERS

In vertebrates, white blood cells (leukocytes, Section 33.4) participate in both innate and adaptive immune responses. Almost all spend at least part of their life circulating in blood and lymph. Many populate the lymph nodes, spleen, and other solid tissues. White blood cells communicate by secreting and responding to chemical signaling molecules. These molecules, which include proteins and polypeptides called **cytokines**, allow cells throughout the body to coordinate activities during an immune response. Interleukins, interferons, and tumor necrosis factors are examples of cytokines.

All white blood cells are phagocytic—they carry out phagocytosis (Section 5.8) to some degree. A phagocyte engulfs and digests pathogens, dead cells, or other particles. All white blood cells also have the ability to process and present antigen, a mechanism that alerts the adaptive immune system to threats (Section 34.5 returns to antigen processing). Aside from these similarities, different types of white blood cells specialize in different tasks (**Figure 34.2**). **Neutrophils** are the most abundant phagocytes in blood, so they are often the first responders at a site of infection or tissue damage. Monocytes also circulate in blood, but during an infection they migrate into solid tissue, then differentiate into macrophages and dendritic cells. **Macrophages** are professional phatocytes that engulf essentially everything except undamaged body cells. **Dendritic cells** specialize in antigen presentation.

Some white blood cells have secretory vesicles called granules, which can contain cytokines, local signaling molecules, destructive enzymes, and toxins such as hydrogen peroxide. A cell degranulates (releases the contents of its granules) in response to a trigger such as antigen binding. Neutrophils have granules, as do **eosinophils** that can target parasites too big for phagocytosis. **Basophils** and **mast cells** degranulate in response to injury or antigen. Mast cells, unlike other white blood cells, are permanently anchored in tissues. These cells also degranulate in response to signaling molecules secreted by cells of the endocrine and nervous systems.

Lymphocytes are a special category of white blood cells that carry out adaptive immune responses. **B cells** (B lymphocytes) make antibodies (more about these proteins in Section 34.4). **T cells** (T lymphocytes) play a central role in all adaptive immune responses. Cytotoxic T cells are specialized to kill infected or cancerous body cells. NK cells (natural killer cells) are lymphocytes that kill cancerous body cells that cytotoxic T cells cannot detect.

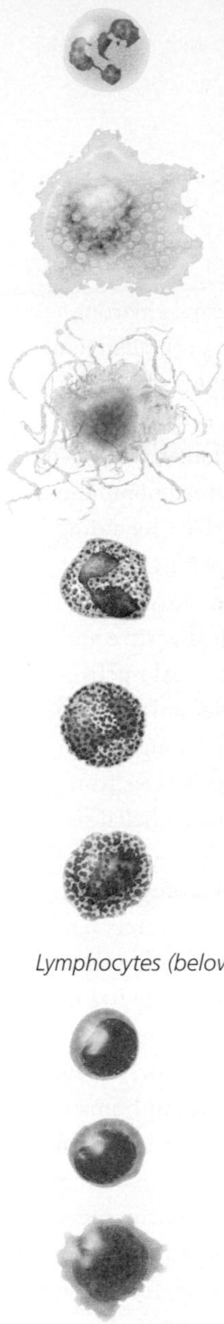

Neutrophil Most abundant white blood cell. Phagocyte that circulates in blood and migrates into damaged or infected tissues, where it participates in inflammatory response by degranulating.

Macrophage Starts out as a monocyte in blood, then enters damaged tissue and differentiates. Major phagocytic cell that also presents antigen.

Dendritic cell Starts out as a monocyte in blood, then enters damaged tissue and differentiates. Major antigen-presenting cell that is also phagocytic.

Eosinophil Circulates in blood; migrates into damaged or infected tissues, where it participates in inflammatory response. Granules contain enzymes effective against multicellular parasites.

Basophil Circulates in blood; migrates into damaged or infected tissues, where it participates in inflammatory response by degranulating.

Mast cell Anchored in tissues. Participates in inflammation and allergies by degranulating.

Lymphocytes (below) carry out adaptive immune responses.

B cell Only type of cell that produces antibodies in antibody-mediated adaptive responses.

T cell Several kinds play a central role in all adaptive immune responses.

Natural killer (NK) cell Kills cancer cells undetectable by other white blood cells.

Figure 34.2 Preview of white blood cells. Details such as lobed nuclei and cytoplasmic granules are revealed by staining.

Invertebrates have no white blood cells and no adaptive immunity, but they do have other specialized immune cells and robust innate immunity. In these organisms, PAMP receptor binding triggers immune cell phagocytosis and the release of antimicrobial compounds. Some invertebrate immune cells also surround and encapsulate pathogens too big for phagocytosis. Immune mechanisms are complex and they vary widely, as this example illustrates. For clarity, the remainder of the chapter focuses on immunity in humans and other mammals.

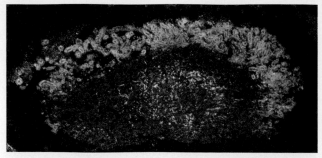

A Biofilm community in a chunk of human dental plaque. Six different genera of oral bacteria are indicated in different colors; each tiny dot or rod is a single bacterial cell.

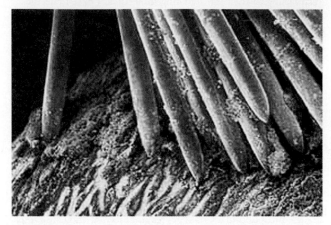

B Micrograph of toothbrush bristles scrubbing plaque on the surface of a tooth.

Figure 34.3 Dental plaque.

BIOLOGICAL BARRIERS

Your skin is in constant contact with the external environment, so it picks up many microorganisms. It normally teems with more than 4,000 species of bacteria, fungi, viruses, and archaea: up to 25 million microorganisms on every square inch of your external surfaces. They tend to flourish in warm, moist areas, such as the crease of the elbow. Huge populations inhabit structures that open on the body's surface, including the eyes, nose, mouth, and anal and genital openings.

Microorganisms that typically live on human surfaces, including the interior tubes and cavities of the digestive and respiratory tracts, are called normal flora or normal microbiota (Section 19.5). Our body surfaces provide microorganisms with a stable environment and nutrients. Competition for attachment sites can deter more dangerous species from colonizing—and potentially penetrating—our epithelial tissues (Section 28.2).

The human body has at least as many microorganisms as human cells. Most of them inhabit the digestive tract, and normally they make up a large and diverse community there. Gut microbiota are not just random passengers; they live in symbiosis with us (Section 18.4). These microorganisms help us digest food, and they make essential nutrients such as vitamins K and B$_{12}$. They also play a significant role in human health that goes far beyond nutrition. Mounting evidence suggests that species composition and diversity of gut microbiota have a major influence on our physiology, metabolism, mood, and immune function. Imbalances in this community have been associated with conditions that include diabetes, obesity, inflammatory bowel disease, and cancer.

Microbiota are helpful only on body surfaces. Consider a major constituent of microbiota, *Propionibacterium acnes* (below). This bacterium feeds on sebum, a greasy mixture of

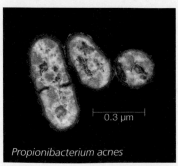

0.3 μm

Propionibacterium acnes

fats, waxes, and glycerides that lubricates hair and skin. Glands in the skin secrete sebum into hair follicles. During puberty, an increase in sex hormone production triggers an increase in sebum production. Excess sebum combines with dead, shed skin cells and blocks the openings of hair follicles. *P. acnes* can survive on the surface of the skin, but far prefers anaerobic habitats such as the interior of blocked hair follicles. There, the cells multiply to tremendous numbers. Secretions of their flourishing populations leak into internal tissues of the follicles and initiate inflammation (we return to inflammation in the next section). The resulting pustules are called acne.

About 60 species of microorganisms that inhabit the mouth are part of **dental plaque**, a thick biofilm of various bacteria, fungi, and archaea together with their extracellular products and saliva glycoproteins (Section 4.4). Plaque clings firmly to teeth (**Figure 34.3**). Some of the bacteria in plaque carry out lactate fermentation. The lactate they produce is acidic enough to dissolve minerals that make up the tooth, causing holes called cavities.

In young, healthy people, tight junctions (Section 4.10) normally seal gum epithelium to teeth. The tight seal excludes oral microorganisms. As we age, the connective tissue beneath the epithelium thins, so the seal between gums and

CREDITS: (in text) Kwangshin Kim/Science Source; (3A) Blair Rossetti Forsyth Institute; (3B) www.zahnarzt-stuttgart.com.

teeth weakens. Deep pockets form, and a nasty collection of anaerobic bacteria and archaea tends to accumulate in them. The microorganisms secrete destructive enzymes and acids that cause inflammation of the surrounding gum, a condition called periodontitis. Periodontal wounds are an open door to the circulatory system and its arteries, and all species of oral bacteria associated with periodontitis are also found in atherosclerotic plaque (Section 33.9). Atherosclerosis is now known to be a disease of inflammation. What role oral microorganisms play in atherosclerosis is not yet clear, but one thing is certain: They contribute to the inflammation that fuels coronary artery disease.

Microbiota can cause or worsen disease when they invade tissues. Serious illnesses associated with microbiota include pneumonia; ulcers; colitis; whooping cough; meningitis; abscesses of the lung and brain; and cancers of the colon, stomach, and intestine. The bacterial agent of tetanus, *Clostridium tetani*, is considered a normal inhabitant of human intestines. The bacterial species responsible for diphtheria, *Corynebacterium diphtheriae*, was normal skin flora before widespread use of the vaccine eradicated it. *Staphylococcus aureus*, a resident of human skin and linings of the mouth, nose, throat, and intestines, is also a leading cause of human bacterial disease. A particularly dangerous kind, MRSA (methicillin-resistant *S. aureus*), is resistant to several antibiotics. MRSA causes life-threatening bloodstream infections in medical facilities.

PHYSIOLOGICAL AND ANATOMICAL BARRIERS

In contrast to body surfaces, the blood and tissue fluids of healthy people are typically free of microorganisms (sterile). Surface barriers (**Table 34.2**) usually prevent microorganisms from entering the body's internal environment. The tough outer layer of skin (Section 28.7) is an example. Microorganisms flourish on skin's waterproof, oily surface, but they rarely penetrate its thick layer of dead cells (**Figure 34.4**). The thinner epithelial tissues that line the body's interior tubes and cavities also have surface barriers. Sticky mucus secreted by cells of these linings can trap microorganisms. The mucus contains **lysozyme**, an enzyme that kills bacteria. In the sinuses and respiratory tract, the coordinated beating of cilia sweeps trapped microorganisms away before they have a

---

dental plaque On teeth, a thick biofilm composed of various microorganisms, their extracellular products, and saliva proteins.
lysozyme (LICE-uh-zime) Antibacterial enzyme in body secretions such as saliva and mucus.

TABLE 34.2

Examples of Surface Barriers

Biological	Established populations of microbiota
Physiological	Secretions (sebum, other waxy coatings); low pH of urine, gastric juices, urinary and vaginal tracts; lysozyme
Anatomical	Epithelia that line tubes and cavities such as the gut and eye sockets; skin; mucus; broomlike action of cilia; flushing action of tears, saliva, urination, diarrhea

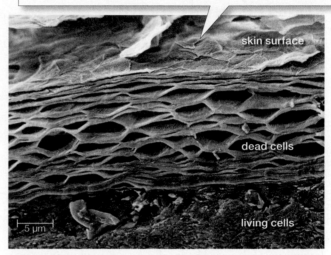

Figure 34.4 One surface barrier to infection: the epidermis of human skin. A thick, waterproof layer of dead skin cells usually keeps microbiota from penetrating internal tissues.

chance to breach the delicate walls of these structures (ciliated epithelium of the respiratory tract is shown in Figure 34.1).

Your mouth is a particularly inviting habitat for microorganisms because it offers plenty of nutrients and warm, moist pockets suitable for colonization. Accordingly, it harbors huge populations of microbiota that can resist lysozyme in saliva. Microorganisms that get swallowed are typically killed in the stomach by gastric fluid, a potent brew of protein-digesting enzymes and acid. Those that survive passage to the small intestine are usually killed by strong salts secreted into the fluid there. The hardy ones that reach the large intestine must compete with about 500 resident species that are specialized to live there and have already established large populations. Any that disrupt this resident microbial community are typically flushed out by diarrhea.

Urination's flushing action usually prevents pathogens from colonizing the urinary tract. Lactic acid produced by *Lactobacillus* bacteria helps keep the vaginal pH outside the range of tolerance of most fungi and other bacteria.

What happens if a microorganism slips by surface barriers and enters body tissues or fluids? Fast-acting, general response mechanisms of innate immunity can stop or slow a pathogen's proliferation for a few days until adaptive immunity peaks.

COMPLEMENT ACTIVATION

Complement is a set of proteins that circulate in inactive form in blood and tissue fluid; when activated, complement is part of responses against extracellular pathogens. Complement was named because the proteins were originally identified by their ability to "complement" the action of antibodies in adaptive immune responses (we return to these responses in Section 34.6). Years later, researchers discovered that some complement proteins function independently of antibodies.

Mammals have about 30 kinds of complement proteins. Some become activated when they bind directly to microorganisms. Others are activated by binding to antibodies clustered on the surface of a cell. Still others become activated when they encounter cytoplasmic or mitochondrial proteins leaking out of damaged body cells. An activated complement protein activates other complement proteins, which activate other complement proteins, and so on. These cascading reactions (a mechanism of positive feedback, Section 29.4) quickly produce huge amounts of activated complement, which diffuses into surrounding tissues to form a gradient around an affected site.

Activated complement forms a coating on microorganisms, damaged cells, and debris that enhances their uptake by phagocytic white blood cells (**Figure 34.5**). Activated complement proteins also assemble into structures that puncture plasma membranes. Ions that flow through these channels cause the cell to burst (lysis, Section 19.1). Normal body cells continuously produce proteins that inactivate complement, thus preventing a complement cascade from spreading too far into healthy tissue. Microorganisms do not make complement-inactivating proteins, so they are singled out for destruction.

PHAGOCYTOSIS

Receptors for cytokines and activated complement allow neutrophils, macrophages, and dendritic cells to follow a chemical trail leading to an affected tissue. There, the phagocytes engulf complement-coated cells and other particles.

Neutrophils are among the first responders to injury or infection, collecting within minutes at a site of tissue damage. After a neutrophil engulfs a microorganism, it releases the contents of its granules into the vesicle containing the engulfed pathogen (Section 5.8), and also to the exterior of the cell. Enzymes and toxins released into extracellular fluid destroy all cells in the vicinity (including healthy body cells). Neutrophils literally explode in response to a certain combination of signaling molecules and complement, in the process ejecting their nuclear DNA and associated proteins along with the contents of their granules. The mixture solidifies into a net, trapping pathogens near the released antimicrobial compounds. Neutrophil nets are very effective at killing bacteria.

Macrophages in interstitial fluid engulf microorganisms, cell debris, and other particles, but they are more than just scavengers. Upon engulfing antigen, a macrophage secretes cytokines that alert other white blood cells to the presence of invading pathogens.

Dendritic cells patrol tissues that contact the external environment, such as the lining of respiratory airways. Phagocytosis by these cells protects the lungs from pathogens and harmful particles. However, the main function of dendritic cells is to present antigen to T cells in an adaptive immune response (as described in Section 34.5).

Figure 34.5 Complement function.
Activated complement coats pathogens, recruits phagocytic white blood cells, and destroys microorganisms directly.

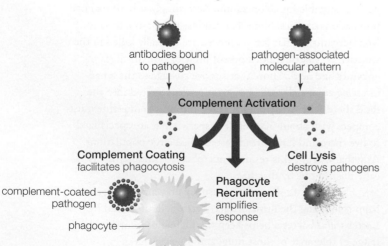

antibodies bound to pathogen

pathogen-associated molecular pattern

Complement Activation

Complement Coating facilitates phagocytosis

Phagocyte Recruitment amplifies response

Cell Lysis destroys pathogens

complement-coated pathogen

phagocyte

complement A set of proteins that circulate in inactive form, and when activated play a role in immune responses.

fever A temporary, internally induced rise in core body temperature above the normal set point.

inflammation A local response to tissue damage or infection; characterized by redness, warmth, swelling, and pain.

INFLAMMATION

Complement activation and cytokines released by phagocytes trigger two hallmarks of infection: inflammation and fever. **Inflammation** is a fast, local response that simultaneously destroys affected tissues and jump-starts the healing process (**Figure 34.6**). Inflammation begins when basophils, mast cells, or neutrophils degranulate, releasing the contents of their granules into an affected tissue. Degranulation occurs in response to a number of stimuli, including (for example) PAMP receptor binding to a bacterial antigen ❶.

A degranulating white blood cell releases cytokines. It also releases prostaglandins and histamines ❷, local signaling molecules that have two effects. First, they cause nearby arterioles to widen, so blood flow to the area increases. The increased flow speeds the arrival of more phagocytes, which are attracted to the cytokines. Second, the local signaling molecules cause spaces to open up between the cells that make up the walls of nearby capillaries. Phagocytes arriving in the bloodstream can move quickly into tissues by squeezing through the spaces ❸. By the time these phagocytes have entered an affected tissue, any invading cells have become coated with activated complement ❹. The coating makes them easy targets for phagocytosis ❺.

The increased permeability of capillary walls allows plasma proteins to escape into interstitial fluid. The fluid becomes hypertonic with respect to blood, and water follows by osmosis. The tissue swells with excess fluid, putting pressure on nerves and thus causing pain. Other outward symptoms of inflammation include redness and warmth, which are both indications of the area's increased blood flow.

Inflammation continues as long as its triggers do. When these stimuli subside, for example after invading bacteria have been cleared from an infected tissue, macrophages produce compounds that suppress inflammation and promote tissue repair. If the stimulus persists, inflammation becomes chronic. Chronic inflammation is not a normal condition. It does not benefit the body; rather, it causes or contributes to several diseases, including asthma, Crohn's disease, rheumatoid arthritis, atherosclerosis, diabetes, and cancer.

FEVER

Fever is a temporary rise in body temperature above the normal 37°C (98.6°F) that occurs in response to infection or serious injury. Some cytokines stimulate brain cells to release prostaglandins that act on the hypothalamus (Section 29.9) to raise the body's internal (core) temperature set point. As long as core temperature remains below the new set point, the hypothalamus sends out signals that cause blood vessels in the skin to constrict, thus reducing heat loss from the skin.

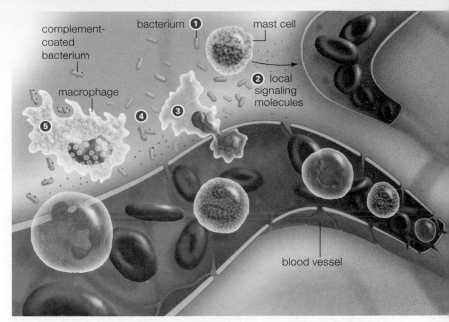

Figure 34.6 Example of inflammation as a response to bacterial infection.

❶ PAMP receptors on mast cells in the tissue recognize and bind to bacteria.

❷ The mast cells release prostaglandins and histamines (blue dots) that cause arterioles to widen. The resulting increase in blood flow, which reddens and warms the tissue, hastens the arrival of phagocytes.

❸ The signaling molecules also increase capillary permeability, which allows phagocytes arriving in the bloodstream to move quickly through the vessel walls into the tissue. The tissue swells with fluid as plasma proteins seep out of the leaky capillaries.

❹ Meanwhile, bacterial antigens have triggered complement cascades, and invading bacteria have become coated with complement (purple dots).

❺ Phagocytes in the tissue recognize and engulf the complement-coated bacteria.

The signals also increase the metabolic rate, quickening heartbeat and respiration, and causing reflexive movements called shivering, or "chills," that boost the metabolic heat output of muscles. Together, these responses raise the body's internal temperature. If core temperature rises too much, sweating and flushing quickly lower it, thus maintaining the body's new temperature set point. A fever is a sign that the body is fighting something, so it should never be ignored. However, a fever of 40.6°C (105°F) or less does not require treatment in an otherwise healthy adult. Body temperature usually will not rise above that value. Brain damage or death can occur if the core temperature reaches 42°C (107.6°F).

Fever enhances almost every step in innate and adaptive immunity. For example, fever speeds the production of white blood cells while stressing infected and cancerous body cells. It enhances the phagocytic activity of white blood cells and boosts their production of antigen receptors, and also enhances their recruitment to an affected area. Fever even increases the effectiveness of chemicals released by degranulating white blood cells.

If innate immune mechanisms do not quickly rid the body of an invading pathogen, an infection may become established in internal tissues. By that time, long-lasting mechanisms of adaptive immunity have already begun to target the invaders specifically. These mechanisms were triggered by cells that detected antigen via their antigen receptors. All body cells have one type of antigen receptor: plasma membrane proteins that recognize PAMPs. T cells also have antigen receptors called **T cell receptors**, or TCRs. Part of a TCR recognizes antigen as nonself. Another part recognizes certain proteins in the plasma membrane of body cells as self. These self-proteins are called **MHC markers** (left) after the major histocompatibility complex genes that encode them. MHC genes have thousands of alleles, so the cells of even closely related individuals rarely bear the same MHC markers.

—an MHC marker

Antibodies are another type of antigen receptor. **Antibodies** are Y-shaped antigen receptor proteins made by B cells. Many antibodies circulate in blood, and they can enter interstitial fluid during inflammation, but they do not kill pathogens directly. Instead, they activate complement and facilitate phagocytosis. Antibody binding also can prevent pathogens and toxins from attaching to body cells.

An antibody molecule consists of four polypeptides: two identical "light" chains and two identical "heavy" chains (**Figure 34.7A**). Each chain has a region that can vary (the variable region) and one that does not (the constant region).

TABLE 34.3

Structural Classes of Antibodies

Secreted antibodies

IgG		Main antibody in blood; binds pathogens, neutralizes toxins, activates complement. Can cross placenta to protect fetus.
IgA		Abundant in exocrine gland secretions (e.g., tears, saliva, milk, mucus), where it occurs in dimeric form (shown). Interferes with binding of pathogens to body cells; interacts with white blood cells to initiate inflammation.

Membrane-bound antibodies

IgE		Anchored to surface of basophils, mast cells, eosinophils, and some dendritic cells. IgE binding to antigen induces anchoring cell to release histamines and cytokines. Factor in allergies and asthma.
IgD		B cell receptor.
IgM		B cell receptor, as a monomer. Also secreted as polymers (left) that are especially effective at binding antigen and activating complement.

When the chains fold up together as an intact antibody, the variable regions form two antigen-binding sites at the tips of the Y. These sites have a specific distribution of bumps, grooves, and charge, and they bind only to antigen that has a complementary distribution of bumps, grooves, and charge (**Figure 34.7B**). Antigen-binding sites vary greatly among antibodies, so they are called hypervariable regions.

An antibody's constant region determines its structural identity, or class: IgG, IgA, IgE, IgM, or IgD (Ig stands for

Figure 34.7 Antibody structure.

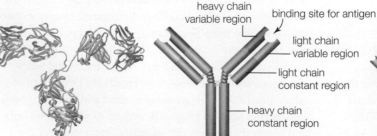

heavy chain variable region
binding site for antigen
light chain variable region
light chain constant region
heavy chain constant region

A An antibody molecule consists of four polypeptide chains joined in a Y-shaped configuration. The chains fold up so two antigen-binding sites form at the tips of the Y.

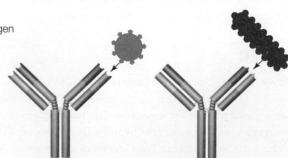

B The antigen-binding sites of each antibody bind only to an antigen that has a complementary distribution of bumps, grooves, and charge.

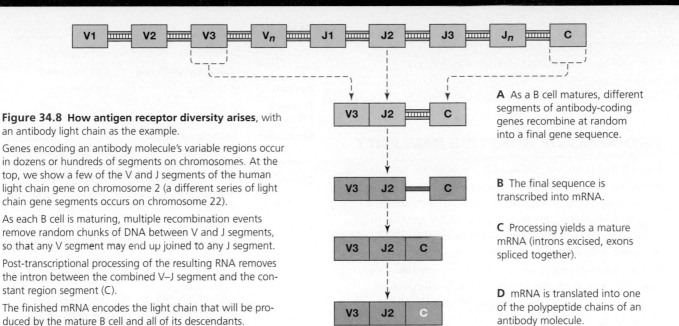

Figure 34.8 How antigen receptor diversity arises, with an antibody light chain as the example.

Genes encoding an antibody molecule's variable regions occur in dozens or hundreds of segments on chromosomes. At the top, we show a few of the V and J segments of the human light chain gene on chromosome 2 (a different series of light chain gene segments occurs on chromosome 22).

As each B cell is maturing, multiple recombination events remove random chunks of DNA between V and J segments, so that any V segment may end up joined to any J segment.

Post-transcriptional processing of the resulting RNA removes the intron between the combined V–J segment and the constant region segment (C).

The finished mRNA encodes the light chain that will be produced by the mature B cell and all of its descendants.

A As a B cell matures, different segments of antibody-coding genes recombine at random into a final gene sequence.

B The final sequence is transcribed into mRNA.

C Processing yields a mature mRNA (introns excised, exons spliced together).

D mRNA is translated into one of the polypeptide chains of an antibody molecule.

immunoglobulin, another name for antibody). The different classes serve different functions (**Table 34.3**). Most of the antibodies circulating in the bloodstream and tissue fluids are IgG, which binds pathogens, neutralizes toxins, and activates complement. IgG is the only antibody that can cross the placenta to protect a fetus before its own immune system is active.

IgA is the main antibody in mucus and other exocrine gland secretions (Section 28.2). IgA is secreted as a dimer (two antibodies bound together), which makes the molecule stable enough to function in harsh environments such as the interior of the digestive tract. There, IgA encounters pathogens before they contact body cells. Bound to antigen, IgA interacts with mast cells, basophils, macrophages, and NK cells to initiate inflammation.

IgE made and secreted by B cells is taken up by mast cells and basophils, and then incorporated into their plasma membranes. Binding of antigen to membrane-bound IgE causes a mast cell or basophil to degranulate.

IgM or IgD antibodies produced by a new B cell stay attached to the cell's plasma membrane as **B cell receptors.** The base of each receptor is embedded in the lipid bilayer of the cell's plasma membrane, and the two arms of the Y project into the extracellular environment (right). A mature B cell bristles with more than 100,000 receptors. IgM is also secreted as polymers of five or six antibodies. The polymers are especially efficient at binding antigen and activating complement.

B cell

B cell receptor

ANTIGEN RECEPTOR DIVERSITY

Humans can make billions of unique B and T cell receptors. The diversity arises because the genes that encode these receptors occur in segments, and there are dozens to hundreds of different versions of segments that encode the variable regions (**Figure 34.8**). The gene segments become spliced together during B and T cell differentiation, but which version of each segment gets spliced into the antigen receptor gene of a particular cell is random. As a B cell or T cell differentiates, it ends up with one of about 2.5 billion different combinations of gene segments. When the resulting gene is expressed, the cell produces thousands of receptors, all of which can recognize the same specific antigen.

Like all other blood cells, lymphocytes form in bone marrow. New B cells begin making antigen receptors before leaving bone, then migrate to the spleen or a lymph node to mature. T cells also form in bone marrow, but they mature in the thymus gland (another lymphatic organ, Section 33.10). There, they encounter hormones that stimulate them to make T cell receptors.

---

antibody (AN-tee-baa-dee) Y-shaped antigen receptor protein made by B cells.
B cell receptor Antigen receptor on the surface of a B cell; an antibody that stays anchored in the B cell's plasma membrane.
MHC markers Cell surface self-proteins of vertebrates.
T cell receptor (TCR) Special antigen receptor that only T cells have; recognizes antigen as nonself, and MHC markers as self.

LEARNING OBJECTIVES

Distinguish between effector cells and memory cells.

Explain the four defining characteristics of adaptive immunity.

Describe antigen processing.

TWO ARMS OF ADAPTIVE IMMUNITY

Like a boxer's one-two punch, adaptive immunity has two separate arms: two types of responses that work together to eliminate diverse threats. Why two arms? Not all threats present themselves in the same way. Consider bacteria, fungi, and toxins that circulate in blood or interstitial fluid. These and other extracellular threats are intercepted by phagocytes that initiate an antibody-mediated immune response. In an **antibody-mediated immune response**, B cells produce antibodies that bind to the pathogen or toxin that triggered the response.

An antibody-mediated immune response is not the most effective way of countering all threats, however. Viruses and some bacteria, fungi, and protists reproduce inside body cells. These intracellular pathogens may be vulnerable to an antibody-mediated response only when they exit one cell to infect another. Intracellular threats are targeted by the **cell-mediated immune response**, in which cytotoxic T cells and NK cells detect and destroy infected or cancerous body cells. **Table 34.4** previews and compares features of the two types of adaptive responses.

Effector and memory cells form during adaptive immune responses. **Effector cells** are lymphocytes that act in an immune response as soon as they form. **Memory cells** are long-lived lymphocytes reserved for possible future encounters with the same threat. Effector cells die after a few days or weeks; memory cells can persist for decades. If the same antigen is detected in the body later, memory cells carry out a faster, stronger secondary response (**Figure 34.9**).

Lymphocytes and other phagocytic white blood cells interact to bring about the four defining characteristics of adaptive immunity:

1. *Self/nonself recognition*, based on the ability of T cell receptors to recognize self (in the form of MHC markers), and that of all antigen receptors to recognize nonself (in the form of antigen).

2. *Specificity*, which means that adaptive immune responses are tailored to combat specific antigens.

3. *Diversity*, which refers to the diversity of antigen receptors that an individual can make. The ability to produce billions of different antigen receptors offers the ability to counter billions of different threats.

TABLE 34.4

Antibody-Mediated and Cell-Mediated Responses

	Antibody-Mediated	Cell-Mediated
Antigen	Extracellular	Intracellular
Primary Effector Cells	B cells	Cytotoxic T cells, NK cells
Mechanism	Secreted antibodies activate complement, facilitate phagocytosis	Activated lymphocytes kill infected/cancerous body cells
Outcome(s)	Pathogen or toxin removed from blood and tissue fluids	Pathogen eliminated along with ailing cells; cancer cells destroyed

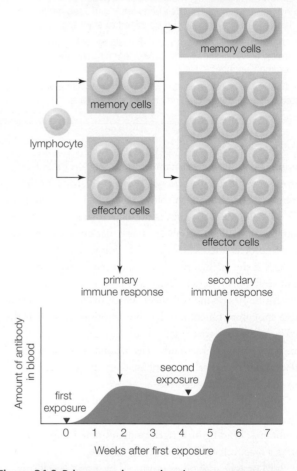

Figure 34.9 Primary and secondary immune responses.
A first exposure to an antigen causes a primary immune response in which effector cells fight the infection. Memory cells that also form initiate a faster, stronger secondary response if the antigen returns at a later time.

4. *Memory*, the capacity of the adaptive immune system to "remember" an antigen via memory cells. It takes about a week for B and T cells to respond in force the first time an antigen is detected. If the same antigen shows up later, the secondary response is faster and stronger.

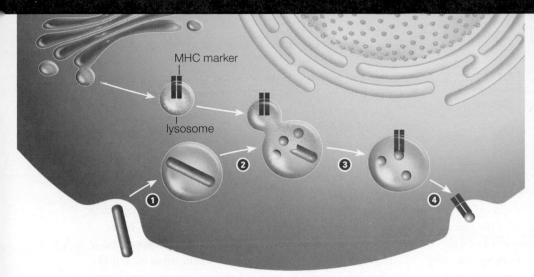

Figure 34.10 Antigen processing.
This drawing shows what happens in a dendritic cell or other phagocytic white blood cell after it engulfs an antigenic particle.

❶ An endocytic vesicle forms around a bacterium as it is engulfed by a phagocytic cell.

❷ The vesicle fuses with a lysosome, which contains enzymes and MHC markers.

❸ Lysosomal enzymes digest the bacterium to molecular bits. Fragments of proteins and polysaccharides bind to MHC markers.

❹ The vesicle fuses with the cell's plasma membrane by exocytosis. When it does, the antigen–MHC complexes become displayed on the cell's surface.

ANTIGEN PROCESSING

A new lymphocyte is "naive," which means that no antigen has bound to its receptors yet. Receptors of a naive B cell can bind directly to antigen, but receptors of a naive T cell cannot. T cell receptors recognize and bind only to antigen that is displayed on the surface of another cell. All body cells can display antigen. White blood cells that display antigen are called antigen-presenting cells, and these have a special role in adaptive immune responses.

The mechanism by which a white blood cell prepares antigen for display is called antigen processing. It begins when a phagocyte engulfs a pathogen, ailing cell, or other antigen-bearing particle (**Figure 34.10**). A vesicle that contains the particle forms in the phagocyte's cytoplasm ❶ and fuses with a lysosome ❷. Lysosomal enzymes then break down the ingested particle into molecular bits. Fragments of proteins and polysaccharides bind to MHC markers embedded in the lysosome's membrane, forming antigen–MHC complexes ❸. The vesicle moves to the plasma membrane, and its membrane fuses with (becomes part of) the plasma membrane ❹. As this occurs, the antigen–MHC complexes become displayed (presented) at the cell's surface.

Naive T cells often encounter antigen-presenting cells in the spleen or lymph nodes. Phagocytes engulf antigen-bearing particles, and then migrate to these lymphatic organs (Section 33.10) as they start to present antigen. Each day, billions of naive T cells filter through the lymph nodes and spleen, where they come into close contact with antigen-presenting cells that have taken up residence there (**Figure 34.11**). As you will see, a T cell with receptors that bind to an antigen–MHC complex stimulates other white blood cells to divide and differentiate in an adaptive immune response.

During an infection, the lymph nodes swell as lymphocytes accumulate and proliferate inside them. When you are ill, you may notice your swollen lymph nodes as tender lumps under the jaw or elsewhere in your body.

---

antibody-mediated immune response Immune response in which antibodies are produced that target the pathogen or toxin that triggered the response.

cell-mediated immune response Immune response in which cytotoxic T cells and NK cells kill infected or cancerous body cells.

effector cell Antigen-sensitized lymphocyte that forms in an immune response and acts immediately.

memory cell Long-lived, antigen-sensitized lymphocyte that can carry out a secondary immune response.

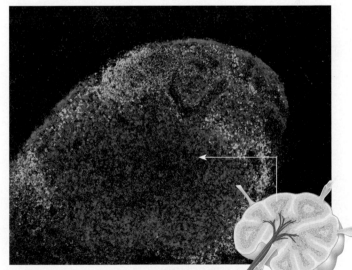

Figure 34.11 Lymph is filtered through at least one lymph node before it merges with the bloodstream. The fluorescence micrograph shows naive T cells (blue) that are passing through a lymph node and interacting with populations of resident dendritic cells (red) and B cells (green).

An antibody-mediated immune response is often called a humoral response because it involves mainly elements of blood and tissue fluid (from Latin *umor*, body fluid). B cells make antibodies to specific antigens in this adaptive response. If we liken B cells to assassins, then each one has a genetic assignment to liquidate a particular target: an antigen-bearing extracellular pathogen or toxin. Antibodies are B cells' molecular bullets, as the following example illustrates.

Suppose that you accidentally nick your finger. Being opportunists, some *Staphylococcus aureus* bacteria on your skin immediately enter the cut, invading your internal environment. Complement in interstitial fluid quickly attaches to peptidoglycan in the cell walls of the bacteria, and complement activation cascades begin.

Within an hour, complement-coated bacteria tumbling along in lymph vessels have become trapped by a lymph node. There, they encounter millions of naive B cells maturing in the node. One of the B cells makes antigen receptors that recognize an *S. aureus* protein called ClfA. The B cell's receptors bind to one of the bacteria, and the complement coating stimulates the B cell to engulf it (**Figure 34.12 ❶**). The B cell is now activated (it is no longer naive).

Meanwhile, more *S. aureus* bacteria have been secreting metabolic products into interstitial fluid around your cut. The secretions are attracting phagocytic white blood cells. A dendritic cell engulfs several bacteria, then migrates to the lymph node in your elbow. By the time it reaches its destination, the dendritic cell has digested the bacteria and is displaying their fragments as antigens bound to MHC markers ❷. In the lymph node, one of your T cells recognizes and binds to a fragment of ClfA protein displayed by the dendritic cell ❸. This T cell is called a **helper T cell** because it activates other lymphocytes in an adaptive immune response.

The helper T cell and the dendritic cell interact in the lymph node for about 24 hours and then disengage. The helper T cell returns to the circulatory system and undergoes repeated mitotic divisions, forming a huge population of identical T cells. These clones mature into effector and memory cells ❹, each with receptors that recognize the ClfA protein.

Let's now go back to that B cell in the lymph node. By now, it has digested the engulfed bacterium, and it is displaying bits of *S. aureus* bound to MHC markers on its plasma

Figure 34.12 An example of an antibody-mediated immune response.

❶ The B cell receptors (green) on a naive B cell bind to an antigen on the surface of a bacterium (red). The bacterium's complement coating (purple dots) triggers the B cell to engulf and digest it. Bacterial fragments (red) bound to MHC markers (blue) become displayed at the surface of the B cell.

❷ A dendritic cell engulfs and digests the same kind of bacterium that the B cell encountered. Bacterial fragments (red) bound to MHC markers (blue) become displayed at the surface of the dendritic cell.

❸ The receptors of a naive helper T cell recognize bacterial antigen displayed by the dendritic cell. The two cells interact and disengage.

❹ The T cell begins to divide. Its descendants differentiate into effector helper T cells and memory helper T cells.

❺ The receptors of one of the effector helper T cells recognize and bind to antigen displayed by the B cell. Binding causes the T cell to secrete cytokines (blue dots).

❻ The cytokines induce the B cell to undergo repeated mitotic divisions. Its many descendants differentiate into effector B cells and memory B cells.

❼ The effector B cells begin making and secreting huge numbers of antibodies, all of which recognize the same antigen as the original B cell receptor. The new antibodies circulate throughout the body and bind to any remaining bacteria.

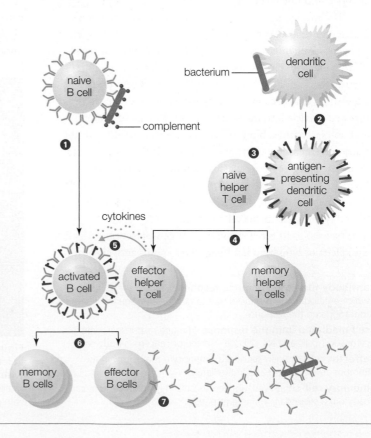

membrane. The new helper T cells recognize a fragment of ClfA protein displayed by the B cell. One of these helper T cells binds to the B cell. Like long-lost friends, the two cells stay together for a while and communicate ❺. One of the messages that is communicated consists of cytokines secreted by the helper T cell. The cytokines stimulate the B cell to undergo repeated mitotic divisions after the two cells disengage. A huge clonal population of descendant cells forms, and these B cells mature as effector and memory cells ❻. The effector B cells start releasing antibodies ❼. These antibodies are a secreted version of the original B cell's receptors, so they can recognize and bind to *S. aureus* ClfA protein.

By a theory called clonal selection, lymphocytes that give rise to huge clonal populations are "selected" during an immune response because antigen binds most tightly to their receptors. B cells or T cells with receptors that do not bind antigen do not divide; those that bind antigen loosely do not form huge populations (**Figure 34.13**).

Huge numbers of antibodies now circulate throughout the body and bind to any *S. aureus* cells. An antibody coating prevents the bacteria from attaching to body cells, and it facilitates phagocytosis to speed disposal. Antibodies also glue the foreign cells together into clumps, a process called **agglutination**. The clumps are easily removed from the circulatory system by the liver and spleen.

ANTIBODIES IN ABO BLOOD TYPING

As you learned in Section 13.4, a carbohydrate on red blood cell membranes determines your blood type. This carbohydrate is called H antigen, and it occurs in two forms. People with one form of the antigen have type A blood; people with the other form have type B blood. People with both forms have type AB blood; those with neither are type O.

Early in life, each individual starts making antibodies that recognize molecules foreign to the body, including any nonself form of the H antigen (**Table 34.5**). If a blood transfusion becomes necessary, it is especially important to know which H antigens your blood cells carry. A transfusion of incompatible red blood cells can cause a potentially fatal transfusion reaction in which the recipient's antibodies recognize and bind to H antigens on the transfused cells. The binding activates complement, which punctures the membranes of the foreign cells, thus releasing a massive amount of hemoglobin that can very quickly cause the kidneys to fail.

Identifying red blood cell surface antigens helps prevent pairing of incompatible transfusion donors and recipients, and also alerts physicians to blood incompatibility problems that may arise during pregnancy. A typical blood typing test involves mixing drops of a patient's blood with antibodies to

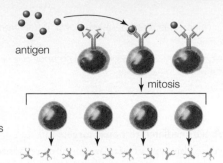

Antigen binds only to a matching B cell receptor.

B cell that binds antigen undergoes mitosis.

Many effector B cells form and secrete many antibodies.

Figure 34.13 Example of clonal selection, with B cells. Only lymphocytes with receptors that bind to antigen divide and mature. Lymphocytes with receptors that best fit an antigen are favored in this process.

TABLE 34.5

Antibodies to H Antigen

ABO Blood Type	H Antigen on Red Blood Cells	H Antigen Antibodies
A	A	anti-B
B	B	anti-A
AB	Both A and B	none
O	Neither A nor B	anti-A, anti-B

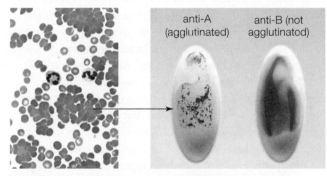

Figure 34.14 Agglutination of red blood cells. In an ABO blood typing test, samples of a patient's blood are mixed with antibodies to the different types of H antigens (A or B). Antibodies that recognize H antigen will agglutinate (clump) the red blood cells.

FIGURE IT OUT
Which ABO blood type does this patient have?

Answer: Type B

the different forms of red blood cell antigens. Agglutination occurs when the antibodies recognize antigen on the patient's cells (**Figure 34.14**).

---

agglutination (uh-glue-tin-A-shun) The clumping together of antigen-bearing cells or particles by antibodies.
helper T cell T lymphocyte that activates other lymphocytes; required for all adaptive immune responses.

CREDITS: (14) left, Courtesy Kristine Krafts, M.D.; right, lkordela/Shutterstock.com.

CHAPTER 34 IMMUNITY   **603**

A cell-mediated response targets infected or cancerous body cells. This type of adaptive immune response does not involve antibodies; it is carried out by lymphocytes that can recognize specific ailing body cells. Most ailing body cells display molecules not found on healthy cells. For example, cancer cells display abnormal proteins; body cells infected with viruses or other intracellular pathogens display polypeptides of the infecting agent. Both types of cell are killed in a cell-mediated response. The killers are antigen-sensitized **cytotoxic T cells**: If B cells are like assassins, then cytotoxic T cells specialize in cell-to-cell combat.

CYTOTOXIC T CELLS: ACTIVATION AND ACTION

A typical cell-mediated immune response starts in interstitial fluid during inflammation, when a dendritic cell engulfs a sick body cell or the remains of one (**Figure 34.15**), then migrates to a lymph node or the spleen. By the time it get there, the dendritic cell is displaying antigen (abnormal or foreign protein fragments from the ingested cell) together with MHC markers ❶. These antigen-MHC complexes are inspected by millions of naive T cells ❷, ❸ passing through the lymphatic organ. Some T cells have receptors that recognize antigen presented by the dendritic cell. These cells interact with the dendritic cell, then disengage. The T cells are now activated.

Helper T cells that have been activated by interacting with the dendritic cell return to the circulatory system and undergo repeated mitotic divisions. Their many descendants mature as effector and memory helper T cells ❹. The new effector cells recognize and interact with antigen-presenting macrophages. Any macrophage that interacts with an activated helper T cell increases its production of cytokines and pathogen-busting enzymes and toxins.

The new effector T cells also secrete cytokines ❺. Any cytotoxic T cells that have been activated by interacting with the dendritic cell (❷) recognize these cytokines as a signal to divide repeatedly, and their many descendants mature as effector and memory cells ❻.

All of the new effector cytotoxic T cells have receptors that recognize abnormal or foreign proteins displayed by that first ailing body cell. These cytotoxic T cells now circulate throughout blood and interstitial fluid, and bind to any

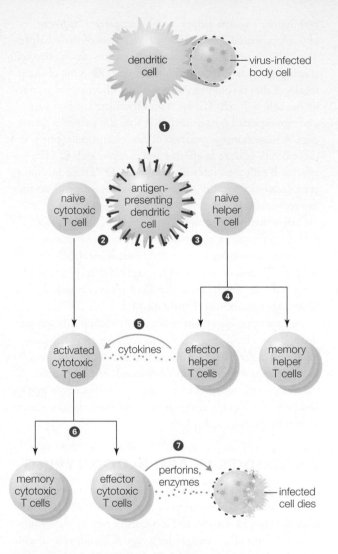

Figure 34.15 Example of a cell-mediated immune response targeting virus-infected body cells.

❶ A dendritic cell engulfs a virus-infected cell. Protein fragments of the virus (red) bind to MHC markers (blue), and the complexes become displayed at the dendritic cell's surface. The dendritic cell, now an antigen-presenting cell, migrates to a lymph node.

❷ Receptors on a naive cytotoxic T cell bind to antigen displayed by the dendritic cell. The interaction activates the cytotoxic T cell.

❸ Receptors on a naive helper T cell bind to antigen displayed by the dendritic cell. The interaction activates the helper T cell.

❹ The activated helper T cell divides again and again. Its many descendants mature as effector and memory cells, each with T cell receptors that recognize the same antigen.

❺ The effector helper T cells secrete cytokines.

❻ The cytokines induce the activated cytotoxic T cell to divide again and again. Its many descendants differentiate into effector and memory cells. Each cell bears T cell receptors that recognize the same antigen.

❼ The new effector cytotoxic T cells circulate throughout the body. They kill any body cell that displays viral antigen on its surface.

other body cell displaying the same molecules (**Figure 34.16**). Binding causes a cytotoxic T cell to release protein-digesting enzymes and small molecules called perforins. Perforins, like complement proteins, assemble into structures that penetrate a cell's plasma membrane and form large channels through it. The channels allow the enzymes to enter the body cell and cause it to lyse (burst) or commit suicide ❼.

THE ROLE OF NK CELLS

Remember that part of a T cell receptor recognizes antigen, and the other recognizes the MHC markers on body cells as "self." Only ailing cells that have intact MHC markers can be targeted and killed by cytotoxic T cells. However, some infections and cancers alter body cells so much that part or all of their MHC markers are missing. **NK cells** (natural killer cells) are crucial for eliminating such cells from the body. Unlike cytotoxic T cells, NK cells can kill body cells that lack MHC markers. Cytokines secreted by effector helper T cells also stimulate NK cell division, and the resulting populations of NK cells recognize and attack body cells that have antibodies bound to them. They also recognize certain proteins displayed by body cells that are under stress. Stressed body cells with normal MHC markers are ignored; only those with altered or missing MHC markers are killed. Because NK cells can operate early in immune defenses, they are often considered to be part of innate immunity. However, they also have features associated with lymphocytes of adaptive immunity: activation by cytokines, for example, and memory.

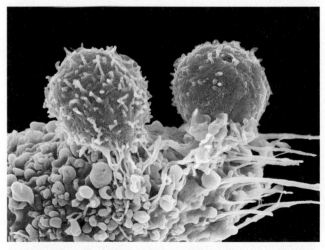

Figure 34.16 Cytotoxic T cells (pink) **killing a cancer cell.**

cytotoxic T cell T cell that kills ailing or cancerous body cells.
NK cell Lymphocyte that can kill stressed body cells lacking MHC markers.

Dr. Mark Merchant

Crocodiles are notorious for violent territorial disputes, leaving other crocodiles with large, jagged wounds and sometimes lost limbs. Loss of a limb in swampy, bacteria-infested water would probably have a fatal outcome for a human, but severe wounds in a crocodile typically heal very rapidly and without infection.

The human immune system depends largely on adaptive immunity to combat specific infectious agents. An adaptive response can take days to fully develop. Crocodiles also have adaptive immunity, but their innate immunity is much more robust than ours. Unique antimicrobial peptides made by crocodilians are very effective at preventing infection. National Geographic Explorer Mark Merchant has been investigating these peptides for use as drugs in humans. Most conventional antibiotics fight infection by targeting bacterial enzymes, but the fast mutation rate in bacteria can result in antibiotic-resistant strains (Chapter 17 Application). Crocodilian peptides interact ionically with bacterial cell membranes and cause them to rupture. This mechanical strategy makes it much more difficult for bacteria to become resistant.

Merchant compared the effects of cell-free serum from humans and alligators on 23 strains of bacteria, including the antibiotic-resistant MRSA. Human serum outright killed 8 of the bacterial strains; alligator serum killed all 23. In addition, the alligator serum was able to destroy HIV, the virus that causes AIDS.

LEARNING OBJECTIVES

Describe the symptoms of allergies and how they arise.

Using an appropriate example, explain autoimmunity.

Give some examples of the ways in which pathogens can evade mechanisms of innate and adaptive immunity.

Despite built-in quality controls and redundancies in function, immunity does not always work as well as it should. Its complexity is part of the problem, because there are simply more opportunities for failure to occur in systems with many components. Even a small failure in immune function can have a major effect on health.

OVERLY VIGOROUS RESPONSES

Allergies An **allergen** is a normally harmless substance that stimulates an immune response in some people. Common allergens include drugs, foods, pollen, dust mite feces, fungal spores, and insect venom. Sensitivity to an allergen is called an **allergy**.

Some people are genetically predisposed to have allergies, but factors such as infections, emotional stress, exercise, and changes in air temperature can trigger or worsen them. Typically, a first exposure to an allergen stimulates an antibody-mediated response targeting the allergen. Some IgE produced during this response becomes anchored to mast cells and basophils. Upon a later exposure, the anchored IgE binds to the allergen. Binding triggers the anchoring cell to degranulate and release histamines that initiate inflammation. If an allergen is detected by mast cells in the lining of the respiratory tract, the resulting inflammation constricts the airways and increases mucus secretion; sneezing, stuffed-up sinuses, and a drippy nose result (**Figure 34.17A**). Antihistamines relieve these symptoms by dampening the effects of histamines. Other drugs can inhibit mast cell degranulation, thereby preventing histamine release.

Physical contact with some allergens can provoke an inflammatory response characterized by an intensely itchy, irritated patch of skin (**Figure 34.17B**). Oils produced by poison oak and similar plants commonly trigger contact allergies, for example. Metal ions or small compounds such as these oils penetrate skin and covalently bond to molecules normally found in epidermal tissue. In some people, the resulting hybrid compounds bind to PAMP receptors on mast cells and keratinocytes (the main cellular constituents of epidermis, Section 28.7). Recent research has revealed that keratinocytes function as immune cells: When antigen binds to their PAMP receptors, keratinocytes release cytokines that trigger inflammation and a cell-mediated response. Effector

A Hay fever is caused by allergy to grass pollen. Coldlike symptoms (such as sneezing and a runny nose) are a result of inflammation of the mucous membranes in respiratory airways.

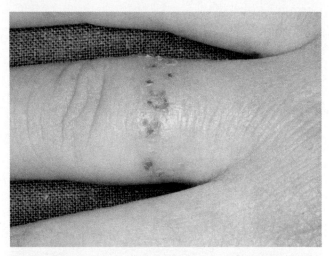

B Direct contact with an allergen can cause an itchy, irritated patch of skin. In this case, the offending substance was nickel (a metal) in a ring. A nickel allergy is common in humans because many of us have an allele that specifies a PAMP receptor with two amino acid substitutions. Nickel normally binds to a carbohydrate in skin, and the resulting compound is recognized by the altered receptor.

Figure 34.17 Allergies.

and memory T cells that form in the cell-mediated response migrate into dermis all over the body, a process that takes about two weeks. Upon later contact with the allergen, these cells release cytokines that trigger an enhanced local inflammatory reaction peaking a day or two after exposure. Each subsequent exposure triggers a faster and greater response.

Acute Illnesses Immune defenses that eliminate a threat can also damage body tissues. Thus, mechanisms that limit these defenses are always in play. Consider that some complement proteins activate spontaneously, even in the absence of infection or tissue damage. Without inhibitory molecules that inactivate complement, complement cascades would occur constantly, with disastrous effects on body tissues.

Acute illnesses can arise when mechanisms that limit immune responses fail. Exposure to an allergen sometimes causes a rapid and severe allergic reaction called anaphylaxis.

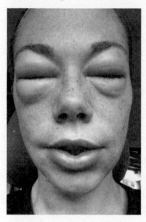

Huge amounts of inflammatory molecules, including histamines and prostaglandins, are released all at once in all parts of the body. Too much fluid leaks from blood into tissues, causing a sudden and dramatic drop in blood pressure (a reaction called shock). Rapidly swelling tissues (left) constrict the airways and may block them. Anaphylaxis is rare but life-threatening and requires immediate treatment. It may occur at any time upon exposure to even a tiny amount of allergen. Risks include any prior allergic reaction.

Severe episodes of asthma or septic shock occur when too many neutrophils degranulate at once. A "cytokine storm" occurs when too many white blood cells release cytokines at the same time. The cytokine overdose activates more white blood cells, which release more cytokines, and so on—a positive feedback loop that results in an exaggerated immune response. Massive inflammation and organ failure can occur, with potentially fatal results. Viruses such as H5N1 and Ebola (Section 19.2) have an unusually high mortality rate because they trigger cytokine storm in their hosts.

Autoimmunity People usually do not make antibodies to molecules that occur on their own healthy body cells. Quality control mechanisms built into the thymus and other lymphoid organs weed out T cells and B cells with receptors that recognize one's own body proteins. If these mechanisms fail, lymphocytes that do not discriminate between self and nonself may be produced. Such lymphocytes can mount an **autoimmune response**, which is an immune response that targets one's own body tissues. Autoimmunity is beneficial when a cell-mediated response targets cancer cells, but in most other cases it is not.

---

allergen (AL-er-jen) A normally harmless substance that provokes an immune response in some people.
allergy Sensitivity to an allergen.
autoimmune response Immune response that targets one's own body tissues.

The neurological disorder called multiple sclerosis occurs when self-reactive T cells mount a cell-mediated response targeting myelin in the brain and spinal cord (Section 29.4). Symptoms include weakness, loss of balance, paralysis, and blindness. Some alleles of MHC genes increase susceptibility, but a bacterial or viral infection may trigger the disorder.

Autoantibodies (antibodies that recognize self-proteins) produced by self-reactive B cells can damage the body or disrupt its functioning. In Graves' disease, autoantibodies that bind stimulatory receptors on the thyroid gland cause it to produce excess thyroid hormone, which quickens the body's overall metabolic rate (Section 31.4). Antibodies are not part of the feedback loops that normally regulate thyroid hormone production. So, antibody binding continues unchecked, the thyroid continues to release too much hormone, and the metabolic rate spins out of control. Symptoms of Graves' disease include uncontrollable weight loss; rapid, irregular heartbeat; sleeplessness; pronounced mood swings; and bulging eyes.

IMMUNE EVASION

Sometimes, the immune system fails to eliminate a pathogen from the body. In most cases, the failure occurs because evolved defenses allow a persistent pathogen to evade or suppress its host's immune system. Consider how antigen recognition is a key factor in adaptive immunity. Fast replication coupled with a high mutation rate allow pathogenic bacteria and viruses to evolve quickly, and a normal adaptive immune response provides plenty of selection pressure to drive the change. Recombination among different strains of viruses (or bacteria) adds even more evolutionary flexibility.

Antigen receptors, as you know, are quite specific, so even a small structural change in a viral envelope protein or bacterial carbohydrate can render a whole army of lymphocytes useless. In the few days it takes to make a new army of lymphocytes that recognize the changed molecule, a pathogen may be free to multiply unchecked.

Interfering with host adaptive immune responses is a common pathogen strategy. Some bacteria produce molecules that disrupt complement function or prevent antibody binding; extra-slimy capsules and biofilms can discourage phagocytosis by white blood cells. A toxin produced by *Staphyloccocus aureus* inhibits phagocyte movement. Herpes simplex viruses, which cause cold sores and genital herpes, are persistent because they inhibit antigen presentation. The bacterial species that causes tuberculosis disrupts T cell receptor signaling, as does the one that causes leprosy. HIV (the virus that causes AIDS) and Epstein-Barr virus (which causes mononucleosis) suppress expression of MHC genes, so cytotoxic T cells cannot recognize infected cells.

LEARNING OBJECTIVES

Describe the progress of a typical infection with HIV and how it causes AIDS.

Explain how HIV is transmitted.

Insufficient immune function—immune deficiency—makes an individual vulnerable to infections by opportunistic agents that are typically harmless to those in good health. Primary immune deficiencies, which are present at birth, are the outcome of mutations. Severe combined immunodeficiency (SCID) is an example (Section 15.6). Secondary immune deficiency is the loss of immune function after exposure to a virus or other outside agent. **AIDS** (acquired immune deficiency syndrome), the most common secondary immune deficiency, occurs as a result of infection with HIV (Section 19.1). Worldwide, almost 37 million individuals are infected with this virus (**Table 34.6** and **Figure 34.18**).

AIDS is a syndrome of disorders that occurs because HIV cripples the immune system, thus making the body susceptible to infections by other pathogens and to rare forms of cancer. A person newly infected with HIV appears to be in good health, perhaps fighting a cold or "the flu." If the infection is untreated, symptoms eventually emerge that foreshadow AIDS: fever, many enlarged lymph nodes, chronic fatigue and weight loss, and drenching night sweats. Then, infections caused by normally harmless microorganisms strike. Yeast infections of the mouth, esophagus, and vagina often occur, as well as a form of pneumonia caused by a fungus. Gastrointestinal inflammation due to infection by a yeast or virus causes diarrhea. Colored lesions that erupt are evidence of Kaposi's sarcoma, a type of cancer that is common among AIDS patients but rare among the general population. Other cancers are also common, as are infections by cancer-causing viruses such as Epstein–Barr virus. These medical problems are relatively uncommon in people with healthy immune systems.

HIV REVISITED

HIV mainly infects macrophages, dendritic cells, and helper T cells. When the virus particles enter the body, dendritic cells engulf them. The dendritic cells then migrate to lymph nodes, where they present processed HIV antigen to naive T cells. Armies of HIV-neutralizing antibodies and HIV-specific cytotoxic T cells form.

We have just described a typical adaptive immune response. It rids the body of most—but not all—of the virus. HIV persists in a few helper T cells in a few lymph nodes. For years or even decades, antibodies keep the level of HIV in the blood low, and cytotoxic T cells kill most of the HIV-

TABLE 34.6

Global HIV Cases

Region	People Living With HIV	New HIV Infections
Sub-Saharan Africa	25,800,000	1,400,000
Asia and the Pacific	5,000,000	340,000
Western/Central Europe and North America	2,400,000	85,000
Latin America	1,700,000	87,000
Central Asia/Eastern Europe	1,500,000	140,000
Caribbean Islands	280,000	13,000
Middle East/North Africa	240,000	22,000
Approx. worldwide total	36,900,000	2,087,000

Source: Joint United Nations Programme HIV/AIDS, 2015 report

infected cells. During this stage, infected people often have no symptoms of AIDS, but they can pass the virus to others.

Eventually, the level of virus-neutralizing antibodies plummets, and the production of T cells slows. Why this occurs is still a major topic of research, but its effect is certain: The immune system becomes progressively less effective at fighting the virus. The number of virus particles rises, and more and more helper T cells become infected. Lymph nodes begin to swell with infected T cells, and the battle tilts as the body makes fewer replacement helper T cells and immunity fails. Other types of viruses replicate more quickly than HIV, but the immune system eventually demolishes them. HIV demolishes the immune system. Secondary infections and tumors kill the patient.

Transmission HIV is not transmitted by casual contact; most infections are the result of having unprotected sex with an infected partner. The virus occurs in semen and vaginal secretions, and it can enter a sexual partner through epithelial linings of the penis, vagina, rectum, and mouth. The risk of transmission increases by the type of sexual act; for example, anal sex carries 50 times the risk of oral sex. Infected mothers can transmit HIV to a child during pregnancy, labor, delivery, or breast-feeding. HIV also travels in tiny amounts of infected blood in the syringes shared by intravenous drug abusers, or by hospital patients in less-developed countries. Many people have become infected via blood transfusions, but this transmission route is becoming rarer because blood is now tested prior to use for transfusions.

Testing Most AIDS tests check blood, saliva, or urine for antibodies that bind to HIV antigens. These antibodies are detectable in 99 percent of infected people within three months of exposure to the virus. One test that cross checks

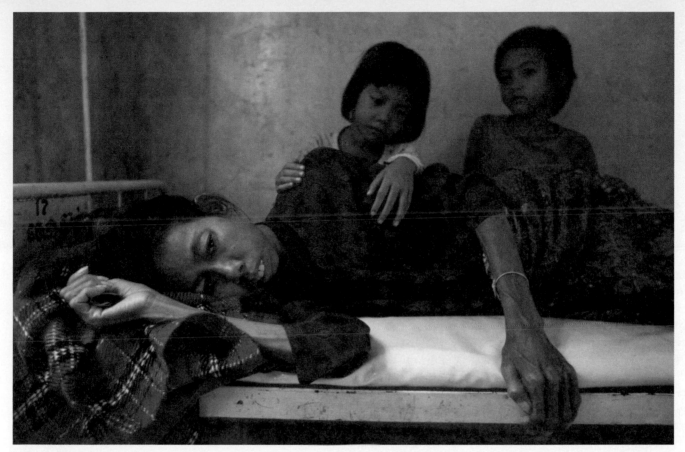

Figure 34.18 In Cambodia, a mother lies dying of AIDS in front of her children. This photo was taken in 2002. Today, the country has the highest incidence of AIDS in Southeast Asia.

antibodies with viral antigen can detect infection 3 weeks after exposure. Currently, the only reliable tests are performed in clinical settings; home test kits are generally less accurate than clinical tests. A false negative result may cause an infected person to unknowingly transmit the virus.

Treatments There is no way to rid the body of HIV, no cure for those already infected. Of the twenty or so FDA-approved drugs to treat AIDS, most target processes unique to retroviral replication. HIV injects its RNA genome and viral enzymes into a host cell (Section 19.1). One of the enzymes, reverse transcriptase, uses the viral RNA as a template to assemble DNA (Section 15.1). Modified nucleotides such as AZT and dideoxynucleotides (Section 15.3) stop the DNA synthesis reaction when they substitute for normal nucleotides, so these molecules can be used as drugs to slow the progression of AIDS.

Protease inhibitors used as HIV drugs prevent viral enzymes from carrying out post-translational cleavage of viral proteins; the uncut proteins cannot assemble into new viruses. A three-drug "cocktail" of one protease inhibitor plus two

reverse transcriptase inhibitors is currently the most successful AIDS therapy. The drugs are somewhat toxic, but they have changed the typical course of the disease from a short-term death sentence to a long-term, often manageable illness. Unfortunately, strains of the virus resistant to the most common drugs are becoming common. HIV has a very high mutation rate—the highest known for any biological entity.

Prevention Preventive use of a drug that contains two reverse transcriptase inhibitors has recently been shown to greatly reduce the HIV infection rate in high-risk populations. However, education is still our best option for halting the global spread of AIDS. In most circumstances, HIV infection is the consequence of unprotected sex or use of a shared needle for intravenous drugs. Programs that teach people how to avoid these unsafe behaviors are having an effect on the spread of the virus, but overall, our global battle against AIDS is not being won.

AIDS Acquired immune deficiency syndrome. A secondary immune deficiency that develops as the result of infection by the HIV virus.

Figure 34.19 Five-year-old Aliana is critically ill with encephalitis caused by measles. She is not expected to survive much longer.

COMMUNITY IMMUNITY

Immunization refers to procedures designed to induce immunity. In active immunization, a preparation that contains antigen—a **vaccine**—is injected or given orally. This immunization elicits a primary immune response, just as an infection would. A second immunization, or booster, elicits a secondary immune response for enhanced protection.

The first vaccine was developed as a result of desperate attempts to survive epidemics of smallpox, a severe disease that kills up to one-third of the people it infects. Before 1880, no one knew what caused infectious diseases or how to protect anyone from getting them, but there were clues. In the case of smallpox, survivors seldom contracted the disease a second time. They were said to be immune—protected from infection.

At the time, the idea of acquiring immunity to smallpox was extremely appealing. People had been risking their lives on it by poking into their skin bits of smallpox scabs or threads soaked in pus from smallpox sores. Some survived the crude practices, but many others did not.

By 1774, it was known that dairymaids usually did not get smallpox after they had contracted cowpox, a mild disease that affects humans as well as cattle. An English farmer collected pus from a cowpox sore on a cow's udder, and poked it into the arm of his pregnant wife and two small children. All survived the smallpox epidemic, but were from that time on subject to rock peltings by neighbors convinced they would turn into cows.

Twenty years later, the English physician Edward Jenner injected liquid from a cowpox sore into the arm of a healthy boy. Six weeks later, Jenner injected the boy with liquid from a smallpox sore. Luckily, the boy did not get smallpox. Jenner's experiment showed directly that the agent of cowpox elicits immunity to smallpox. Jenner named his procedure "vaccination," after the Latin word for cowpox (*vaccinia*). Though still controversial, Jenner's vaccine spread quickly through Europe, then to the rest of the world. The last known case of naturally occurring smallpox was in 1977. Use of the vaccine eradicated the disease.

Today, we know that immunization with cowpox is effective against smallpox because the two diseases are caused by closely related viruses; antibodies produced during an infection with one of them recognize antigens of both.

Vaccines for many other infectious diseases have since reduced suffering and deaths overwhelmingly, but only where they are available—and used. Public confidence is a necessary part of their success. When enough individuals refuse vaccination, outbreaks of preventable and sometimes fatal diseases occur. More and more parents are choosing not to vaccinate their children, most citing fears about vaccine safety.

Is this concern justified? Like all medical treatments, vaccines are not risk-free. Consider the MMR vaccine, which prevents infection from viruses that cause measles, mumps, and rubella. A rash or mild fever are relatively common side effects. However, one frequently cited concern, an increased risk of the neurological disorder autism, has no scientific basis. The 1998 scientific paper that first proposed this risk was later shown to contain falsified data, and the lead author of the paper lost his medical license as a result of this fraud. Rigorous studies of thousands of individuals have repeatedly failed to show a link between vaccinations and autism.

Compare the risk of mild side effects from a vaccination to the risk posed by contracting and spreading measles. Measles is a very serious and sometimes fatal disease. The virus spreads easily from person to person in air, and it remains infectious for hours in air or on solid surfaces. Measles is also highly contagious: An unvaccinated person exposed to measles has a 90 percent chance of contracting the disease; vaccination lowers the chance to 5 percent. An infected person is contagious for four days before the telltale rash appears, so there is no way to avoid infection by being vigilant.

Lola-May was three when she attended a birthday party for a child who later became ill with measles. Her mother Rachel said, "I'd chosen not to vaccinate Lola-May or perhaps not put as much thought into it as I should've done." Lola became ill a few days later. Cold-like symptoms gave way to a barking cough, difficulty breathing, agonizing ear pain, and uncontrollable fever. Then the rash appeared. Worsening symptoms required hospitalization, intravenous antibiotics, and other treatments to keep her lungs working. "When you're looking at a child who's sick and knowing…that the decision you made inevitably has caused this to happen…the guilt…is awful," said Rachel. Lola survived, but measles left her with permanent hearing loss and a perforated eardrum that prevents the use of hearing aids. "She just has to struggle. It's affected her speech and language, which has affected her confidence…I kind of thought that if everyone else had done it, I perhaps didn't need to. And that complacency cost us so much as a family."

Other children who contract measles are not nearly as lucky as Lola-May. The virus outright kills one or two of every thousand children it infects. Optic nerve infection results in blindness; encephalitis (dangerous swelling of the brain) can cause permanent intellectual disability or death (**Figure 34.19**). Rarely, a fatal central nervous system disease appears decades after a measles infection has apparently resolved.

Those of us who live in developed countries are fortunate to have access to vaccines that can prevent terrible diseases. Decades of widespread vaccination have all but removed the immediate threat of infection, so it is perhaps easy to take our collective immunity for granted. Refusing vaccination attaches a serious risk to an individual. People who cannot be vaccinated (children who are too young, for example, or those with immune deficiencies) have no choice but to rely on community immunity, which is effective only if 92 percent of the population are vaccinated. Parents who choose not to vaccinate their children also choose to risk the lives of others in their community.

immunization Any procedure designed to induce immunity to a specific disease.
vaccine (vax-SEEN) A preparation introduced into the body in order to elicit immunity to a specific antigen.

Section 34.1 The body's ability to resist and fight infections is called **immunity. Antigens** such as **PAMPs** trigger **innate immunity**, a set of general defenses that can prevent pathogens from becoming established inside the body. **Adaptive immunity**, which can specifically target billions of different antigens, follows. Signaling molecules called **cytokines** help coordinate the activities of white blood cells such as **dendritic cells, macrophages, neutrophils, basophils, mast cells**, and **eosinophils**. Lymphocytes (**B cells, T cells**, and NK cells) are white blood cells with special roles in immune responses.

Section 34.2 Microbiota (microorganisms that normally colonize body surfaces, including the linings of tubes and cavities) offer a biological barrier to infection. Anatomical and physiological barriers (such as **lysozyme**) also help keep microorganisms such as the ones in **dental plaque** on the outside of the body.

Section 34.3 Microorganisms that breach surface barriers trigger innate immunity. The presence of antigen in the body activates **complement** in cascading reactions. White blood cells can follow activated complement back to an affected tissue. Activated complement proteins also kill cells by puncturing their plasma membrane, and coat cells and particles to enhance their uptake by phagocytes. Phagocytes that engulf a microorganism or encounter damaged tissue release cytokines and local signaling molecules that initiate **inflammation**. Increased blood flow and capillary permeability speed delivery of more phagocytes to the affected area. Cytokines trigger **fever** that enhances almost every step of innate immune responses.

Section 34.4 Vertebrate antigen receptors collectively have the ability to recognize billions of specific antigens, a diversity that arises from random splicing of antigen receptor genes. **T cell receptors** are the basis of self/nonself discrimination; they recognize antigen only when it is displayed together with **MHC markers. B cell receptors** are **antibodies** that have not been released from a B cell.

Section 34.5 B cells and T cells carry out adaptive immune responses. The four main characteristics of these responses are self/nonself recognition, specificity (the ability to be customized for specific antigens), diversity (the potential to intercept a tremendous variety of pathogens), and memory. **Antibody-mediated** and **cell-mediated immune responses** work together to rid the body of a specific pathogen. **Effector cells** act in an immune response as soon as they form. **Memory cells** that also form are reserved for a later encounter with the same antigen, in which case they trigger a

faster, stronger secondary response. In all adaptive responses, phagocytic white blood cells engulf, process, and present antigen to T cells in the spleen or lymph nodes.

Section 34.6 **Helper T cells**, which activate other lymphocytes, are required in all adaptive immune responses. Antibodies that recognize a specific antigen are secreted by B cells during an antibody-mediated immune response. Antibody binding facilitates phagocytosis of antigen-bearing particles, and activates complement. ABO blood typing is an **agglutination** test that reveals the form of H antigen on a person's red blood cells.

Section 34.7 In a cell-mediated immune response, **cytotoxic T cells** kill body cells that have been altered by infection or cancer. **NK cells** can kill ailing body cells that cannot be detected by cytotoxic T cells.

Section 34.8 **Allergens** are normally harmless substances that trigger immune responses; sensitivity to an allergen is called an **allergy**. A malfunction in the immune system or in its checks and balances can cause dangerous acute illnesses, or chronic and sometimes deadly disorders. In an **auto-immune response**, a body's own cells are recognized as foreign and attacked.

Section 34.9 Immune deficiency is a reduced capacity to mount an immune response. **AIDS** is caused by the human immunodeficiency virus (HIV). The virus infects white blood cells, so it eventually cripples the immune system.

Application **Immunizations** with **vaccines** designed to elicit immunity to specific diseases are a critical part of worldwide health programs that save millions of lives. Effective vaccination programs require widespread participation.

SELF-ASSESSMENT
ANSWERS IN APPENDIX VII

1. _____ trigger(s) immune responses.
   - a. Cytokines
   - b. Lysozyme
   - c. Antibodies
   - d. Antigens
   - e. Histamines
   - f. MHC markers

2. Which of the following is *not* a surface barrier to infection?
   - a. skin
   - b. acidic gastric fluid
   - c. lysozyme in saliva
   - d. resident bacterial populations
   - e. complement activation
   - f. flushing action of diarrhea

3. Which of the following is *not* part of innate immunity?
   - a. phagocytic cells
   - b. fever
   - c. histamines
   - d. cytokines
   - e. inflammation
   - f. complement activation
   - g. presenting antigen
   - h. all take part

4. Which does *not* take part in adaptive immunity?
   - a. phagocytic cells
   - b. antigen-presenting cells
   - c. MHC markers
   - d. cytokines
   - e. antigen receptors
   - f. complement activation
   - g. antibodies
   - h. all take part

5. Activated complement proteins _____ .
   - a. puncture cells
   - b. promote inflammation
   - c. attract macrophages
   - d. all of the above

6. Choose the defining characteristics of innate immunity.
   - a. self/nonself recognition
   - b. immediate response
   - c. set of general defenses
   - d. antigen memory
   - e. diverse antigen receptors
   - f. fixed number of PAMPs
   - g. tailored for specific antigens

7. Choose the defining characteristics of adaptive immunity.
   - a. self/nonself recognition
   - b. immediate response
   - c. set of general defenses
   - d. antigen memory
   - e. diverse antigen receptors
   - f. fixed number of PAMPs
   - g. tailored for specific antigens

8. Antibodies are _____ .
   - a. antigen receptors
   - b. made only by B cells
   - c. proteins
   - d. all of the above

9. A dendritic cell engulfs a bacterium, then presents bacterial bits on its surface along with a(n) _____ .
   - a. MHC marker
   - b. antibody
   - c. T cell receptor
   - d. antigen

10. Antibody-mediated responses are most effective against _____ .
    - a. intracellular pathogens
    - b. extracellular pathogens
    - c. cancerous cells
    - d. a and c

11. Cell-mediated responses are most effective against _____ .
    - a. intracellular pathogens
    - b. extracellular pathogens
    - c. cancerous cells
    - d. a and c

12. _____ are targets of cytotoxic T cells.
    - a. Extracellular virus particles in blood
    - b. Virus-infected body cells or tumor cells
    - c. Parasitic worms in the liver
    - d. Bacterial cells in pus
    - e. Pollen grains in nasal mucus

13. Which combination of the following types of antibodies and immune cells is central to hay fever?
    - a. IgE and mast cells
    - b. IgG and basophils
    - c. IgA and lymphocytes
    - d. IgM and macrophages

14. Match the immune cell with its main function.
    - ___ dendritic cell
    - ___ B cell
    - ___ helper T cell
    - ___ NK cell
    - ___ macrophage
    - a. professional phagocyte
    - b. antigen-presenter
    - c. activates other lymphocytes
    - d. makes antibodies
    - e. kills ailing body cells that lack MHC markers

CERVICAL CANCER INCIDENCE IN HPV-POSITIVE WOMEN

A persistent infection with one of about 10 strains of genital HPV (human papillomavirus) is the main risk factor for cervical cancer. The virus spreads easily by sexual contact, but vaccines that prevent infection have been available since 2006. The vaccines consist of viral proteins that self-assemble into virus-like particles. The particles are not infectious (they contain no viral DNA), but their component proteins trigger an immune response that can prevent HPV infection and the cervical cancer it causes.

In 2003, Michelle Khan and her coworkers published results of their 10-year study correlating HPV status with cervical cancer incidence in women (**Figure 34.20**). All 20,514 participants were free of cervical cancer when the study began.

1. At 110 months into the study, what percentage of women who were not infected with any type of cancer-causing HPV had cervical cancer? What percentage of women who were infected with HPV16 also had cancer?

2. In which group would women infected with both HPV16 and HPV18 fall?

3. Is it possible to estimate from this graph the overall risk of cervical cancer that is associated with infection of cancer-causing HPV of any type?

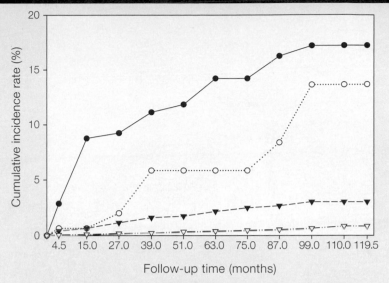

Figure 34.20 Cumulative incidence rate of cervical cancer correlated with HPV status. The data were grouped as follows:

● HPV16 positive
○ HPV16 negative and HPV18 positive
▼ All other cancer-causing HPV types combined
▽ No cancer-causing HPV type was detected.

---

15. Allergies occur when the immune system responds to _____ .
    - a. pathogens
    - b. toxins
    - c. normally harmless substances
    - d. all of the above

16. Match the immunity concept with the best description.
    - ___ anaphylactic shock
    - ___ immune memory
    - ___ autoimmunity
    - ___ inflammation
    - ___ immune deficiency
    - ___ antigen receptor
    - ___ antigen processing

    - a. recognizes antigen
    - b. inadequate immune response
    - c. general defense mechanism
    - d. immune response against one's own body
    - e. secondary response
    - f. acute allergic reaction
    - g. presenting antigen together with MHC markers

CLASS ACTIVITY
CRITICAL THINKING

1. Elena developed chicken pox when she was in first grade. Later in life, when her children developed chicken pox, she remained healthy even though she was exposed to countless virus particles daily. Explain why.

2. Monoclonal antibodies are produced by immunizing a mouse with a particular antigen, then removing its spleen. Individual B cells producing antibodies specific for the antigen are isolated from the mouse's spleen and fused with cancerous B cells from a myeloma cell line. The resulting hybrid myeloma ("hybridoma") cells are cloned: Individual cells are grown in tissue culture as separate cell lines. Each line produces and secretes antibodies that recognize the antigen to which the mouse was immunized. These antibodies are called monoclonal antibodies, and they can be purified and used for research or other purposes.

Monoclonal antibodies are sometimes used in passive immunization. With this procedure, the patient receives antibodies purified from the blood of another. Passive immunization offers immediate benefit for someone who has been exposed to a lethal agent such as tetanus, rabies, Ebola virus, or a venom or toxin. Because the antibodies were not made by the recipient's lymphocytes, effector and memory cells do not form, so benefits last only as long as the injected antibodies do.

Monoclonal antibodies are effective for passive immunization, but only in the immediate term. Antibodies produced by one's own immune system can last up to about six months in the bloodstream, but monoclonals delivered in passive immunization often last for less than a week. Why the difference?

3. The immune system immediately attacks a transplanted tissue or organ, so transplant recipients require immunosuppressive drugs. What type of lymphocyte is responsible for this response, which is called transplant rejection?

4. A flu shot consists of a vaccine against several strains of influenza virus. This year, you get the shot and "the flu." What happened? (There are at least three explanations.)

for additional quizzes, flashcards, and other study materials
▶ ACCESS MINDTAP AT WWW.CENGAGEBRAIN.COM

CREDIT: (20) Michelle Khan et al., "The Elevated 10-year Risk of Cervical Precancer and Cancer in Women with Human Papillomavirus (HPV) Type 16 or 18 and the Possible Utility of Type Specific HPV Testing in Clinical Practice; Journal of the National Cancer Institute, Vol. 97, No. 14, July 20, 2005.

CHAPTER 34 IMMUNITY   613

35

Respiration

Links to Earlier Concepts

This chapter draws on your knowledge of diffusion and membrane transport (Sections 5.6, 5.7), and aerobic respiration (7.1). You will learn about respiration in invertebrates (Chapter 23) and vertebrate gills and lungs (24.2, 24.3). You will see how hemoglobin and red blood cells (9.5, 33.4) function in gas exchange, and how chemoreceptors (30.1) and the brain stem (29.9) interact to regulate breathing.

Core Concepts

Organisms must exchange matter with the environment in order to grow, maintain themselves, and reproduce.

Respiration is a physiological process that moves oxygen and carbon dioxide between body tissues and the external environment. Aquatic animals exchange gases with water, as by way of gills. Air-breathing invertebrates have a system of tubes that deliver air into the body. Air-breathing vertebrates exchange gases inside their lungs.

Evolution underlies the unity and diversity of life.

Shared core processes and features provide evidence that living things are related. Vertebrate lungs evolved from outpouchings of the gut wall in fishes. Evolutionary modifications increased the efficiency of lungs, especially in mammals and birds. Birds have the most efficient lungs, which they require to support their high metabolism.

Living things sense and respond appropriately to their internal and external environments.

Interactions between the respiratory, circulatory, and nervous systems involve feedback control mechanisms. These homeostatic mechanisms allow a human body to respond dynamically to internal and external conditions. The brain monitors the gas content of the blood and adjusts the rate and depth of breathing accordingly, thus matching oxygen delivery to cells with the cellular activity level.

A manatee surfaces to breathe through nostrils on the top of its snout. Like other marine mammals, manatees have lungs and must hold their breath when underwater.

Photograph by Brian J. Skerry, National Geographic Creative.

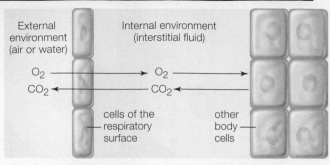

A Cells of the respiratory surface exchange gases with both the external and internal environment.

B Other body cells exchange gases with the internal environment.

Figure 35.1 Two sites of gas exchange.

SITES OF GAS EXCHANGE

In Chapter 7 you learned about aerobic respiration, an energy-releasing pathway that requires oxygen (O_2) and produces carbon dioxide (CO_2) as summarized below:

$$C_6H_{12}O_6 + O_2 \longrightarrow CO_2 + H_2O$$

glucose   oxygen   carbon dioxide   water

This chapter focuses on **respiration**, the physiological processes that collectively supply an animal's body cells with oxygen from the environment, and deliver waste carbon dioxide to the environment.

Gases enter and leave an animal body by diffusing across a thin, moist **respiratory surface** (**Figure 35.1A**). A typical respiratory surface is no more than one or two cell layers thick. The respiratory surface must be thin because gases must diffuse quickly through it. It must be moist because gases must dissolve in fluid to diffuse across the cells of the respiratory surface and enter the internal environment. Depending on the animal, the respiratory surface may be the outer body surface or the surface of an respiratory organ.

A second exchange of gases occurs internally, at the plasma membrane of body cells (**Figure 35.1B**). Here, oxygen diffuses from the interstitial fluid into a cell, and carbon dioxide diffuses in the opposite direction. In invertebrates that do not have a circulatory system, oxygen that crosses the respiratory surface reaches body cells by diffusion. In most invertebrates and all vertebrates, a circulatory system facilitates the movement of gases between the two sites of gas exchange.

FACTORS AFFECTING GAS EXCHANGE

The larger the surface area of a respiratory surface, the more molecules can cross it in any given time. This is why the surface area of a respiratory surface is often surprisingly large relative to the animal's body size. Branches and folds allow a large respiratory surface to fit in a small volume.

The concentrations of gases on either side of the respiratory surface also affect the rate of gas exchange. The steeper the concentration gradient across this surface, the faster diffusion proceeds. As a result, many animals have mechanisms that keep oxygen-rich air or water flowing over their respiratory surface. Your inhalations and exhalations move air rich in carbon dioxide away from the respiratory surface in your lungs and replace it with oxygen-rich air.

In many animals, respiratory proteins steepen the oxygen concentration gradient at a respiratory surface. A **respiratory protein** contains one or more metal ions that bind oxygen when the oxygen concentration is high, and release it when the oxygen concentration is low. Hemoglobins are the main respiratory protein in vertebrates and some invertebrates, including annelids such as earthworms. When an oxygen atom binds to hemoglobin, that atom no longer contributes to the oxygen concentration of blood or hemolymph. Thus, the presence of hemoglobin lowers the effective oxygen concentration in the blood and encourages faster diffusion of oxygen from the air into the blood.

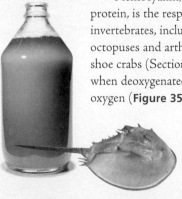

Hemocyanin, a copper-containing protein, is the respiratory protein in some invertebrates, including mollusks such as octopuses and arthropods such as horseshoe crabs (Section 23.10). It is colorless when deoxygenated, and blue when carrying oxygen (**Figure 35.2**).

Figure 35.2 Respiratory protein. Horseshoe crab hemolymph is colored blue by hemocyanin with bound oxygen.

respiration Physiological processes that collectively supply animal cells with oxygen and dispose of their waste carbon dioxide.

respiratory protein A protein that reversibly binds oxygen when the oxygen concentration is high and releases it when the oxygen concentration is low. Hemoglobin is an example.

respiratory surface Thin, moist layer of cells across which gases are exchanged between animal cells and the external environment.

35.2 Invertebrate Respiratory Organs

LEARNING OBJECTIVES

Describe gas exchange in a flatworm.

Explain how invertebrate circulatory systems interact with respiratory systems.

Compare the respiratory systems of a clam, a sea slug, and a land snail.

Describe the respiratory system of an insect.

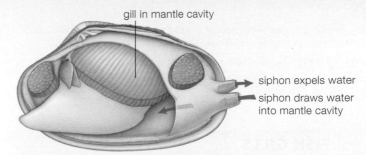

A Clam with internal gills. (Drawn with upper half of shell removed.)

B Sea slug with external gills.   **C** Land snail with a lung inside shell.

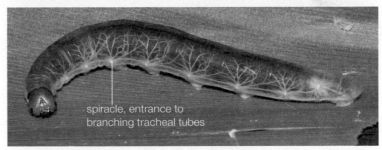

D Insect with a tracheal tube respiratory system. Spiracles at the body surface are the entrance to the branching air-filled tracheal tubes that are visible through the transparent cuticle of this butterfly larva.

Figure 35.3 Examples of invertebrate respiratory mechanisms.

Some invertebrates that live in aquatic or continually damp land environments have neither respiratory nor circulatory organs. For example, in flatworms (Section 23.5), all gas exchange occurs across the body surface and the lining of the gastrovascular cavity. Animals that lack both respiratory and circulatory organs are either small and flat like flatworms or, when larger, have cells arranged in thin layers.

Evolution of a circulatory system increased the efficiency of oxygen distribution through the body. In earthworms, a closed circulatory system carries hemoglobin-containing blood to and from a moist body wall that serves as the respiratory surface. Earthworms do not have red blood cells; their hemoglobin is in the fluid portion of their blood.

Aquatic invertebrates that have a circulatory system usually have **gills**—folded or filamentous respiratory organs that exchange gases with water. As blood or hemolymph moves through gills, it picks up oxygen from the water and gives up carbon dioxide. Crabs have gills in a chamber beneath the exoskeleton of their thorax. Among mollusks, most have gills in their mantle cavity (**Figure 35.3A**), but some sea slugs have external gills (**Figure 35.3B**).

A **lung** is an air-filled internal respiratory organ that functions in gas exchange. Some land snails and slugs have a lung (**Figure 35.3C**). Air that enters their lung exchanges gases with the animal's hemolymph.

The most successful air-breathing land invertebrates are insects. An exoskeleton helps these animals conserve water, but it also greatly slows gas exchange across the body's surface. Insects overcome this limitation with a **tracheal system**, which is a set of branching, chitin-reinforced tubes that convey air directly to tissues deep in the body.

The tracheal tubes start at spiracles, which are small openings across the exoskeleton (**Figure 35.3D**). There is usually a pair of spiracles per segment: one on each side of the body. Spiracles can be opened or closed to regulate the amount of oxygen that enters the body. Substances that clog spiracles are used as insecticides. For example, horticultural oils sprayed on fruit trees kill scale bugs, aphids, and mites by smothering them in this manner.

Inside the insect body, tracheal tubes divide into finer and finer branches. At the tips of the finest branches, oxygen from air in the tube dissolves in interstitial fluid, from which it diffuses into cells. Carbon dioxide diffuses in the opposite direction, from the cells into the tracheal tubes. Tracheal tubes deliver oxygen to fluid surrounding cells, so insects have no need for a respiratory protein such as hemoglobin or hemocyanin to transport gases over a long distance in fluid.

---

gill Of an aquatic animal, a folded or filamentous respiratory organ in which blood or hemolymph exchanges gases with water.

lung Internal respiratory organ in which blood exchanges gases with the air.

tracheal system (TRAY-key-ul) Of insects, a system of air-filled tubes that convey gases between openings at the body surface and internal tissues.

FISH GILLS

In nearly all fishes, gas exchange occurs across a pair of internal gills. Gill slits that open across the pharynx (throat) are visible at the body surface of jawless fishes and cartilaginous fishes, but most bony fishes have a movable gill cover. Water flows into a fish's mouth, continues into the pharynx (throat region), then exits the body through the gill slits (**Figure 35.4A**). A bony fish sucks water inward by opening its mouth, closing the cover over each gill, and contracting muscles that enlarge the oral cavity. Water is forced out over the gills when the fish closes its mouth, opens its gill covers, and contracts muscles that reduce the size of the oral cavity.

If you could remove the gill covers of a bony fish, you would see that the gills themselves consist of bony gill arches,

each with many gill filaments attached (**Figure 35.4B**). Each gill filament contains numerous capillary beds where gases in water are exchanged with gases in the blood.

Countercurrent exchange maximizes the efficiency of gas exchange across fish gills. **Countercurrent exchange** is a mechanism by which a substance is transferred between two fluids that are flowing in opposite directions on either side of a semipermeable membrane. Water flowing over gills and blood flowing through gill capillaries travel in opposite directions (**Figure 35.4C**). As a result, the water adjacent to a gill capillary always holds more oxygen than the blood flowing inside that capillary (**Figure 35.4D**). This persistent concentration gradient causes oxygen to diffuse from water to blood along the entire length of the capillary.

VERTEBRATE LUNGS

The earliest vertebrate lungs evolved from outpouchings of the gut wall in a lineage of bony fishes. Small, simple lungs probably helped these fishes survive in oxygen-poor water or when they made short trips over land. Over the course of vertebrate evolution, anatomical changes that accompanied the transition to land (and full-time air breathing) increased lung efficiency and improved the blood supply to the lungs (Section 33.1). The shift also involved changes in skull anatomy. Unlike most fishes, tetrapods (amphibians, reptiles, birds, and mammals) have internal nostrils that connect the sinus cavity with the oral cavity (**Figure 35.5A**).

An adult amphibian such as a frog draws air into its oral cavity by opening its nostrils, closing the entrance to its throat, and lowering the floor of its mouth (**Figure 35.5B**). The resulting negative pressure draws air into the oral cavity. To fill its lungs, the frog then shuts its nostrils and lifts the floor of the mouth to push air from the oral cavity through its now open pharynx into the lungs (**Figure 35.5C**). Amphibians also exchange gases across their scaleless, thin-skinned body surface. In fact, some salamanders rely entirely on such exchanges; over evolutionary time, they lost their lungs.

Figure 35.4 Structure and function of the gills of a bony fish.

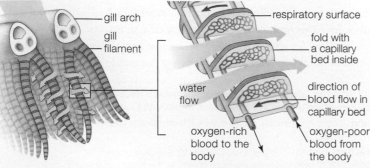

A Bony fish with its gill cover removed. Water flows in through the mouth, over gills, then out through gill slits.

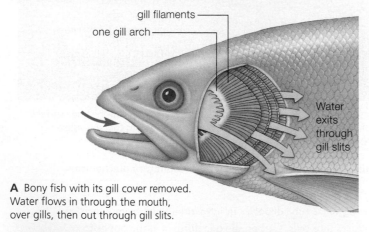

B Two gill arches with filaments.   **C** Countercurrent flow of water and blood.

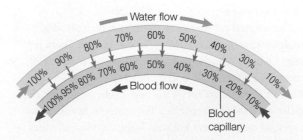

D Oxygen flow from water into a capillary. Percentages indicate the degree of oxygenation of water (blue) and blood (red). All along the capillary, oxygen diffuses from water into blood.

CREDITS: (4D) From Russell/Wolfe/Hertz/Starr. *Biology,* 1e. © 2008 Cengage Learning Inc.

A Nostril modifications. In frogs and other tetrapods, a pair of external nostrils opens from the nasal cavity to the external environment and a pair of internal nostrils connects the nasal cavity to the oral cavity. As a result, these animals can draw air into their oral cavity through their nose. By contrast, most bony fishes have two sets of external nostrils that are used to detect odors, but do not function in respiration. Water enters one set of nostrils and exits through another; the nasal and oral cavities are not connected.

B The frog lowers the floor of its mouth, pulling air (blue) through its nostrils and into the oral cavity.

C Closing the nostrils and elevating the floor of the mouth pushes air into the lungs.

Figure 35.5 Frog respiration.

Reptiles, birds, and mammals have a thick, dry, keratin-rich skin, so the lining of their lungs is their only respiratory surface. Unlike amphibians, these animals pull air into their lungs. By expanding their thoracic (chest) cavity, they decrease the pressure inside their lungs, creating a pressure gradient that draws air inward.

In mammals, inhaled air flows through increasingly smaller airways until it reaches tiny sacs called **alveoli** (singular, alveolus) where gas exchange with the blood occurs. These sacs are like a dead-end street, so air flows out the same way it came in. As a result of this two-way air flow, the oxygen content of air in an alveolus fluctuates. It peaks at the end of each inhalation when fresh air arrives, then declines as oxygen diffuses from the lungs into the blood. Lungs never deflate completely, so some stale air remains even after exhalation.

Birds have a more efficient respiratory system than mammals, meaning they are able to take up more oxygen from any given volume of air. A bird's small, inelastic lungs do not expand and contract (**Figure 35.6A**). Instead, a series of large, elastic air sacs attached to the lungs inflate and deflate. No gas exchange occurs across these air sacs. Instead, their function is to provide a continual one-way flow of fresh, oxygen-rich air through a bird's lungs. Bird lungs never contain stale air.

It takes two breaths for an inhaled volume of air to travel through a bird's respiratory system (**Figure 35.6B**). The first inhalation draws fresh air into posterior air sacs. During the first exhalation, this fresh air moves from posterior air sacs through tiny tubes in the lungs. The interior of these tubes is the bird's respiratory surface. During the second inhalation, the air that was in the lungs moves into anterior air sacs. It exits these air sacs during the second exhalation.

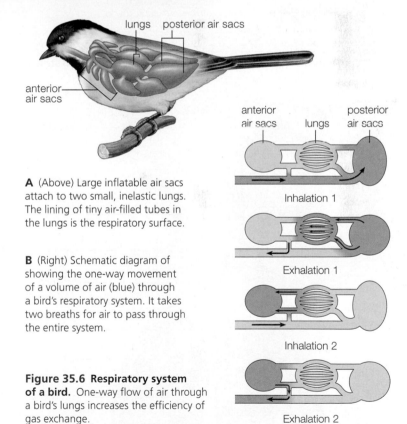

A (Above) Large inflatable air sacs attach to two small, inelastic lungs. The lining of tiny air-filled tubes in the lungs is the respiratory surface.

B (Right) Schematic diagram of showing the one-way movement of a volume of air (blue) through a bird's respiratory system. It takes two breaths for air to pass through the entire system.

Figure 35.6 Respiratory system of a bird. One-way flow of air through a bird's lungs increases the efficiency of gas exchange.

---

alveolus (al-VEE-oh-luss) In a mammalian lung, one of many tiny air sacs where gas exchange with the blood occurs.

countercurrent exchange Exchange of substances between two fluids moving in opposite directions.

LEARNING OBJECTIVES

List the structures that inhaled air passes through as it flows from the human nasal cavity to an alveolus.

Describe the location of the epiglottis and explain its function.

Explain the role of the respiratory system in giving rise to the voice.

Describe the location and structure of the human lungs.

The human respiratory system functions primarily in gas exchange, but it also has other roles. We speak as air moves past our vocal cords. We have a sense of smell because inhaled molecules stimulate olfactory receptors in the nose. Cells in nasal passages and other airways intercept and neutralize airborne pathogens. The respiratory system contributes to acid–base balance by expelling waste CO_2 that can make the blood too acidic. Controls over breathing also help maintain body temperature; water evaporating from airways has a cooling effect.

In this section, we will look at the anatomy of our respiratory system and associated structures. The sections that follow examine this system's respiratory function in more detail.

THE AIRWAYS

Take a deep breath. Now look at **Figure 35.7A** to see where the air went. If you are healthy and sitting quietly, air usually enters through your nose, rather than your mouth. As air moves through your nostrils, tiny hairs filter out large particles. Mucus secreted by some cells of the nasal lining captures most fine particles and airborne chemicals. Waving cilia on other cells in the lining sweep away captured contaminants.

Air from the nostrils enters the nasal cavity, where it is warmed and moistened. It flows next to the **pharynx**, or throat. It continues to the **larynx**, a short airway commonly known as the voice box because of the pair of vocal cords that span it (**Figure 35.8**). Each vocal cord consists of skeletal muscle with a cover of mucus-secreting epithelium. Contrac-

Figure 35.7 Human respiratory system and associated structures.

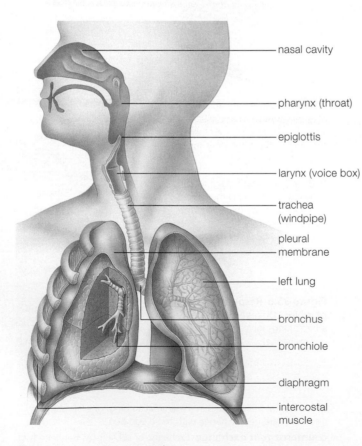

- nasal cavity
- pharynx (throat)
- epiglottis
- larynx (voice box)
- trachea (windpipe)
- pleural membrane
- left lung
- bronchus
- bronchiole
- diaphragm
- intercostal muscle

A Airways and respiratory muscles.

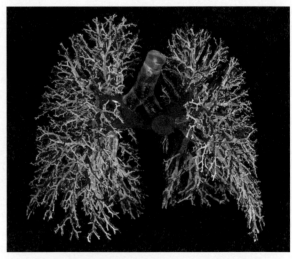

C Cast of airways (white) and blood vessels (red) in lungs.

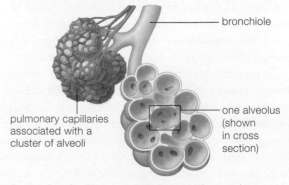

- bronchiole
- pulmonary capillaries associated with a cluster of alveoli
- one alveolus (shown in cross section)

B Alveoli and pulmonary capillaries; site of gas exchange.

tion of the vocal cords changes the size of the **glottis**, the gap between them.

When the glottis is wide open, air flows through it silently. When muscle contraction narrows the glottis, outgoing air flowing through the tighter gap makes the vocal cords vibrate, giving rise to sounds. The tension on the cords and changes in the position of the larynx alter the sound's pitch. To get a feel for how this mechanism works, place one finger on your "Adam's apple," the laryngeal cartilage that sticks out at the front of your neck. Hum a low note, then a high one. You will feel your vocal cords vibrate and notice how laryngeal muscles shift the position of the larynx as you change the pitch of your hum.

Passage of material out of the throat is controlled by the **epiglottis**, a leaf-shaped flap of membrane-covered elastic cartilage. When you swallow, the vocal cords come together to close the glottis, and the epiglottis folds down over the entrance to the larynx. As a result of these actions, materials move into the esophagus, the muscular tube that connects the pharynx to the stomach. When you are breathing, the epiglottis is tucked upward so the entrance to the larynx is clear. Air flows through the larynx into the **trachea** (windpipe).

The cartilage-reinforced trachea branches into two similarly reinforced **bronchi** (singular, bronchus). Each bronchus delivers air to a lung. Ciliated and mucus-secreting cells in the epithelial lining of the trachea and bronchi help fend off respiratory tract infections. Bacteria and airborne particles get caught in the secreted mucus, then cilia sweep the mucus toward the throat for disposal.

THE LUNGS

Human lungs are conical, spongy organs that reside in the chest, one on either side of the heart. The rib cage encloses and protects the lungs, and a two-layer-thick pleural membrane covers them. The pleural membrane's outer layer adheres to the wall of the thoracic cavity and its inner layer attaches to the outer surface of the lungs. A secreted fluid fills the small space between the pleural membrane's inner and outer layers. This fluid serves as a lubricant that reduces friction between the membrane's two layers when the lungs expand and contract during respiration.

Air that enters a lung through a bronchus flows through finer and finer branchings of a "bronchial tree." All of these branches are **bronchioles**. Unlike the airways that lead to them, bronchioles do not have cartilage-reinforced walls. Contraction and relaxation of smooth muscle in the walls of bronchioles can alter the bronchioles' internal diameter.

The smallest bronchioles have clusters of alveoli at their tips (**Figure 35.7B**). These tiny air sacs are the site of gas

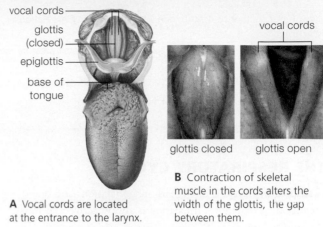

vocal cords
glottis (closed)
epiglottis
base of tongue

vocal cords

glottis closed glottis open

A Vocal cords are located at the entrance to the larynx.

B Contraction of skeletal muscle in the cords alters the width of the glottis, the gap between them.

Figure 35.8 Vocal cords and the glottis.

exchange. Each human lung has hundreds of million of alveoli, and both lungs together provide about 50 square meters (540 square feet) of surface area for gas exchange.

Branching blood vessels associate with the branching airways (**Figure 35.7C**). The smallest of these vessels are pulmonary capillaries that wrap in a meshlike manner around the alveoli. As blood flows through a pulmonary capillary, it exchanges gases with the air inside an alveolus. We will look at this exchange in detail in Section 35.6.

MUSCLES OF RESPIRATION

Like other mammals, humans draw air into their lungs by expanding the thoracic cavity. The **diaphragm**, a broad sheet of skeletal muscle beneath the lungs, separates the thoracic cavity from the abdominal cavity. Together with the intercostal muscles—skeletal muscles between the ribs—the diaphragm alters the volume of the thoracic cavity during breathing. The next section discusses in detail exactly how these muscles allow us to breathe.

---

bronchiole (BRON-key-ole) A small airway that leads from a bronchus to alveoli.

bronchus (BRON-cuss) Airway connecting the trachea to a lung.

diaphragm (DIE-uh-fram) Sheet of smooth muscle between the thoracic and abdominal cavities; contracts during inhalation.

epiglottis (ep-uh-GLOT-iss) Tissue flap that folds down to prevent food from entering the airways during swallowing.

glottis Opening formed when the vocal cords relax.

larynx (LAIR-inks) Short airway that contains the vocal cords; the "voice box."

pharynx (FAIR-inks) Throat; opens to airways and digestive tract.

trachea (TRAY-key-uh) Airway to the lungs; windpipe.

LEARNING OBJECTIVES

List the types of muscles that contract during inhalation, passive exhalation, and forced exhalation.

Explain the effect that contraction of the diaphragm has on the air pressure inside the lungs.

Describe the mechanisms that control the rate and depth of breathing.

Describe the recommended first aid for choking.

THE RESPIRATORY CYCLE

A **respiratory cycle** is one breath in (inhalation) and one breath out (exhalation). Inhalation is always active, and muscle contractions drive it.

Changes in the volume of the lungs and thoracic cavity during a respiratory cycle alter pressure gradients between air inside and outside the respiratory tract. When you start to inhale, the diaphragm contracts, flattening and moving downward (**Figure 35.9A**). At the same time, intercostal muscles on the outside of the rib cage contract, lifting the rib cage up and expanding it outward. As the thoracic cavity expands, so do the lungs. When pressure in the alveoli falls below atmospheric pressure, air follows the pressure gradient and flows into the airways.

Exhalation is usually passive. When intercostal muscles and the diaphragm relax, the lungs passively recoil and lung volume decreases. This compresses the alveoli, causing the air pressure inside them to increase above atmospheric pressure. As a result of this increase in pressure, air moves out of the lungs (**Figure 35.9B**).

Exhalation becomes active when you exercise vigorously or consciously attempt to expel extra air. During active exhalation, muscles of the abdominal wall contract. Pressure in the abdominal cavity increases, exerting an upward-directed force on the diaphragm. At the same time, contraction of intercostal muscles inside the rib cage pulls the thoracic wall inward and downward. As a result, the volume of the thoracic cavity decreases more than usual and additional air is forced out of the lungs.

The maximum amount of air that the lungs can hold averages 5.7 liters in men and 4.2 liters in women. **Vital capacity**, the maximum volume of air that can be moved in and out with forced inhalation and exhalation, is a measure of lung health. People who exercise regularly can increase their vital capacity. A person typically fills his or her lungs only halfway during a normal resting inhalation. Lungs never deflate completely. As you exhale, the smallest airways collapse, temporarily trapping some air. As a result, air in alveoli is always a mix of freshly inhaled air and air left behind during the previous exhalation.

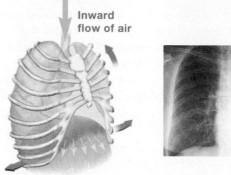

A Inhalation.
The thoracic cavity and lungs expand as the diaphragm contracts and moves downward, while the external intercostal muscles contract, lifting the rib cage.

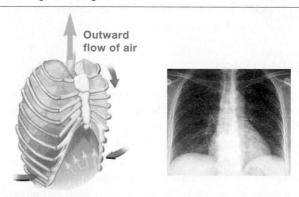

B Exhalation.
The thoracic cavity and lungs shrink as the diaphragm relaxes and moves upward and the external intercostal muscles relax.

Figure 35.9 Changes in the size of the thoracic cavity during a single respiratory cycle. The x-ray images reveal how inhalation and expiration change the lung volume.

FIGURE IT OUT
What effect does contraction of the diaphragm have on the volume of the thoracic cavity?

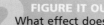
Answer: It increases the volume

CONTROL OF BREATHING

You do not have to think about breathing. Neurons in the medulla oblongata of your brain stem (Section 29.9) function as the pacemaker for inhalation. These neurons trigger 10–20 action potentials per minute. Nerves relay the action potentials to the diaphragm and intercostal muscles, where the stimulation results in contractions that cause inhalation. Between action potentials, the muscles relax and you exhale.

Your rate of breathing varies with your activity level (**Figure 35.10**). During exercise, your muscle cells increase their rate of aerobic respiration, so they produce more CO_2. This CO_2 diffuses into the blood, where it excites chemoreceptors in the walls of the carotid arteries and aorta. These receptors respond by signaling the brain. The brain itself also has chemoreceptors for CO_2. In response to signals that CO_2 is rising, the brain alters the breathing pattern, so breaths become more frequent and deeper. Sympathetic nervous stimulation also increases the respiratory rate during fright or excitement (Section 29.6). In many birds and mammals, a rise in body temperature triggers shallow, rapid breaths (panting). Panting helps cool a body by increasing the rate of evaporative cooling across the respiratory tract. Humans cool themselves primarily by sweating, and do not pant in response to heat.

Reflexes such as swallowing or coughing can temporarily halt breathing, but humans rarely hold their breath for more than a minute. When breathing stops, the resulting rise in CO_2 usually triggers a gasp for air within a minute or so.

CHOKING—A BLOCKED AIRWAY

Choking occurs when food or another object enters and blocks the larynx. A person who is choking cannot move air through the larynx, and so cannot breathe, cough, or speak. If you suspect someone is choking, encourage the person to cough. Being able to cough means the airway is not fully obstructed, and coughing will probably clear it.

If someone is choking, a rescuer can help clear the person's airway and keep the person from dying of lack of air. First aid for choking involves two types of maneuvers. Blows to the back can help dislodge the foreign material with little risk of injury to internal organs (**Figure 35.11A**). If back blows are ineffective, thrusts to the choker's abdomen can raise intra-abdominal pressure, forcing the diaphragm upward. Raising air pressure inside the lungs by this maneuver can dislodge the object, allowing the victim to resume breathing normally (**Figure 35.11B**).

respiratory cycle One inhalation and one exhalation.
vital capacity Maximum amount of air moved in and out of lungs with forced inhalation and exhalation.

STIMULUS

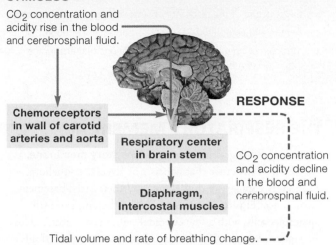

CO_2 concentration and acidity rise in the blood and cerebrospinal fluid.

Chemoreceptors in wall of carotid arteries and aorta

RESPONSE

Respiratory center in brain stem

CO_2 concentration and acidity decline in the blood and cerebrospinal fluid.

Diaphragm, Intercostal muscles

Tidal volume and rate of breathing change.

Figure 35.10 Respiratory response to CO_2 production during exercise. When chemoreceptors in arteries and the brain are excited by increased CO_2, they signal brain regions that govern breathing. As a result, breaths becomes deeper and more frequent, so more CO_2 is expelled.

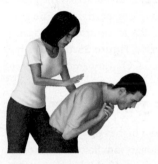

A Back blows. Have the person lean forward. Use the heel of your hand to strike between the shoulder blades.

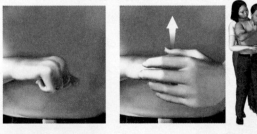

B Abdominal thrusts. Stand behind the person and place one fist below the rib cage, just above the navel, with your thumb facing inward. Cover the fist with your other hand and thrust inward and upward with both fists.

Figure 35.11 Maneuvers to assist a conscious adult who is choking. The Red Cross recommends that rescuers ask if a victim is choking and wants help. If the person nods, the rescuer should alternate between a series of five back blows (**A**) and five abdominal thrusts (**B**).

CREDIT: (10) photo, © C. Yokochi and J. Rohen, *Photographic Anatomy of the Human Body*, 2nd Ed., Igaku-Shoin, Ltd., 1979.

35.6 Gas Exchange and Transport

LEARNING OBJECTIVES

Explain how the structure of the respiratory membrane enhances its function.

Explain how oxygen and carbon dioxide are transported in the blood.

Describe the structure of hemoglobin and the conditions that encourage it to bind or to release oxygen.

THE RESPIRATORY MEMBRANE

Gases are exchanged at the lung's **respiratory membrane**, a multilayered structure that consists of alveolar epithelium, pulmonary capillary endothelium, and their fused basement membranes (**Figure 35.12**). Alveolar epithelium is mostly squamous cells, with a few cuboidal cells. Gases diffuse across the squamous cells. The thicker cuboidal cells secrete fluid that lubricates the walls of the alveolus, so these walls do not stick together when the alveolus has to inflate.

Oxygen from air in an alveolus dissolves in the secreted fluid, then diffuses passively across the respiratory membrane. The membrane is very thin, so gases do not need to diffuse far. The net direction of movement for each gas depends on its concentration gradient across the respiratory membrane, or as we say for gases, partial pressure gradient. The partial pressure of a gas is its contribution to the pressure exerted by a mixture of gases. **Figure 35.13** shows how the partial pressures of O_2 and CO_2 vary among air, blood, and body tissues.

OXYGEN TRANSPORT

Inhaled air has a higher partial pressure of O_2 than does blood in pulmonary capillaries. As a result, O_2 tends to diffuse from alveoli into these capillaries. Once in the blood, most O_2 diffuses into red blood cells, where it binds to the respiratory protein hemoglobin. Hemoglobin consists of four polypeptide (globin) subunits, each with an associated iron-containing heme group (**Figure 35.14**). When O_2 is bound to one or more of hemoglobin's heme groups, the molecule is called oxyhemoglobin.

Binding of O_2 to heme is reversible. Heme picks up O_2 in pulmonary capillaries, where the temperature is relatively cool, O_2 partial pressure is high, and CO_2 partial pressure is low. The oxygen-rich blood then flows to a systemic capillary bed in a tissue. Here, the temperature and partial pressure of CO_2 are high and the partial pressure of O_2 is low. These conditions cause the hemes to release O_2.

CARBON DIOXIDE TRANSPORT

CO_2 diffuses into capillaries from metabolically active tissues, which have higher CO_2 partial pressure than blood. Once in the blood, CO_2 is transported to the lungs in three forms. About 10 percent remains dissolved in plasma. Another 30 percent reversibly binds with amino groups of hemoglobin and forms carbaminohemoglobin. However, the majority of CO_2 is transported in plasma as bicarbonate (HCO_3^-).

An enzyme in red blood cells (carbonic anhydrase) catalyzes formation of carbonic acid from CO_2 and water. Carbonic acid then dissociates into bicarbonate and H^+:

$$CO_2 + H_2O \rightleftharpoons \underset{\text{carbonic acid}}{H_2CO_3} \rightleftharpoons \underset{\text{bicarbonate}}{HCO_3^- + H^+}$$

Being ions, bicarbonate and H^+ cannot diffuse across a red blood cell's plasma membrane. The H^+ binds to hemoglobin, and the bicarbonate is transported into the plasma by proteins in the red blood cell's plasma membrane. When the blood reaches pulmonary capillaries, where the CO_2 partial pressure is relatively low, bicarbonate reenters the red blood cell and the above reaction is reversed. The CO_2 produced by this reverse reaction then diffuses into alveoli and leaves the body in exhalations.

Figure 35.12 Zooming in on the site of gas exchange.

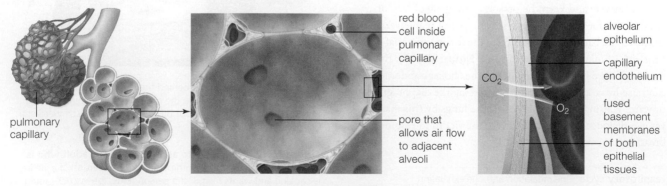

A Each alveolus is surrounded by a mesh of pulmonary capillaries.

B Cutaway view of an alveolus and associated pulmonary capillaries. Pores connect adjacent alveoli.

C Components of the respiratory membrane.

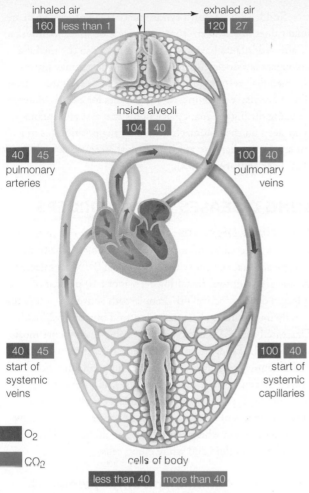

inhaled air → exhaled air

160 | less than 1   120 | 27

inside alveoli
104 | 40

40 | 45
pulmonary
arteries

100 | 40
pulmonary
veins

40 | 45
start of
systemic
veins

100 | 40
start of
systemic
capillaries

■ O₂

■ CO₂

cells of body

less than 40 | more than 40

Figure 35.13 Partial pressures (mm Hg) for O₂ (red boxes) and CO₂ (blue boxes) in air, blood, and tissues.

FIGURE IT OUT
Where is the biggest drop in partial pressure of O₂?

Answer: Between the start of the systemic capillaries and systemic veins.

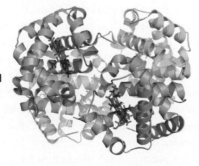

Figure 35.14 Hemoglobin, the oxygen-carrying protein in red blood cells. Heme groups are shown in red, globin chains in blue and green.

---

respiratory membrane Membrane consisting of alveolar epithelium, capillary endothelium, and their fused basement membranes.

Dr. Cynthia Beall

Air pressure decreases with altitude, so air becomes "thinner." At 5,500 meters (about 18,000 feet), a given volume of air contains only half as many O₂ molecules as air at sea level. This is why people who normally live at sea level often feel dizzy, disoriented, and breathless when they visit a high elevation. Typically, members of long-established, high-altitude populations do not experience these ill effects, because natural selection has adapted these populations to their low-oxygen environment.

National Geographic Explorer Dr. Cynthia Beall (at the center in the photo above) studies the adaptations of high-altitude human populations. She began her work in the 1970s, when people of the Andean plateau were the only high-altitude group whose physiology had been studied. The Andeans have an unusually high concentration of hemoglobin in their blood, a trait that maximizes the amount of oxygen carried in the blood. Beall's comparative studies revealed that high-altitude populations in other places adapted in other ways. For example, Tibetan plateau dwellers maintain a low hemoglobin level, but take more breaths per minute. Tibetans also make more nitric oxide. This gas encourages blood vessels to widen, increasing blood flow through lung capillaries. Although the Andean and Tibetan populations both face the same selective pressure (low oxygen), a different mechanism of countering that pressure evolved in each. In Andeans, the blood composition was altered; in Tibetans, respiratory rate and blood flow were affected.

35.7 Respiratory Diseases and Disorders

LEARNING OBJECTIVES

Explain the possible causes of sleep apnea, and why it is dangerous.

Describe some respiratory disorders caused by infectious organisms.

Explain how emphysema affects the lungs, and describe its symptoms.

Genetic disorders, infectious disease, and lifestyle choices can increase the risk of respiratory problems.

INTERRUPTED BREATHING

Apnea is a disorder in which breathing repeatedly stops and restarts spontaneously, especially during sleep. A tumor or damage to the brain stem's medulla oblongata can affect respiratory controls and result in apnea. However, sleep apnea more often occurs because the tongue or other soft tissue obstructs the upper airways. Breathing may stop for up to several seconds many times each night, interrupting normal sleep and causing daytime fatigue. Risk for heart attacks and strokes rises with sleep apnea because blood pressure soars when breathing stops. Sleep apnea caused by obstructing tissues can be reduced by changes in sleeping position or by wearing a mask that delivers pressurized air (**Figure 35.15A**).

Figure 35.15 Preventing apnea.

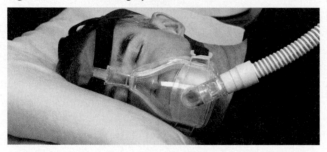

A Wearing a mask that delivers pressurized air can prevent sleep apnea caused by soft tissue obstruction.

B Putting infants to sleep on their back reduces the risk of apnea and SIDS (sudden infant death syndrome).

Sudden infant death syndrome (SIDS) occurs when an infant does not awaken from an episode of apnea. Autopsies of affected infants have shown that a defect in the medulla oblongata is associated with SIDS. The defect may impair the medulla's response to potentially deadly respiratory stress. There are also environmental risk factors for SIDS. Maternal smoking during pregnancy heightens the risk, and infants who sleep on their stomach or in a sitting position (as in a car seat or stroller) are at higher risk than those who sleep on their back (**Figure 35.15B**).

LUNG DISEASES AND DISORDERS

Infectious Diseases Worldwide, about one in three people is infected by *Mycobacterium tuberculosis*, the bacterial species that can cause tuberculosis (TB). Most infected people are symptom-free. However, about 10 percent of those infected develop "active TB." People with active TB cough up bloody mucus, have chest pain, and find breathing difficult. Untreated active TB can be fatal. Antibiotics can cure most infections, but they must be taken for at least six months.

Pneumonia is a general term for lung inflammation caused by an infection. Bacteria, viruses, and fungi can cause pneumonia. Typical symptoms include a cough, an aching chest, shortness of breath, and fever. A chest x-ray will show infected tissues in which spaces that should be filled with air instead contain fluid and white blood cells.

Bronchitis, Asthma, and Emphysema Inflammation of bronchial epithelium is called bronchitis. Inflamed epithelial cells secrete extra mucus that triggers coughing. Bacteria colonize the mucus, leading to more inflammation, more mucus, and more coughing. Bronchitis often arises after an upper respiratory infection. Inhaling irritants such as cigarette smoke can cause chronic bronchitis.

Asthma is a disorder in which inhaling an allergen or irritant triggers constriction of airways and makes breathing difficult. A tendency to have asthma is inherited, but exposure to irritants such as smoke raises the frequency of asthma attacks (episodes of constriction). A severe asthma attack is treated with inhaled drugs that dilate the airways.

With emphysema, tissue-destroying bacterial enzymes digest the thin, elastic alveolar wall. As these walls disappear, the area of the respiratory surface declines. Lungs become distended and inelastic like an over-inflated balloon, leaving the person constantly feeling short of breath. Some people inherit a predisposition to emphysema. They do not have a functioning gene for an enzyme that inhibits bacterial attacks on alveoli. However, tobacco smoking is by far the main risk factor for emphysema.

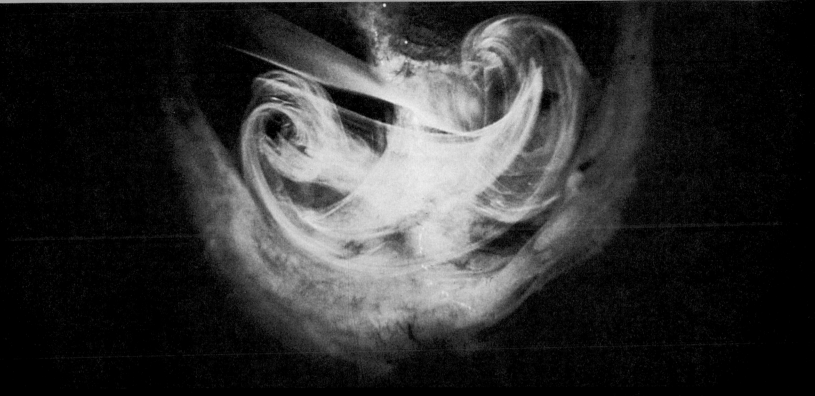

Figure 35.16 Cigarette smoke at the junction between the trachea and the bronchi. Smoke immobilizes cilia of cells lining the airways.

EFFECTS OF SMOKING

By now, everyone knows that smoking tobacco raises the risk of lung cancer. Tobacco smoke contains more than forty carcinogens (cancer-causing chemicals), and more than 80 percent of lung cancers occur in smokers. Once diagnosed with lung cancer, the majority of smokers die within a year. Female smokers are at an especially high risk of developing cancer. On average, they develop cancers at a younger age than men, and with lower exposure to tobacco.

Lung cancer takes years to become large enough to be diagnosed, but tobacco smoke impairs respiratory function immediately. When the smoke enters upper airways (**Figure 35.16**), it immobilizes cilia that help keep pathogens and pollutants from reaching the lungs. Smoke from a single cigarette can paralyze these cilia for hours. Smoke also kills white blood cells that patrol and defend respiratory tissues, allowing pathogens to multiply in the undefended airways. The result is a heightened susceptibility to colds, asthma attacks, bronchitis, and emphysema.

Tobacco smoke also delivers a hefty dose of carbon monoxide (CO), a colorless, odorless, poisonous gas. Inhaled CO interferes with gas exchange. It binds to hemoglobin at the hemes that normally bind oxygen, forming COHb. CO blocks these oxygen-binding sites more or less permanently, because heme binds CO 200 times more tightly than it binds oxygen. Binding of CO also makes hemoglobin hold more tightly to any oxygen it does bind. Thus, when CO is present, blood holds less oxygen and delivers less of what oxygen it does have to tissues.

Smoking cigarettes increases blood COHb by about 5 percent for each pack smoked per day. Over the short term, a blood COHb level as low as 4–6 percent decreases a healthy young person's capacity to exercise. Over the longer term, the increased COHb level harms the heart, which must work overtime to make up for CO's suffocating effects on normal respiratory function.

Respiratory effects of marijuana smoke are less well studied than those of tobacco smoke. We do know that marijuana smoke contains carbon monoxide and an assortment of carcinogens, including arsenic and ammonia. Nevertheless, the few studies that have been done show no increased risk for lung cancer with marijuana smoking only. On the other hand, people who smoke both marijuana and tobacco have more respiratory problems than tobacco-only smokers.

Vaping—using an electronic device to produce an inhalable aerosol—is a relatively new phenomenon, so its health effects are even less well studied. We do know that vaping delivers far lower doses of carcinogens than smoking standard cigarettes, and no carbon monoxide. However, vaping is not harmless. Users still inhale toxins such as formaldehyde, as well as tiny lung-damaging particles.

CREDIT: (16) © Lennart Nilsson/Bonnierforlagen AB.

Section 35.1 Aerobic respiration requires O_2 and releases CO_2 as a product. **Respiration** is a physiological process by which O_2 enters a body's internal environment and CO_2 leaves it. Both gases diffuse across a **respiratory surface**. Hemoglobin and other **respiratory proteins** facilitate gas exchange by maintaining steep concentration gradients for the two gases.

Section 35.2 Some invertebrates in aquatic or damp habitats exchange gases mainly across the body surface. In many aquatic invertebrates gas exchange occurs at **gills**. Land snails and slugs have a simple **lung**. In insects, a **tracheal system** delivers air to cells deep inside the body.

Section 35.3 In fish gills, the opposing direction of water flow and blood flow allows a highly efficient **countercurrent exchange** of gases. Most land vertebrates have paired lungs; amphibians also exchange gases across the skin. Frogs pull air into their mouth through their nostrils, then push it into their lungs. Other tetrapods pull air into their lungs by expanding their thoracic cavity and lungs. In mammals, gas exchange occurs in **alveoli** at the ends of airways. Birds have a more efficient system than other vertebrates; gas exchange occurs as air flows through tubes in their lungs.

Section 35.4 In humans, air flows through the nose and the mouth into the **pharynx**, then the **larynx**, then the **trachea** (windpipe). The larynx contains the vocal cords, whose movements alter the size of the opening (the **glottis**) between them. When you swallow, the position of the **epiglottis** at the entrance to the larynx shifts, keeping food out of the trachea. The trachea branches into two **bronchi** that enter the lungs. These large airways connect to narrower **bronchioles**. At the ends of the finest branches are the thin-walled alveoli, where gas exchange occurs. Contraction of the **diaphragm** at the base of the thoracic cavity and of muscles attached to ribs alters the size of the thoracic cavity during breathing.

Section 35.5 A **respiratory cycle** is one inhalation and one exhalation. Inhalation is active. As muscle contraction expands the chest cavity, pressure in the lungs decreases below atmospheric pressure, so air flows into the lungs. These events are reversed during exhalation, which is normally passive. The most air that can move in and out in one cycle is the **vital capacity**. Commands from the brain stem adjust the rate and depth of breathing.

Section 35.6 Gas exchange occurs at the alveoli. A meshlike array of pulmonary capillaries surrounds each alveolus. The wall of an alveolus, the wall of a pulmonary capillary, and their fused basement membranes constitute a thin **respiratory membrane** between air inside an alveolus and the

blood. O_2 following its partial pressure gradient diffuses across the respiratory membrane, into the plasma of the blood, and finally into red blood cells.

Where O_2 partial pressure is high (in lungs), heme in hemoglobin of red blood cells binds O_2 and forms oxyhemoglobin. Heme groups release O_2 where its partial pressure is low (in respiring tissues).

CO_2 follows its partial pressure gradient and diffuses from cells to interstitial fluid, to blood. Most CO_2 reacts with water in red blood cells to form bicarbonate. The enzyme carbonic anhydrase speeds this reaction. The reaction is reversed in the lungs. There, CO_2 diffuses out of blood into air inside alveoli. It is expelled, along with water vapor, in exhalations.

Section 35.7 With apnea, breathing is interrupted by an obstructing tissue or as the result of a problem in the brain's respiratory center. SIDS occurs when an infant does not recover from an episode of apnea. Tuberculosis, pneumonia, bronchitis, and emphysema are respiratory diseases.

Application Smoking tobacco raises the risk of lung cancer and damages cells that protect airways from pathogens. It also delivers carbon monoxide, a gas that interferes with oxygen delivery by binding to hemoglobin more strongly than oxygen does. Smoking marijuana also delivers carbon monoxide. Vaping also introduces toxins into the lungs, and its health risks are not well studied.

SELF-ASSESSMENT

ANSWERS IN APPENDIX VII

1. Respiratory proteins such as hemoglobin _____ .
   a. contain metal ions
   b. occur only in vertebrates
   c. permanently bind oxygen
   d. strengthen the walls of lung cells

2. In _____ , gas exchange occurs at the body surface and gas is distributed by diffusion alone.
   a. earthworms   c. frogs
   b. flatworms   d. insects

3. Countercurrent flow of water and blood increases the efficiency of gas exchange in _____ .
   a. fishes   c. birds
   b. amphibians   d. all of the above

4. In human lungs, gas exchange occurs at the _____ .
   a. two bronchi   c. alveoli
   b. pleural sacs   d. trachea

5. When you breathe quietly, inhalation is _____ and exhalation is _____ .
   a. passive; passive   c. passive; active
   b. active; active   d. active; passive

RISKS OF RADON Radon is a colorless, odorless gas emitted by many rocks and soils. It is formed by the radioactive decay of uranium and is itself radioactive. There is some radon in the air almost everywhere, but routinely inhaling radon raises the risk of lung cancer, especially in people who smoke tobacco. **Figure 35.17** is an estimate of how radon in homes influences the risk of lung cancer mortality.

Radon Level (pCi/L)	Risk of Cancer Death From Lifetime Radon Exposure	
	Never Smoked	**Current Smokers**
20	36 out of 1,000	260 out of 1,000
10	18 out of 1,000	150 out of 1,000
8	15 out of 1,000	120 out of 1,000
4	7 out of 1,000	62 out of 1,000
2	4 out of 1,000	32 out of 1,000
1.3	2 out of 1,000	20 out of 1,000
0.4	<1 out of 1,000	6 out of 1,000

Note that these data show only the risk of death for radon-induced cancers. Smokers are also at risk from lung cancers that are caused by tobacco alone.

1. If 1,000 smokers were exposed to a radon level of 1.3 pCi/L over a lifetime (the average indoor radon level), how many would die of a radon-induced lung cancer?

2. How high would the radon level have to be to cause approximately the same number of cancers among 1,000 nonsmokers?

3. The risk of dying in a car crash is about 7 out of 1,000. Is a smoker in a home with an average radon level (1.3 pCi/L) more likely to die in a car crash or of radon-induced cancer?

Figure 35.17 Estimated risk of lung cancer death as a result of lifetime radon exposure. Radon levels are in picocuries per liter (pCi/L). A curie is a measure of radiation. The Environmental Protection Agency considers a radon level above 4 pCi/L unsafe. To learn about testing for radon and what to do if the radon level is high, visit the EPA's Radon Information Site at www.epa.gov/radon.

6. During inhalation, _____ .
    a. the thoracic cavity expands
    b. the diaphragm relaxes
    c. atmospheric pressure declines
    d. the lungs deflate

7. What type of metal associates with hemoglobin?

8. _____ binds to hemoglobin more strongly than O_2 does.
    a. Carbon dioxide   c. Oxyhemoglobin
    b. Carbon monoxide   d. Carbonic anhydrase

9. Carbonic anhydrase in red blood cells catalyzes formation of bicarbonate from water and _____ .
    a. oxygen   c. oxyhemoglobin
    b. hemoglobin   d. carbon dioxide

10. The medulla oblongata _____ .
    a. produces hemoglobin   c. regulates breathing rate
    b. is in the forebrain   d. detects carbon monoxide

11. _____ in some arteries detect changes in the CO_2 concentration of the blood.
    a. Mechanoreceptors   c. Photoreceptors
    b. Neurotransmitters   d. Chemoreceptors

12. True or false? Human lungs hold some air even after forced exhalation.

13. Vital capacity measures the amount of _____ .
    a. air moved in a breath   c. CO_2 in blood
    b. hemoglobin in red cells d. blood flow to the lungs

14. What type of organism causes tuberculosis?
    a. Bacteria   c. Fungi
    b. Viruses   d. Worms

15. Match the structures with their descriptions.
    ___ trachea   a. muscle of respiration
    ___ diaphragm   b. gap between vocal cords
    ___ alveolus   c. between bronchi and alveoli
    ___ bronchus   d. windpipe
    ___ bronchiole   e. site of gas exchange
    ___ glottis   f. airway leading to lung

CRITICAL THINKING

1. The red blood cell enzyme carbonic anhydrase contains the metal zinc. Humans obtain zinc from their diet, especially from red meat and some seafoods. A zinc deficiency does not reduce the number of red blood cells, but it does impair respiratory function by reducing carbon dioxide output. Explain why a zinc deficiency has this effect.

2. Look again at **Figure 35.13**. Notice that the oxygen and carbon dioxide content of blood in pulmonary veins is the same as that at the start of the systemic capillaries. Notice also that systemic veins and pulmonary arteries have equal partial pressures. Explain the reason for these similarities.

3. Respiration supplies cells with the oxygen they need for aerobic respiration. Explain the role of oxygen in this energy-releasing metabolic pathway. Where is it used and what is its function?

4. A developing fetus gets oxygen from its mother's blood. Fetal capillaries run through pools of maternal blood in an organ called the placenta. As fetal blood runs through these capillaries, it exchanges substances with the maternal blood around the capillary. The hemoglobin made by a fetus is different than that made after birth. Fetal hemoglobin binds oxygen more strongly at low oxygen levels than normal hemoglobin. How would fetal hemoglobin's somewhat higher affinity for oxygen benefit the fetus?

36

Digestion and Human Nutrition

Links to Earlier Concepts

This chapter expands the discussion of digestive systems in Chapters 23 and 24. You will consider dietary aspects of organic compounds (Sections 3.2–3.4) and the use and storage of glucose (7.6). This chapter revisits diffusion, transport mechanisms, and osmosis (5.6–5.8); pH and cofactors (2.5, 5.5); and epithelium (28.2) and adipose tissue (28.3).

Core Concepts

Organisms exchange matter and energy with the environment in order to grow, maintain themselves, and reproduce.

All animals meet their energy needs and nutritional requirements by acquiring food. Food supplies all of the nutrients necessary to grow and maintain a body. Organic molecules in food provide a source of energy and serve as chemical building blocks for new organic molecules.

Evolution underlies the unity and diversity of life.

Shared core processes and features provide evidence that all living things are descended from a common ancestor. All animals break food down inside their body. In all vertebrates, digestion takes place inside a tubular digestive tract that runs through the body. Specialized anatomical and physiological traits, including variations in structure, size, and shape of organs along the digestive tract, are adaptations to different types of diet.

Interactions among the components of each biological system give rise to complex properties.

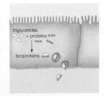

All animals have body regions in which functions related to the breakdown of organic molecules occur. Specialized components of the human digestive system interact to support functioning of the entire body. The body's efficient use of food arises from cooperative interactions among these components.

◀ An aye-aye feeding on a coconut. Constantly growing incisors allow these unusual primates to gnaw through woody nuts and grub-containing wood without wearing down their teeth.

631

36.1 Evolution of Digestive Systems

LEARNING OBJECTIVES

Explain how a sponge digests food and distributes nutrients.

Compare the structure of a gastrovascular cavity and a complete digestive tract, and explain the advantages of the latter.

Provide examples of regional specializations in the digestive tract.

All animals are heterotrophs that take food into their body and digest it. In this context, **digestion** means breaking food down into chemical components that cells of an animal body can use as sources of energy and as building blocks. Essential molecules and elements that an animal must obtain from its food are referred to as nutrients.

SITES OF DIGESTION

Intracellular Digestion Animals evolved from a colonial protist (Section 23.2). In such protists, individual cells take in food by phagocytosis, break it down inside food vacuoles, then expel wastes by exocytosis. Sponges, which do not have tissues or organs, feed in a similar manner (**Figure 36.1**). Flagellated collar cells at the surface of the sponge capture food by waving their flagellum. The movement draws water through a sieve-like collar of microvilli that rings the flagellum. Food particles that collect on the collar are taken into the cell, and either digested or passed along to motile

amoeboid cells. The amoeboid cells digest food and distribute nutrients to cells throughout the sponge body.

Extracellular Digestion In most animals, digestion is extracellular. Food is broken down within the body, but inside a hollow sac or tube that opens to the external environment. In animals with tissues, nutrition involves four tasks:

1. *Ingestion*. Taking food into the body.
2. *Digestion*. Breaking down food. Mechanical digestion smashes food into smaller fragments. Chemical digestion is the enzyme-facilitated breakdown of large polymers into smaller, absorbable subunits.
3. *Absorption*. Moving nutrients across the lining of the digestive region and into the internal environment (the fluid around body cells).
4. *Elimination*. Expelling any leftover material that was not digested and absorbed.

SAC OR TUBE?

Flatworms such as planarians digest their food inside a saclike **gastrovascular cavity** that also functions in gas exchange. A gastrovascular cavity has a single opening through which food enters and wastes leave. In flatworms, the opening is at the tip of a muscular tube called the pharynx (**Figure 36.2A**). The two-way traffic in a gastrovascular cavity makes digestion relatively inefficient. Food must be broken down, its nutrients absorbed, and wastes eliminated before new food can enter.

A gastrovascular cavity is sometimes referred to as an incomplete digestive tract. By contrast, a **complete digestive tract** is a tubular gut, with a mouth at one end and an opening for expelling waste at the other end. Some invertebrates such a earthworms (**Figure 36.2B**) have a complete digestive tract, as do all vertebrates. A coelomate animal (Section 23.1) with a complete digestive tract has a tube-within-a-tube body plan. The body wall is the outer tube, the digestive tract is the inner tube, and the coelom lies between them (**Figure 36.2C**).

Figure 36.1 Food-processing cells of a sponge. Digestion is intracellular.

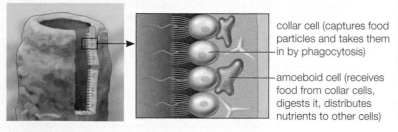

collar cell (captures food particles and takes them in by phagocytosis)

amoeboid cell (receives food from collar cells, digests it, distributes nutrients to other cells)

Figure 36.2 Complete versus incomplete digestive tracts. In both, digestion occurs within the body, but outside of cells.

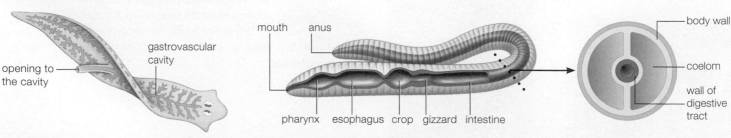

opening to the cavity

gastrovascular cavity

mouth   anus

pharynx   esophagus   crop   gizzard   intestine

body wall

coelom

wall of digestive tract

A Incomplete digestive tract of a flatworm. The branching gastrovascular cavity has a single opening.

B Complete digestive tract of an earthworm. There are two openings (mouth and anus) and specialized regions in between.

C Cross-section illustrating the tube-within-a-tube body plan.

CREDIT: (2B) From Russel/Wolfe/Hertz/Starr, Biology, 2E. © 2011 Cengage Learning, Inc.

The evolution of a complete digestive tract was an important innovation. As food travels the length of a complete digestive tract, it passes through regions specialized for food storage, food breakdown, nutrient absorption, and waste elimination. Unlike a gastrovascular cavity, a tubular gut can carry out all of these tasks simultaneously. Like you, an earthworm can take in new food even when food it ingested earlier has not yet exited the digestive tract.

SPECIALIZED REGIONS

In a complete digestive tract, a tubular organ called the **esophagus** receives food from the pharynx (throat). In some animals, including earthworms and many birds, part of the esophagus has become enlarged to serve as a food-storing pouch called the **crop** (**Figure 36.3**).

The esophagus delivers food to the **stomach**, a muscular digestive organ that functions in mechanical and chemical digestion. In earthworms, birds, and other animals that lack teeth, a portion of the stomach is specialized as a **gizzard**, an organ that grinds up food and begins the process of mechanical digestion. In the earthworm gizzard, hard bits of dirt ingested by the worm break up food. In birds, the gizzard is a muscular chamber with a hardened lining of glycoprotein. Seeds are difficult to grind up, so seed-eating birds have larger gizzards than those with other diets.

Ruminants such as cattle, goats, sheep, antelope, and deer are hoofed grazers that have multiple stomach chambers (**Figure 36.4**). Their digestive system adapts them to a diet of cellulose-rich plant foods. Like other animals, ruminants do not make enzymes that can break down cellulose. However, microbes living in the first two chambers of their stomach do make these enzymes. Solids from food accumulate in the second chamber, forming a "cud" that is regurgitated—moved back into the mouth for a second round of chewing. Fluid rich in nutrients moves from the second chamber to the third and fourth chambers, and finally to the intestine.

In a complete digestive tract, the **intestine** is the site of most chemical digestion and nutrient absorption. Meat is easier for animals to digest than plant material, so carnivores (animals that eat other animals) typically have a shorter intestine than herbivores (animals that eat plants).

In some animals with a complete digestive tract, digestive wastes exit through an **anus**, an opening that serves this purpose alone. In others, digestive wastes leave the body through a **cloaca**, a multipurpose opening that also releases urinary waste and functions in reproduction. Amphibians, reptiles, and birds have a cloaca, whereas humans and other placental mammals have an anus for digestive waste and a separate opening for urinary waste.

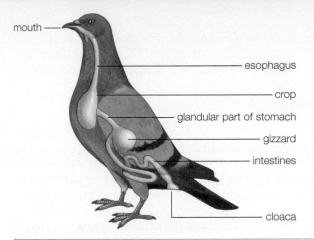

Figure 36.3 Complete digestive tract of a pigeon. Food enters through the mouth and wastes exit through the cloaca.

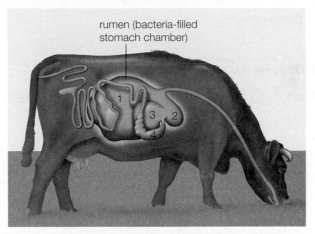

Figure 36.4 Multiple-chambered stomach of a ruminant (cow). There are four chambers. The large first chamber, the rumen (outlined in yellow), contains microbes that break down cellulose.

---

anus Body opening that serves solely as the exit for wastes from a complete digestive tract.

cloaca (cloh-AY-kuh) Body opening that serves as the exit for digestive and urinary waste; also functions in reproduction.

complete digestive tract Tubelike digestive system; food enters through one opening and wastes leave through another.

crop Of birds and some invertebrates, an enlarged portion of the esophagus that stores food.

digestion Breakdown of food into its chemical components.

esophagus (eh-SOFF-ah-gus) Tubular organ that connects the pharynx to the stomach.

gastrovascular cavity Saclike gut that also functions in gas exchange.

gizzard Of birds and some other animals, a digestive chamber lined with a hard material that grinds up food.

intestine Tubular organ in which most chemical digestion occurs and from which nutrients are absorbed.

stomach Tubular, muscular organ that functions in chemical and mechanical digestion.

Like other vertebrates, humans have a complete digestive tract (**Figure 36.5**). If the human digestive tract were laid out in a straight line, it would extend 6.5 to 9 meters (21 to 30 feet). Its various regions specialize in digesting food, absorbing nutrients, and concentrating and storing unabsorbed wastes. In this section we introduce the digestive tract and the accessory organs that secrete materials into it. Sections that follow consider the specialized regions of the human digestive tract in more detail.

Figure 36.5 Components of the human digestive system.

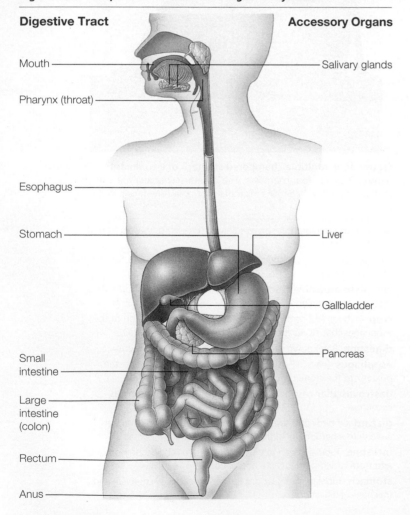

Digestive Tract

Mouth

Pharynx (throat)

Esophagus

Stomach

Small intestine

Large intestine (colon)

Rectum

Anus

Accessory Organs

Salivary glands

Liver

Gallbladder

Pancreas

Figure 36.6 Peristalsis.
Rings of smooth muscle contraction (indicated by the inward pointing arrows) travel in the mouth-to-anus direction along the digestive tract. The contraction narrows the interior of the tract, pushing material inside it (green) through the tube.

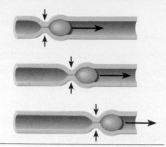

Processing of food begins inside the mouth (oral cavity.) **Salivary glands**, which are exocrine glands that open into the mouth, produce saliva that moistens food and begins the process of chemical digestion. Swallowing moves the food–saliva mixture into the pharynx (throat), from which it proceeds into the esophagus. The esophagus extends down through the thoracic cavity parallel to and behind the trachea. It then passes through an opening in the diaphragm to extend into the abdominal cavity, where it connects to the stomach.

The wall of the esophagus and all portions of the digestive tract beyond it contain smooth muscle. By a process called **peristalsis**, waves of smooth muscle contraction propel material through the digestive tract (**Figure 36.6**).

Sphincters at various points along the digestive tract control passage through the tract. A sphincter is a ring of muscle. When it relaxes, it opens to allow substances through a passageway. When it contracts, it closes off the passage. For example, when you swallow, a sphincter in the upper esophagus opens to allow food into the esophagus, then closes behind it. The sound made during a burp, or belch, arises when previously swallowed air flows outward through this sphincter, causing it to vibrate.

The stomach empties into the intestine, which has two regions that differ in their structure and function. The first region, the **small intestine**, completes the process of chemical digestion and absorbs most nutrients. Secretions from the liver, gallbladder, and pancreas assist the small intestine in these tasks. The second region of the intestine, the **large intestine**, absorbs fluid and concentrates waste. The terminal portion of the large intestine, the **rectum**, stores waste until it is expelled from the body through the anus.

large intestine Organ that receives digestive waste from the small intestine and concentrates it as feces.

peristalsis (pehr-ih-STAL-sis) Wavelike smooth muscle contractions that propel food through the digestive tract.

rectum Terminal region of the large intestine; stores digestive waste.

salivary gland Exocrine gland that secretes saliva into the mouth.

small intestine Longest portion of the digestive tract, and the site of most digestion and absorption.

36.3 Taking in Food

LEARNING OBJECTIVES

Compare the teeth of mammalian herbivores and carnivores.

List the components of saliva and describe their functions.

Describe how variations in beak shape among birds reflect differences in their diets.

MAMMALIAN MOUTHS

In mammals, mechanical digestion begins when our teeth rip and crush food. Each tooth consists mostly of dentin, a bone-like material. Enamel, the hardest material in the body, covers the tooth's exposed portion, which is called the crown. The lower portion of a tooth (the "root") is embedded in the bone of the jaw at a fibrous joint (Section 32.3).

As a species, humans are omnivores, meaning that we are adapted to feed on both plant and animal material. As a group, mammals have four types of teeth (Section 24.7). Being omnivores, we have all four tooth types and all are equally well developed (**Figure 36.7A**). By comparison, carnivorous mammals typically have enlarged canine teeth for piercing and tearing flesh (**Figure 36.7B**), and herbivores tend to have reduced canine teeth (**Figure 36.7C**). Premolars of carnivores have a narrow, bladelike surface that helps them shear through meat. By contrast, premolars of herbivores and humans are flat and broad, with bumpy crowns that function as platforms for grinding and crushing plant material.

Like most mammals, humans have two sets of teeth over the course of a lifetime. Their deciduous teeth or "baby teeth," are replaced by adult teeth. Human adult teeth do not grow. Mammals that routinely chew on rough material have teeth adapted to their diet; their teeth grow continually so they do not wear down. The aye-aye shown at the beginning of this chapter has continually growing incisors, as do rodents.

A mammal's tongue is a bundle of membrane-covered skeletal muscle attached to the floor of the mouth. Tongue movements help position food where teeth can chop or shred it. They also mix food with saliva from salivary glands. These glands open into the mouth beneath the tongue and on the inner surface of the cheeks next to the upper molars. Salivary amylase, an enzyme in saliva, begins the process of chemical digestion by breaking starch into disaccharides (two-sugar units). Like many other reactions of chemical digestion, this is an example of hydrolysis (Section 3.1). Saliva also contains glycoproteins that combine with water to form mucus. The mucus helps small bits of food stick together in moist, easy-to-swallow clumps that will slide down the throat and through the esophagus.

The presence of food at the back of the throat triggers a swallowing reflex. When you swallow, the larynx rises, causing

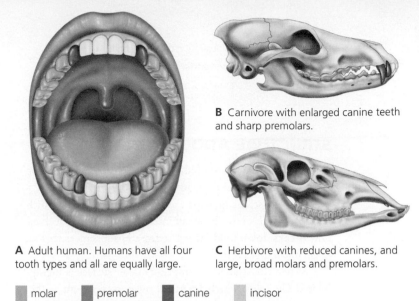

B Carnivore with enlarged canine teeth and sharp premolars.

A Adult human. Humans have all four tooth types and all are equally large.

C Herbivore with reduced canines, and large, broad molars and premolars.

■ molar   ■ premolar   ■ canine   ■ incisor

Figure 36.7 Variations in mammalian tooth patterns.

Meat eater (eagle)   Seed eater (finch)   Nectar eater (hummingbird)

Figure 36.8 Variations in beak shape.

the epiglottis to fold over and cover the entrance to the larynx (Section 35.4). At the same time, the vocal cords constrict. These actions block the route between the pharynx and larynx so food cannot enter the airway and choke you.

BIRD BEAKS

Unlike mammals, birds do not have teeth; their jawbones are covered with a layer of the structural protein keratin to form a beak. The size and shape of the beak vary among birds and adapt them to different diets. Hawks and other predatory birds have a sharp, curved beak that they use to tear flesh, whereas some other birds have a short, thick beak that can open seeds, or a long, narrow beak for reaching into flowers to sip nectar (**Figure 36.8**).

Lacking teeth, birds cannot chew food. They swallow large pieces of food and any necessary mechanical digestion takes place in the gizzard.

CREDITS: (8) left, Michal Ninger/Shutterstock.com; middle, Charles Brutlag/Shutterstock.com; right, Steve Byland/Shutterstock.com.

CHAPTER 36 DIGESTION AND HUMAN NUTRITION   **635**

36.4 The Stomach

LEARNING OBJECTIVES

List the functions of the stomach.

Explain the process of protein digestion in the stomach.

Explain what triggers secretion of gastrin and describe gastrin effects.

Describe the main cause of stomach ulcers.

STRUCTURE AND FUNCTION

The human stomach is a J-shaped, muscular, stretchable sac with a sphincter at either end (**Figure 36.9**). It has three functions. First, it stores food and controls the rate at which food continues on to the small intestine. Second, it mechanically breaks down food. Third, it secretes substances that aid in chemical digestion.

When the stomach is empty, its inner surface is highly folded. The folds, which are called rugae, smooth out as the stomach fills with food, thus increasing the stomach's capacity. In an average adult, the stomach can expand enough to hold about 1 liter of fluid (a little more than a quart).

Glandular epithelium, also called mucosa, lines the stomach's inner wall. Cells of this lining secrete about 2 liters of **gastric fluid** each day. Gastric fluid includes mucus, hydrochloric acid, and the protein-splitting enzyme pepsin.

Chemical digestion of proteins begins in the stomach. The acidity denatures most proteins (unfolds them), exposing their peptide bonds. High acidity also activates the pepsin. Activated pepsin breaks peptide bonds, snipping proteins into smaller polypeptides.

The pH of the stomach varies between 1.5 and 3 depending on how much acid has been secreted, which in turn depends on what has been eaten and when. Arrival of food in the stomach, especially protein-rich food, triggers endocrine cells in the stomach lining to secrete the hormone gastrin into the blood. Gastrin acts on acid-secreting cells of the stomach lining, causing them to increase their output. Gastrin also stimulates peristalsis in the stomach. The resulting smooth muscle contractions mix gastric fluid with food to form a semiliquid mass called **chyme**. They also propel chyme out through the pyloric sphincter and into the first segment of the small intestine.

When the stomach is empty, contraction of stomach muscle slows and gastrin secretion declines. The resulting decrease in acid secretion minimizes the likelihood that excess acid will damage the stomach wall.

Nutrients are not absorbed across the stomach wall, although some water can be. Ethanol and some other drugs such as aspirin are also absorbed from the stomach.

STOMACH DISORDERS

In some people, the sphincter at the entrance to the stomach (the gastroesophageal sphincter) does not close properly or opens when it should be closed. The result is gastroesophageal reflux, or acid reflux. Acidic chyme splashes into the esophagus, causing a burning pain commonly called heartburn or acid indigestion. Occasional acid reflux can be treated with over-the-counter antacids, but a chronic problem should be discussed with a doctor. Repeated exposure to acid can lead to chronic inflammation of the esophagus and raise the risk of esophageal cancer.

A protective layer of secreted mucus prevents acid and enzymes in gastric fluid from damaging the stomach lining. When this protection fails, the result is an ulcer (a craterlike sore). Most stomach ulcers arise after acid-tolerant *Helicobacter pylori* bacteria infect cells of the stomach lining. *H. pylori* degrades the protective mucus and makes chemicals that increase gastrin secretion, causing the stomach to secrete extra acid. The heightened acidity causes chronic inflammation of the stomach lining. If untreated, infection with *H. pylori* raises the risk of stomach cancer. Antibiotics can halt the infection and allow the ulcer to heal.

Overuse of nonsteroidal anti-inflammatory drugs such as ibuprofen or aspirin can also cause a stomach ulcer. These drugs inhibit secretion of the stomach's protective mucus, allowing acid to contact cells of the stomach lining.

Figure 36.9 Location and structure of the stomach. The folds shown on the inner surface (the rugae) smooth out when the stomach fills with food.

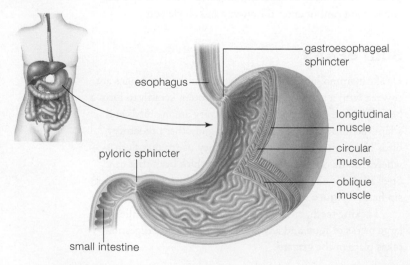

esophagus

gastroesophageal sphincter

longitudinal muscle

circular muscle

oblique muscle

pyloric sphincter

small intestine

---

chyme (kime) Mix of food and gastric fluid.

gastric fluid Fluid secreted by the stomach lining; contains digestive enzymes, acid, and mucus.

36.5 The Small Intestine

Explain the roles of the small intestine.

Describe the structural features that contribute to the large surface area of the small intestine.

Chyme that is forced out of the stomach through the pyloric sphincter enters the small intestine. The small intestine is "small" only in terms of its diameter—about 2.5 cm (1 inch). It is the longest segment of the gut. Uncoiled, the adult small intestine would extend about 5 to 7 meters (16 to 23 feet). There are three regions of the small intestine: the duodenum (which receives food from the stomach), the jejunum, and the ileum (which empties into the large intestine).

Most digestion and absorption takes place at the lining of the small intestine. This lining is highly folded (**Figure 36.10A**). Unlike the folds of an empty stomach, those of the small intestine are permanent. The surface of each fold has many **villi** (singular, villus). Each villus is a hairlike multi-celled projection about 1 millimeter long, with blood vessels and lymph vessels running through its interior (**Figure 36.10B,C**). Millions of villi on the intestinal lining give it a furry or velvety appearance.

Most of the cells at the surface of a villus have even tinier cylindrical protrusions called **microvilli** (singular, microvillus). These cells are sometimes called **brush border cells**

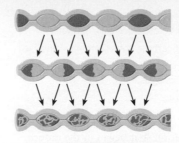

Figure 36.11
Segmentation contraction in the small intestine. Contractions of rings of muscle in the wall of the small intestine cause chyme to slosh back and forth, mixing it with digestive enzymes and forcing it against the wall of the small intestine.

because the many microvilli at each cell's free surface make its outer edge resemble a brush (**Figure 36.10D,E**). Collectively, the many folds and projections of the small intestinal lining increase its surface area by hundreds of times. As a result, the surface area of the small intestine is comparable to that of a tennis court. Having such a large surface area maximizes the number of membrane-embedded digestive enzymes and transport proteins that come into contact with the chyme.

Two patterns of muscle contraction occur in the small intestine. The main pattern is segmentation, in which smooth muscle contractions squeeze the chyme so that it sloshes alternatively forward and backward (**Figure 36.11**). Segmentation enhances digestion and absorption by mixing chyme with digestive enzymes, and sloshing the mixture against the small intestine's interior wall. The second pattern is peristalsis, which propels chyme through the small intestine.

brush border cell In the lining of the small intestine, an epithelial cell with microvilli at its surface.

microvilli (my-croh-VILL-lie) Thin projections that increase the surface area of some epithelial cells.

villi Multicelled projections from the lining of the small intestine.

FIGURE IT OUT
Are microvilli multicelled or smaller than a cell?

Answer: Microvilli are smaller than a cell. Villi are multicellular.

Figure 36.10 Structure of the small intestine.

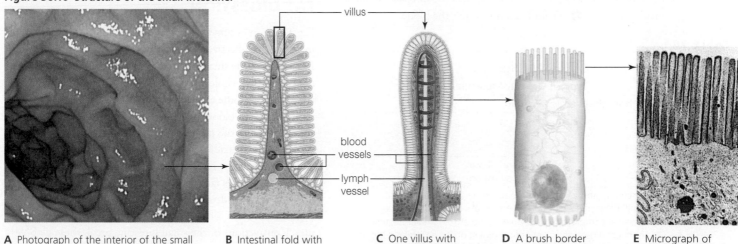

A Photograph of the interior of the small intestine showing its folded lining.

B Intestinal fold with villi at its surface.

C One villus with brush border cells at its surface.

D A brush border cell with microvilli at its free surface.

E Micrograph of microvilli on a brush border cell.

The process of chemical digestion that began in the mouth and continued in the stomach is completed in the small intestine (**Table 36.1**). The lumen of the small intestine (the region inside the tube) receives chyme from the stomach. It also receives digestive enzymes and bicarbonate from the pancreas, and bile from the gallbladder (**Figure 36.12**). Pancreatic enzymes work in concert with enzymes at the surface of brush border cells to complete the breakdown of large organic compounds into absorbable subunits. The bicarbonate secreted by the pancreas raises the pH of the chyme enough for digestive enzymes to function properly (Section 5.3).

CARBOHYDRATE DIGESTION

In the small intestine, carbohydrates are broken down into monosaccharides, or simple sugars (**Figure 36.13 ❶**). This process began in the mouth, where salivary amylase broke some bonds between sugar units in polysaccharides to release disaccharides (two-unit sugars). A pancreatic amylase continues these reactions in the small intestine. In addition, enzymes embedded in the plasma membrane of brush border cells split disaccharides into monosaccharides. For example, sucrase breaks sucrose into glucose and fructose subunits, and lactase splits lactose into glucose and galactose. Transport proteins move the resulting monosaccharides from the lumen into a brush border cell. At the other side of the cell, the monosaccharides are transported into the interstitial fluid inside the villus ❷. From here, they enter the blood.

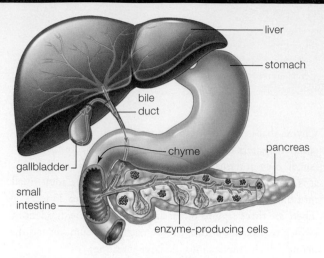

Figure 36.12 Organs that contribute to the contents of the small intestine. Chyme from the stomach is joined by bile and pancreatic enzymes. Bile is produced by the liver, stored in the gallbladder, and enters the small intestine through the bile duct.

PROTEIN DIGESTION

In the small intestine, polypeptides are broken down into amino acids. Protein digestion began in the stomach, where pepsin broke some of the peptide bonds between amino acid subunits in proteins to release polypeptides. It is completed in the small intestine, where pancreatic proteases such as trypsin and chymotrypsin break the polypeptides into smaller fragments ❸. Enzymes at the surface of the brush border cell then break these fragments first into smaller peptides, then into amino acids. Like monosaccharides, amino acids are transported into brush border cells, then into the interstitial fluid by membrane proteins ❹. From here, the amino acids enter the blood.

FAT DIGESTION

Fat digestion occurs entirely in the small intestine. Here, bile increases the effectiveness of lipases (fat-digesting enzymes) secreted into the small intestine by the pancreas. **Bile** contains salts, pigments, cholesterol, and lipids. It is made in the **liver**, a large organ in the abdomen that also produces plasma proteins, filters the blood, and stores glycogen. Bile is stored and concentrated in the **gallbladder**. A fatty meal causes the gallbladder to contract, forcing bile out through a short duct (the common bile duct) into the small intestine.

Bile enhances fat digestion by aiding in **emulsification**, the dispersion of droplets of fat in a fluid. Water-insoluble triglycerides from food tend to clump together as fat globules. Contractions of the small intestine break these globules into smaller droplets, then bile salts coat the droplets so they remain separated ❺. Compared to big globules, the many

TABLE 36.1

Locations and Products of Chemical Digestion

Organ	Type of Organic Molecule		
	Carbohydrate	Protein	Fat
Mouth	Salivary amylase begins chemical digestion.		
Stomach		Acid and pepsin begin digestion.	
Small intestine	Pancreatic and small intestinal enzymes complete chemical digestion.	Pancreatic and small intestinal enzymes complete chemical digestion.	Bile emulsifies fats; pancreatic, small intestinal enzymes digest it.
Absorbable Products	Monosaccharides (simple sugars)	Amino acids	Fatty acids, monoglycerides

CREDIT: (12) From Russell/Wolfe/Hertz/Starr, *Biology*, 2E, © 2011 Cengage Learning, Inc.

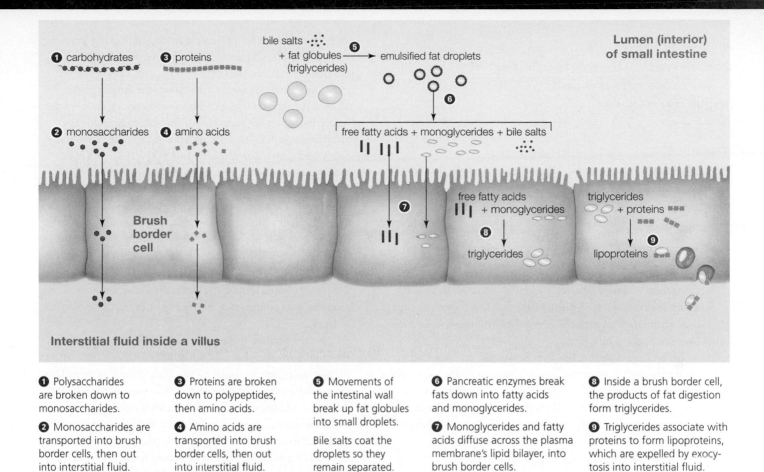

1 carbohydrates

3 proteins

bile salts + fat globules (triglycerides) **5** → emulsified fat droplets

Lumen (interior) of small intestine

2 monosaccharides

4 amino acids

6

free fatty acids + monoglycerides + bile salts

Brush border cell

7

free fatty acids + monoglycerides

triglycerides + proteins

8

triglycerides

9

lipoproteins

Interstitial fluid inside a villus

1 Polysaccharides are broken down to monosaccharides.

2 Monosaccharides are transported into brush border cells, then out into interstitial fluid.

3 Proteins are broken down to polypeptides, then amino acids.

4 Amino acids are transported into brush border cells, then out into interstitial fluid.

5 Movements of the intestinal wall break up fat globules into small droplets.

Bile salts coat the droplets so they remain separated.

6 Pancreatic enzymes break fats down into fatty acids and monoglycerides.

7 Monoglycerides and fatty acids diffuse across the plasma membrane's lipid bilayer, into brush border cells.

8 Inside a brush border cell, the products of fat digestion form triglycerides.

9 Triglycerides associate with proteins to form lipoproteins, which are expelled by exocytosis into interstitial fluid.

Figure 36.13 Summary of digestion and absorption in the small intestine.

small droplets present a greater surface area to lipases. Only at the surfaces of fat droplets can lipases break the glycerol backbone of triglycerides to release fatty acids and monoglycerides **6**. Being lipid soluble, fatty acids and monoglycerides can enter a villus by diffusing across the lipid bilayer of brush border cells **7**. Inside the cells, these compounds are reassembled into triglycerides **8**. The triglycerides become coated with proteins, and the resulting lipoproteins are moved by exocytosis into interstitial fluid inside a villus **9**. From interstitial fluid, triglycerides enter lymph vessels (Section 33.10) that eventually drain into the blood.

Bile salts used in fat digestion are recycled. They are absorbed from the final portion of the small intestine (the ileum) and returned to the liver by the blood.

In some people, components of bile accumulate to form pebble-like gallstones in the gallbladder. Most gallstones are harmless, but some block or become lodged in the bile duct. In this case, the gallbladder or gallstones are usually removed surgically. After removal of a gallbladder, all bile from the liver drains directly into the small intestine.

WATER UPTAKE

Each day, eating and drinking puts 1 to 2 liters of fluid into the lumen of your small intestine. Secretions from your stomach, the liver and pancreas, and the intestinal lining add another 6 to 7 liters. About 80 percent of the water that enters the small intestine moves across the intestinal lining and into the internal environment by osmosis. Transport of salts, sugars, and amino acids across brush border cells creates an osmotic gradient, and water follows that gradient from chyme into the interstitial fluid.

bile Mixture of salts, pigments, and cholesterol produced in the liver, then stored and concentrated in the gallbladder; emulsifies fats when secreted into the small intestine.

emulsification Dispersion of fat droplets in a fluid.

gallbladder Organ that stores and concentrates bile.

liver Abdominal organ that produces plasma proteins and bile, filters the blood, and stores glycogen.

36.7 The Large Intestine

Not everything that enters the small intestine can or should be absorbed. Contraction of the small intestine propels indigestible material, dead bacteria and mucosal cells, inorganic substances, and some water into the large intestine. The large intestine is wider than the small intestine, but shorter—only about 1.5 meters (5 feet) long.

As wastes travel through the large intestine, they become compacted as **feces**. The large intestine concentrates wastes as active transport proteins in cells of its lining pump sodium ions across its wall, into the internal environment. Water follows the sodium by osmosis.

The large intestine has a moderate pH and material moves through it relatively slowly. These conditions favor the growth of bacteria and archaea that are part of our normal microbiota (Section 19.5). Bacteria that live in the large intestine produce vitamin B_{12} that we absorb across the lining of this organ.

The first region of the large intestine, the part that connects to the ileum of the small intestine, is a cup-shaped pouch called the cecum (**Figure 36.14**). Herbivores such as rabbits have a large cecum containing many bacteria that can break down cellulose. The human cecum is comparatively small. In humans and many other mammals, a short, tubular **appendix** projects from the cecum. The appendix serves as a reservoir for beneficial bacteria.

Appendicitis—an inflammation of the appendix— often occurs after a bit of feces lodges in the appendix and infection sets in. It requires prompt surgical treatment. Removing an inflamed appendix prevents it from bursting and releasing bacteria into the abdominal cavity, where they could cause a life-threatening infection.

The cecum connects to the **colon**, the longest portion of the large intestine. Material that enters the colon from the cecum moves up the right side of the abdomen inside the ascending colon, crosses the abdomen inside the transverse colon, and moves down the left side of the abdomen inside the descending colon. It then enters the sigmoid colon, an S-shaped segment that delivers feces to the rectum.

Peristalsis propels the contents of the colon along. After a meal, signals from autonomic nerves cause strong peristaltic contractions that propel feces into the rectum. The resulting stretching of the rectum activates a defecation reflex that opens a sphincter of smooth muscle at the base of the rectum. Voluntary contraction of a sphincter of skeletal muscle at the anus provides control over the timing of defecation (expulsion of feces).

Healthy adults typically defecate once a day, on average. Emotional stress, a diet low in fiber, minimal exercise, dehydration, and some medications can lead to constipation. This means defecation occurs fewer than three times a week, is difficult, and yields small, hardened, dry feces. Chronic constipation should be discussed with a physician. Infection by a viral, bacterial, or protozoal pathogen can cause an episode of diarrhea, the frequent passing of watery feces. Diarrhea can be dangerous because it causes dehydration, disrupting the normal concentration of solutes in the body.

Flatulence, passing gas from the anus, is normal. Gases that escape the digestive tract in this manner are by-products of carbohydrate breakdown by bacteria and archaea in the colon. A high-fiber diet provides more raw material for these microbes to dine on and thus can increase flatulence.

Colon cancer is the third most common cause of cancer deaths in the United States. In 2015, it killed about 50,000 people. Colon cancers start out as small growths called polyps on the colon wall. Most colon polyps are benign, but some become cancerous. People with no family history of colon cancer should begin getting screened for this cancer at age 50. People with a family history of this cancer should be screened earlier. Colon cancer screening involves examination of feces samples for cancer cells or colonoscopy, a procedure in which a physician uses a camera to examine the colon. If detected early enough, colon cancer can be removed before it spreads.

Figure 36.14 The large intestine.

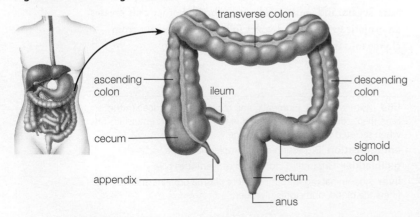

appendix Wormlike projection from the first part of the large intestine.

colon Main portion of the human large intestine.

feces Unabsorbed food material and cellular waste that is expelled from the digestive tract.

LEARNING OBJECTIVES

Explain how the body uses the breakdown products of carbohydrates, proteins, and fats.

Describe the role of the liver in sugar storage and amino acid breakdown.

Explain the difference between complete and incomplete proteins and provide examples.

Dietary carbohydrates, fats, and proteins are macronutrients, meaning that we require them in large amounts. Macronutrients serve as sources of energy and as building blocks. As Section 7.6 explained, these molecules can be broken down in aerobic respiration to produce ATP. **Figure 36.15** rounds out this picture by illustrating the major routes by which the body uses organic compounds obtained from food.

Cells break down starch and sugars to release glucose, their primary source of energy. When glucose absorption exceeds the body's immediate needs, the excess is stored for later use. Blood that flows through capillaries in the small intestine travels directly to the liver, carrying glucose that the liver stores as glycogen. The liver and adipose cells also use glucose to build fats.

In addition to their function as an energy source, fats are used to build components of cell membranes and to make steroid hormones. They also help you take up fat-soluble vitamins. Your body can remodel carbohydrates to make most of the fats that it requires, but those it cannot make must come from foods. **Essential fatty acids** are those that cannot be synthesized by your cells, so these components of fats must be obtained from a dietary source such as nuts, seeds,

and vegetable oils. We discuss the two types of essential fatty acids (omega-3 fatty acids and omega-6 fatty acids) in more detail in Section 36.10.

Your body uses amino acids from dietary proteins to build peptides and proteins and to make nucleotides. Although most organs do not routinely break down amino acids for a source of energy, the liver does. Liver enzymes separate the amino group from its carbon skeleton, which is used to fuel the citric acid cycle. The liver then converts the toxic ammonia (NH_3) produced by this reaction into urea, a somewhat less toxic compound that is excreted in urine.

There are nine **essential amino acids** that your body cannot synthesize and must obtain from food in order to build its own proteins. Proteins from animal sources (meat, fish, etc.) are complete, meaning they contain all of the essential amino acids in the ratio that a human needs. By contrast, most plant proteins are incomplete, meaning they lack or have a low amount of one or more essential amino acids. You can meet your amino acid requirements with plant-based foods, but amino acids missing from one food must be provided by others. As an example, rice and beans together provide all necessary amino acids, but rice alone or beans alone do not. Complementary sources of amino acids do not need to be eaten simultaneously, but can instead be "combined" over the course of a day. Soybeans and quinoa seeds are two of the few plant foods that supply all essential amino acids on their own.

essential amino acid Amino acid that the body cannot make and must obtain from food.

essential fatty acid Fatty acid that the body cannot make and must obtain from food.

Figure 36.15 Main ways in which cells use and reuse dietary carbohydrates, fats, and proteins.

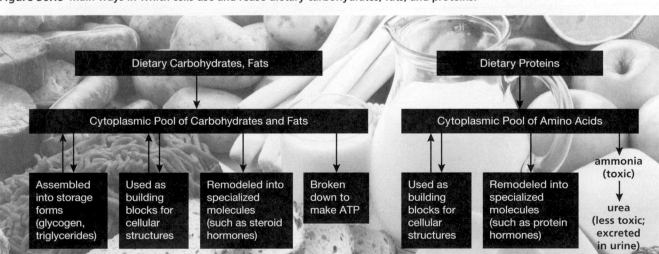

36.9 Vitamins and Minerals

Explain the difference between vitamins and minerals.

Provide an example of a water-soluble vitamin and a fat-soluble vitamin and explain why each is essential.

Give an example of an essential mineral and explain its role in the body.

In addition to macronutrients, normal metabolism requires intake of vitamins and minerals, which are micronutrients.

Vitamins are organic substances that are required in the diet in very small amounts. **Table 36.2** lists the major vitamins humans need. Vitamins A, D, E, and K are fat soluble. Heat has little effect on these vitamins, which are abundant in both cooked and fresh foods. Fat-soluble vitamins can be stored in the body's own fat, so it is not necessary to eat them every day. By contrast, water-soluble vitamins (such as vitamins B and C) are generally not stored in the body, so they must be eaten more or less daily. Water-soluble vitamins are also more sensitive to heat than fat-soluble vitamins, so they are more easily destroyed by cooking.

The body remodels vitamin A into the visual pigment made by rod cells (Section 30.5). In the United States, milk and soy milk products are typically fortified with vitamin A, so a deficiency in this vitamin is rare. In less-developed countries, vitamin A deficiency is the most common cause of childhood blindness.

Vitamin D increases calcium uptake from the gut and encourages retention of calcium in bone. It can be obtained from dietary sources, or made in the skin. Vitamin D production by skin requires exposure to sunlight. People with dark skin make less vitamin D than lighter-skinned people, so they are more likely to be deficient in this vitamin. A vitamin D deficiency that occurs while bones are growing impairs bone growth and can result in rickets. Affected individuals have unusually soft, weak bones, short stature, and, in extreme cases, misshapen bones.

TABLE 36.2

Major Vitamins: Sources and Functions

Vitamin	Main Dietary Sources	Main Functions	Symptoms of Deficiency
Fat-Soluble Vitamins			
A	Orange fruits and vegetables, leafy greens, egg yolk	Component of visual pigments; maintains epithelia	Night blindness, skin problems
D	Fatty fish, egg yolk	Aids uptake and use of calcium	Weak/soft bones, rickets in children
E	Nuts, seeds, vegetable oils	Antioxidant, aids fat absorption, helps maintain cell membranes	Muscle weakness, nerve damage
K	Green vegetables	Needed for blood clotting	Impaired blood clotting
Water-Soluble Vitamins			
B_1 (thiamine)	Meats, nuts, legumes	Coenzyme in carbohydrate metabolism	Beri beri (neurological and heart problems)
B_2 (riboflavin)	Meats, eggs, nuts, milk, green vegetables	Component of the coenzyme FAD	Anemia, sores in mouth, sore throat
B_3 (niacin)	Meats, fish, dairy products, nuts, legumes	Component of the coenzyme NAD	Skin and mucous membrane sores, gut pain, diarrhea, psychosis, dementia
B_6	Meats, fish, starchy vegetables, noncitrus fruits	Coenzyme in protein metabolism	Anemia, sores on lips, depression, impaired immune function
B_7 (biotin)	Meats, fish, nuts, legumes, whole grains	Coenzyme in many reactions	Hair loss, dry skin, dry eyes, fatigue, insomnia, depression
B_9 (folic acid)	Meats, fruits, green vegetables, whole grains	Coenzyme in nucleic acid synthesis and amino acid metabolism	Anemia, sores in mouth; deficiency during pregnancy causes neurological birth defects
B_{12}	Meats, seafood, dairy products	Coenzyme in amino acid synthesis	Anemia; fatigue, neurological problems
C (ascorbic acid)	Fruits (especially citrus) and vegetables	Required for collagen synthesis, antioxidant	Scurvy (anemia, bleeding gums, impaired wound healing, swollen joints)

Vitamin E is an antioxidant: a molecule that prevents oxidation of other molecules by free radicals (Section 5.5). Being fat soluble, vitamin E plays an important role in protecting the lipids of cell membranes from oxidation.

Vitamin K is a coenzyme that assists enzymes involved in blood clotting. (K denotes the German word *Koagulation*.) Absorption of vitamin K produced by bacteria in the large intestine supplements vitamin K extraction from food.

B vitamins are remodeled into coenzymes that function in a variety of essential pathways. For example, vitamin B_3 (niacin) is used to make NAD, and vitamin B_2 (riboflavin) is used to make FAD. Both of these coenzymes play vital roles in aerobic respiration, as summarized in Section 7.4.

Vitamin C is needed to make collagen, the body's most abundant protein. A deficiency causes scurvy, a disorder in which skin and bones deteriorate and wounds are slow to heal. Vitamin C also functions as an antioxidant.

Minerals are elements that are essential in the diet in small amounts. The seven minerals required in largest quantities are calcium, phosphorus, potassium, sulfur, sodium, chlorine, and magnesium. Calcium and phosphorus, the body's most abundant minerals, are components of teeth and bones. Calcium also plays a role in intercellular signaling, and phosphorus is also used to make ATP and nucleic acids. Sodium and potassium are important in nerve and muscle function, sulfur is a component of some proteins, and chlorine is a component of hydrochloric acid (HCl) in gastric fluid. Magnesium is a cofactor for many reactions, including most reactions in glycolyis. Deficiencies in these major elements are generally rare, although calcium levels may be low in people who eat a solely plant-based diet.

Iodine, another essential mineral, is necessary to produce thyroid hormone, which has roles in development and metabolism (Section 31.4). Iodine is abundant in fish, shellfish, and seaweeds. A deficiency in this mineral causes goiter, an enlarged thyroid gland, which is a symptom of disrupted metabolism. In the United States, general use of iodized salt makes iodine deficiency rare.

Iron is an essential component of heme, the chemical group that allows hemoglobin to bind oxygen (Section 35.6) and myoglobin. Heme is also a cofactor for many enzymes, including critical components of electron transfer chains (Section 5.4) and the antioxidant catalase (Section 5.5). Worldwide, iron is the mineral most commonly deficient in the diet. Symptoms of iron-deficiency anemia include fatigue, shortness of breath, dizziness, and chest pain.

---

mineral In the diet, an element that is required in small amounts for normal metabolism.

vitamin Organic substance required in the diet in small amounts for normal metabolism.

Dr. Christopher Golden

To conservationists, the aye-aye shown in this chapter's opening photo is an endangered lemur in need of improved protection. However, to some of the aye-aye's human neighbors, it and other endangered animals are a valuable source of food. National Geographic Explorer Christopher Golden, who studies public health and ecology in rural Madagascar, has found that bushmeat is an important source of dietary iron and other nutrients for the local people. He says, "Everyone can get adequate protein from vegetables. However, in areas without fortification and supplementation programs, it is extraordinarily difficult to obtain adequate iron without eating meat." Golden estimates that cutting off access to bushmeat would triple the incidence of iron-deficiency anemia among the poorest Malagasy children. Among other ill effects, childhood anemia stunts growth and decreases mental capacity.

Golden fears that the people in his study area will eventually lose access to bushmeat, regardless of whether conservation laws are better enforced. Improved law enforcement would cut off this food supply quickly. Allowing unsustainable hunting to continue would do the same more slowly, by decimating wildlife populations. Golden argues that it is possible to protect both human health and biodiversity. He advocates ramping up efforts to prevent illegal hunting, while also providing alternative sources of meat, such as chickens. In rural Madagascar, unlike some other regions, the vast majority of bushmeat is consumed, not sold.

In addition to protecting wildlife, reducing bushmeat consumption would benefit public health by minimizing people's exposure to potentially dangerous viruses that commonly infect wild animals. Golden is collaborating with Nathan Wolfe, also a National Geographic Explorer (Section 19.2), to determine what types of viruses lemurs, bats, and the other animals hunted for bushmeat carry.

CREDIT: Courtesy of Dr. Christopher Golden.

LEARNING OBJECTIVES

Compare the two types of dietary fiber and explain the health benefits of each.

List the types of fats in foods and explain which are considered healthy choices, and which should be avoided.

Explain why current nutritional guidelines call for lowering intake of salt and sugar.

The United States government issues dietary guidelines designed to help people maintain a healthy weight, promote health, and prevent disease. **Figure 36.16** shows an example of their recommendations. You can generate your own healthy eating plan by visiting the USDA website: www.choosemyplate.gov. Food labels (**Figure 36.17**) can also help guide your dietary choices.

FRUITS, VEGETABLES, WHOLE GRAINS

Fruits, vegetables, and grains should make up the largest proportion of your diet. These foods provide sugars and starches, your primary sources of energy. The energy stored in food is measured in kilocalories, or as written on food labels, "Calories" (with a capital C). Fruits, vegetables, and grains also provide vitamins, minerals, and dietary fiber.

There are two types of fiber. Soluble fiber consists of polysaccharides that form a gel when mixed with water. Eating foods high in soluble fiber helps lower one's cholesterol level and may reduce the risk of heart disease. Insoluble fiber such as cellulose does not dissolve, so it passes through the human digestive tract more or less intact. Eating insoluble fiber helps prevent constipation.

Whole grains provide more vitamins and fiber than their processed counterparts. A whole grain includes all components of a grain seed. For example, whole wheat includes bran (the fiber-rich seed coat) and wheat germ (the protein- and vitamin-rich plant embryo), as well as starchy endosperm. By contrast, white wheat flour is made solely from endosperm.

You may have noticed breads and other grain-based foods labeled as "gluten-free." Gluten is a protein found in wheat and many other grains. An estimated 1 percent of the population has a genetic disorder called celiac disease, in which gluten causes an autoimmune reaction that harms the small intestine's villi. Celiac disease is treated by eliminating gluten from the diet.

HEART-HEALTHY OILS

A healthy diet includes fats that provide energy and meet your need for essential fatty acids (**Table 36.3**). Essential

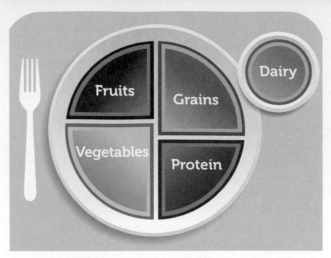

USDA Nutritional Guidelines

Food Group	Amount Recommended
Grains	6 ounces/day
Vegetables	2.5 cups/day
Fruits	2 cups/day
Dairy products	3 cups/day
Meat and beans	5.5 ounces/day

Figure 36.16 Example of nutritional guidelines from the United States Department of Agriculture (USDA).
These recommendations are for females ages ten to thirty who get less than 30 minutes of vigorous exercise daily. Portions add up to a 2,000-kilocalorie daily intake.

TABLE 36.3

Main Types of Dietary Fats

Polyunsaturated Fatty Acids: Liquid at room temperature; essential for health.
    Omega-3 fatty acids
        Alpha-linolenic acid and its derivatives
        Sources: Nut oils, vegetable oils, oily fish
    Omega-6 fatty acids
        Linoleic acid and its derivatives
        Sources: Nut oils, vegetable oils, meat

Monounsaturated Fatty Acids: Liquid at room temperature. Main dietary source is olive oil. Beneficial in moderation.

Saturated Fatty Acids: Solid at room temperature. Main sources are meat and dairy products and palm and coconut oils. Excessive intake may raise risk of heart disease.

Trans Fatty Acids (Hydrogenated Fats): Solid at room temperature. Made from vegetable oils and used in some processed foods. Excessive intake raises risk of heart disease.

fatty acids are polyunsaturated (Section 3.3), meaning their tails have two or more double bonds. There are two types of essential fatty acids: omega-3 fatty acids and omega-6 fatty acids. Omega-3 fatty acids, the main fat in oily fish, seem to have special health benefits. Studies suggest that a diet high in omega-3 fatty acids can reduce the risk of cardiovascular disease, lessen the inflammation associated with rheumatoid arthritis, and help diabetics control their blood glucose.

Oleic acid, the main fat in olive oil, is monounsaturated, which means its carbon tails have only one double bond. While monounsaturated fats have not been shown to have the same benefits as polyunsaturated fats, they can be beneficial if substituted for less healthy fats.

Trans fats (Chapter 3 Application) should be avoided. A diet rich in these artificially modified fats increases the risk of heart disease. In recognition of the harmful effects of *trans* fats, the FDA has ruled that food manufacturers must eliminate these fats from most products by 2018.

Most nutritionists also recommend moderating one's intake of foods rich in saturated fat in order to minimize the risk of cardiovascular disease. Such foods include eggs, red meat, and whole-fat dairy products.

LEAN MEAT AND LOW-FAT DAIRY

Meat (including poultry and fish) is the richest source of protein and a rich source of essential iron. Choosing lean meats or fish helps minimize intake of saturated fats and cholesterol. Eating soybean products such as tofu provides complete protein without harmful fats or cholesterol, as does eating the proper combinations of plant foods such as rice and beans.

Milk and dairy products such as yogurt are good sources of protein, vitamins, and minerals, but whole milk is rich in saturated fats. Low-fat or skim alternatives may be healthier options. People who are lactose intolerant or who wish to avoid animal products can substitute "milks" made from soybeans, rice, almonds, or other plants. Such products have the advantage of being free of saturated fats and cholesterol. Many are fortified to provide the same amount of calcium and vitamins as dairy milk.

MINIMAL ADDED SALT AND SUGAR

Salt contains two essential minerals, sodium and chloride, but both are easily obtained in sufficient quantity from unsalted foods. Eating foods with added salt elevates the body's sodium level, which raises the risk of high blood pressure in some people. Most canned and otherwise processed foods are high in salt, so replacing these items with fresh foods lowers sodium intake. Food labels show sodium content as a percentage of the recommended daily maximum for a person who has normal blood pressure.

Commercially prepared foods and drinks are often high in added sugar. The USDA estimates that added sugars,

which provide no nutritional benefits, contribute an astounding 16 percent of the total calories in American diets. Sodas, energy drinks, and sport drinks contribute the greatest proportion of calories. The American Heart Association recommends that women consume no more than 24 grams of sugar per day and men no more than 36 grams, from all sources. A typical can of soda has more than 30 grams of sugar.

Nutrition Facts

Serving Size 1 cup (228g)
Servings Per Container 2

Amount Per Serving

Calories 250	Calories from Fat 110

	% Daily Value*
Total Fat 12g	18%
Saturated Fat 3g	15%
Trans Fat 1.5g	
Cholesterol 30mg	10%
Sodium 470mg	20%
Total Carbohydrate 31g	10%
Dietary Fiber 0g	0%
Sugars 5g	
Protein 5g	

Vitamin A	4%
Vitamin C	2%
Calcium	20%
Iron	4%

* Percent Daily Values are based on a 2,000 calorie diet. Your Daily Values may be higher or lower depending on your calorie needs:

	Calories:	2,000	2,500
Total Fat	Less than	65g	80g
Sat Fat	Less than	20g	25g
Cholesterol	Less than	300mg	300mg
Sodium	Less than	2,400mg	2,400mg
Total Carbohydrate		300g	375g
Dietary Fiber		25g	30g

Check serving size. A package often holds more than one serving, but nutritional information is given per serving.

Avoid foods in which a large proportion of calories comes from fat.

Choose foods that provide a low percentage of the recommended maximum of saturated fat, *trans* fat, cholesterol, and sodium. 20% or more is high.

Choose foods that are high in dietary fiber and low in sugar.

If you eat meat, you probably get more than enough protein.

Choose foods that provide a high percentage of your daily vitamin and mineral requirements.

This part of the label shows recommended intake of nutrients for two levels of calorie intake. Keeping fat and salt intake below recommended levels and dietary fiber above recommended levels decreases risk of some chronic health problems.

Figure 36.17 How to read a food label. Information on a food label can be used to ensure that you get the nutrients you need without exceeding recommended limits on less healthy substances such as salt and *trans* fats. This hypothetical label is for a ready-to-eat macaroni and cheese product.

FIGURE IT OUT
What proportion of the fat in a serving of this product comes from the least healthy forms of fat (saturated fat and *trans* fat)?

Answer: Of the total fat content in a serving (12 grams), 3 g are saturated fat and 1.5 g are *trans* fat. Thus 4.5 g, or more than a third of the fat, is from unhealthy sources.

LEARNING OBJECTIVES

Explain how BMI is measured, what it measures, and why it is only a crude measure of health risks.

Explain what resting metabolic rate it is, factors that affect it, and how it influences the risk of obesity.

When the food you eat contains more energy than you need at the time, you store the excess as bond energy in organic compounds. The body's largest energy store is fat in adipose tissue. For most of our species' history, an ability to store energy as fat in adipose tissue was selectively advantageous. Putting on fat when food was abundant increased the chances of survival in the event of a later famine. However, most people in the United States now have more than enough food all of the time. As a result, about two-thirds of adults are overweight or obese.

Body mass index (BMI) is a measurement designed to help assess the health risk associated with body weight. You can calculate your body mass index with this formula:

$$\text{BMI} = \frac{\text{weight (pounds)} \times 703}{\text{height (inches)} \times \text{height (inches)}}$$

Generally, individuals with a BMI of 25 to 29.9 are said to be overweight. A score of 30 or more indicates obesity: an overabundance of fat in adipose tissue that may lead to severe health problems. Conversely, a BMI of 18.5 or lower is considered dangerously underweight.

Keep in mind that BMI is a rather inexact tool for determining whether or not a person is healthy. Some people, such as highly muscular athletes, have a high BMI but do not have excessive body fat. In addition, the location of fat deposits affects health. Fat deposits in the abdomen pose a far greater threat than fat deposits elsewhere in the body.

To maintain a given weight, you must balance the amount of energy in the food you eat with the energy you expend in your activities. Here is a way to estimate how many kilocalories you should take in daily to maintain a given weight. First, multiply the weight (in pounds) by 10 if you are not physically active, by 15 if you are moderately active, and by 20 if you are highly active. Next, subtract one of the following amounts from the multiplication result:

Age:	25–34	Subtract:	0
	35–44		100
	45–54		200
	55–64		300
	Over 65		400

For example, if you are 25 years old, are highly active, and weigh 120 pounds, you will require $120 \times 20 = 2,400$

TABLE 36.4

Energy Expended in Activities

Activity	Energy expended (Calories per kilogram of body weight per hour)
Sitting quietly	1
Standing	2
Walking slowly	3
Bicycling to class	4–7
Swimming	6–10
Basketball game	8
Running (5 mph)	8

kilocalories daily to maintain weight. If you want to gain weight you will require more; to lose, you will require less.

You can also lose weight by increasing the rate at which your body uses energy. Even when you are at rest, your body is using energy in essential processes such as breathing, generating body heat, fighting off pathogens, moving material though your digestive tract, and circulating your blood. The amount of energy you use in these activities is your **basal metabolic rate**. Basal metabolic rate varies with lean body weight. Larger, more muscular bodies use more energy than smaller, less muscular ones. Men tend to be more muscular than women of the same weight, so their metabolic rate is higher. In both sexes, metabolic rate slows with age. Thyroid hormone level influences basal metabolic rate too. People with a high level of thyroid hormone use more energy at rest than those with a lower level.

There are also genetic differences in metabolic rate. For example, several mutant alleles of the human gene *KSR2* are associated with a lower than normal metabolic rate and a greater than normal appetite. The result is a predisposition to early-onset obesity.

Dieting can influence one's basal metabolic rate. Often, when people eat less, their resting metabolic rate declines. This mechanism evolved to promote survival when food became scarce. However, it is a source of great frustration to dieters who have chosen to limit their food intake in the hope of losing weight.

Expending energy by exercising can help you lose weight. **Table 36.4** shows energy expenditures during various activities. Exercise that increases the body's muscle mass also has the benefit of increasing the resting metabolic rate.

basal metabolic rate Rate at which the body uses energy when you are at rest.

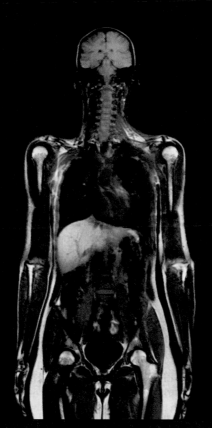
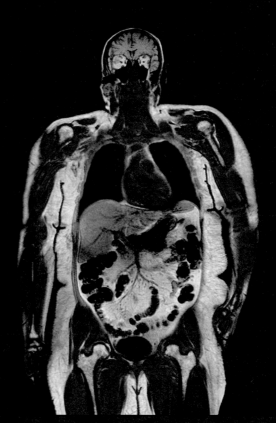

Figure 36.18 MRIs of an obese woman (right) and a woman of normal body weight (left). With obesity, the abdomen fills with fat that squashes internal organs.

THE OBESITY EPIDEMIC

The increasing prevalence of obesity has dire implications for public health. Obesity increases the risk for a long list of diseases including high blood pressure, type 2 diabetes, heart disease, stroke, gallbladder disease, osteoarthritis, sleep apnea, and cancers of the uterus, breast, prostate gland, kidney, and colon.

An obese person's internal organs are hemmed in by fat (**Figure 36.18**), and the pressure can impair the organs' function. For example, breathing can become difficult because fat in the abdomen impairs the ability of the diaphragm to descend downward during inhalation.

Fat also impairs function at the cellular level. Adipose cells of people who are at a healthy weight hold a moderate amount of triglycerides. In obese people,

adipose cells are overstuffed with triglycerides. Like cells that are stressed in other ways, the overstuffed adipose cells respond by sending out chemical signals that summon up an inflammatory response (Section 34.3). The resulting chronic inflammation harms organs throughout the body and increases the risk of cancer. Overstuffed adipose cells also increase their secretion of chemical messages that interfere with the effect of insulin. Remember that this hormone encourages cells to take up sugar from the blood (Section 31.8). When insulin becomes ineffective, the result is type 2 diabetes.

In 1960, the obesity rate in the United States was 15 percent. Today it is more than double that. Multiple factors contributed to this increase. The proportion

of meals consumed outside the home increased, as did the average portion sizes of restaurant meals. Soda consumption rose, and physical activity decreased. We spend more time in front of televisions and computers, and fewer of us have jobs that require physical exertion.

Preventing obesity is important. Once a person becomes obese, dieting alone is seldom effective in restoring a normal body weight. A variety of existing drugs can reduce weight somewhat, but they have negative side effects and must be taken continually to prevent rebound weight gain. Surgical procedures that reduce stomach volume can produce a dramatic sustainable weight loss, but these drastic interventions are expensive and can result in serious complications.

37

Maintaining the Internal Environment

Links to Earlier Concepts

This chapter's discussion of fluid regulation touches on osmosis (Section 5.6), transport proteins (5.7) and wastes produced by protein breakdown (36.8). The discussion of urine formation refers to epithelium (28.2), blood pressure (33.6), and adrenal glands (31.6). We also reconsider sweat glands (28.2) and regulation of body temperature (28.8).

Core Concepts

Organisms must exchange matter with the environment in order to grow, maintain themselves, and reproduce.

Maintaining an internal enviroment that differs from an external environment requires constant movement of substances across cell membranes via osmosis, diffusion, and membrane transport proteins. This movement is central to processes that rid the animal body of metabolic wastes.

Interactions among the components of a biological system give rise to complex properties.

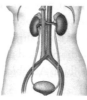

Interactions and coordination among specialized organs and organ systems contribute to the overall functioning of the animal body. These interactions include various strategies for regulating solute balance and body temperature. Internal and external factors trigger homeostatic responses in organs; in turn, these responses affect the entire body and contribute to the overall functioning of the individual.

Living things sense and respond appropriately to their internal and external environments.

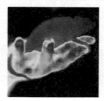

Disturbances to homeostasis can negatively affect cellular processes. Animals respond to some types of changes in the environment through physiological mechanisms. Negative feedback mechanisms that are triggered by change (in temperature and solute balance, for example) return conditions in the internal environment back to target set points. These mechanisms maintain the internal environment's conditions within optimal ranges.

◀ A dromedary camel's ability to tolerate dehydration and temperature extremes adapts it to a desert habitat.

KEEPING COOL

As Section 28.8 explained, a negative feedback mechanism maintains the human core body temperature at about 98.6° Fahrenheit (37°C). When our core temperature rises above this point, blood flow to our skin increases and we, like other primates, begin to sweat. Humans have a denser array of sweat glands than other primates. For example, we have about ten times the number of sweat glands of a chimpanzee. Our

Figure 37.11 Metabolic response of brown adipose tissue to cold. PET/CT scans showing metabolic activity of one individual at two temperatures. Dark regions indicate areas of highest metabolic activity. The scans show that brown adipose tissue in the neck and clavicle region increased its metabolic activity in response to cold.

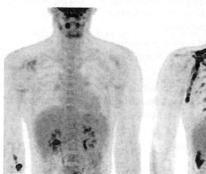 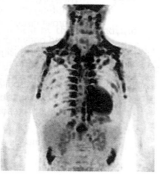

| Metabolic activity at 22°C (72°F) | Metabolic activity after two hours at 16°C (61°F) |

TABLE 37.2

Impact of Increases in Cold Stress

Core Temperature	Physiological Responses
36–34°C (about 95°F)	Shivering response; faster breathing. Peripheral vasoconstriction moves blood away from body surface. Dizziness, nausea.
33–32°C (about 91°F)	Shivering response ends. Metabolic heat output declines.
31–30°C (about 86°F)	Capacity for voluntary motion is lost. Eye and tendon reflexes inhibited. Consciousness lost. Cardiac muscle action becomes irregular.
26–24°C (about 77°F)	Ventricular fibrillation sets in (Section 33.9). Death follows.

impressive sweating capacity is thought to have arisen when our early ancestors spent much of their time walking upright across the hot, sunny savannas of equatorial Africa. Under these conditions, an increased capacity to cool the body by sweating is selectively advantageous.

The sparse body hair of humans relative to other primates is also considered a heat-related adaptation. A thick coat of hair slows evaporative cooling and insulates the body, thus reducing the heat lost by convection.

Sweating depletes the body of water and salts, a loss that can be dangerous if not remedied. Most heat-related problems in otherwise healthy people occur when overexertion is coupled with dehydration. A rise in core body temperature above 104°F (40°C) results in heat stroke, a potentially fatal condition. Symptoms of heat stroke include nausea, headache, a racing heart, confusion, and a decrease in the capacity to sweat despite ongoing heat.

To stay safe outside on a hot day, drink plenty of water and avoid excessive exercise. Avoid direct sunlight, wear a hat and light-colored clothing, and use sunscreen. Sunburn impairs the skin's ability to transfer heat to the air. Also keep in mind that the sweat glands of a person who is not normally exposed to heat produce less sweat than those of a person who has routine heat exposure.

STAYING WARM

In a cold environment, humans warm themselves by shivering and, in some cases, by nonshivering heat production (**Figure 37.11**). Exposure to cold increases secretion of thyroid hormone (Section 31.4), which encourages nonshivering heat production. People who make too little thyroid hormone cannot increase their nonshivering heat production, so they often feel cold. They also tend to be overweight, because nonshivering heat production burns a lot of calories.

Hypothermia occurs when normal mechanisms fail to keep core temperature from dropping so low that normal function is disrupted (**Table 37.2**). Brain activity becomes impaired when core body temperature falls to 95°F (35°C). Thus, "stumbles, mumbles, and fumbles" are symptoms of early hypothermia. Severe hypothermia causes loss of consciousness, disrupts heart rhythm, and can be fatal.

A person who has even moderate hypothermia may also suffer from frostbite. When body temperature declines, blood flow shifts away from peripheral tissues and toward the body's core. As peripheral tissues cool, interstitial fluid inside them begins to freeze. Fingers and toes are the most common sites of frostbite. In severe cases, frostbite kills these tissues. Any frostbite-killed tissue must be surgically removed to prevent infection of other body parts.

CREDITS: (11) From Cold-Activated Brown Adipose Tissue in Healthy Men. van Marken Lichtenbelt, Wouter D. et al. Article DOI: 10.1056/NEJMoa0808718. © 2009 *The New England Journal of Medicine*. Reprinted with permission.

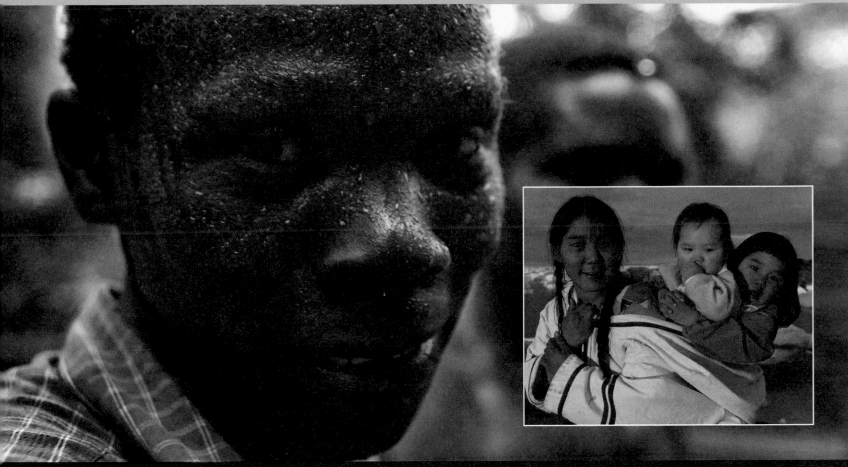

Figure 37.12 Climatic challenges. Ituri pygmies in the Congo (above) are adapted to heat and Inuit people in the Canadian Arctic (inset) to cold.

GENETIC ADAPTATION TO DIFFERING CLIMATES

As humans dispersed out of Africa, the climatic challenges they faced changed. In equatorial Africa, where humans evolved, hot days are common, so people often sweat (**Figure 37.12**). Sweating lowers the body temperature, but also results in the loss of water and sodium.

Sodium is an essential nutrient, and was a limited resource for many early populations, especially those living inland. In these populations, individuals with kidneys that reabsorbed a large amount of sodium were at a selective advantage. Later, some humans migrated to colder regions. In cooler environments, a tendency to reabsorb a large amount of sodium from filtrate would have been less advantageous. It could even be maladaptive, because high blood sodium increases

risk of hypertension and cardiovascular disorders.

Variations in the allele frequencies of some genes involved in sodium reabsorption provide evidence of climate-related selective pressure. Consider the *ACE* gene, which encodes an enzyme needed to make aldosterone. Aldosterone, remember, is the hormone that promotes sodium reabsorption from urine. The ancestral allele for the *ACE* gene (*D*) is most common in modern African and Middle Eastern populations. A different allele is most common in people of European or Asian descent. This allele has an insertion mutation that results in a less active form of the aldosterone-making enzyme.

Distribution of *ACE* alleles in modern populations may be of medical importance.

People homozygous for the *D* allele (*DD*) reabsorb more sodium than those with other genotypes. In modern societies, where food often contains a large amount of sodium, a high capacity for sodium reabsorption may contribute to health problems. Some studies have found an increased risk of hypertension or heart attack with the *DD* genotype.

Humans who settled in the coldest climates also show evidence of climate-related selection. In these regions, an ability to maximize metabolic heat provided a selective advantage. We see results of this selection in the distribution of alleles for genes involved in nonshivering heat production. Alleles that result in increased nonshivering heat production tend to be most frequent in arctic populations.

Section 37.1 Extracellular fluid, the body fluid outside of cells, is the body's internal environment. Maintaining the volume and composition of this fluid is an essential aspect of homeostasis. Organisms must balance solute and fluid gains with solute and fluid losses. They also must eliminate metabolic wastes such as **ammonia** produced by protein breakdown. Some animals excrete ammonia; others conserve water by converting ammonia to **urea** or **uric acid**. Excretory organs include **nephridia** of earthworms, **Malpighian tubules** of insects, and the paired **kidneys** of vertebrates, which excrete **urine**.

Section 37.2 A kidney has more than a million **nephrons**, each consisting of a tubule and some associated capillaries. A **Bowman's capsule** in the renal cortex is the entrance to a nephron tubule. It receives fluid filtered out of the capillaries of the **glomerulus**. This fluid continues through a **proximal tubule**, a **loop of Henle** that descends into and ascends from the renal medulla, and a **distal tubule** that drains into a **collecting tubule**. **Peritubular capillaries** lie in close proximity to the tubular portion of a nephron and exchange substances with it.

Fluid from collecting tubules drains into the renal pelvis, where it is called urine. **Ureters** carry urine formed in the kidneys to the **urinary bladder**. The bladder stores urine until it is expelled through the **urethra**. Urination is a reflex that can be overridden by voluntary control.

Section 37.3 Urine formation begins when **glomerular filtration** produces protein-free plasma that enters the tubular part of a nephron. Most water and solutes are returned to the blood by **tubular reabsorption**. Substances that are not reabsorbed, and substances added to the filtrate by **tubular secretion**, end up in the urine. Hormones regulate the urine's concentration and composition. **Antidiuretic hormone** makes tubules more permeable to water, so more is reabsorbed and urine is more concentrated. **Aldosterone** increases sodium reabsorption. Water follows sodium into the blood, so aldosterone indirectly concentrates the urine.

Sections 37.4, 37.5 Kidneys can be harmed by chronic disease, genetic factors, infections, or drugs. When kidneys fail, frequent dialysis or a kidney transplant is required to sustain life. Analysis of urine provides information about health and hormonal status. Drugs and toxins such as pesticides are excreted in urine and can be detected by urine tests.

Section 37.6 All animals produce metabolic heat. **Thermoregulation** requires that heat produced by metabolism and gained from the environment must balance heat lost to the environment. Animals also gain or lose heat by **thermal radiation**, **conduction**, and **convection** and lose heat by **evaporation** of water from body surfaces.

For **ectotherms** such as reptiles, core temperature depends more on heat exchanges with the environment than on metabolic heat production. In such animals, core temperature is controlled mainly by behavioral changes. For **endotherms** (most birds and mammals), a high metabolic rate is the primary source of heat. In endotherms, core temperature is controlled by regulating the production and loss of metabolic heat. In **heterotherms**, core temperature is tightly controlled part of the time, and allowed to fluctuate with the external temperature at other times. In some animals, temperature fluctuates daily. Those that hibernate allow it to fall during hibernation.

Dilating blood vessels in the skin, sweating, and panting can cool a body. Shivering and **nonshivering heat production** by brown adipose tissue generate heat. Constricting blood vessels in the skin retain heat in the body core.

Section 37.7 Sweating is the main cooling mechanism in humans. We have less hair and more sweat glands than other primates. Sweating without restoring lost water and solutes raises the risk of hyperthermia. Humans warm themselves by shivering and nonshivering heat production. A decline in core body temperature can be fatal. Frostbite is damage to extremities that occurs when tissue freezes.

Application Humans evolved in tropical Africa, where individuals who maintained their sodium level despite sweating were at a selective advantage. Reabsorbing extra sodium can cause health problems where salt is plentiful. Mechanisms that enhanced metabolic heat production evolved in humans who migrated to cold regions.

SELF-ASSESSMENT
ANSWERS IN APPENDIX VII

1. Malpighian tubules of _____ eliminate uric acid.
    a. birds
    c. insects
    b. marine fishes
    d. earthworms

2. Breakdown of _____ produces ammonia.
    a. sugars
    c. starches
    b. fats
    d. proteins

3. Urine of a healthy person contains _____ .
    a. ammonia
    c. uric acid
    b. urea
    d. proteins

4. Bowman's capsule, the start of the tubular part of a nephron, is located in the _____ .
    a. renal cortex
    c. renal pelvis
    b. renal medulla
    d. renal artery

5. Plasma fluid filtered into Bowman's capsule flows directly into the _____ .
    a. renal artery
    c. distal tubule
    b. proximal tubule
    d. loop of Henle

PESTICIDES IN URINE Food that carries the USDA's organic label (left) is produced without pesticides such as malathion and chlorpyrifos that are used on conventionally grown crops. Chensheng Lu of Emory University tested whether eating organic food significantly affects the level of pesticide residues in a child's body (**Figure 37.13**). He tested urine of 23 children for metabolites (breakdown products) of pesticides. During the first five days, children ate their standard, nonorganic diet. For the next five days, they ate organic versions of the same foods and drinks. For the final five days, they returned to their standard diet.

1. During which phase of the experiment did the children's urine contain the lowest level of the malathion metabolite?

2. In which phase of the experiment was the most chlorpyrifos metabolite detected?

3. Did switching to an organic diet lower the amount of pesticide residues excreted by the children?

		Malathion Metabolite		Chlorpyrifos Metabolite	
Study Phase	No. of Samples	Mean (μg/liter)	Maximum (μg/liter)	Mean (μg/liter)	Maximum (μg/liter)
1. Conventional	87	2.9	96.5	7.2	31.1
2. Organic	116	0.3	7.4	1.7	17.1
3. Conventional	156	4.4	263.1	5.8	25.3

Figure 37.13 Levels of metabolites (breakdown products) of malathion and chlorpyrifos in the urine of children taking part in a study of effects of an organic diet. The difference between the mean level of metabolites in the organic and inorganic phases of the study was statistically significant.

4. Even in the nonorganic phases of this experiment, the highest pesticide metabolite levels detected were far below those known to be harmful. Given these data, would you spend more to buy organic foods?

6. Blood pressure forces water and small solutes out of blood and into nephrons during _____ .
   a. glomerular filtration
   b. tubular reabsorption
   c. tubular secretion
   d. both a and c

7. Kidneys return most of the water and small solutes back to blood by way of _____ .
   a. glomerular filtration
   b. tubular reabsorption
   c. tubular secretion
   d. both a and b

8. Tubular secretion moves _____ into kidney tubules.
   a. H^+ b. glucose   c. water   d. protein

9. Antidiuretic hormone makes distal tubules and collecting tubules more permeable to _____ .

10. _____ can keep people with kidney failure alive, but it cannot cure them.

11. Match each structure with a function.
    ___ ureter
    ___ Bowman's capsule
    ___ urethra
    ___ collecting tubule
    ___ pituitary gland

    a. start of nephron
    b. delivers urine to body surface
    c. carries urine from kidney to bladder
    d. secretes ADH
    e. target of aldosterone

12. Which of the following is an endotherm?
    a. a shark   b. a frog   c. a monkey   d. a snake

13. Match each term with the most suitable description.
    ___ endotherm
    ___ ectotherm
    ___ convection
    ___ conduction
    ___ thermal radiation

    a. environment dictates core temperature
    b. metabolism dictates core temperature
    c. heat transfer between objects that are in direct contact
    d. water or air current transfers heat
    e. emission of radiant energy

14. Nonshivering heat production involves the mitochondria of _____ .
    a. the extremities
    b. the kidneys
    c. brown adipose tissue
    d. epithelial tissue

CLASS ACTIVITY
CRITICAL THINKING

1. Like desert rodents such as the kangaroo rat (**Figure 37.3**), marine mammals have highly efficient kidneys that produce only a tiny amount of very concentrated urine. What common selective pressure shaped this trait in both animals?

2. When marathoners or other endurance athletes sweat heavily and drink a large amount of pure water, their sodium level drops. The resulting condition, called "water intoxication," can be fatal. Why is maintaining the sodium level so important?

3. In cold habitats, ectotherms are few and endotherms often show morphological adaptations to cold. Compared to closely related species that live in warmer areas, cold dwellers tend to have smaller appendages. Also, animals adapted to cool climates tend to be larger than closely related species that dwell in warmer climates. For example, the largest bear species is the polar bear and the largest penguin is Antarctica's emperor penguin. Think about heat transfers between animals and their habitat, then explain why smaller appendages and larger overall body size are advantageous in very cold places.

4. Drinking alcohol inhibits ADH secretion. What effect will drinking a beer have on the permeability of kidney tubules to sodium? To water?

for additional quizzes, flashcards, and other study materials
▶ ACCESS MINDTAP AT WWW.CENGAGEBRAIN.COM

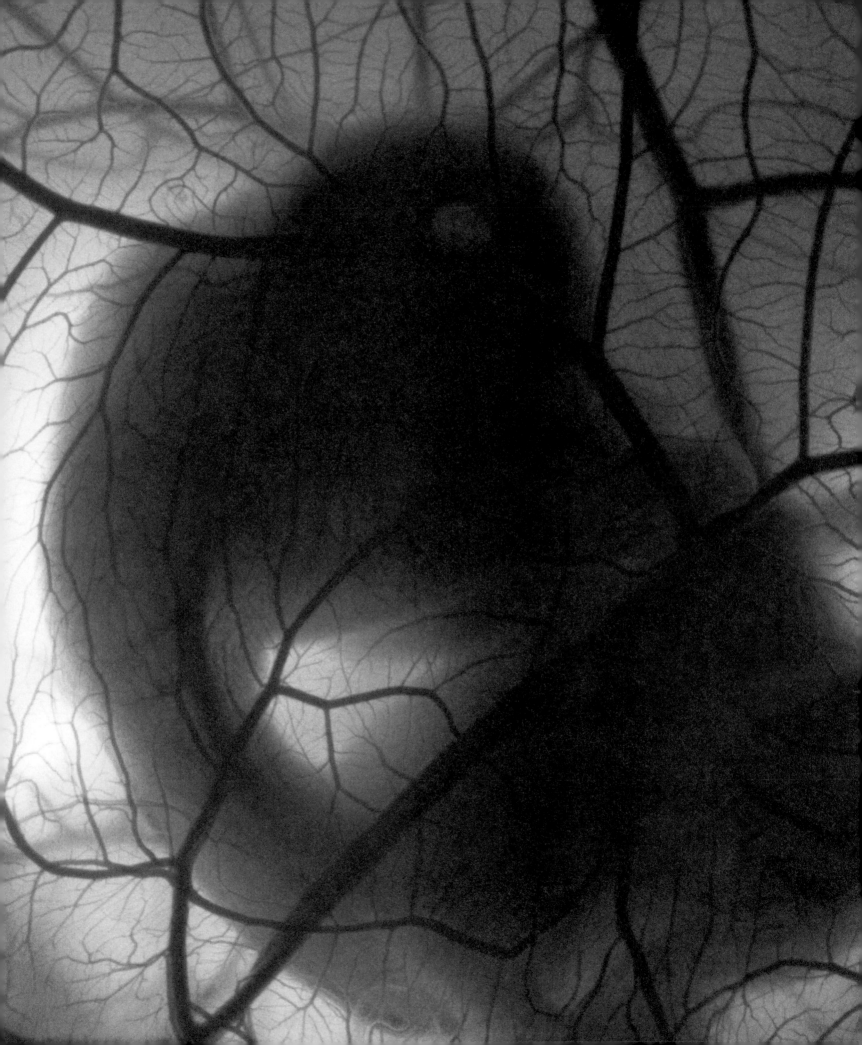

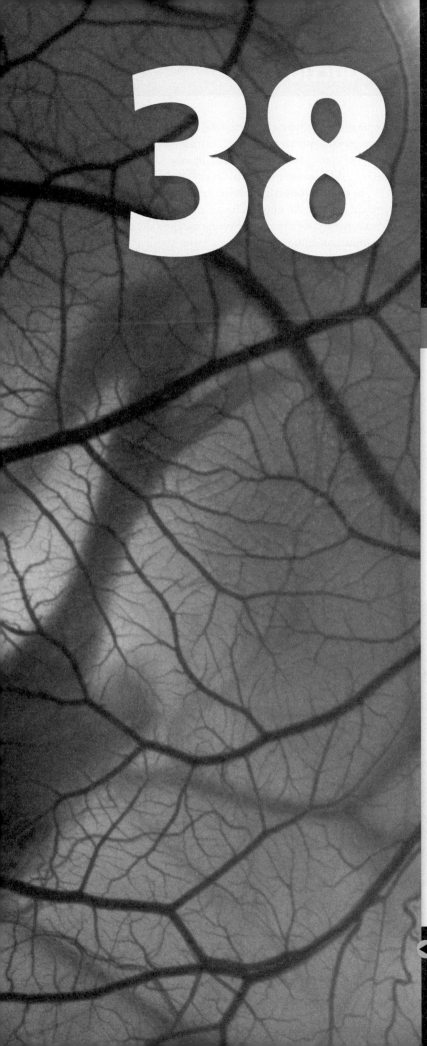

38 Reproduction and Development

Links to Earlier Concepts

This chapter returns to sexual reproduction (Sections 12.1 and 12.2) and the role of master genes (10.2) in cell differentiation (8.6). You will learn more about stem cells (Chapter 28 Application), primary tissue layers (23.1), the mammalian placenta (24.7), and sex hormones (31.7). We also reconsider the harmful effects of HIV (19.1, 34.9), mercury, alcohol, and endocrine disruptors (Chapters 2, 5, and 31 Applications).

Core Concepts

All organisms alive today are linked by lines of descent from shared ancestors.

Shared core processes and features provide evidence that all living things are descended from a common ancestor: Most animals reproduce sexually, thus producing offspring that are genetically different from one another and from their parents. Many of the master genes that direct embryonic development are evolutionarily conserved across major animal groups.

Interactions among the components of a biological system give rise to complex properties.

During development, cascades of master gene expression shape a complex, multicelled body from a single-celled zygote. Localized patterns of master gene expression trigger responses in embryonic cells. These responses, which include cell differentiation, cell movement, and cell death, ultimately result in the formation of tissues and organs in specific regions at specific times.

Growth, reproduction, and homeostasis depend on proper timing and coordination of specific molecular events.

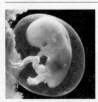

The development of a complex animal body depends on the presence of necessary genetic information as well as the correct and timely expression of this information. Because environmental stimuli influence master gene expression, nutritional deficiencies, pathogens, and exposure to harmful chemicals can interfere with development.

◀ A two-week-old turkey embryo. Similar developmental processes among vertebrates are evidence of shared ancestry.

Using appropriate examples, describe some of the ways that animals reproduce asexually.

Explain the genetic costs and benefits of sexual reproduction.

Describe the different types of hermaphrodites.

Compare internal and external fertilization.

Using appropriate examples, describe the ways in which developing animal offspring are nourished.

ASEXUAL OR SEXUAL?

With **asexual reproduction**, a single individual produces offspring. All offspring are genetic replicas (clones) of the parent and identical to one another. Such genetic uniformity is advantageous in a stable environment where the combination of alleles that makes a parent successful is likely to do the same for its offspring (Section 11.1).

Invertebrates reproduce by a variety of asexual mechanisms. A new individual may grow on the body of its parent, a process called budding (**Figure 38.1A**). Alternatively, offspring may arise by fragmentation, in which a piece of the parent (such as a coral) breaks off and develops into a new animal. Some flatworms reproduce by transverse fission: The worm divides in two, leaving one piece headless and one tailless. Each piece then grows the missing body parts.

Parthenogenesis is a mechanism of asexual reproduction in which female offspring develop from unfertilized eggs.

Some rotifers (a type of invertebrate) reproduce solely by parthenogenesis. Among vertebrates, parthenogenesis has been observed among fishes, amphibians, lizards, and one bird (the turkey). By contrast, mammals can only reproduce sexually.

Sexual reproduction involves two parents that produce haploid gametes. An **egg** is a female gamete and a **sperm** is a male gamete. As a result of crossing over and the random assortment of chromosomes during meiosis, each gamete receives a different combination of maternal and paternal alleles (Section 12.4). An egg and a sperm combine to produce a genetically unique individual with alleles from both parents.

Producing offspring that differ from both parents and from one another can be advantageous in a changing environment (Section 12.1). By reproducing sexually, a parent increases the likelihood that some of its offspring will have a combination of alleles that suits them to conditions in a changed environment.

Some animals reproduce asexually when conditions are stable and favorable, but switch to sexual reproduction when conditions begin to change. Consider aphids, a type of plant-sucking insect. In early spring, when tender plant parts are plentiful and aphids are not, a female aphid can give birth to several smaller clones of herself every day by parthenogenesis. In autumn, when plants prepare for dormancy, tender plant parts become scarce and competition for them increases. At this time, the aphids switch reproductive modes to produce offspring by sexual means.

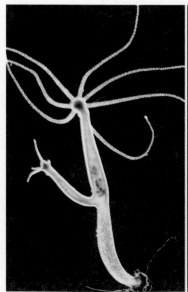

A Asexual reproduction in hydra (a cnidarian). A new individual (left) is budding from its parent.

B Sexual reproduction in the barred hamlet, a simultaneous hermaphrodite. Each fish lays eggs and also fertilizes its partner's eggs.

C Sexual reproduction in elephants. The male is inserting his penis into the female. Eggs will be fertilized and the offspring will develop inside the female's body.

Figure 38.1 Examples of animal reproduction.

VARIATION AMONG SEXUAL REPRODUCERS

Animal gametes form in **gonads**, which are primary reproductive organs. Most vertebrates have separate sexes that remain fixed for life; each individual has either male or female gonads. Some animals are **hermaphrodites**—they have both male and female gonads during their life span. Tapeworms and some roundworms are simultaneous hermaphrodites; they produce eggs and sperm at the same time, and can fertilize themselves. Earthworms, land snails, and slugs are simultaneous hermaphrodites too, but they require a partner. So do hamlets, a type of marine fish (**Figure 38.1B**). When hamlets mate, they take turns donating and receiving sperm. Some other fishes are sequential hermaphrodites, which switch from one sex to another during a lifetime.

Most aquatic invertebrates, fishes, and amphibians release gametes into the water, where the gametes combine during **external fertilization**. With **internal fertilization**, sperm fertilize an egg inside the female's body (**Figure 38.1C**). Internal fertilization evolved in most land animals, including insects and the amniotes.

Depending on the species, eggs that have been fertilized internally are either released to develop in the environment, or retained inside the female's body for some portion of their development. Most insects (**Figure 38.2A**), all birds, and one group of mammals (the monotremes) lay eggs. By contrast, the eggs of many sharks, snakes, and lizards develop while enclosed by an egg sac inside the mother's body. The eggs hatch inside the mother's body shortly before she gives birth to well-developed young (**Figure 38.2B**).

A developing animal requires nutrients. Most animals are nourished by **yolk**, a thick fluid rich in proteins and lipids that is deposited in the egg during its formation. In birds, the proportion of yolk in an egg varies with the incubation time characteristic of the species. The average bird egg, which is about one-third yolk, takes 21 days to hatch. Kiwi birds have the longest incubation period of any bird—11 weeks. Their eggs are about two-thirds yolk by volume.

Humans and other placental mammals (Section 24.7) have almost yolkless eggs. An organ called the **placenta** that forms during pregnancy allows exchange of substances between a mother's blood and that of her developing offspring (**Figure 38.2C**). Nutrients in the maternal blood diffuse across the placenta into an offspring's blood and support the offspring's development. We discuss the structure and function of the mammalian placenta in detail in Section 38.14.

Structures analogous to the mammalian placenta evolved independently in several groups of live-bearing fishes and one group of live-bearing lizards (skinks).

A Bug laying eggs. Yolk in the eggs will provide offspring of this bug with all the nutrients they need to develop.

B Snake giving birth to young that developed in an egg sac inside her body. Yolk nourished the developing young.

C Elk is examining her newborn calf. The placenta that allowed nutrients to diffuse from her blood into the calf's blood dangles at the left. It is expelled after birth.

Figure 38.2 Variations in where offspring develop.

asexual reproduction Reproductive mode by which offspring arise from one parent only.

egg Female gamete.

external fertilization Sperm fertilize eggs after gametes are released into the environment.

gonad Gamete-forming organ of an animal.

hermaphrodite (herm-AFF-roh-dyte) Animal that has both male and female gonads, either simultaneously or at different times in its life cycle.

internal fertilization Sperm fertilize eggs inside a female's body.

placenta (pluh-SEN-tah) Organ formed during pregnancy; allows diffusion of material between maternal and embryonic bloods.

sexual reproduction Reproductive mode by which offspring arise from two parents and inherit genes from both.

sperm Male gamete.

yolk Nutritious material in many animal eggs.

CREDITS: (2A) R. Scott Cameron, Advanced Forest Protection, Inc., Bugwood.org; (2B) © Tony Phelps/naturepl.com; (2C) NPS Yellowstone/Becky Wyman.

In all animals, a new individual produced by sexual reproduction develops through a series of predictable stages to adulthood. **Figure 38.3** illustrates the stages of development in one vertebrate, the leopard frog.

GAMETES FORM AND UNITE

Male gonads, which are called **testes** (singular, testis), produce sperm. Female gonads, which are called **ovaries**, produce eggs. The number, structure, and location of gonads vary among animal groups. For example, a sea star has a pair of gonads in each arm. An octopus has a single gonad near the rear of its head, inside the mantle cavity. A frog has a pair of gonads in its abdominal cavity.

An egg is always much larger than a sperm of the same species. Meiosis of a male germ cell is accompanied by two equal cytoplasmic divisions, so four sperm form from each male germ cell. By contrast, meiosis of female germ cells involves unequal cytoplasmic divisions. After each meiotic division, one cell receives the bulk of the cytoplasm. As a result, meiosis of a female germ cell yields a single egg, with a large amount of cytoplasm. (We discuss how human sperm and eggs form in more detail in Section 38.6 and 38.8.)

During fertilization, a haploid egg and a haploid sperm unite to form a diploid zygote. (We discuss the process of fertilization in more detail in Section 38.9.)

CLEAVAGE

New cells arise when **cleavage** carves up a zygote by repeated mitotic cell divisions ❶. During cleavage, the number of cells increases, but the zygote's original volume remains unchanged. Thus, cells become more numerous but smaller in size. Cleavage produces a **blastula**: a ball of cells (called blastomeres) surrounding a fluid-filled cavity (the blastocoel) ❷. Tight junctions hold the cells of the blastula together.

Although all the cells that make up a blastula are descended from the same zygote, some contents of their cytoplasm vary. The differences among the cells arises because maternal mRNAs are not distributed evenly throughout the egg cytoplasm. During cleavage, different cells receive different amounts and types of these maternal mRNAs. This process

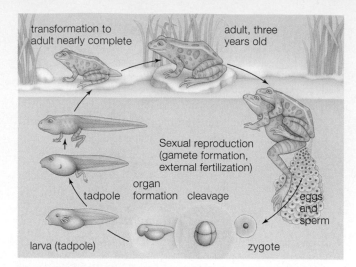

Figure 38.3 Vertebrate development. Above, overview of reproduction and development in the leopard frog. Opposite, a closer look at some stages.

TABLE 38.1

Fates of Vertebrate Germ Layers

Germ Layer	Organs and Tissues in Adult
Ectoderm (outer layer)	Outer layer (epidermis) of skin; nervous tissue
Mesoderm (middle layer)	Connective tissue of skin; skeletal, cardiac, smooth muscle; bone; cartilage; blood vessels; urinary system; gut organs; reproductive tract; lining of coelom
Endoderm (inner layer)	Lining of gut and respiratory tract, and organs derived from these linings

distributes different maternal mRNAs into different parts of the embryo, and the localization will determine which master genes are turned on where.

GASTRULATION

During **gastrulation**, cells of the blastula move about and organize themselves as the layers of the **gastrula** ❸. In most animals and all vertebrates, a gastrula consists of three embryonic tissue layers called **germ layers**. Each germ layer gives rise to particular tissues (**Table 38.1**). **Ectoderm**, the outer germ layer, gives rise to nervous tissue and to the outer layer of skin or other body covering. **Endoderm**, the inner germ layer, is the start of the respiratory tract and gut linings. **Mesoderm** forms between the ectoderm and the endoderm. Mesoderm is the source of all muscles, connective tissues, and the circulatory system. The same tissues and organs form from the same germ layers in all vertebrates.

blastocoel

blastula

❶ Here we show the first three divisions of cleavage, a process that carves up a zygote's cytoplasm. In this species, cleavage results in a blastula, a ball of cells enclosing a fluid-filled cavity.

❷ Cleavage is over when the blastula forms.

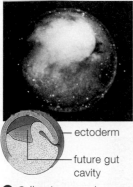

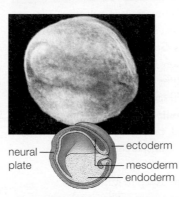
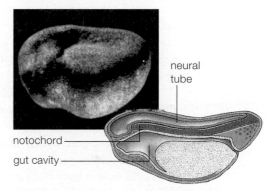

ectoderm

future gut cavity

neural plate

ectoderm

mesoderm

endoderm

neural tube

notochord

gut cavity

❸ Cells migrate and reorganize during gastrulation. In this species, gastrulation produces a gastrula that has an outer layer of ectoderm, a middle layer of mesoderm, and an inner layer of endoderm.

❹ A primitive gut cavity opens up. A neural tube, then a notochord, and other organs form.

Tadpole, a swimming larva with segmented muscles and a notochord extending into a tail.

Limbs grow and the tail is absorbed during metamorphosis to the adult form.

Sexually mature, four-legged adult leopard frog.

❺ The frog's body form changes as it grows and its tissues specialize.

TISSUES AND ORGANS FORM

An embryo's tissues and organs begin to form after gastrulation ❹. Many organs incorporate tissues derived from more than one germ layer. For example, the stomach's epithelial lining is derived from endoderm, and the smooth muscle that makes up the stomach wall develops from mesoderm. In most animals, a newly hatched or born individual continues to grow and develop. In frogs, a larva (a tadpole) grows, then undergoes metamorphosis, which is a drastic remodeling of tissues into the adult form ❺.

blastula (BLAST-yoo-luh) Hollow ball of cells that forms as a result of cleavage.

cleavage In animal development, stage during which repeated mitotic divisions produce a blastula.

ectoderm Outermost tissue layer of an animal embryo.

endoderm Innermost tissue layer of an animal embryo.

gastrula (GAS-true-luh) Three-layered developmental stage formed by gastrulation in an animal.

gastrulation (gas-true-LAY-shun) Animal developmental process by which cell movements produce a three-layered gastrula.

germ layer One of three primary layers in an early embryo.

mesoderm Middle tissue layer of a three-layered animal embryo.

ovary Egg-producing animal gonad.

testis Sperm-producing animal gonad.

CREDITS: (3) 1–4: Carolina Biological Supply Company; 5: left and center, © David M. Dennis/Tom Stack & Associates, Inc.; right, © John Shaw/Tom Stack & Associates.

38.3 Tissue and Organ Formation

LEARNING OBJECTIVES

Explain the role of selective gene expression in development.

Describe the process of embryonic induction and provide an example.

Give an example of how cell shape changes contribute to organ formation.

Explain the process of apoptosis and its role in development.

Describe how cells can shift their position during development.

CELL DIFFERENTIATION

All cells in a developing animal are descended from the same zygote, so all have the same genes. How, then, do specialized tissues and organs form? From gastrulation onward, selective gene expression occurs, which means that different cell lineages express different subsets of genes. Selective gene expression is the key to differentiation (Section 10.1), the process by which cells become specialized as different tissues.

A human body has about 200 differentiated cell types, all descended from pluripotent embryonic stem cells (Chapter 28 Application). As your eye developed, cells of one lineage began expressing genes for crystallin, a transparent protein. These differentiating cells formed the lens of your eye. No other cells in your body make crystallin. Keep in mind that a differentiated cell still retains the entire genome. That is why it is possible to clone an adult animal from one of its differentiated cells (Section 8.6).

EMBRYONIC INDUCTION

Intercellular signals encourage selective gene expression. By a process called **embryonic induction**, one subset of cells affects the developmental pathway of other cells. Induction often involves **morphogens**, which are substances (usually proteins) that are produced in a specific region and affect gene expression in a concentration-dependent manner. Cells near the morphogen-producing region are exposed to a high concentration of morphogen and express one set of genes. Cells farther from the morphogen source are exposed to less morphogen, and therefore express different genes.

The bicoid protein of fruit flies is an example of a morphogen. *Bicoid* is a maternal effect gene, meaning it is expressed during egg production and its product influences development. During egg production, bicoid mRNA accumulates at one end of an egg. After fertilization, this mRNA is translated into bicoid protein. The bicoid protein diffuses away from the site of translation, thus forming a gradient from one end of the zygote to the other. This gradient determines the front-to-back axis of the developing individual: The head will form at the end where bicoid protein concentration is highest, and the abdomen will form at the end

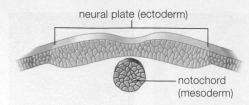

❶ Morphogens produced by the notochord induce the ectoderm above it to thicken and form a neural plate.

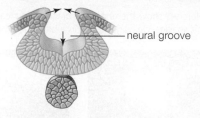

❷ As cells in the neural plate change shape, the plate's edges fold toward the plate center to form a neural groove.

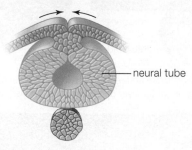

❸ As inward folding continues, the edges of the neural plate meet and fuse to form the neural tube.

Figure 38.4 Neural tube formation.

where the bicoid protein concentration is lowest. *Bicoid* is one of the master genes described in Section 10.2; it encodes a transcription factor that affects expression of many other master genes. The resulting cascade of master gene expression culminates in the expression of homeotic genes in particular regions of the embryo. Expression of a homeotic gene results in the formation of a specific body part.

As another example of embryonic induction, consider the formation of the vertebrate neural tube, the embryonic forerunner of the spinal cord and brain (Section 29.9). Formation of the neural tube from ectoderm is regulated by morphogens secreted by the notochord, which lies directly beneath it (**Figure 38.4**).

Development of the neural tube begins when the region of ectoderm directly above the notochord thickens, forming a neural plate **❶**. Next, cells at the edges of the neural plate become wedge-shaped as actin microfilaments at one end of

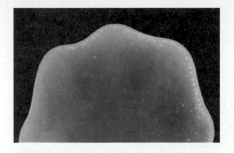 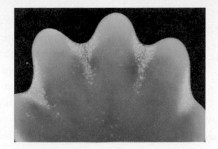 

Figure 38.5 Paw development in a mouse embryo. The yellow stain indicates regions where cells are undergoing apoptosis.

each cell shorten. These changes in cell shape cause the edges of the neural plate to fold inward, forming a depression called the neural groove ❷. With continued folding, the two edges of the neural plate meet at the midline and fuse to form the neural tube ❸.

Cell shape changes are one mechanism of morphogenesis, the developmental process that shapes tissues and organs and determines their position within the animal body. We turn next to two additional mechanisms.

APOPTOSIS

Apoptosis, or programmed cell death, is a proccess by which a cell self-destructs in a controlled, predictable manner. During apoptosis, a cell activates enzymes that destroy its component parts. Some enzymes chop up cytoskeletal proteins and the histones that organize DNA. Others snip apart nucleic acids. As a result of this self-inflicted damage, the cell shrivels up and dies without spewing out its contents. The cell remains can then be engulfed and disposed of by other body cells. In a mature body, apoptosis plays a role in normal cell turnover and in defense against cancer.

During development, apoptosis shapes structures by eliminating excess cells, removing those cells that do not contribute to the expected final form. Consider its role in shaping a mouse paw. At about two weeks of development, the paw of a developing mouse is a fleshy, paddle-like structure with the bones hidden inside (**Figure 38.5**). By day 16, apoptosis has removed many cells, leaving only thin webs of tissue between

individual digits. By day 18, this webbing is completely gone and the individual digits are free. Apoptosis also makes the tail of a tadpole disappear during metamorphosis, and eliminates the tail that forms early in human development.

Figure 38.6 Cell migration in development.

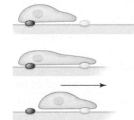

A cell adheres to one point and protrudes forward.

The protruding portion of the cell attaches to a new point.

The trailing portion of the cell releases its grip.

CELL MIGRATIONS

Cell migrations also play a role in morphogenesis. For example, ectodermal cells that form at the tip of the neural tube (a region called the neural crest) migrate outward to positions throughout the body. The descendants of neural crest cells include the neurons and glial cells of the peripheral nervous system, and the melanocytes in skin.

Cells travel by inching along in an amoeba-like fashion (**Figure 38.6**). A portion of the cell protrudes forward when actin microfilaments assemble along the forward edge. Adhesion proteins in the plasma membrane then anchor the advancing portion of the cell to a protein in the membrane of another cell or in the extracellular matrix. Once the front of a cell is thus anchored, adhesion proteins in the trailing part of the cell release their grip, and the rear of the cell is drawn forward. How does the cell know where to go? Some cells respond to a chemical signal by following its concentration gradient, moving either toward or away from a particular signal. Others cells follow a "trail" of molecules that its adhesion proteins recognize.

apoptosis (ap-op-TOE-sis or ay-poe-TOE-sis) Programmed cell death; a predictable, controlled process of cell self-destruction.
embryonic induction Embryonic cells in one region produce signals that alter the behavior of neighboring cells.
morphogen (MOR-foe-jen) Substance that regulates development; affects cells in a concentration-dependent manner.

38.4 Evolutionary Developmental Biology

LEARNING OBJECTIVES

Describe the three types of factors that constrain animal body plans.

Using appropriate examples, explain how mutations that affect development can alter a body plan.

Evolutionary developmental biology, commonly known as evo-devo, looks at the genetic basis for the developmental pathways common to all animals, how evolved variations in these pathways can give rise to diverse body plans, and how developmental factors constrain this variation.

CONSTRAINTS ON BODY PLANS

Let us first consider developmental constraints. These are factors that limit the types of body plans that can evolve. There are three general categories of developmental constraints.

First, physical constraints limit body form. For example, large body size cannot evolve in an animal that does not have circulatory and respiratory mechanisms sufficient to service cells far from the body surface. Also, an open circulatory system, in which blood is not constrained within vessels, cannot move fluid against gravity to the same extent that a closed system can. Thus, insects the size of hippos cannot evolve.

Second, an existing body framework imposes architectural constraints. For example, the common ancestor of all living land vertebrates had two pairs of limbs. Some vertebrates have lost one or both pairs of limbs, but none have gained an extra pair. The evolution of wings in birds and bats occurred through modification of existing forelimbs, not by sprouting new limbs. Although it would seem advantageous to have both wings and arms, no living or fossil vertebrate with such a body plan has ever been discovered.

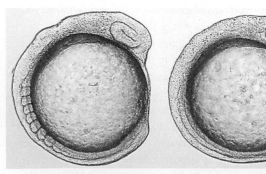

A Normal zebrafish embryo with somites (bumps of mesoderm that give rise to bone and muscle).

B Embryo with a mutation that prevents somite formation. It will die in early development.

Figure 38.7 Lethal effect of a mutation in a zebrafish gene called *fused somites* that functions in early development.

Finally, there are phyletic constraints on body plans. These are constraints imposed by interactions among the genes that regulate development. Once master genes evolved, their interactions determined the basic body form. Mutations that dramatically alter these interactions are usually lethal.

Consider how paired bones and skeletal muscles become arrayed along a vertebrate's head-to-tail axis. This pattern arises early in development, when the mesoderm on either side of the embryo's neural tube becomes divided into blocks of cells called somites (**Figure 38.7A**). The somites will later develop into bones and skeletal muscles. A complex pathway involving many genes governs somite formation. Any mutation that disrupts this pathway so that somites do not form is lethal to the embryo (**Figure 38.7B**).

DEVELOPMENTAL MUTATIONS

Evolutionary changes in body plans arise by modifications to existing developmental pathways rather than blazing entirely new genetic trails.

Modifications in the rate and/or the timing of developmental process are one mechanism for changing body form. Consider how the length of a vertebrate body varies among species. All vertebrates develop somites, but their number of somites, and hence their number of vertebrae, varies. A typical lizard embryo has about 61 somites, whereas a corn-snake embryo has more than 300. Differences in body length between lizards and snakes arise from differences in the rate at which somites form during early development. In snakes, somites form much more rapidly than they do in other vertebrates. As a result, the mesoderm is divided into more somites, which give rise to more vertebrae.

Changes in gene regulation can also affect development. Consider the variation in beak sizes and shapes among the finches that Darwin observed on the Galápagos Islands (Section 16.2). All 14 species of these birds share a common ancestor, but different types of beaks adapt them to different food sources. Whether a species has a blunt beak or a pointed one is determined in large part by the genotype at the *ALX1* locus. The *ALX1* encodes a protein that acts as a transcription factor (Section 10.1) during vertebrate facial development. In humans, mutations in the *ALX1* gene are associated with facial abnormalities such as cleft lip (a defect that we discuss in Section 38.14).

Among Galápagos finches, there are two alleles for *ALX1*; one produces a blunt beak and the other a pointy beak. Heterozygotes have a beak of intermediate shape. In most Galápagos finch species, either one or the other of these alleles has become fixed (Section 17.6); all individuals of the species are homozygous for the same allele.

38.5 Overview of Human Development

Now that you have a general understanding of animal reproduction and development, we will turn our attention to the details of these processes in humans. In this section, we provide an overview of human development and define its stages. Prenatal (before birth) and postnatal (after birth) stages are listed in **Table 38.2**. The remainder of this chapter describes human reproductive anatomy and discusses human reproductive function—from gamete formation to birth.

Human prenatal development normally lasts 38 to 40 weeks from the time of fertilization. It is sometimes discussed in terms of trimesters. The first trimester includes months one through three, the second trimester, months four through six, and the third trimester, months seven through nine.

In discussing animals, the term **embryo** generally refers to an individual from its first cleavage to the time when it

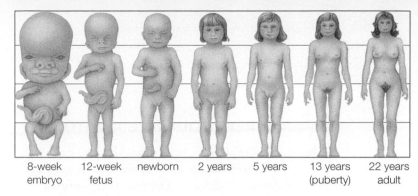

| 8-week embryo | 12-week fetus | newborn | 2 years | 5 years | 13 years (puberty) | 22 years adult |

Figure 38.8 Proportional changes during human development.

hatches or is born. With regard to humans, the term embryo is usually reserved for the period from 2 weeks (when implantation in the uterus occurs) to 8 weeks after fertilization. From week nine until birth, the individual is referred to as a **fetus**. During embryonic development, the human body plan is set in place and organs form. During fetal development, organs mature and begin to function.

In humans, birth before 37 weeks is considered premature. To date, the earliest a child has been born and survived is 21 weeks and six days. However, most infants born before 23 weeks die before leaving the hospital. A child born before 28 weeks will not have fully formed lungs and so will require a respirator. Premature birth also raises the risk of infections, heart problems, bleeding into the brain, and severe anemia.

After birth, the human body continues to grow and its body parts continue to change in proportion (**Figure 38.8**). The head grows much more slowly than other body parts, and the legs grow much faster. Compared to other primates, humans have long legs for their body size. Our long-legged body is an adaptation to bipedal locomotion (upright walking). By about age 7, human legs make up about 50 percent of the body's total height, which is the proportion required for efficient upright walking.

Sexual maturation occurs at **puberty**. At this time, the gonads increase their production of sex hormones and begin to produce gametes. Secondary sexual characteristics appear. Bones stop growing shortly after completion of puberty. The brain is the last organ to become fully mature: Parts of the brain continue to develop until age 20.

TABLE 38.2

Stages of Human Development

Prenatal period

Zygote	Single cell resulting from fusion of sperm nucleus and egg nucleus at fertilization.
Blastocyst (blastula)	Ball of cells with surface layer, fluid-filled cavity, and inner cell mass.
Embryo	Individual from the completion of implantation (2 weeks) until the end of week 8.
Fetus	Individual from week 9 until birth.

Postnatal period

Newborn	Individual during the first two weeks after birth.
Infant	Individual from two weeks to fifteen months.
Child	Individual from infancy to about twelve years.
Pubescent	Individual at puberty, when secondary sexual traits develop.
Adolescent	Individual from puberty until adulthood.
Adult	Begins between 18 and 25 years; bone formation and growth cease. Changes proceed slowly after this.
Old age	Aging processes result in tissue deterioration.

embryo In animals, a developing individual from first cleavage until hatching or birth; in humans, usually refers to an individual during weeks 2 to 8 of development.

fetus Developing human from about 9 weeks until birth.

puberty Period when reproductive organs begin to mature.

Describe the location and functions of human ovaries.

Explain the functions of the oviducts, uterus, and vagina.

Describe the events of one ovarian cycle.

FEMALE REPRODUCTIVE ANATOMY

A human female's gonads—her ovaries—lie deep inside her pelvic cavity (**Figure 38.9A**, **B**). They are about the size and shape of almonds. Ovaries produce and release **oocytes**, which are immature eggs. They also secrete estrogens and progesterone, the main sex hormones in females. **Estrogens** maintain the female reproductive tract and trigger the development of female sexual traits. **Progesterone** prepares the reproductive tract for pregnancy.

Adjacent to each ovary is an **oviduct**, a hollow tube that connects the ovary to the uterus. (Mammalian oviducts are sometimes referred to as Fallopian tubes.) An oocyte released from an ovary is drawn into an oviduct by the motion of cilia on fingerlike projections at the oviduct's entrance. Cilia in the oviduct lining then propel the oocyte along the length of the tube. Fertilization usually occurs in an oviduct.

Each oviduct opens into the **uterus**, a hollow, pear-shaped organ commonly called the womb. Myometrium, a thick layer of smooth muscle, makes up the bulk of the uterine wall. The uterine lining, or endometrium, consists of glandular epithelium, connective tissues, and blood vessels. A woman becomes pregnant when a blastula attaches to and then implants in the endometrium.

The lowest portion of the uterus is a narrowed region called the **cervix**, which opens into the vagina. The **vagina**,

Figure 38.9 Reproductive system of the human female.

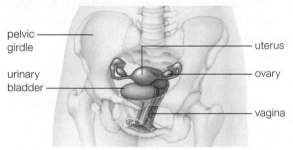

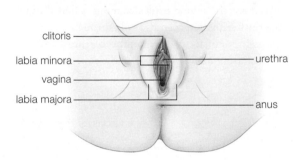

A Location of reproductive organs in pelvic cavity.

C External view of reproductive organs.

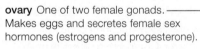

ovary One of two female gonads. Makes eggs and secretes female sex hormones (estrogens and progesterone).

oviduct One of a pair of ducts through which oocytes are propelled from an ovary to the uterus; usual site of fertilization.

uterus Womb, chamber in which an embryo develops. Includes myometrium (smooth muscle layer) and endometrium (epithelial lining). Narrowed lower portion (the cervix) secretes mucus into the vagina.

vagina Organ of sexual intercourse; birth canal.

clitoris Highly sensitive erectile organ. Only the tip is externally visible; bulk of the organ extends internally on either side of the vagina.

labium minus One of a pair of inner skin folds (the labia minora).

labium majus One of a pair of fatty outer skin folds (the labia majora).

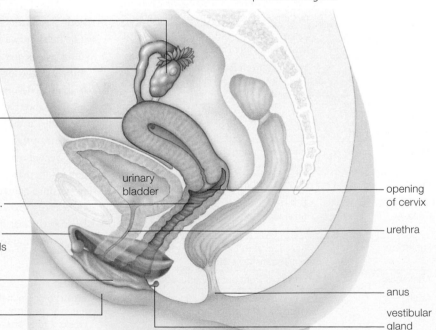

B Components of the system and their functions.

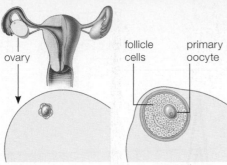 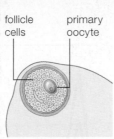 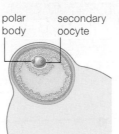 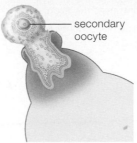 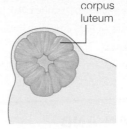 

ovary — follicle cells — primary oocyte — polar body — secondary oocyte — secondary oocyte — corpus luteum

❶ One of many immature follicles in an ovary. Each consists of a primary oocyte and the surrounding follicle cells.

❷ A fluid-filled cavity begins to form in the follicle's cell layer.

❸ The primary oocyte completes meiosis I and divides unequally, forming a secondary oocyte and a polar body.

❹ Ovulation. Rupture of the mature follicle releases a secondary oocyte coated with secreted protein and follicle cells.

❺ A corpus luteum develops from follicle cells left behind after ovulation.

❻ If pregnancy does not occur, the corpus luteum degenerates.

Figure 38.10 Events during one ovarian cycle.

which extends from the cervix to the body's surface, functions as the organ of intercourse and the birth canal. Typically, when a girl is born, a membrane called the hymen partially covers the external entrance to her vagina. It is usually stretched open during the first episode of intercourse, if it has not previously been broken by other physical activity.

Genitals are the externally visible parts of the reproductive tract (**Figure 38.9C**). Female genitals include two pairs of liplike skin folds that enclose the openings of the vagina and urethra. Adipose tissue fills the labia majora, the thick outer folds. The thin inner folds are the labia minora. An erectile organ called the clitoris lies near the anterior junction of the labia minora. The clitoris and penis develop from the same embryonic tissue and both are sensitive to tactile stimulation.

EGG PRODUCTION AND RELEASE

In humans, all production of oocytes from germ cells occurs before birth. A girl is born with about 2 million primary oocytes—oocytes that entered meiosis but stopped in prophase I (Section 12.3), rather than completing meiosis. At puberty, hormonal changes prompt oocytes to mature, one at a time, in an approximately twenty-eight-day ovarian cycle.

Figure 38.10 shows one round of this cycle. A maturing oocyte and the cells around it constitute an **ovarian follicle** ❶. In the first part of the ovarian cycle, the primary oocyte enlarges and secretes a layer of proteins. Follicle cells surrounding the primary oocyte divide repeatedly and a fluid-filled cavity forms around the oocyte ❷.

Often, more than one follicle starts to develop, but usually only one becomes fully mature. In that follicle, the primary oocyte completes meiosis I and undergoes unequal cytoplasmic division. This division produces a large secondary oocyte and a tiny **polar body** ❸, which will eventually dis-

integrate. The secondary oocyte begins meiosis II, then halts in metaphase II. It will not complete meiosis and become a mature egg until fertilization. At that time, a second polar body will form. It too will disintegrate.

About two weeks after the follicle began to mature, its wall ruptures and **ovulation** occurs: The secondary oocyte, polar body, and some surrounding follicle cells are ejected into the adjacent oviduct ❹. The oocyte must meet up with sperm within 24 hours for fertilization to occur. Meanwhile, in the ovary, the cells of the ruptured follicle develop into a hormone-secreting **corpus luteum** ❺. If pregnancy does not occur, the corpus luteum will break down ❻, and another follicle will begin to mature.

Occasionally, two oocytes mature and are released at the same time. If each egg is later fertilized by a different sperm, the result is fraternal twins. **Fraternal twins** are genetically nonidentical; they may be the same sex or different sexes.

cervix Narrow part of uterus that connects to the vagina.
corpus luteum (CORE-pus LOO-tee-um) Hormone-secreting structure that forms from follicle cells left behind after ovulation.
estrogens Hormones secreted by ovaries; cause development of female sexual traits and maintain the reproductive tract.
fraternal twins Genetically nonidentical twins.
oocyte (OH-uh-cite) Immature egg.
ovarian follicle In animals, oocyte and surrounding follicle cells.
oviduct Duct between an ovary and the uterus.
ovulation Release of a secondary oocyte from an ovary.
polar body Tiny cell produced by unequal cytoplasmic division during egg production.
progesterone Hormone secreted by ovaries; prepares the uterus for pregnancy.
uterus Muscular chamber where offspring develop; womb.
vagina Female organ of intercourse and the birth canal.

OVARIAN AND MENSTRUAL CYCLES

The ovarian cycle described in the previous section is coordinated with cyclic changes in the uterus. The approximately monthly changes in the uterus are referred to as the **menstrual cycle**. Each menstrual cycle begins with **menstruation**, an interval during which bits of uterine lining and a small amount of blood from the uterus flow through the cervix and out of the vagina.

Hormones control the ovarian and menstrual cycles (**Table 38.3** and **Figure 38.11**). When the cycles begin, secretion of gonadotropin-releasing hormone (GnRH) by the hypothalamus is stimulating the pituitary to release FSH and LH ❶. As its name would suggest, **follicle-stimulating hormone** stimulates an ovarian follicle to begin maturing ❷. The interval of follicle maturation that precedes ovulation is called the follicular phase of the menstrual cycle. During follicle maturation, cells around the oocyte secrete estrogens ❸. Estrogens bind to cells of the endometrium and encourage them to begin mitosis. The resulting cell divisions thicken the endometrium ❹.

A small increase in the level of estrogens triggers a positive feedback loop by stimulating the hypothalamus to increase its secretion of GnRH. In response to the rise in GnRH, the pituitary releases more FSH, which in turn stimulates the follicle to grow and produce more estrogens. The high level of estrogens that result from this feedback loop triggers the anterior pituitary to release **luteinizing hormone** (LH), which in females has roles in ovulation and formation of the corpus luteum.

The anterior pituitary releases the LH in a surge around the midpoint in the cycle ❺. The primary oocyte responds to the surge in LH by completing meiosis I and undergoing cytoplasmic division to become a secondary oocyte. LH also causes the follicle to swell and burst. Thus, the mid-cycle surge of LH is the trigger for ovulation ❻.

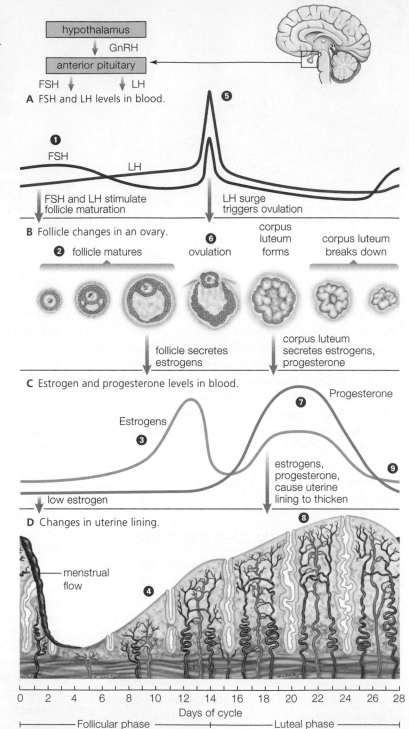

Figure 38.11 Reproductive cycle of a human female.
Day 1 of the cycle is the onset of menstruation.

FIGURE IT OUT

What is the source of the progesterone secreted after ovulation?

Answer: The corpus luteum

TABLE 38.3

Events of a Menstrual Cycle Lasting Twenty-Eight Days

Phase	Events	Day of Cycle
Pre-ovulation (follicular phase)	Menstruation; endometrium breaks down	1–5
	FSH stimulates follicle growth, estrogen level rises, endometrium is rebuilt	6–13
Ovulation	High level of estrogens causes LH surge that triggers completion of meiosis I in oocyte and rupture of the follicle	14
Post-ovulation (luteal phase)	Corpus luteum forms, secretes estrogens and progesterone; endometrium thickens; FSH and LH secretion are inhibited.	15–25
	Corpus luteum degenerates; estrogens and progesterone decline, allowing FSH and LH to begin to rise	26–28

The luteal phase of the menstrual cycle begins immediately after ovulation. At first the concentration of estrogens declines a bit, then LH stimulates corpus luteum formation. As the luteal phase continues, the corpus luteum secretes estrogens and a lot of progesterone ❼. Estrogens and progesterone cause the uterine lining to thicken and encourage blood vessels to grow through it. The uterus is now ready for pregnancy ❽.

Secretion of progesterone by the corpus luteum has a negative feedback effect on the hypothalamus and pituitary. The high level of progesterone causes a reduction in the secretion of FSH and LH, thus preventing maturation of other follicles and inhibiting additional ovulations.

If fertilization does not occur, the level of LH continues to decline, allowing the corpus luteum to degenerate. Breakdown of the corpus luteum causes estrogen and progesterone levels to plummet ❾. In the uterus, the decline in estrogens and progesterone results in degeneration of the uterine lining, and menstruation begins. Blood and endometrial tissue flow out of the vagina for three to six days. At the same time, the pituitary increases secretion of FSH and LH once again.

FROM PUBERTY TO MENOPAUSE

When a woman enters puberty, increased estrogen secretion by her ovaries results in the development of female secondary sexual traits such as breasts and pubic hair. She also begins to menstruate. Many women regularly experience discomfort a week or two before they menstruate. Estrogens and progesterone released during the cycle cause milk ducts to widen, making breasts feel tender. Other tissues swell too, because aldosterone secretion increases. This hormone stimulates reabsorption of sodium and, indirectly, water (Section 37.3). Cycle-associated hormonal changes can also cause depression, irritability, anxiety, and headaches, and can disrupt sleep. Regular recurrence of these symptoms is known as premenstrual syndrome (PMS). Use of oral contraceptives minimizes hormone swings and therefore PMS.

As the uterine lining breaks down, it releases local signaling molecules (prostaglandins) that stimulate contraction of smooth muscle in the uterine wall. Many women do not feel the muscle contractions, but others experience a dull ache or sharp pains known as menstrual cramps. Severe pain and heavy bleeding during menstruation are not normal and may be caused by benign tumors in the uterus called fibroids.

A woman enters **menopause** when all the follicles in her ovaries have either been released during menstrual cycles or have disintegrated as a result of normal aging. With no follicles left to mature, production of estrogen and progesterone is dramatically diminished and menstrual cycles cease.

Menopause is rare among animals. Female fertility declines during old age in many species. However, humans and killer whales are the only animals in which the females have a long post-reproductive interval in their life cycle.

ESTROUS CYCLES

All female placental mammals replace their uterine lining on a cyclic basis. However, most have an estrous cycle, rather than a menstrual cycle. In an **estrous cycle**, the endometrial lining is reabsorbed rather than shed, so the female never menstruates. Females with an estrous cycle are sexually receptive only when they can conceive, a period known as estrus or heat. The length of estrous cycles varies among species. For example, in dogs it is about six months; in mice it is six days.

In many species, visual and chemical signals alert males when a female enters estrus. Often a female's labia swell and she produces a discharge containing pheromones that help attract potential mates. For example, male dogs are attracted by the odor of the thin, bloody discharge released from the vagina of a female dog who is in heat.

estrous cycle (ESS-truss) Reproductive cycle in which the uterine lining thickens, and, if pregnancy does not occur, is reabsorbed.
follicle-stimulating hormone (FSH) Anterior pituitary hormone with roles in ovarian follicle maturation and sperm production.
luteinizing hormone (LH) Anterior pituitary hormone with roles in ovulation, corpus luteum formation, and sperm production.
menopause Permanent cessation of menstrual cycles.
menstrual cycle Reproductive cycle in which the uterus lining thickens and then, if pregnancy does not occur, is shed.
menstruation Flow of shed uterine tissue out of the vagina.

FUNCTIONS OF THE TESTES

Human testes, like ovaries, initially form in the pelvic cavity. However, before a male is born, his testes descend into the scrotum, a pouch of skin and smooth muscle suspended below the bones of the pelvic girdle (**Figure 38.12**). When a male is cold or frightened, reflexive muscle contractions draw the testes closer to his body. When he feels warm, relaxation of smooth muscle in the scrotum allows his testes to hang lower, so sperm-making cells do not overheat. These cells function best just below normal body temperature.

At puberty, a man's testes enlarge and begin producing a large amount of **testosterone**, the main sex hormone in males. The increase in testosterone stimulates development of secondary sex characteristics such as facial hair. It also triggers formation of sperm inside the **seminiferous tubules** that make up the bulk of the testes.

Diploid male germ cells line the inner wall of each seminiferous tubule (**Figure 38.13 ❶**). These cells divide by mitosis to produce primary spermatocytes ❷. The spermatocytes undergo meiosis to produce round, haploid cells called spermatids, which then differentiate into specialized cells with a long flagellum—sperm ❸. Also inside the seminiferous tubule are large, elongated cells called nurse cells. These cells support the developing sperm. The testes' testosterone-secreting cells (Leydig cells) reside between the seminiferous tubules.

FSH and LH regulate sperm formation. LH stimulates Leydig cells to secrete testosterone. FSH targets nurse cells and, in combination with testosterone, encourages them to produce chemicals essential to sperm development.

DUCTS AND ACCESSORY GLANDS

After sperm form, cilia push them into a coiled duct called the **epididymis** (plural, epididymides), where they mature. The final portion of each epididymis is continuous with a **vas deferens** (plural, vasa deferentia). The vasa deferentia, the longest ducts of the male reproductive tract, convey sperm to a short ejaculatory duct. This duct connects to the urethra, which extends through the penis to the body surface.

The **penis**, the male organ of intercourse, has a rounded head (the glans) at the end of a narrower shaft. Nerve endings in the glans make it highly sensitive to touch. The foreskin, a retractable tube of skin, covers the glans when a man is not sexually excited. In some cultures, males undergo circumci-

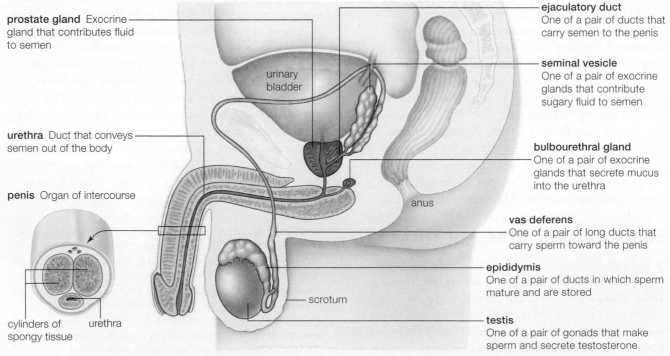

prostate gland Exocrine gland that contributes fluid to semen

urinary bladder

urethra Duct that conveys semen out of the body

penis Organ of intercourse

cylinders of spongy tissue   urethra

ejaculatory duct One of a pair of ducts that carry semen to the penis

seminal vesicle One of a pair of exocrine glands that contribute sugary fluid to semen

bulbourethral gland One of a pair of exocrine glands that secrete mucus into the urethra

anus

vas deferens One of a pair of long ducts that carry sperm toward the penis

epididymis One of a pair of ducts in which sperm mature and are stored

scrotum

testis One of a pair of gonads that make sperm and secrete testosterone

Figure 38.12 Components of the human male reproductive system and their reproductive functions.

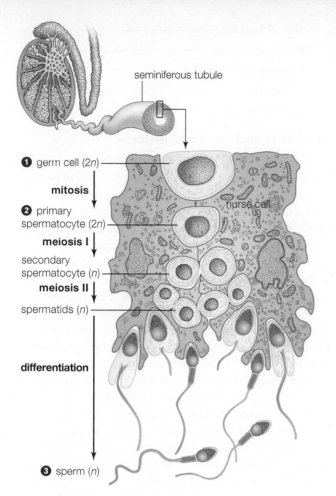

seminiferous tubule

❶ germ cell (2n)

mitosis

❷ primary spermatocyte (2n)

meiosis I

secondary spermatocyte (n)

meiosis II

spermatids (n)

nurse cell

differentiation

❸ sperm (n)

Figure 38.13 Sperm formation in the testes.

sion, which is surgical removal of the foreskin. Inside the penis, connective tissue encloses three elongated cylinders of spongy tissue. When a male is sexually excited, blood flows into the spongy tissue faster than it flows out, so fluid pressure rises inside the penis and it stiffens.

Sperm travel to the body surface only when a male reaches the peak of sexual excitement. During ejaculation, smooth muscle contractions propel sperm and exocrine gland secretions out of the body as a thick, white fluid called **semen**. Sperm make up less than 5 percent of the semen's volume. The main component of semen is fluid secreted by exocrine glands called **seminal vesicles**. Secretions from the **prostate gland**, a walnut-sized exocrine gland that encircles the urethra, are the other major component of semen volume. Prostate secretions help raise the pH of the female reproductive tract, making it more hospitable to sperm. Other exocrine glands, called bulbourethral glands, secrete an alkaline mucus that neutralizes any residue of acidic urine in the urethra.

Dr. Stewart C. Nicol

National Geographic Explorer Stewart Nicol studies spiny, ant-eating monotremes called echidnas. Monotreme means "one opening," and refers to the cloaca of these egg-laying mammals. A male monotreme expels urine through his cloaca, so his penis functions solely in reproduction. He extends it out through his cloaca only when ready to mate.

In a short-beaked echidna (left), the male's penis is large for his size and forked, with two rosette-shaped openings at the tip of each fork. Like reptiles, most monotremes have a forked penis with two openings, but the short-beaked echidna's four-barreled penis is unique.

Nicol's research has revealed that both sexes of short-beaked echidnas have multiple mates. This may explain why short-beaked echidna males have very large testes relative to their body size. During the mating season, a male's echidna's testes can account for 1 percent of his body weight. Bigger testes can make more sperm, and the more sperm a male can put into a female, the greater the chance that one of his sperm will fertilize her egg.

Nicol has found evidence of sexual competition among males. In Tasmania, some impatient males mate with females who are still hibernating. Echidnas are one of the few mammals that can sustain a pregnancy during hibernation, and by mating early with the fittest males and then re-entering hibernation, they can delay egg-laying until conditions are more suitable for egg-laying.

epididymis (epp-ih-DID-ih-muss) Duct where sperm mature; empties into a vas deferens.

penis Male organ of intercourse.

prostate gland (PRAH-state) Exocrine gland that contributes to semen.

semen (SEE-men) Sperm and secretions from exocrine glands.

seminal vesicles (SEHM-in-al) Exocrine glands that add sugary fluid to semen.

seminiferous tubules (sehm-in-IF-er-us) In testes, tiny tubes where sperm form.

testosterone Hormone secreted by testes; functions in sperm formation and development of male secondary sex characteristics.

vas deferens (vas DEF-er-ens) One of a pair of long ducts that carry mature sperm to the ejaculatory duct.

LEARNING OBJECTIVES
Describe the mechanism by which the penis becomes erect.
Describe the structure of a sperm.
Explain how a sperm reaches an oviduct.
Describe what occurs during fertilization.

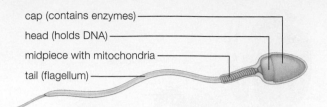

cap (contains enzymes)
head (holds DNA)
midpiece with mitochondria
tail (flagellum)

Figure 38.14 Structure of a human sperm.

SEXUAL INTERCOURSE

For males, intercourse requires an erection. The penis consists mostly of long cylinders of spongy tissue. When a male is not sexually aroused, his penis is limp because arteries that transport blood to its spongy tissue remain constricted. With arousal, increased activity of sympathetic nerves (Section 29.6) causes these arteries to dilate. The resulting inflow of blood expands the spongy tissue and compresses veins that carry blood out of the penis. Inward blood flow now exceeds outward flow, causing an increase in internal fluid pressure. This pressure enlarges and stiffens the penis so it can be inserted into a female's vagina.

The ability to obtain and sustain an erection peaks during the late teens. As a male ages, he may have episodes of erectile dysfunction. With this disorder, the penis does not stiffen enough for intercourse. Drugs prescribed for erectile dysfunction act by relaxing smooth muscle in the walls of arterioles supplying the penis.

When a woman becomes sexually excited, blood flow to the vaginal wall, labia, and clitoris increases. Glands in the cervix secrete mucus, and glands on the labia (the equivalent of a male's bulbourethral glands) produce a lubricating fluid. The vagina itself does not have any glandular tissue. It is moistened by mucus from the cervix and by plasma fluid that seeps out between epithelial cells of the vaginal lining.

During intercourse, increased signals from sympathetic nerves raise the heart rate and breathing rate in both partners. The posterior pituitary steps up its oxytocin secretion. The oxytocin acts in the brain, inhibiting signals from the amygdala, the brain region that controls fear (Section 29.11).

Continued mechanical stimulation of the penis or clitoris can lead to orgasm. During orgasm, endorphins flood the brain and evoke feelings of pleasure. At the same time, release of oxytocin causes rhythmic contractions of smooth muscle in both the male and female reproductive tract. In males, orgasm is usually accompanied by ejaculation, in which contracting muscles force the semen out of the penis.

THE SPERM'S JOURNEY

An ejaculation can put 300 million sperm into the vagina. **Figure 38.14** shows the structure of a mature sperm. It is a haploid cell with a "head" packed full of DNA and tipped by an enzyme-containing cap. The enzymes will help a sperm penetrate an oocyte by breaking down the layer of proteins that encases the oocyte. At its other end, the sperm has a flagellum that allows it to swim toward an egg. The sperm does not have ribosomes, endoplasmic reticulum, or Golgi bodies. However, its midsection contains many mitochondria that supply the ATP required for flagellar movement. While in a male's body, sperm produce ATP by breaking down the fructose in seminal vesicle secretions. Once sperm enter the female reproductive tract, they take up carbohydrates from their environment to fuel their movement.

To travel from the vagina into the uterus, sperm must swim through a canal in the center of the cervix. During parts of the reproductive cycle when a woman's estrogen level is low, a thick plug of acidic mucus bars the passage of sperm through this canal. As the estrogen level rises before ovulation, the cervix becomes more sperm-friendly. It begins secreting a thinner, more alkaline mucus containing glucose that the sperm can use as fuel. Even so, passing through cervical mucus is challenging, so only the strongest sperm pass through the cervix into the chamber of the uterus.

Once sperm are in the uterus, contractions of smooth muscle in the uterus wall help move them upwards, toward the oviducts. Sperm enter both oviducts, but because ovulation occurs in only one ovary at a time, about half of them will continue onward without any possibility of ever encountering an egg. An egg does produce chemicals that attract sperm, but this attraction is effective only at close range.

FERTILIZATION

Fertilization usually happens in the upper part of an oviduct (**Figure 38.15 ❶**). Of the millions of sperm ejaculated into the vagina, only a few hundred make it this far. Sperm live for three days after ejaculation, so fertilization can occur even if intercourse takes place a few days before ovulation.

A secondary oocyte released at ovulation retains a wrapping of follicle cells. Beneath those cells is a jelly coat composed of secreted glycoproteins ❷. The sperm makes its way between the follicle cells to the jelly coat. The plasma

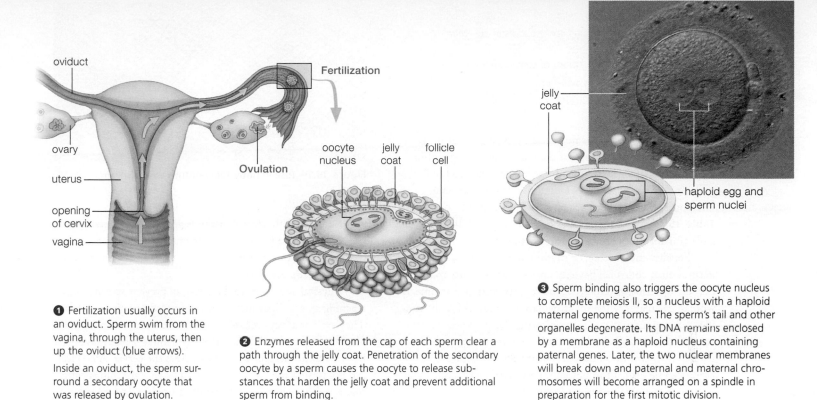

oviduct

ovary

uterus

opening
of cervix

vagina

Fertilization

Ovulation

oocyte
nucleus

jelly
coat

follicle
cell

jelly
coat

haploid egg and
sperm nuclei

❶ Fertilization usually occurs in an oviduct. Sperm swim from the vagina, through the uterus, then up the oviduct (blue arrows).

Inside an oviduct, the sperm surround a secondary oocyte that was released by ovulation.

❷ Enzymes released from the cap of each sperm clear a path through the jelly coat. Penetration of the secondary oocyte by a sperm causes the oocyte to release substances that harden the jelly coat and prevent additional sperm from binding.

❸ Sperm binding also triggers the oocyte nucleus to complete meiosis II, so a nucleus with a haploid maternal genome forms. The sperm's tail and other organelles degenerate. Its DNA remains enclosed by a membrane as a haploid nucleus containing paternal genes. Later, the two nuclear membranes will break down and paternal and maternal chromosomes will become arranged on a spindle in preparation for the first mitotic division.

Figure 38.15 Human fertilization. The photo at the upper right is a light micrograph taken one day after fertilization.

membrane of the sperm's head has receptors that can bind species-specific proteins in the jelly. Binding these proteins triggers the release of protein-digesting enzymes from the cap on the sperm's head. The collective effect of enzyme release from many sperm clears a passage through the jelly coat to the oocyte's plasma membrane. Receptors in this membrane bind a sperm's plasma membrane and the two membranes fuse. The sperm is then drawn into the oocyte. Usually only one sperm enters the secondary oocyte. Its entry causes the egg to release substances that change the consistency of the jelly coat, making it difficult for other sperm to bind.

It only takes one sperm to fertilize an egg, but many must bind to release enough enzyme to clear a way to the egg's plasma membrane. This is why a man who has healthy sperm but a low sperm count can be functionally infertile.

Remember that the secondary oocyte released at ovulation was halted in the middle of meiosis II. Binding of the sperm membrane to the oocyte membrane triggers the oocyte to complete meiosis. The unequal cytoplasmic division that follows produces a single mature egg—an **ovum** (plural, ova)—and the second polar body.

After a sperm penetrates an oocyte, its tail breaks down, leaving the nucleus ❸. Chromosomes in the haploid egg and

sperm nuclei become the genetic material of the new zygote. The sperm also supplies a single centriole (Section 4.9). The sperm's centriole replicates to form the pair of centrioles that organize the spindle for the zygote's first mitotic division. Sperm mitochondria enter an oocyte too, but they are typically broken down. Thus, mitochondria are inherited only from the mother.

An egg or a sperm is a fully differentiated cell, so it expresses only those genes required for its function as a gamete. However, a zygote must be genetically flexible enough to give rise to all types of body cells. Thus, fertilization begins a process of dedifferentiation. Producing an undifferentiated diploid genome from two haploid gamete genomes requires reversing the changes that previously shut down specific genes in each gamete. This dedifferentiation process involves removing methyl groups from DNA and reversing modifications to histone proteins. Most of these epigenetic modifications are reset, but some are not. This is why some environmentally induced changes in gene expression can be passed from one generation to the next (Section 10.5).

ovum Mature animal egg.

38.10 Birth Control

LEARNING OBJECTIVES

List some types of information a woman can use to estimate when she is ovulating.

Describe some surgical and barrier methods of contraception.

Explain which contraceptives prevent ovulation.

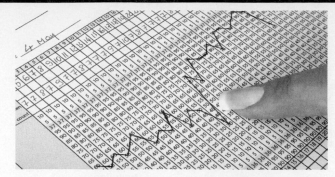

Figure 38.16 Rise in body temperature at ovulation.

A fertile couple who takes no precautions to prevent pregnancy has about an 85 percent chance of conceiving within a year. A variety of methods can lower the likelihood of pregnancy. The most effective option is abstinence—no intercourse. It is 100 percent effective, but difficult to sustain. **Table 38.4** lists common types of contraception (birth control) available to people who are sexually active.

Fertility-monitoring, or rhythm, methods are a type of natural birth control. They rely on a woman's ability to detect her fertile period and to avoid intercourse during that time. A woman calculates when she is ovulating by recording the timing of her menstrual cycles, monitoring the thickness of her cervical mucus, and checking her body temperature, which usually rises around ovulation (**Figure 38.16**). However, ovulation can be difficult to pinpoint even with meticulous monitoring, so miscalculations are common.

Withdrawal, the removal of the penis from the vagina before ejaculation, is often ineffective because sperm can leave the urethra prior to ejaculation. Similarly, rinsing the vagina (douching) after intercourse is unreliable because sperm can swim through the cervical canal within seconds of ejaculation.

Surgical birth control methods are highly effective, but they are meant to result in permanent sterility. A vasectomy blocks or cuts a man's vas deferens. A tubal ligation blocks or cuts a woman's oviducts.

Physical and chemical barriers can prevent sperm from reaching an egg. Spermicidal foams and jellies poison sperm. They are not always reliable, but their use with a condom or diaphragm reduces the chance of pregnancy. A diaphragm is a flexible, dome-shaped device that covers the cervix. A cervical cap is a similar but smaller device. Condoms are thin sheaths that are either worn over the penis or used to line the vagina during intercourse.

An intrauterine device, or IUD, must be inserted into the uterus by a physician. Some IUDs thicken cervical mucus so much that sperm cannot swim through it. Others release copper ions that prevent an early embryo from implanting.

The birth control pill is the most commonly used fertility control method in developed countries. "The Pill" is a combination of synthetic estrogens and progesterone-like hormones that prevents both maturation of oocytes and ovulation. The birth control patch delivers the same mixture of hormones and it blocks ovulation the same way, as do hormone injections or hormone-releasing implants.

Emergency contraception can prevent pregnancy if used within days of unprotected intercourse. The most commonly used type of emergency contraceptive pill delivers a dose of synthetic progesterone that prevents ovulation and interferes with fertilization up to three days after intercourse. An alternative pill blocks progesterone receptors. Placement of a copper IUD by a physician can also be used to prevent embryonic implantation as late as 5 days after intercourse.

The drug mifepristone (RU-486) in combination with a synthetic prostaglandin can be taken orally to terminate a pregnancy up to 9 weeks after a woman's last menstrual period. The drugs must be prescribed by a physician. Mifepristone interferes with progesterone's ability to sustain the pregnancy, and the synthetic prostaglandin encourages expulsion of the contents of the uterus through the vagina.

TABLE 38.4

Common Methods of Contraception

Method	Mechanism of Action	Pregnancy*
Abstinence	Avoid intercourse entirely	0% per year
Rhythm method	Avoid intercourse when female is fertile	25% per year
Withdrawal	End intercourse before male ejaculates	27% per year
Vasectomy	Cut or close off male's vasa deferentia	<1% per year
Tubal ligation	Cut or close off female's oviducts	<1% per year
Condom	Enclose penis, block sperm entry to vagina	15% per year
Diaphragm, cervical cap	Cover cervix, block sperm entry to uterus	16% per year
Spermicides	Kill sperm	29% per year
Intrauterine device	Prevent sperm entry to uterus or prevent implantation	<1% per year
Oral contraceptives	Prevent ovulation	<1% per year
Hormone patches, implants, or injections	Prevent ovulation	<1% per year
Emergency contraception pill	Prevent ovulation	15–25% per use**

* Percent of pregnancies with consistent, correct use
** Not meant for regular use

38.11 Sexually Transmitted Diseases

Give examples of protozoal, bacterial, and viral STDs.

Describe some negative health effects of untreated STDs.

Explain how to reduce the risk of acquiring an STD.

Sexually transmitted diseases (STDs) are contagious diseases that spread through sexual contact (**Table 38.5**). For all STDs, women are generally more easily infected than men, and have more complications. Women can also infect their offspring during childbirth, so sexually active women who intend to become pregnant should be tested for STDs.

To reduce the risk of STD transmission, physicians recommend use of a latex condom and a lubricant during sexual encounters. The condom serves as a barrier to the pathogen and the lubricant helps prevent small abrasions that would make it easier for the pathogen to enter the body.

Trichomoniasis (also called trich) is a common STD caused by the flagellated protozoan *Trichomonas vaginalis*. Infected women have a yellowish discharge and a sore, itchy vagina. Men usually have no symptoms, but untreated trichomoniasis may increase their risk of prostate enlargement and aggressive prostate cancer. In both sexes, untreated infection can cause infertility. A single dose of an antiprotozoal drug can quickly cure the infection.

Bacterial STDs cause a discharge from the penis and painful urination in most infected men. Women often have no initial symptoms, but they can develop pelvic inflammatory disease (PID). This secondary outcome of bacterial STDs scars the female reproductive tract and can cause infertility and chronic pain. PID also raises the risk of tubal pregnancies, in which an embryo begins to develop in the oviduct. A tubal pregnancy cannot be carried to term and must be terminated to protect the mother.

The most common bacterial STD is chlamydia, caused by *Chlamydia trachomatis*. Bacteria are also responsible for gonorrhea and syphilis. An untreated infection by the spiral-shaped bacteria (*Treponema pallidum*) that cause syphilis is especially dangerous. Such an infection produces skin ulcers (**Figure 38.17**) and can damage the liver, bones, and the brain. Fortunately, most bacterial STDs can be cured with antibiotics, although antibiotic-resistant strains of syphilis and gonorrhea are becoming increasingly common.

Infection by human papillomavirus (HPV) is the most common viral STD worldwide. Most HPV strains cause benign growths called warts and are cleared from the body within two years. However, a few strains cause cancer. Depending on where the virus is introduced, it can cause cancer of the throat, cervix, penis, or anus. Vaccinations and screenings are a key part of the fight against HPV-associated

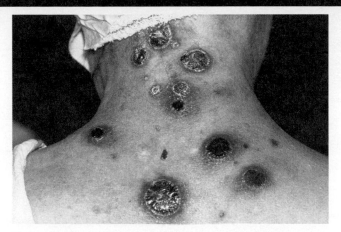

Figure 38.17 Skin ulcers caused by untreated syphilis.

TABLE 38.5

Prevalence and Causes of STDs in the United States

STD	Cases in U.S.	Pathogen
Trichomoniasis	>8,000,000	Protozoan
Chlamydia	1,800,000	Bacteria
Gonorrhea	350,000	Bacteria
Syphilis	63,000	Bacteria
HPV infection	79,100,000	Virus
HSV-2 infection	24,000,000	Virus
HIV infection	1,200,000	Virus

cancers. A vaccine can prevent HPV infection in both males and females, if given before viral exposure occurs. A blood test can determine whether a person has been infected by one of the two strains that most commonly cause cancer, and a Pap test can detect early signs of cervical cancer.

Genital herpes is caused by type 2 herpes simplex (HSV-2) virus. An initial infection commonly causes small sores at the site of infection. The sores heal, but the virus remains dormant and may be reactivated periodically. Tingling, itching, and more sores are symptoms of a reactivated infection. Antiviral drugs help sores heal and lessen the likelihood of viral reactivation.

Infection by HIV (or human immunodeficiency virus) can cause AIDS (acquired immunodeficiency syndrome). Sections 19.1 and 19.2 described the HIV virus and Section 34.9 explained how its effects on the immune system lead to AIDS. With regard to sexual transmission, oral sex is least likely to transmit an infection. Unprotected anal sex is 5 times more dangerous than unprotected vaginal sex and 50 times more dangerous than oral sex. If you think you have been exposed to HIV, get tested as soon as possible. Treatment with antiviral drugs can prevent an HIV-infected person from developing AIDS and reduce his or her likelihood of passing the infection to others.

Describe the formation and implantation of a human blastocyst.

Explain the function of the extraembryonic membranes.

Describe formation of the neural tube and explain its function.

Human development starts with cleavage. Cell divisions begin within a day of fertilization, as cilia propel the zygote away from the upper portion of the oviduct where fertilization occurred (**Figure 38.18**). The zygote divides by mitosis to form two cells, which become four, which become eight, and so on ❶. Typically, all cells adhere tightly to one another as divisions continue.

By 5 days after fertilization, the dividing cluster of cells has reached the uterus and formed a **blastocyst**, the mamma-

Figure 38.18 Early embryonic development. Graphics 3–5 show a cross-section through the uterus.

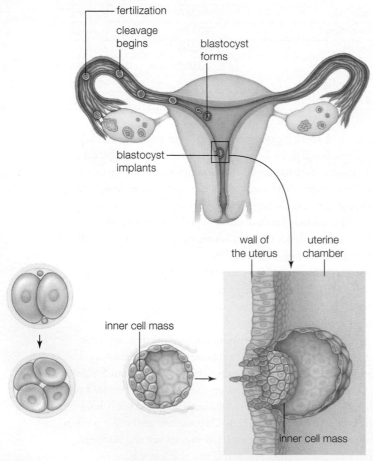

- fertilization
- cleavage begins
- blastocyst forms
- blastocyst implants
- wall of the uterus
- uterine chamber

lian version of a blastula ❷. The blastocyst consists of an outer layer of cells, a cavity (the blastocoel) filled with their fluid secretions, and an inner cell mass. Of all cells in the human blastocyst, only those in the inner cell mass will give rise to the developing individual. The other cells will give rise to protective and supportive extraembryonic membranes.

Up to this point, the developing individual has been nourished by nutrients it absorbed from maternal secretions in the oviduct and uterus. For development to continue, the blastocyst must implant itself in the wall of the uterus, where it can obtain nutrients from the maternal bloodstream. Implantation requires direct contact between the surface of the blastocyst and cells in the lining of the uterus. Thus, before a blastocyst can implant, it must shed the protein coat that surrounds it.

Once free of its protein coat, the blastocyst attaches to the lining of the uterus and implantation begins ❸. As the blastocyst burrows into the uterine wall, the inner cell mass develops into two flattened layers of cells that are collectively called an embryonic disk ❹. At the same time, the extraembryonic membranes typical of amniotes begin to form. A membrane called the **amnion** encloses a fluid-filled amniotic cavity between the embryonic disk and the blastocyst surface. Fluid in this cavity (amniotic fluid) acts as a buoyant cradle in which an embryo grows and moves freely, protected from temperature changes and mechanical impacts. As the amnion forms, other cells move around the inner wall of the blastocyst to form the lining of a yolk sac. In reptiles and birds, this sac holds yolk that nourishes the developing embryo. In humans, it has no nutritional role. Some cells that form in a human yolk sac give rise to the embryo's first blood cells.

As implantation continues, spaces in the uterine tissue around the blastocyst become filled with blood seeping

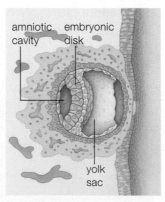

amniotic cavity · embryonic disk

yolk sac

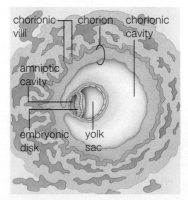

chorionic villi · chorion · chorionic cavity

amniotic cavity

embryonic disk · yolk sac

inner cell mass

inner cell mass

❶ Days 1–4
Cleavage divides the zygote into an increasing number of smaller cells.

❷ Days 5–7
The blastocyst forms, expands, then sheds its protein coat.

❸ Days 8–9
Implantation begins. The blastocyst attaches to the wall of the uterus and begins to embed itself in it.

❹ Days 10–11
The inner cell mass develops into a two-layered embryonic disk. Extraembryonic membranes form.

❺ Day 14
Chorionic villi (fingerlike extensions of the chorion) begin to grow into blood-filled spaces in the lining of the uterus.

in from ruptured maternal capillaries. The extraembryonic membrane called the **chorion** sprouts many tiny fingerlike projections (chorionic villi) that extend into blood-filled maternal tissues ❺. The chorion will become part of the placenta. An outpouching of the yolk sac will become the fourth extraembryonic membrane, the **allantois**. In humans, the allantois forms blood vessels of the umbilical cord that connects the fetus to the placenta.

Implantation of a blastocyst in the uterus prevents menstruation because the chorion secretes human chorionic gonadotropin (HCG). This is the first hormone that a new human produces, and it prevents degeneration of the corpus luteum. By the beginning of the third week, HCG can be detected in a mother's blood or urine. At-home pregnancy tests have a "dipstick" with a region that changes color when exposed to urine containing HCG (**Figure 38.19**).

Gastrulation occurs on about day 15. Cells migrate inward along the primitive streak, a groove on the disk's surface ❻. Shortly thereafter, the typical chordate features begin to emerge. The embryonic disk develops two folds that will merge to form a neural tube, the precursor of the spinal cord and brain ❼. Beneath the neural tube, mesoderm folds into another tube that develops into the notochord. The human notochord acts as a structural model for the backbone.

---

allantois (al-LAN-toh-is) Extraembryonic membrane that, in mammals, becomes part of the umbilical cord.

amnion (AM-nee-on) Extraembryonic membrane that encloses an amniote embryo and the amniotic fluid.

blastocyst Mammalian blastula.

chorion (KOR-ee-on) Outermost extraembryonic membrane of amniotes; major component of the placenta in placental mammals.

identical twins Genetically identical individuals that develop after an embryo splits in two.

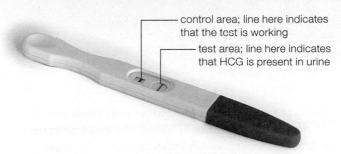

control area; line here indicates that the test is working

test area; line here indicates that HCG is present in urine

Figure 38.19 One type of urine test for pregnancy. A dipstick containing an absorbent material is dipped in a woman's urine. If the hormone HCG is present in the urine, it reacts with antibodies and dye in the stick, and a colored line appears in the test area. If HCG is not present, a colored line appears only in the control area.

Somites form by the end of the third week ❽. These paired blocks of mesoderm on either side of the notochord will develop into bones and skeletal muscles of the head and trunk, and the dermis of the skin that overlies them. Pharyngeal arches develop that will form structures in the head and neck ❾. In fishes, pharyngeal arches develop into gills, but a human embryo never has gills. Instead, it receives oxygen and other essential material by way of the placenta.

Sometimes a pre-implantation embryo splits in two, giving rise to **identical twins**, two individuals that have the same genome. If the split occurs at the two-cell stage, each twin will implant separately and will have its own placenta. If the split occurs just before implantation, the twins will share a placenta. By contrast, fraternal twins always have their own individual amnion sacs and placentas. Conjoined twins arise when an embryo does not split completely, so the two identical individuals that develop share some skin and organs.

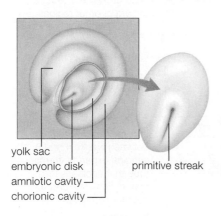

yolk sac
embryonic disk
amniotic cavity
chorionic cavity
primitive streak

❻ **Day 15** Gastrulation begins at a depression called the primitive streak.

paired neural folds

neural groove (below, notochord is forming)

❼ **Day 16** Neural tube begins to form.

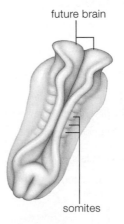

future brain

somites

❽ **Days 18–23** Somites develop.

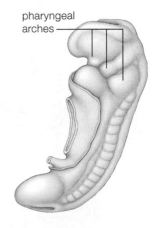

pharyngeal arches

❾ **Days 24–25** Pharyngeal arches appear.

LEARNING OBJECTIVES

Describe the size and appearance of a human at the beginning of the fetal period.

Describe the developmental events of the second and third trimester.

When the fourth week of development ends, the embryo is 500 times the size of a zygote, but still less than 1 centimeter long. Like other chordate embryos, it has a tail (**Figure**

38.20). Growth slows as details of organs begin to fill in. Limbs form, and paddles are sculpted into fingers and toes. Growth of the head surpasses that of all other regions. At the end of the eighth week, the heart has begun beating, eyelids and external ears have formed, and apoptosis has eliminated most of the tail. At this point, we define the individual as a human fetus.

In the second trimester (months 3–6), obstetric sonography (ultrasound) can reveal the sex of a fetus. By four

Figure 38.20 Human organ development and growth.

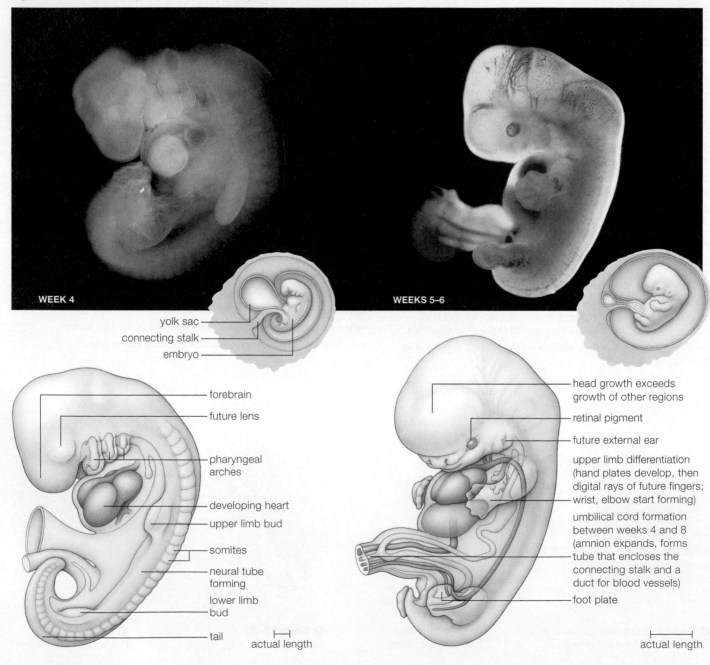

WEEK 4

yolk sac
connecting stalk
embryo

forebrain

future lens

pharyngeal arches

developing heart

upper limb bud

somites

neural tube forming

lower limb bud

tail

actual length

WEEKS 5–6

head growth exceeds growth of other regions

retinal pigment

future external ear

upper limb differentiation (hand plates develop, then digital rays of future fingers; wrist, elbow start forming)

umbilical cord formation between weeks 4 and 8 (amnion expands, forms tube that encloses the connecting stalk and a duct for blood vessels)

foot plate

actual length

months, most neurons have formed. Reflexive movements begin as developing nerves and muscles connect. Legs kick, arms wave about, and fingers grasp. The fetus frowns, squints, puckers its lips, sucks, and hiccups. Soft fetal hair (lanugo) and a thick, cheesy coating (vernix) cover the skin. Most of the hair will be shed before birth.

Beginning at about five months, the fetal heartbeat can be heard clearly through a stethoscope positioned on the mother's abdomen.

By the seventh month, the fetus has begun not only to detect sounds, but also to remember some of them. Exposure to human singing and speech during this period may enhance language acquisition after birth.

In this final trimester, the fetus opens its eyes, loses most of its fetal hair, and puts on a layer of fat. In preparation for birth, it practices muscle movements involved in breathing, inhaling amniotic fluid in place of air. Its lungs will later inflate with its first breath after birth.

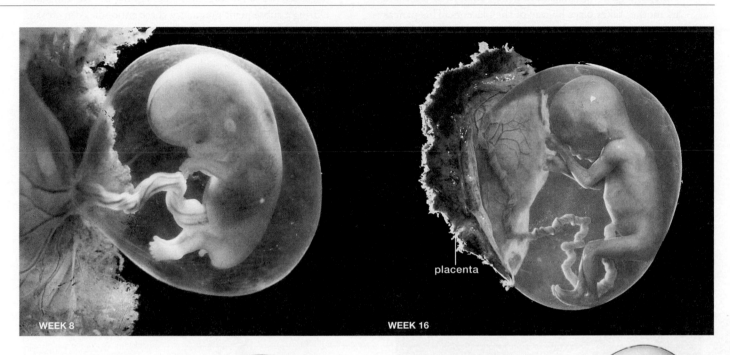

WEEK 8

WEEK 16

placenta

final week of embryonic period; embryo looks distinctly human compared to other vertebrate embryos

upper and lower limbs well formed; fingers and then toes have separated

primordial tissues of all internal, external structures now developed

tail has become stubby

actual length

| Length: | 16 centimeters (6.4 inches) |
| Weight: | 200 grams (7 ounces) |

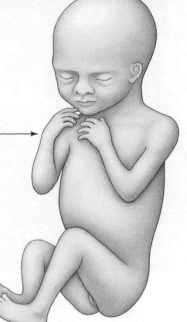

WEEK 38 (full term)

| Length: | 50 centimeters (20 inches) |
| Weight: | 3,400 grams (7.5 pounds) |

During fetal period, length measurement extends from crown to heel (for embryos, it is the longest measurable dimension, as from crown to rump).

Describe the structure of the placenta.

Explain how materials are exchanged across the placenta.

Using appropriate examples, explain how maternal behavior can increase the risk of miscarriage or birth defects.

FUNCTION OF THE PLACENTA

All exchange of materials between an embryo or fetus and its mother takes place by way of the placenta. This pancake-shaped organ consists of uterine lining, extraembryonic membranes, and embryonic blood vessels (**Figure 38.21**). At full term, the placenta covers about a quarter of the inner surface of the uterus.

A coiled umbilical cord connects the embryo to the placenta. Embryonic blood vessels extend through this cord to the placenta, and into the chorionic villi (fingerlike projections of the chorion). These villi are surrounded by pools of maternal blood.

Maternal and embryonic bloodstreams never mix. Instead, substances move between maternal and embryonic blood by diffusing across the walls of the embryonic vessels in the chorionic villi. Oxygen diffuses from maternal blood into embryonic blood, and carbon dioxide diffuses in the opposite direction. Transport proteins assist in the movement of essential nutrients from the maternal blood into embryonic blood vessels inside the villi.

One type of maternal antibody (IgG) is also transferred from maternal blood to the fetal bloodstream. This transfer can cause problems if the mother and her child differ in their Rh blood type. An Rh-negative woman lacks the Rh antigen on her red blood cells. When she gives birth to her first Rh-positive child, exposure to the child's blood during childbirth can cause her to develop antibodies against the Rh antigen. During a subsequent pregnancy, these antibodies can cross the placenta and cause the mother's immune system to attack an Rh-positive fetus.

In addition to providing essential substances to the embryo, the placenta produces hormones that sustain the pregnancy. From the third month of pregnancy on, it secretes the HCG, progesterone, and estrogens that encourage the ongoing maintenance of the uterine lining.

EFFECTS OF MATERNAL HEALTH AND BEHAVIOR

Development of a human infant from a single-celled zygote is a complicated process, so a lot can go wrong along the way. When it does, a pregnancy can end in miscarriage or

Figure 38.21 **Life support system of a developing human.**

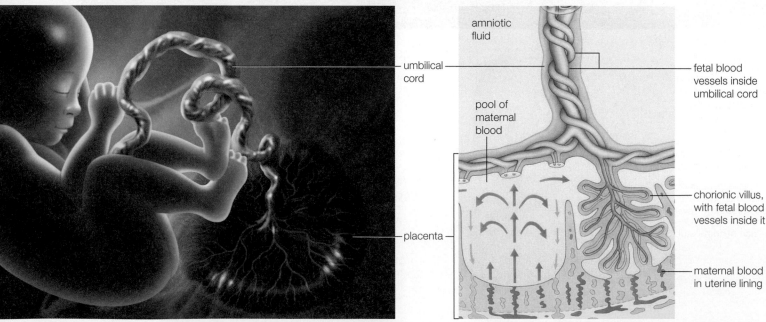

amniotic fluid

umbilical cord

pool of maternal blood

placenta

fetal blood vessels inside umbilical cord

chorionic villus, with fetal blood vessels inside it

maternal blood in uterine lining

Artist's depiction of the view inside the uterus, showing a fetus connected by an umbilical cord to the pancake-shaped placenta.

The placenta consists of maternal and fetal tissue. Fetal blood flowing in vessels of chorionic villi exchanges substances by diffusion with maternal blood around the villi. The bloodstreams do not mix.

stillbirth, or a child can be born with a birth defect. With miscarriage, death occurs before 20 weeks. Miscarriages that occur in the first month or two of development often go undetected, so their frequency is not known. The risk of miscarriage for recognized pregnancies is 12 to 15 percent. Most of these occur early in pregnancy. Death of a fetus after 20 weeks is less common, and is called a stillbirth. Some birth defects have a genetic basis, whereas others are caused by poor maternal nutrition or other environmental factors.

Maternal Nutrition A pregnant woman's health and behavior affect development of the child she carries. Women who are underweight or highly stressed are more likely to miscarry than those who are a bit heavier and happier. However, excessive weight is also risky; obese women have more stillbirths. Deficiencies in specific nutrients can disrupt prenatal development. If a mother does not eat an adequate amount of iodine, her newborn may have cretinism, a disorder that affects brain function and motor skills (Section 31.4). A maternal folate (folic acid) deficiency puts the child at risk for neural tube defects. A deficiency in zinc or folate increases the risk of a cleft lip or palate (**Figure 38.22**).

Teratogens Environmental factors that disrupt development and cause birth defects are called **teratogens**. X-ray radiation is a teratogen, as are some pathogens that cross the placenta. Maternal infection by the rubella virus during the first eight weeks of pregnancy nearly always causes severe birth defects or miscarriage. Maternal infection by the Zika virus can result in microcephaly (an abnormally small brain and head) and other neurological defects. A recent rise in the number of microcephaly cases in South America has been attributed to the introduction of Zika virus to this region.

Maternal toxoplasmosis can lead to developmental problems, a miscarriage, or stillbirth. Toxoplasmosis is caused by an apicomplexan protist (Section 20.4). The protists can be found in garden soil, cat feces, and undercooked meat. To avoid contracting toxoplasmosis, pregnant women should eat well-cooked meat and avoid cats that have been outdoors.

Alcohol crosses the placenta; its use during pregnancy is the leading preventable cause of inborn mental impairment. Lacking a fully developed liver, a developing child cannot detoxify alcohol (or other toxins) as effectively as an adult. The most dramatic effects of prenatal alcohol exposure are seen in children with fetal alcohol syndrome (FAS). Affected individuals have a small head and facial abnormalities, heart malfunctions, skeletal abnormalities, mental impairment, behavioral problems, and poor coordination. Likelihood of FAS increases with the frequency of maternal drinking and with the amount of alcohol consumed at one time. Most

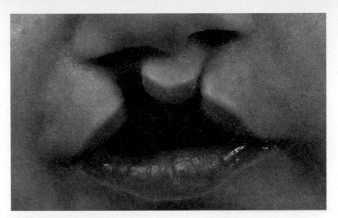

Figure 38.22 Cleft lip in an infant. This defect arises when tissues fail to fuse properly during early development. To learn more about the causes and prevention of birth defects, visit the National Birth Defects Prevention Study website: www.nbdps.org.

doctors advise women who are pregnant or attempting to become pregnant to avoid alcohol entirely.

Some commonly used medications cause birth defects. Isotretinoin (Accutane), a highly effective treatment for severe acne, is often prescribed for young women. If taken early in a pregnancy, it can cause heart problems or facial and cranial deformities in the embryo. Some antidepressants also increase the risk of birth defects. Paroxetine (Paxil) and related drugs inhibit the reuptake of serotonin. Use of these drugs during early pregnancy increases the risk of fetal heart malformations. Taking them later in pregnancy increases the fetal risk of fatal heart and lung disorders.

Environmental pollutants are another risk. Prenatal exposure to methyl mercury impairs brain development, which is why pregnant women should avoid eating large predatory fish such as albacore tuna. These fish often contain high levels of methylmercury. Prenatal exposure to endocrine-disrupting phthalates increases the risk of reproductive abnormalities in male offspring (Chapter 31 Application).

Investigating Birth Defects Scientists have identified many risk factors for birth defects, but most such defects arise from unknown causes. The federally funded National Birth Defects Prevention Study (NBDPS) is designed to further illuminate causes of birth defects. Since 1997, participating researchers have been interviewing and collecting genetic material from women who gave birth to a child with a birth defect. Women who gave birth to children without birth defects comprise a control group. So far, over 35,000 women have participated, and the study is ongoing.

teratogen (teh-RAT-uh-jenn) Environmental factor that disrupts normal development and causes birth defects.

38.15 Birth and Lactation

LEARNING OBJECTIVES

Describe normal childbirth and compare it with a surgical delivery.

List the components of milk and describe their functions.

CHILDBIRTH

A mother's body changes as her fetus nears full term, at about 38 weeks after fertilization. Until the last few weeks, her firm cervix helped prevent the fetus from slipping out of her uterus prematurely. Now, connective tissue in the cervix becomes thinner, softer, and more flexible so it can stretch during childbirth.

The birth process is known as **labor**. Typically, the amnion ruptures right before birth, so amniotic fluid drains out from the vagina. The cervical canal dilates. Strong contractions propel the fetus through the cervix, then out through the vagina (**Figure 38.23**).

A positive feedback mechanism operates during labor. As the fetus nears full term, it typically shifts position so that its head touches the mother's cervix. Receptors in the cervix respond to pressure by sending signals along nerves to the hypothalamus, which in turn tells the posterior lobe of the pituitary to secrete the hormone oxytocin. In a positive feedback loop, oxytocin binds to receptors on smooth muscle cells of the uterus, causing stronger uterine contractions that put additional pressure on the cervix. The added pressure triggers more oxytocin secretion, causing more cervical stretching, and so on, until the fetus is expelled. Synthetic oxytocin is often given to induce or speed labor.

Strong muscle contractions also detach and expel the placenta from the uterus as the "afterbirth." The umbilical cord that connects the newborn to this mass of expelled tissue is clamped, cut short, and tied. The body's navel marks the former attachment site of this cord.

In the United States, more than 30 percent of births involve a cesarean section. During this surgical procedure, a doctor makes an incision in the mother's abdominal wall, then cuts open her uterus to remove the fetus.

A variety of conditions can necessitate a surgical delivery. A placenta that grows in the lower portion of the uterus and covers the cervix can make it impossible for the fetus to exit safely through the vagina. A fetus that is positioned with its feet, rather than its head, at the cervix may also have to be extracted surgically. Surgery may also be necessary if the placenta dislodges too early in labor or the umbilical cord becomes kinked. A cesarean section can also be used to prevent transmission of herpes or another sexually transmitted disease to an infant.

LACTATION

Before a pregnancy, a woman's breast tissue is largely adipose tissue. During pregnancy, milk ducts and mammary glands enlarge in preparation for **lactation**, or milk production. Prolactin, a hormone secreted by the mother's anterior pituitary, encourages milk synthesis. When a newborn suckles, neural signals cause the release of oxytocin. The hormone stimulates muscles around the milk glands to contract and force milk into the ducts.

Human milk includes sugar (lactose), easily digested fats and proteins, essential vitamins and minerals, and enzymes that aid in digestion. It has proteins that encourage growth of beneficial gut bacteria, and lysozyme (Section 34.2) to kill harmful bacteria. In addition, milk contains maternal antibodies that coat the lining of the newborn's throat and gut lining, lessening the risk of dangerous infections.

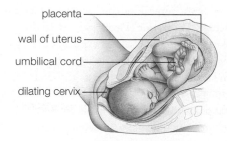

placenta
wall of uterus
umbilical cord
dilating cervix

A Fetus positioned for childbirth; its head is against the mother's cervix, which is dilating (widening).

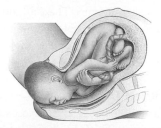

B Muscle contractions stimulated by oxytocin force the fetus out through the vagina.

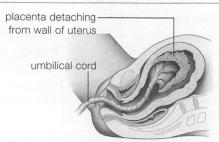

placenta detaching from wall of uterus

umbilical cord

C The placenta detaches from the wall of the uterus and is expelled.

Figure 38.23 Expulsion of fetus and afterbirth during labor. The afterbirth consists of the placenta, tissue fluid, and blood.

labor Expulsion of a placental mammal from its mother's uterus by muscle contractions.

lactation Milk production by a female mammal.

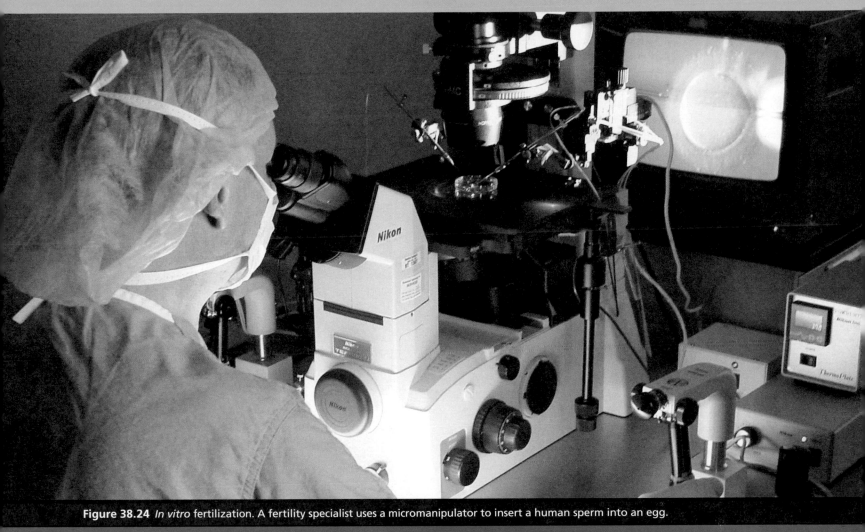

Figure 38.24 *In vitro* fertilization. A fertility specialist uses a micromanipulator to insert a human sperm into an egg.

REPRODUCTIVE TECHNOLOGY

A couple that cannot conceive naturally may turn to assisted reproduction. The oldest and most common form of assisted reproduction is the use of donor sperm. In the United States, the Food and Drug Administration regulates sperm banks, requiring them to assess donors' health, test donors for sexually transmitted diseases, record the donors' family history, and test semen for certain pathogens.

With *in vitro* fertilization (IVF), an egg and sperm are combined outside the body. After fertilization, the zygote develops into a blastocyt that can be placed in a woman's uterus to develop.

The first child conceived by IVF was born in 1976. Scientists worried that this new, unnatural procedure would produce children with psychological and genetic defects. Ethicists warned about the societal implications of manipulating human embryos. Despite the initial reservations, IVF is now widely accepted and practiced. Worldwide, the technique has produced more than 3 million children. The earliest test tube babies have reached adulthood and begun having children of their own.

Research into IVF opened the way to a variety of new reproductive technologies. If a man makes sperm, but does not ejaculate them properly or does not ejaculate a large enough number to allow fertilization by normal means, intracytoplasmic sperm injection can put his sperm inside his partner's egg (**Figure 38.24**). If a woman cannot conceive, she can now obtain donated eggs and use IVF to produce a blastocyst that she can carry to term. A woman who has fertile eggs but cannot or does not want to carry them herself can have a blastocyst conceived by IVF implanted in a surrogate mother. Blastocysts created by IVF can be frozen away for later use.

Eggs can now be frozen too. Human sperm have been stored this way for decades, but the Food and Drug Administration first approved an egg-freezing procedure in 2012. The ability to freeze eggs decreases the cost of using donated eggs for IVF. It also gives women who face the loss of fertility as a result of a medical condition or aging a way to retain the option of reproducing later in life.

Section 38.1 **Asexual reproduction** produces genetic copies of a parent; **sexual reproduction** yields varied offspring. Most animals reproduce sexually and have separate sexes. **Hermaphrodites** have both male and female **gonads**, so they make **sperm** and **eggs**. With **external fertilization**, gametes are released into water. With **internal fertilization**, which occurs in most land animals, gametes meet inside the female's body. Offspring may develop inside or outside the maternal body. **Yolk** helps nourish developing young of most animals. In placental mammals, nutrients are delivered via the **placenta**.

Sections 38.2–38.4 All animals go through similar developmental stages. Gametes are produced in gonads: a male's **testes** or a female's **ovaries**. After fertilization, **cleavage** divides the zygote into many smaller cells, each containing different components of the maternal cytoplasm. Cleavage ends with the formation of the **blastula**. Cellular rearrangements during **gastrulation** yield a **gastrula** with three **germ layers** (**ectoderm**, **mesoderm**, and **endoderm**).

Cells become specialized by selective gene expression. With **embryonic induction,** some cells affect gene expression in other cells, as by secreting **morphogens**. Organs take shape as cells migrate, tissue layers fold, and cells self-destruct by the process of **apoptosis**. Genes that regulate development are highly conserved. Mutations that dramatically alter development are usually fatal.

Section 38.5 Human prenatal development lasts about 40 weeks. From weeks 2 to 8, the individual is called an **embryo**. From week 9 to birth it is a **fetus**. An individual's sex organs mature at **puberty**.

Sections 38.6, 38.7 Ovaries contain **ovarian follicles** that make **oocytes** and secrete **estrogens** and **progesterone**. Two **oviducts** convey oocytes to the **uterus**, which opens into the **vagina** via the **cervix**. Most mammals have an **estrous cycle**. Humans are unusual in that they have a **menstrual cycle** and undergo **menopause**. During each menstrual cycle, **follicle-stimulating hormone** (FSH) stimulates the formation of a secondary oocyte and a **polar body**. FSH also causes the follicle to secrete estrogen. The rise in estrogen triggers a surge of **luteinizing hormone** that in turn triggers **ovulation**. After ovulation, the **corpus luteum** secretes progesterone that primes the uterus for pregnancy. When the corpus luteum breaks down, the uterine lining is shed and **menstruation** occurs. Usually, only one oocyte is released during a menstrual cycle; **fraternal twins** result when two eggs are released and each is fertilized by a different sperm.

Section 38.8 A human male's testes produce sperm and the sex hormone **testosterone**. Hormones from the pituitary regulate testosterone secretion and sperm production. Sperm form in **seminiferous tubules** and mature in an **epididymis** that opens into a **vas deferens**. Secretions from the **seminal vesicles** and **prostate gland** join with sperm to form **semen**. Semen leaves the body through the **penis**.

Sections 38.9–38.11 Intercourse delivers sperm into the vagina. It can also transfer sexually transmitted diseases. Fertilization occurs when a sperm penetrates a secondary oocyte. The nucleus of the resulting **ovum** and that of the sperm supply the genetic material of the zygote. Pregnancy can be prevented by abstinence; with surgical, physical, or chemical barriers; and by manipulating hormones.

Sections 38.12–38.13 Human fertilization usually occurs in an oviduct. Cleavage produces a **blastocyst** that implants in the uterine wall. An embryo that splits in two before implantation becomes **identical twins**. After implantation, an **amnion** forms and encloses the embryo in fluid. The **chorion** becomes part of the placenta, and the **allantois** becomes part of the umbilical cord. Gastrulation occurs at about 2 weeks; then a neural tube, pharyngeal arches, and a tail form in the early embryo. The tail disappears and organ formation ends before an individual becomes a fetus after 8 weeks. Organs mature and body size increases during the fetal period.

Section 38.14 The placenta allows exchanges between embryonic and maternal blood without directly mixing the two bloodstreams. A mother's health, nutrition, and exposure to **teratogens** can affect the growth and development of her future child.

Section 38.15 Hormones induce **labor** at about 38 weeks. Positive feedback governs contractions that expel a fetus and then the afterbirth. Hormones control maturation of the mammary glands and **lactation**.

Application A variety of methods can help couples with fertility problems have children. *In vitro* fertilization combines gametes outside the body. Human sperm, eggs, and early embryos can now be frozen for later use.

1. Most land animals have _____ fertilization.
   a. internal   b. external

2. An individual that makes eggs and sperm is a _____ .
   a. zygote   c. gastrula
   b. hermaphrodite   d. gamete

3. Testosterone is secreted by the _____ .
   a. testes   c. prostate gland
   b. hypothalamus   d. pituitary gland

DATA ANALYSIS
CLASS ACTIVITY

SPERM COUNTS DOWN ON THE FARM Contamination of water by agricultural chemicals affects the reproductive function of some animals. To find out if these chemicals have similar effects in humans, epidemiologist Shanna Swan and her colleagues studied sperm collected from men in four cities in the United States (**Figure 38.25**). The men were partners of women who had become pregnant and were visiting a prenatal clinic, so all were fertile. Of the four cities, Columbia, Missouri, is located in the county with the most farmlands. New York City in New York represents an area with no agriculture.

1. Where did researchers record the highest and lowest sperm counts?

2. In which cities did samples show the highest and lowest sperm motility (ability to move)?

3. Aging, smoking, and sexually transmitted diseases harm sperm. Do regional differences in these variables explain differences in sperm count?

4. Do these data support the hypothesis that living near farmlands can adversely affect male reproductive function?

	Location of Clinic			
	Columbia, Missouri	Los Angeles, California	Minneapolis, Minnesota	New York, New York
Average age	30.7	29.8	32.2	36.1
Percent nonsmokers	79.5	70.5	85.8	81.6
Percent with history of STD	11.4	12.9	13.6	15.8
Sperm count (million/ml)	58.7	80.8	98.6	102.9
Percent motile sperm	48.2	54.5	52.1	56.4

Figure 38.25 Data from a study of sperm collected from men who were partners of pregnant women that visited prenatal health clinics in one of four cities. STD stands for sexually transmitted disease.

4. A midcycle surge of _____ triggers ovulation.
   a. estrogens   b. progesterone   c. LH   d. FSH

5. The corpus luteum develops from _____ .
   a. a polar body   c. a secondary oocyte
   b. follicle cells   d. spermatogonia

6. Sexually transmitted bacteria cause _____ .
   a. trichomoniasis   c. syphilis
   b. genital herpes   d. all of the above

7. Match each term with the most suitable description.
   ___ epididymis   a. conveys oocyte to uterus
   ___ vas deferens   b. can be inflated by blood
   ___ vagina   c. where sperm mature
   ___ seminal vesicle d. target of vasectomy
   ___ penis   e. main contributor to semen
   ___ oviduct   f. sheds lining monthly
   ___ uterus   g. birth canal

8. A homeotic gene regulates _____ .
   a. body part formation   c. secondary sexual traits
   b. milk production   d. sperm formation

9. A human blastocyst normally implants in _____ .
   a. an oviduct   c. the uterus
   b. a seminiferous tubule d. the vagina

10. Put these human developmental events in order, from earliest to latest.
    a. blastocyst forms   d. implantation
    b. heart begins beating   e. breathing practice begins
    c. gastrulation   f. neural tube forms

11. Most miscarriages _____ .
    a. occur in the first trimester
    c. are caused by exposure to a pathogen
    b. can be prevented by a cesarean section
    d. threaten the life of the mother

12. _____ removes webs between developing digits.
    a. Gastrulation   c. Apoptosis
    b. Lactation   d. Implantation

13. Match each hormone with its effect.
    ___ oxytocin   a. growth of facial hair
    ___ testosterone   b. uterus contracts
    ___ LH   c. LH, FSH are released
    ___ GnRH   d. surge causes ovulation
    ___ FSH   e. follicle develops
    ___ HCG   f. milk is produced
    ___ prolactin   g. corpus luteum is maintained

CLASS ACTIVITY
CRITICAL THINKING

1. At no point in human development does the embryo have functional gills. Explain how the embryo obtains the oxygen it requires for aerobic respiration.

2. Fraternal twins are nonidentical siblings that form when two eggs mature and are released and fertilized at the same time. Explain why an increased level of FSH raises the likelihood of fraternal twins.

3. The most common ovarian tumors in young women are teratomas. The name comes from the Greek word *teraton*, which means monster. The "monstrous" feature of these tumors is the presence of a variety of tissues, most commonly bones, teeth, fat, and hair. Explain why a tumor that arises from a germ cell can contain a wider assortment of tissues than one derived from a differentiated body cell.

for additional quizzes, flashcards, and other study materials
▶ ACCESS MINDTAP AT WWW.CENGAGEBRAIN.COM

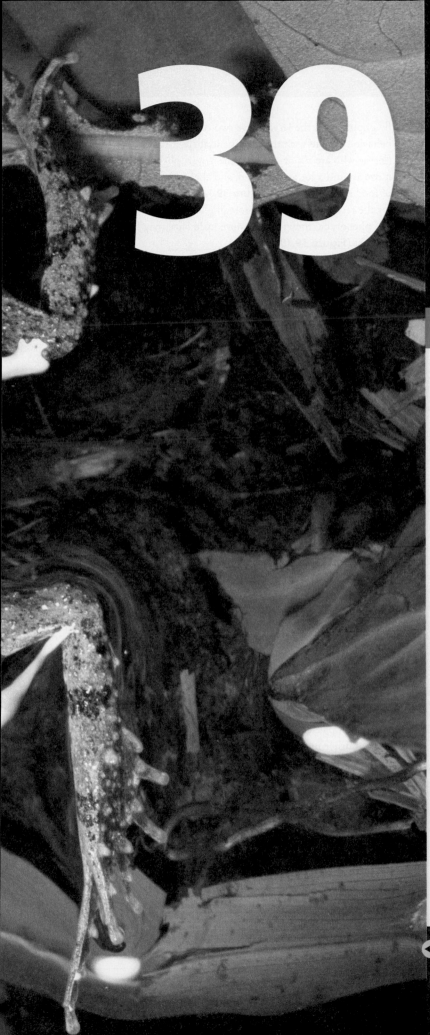

39

Animal Behavior

Links to Earlier Concepts

This chapter builds on your knowledge of sensory and endocrine systems (Chapters 30, 31). You will revisit pheromones (30.3) and learn about additional effects of the hormone oxytocin (31.3). Be sure that you understand the concepts of sexual selection (17.5) and adaptation (16.2). The chapter has many examples of scientific experiments (1.5, 1.6).

Core Concepts

Interactions among the components of a biological system give rise to complex properties.

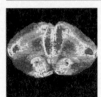 Genes and epigenetic mechanisms are the basis of innate (inherited) behaviors and the capacity for learning through experience. Natural selection acts on these traits, so it shapes behaviors as well as their timing and coordination. A population's overall survival depends on communication and cooperation among its members, and is influenced by individual behavior.

Adaptation of organisms to a variety of environments has resulted in diverse structures and physiological processes.

 Many animal behaviors occur in response to environmental cues and interactions with other organisms. Different species communicate information via visual, audible, tactile, electrical, or chemical signals that affect the behavior of other organisms. Genetic variation in these traits allows them to evolve under selection.

The continuity of life arises from and depends on genetic information in DNA.

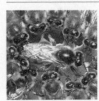 Behavioral traits, like other traits, have a basis in genes. When animals reproduce sexually, they pass about half of their alleles to offspring. Alleles underlying behaviors that increase the chances of survival of offspring (or of close relatives) are adaptive, so they are passed preferentially to successive generations.

◀ A male tungara frog inflates his vocal sac to make his species-specific mating call. Female frogs are attracted by the sound of this call.

Photograph by George Grall, National Geographic Creative.

LEARNING OBJECTIVES

Explain the difference between proximate and ultimate causes of behavior.

Using appropriate examples, explain how genetic and epigenetic mechanisms can give rise to differences in behavior.

Why do animals behave the way they do? In thinking about what drives animal behavior, biologists distinguish between "proximate" and "ultimate" causes. "Proximate" means nearby or immediate; **proximate causes** of a behavior include genes and epigenetic factors that specify the behavior, the physiological mechanisms that bring about the behavior, and environmental factors that influence and elicit the behavior. The **ultimate causes** of a behavior are the evolutionary factors that shaped it. Variations in behavior affect fitness, so heritable behavioral traits will evolve under natural selection.

Studying variations in behavior, either among members of a species or among closely related species, can help uncover the underlying causes of the behavior. A few examples will illustrate this approach.

FORAGING IN FRUIT FLIES

Most behavior has a polygenic basis. However, there are some cases in which behavioral differences can be traced to allele differences at a single gene. The *foraging* gene, which determines the feeding behavior of fruit fly larvae, is one such gene. Fruit fly larvae are wingless, with a wormlike body, and they feed on yeast cells that grow on decaying fruit. In wild fruit

fly populations, the larvae have two distinct types of feeding behavior. Larvae with the dominant allele at the *foraging* locus (larvae who are *FF* or *Ff*) have the "rover" phenotype. Rovers move around a lot as they feed, and they often leave one patch of food to seek another (**Figure 39.1A**). Larvae homozygous for the recessive allele at the foraging locus (*ff*) have the "sitter" phenotype; they tend to move little once they find a patch of yeast (**Figure 39.1B**).

The ultimate cause of this behavioral variation in larval foraging behavior is selection related to competition for food. In experimental populations with limited food, both rovers and sitters are most likely to survive to adulthood when their foraging type is rare. A rover does best when surrounded by sitters, and vice versa. Presumably, when there are lots of larvae of one type, they all compete for food in the same way. Under these circumstances, a fly larva that behaves differently than the majority is at an advantage. Thus, natural selection maintains both alleles of the *foraging* gene.

PAIR-BONDING IN VOLES

Comparisons among closely related species can sometimes reveal proximate causes of a behavior. For example, studies of small rodents called voles (**Figure 39.2A**) showed that inherited differences in the number and distribution of certain hormone receptors influence mating and bonding behavior. Most voles, like most mammals, are promiscuous; they have multiple mates. However, some voles form lifelong, largely monogamous relationships. For example, in prairie voles, a permanent social bond forms after a night of repeated matings. Males and females continue to associate and collaborate in care of their offspring.

Oxytocin, a peptide that functions both as a hormone and a neurotransmitter, plays a central role in a female prairie vole's bonding behavior. Experimentally disrupting oxytocin function in the brain of a female who is a member of an established pair causes her to dump her long-term partner in favor of other males. Females of promiscuous vole species are less influenced by oxytocin than prairie voles. When researchers compared the brains of promiscuous and monogamous species, they found a striking difference in the number and distribution of oxytocin receptors (**Figure 39.2B,C**). Monogamous prairie vole females have many oxytocin receptors in the part of the brain associated with social behavior. Promiscuous mountain vole females have far fewer oxytocin receptors.

Similarly, pair-bonding prairie vole males have more brain receptors for arginine vasopressin (AVP) than males of promiscuous species. AVP is a hormone that is similar to oxytocin. Additional evidence that AVP plays a role in male pair-bonding comes from an experiment in which the AVP

Figure 39.1 Genetic polymorphism for foraging behavior in fruit fly larvae. When a larva is placed in the center of a yeast-filled plate, its genotype at the *foraging* locus influences how much it moves while it feeds. Lines show a representative larva's path.

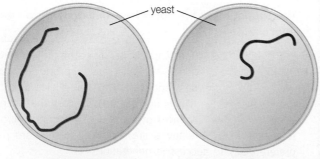

A Rovers (genotype *FF* or *Ff*) move often as they feed. When a rover's movements on a petri dish filled with yeast are traced for 5 minutes, the trail is relatively long.

B Sitters (genotype *ff*) move little as they feed. When a sitter's movements on a petri dish filled with yeast are traced for 5 minutes, the trail is relatively short.

A Vole (*Microtus*), a type of small rodent.

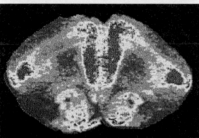

B PET scan of a monogamous prairie vole's brain with many oxytocin receptors (red).

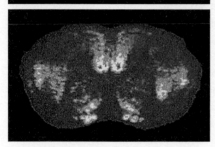

C PET scan of a promiscuous mountain vole's brain with fewer oxytocin receptors.

Figure 39.2 Genetic roots of mating and bonding behavior in voles. Closely related vole species vary in their behavior, and in the number and distribution of receptors for the hormone oxytocin.

receptor gene from a prairie vole was experimentally transferred into the forebrain of male mice. Mice are naturally promiscuous. However, male mice who received the prairie vole gene gave up their playboy ways. After the procedure, they preferred a female with whom they had mated before.

Formation of a pair bond that lasts a vole's lifetime occurs by way of an epigenetic mechanism. As Section 10.5 explained, epigenetic modifications alter gene expression without changing DNA sequence. In vole species that form a pair bond, mating triggers modification of histone proteins at promoters for the oxytocin receptor gene and vasopressin receptor gene. This modification (removal of an acetyl group) increases expression of these genes.

Insights from studies of vole behavior may help researchers understand human behavior. Evidence of oxytocin's role in animal bonding led researchers to examine this hormone's role in autism, a condition in which a person has trouble making social attachments. There are many alleles at the human oxytocin receptor locus and some of them are associated with an elevated risk of autism.

Figure 39.3 Social dominance. Dominant male cichlids with distinctive eye bars face off.

SOCIAL DOMINANCE IN CICHLIDS

Differences in behavior within a species can arise as a result of environmental factors. In cichlids (a type of fish) a male's social environment affects his behavior. During the course of a male cichlid's life, he can switch back and forth between two social roles. When in a dominant role, he courts and breeds with females, and engages in confrontations with other males. Dominant males have bright coloration and a dark eye bar (**Figure 39.3**). When in a subordinate role, a male cichlid flees from dominant males, does not court or breed, and has a drab, pale gray coloration.

Whether a male cichlid exhibits the dominant or subordinate phenotype depends on a variety of factors, most importantly the presence of a larger dominant male. Remove all larger dominant males from an environment, and a subordinate male can switch to the dominant phenotype. Within minutes, he becomes more colorful and begins chasing other males. Within hours, his previously shrunken testes enlarge and produce viable sperm.

Researchers suspected that the dramatic transformation between subordinate and dominant phenotypes involves epigenetic changes. To test this hypothesis, they injected some male cichlids who had been lifelong subordinates with methionine, which provides the methyl groups used for methylating DNA. They injected other lifelong subordinates with a chemical that inhibits DNA methylation. Pairs of similarly sized, former subordinates—one injected with methionine, the other with the methylation inhibitor—were then placed in a tank without other males. Within a day, one of the males had become dominant, and in the vast majority of cases it was the methionine-injected individual. An epigenetic effect, specifically methylation, seems to facilitate this change.

proximate cause Of a behavior; the genes, physiology, or environmental conditions that trigger the behavior.
ultimate cause Of a behavior; the evolutionary factors that shaped the behavior.

CREDITS: (2A) © Robert M. Timm & Barbara L. Clauson, University of Kansas; (2B–C) Reprinted from *Trends in Neuroscience*, Vol. 21, Issue 2, 1998, L.J.Young, W. Zuoxin, T.R. Insel, "Neuroendocrine bases of monogamy", Pages 71–75, ©1998, with permission from Elsevier Science; (3) © Lemi Hacıoğlu Photography, lemihacioglu.net/photography.

LEARNING OBJECTIVES

Provide an example of instinctive behavior.

Explain how instinct and learning interact in shaping bird song.

Describe five types of learning and give an example of each.

INSTINCTIVE BEHAVIOR

All animals have some **instinctive behaviors**—inborn responses to a specific and usually simple stimulus.

The life cycle of the cuckoo bird provides several examples of instinct at work. The European cuckoo is a "brood parasite," meaning it lays its eggs in the nest of other birds. The female cuckoo scatters her eggs among foster nests, leaving just one egg in each. A newly hatched cuckoo is blind, but contact with an object beside it stimulates an instinctive response. The cuckoo maneuvers the object, which is usually one of its foster parents' eggs, onto its back, then shoves it out of the nest (**Figure 39.4A**). The cuckoo will also shove its foster siblings over the side, if they hatched before it did. Instinctively shoving these objects out of the nest ensures that the cuckoo receives its foster parents' undivided attention.

Instinctive responses are adaptive when the triggering stimulus almost always signals the same situation and requires the same response. Doing away with an egg or foster sibling always benefits a cuckoo chick because it rids the nest of future competitors for food. However, instinctive responses sometimes open the way to exploitation. For example, most birds instinctively fill any wide-open mouth they see in their nest, a response that is clearly adaptive when those mouths belong to their chicks. However, this parental instinct opens the way to exploitation by cuckoos, which often look nothing like the chicks they displace (**Figure 39.4B**).

LEARNED BEHAVIOR

A **learned behavior** is a behavior that has been altered as a result of experience. Some instinctive behavior can be modified with learning. For example, a newly hatched toad instinctively tries to capture and eat any insect-sized moving object it sees. This behavior is modified after encounters with bees or wasps. The toad quickly learns that insect-sized objects with certain color patterns provide a sting rather than a meal.

Time-Sensitive Learning **Imprinting** is a form of learning that occurs during a genetically determined time period early in life. For example, newly hatched geese and cranes learn to follow the large moving object that looms over them after they hatch. That object is usually their mother, but the young birds will imprint on and follow other animals or

A Young cuckoo shoving its foster parents' egg from the nest. It pushes out any object beside it.

B A foster parent (left) feeds a cuckoo chick (right) in response to a simple cue: a gaping mouth.

Figure 39.4 Instinctive responses to simple cues.

humans if these substitutes replace their mother during the appropriate time (**Figure 39.5A**).

A genetic capacity to learn, combined with experiences in the environment, shapes most forms of behavior. For example, a male songbird has an inborn capacity to recognize his species' song when he hears older males singing it. The young male uses these overheard songs as a model for his own song. A male reared with no model or exposed to songs of other species will sing a simple version of his own species' song.

Many birds can only learn the details of their species-specific song during a limited period early in life. For example, a male white-crowned sparrow must hear a male "tutor" of his own species during his first month, or he will not learn the appropriate song. Hearing such a same-species tutor later in life will not normalize his singing.

Conditioned Responses Nearly all animals learn to associate certain stimuli with rewards and others with negative consequences. With classical conditioning, an animal's involuntary response to a stimulus becomes associated with a stimulus that accompanies it. In the most famous example, Ivan Pavlov always rang a bell just before feeding a dog. Eventually, the dog's reflexive response—increased salivation—was elicited by the sound of the bell alone.

Conditioned taste aversion is a type of classical conditioning in which an animal learns to avoid a food that sickened it. This response protects the animal from repeatedly ingesting toxic substances. The evolved defense of monarch butterflies depends on conditioned taste aversion in their predators. Plants that monarch larvae feed on contain a chemical that induces vomiting in birds, and the larvae store the chemical in their body. Once a bird has been sickened by a larval or adult monarch, it avoids these insects in the future.

Operant conditioning is a learning process in which an animal modifies its voluntary behavior in response to the behavior's consequences. This type of learning was first described in the context of learning experiments in the lab. A rat that presses a lever and is rewarded with a food pellet becomes more likely to press the lever again. A rat that receives a shock when it enters a particular area will quickly learn to avoid that area. In nature as well, animals learn to repeat behaviors that provide food or mating opportunities and to avoid those that cause pain or stress.

Habituation With **habituation**, an animal learns by experience not to respond to a stimulus that is irrelevant to its well-being. Pigeons in crowded cities often become habituated to people: They learn not to flee from the throngs who walk past them. Habituation is considered the simplest type of learning. Animals as simple as jellies can learn to ignore a repeated harmless touch.

Spatial Learning Many animals learn the landmarks in their environment, creating a mental map. Such spatial learning allows a fiddler crab foraging up to 10 meters (30 feet) from its burrow to scurry straight home in response to a threat. Learned landmarks also play a role in some migrations. Young whooping cranes follow older birds to learn the route of their long seasonal migration. A bird's ability to remember the route improves with repeated trips.

Social Learning Animals also learn the details of their social landscape. Many animals learn to recognize mates, offspring, or competitors by their appearance, calls, odor, or some combination of cues. For example, two male lobsters that meet up for the first time will fight (**Figure 39.5B**). Later, they will recognize one another by scent and behave accordingly, with the loser actively avoiding the winner.

Imitative Learning With imitative learning, an animal copies behavior it observes in another individual. Consider how a chimpanzee strips leaves from branches to make a "fishing stick" for use in capturing termites as food. The chimpanzee inserts a fishing stick into a termite mound, then withdraws the stick and eats the termites that cling to it. Different groups of chimpanzees use different methods of tool-shaping and insect-fishing. In each group, young individuals learn by imitating behavior of others (**Figure 39.5C**).

---

habituation Learning not to respond to a repeated neutral stimulus.
imprinting Learning that can occur only during a specific interval in an animal's life.
instinctive behavior An innate response to a simple stimulus.
learned behavior Behavior that is modified by experience.

A Imprinting. Nobel laureate and animal behaviorist Konrad Lorenz with geese that imprinted on him. The inset photos shows a more common result of imprinting.

B Social learning. Captive male lobsters fight at their first meeting. The loser remembers the winner's scent and avoids him. Without another meeting, memory of the defeat lasts up to two weeks.

C Imitative learning. Chimpanzees use sticks as tools for extracting tasty termites from a nest. The method of shaping the stick and catching termites is learned by imitation.

Figure 39.5 Examples of learned behavior.

LEARNING OBJECTIVES

List the different types of communication signals and give examples of each.

Describe some ways that predators can take advantage of prey communication systems.

Communication signals are evolved cues that transmit information from one member of a species to another. A communication signal arises and persists only if it benefits both the signal sender and the signal receiver. If signaling is disadvantageous for either party, then natural selection will favor individuals that do not send or respond to it.

TYPES OF COMMUNICATION SIGNALS

Pheromones are chemical signals that convey information among members of a species. Signal pheromones cause a rapid shift in the receiver's behavior. Sex attractants that help males and females of many species find each other are one example. Priming pheromones cause longer-term responses, as when a chemical in the urine of male mice triggers ovulation in females.

A Ground squirrel giving an alarm call.　**B** Courtship display in grebes; the birds move in unison.

C Threat display of a male collared lizard. This display allows rival males to assess one another's strengths without engaging in a potentially damaging fight.

Figure 39.6 Examples of vertebrate communication.

Producing a pheromone requires less energy than calling or gesturing. However, the amount of information a pheromone can convey is limited; it is either released or not. Properties of acoustical, visual, and tactile signals vary continuously, so they can convey more information.

Acoustical signals often advertise the presence of an animal or group of animals. Many male vertebrates, including songbirds, whales, frogs, some fish, and many insects, make sounds to attract prospective mates. In many species, sounds made by males also function in territoriality. Some birds and mammals give alarm calls that inform others of potential threats (**Figure 39.6A**). In many cases, the calls convey more than simply "Danger!" A ground squirrel emits one type of bark when it detects an eagle, and another when it sees a coyote. Upon hearing the call, other ground squirrels respond appropriately: They either dive into burrows (to escape an eagle's attack) or stand erect (to spot the coyote).

Natural selection shapes the properties of acoustic signals. Thus, aspects of these signals differ depending on whether or not the signaler benefits by revealing its position. Sounds that lure mates or offspring are easily localized, whereas alarm calls are not.

Visual communication is most widespread in animals that have good eyesight and are active during the day. In most birds, sex is preceded by a courtship display that involves coordinated visual signaling (**Figure 39.6B**). Natural selection favors production of clear signals that are easily seen and understood by prospective mates, so movements that serve as signals often become exaggerated. Body form may evolve in concert with movements, as when brightly colored feathers enhance a courtship display. A courtship display assures a prospective mate that the displayer is of the correct species and in good health.

Threat displays advertise good health too, but they serve a different purpose. When two potential rivals meet, a threat display demonstrates each individual's strength and how well armed it is (**Figure 39.6C**). If the rivals are not evenly matched, the weaker individual retreats. Threat displays benefit both participants by allowing them to avoid an energetically costly fight that could lead to injury.

With tactile displays, touch transmits information. For example, after discovering food, a foraging honeybee worker returns to the hive and dances on the vertical honeycomb surrounded by a crowd of workers. The interior of the hive is dark, so the workers follow the dancer's progress by touch. The speed and orientation of a successful forager's dance convey information about the location of the food (**Figure 39.7**).

pheromone (FAIR-uh-moan) Chemical that serves as a communication signal between members of an animal species.

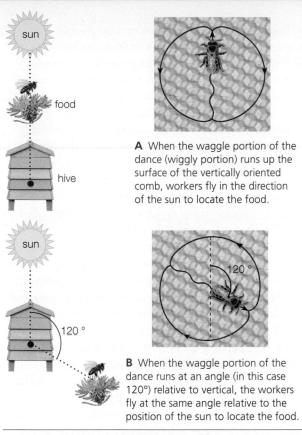

A When the waggle portion of the dance (wiggly portion) runs up the surface of the vertically oriented comb, workers fly in the direction of the sun to locate the food.

B When the waggle portion of the dance runs at an angle (in this case 120°) relative to vertical, the workers fly at the same angle relative to the position of the sun to locate the food.

Figure 39.7 Honeybee waggle dance. The angle of the waggle run, in which the bee wiggles back and forth, gives the direction of the food. The faster the dance is performed, the closer the food.

FIGURE IT OUT
How will a waggle run be oriented if food is located in the opposite direction from the sun?

Answer: Straight down the comb

EAVESDROPPERS AND FAKE SIGNALS

Predators can benefit by intercepting signals sent by their prey. For example, frog-eating bats locate a male tungara frog (such as the one shown in the chapter opening photo) by listening for his mating call. This poses a dilemma for the male frog. Female tungara frogs prefer complex calls, but complex calls are easier for bats to localize. Thus, when bats are near, male frogs call less, and with less flair. This subdued signal is a trade-off between competing pressures to attract a mate and to avoid being eaten.

Other predators lure prey with counterfeit signals. Fireflies are nocturnal beetles that attract mates with flashes of light. When a predatory female firefly sees the flash from a male of the prey species, she flashes back as if she were a female of his own species. If she lures him close enough, she captures and eats him.

Dr. Isabelle Charrier

Some animals are able to recognize individuals by their voice. National Geographic Explorer Isabelle Charrier says, "My scientific interest is in understanding how animals are able to identify each other individually by voice, especially in species showing strong environmental and ecological constraints. Colonial birds and mammals are good models for this problem: Vocal recognition between mates or between parents and offspring is effective and reliable in spite of the high background noise and the high risk of confusion between individuals."

Much of Charrier's research focuses on seals and related marine mammals. These animals spend most of their life at sea, but come ashore to give birth and nurse their young. Charrier's field studies span the globe, from walrus colonies in the Arctic to sea lion colonies in Australia. She has studied the features of vocal signals and their role in recognition in interactions between mothers and pups, between males and females, and between territorial males. She says, "My favorite field experience was when I did my Ph.D.: I stayed nearly nine months on Amsterdam Island [in the southern Indian Ocean] to study subantarctic fur seals. This was so great, to study mother–pup vocal recognition from birth until weaning. You can really develop special interactions with the animals by being on the colony for hours every day." Her greatest challenge has been studying walruses in the Arctic. She says, "You fight every day with the weather and sea conditions, and walruses are quite difficult to find and approach. It results in a great amount of frustration, but walruses are amazing to study."

As for why she does her research, she says, "Learning more about animal communication systems is not only important for our general knowledge, but also an essential approach and a powerful tool to protect some species and their environment."

Explain why genetic monogamy is rare among animals.

Describe the mating system typically associated with sexual dimorphism.

Using appropriate examples, explain some factors that determine whether parental care will evolve in a species and which parent will provide it.

MATING SYSTEMS

Animal mating systems have traditionally been categorized as promiscuous (multiple mates for both sexes), polygynous (multiple mates for males only), polyandrous (multiple mates for females only), or monogamous (a single mate for both sexes). In recent years, studies that integrate paternity analysis with behavioral observations have shown that species do not always fit cleanly into these categories. Members of a species often vary in their behavior.

Some animals such as prairie voles form a pair bond, which means two individuals mate, preferentially spend time together, and cooperate in rearing offspring. However, the participants in a pair bond may also seize opportunities for matings with other individuals. A paternity study of the prairie voles discussed in Section 39.1 found that 7 percent of pair-bonded females produced litters of mixed male parentage, and 15 percent of pair-bonded males sired offspring with females other than their partner. Thus, prairie voles are now described as "socially monogamous, but genetically promiscuous." As a species, prairie voles tend to form an exclusive social relationship with a single partner. However, the genetic composition of their offspring reflects a tendency to sometimes mate with multiple partners.

Even social monogamy is rare in most animals, with the exception of birds. An estimated 90 percent of birds are socially monogamous. Among this group, the vast majority of species that have been examined for paternity patterns are genetically promiscuous. Promiscuity benefits both males and females by increasing the genetic diversity among the offspring they produce. Multiple matings can also provide another benefit: more offspring. In most cases, this benefit applies mainly to males. Sperm are energetically inexpensive to produce, so a male's reproductive success is usually limited by access to mates. By contrast, the main limit on a female's reproductive success is her capacity to produce large, yolk-rich eggs or, in mammals, to carry developing young.

The selective benefits of promiscuity are offset somewhat when two parents must cooperate to rear the young. Promiscuity among females reduces males' probability of paternity, which in turn selects for them to contribute less parental care.

Selection resulting from differential mating success is called sexual selection (Section 17.5). In many species, females choose among males on the basis of their ability to provide necessary resources. For example, a female hangingfly will only mate with a male while feeding on prey that he has provided as a "nuptial gift" (**Figure 39.8A**).

In other species, males establish a mating territory that includes resources required by females for reproduction. A **territory** is an area from which an animal or group of animals actively excludes others. A male fiddler crab's territory is a stretch of shoreline in which he excavates a burrow. Fiddler crabs have **sexual dimorphism**, in which traits of males and females differ. A male has one oversized claw (**Figure 39.8B**) and during mating season, he stands outside his burrow

A Male hangingfly dangling a moth as a gift for a potential mate.

B Male fiddler crab waving his one enlarged claw to attract a female to his burrow.

C Male sage grouse on his portion of a communal display ground (a lek), where he will dance, inflate the sacs on his chest, and make gurgling calls.

Figure 39.8 Male courtship behavior.

waving it. Females attracted by a male's display inspect the location and dimensions of his burrow before mating with him. Burrow location and size are important because they affect survival of the larvae.

In mammals such as elephant seals and elk, the strongest males hold large territories, and mate with all of the females inside the area. Other males may never get to mate at all unless they sneak into an area under another male's control.

In some birds such as sage grouse, males converge at a communal display ground called a **lek**. Each male performs a courtship display for females (**Figure 39.8C**). A few males attract and mate with most females. It may seem strange for females to base their choice on a male's ability to sing and dance, but the healthiest males perform the most stunning displays. The quality of the display may thus indicate other traits, not directly visible, that make a male a desirable mate. In addition, males whose courtship behavior attracts many females pass female-pleasing traits to their sons.

PARENTAL CARE

Parental care requires time and energy that an individual could otherwise invest in reproducing again. It arises only if the genetic benefit of providing care (increased offspring survival now) offsets the cost (reduced opportunity to produce offspring later). If the parental care provided by a single parent is sufficient to successfully rear an offspring, individuals who spare themselves this cost by leaving parental duties to their mate are at a selective advantage. Thus, whether parental care occurs and, if so, who provides it vary among species.

Most fishes provide no parental care, but in those that do, this duty usually falls to the male. In many species, males guard eggs until they hatch (**Figure 39.9A**). In some cases, males retain eggs or young on or in their body. Male seahorses have a special brood pouch. Mouth brooding (housing developing eggs in the mouth) has also evolved in a variety of fish species. The prevalence of male parental care in fishes may be related to sex differences in how fertility changes with age. A female fish's fertility increases with age, whereas a male's does not. Thus, a female fish who invests energy in care forgoes more future reproduction than a male fish does.

In birds, two-parent care is most common (**Figure 39.9B**). Chicks of birds that cooperate in care of their young tend to hatch while in a relatively helpless state, so the chicks require the care of two parents to survive. Chicks of sage grouse and other birds in which females alone provide care typically hatch when more fully developed.

In about 90 percent of mammals, the female rears young (**Figure 39.9C**). In the remaining 10 percent, both sexes participate. No mammal relies solely on male parental care.

A Male sergeant majors guard eggs (pink) until they hatch.

B Male and female terns cooperate in the care of their chicks.

C A female grizzly bear cares for her cub for as long as two years.
Figure 39.9 Parental care.

Female mammals sustain developing young in their body, so they do not have the option of pursuing a new mate rather than providing care. Females also have a greater investment in newborns than males, and this investment continues with lactation. Most male mammals cannot lactate, although in two fruit bat species, both males and females nurse the young.

lek Of some birds, a communal mating display area for males.
sexual dimorphism Distinct male and female traits.
territory Area from which an animal or group of animals actively excludes others.

LEARNING OBJECTIVES

Describe some benefits and costs of living in a group.

Explain how altruistic behavior can evolve.

Some animals live solitary lives, coming together only to mate. However, individuals of many species spend time in groups. Some groups are temporary, like schools of fish; others such as prairie dog colonies or wolf packs are more permanent. A tendency to group together, whether briefly or for the longer term, will evolve only if the benefits of being near others outweigh the costs.

BENEFITS OF GROUPING

Whenever animals cluster together, individuals at the margins of the group inadvertently shield others from predators. The **selfish herd effect** describes how groups can emerge from the collective tendencies of individuals to reduce their risk of danger by drawing near one another. For example, in the presence of a predator, small fish form a tight school (**Figure 39.10A**) because each fish tries to hide behind others.

Once a group forms, multiple individuals can be on the alert for predators. In some cases, an animal that spots a threat will warn others of its approach. Birds, monkeys, prairie dogs, and ground squirrels are among the animals that make alarm calls in response to a predator. Even in species that do not make alarm calls, individuals can benefit from the vigilance of others. Often, when an animal notices another group member beginning to flee, it will do likewise. In what is called the "confusion effect," a predator has more difficulty choosing an individual to pursue when a group of prey is

scattering. Some animals stand their ground to present a united defense. For example, a group of threatened musk oxen forms a tight circle so a predator faces their sharp horns.

Eastern tent caterpillars (**Figure 39.10B**) are moth larvae that collaborate to spin a protective tent into which the group retreats when not feeding. To fend off a predator or parasite, the caterpillars thrash about as a group and regurgitate fluid laced with cyanide. (The cyanide is from their diet of leaves.) The caterpillars also benefit by sharing information about food. A caterpillar who finds a good feeding site leaves a pheromone trail as it returns to the communal tent. Other caterpillars follow this trail, feed, then add more pheromone to the trail. The more pheromone a trail has, the more likely hungry caterpillars are to follow it.

The eastern tent caterpillar's group defense and information sharing capacities have led some scientists to label them "social caterpillars." However, biologists have traditionally defined **social animals** as species in which individuals benefit by cooperating in a permanent multigenerational group. Tent caterpillars do not fit this definition because there is no generational overlap; all members of the group are larvae that hatched at about the same time. As this example shows, there is a range of what is considered social behavior. The most extreme social behavior occurs among ants, bees, and termites (we return to this topic in the next section).

Wolves and lions are social animals who live in multigenerational groups and cooperate in hunting prey. Cooperative hunting allows a group of predators to capture larger or faster prey. However, on a per-hunter basis, cooperative hunting is usually no more efficient than hunting alone. In one study, researchers observed that a solitary lion catches prey about 15 percent of the time. Two lions hunting together catch prey

A Sardines attempting to hide behind one another during an attack by predatory sailfish.

B Tent caterpillars on the surface of their communally produced tent. If threatened, the group wriggles and produces cyanide-laced vomit.

Figure 39.10 Safety in numbers. Clustering with others provides defense against parasites and predators.

twice as often as a solitary lion, but having to share the spoils of the hunt means the amount of food per lion is the same. Among carnivores that hunt cooperatively, hunting success is not the major advantage of group living. Individuals hunt together, but also cooperate in fending off scavengers (**Figure 39.11**), caring for one another's young, and defending the group's territory.

Most primates live in social groups. Primates that spend most of their time on the ground, including chimpanzees and baboons, live in multi-male, multi-female groups. Members of these groups benefit from collective defense against predators and shared parental care. Individuals also benefit by sharing information about food and receiving grooming to remove parasites. Group living also facilitates imitation learning.

COSTS OF GROUP LIVING

Individuals that group together are easier for predators to locate. In addition, parasites and contagious diseases spread more quickly when individuals come into frequent close contact. Grouping also increases competition for food and mates, and the resulting conflict increases stress.

In most social groups, costs and benefits are not equally distributed among the group members. Many animals that live in groups form a **dominance hierarchy**, a social system in which dominant animals receive a greater share of resources and breeding opportunities than subordinates. Typically, dominance is established by physical confrontation. In wolf packs, the dominant male and dominant female are usually the only breeders. All members of the pack hunt and carry food back to individuals that guard the young in their den. Why would a wolf stay in a pack where it is subordinate? Belonging to a group enhances the subordinate's chance of survival, and some subordinates do get a chance to reproduce in periods when food is abundant. A subordinate may also one day ascend to the dominant reproductive role.

EVOLUTION OF COOPERATION

Why do animals assist other members of their social group? In many animal groups, individuals display **altruism**, which is behavior that decreases the altruist's fitness but increases the fitness of other group members. Giving an alarm call is an example of an altruistic act; it puts the caller at an increased risk of predation but reduces the risk to others.

One way that altruistic behavior can evolve is through **kin selection**, a type of selection that favors individuals who have the highest inclusive fitness. An individual's **inclusive fitness** is the number of offspring that it could produce with no help from others, plus a fraction of the number of extra

Figure 39.11 Cooperating to obtain food. Lions enjoy the results of a communal hunt while a group of hyenas collaborate in attempting to steal a share.

offspring produced by other individuals because of help from it. The fraction mentioned in the previous sentence is determined by the individual's relatedness to those whom it helps.

Field studies of social animals confirm that many altruistic behaviors are preferentially directed at relatives. For example, ground squirrels and prairie dogs are more likely to give an alarm call if they have close relatives living nearby. In hyenas, which often have intragroup fights, dominant females are more likely to risk injury by joining a fight to assist a relative than to assist a nonrelative.

Cooperative behavior can also arise through reciprocity, in which an individual assists group members who have previously assisted it or may assist it in the future. Consider vampire bats, which roost during the day in large groups. Hungry bats who have not successfully fed the night before are offered food (regurgitated blood) by other bats. Close kinship increases the likelihood of food sharing, but sharing also occurs among unrelated bats. Bats in need who previously offered food to many different recipients receive more offers of food than bats who were less generous.

---

altruism (AL-true-izm) Behavior that benefits others at the expense of the individual that performs it.

dominance hierarchy Social system in which resources and mating opportunities are unequally distributed within a group.

inclusive fitness Genetic contribution an individual makes by reproducing, plus a fraction of the contribution it makes by facilitating reproduction of relatives.

kin selection Type of selection that favors individuals with the highest inclusive fitness.

selfish herd effect Animal groups arise from the collective tendency of individuals to reduce their risk of danger by drawing near one another.

social animal Animal that lives in a multigenerational group in which members benefit by cooperating in some tasks.

39.6 Eusocial Animals

From a genetic standpoint, the greatest cost an individual can pay for living in a group is the failure to breed. Yet, in some animals, a social system has evolved in which permanently sterile workers care cooperatively for the offspring of just a few breeding individuals. These species are said to be eusocial. **Eusocial animals** live in a multigenerational family group in which sterile workers carry out tasks essential to the group's welfare, while other members of the group reproduce.

Many eusocial species are members of the order Hymenoptera, which includes ants, bees, and wasps. All ants are eusocial, as are are some bees and wasps. In eusocial hymenoptera all workers are females. Workers forage for food, maintain the hive, care for young, and defend the colony.

Honeybees are eusocial. The only egg-laying female in a honeybee colony is the queen, who is larger than the workers and anatomically distinct (**Figure 39.12A**). A female's diet as a larva determines whether the individual becomes a queen or a worker. A larva destined to be a queen eats only a special glandular secretion called royal jelly. Larva that will become workers get a bit of royal jelly, plus pollen and honey. Components of the pollen and honey trigger epigenetic changes that result in the worker phenotype. Queen and worker methylation patterns differ, with more than 500 genes involved.

The thousands of species of termites are also eusocial. The termite queen is relatively enormous (**Figure 39.12B**), far larger than her mate, the king, who also resides in the colony, or the male and female workers who tend to her.

Only two species of eusocial vertebrates have been discovered. Both are African mole-rats: mouse-sized rodents that live in underground burrows. The best studied is *Heterocephalus glaber*, the naked mole-rat (**Figure 39.12C**). Clans of this nearly hairless rodent live in burrows in dry parts of East Africa. A reproducing female (the queen) dominates the clan and mates with one to three males (the king or kings). Nonbreeding members of the clan protect and care for the queen, the king(s), and their offspring. Sterile diggers excavate tunnels and chambers. When a digger finds an edible root, it hauls a bit back to the main chamber and chirps to alert other

eusocial animal Animal species that lives in a multigenerational group in which many sterile workers cooperate in all tasks essential to the group's welfare; only a few members of the group produce offspring.

A Honeybee queen surrounded by sterile worker females.

B Termite queen and workers.

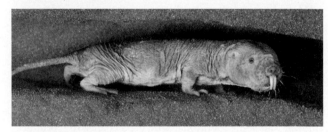

C Naked mole rat worker.

Figure 39.12 Examples of eusocial animals.

workers about the food source. Other sterile workers guard the colony. When a predator appears, they chase and attack it at great risk to themselves.

Inbreeding, which increases genetic similarity among relatives, may have played a role in the evolution of naked mole-rat eusociality. Some colonies are highly inbred, so workers share an unusually high number of genes with the king and queen. However, ecological factors probably also play a role. According to one hypothesis, the mole-rat's arid habitat and patchy food supply favor a genetic predisposition to cooperate in digging burrows, searching for food, and fending off competitors. Individuals who strike out on their own are unlikely to have high reproductive success.

Figure 39.13 Whales in a shipping channel. Ship noise can impair whales' ability to communicate.

CAN YOU HEAR ME NOW?

Right whales, like other whales, communicate with one another through sound. The sound they make most frequently is a whooplike contact call that informs other right whales of their presence. Other calls sound like screams, moans, or gunshots.

In the past, right whale populations were driven to the brink of extinction throughout the world. The North Atlantic population declined to a low of about 100 animals; today it includes about 450. Unfortunately, this population lives near the eastern coasts of Canada and the United States, where commercial ship traffic is heavy. For whales, potentially fatal collisions with ships are a problem, as is ship-related noise pollution (**Figure 39.13**).

Marine biologist Susan Parks studies how North Atlantic right whales communicate, the effects of man-made noise on the whales, and ways to minimize any negative effects. Sound-recording devices attached to individual right whales revealed that the level of shipping noise influences the whales' calling patterns.

Whales call more loudly in the presence of a ship, and their calls become more shrill. The louder the ship noise, the louder the whales call. They behave like humans at a noisy restaurant, shouting at one another in high-pitched voices to be heard above the din. In the sea, as in a restaurant, such vocal alterations improve sound transmission, but vital information can still be drowned out. We do not know if the modifications the whales make to their voices are sufficient to allow their calls to be understood despite the noise.

For humans, chronic exposure to noise is stressful, and the same may be true for whales. It would be impossible to ask tankers to stop their engines for days to monitor how noise affects whales, but a quiet period did occur immediately after the terrorist attacks of September 11, 2001. During this time, another scientist, Rosalind Rolland, was collecting whales' feces to test their level of cortisol, a hormone whose secretion increases with stress (Section 31.6). Parks and Rolland

pooled data regarding ship-related noise and cortisol secretion, and found that the brief decline in shipping noise was accompanied by a significant drop in the whales' cortisol level. There was no drop in cortisol immediately after September 11 of other years. These data suggest that the normally high noise level related to shipping causes a chronic stress response in right whales. We know that, in other mammals, chronic stress dampens the immune response and impairs reproduction. Thus, noise-induced stress may slow the recovery of this highly endangered species.

As this example shows, humans can unknowingly interfere with animal signaling. The communication systems of whales and other animals evolved over countless generations in an environment free of human-generated noise, light, and other sensory distractions. Finding ways to meet human needs without unnecessarily disrupting the communication mechanisms of Earth's other species is an ongoing challenge.

Section 39.1 Studies of behavioral differences within a species or among closely related species can shed light on the **proximate causes** of a behavior (what triggers it) and the **ultimate causes** (why it evolved). Some behavioral differences such as variations in fruit fly foraging behavior arise from alleles of a single gene. Most behavior is governed by multiple genes. Behaviors also arise as a result of epigenetic effects.

Section 39.2 **Instinctive behavior** is inborn; it occurs without any prior experience. Instinctive responses are often triggered by a simple stimulus. **Learned behavior** arises in response to experience. Most behavior has a learned component. **Imprinting** is time-sensitive learning. With **habituation**, an animal learns to disregard frequently encountered neutral stimuli. Animals learn landmarks, learn the identity of other individuals, develop conditioned responses, and imitate observed behaviors.

Section 39.3 Animals communicate by means of chemical signals (**pheromones**) as well as visual, acoustical, or tactile signals. Signaling systems are adaptive only when they benefit both the sender and the receiver. Natural selection influences properties of signals, as when courtship calls are easily localized but alarm calls are not. Predators sometimes take advantage of the communication system of their prey.

Section 39.4 Animal mating systems vary among species. Monogamy is rare. Even among species that form pair bonds, either or both partners may mate with others. Both males and females benefit by mating with multiple partners, but males tend to be less choosy about mates than females. Some male birds display for females at a **lek**. In other animals, males establish and defend a **territory**. With territorial systems, some males have many mates, and others do not mate at all. Natural selection operating on traits related to mating may result in **sexual dimorphism.**

Parental care evolves when its cost (in terms of lost opportunities to produce additional offspring later) is offset by its benefit (of increased survival among current offspring). Which parent or parents care for offspring varies among animal groups. Two-parent care is common in birds; in mammals, females are usually the sole care providers.

Section 39.5, 39.6 Some animal species form temporary or long-term groups. The **selfish herd effect** describes how animals group for protection. Animals that live in groups can also benefit by cooperating in tasks such as defense, obtaining food, or rearing the young. **Social animals** are multigenerational groups of cooperating individuals. Costs of group living include increased disease and parasitism, and more intense competition for resources. With a **dominance hierarchy**, resource and mating opportunities are distributed unequally among members of the group.

Altruism can evolve through **kin selection**, in which individuals whose actions result in increased reproduction by their relatives benefit from increased **inclusive fitness**. Cooperative behavior can also evolve through reciprocity: Individuals help those who helped or may help them.

Bees, ants, termites, and some rodents called mole-rats are **eusocial animals**: They live in colonies with overlapping generations and have a reproductive division of labor. Most colony members do not reproduce; they assist their relatives instead. In honeybees, an epigenetic mechanism determines whether a female becomes a worker or a queen.

Application Whales, which rely on vocal communication, alter their behavior in response to noise generated by human activities. Exposure to human-generated noise can stress animals and lower their reproductive output.

SELF-ASSESSMENT
ANSWERS IN APPENDIX VII

1. _____ , a hormone that plays a role in vole mating behavior, also affects human social behavior.
   a. Testosterone   c. FSH
   b. Oxytocin   d. Estrogen

2. In cichlid fish, the transformation from subordinate to dominant phenotype _____ .
   a. involves epigenetic changes
   b. depends on diet
   c. occurs over several years
   d. is an example of altruistic behavior

3. The proximate cause of feeding differences between sitter and rover fruit flies is _____ .
   a. kin selection
   b. natural selection related to competition for food
   c. a difference in alleles at one locus
   d. an environmentally stimulated epigenetic change

4. A honeybee dance transmits information about _____ .
   a. predators   c. the location of food
   b. mating opportunities   d. the amount of honey

5. A _____ is a chemical signal that conveys information between individuals of the same species.
   a. pheromone   c. hormone
   b. neurotransmitter d. releaser

6. In what group of vertebrates are monogamy and cooperative care of the young by two parents most common?
   a. fishes   c. birds
   b. amphibians   d. mammals

7. The reproductive success of _____ is most often limited by access to mates.
   a. males   b. females

NESTLING RESPONSES TO ALARM CALLS In many animals, alarm calls provide information about the type of threat. For example, as adults, Australian songbirds known as scrubwrens give different alarm calls depending on whether a predatory bird is spotted on the ground or in flight. The length of the call indicates the urgency of the threat. Scrubwrens make well-concealed nests on the ground, so predators on the ground pose more of a threat to nestlings than those flying above. By contrast, flying predators pose more of a threat to adults. To gauge the response of scrubwren nestlings to parental alarm calls, Dirk Platzen recorded the calls and played them to 5- and 11-day-old nestlings. **Figure 39.14** shows how playback of the recordings affected the rate at which chicks made peeping sounds.

1. Which type of alarm call reduced peeping most in 11-day-old nestlings? Which had the least effect?

2. Given that a predator on the ground poses the greatest threat to young birds, which response would you expect to be greatest in 5-day-old nestlings? Is it?

3. Which types of calls did chicks become increasingly responsive to as they aged?

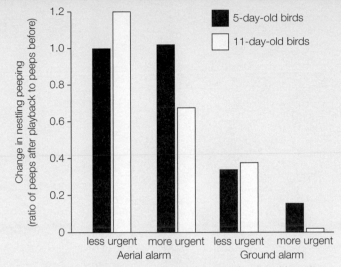

Figure 39.14 Change in peeping behavior of 5-day-old nestlings (black bars) and 11-day-old nestlings (white bars) in response to playback of alarm calls by their parents. A value of 0 indicates complete suppression of nestling peeps, 1 indicates no change, and a value above 1 indicates an increase in peeping.

8. In _____ , all workers are sterile females.
   a. termites
   b. honeybees
   c. naked mole-rats
   d. all of the above

9. Eusocial insects _____ .
   a. live in extended family groups
   b. include termites, honeybees, and ants
   c. show a reproductive division of labor
   d. all of the above

10. Helping other individuals at a reproductive cost to oneself can be adaptive if those helped are _____ .
    a. members of another species
    b. competitors for mates
    c. close relatives

11. A vampire bat sharing food with an unrelated bat that previously fed her is an example of _____ .
    a. kin selection
    b. imprinting
    c. reciprocity
    d. conditioning

12. Tent caterpillars _____ .
    a. live in a multigenerational group
    b. can recruit others to a food source
    c. are all female
    d. are eusocial insects

13. A female lion who shares food with her sister's offspring increases her own _____ .
    a. dominance
    b. reproductive success
    c. sexual dimorphism
    d. inclusive fitness

14. Match the terms with their most suitable description.
    ___ altruism
    ___ selfish herd
    ___ habituation
    ___ lek
    ___ territory
    ___ imprinting
    ___ dominance hierarchy
    ___ classical conditioning

    a. time-dependent learning
    b. communal display area
    c. learning to ignore a stimulus
    d. defended area with resources
    e. assisting another individual at one's own expense
    f. involuntary response becomes tied to a stimulus
    g. individuals hide behind others
    h. unequal distribution of benefits

CLASS ACTIVITY
CRITICAL THINKING

1. For billions of years, the only bright objects in the night sky were stars and the moon. Night-flying moths use these light sources to navigate a straight line. Today, the instinct to fly toward bright objects causes moths to exhaust themselves fluttering around streetlights and banging against brightly lit windowpanes. This behavior clearly is not adaptive, so why does it persist?

2. Many animals preferentially assist individuals with whom they are most familiar. Explain how such a preference could evolve through kin selection.

for additional quizzes, flashcards, and other study materials
▶ **ACCESS MINDTAP AT WWW.CENGAGEBRAIN.COM**

CREDIT: (14) Platzen, D. & Magrath, R., "Adaptive differences in response to two types of parental alarm call in altricial nestlings," Proceedings of the Royal Society of London 2005, Series B; Biological Sciences, vol. 272. pp. 1101–1106, by permission of The Royal Society.

40

Population Ecology

Links to Earlier Concepts

The evolutionary history and genetics of populations were discussed earlier (Sections 1.1, 17.1), as was the expansion of the human population (24.12). You will learn here about experiments (1.5, 1.6) that showed how predation can result in directional selection (17.3) that affects traits of a population.

Core Concepts

Interactions among the components of a biological system give rise to complex properties.

Properties of a population emerge from the characteristics of its members, change over time, and are affected by resource availability. Competition, cooperation, and other interactions among members of a population affect individuals and the population as a whole, as do interactions with the living and nonliving environment.

Organisms exchange matter and energy with the environment in order to grow, maintain themselves, and reproduce.

A population's interactions with its living and nonliving environment involve the exchange of matter and energy. The finite availability of natural resources limits a population's size and influences its other characteristics. Competition for a limited resource can result in natural selection that changes the genetic makeup of a population. Such change is evolution.

The field of biology relies upon experimentation and the collection and analysis of scientific evidence.

Carefully designing experiments helps researchers unravel cause-and-effect relationships in complex natural systems. Observations of natural populations illustrated the interplay of predictable and random events in population growth. Field experiments involving natural populations clarified the role of predation in shaping population characteristics.

◄ This red knot sandpiper is marked with leg bands for an ongoing study of its rapidly declining population.

Photograph by © Kevin Elsby/iStock Images.

713

Ecology is the study of interactions between organisms and their environments. Ecology is not the same as environmentalism, which is advocacy for protection of the environment. However, environmentalists sometimes cite the results of ecological studies when drawing attention to their concerns.

Population ecology investigates the factors that influence the size, distribution, and other properties of natural populations. Data from population ecology studies can be used to make decisions about how to manage a species. For example, information about a specific population can help wildlife managers determine the number of individuals that can be hunted or fished without driving the population into decline. Population ecology studies are also used to assess the effects of other human activities on wild species.

SIZE, DENSITY, AND DISTRIBUTION

A population is a group of organisms that interbreed with one another more often than they interbreed with other members of their species. **Demographics** are statistical characteristics that can be used to describe a population. They include population size, density, and distribution, which we discuss here, as well as age structure and fertility rates, which we consider in Section 40.6. **Population size** is the total number of individuals in a population. **Population density** is the number of individuals per unit area or volume. Examples of population density include the number of dandelions per square meter of lawn or the number of amoebas per milliliter of pond water. **Population distribution** describes the location of individuals relative to one another. Members of

a population may be clumped together, be an equal distance apart, or be distributed randomly. The most common population distribution pattern is a clumped one, in which members of the population are closer to one another than would be predicted by chance alone. A patchy distribution of resources encourages clumping. Hippopotamuses clump in muddy river shallows, for example (**Figure 40.1**). Moisture-loving ferns may cover a damp, north-facing slope and be absent from an adjacent drier and sunnier south-facing slope. A limited ability to disperse increases the likelihood of a clumped distribution too: The nut really does not fall far from the tree. Asexual reproduction is another source of clusters. It results in colonies of coral and vast stands of poplar trees. Finally, some animals have a clumped distribution because they benefit by living in groups.

Intense competition for limited resources can produce a near-uniform distribution, with individuals more evenly spaced than would be expected by chance. Creosote bushes in deserts of the American Southwest grow in this pattern. Competition for limited water among the root systems keeps the plants from growing in close proximity. Similarly, seabirds in breeding colonies often show a near-uniform distribution. Each nesting bird aggressively repels others that get within reach of its beak (**Figure 40.1B**).

A random distribution occurs when resources are distributed uniformly through the environment, and proximity to others neither benefits nor harms individuals. For example, when the wind-dispersed seeds of dandelions land on the uniform environment of a suburban lawn, dandelion plants grow in a random pattern (**Figure 40.1C**).

The scale of the area sampled and timing of a study can influence observed demographics. For example, seabirds are spaced almost uniformly at a nesting site, but the nesting sites are clumped along a shoreline. The birds crowd together during the breeding season, but disperse when breeding is over.

Figure 40.1 Population distribution patterns.

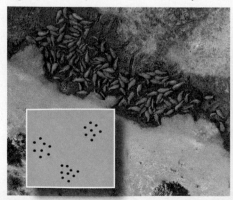  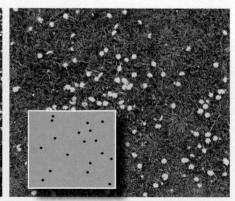

A Clumped distribution of hippopotamuses.   **B** Near-uniform distribution of nesting seabirds.   **C** Random distribution of dandelions.

SAMPLING A POPULATION

It is often impractical to count all members of a population, so biologists frequently use sampling techniques to estimate population size. **Plot sampling** is a method of estimating the total number of individuals in an area based on data from direct counts in some portion of the area. For example, ecologists might estimate the number of grass plants in a grassland, or the number of clams in a mudflat, by measuring the number of individuals in each of several 1-meter by 1-meter square plots. To estimate total population size, scientists first determine the average number of individuals per sample plot. They then multiply that average by the number of plots that would fit in the population's range. Estimates derived from plot sampling are most accurate when species are not very mobile and conditions across their habitat are uniform.

Mark–recapture sampling is used to estimate the population size of mobile animals. With this technique, animals are captured, marked with a unique identifier of some sort, then released. Some time later, scientists capture another group of individuals from the same population. The proportion of marked animals in the second sample is taken to be representative of the proportion marked in the population as a whole. Suppose 100 deer are captured, marked, and released. Later, 50 of these deer are recaptured along with 50 unmarked deer. Marked deer constitute half the recaptured group, so the group previously caught and marked (100 deer) must have been half of the population. Thus the total population is estimated at 200.

Information about the traits of individuals in a sample plot or capture group can be used to infer properties of the whole population. For example, if half the recaptured deer are of reproductive age, half of the population is assumed to share this trait.

Giving each captured animal a unique mark (such as the bands on the sandpiper in the photo that opens the chapter) allows individuals to be followed over time.

---

demographics Statistics that describe a population.

ecology The study of interactions between organisms and their environments.

mark–recapture sampling Method of estimating population size of mobile animals by marking individuals, releasing them, then checking the proportion of marks among individuals recaptured at a later time.

plot sampling Using demographics observed in sample plots to estimate demographics of a population as a whole.

population density Number of individuals per unit area.

population distribution Location of population members relative to one another: clumped, uniformly dispersed, or randomly dispersed.

population size Total number of individuals in a population.

Dr. Karen DeMatteo

National Geographic Explorer Karen DeMatteo has an unusual field assistant. A Chesapeake Bay retriever named Train helps her study the population structure and movements of jaguars, pumas, ocelots, oncillas, and bush dogs. These predators occur at low density in the dense South American forest, so locating and tracking them is a challenge. DeMatteo could have used standard survey technology (such as radio telemetry or camera traps) to get information, but those methods have some weaknesses. For example, an animal could avoid the part of the forest where a camera is set up. So, DeMatteo decided to look for what the animals leave behind in the forest—their scat (feces).

This is where Train plays a role. Like other dogs, he has a terrific sense of smell, and he has been trained to use it to find specific types of scat. He behaves like a drug-sniffing dog, except instead of alerting his handler when he finds a drug, he alerts DeMatteo when he finds predator scat. DeMatteo collects the scat that Train finds, and later analyzes it in the laboratory. Analysis of material from scat can reveal the species of the individual that produced it, as well as that individual's sex. Results of the scat analysis, along with GPS and spatial mapping locations of where scat was found, allow DeMatteo to determine the predators' distribution and understand differences in habitat use. DeMatteo says, "The goal is to expand our knowledge of how these species move through the landscape so we can determine locations for biological corridors/wildlife crossings that maximize animal movement and minimize human–wildlife conflict."

List the factors that affect the number of individuals in a population.

Provide the equation that describes exponential growth and explain what the terms in the equation mean.

Define biotic potential and give examples of organisms that differ in their biotic potential.

IMMIGRATION AND EMIGRATION

In nature, populations continually change in size. Individuals are added to a population by births and **immigration**, the arrival of new residents that previously belonged to another population. Individuals are removed from it by deaths and **emigration**, the departure of individuals who take up permanent residence elsewhere.

In many animal species, young of one or both sexes leave the area where they were born to breed elsewhere. For example, young freshwater turtles typically emigrate from their parental population and become immigrants at another pond some distance away. By contrast, seabirds typically breed where they were born. However, some individuals may emigrate and end up at breeding sites more than a thousand kilometers away. In most species, the tendency of individuals to emigrate to a new breeding site is related to resource availability and crowding. As resources decline and crowding increases, the likelihood of emigration rises.

ZERO TO EXPONENTIAL GROWTH

If we set aside the effects of immigration and emigration, we can define **zero population growth** as an interval during which the number of births is balanced by an equal number of deaths. As a result, population size remains unchanged, with no net increase or decrease in the number of individuals.

We can measure births and deaths in terms of rates per individual, or per capita. *Capita* means head, as in a head count. Subtract a population's per capita death rate (d) from its per capita birth rate (b) and you have the **per capita growth rate**, or r:

$$\underset{\substack{\text{(per capita} \\ \text{birth rate)}}}{b} - \underset{\substack{\text{(per capita} \\ \text{death rate)}}}{d} = \underset{\substack{\text{(per capita} \\ \text{growth rate)}}}{r}$$

Imagine 2,000 mice living in the same field. If 1,000 mice are born each month, then the birth rate is 0.5 births per mouse per month (1,000 births/2,000 mice). If 200 mice die one way or another each month, then the death rate is 200/2,000 or 0.1 deaths per mouse per month. Thus, r is 0.5 − 0.1, or 0.4 per mouse per month.

Starting Size of Population		Net Monthly Increase	New Size of Population
2,000	× r =	800	2,800
2,800	× r =	1,120	3,920
3,920	× r =	1,568	5,488
5,488	× r =	2,195	7,683
7,683	× r =	3,073	10,756
10,756	× r =	4,302	15,058
15,058	× r =	6,023	21,081
21,081	× r =	8,432	29,513
29,513	× r =	11,805	41,318
41,318	× r =	16,527	57,845
57,845	× r =	23,138	80,983
80,983	× r =	32,393	113,376
113,376	× r =	45,350	158,726
158,726	× r =	63,490	222,216
222,216	× r =	88,887	311,103
311,103	× r =	124,441	435,544
435,544	× r =	174,218	609,762
609,762	× r =	243,905	853,667
853,667	× r =	341,467	1,195,134

A Increases in size over time. Note that the net increase becomes larger with each generation.

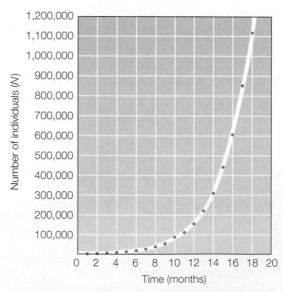

B Graphing numbers over time produces a J-shaped curve.

Figure 40.2 Exponential growth in a hypothetical population of mice with a per capita rate of growth (r) of 0.4 per mouse per month and a population size of 2,000.

As long as r remains constant and greater than zero, **exponential growth** will occur, which means that the population's size will increase by the same proportion of its total in every successive time interval. We can calculate population growth (G) for each interval based on the number of individuals (N) and the per capita growth rate:

$$
\begin{array}{ccccc}
N & \times & r & = & G \\
\text{(number of} & & \text{(per capita} & & \text{(population growth} \\
\text{individuals)} & & \text{growth rate)} & & \text{per unit time)}
\end{array}
$$

Let's use this equation to study growth of our hypothetical population of field mice. In one month, the population will expand from 2,000 to 2,800 individuals (**Figure 40.2A**). A net increase of 800 mice has increased the number of breeders. If all of the mice reproduce, the population size will expand by 1,120 individuals (2,800 × 0.4) in the next month. The total population size is now 3,920. At this growth rate, the number of mice would rise from 2,000 to more than 1 million in under two years! Graphing the increases against time results in a J-shaped curve, which is characteristic of exponential population growth (**Figure 40.2B**).

With exponential growth, the number of new individuals increases with each generation, even though the per capita growth rate stays the same. Exponential population growth is analogous to compound interest in a bank account. The annual interest *rate* stays fixed, yet every year the *amount* of interest paid increases. The annual interest paid into the account adds to the size of the balance, so the next interest payment will be based on the increased balance. Similarly, with exponential growth, the number of new individuals increases in each generation.

Exponential growth will occur in any population in which the birth rate exceeds the death rate—in other words, as long as r is greater than zero. Imagine a single bacterium in a culture flask. After thirty minutes, the cell divides in two. Those two cells divide, and so on every thirty minutes. If no cells die between divisions, then the population size will double in every interval—from 1 to 2, then 4, 8, 16, 32, and so

---

biotic potential (by-AH-tick) Maximum possible population growth rate under optimal conditions.

emigration Movement of individuals out of a population.

exponential growth A population grows by a fixed percentage in successive time intervals; the size of each increase is determined by the current population size.

immigration Movement of individuals into a population.

per capita growth rate (r) Of a population, per capita (per individual) birth rate minus per capita death rate.

zero population growth Interval during which the number of births equals the number of deaths.

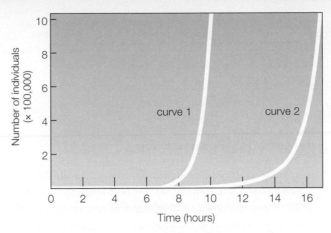

Figure 40.3 Effect of death rate on population increase.
Consider two hypothetical populations of bacteria; both start out with one cell and both reproduce asexually every half hour.

Curve 1 shows how population size increases when no cells die between divisions.

Curve 2 shows how population size increases when 25 percent of the cells die between divisions. Note that these deaths slow the rate of increase, but as long as the birth rate exceeds the death rate, exponential growth will continue.

on. After 9–1/2 hours, there have been nineteen doublings, so the population now consists of more than 500,000 cells. Ten hours (twenty doublings) later, there are more than a million. Curve 1 in **Figure 40.3** is a plot of this increase.

Now suppose that 25 percent of the descendant cells die every thirty minutes in our hypothetical population of bacteria. In this scenario, it takes seventeen hours, not ten, for that population to reach 1 million. Thus, deaths slow the rate of increase but do not stop exponential growth (curve 2 in **Figure 40.3**).

BIOTIC POTENTIAL

The growth rate for a population under ideal conditions is its **biotic potential**. This is a theoretical rate at which the population would grow if shelter, food, and other essential resources were unlimited and there were no predators or pathogens. Microbes such as bacteria have some of the highest biotic potentials, whereas large-bodied mammals have some of the lowest. Factors that affect biotic potential include the age at which reproduction typically begins, how long individuals remain reproductive, and the number of offspring that are produced each time an individual reproduces. Section 40.4 considers how natural selection influences these factors.

Regardless of the species, populations seldom reach their biotic potential because of the effects of limiting factors, a topic we discuss in detail in the next section.

DENSITY-DEPENDENT FACTORS

No population can grow exponentially forever. As the degree of crowding increases, **density-dependent limiting factors** cause birth rates to slow and death rates to rise, so the rate of population growth decreases. These factors include predation, parasitism and disease, and competition for a limited resource.

Any natural area has limited resources. Thus, an increase in the number of individuals in an area leads to increased **intraspecific competition**—competition for a resource among members of the same species. As a result of increased competition, some individuals may fail to secure what they need to survive and reproduce. Competition has a detrimental effect even on winners, because energy they use in competition for resources is not available for reproduction. Essential resources for which animals might compete include food, water, hiding places, and nesting sites (**Figure 40.4**). Plants compete for nutrients, water, and access to sunlight.

The negative effects of parasites and contagious disease increase with crowding because the closer individuals are to one another, the more easily parasites and pathogens can spread. Predation increases with density too, because predators often concentrate their efforts on the most abundant prey.

LOGISTIC GROWTH

Logistic growth occurs when density-dependent factors affect population size over time, so that a plot of numbers versus time yields an S-shaped curve (**Figure 40.5**). When the population is small, density-dependent limiting factors have little effect and the population grows exponentially ❶. Then, as population size and the degree of crowding rise, these factors begin to limit growth ❷. Eventually, the population size levels off at the carrying capacity ❸. **Carrying capacity (K)** is the maximum number of individuals of a particular species that a population's environment can support indefinitely.

The equation that describes logistic growth is:

$$(r \times N)\frac{(K - N)}{K} = G$$

As with the exponential growth equation, G is growth per unit time, r is the per capita growth rate, and N is current population size. K is carrying capacity. The $(K-N)/K$ part of the equation represents the proportion of carrying capacity not yet used. As a population grows, this proportion decreases, so G becomes smaller and smaller. At carrying capacity, the equation become $G = r \times N \times 0$, which means the size of the population cannot increase.

Figure 40.4 Example of a limiting factor. Wood ducks build nests only inside tree hollows of specific dimensions (left). In some places, lack of access to appropriate hollows now limits the size of the wood duck population. Adding artificial nesting boxes (right) to these environments can help increase the size of the duck populations.

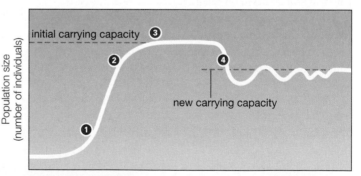

Figure 40.5 Logistic growth. Note the initial S-shaped curve.

❶ At low density, the population grows exponentially.

❷ As crowding increases, density-dependent limiting factors slow the rate of growth.

❸ Population size levels off at the carrying capacity.

❹ Any change in the carrying capacity will result in a corresponding shift in population size.

FIGURE IT OUT
How does adding nest boxes affect the carrying capacity for wood duck populations?

Answer: It increases carrying capacity

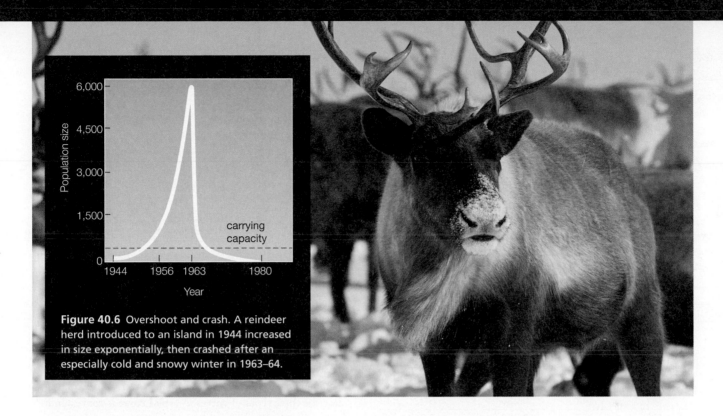

Figure 40.6 Overshoot and crash. A reindeer herd introduced to an island in 1944 increased in size exponentially, then crashed after an especially cold and snowy winter in 1963–64.

Carrying capacity is species-specific, environment-specific, and can change over time ❹. For example, the carrying capacity for a plant species decreases when nutrients in the soil become depleted. Human activities can affect carrying capacity. For example, human harvesting of horseshoe crabs has decreased the carrying capacity for red knot sandpipers (shown in the chapter opening photo). Horseshoe crab eggs are the birds' main food during their long-distance migration.

DENSITY-INDEPENDENT FACTORS

Sometimes, natural disasters or weather-related events affect population size. A volcanic eruption, hurricane, or flood can decrease population size. So can human-caused events such as oil spills. These events are called **density-independent limiting factors**, because crowding does not influence the likelihood of their occurrence or the magnitude of their effect.

In nature, density-dependent and density-independent factors often interact to determine a population's size. Consider what happened after the 1944 introduction of 29 reindeer to St. Matthew Island, an uninhabited island off the coast of Alaska. When biologist David Klein visited the island in 1957, he found 1,350 well-fed reindeer (**Figure 40.6**). Klein returned in 1963 and counted 6,000 reindeer. The population had soared far above the island's carrying capacity. A population can temporarily overshoot an environment's carrying capacity, but the high density cannot be sus-

tained. Klein observed that some effects of density-dependent limiting factors were already apparent. For example, the average body size of the reindeer had decreased. When Klein returned in 1966, only 42 reindeer survived, and only one was male. There were no fawns. Thousands of reindeer had starved to death during the winter of 1963–1964. That winter was unusually harsh, with low temperatures, high winds, and 140 inches of snow. Most reindeer, already in poor condition as a result of increased competition, starved when deep snow covered their food. A decline in the number of reindeer in this population had been expected—populations that overshoot their environment's carrying capacity will necessarily shrink—but bad weather magnified the extent of the crash. By the 1980s, there were no reindeer left on the island.

carrying capacity (K) Maximum number of individuals of a species that a particular environment can sustain; can change over time.

density-dependent limiting factor Factor that limits population growth and has a greater effect in denser populations; for example, competition for a limited resource.

density-independent limiting factor Factor that limits population growth and arises regardless of population size; for example, a flood.

intraspecific competition Competition for resources among members of the same species.

logistic growth A population grows exponentially at first, then growth slows as population size approaches the environment's carrying capacity for that species.

TABLE 40.1

Life Table for a Cohort of an Annual Plant Species*

Age Interval (days)	Survivorship (number surviving at start of interval)	Number Dying During Interval	Death Rate (number dying/ number surviving)	"Birth" Rate During Interval (number of seeds from each plant)
0–63	996	328	0.329	0
63–124	668	373	0.558	0
124–184	295	105	0.356	0
184–215	190	14	0.074	0
215–264	176	4	0.023	0
264–278	172	5	0.029	0
278–292	167	8	0.048	0
292–306	159	5	0.031	0.33
306–320	154	7	0.045	3.13
320–334	147	42	0.286	5.42
334–348	105	83	0.790	9.26
348–362	22	22	1.000	4.31
362–	0	0	0	0
		996		

* *Phlox drummondii*; data from W. J. Leverich and D. A. Levin, 1979.

A population's growth rate is affected by its members' **life history**—the manner in which individuals allocate resources to growth, survival, and reproduction over the course of their lifetimes. Survival-related traits include the probability of surviving to a given age, and of dying at specific ages. Traits related to reproduction include the age at which reproduction begins, the frequency of reproduction, the number of offspring produced by each reproductive event, and the extent of parental investment in each offspring.

TIMING OF BIRTHS AND DEATHS

One way to investigate life history traits is to focus on a **cohort**—a group of individuals born during the same interval—from their time of birth until the last one dies. Ecologists often divide a natural population into age classes and record the age-specific birth rates and mortality. The resulting data is summarized in a life table (**Table 40.1**).

Information about age-specific death rates can also be illustrated by a **survivorship curve**, a plot that shows how many members of a cohort remain alive over time. Ecologists describe three types of curves. A type I curve is convex, indicating that the death rate remains low until relatively late in life (**Figure 40.7A**). Humans and other large mammals that produce and care for one or two offspring at a time have this pattern. A diagonal, type II curve indicates that the death rate of the population does not vary much with age (**Figure**

Figure 40.7 Survivorship curves. Gray lines are theoretical curves. Red dots are data from field studies.

? FIGURE IT OUT
Based on the death rates listed for the annual plant in Table 40.1, what type of survivorship curve does that plant have?

Answer: A type III curve, with mortality highest early in life.

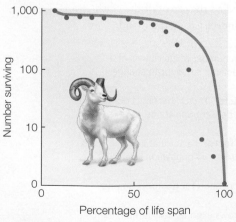

A Type I curve. Mortality is highest late in life. Data for Dall sheep (*Ovis dalli*).

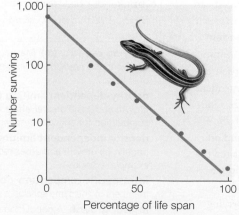

B Type II curve. Mortality varies little with age. Data for five-lined skink (*Eumeces fasciatus*).

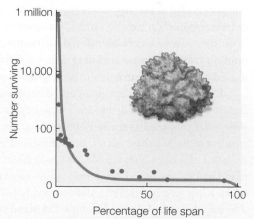

C Type III curve. Mortality is highest early in life. Data for a desert shrub (*Cleome droserifolia*).

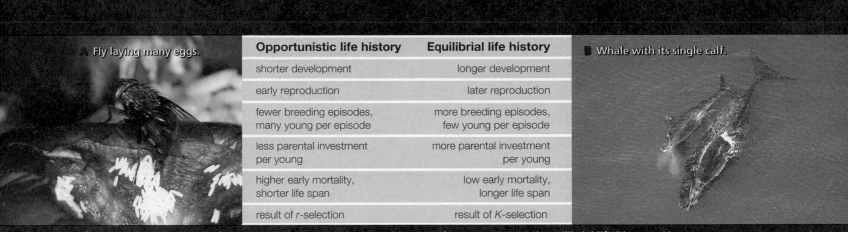

Opportunistic life history	Equilibrial life history
shorter development	longer development
early reproduction	later reproduction
fewer breeding episodes, many young per episode	more breeding episodes, few young per episode
less parental investment per young	more parental investment per young
higher early mortality, shorter life span	low early mortality, longer life span
result of *r*-selection	result of *K*-selection

A Fly laying many eggs.

B Whale with its single calf.

Figure 40.8 Two types of life history. Most species have a combination of opportunistic and equilibrial life history traits.

40.7B). In lizards, small mammals, and large birds, old individuals are about as likely to die of disease or predation as young ones. A type III curve is concave, indicating that the death rate for a population peaks early in life (**Figure 40.7C**). Marine animals that release eggs into water have this type of curve, as do plants that release huge numbers of tiny seeds.

r-SELECTION AND *K*-SELECTION

To produce offspring, an individual must invest resources that it could otherwise use to grow and maintain itself. Species differ in the ways which they distribute parental investment among offspring. Life history patterns vary continuously among species, but ecologists have described two theoretical extremes at either end of this continuum. Which pattern evolves will depend on what particular allocation of resources to growth, survivorship, and reproduction will maximize the number of offspring that survive to adulthood.

When a species lives where conditions vary in an unpredictable manner, its populations seldom reach the carrying capacity of their environment. As a result, there is little competition for resources, and deaths occur mainly as a result of density-independent factors. These conditions favor an opportunistic life history, in which individuals produce as many offspring as possible, as quickly as possible. Opportunistic species are said to be subject to *r*-selection, because they maximize *r*, the per capita growth rate. They tend to have a short generation time and small body size. Opportunistic species usually have a type III survivorship curve, with mortality heaviest early in life. For example, weedy plants such as dandelions have an opportunistic life history. They mature within weeks, produce many tiny seeds, then die. Flies are opportunistic animals. A female fly can lay hundreds of eggs in a temporary food source such as a rotting tomato (**Figure 40.8A**).

When a species lives in a more stable environment, its populations often approach carrying capacity. Under these circumstances, the ability to successfully compete for resources has a major influence on reproductive success. Thus, an equilibrial life history, in which parents produce a few, high-quality offspring, is adaptive. Equilibrial species are shaped by *K*-selection, in which adaptive traits provide a competitive advantage when population size is near carrying capacity (*K*). Such species tend to have a large body and a long generation time. This type of life history is typical of large mammals that take years to reach adulthood and begin reproducing. For example, a female blue whale reaches maturity at the age of 6 to 10 years. She then produces only one large calf at a time, and continues to invest in the calf by nursing it after its birth (**Figure 40.8B**). Similarly, a coconut palm grows for years before beginning to produce a few coconuts at a time. In both whales and coconut palms, a mature individual produces young for many years. Many equilibrial species have a type I survivorship curve, with mortality heaviest later in life.

Some species have combinations of traits that cannot be explained by *r*-selection or *K*-selection alone. For example, Atlantic eels and Pacific salmon are among the few vertebrates that have a one-shot reproductive strategy. They breed once, then die. This strategy can evolve when opportunities for reproduction are unlikely to be repeated. In the eels and salmon, physiological changes related to migration between fresh water and salt water make a repeat journey impossible.

cohort Group of individuals born during the same interval.

K-selection Pattern of selection in which adaptive traits allow their bearers to outcompete others for limited resources; occurs when a population is near its environment's carrying capacity.

life history Schedule of how resources are allocated to growth, survival, and reproduction over a lifetime.

r-selection Pattern of natural selection in which producing the most offspring the most quickly is adaptive; occurs when population density is low and resources are abundant.

survivorship curve Graph showing how many members of a cohort remain alive over time.

40.5 Predation Effects on Life History

LEARNING OBJECTIVES

Using the example of guppies, explain how predators act as selective agents on the life history patterns of prey species.

Explain how the study of life history data can help in conservation.

Many predators prefer prey of a specific size, and individuals of most prey species change in size over their lifetime. Thus, predation can affect life history traits of prey. When predators prefer large prey, prey individuals who reproduce when still small and young are at a selective advantage. When predators focus on small prey, fast-growing individuals have the selective advantage.

AN EXPERIMENTAL STUDY

A long-term study by the evolutionary biologists John Endler and David Reznick illustrates the effect of predation on life history traits. Endler and Reznick studied populations of guppies, small fishes that are native to shallow freshwater streams in the mountains of Trinidad (**Figure 40.9**). The scientists focused their attention on a region where many small waterfalls prevent guppies in one part of a stream from moving easily to another. As a result of these natural barriers, each stream holds several populations of guppies that have very little gene flow between them (Section 17.6).

The waterfalls also keep guppy predators from moving from one part of the stream to another. The main guppy

Figure 40.9 How predation affects life history traits in guppies. David Reznick (shown above) and John Endler carried out an experiment using wild guppies that had evolved in the presence of a predator (cichlid) that preferentially preys on large guppies. Control group guppies remained in their home pool with cichlids. Other guppies were moved to a guppy-free pool with a predator (killifish) that preferentially eats small guppies. Data at left show the average age and weight at maturity for descendants of both groups 11 years after the study began.

CREDITS: (9) photo, © Helen Rodd; art, Based on data from Reznick D.A., Bryga H., and Endler J.A. (1990) *Nature* 346: 357–359.

predators, killifishes and cichlids, differ in their size and prey preferences. The relatively small killifish preys mostly on immature guppies, and ignores the larger adults. The cichlids are larger fish. They tend to pursue large, mature guppies and ignore small ones.

In some regions of the stream, guppies are exposed to one type of predator but not the other. Reznick and Endler looked at the life history traits of guppies in different regions to see if they varied. They discovered that guppies in regions where cichlids are the only predators grow faster and are smaller at maturity than guppies in regions where killifish are the only threat. Guppies in populations preyed on by cichlids alone also reproduce earlier, produce more offspring each time they breed, and breed more frequently.

The researchers were not sure whether the observed differences in life history traits were genetic, or a result of some environmental variation between regions. To find out, they collected guppies from both cichlid- and killifish-dominated streams. They reared the groups in separate aquariums under identical predator-free conditions. Two generations later, the guppy groups continued to have the differences observed in natural populations. The researchers concluded that the predator-associated life history differences they had observed between guppies have a genetic basis.

Reznick and Endler hypothesized that the predators act as selective agents on guppy life history patterns. They made a prediction: If life history traits evolve in response to predation, then these traits will change in a population after a new predator that favors different prey traits is introduced. To test their prediction, they carried out an experiment. They located a stream region that was above a waterfall and had killifish but no guppies or cichlids. To this region, they introduced guppies from a site below the waterfall, where there were cichlids but no killifish. Thus, at the experimental site, guppies that had previously lived only with cichlids (which eat large guppies) were now exposed to killifish (which eat smaller ones). The control site was the downstream region below the waterfall, where relatives of the transplanted guppies still coexisted with cichlids.

Reznick and Endler revisited the stream over the course of eleven years and thirty-six generations of guppies. They monitored traits of guppies above and below the waterfall. The recorded data showed that guppies at the upstream experimental site were evolving. Exposure to a previously unfamiliar predator caused changes in the guppies' rate of growth, age at first reproduction, and other life history traits. By contrast, guppies at the control site showed no such changes. Reznick and Endler concluded that life history traits in guppies can evolve rapidly in response to the selective pressure exerted by predation.

Figure 40.10 Fishermen with a prized catch, a large Atlantic codfish. Both sport fishermen and commercial fisherman preferentially harvested the largest codfish.

COLLAPSE OF A FISHERY

The evolution of life history traits in response to predation is not merely of theoretical interest. It has economic importance. Just as guppies evolved in response to predators, a population of Atlantic codfish (*Gadus morhua*) evolved in response to human fishing pressure. From the mid-1980s to early 1990s, the number of fishing boats targeting the North Atlantic population of codfish increased. As the yearly catch rose, the age at which codfish become sexually mature shifted; fishes that reproduce while young and small increased in frequency in the population. These early-reproducing individuals were at an advantage because both commercial fisherman and sports fishermen preferentially caught and kept larger fish (**Figure 40.10**). Fishing pressure continued to rise until 1992, when declining cod numbers caused the Canadian government to ban cod fishing in some areas. That ban, and later restrictions, came too late to stop the Atlantic cod population from crashing. In some areas, the population declined by 97 percent and still shows no signs of recovery.

Looking back, it is clear that life history changes were an early sign that the North Atlantic cod population was in trouble. Had biologists recognized what was happening, they might have been able to save the fishery and protect the livelihood of more than 35,000 fishers and associated workers. Ongoing monitoring of the life history data for other economically important fishes may help prevent similar disastrous crashes in the future.

41

Community Ecology

Links to Earlier Concepts

In this chapter you will revisit biogeography and take a closer look at global patterns in species distribution (Section 16.1). You will revisit effects of pathogens (19.2) and how species interactions lead to natural selection (17.10). You will see examples of field experiments (1.6) and apply what you know about populations and their growth (40.1–40.3).

Core Concepts

Interactions among the components of a biological system give rise to complex properties.

Interactions among organisms in a community involve the exchange of energy and matter. These interactions are complex: A change in any one component can affect the entire community. Community structure, including species variety, abundance, and patterns of distribution, changes over time as a result of disturbances, environmental change, and random factors.

Adaptation of organisms to a variety of environments has resulted in diverse structures and physiological processes.

Each species is adapted to a specific habitat, and other species in its biological community are one component of that habitat. Cooperative behavior and symbiotic relationships contribute to the survival of participating populations. Competition between species is a selective pressure that sometimes leads to changes in resource use.

The field of biology relies upon experimentation and the collection and analysis of scientific evidence.

Carefully designing experiments helps researchers unravel cause-and-effect relationships in complex natural systems. The interactions among populations of a community, and the effects of disturbances or environmental change, can be modeled mathematically. However, random factors and disturbances that affect community structure make future change difficult to predict.

◀ A red-billed oxpecker perches on an impala. Oxpeckers benefit impalas and other grazing animals by removing and eating the ticks that parasitize the grazers.

41.1 Factors That Shape Communities

LEARNING OBJECTIVES

Using appropriate examples, describe some factors that influence community structure.

Distinguish between the two components of species diversity.

Explain symbiosis and give an example.

TABLE 41.1

Direct Two-Species Interactions

Types of Interaction	Effect on Species 1	Effect on Species 2
Commensalism	Helpful	None
Mutualism	Helpful	Helpful
Interspecific competition	Harmful	Harmful
Predation, herbivory, parasitism, parasitoidism	Helpful	Harmful

A biological **community** includes all the populations of all species that live in a particular place at the same time. The scale of a community is defined by a human observer, so one community can contain smaller communities or be a component of a larger one. Consider the community of microbial organisms that live inside the digestive tract of a termite. That termite is part of a larger community of organisms that live in and on a fallen log. The many log-dwellers are part of a still larger forest community.

Communities can differ in species diversity even if they are of similar size. There are two components to diversity. The first, **species richness**, refers to the number of different species in a community. The second is **species evenness**, or the relative abundance of each species. For example, a pond that has five fish species in nearly equal numbers has a higher species diversity than a pond with one abundant fish species and four rare ones.

Community structure is dynamic, which means that in any community, the array of species and their relative abundances tend to change over time. Communities change over a long time span as they form and then age. They also change over the short term as a result of disturbances.

Each species can only live in a specific **habitat**, a place with resources it needs and conditions it can tolerate. Thus, geography and climate affect community structure. Factors such as soil quality, sunlight intensity, rainfall, and temperature vary with latitude and elevation. Tropical (low latitude) regions receive the most sunlight energy and have the most even temperature. For most plant and animal groups, the number of species is greatest near the equator, and declines as you move toward the poles. Tropical forest communities have more types of trees than temperate ones. Similarly, tropical reef communities are more diverse than marine communities farther from the equator.

Species interactions also influence community structure. In some cases, the effect is indirect. For example, when songbirds eat caterpillars, the birds indirectly benefit the trees that the caterpillars feed on, while directly reducing the abundance of caterpillars.

Biologists categorize direct species interactions by their effects on both participating species (**Table 41.1**). For example, **commensalism** helps one species and has no effect on the other. Commensal orchids live attached to the trunk or branches of a tree (**Figure 41.1**). Having a perch in the light benefits the orchid, and the tree is unaffected. Relationships are considered commensal only when one species benefits and the other neither benefits nor is harmed by the relationship. If evidence of helpful or harmful effects is discovered, the relationship is reclassified.

Species interactions may be fleeting or longer term. **Symbiosis**, which means "living together," refers to a relationship in which two species have a prolonged close association. Mutualism, parasitism, and commensalism can be symbioses. Two species that interact closely for generations often coevolve. As Section 17.10 explained, coevolution is an evolutionary process in which each species acts as a selective agent that shifts the range of variation in the other.

---

commensalism Species interaction that benefits one species and neither helps nor harms the other.
community All populations of all species living in a particular area.
habitat Type of environment in which a species typically lives.
species evenness Of a community, the relative abundance of species.
species richness Of a community, the number of species.
symbiosis (sim-by-OH-sis) One species lives in or on another in a commensal, mutualistic, or parasitic relationship.

Figure 41.1 Commensal orchids on a tree trunk. The orchids benefit by growing on the tree, which is unaffected by their presence.

41.2 Mutualism

LEARNING OBJECTIVES

Using appropriate examples, describe the types of benefits that can be exchanged by partners in a mutualism.

Explain why mutualism is considered reciprocal exploitation.

Mutualism is an interspecific interaction that benefits both participants. Flowering plants and their pollinators are a familiar example. In the most extreme cases, two species coevolve and become mutually dependent. For example, each species of yucca plant is pollinated by one species of yucca moth, whose larvae develop only on that plant (**Figure 41.2**). Mutualistic relationships are typically less exclusive: Most flowering plants have more than one pollinator, for example, and most pollinators service more than one species of plant.

Photosynthetic organisms often supply food for nonphotosynthetic partners, as when plants lure pollinators with sugary nectar. In addition, many plants make fruits that attract seed-dispersing animals. Plants also provide sugars to mycorrhizal fungi and nitrogen-fixing bacteria. The plants' fungal or bacterial symbionts return the favor by supplying their host with other essential nutrients. Similarly, photosynthetic dinoflagellates provide sugars to reef-building corals, and photosynthetic bacteria or algae in a lichen feed their fungal partner. All are examples of mutualisms.

Many animals have mutualistic microorganisms living in their digestive tract. For example, *Escherichia coli* bacteria in your colon provide you with vitamin K, and you provide them with a steady food supply and a warm habitat.

Other mutualisms involve protection. For example, an anemonefish and a sea anemone fend off one another's preda-

Figure 41.3 Mutual protection. The stinging tentacles of this sea anemone protect its partner (a pink anemonefish) from fish-eating predators. In return, the anemonefish chases away fish that eat sea anemone tentacles. The anemonefish secretes a special mucus that prevents the anemone from stinging it.

tors (**Figure 41.3**). Ants protect bull acacia trees from leaf-eating insects, and in return the tree houses the ants in special hollow thorns and provides them with sugar-rich food. The oxpecker bird shown in the chapter's opening photo protects its partner from ticks, which it eats.

From an evolutionary standpoint, mutualism is best described as reciprocal exploitation. Each individual increases its fitness by extracting a resource, such as protection or food, from its partner. If taking part in the mutualism has a cost, then minimizing that cost is evolutionarily advantageous. Most flowering plants produce the minimum amount of nectar necessary to attract pollinators. Producing more nectar than necessary would waste resources that could otherwise be used in growth and reproduction.

mutualism Species interaction that benefits both species.

Figure 41.2 Obligate mutualists: a yucca moth on a yucca flower. Yucca moths mate on these flowers. Afterward, the female gathers pollen from one flower and carries it with her to another flower that has not yet begun to produce pollen. She lays her eggs in the second flower's ovary, then uses the pollen she brought along to fertilize it. By pollinating the flower, she ensures that seeds will form to feed her larvae. Moth larvae eat some seeds, but plenty survive to give rise to new yucca plants.

41.3 Interspecific Competition

LEARNING OBJECTIVES

Using appropriate examples, describe the two types of interspecific competition.

Describe the process of competitive exclusion.

Explain the role of character displacement in resource partitioning.

TYPES OF COMPETITION

Interspecific competition is competition among members of different species. It is not usually as intense as intraspecific competition (competition within a species). The requirements of two species might be similar, but they are never as close as they are for members of the same species.

Each species has a unique set of ecological requirements and roles that we refer to as its **ecological niche**. Both physical and biological factors define the niche. Aspects of an animal's niche include the temperature range it can tolerate, the species it eats, and the places it can breed. A description of a

A Interference competition between scavengers.
A golden eagle attacks a fox with its talons to drive the fox away from a resource—the moose carcass in the foreground.

B Exploitative competition between insect eaters.
Sundew plants (left) and wolf spiders (right) both feed on insects. The presence of one species reduces the food available to the other.

Figure 41.4 Interspecific competition.

flowering plant's niche would include its soil, water, light, and pollinator requirements. The more similar the niches of two species are, the more intensely those species will compete.

Competition generally takes one of two forms. With interference competition, one species actively prevents the other from accessing some limited resource. As an example, one species of scavenger will often chase another away from a carcass (**Figure 41.4A**). Plants engage in interference competition too, although it is less obvious. For example, some plants use chemicals to fend off potential competitors. Aromatic compounds that ooze from tissues of sagebrush plants, black walnut trees, and eucalyptus trees seep into the soil and prevent most plants from germinating or growing nearby.

In exploitative competition, species do not interact directly, but by using the same resource, each reduces the amount of that resource available to the other. For example, wolf spiders and carnivorous plants called sundews both feed on insects in Florida swamps (**Figure 41.4B**). These two very different organisms do not fight over food, but they do compete. By catching insects, each reduces the number of insects available to the other.

COMPETITIVE EXCLUSION

Competition between species will be most intense when the supply of a shared resource is an important limiting factor for both species.

In the 1930s, G. Gause conducted experiments with two species of ciliated protozoans (*Paramecium*) that compete for the same prey: bacteria. Gause cultured the *Paramecium* species separately and together (**Figure 41.5**). Within weeks, population growth of one species outpaced the other, which went extinct. This and other experiments are the basis for the concept of **competitive exclusion**: Whenever two species

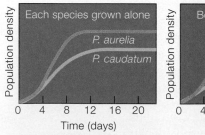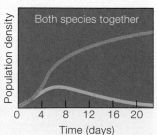

Figure 41.5 Competitive exclusion.
Growth curves for *Paramecium aurelia* (orange) and *P. caudatum* (green) when grown separately (left) and together (right).

? **FIGURE IT OUT**
Which *Paramecium* species is the better competitor?

Answer: *P. aurelia*

CREDITS: (4A) © Pekka Komi; (4B) left, scaners3d/Shutterstock.com; right, Cathy Keifer/Shutterstock.com.

require the exact same limited resource to survive or reproduce, the better competitor will drive the less competitive species to extinction in that habitat.

When resource needs of competitors are not exactly the same, competing species can coexist, but the presence of each reduces the carrying capacity (Section 40.3) of the habitat for the other. For example, the reproductive success of a flowering plant is decreased by competition from other species that flower at the same time and share the same pollinators. As another example, consider again those fly-eating sundew plants. A sundew grown in the presence of wolf spiders makes fewer flowers than one grown in a spider-free area. Competition for insect prey reduces the availability of resources that a sundew could otherwise use to make flowers.

RESOURCE PARTITIONING

Resource partitioning is an evolutionary process by which different species become adapted to share a limiting resource in a way that minimizes competition. Consider three species of annual plants that commonly coexist in abandoned fields. All require water and nutrients, but the roots of each species extend to and take resources up from soil at a different depth. This variation in root form allows the plants to coexist.

Resource partitioning arises as a result of directional selection that occurs when species with similar requirements share a habitat and compete for a limiting resource. In each species, those individuals whose traits minimize their need to compete with the other species for the resource will be at a selective advantage. Thus, competition between species can lead to **character displacement**: The range of variation for one or more characters shifts in a direction that lessens the intensity of competition for a limiting resource. In short, the resource needs of the competing species become less similar.

Results from a long-term study of Galápagos finches demonstrate how character displacement can occur. In 1982, an established population of medium ground finches on one of the Galápagos islands was joined by a new competitor, the large ground finch. Both types of ground finches eat seeds, but the large ground finch (which has a bigger beak) is better at opening big seeds. From 2003 to 2005 a drought decreased the availability of all seeds, competition for food increased, and populations of both finch species declined. By the end of this period, the distribution of beak size among the medium ground finches had shifted; smaller beaks had become more common. Smaller-beaked individuals had survived the drought-induced seed shortage better than larger-beaked members of their species because they concentrated their efforts on smaller seeds. Thus, they faced less competition for food from the large ground finches.

ENGAGE: PROCESS OF SCIENCE

Dr. Nayuta Yamashita

In animals, resource partitioning often involves evolved differences in feeding behavior. National Geographic Explorer Nayuta Yamashita (above left) and her colleagues Chia Tan (center) and Chris Vinyard (right) study how differences in body form influence food usage among primates. They are especially interested in three species of lemurs (Section 24.8) that coexist in Madagascar. Yamashita says, "The lemurs are all bamboo specialists that feed on the same species of giant bamboo, though they eat different parts."

To find out the mechanisms by which the lemurs divvy up their food supply, Yamashita measured physical properties of the bamboo they eat and compared the shapes of the lemurs' teeth and jaws. She also measured how much force their jaws could exert. Results of these studies revealed that differences in bite force reflect resource partitioning. The greater bamboo lemur (shown in the photo above) has the most powerful bite. It is the only bamboo lemur that can access the toughest part of the bamboo plant—the pith inside mature shoots.

character displacement As a result of competition, two species become less similar in their resource requirements.

competitive exclusion Process whereby two species compete for a limiting resource, and one drives the other to local extinction.

ecological niche Of a species, unique set of requirements and roles in an ecosystem.

interspecific competition Competition between members of two species.

resource partitioning Evolutionary process whereby species become adapted in different ways to access different portions of a limited resource; allows species with similar needs to coexist.

41.4 Predation and Herbivory

LEARNING OBJECTIVES

Explain why predator abundance and prey abundance affect one another.

Give examples of evolved defenses against predation and herbivory.

Distinguish between the two types of mimicry.

PREDATOR AND PREY ABUNDANCE

With **predation**, one species (the predator) captures, kills, and eats another species (the prey). Predation removes a prey individual from the population immediately.

In any community, predator abundance and prey abundance are interconnected. Generally, an increase in the size of a predator population results in a decrease in the abundance of its prey. Understanding how a predator population responds to changes in prey density is important because it helps ecologists predict long-term effects of predation.

For passive predators, such as web-spinning spiders, the proportion of prey killed is constant, so the number killed in any given interval depends solely on prey density. The number of flies caught in spider webs is proportional to the total number of flies: The more flies there are, the more get caught in webs. More typically, the rate at which prey are killed depends in part on the time it takes predators to process prey. For example, a wolf that just killed a caribou will not hunt another until it has eaten and digested the first one. As prey density increases, the rate of kills rises steeply at first because there are more prey to catch. The rate slows when predators are exposed to more prey than they can handle at one time.

Predator and prey populations sometimes rise and fall in a cyclical fashion. **Figure 41.6** shows historical data for the numbers of lynx and their main prey, the snowshoe hare. Both populations rise and fall over an approximately ten-year cycle, with predator abundance lagging behind prey abundance. Field studies indicate that lynx numbers fluctuate mainly in response to hare numbers. However, the size of the hare population is affected by the abundance of the hare's food as well as the number of lynx. Hare populations continue to rise and fall even when predators are experimentally excluded from areas.

PREDATOR–PREY ARMS RACES

Predator and prey exert selective pressure on one another. Suppose a mutation arises that gives a prey species a more effective defense. Over generations, this mutation will tend to spread through the prey population as a result of natural selection. As this occurs, traits that make individual predators better at thwarting the improved prey defense become adaptive in predator populations. Thus, predators exert selective

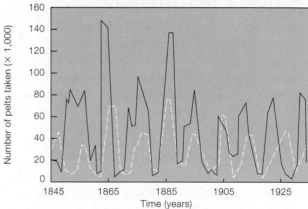

Figure 41.6 Predator–prey cycles. The graph shows the abundance of the Canadian lynx (dashed line) and snowshoe hares (solid line), based on counts of pelts sold by trappers to Hudson's Bay Company from 1845 to 1925.

pressure that favors improved prey defense, which in turn exerts selective pressure on predators, and so it goes over many generations.

You have already learned about some defensive adaptations. Many prey species have hard or sharp parts that make them difficult to eat. Think of a snail's shell or a sea urchin's spines. Other prey contain chemicals that taste bad or sicken predators. Most such defensive compounds in animals come from the plants they eat. For example, a monarch butterfly caterpillar takes up chemicals from the milkweed that it feeds on. A bird that eats a monarch butterfly will be sickened by these chemicals and avoid similar butterflies later.

Prey animals use a variety of mechanisms to fend off a predator. Section 1.6 described how eyespots and a hissing sound protect some butterflies. A lizard's tail may detach from the body and wiggle a bit as a distraction. Many ani-

CREDITS: (6) top, Ed Cesar/Science Source; bottom, Precision Graphics.

A Wasp that can inflict a painful sting. Like many stinging bees and wasps, it has a yellow and black pattern.

B Fly, which lacks a stinger, mimics the color pattern of stinging insects.

Figure 41.7 Mimicry.

mals, including skunks, exude or squirt a foul-smelling, irritating repellent when frightened. Bees and wasps sting.

Well-defended prey often have warning coloration, a conspicuous color pattern that predators learn to avoid. For example, many species of stinging wasps and bees have a pattern of black and yellow stripes (**Figure 41.7A**). The similar appearance of bees and wasps is an example of one type of **mimicry**, an evolutionary pattern in which one species comes to resemble another. Bees and wasps benefit from their similar appearance. The more often a predator is stung by a black-and-yellow-striped insect, the less likely it is to attack any similar-looking insect.

In another type of mimicry, an undefended prey species masquerades as a well-defended species. For example, some flies that cannot sting resemble bees or wasps that can (**Figure 41.7B**). The fly benefits when predators avoid it after an encounter with the better-defended look-alike species.

Camouflage is a body form and coloration pattern that allows an animal to blend into its surroundings, and thus avoid detection. For example, snowshoe hares such as the one in **Figure 41.6** turn white in winter, making it harder for predators to spot them in a snowy landscape.

Predators benefit from camouflage that hides them from their prey (**Figure 41.8**). Other predator adaptations include sharp teeth and claws that can pierce protective hard parts. Speedy prey select for faster predators. For example, the cheetah, the fastest land animal, can run 114 kilometers per hour (70 mph). Its preferred prey, Thomson's gazelles, run 80 kilometers per hour (50 mph).

PLANT–HERBIVORE ARMS RACE

With **herbivory**, an animal eats a plant or plant parts. The number and type of plants in a community can influence the abundance and type of herbivores present. Two types of defenses have evolved in response to herbivory. Some plants can withstand and recover quickly from the loss of some parts. For example, prairie grasses store enough resources in

A Pink body parts of a flower mantis help hide it from insect prey attracted to the real flowers.

B Fleshy protrusions give a scorpionfish the appearance of an algae-covered rock. Fish that come close for a nibble end up as prey.

Figure 41.8 Camouflaged predators.

their roots to grow back lost shoots, so they are seldom killed by grazers such as bison. Other plants have traits that deter herbivory: spines or thorns, leaves that are hard to chew or digest, or chemicals that taste bad or sicken herbivores. Ricin, the toxin made by castor bean plants, is an example.

Plant defenses select for traits that allow herbivores to overcome those defenses. For example, eucalyptus leaves contain toxins that make them poisonous to most mammals, but not to koalas. Koalas have specialized liver enzymes that break down these toxins.

---

camouflage Coloration or body form that helps an organism blend in with its surroundings and escape detection.
herbivory An animal feeds on plants or plant parts.
mimicry An evolutionary pattern in which one species becomes more similar in appearance to another.
predation One species captures, kills, and eats another.

Distinguish between parasites and parasitoids.

Describe how parasites can reduce the population size of their hosts.

Using an appropriate example, explain brood parasitism.

Explain the benefits and risks of biological pest control.

A Endoparasitic roundworms in the intestine of a host pig.

B Ectoparasitic ticks attached to and sucking blood from a finch.

C Ectoparasitic dodder (*Cuscuta*), also known as strangleweed or devil's hair. This parasitic flowering plant has almost no chlorophyll. Leafless stems wrap around a host plant, and modified roots absorb water and nutrients from the host plant's vascular tissue.

Figure 41.9 Parasites inside and out.

PARASITISM

With **parasitism**, one species (the parasite) benefits by feeding on another (the host), without immediately killing it. Endoparasites such as parasitic roundworms live and feed in their host (**Figure 41.9A**). An ectoparasite such as a tick feeds while attached to a host's external surface (**Figure 41.9B**).

A parasitic way of life has evolved in a diverse variety of groups. Bacterial, fungal, protistan, and invertebrate parasites feed on vertebrates. Lampreys (Section 24.2) attach to other fish and feed on them. There are even a few parasitic plants that withdraw nutrients from other plants (**Figure 41.9C**).

In terms of evolutionary fitness, killing a host too fast is bad for the parasite. The longer the host survives, the more parasite offspring can be produced. Thus, parasites with less-than-fatal effects on hosts are at a selective advantage.

Although parasites typically do not kill their hosts, many still have an important impact on a host population. Many parasites are pathogens that cause disease in their hosts. Even when a parasite does not cause obvious symptoms, its presence can weaken a host, making it more vulnerable to predation or less attractive to potential mates. Some parasites cause their host to become sterile or limit its number of offspring.

Adaptations to a parasitic lifestyle include traits that allow the parasite to locate hosts and to feed undetected. Ticks that feed on mammals or birds move toward a source of heat and carbon dioxide, which may be a potential host. A chemical in tick saliva acts as a local anesthetic; it prevents an animal from feeling a tick that is feeding on it. Endoparasites often have traits that help them evade a host's immune defenses. For example, some parasitic worms that live in the human digestive tract turn down our inflammatory response. By one hypothesis, the increasing rate of autoimmune disease in highly developed countries is an unexpected consequence of a reduction in inflammation-reducing worm infections.

Among hosts, an ability to fend off parasites or reduce the toll that a parasite takes on fitness is adaptive. The crested auklet (a type of seabird) secretes a citrus-scented compound that repels lice and ticks. In many other birds and mammals, preening or grooming behavior removes such ectoparasites. Chimpanzees and other primates sometimes fold up tough, indigestible leaves and swallow them whole, a practice that helps rid their gut of parasitic worms.

biological pest control Use of a pest's natural enemies to control its population size.

brood parasitism One egg-laying species benefits by having another raise its offspring.

parasitism Relationship in which one species withdraws nutrients from another species, without immediately killing it.

parasitoid (PAIR-uh-sit-oyd) An insect that lays eggs in another insect, and whose young devour their host from the inside.

Figure 41.10 Biological pest control. A commercially raised parasitoid wasp about to deposit a fertilized egg in an aphid. This wasp is used to reduce aphid populations. After the egg hatches, a wasp larva devours the aphid from the inside.

Figure 41.11 Brood parasitism. Ants tending a larva of the Alcon blue butterfly (*Phengaris alcon*). Worker ants feed the butterfly larva more eagerly than they feed ant larvae and will even feed it ant larvae if food runs short.

PARASITOIDS

An estimated 15 percent of insects are **parasitoids**, meaning their larvae develop in or on a host insect. Typically, the larvae hatch from the eggs laid in or on the host. As the larvae develop, they feed on the host's tissues. Parasitoids reduce the size of a host population in two ways. First, as the parasitoid larvae grow inside their host, they withdraw nutrients and prevent the host from reproducing. Second, the presence of parasitoid larvae often leads to the death of the host.

BIOLOGICAL PEST CONTROL

Biological pest control is the practice of using a pest's natural enemies to reduce its numbers. Commercially raised parasites and parasitoids are often used as agents of biological pest control (**Figure 41.10**). The method of pest control has some advantages over pesticides. Most chemical insecticides kill a wide variety of insects, including helpful species. Insecticides also have negative effects on human health. By contrast, the parasites and parasitoids usually used as biological control agents target only a limited number of species.

For a species to be an effective biological control agent, it must target only one specific host species and be able to survive in that host's habitat. The ideal biological control agent excels at finding the target host species, has a population growth rate comparable to the host's, and has offspring that disperse widely.

Introducing a species into a community as a biological control agent always entails some risks. The introduced parasites sometimes attack nontargeted species in addition to, or instead of, those that they were expected to control. Consider the parasitoid wasps that were introduced to Hawaii to con-

trol stinkbugs that feed on some Hawaiian crops. Instead, the parasitoids decimated the population of koa bugs, Hawaii's largest native bug. Introduced parasitoids have also been implicated in ongoing declines of many native Hawaiian butterfly and moth populations.

BROOD PARASITISM

With **brood parasitism**, one egg-laying species benefits by having another raise its offspring. The host species suffers by squandering care on unrelated individuals. The European cuckoos described in Section 39.2 are brood parasites, as are North American cowbirds. Not having to invest in parental care allows a female cowbird to produce a large number of eggs—as many as thirty in a single reproductive season.

The presence of brood parasites decreases the reproductive rate of the host species and favors an ability to detect and eject foreign young. Some avian brood parasites counter this host defense by producing eggs that closely resemble those of their host species. In cuckoos, different subpopulations have different host preferences and egg coloration. Females of each subpopulation lay eggs that closely resemble those of their preferred host.

Some butterflies are brood parasites that outsource care of their larvae to ants. Caterpillars (larvae) of the Alcon blue butterfly smell like an ant worker and mimic the sounds made by an ant queen. Worker ants, fooled by these false cues, carry the caterpillars into their nest, where they care for them as if they were members of the colony, feeding them and protecting them from predators (**Figure 41.11**). The large blue butterfly described in Section 17.10 is a relative of the Alcon blue butterfly, and it too is a brood parasite of ants.

LEARNING OBJECTIVES

Distinguish between primary and secondary succession.

Explain why the exact course of succession can be unpredictable.

Describe how the frequency and magnitude of disturbance influences the species richness of a community.

Explain how indicator species can be used to monitor the environment.

ECOLOGICAL SUCCESSION

The species composition of a community will change over time. Often, some species alter the habitat in ways that allow others to come in and replace them. This type of change, which takes place over a long interval, is called ecological succession. (Succession refers to a series of things, one following after the other.)

Ecological succession begins with the arrival of **pioneer species**, which are species whose traits allow them to colonize new or newly vacated habitats. Pioneer species have an opportunistic life history (Section 40.4): They grow and mature quickly, and they produce many offspring capable of dispersing. Later, other species replace the pioneers. Then the replacements are replaced, and so on.

There are two types of succession: primary succession and secondary succession.

Primary succession takes place in a barren habitat that lacks soil, such as land exposed by the retreat of a glacier, a newly formed volcanic island, or a region where volcanic material has buried existing soil. The earliest pioneers to colonize such environments are often mosses and lichens (Sections 21.2 and 22.3), which are small, have a brief life cycle, and can tolerate intense sunlight, extreme temperature changes, and little or no soil. Some hardy annual flowering plants with wind-dispersed seeds are also frequent pioneers. Pioneer species help build and improve the soil. In doing so, they often set the stage for their own replacement. Many pioneer species partner with nitrogen-fixing bacteria, so they can grow in nitrogen-poor habitats. Seeds of later species can take root inside mats of the pioneers. Organic wastes and remains accumulate and, by adding volume and nutrients to soil, this material helps other species take hold. Later successional species often shade and eventually displace earlier ones.

In **secondary succession**, a disturbed area within a community recovers. It commonly occurs in abandoned agricultural fields and burned forests. Because improved soil is present from the start, secondary succession usually occurs faster than primary succession.

When the concept of ecological succession was first developed in the late 1800s, it was thought to be a predictable and directional process. Which species are present at each stage in succession was thought to be determined primarily by physical factors such as climate, altitude, and soil type. In this view, succession culminates in a "climax community," an array of species that persists over time and will be reconstituted in the event of a disturbance.

Ecologists now know that the species composition of a community changes in unpredictable ways. Communities do not journey along a well-worn path to a predetermined climax state. Random events determine the order in which species arrive in a habitat, and thus affect the course of succession.

A The community at the base of Mount Saint Helens was wiped out by a volcanic eruption in 1980.

B In less than a decade, pioneer species had arrived and the process of primary succession had begun.

C Twelve years later, seedlings of Douglas firs had taken hold.

Figure 41.12 An example of succession.

Ecologists had an opportunity to investigate how random factors influence succession after the 1980 eruption of Mount Saint Helens leveled about 600 square kilometers (235 square miles) of forest in Washington State (**Figure 41.12**). They recorded the natural pattern of colonization, and carried out experiments in plots inside the blast zone. The results of these and other studies showed that the types of pioneer species that arrive first were most important. Which species arrived first in a particular area was a random event that had a major influence on which species followed.

EFFECTS OF DISTURBANCE

The magnitude and frequency of disturbances affect communities. A variety of field studies support the **intermediate disturbance hypothesis**, which states that species richness is greatest when physical and biological disturbances are moderate in their intensity or frequency (**Figure 41.13**).

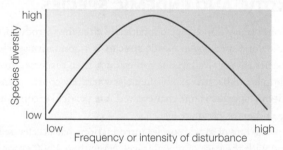

Figure 41.13 Intermediate disturbance hypothesis.

When disturbance is infrequent and of low intensity (does not remove many individuals), the most competitive species will exclude others, so diversity is low. By contrast, when disturbance occurs often or is of high intensity, most species present will be colonizers. With a moderate level of disturbance, the community will contain a mix of early and late successional species, and so will be most diverse.

In communities that repeatedly experience a particular type of disturbance, individuals that withstand or benefit from that disturbance have a selective advantage. For example, some plants in areas subject to periodic fires produce seeds that germinate only after a fire. Seedlings of these plants benefit from the lack of competition for resources in newly burned areas. Other plants have an ability to resprout fast after a fire (**Figure 41.14**). Because different species respond differently to fire, the frequency of this disturbance affects competitive interactions. Human suppression of naturally occurring fires can alter the composition of a biological community. In the absence of fires, species whose numbers would otherwise be suppressed by fire can become dominant.

Figure 41.14 Adapted to disturbance. Some woody shrubs, such as this toyon, resprout from their roots after a fire. In the absence of occasional fire, toyons are outcompeted and displaced by species that grow faster but are less fire resistant.

Figure 41.15 Indicator species. Lichens take up nutrients—and pollutants—from airborne dust, so many lichens cannot survive where the level of air pollution is high. The U.S. Forest Service monitors the types and numbers of lichens on tree trunks as part of its program to assess the effects of air pollution on forest communities.

Some species are especially intolerant of physical disturbance of their environment. These **indicator species** are the first to decline or disappear when conditions change, so they can provide an early warning of environmental degradation. For example, a decline in a trout population can be an early sign of problems in a stream, because trout are highly sensitive to pollutants and they cannot tolerate low oxygen levels. Some lichens that are intolerant of air pollution serve as indicators of air quality (**Figure 41.15**).

indicator species Species whose presence and abundance in a community provide information about conditions in the community.
intermediate disturbance hypothesis Species richness is greatest in communities in which disturbances are moderate in their magnitude and frequency.
pioneer species Species that can colonize a new habitat.
primary succession A new community becomes established in an area where there was previously no soil.
secondary succession A new community becomes established in a site where a community previously existed.

The loss or addition of one species sometimes alters the abundance of many other species in a community.

KEYSTONE SPECIES

A **keystone species** is a species that has a disproportionately large effect on a community relative to its size and abundance. It is named after a central, wedge-shaped stone that holds others stones of a archway in place. Robert Paine coined the term "keystone species" to describe the results of his experiments on the rocky shores of Washington state. Species in the rocky intertidal zone withstand pounding surf by clinging to rocks. A rock to cling to is a limiting factor. Paine set

Figure 41.16 The effect of sea star removal on species richness in tide pools. The sea star *Pisaster ochraceus* (*right*) was removed from experimental plots, but left in place in control plots. The graph below shows the species richness of control and experimental plots over time. These results indicate that the sea star is a keystone species.

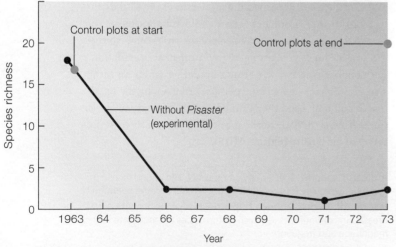

FIGURE IT OUT

During which interval was the decline in species richness of experimental plots the greatest?

Answer: 1963–1966

up control plots with the sea star *Pisaster ochraceus* and its main prey—chitons, limpets, barnacles, and mussels. Then he removed all sea stars from his experimental plots.

Sea stars prey mainly on mussels. With sea stars missing from experimental plots, mussels took over, crowding out seven other species of invertebrates (**Figure 41.16**). Paine concluded that sea stars normally prevent competitive exclusion by mussels in the intertidal zone, so they keep the number of species here high.

Keystone species need not be predators. For example, beavers are keystone species in some communities. These large, herbivorous rodents cut down trees by gnawing through their trunks. The beaver then uses the felled trees to build a dam, thus creating a deep pool where a shallow stream would otherwise exist. By altering the physical conditions in a section of the stream, the beaver changes which species of fish and aquatic invertebrates can live there.

EXOTIC AND ENDEMIC SPECIES

A community can change dramatically after the introduction of an exotic species. An **exotic species** is a nonnative species, a species that evolved in one community, then dispersed from its home and became established elsewhere. By contrast, an **endemic species** is one that evolved in a particular community and exists nowhere else.

The rate of exotic species introductions has accelerated along with the pace of global travel. More than 4,500 exotic species have become established in the United States. An estimated 25 percent of Florida's plant and animal species are exotics. In Hawaii, 45 percent are exotic. Some of these species were brought in as food crops, as ornamentals for gardens, or as a source of textiles. Other species arrived as stowaways along with cargo from distant regions.

Many exotic species have little impact on their adoptive community, but some of them are invasive. An invasive species is an exotic species that harms members of its new community. In some cases, the arrival of an exotic species can lead to the extinction of endemic species.

Invasive species often have a far greater impact in their new community than they did in the region in which they originated. When a species leaves its community of origin behind, it also leaves behind the competitors, predators, and parasites with which it coevolved and which helped to keep its population in check. If the invasive species is a parasite, predator, or herbivore, it also leaves behind hosts or prey that had coevolved with it and had defenses against it. Its new hosts are often defenseless against it. As a result, an invasive species often reaches a higher population density in its new home than it achieved in its old one.

Gypsy moth caterpillars native to Europe and Asia now feed on oaks in much of the United States.

Nutrias native to South America now abound in freshwater marshes of the Gulf coast states.

Kudzu native to Asia is overgrowing trees across the southeastern United States.

Figure 41.17 Three exotic species that are altering natural communities in the United States. To learn more about invasive species in the United States, visit the National Invasive Species Information Center online at www.invasivespeciesinfo.gov.

The water mold that is currently killing off oaks in the Pacific Northwest (Section 20.4) is an invasive species, as are the fungi that currently threaten North America's bats (Section 22.3) and frogs (Chapter 22 Application). Sea lampreys are an invasive species in the Great Lakes (Section 24.2).

As another example, consider the woody perennial vine called kudzu (*Pueraria lobata*). Native to Asia, kudzu was introduced to the American Southeast as a food for grazers and to control erosion, but it quickly became an invasive weed. It is now present in 31 states, and continues to spread. Kudzu decreases the diversity of natural communities by overgrowing native trees and shrubs, and outcompeting them for sunlight (**Figure 41.17A**). Kudzu also poses a threat to agriculture; it serves as the main winter host for soybean rust, a fungal pathogen that can drastically reduce soybean yields.

Gypsy moths (*Lymantria dispar*) are an invasive species too. They entered the northeastern United States in the mid-1700s and now range into the Southeast, Midwest, and Canada. Gypsy moth caterpillars (**Figure 41.17B**) preferentially feed on oaks. Loss of leaves to gypsy moths weakens the trees, making them less efficient competitors and more susceptible to pathogens.

Large semiaquatic rodents called nutrias (*Myocastor coypus*) were imported from South America for their fur. Today, descendants of animals that either escaped or were intentionally released thrive in freshwater marshes and along rivers in twenty states (**Figure 41.17C**). Their appetite for plants threatens native vegetation and crop plants, and their burrowing contributes to marsh erosion and damages levees, increasing the risk of flooding.

endemic species (en-DEM-ick) A species found in the region where it evolved and nowhere else.

exotic species A species that evolved in one community and later became established in a different one.

keystone species A species that has a disproportionately large effect on community structure relative to its abundance.

CREDITS: (17A) Danny E Hooks/Shutterstock.com; (17B) Photo by Scott Bauer, USDA/ARS; (17C) © Greg Lasley Nature Photography, www.greglasley.net.

LEARNING OBJECTIVES

Describe the formation and colonization of the island of Surtsey.

Explain how island size and area affect the number of species on the island.

Islands are natural laboratories for population studies. They have also been laboratories for community studies. Consider Surtsey, an island that appeared in the mid-1960s when an undersea volcano erupted 33 kilometers (21 miles) from the coast of Iceland (**Figure 41.18**). Bacteria and fungi were early colonists. Vascular plants followed in 1965, and mosses appeared two years later. The first lichens were found five years after that.

The rate at which vascular plants were introduced to the island increased dramatically after a seagull colony became established in 1986, but the number of species on Surtsey can not continue increasing forever. How many species will there be when the number levels off? According to the **equilibrium model of island biogeography**, the eventual number of species living on any island reflects a balance (or equilibrium) between immigration of new species and extinction of established ones.

Two factors determine the equilibrium number of species (**Figure 41.19**). First, there is a **distance effect**: Islands far from a source of colonists receive fewer immigrants than those closer to a source. Most species cannot disperse very far, so they will not turn up far from a mainland. Second, there is an **area effect**: An island's size affects both immigration rates and extinction rates. More colonists will happen upon a larger island simply by virtue of its size. Also, big islands are more likely to offer a variety of habitats, such as high and low elevations. These options make it more likely that a new arrival will find a suitable habitat. Finally, big islands can support larger populations of species than small islands. The larger a population, the less likely it is to become locally extinct as the result of some random event. For example, a fire large enough to wipe out the population of a small island might leave survivors on a larger island.

Robert H. MacArthur and Edward O. Wilson developed the equilibrium model of island biogeography in the late 1960s. Since then, the model has been modified and its use has been expanded to help scientists think about habitat islands, which are natural settings surrounded by a "sea" of habitat that has been disturbed by humans. Many parks and wildlife preserves fit this description. Island-based models help ecologists estimate the size of an area needed to ensure survival of a species in a region.

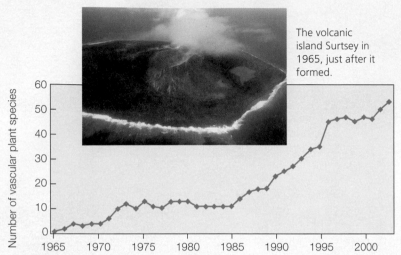

The volcanic island Surtsey in 1965, just after it formed.

Figure 41.18 Vascular plant colonization of Surtsey. Seagulls began nesting on the island in 1986.

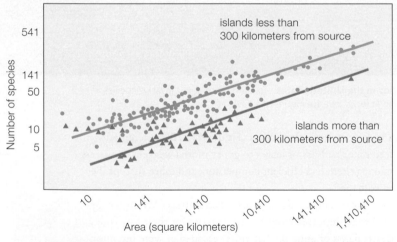

Figure 41.19 Island biodiversity patterns.

Distance effect: Species richness on islands of a given size declines as distance from a source of colonists rises. *Green* circles are values for islands less than 300 kilometers from the colonizing source. *Orange* triangles are values for islands more than 300 kilometers (190 miles) from a source of colonists.

Area effect: Among islands the same distance from a source of colonists, larger islands tend to support more species than smaller ones.

area effect Larger islands receive more immigrants and have a lower extinction rate than small ones.

distance effect Islands close to a mainland receive more immigrants than those farther away.

equilibrium model of island biogeography Model that predicts the number of species on an island based on the island's area and distance from a colonizing source.

? FIGURE IT OUT

Which has more species, a 100-km² island more than 300 km from a colonizing source or a 500-km² island less than 300 km from a colonist source?

Answer: The 500-km² island

Figure 41.20 Nest mounds of red imported fire ants in a Texas field. The inset photo shows the ants attacking a quail egg.

FIGHTING FOREIGN FIRE ANTS

Like most ants, red imported fire ants (*Solenopsis invicta*) nest in the ground (**Figure 41.20**). Accidentally step on one of their nests, and you will quickly realize your mistake. *S. invicta* defend their nest by stinging, and their venom causes a burning sensation that gives the ants their common name.

S. invicta is native to South America. The species first arrived in the southeast United States in the 1930s when some of the ants traveled as stowaways on a cargo ship. Since then, *S. invicta* has gradually expanded its range across the south, and has been accidentally introduced to California and New Mexico. More recently, it became established in the Caribbean, Australia, New Zealand, and several countries in Asia. Genetic comparisons among *S. invicta* populations revealed that the ants involved in these recent introductions originated in the southeastern United States, rather than South America.

Increased dispersal of pest species is an unanticipated side effect of increased global trade and improvements in shipping. Speedier ships make quicker trips, increasing the likelihood that pests hidden away in cargo holds will survive a journey.

The spread of *S. invicta* concerns ecologists because this species has a negative impact on communities where it is introduced. Competition from *S. invicta* typically causes a region's native ant populations to decline, and the resulting change in species composition can harm ant-eating animals. For example, the Texas horned lizard feeds mainly on native harvester ants, and cannot eat the red imported fire ants that have now largely replaced its natural prey. Red imported fire ants also harm native species by feeding on their eggs, and feeding on or stinging their young. Ground-nesting animals such as quail are especially vulnerable to fire ant predation.

The arrival of red imported fire ants can even affect native plants. The ants interfere with pollination by displacing or preying on native pollinators such as ground-nesting bees. They also impede dispersal of native plants whose seeds would normally be spread by native ants.

Given the problems that the imported ants are causing in the United States, you might wonder what things are like in their native South America. The ants are not considered much of a concern there,

in part because they are far less common. Coevolved parasites, predators, and diseases keep the ants' numbers in check.

Invicta means "invincible" in Latin, and *S. invicta* lives up to its name. Pesticides have not halted its spread, so scientists are now turning to some of the ants' coevolved South American enemies.

Phorid flies are one such natural enemy. In South America, some species of these parasitoids target red imported fire ants. A female phorid fly pierces the cuticle of an adult ant and lays an egg in the ant's soft tissues. The egg hatches into a larva, which grows and eats its way through the tissues to the ant's head. When the larva is ready to undergo metamorphosis, it secretes an enzyme that makes the ant's head fall off. The fly larva then develops into an adult within the shelter of the detached ant head.

Several South American phorid fly species have been introduced to southern states. The flies are surviving, reproducing, and, in some cases, increasing their range. The imported flies are not expected to kill off all *S. invicta* in affected areas. Rather, the hope is that these flies will reduce the density of invading colonies.

CREDITS: (20) background, Photography by B. M. Drees, Texas A&M University, http://fireant.tamu; inset, © James Mueller.

Section 41.1 Each species in a **community** is adapted to a certain **habitat**. Community diversity is described in terms of **species richness** and **species evenness**. Species interactions affect community structure. A **symbiosis** is an interaction in which one species lives in or on another; the interaction can be beneficial, neutral, or harmful. With **commensalism**, one species benefits and the other is unaffected.

Section 41.2 In a **mutualism**, both species benefit from an interaction. Some mutualists cannot complete their life cycle without the interaction. Mutualists who maximize their own benefits while limiting the cost of cooperating are at a selective advantage.

Section 41.3 A species' roles and requirements define its unique **ecological niche**. With **interspecific competition**, two species with similar resource requirements are harmed by one another's presence. With **competitive exclusion**, one competitor drives the other to local extinction. Competition between species with similar requirements results in directional selection; the species become less similar, an effect called **character displacement**. The change in traits allows **resource partitioning**.

Section 41.4 **Predation** benefits a predator at the expense of the prey it captures, kills, and eats. Predators and their prey exert selective pressure on one another. Evolved prey defenses include **camouflage** and **mimicry**. Predators have traits that allow them to overcome prey defenses. With **herbivory**, an animal eats a plant or plant parts. Some plants have traits that discourage herbivory. Others withstand herbivory by making a quick recovery from the loss of shoots.

Section 41.5 **Parasitism** involves feeding on a host without killing it. **Parasitoids** are insects that lay eggs on a host insect, then larvae devour the host. Parasites and parasitoids are often used in **biological pest control**. A **brood parasite** steals parental care from another species.

Section 41.6 Ecological succession is the sequential replacement of one array of species by another over time. **Primary succession** happens in newly formed habitats that lack soil. **Secondary succession** occurs in disturbed habitats. In both types of succession, the first species in a community are **pioneer species**. Their presence may help, hinder, or not affect later colonists.

Modern models of succession emphasize the unpredictability of the outcome as a result of chance events, ongoing changes, and disturbance. The **intermediate disturbance hypothesis** predicts that a moderate level of disturbance keeps a community diverse. An **indicator species** is highly sensitive to disturbance and can provide information about the health of the environment.

Section 41.7 **Keystone species** play a major role in determining the composition of a community. Removing a keystone species or introducing an **exotic species** can alter community structure, as by harming **endemic species**.

Section 41.8 A community supports a finite number of species. The **equilibrium model of island biogeography** predicts the number of species that an island will sustain based on the **area effect** and the **distance effect**. Scientists can use this model to predict the number of species that habitat islands such as parks can sustain.

Application Spread of the red imported fire ant has harmed populations of native ants that compete for the same resources, as well as species that depend on the native ants. Parasitoids that target ants are now being used as biological controls to reduce ant numbers.

SELF-ASSESSMENT
ANSWERS IN APPENDIX VII

1. The type of place where a species typically lives is called its _____ .
   a. niche   c. community
   b. habitat   d. population

2. Which cannot be a symbiosis?
   a. mutualism   c. commensalism
   b. parasitism   d. interspecific competition

3. Lizards that eat flies they capture on the ground and birds that eat flies they catch in the air are engaged in _____ competition.
   a. exploitative   b. interference

4. _____ can lead to resource partitioning.
   a. Mutualism   c. Commensalism
   b. Parasitism   d. Interspecific competition

5. Match the terms with the most suitable descriptions.
   ___ mutualism   a. one free-living species kills and eats another
   ___ parasitism
   ___ commensalism   b. two species interact and both benefit by the interaction
   ___ predation
   ___ interspecific   c. two species interact and one competition   benefits while the other is neither helped nor harmed
                          d. one species feeds on another but usually does not kill it
                          e. two species access the same resource

6. A tick is a(n) _____ .
   a. brood parasite   b. ectoparasite   c. endoparasite

7. _____ species are the first to colonize a new habitat.
   a. Pioneer   b. Endemic   c. Climax   d. Exotic

BIOLOGICAL CONTROL OF FIRE ANTS Ant-decapitating phorid flies are just one of the biological control agents used to battle imported fire ants. Researchers have also enlisted the help of a fungal parasite that infects the ants and slows their production of offspring. An infected colony dwindles in numbers and eventually dies out.

Is this biological control useful against imported fire ants? To find out, USDA scientists treated infested areas with either traditional pesticides or pesticides plus biological controls (both flies and the parasite). The scientists left some plots untreated as controls. **Figure 41.21** shows the results.

1. How did population size in the control plots change during the first four months of the study?

2. How did population size in the two types of treated plots change during this same interval?

3. If this study had ended after the first year, would you conclude that biological controls had a major effect?

4. How did the two types of treatment (pesticide alone versus pesticide plus biological controls) differ in their longer-term effects? Which is most effective?

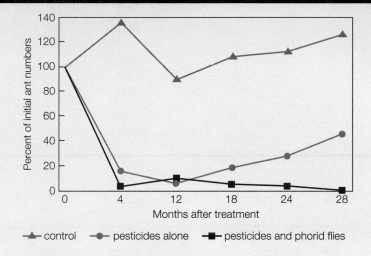

Figure 41.21 Effects of two methods of controlling red imported fire ants. The graph shows the numbers of red imported fire ants over a 28-month period. Orange triangles represent untreated control plots. Green circles are plots treated with pesticides alone. Black squares are plots treated with pesticide and biological control agents (phorid flies and a fungal parasite).

8. By the currently favored hypothesis, species richness of a community is greatest when disturbances are of a(n) _____ intensity and frequency.
   - a. low
   - b. intermediate
   - c. high
   - d. variable

9. Growth of a forest in an abandoned corn field is an example of _____ .
   - a. primary succession
   - b. resource partitioning
   - c. secondary succession
   - d. competitive exclusion

10. Species richness is greatest in communities _____ .
    - a. near the equator
    - b. in temperate regions
    - c. near the poles
    - d. at high elevations

11. A(n) _____ has another species rear its young.
    - a. mutualist
    - b. pioneer species
    - c. brood parasite
    - d. exotic species

12. Herbivory benefits _____ .
    - a. the herbivore
    - b. the plant
    - c. both a and b
    - d. neither a nor b

13. Match the terms with the most suitable descriptions.
    - ___ area effect
    - ___ endemic species
    - ___ indicator species
    - ___ keystone species
    - ___ exotic species
    - ___ resource partitioning

    - a. greatly affects other species
    - b. species that lives only where it evolved
    - c. more species on large islands than small ones at same distance from the source of colonists
    - d. species that is especially sensitive to changes in the environment
    - e. allows competitors to coexist
    - f. often outcompete, displace native species of established community

1. With antibiotic resistance rising, researchers are looking for ways to reduce the use of antibiotics. Instead of antibiotic-laced food, some cattle now get probiotic feed with cultured bacteria that can establish or bolster populations of helpful bacteria in the animal's gut. The idea is that if a large population of beneficial bacteria is in place, then harmful bacteria cannot become established or thrive. Explain the ecological principle that is guiding this practice.

2. Flightless birds on islands often have relatives on the mainland that can fly. The island species presumably evolved from fliers that, in the absences of predators, lost their ability to fly. Many island populations of flightless birds are in decline because rats have been introduced to their previously isolated island habitats. Despite current selection pressure in favor of flight, no flightless island bird species has regained the ability to fly. Why is this unlikely to happen?

3. Some attempts to use biological controls prove ineffective. Although the introduced control agent targets only the expected pest species, the introduced control agent does not take hold and spread in its new habitat. Describe some community interactions that could slow or prevent an introduced biological control agent from becoming established.

42

Ecosystems

Links to Earlier Concepts

This chapter reconsiders the one-way flow of energy (Sections 1.2, 5.1) and looks at nitrogen fixation (19.4), algal blooms (Chapter 20 Application), and erosion and leaching (26.1) in the context of nutrient cycles. You will revisit properties of chemical bonds (2.3) and the pathways of photosynthesis (6.3) and aerobic respiration (7.1). You will also see how slow movements of Earth's crust (16.5) influence the cycling of some nutrients.

Core Concepts

Organisms exchange matter and energy with the environment in order to grow, maintain themselves, and reproduce.

Organisms have various strategies to capture and store energy for use in biological processes. In an ecosystem, nutrients and energy are transferred through food chains that connect as food webs. With each transfer or conversion, some energy disperses.

Interactions among the components of each biological system give rise to complex properties.

The complexity and diversity of every biological system arise from interactions among its component parts. Ecosystems and their distribution can be affected by environmental catastrophes and geological events; ecosystems with the greatest species diversity are also the most stable and resilient to change. Human activities accelerate change in ecosystems at the local and global level.

Interactions between organisms and their environment result in the movement of matter and energy.

Human activities are altering biogeochemical cycles that sustain ecosystems. Diversion of water for human use drains rivers and underground reservoirs. Addition of carbon to the atmosphere alters the climate. Nitrogen and phosphorus pollution of aquatic habitats encourages algal growth and results in algal blooms.

◄ An energy and nutrient transfer. The caterpillar is tapping into the sunlight energy that this leaf captured and stored in the chemical bonds of sugars. The caterpillar also obtains nutrients such as nitrogen that the plant took up from the soil and used to build its leaves.

749

42.7 The Phosphorus Cycle

LEARNING OBJECTIVES

Describe the geochemical portion of the phosphorus cycle.

Describe how phosphorus enters food webs.

Distinguish between a sedimentary cycle and an atmospheric cycle.

List the sources of phosphate pollution and describe the effect of excess phosphorus on aquatic habitats.

Atoms of phosphorus are highly reactive, so phosphorus does not occur naturally in its elemental form. Most of Earth's phosphorus is bonded to oxygen as phosphate, which occurs in rocks and sediments. In the **phosphorus cycle**, phosphorus passes quickly through food webs as it moves from land to ocean sediments, then slowly back to land (**Figure 42.12**). Little phosphorus exists in a gaseous form. Because the atmosphere plays little role in the phosphorus cycle and the major reservoir for phosphorus is sedimentary rock, the phosphorus cycle is called a **sedimentary cycle**.

In the geochemical portion of the phosphorus cycle, weathering and erosion move phosphates from rocks into soil, lakes, and rivers ❶. Leaching and runoff carry dissolved phosphate (PO_4^{3-}) to the ocean ❷. Here, most dissolved phosphorus comes out of solution and settles as rocky deposits along continental margins ❸. Tectonic movements of Earth's crust can uplift these deposits onto land ❹, where weathering releases phosphates from rocks once again.

All organisms use phosphorus to build nucleic acids and phospholipids. The biological portion of the phosphorus cycle begins when producers take up phosphate. Land plants take up dissolved phosphate from soil water ❺. Animals on land get phosphates by eating the plants or one another. Phosphorus returns to the soil in the wastes and remains of organisms ❻. In the seas, phosphorus enters food webs when producers take up phosphate dissolved in seawater ❼. As on land, wastes and remains continually replenish the phosphorus supply ❽.

Like nitrogen, phosphorus is often a limiting factor for plant growth, so most fertilizers contain phosphorus as well as nitrogen. Phosphate-rich rocks are mined for use in the industrial production of fertilizer. Guano, which is phosphate-rich droppings from seabird or bat colonies, is also mined and used as fertilizer. Phosphates from fertilizers often run off from the site where they are applied and enter aquatic habitats. Other sources of aquatic phosphate pollution include animal waste from farms, sewage released from cities, and phosphate-rich detergents used to wash laundry and dishes. An influx of phosphorus can encourage the growth of aquatic producers, resulting in an algal bloom.

phosphorus cycle Movement of phosphorus among Earth's rocks and waters, and into and out of food webs.

sedimentary cycle Biochemical cycle in which the atmosphere plays little role and rocks are the major reservoir.

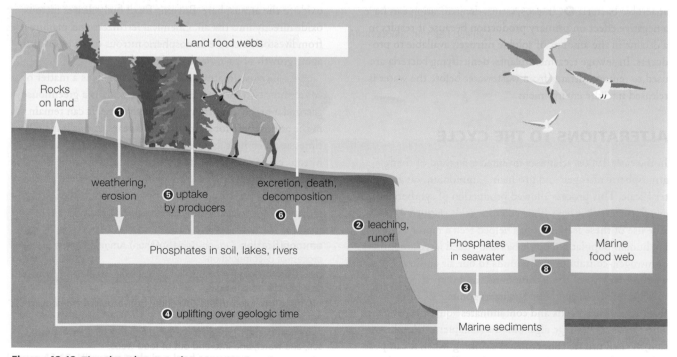

Figure 42.12 The phosphorus cycle.

Figure 42.13 Loon with its fish prey. Along with nutrients, the loon takes in any toxic pollutants that accumulated in the fish's body.

TOXIC TRANSFERS

Nutrients are not the only substances that move up food chains. Pollutants also enter food chains and pass from one trophic level to the next.

By the process of **bioaccumulation**, an organism's tissues store a pollutant taken up from the environment, so that the concentration of pollutant in the body increases over time. The ability of some plants to bioaccumulate toxic substances makes them useful in phytoremediation of polluted soils (Chapter 26 Application).

In animals, fat-soluble chemical pollutants ingested or absorbed across skin tend to accumulate in fatty tissues. Because the concentration of these pollutants in an animal's body increases over time, longer-lived species tend to be more affected by fat-soluble pollutants than shorter-lived ones. Within a species, old individuals tend to have a higher pollutant load than younger ones.

The concentration of a pollutant in organisms also increases as the pollutant moves up a food chain, a process known as **biological magnification**. As a result of bioaccumulation and biological magnification, even seemingly low concentrations of pollutants in an environment can cause harm, with long-lived animals at the top of food chains being most affected.

Consider the effects of pollution on a fish-eating bird such as a loon (**Figure 42.13**). A loon's tissues can contain a high concentration of contaminants, all derived from the bodies of the many fish the bird ate. In many regions, loons contain high levels of methylmercury, a neurotoxin that can form from mercury emitted by coal-burning power plants (Chapter 2 Application). The higher the mercury level in a lake where a loon lives, the lower the bird's reproductive success. Loons with a high mercury level show impaired parental behavior, so they raise fewer offspring.

bioaccumulation The concentration of a chemical pollutant in the tissues of an organism rises over the course of the organism's lifetime.
biological magnification A chemical pollutant becomes increasingly concentrated as it moves up through food chains.

Section 42.1 There is a one-way flow of energy into and out of an **ecosystem**, and a cycling of materials among the organisms within it. Most **producers** convert light energy into chemical bond energy. **Primary production**, the rate at which producers capture and store energy, can vary over time and between locations. **Consumers** eat producers or one another. **Detritivores** are consumers that eat small bits of organic remains, as are **decomposers** that break wastes and remains into inorganic components.

A **food chain** shows one path of energy and nutrient flow among organisms. Each organism in a food chain is at a different **trophic level**, with the primary producer being the first level and consumers at higher levels. The inefficiency of energy transfers from one trophic level to the next limits most food chains to four or five links.

Section 42.2 The many food chains within an ecosystem interconnect to form a **food web**. Most food webs include both **grazing food chains**, in which herbivores eat producers, and **detrital food chains**, in which producers die and are then consumed by detritivores. As a result of the multiple connections through food webs, a change that affects one species in an ecosystem will have effects on many others.

Energy pyramids and **biomass pyramids** are graphics that show how energy and organic compounds are distributed among organisms within an ecosystem. All energy pyramids are largest at their base.

Section 42.3 In the nonbiological portion of a **biogeochemical cycle**, an element moves among environmental reservoirs such as Earth's atmosphere, rocks, and waters. In the biological portion, elements move through a food web, then return to the environment.

Section 42.4 In the **water cycle**, evaporation, condensation, and precipitation move water from its main reservoir—the oceans—into the atmosphere, onto land, then back to oceans. Water that falls onto land may become part of the **groundwater**, which means it may be become **soil water** or be stored in an **aquifer**. Alternatively, it may become **runoff**. The water cycle helps move soluble forms of other nutrients. Earth has a limited amount of freshwater, most of which is ice. Overdrafts of water from rivers or aquifers to meet human needs can harm other species.

Section 42.5 The main reservoir for carbon is rocks, but the **carbon cycle** moves carbon mainly among seawater, the air, soils, and living organisms in an **atmospheric cycle**. Carbon dioxide contributes to the **greenhouse effect**. It is one of the **greenhouse gases**, which keep Earth's surface warm enough to support life. However, as a result of rising fossil fuel consumption the level of atmospheric carbon dioxide is increasing. Its increase correlates with and is considered the most likely cause of current **global climate change**.

Section 42.6 The **nitrogen cycle** is an atmospheric cycle. Air is the main reservoir for N_2, a gaseous form of nitrogen. **Nitrogen fixation** by bacteria converts N_2 to ammonium. **Ammonification** by bacterial and fungal decomposers also adds ammonium to the soil. Plants take up ammonium and nitrates, which are produced by **nitrification** of ammonium. Bacterial **denitrification** of nitrates returns nitrogen to the air. Humans add extra nitrogen to ecosystems by using synthetic fertilizer and by burning fossil fuels, which releases nitrous oxide. Excess nitrogen in aquatic habitats encourages algal blooms.

Section 42.7 The **phosphorus cycle** is a **sedimentary cycle** with no significant atmospheric component. Phosphorus from rocks dissolves in water and is taken up by producers. Phosphate-rich rocks and deposits of bird droppings are mined for use as fertilizer. Phosphorus from fertilizers, sewage, and detergents can pollute aquatic habitats and encourage algal blooms.

Application Toxic substances move up food chains. **Bioaccumulation** of fat-soluble toxins increases their concentration with age. Predators end up with high levels of these toxins because of **biological magnification**; each time they eat prey, they ingest all the toxins that accumulated in the prey's body over its lifetime.

SELF-ASSESSMENT
ANSWERS IN APPENDIX VII

1. In most ecosystems, producers use energy from _____ to build organic compounds.
   a. inorganic chemicals c. heat
   b. sunlight d. lower trophic levels

2. Decomposers are commonly _____ .
   a. fungi and bacteria c. animals
   b. plants d. green algae

3. Organisms at the first trophic level _____ .
   a. capture energy from a nonliving source
   b. are eaten by organisms at lower trophic levels
   c. are shown at the top of an energy pyramid
   d. are consumers

4. Primary productivity on land is affected by _____ .
   a. nutrient availability c. temperature
   b. amount of sunlight d. all of the above

RISING ATMOSPHERIC CARBON To assess the impact of human activity on the carbon dioxide level in Earth's atmosphere, it helps to take a long view. One useful data set comes from deep core samples of Antarctic ice. The oldest ice core that has been fully analyzed dates back a bit more than 400,000 years. Air bubbles trapped in the ice are a record of the atmosphere's composition at the time the ice formed. Combining ice core data with more recent direct measurements of atmospheric carbon dioxide—as in **Figure 42.14**—can help scientists put current changes in the atmospheric carbon dioxide into historical perspective.

1. What was the highest carbon dioxide level between 400,000 B.C. and 0 A.D.?

2. During this period, how many times did carbon dioxide reach a level comparable to that measured in 1980?

3. The industrial revolution occurred around 1850. How much did carbon dioxide levels change in the 800 years prior to the event? In 175 years after it?

4. Did carbon dioxide levels rise more between 1800 and 1975 or between 1980 and 2013?

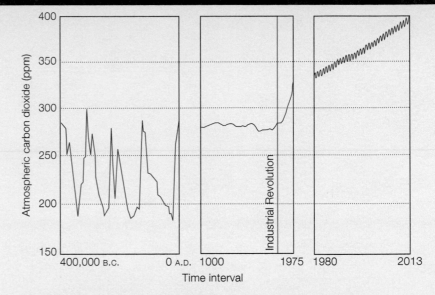

Figure 42.14 Changes in atmospheric carbon dioxide levels. Direct measurements began in 1958. Earlier data are based on air bubbles in ice cores. The oscillations in the directly measured data result from seasonal differences in photosynthesis.

5. A(n) _____ is an autotroph.
    a. producer
    b. herbivore
    c. detritivore
    d. top carnivore

6. Most of Earth's fresh water is _____ .
    a. in lakes and streams
    b. in aquifers and soil
    c. frozen as ice
    d. in bodies of organisms

7. Earth's largest carbon reservoir is _____ .
    a. the atmosphere
    b. sediments and rocks
    c. seawater
    d. living organisms

8. Carbon is released into the atmosphere by _____ .
    a. photosynthesis
    b. aerobic respiration
    c. burning fossil fuels
    d. b and c

9. Greenhouse gases _____ .
    a. help keep Earth's surface warm enough for life
    b. are released only by human activities
    c. have decreased since the industrial revolution

10. The _____ cycle is a sedimentary cycle.
    a. phosphorus
    b. carbon
    c. nitrogen
    d. water

11. Earth's largest phosphorus reservoir is _____ .
    a. the atmosphere
    b. bird droppings
    c. sediments and rocks
    d. living organisms

12. Plants obtain _____ by taking it up from the air.
    a. nitrogen
    b. carbon
    c. phosphorus
    d. water

13. Nitrogen fixation converts _____ to _____ .
    a. nitrogen gas; ammonium
    b. nitrates; nitrites
    c. ammonia; nitrates
    d. nitrites; nitrogen oxides

14. Soil in _____ is richest in carbon.
    a. the arctic
    b. the tropics
    c. temperate zones

15. Match each term with its most suitable description.
    ___ carbon dioxide
    ___ bicarbonate
    ___ nitrate
    ___ nitrogen gas
    a. contains triple bond
    b. nutrient taken up by plant roots
    c. marine carbon source
    d. greenhouse gas

CLASS ACTIVITY
CRITICAL THINKING

1. Marguerite has a vegetable garden in Maine. Eduardo has one in Florida. List the variables that could cause differences in the primary production of these gardens.

2. A watershed is an area in which all rainfall drains into a particular river. Find out which watershed you live in at the *Science in Your Watershed* website, http://water.usgs.gov/wsc.

3. The sulfur cycle is another important biogeochemical cycle. Rocks are the main reservoir for sulfur, but some sulfur compounds are dissolved in water, and sulfur oxide occurs in the atmosphere. Which of these sources do you think plants tap to meet their need for sulfur? Which essential biological compounds contain sulfur?

4. Rather than using fertilizer, a farmer may rotate crops, planting legumes one year, then another crop, then legumes again. Explain how such crop rotation helps keep soil fertile.

for additional quizzes, flashcards, and other study materials
▶ ACCESS MINDTAP AT WWW.CENGAGEBRAIN.COM

43

The Biosphere

Links to Earlier Concepts

In this chapter, you will consider the biosphere (Section 1.1) as a whole, revisit properties of water (2.4), and see how carbon-fixing pathways (6.5), soils (26.1), herbivory (41.4), and fire (41.6) affect plant distribution. We compare regional primary production (42.1) and learn more about the effects of global climate change (42.5), succession (41.6), and morphological convergence (16.7).

Core Concepts

Organisms exchange matter and energy with the environment in order to grow, maintain themselves, and reproduce.

The main flow of energy through the biosphere starts with the capture of solar energy by way of photosynthesis. Regional differences in the amount of solar energy that reaches the Earth's surface, and the extent to which that energy varies with the seasons, give rise to regional differences in primary production.

Interactions among the components of a biological system give rise to complex properties.

A biome consists of geographically separated regions that have a similar climate and soils. These regions support producers that are similarly adapted to the conditions within them. The distribution of different types of producers, in combination with differences in climate, determines the types of consumers that each biome includes.

Adaptation of organisms to a variety of environments has resulted in diverse structures and physiological processes.

Evolved differences in the ability to withstand cold, heat, drought, fire, and salinity allow different species to thrive in different environments. Organisms found in the same biome in different parts of the world often have independently evolved adaptations to the environmental stresses that occur in that biome.

◄ The Congo rain forest in central Africa, Earth's second largest tropical rain forest. Year-round warmth and rainfall support a diverse variety of evergreen, broadleaf trees. Their continual photosynthesis removes an enormous amount of carbon dioxide from the atmosphere.

765

Photograph by Frans Lanting, National Geographic Creative.

LEARNING OBJECTIVES

Explain why the amount of sunlight that reaches the ground varies with latitude (distance from the equator).

Describe how latitudinal differences in sunlight energy lead to global air circulation patterns that affect climate.

Explain why winds trace a curved path relative to Earth's surface.

The **biosphere** includes all places where life exists on Earth (Section 1.1). The distribution of species within the biosphere depends mainly on climate. **Climate** refers to average weather conditions in a region, including cloud cover, temperature, humidity, and wind speed. Climates differ among

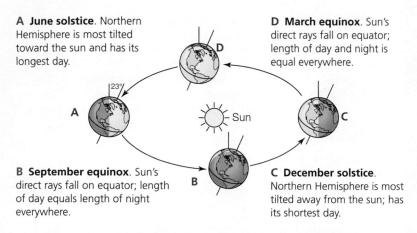

A June solstice. Northern Hemisphere is most tilted toward the sun and has its longest day.

D March equinox. Sun's direct rays fall on equator; length of day and night is equal everywhere.

23°

A

Sun

C

B September equinox. Sun's direct rays fall on equator; length of day equals length of night everywhere.

B

C December solstice. Northern Hemisphere is most tilted away from the sun; has its shortest day.

Figure 43.1 Effects of Earth's tilt and yearly rotation around the sun. The 23° tilt of Earth's axis causes the Northern Hemisphere to receive more intense sunlight and have longer days in summer than in winter.

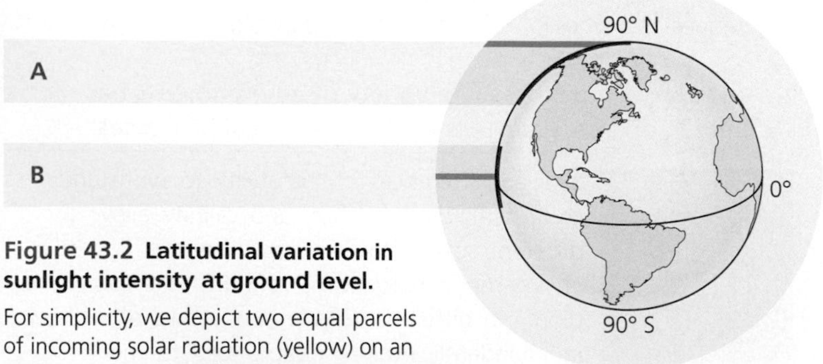

Figure 43.2 Latitudinal variation in sunlight intensity at ground level.

90° N

A

B

0°

90° S

For simplicity, we depict two equal parcels of incoming solar radiation (yellow) on an equinox, when incoming rays are perpendicular to Earth's axis. Rays that fall on high latitudes **A** have passed through more atmosphere (blue) than those that fall near the equator **B**. Compare the length of the green lines. (Atmosphere is not to scale.)

In addition, energy in the rays that fall at the high latitude is spread over a greater area than energy that falls on the equator. Compare the length of the red lines.

As a result of these two factors, the amount of solar energy that reaches the Earth's surface decreases with increasing latitude.

regions because many factors that influence winds and ocean currents vary from place to place.

SEASONAL EFFECTS

Primary production is influenced by the amount of incoming solar energy, which varies seasonally by region. Each year, Earth rotates around the sun in an elliptical path. Seasonal changes in daylight and temperature arise because Earth's axis is not perpendicular to the plane of this ellipse, but rather tilts at a 23-degree angle (**Figure 43.1**). In June, the Northern Hemisphere is tipped toward the sun, so it receives more direct sunlight and more daylight than the Southern Hemisphere, which is tipped away from the sun. The June solstice is the day when the Northern Hemisphere receives the most sunlight (**Figure 43.1A**), and the Southern Hemisphere receives the least. On the December solstice, the opposite occurs (**Figure 43.1C**). Twice a year—on the September and March equinoxes—Earth's axis is perpendicular to incoming sunlight. On these days, every place on Earth has 12 hours of daylight and 12 hours of darkness (**Figure 43.1B,D**).

In each hemisphere, the extent of seasonal change in daylight increases with latitude. Latitude, the distance from the equator, is measured in degrees (°), with the equator being a latitude of 0° and the poles being 90°. In Honolulu, Hawaii (21° north of the equator) the longest day is about 13 hours and the shortest is about 10 hours. In Anchorage, Alaska (61° north of the equator) the longest day is 19 hours and the shortest is 5.5 hours.

AIR CIRCULATION AND RAINFALL

Every day, regions near the equator receive more solar energy than higher latitudes. This occurs for two reasons. First, dust particles, water vapor, and gases in the atmosphere absorb or reflect some solar radiation. Sunlight traveling to high latitudes passes through more atmosphere to reach Earth's surface than light traveling to the equator, so less of its energy reaches the ground (**Figure 43.2A**). Second, energy in an incoming parcel of sunlight is spread out over a smaller surface area at the equator than at the higher latitudes (**Figure 43.2B**). As a result of this difference in incoming energy, Earth's surface warms more at the equator than at the poles.

Knowing about two properties of air can help you understand how regional differences in surface warming give rise to global air circulation and rainfall patterns. First, as air warms, it becomes less dense and rises. Hot-air balloon-

biosphere All regions of Earth where life exists.
climate Average weather conditions in a region.

Idealized Pattern of Air Circulation

D At the poles, cold air sinks and moves toward lower latitudes.

C Air rises again at 60° north and south, where air flowing poleward meets air coming from the poles.

B At around 30° north and south latitude, the now cool, dry air sinks and flows north and south along Earth's surface.

A Warmed by energy from the sun, air at the equator picks up moisture and rises. It reaches a high altitude, and spreads north and south. As the air flows toward higher latitudes, it cools and loses moisture as rain.

Major Winds Near Earth's Surface

E Air masses do not travel directly north and south across Earth's surface. As a result of the rotation of the Earth, winds are deflected to the right of their initial direction in the Northern Hemisphere, and to the left in the Southern Hemisphere.

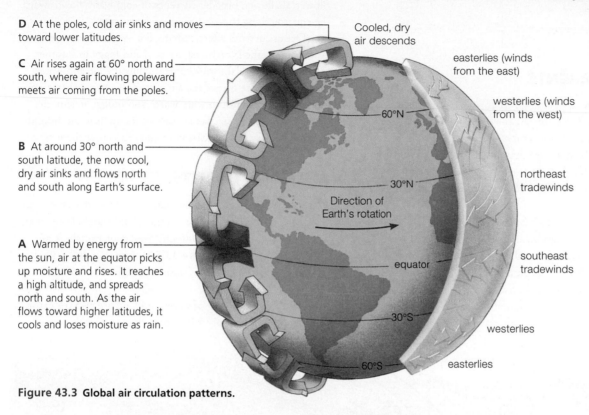

Cooled, dry air descends

easterlies (winds from the east)

westerlies (winds from the west)

60°N

30°N

Direction of Earth's rotation

northeast tradewinds

equator

southeast tradewinds

30°S

westerlies

60°S

easterlies

Figure 43.3 Global air circulation patterns.

ists take advantage of this effect when they take off from the ground by heating the air inside their balloon. Second, warm air holds more water than cooler air. You "see your breath" in cold weather because water vapor in warm exhaled air condenses into droplets as it cools outside of the body.

Global air circulation patterns are set in motion at the equator, where intense sunlight heats air and causes evaporation from the ocean. The result is an upward movement of warm, moist air (**Figure 43.3A**). As this air rises, it cools and flows away from the equator (north and south), releasing moisture as rain. This rain supports tropical rain forests such as the one shown in the chapter opening photo. By the time the air reaches 30° north or south of the equator, it has given up most of its moisture, so little rain falls here. Many of the world's great deserts are about 30° from the equator. The air has also cooled, so it sinks toward Earth's surface (**Figure 43.3B**). This air flows along Earth's surface toward the poles, again picking up heat and moisture. The warm, moist air rises at latitudes around 60° north and south, losing moisture as it does so (**Figure 43.3C**). The resulting rains support temperate zone forests. Near the poles, cold, dry air descends (**Figure 43.3D**). Precipitation is sparse, and polar deserts exist.

SURFACE WIND PATTERNS

The rising and falling of air masses set in motion the pattern of prevailing winds (**Figure 43.3E**). Air flows as wind across Earth's surface, moving from areas where air is sinking downward (at the Poles and at 30° north and south latitude) toward areas where it is rising (at the equator and at 60° north and south latitude). Thus, winds in the tropics tend to blow toward the equator, whereas those in temperate regions tend to blow toward the poles.

Winds do not blow directly north and south because of the effects of Earth's rotation. Earth rotates on its axis from west to east. As it rotates, points on the equator (where Earth is widest) travel faster than points nearer the poles (where Earth is narrower). These regional differences in the speed of rotation result in an easterly or westerly deflection of winds. In the tropics, the prevailing winds (the trade winds) blow from the east toward the equator. In temperate zones, prevailing winds blow from west to east toward the poles. Winds are named for the direction from which they blow, so prevailing winds in the United States are westerlies. They blow from the west toward the east.

OCEAN CURRENTS

Latitudinal variations in sunlight that reaches Earth set the oceans' waters in motion. At the equator, a vast volume of water warms and expands, making the sea level about 8 centimeters (3 inches) higher than it is at either pole. The existence of this "slope" starts sea surface water moving toward the poles. As the water moves toward cooler latitudes, it gives up heat to the air above it.

The direction of surface currents is influenced by the major winds, Earth's rotation, and the distribution of land masses. Surface currents generally circulate clockwise in the Northern Hemisphere, and counterclockwise in the Southern Hemisphere (**Figure 43.4**). Swift, deep, narrow currents of water flow away from the equator along the east coast of continents. The Gulf Stream is an example; this warm water current flows north along the east coast of North America. Slower, shallower, broader currents of cold water flow toward the equator along the west coast of continents.

Ocean currents affect climate. For example, coasts of the American Pacific Northwest are cool and foggy in summer because the cold California current chills the air, and water droplets condense out of the air as it cools. As another example, Boston and Baltimore are warm and muggy in summer because air over these coastal cities picks up heat and humidity from the warm Gulf Stream, which runs past them.

REGIONAL EFFECTS

Water can hold more heat than land, and this difference gives rise to coastal breezes. In the daytime, land warms faster than water. As air over land warms and rises, cooler offshore air moves in to replace it (**Figure 43.5A**). After sundown, land cools more quickly than the water, so the breezes reverse direction (**Figure 43.5B**).

Differential heating of water and land are also the cause of **monsoons**, which are winds that change direction sea-

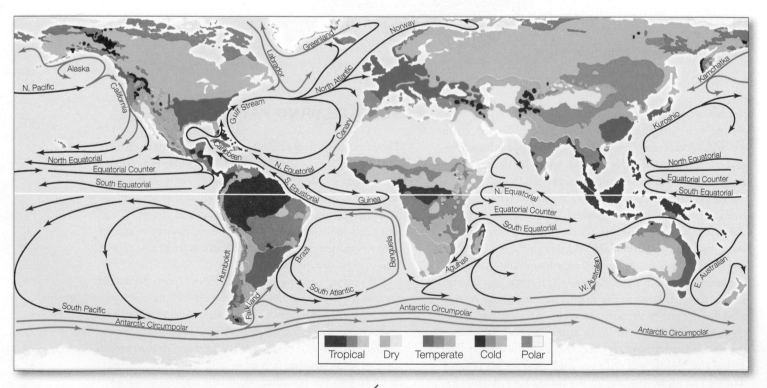

Figure 43.4 Major climate zones correlated with surface currents. Warm surface currents start moving from the equator toward the poles, but winds, Earth's rotation, gravity, the shape of ocean basins, and landforms influence the direction of flow. Water temperatures, which differ with latitude and depth, contribute to regional differences in air temperature and rainfall.

warm current

cold current

CREDIT: (4) NASA.

sonally. Consider how the continental interior of Asia heats up in the summer, so air rises above it. Moist air from over the warm Indian Ocean, which is to the south, moves in to replace the rising air, and this north-blowing wind delivers heavy rains. In the winter, the continental interior is cooler than the ocean. Thus, a cool, dry wind blows from the north toward warm southern coasts, causing a seasonal drought.

Proximity to an ocean moderates climate. Seattle, Washington has much milder winters than Minneapolis, Minnesota, even though Seattle is a bit farther north. Air over Seattle draws heat from the adjacent Pacific Ocean, a heat source not available to Minneapolis.

Mountains, valleys, and other surface features of the land affect climate too. Suppose you track a warm, moisture-laden air mass as it travels eastward from the Pacific Ocean into California (**Figure 43.6A**). When the air moves inland as wind from the west, it runs into the Sierra Nevada, a high mountain range that parallels the coast. The air cools as it rises in altitude to cross these mountains, so it can hold less water. As a result, clouds form and rain falls on the *windward* side of the mountain—the side facing the wind (**Figure 43.6B**). The air gives up most of its moisture before crossing the peaks, and moving down the leeward side the mountain (the side facing away from the wind). As the

A In afternoons, land is warmer than the sea, so a breeze blows onto shore.

B In evenings, the sea is warmer than land, so the breeze blows out to sea.

Figure 43.5 Coastal breezes.

cool air descends, it picks up heat from the ground, so any clouds dissipate and little or no rain falls. The result is a **rain shadow**, a region of sparse rainfall on the leeward side of the coastal mountains (**Figure 43.6C**). The Himalayas, Andes, Rockies, and other mountain ranges cast similar rain shadows.

monsoon (mon-SOON) Wind that reverses direction seasonally.
rain shadow Dry region downwind of a coastal mountain range.

A Prevailing winds move moisture inland from the Pacific Ocean.

B Clouds pile up and rain forms on side of mountain range facing prevailing winds.

C Rain shadow on side facing away from prevailing winds makes arid conditions.

moist habitats

4,000/ 75
3,000/ 85   2,000/ 25
1,800/ 125   1,000/ 25
1,000/ 85
15/ 25

Figure 43.6 Rain shadow caused by the Sierra Nevada mountains. Black numbers are average annual precipitation, in centimeters. White numbers are the elevation, in meters. Photos show the vegetation on the two sides of the mountain range.

LEARNING OBJECTIVES

List the factors that determine the distribution of biomes.

Explain why organisms in different regions of a biome often have similar characteristics.

DIFFERENCES AMONG BIOMES

Biomes are regions of land characterized by their main type of vegetation (**Figure 43.7**). Most biomes are geographically discontinuous, meaning they consist of widely separated areas on different continents. For example, the temperate grassland biome includes areas of North American prairie, South African veld, South American pampa, and Eurasian steppe. Grasses and other nonwoody flowering plants constitute the bulk of the vegetation in all of these regions.

Rainfall and temperature are the main determinants of the type of biome in a given region. Desert biomes get the least annual rainfall, grasslands and shrublands get more, and forests get the most. Deserts occur where temperatures soar the highest and tundra where they drop the lowest.

Soils also influence biome distribution. Soils consist of a mixture of mineral particles and varying amounts of humus (Section 26.1). Water and air fill spaces between soil particles. Soil properties vary depending on the types, proportions, and compaction of particles. Deserts have sandy or gravelly, fast-draining soil with little topsoil. Topsoil tends to be deepest in natural grasslands, and it can be more than one meter thick in these biomes. For this reason, grasslands are often converted to cropland. Climate and soils affect primary production (Section 42.1), so primary production varies among biomes.

SIMILARITIES WITHIN A BIOME

Different parts of a biome often have unrelated species that appear similar because they have adapted to similar conditions, an example of convergent evolution (Section 16.7). Cacti (North American desert plants with water-storing stems) appear similar to euphorbs (African desert plants with water-storing stems). Cacti and euphorbs did not both evolve from a shared ancestor with a water-storing stem. Rather, this feature arose independently in the two groups as a result of similar environmental pressures. Similarly, an ability to carry out C4 photosynthesis evolved independently in grasses growing in warm grasslands on different continents. Under hot, dry conditions, C4 photosynthesis is more efficient than the more common C3 pathway (Section 6.5).

biome (BY-ohm) A region (often discontinuous) characterized by its climate and dominant vegetation.

Figure 43.7 Major biomes. Climate is the main factor determining the distribution of biomes.

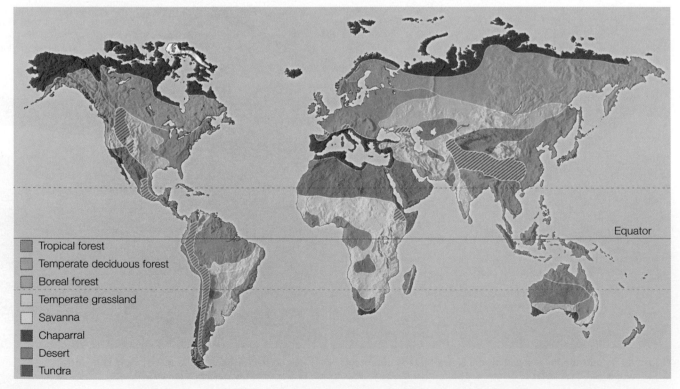

- Tropical forest
- Temperate deciduous forest
- Boreal forest
- Temperate grassland
- Savanna
- Chaparral
- Desert
- Tundra

Equator

43.4 Tropical Rain Forests

LEARNING OBJECTIVES

Explain why primary production and species diversity are high in tropical rain forests.

Describe rain forest soil and explain why it is not well suited to agriculture.

Tropical rain forests dominated by evergreen broadleaf trees form mainly between latitudes 10° north and south in equatorial Africa, the East Indies, Southeast Asia, South America, and Central America. Rain falls throughout the year, adding up to an annual total of 130 to 200 centimeters (50 to 80 inches). The regular rains, combined with an average warm temperature of 25°C (77°F) and little variation in day length, allow photosynthesis to continue year-round. Of all land biomes, tropical forests have the greatest primary production. Per unit area, they remove more carbon from the atmosphere than other forests or grasslands.

The high primary production supports an enormous variety of species. Tropical rain forest is the most species-rich biome, as well as the oldest. The biome's age may contribute to its diversity. Some rain forests have existed for more than 50 million years, a long interval in which many opportunities for speciation could have arisen. Compared to other land biomes, tropical rain forest has the greatest variety of plants, insects, birds, and primates.

Rain forests have a multilayer structure (**Figure 43.8**). Broadleaf trees standing up to 30 meters (100 feet) tall often form a closed canopy that prevents most sunlight from reaching the forest floor. Epiphytes (plants that grow on another plant, but do not withdraw nutrients from it) and vines are adapted to life in the shade beneath the canopy.

Trees in tropical rain forests shed leaves continually, but litter does not accumulate because decomposition and mineral cycling happen fast in this warm, moist environment. Rapid decay of leaf litter provides the nutrients that sustain the forests' high primary production. The soil itself is highly weathered and heavily leached. Being a poor nutrient reservoir, this soil is not well suited to agriculture if the forest is removed. Nevertheless, deforestation is an ongoing threat to tropical rain forests. Most are located in developing countries with a rapidly growing human population that looks to the forest as a source of lumber, fuel, and potential cropland.

Deforestation in any region leaves fewer trees to remove carbon dioxide from the atmosphere. In tropical rain forests, it also causes the extinction of species found nowhere else in the world. Among the potential losses are plants that make potentially life-saving chemicals. Two chemotherapy drugs, vincristine and vinblastine, were discovered in rosy periwinkle, a low-growing plant native to Madagascar's rain forests. Today, these drugs help fight leukemia, lymphoma, breast cancer, and testicular cancer.

---

tropical rain forest Highly productive and species-rich biome in which year-round rains and warmth support continuous growth of evergreen broadleaf trees.

Figure 43.8 Tropical rain forest. The graphic to the right shows the soil profile.

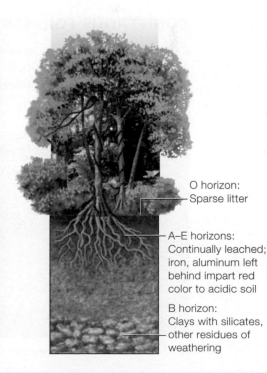

O horizon:
Sparse litter

A–E horizons:
Continually leached;
iron, aluminum left
behind impart red
color to acidic soil

B horizon:
Clays with silicates,
other residues of
weathering

Compare the conditions that prevail in temperate deciduous forests and boreal forests.

Describe some traits that adapt conifers to cold climates.

TEMPERATE DECIDUOUS FORESTS

Temperate deciduous forests are dominated by broadleaf (angiosperm) trees that lose all their leaves seasonally before a cold winter. Leaves often change color before dropping (**Figure 43.9A**). During winter, deciduous broadleaf trees remain dormant while water is locked in snow and ice. In the spring, when conditions again favor growth, these trees flower and put out new leaves. Also during the spring, leaves that were shed the prior autumn decay to form a rich humus. Nutrient-rich soil and a somewhat open canopy that lets some sunlight through allows shorter plants to grow beneath the trees.

Temperate deciduous forests are limited to the Northern Hemisphere, occurring in parts of eastern North America, western and central Europe, and areas of Asia, including Japan. In all these regions, 50 to 150 centimeters (about 20–60 inches) of precipitation falls throughout the year. Winters are cool and summers are warm.

North America has the most species-rich examples of the temperate deciduous forest biome, with different species characterizing forests in different areas. For example, Appalachian forests include mainly oaks, whereas beeches and maples dominate Ohio's forests.

CONIFEROUS FORESTS

Conifers (gymnosperms with seed-bearing cones) are the main plants in coniferous forests. Conifers shed and replace their leaves, but they do so continually, not seasonally as deciduous trees do. As a group, conifers tolerate drought, cold, and nutrient-poor soils better than broadleaf trees. Conifer leaves are typically needle-shaped, with a thick, waxy cuticle and stomata that are sunk below the leaf surface. These adaptations help conifers conserve water during drought and at times when the ground is frozen. The shape of conifer leaves also helps the tree shed snow, so heavy snowfall does not break limbs.

The most extensive land biome is the coniferous forest that sweeps across northern Asia, Europe, and North America (**Figure 43.9B**). It is known as **boreal forest**, or taiga, which means "swamp forest" in Russian. Pine, fir, and spruce predominate. Most rain falls in the summer, and little evaporates into the cool summer air. Winters are long, cold, and dry. Moose are the main grazers.

Other conifer-dominated ecosystems are less extensive. In the Northern Hemisphere, montane coniferous forests extend southward through great mountain ranges. Spruce and fir dominate at the highest elevations. At lower elevations, the mix becomes firs and pines. Conifers also dominate temperate lowlands along the Pacific coast from Alaska into northern California. These coastal forests hold the world's tallest trees: Sitka spruce to the north; coast redwoods to the south. In the eastern United States, about a quarter of New Jersey is covered by pine barrens—a mixed forest of pitch pines and scrub oaks that grow in sandy, acidic soil. Pine forest covers about one-third of the Southeast. The fast-growing loblolly pines that dominate this forest are a major source of lumber.

boreal forest (BORE-ee-uhl) Extensive high-latitude forest of the Northern Hemisphere; conifers are the predominant vegetation.

temperate deciduous forest Northern Hemisphere biome in which the main plants are broadleaf trees that lose their leaves in fall and become dormant during cold winters.

Figure 43.9 Cool forest biomes.

A North American temperate deciduous forest in fall.

B Siberian boreal forest (taiga) in summer.

LEARNING OBJECTIVES

Explain the factors that can prevent grasslands from becoming woodlands.

Compare the climate and dominant plants in prairies, savannas, and chaparral.

Plants adapted to periodic lightning-ignited fires dominate grasslands, savanna, and chaparral.

Grasslands form in the interior of continents between deserts and temperate forests. Their soils are rich, with a deep layer of topsoil that has been enriched over a long period of time by the decay of many grass roots. Annual rainfall is enough to keep desert from forming, but not enough to support woodlands. Low-growing grasses and other nonwoody plants tolerate strong winds, sparse and infrequent rain, and intervals of drought. Constant trimming by grazers, along with periodic fires, prevents trees and most shrubs from taking hold. When humans suppress fire, the low-growing grasses may be overgrown by woody plants.

Temperate grasslands are warm in summer, but cold in winter. Annual rainfall is 25 to 100 centimeters (10 to 40 inches), with rains throughout the year. In North America, temperate grasslands called prairies once covered much of the continent's interior, where summers are hot and winters are cold and snowy. The prairies supported herds of elk, pronghorn antelope, and bison (**Figure 43.10A**) that were prey to wolves. Today, these predators and prey are largely absent from most of their former range. The majority of prairies have been plowed under and now sustain production of wheat and other crops.

Savannas are broad belts of tropical grasslands with a few scattered shrubs and trees. These regions remain warm to hot year-round and receive 50 to 125 cm (about 20–50 inches) of rain during a rainy season. Savannas separate the tropical forests and hot deserts of Africa, India, and Australia. Africa's savannas are famous for their abundant wildlife. Grazers include giraffes, zebras, a variety of antelopes, and immense herds of wildebeests (**Figure 43.10B**).

Chaparral is a biome dominated by drought-resistant, fire-adapted shrubs whose small, leathery leaves help them withstand drought. Chaparral occurs along the western coast of continents, between 30 and 40 degrees north or south lati-

A In Kansas, tallgrass prairie with grazing bison.

B African savanna with grazing wildebeest.

C In California, chaparral with grazing mule deer.

Figure 43.10 Fire-adapted biomes.

tude. Mild winters bring up to 75 centimeters (30 inches) of rain, but summers are warm and dry. Chaparral is California's most extensive ecosystem, and deer are its main grazers (**Figure 43.10C**).

Chaparral also occurs in regions that border the Mediterranean, and in Chile, Australia, and South Africa. Many chaparral plants produce aromatic oils that help fend off insects, but also make them highly flammable. After a fire, plants resprout from roots and fire-resistant seeds germinate. Seeds of some species germinate only after a fire, when competition with other plants will be minimal (Section 41.6).

chaparral (shap-ur-RAHL) Biome of dry shrubland in regions with hot, dry summers and cool, rainy winters.

savanna (suh-VAN-uh) Tropical biome dominated by grasses with a few scattered shrubs and trees.

temperate grassland Biome in the interior of continents; perennial grasses and other nonwoody plants adapted to grazing and fire predominate.

Describe the climate of desert biomes.

List some traits that adapt plants to life in the desert.

Describe the composition of the desert crust and explain its ecological importance.

Figure 43.11 Sonoran Desert after a rain. Perennial saguaro cacti and drough-resistant shrubs grow beside annual wildflowers.

Figure 43.12 Desert crust.

Deserts receive an average of less than 10 centimeters (4 inches) of rain per year. They cover about one-fifth of Earth's land surface and many are located at about 30° north and south latitude, where global air circulation patterns cause dry air to sink. Rain shadows can also produce deserts. Chile's Atacama Desert is on the leeward side of the Andes, for example, and the Himalayas prevent rain from falling in China's Gobi desert.

Minimal rainfall keeps the humidity in deserts low. With little water vapor to block the sun's rays, intense sunlight reaches and heats the ground. At night, the lack of insulating water vapor in the air allows the temperature to fall fast. As a result, deserts tend to have larger daily temperature shifts than other biomes. Desert soils have very little topsoil. They also tend to be somewhat salty, because rain that falls usually evaporates before seeping into the ground. This rapid evaporation allows any salt that is in the rainwater to accumulate at the soil surface.

PLANT ADAPTATIONS TO DROUGHT

Despite their harsh conditions, most deserts support some plant life (**Figure 43.11**). Desert annuals avoid drought—their seeds sprout only when the soil is moist and the new plants produce seeds quickly. Desert perennials, which produce seeds over several seasons, have many structural adaptations. Spines or fuzz at their surface both deter herbivores and reduce the evaporative water loss. In deserts where rains fall seasonally, perennials can reduce water loss by producing leaves only after a rain, then shedding them when dry conditions return. Other desert-adapted perennials store water in their tissues. For example, the stem of a barrel cactus has a spongy pulp that holds water. The stem swells after a rain, then shrinks as the plant uses stored water.

Woody desert shrubs such as mesquite and creosote have extensive, efficient root systems that take up the little water that is available in soil. Mesquite roots can tap into water that lies as much as 60 meters (197 feet) beneath the soil surface.

Alternative carbon-fixing pathways also help desert plants minimize water loss. Cacti, agaves, and euphorbs are CAM plants, which open their stomata only at night when temperature typically declines (Section 6.5). This adaptation reduces the amount of water lost to evaporation.

DESERT CRUST

A desert crust forms at the surface of many desert soils. The crust is a community that can include cyanobacteria, lichens, mosses, and fungi (**Figure 43.12**). These organisms secrete organic molecules that glue them and the surrounding soil particles together. The resulting crust benefits members of the larger desert community in several ways. Bacteria in the crust fix nitrogen (Section 19.4), making this nutrient available to plants. The crust also holds soil particles in place. When the fragile connections within the desert crust are broken, the soil can blow away. Negative effects of such disturbance increase when windblown soil buries healthy crust in an undisturbed area. Once buried, crust organisms in the previously undisturbed area die off, allowing still more soil to take flight.

---

desert Biome with little rain and low humidity; plants that have water-storing and water-conserving adaptations predominate.

43.8 Arctic Tundra

LEARNING OBJECTIVES

Describe the climate and location of arctic tundra.

Explain why the soil of arctic tundra contains a large amount of carbon.

Explain why arctic tundra communities are relatively young.

Arctic tundra extends between the ice caps of the North Pole and the belts of boreal forests in the Northern Hemisphere. Most of this biome is in northern Russia and Canada. Arctic tundra is Earth's youngest modern biome. Arctic tundra communities first became established about 10,000 years ago, when glaciers retreated at the end of the last ice age.

Conditions in this biome are harsh; snow blankets the ground for as long as nine months of the year. Annual precipitation is usually less than 25 centimeters (10 inches), and cold temperature keeps the snow that does fall from melting. During a brief summer, plants grow fast under the nearly continuous sunlight (**Figure 43.13**). Lichens and shallow-rooted, low-growing plants are the main producers.

Even at midsummer, only the surface layer of tundra soil thaws. Beneath the thawed layer is **permafrost**, a layer of frozen soil that is as much as 500 meters (1,600 feet) thick in places. Permafrost acts as a barrier that prevents drainage, so the soil above it remains perpetually waterlogged. Decay is slow in the cool, anaerobic conditions of this soil, so organic remains accumulate. Organic matter in permafrost makes the arctic tundra one of Earth's greatest stores of carbon.

Figure 43.13 Arctic tundra in the summer.

arctic tundra Highest-latitude Northern Hemisphere biome, where low, cold-tolerant plants survive with only a brief growing season.

permafrost Of arctic tundra, a perpetually frozen layer of soil that prevents water from draining.

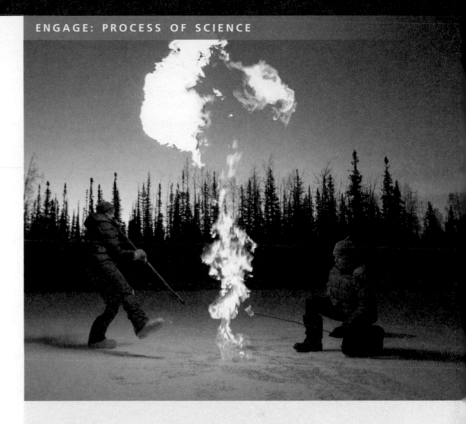

Dr. Katey Walter Anthony

Thawing permafrost releases methane, the same flammable gas used to heat homes and cook meals. In arctic lakes, the released methane bubbles up from the depths in a way that is hard to quantify—until the first clear ice of fall captures a snapshot of emissions from the lake in bubbles at its surface. To test whether a frozen bubble contains methane, National Geographic Explorer Katey Walter Anthony has an assistant plunge a pick into a bubble while she holds a match near it. If the bubble is methane, the gas ignites, as shown in the photo above.

Walter Anthony says the amount of methane bubbling up out of arctic lakes is troubling, because some of it seems to be coming not from bottom mud but from deeper geologic reservoirs that had previously been securely capped by permafrost. These reservoirs contain hundreds of times more methane than is in the atmosphere now. By venting methane into the atmosphere, the lakes are amplifying the global warming that created them: Methane is a greenhouse gas that traps 25 times more heat than carbon dioxide. Release of arctic methane results in warming that encourages the release of still more methane.

"If we could only capture the gas, it would make a great energy source," Walter Anthony says. Unlike coal, methane burns without leaving ash behind or releasing sulfur dioxide and mercury.

CREDITS: (13) Fexel/Shutterstock.com; (in text) Mark Thiessen/National Geographic Creative.

CHAPTER 43 THE BIOSPHERE 775

LEARNING OBJECTIVES

List some of the factors that affect a lake's primary production.

Describe the zonation of a lake.

Explain why a temperate zone lake turns over in spring and fall, and the effect that this turnover has on primary production.

Explain the factors that affect the oxygen content of a river.

With this section, we turn our attention to Earth's waters. We begin here with freshwater (non-saline) ecosystems, continue to coasts in the next section, then dive into the oceans. Note that the term biomes was originally coined to apply only to regions of land, but many people today refer to aquatic ecosystems as "aquatic biomes."

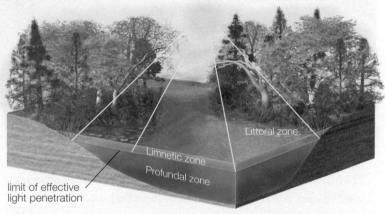

Figure 43.14 Lake zonation. A lake's littoral zone extends all around the shore to a depth where rooted aquatic plants stop growing. Its limnetic zone is the open water where light penetrates and photosynthesis occurs. Below the limnetic zone is the cooler, dark water of the profundal zone.

Figure 43.15 A young lake. Crater Lake in Oregon is a collapsed volcano that filled with snowmelt. It began filling about 7,700 years ago; from a geologic standpoint, it is a young lake.

LAKES

A lake is a body of standing fresh water. If sufficiently deep, it will have zones that differ in their physical characteristics and species composition (**Figure 43.14**). Nearest shore is the littoral zone. Here, sunlight penetrates all the way to the lake bottom, and aquatic plants are the main producers. A lake's open waters include an upper, well-lit limnetic zone, and a profundal zone where light does not penetrate. The main producers in the limnetic zone are members of the phytoplankton, a group of photosynthetic microorganisms that includes green algae, diatoms, and cyanobacteria. They serve as food for zooplankton, which are tiny consumers such as copepods. In the profundal zone, there is not enough light for photosynthesis, so consumers depend on food produced above. Debris that drifts down feeds detritivores and decomposers.

Nutrient Content and Succession Like a habitat on land, a lake undergoes succession (it changes over time, Section 41.6). A newly formed lake is deep, clear, and nutrient-poor, with low primary production (**Figure 43.15**). Over time, the lake becomes eutrophic. **Eutrophication** refers to processes, either natural or artificial, that enrich a body of water with nutrients. The increase in nutrients allows more producer growth, so primary production rises.

Seasonal Changes Temperate zone lakes undergo seasonal changes that affect primary production. Unlike most other substances, water is not most dense in its solid state (ice). As water cools, its density increases until it reaches 4°C (39°F). Below this temperature, water becomes less dense until it eventually freezes as ice. This is why ice floats on water (Section 2.4). In an ice-covered lake, there is a thin layer of water that is between 0°C and 4°C just beneath the ice. Below that is the densest (4°C) water (**Figure 43.16A**).

In spring, winds and rising temperature cause currents that lead to a spring overturn, during which oxygen-rich water in the surface layers moves down and nutrient-rich water from the lake's depths moves up (**Figure 43.16B**). After the spring overturn, longer days and the dispersion of nutrients through the water encourage primary production.

In summer, a lake has three layers (**Figure 43.16C**). The upper layer is warm and oxygen-rich. Below this is a thermocline, a thin boundary layer across which temperature falls rapidly with depth. The coldest, densest water is below the thermocline. The temperature and density difference across a thermocline prevents the lake's upper and lower layers from mixing. With no mixing, decomposers deplete the oxygen near the lake bottom, and nutrients released by decomposition at the bottom of the lake cannot reach surface waters. Thus, primary production declines.

In autumn, the lake's upper waters cool, the thermocline vanishes, and a fall overturn occurs (**Figure 43.16D**). Oxygen-rich water moves down while nutrient-rich water moves up. This overturn brings nutrients to the surface and favors a brief burst of primary production. However, unlike the spring overturn, it does not lead to sustained production because decreasing light and temperature slow photosynthesis. Primary production will not peak again until after the next spring overturn.

STREAMS AND RIVERS

Flowing-water ecosystems start as freshwater springs or seeps. Streams grow and merge as they flow downhill. Rainfall, snowmelt, geography, altitude, and shade affect flow volume and temperature. Minerals in submerged rocks dissolve in the water and affect its solute concentrations. Water in different parts of a river moves at different speeds, contains different solutes, and differs in temperature, and so the species composition of a river varies along its length.

THE ROLE OF DISSOLVED OXYGEN

The amount of oxygen dissolved in water is one of the most important factors affecting aquatic organisms. More oxygen dissolves in cooler, fast-flowing, turbulent water than in warmer, smooth-flowing or still water (**Figure 43.17**). Thus, an increase in water temperature or decrease in its flow rate can cause aquatic species with high oxygen needs to suffocate.

In freshwater habitats, aquatic larvae of mayflies and stoneflies are the first invertebrates to disappear when the oxygen content of the water decreases. These insect larvae are active predators that demand considerable oxygen, so they serve as indicator species (Section 41.6). Aquatic snails disappear, too. Declines in these invertebrates can have cascading effects on the fishes that feed on them. Fishes can also be more directly affected. Trout and salmon are especially intolerant of low oxygen. Carp (including goldfish) are among the most tolerant; they survive tepid, stagnant water in ponds.

No fishes survive when the oxygen content of water falls below 4 parts per million. Leeches persist, but most other invertebrates disappear. At the lowest oxygen concentration, annelids called sludge worms (*Tubifex*) often are the only animals. The worms are colored red by their large amount of hemoglobin. This abundance of hemoglobin adapts them to their low-oxygen habitat, where predators and competition for food are scarce.

eutrophication (you-truh-fih-CAY-shun) Nutrient enrichment of an aquatic habitat.

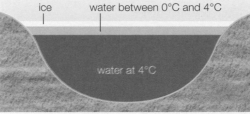

A Winter. Ice covers a thin layer of slightly warmer water just below it. Densest (4°C) water is at the bottom. Winds do not affect water under the ice, so there is little circulation.

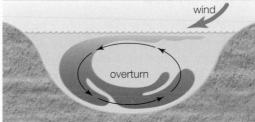

B Spring. Ice thaws. Upper water warms to 4°C and sinks. Winds blowing across water create currents that help overturn water, bringing nutrients up from the bottom.

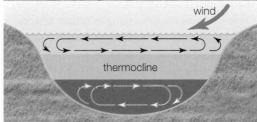

C Summer. Sun-warmed water floats on a thermocline, a layer across which temperature declines abruptly with depth. Upper and lower water do not mix because of this thermal boundary.

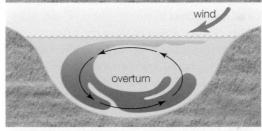

D Fall. Upper water cools and sinks downward, eliminating the thermocline. Vertical currents mix water that was separated during the summer.

Figure 43.16 Seasonal changes in a temperate zone lake.

FIGURE IT OUT
Which overturn results in the greatest rise in primary production?

Answer: The spring overturn, because increased nutrient availability is accompanied by increased light.

Figure 43.17 Effect of turbulence. As water flows over rocks, it picks up soluble minerals and becomes aerated, so it contains more oxygen.

LEARNING OBJECTIVES

Explain why estuaries are especially nutrient-rich.

Compare the main types of food chains on rocky and sandy shores.

Describe a coral reef.

Explain what causes coral bleaching.

COASTAL WETLANDS

An **estuary** is a partly enclosed body of water where fresh water from a river or rivers mixes with seawater. The size and shape of the estuary, and how fast freshwater flows into it, determine how quickly the saltwater and fresh water mix. In all estuaries, water from upstream constantly delivers the nutrients necessary to sustain a high level of primary production. Incoming fresh water also carries silt. Where water flow slows, the silt falls to the bottom, forming mudflats. Photosynthetic bacteria and protists in biofilms on mudflats often account for a large portion of an estuary's primary production.

Plants that live in estuaries are adapted to survive in water-logged, oxygen-poor, highly salty soil. These plants have a high salt content that makes them unpalatable to most herbivores, so detrital food chains (Section 42.2) usually predominate. Consider cordgrass (*Spartina*), the dominant plant in the salt marshes of many estuaries along the Atlantic coast (**Figure 43.18A**). Oxygen taken in by cordgrass shoots above water diffuses through long, hollow spaces in the stems to support roots buried in anaerobic soil. (All plant roots need oxygen to carry out aerobic respiration.) Salt that roots take up with water is excreted by specialized glands on the leaves.

Estuaries and tidal flats of tropical and subtropical latitudes often support nutrient-rich mangrove wetlands (**Figure 43.18B**). "Mangrove" is the common term for salt-tolerant woody plants that live in sheltered areas along tropical coasts. The plants have many prop roots that extend out from their trunk and help the plant stay upright in the soft sediments. Specialized cells at the surface of some exposed roots allow gas exchange necessary for aerobic respiration.

ROCKY AND SANDY SEASHORES

As with lakes, an ocean's shoreline is the littoral zone. This zone can be divided into three vertical regions that differ in their physical characteristics and species diversity (**Figure 43.19**). The most obvious difference between the regions is the amount of time they spend underwater. The upper littoral zone, also called the splash zone, regularly receives ocean spray but is completely submerged only during the highest of high tides. This zone gets the most sun, but has the fewest species. The midlittoral zone is covered by water during an average high tide and dry during an average low tide. The

A Cordgrass (*Spartina*) in a salt marsh. Salt taken up in water by roots is excreted by glands on the leaves.

B Mangroves along a tropical coast. Specialized cells at the surface of exposed prop roots allow gas exchange with the air.

Figure 43.18 Two types of coastal wetlands.

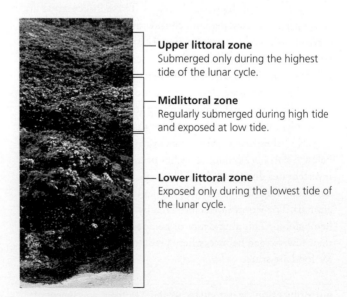

Upper littoral zone
Submerged only during the highest tide of the lunar cycle.

Midlittoral zone
Regularly submerged during high tide and exposed at low tide.

Lower littoral zone
Exposed only during the lowest tide of the lunar cycle.

Figure 43.19 Vertical zonation in the intertidal zone.

A Healthy coral reef near Fiji. The coral gets its color from the pigments of symbiotic dinoflagellates that live in its tissues.

B Bleached coral. The coral skeletons are mostly staghorn coral (*Acropora*), a genus especially likely to undergo coral bleaching.

Figure 43.20 Coral reefs.

lower littoral zone, which is exposed only during the lowest of low tides, is home to the most species.

You can easily see the zonation along a rocky shore. Multicelled algae ("seaweeds") that cling to rocks in the lower littoral zone are the main producers, and grazing food chains predominate. Snails are primary consumers. Zonation is less obvious on sandy shores where detrital food chains start with material washed ashore. Some crustaceans eat detritus in the upper littoral zone. In the lower littoral zone, other invertebrates feed as they burrow through the sand.

CORAL REEFS

Coral reefs are wave-resistant formations that consist primarily of calcium carbonate secreted by many generations of coral polyps. Reef-forming corals live mainly in shallow, clear, warm waters between latitudes 25° north and 25° south, so about 75 percent of all coral reefs are in the Indian and Pacific Oceans. Reef-building corals are keystone species (Section 41.7). A healthy reef is home to both corals and a huge number of other species (**Figure 43.20A**). Biologists estimate that coral reefs support about a quarter of all marine fish species.

Photosynthetic protists called dinoflagellates live inside the tissues of all reef-building corals (Section 23.4). The dinoflagellates are protected within the coral's tissues, and they provide the coral with oxygen and sugars that it depends on.

When a coral is stressed, as by a change in water temperature, it expels its dinoflagellates. These protists give the coral its color, so expelling them removes that color, an event called **coral bleaching**. If the stress subsides quickly, a few dinoflagellates may repopulate the coral's tissues or it may take up new ones. If the stress persists, the coral dies, leaving

only its white skeleton behind (**Figure 43.20B**). Coral bleaching events are increasingly common, in part because of the rise in sea temperature associated with global climate change (Section 42.5). Coral reefs are also under threat from many other factors such as discharge of sewage and other pollutants into coastal waters, human-induced erosion that clouds water with sediments, destructive fishing practices, and introduced invasive species. For example, some Hawaiian reefs are now threatened by exotic algal species that were imported for cultivation in the 1970s.

Another threat to coral is **ocean acidification**, a decrease in seawater pH caused by the human-induced rise in atmospheric carbon dioxide (CO_2). When CO_2 from the atmosphere dissolves in the ocean, seawater becomes more acidic. Since the onset of the industrial revolution, acidity of the ocean has risen by about 30 percent. The increased acidity makes it more difficult for corals to take up the calcium carbonate they need to build their skeleton. It also makes it more difficult for other marine organisms to make calcium carbonate parts such as shells.

As a result of these various threats, 25 species of coral are now considered threatened or endangered. This means their population has declined so much that they are now at risk of extinction.

coral bleaching A coral expels its photosynthetic dinoflagellate symbionts in response to stress and becomes colorless.

coral reef Highly diverse marine ecosystem centered around reefs built by living corals that secrete calcium carbonate.

estuary (ESS-chew-erry) A highly productive ecosystem where nutrient-rich water from a river mixes with seawater.

ocean acidification Decrease in seawater pH caused by the rise in atmospheric carbon dioxide.

LEARNING OBJECTIVES

Distinguish between the pelagic and benthic provinces.

Describe the producers at hydrothermal vents.

Describe a seamount.

Regional variations in light, nutrient availability, temperature, and oxygen concentration occur in oceans as well as lakes (**Figure 43.21**). The ocean's open waters are the **pelagic province**. This province includes the water over continental shelves and the more extensive waters farther offshore. In the ocean's upper, bright waters, phytoplankton such as single-celled algae and bacteria are the primary producers, and grazing food chains predominate. Even in clear water, light sufficient for photosythesis can penetrate only 200 meters beneath the sea surface. In deeper waters, organisms live in continual darkness. Here, organic material that drifts down from above serves as the basis of detrital food chains.

The **benthic province** is the ocean bottom, its rocks, and sediments. Species richness is greatest on continental shelves (the underwater edges of continents). The benthic province also includes mostly unexplored species-rich regions such as seamounts and hydrothermal vents. **Seamounts** are undersea mountains that stand 1,000 meters or more above the ocean floor, but are still below the sea surface. They attract large

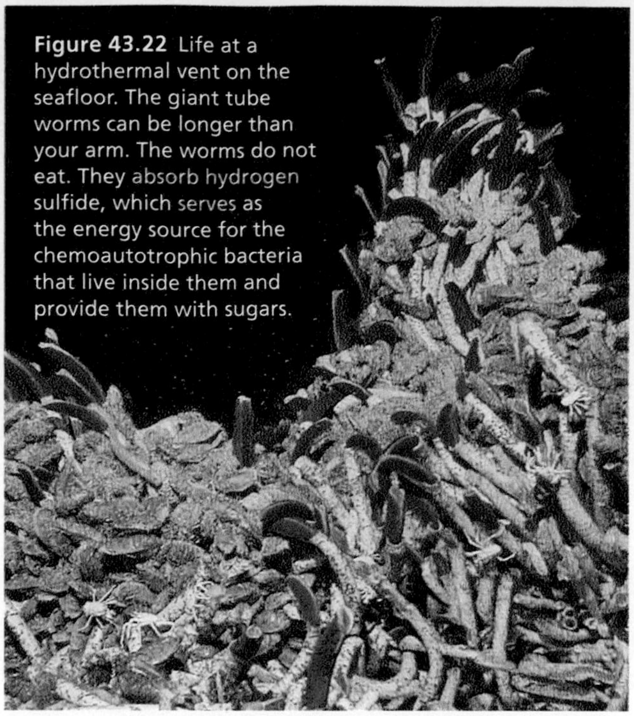

Figure 43.22 Life at a hydrothermal vent on the seafloor. The giant tube worms can be longer than your arm. The worms do not eat. They absorb hydrogen sulfide, which serves as the energy source for the chemoautotrophic bacteria that live inside them and provide them with sugars.

numbers of fishes and are home to many marine invertebrates. Many invertebrates present in large numbers on sea mounts are far less common elsewhere in the ocean.

At **hydrothermal vents**, mineral-rich seawater heated by geothermal energy spews out from an opening on the ocean floor. When this hot water mixes with cold deep-sea water, minerals come out of solution, forming rocky deposits. Chemoautotrophic bacteria and archaea that obtain energy by removing electrons from the minerals are the producers in hydrothermal vent communities. The energy they capture supports a food web that includes diverse invertebrates, such as crabs and tube worms (**Figure 43.22**).

Life exists even in the deepest sea, where it is cold, dark, and the pressure exerted by the water above is extremely high. A remote-controlled submersible that sampled sediments in the deepest part of the ocean (the Mariana Trench) brought up foraminifera that live 11 kilometers (7 miles) below the surface. Another submersible recorded video of eyeless, scaleless fish called snailfish living 8 kilometers (5 miles) down. There may be fish in even deeper water; the biodiversity of the deep sea remains largely unknown.

Figure 43.21 Ocean zonation. Zone dimensions are not drawn to scale.

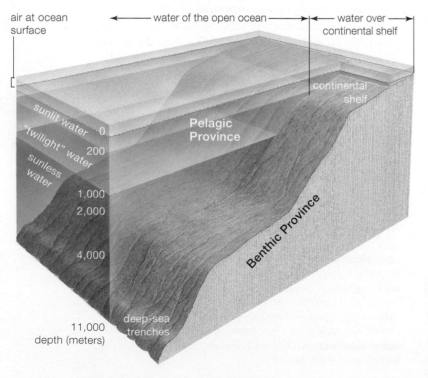

air at ocean surface

water of the open ocean

water over continental shelf

continental shelf

Pelagic Province

sunlit water — 0

"twilight" water — 200

sunless water

1,000
2,000

Benthic Province

4,000

deep-sea trenches

11,000
depth (meters)

---

benthic province (BEN-thick) The ocean's sediments and rocks.
hydrothermal vent Place where hot, mineral-rich water streams out from an underwater opening in Earth's crust.
pelagic province (peh-LAJ-ic) The ocean's open waters.
seamount An undersea mountain.

Figure 43.23 Satellite photos showing the effect of El Niño on primary production in the Pacific Ocean.

EFFECTS OF EL NIÑO

The El Niño Southern Oscillation, or ENSO, is a naturally occurring, irregularly timed fluctuation in sea surface temperature and wind patterns in the equatorial Pacific. The two extremes of this oscillation are referred to as El Niño and La Niña. Their influence is felt most strongly during the winter and along the western coast of South America, but they affect weather patterns year-round and worldwide.

The term El Niño means "baby boy" and refers to Jesus Christ; it was first used by Peruvian fishermen to describe local weather changes and a shortage of fishes that occurred in some winters and began around the time of Christmas. During an El Niño, unusually warm water flows toward eastern Pacific coasts, displacing the Humboldt Current that would otherwise bring up cooler, nutrient-rich water from the deep (Figure 43.4). Without this input of nutrients, marine primary producers decline in numbers. Dwindling producer populations and warming water decrease the number of small, cold-water fishes, as well as the larger fishes that eat them. This is why Peruvian fishermen catch fewer fishes during an El Niño.

El Niño episodes persist for 6 to 18 months. Often they are followed by a La Niña, in which Pacific waters become cooler than usual. During a La Niña, cold, nutrient-rich water flows toward the western coast of South America, phytoplankton populations rebound, and so do populations of fishes that feed on phytoplankton. At other times, waters of the Pacific are not much warmer or colder than average.

An extreme El Niño occurred in the winter of 1997–1998. Average sea surface temperatures in the eastern Pacific rose by 5°C (9°F) and a plume of warm water extended 9,660 kilometers (6,000 miles) west from the coast of Peru. These changes had extraordinary effects on the primary production in the equatorial Pacific. With the massive eastward flow of nutrient-poor warm water, photoautotrophs were almost undetectable in satellite photos that measure primary production (**Figure 43.23A**). Dwindling populations of producers and the warming water caused decreases in populations of cold-water fishes. Animals that feed on these fishes declined too. Primary production rebounded during the La Niña that followed (**Figure 43.23B**).

Rainfall patterns also shift during an El Niño. An El Niño brings heavy rains to eastern Pacific coasts, while Australia and Indonesia suffer from drought. An El Niño typically brings cooler, wetter weather to the American Gulf states, and reduces their likelihood of hurricanes.

Outbreaks of human disease often occur during an El Niño. For example, the increased ocean temperature in the Pacific leads to an increased incidence of cholera. The rise in the ocean surface temperature causes a population explosion copepods, a type of small crustacean that can carry cholera-causing bacteria. People become infected when they drink water from an estuary that contains the copepods. An El Niño also increases the incidence of malaria in southern Asia and Latin America. Heavy rains produce many puddles of freshwater in which the mosquitoes that carry malaria can breed.

Among its other duties, the United States National Oceanographic and Atmospheric Administration (NOAA) monitors sea surface temperature and studies El Niño events. NOAA's goal is to determine how El Niño affects global weather patterns and the extent of its effects. The results of these studies will help them predict when an El Niño or La Niña is likely to occur, and which regions are at a heightened risk for flooding, drought, hurricanes, or epidemics. Being able to plan for these events can help minimize their harmful effects. Current data on sea surface temperature, as well as information about the monitoring program, are available on NOAA's website at www.elnino.noaa.gov.

CREDIT: (23) NASA Goddard Space Flight Center Scientific Visualization Studio.

Sections 43.1, 43.2 Global air circulation patterns affect **climate** and the distribution of communities within the **biosphere**. Air is set in motion when sunlight heats tropical regions more than higher latitudes. Ocean currents distribute heat worldwide and influence weather patterns. Interactions between ocean currents, air currents, and landforms determine where regional phenomena such as **rain shadows** or **monsoons** occur.

Section 43.3 **Biomes** are characterized by a particular type of vegetation. Many biomes include multiple discontinuous areas. Climate and soil properties affect the distribution of biomes.

Sections 43.4, 43.5 Broadleaf evergreen trees that predominate in **tropical rain forests** grow all year. These highly productive forests are home to a large number of species. The tropical rain forest biome is the oldest that exists. Trees in **temperate deciduous forests** shed their leaves all at once just before a cold winter that prevents growth. Conifers that dominate Northern Hemisphere high-latitude **boreal forests** withstand cold and drought better than broadleaf trees.

Section 43.6 **Temperate grasslands** dominated by plants adapted to fire and grazing form in the somewhat moist interior of midlatitude continents. **Savannas** include fire-adapted grasses and scattered shrubs. Shrubby, fire-adapted **chaparral** is common in California and other coastal regions with hot, dry summers and cool, wet winters.

Section 43.7 **Deserts** form in regions with little precipitation and widely fluctuating temperature. Drought-adapted plants dominate. Desert crust holds soil particles in place and provides plants with nutrients.

Section 43.8 The Northern Hemisphere's **arctic tundra** is dominated by short plants that grow during the brief, cool summer when daylight is abundant. Tundra is the youngest biome, and its **permafrost** is a great reservoir of carbon.

Section 43.9 Variations in light, temperature, dissolved gases, and nutrients affect the distribution of species in aquatic ecosystems. Lakes undergo a natural process of **eutrophication**, becoming more nutrient-rich and productive over time. In temperate zone lakes, a spring overturn and a fall overturn cause vertical mixing of waters.

Sections 43.10, 43.11 Nutrient-rich fresh water mixes with seawater in an **estuary**. Plants in estuaries are adapted to high salinity and water-logged soils. Along rocky shores, multicellular algae form the base for grazing food chains. Detrital food chains predominate on sandy shores. In shallow tropical waters, accumulated calcium carbonate skeletons of coral polyps form the foundation of **coral reefs**. Photosynthetic dinoflagellates that live in the corals are producers for this species-rich ecosystem. When stressed, a coral may eject its photosynthetic symbionts, an event called **coral bleaching**. Coral reefs are also threatened by **ocean acidification**.

Life exists throughout the ocean. In the **pelagic province**, grazing food chains predominate. Detritus is the base for food chains in deeper, darker waters. **Seamounts** are regions of high diversity in the **benthic province**. At **hydrothermal vents**, chemoautotrophic bacteria and archaea are the producers.

Application Interactions among Earth's air and waters affect weather worldwide and have effects on human health, as when the Pacific warms during an El Niño and then cools during a La Niña.

SELF-ASSESSMENT
ANSWERS IN APPENDIX VII

1. The Northern Hemisphere is most tilted toward the sun in _____ .
   - a. spring
   - b. summer
   - c. autumn
   - d. winter

2. Which latitude will have the most hours of daylight on the summer solstice?
   - a. 0° (the equator)
   - b. 30° north
   - c. 45° north
   - d. 60° north

3. Warm air _____ and it holds _____ water than cold air.
   - a. sinks; less
   - b. rises; less
   - c. sinks; more
   - d. rises; more

4. A rain shadow is a reduction in rainfall _____.
   - a. on the inland side of a coastal mountain range
   - b. during an El Niño event
   - c. that results from global warming

5. The presence of the Gulf Stream _____ areas along the eastern coast of the United States.
   - a. warms
   - b. cools

6. _____ have a deep layer of humus-rich topsoil.
   - a. Deserts
   - b. Grasslands
   - c. Rain forests
   - d. Seamounts

7. Biomes differ in their _____.
   - a. climate
   - b. dominant plants
   - c. soils
   - d. all of the above

8. Grasslands most often are found _____.
   - a. near the equator
   - b. at high altitudes
   - c. in the interior of continents
   - d. in rain shadows

9. Permafrost is part of the soil in _____.
   - a. arctic tundra
   - b. temperate forest
   - c. boreal forest
   - d. grasslands

DATA ANALYSIS

CLASS ACTIVITY

CHANGING SEA TEMPERATURES In an effort to predict upcoming El Niño or La Niña events, the National Oceanographic and Atmospheric Administration collects information about sea surface temperature (SST) and atmospheric conditions. Scientists compare monthly temperature averages in the eastern equatorial Pacific Ocean to historical data, then calculate the difference (degree of anomaly) to see if El Niño conditions, La Niña conditions, or neutral conditions are developing. El Niño is a rise in the average SST above 0.5°C. La Niña is a decline of the same amount. **Figure 43.24** shows data for more than 40 years.

1. When did the greatest positive temperature deviation occur during this time period?

2. What type of conditions occurred during the winter of 1982–1983? What about the winter of 2001–2002?

3. During a La Niña, less rain than normal falls in the American West and Southwest. In the time interval shown, what was the longest interval without a La Niña?

4. What type of conditions were in effect in the fall of 2007 when California suffered severe wildfires?

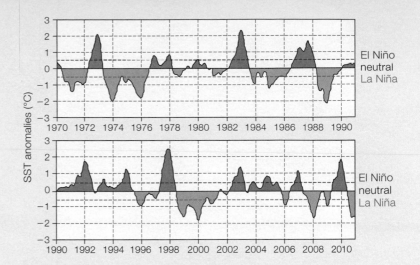

Figure 43.24 Sea surface temperature anomalies (differences from the historical mean) in the eastern equatorial Pacific Ocean. A rise above the dashed red line is an El Niño; a decline below the blue line is a La Niña.

---

10. The warmer water is, the _____ oxygen it can hold.
    a. more
    b. less

11. Chemoautotrophic bacteria and archaea are the primary producers for food webs _____ .
    a. in mangrove wetlands
    b. at seamounts
    c. on coral reefs
    d. at hydrothermal vents

12. Corals rely on symbiotic _____ for sugars.
    a. fungi
    b. bacteria
    c. dinoflagellates
    d. green algae

13. Which of the following biomes borders on boreal forest?
    a. savanna
    b. taiga
    c. tundra
    d. chaparral

14. Unrelated species living in geographically separated parts of a biome may resemble one another as a result of _____ .
    a. competitive interactions
    b. morphological convergence
    c. morphological divergence
    d. coevolution

15. Match the terms with the most suitable description.
    ___ tundra
    ___ chaparral
    ___ desert
    ___ savanna
    ___ estuary
    ___ boreal forest
    ___ pelagic province
    ___ tropical rain forest
    ___ hydrothermal vents

    a. broadleaf forest near equator
    b. partly enclosed by land; where fresh water and seawater mix
    c. African grassland with trees
    d. low-growing plants at high latitudes or elevations
    e. dry shrubland
    f. at latitudes 30° north and south
    g. mineral-rich, superheated water supports communities
    h. conifers dominate
    i. deep ocean

CLASS ACTIVITY

CRITICAL THINKING

1. In March of 2011, a tsunami (tidal wave) damaged a nuclear power plant in Fukushima, Japan. Radioisotopes released into the air at Fukushima were distributed by the prevailing winds. In what direction did these winds carry the isotopes?

2. Owners of off-road recreational vehicles would like increased access to government-owned deserts. Some argue that it is the perfect place for off-roaders because "There's nothing there." Why do biologists disagree?

3. Rita Colwell, the scientist who discovered why cholera outbreaks often occur during an El Niño, is concerned that global climate change could increase the incidence of this disease. By what mechanism might global warming cause an increase in cholera outbreaks?

4. London, England, is at the same latitude as Calgary in Canada's province of Alberta. However, the mean January temperature in London is 5.5°C (42°F), whereas in Calgary it is minus 10°C (14°F). Review Section 43.2, then suggest a reason for the difference in climate.

5. As global temperature rises as a result of increased atmospheric carbon dioxide, temperate zone lakes are freezing later in the winter and defrosting earlier in spring. How will these changes affect the timing of nutrient availability in these lakes? What effect might this change in timing have on life in the lakes?

for additional quizzes, flashcards, and other study materials
▶ ACCESS MINDTAP AT WWW.CENGAGEBRAIN.COM

44 Human Effects on the Biosphere

Links to Earlier Concepts

This chapter considers the causes of an ongoing mass extinction (Section 16.6) in light of human population growth (40.6). We look again at effects of species introductions and will draw on your knowledge of pH (2.5), the ozone layer (18.3), plant nutrition (26.1), and water and nutrient cycles (42.4–42.7).

Core Concepts

Adaptation of organisms to a variety of environments has resulted in diverse structures and physiological processes.

All living species are products of an ongoing evolutionary process that stretches back billions of years and each has a unique combination of evolved traits that suit it to its habitat. Extinction removes that collection of traits from the world forever. Humans have increased the frequency of extinctions and the extent of the losses is not fully known.

Interactions among the components of each biological system give rise to complex properties.

Human activities that alter one component of a habitat can have long-range and long-term effects. Plowing grasslands and cutting forests allow soil erosion that affects rainfall patterns, thus altering the habitat in a way that is difficult to reverse. Raising the temperature of the land and seas causes cascading effects by altering feeding, migration, and breeding patterns.

The field of biology relies upon experimentation and the collection and analysis of scientific evidence.

Carefully designing experiments helps researchers unravel cause-and-effect relationships in complex natural systems. Long-term observations can reveal patterns such as decline in the ozone level or a rise in global temperature. Documenting biodiversity and how it changes over time is an enormous and ongoing process.

◀ Young elephants and their keepers at David Sheldrick Wildlife Trust in Kenya. Poaching and human–wildlife conflicts left the elephants motherless at an early age. They will need human care for eight to ten years before they can survive in the wild.

Photograph by Michael Nichols, National Geographic Creative.

Extinction, like speciation, is a natural process; species arise and become extinct on an ongoing basis. The overwhelming majority of all species that have ever lived are now extinct.

The rate of extinction increases dramatically during a mass extinction, when many kinds of organisms in many different habitats become extinct in a relatively short period. Such an event is happening now. Scientists estimate that the current extinction rate is about 1,000 times that of the rate when a mass extinction is not occurring. Unlike most previous mass extinctions, this one is not the inevitable result of a physical catastrophe such as a volcanic eruption or asteroid impact. Humans are driving the current rise in extinctions, and our actions will determine the extent of the losses.

For purposes of conservation, a species is considered extinct if repeated, extensive surveys of its known range fail to turn up signs of any individuals. It is "extinct in the wild" if the only known members of the species are in captivity. Some species are difficult to find, so occasionally a population of species previously thought to be extinct or extinct in the wild turns up, but this does not happen very often.

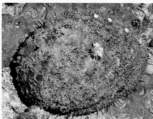

A White abalone. **B** Pyne's ground plum.

C Texas blind salamander.

Figure 44.1 Some endangered species of the United States. To learn more about these and other endangered species, visit the United States Fish and Wildlife Service's website at www.fws.gov/endangered/.

An **endangered species** faces extinction in all or part of its range. A **threatened species** is one that is likely to become endangered in the near future. Keep in mind that not all rare species are threatened or endangered. Some species have always been uncommon. In the United States, species are listed as endangered or threatened by the United States Fish and Wildlife Service (USFWS).

CAUSES OF SPECIES DECLINE

Overharvesting In some cases, people directly reduce a species' number. Consider that when European settlers first arrived in North America, they found between 3 and 5 billion passenger pigeons. In the 1800s, commercial hunting of these birds for food caused a steep decline in their number. The last time anyone saw a wild passenger pigeon was 1900, and he shot it. The last captive member of the species died in 1914.

We continue to overharvest species. The overfishing of the Atlantic codfish population, described in Section 40.5, is one recent example. Consider also the fate of the white abalone, a gastropod mollusk native to kelp forests off the coast of California (**Figure 44.1A**). Heavy harvesting of this species during the 1970s reduced the population to about 1 percent of its original size. In 2001, it became the first invertebrate to be listed as endangered by the USFWS. Although some white abalone remain in the wild, their population density remains too low for effective reproduction. The species' only hope for survival is a program of captive breeding. If this program succeeds, individuals will be reintroduced to the wild.

Species are overharvested not only as food, but also for use in traditional medicine, for the pet trade, and for ornamentation. Some orchids prized by collectors have become nearly extinct in the wild. Most of the orphan elephants shown in the chapter opening photo lost their mother to poachers who kill to obtain ivory tusks. The majority of elephant tusks harvested this way end up in China, in the form of decorative carved objects.

Habitat Degradation and Fragmentation Humans also harm species indirectly by altering the species' habitat. Each species requires a specific type of habitat, and any degradation, fragmentation, or destruction of that habitat reduces population numbers.

An **endemic species** occurs only in the area in which it evolved. Endemic species are more likely to go extinct as a result of habitat degradation than species that have spread to many areas. Consider Pyne's ground plum (**Figure 44.1B**), a flowering plant that lives only in cedar forest near a rapidly growing city in Tennessee. The plant is threatened by conversion of its habitat to homes and industrial use. Texas blind

CREDITS: (1A) John Butler, NOAA; (1B) Joel Sartore/National Geographic Creative; (1C) Joe Fries, U.S. Fish & Wildlife Service.

salamanders (**Figure 44.1C**) are among the species endemic to Edwards Aquifer, a series of water-filled, underground limestone formations. Excessive withdrawals of water from the aquifer, in combination with water pollution, threaten the salamander and other species in the aquifer.

Deliberate or accidental species introductions can be another type of habitat degradation (Section 41.7). Rats that reached islands by stowing away on ships endanger many ground-nesting birds that evolved in the absence of egg-eating ground predators. Exotic species can outcompete native ones. For example, California's endemic golden trout declined after European brown trout and eastern brook trout were introduced into California's mountain streams for sport fishing. The introduced trout outcompete the native trout for food.

Habitat fragmentation creates islands of suitable habitat scattered across what was once a continuous range. Buildings, roads, and fences can fragment a habitat, thus breaking a large population into smaller populations in which inbreeding is more likely. Inbreeding can have deleterious genetic effects (Section 17.6). Fragmentation also prevents individuals from accessing essential resources. Roads, fences, and houses prevent spotted turtles endemic to the eastern United States from moving among wetlands to feed and hibernate. The wetlands still exist, but turtles cannot move between them.

THE UNKNOWN LOSSES

Estimates of the number of currently living species range as high as one trillion, but only about 2 million species have been described and named. Simply put, we do not yet know what we have to lose. Scientists racing to document Earth's diversity are attempting to identify the species most in need of protection before it is too late.

The International Union for Conservation of Nature and Natural Resources (IUCN) monitors threats to species worldwide. However, its species listings have historically focused on vertebrates. Threats to invertebrates and plants have only recently been considered. Our impact on protists and fungi is largely unknown, and the IUCN does not address threats to bacteria or archaea. These microorganisms are essential to nutrient cycles that humans rely upon, yet we have almost no idea whether or how our activities affect their numbers and distribution.

---

endangered species Species that faces extinction in all or a part of its range.

endemic species (en-DEM-ick) Species that remains restricted to the area where it evolved.

threatened species Species likely to become endangered in the near future.

Dr. Paula Kahumbu

Linking conservationists with members of the public and with one another is one of Paula Kahumbu's main goals. This National Geographic Explorer says, "Conservationists do crucial work on a shoestring, cut off from the rest of the world. They're in remote, isolated places, some even risking their lives, with no chance of getting on the international radar screen. Meanwhile, millions of people who care about the catastrophic loss of wildlife and habitats aren't sure how to help." Kahumbu is executive director of WildlifeDirect (http://wildlifedirect.org), an online platform that gives conservationists a way to share day-by-day challenges and victories via blogs, diaries, videos, photos, and podcasts. Thanks to her efforts, people concerned about wildlife and wild places can view problems in real time and track the impact of their own contributions. The site attracts thousands of visitors daily, with online donations going directly to projects across Africa, Asia, and South America.

Information from conservationists involved with WildlifeDirect helped Kahumbu recognize an important conservation issue in Africa—misuse of the insecticide carbofuran. This chemical, which is banned for use on food crops in the United States and for all uses in Europe, remains widely available in Africa. Reports flowing into WildlifeDirect revealed that some farmers were using carbofuran-laced carcasses to poison lions that they believed threatened livestock. This practice caused declines in endangered lions, hyenas, and vultures. People were also putting carbofuran into irrigated rice fields to deliberately kill water birds for consumption as human food. Although both harmful practices have declined as a result of the efforts of Kahumbu and other conservationists, she continues to work toward a ban on the sale of carbofuran in Africa.

Kahumbu began her conservation work with research on African elephants and she remains a strong advocate for their protection. She favors a global ban on all trade in ivory, increased efforts to detect smuggled ivory, and stiffer penalties for those who take part in any aspect of the ivory trade.

With this section, we begin a survey of ways that human activities threaten species by destroying or degrading habitats.

DESERTIFICATION

Deserts naturally expand and contract over geological time as climate conditions vary. However, human activities sometimes result in the rapid conversion of a grassland or woodland to desert, a process called **desertification**. As human populations increase, greater numbers of people are forced to farm in areas that are ill suited to agriculture. In other places, people allow livestock to overgraze in grasslands. In both cases, the result can be habitat degradation through desertification.

A well-documented instance of desertification occurred in the United States during the mid-1930s, when large portions of prairie on the southern Great Plains were plowed under to plant crops. Plowing exposed deep prairie topsoil to the force of the region's constant winds. Then came a drought, and the result was an economic and ecological disaster. Winds carried more than a billion tons of topsoil aloft as sky-darkening dust clouds (**Figure 44.2A**), turning the region into what came to be known as the Dust Bowl. Millions of tons of soil was displaced, and some fell to earth as far away as New York City and Washington, D.C.

Today, Africa's Sahara desert is expanding south into the Sahel region. Overgrazing in this region strips grasslands of their vegetation, exposing the soil to erosion by wind. Wind carries billions of tons of soil aloft and westward (**Figure 44.2B**). Soil particles land as far away as the southern United States and the Caribbean. In China's northwestern regions, overplowing and overgrazing have expanded the Gobi desert so that dust clouds periodically darken skies above Beijing. Winds carry some of this soil across the Pacific to the west coast of the United States.

Drought encourages desertification, which results in more drought (a positive feedback cycle). Plants cannot thrive in a region where the topsoil has blown away. Fewer plants means less transpiration (Section 26.3), so less water enters the atmosphere and local rainfall decreases.

The best way to prevent desertification is to avoid farming in areas subject to high winds and periodic drought. If these areas must be used, methods that do not repeatedly disturb the soil can minimize risk of desertification.

DEFORESTATION

The amount of forested land is currently stable or increasing in North America, Europe, and China, but tropical forests continue to disappear at an alarming rate. In Brazil, increases in the export of soybeans and free-range beef have helped make the country the world's seventh-largest economy. However, this economic expansion has come at the expense of the country's woodlands and forests (**Figure 44.3**).

Deforestation has detrimental effects beyond the immediate destruction of forest organisms. For example, deforestation encourages flooding, because water that no longer is taken up by tree roots runs off instead. Deforestation in hilly

Figure 44.2 Dust storms, one outcome of desertification.

A Dust cloud in the Great Plains during the 1930s.

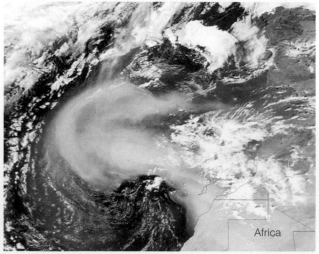

B Dust blows across the Atlantic from North Africa.

Figure 44.3 Deforestation. Cropland that once was rainforest extends right up to the edge of Iguaçu National Park, a World Heritage Site that straddles the border between Brazil and Argentina. Threatened primates and large cats are among the species that rely on this forest habitat.

areas raises the risk of landslides. Tree roots tend to stabilize the soil. When they are removed, waterlogged soil becomes more likely to slide.

Soils of deforested areas become nutrient-poor because of the increased loss of nutrient ions in runoff. **Figure 44.4** shows results of an experiment in which scientists deforested a region in New Hampshire and monitored the nutrient content of runoff. Deforestation caused a spike in loss of essential soil nutrients such as calcium.

Like desertification, deforestation affects local climate. The loss of plants means reduced transpiration, so the amount of local rainfall declines. In shady forests, transpiration also results in evaporative cooling. When a forest is cut down, shade disappears and the evaporative cooling ceases. Thus, the temperature in a deforested area is typically higher than in an adjacent forested area.

Once a tropical forest has been logged, the resulting nutrient losses and drier, hotter conditions can make it impossible for tree seeds to germinate or for seedlings to survive. Thus, deforestation can be difficult to reverse.

Forests take up and store huge amounts of carbon dioxide (Chapter 25 Application), so forest losses also contribute to global climate change. With fewer trees, less carbon dioxide is taken up from the atmosphere and stored in wood.

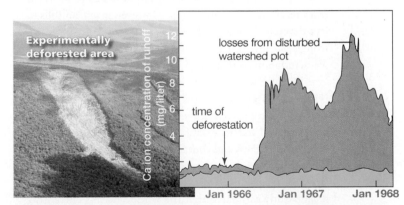

Figure 44.4 Effect of experimental deforestation on nutrient losses from soil. After deforestation, calcium (Ca) levels in runoff increased sixfold (gray). An undisturbed plot in the same forest showed no increase during this time (green).

FIGURE IT OUT
In which January was the Ca ion concentration in runoff from the disturbed forest the highest?　　Answer: January 1967

---

desertification (dih-zert-ih-ifh-KAY-shun) Conversion of grassland or woodland to desert.

LEARNING OBJECTIVES

Explain the negative impact of burying trash on land.

Describe how plastic in the sea harms marine organisms.

Compare the negative effects of acid, ammonium, and mercury deposition.

Pollutants are natural or synthetic substances released into soil, air, or water in greater than natural amounts. The presence of a pollutant disrupts the physiological processes of organisms that evolved in its absence, or that are adapted to lower levels of it. Some pollutants come from a few easily identifiable sites, or point sources. A factory that discharges pollutants into the air or a landfill that releases pollutants into water is a point source. Pollutants that come from point sources are easiest to control: Identify the few specific sources, and you can take action there. Eradicating pollution from nonpoint sources is more challenging. Pollution from nonpoint sources stems from widespread release of a pollutant. For example, leaked motor oil that pollutes waterways comes from vehicles in many roads and driveways.

TALKING TRASH

Seven billion people use and discard a lot of stuff. Where does all the waste go? Historically, much of it was buried in the ground or dumped out at sea. Trash was out of sight, and also out of mind. We now know that these practices have negative impacts on both biodiversity and human health.

Piling up or burying garbage on land can contaminate groundwater, as when lead from discarded batteries seeps into aquifers. Modern landfills in developed countries are lined with clay and/or plastic to prevent toxins in the trash from reaching the groundwater. However, many landfills in less-developed countries have no such barrier. Toxins leached from trash enter the groundwater and can turn up in waterways and in wells that supply drinking water.

Discarding solid waste in oceans threatens marine life (**Figure 44.5**). In the United States, solid waste can no longer legally be dumped at sea. Nevertheless, plastic constantly enters coastal waters. Plastic shopping bags, plastic water bottles, foam cups and containers from fast-food outlets, and other litter often enters storm drains. From there it is carried to streams and rivers that can convey it to the sea. A recent study estimated that the world's oceans currently contain 268,940 tons of floating plastic.

Ocean currents can carry bits of plastic for thousands of miles. These plastic bits accumulate in some areas of the ocean. Consider the Great Pacific Garbage Patch, a region of the north central Pacific that the media often describe as an "island of trash." In fact, the plastic is not easily visible.

A Juvenile sea lion with ring of discarded plastic around its neck. As the animal grows, the plastic will cut into its neck, causing a wound and impairing its ability to feed.

B Recently deceased Laysan albatross chick, dissected to reveal the contents of its gut. Scientists found more than 300 pieces of plastic inside this bird. It died because one of the pieces had punctured its gut wall. The chick was fed the plastic by its parents, who gathered the material from the ocean surface, mistaking it for food.

Figure 44.5 Perils of plastic.

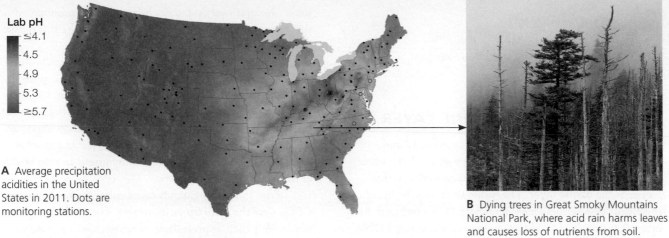

A Average precipitation acidities in the United States in 2011. Dots are monitoring stations.

Lab pH
- ≤4.1
- 4.5
- 4.9
- 5.3
- ≥5.7

B Dying trees in Great Smoky Mountains National Park, where acid rain harms leaves and causes loss of nutrients from soil.

Figure 44.6 Acid precipitation in the United States.

Rather, the garbage patch is a region where a high concentration of confetti-like plastic particles swirl slowly around an area as large as the state of Texas. The small bits of plastic absorb and concentrate toxic compounds such as pesticides and industrial chemicals from the seawater around them, making the plastic all the more harmful to marine organisms that mistakenly eat it.

ATMOSPHERIC DEPOSITION

Pollutants that enter the atmosphere can harm organisms when they fall to Earth as dust or in precipitation.

Acid Deposition Common air pollutants include sulfur dioxides (released by coal-burning power plants) and nitrogen oxides (released by combustion of gasoline and oil). In dry weather, airborne sulfur and nitrogen oxides coat dust particles that fall to the ground as dry acid deposition. Wet acid deposition, or **acid rain**, occurs when pollutants combine with water and fall as acidic precipitation. The pH of acid rain can be as low as 2; normal rainwater has a pH above 5 (Section 2.5). In the United States, federal regulations limiting sulfur dioxide emissions from coal-burning power plants have made precipitation less acidic (**Figure 44.6A**). The world's main sulfur dioxide emitters are now China and India, where industrialization and coal use continue to increase.

Acid rain that falls on or drains into waterways, ponds, and lakes can harm aquatic organisms. Acidic water can kill adult fish and prevent fish eggs from developing. Acid rain that falls on a forest burns tree leaves and alters the nutrient content of forest soil. As acidic water drains through the soil, positively charged hydrogen ions displace positively charged nutrient ions such as calcium, which leach from the soil in runoff. The acidity also causes soil particles to release metals such as aluminum that can harm plants. A combination of fewer nutrients and greater exposure to toxic metals weakens trees, making them more susceptible to insects and pathogens, and thus more likely to die. Trees at high elevations, where exposure to clouds of acidic droplets occurs more frequently, are most at risk of harm from acid rain (**Figure 44.6B**).

Ammonium Deposition Sulfur dioxides and nitrogen oxides that enter the air can also combine with gaseous ammonia released by fertilizers and animal wastes. When the resulting ammonium salts fall to Earth, they act like fertilizer. Addition of these salts to soil allows plants with high nitrogen needs to thrive where they would otherwise be excluded, thus altering the array of species in an ecosystem.

Mercury Deposition Emissions from coal-burning power plants and factories are the main source of mercury in the atmosphere. When mercury falls to Earth, microorganisms combine it with carbon to form methylmercury, which is toxic to animals. As explained in the Chapter 2 and 42 Applications, methylmercury commonly enters aquatic food chains and undergoes biological magnification—its concentration in organisms increases as it passes from one trophic level to the next. Thus upper-level consumers of aquatic organisms are most affected. Mercury does not harm plants; in fact, plants are sometimes used to bioremediate mercury pollution.

---

acid rain Low-pH rain that forms when sulfur dioxide and nitrogen oxides mix with water vapor in the atmosphere.

pollutant A substance released into the environment by human activities; interferes with the function of organisms that evolved in the absence of the substance or with lower levels.

LEARNING OBJECTIVES

Explain the importance of the ozone layer.

List the pollutants that affect the ozone layer.

Describe what has been done to slow ozone layer thinning.

DEPLETION OF THE OZONE LAYER

In the upper layers of the atmosphere, between 17 and 27 kilometers (10.5 and 17 miles) above sea level, the ozone (O_3) concentration is so great that scientists refer to this region as the **ozone layer**. The ozone layer benefits living organisms by absorbing most ultraviolet (UV) radiation from incoming sunlight. UV radiation, remember, damages DNA and causes mutations (Section 8.5).

In the mid-1970s, scientists noticed that Earth's ozone layer was thinning. Its thickness had always varied a bit with the season, but now the average level was declining steadily from year to year. By the mid-1980s, the spring thinning of

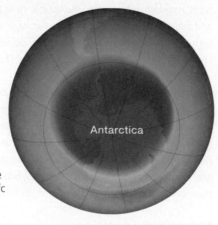

A Ozone levels in the upper atmosphere in September 2012, the Antarctic spring.

Purple indicates the least ozone; blue, green, and yellow indicate increasingly higher levels.

Check the current status of the ozone hole at NASA's website (http://ozonewatch.gsfc.nasa.gov/).

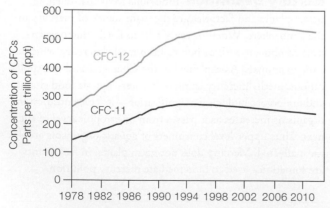

B Concentration of two CFCs in the upper atmosphere. These pollutants destroy ozone. A worldwide ban on CFCs has successfully halted the rise in atmospheric CFC concentration.

Figure 44.7 Destruction of the ozone layer.

the ozone layer over Antarctica had become so pronounced that people began referring to an "ozone hole" over this region (**Figure 44.7A**).

Declining ozone quickly became an international concern. With a thinner ozone layer, people would be exposed to more UV radiation, the main cause of skin cancers. Higher UV levels also harm animals, which do not have the option of using sunscreeens or avoiding sunlight. In addition, exposure to higher-than-normal UV levels damages plants and other producers, slowing their rate of photosynthesis and the release of oxygen into the atmosphere.

In the 1980s, compounds called chlorofluorocarbons, or CFCs, were the main ozone destroyers. These odorless gases were widely used as propellants in aerosol cans, as coolants, and in solvents and plastic foam. In response to the potential threat posed by the thinning ozone layer, countries worldwide agreed in 1987 to phase out the production of CFCs and other ozone-destroying chemicals. As a result of that agreement, known as the Montreal Protocol, the concentrations of CFCs in the atmosphere are no longer rising dramatically (**Figure 44.7B**). However, CFCs break down quite slowly, so they are expected to remain at a level that significantly thins the ozone layer for several decades.

Today, the main ozone-depleting pollutant entering the atmosphere is nitrous oxide (N_2O). This gas is released when people burn fossil fuels and when bacteria feed on fertilizers and animal waste (Section 42.6).

NEAR-GROUND OZONE POLLUTION

Ozone does not naturally occur in the lower atmosphere, but it can form when compounds released by burning or evaporating fossil fuels are exposed to sunlight. Warm temperature speeds this reaction, so ground-level ozone varies daily (it is higher in the daytime) and seasonally (it is higher during the summer). The ozone that forms in the lower atmosphere never reaches the ozone layer where it would be useful. Instead, ozone that forms in the lower atmosphere acts as a harmful pollutant. It irritates the eyes and respiratory tracts of animals and interferes with plant growth.

To help reduce ozone pollution, avoid actions that put fossil fuels or their combustion products into the air at times that favor ozone production. During hot, sunny, still weather, postpone filling your gas tank or using gasoline-powered appliances until the evening, when there is less sunlight to power the conversion of pollutants to ozone.

ozone layer High atmospheric layer with a high concentration of ozone (O_3) that prevents much ultraviolet radiation from reaching Earth's surface.

CREDITS: (7A) NASA Ozone Watch; (7B) From www.esrl.noaa.gov.

44.5 Global Climate Change

Explain why a rise in global temperature raises sea level.

Describe some ways in which global climate change can alter biological communities.

Ongoing climate change affects ecosystems worldwide. Average temperatures are increasing (**Figure 44.8**), with the more pronounced rise at temperate and polar latitudes. This rising temperature is elevating sea level by two mechanisms. Water expands as it is heated, and heating also melts sea ice and glaciers (**Figure 44.9**), adding meltwater to the sea. In the past century, sea level has risen about 20 centimeters (8 inches). As a result, some coastal wetlands and low-lying islands are disappearing underwater.

Temperature changes are important cues for many temperate zone species. Abnormally warm spring temperatures are causing deciduous trees to put leaves out earlier, and spring-blooming plants to flower earlier. Animal migration times and breeding seasons are also shifting. Species arrays in biological communities are changing as warmer temperatures allow some species to expand their range to higher latitudes or elevations that were previously too cold for them.

Not all species can move or disperse quickly to a new location, and warmer temperatures are expected to drive some of these species to extinction. In tropical seas, warming water is stressing reef-building corals and increasing the frequency of coral bleaching events (Section 43.10). In the arctic, seasonal sea ice now forms later and melts earlier, stressing polar bears. During late spring and early fall, regions where these bears previously would have walked across sea ice to gain access to their seal prey are now open water.

Global warming is just one aspect of global climate change. Temperature affects evaporation, winds, and currents, so many weather patterns are expected to change as the world continues to warm. For example, warmer temperatures are correlated with extremes in rainfall patterns: periods of drought interrupted by unusually heavy rains. In addition, warmer seas tend to increase the intensity of hurricanes.

As Section 42.5 explained, the main cause of global climate change is a rise in the atmospheric concentrations of greenhouse gases such as carbon dioxide. Fossil fuel combustion is the single biggest source of greenhouse gas emissions, and the use of these fuels continues to rise as China and India become increasingly industrialized. Reducing greenhouse gas emissions will be a challenge, but international efforts are under way to increase the efficiency of processes that require fossil fuels, to shift to alternative energy sources such as solar and wind power, and to develop innovative ways to store carbon dioxide.

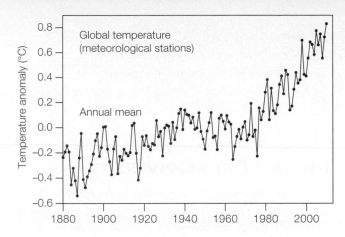

Figure 44.8 Rising global temperature. The temperature anomaly is the amount a temperature deviates from average temperature during the interval from 1951 to 1980. Temperatures were measured at weather stations on land.

Figure 44.9 Melting glaciers, a sign of a warming world. A recent study estimates that Alaska's glaciers lost an average of 75 million tons of ice per year between 1994 and 2013.

CREDITS: (8) NASA Goddard Institute for Space Studies; (9) top, National Snow and Ice Data Center, W. O. Field; bottom, National Snow and Ice Data Center, B. F. Molnia.

LEARNING OBJECTIVES

List the three levels of biodiversity.

Explain the benefits of protecting biodiversity.

Explain what an ecological hot spot is and give an example.

Using an appropriate example, explain the goal of ecological restoration.

THE VALUE OF BIODIVERSITY

Biodiversity (biological diversity) refers to the variety and variability of life forms. A region's **biodiversity** is measured at three levels: the genetic diversity within species, species diversity, and ecosystem diversity. Biodiversity is currently declining at all three levels, in all regions.

Conservation biology addresses these declines. The goals of this relatively new field of biology are (1) to survey the range of biodiversity, and (2) to find ways to maintain and use biodiversity to benefit human populations by encouraging people to value their region's natural resources and use those resources in nondestructive ways.

Why should we protect biodiversity? From a selfish standpoint, doing so is an investment in our future. Healthy ecosystems are essential to the survival of our species. Other organisms produce the oxygen we breathe and the food we eat. They remove waste carbon dioxide from the air and decompose and detoxify wastes. Plants take up rain and hold soil in place, preventing erosion and reducing the risk of flooding. Wild species produce medically valuable compounds that we are still discovering; wild relatives of crop plants are reservoirs of genetic diversity that plant breeders draw on to protect and improve crops.

There are ethical reasons to preserve biodiversity too. All living species are the result of an ongoing evolutionary process that stretches back billions of years. Each species has a unique combination of traits, and extinction removes that collection of traits from the world forever. The more diverse a biological system is, the greater its capacity to recover from damage. This applies at all levels of life. Thus, reducing the biodiversity of ecosystems makes them less resilient to disturbances.

SETTING PRIORITIES

Protecting biological diversity is often a tricky proposition. Even in developed countries, people often oppose environmental protections because they fear adverse economic consequences. However, taking care of the environment can make good economic sense. With a bit of planning, people can both preserve and profit from their biological wealth.

The resources available for conserving areas are limited, so conservation biologists must often make difficult choices about which areas should be targeted for protection first. These biologists identify **biodiversity hot spots**, places that are home to species found nowhere else and are under great threat of destruction. Once identified, hot spots can take priority in worldwide conservation efforts.

Figure 44.10 The location and conservation status of the land ecoregions deemed most important by the World Wildlife Fund.

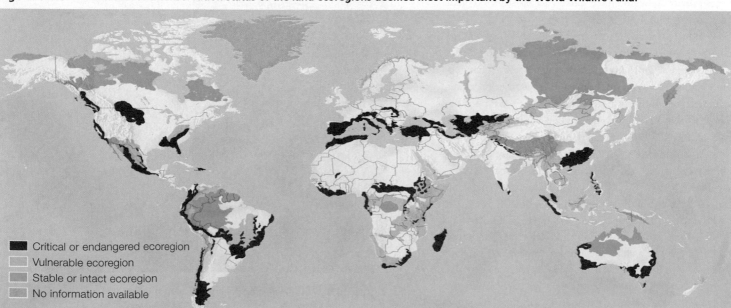

- Critical or endangered ecoregion
- Vulnerable ecoregion
- Stable or intact ecoregion
- No information available

On a broader scale, conservation biologists define eco-regions, which are land or aquatic regions characterized by climate, geography, and the species found within them. The most widely used system of ecoregion definitions was developed by scientists of the World Wildlife Fund. This system defines 867 distinctive land ecoregions, each with a distinctive arrays of species. **Figure 44.10** shows the locations and conservation status of major ecoregions considered the top priority for conservation efforts.

The Klamath–Siskiyou forest in southwestern Oregon and northwestern California is one of North America's endangered ecoregions (**Figure 44.11**). It is home to many rare conifers. Two endangered birds, the northern spotted owl and the marbled murrelet, nest in old-growth parts of the forest, and endangered coho salmon breed in streams that run through the forest. Logging threatens all of these species.

By focusing on protecting biodiversity hot spots and critical ecoregions rather than on individual endangered species, scientists hope to maintain the ecosystem interactions that naturally sustain biological diversity.

PRESERVATION AND RESTORATION

Worldwide, many biodiverse regions have been protected in ways that benefit local people. The Monteverde Cloud Forest in Costa Rica is one example. In the 1970s, George Powell was studying birds in this forest, which was rapidly being cleared. Powell decided to buy part of the forest as a nature sanctuary. His efforts inspired individuals and conservation groups to donate funds, and much of the forest is now a nature reserve. The reserve's plants and animals include more than 100 mammal species, 400 bird species, and 120 species of amphibians and reptiles. It is one of the few habitats left for jaguars and ocelots. A tourism industry centered on the reserve provides economic benefits to local people.

Sometimes an ecosystem is so damaged, or so little of it is left, that conservation alone is not enough to sustain biodiversity. **Ecological restoration** is the process of renewing a natural ecosystem that has been degraded or destroyed. Consider the ecological restoration occurring in Louisiana's coastal wetlands. More than 40 percent of the coastal wetlands in the United States are in Louisiana. These marshes are an ecological and economic treasure, but they are in trouble. Dams and levees built upstream of the marshes hold back sediments that should replace sediment the marshes lose to the sea. Channels cut through the marshes for oil exploration and extraction have encouraged erosion, and the rising sea level threatens to drown existing plants. Restoration efforts now under way aim to reverse some of those losses by restoring marsh grass to regions that have become open water (**Figure 44.12**).

Figure 44.11 Klamath–Siskiyou forest, one of North America's critical ecoregions. Endangered northern spotted owls (inset) are endemic to this coniferous forest.

Figure 44.12 Ecological restoration in Sabine National Wildlife Refuge. The squares contain sediment that was brought in and planted with marsh grass. Restoring marsh in areas that had become open water increases animal abundance and overall species richness of the area.

biodiversity Of a region, the genetic variation within species, variety of species, and variety of ecosystems.

conservation biology Field of applied biology that surveys biodiversity and seeks ways to maintain and use it nondestructively.

ecological restoration Actively altering an area in an effort to restore an ecosystem that has been damaged or destroyed.

biodiversity hot spot Threatened region with many species not found elsewhere; a high priority for conservation efforts.

CREDITS: (11) background, David Patte, USFWS; inset, USFWS; (12) Diane Borden-Bilot, U.S. Fish and Wildlife Service.

Explain what is meant by living sustainably.

Describe some ways that people can reduce their negative impact on other species.

An understanding that all life on Earth draws upon the same limited resources and that the health of the environment affects human well-being has given rise to call for more sustainable practices. In this context, sustainability means using resources in a way that meets the needs of the current human population without degrading the environment so that future generations will be unable to meet their own needs.

Living sustainably begins with recognizing the environmental consequences of one's own lifestyle. People in industrial nations use enormous quantities of resources, and the extraction, delivery, and use of these resources has negative effects on biodiversity. In the United States, the size of the average family has declined since the 1950s, while the size of the average home has doubled. All of the materials used to build and furnish those larger homes come from the environment. For example, an average new home contains about 500 pounds of copper in its wiring and plumbing.

Where does that copper come from? Like most other nonrenewable mineral elements used in manufacturing, most copper is mined (**Figure 44.13A**). Surface mining strips an area of vegetation and soil, creating an ecological dead zone. Like other types of mining, surface mining can put particulate matter into the atmosphere, creates mountains of rocky waste, and can contaminate nearby waterways.

Minerals are mined worldwide, and global trade makes it difficult to know the source of the raw materials in products you buy. Keep in mind that resource extraction in developing countries is often carried out under regulations that are less strict or less stringently enforced than those in the United States. As a result, the environmental impact of mining is even greater in these countries.

Nonrenewable mineral resources are used in electronic devices such as phones, computers, and televisions. Constantly trading up to the newest device may be good for the ego and for the economy, but it is bad for the environment. Reducing consumption by fixing existing products is a sustainable resource use, as is recycling. Reuse and recycling reduce the need for extracting nonrenewable resources, and they also help keep material out of landfills. In 2013, the United States recycled 87 million tons of material that would otherwise have ended up in landfills. Even so, there is plenty of room for improvement. That 87 million tons was only one-third of the municipal solid waste produced.

Minimizing energy use is also part of living sustainably. Fossil fuels such as petroleum, natural gas, and coal supply most of the energy used by developed countries. You already know that burning these nonrenewable fuels contributes to global warming and acid rain. Extracting and transporting fossil fuels add environmental costs. For example, oil that escapes from ocean-drilling operations harms aquatic species

Figure 44.13 Environmental costs of resource extraction.

A Bingham copper mine near Salt Lake City, Utah. This open pit mine is 4 kilometers (2.5 miles) wide and 1,200 meters (0.75 miles) deep, the largest man-made excavation on Earth.

B Pelican covered with oil that accidentally escaped from a deep-sea drilling platform in the Gulf of Mexico.

(**Figure 44.13B**). Natural gas that leaks from wells and pipes contributes to global warming. For example, between October 2015 and February 2016, a natural gas well in Southern California leaked more than 2 million tons of methane into the air. Methane is one of the greenhouse gases.

Nuclear energy produces no carbon dioxide, but occasional accidents allow dangerous radioactive material to escape into the environment.

Renewable energy sources have their own drawbacks. For example, dams in rivers can generate renewable hydroelectric power, but they can also alter the array of species in the river by slowing its flow. Wind turbines kill birds and bats. This occurs most often if turbines are installed without regard for migration routes and/or the structure supporting the turbine provides a place to perch. Industrial-scale solar farms can pose a threat to birds too. Birds sometimes mistake the array of solar panels or mirrors at such facilities for a lake. As a result, the birds collide with the panels or are burned by intense sunlight concentrated by the mirrors. Placing a solar farm in an undisturbed area also displaces native species that lived in or used the area where the facility is situated.

In short, all energy use has some negative environmental impacts, so the best way to minimize your impact is to use less energy.

If you want to make more of a difference, learn about the threats to ecosystems in your own area. Support efforts to document, preserve, and restore local biodiversity. Many ecological restoration projects are supervised by trained biologists but carried out primarily by volunteers (**Figure 44.14**).

Figure 44.14 Volunteers restoring the Little Salmon River in Idaho. Planting native vegetation along the river's banks will reduce the amount of sediment that enters the river and provide shade to cool the water. These alterations will improve the habitat for threatened salmon species that breed in this river.

Figure 44.15 Documenting biodiversity.

CITIZEN SCIENTISTS

Documenting Earth's existing biodiversity and monitoring the effects of climate change and other human-caused factors are enormous undertakings. Scientists need to know the locations and sizes of a species' populations in order to monitor changes in its range and numbers. Given the vast number of existing species, gathering such information is a task too big to be accomplished by scientists alone.

Fortunately, modern technology makes it easy to crowdsource data—to gather and organize information collected independently by a large number of individuals. Anyone with a camera can photograph organisms that they encounter (**Figure 44.15**), then upload information about their sighting to an online database. Researchers then access this information.

 Consider the iNaturalist project, an online social network for sharing information about biodiversity. The goal of iNaturalist is to foster a public appreciation of biodiversity, while also gathering scientifically valuable biodiversity data. The iNaturalist app (available for iOS and Android) allows users to upload photos of organisms they observe, along with the time and location of the observation. Users do not need to know exactly what they saw; in fact, they are encouraged to post unknown organisms. Members of the iNaturalist community, which includes individuals with specialized expertise, collaborate in identifying the species in the uploaded photos, and the resulting data is made available to anyone who is interested.

By the beginning of 2016, iNaturalist had amassed more than 1.8 million observations documenting sightings of over 80,000 different species from locations throughout the world. To learn more about iNaturalist, visit iNaturalist.org, check out the iNaturalist page on Facebook, or download the iNaturalist app.

CREDITS: (14) Mountain Visions/NOAA; (15) iStockphoto.com/lzf; (in text) iNaturalist.org.

CHAPTER 44 HUMAN EFFECTS ON THE BIOSPHERE   **797**

Section 44.1 Extinction is a natural process, but we are in the midst of a human-caused mass extinction with rising numbers of **threatened species** and **endangered species**. **Endemic species** are especially likely to be at risk. Overharvesting, species introductions, and habitat destruction, degradation, and fragmentation can push species toward extinction. We know about only a tiny fraction of the species that are currently under threat.

Section 44.2 Overplowing or overgrazing of grassland can cause **desertification**. Both desertification and deforestation affect soil properties and can alter rainfall patterns. Changes caused by deforestation are especially difficult to reverse.

Section 44.3 **Pollutants** are released by both point sources (such as landfills) and non-point sources (such as roadways). Chemicals that seep out of buried trash in landfills can pollute groundwater. Trash, especially plastic, that washes into or is dumped into oceans poses a threat to marine life.

Burning coal releases sulfur dioxide, and burning oil or gas releases nitrogen oxides. These pollutants react with gases and water vapor in air, then fall to earth as **acid rain**. Deposition of mercury and ammonia that entered the atmosphere also has disruptive effects on ecosystems.

Section 44.4 The **ozone layer** in the upper atmosphere reduces the amount of incoming UV radiation that reaches Earth's surface. CFCs were banned after they were found to cause thinning of the ozone layer. Nitrous oxide also destroys ozone. It enters the air as a result of fossil fuel combustion and use of nitrogen fertilizers. Near the ground, where the ozone concentration is naturally low, ozone produced as a result of fossil fuel use is a pollutant.

Section 44.5 Global climate change resulting from an increase in greenhouse gases is causing glaciers to melt, thus raising the sea level. It also is affecting the range of species, allowing some to move into higher elevations or latitudes. Other species such as corals are showing signs of temperature-related stress. In addition, global climate change is expected to alter rainfall patterns and the intensity of hurricanes.

Section 44.6 Genetic diversity, species diversity, and ecosystem diversity are components of **biodiversity**, which is declining in all regions. **Conservation biology** involves surveying biodiversity and developing strategies to protect it and allow its sustainable use. Because resources are limited, biologists often focus on **biodiversity hot spots**, where many

unique species are under threat. They also attempt to ensure that portions of all ecoregions are protected. When an ecosystem has been totally or partially degraded, **ecological restoration** can help restore biodiversity.

Section 44.7 Extraction of nonrenewable mineral resources and fuels have detrimental effects on an ecosystem. Individuals can help sustain biodiversity by limiting their energy use and by reducing resource consumption through reuse and recycling.

Application Documenting existing biodiversity and monitoring changes are enormous tasks. Technology for crowdsourcing information now makes it possible for the public to assist in this undertaking.

SELF-ASSESSMENT

1. A(n) _____ species has population levels so low it is at great risk of extinction in the near future.
   - a. endemic
   - c. threatened
   - b. endangered
   - d. indicator

2. Species are threatened by habitat _____ .
   - a. fragmentation
   - c. destruction
   - b. degradation
   - d. all of the above

3. Deforestation _____ .
   - a. increases mineral runoff from soil
   - b. decreases local temperature
   - c. increases local rainfall
   - d. all of the above

4. Sulfur dioxide released by coal-burning power plants contributes to _____ .
   - a. ozone destruction
   - c. acid rain
   - b. sea level rise
   - d. desertification

5. The "hole" in the ozone layer is most pronounced in _____ over _____ .
   - a. fall; the Arctic
   - c. spring; Antarctica
   - b. fall; the equator
   - d. spring; the equator

6. An increase in the size of the ozone hole would be expected to _____ .
   - a. increase skin cancers
   - c. melt more glaciers
   - b. acidify the sea
   - d. increase atmospheric CO_2

7. Ocean currents have caused a large amount of plastic to accumulate in the _____ .
   - a. North Pacific Ocean
   - c. Gulf of Mexico
   - b. South Atlantic Ocean
   - d. Indian Ocean

8. Global climate change is causing _____ .
   - a. a decrease in sea level
   - c. acid rain
   - b. glacial melting
   - d. all of the above

ARCTIC PCB POLLUTION Winds can carry chemical pollutants long distances. For example, pollutants produced and released at temperate latitudes are carried to the Arctic, where the chemicals enter food webs. As a result of biological magnification, top carnivores in arctic food webs, including people and polar bears, can have high concentrations of these chemicals in their body. Arctic people who consume a lot of local wildlife tend to have unusually high levels of industrial chemicals called polychlorinated biphenyls, or PCBs, in their body. The Arctic Monitoring and Assessment Programme studies the effects of these industrial chemicals on the health and reproduction of Arctic people. **Figure 44.16** shows how sex ratio at birth varies with average maternal PCB levels among people native to the Russian Arctic.

1. Which sex was more common in offspring of women with less than 1 microgram per milliliter of PCB in serum?

2. At what PCB concentrations were women more likely to have daughters?

3. In some villages in Greenland, nearly all recent newborns are female. Would you expect PCB levels in those villages to be above or under 4 micrograms per milliliter?

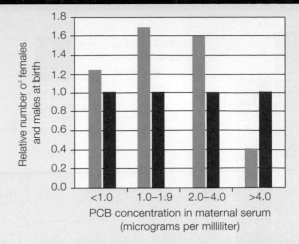

Figure 44.16 Effect of maternal PCB concentration on sex ratio of newborns in human populations native to the Russian Arctic. Blue bars indicate the relative number of males born per one female (pink bars).

9. The Montreal Protocol banned use of _____ , which contribute(s) to destruction of the ozone layer.
   a. DDT
   b. CFCs
   c. fossil fuels
   d. sulfur dioxides

10. A highly threatened region that is home to unique species is a(n) _____ .
   a. ecoregion   b. biome   c. hot spot   d. community

11. Biodiversity refers to all except _____ .
   a. genetic diversity
   b. species diversity
   c. ecosystem diversity
   d. geographic diversity

12. Restoring a marsh that has been damaged by human activities is an example of _____ .
   a. biological magnification   c. ecological restoration
   b. bioaccumulation   d. globalization

13. Individuals help sustain biodiversity by _____ .
   a. reducing resource consumption
   b. reusing materials
   c. recycling materials
   d. all of the above

14. Match the terms with the most suitable description.
   ___ hot spot
   ___ ozone
   ___ biodiversity
   ___ acid rain
   ___ endemic species
   ___ nonpoint source of pollution
   ___ global climate change
   ___ deforestation
   ___ desertification

   a. good up high; bad nearby
   b. tree loss alters rainfall pattern and is difficult to reverse
   c. can increase dust storms
   d. evolved in one region and remains there alone
   e. coal-burning is major cause
   f. release of pollutant in many areas
   g. has unique threatened species
   h. cause of rising sea level
   i. genetic, species, and ecosystem diversity

1. In one seaside community in New Jersey, the U.S. Fish and Wildlife Service suggested trapping and removing feral cats (domestic cats that live in the wild). The goal was to protect some endangered wild birds (plovers) that nested on the town's beaches. Many residents were angered by the proposal, arguing that the cats have as much right to be there as the birds. Do you agree? Why or why not?

2. Burning fossil fuel puts excess carbon dioxide into the atmosphere, but deforestation and desertification also affect the atmospheric carbon dioxide concentration. Explain why a global decrease in the amount of vegetation is contributing to the rise in the atmospheric content of carbon dioxide.

3. The magnitude of acid rain's effects can be influenced by the properties of the rock that the rain runs over. Acid rain is least likely to significantly acidify lakes in regions where there is a large amount of calcium carbonate–rich rock. How does the presence of these rocks mitigate the effects of acid rain?

4. In some areas, the barrier built to prevent uncontrolled movement of people across the U.S.–Mexico border consists of miles of 6-meter-tall steel beams placed four inches apart. Discuss the ecological effects of such a barrier. What types of organisms would be most adversely affected by this habitat fragmentation?

5. Restoring river ecosystems sometimes involves removing dams. Explain how changing the speed of water flow alters which species can live in a river.

for additional quizzes, flashcards, and other study materials
▶ ACCESS MINDTAP AT WWW.CENGAGEBRAIN.COM

Atomic number →	**11**
Symbol →	**Na**
Atomic weight →	22.99

(Atomic weight is a ratio: the mass of an atom relative to 1/12 the mass of a carbon 12 atom. Atomic weights given in parentheses are those of the most stable or best known isotopes of radioactive elements.)

1	2	3	4	5	6	7	8	9	10	11	12	13	14	15	16	17	18
1 H 1.008																	**2** He 4.0026
3 Li 6.94	**4** Be 9.0122											**5** B 10.81	**6** C 12.011	**7** N 14.007	**8** O 15.999	**9** F 18.998	**10** Ne 20.180
11 Na 22.990	**12** Mg 24.305											**13** Al 26.982	**14** Si 28.085	**15** P 30.974	**16** S 32.06	**17** Cl 35.45	**18** Ar 39.948
19 K 39.098	**20** Ca 40.078	**21** Sc 44.956	**22** Ti 47.867	**23** V 50.942	**24** Cr 51.996	**25** Mn 54.938	**26** Fe 55.845	**27** Co 58.933	**28** Ni 58.693	**29** Cu 63.546	**30** Zn 65.38	**31** Ga 69.723	**32** Ge 72.63	**33** As 74.922	**34** Se 78.96	**35** Br 79.904	**36** Kr 83.798
37 Rb 85.468	**38** Sr 87.62	**39** Y 88.906	**40** Zr 91.224	**41** Nb 92.906	**42** Mo 95.96	**43** Tc (97.91)	**44** Ru 101.07	**45** Rh 102.91	**46** Pd 106.42	**47** Ag 107.87	**48** Cd 112.41	**49** In 114.82	**50** Sn 118.71	**51** Sb 121.76	**52** Te 127.60	**53** I 126.90	**54** Xe 131.29
55 Cs 132.91	**56** Ba 137.33	*	**71** Lu 174.97	**72** Hf 178.49	**73** Ta 180.95	**74** W 183.84	**75** Re 186.21	**76** Os 190.23	**77** Ir 192.22	**78** Pt 195.08	**79** Au 196.97	**80** Hg 200.59	**81** Tl 204.38	**82** Pb 207.2	**83** B 208.98	**84** Po (208.98)	**85** At (209.99)	**86** Rn (222.02)
87 Fr (223.02)	**88** Ra (226.03)	**	**103** Lr (262.11)	**104** Rf (265.12)	**105** Db (268.13)	**106** Sg (271.13)	**107** Bh (270)	**108** Hs (277.15)	**109** Mt (276.15)	**110** Ds (281.16)	**111** Rg (280.16)	**112** Cn (285.17)	**113** Nh (284.18)	**114** Fl (289.19)	**115** Mc (288.19)	**116** Lv (293)	**117** Ts (294)	**118** Og (294)

*Lanthanoids

57 La 138.91	**58** Ce 140.12	**59** Pr 140.91	**60** Nd 144.24	**61** Pm (144.91)	**62** Sm 150.36	**63** Eu 151.96	**64** Gd 157.25	**65** Tb 158.93	**66** Dy 162.50	**67** Ho 164.93	**68** Er 167.26	**69** Tm 168.93	**70** Yb 173.05

**Actinoids

89 Ac (227.03)	**90** Th 232.04	**91** Pa 231.04	**92** U 238.03	**93** Np (237.05)	**94** Pu (244.06)	**95** Am (243.06)	**96** Cm (247.07)	**97** Bk (247.07)	**98** Cf (251.08)	**99** Es (252.08)	**100** Fm (257.10)	**101** Md (258.10)	**102** No (259.10)

1	Hydrogen	H	25	Manganese	Mn	49	Indium	In	73	Tantalum	Ta	97	Berkelium	Bk
2	Helium	He	26	Iron	Fe	50	Tin	Sn	74	Tungsten	W	98	Californium	Cf
3	Lithium	Li	27	Cobalt	Co	51	Antimony	Sb	75	Rhenium	Re	99	Einsteinium	Es
4	Beryllium	Be	28	Nickel	Ni	52	Tellurium	Te	76	Osmium	Os	100	Fermium	Fm
5	Boron	B	29	Copper	Cu	53	Iodine	I	77	Iridium	Ir	101	Mendelevium	Md
6	Carbon	C	30	Zinc	Zn	54	Xenon	Xe	78	Platinum	Pt	102	Nobelium	No
7	Nitrogen	N	31	Gallium	Ga	55	Cesium	Cs	79	Gold	Au	103	Lawrencium	Lr
8	Oxygen	O	32	Germanium	Ge	56	Barium	Ba	80	Mercury	Hg	104	Rutherfordium	Rf
9	Fluorine	F	33	Arsenic	As	57	Lanthanum	La	81	Thallium	Tl	105	Dubnium	Db
10	Neon	Ne	34	Selenium	Se	58	Cerium	Ce	82	Lead	Pb	106	Seaborgium	Sg
11	Sodium	Na	35	Bromine	Br	59	Praseodymium	Pr	83	Bismuth	Bi	107	Bohrium	Bh
12	Magnesium	Mg	36	Krypton	Kr	60	Neodymium	Nd	84	Polonium	Po	108	Hassium	Hs
13	Aluminum	Al	37	Rubidium	Rb	61	Promethium	Pm	85	Astatine	At	109	Meitnerium	Mt
14	Silicon	Si	38	Strontium	Sr	62	Samarium	Sm	86	Radon	Rn	110	Darmstadtium	Ds
15	Phosphorus	P	39	Yttrium	Y	63	Europium	Eu	87	Francium	Fr	111	Roentgenium	Rg
16	Sulfur	S	40	Zirconium	Zr	64	Gadolinium	Gd	88	Radium	Ra	112	Copernicium	Cn
17	Chlorine	Cl	41	Niobium	Nb	65	Terbium	Tb	89	Actinium	Ac	113	Nihonium	Nh
18	Argon	Ar	42	Molybdenum	Mo	66	Dysprosium	Dy	90	Thorium	Th	114	Flerovium	Fl
19	Potassium	K	43	Technetium	Tc	67	Holmium	Ho	91	Protactinium	Pa	115	Moscovium	Mc
20	Calcium	Ca	44	Ruthenium	Ru	68	Erbium	Er	92	Uranium	U	116	Livermorium	Lv
21	Scandium	Sc	45	Rhodium	Rh	69	Thulium	Tm	93	Neptunium	Np	117	Tennessine	Ts
22	Titanium	Ti	46	Palladium	Pd	70	Ytterbium	Yb	94	Plutonium	Pu	118	Oganesson	Og
23	Vanadium	V	47	Silver	Ag	71	Lutetium	Lu	95	Americium	Am			
24	Chromium	Cr	48	Cadmium	Cd	72	Hafnium	Hf	96	Curium	Cm			

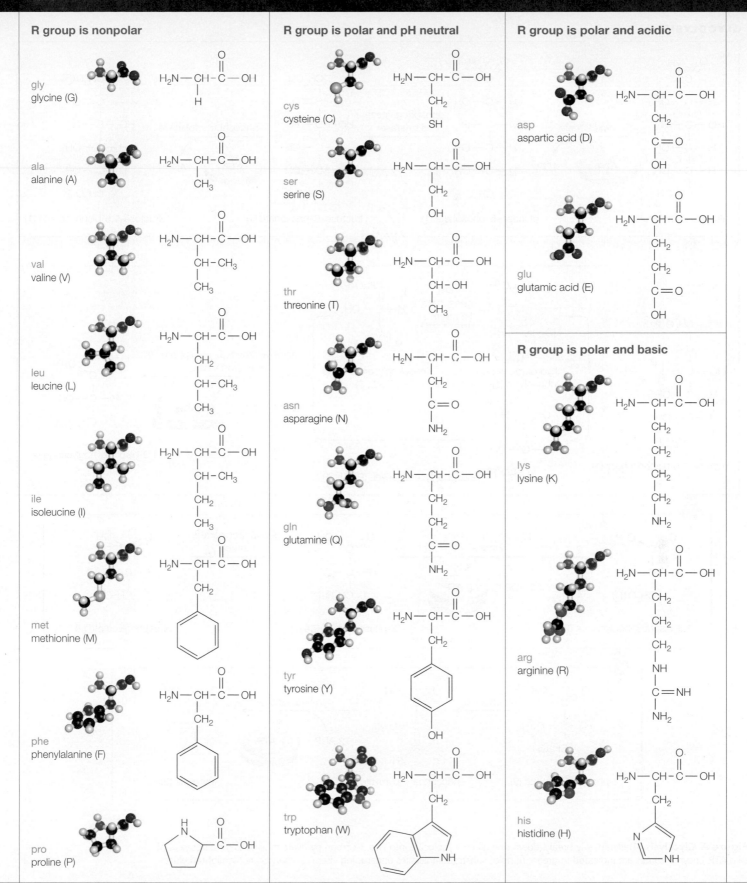

R group is nonpolar

gly
glycine (G)

ala
alanine (A)

val
valine (V)

leu
leucine (L)

ile
isoleucine (I)

met
methionine (M)

phe
phenylalanine (F)

pro
proline (P)

R group is polar and pH neutral

cys
cysteine (C)

ser
serine (S)

thr
threonine (T)

asn
asparagine (N)

gln
glutamine (Q)

tyr
tyrosine (Y)

trp
tryptophan (W)

R group is polar and acidic

asp
aspartic acid (D)

glu
glutamic acid (E)

R group is polar and basic

lys
lysine (K)

arg
arginine (R)

his
histidine (H)

GLYCOLYSIS

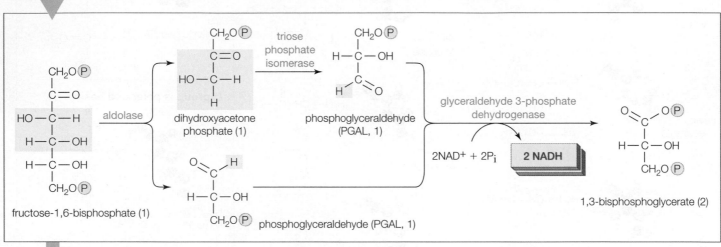

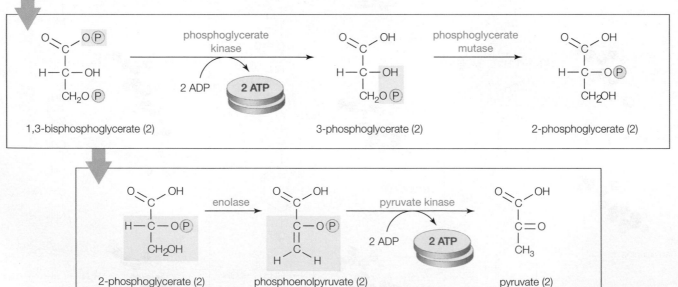

Figure A Glycolysis. This pathway breaks down one glucose molecule into two 3-carbon pyruvate molecules for a net yield of two ATP. Enzyme names are indicated in green; parts of substrate molecules undergoing chemical change are highlighted blue.

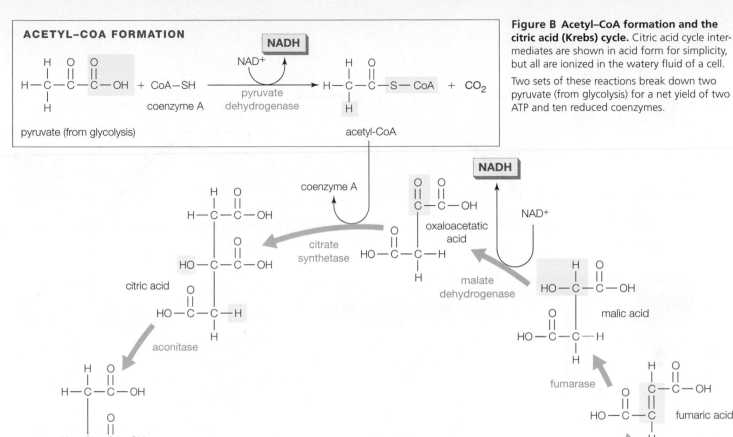

ACETYL–COA FORMATION

Figure B Acetyl–CoA formation and the citric acid (Krebs) cycle. Citric acid cycle intermediates are shown in acid form for simplicity, but all are ionized in the watery fluid of a cell.

Two sets of these reactions break down two pyruvate (from glycolysis) for a net yield of two ATP and ten reduced coenzymes.

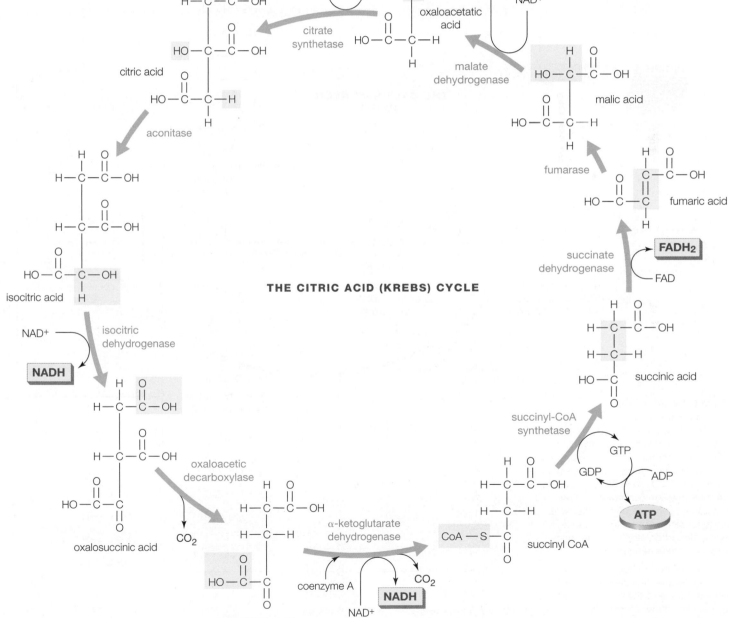

THE CITRIC ACID (KREBS) CYCLE

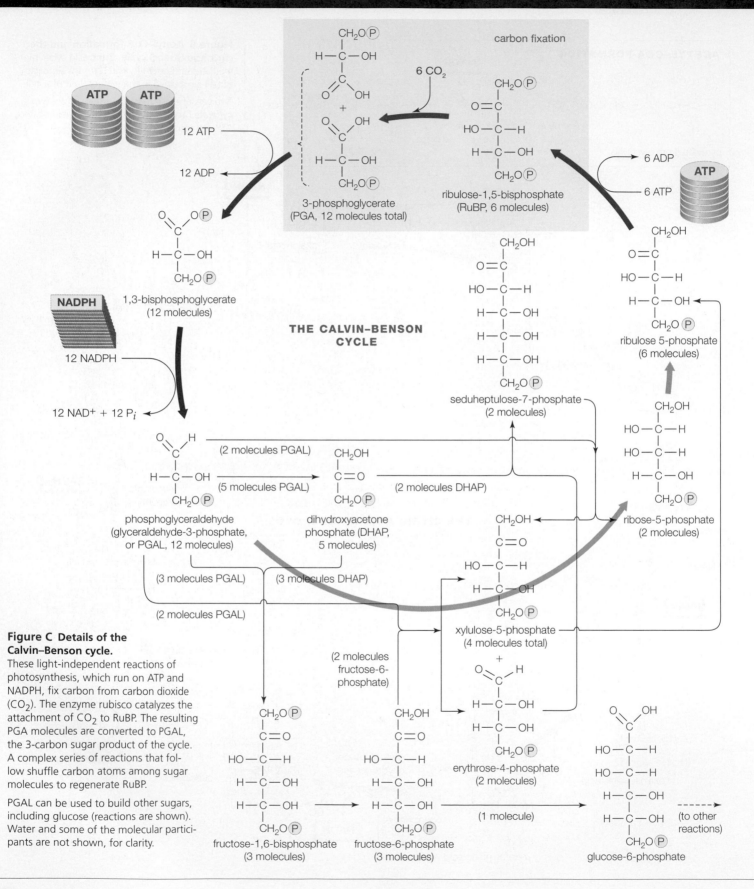

Figure C Details of the Calvin–Benson cycle.

These light-independent reactions of photosynthesis, which run on ATP and NADPH, fix carbon from carbon dioxide (CO_2). The enzyme rubisco catalyzes the attachment of CO_2 to RuBP. The resulting PGA molecules are converted to PGAL, the 3-carbon sugar product of the cycle. A complex series of reactions that follow shuffle carbon atoms among sugar molecules to regenerate RuBP.

PGAL can be used to build other sugars, including glucose (reactions are shown). Water and some of the molecular participants are not shown, for clarity.

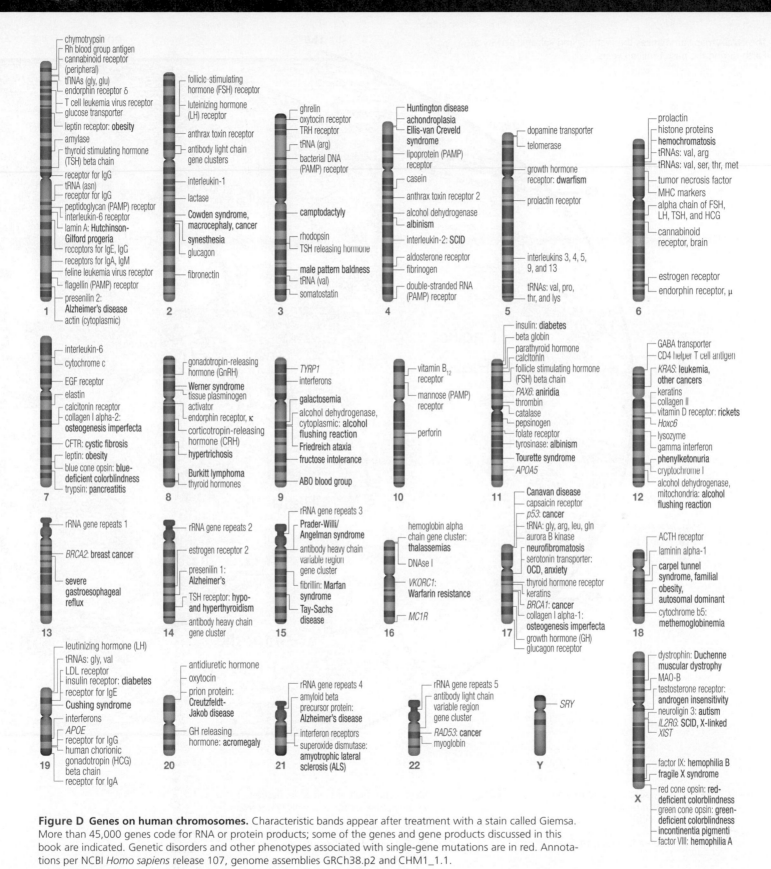

Figure D Genes on human chromosomes. Characteristic bands appear after treatment with a stain called Giemsa. More than 45,000 genes code for RNA or protein products; some of the genes and gene products discussed in this book are indicated. Genetic disorders and other phenotypes associated with single-gene mutations are in red. Annotations per NCBI *Homo sapiens* release 107, genome assemblies GRCh38.p2 and CHM1_1.1.

This NASA map summarizes the tectonic and volcanic activity of Earth during the past 1 million years.

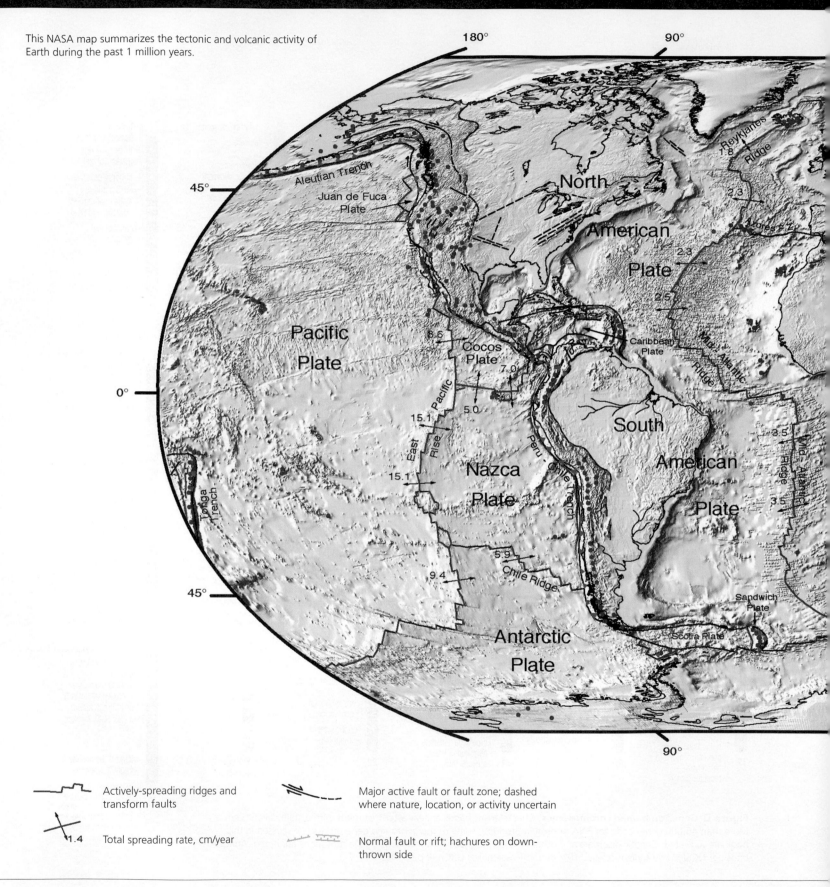

Actively-spreading ridges and transform faults

Total spreading rate, cm/year

Major active fault or fault zone; dashed where nature, location, or activity uncertain

Normal fault or rift; hachures on down-thrown side

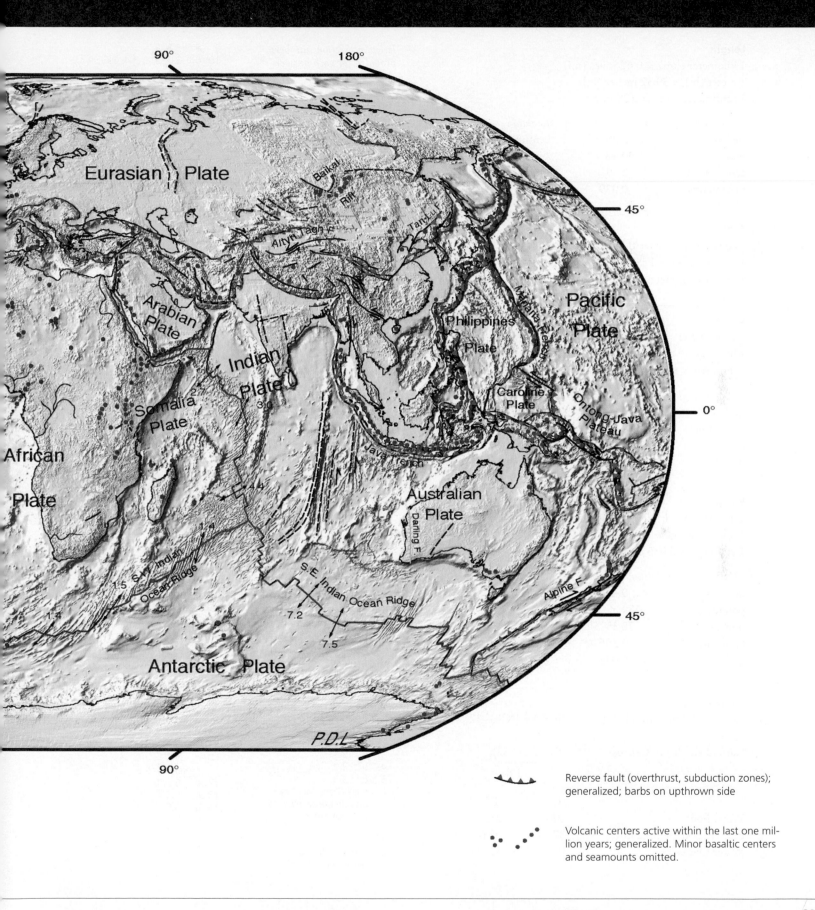

Reverse fault (overthrust, subduction zones); generalized; barbs on upthrown side

Volcanic centers active within the last one million years; generalized. Minor basaltic centers and seamounts omitted.

Appendix VI. Units of Measure

Length

1 kilometer (km) = 0.62 miles (mi)
1 meter (m) = 39.37 inches (in)
1 centimeter (cm) = 0.39 inches

To convert	multiply by	to obtain
inches	2.25	centimeters
feet	30.48	centimeters
centimeters	0.39	inches
millimeters	0.039	inches

Area

1 square kilometer = 0.386 square miles
1 square meter = 1.196 square yards
1 square centimeter = 0.155 square inches

Volume

1 cubic meter = 35.31 cubic feet
1 liter = 1.06 quarts
1 milliliter = 0.034 fluid ounces = 1/5 teaspoon

To convert	multiply by	to obtain
quarts	0.95	liters
fluid ounces	28.41	milliliters
liters	1.06	quarts
milliliters	0.03	fluid ounces

Weight

1 metric ton (mt) = 2,205 pounds (lb) = 1.1 tons (t)
1 kilogram (kg) = 2.205 pounds (lb)
1 gram (g) = 0.035 ounces (oz)

To convert	multiply by	to obtain
pounds	0.454	kilograms
pounds	454	grams
ounces	28.35	grams
kilograms	2.205	pounds
grams	0.035	ounces

Temperature

Celsius (°C) to Fahrenheit (°F):   °F = 1.8 (°C) + 32

Fahrenheit (°F) to Celsius: $°C = \dfrac{(°F - 32)}{1.8}$

	°C	°F
Water boils	100	212
Human body temperature	37	98.6
Water freezes	0	32

Appendix VII. Answers to Self-Assessments/Genetics Problems

Relevant Section numbers provided for each answer

Chapter 1

1. a 1.1
2. c 1.1
3. energy, nutrients 1.2
4. homeostasis 1.2
5. d 1.2
6. domains 1.4
7. reproduction 1.2
8. d 1.2
9. a 1.1
   d 1.3
   e 1.2
10. a 1.1
    b 1.3
11. b 1.5
12. b 1.7
13. b 1.7
14. a 1.8
15. h 1.5
    e 1.7
    b 1.1
    f 1.8
    a 1.5
    g 1.2
    d 1.5
    c 1.1

Chapter 2

1. a 2.2
2. b 2.1
3. d 2.2
4. d 2.1
5. a 2.2
6. a, c, b 2.3
7. a 2.3
8. c 2.3
9. c 2.4
10. b 2.4
11. d 2.5
12. a 2.5
13. c 2.5
14. b 2.4
15. c 2.4
    b 2.1
    d 2.4
    a 2.1, 2.2
    f 2.4
    e 2.1, 2.2

Chapter 3

1. c 3.1
2. four 3.1
3. b 3.1
4. a 3.2
5. c 3.3
6. false (trans fats have double bonds that do not kink the chain) 3.3, Application
7. b 3.3
8. starch, cellulose, glycogen 3.2

9. e 3.3
10. d 3.1, 3.2, 3.4, 3.6
11. d 3.5
12. d 3.4
13. a. polypeptide 3.4
    b. carbohydrate 3.2
    c. amino acid 3.4
    d. fatty acid 3.3
14. c 3.3
    a 3.2
    b 3.3
    e 3.2
    d 3.2
15. g 3.4
    a 3.3
    b 3.4
    c 3.3
    d 3.6
    f 3.4
    e 3.2

Chapter 4

1. c 4.1
2. c 4.1
3. b 4.1
4. b 4.3
5. c 4.3
6. b 4.5
7. a 4.3
8. a 4.6
9. c 4.5, 4.7
10. a 4.6
11. c, b, d, a 4.6
12. d 4.10
13. b 4.4, 4.5-4.8
14. a 4.10
15. c 4.7
    f 4.8
    e 4.4, 4.5
    d 4.5
    a 4.10
    g 4.9
    b 4.10

Chapter 5

1. b 5.1
2. c 5.1
3. b 5.1
4. c 5.2
5. c 5.3
6. Temperature, pH, salt concentration, pressure 5.3
7. d 5.4
8. b 5.6
9. a 5.5
10. c 5.4
11. more/less 5.6
12. c 5.6, 5.7
13. b 5.6
14. a 5.6
15. e 5.2
    d 5.8
    c 5.1
    b 5.2
    a 5.5

g 5.6
h 5.7
f 5.7
k 5.4
j 5.4
i 5.7

Chapter 6
1. autotroph: weed; heterotrophs: cat, bird, caterpillar 6.1
2. a 6.1
3. a 6.1, 6.3
4. b 6.1
5. c 6.2
6. c 6.3
7. b 6.3
8. b 6.3
9. c 6.3
10. c 6.1, 6.4
11. b 6.4
12. b 6.4
13. a 6.4
14. c 6.4
15. f 6.4
   h 6.4
   c 6.1
   d 6.3
   e 6.4
   b 6.1, 6.3
   a 6.2
   g 6.3

Chapter 7
1. False 7.1
2. d 7.2
3. a 7.1
4. c 7.2
5. b 7.3, 7.4
6. d 7.5
7. b 7.3
8. c 7.4
9. c 7.4
10. b 7.5
11. d, b, c, a 7.1, 7.4
12. d 7.6
13. f 7.5
14. c 7.4
15. b 7.3
   d 7.2
   a 7.2, 7.5
   c 7.3, 7.4
   e 7.2
   f 7.1

Chapter 8
1. c 8.2
2. c 8.2
3. b 8.2
4. b 8.2
5. a 8.3
6. b 8.3
7. b 8.3
8. d 8.4

9. a 8.4
10. a 8.4
11. b 8.4
12. d 8.3
13. d 8.5
14. a 8.6
15. a 8.2
   b 8.6
   c 8.3
   e 8.4
   f 8.5
   d 8.1
   g 8.4
   h 8.4

Chapter 9
1. c 9.1
2. b 9.2
3. a 9.1
4. c 9.1
5. a 9.1
6. b 9.1, 9.3
7. b 9.2
8. c 9.3
9. a 9.3
10. a 9.3
11. a 9.2
12. nucleus 9.2, 9.4
13. cytoplasm 9.4
14. c 9.4
15. c, a, d, b 9.2, 9.4
16. c 9.3
   b 9.2
   e 9.4
   a 9.2
   f 9.3
   d 9.2
   j 9.3
   g Application
   h 9.5
   i 9.3

Chapter 10
1. d 10.1
2. b 10.1
3. b 10.1
4. d 10.1
5. b 10.1
6. c 10.2
7. c 10.2
8. b 10.2
9. b 10.2
10. b 10.3
11. b 10.3
12. c 10.3
13. b 10.4
14. True 10.5
15. d 10.5
   a 10.3
   b 10.4
   e 10.3
   c 10.2

Chapter 11
1. b 11.1
2. b 11.1
3. a 11.1
4. e 11.1
5. c 11.2
6. c 11.1
7. a 11.1
8. c 11.1
9. a 11.1
10. interphase, prophase, metaphase, anaphase, telophase 11.2
11. d 11.3
12. d 11.5
13. a 11.5
14. c 11.3
   f 11.2
   a 11.5
   g 11.3
   b 11.5
   e 11.5
   d 11.2
   h 11.4
15. d 11.2
   b 11.2
   c 11.2
   e 11.1
   a 11.2
   f 11.3

Chapter 12
1. b 12.1
2. c 12.1
3. b 12.2
4. c 12.2
5. d 12.2
6. a 12.2
7. Sister chromatids are attached 12.3
8. c 12.3
9. b 12.4
10. a 12.4
11. d 12.4
12. b 12.5
13. c 12.3
   d 12.3
   a 12.1
   f 12.2
   e 12.2
   b 12.1
   g 12.4

Chapter 13
1. b 13.1
2. a 13.1
3. b 13.2
4. c 13.2
5. b 13.2
6. a 13.2
7. b 13.3
8. b 13.2
9. d 13.3
10. c 13.3

11. false 13.4–13.6
12. c 13.4
13. Continuous variation 13.6
14. b 13.4
15. b 13.3
   d 13.2
   a 13.1
   c 13.1

Chapter 14
1. b 14.1
2. b 14.1
3. a 14.1
4. False 14.1
5. b 14.2
6. d 14.1, 14.2, 14.3 (Outcome could be an effect of both parents carrying an autosomal recessive allele, or the mother carrying an X-linked recessive allele)
7. d 14.3
8. X from mom, Y from dad 14.3
9. d 14.1, 14.2
10. Y-linked 14.3
11. True 14.4
12. c 14.5
13. b 14.5
14. c 14.5
15. c 14.5
   e 14.4
   f 14.5
   b 14.5
   a 14.1
   d 14.4
   g 14.3

Chapter 15
1. c 15.1
2. a 15.1
3. b 15.1
4. b 15.2
5. c 15.2
6. b 15.3
7. b 15.3
8. b 15.4
9. d 15.4
10. a 15.2
11. e, a, d, b, c 15.1–15.3
12. True 15.5, 15.6
13. b 15.6
14. True 15.6
15. c 15.4
   f 15.5
   d 15.1
   b 15.4
   a 15.5
   e 15.1

Chapter 16
1. b 16.1
2. c 16.1, 16.7
3. b 16.2
4. b 16.2
5. b 16.3
6. True 16.3
7. d 16.4
8. Gondwana 16.5
9. b 16.2, 16.3, 16.5, 16.6
10. d 16.7
11. b 16.7
12. b 16.8
13. 66 Application
14. e 16.7, 16.8
15. i 16.2
   f 16.1, 16.3
   c 16.2
   e 16.8
   d 16.4
   g 16.7
   h 16.7
   b 16.3
   a 16.8

Chapter 17
1. a 17.1
2. d 17.2
3. a, c, b 17.3–17.4
4. c 17.5
5. d 17.5
6. b 17.6
7. d 17.5, 17.7
8. a 17.7
9. f 17.3–17.6
10. allopatric speciation 17.8
11. d 17.11
12. species 17.11
13. c 17.11
14. c 17.6
   f 17.5
   g 17.10
   b 17.6
   d 17.11
   i 17.10
   j 17.11
   a 17.10
   e 17.1
   h 17.11
15. d 17.10, 17.11

Chapter 18
1. c 18.1
2. b 18.1
3. c 18.2
4. a 18.1
5. c 18.2
6. a 18.2
7. b 18.3
8. c 18.4
9. b 18.3
10. d 18.2

11. a 18.3
12. d 18.5
13. c 18.4
14. h 18.5
15. a 18.1
   c 18.2
   b 18.2
   e 18.4
   f 18.5
   d 18.5

Chapter 19
1. c 19.1
2. d 19.3
3. c 19.1
4. b 19.1
5. d 19.2
6. c 19.3
7. one 19.6
8. c 19.4
9. d 19.4
10. d 19.3–19.5
11. c 19.5
12. a 19.5
13. a 19.7
14. b 19.5
   a 19.2
   a 19.2
   b 19.5
   b 19.5
15. d 19.5
   a 19.1
   e 19.6
   f 19.1
   b 19.6
   c 19.6

Chapter 20
1. d 20.1
2. d 20.4
3. b 20.2
4. c 20.3
5. b 20.5
6. a 20.6
7. d 20.6
8. c 20.4
9. c 20.4
10. d 20.4
11. c 20.6
12. a 20.7
13. a 20.2
14. a 20.7
15. d 20.8
   g 20.4
   a 20.4
   b 20.4
   f 20.4
   c 20.6
   e 20.4

Chapter 21

1. c 21.2
2. a 21.3
3. b 21.2
4. c 21.3
5. d 21.4
6. a 21.2
7. b 21.4–21.6
8. a 21.3
9. c 21.6
10. a 21.1
11. a 21.1
12. c 21.7
13. b 21.5
14. c 21.2
    d 21.3
    a 21.5
    b 21.6
15. c 21.4
    a 21.6
    d 21.2
    g 21.3
    e 21.1, 21.6
    f 21.6
    b 21.3

Chapter 22

1. c 22.1
2. a 22.1
3. b 22.2
4. a 22.3
5. d 22.2
6. c 22.2
7. c 22.2
8. d 22.2
9. b 22.4
10. d 22.3
11. a 22.1, 22.3
12. c 22.4
13. b
Application
14. b 22.3
15. c 22.1
    b 22.1
    a 22.1
    e 22.2
    f 22.1
    h 22.3
    d 22.3
    g 22.1

Chapter 23

1. b 23.1, 23.4
2. c 23.2
3. d 23.1
4. a 23.4
5. a 23.5
6. b 23.11
7. a 23.7
8. a 23.1, 23.9
9. c 23.9,
23.11
10. b 23.12
11. b 23.6, 23.7
12. d 23.11
13. b 23.12
    f 23.7
    d 23.3

h 23.4
c 23.5
a 23.8
g 23.6
e 23.9

Chapter 24

1. c 24.1
2. c 24.1
3. a 24.2
4. b 24.2
5. d 24.3
6. c 24.4
7. f 24.4
8. a 24.5
9. c 24.6
10. b 24.9
11. Africa 24.12
12. b 24.1
    e 24.2
    g 24.3
    f 24.8
    c 24.6
    d 24.7
    a 24.7
    h 24.7
13. b 24.4
    e 24.2
    d 24.4
    c 24.4
    f 24.7
    a 24.6

Chapter 25

1. b 25.2
2. a 25.2
3. b 25.2
4. b 25.3
5. false (roots
   do not have
   nodes) 25.5
6. c, d 25.2
7. c 25.4
8. a 25.3, 25.5
9. It is a (modi-
   fied) stem.
   Roots grow
   from a basal
   plate. 25.3
10. d 25.5
11. b 25.5
12. a 25.6
13. b 25.6
14. b 25.7
15. c 25.6
    d 25.2, 25.4
    e 25.6
    a 25.3, 25.5
    b 25.3
    g 25.1
    f 25.4

Chapter 26

1. d 26.1
2. b 26.1
3. c 26.2
4. b 26.2

5. b 26.2
6. c 26.3
7. d 26.3
8. d 26.3
9. c 26.5
10. a 26.4
11. c 26.4
12. b 26.5
13. a 26.5
14. i 26.4
    c 26.1
    e 26.5
    b 26.2
    g 26.5
    h 26.3
    f 26.5
    a 26.3
    d 26.3

Chapter 27

1. a, b 27.1
2. c 27.1
3. b 27.1
4. b 27.1
5. c 27.3
6. c 27.3
7. d 27.3
8. c 27.4
9. a 27.4
10. d 27.5
11. d 27.9
12. c 27.11
13. b 27.10
    d 27.10
    a 27.10
    f 27.11
    c 27.11
    e 27.10
14. c 27.8
15. e 27.8
    c 27.6
    b 27.10
    a 27.7
    d 27.8
    f 27.12

Chapter 28

1. a 28.2
2. c 28.4
3. a 28.2
4. b 28.3
5. b 28.3
6. c 28.3
7. c 28.4
8. a 28.4
9. d 28.5
10. a 28.7
11. a 28.6
12. b 28.8
13. d 28.3
14. c 28.7
15. b 28.2
    j 28.2
    a 28.3
    c 28.7
    d 28.4
    h 28.3
    f 28.7

i 28.3
e 28.5
k 28.3
g 28.2

Chapter 29

1. a 29.2
2. c 29.3
3. b 29.4
4. a 29.5
5. a 29.4
6. c 29.6
7. a 29.6
8. b 29.11
9. c 29.12
10. c 29.1
11. b 29.9
12. a 29.7
13. j 29.9
    d 29.5
    g 29.11
    b 29.9
    h 29.10
    a 29.9
    e 29.2
    i 29.1
    c 29.7
    f 29.7

Chapter 30

1. b 30.2
2. c 30.1
3. a 30.2
4. c 30.3
5. d 30.2
6. b 30.8
7. b 30.7
8. a 30.2
9. b 30.7
10. c 30.5
11. d 30.4
12. d 30.5
13. a 30.8
14. c 30.3
15. d 30.5
    g 30.7
    f 30.4
    a 304.
    h 30.5
    e 30.3
    b 30.8
    i 30.7
    c 30.3

Chapter 31

1. a 31.1
2. b 31.3
3. a 31.3
4. d 31.1
5. c 31.8
6. a 31.4
7. b 31.8
8. d 31.4
9. c 31.4
10. c 31.4
11. c 31.7
12. d 31.6

g 31.4
c 31.4
e 31.8
a 31.5
b 31.3
f 31.7
13. b 31.9
14. d 31.5
    c 31.4
    b 31.6
    g 31.7
    f 31.7
    e 31.8
    a 31.8

Chapter 32

1. a 32.2
2. c 32.7
3. b 32.4
4. a 32.3
5. a 32.4
6. c 32.4
7. a 32.3
8. b 32.5
9. d 32.6
10. d 32.8
11. c 32.8
12. d 32.7
13. a 32.8
14. c 32.1
15. h 32.5
    f 32.5
    g 32.8
    e 32.3
    i 32.5
    c 32.3
    b 32.2
    d 32.5
    a 32.7

Chapter 33

1. a 33.1
2. c 33.1
3. pulmonary
33.1
4. c 33.4
5. b 33.4
6. a 33.4
7. b 33.3
8. d 33.3
9. a 33.6
10. b 33.8
11. d 33.3
12. a 33.2
13. b 33.10
14. b 33.9
15. f 33.1
    a 33.10
    e 33.2
    g 33.3
    b 33.3
    c 33.8
    d 33.2

Chapter 34

1. d 34.1
2. e 34.2

3. g 34.3, 34.4
4. h 34.3–34.6
5. d 34.3
6. b 34.1, 34.3
    c 34.1, 34.3
    f 34.1
7. a, d, e, g
34.5
8. d 34.4
9. a 34.5
10. b 34.6, 34.7
11. d 34.7
12. b 34.7
13. a 34.8
14. b 34.5
    d 34.4
    c 34.6
    e 34.7
    a 34.1
15. c 34.8
16. f 34.8
    e 34.5
    d 34.8
    c 34.3
    b 34.9
    a 34.5
    g 34.5

Chapter 35

1. a 35.1
2. b 35.2
3. a 35.3
4. c 35.4
5. d 35.5
6. a 35.5
7. iron 35.6
8. b Application
9. d 35.6
10. c 35.5
11. d 35.5
12. True 35.5
13. a 35.5
14. a 35.7
15. d 35.4
    a 35.4
    e 35.3
    f 35.4
    c 35.4
    b 35.4

Chapter 36

1. c 36.1
2. b 36.4
3. c 36.6
4. b 36.6
5. a 36.6
6. c 36.7, 36.9
7. a 36.4,
36.6, 36.7
8. b 36.6
9. d 36.9
10. d 36.8
11. c 36.5
12. c 36.10
13. b 36.11
14. c 36.9
15. g 36.6
    b 36.7

a	36.6	7. a	39.4	6. c	42.4		
d	36.6	8. b	39.6	7. b	42.5		
h	36.4	9. d	39.6	8. d	42.5		
c	36.6	10. c	39.5	9. a	42.5		
f	36.2	11. c	39.5	10. a	42.7		
e	36.7	12. b	39.5	11. c	42.7		
		13. d	39.5	12. b	42.5		
		14. e	39.6	13. a	42.6		

Chapter 37

1. c 37.1
2. d 37.1
3. b 37.1
4. a 37.2
5. b 37.3
6. a 37.3
7. b 37.3
8. a 37.3
9. water 37.3
10. dialysis 37.4
11. c 37.2
   a 37.2
   b 37.2
   e 37.3
   d 37.3
12. c 37.6
13. b 37.6
   a 37.6
   d 37.6
   c 37.6
   e 37.6
14. c 37.7

Chapter 38

1. a 38.1
2. b 38.1
3. a 38.8
4. c 38.7
5. b 38.6
6. c 38.11
7. c 38.6
   d 38.10
   g 38.7
   e 38.6
   b 38.6
   a 38.7
   f 38.8
8. a 38.3
9. c 38.12
10. a, d, c, f 38.12
    b, e 38.13
11. a 38.14
12. c 38.3
13. b 38.15
    a 38.6
    d 38.8
    c 38.8
    e 38.8
    g 38.12
    f 38.15

Chapter 39

1. b 39.1
2. a 39.1
3. c 39.1
4. c 39.3
5. a 39.3
6. c 39.4

(continued)
   g 39.5
   c 39.2
   b 39.4
   d 39.4
   a 39.2
   h 39.5
   f 39.2

Chapter 40

1. b 40.1
2. f 40.2
3. 400 40.1
4. 1,600 40.2
5. a 40.2
6. d 40.3
7. a 40.5
8. b 40.6
9. d 40.7
10. a 40.6
11. a 40.4
12. c 40.4
13. c 40.3
    d 40.2
    a 40.2
    e 40.3
    b 40.3
14. d 40.7

Chapter 41

1. b 41.1
2. d 41.1, 41.3
3. a 41.3
4. d 41.3
5. b 41.2
   d 41.5
   c 41.1
   a 41.4
   e 41.3
6. b 41.5
7. a 41.6
8. b 41.6
9. c 41.6
10. a 41.1
11. c 41.5
12. a 41.4
13. c 41.8
    b 41.7
    d 41.6
    a 41.7
    f 41.7
    e 41.3

Chapter 42

1. b 42.1
2. a 42.1
3. a 42.1
4. d 42.1
5. a 42.1

14. a 42.5
15. d 42.5
    c 42.5
    b 42.6
    a 42.6

Chapter 43

1. b 43.1
2. d 43.1
3. d 43.1
4. a 43.2
5. a 43.2
6. b 43.6
7. d 43.3
8. c 43.6
9. a 43.8
10. b 43.9
11. d 43.11
12. c 43.10
13. c 43.3
14. b 43.3
15. d 43.8
    e 43.6
    f 43.7
    c 43.6
    b 43.10
    h 43.5
    i 43.11
    a 43.4
    g 43.1

Chapter 44

1. b 44.1
2. d 44.1
3. a 44.2
4. c 44.3
5. c 44.4
6. a 44.4
7. a 44.3
8. b 44.5
9. b 44.4
10. c 44.6
11. d 44.6
12. c 44.6
13. d 44.7
14. g 44.6
    a 44.4
    i 44.6
    e 44.3
    d 44.1
    f 44.3
    h 44.5
    b 44.2
    c 44.2

Chapter 13: Genetics

1. Yellow is recessive. Because F_1 plants have a green phenotype and must be heterozygous, green must be dominant over the recessive yellow.
2. a. *AB*
   b. *AB*, *aB*
   c. *Ab*, *ab*
   d. *AB*, *Ab*, *aB*, *ab*
3. a. All offspring will be *AaBB*.
4. A mating of two M^L cats yields 1/4 *MM*, 1/2 M^LM, and 1/4 M^LM^L. Because M^LM^L is lethal, the probability that any one kitten among the survivors will be heterozygous is 2/3.
5. Because both parents are heterozygous (*HbA/HbS*), each child has a 1 in 4 possibility of inheriting two *HbS* alleles and having sickle cell anemia.
6. The data reveal that these genes do not assort independently because the observed ratio is very far from the 9:3:3:1 ratio expected with independent assortment. Instead, the results can be explained if the genes are located close to each other on the same chromosome.

Chapter 14: Genetics

1. Autosomal recessive. If the allele was inherited in a dominant pattern, individuals in the last generation would all have the phenotype. If it was X-linked, offspring of the first generation would all have the phenotype.
2. a. A male with an X-linked allele produces two kinds of gametes: one with an X chromosome (and the X-linked allele), and the other with a Y chromosome.
   b. A female homozygous for an X-linked allele produces one type of gamete, which carries the X-linked allele.
   c. A female heterozygous for an X-linked allele produces two types of gametes: one that carries the X-linked allele, and another that carries the partnered allele on the homologous chromosome.
3. As a result of translocation, chromosome 21 may get attached to the end of chromosome 14. The new individual's chromosome number would still be 46, but its somatic cells would have the translocated chromosome 21 in addition to two normal chromosomes 21.
4. 50 percent.
5. A daughter could develop this muscular dystrophy only if she inherited two X-linked recessive alleles—one from each parent. Males who carry the allele are unlikely to father children because they develop the disorder and die early in life.

Glossary

ABA *See* abscisic acid.

abscisic acid (ABA) (ab-SIH-sick) Plant hormone with a major role in stress responses such as stomata closure; also inhibits seed germination. **466**

abscission (ab-SIH-zhun) Process by which plant parts are shed. **466**

acid Substance that releases hydrogen ions in water. **32**

acid rain Low-pH rain that forms when sulfur dioxide and nitrogen oxides mix with water vapor in the atmosphere. **791**

ACTH *See* adrenocorticotropic hormone.

actin Spherical protein that plays a role in cell movements; the main component of thin filaments in a sarcomere. **563**

action potential Brief reversal of the charge difference across a neuron membrane. **503**

activation energy Minimum amount of energy required to start a reaction. **81**

activator Transcription factor that increases the rate of transcription when it binds to a promoter or enhancer. **165**

active site Pocket in an enzyme where substrates react and are converted to products. **82**

active transport Energy-requiring mechanism in which a transport protein pumps a solute across a cell membrane against the solute's concentration gradient. **91**

adaptation *See* adaptive trait.

adaptive immunity In vertebrates, a set of immune defenses that can be tailored to specific pathogens as an organism encounters them during its lifetime. Characterized by self/nonself recognition, specificity, diversity, and memory. **592**

adaptive radiation Macroevolutionary pattern in which a lineage undergoes a burst of genetic divergences that gives rise to many species. **292**

adaptive trait (adaptation) A form of a heritable trait that enhances an individual's fitness. **257**

ADH *See* antidiuretic hormone.

adhering junction Cell junction that fastens an animal cell to another cell, or to basement membrane. Connects to cytoskeletal elements inside the cell; composed of adhesion proteins. **69, 484**

adhesion protein Plasma membrane protein that helps cells stick together in animal tissues. **57**

adipose tissue Connective tissue specializing in fat storage. **486**

adrenal cortex (uh-DREE-null) Outer portion of adrenal gland; secretes the hormones aldosterone and cortisol. **546**

adrenal gland Endocrine gland that is located atop the kidney. **546**

adrenal medulla Inner portion of adrenal gland; secretes epinephrine and norepinephrine. **546**

adrenocorticotropic hormone (ACTH) Anterior pituitary hormone that stimulates cortisol secretion by the adrenal glands. **543**

aerobic (air-OH-bick) Involving or occurring in the presence of oxygen. **114**

aerobic respiration Oxygen-requiring cellular respiration; breaks down organic molecules (particularly glucose) and produces ATP, carbon dioxide, and water. Includes glycolysis, acetyl–CoA formation, the Krebs cycle, and electron transfer phosphorylation. **115**

age structure Distribution of population members among various age categories. **725**

agglutination (uh-glue-tin-A-shun) The clumping together of antigen-bearing cells or particles by antibodies. Basis of blood typing tests. **603**

AIDS Acquired immune deficiency syndrome. A secondary immune deficiency that develops as the result of infection by the HIV virus. **608**

alcoholic fermentation Fermentation pathway that produces ATP, carbon dioxide, and ethanol. **122**

aldosterone (al-DOSS-ter-ohn) Adrenal hormone that encourages sodium reabsorption by the kidney; concentrates the urine. **657**

allantois (al-LAN-toh-is) Extraembryonic membrane that, in mammals, becomes part of the umbilical cord. **687**

allele frequency Proportion of one allele relative to all alleles at a particular locus in a population's chromosomes. **275**

alleles (uh-LEELZ) Forms of a gene with slightly different DNA sequences; may encode different versions of the gene's product. **192**

allergen (AL-er-jen) A normally harmless substance that provokes an immune response in some people. **606**

allergy Sensitivity to an allergen. **606**

allopatric speciation Speciation pattern in which a physical barrier arises and separates populations, ending gene flow between them. **288**

allosteric regulation (al-oh-STARE-ick) Control of an enzyme's activity by a regulatory molecule that binds outside the active site. **84**

alternation of generations Life cycle that alternates between haploid and diploid multicelled bodies. **340**

alternative splicing Post-translational RNA modification process in which exons are joined in different combinations. **151**

altruism (AL-true-izm) Behavior that benefits others at the expense of the individual performing it. **707**

alveolus (al-VEE-oh-luss) In a mammalian lung, one of many tiny air sacs where gas exchange with the blood occurs. **619**

amino acid (uh-ME-no) Small organic compound that consists of a carboxyl group, an amine group, and a characteristic side group (R), all typically bonded to the same carbon atom; monomer of proteins. **44**

amino acid–derived hormone An amine (modified amino acid), peptide, or protein that functions as a hormone. **538**

ammonia Nitrogen-containing compound (NH_3) that is a toxic product of amino acid breakdown; a large amount of water is required to excrete it. **652**

ammonification (am-on-if-ih-CAY-shun) Ammonium-producing process of decomposition. **758**

amnion (AM-nee-on) Extraembryonic membrane that encloses an amniote embryo and the amniotic fluid. **686**

amniote (AM-nee-oat) Vertebrate whose egg has waterproof membranes that allow the embryo to develop away from water; a reptile, bird, or mammal. **403**

amoeba (uh-ME-buh) Single-celled protist that extends pseudopods to move and to capture prey. **342**

amphibian (am-FIB-ee-un) Tetrapod with scaleless skin; it typically develops in water, then lives on land as a carnivore with lungs. For example, a frog or salamander. **406**

anaerobic (an-air-OH-bick) Occurring in the absence of oxygen. **114**

analogous structures Similar body structures that evolved separately in different lineages (by morphological convergence). **267**

anaphase (ANN-uh-faze) Stage of mitosis during which sister chromatids separate and move toward opposite spindle poles. **181**

aneuploidy (AN-you-ploy-dee) Condition of having too many or too few copies of a particular chromosome. **230**

angiosperms (AN-jee-oh-sperms) Highly diverse seed plant lineage; only plants that make flowers and fruits. **358**

animal Multicelled consumer that breaks down food inside its body, develops through a series of stages, and moves about during part or all of its life. **8, 378**

animal hormone Intercellular signaling molecule that is secreted by an endocrine gland or cell and travels in the blood. **538**

annelid (ANN-uh-lid) Segmented worm with a coelom, tubular digestive tract, and closed circulatory system; a polychaete, oligochaete, or leech. **386**

antenna Of some arthropods, sensory structure on the head that detects touch and odors. **391**

anther (AN-thur) Of a flower, part of the stamen that produces pollen grains. **456**

anthropoid primate Humanlike primate; monkey, ape, or human. **413**

antibody (AN-tee-baa-dee) Y-shaped antigen receptor protein made by B cells; immunoglobulin. E.g., IgA, IgG, IgE, IgM, IgD. **598**

antibody-mediated immune response Immune response in which antibodies are produced that target the pathogen or toxin that triggered the response. **600**

anticodon In a tRNA, set of three nucleotides that base-pairs with an mRNA codon. **153**

antidiuretic hormone (an-tee-die-er-ET-ick) (**ADH**). Posterior pituitary hormone that affects urine formation and blood pressure; encourages water reabsorption in the kidney. **543, 657**

antigen (AN-tih-jenn) A molecule or particle that the immune system recognizes as nonself. Its presence in the body triggers an immune response. **592**

antioxidant (ann-tee-OX-ih-dunt) Substance that interferes with the oxidation of other molecules. **87**

anus Body opening that serves solely as the exit for wastes from a complete digestive tract. **633**

aorta (ay-OR-tuh) Large artery that receives oxygenated blood pumped out of the heart's left ventricle. **575**

ape Common name for a tailless nonhuman primate; a gibbon, orangutan, gorilla, chimpanzee, or bonobo. **413**

apical dominance Effect in which a lengthening shoot tip inhibits the growth of lateral buds. **464**

apical meristem (A-pih-cull) Meristem in the tip of a shoot or root; gives rise to primary growth (lengthening) in plants. **434**

apicomplexans Parasitic protists that reproduce in animal cells. **337**

apoptosis (ap-op-TOE-sis or ay-poe-TOE-sis) Programmed cell death; a predictable, controlled process of cell self-destruction. **673**

appendicular skeleton Bones of the limbs or fins and the bones that connect them to the axial skeleton. **559**

appendix Wormlike projection from the first part of the large intestine. **640**

aquifer (AH-qwih-fur) Porous rock layer that holds some groundwater. **754**

arachnids (uh-RAK-nidz) Land-dwelling arthropods with no antennae and four pairs of walking legs; spiders, scorpions, mites, and ticks. **392**

archaea (are-KEY-uh) Singular **archaean**. Group of single-celled organisms that lack a nucleus but are more closely related to eukaryotes than to bacteria. **8, 325**

arctic tundra High-latitude Northern Hemisphere biome, where low, cold-tolerant plants survive with only a brief growing season. **775**

area effect Larger islands receive more immigrants and have a lower extinction rate than small ones. **744**

arteriole Blood vessel that conveys blood from an artery to capillaries. **580**

artery Large-diameter blood vessel that carries blood away from the heart. **574**

arthropod Invertebrate with jointed legs and a hard exoskeleton that is periodically molted; for example, an insect or crustacean. **391**

asexual reproduction Reproductive mode of eukaryotes by which offspring arise from a single parent only. **179, 668**

atherosclerosis (ath-err-oh-scler-OH-sis) Cardiovascular disorder in which lipid-rich plaques form on the interior walls of blood vessels and narrow the vessels. **584**

atmospheric cycle Biogeochemical cycle in which a gaseous form of an element plays a significant role. **756**

atom Smallest unit of a substance; consists of varying numbers of protons, neutrons, and electrons. **4**

atomic number Number of protons in the atomic nucleus; determines the element. **24**

ATP Adenosine triphosphate. (uh-DEN-uh-seen) Nucleotide that consists of an adenine base, a ribose sugar, and three phosphate groups. Functions as a subunit of RNA and as a coenzyme in many reactions. Important energy carrier. **47**

ATP/ADP cycle Process by which cells regenerate ATP. ADP forms when a phosphate group is removed from ATP, then ATP forms again as ADP gains a phosphate group. **87**

atrioventricular (**AV**) **node** (AY-tree-oh-ven-TRICK-you-lur) Clump of cells that conveys excitatory signals between the atria and ventricles. **577**

atrium (AY-tree-um) Heart chamber that receives blood and pumps it into a ventricle. **572**

autoimmune response Immune response that targets one's own body tissues. **607**

autonomic nervous system Division of the peripheral nervous system that relays signals to and from internal organs and glands. **509**

autosome Chromosome of a pair that is the same in males and females; a chromosome that is not a sex chromosome. **137**

autotroph (AH-toe-trof) Producer. Organism that makes its own food using energy from the environment and carbon from inorganic molecules such as carbon dioxide. **100**

auxin (OX-in) Plant hormone that causes cell enlargement; also has a central role in growth and development by coordinating the effects of other hormones. **464**

AV node *See* atrioventricular node.

axial skeleton Bones of the main body axis, such as skull, backbone, and ribs. **559**

axon Of a neuron, a cytoplasmic extension that transmits electrical signals along its length and secretes chemical signals at its endings. **502**

bacteria Singular **bacterium**. The most diverse and well-known group of prokaryotes. **8, 320**

bacteriophage (bac-TEER-ee-oh-fayj) Type of virus that infects bacteria. **133, 326**

balanced polymorphism Maintenance of two or more alleles of a gene at high frequency in a population. **283**

bark Informal term for all living and dead tissues outside the ring of vascular cambium in woody plants. **435**

Barr body Inactivated (condensed) X chromosome in a cell of a female mammal. **168**

basal metabolic rate Rate at which the body uses energy when at rest. **646**

base Substance that accepts hydrogen ions in water. **32**

basement membrane Secreted protein layer that attaches an epithelium to an underlying tissue. **484**

base-pair substitution Type of mutation in which a single base pair changes. **156**

basophil (BASE-uh-fill) Circulating white blood cell that migrates into damaged or infected tissues and degranulates in response to injury or antigen. **593**

B cell B lymphocyte. White blood cell that makes antibodies in antibody-mediated responses. **593**

B cell receptor Antigen receptor (IgG or IgM) on the surface of a B cell. **599**

bell curve Bell-shaped curve; typically results from graphing frequency versus distribution for a trait that varies continuously. **216**

benthic province (BEN-thick) The ocean's sediments and rocks. **780**

bilateral symmetry Having paired structures so the right and left halves of an object are mirror images. **379**

bile Mixture of salts, pigments, and cholesterol produced in the liver, then stored and concentrated in the gallbladder; emulsifies fats when secreted into the small intestine. **638**

binary fission Method of asexual reproduction in which one prokaryotic cell divides into two identical descendant cells. **320**

bioaccumulation The concentration of a chemical pollutant in the tissues of an organism rises over the course of the organism's lifetime. **761**

biodiversity Of a region, the genetic variation within species, variety of species, and variety of ecosystems. **794**

biodiversity hot spot Threatened region with many species not found elsewhere; a high priority for conservation efforts. **794**

biofilm Community of microorganisms living within a shared mass of secreted slime. **59**

biogeochemical cycle A nutrient moves among environmental reservoirs and into and out of food webs. **754**

biogeography Study of patterns in the geographic distribution of species and communities. **254**

biological magnification A chemical pollutant becomes increasingly concentrated as it moves up through food chains. **761**

biological pest control Use of a pest's natural enemies to control its population size. **739**

biology The scientific study of life. **12**

biomass pyramid For an ecosystem, a graphic depicting the mass of organic material in the bodies of organisms at each trophic level at a specific time. **753**

biome (BY-ohm) A region (often discontinuous) characterized by its climate and dominant vegetation. **770**

biosphere (BY-oh-sfeer) All regions of Earth where organisms live. **4, 766**

biotic potential (by-AH-tick) Maximum possible population growth rate under optimal conditions. **717**

bipedalism Habitual upright walking. **414**

bird Feathered amniotes; most have a body adapted for flight. **410**

bivalve (BYE-valv) Mollusk with a hinged, two-part shell. **388**

blastocyst Mammalian blastula. **686**

blastula (BLAST-yoo-luh) Hollow ball of cells that forms as a result of cleavage. **670**

blood Circulatory fluid of a closed circulatory system. Of vertebrates, fluid connective tissue consisting of plasma and cellular components (red cells, white cells, platelets) that form in bones. **487, 572**

blood–brain barrier Protective mechanism that prevents unwanted substances from entering cerebrospinal fluid. **510**

blood pressure Pressure exerted by blood against a vessel wall. **581**

B lymphocyte See B cell.

bone tissue Connective tissue that consists of cells surrounded by a mineral-hardened matrix. **487**

boreal forest (BORE-ee-uhl) Extensive high-latitude forest of the Northern Hemisphere; conifers are the predominant vegetation. **772**

bottleneck Reduction in population size so severe that it reduces genetic diversity. **284**

Bowman's capsule Region of a nephron that forms a cup around the glomerulus; connects to the proximal tubule. **655**

brain Central control organ of a nervous system; receives and integrates sensory information, regulates internal conditions, and sends out signals that result in movements. **501**

bronchiole (BRON-key-ole) A small airway that leads from a bronchus to alveoli. **621**

bronchus (BRON-cuss) Airway connecting the trachea to a lung. **621**

brood parasitism One egg-laying species benefits by having another raise its offspring. **739**

brown algae Multicelled marine protists with a brown accessory pigment in their chloroplasts. **336**

brush border cell In the lining of the small intestine, an epithelial cell with microvilli at its surface. **637**

bryophyte (BRY-oh-fight) Nonvascular plant; a moss, liverwort, or hornwort. **350**

buffer Set of chemicals that can keep the pH of a solution stable by alternately donating and accepting ions that contribute to pH. **33**

C3 plant Type of plant that fixes carbon by the Calvin–Benson cycle alone. **107**

C4 plant Type of plant that minimizes photorespiration by fixing carbon twice, in two cell types. **108**

calcitonin (cal-sih-TOE-nin) Thyroid hormone that encourages bones to take up and incorporate calcium. **545**

Calvin–Benson cycle Light-independent reactions of photosynthesis; cyclic carbon-fixing pathway that builds sugars from CO_2. **106**

camera eye Eye with an adjustable opening and a single lens that focuses light on a retina. **526**

camouflage Coloration or body form that helps an organism blend in with its surroundings and escape detection. **737**

CAM plant Type of plant that conserves water by fixing carbon twice, at different times of day. **108**

cancer Disease that occurs when malignant cells physically and metabolically disrupt body tissues. **185**

capillary Smallest diameter blood vessel; exchanges with interstitial fluid occur across its wall. **572**

carbohydrate (car-bow-HI-drait) Molecule that consists mainly of carbon, hydrogen, and oxygen atoms in a 1:2:1 ratio. A sugar, or a polymer of sugars (e.g., cellulose, starch, glycogen). **40**

carbon cycle Biogeochemical cycle in which carbon is exchanged between oceans, soils, atmosphere, and living organisms. **756**

carbon fixation Process by which carbon from an inorganic source such as carbon dioxide is incorporated (fixed) into an organic molecule. **106**

cardiac cycle Sequence of contraction and relaxation of heart chambers that occurs with each heartbeat. **576**

cardiac muscle tissue Muscle of the heart wall. **488**

carpel (CAR-pull) Of a flower, reproductive organ that produces the female gametophyte; consists of an ovary, stigma, and usually a style. **359, 456**

carrying capacity (K) Maximum number of individuals of a species that a particular environment can sustain; can change over time. **718**

cartilage Connective tissue that consists of cells surrounded by a rubbery matrix. **487**

cartilaginous fish (car-tih-LAJ-ih-nuss) Jawed fish with a skeleton of cartilage; a shark, ray, or skate. **404**

Casparian strip Waxy band between the plasma membranes of abutting root endodermal cells; forms a seal that prevents soil water from seeping through cell walls into the vascular cylinder. **444**

catalysis (cut-AL-ih-sis) The acceleration of a reaction rate by a molecule that is unchanged by participating in the reaction. **82**

cDNA Complementary strand of DNA synthesized from an RNA template by the enzyme reverse transcriptase. **239**

cell Smallest unit of life; at minimum, consists of a plasma membrane, cytoplasm, and DNA. **4**

cell cortex Reinforcing mesh of cytoskeletal elements under a plasma membrane. **66**

cell cycle Collective series of intervals and events of a cell's life, from the time it forms until its cytoplasm divides. **178**

cell junction Structure that connects a cell directly to another cell or to extracellular matrix. See tight junction, adhering junction, gap junction, plasmodesma. **69**

cell-mediated immune response Immune response in which cytotoxic T cells and NK cells detect and destroy infected or cancerous body cells. **600**

cell plate Of plants, disk-shaped structure that forms and partitions descendant cells during cytokinesis. **182**

cell theory Theory that all organisms consist of one or more cells, which are the basic unit of life; all cells arise by division of preexisting cells; and all cells pass hereditary material to offspring. **52**

cell wall Rigid but permeable layer of extracellular matrix that surrounds the plasma membrane of some cells. **58**

cellular respiration Pathway that uses an electron transfer chain to harvest energy from an organic molecule and make ATP. See aerobic respiration. **114**

cellular slime mold Amoeba-like protist that feeds as a single predatory cell; joins with others to form a multicellular spore-bearing structure under unfavorable conditions. **342**

cellulose (SELL-you-low-ss) Tough, insoluble polysaccharide of glucose monomers; the major structural material in plants. **40**

central nervous system Of vertebrates, a brain and spinal cord. **501**

central vacuole Very large vacuole that makes up most of the volume of a plant cell. **62**

centriole (SEN-tree-ole) In animal cells, a barrel-shaped organelle from which microtubules grow. **67**

centromere (SEN-truh-meer) Of a duplicated eukaryotic chromosome, constricted region where sister chromatids attach to each other. **136**

cephalization Evolutionary trend whereby neurons that detect stimuli and process information become concentrated in the head of a bilateral animal. **500**

cephalopod (CEFF-uh-luh-pod) Predatory mollusk with a closed circulatory system; moves by jet propulsion. **389**

cerebellum (cer-uh-BELL-um) Hindbrain region that coordinates voluntary movements. **512**

cerebral cortex Outer gray matter layer of the cerebrum; region responsible for most complex behavior. **514**

cerebrospinal fluid Clear fluid that surrounds the brain and spinal cord and fills ventricles within the brain. **510**

cerebrum (suh-REE-brum) Forebrain region that controls higher functions. **513**

cervix Narrow part of uterus that connects to the vagina. **676**

chaparral (shap-ur-RAHL) Biome of dry shrubland in regions with hot, dry summers and cool, rainy winters. **773**

character Quantifiable, heritable trait such as the number of segments in a backbone or the nucleotide sequence of ribosomal RNA. **294**

character displacement As a result of competition, two species become less similar in their resource requirements. **735**

chelicerates (kell-ISS-er-ates) Arthropod group with specialized feeding structures (chelicerae) and no antennae; arachnids and horseshoe crabs. **392**

chemical bond Attractive force that arises between two atoms when their electrons interact; links atoms into molecules. *See* covalent bond, ionic bond. **28**

chemoautotroph (keem-oh-AWE-toe-trof) Organism that uses carbon dioxide as its carbon source and obtains energy by oxidizing inorganic molecules. **321**

chemoheterotroph (keem-oh-HET-ur-o-trof) Organism that obtains energy and carbon by breaking down organic compounds. **321**

chemoreceptor Sensory receptor that responds to a chemical. **522**

chitin (KIE-tin) Nitrogen-containing polysaccharide that composes fungal cell walls and arthropod exoskeletons. **41**

chlorophyll *a* (KLOR-uh-fil) Main photosynthetic pigment in plants. **102**

chloroplast (KLOR-uh-plast) Organelle of photosynthesis in plants and photosynthetic protists. Light-dependent reactions occur at its inner thylakoid membrane; light-independent reactions, in the stroma. **65**

choanoflagellates (ko-an-oh-FLAJ-el-ets) Flagellated heterotrophic protists closely related to animals. **343**

chordate (CORE-date) Animal with an embryo that has a notochord, dorsal nerve cord, pharyngeal gill slits, and a tail that extends beyond the anus. For example, a lancelet or a vertebrate. **402**

chorion (KOR-ee-on) Outermost extraembryonic membrane of amniotes; major component of the placenta in placental mammals. **687**

chromosome A structure that consists of DNA together with associated proteins; carries part or all of a cell's genetic information. **136**

chromosome number The total number of chromosomes in a cell of a given species. **137**

chyme (kime) Mix of food and gastric fluid. **636**

chytrids (KIH-trids) Fungi with flagellated spores. **367**

cilia (SILL-ee-uh) Singular **cilium**. Short, hairlike structures that project from the plasma membrane of some eukaryotic cells. Motile types often occur in clumps that beat in unison. **67**

ciliates (SIL-ee-ets) Single-celled or colonial protists with cilia. **336**

circadian rhythm (sir-KAY-dee-un) A biological activity that is repeated about every 24 hours. **171, 472**

circulatory system Organ system consisting of a heart or hearts and vessels that distribute circulatory fluid through a body. **386**

citric acid cycle Cyclic pathway that harvests energy from acetyl–CoA; part of aerobic respiration. Also called the Krebs cycle. **118**

clade (CLAYD) A group of taxa whose members share one or more defining derived traits. **294**

cladistics (cluh-DISS-ticks) Making hypotheses about evolutionary relationships among clades. **295**

cladogram (CLAD-oh-gram) Evolutionary tree diagram that visually summarizes a hypothesis about how a group of clades are related. **295**

cleavage In animal development, stage during which repeated mitotic divisions produce a blastula. **670**

cleavage furrow In a dividing animal cell, the indentation where cytoplasmic division will occur. **182**

climate Average weather conditions in a region. **766**

cloaca (klo-AY-kuh) Body opening that serves as the exit for digestive waste and urine; also functions in reproduction. **404, 633**

clone Genetic copy of an organism. **142**

cloning vector *See* vector.

closed circulatory system Circulatory system in which blood flows through a continuous system of vessels, and substances are exchanged across the walls of the smallest vessels. **386, 572**

club fungi Fungi that produce spores on club-shaped structures during sexual reproduction. **367**

cnidarian (nigh-DARE-ee-en) Radially symmetrical invertebrate with two tissue layers; uses tentacles with stinging cells to capture food. **382**

cnidocyte (NIGH-duh-site) Stinging cell unique to cnidarians. **382**

coal Fossil fuel formed over millions of years by compaction and heating of plant remains. **354**

cochlea (COCK-lee-uh) Coiled, fluid-filled structure in the inner ear that holds the mechanoreceptors involved in hearing. **530**

codominance Effect in which the full and separate phenotypic effects of two alleles are apparent in heterozygous individuals. **212**

codon (CO-don) In an mRNA, a nucleotide base triplet that codes for an amino acid or stop signal during translation. **152**

coelom (SEE-lum) A fluid-filled cavity between the gut and body wall that is lined by tissue derived from mesoderm; the main body cavity of most animals. **379**

coenzyme An organic cofactor. **86**

coevolution The joint evolution of two closely interacting species; macroevolutionary pattern in which each species is a selective agent for traits of the other. **292**

cofactor (KO-fack-ter) A coenzyme or metal ion that associates with an enzyme and is necessary for its function. **86**

cohesion (ko-HE-zhun) Property of a substance (such as water) that arises from the tendency of its molecules to resist separating from one another. **31**

cohesion–tension theory Explanation of how transpiration creates a tension that pulls a cohesive column of water upward through xylem. **447**

cohort Group of individuals born during the same interval. **720**

Glossary (continued)

coleoptile (coal-ee-OPP-till) Of monocots, a rigid sheath that protects the plumule (embryonic shoot) as it grows up to the soil surface. **468**

collecting tubule Kidney tubule that receives fluid from several nephrons and delivers it to the renal pelvis. **655**

collenchyma (coal-EN-kuh-muh) In plants, simple tissue composed of living cells with unevenly thickened walls; provides flexible support. **426**

colon Main portion of the human large intestine. **640**

colonial organism Organism composed of many similar cells, each capable of living and reproducing on its own. **332**

colonial theory of animal origins Hypothesis that the first animals evolved from a colonial protist. **380**

commensalism Species interaction that benefits one species and neither helps nor harms the other. **732**

community All populations of all species in a defined area. **4, 732**

compact bone Dense bone; makes up the shaft of long bones. **560**

companion cell In phloem, cell that provides metabolic support to its partnered sieve element. **448**

comparative morphology (more-FALL-uh-jee) Scientific study of similarities and differences in body plans. **254**

competitive exclusion Process whereby two species compete for a limiting resource, and one drives the other to local extinction. **734**

complement Set of proteins that circulate in inactive form in the blood; activated complement attracts phagocytes, coats antigenic particles, and punctures cells. **596**

complete digestive tract Tubelike digestive system; food enters through one opening and wastes leave through another. **632**

compound Molecule that has atoms of more than one element. **28**

compound eye Of some arthropods, a motion-sensitive eye made up of many image-forming units. **391, 526**

concentration Amount of solute per unit volume of a solution. **30**

condensation Chemical reaction in which a large molecule is assembled from smaller subunits; water also forms. **39**

conduction Of heat, the transfer of heat within an object or between two objects in contact with one another. **660**

cone cell Photoreceptor involved in sharp daytime vision and in color vision. **528**

conjugation (con-juh-GAY-shun) Mechanism of horizontal gene transfer in which one prokaryote passes a plasmid to another. **321**

connective tissue Animal tissue with an extensive extracellular matrix; provides structural and functional support. **486**

conservation biology Field of applied biology that surveys biodiversity and seeks ways to maintain and use it nondestructively. **794**

consumer (kun-SUE-murr) Organism that gets energy and nutrients by feeding on tissues, wastes, or remains of other organisms. **6, 750**

continuous variation Range of small differences in a trait. **216**

contractile vacuole In freshwater protists, an organelle that collects and expels excess water. **334**

control group Of an experiment, group of individuals identical to an experimental group except for the independent variable under investigation. **13**

convection Transfer of heat by moving molecules of air or water. **660**

coral bleaching A coral expels its photosynthetic dinoflagellate symbionts in response to stress and becomes colorless. **779**

coral reef Highly diverse marine ecosystem centered around reefs built by living corals that secrete calcium carbonate. **779**

cork Plant tissue that waterproofs, insulates, and protects the surfaces of woody stems and roots. Produced by cork cambium. **435**

cork cambium (CAM-bee-um) Lateral meristem that produces cork. **435**

cornea (CORE-nee-uh) Clear, protective covering at the front of a vertebrate eye; helps focus light on the retina. **527**

corpus luteum (CORE-pus LOO-tee-um) Hormone-secreting structure that forms from follicle cells left behind after ovulation. **677**

cortisol (CORT-is-all) Adrenal cortex hormone that influences metabolism and immunity; secretions rise with stress. **546**

cotyledon (cot-uh-LEE-dun) Seed leaf of a flowering plant embryo. **359, 424**

countercurrent exchange Exchange of substances between two fluids moving in opposite directions. **618**

covalent bond (co-VAY-lent) Type of chemical bond in which two atoms share a pair of electrons. Can be polar or nonpolar. **29**

critical thinking The deliberate process of judging the quality of information before accepting it. **12**

crop Of birds and some invertebrates, an enlarged portion of the esophagus that stores food. **633**

crossing over Process by which homologous chromosomes exchange corresponding segments of DNA during prophase I of meiosis. **198**

crustaceans (krus-TAY-shuns) Mostly marine arthropods with a calcium-hardened cuticle and two pairs of antennae; for example lobsters, crabs, krill, and barnacles. **393**

cuticle (KEW-tih-cull) Secreted covering at a body surface. **68, 348**

cyanobacteria (sigh-an-o-bac-TEER-ee-uh) Photosynthetic, oxygen-producing bacteria. **322**

cytokines (SITE-uh-kynes) Cell-to-cell signaling molecules secreted by white blood cells to coordinate activities during immune responses. **593**

cytokinesis (site-oh-kye-KNEE-sis) Cytoplasmic division; process in which a eukaryotic cell divides in two after mitosis or meiosis. **182**

cytokinin (site-oh-KINE-in) Plant hormone that promotes cell division in shoot apical meristem and cell differentiation in root apical meristem. Often opposes auxin's effects. **464**

cytoplasm (SITE-uh-plaz-um) Semifluid substance enclosed by a cell's plasma membrane. **52**

cytoskeleton (sigh-toe-SKEL-uh-ton) Network of microtubules, microfilaments, and intermediate filaments that support, organize, and move eukaryotic cells and their internal structures. **66**

cytotoxic T cell White blood cell that can kill ailing or cancerous body cells during a cell-mediated adaptive immune response. **604**

data (DAY-tuh) Experimental results. **13**

decomposer Heterotroph that feeds on organic remains by breaking them down into smaller, absorbable subunits. **322, 366, 750**

deductive reasoning Using a general idea to make a conclusion about a specific case. **13**

deletion Mutation in which one or more nucleotides are lost from DNA. **157**

demographics Statistics that describe a population. **714**

demographic transition model Model describing the changes in human birth and death rates that occur as a region becomes industrialized. **726**

denaturation (dee-nay-turr-AY-shun) Loss of a protein's shape and therefore its function. **46**

dendrite Of a neuron, a cytoplasmic extension that receives chemical signals sent by other neurons. **502**

dendritic cell (den-DRIH-tick) Phagocytic white blood cell that specializes in antigen presentation during adaptive immune responses. **593**

denitrification (dee-ny-triff-ih-CAY-shun) Conversion of nitrates or nitrites to nitrogen gas. **759**

dense, irregular connective tissue Connective tissue consisting of randomly arranged fibers and scattered fibroblasts. **486**

dense, regular connective tissue Connective tissue that consists of fibroblasts arrayed between parallel arrangements of fibers. **486**

density-dependent limiting factor Factor that limits population growth and has a greater effect in denser populations; for example, competition for a limited resource. **718**

density-independent limiting factor Factor that limits population growth and arises regardless of population size; for example, a flood. **719**

dental plaque On teeth, a thick biofilm composed of various microorganisms, their extracellular products, and saliva glycoproteins. **594**

deoxyribonucleic acid *See* DNA.

dependent variable In an experiment, a variable presumed to be influenced by the independent variable being tested. **13**

derived trait A character that is present in a clade under consideration, but not in any of the clade's ancestors. **294**

dermal tissue system Tissues that cover and protect a plant. *See* epidermis, periderm. **424**

dermis Deep layer of skin that consists of connective tissue with nerves and blood vessels running through it. **493**

desert Biome with little rain and low humidity; plants that have water-storing and water-conserving adaptations predominate. **774**

desertification (dih-zert-ih-ifh-KAY-shum) Conversion of grassland or woodland to desert. **788**

detrital food chain Food chain in which energy is transferred directly from producers to detritivores. **752**

detritivore (dih-TRITE-ih-vore) Animal that feeds on small bits of organic material. **750**

deuterostomes (DUE-ter-oh-stomes) Lineage of bilateral animals in which the second opening on the embryo surface develops into a mouth; for example, echinoderms and chordates. **379**

development (dih-VELL-up-ment) Process by which the first cell of a multicelled organism gives rise to an adult. **7**

diaphragm Sheet of smooth muscle between the thoracic and abdominal cavities; contracts during inhalation. **621**

diastole (die-ASS-toll-ee) Relaxation phase of the cardiac cycle. **576**

diastolic pressure (die-ah-STAHL-ic) Blood pressure when the left ventricle is most relaxed; lowest pressure of the cardiac cycle. **581**

diatoms (DIE-ah-tom) Single-celled, photosynthetic protists with a two-part silica shell. **336**

differentiation Process by which cells become specialized during development; occurs as different cells in an embryo begin to use different subsets of their DNA. **143**

diffusion (dif-YOU-zhun) Spontaneous spreading of molecules or atoms through a fluid or gas. **88**

digestion Breakdown of food into its chemical components. **632**

dihybrid cross Cross between two individuals identically heterozygous for two genes; for example, *AaBb* × *AaBb*. **210**

dikaryotic (die-kary-OTT-ik) Having two genetically distinct nuclei (*n*+*n*). **368**

dinoflagellates (di-no-FLAJ-el-lets) Single-celled, aquatic protists that move with a whirling motion; may be heterotrophic or photosynthetic. **337**

dinosaurs Group of reptiles that include the ancestors of birds; became extinct at the end of the Cretaceous. **408**

diploid (DIP-loyd) Having two of each type of chromosome characteristic of the species (2*n*). **137**

directional selection Mode of natural selection in which forms of a trait at one end of a range of variation are adaptive. **278**

disruptive selection Mode of natural selection in which forms of a trait at both ends of a range of variation are adaptive, and intermediate forms are selected against. **281**

distal tubule Region of a kidney tubule that connects to a collecting tubule. **655**

distance effect Islands close to a mainland receive more immigrants than those farther away. **744**

DNA Deoxyribonucleic (dee-ox-ee-rye-bo-new-KLAY-ick) acid; nucleic acid that carries hereditary information in its sequence. Double helix structure consists of two chains (strands) of deoxyribonucleotides (adenine, guanine, thymine, cytosine). **7**, **47**

DNA cloning Set of methods that uses living cells to mass-produce targeted DNA fragments. **238**

DNA library Collection of cells that host different fragments of foreign DNA, often representing an organism's entire genome. **240**

DNA ligase (LIE-gayce) Enzyme that seals strand breaks or gaps in double-stranded DNA. **139**

DNA polymerase Enzyme that carries out DNA replication. Uses a DNA template to assemble a complementary strand of DNA. **138**

DNA profiling Identifying an individual by analyzing the unique parts of his or her DNA. **245**

DNA replication Process by which a cell duplicates all of its DNA before it divides. **138**

DNA sequence Order of nucleotide bases in a strand of DNA; genetic information. **135**

DNA sequencing *See* sequencing.

dominance hierarchy Social system in which resources and mating opportunities are unequally distributed within a group. **707**

dominant Refers to an allele that masks the effect of a recessive allele paired with it in heterozygous individuals. **207**

dormancy Period of temporarily suspended metabolism. **458**

dosage compensation Any mechanism of equalizing gene expression between males and females. **168**

double fertilization Fertilization as it occurs in flowering plants; one sperm cell fuses with an egg, and a second sperm cell fuses with the endosperm mother cell. **458**

duplication Repeated section of DNA in a chromosome. **228**

ecdysone (eck-DYE-sohn) Steroid hormone that controls molting in arthropods. **549**

echinoderms (eh-KYE-nuh-derms) Invertebrates with a water–vascular system, and hardened plates and spines in the skin or body. **396**

ECM *See* extracellular matrix.

ecological footprint Area of Earth's surface required to sustainably support a particular level of development and consumption. **727**

ecological niche Of a species, unique set of requirements and roles in an ecosystem. **734**

ecological restoration Actively altering an area in an effort to restore an ecosystem that has been damaged or destroyed. **795**

ecology The study of interactions between organisms and their environments. **714**

ecosystem A biological community interacting with its environment. **4**, **750**

ectoderm (EK-toe-derm) Outermost tissue layer of an animal embryo. **378**, **670**

ectotherm Animal in which internal temperature is regulated by changes in behavior; for example, a fish or a lizard. **408**, **660**

effector cell Antigen-sensitized white blood cell (T cell, B cell, or NK cell) that forms in an immune response and acts immediately. **600**

egg Female gamete. **668**

electron Negatively charged subatomic particle. **24**

electronegativity Measure of the ability of an atom to pull electrons away from other atoms. **28**

electron transfer chain In a cell membrane, series of enzymes and other molecules that accept and give up electrons, thus releasing the energy of the electrons in small, usable steps. **85**

electron transfer phosphorylation Process in which electron flow through electron transfer chains sets up a hydrogen ion gradient that drives ATP formation. **104**

electrophoresis (eh-lek-troh-fur-EE-sis) Laboratory technique that separates DNA fragments by size. **242**

element A pure substance that consists only of atoms with the same number of protons. **24**

embryo In animals, a developing individual from first cleavage until hatching or birth; in humans, usually refers to an individual during weeks 2 to 8 of development. **675**

embryonic induction Embryonic cells in one region produce signals that alter the behavior of neighboring cells. **672**

emergent property (ee-MERGE-ent) A characteristic of a system that does not appear in any of the system's component parts. **4**

emerging disease A disease that was previously unknown or has recently begun spreading to a new region. **318**

emigration Movement of individuals out of a population. **716**

emulsification Dispersion of fat droplets in a fluid. **638**

endangered species Species that faces extinction in all or a part of its range. **786**

endemic species (en-DEM-ick) A species found only in the region where it evolved. **742, 786**

endergonic (end-er-GON-ick) Describes a reaction that requires a net input of free energy. **80**

endocrine gland Ductless gland; aggregation of hormone-secreting epithelial cells with associated connective tissue and blood vessels. **485, 540**

endocrine system Organ system that coordinates responses among cells of an animal body by means of hormones. **538**

endocytosis (en-doe-sigh-TOE-sis) Process by which a cell takes in a small amount of extracellular fluid (and its contents) by the ballooning inward of the plasma membrane. **92**

endoderm (EN-doe-derm) Innermost tissue layer of an animal embryo. **378, 670**

endodermis In a plant root, a sheet of cells just outside the vascular cylinder. A Casparian strip between these cells prevents water from seeping through their walls. **433**

endomembrane system Multifunctional network of membrane-enclosed organelles (endoplasmic reticulum, Golgi bodies, vesicles). **62**

endoplasmic reticulum (ER) (en-doh-PLAZ-mick ruh-TICK-you-lum) Membrane-enclosed organelle that is a system of sacs and tubes extending from the nuclear envelope. Smooth ER makes phospholipids, stores calcium, and has additional functions in some cells; ribosomes on the surface of rough ER make proteins. **62**

endorphin (en-DOR-fin) Pain-relieving chemical made by the body. **523**

endoskeleton Internal skeleton made up of hardened components such as bones. **403, 558**

endosperm Nutritive tissue in the seeds of flowering plants. **359, 458**

endospore Resistant resting stage of some soil bacteria. **323**

endosymbiont (en-doh-SIM-bee-ont) Organism that lives inside another organism and does not harm it. **309**

endosymbiont theory Theory that mitochondria and chloroplasts evolved from bacteria that entered and lived inside a host cell. **309**

endotherm Animal that maintains its temperature by adjusting its production of metabolic heat; for example, a bird or mammal. **408, 660**

energy The capacity to do work. **78**

energy pyramid For an ecosystem, a graphic depicting the energy that flows through each trophic level in a given interval. **753**

entropy (EN-truh-pee) Measure of how much the energy of a system is dispersed. **78**

enzyme Protein or RNA that speeds up a reaction without being changed by it. **39**

eosinophil (ee-uh-SIN-uh-fill) Granular, circulating white blood cell that targets multicelled parasites that are too large for phagocytosis. **593**

epidermis (epp-ih-DUR-miss) Outermost tissue layer. In animals, the epithelial layer of skin. In plants, the dermal tissue. **427, 492**

epididymis (epp-ih-DID-ih-muss) Duct where sperm mature; empties into a vas deferens. **680**

epigenetic Refers to potentially heritable mechanisms that stably alter gene expression without changing the DNA sequence. **172**

epiglottis Tissue flap that folds down to prevent food from entering the trachea during swallowing. **621**

epistasis (epp-ih-STAY-sis) Polygenic inheritance, in which a trait is influenced by multiple genes. **213**

epithelial tissue Sheetlike animal tissue that covers outer body surfaces and lines internal tubes and cavities. **484**

equilibrium model of island biogeography Model that predicts the number of species on an island based on the island's area and distance from a colonizing source. **744**

ER *See* endoplasmic reticulum.

erosion *See* soil erosion.

esophagus (eh-SOFF-ah-gus) Tubular organ that connects the pharynx to the stomach. **633**

essential amino acid Amino acid that the body cannot make and must obtain from food. **641**

essential fatty acid Fatty acid that the body cannot make and must obtain from food. **641**

estrogens Sex hormones secreted by ovaries; function in reproduction and cause development of female secondary sexual characteristics. **547, 676**

estrous cycle (ESS-truss) Reproductive cycle in which the uterine lining thickens, and, if pregnancy does not occur, is reabsorbed. **679**

estuary (ESS-chew-erry) A highly productive ecosystem where nutrient-rich water from a river mixes with seawater. **778**

ethylene (ETH-ill-een) Gaseous plant hormone that participates in germination, abscission, ripening, and stress responses. **467**

eudicot (YOU-dih-cot) Flowering plant in which the embryo has two cotyledons. **359**

euglenoid (you-GLEEN-oyd) Free-living, flagellated protist in the supergroup Excavata; some have chloroplasts descended from a green alga. **335**

eukaryote (you-CARE-ee-oat) Single- or multi-celled organism whose cells characteristically have a nucleus; a protist, fungus, plant, or animal. **8**

eusocial animal Animal species that lives in a multigenerational group in which many sterile workers cooperate in all tasks essential to the group's welfare; only a few members of the group produce offspring. **708**

eutrophication (you-truh-fih-CAY-shun) Nutrient enrichment of an aquatic habitat. **776**

evaporation Transition of a liquid to a vapor. **31, 660**

evolution Change in a line of descent. **256**

evolutionary tree Diagram showing evolutionary connections. **295**

exaptation (eggs-app-TAY-shun) A trait that becomes used for an entirely new purpose during evolution. **292**

exergonic (ex-er-GON-ick) Describes a reaction that ends with a net release of free energy. **80**

exocrine gland Gland that secretes milk, sweat, saliva, or some other substance through a duct. **485**

exocytosis (ex-oh-sigh-TOE-sis) Process by which a cell expels a vesicle's contents to extracellular fluid. **92**

exon Nucleotide sequence that remains in an RNA after post-transcriptional modification. **151**

exoskeleton External skeleton of some invertebrates; hard external parts that muscles attach to and move. **391, 558**

exotic species A species that evolved in one community and later became established in a different one. **742**

experiment A test that is designed to evaluate a prediction. **13**

experimental group In an experiment, group of individuals who have a certain characteristic or receive a certain treatment as compared with a control group. **13**

exponential growth A population grows by a fixed percentage in successive time intervals; the size of each increase is determined by the current population size. **717**

external fertilization Sperm fertilize eggs after gametes are released into the environment. **669**

extinct Refers to a species that no longer has living members. **292**

extracellular fluid Of a multicelled organism, body fluid outside of cells; serves as the body's internal environment. **482**

extracellular matrix (ECM) Complex mixture of cell secretions; its composition and function vary by the type of cell that secretes it. **68**

extreme halophile (HAY-lo-file) Organism adapted to life in a highly salty environment. **325**

extreme thermophile (THERM-o-file) Organism adapted to life at a very high temperature. **325**

facilitated diffusion Passive transport mechanism in which a solute follows its concentration gradient across a membrane by moving through a transport protein. **90**

fat Substance that consists mainly of triglycerides. *See* saturated fat, unsaturated fat. **42**

fatty acid Organic compound that consists of an acidic carboxyl group "head" and a long hydrocarbon "tail." Can be saturated or unsaturated. **42**

feces Unabsorbed food material and cellular waste that is expelled from the digestive tract. **640**

feedback inhibition Regulatory mechanism in which a change that results from some activity decreases or stops the activity. **84**

fermentation Anaerobic glucose-breakdown pathway that produces ATP without the use of electron transfer chains. **115**

fertilization Fusion of two gametes to form a zygote; part of sexual reproduction. **195**

fetus Developing human from about 9 weeks until birth. **675**

fever A temporary, internally induced rise in core body temperature above the normal set point. **597**

fibrin Threadlike protein that is formed from the soluble plasma protein fibrinogen during blood clotting. **579**

fibrous root system In plants, root system composed of an extensive mass of similar-sized adventitious roots; typical of monocots. **432**

first law of thermodynamics Energy cannot be created or destroyed. **78**

fitness Degree of adaptation to an environment, as measured by the individual's relative genetic contribution to future generations. **257**

fixed Refers to an allele for which all members of a population are homozygous. **284**

flagellum (fluh-JEL-um) Long, slender cellular structure used for motility. **59**

flatworm Bilaterally symmetrical invertebrate with organs but no body cavity; for example, a planarian or tapeworm. **384**

flower Specialized reproductive shoot of a flowering plant (angiosperm). **358, 456**

fluid mosaic model Model of a cell membrane as a two-dimensional fluid of mixed composition. **56**

follicle-stimulating hormone (FSH) Anterior pituitary hormone with roles in ovarian follicle maturation and sperm production. **543, 678**

food chain Description of who eats whom in one path of energy flow through an ecosystem. **751**

food web Set of cross-connecting food chains. **752**

foraminifera (for-am-in-IF-er-ah) Heterotrophic single-celled protists with a porous calcium carbonate shell and cytoplasmic extensions. **337**

fossil Physical evidence of an organism that lived in the ancient past. **255**

founder effect After a small group of individuals found a new population, allele frequencies in the new population differ from those in the original population. **285**

fovea (FOE-vee-uh) Retina region with the greatest density of cone cells; has no rod cells. **528**

fraternal twins Genetically nonidentical twins. **677**

free radical Atom with an unpaired electron. Most are highly reactive. **27**

frequency-dependent selection Natural selection in which a trait's adaptive value depends on its frequency among members of a population. **283**

fruit Mature ovary of a flowering plant, often with accessory parts; encloses a seed or seeds. **359, 460**

FSH *See* follicle-stimulating hormone.

functional group An atom (other than hydrogen) or a small molecular group bonded to a carbon of an organic compound; imparts a specific chemical property. **38**

fungus Single- or multicelled eukaryotic consumer that digests food outside the body, then absorbs the resulting breakdown products. Has chitin-containing cell walls. **8, 366**

gallbladder Organ that stores and concentrates bile. **638**

gamete (GAM-eat) Mature, haploid reproductive cell; e.g., a sperm. **194**

gametophyte (gam-EAT-uh-fight) Gamete-producing haploid body that forms in the life cycle of plants and some multicelled algae. **340, 348**

ganglion (GANG-lee-on) Cluster of nerve cell bodies. **500**

gap junction Cell junction that forms a closable channel across the plasma membranes of adjoining animal cells. **69**

gastric fluid Fluid secreted by the stomach lining; contains digestive enzymes, acid, and mucus. **636**

gastropod Mollusk in which the lower body is a broad "foot;" for example, a snail or slug. **388**

gastrovascular cavity A saclike gut that also functions in gas exchange. **382, 632**

gastrula (GAS-true-luh) Three-layered developmental stage formed by gastrulation. **670**

gastrulation (gas-true-LAY-shun) Animal developmental process by which cell movements produce a three-layered gastrula. **670**

gene (JEEN) Chromosomal DNA sequence that encodes an RNA or protein product. **148**

gene expression Multistep process by which the information in a gene guides assembly of an RNA or protein product. Includes transcription and translation. **148**

gene flow The movement of alleles between populations. **285**

gene pool All the alleles of all the genes in a population; a pool of genetic resources. **275**

gene therapy Treating a genetic defect or disorder by transferring a normal or modified gene into the affected individual. **248**

genetically modified organism (GMO) Organism whose genome has been modified by genetic engineering. **246**

genetic code Complete set of sixty-four mRNA codons. **152**

genetic drift Change in allele frequency due to chance alone. **284**

genetic engineering Process by which an individual's genome is altered deliberately, in order to change its phenotype. **246**

genetic equilibrium Theoretical state in which an allele's frequency never changes in a population's gene pool. **276**

genome An organism's complete set of genetic material. **240**

genomics The study of whole-genome structure and function. **244**

genotype (JEEN-oh-type) The particular set of alleles that is carried by an individual's chromosomes. **207**

genus (JEE-nuss) Plural **genera**. A group of species that share a unique set of traits; also the first part of a species name. **10**

geologic time scale Chronology of Earth's history; correlates geologic and evolutionary events. **264**

germ cell Immature reproductive cell; gives rise to haploid gametes when it divides. **194**

germinate To resume metabolic activity after dormancy (in pollen and seeds). **458**

germ layer One of three primary layers in an early animal embryo. **670**

GH *See* growth hormone.

gibberellin (jib-er-ELL-in) Plant hormone that induces stem elongation; also helps seeds break dormancy. **465**

gill Of an aquatic animal, a folded or filamentous respiratory organ in which blood or hemolymph exchanges gases with water. **388, 617**

gizzard Of birds and some other animals, a digestive chamber lined with a hard material that grinds up food. **633**

global climate change An ongoing trend toward warmer temperatures and shifts in climate patterns worldwide; caused by a rise in greenhouse gases. **757**

glomeromycetes (gloh-mere-oh-MY-seats) Fungi that partner with plant roots; their hyphae grow inside cell walls of root cells. **367**

glomerular filtration (glow-MER-you-lar) Protein-free plasma forced out of glomerular capillaries by blood pressure enters Bowman's capsule. **656**

glomerulus (gloh-MAIR-yuh-lus) Cluster of leaky capillaries enclosed by a Bowman's capsule. **655**

glottis Opening formed when the vocal cords relax. **621**

glucagon (GLUE-kah-gon) Pancreatic hormone that encourages glycogen breakdown; increases blood glucose level. **548**

glycogen (GLY-ko-jen) Highly branched polysaccharide of glucose monomers. Principal form of stored sugars in animals. **41**

glycolysis (gly-COLL-ih-sis) Set of reactions that convert glucose to two pyruvate for a net yield of two ATP and two NADH. **116**

GMO *See* genetically modified organism.

Golgi body (GOAL-jee) Membrane-enclosed organelle that modifies proteins and lipids, then sorts the finished products into vesicles. Shaped like a stack of pancakes. **63**

gonads Primary sex organ; produce gametes and sex hormones. **547, 669**

Gondwana (gond-WAN-uh) Supercontinent that existed before Pangea, more than 500 million years ago. **263**

Gram-positive bacteria Bacteria with a thick cell wall that stains purple when prepared for microscopy by Gram staining. **322**

gravitropism (grah-vih-TRO-pizm) Directional response to gravity. **470**

gray matter Central nervous system tissue consisting of neuron axon terminals, cell bodies, and dendrites, and some neuroglia. **510**

grazing food chain Food chain in which energy is transferred from producers to grazers (herbivores). **752**

green alga Single-celled, colonial, or multicelled photosynthetic protist that has chloroplasts containing chlorophylls *a* and *b*. **341**

greenhouse effect Warming of Earth's lower atmosphere and surface as a result of heat trapped by greenhouse gases. **757**

greenhouse gas Atmospheric gas, such as carbon dioxide, that absorbs and reradiates heat, warming Earth's surface. **757**

ground tissue system Of plants, all plant tissues other than vascular and dermal tissues. Carries out basic processes of survival such as photosynthesis, storage, and growth. **424**

groundwater Soil water and water in aquifers. **754**

growth In multicelled species, an increase in the number, size, and volume of cells. **7**

growth factor Molecule that stimulates mitosis and differentiation. **184**

growth hormone (**GH**) Anterior pituitary hormone that regulates growth and metabolism. **543**

guard cell One of a pair of cells that define a stoma across plant epidermis. **448**

gymnosperm (JIM-no-sperm) Seed plant whose seeds are not enclosed within a fruit; a conifer, cycad, ginkgo, or gnetophyte. **356**

habitat Type of environment in which a species typically lives. **732**

habituation Learning not to respond to a repeated neutral stimulus. **701**

half-life Characteristic time it takes for half of a quantity of a radioisotope to decay. **260**

haploid (HAP-loyd) Having one of each type of chromosome characteristic of the species. **194**

heart Muscular organ that pumps blood through a body. **572**

helper T cell Type of T cell that activates other white blood cells; required for all adaptive immune responses. **602**

hemolymph (HEEM-oh-limf) Fluid that circulates in an open circulatory system. **572**

hemostasis Process by which blood clots in response to injury. **579**

herbivory An animal feeds on plants or plant parts. **737**

hermaphrodite (herm-AFF-roh-dyte) Animal that has both male and female gonads, either simultaneously or at different times in its life cycle. **381, 669**

heterotherm Animal in which temperature is at times tightly controlled by changes in metabolism, and at other times fluctuates with the environment. **660**

heterotroph (HET-er-oh-trof) Consumer. Organism that obtains carbon from organic compounds assembled by other organisms. **100**

heterozygous (het-er-uh-ZYE-guss) Having two different alleles at the same locus on homologous chromosomes. **207**

hippocampus Brain region essential to formation of declarative memories. **515**

histone (HISS-tone) Type of protein that associates with DNA and structurally organizes eukaryotic chromosomes. **136**

HIV (**human immunodeficiency virus**) Virus that causes AIDS. **316**

homeostasis (home-ee-oh-STAY-sis) Process in which organisms keep their internal conditions within tolerable ranges by sensing and responding appropriately to change. **7**

homeotic gene (home-ee-OTT-ic) Type of master gene; its expression triggers formation of a specific body part during development. **167**

hominin Human or an extinct primate species more closely related to humans than to any other primates. **414**

Homo erectus Extinct hominin that arose about 1.8 million years ago in East Africa; migrated out of Africa. **416**

Homo habilis Extinct hominin; earliest named *Homo* species; known only from Africa, where it arose 2.3 million years ago. **416**

homologous chromosomes In a nucleus, pair of chromosomes that have the same length, shape, and genes. **178**

homologous structures Body structures that are similar in different lineages because they evolved in a common ancestor. **266**

Homo neanderthalensis Extinct hominin; closest known relative of *H. sapiens*; lived in Africa, Europe, Asia. **417**

Homo sapiens Species name of modern humans. **416**

homozygous (ho-mo-ZYE-guss) Having two identical alleles at the same locus on homologous chromosomes. **207**

horizontal gene transfer Transfer of genetic material between existing individuals. **321**

hormone *See* animal hormone or plant hormone.

human immunodeficiency virus *See* HIV.

humus (HUE-muss) Partially decayed organic material in soil. **443**

hybrid (HI-brid) A heterozygous individual. **207**

hybridization (hi-brih-die-ZAY-shun) Spontaneous establishment of base pairing between two nucleic acid strands. **138**

hydrocarbon Compound that consists only of carbon and hydrogen atoms. **38**

hydrogen bond Attraction between a covalently bonded hydrogen atom and another atom taking part in a separate polar covalent bond. **30**

hydrolysis (hy-DRAWL-uh-sis) Water-requiring chemical reaction that breaks a molecule into smaller subunits. **39**

hydrophilic (hi-druh-FILL-ick) Describes a substance that dissolves easily in water. **30**

hydrophobic (hi-druh-FOE-bick) Describes a substance that resists dissolving in water. **31**

hydrostatic skeleton Of soft-bodied invertebrates, a fluid-filled chamber that contractile cells exert force on. **382, 558**

hydrothermal vent Place where hot, mineral-rich water streams out from an underwater opening in Earth's crust. **303, 780**

hypertonic Describes a fluid that has a high overall solute concentration relative to another fluid. **88**

hypha (HI-fah) One filament in a fungal mycelium. **366**

hypothalamus Forebrain region that controls processes related to homeostasis; produces and controls release of some hormones. **513, 542**

hypothesis (hi-POTH-uh-sis) Testable explanation of a natural phenomenon. **12**

hypotonic Describes a fluid that has a low overall solute concentration relative to another fluid. **88**

identical twins Genetically identical individuals that develop after an embryo splits in two. **687**

immigration Movement of individuals into a population. **716**

immunity The body's ability to resist and fight infections. **592**

immunization Any procedure designed to induce immunity to a specific disease. **610**

imprinting Learning that can occur only during a specific interval in an animal's life. **700**

inbreeding Mating among close relatives. **285**

inclusive fitness Genetic contribution an individual makes by reproducing, plus a fraction of the contribution it makes by facilitating reproduction of relatives. **707**

incomplete dominance Inheritance pattern in which one allele is not fully dominant over another, so the heterozygous phenotype is an intermediate blend between the two homozygous phenotypes. **212**

independent variable Variable that is controlled by an experimenter in order to explore its relationship to a dependent variable. **13**

indicator species Species whose presence and abundance in a community provide information about conditions in the community. **741**

induced-fit model Of enzyme activity, interacting with a substrate causes the active site to change shape so that the fit between them improves and catalysis occurs. **82**

inductive reasoning Drawing a conclusion based on one's observations. **13**

inferior vena cava (VEE-nah CAY-vuh) Vein that delivers oxygen-poor blood from the lower body to the heart. **575**

inflammation Part of innate immunity: a local response to tissue damage or infection; characterized by redness, warmth, swelling, and pain. **597**

inheritance (in-HAIR-ih-tunce) Transmission of DNA to offspring. **7**

inhibiting hormone Animal hormone that discourages secretion of another hormone by its target cells. **538**

innate immunity Of multicelled organisms, a set of immediate, general responses (complement activation, phagocytosis, inflammation, fever) that defend the body from infection. **592**

inner ear Of the mammalian ear, region that contains the fluid-filled vestibular apparatus and cochlea. **530**

insects Most diverse arthropod group; have six legs, two antennae, and, in most subgroups, wings. **394**

insertion Mutation in which one or more nucleotides become inserted into DNA. **157**

instinctive behavior An innate response to a simple stimulus. **700**

insulin Pancreatic hormone that encourages uptake of glucose by target cells; lowers level of glucose in the blood. **548**

intermediate disturbance hypothesis Species richness is greatest in communities in which disturbances are moderate in their magnitude and frequency. **741**

intermediate filament Stable cytoskeletal element of animals and some protists; structurally supports cell membranes and tissues; also forms external structures such as hair. **66**

internal fertilization Sperm fertilize eggs inside a female's body. **669**

interneuron Neuron that integrates information; both receives signals from and sends signals to other neurons. **502**

interphase In a eukaryotic cell cycle, the interval during which a cell grows, roughly doubles the number of its cytoplasmic components, and replicates its DNA in preparation for division. **178**

interspecific competition Competition between members of different species. **734**

intestine Tubular organ in which most chemical digestion occurs and from which nutrients are absorbed. *See* small intestine, large intestine. **633**

interstitial fluid Portion of the extracellular fluid that fills spaces between body cells. **482**

intervertebral disk Cartilage disk between two vertebrae. **559**

intraspecific competition Competition for resources among members of the same species. **718**

intron Nucleotide sequence that intervenes between exons and is removed during post-transcriptional modification. **151**

inversion Structural rearrangement of a chromosome in which part of the DNA has become oriented in the reverse direction. **228**

invertebrate Animal that does not have a backbone. **378**

ion (EYE-on) Atom or molecule that carries a net charge. **27**

ionic bond (eye-ON-ick) Type of chemical bond in which a strong mutual attraction links ions of opposite charge. **28**

iris Circular muscle that adjusts the shape of the pupil to regulate how much light enters the eye. **527**

isotonic Describes two fluids that have identical solute concentrations. **89**

isotopes (ICE-uh-topes) Forms of an element that differ in the number of neutrons their atoms carry. **24**

jawless fish Fish with a skeleton of cartilage, no fins or jaws; a lamprey or hagfish. **404**

joint Region where bones meet. **561**

karyotype (CARE-ee-uh-type) Image of a cell's chromosomes arranged by size, length, shape, and centromere location. **137**

key innovation An evolutionary adaptation that gives its bearer the opportunity to exploit a particular environment more efficiently or in a new way. **292**

keystone species A species that has a disproportionately large effect on community structure relative to its abundance. **742**

kidney Organ of the vertebrate urinary system that filters and adjusts the composition of blood, and produces urine. **653**

kinetic energy (kih-NEH-tick) The energy of motion. **78**

kin selection Type of natural selection in which individuals with the highest inclusive fitness are favored. **707**

Glossary (continued)

knockout An experiment in which a gene is deliberately inactivated in a living organism; also, an organism that carries a knocked-out gene. **167**

Krebs cycle *See* citric acid cycle.

K-selection Pattern of selection in which adaptive traits allow their bearers to outcompete others for limited resources; occurs when a population is near its environment's carrying capacity. **721**

labor Expulsion of a placental mammal from its mother's uterus by muscle contractions. **692**

lactate fermentation Fermentation pathway that produces ATP and lactate. **123**

lactation Milk production by a female mammal. **692**

lancelet (LANCE-uh-lit) Invertebrate chordate that has a fishlike shape and retains the defining chordate traits into adulthood. **402**

large intestine Organ that receives digestive waste from the small intestine and concentrates it as feces. **634**

larva In some animal life cycles, a sexually immature form that differs from the adult in its body plan. **381**

larynx (LAIR-inks) Short airway that contains the vocal cords; the "voice box." **620**

lateral meristem Vascular cambium or cork cambium; cylindrical sheet of meristem that runs lengthwise through a plant's shoots and roots; gives rise to secondary growth (thickening). **435**

law of independent assortment During meiosis, alleles at one gene locus on homologous chromosomes tend to be distributed into gametes independently of alleles at other loci. **210**

law of nature Generalization that describes a consistent natural phenomenon but does not propose a mechanism. **18**

law of segregation A diploid cell has two copies of every gene that occurs on its homologous chromosomes. Two alleles at any locus are distributed into separate gametes during meiosis. **209**

leaching Process by which water moving through soil removes nutrients from it. **443**

leaf vein A vascular bundle in a leaf. **430**

learned behavior Behavior that is modified by experience. **700**

lek Of some birds, a communal mating display area for males. **705**

lens Disk-shaped structure that bends light rays so they fall on an eye's photoreceptors. **526**

lethal mutation Mutation that alters phenotype so drastically that it causes death. **274**

leukocyte *See* white blood cell.

LH *See* luteinizing hormone.

lichen (LIE-kin) Composite organism consisting of a fungus and green algae or cyanobacteria. **370**

life history Schedule of how resources are allocated to growth, survival, and reproduction over a lifetime. **720**

ligament (LIG-uh-ment) Strap of dense connective tissue that holds bones together at a joint. **561**

light-dependent reactions First stage of photosynthesis; convert light energy to chemical energy. A noncyclic pathway produces ATP, oxygen, and NADPH; a cyclic pathway produces ATP only. **100**

light-independent reactions Carbon-fixing second stage of photosynthesis; use ATP and NADPH to assemble sugars from water and CO_2. **100**

lignin (LIG-nin) Material that strengthens cell walls of vascular plants. **68, 348**

limbic system Group of structures deep in the brain that function in emotion. **515**

lineage (LINN-edge) Line of descent. **256**

linkage group All of the genes on a chromosome. **211**

lipid A fatty, oily, or waxy organic compound; a fatty acid derivative (such as a triglyceride), steroid, or wax. **42**

lipid bilayer Double layer of phospholipids arranged tail-to-tail; structural foundation of all cell membranes. **43**

liver Abdominal organ that produces plasma proteins and bile, filters the blood, and stores glycogen. **638**

loam Soil with roughly equal amounts of sand, silt, and clay. **443**

lobe-finned fish Jawed fish with fins supported by internal bones; for example, a coelacanth or lungfish. **405**

locomotion Self-propelled movement from place to place. **556**

locus (LOW-cuss) A particular location on a chromosome. **207**

logistic growth A population grows exponentially at first, then growth slows as population size approaches the environment's carrying capacity for that species. **718**

loop of Henle (HEN-lee) U-shaped portion of a kidney tubule; connects the proximal and distal regions of the tubule. **655**

loose connective tissue Connective tissue that consists of fibroblasts and fibers scattered in a gel-like matrix. **486**

lung Respiratory organ (an air-filled sac) that exchanges gases with the air. **388, 405, 617**

luteinizing hormone (LH) Anterior pituitary hormone with roles in ovulation, corpus luteum formation, and sperm production. **543, 678**

lymph Fluid in the lymph vascular system. **586**

lymph node Small mass of lymphoid tissue through which lymph filters; contains many B cells and T cells. **586**

lymph vascular system System of vessels that takes up interstitial fluid and carries it to the blood as lymph. **586**

lymphocyte *See* white blood cell, B cell, T cell, NK cell.

lysogenic pathway (lice-oh-JEN-ick) Bacteriophage replication path in which viral DNA becomes integrated into the host's chromosome and is passed to the host's descendants. **316**

lysosome (LICE-uh-sohm) Enzyme-filled vesicle that breaks down cellular wastes and debris. **62**

lysozyme (LICE-uh-zime) Antibacterial enzyme in body secretions such as saliva and mucus. **595**

lytic pathway (LIH-tik) Bacteriophage replication pathway in which a virus immediately replicates in its host and kills it. **316**

macroevolution Evolutionary patterns and trends on a larger scale than microevolution; e.g., adaptive radiation, coevolution. **292**

macrophage (MACK-roe-faje) White blood cell that specializes in phagocytosis. **593**

Malpighian tubule (mal-PIG-ee-an) Of insects and spiders, one of many tubular organs that take up waste solutes and deliver them to the digestive tract for excretion. **653**

mammal Animal with hair or fur; females secrete milk from mammary glands. **411**

mantle Of mollusks, a skirtlike extension of the body wall. **388**

mark–recapture sampling Method of estimating population size of mobile animals by marking individuals, releasing them, then checking the proportion of marks among individuals recaptured at a later time. **715**

marsupial (mar-SUE-pee-uhl) Mammal in which young are born at an early stage and complete development in a pouch on the mother's surface. **411**

mass number Of an isotope, the total number of protons and neutrons in the atomic nucleus. **24**

mast cell Granular white blood cell that stays anchored in tissues. Factor in allergies. **593**

master gene Gene encoding a product that affects the expression of many other genes. **166**

mechanoreceptor Sensory receptor that responds to pressure, position, or acceleration. **522**

medulla oblongata (meh-DOOL-uh ob-long-GAH-tah) Hindbrain region that influences breathing and controls reflexes such as coughing and vomiting. **512**

medusa (meh-DUE-suh) Dome-shaped cnidarian body plan with the mouth on the lower surface. Jellies are examples. **382**

megaspore Of seed plants, haploid spore that forms in an ovule and gives rise to an egg-producing gametophyte. **354, 458**

meiosis (my-OH-sis) Nuclear division process required for sexual reproduction. Two nuclear divisions, meiosis I and II, halve the chromosome number for forthcoming gametes. **194**

melatonin (mell-uh-TOE-nin) Pineal gland hormone that regulates sleep–wake cycles and seasonal changes; secreted only in the dark. **545**

membrane potential Voltage difference across a cell membrane. **503**

memory cell Long-lived, antigen-sensitized lymphocyte (B cell, T cell, or NK cell) that can carry out a secondary immune response. **600**

meninges (meh-NIN-jeez) Layers of connective tissue that enclose the brain and spinal cord. **510**

menopause Permanent cessation of menstrual cycles. **679**

menstrual cycle Reproductive cycle in which the uterus lining thickens and then, if pregnancy does not occur, is shed. **678**

menstruation Flow of shed uterine tissue out of the vagina at the onset of the menstrual cycle. **678**

meristem (MARE-ih-stem) Zone of undifferentiated parenchyma cells, the divisions of which give rise to new tissues during plant growth. **434**

mesoderm (MEEZ-o-derm or MEZ-o-derm) Middle tissue layer of a three-layered animal embryo. **378, 670**

mesophyll (MEZ-uh-fill) Photosynthetic parenchyma of plants. **426**

messenger RNA (mRNA) Type of RNA that carries a protein-building message. **148**

metabolic pathway Series of enzyme-mediated reactions by which cells build, remodel, or break down an organic molecule. **84**

metabolism All of the enzyme-mediated reactions in a cell. **39**

metamorphosis Dramatic remodeling of body form during the transition from larva to adult. **391**

metaphase (MEH-tuh-faze) Stage of mitosis at which all chromosomes are aligned midway between spindle poles. **181**

metastasis (meh-TASS-tuh-sis) The process in which malignant cells spread from one part of the body to another. *See* cancer. **185**

methanogen (meth-ANN-o-gen) Prokaryotic organism that produces methane gas (CH_4) as a metabolic by-product. **325**

MHC markers Self-proteins in the plasma membrane of vertebrate body cells. **598**

microevolution Change in allele frequency. **275**

microfilament Cytoskeletal element composed of actin subunits; reinforces cell membranes; functions in movement and muscle contraction. **66**

microRNA Small noncoding RNA that silences gene expression by hybridizing with an mRNA, thus triggering its destruction. **165**

microspore Of seed plants, haploid spore that gives rise to a pollen grain. **354, 458**

microtubule (my-crow-TUBE-yule) Hollow cytoskeletal element composed of tubulin subunits; involved in cellular movement. **66**

microvilli (my-croh-VIL-lie) Fingerlike projections from the plasma membrane of an epithelial cell; increase surface area of the cell. **484, 637**

middle ear Of the mammalian ear, region that contains the eardrum and the tiny bones that transfer sound to the inner ear. **530**

mimicry An evolutionary pattern in which one species becomes more similar in appearance to another. **737**

mineral In the diet, an element that is required in small amounts for normal metabolism. **643**

mitochondrion (my-tuh-CON-dree-un) Double-membraned organelle that produces ATP by aerobic respiration in eukaryotes. **64**

mitosis (my-TOE-sis) Nuclear division mechanism that maintains the chromosome number; occurs in stages (prophase, metaphase, anaphase, and telophase). Basis of body growth, tissue repair, and (in some organisms) asexual reproduction. **178**

model Analogous system used for testing hypotheses. **13**

mold A mass of fungal hyphae that grows rapidly over organic material and produces spores asexually. **367**

molecule (MAUL-ick-yule) Two or more atoms joined by chemical bonds. **4**

mollusk Invertebrate with a reduced coelom and a mantle. **388**

molt To shed and replace an outer body layer or part. **390**

monocot (MAH-no-cot) Flowering plant in which the embryo has one cotyledon. **359**

monohybrid cross Cross between two individuals identically heterozygous for one gene; for example *Aa × Aa*. **208**

monomer Molecule that is a subunit of a polymer. **39**

monophyletic group (ma-no-fill-EH-tick) An ancestor in which a derived trait evolved, together with all of its descendants. **294**

monosaccharide (mon-oh-SACK-uh-ride) Simple sugar; consists of one sugar unit so it cannot be broken apart into monomers. **40**

monotreme (MON-oh-treem) Egg-laying mammal. **411**

monsoon (mon-SOON) Wind that reverses direction seasonally. **768**

morphogen (MOR-foe-jen) Substance that regulates development; secreted by one cell type and affects other cells in a concentration-dependent manner. **672**

morphological convergence Evolutionary pattern in which similar body parts (analogous structures) evolve separately in different lineages. **267**

morphological divergence Evolutionary pattern in which a body part of an ancestor changes differently in its different descendants. **266**

motor neuron Neuron that controls a muscle or gland. **502**

motor protein Type of energy-using protein that interacts with cytoskeletal elements to move the cell's parts or the whole cell. **66**

motor unit One motor neuron and the muscle fibers it controls. **565**

mRNA *See* messenger RNA.

multicellular organism Organism composed of interdependent cells that vary in their structure and function. **332**

multiple allele system Gene for which three or more alleles persist in a population at relatively high frequency. **212**

muscle tension Force exerted by a contracting muscle. **565**

muscle tissue Tissue that consists mainly of contractile cells. **488**

mutation Permanent change in the DNA sequence of a chromosome. **140**

mutualism Species interaction that benefits both species. **733**

mycelium (my-SEAL-ee-um) Mass of threadlike filaments (hyphae) that make up the body of a multicelled fungus. **366**

mycorrhiza (my-kuh-RYE-zah) Mutually beneficial partnership between a fungus and a plant root. **370**

myelin (MY-uh-lin) Fatty material produced by neuroglial cells; insulates axons and thus speeds conduction of action potentials. **505**

myofibril (my-oh-FIE-bril) Within a muscle fiber, a threadlike array of proteins organized as a series of sarcomeres (contractile units). **563**

myoglobin (MY-oh-glow-bin) Muscle protein that reversibly binds oxygen. **566**

myosin (MY-oh-sin) ATP-dependent motor protein; makes up the thick filaments in a sarcomere. **563**

natural killer cell *See* NK cell.

natural selection A major mechanism of evolution: the differential survival and reproduction of individuals of a population based on differences in shared, heritable traits. Outcome of environmental pressures. **257**

nectar Sweet fluid exuded by some flowers; rewards animal pollinators. **457**

negative feedback mechanism A change causes a response that reverses the change; important mechanism of homeostasis. **494**

neoplasm (KNEE-oh-plaz-um) An accumulation of abnormally dividing cells. **184**

nephridium (nef-RID-ee-um) Of some invertebrates, an organ that takes up body fluid and expels excess water and solutes through a pore at the body surface. **653**

nephron (NEF-ron) A kidney tubule and associated capillaries; filters blood and forms urine. **654**

nerve Neuron axons bundled inside a sheath of connective tissue. **500**

nerve cord Bundle of nerves that runs the length of a body. **500**

nerve net Of cnidarians and echinoderms, a mesh of interacting neurons with no central control organ. **382, 500**

nervous system Organ system that gathers information about the internal and external environment, integrates this information, and issues commands to muscles and glands, all by way of neurons. **500**

nervous tissue Animal tissue composed of neurons and neuroglia; detects stimuli and controls responses to them. **489**

neuroglia (nuhr-oh-GLEE-uh) Nervous system cells that assist neurons and facilitate their function in a variety of ways. **502**

neuromuscular junction Synapse between a neuron and a muscle cell. **506**

neuron (NUHR-on) Specialized signaling cell in nervous tissue; transmits electrical signals along its plasma membrane and sends chemical messages to other cells via its endings. **489, 500**

neurosecretory cell Specialized neuron that secretes a hormone into the blood in response to an action potential. **542**

neurotransmitter Chemical signal released by axon terminals or a neuron. **506**

neutral mutation A mutation that has no effect on survival or reproduction. **274**

neutron (NEW-tron) Uncharged subatomic particle that occurs in the atomic nucleus. **24**

neutrophil (NEW-tra-fill) Circulating phagocytic white blood cell; most abundant leukocyte. **593**

niche *See* ecological niche.

nitrification (nigh-triff-ih-CAY-shun) Conversion of ammonium to nitrate. **758**

nitrogen cycle Movement of nitrogen among the atmosphere, soil, and water, and into and out of food webs. **758**

nitrogen fixation Incorporation of nitrogen gas into ammonia. **322, 758**

NK cell Natural killer cell. Lymphocyte that can kill cancer cells undetectable by cytotoxic T cells. **605**

node A region of stem where leaves attach and new shoots form. **428**

nondisjunction Failure of chromosomes to separate properly during nuclear division. **230**

nonshivering heat production Production of heat, rather than ATP, by mitochondria in brown adipose tissue. **661**

nonvascular plant Plant that does not have xylem and phloem; a bryophyte such as a moss. **348**

normal microbiota Collection of normally harmless or beneficial microorganisms that typically live in or on a body. **324**

notochord (NO-toe-cord) Stiff rod of connective tissue that runs the length of the body in chordate larvae or embryos. **402**

nuclear envelope A double membrane that constitutes the outer boundary of the nucleus. Pores in the membrane control which molecules can cross it. **61**

nucleic acid (new-CLAY-ick) Polymer of nucleotides; DNA or RNA. **47**

nucleoid (NEW-klee-oyd) Of a bacterium or archaeon, irregularly shaped region of cytoplasm where the cell's chromosomal DNA is located. **58**

nucleolus (new-KLEE-oh-luss) In a cell nucleus, a dense, irregularly shaped region where ribosomal subunits are assembled. **61**

nucleoplasm (NEW-klee-oh-plaz-um) Of a nucleus, viscous fluid enclosed by the nuclear envelope. **61**

nucleosome (NEW-klee-uh-sohm) Unit of chromosomal organization: a length of DNA wound twice around histone proteins. **136**

nucleotide (NEW-klee-uh-tide) Monomer of nucleic acids; small molecule with a ribose or deoxyribose sugar, a nitrogen-containing base, and one, two, or three phosphate groups. E.g., adenine, guanine, cytosine, thymine, uracil. **47**

nucleus (NEW-klee-us) Of an atom, central core area occupied by protons and neutrons. **24** Of a eukaryotic cell, organelle with a double membrane that holds the cell's DNA. **53**

nutrient (NEW-tree-unt) Substance that an organism must acquire from the environment for growth and survival. **6**

ocean acidification Decrease in seawater pH caused by the rise in atmospheric carbon dioxide. **779**

olfactory receptor Chemoreceptor involved in the sense of smell. **524**

oncogene (ON-ko-jean) Gene that helps transform a normal cell into a tumor cell. **184**

oocyte (OH-uh-cite) Immature egg. **676**

open circulatory system System in which hemolymph leaves vessels and seeps among tissues before returning to the heart. **388, 572**

operator Part of an operon; a DNA binding site for a repressor. **170**

operon (OPP-er-on) Group of genes together with a promoter–operator DNA sequence that controls their transcription. **170**

optic disk Area of the retina where the optic nerve exits the retina; lacks photoreceptors and so causes a blind spot. **528**

organ In multicelled organisms, a structure that consists of tissues engaged in a collective task. **4, 482**

organelle (or-guh-NEL) Structure that carries out a specialized metabolic function inside a cell. **52**

organic Describes a molecule that consists mainly of carbon and hydrogen atoms. **38**

organism (ORG-uh-niz-um) An individual that consists of one or more cells. **4**

organs of equilibrium Sensory organs that respond to body position and motion; function in sense of balance. **532**

organ system In multicelled organisms, set of interacting organs that carry out a particular body function. **4**

osmosis (oz-MOE-sis) Diffusion of water across a selectively permeable membrane; occurs in response to a difference in solute concentration between the fluids on either side of the membrane. **89**

osmotic pressure (oz-MAH-tick) Amount of turgor that prevents osmosis into cytoplasm or other hypertonic fluid. **89**

outer ear Of the mammalian ear, external pinna and the air-filled auditory canal. **530**

ovarian follicle In animals, oocyte and surrounding follicle cells. **677**

ovary In flowering plants, the enlarged base of a carpel, inside which one or more ovules form and eggs are fertilized. In animals, the egg-producing gonad. **359, 456**

oviduct Duct between an ovary and the uterus. **676**

ovulation Release of a secondary oocyte from an ovary. **677**

ovule (AH-vule) Of seed plants, reproductive structure in which the female gametophyte forms and develops. Matures into a seed after fertilization. **354, 456**

ovum Mature animal egg. **683**

oxytocin (ox-ee-TOE-sin) Posterior pituitary hormone that acts on smooth muscle of the reproductive tract and breasts. **543**

ozone layer (OH-zone) Upper atmospheric layer with a high concentration of ozone (O_3) that prevents much ultraviolet radiation from reaching Earth's surface. **307, 792**

pain Perception of tissue injury. **523**

pain receptor Sensory receptor that responds to tissue damage. **522**

PAMPs Pathogen-associated molecular patterns. Molecules found on or in microorganisms that can be pathogenic. **592**

pancreas (PAN-kree-us) Organ that secretes digestive enzymes into the small intestine, and insulin and glucagon into the blood. **548**

Pangea (pan-JEE-uh) Supercontinent that began to form about 300 million years ago and broke up 100 million years later. **262**

parapatric speciation Speciation pattern in which populations speciate while in contact along a common border. **291**

parasitism Relationship in which one species withdraws nutrients from another species, without immediately killing it. **738**

parasitoid (PAIR-uh-sit-oyd) An insect that lays eggs in another insect, and whose young devour their host from the inside. **739**

parasympathetic neurons Neurons of the autonomic system that encourage digestion and tasks that maintain the body. **509**

parathyroid glands Four small endocrine glands on the rear of the thyroid that secrete parathyroid hormone. **545**

parathyroid hormone (**PTH**) Hormone that regulates the concentration of calcium ions in the blood. **545**

parenchyma (puh-REN-kuh-muh) In plants, simple tissue composed of living, thin-walled cells; has various specialized functions such as photosynthesis. **426**

passive transport Membrane-crossing mechanism that requires no energy input. **90**

pathogen Disease-causing agent (virus or cellular organism). **318**

pathogen-associated molecule patterns *See* PAMPs.

PCR *See* polymerase chain reaction.

pedigree Chart of family connections that shows the appearance of a phenotype through generations. **222**

pelagic province (peh-LAJ-ic) The ocean's open waters. **780**

pellicle An outer layer of plasma membrane and elastic proteins that protects and gives shape to many unwalled, single-celled protists. **334**

penis Male organ of intercourse. **680**

peptide Short chain of amino acids linked by peptide bonds. **44**

peptide bond A bond between the amine group of one amino acid and the carboxyl group of another. Joins amino acids in proteins. **44**

per capita growth rate (*r*) Of a population, per capita (per individual) birth rate minus per capita death rate. **716**

perception The meaning a brain derives from a sensation. **522**

pericycle (PAIR-ih-sigh-cull) Of a root, outer layer of the vascular cylinder; a thin layer of cells that retain the ability to divide and differentiate. **433**

periderm (PAIR-ih-durm) Plant dermal tissue that replaces epidermis during secondary growth of eudicots and gymnosperms. **435**

periodic table Tabular arrangement of all known elements by their atomic number. **24**

peripheral nervous system Of vertebrates, nerves that carry signals between the central nervous system and the rest of the body. **501**

peristalsis (pehr-ih-STAL-sis) Wavelike smooth muscle contractions that propel food through the digestive tract. **634**

peritubular capillaries In a kidney, capillaries that branch from an efferent arteriole; surround and exchange substances with a kidney tubule. **655**

permafrost Of arctic tundra, a perpetually frozen layer of soil that prevents water from draining. **775**

peroxisome (purr-OX-uh-sohm) Enzyme-filled vesicle that breaks down fatty acids, amino acids, and toxic substances. **62**

petal Unit of a flower's corolla; often showy and conspicuous. **456**

pH Measure of the number of hydrogen ions in a water-based fluid. **32**

phagocytosis (fag-oh-sigh-TOE-sis) "Cell eating"; type of receptor-mediated endocytosis in which a cell engulfs a large solid particle such as another cell. **93**

pharynx (FAIR-inks) Throat; opens to airways and digestive tract. **620**

phenotype (FEEN-oh-type) An individual's observable traits. **207**

pheromone (FAIR-uh-moan) Chemical that serves as a communication signal between members of an animal species. **524, 702**

phloem (FLOW-um) Complex vascular tissue of plants; its living sieve elements compose sieve tubes that distribute sugars through the plant body. Each sieve element has an associated companion cell that provides it with metabolic support. **348, 427**

phospholipid (foss-foe-LIP-id) A lipid with two (hydrophobic) fatty acid tails and a (hydrophilic) phosphate group in its head; main constituent of lipid bilayers that form cell membranes. **43**

phosphorus cycle Movement of phosphorus among Earth's rocks and waters, and into and out of food webs. **760**

phosphorylation (foss-for-ih-LAY-shun) Reaction in which a phosphate group is added to a molecule. **87**

photoautotroph (foe-toe-AWE-toe-trof) Organism that obtains carbon from carbon dioxide and energy from light. **321**

photoheterotroph (foe-toe-HET-ur-o-trof) Organism that obtains its carbon from organic compounds and its energy from light. **321**

photolysis (foe-TALL-ih-sis) Process by which light energy breaks down a molecule. **104**

photoperiodism (foe-toe-PEER-ee-ud-is-um) Biological response to seasonal changes in the relative lengths of day and night. **472**

photoreceptor Sensory receptor that responds to light. **522**

photorespiration Inefficient sugar-production pathway initiated when rubisco attaches oxygen instead of carbon dioxide to RuBP (ribulose bisphosphate). **107**

photosynthesis (foe-toe-SIN-thuh-sis) Metabolic pathway by which most autotrophs use sunlight to make sugars from carbon dioxide and water. **6, 100**

photosystem Large protein complex in the thylakoid membrane; consists of pigments and other molecules that collectively convert light energy to chemical energy in photosynthesis. **104**

phototropism (foe-toe-TRO-pizm) Directional response to light. **470**

phylogeny (fie-LA-juh-knee) Evolutionary history of a species or group of species. **294**

phytochrome (FIGHT-uh-krome) A light-sensitive pigment that helps set plant circadian rhythms based on length of night. **472**

phytoplankton (FIGH-toe-plank-ton) Community of tiny drifting or swimming photosynthetic organisms. **336**

pigment An organic molecule that can absorb light of certain wavelengths. Reflected light imparts a characteristic color. **102**

pilus (PIE-luss) A protein filament that projects from the surface of some prokaryotic cells. **59**

pineal gland (pie-KNEE-ul) Endocrine gland in the brain that secretes melatonin under low-light or dark conditions. **545**

pioneer species Species that can colonize a new habitat. **740**

pituitary gland (pit-OO-it-airy) Pea-sized endocrine gland in the forebrain; interacts closely with the adjacent hypothalamus. **542**

placenta (pluh-SEN-tah) Organ formed during pregnancy; allows diffusion of material between maternal and embryonic bloods. **669**

Glossary (continued)

placental mammal Mammal in which maternal and embryonic bloodstreams exchange materials by means of a placenta. **411**

plant Multicelled photosynthetic eukaryote that forms and nurtures embryos on the parental body. **8**, **348**

plant hormone Extracellular signaling molecule of plants; exerts an effect on target cells at very low concentration. E.g., auxin, gibberellin. **463**

plaque *See* dental plaque.

plasma Fluid portion of blood. **578**

plasma membrane (PLAZ-muh) Cell membrane that encloses the cytoplasm; separates a cell from its external environment. **52**

plasmid (PLAZ-mid) Small circle of DNA separate from the chromosome(s) in some bacteria and archaea; carries a few genes. **58**

plasmodesma (plaz-mow-DEZ-muh) Cell junction that forms an open channel between the cytoplasm of adjacent plant cells. **69**

plasmodial slime mold (plaz-MOH-dee-ul) Protist that feeds as a multinucleated mass; forms a spore-bearing structure when environmental conditions become unfavorable. **342**

plastid Double-membraned organelle that functions in photosynthesis, pigmentation, or storage in plants and algal cells; for example, a chloroplast, chromoplast, or amyloplast. **65**

plate tectonics theory Theory that Earth's outermost layer of rock is cracked into plates, the slow movement of which conveys continents to new locations over geologic time. **262**

platelet Cell fragment that helps blood clot. **579**

pleiotropic (ply-uh-TROH-pick) Refers to a gene that affects multiple traits. **213**

plot sampling Using demographics observed in sample plots to estimate demographics of a population as a whole. **715**

polar body Tiny cell produced by unequal cytoplasmic division during egg production. **677**

polarity (pole-AIR-ih-tee) Separation of charge into distinct positive and negative regions. **28**

pollen grain Walled, immature male gametophyte of a seed plant. **349**, **456**

pollen sac Of seed plants, reproductive structure in which pollen grains develop. **354**

pollination The arrival of pollen on a stigma or other pollen-receiving reproductive part of a seed plant. **354**, **457**

pollination vector Animal or environmental agent that moves pollen grains from one plant to another. **457**

pollinator An animal that facilitates pollination by moving pollen; an animal pollination vector. **360**, **457**

pollutant A substance released into the environment by human activities; interferes with the function of organisms that evolved in the absence of the substance or with lower levels. **790**

polygenic inheritance *See* epistasis.

polymer Molecule that consists of repeated monomers. **39**

polymerase chain reaction (**PCR**) Laboratory method that rapidly generates many copies of a specific section of DNA. **240**

polyp (PAH-lup) Cylindrical cnidarian body plan with the mouth on the upper surface. Sea anemones are examples. **382**

polypeptide Long chain of amino acids linked by peptide bonds. **44**

polyploidy (PALL-ee-ploy-dee) Condition of having three or more of each type of chromosome characteristic of the species. **230**

polysaccharide (paul-ee-SACK-uh-ride) Carbohydrate that consists of hundreds or thousands of monosaccharide monomers. **40**

pons (PONZ) Hindbrain region that influences breathing and serves as a bridge to the adjacent midbrain. **512**

population A group of organisms of the same species who live in a specific location and breed with one another more often than they breed with members of other populations. **4**, **274**

population density Number of individuals per unit area. **714**

population distribution Location of population members relative to one another: clumped, uniformly dispersed, or randomly dispersed. **714**

population size Total number of individuals in a population. **714**

positive feedback mechanism A response intensifies the conditions that caused its occurrence. **504**

potential energy Energy stored in the position or arrangement of a system's components. **79**

Precambrian Period of Earth's history from 4.6 billion to 541 million years ago: the Hadean, Archean, and Proterozoic eons. **310**

predation One species captures, kills, and eats another. **736**

prediction Statement, based on a hypothesis, about a condition that should occur if the hypothesis is correct. **13**

pressure flow theory Explanation of how a difference in turgor between sieve elements in source and sink regions pushes sugar-rich fluid through a sieve tube. **449**

primary endosymbiosis An organelle evolves from bacteria that entered a heterotrophic protist. **334**

primary growth Of plants, lengthening of young shoots and roots; originates at apical meristems. **434**

primary motor cortex Region of frontal lobe that controls voluntary movement. **514**

primary production The rate at which an ecosystem's producers capture and store energy. **750**

primary succession A new community becomes established in an area where there was previously no soil. **740**

primary wall The first cell wall of young plant cells. **68**

primate Mammal having grasping hands with nails and a body adapted to climbing; for example, a lemur, monkey, ape, or human. **412**

primer Short, single strand of DNA or RNA that base-pairs with a specific DNA sequence and serves as an attachment point for DNA polymerase. **138**

prion (PREE-on) Infectious protein. **46**

probability The chance that a particular outcome of an event will occur; depends on the total number of outcomes possible. **16**

probe Short fragment of DNA labeled with a tracer; designed to hybridize with a nucleotide sequence of interest. **240**

producer Autotroph. Organism that makes its own food using energy and nonbiological raw materials from the environment. **6**, **750**

product A molecule that is produced by a reaction. **80**

progesterone Sex hormone produced by ovaries; prepares a female body for pregnancy and helps maintain a pregnancy. **547**, **676**

prokaryote (pro-CARE-ee-oat) Informal term for a single-celled organism without a nucleus; a bacterium or archaeon. **8**

prolactin (pro-LAC-tin) Anterior pituitary hormone that stimulates milk production by mammary glands. **543**

promoter In DNA, a sequence that is a site where RNA polymerase binds and begins transcription. **150**

prophase (PRO-faze) Stage of mitosis during which chromosomes condense and become attached to a newly forming spindle. **181**

prostate gland (PRAH-state) Exocrine gland that contributes to semen. **681**

protein (PRO-teen) Organic molecule that consists of one or more polypeptides. **44**

proteobacteria (pro-tee-o-bac-TEER-ee-uh) Largest bacterial lineage; members show a wide range of metabolic diversity. **323**

protist Common term for a eukaryote that is not a plant, animal, or fungus. **8**, **332**

protocell Membranous sac that contains interacting organic molecules; hypothesized to have formed prior to the first life forms. **305**

proton Positively charged subatomic particle that occurs in the nucleus of all atoms. **24**

proto-oncogene Gene that, by mutation, can become an oncogene. **184**

protostomes (PRO-toe-stomes) Lineage of bilateral animals in which the first opening on the embryo surface becomes a mouth; include most invertebrates. **379**

proximal tubule Region of kidney tubule nearest Bowman's capsule. **655**

proximate cause Of a behavior, the genes, physiology, or environmental conditions that trigger the behavior. **698**

pseudocoelom (SUE-doe-see-lum) Fluid-filled cavity between the gut and body wall; partially lined by tissue derived from mesoderm. **379**

pseudopod (SUE-doh-pod) A temporary protrusion that helps some eukaryotic cells move and engulf prey. **67**

pseudoscience A claim, argument, or method that is presented as if it were scientific, but is not. **19**

PTH *See* parathyroid hormone.

puberty Period when reproductive organs begin to mature. **675**

pulmonary artery Artery that carries oxygen-poor blood from the heart to a lung. **574**

pulmonary circuit (PUL-mon-erry) Circuit through which blood flows from the heart to the lungs and back. **573**

pulmonary vein Vein that carries oxygen-rich blood from a lung to the heart. **574**

pulse Brief expansion of artery walls that occurs when ventricles contract. **580**

Punnett square (PUN-it) Diagram used to predict the genotypic and phenotypic outcomes of a cross. **208**

pupil Adjustable opening that allows light into a camera eye. **527**

pyruvate (pie-ROO-vait) Three-carbon product of glycolysis. **116**

radial symmetry Having parts arranged around a central axis, like the spokes of a wheel. **379**

radioactive decay Process in which an atom emits energy and/or subatomic particles when its nucleus spontaneously breaks up. **24**

radioisotope Isotope with an unstable nucleus. **24**

radiolaria (ray-dee-oh-LAIR-ee-ah) Heterotrophic single-celled protists with a porous shell of silica and cytoplasmic extensions. **337**

radiometric dating Method of estimating the age of a rock or fossil by measuring the content

and proportions of a radioisotope and its daughter elements. **260**

rain shadow Dry region downwind of a coastal mountain range. **769**

ray-finned fish Jawed fish with fins supported by thin rays derived from skin; member of most diverse lineage of fishes. **405**

reabsorption *See* tubular reabsorption.

reactant (ree-ACK-tunt) Molecule that enters a reaction and is changed by participating in it. **80**

reaction Process of molecular change. **39**

receptor protein Membrane protein that triggers a change in cell activity in response to a specific stimulus. **57**

recessive Refers to an allele with an effect that is masked by a dominant allele on the homologous chromosome. **207**

recombinant DNA (ree-COM-bih-nent) Composite DNA made in a laboratory by joining DNA fragments from multiple sources. **238**

rectum Terminal region of the large intestine; stores digestive waste. **634**

red alga Photosynthetic protist; typically multicelled, with chloroplasts containing red accessory pigments (phycobilins). **340**

red blood cell Hemoglobin-filled blood cell that carries oxygen (an erythrocyte). **578**

red marrow Bone marrow that makes blood cells. **560**

redox reaction (REE-docks) Oxidation–reduction reaction. A typical redox reaction is an electron transfer, in which one molecule accepts electrons (it becomes reduced) from another molecule (which becomes oxidized). **85**

reflex Automatic response that occurs without conscious thought. **508**

releasing hormone Animal hormone that encourages secretion of another hormone by its target cells. **538**

replacement fertility rate Number of children a woman must bear to replace herself with one daughter of reproductive age. **725**

repressor Transcription factor that slows or stops transcription. **164**

reproduction (ree-pruh-DUCK-shun) Processes by which organisms produce offspring. **7**

reproductive base Of a population, members of the reproductive and pre-reproductive age categories. **725**

reproductive cloning Laboratory procedure such as SCNT that produces animal clones from a single somatic cell. **142**

reproductive isolation The end of gene flow between populations; part of speciation. **286**

reptiles Informal amniote subgroup that includes lizards, snakes, turtles, crocodilians. **408**

resource partitioning Evolutionary process whereby species become adapted in different ways to access different portions of a limited resource; allows species with similar needs to coexist. **735**

respiration Physiological processes that collectively supply animal cells with oxygen and dispose of their waste carbon dioxide. **616**

respiratory cycle One inhalation and one exhalation. **622**

respiratory membrane Membrane consisting of alveolar epithelium, capillary endothelium, and their fused basement membranes. **624**

respiratory protein A protein that reversibly binds oxygen when the oxygen concentration is high and releases it when the oxygen concentration is low. Hemoglobin is an example. **616**

respiratory surface Thin, moist layer of cells across which gases are exchanged between animal cells and the external environment. **616**

resting potential Membrane potential of a neuron at rest. **503**

restriction enzyme Type of enzyme that cuts DNA at a specific sequence. **238**

retina Photoreceptor-containing layer of tissue in an eye. **526**

retrovirus RNA virus that uses the enzyme reverse transcriptase to produce viral DNA in a host cell. **317**

reverse transcriptase (trance-CRYPT-ace) An enzyme that assembles a strand of DNA (cDNA) on an mRNA template. **239**

rhizoid (RYE-zoyd) Threadlike structure that holds a nonvascular plant in place. **350**

rhizome (RYE-zohm) Stem that grows horizontally along or under the ground. **352**

ribonucleic acid *See* RNA.

ribosomal RNA (**rRNA**) RNA component of ribosomes. **148**

ribosome (RYE-buh-sohm) Organelle of protein synthesis. An intact ribosome consists of two subunits, each composed of rRNA and proteins. **58**

ribozyme (RYE-boh-zime) RNA that has enzymatic activity. **305**

RNA Ribonucleic (rye-bo-new-CLAY-ick) acid. Nucleic acid with roles in gene expression. Most are single-stranded chains of ribonucleotides (adenine, guanine, cytosine, and uracil). *See* messenger RNA, transfer RNA, ribosomal RNA. **47**

RNA interference Mechanism in which RNA molecules interfere with gene expression by preventing translation of specific mRNAs. **165**

RNA polymerase Enzyme that carries out transcription. **150**

RNA world Hypothetical early interval when RNA served as genetic material. **304**

Glossary (continued)

rod cell Photoreceptor that is active in dim light; allows coarse perception of image and detection of motion. **528**

root hairs Hairlike, absorptive extensions of epidermis cells on young plant roots. **432**

root nodules Of some plant roots, swellings that contain mutualistic nitrogen-fixing bacteria. **445**

roundworm Cylindrical worm with a pseudocoelom. **390**

rRNA *See* ribosomal RNA.

r-selection Pattern of natural selection in which producing the most offspring the most quickly is adaptive; occurs when population density is low and resources are abundant. **721**

rubisco (roo-BIS-co) Ribulose bisphosphate carboxylase. Carbon-fixing enzyme of the Calvin–Benson cycle. **106**

runoff Water that flows over soil into streams. **754**

sac fungi Fungi that form spores in a sac-shaped structure during sexual reproduction. Most diverse fungal group. **367**

salivary gland Exocrine gland that secretes saliva into the mouth. **634**

salt Compound that releases ions other than H⁺ and OH⁻ when it dissolves in water. **30**

sampling error Difference between results derived from testing an entire group of events or individuals, and results derived from testing a subset of the group. **16**

SA node *See* sinoatrial node.

sarcomere Contractile unit of muscle; an array of actin and myosin filaments arranged in parallel and bounded on either end by an array of cytoskeletal elements (a Z line). **563**

sarcoplasmic reticulum (sar-co-PLAZ-mic reh-TIC-you-lum) Specialized endoplasmic reticulum in muscle cells; stores and releases calcium ions. **565**

saturated fat *See* triglyceride.

saturated fatty acid Fatty acid with only single bonds linking the carbons in its tail. **42**

savanna (suh-VAN-uh) Tropical biome dominated by grasses with a few scattered shrubs and trees. **773**

science Systematic study of the physical universe. **12**

scientific method Making, testing, and evaluating hypotheses about the natural world. **13**

scientific theory Hypothesis that has not been falsified after many years of rigorous testing, and is useful for making predictions about a wide range of phenomena. **18**

sclerenchyma (sklair-EN-kuh-muh) In plants, simple tissue composed of cells that die when mature; their tough walls structurally support plant parts. Includes fibers, sclereids. **426**

SCNT *See* somatic cell nuclear transfer.

seamount An undersea mountain. **780**

secondary endosymbiosis A chloroplast evolves from a red alga or green alga that was engulfed by a heterotrophic protist. **334**

secondary growth Of plants, thickening of older stems and roots; originates at lateral meristems. **435**

secondary sexual traits Characteristics that differ between the sexes but do not play a direct role in reproduction. **547**

secondary succession A new community becomes established in a site where a community previously existed. **740**

secondary wall Lignin-reinforced wall that forms inside the primary wall of a mature plant cell. **68**

second law of thermodynamics Energy tends to disperse spontaneously. **78**

second messenger Molecule that forms inside a cell when an animal hormone binds to a receptor in the plasma membrane; sets in motion reactions that alter the cell's activity. **538**

sedimentary cycle Biochemical cycle such as the phosphorus cycle, in which the atmosphere plays little role and rocks are the major reservoir. **760**

seed Embryo sporophyte of a seed plant packaged with nutritive tissue inside a protective coat. **349, 460**

seedless vascular plant Plant that disperses by releasing spores and has xylem and phloem. Ferns and club mosses are examples. **352**

selfish herd effect Animal groups arise from the collective tendency of individuals to reduce their risk of danger by drawing near one another. **706**

semen (SEE-men) Sperm and secretions from exocrine glands. **681**

semicircular canals Organs of equilibrium that respond to rotation and angular movement of the head; part of the vestibular apparatus. **532**

semiconservative replication Describes the process of DNA replication in which one strand of each copy of a DNA molecule is new, and the other is a strand of the original DNA. **139**

seminal vesicles (SEHM-in-al) Exocrine glands that add sugary fluid to semen. **681**

seminiferous tubules (sehm-in-IF-er-us) In testes, tiny tubes where sperm form. **680**

senescent (suh-NESS-ent) Refers to a metabolically active cell with an arrested cell cycle that is a consequence of some type of damage. **183**

sensation Detection of a stimulus. **522**

sensory adaptation Slowing or cessation of a sensory receptor's response to an ongoing stimulus. **522**

sensory neuron Neuron that is activated when its receptor endings detect a specific stimulus, such as light or pressure. **502**

sensory receptor Cell or cell component that responds to a specific stimulus, such as temperature or light. **494**

sequencing Method of determining the order of nucleotides in DNA. **242**

sex chromosomes Pair of chromosomes that differs between males and females, and the difference determines sex. **137**

sex hormone Steroid hormone produced by gonads; functions in reproduction and may affect secondary sexual characteristics. **547**

sexual dimorphism A trait that differs between males and females of a species. **282, 704**

sexual reproduction Reproductive mode by which offspring arise from two parents and inherit genes from both. **192, 668**

sexual selection Mode of natural selection in which some individuals outreproduce others of a population because they are better at securing mates. **282**

shell model Model of electron distribution in an atom. **26**

short tandem repeat In chromosomal DNA, sequences of a few nucleotides repeated multiple times in a row. Used in DNA profiling. **216**

sieve elements Living cells that compose sugar-conducting sieve tubes of phloem. Each sieve tube consists of a stack of sieve elements that meet end to end at sieve plates. **448**

sieve tube Sugar-conducting tube of phloem; consists of stacked sieve elements. **448**

single-nucleotide polymorphism (**SNP**) A one-nucleotide DNA sequence variation carried by a measurable percentage of a population. **245**

sink Of a plant, region of tissue where sugars are being used or put into storage. **448**

sinoatrial (**SA**) **node** (sigh-no-AY-tree-uhl) Cardiac pacemaker; group of cells that spontaneously emits rhythmic action potentials that result in contraction of cardiac muscle. **577**

sister chromatids (KROME-uh-tid) The two identical DNA molecules of a duplicated eukaryotic chromosome, attached at the centromere. **136**

sister groups The two lineages that emerge from a node on a cladogram. **295**

skeletal muscle fiber Multinucleated contractile cell that runs the length of a skeletal muscle. **563**

skeletal muscle tissue Muscle that interacts with skeletal components to move body parts; under voluntary control. **488**

sliding-filament model Explanation of how interactions among actin and myosin filaments shorten a sarcomere and bring about muscle contraction. **564**

small intestine Longest portion of the digestive tract, and the site of most digestion and absorption. **634**

smooth muscle tissue Muscle that lines blood vessels and forms the wall of hollow organs; not under voluntary control. **488**

SNP *See* single-nucleotide polymorphism.

social animal Animal that lives in a multigenerational group in which members benefit by cooperating in some tasks. **706**

soil erosion Loss of soil under the force of wind and water. **443**

soil water Water between soil particles. **754**

solute (SAWL-yute) A dissolved substance. **30**

solution Uniform mixture of solute completely dissolved in solvent. **30**

solvent Liquid in which other substances dissolve. **30**

somatic cell nuclear transfer (**SCNT**) Reproductive cloning method in which the DNA of a donor's body cell is transferred into an enucleated egg. Produces clones of the donor. **142**

somatic nervous system Division of the peripheral nervous system that controls skeletal muscles and relays sensory signals about movements and external conditions. **508**

somatic sensations Sensations such as touch and pain that arise when sensory neurons in skin, muscle, or joints are activated. **523**

sorus (SORE-us) Cluster of spore-producing capsules on a fern leaf. **352**

source Of a plant, region of tissue where sugars are being produced or released from storage. **448**

speciation (spee-see-A-shun) Emergence of a new species. **286**

species (SPEE-sheez) Unique type of organism designated by genus name and specific epithet. Of sexual reproducers, often defined as one or more groups of individuals that can potentially interbreed, produce fertile offspring, and do not interbreed with other groups. **4**

species evenness Of a community, the relative abundance of species. **732**

species richness Of a community, the number of species. **732**

sperm Male gamete. **668**

sphincter Ring of muscle in a tubular organ or at a body opening. **562**

spinal cord Central nervous system organ that extends through the backbone and connects peripheral nerves with the brain. **511**

spindle (SPIN-dull) Temporary structure that moves chromosomes during nuclear division; consists of microtubules. **181**

spirochetes (SPY-roh-keets) Group of small bacteria shaped like a stretched-out spring. **323**

spleen Large lymphoid organ that functions in immunity and filters pathogens, old red blood cells, and platelets from the blood. **586**

sponge Aquatic invertebrate that has no tissues or organs and filters food from the water. **381**

spongy bone Lightweight bone with many internal spaces; contains red marrow. **560**

sporophyte (SPORE-uh-fight) Spore-forming diploid body that forms in the life cycle of land plants and some multicelled algae. **340**, **348**

stabilizing selection Mode of natural selection in which an intermediate form of a trait is adaptive, and extreme forms are selected against. **280**

stamen (STAY-men) Of a flower, reproductive organ that consists of a pollen-producing anther and, in most species, a filament. **359**, **456**

starch Coiled polysaccharide of glucose monomers. Principal form of stored sugars in plants. **40**

stasis (STAY-sis) Evolutionary pattern in which a lineage persists with little or no change over evolutionary time. **292**

statistically significant Refers to a result that is statistically unlikely to have occurred by chance. **16**

stem cell Cell that can divide to produce more stem cells or differentiate into a specialized cell type. **495**

steroid (STARE-oyd) Lipid with four carbon rings, no fatty acid tails. **43**

steroid hormone Lipid-soluble hormone derived from cholesterol. **538**

stigma Of a flower, the upper part of a carpel; adapted to receive pollen. **456**

stomata Singular **stoma**. Of plants, tiny gaps between pairs of guard cells in epidermis; allow internal tissues to exchange gases with the air. Can be closed to prevent water loss. **107**, **348**

stomach Tubular, muscular organ that functions in chemical and mechanical digestion. **633**

strobilus (STROH-bill-us) Of some plants, a spore-forming, cone-shaped structure composed of modified leaves or branches. **353**

stroma (STROH-muh) Thick, cytoplasm-like fluid between the thylakoid membrane and the two outer membranes of a chloroplast. Site of light-independent reactions of photosynthesis. **101**

stromatolite (stroh-MAT-oh-lite) Rocky structure composed of layers of bacterial cells and sediments. **306**

substance P Chemical that enhances transmission of signals related to pain. **523**

substrate Molecule that an enzyme acts upon and converts to a product; reactant in an enzyme-catalyzed reaction. **82**

substrate-level phosphorylation ATP formation by the transfer of a phosphate group from a phosphorylated molecule to ADP. **116**

superior vena cava (VEE-nah CAY-vuh) Vein that delivers oxygen-poor blood from the upper body to the heart. **575**

surface-to-volume ratio A relationship in which the volume of an object increases with the cube of the diameter, and the surface area increases with the square. Limits cell size. **53**

survivorship curve Graph showing how many members of a cohort remain alive over time. **720**

suspension feeder Animal that filters food from the water around it. **381**

swim bladder Of many ray-finned fishes, a gas-filled organ that adjusts buoyancy. **405**

symbiosis (sim-by-OH-sis) One species lives in or on another in a commensal, mutualistic, or parasitic relationship. **732**

sympathetic neurons Neurons of the autonomic system that are activated during stress and danger. **509**

sympatric speciation Speciation pattern in which speciation occurs within a population, in the absence of a physical barrier to gene flow. **290**

synapse (SIN-apps) Region where a neuron's axon terminals transmit signaling molecules (neurotransmitter) to another cell. **506**

synaptic integration The summing of excitatory and inhibitory signals by a postsynaptic cell. **507**

synovial joint (sin-OH-vee-ul) Capsule-enclosed joint at which cartilage-covered ends of bones are separated by a fluid-filled space. **561**

systemic acquired resistance In plants, inducible whole-body resistance to a wide range of pathogens and environmental stresses. **474**

systemic circuit (sis-TEM-ick) Circuit through which blood flows from the heart to the body tissues and back. **573**

systole (SIS-toll-ee) Contractile phase of the cardiac cycle. **576**

systolic pressure (sis-STAHL-ic) Blood pressure when the left ventricle is most contracted; highest pressure of the cardiac cycle. **581**

T cell T lymphocyte. White blood cell central to adaptive immunity; some kinds target infected or cancerous body cells. **593**

T cell receptor (**TCR**) Special antigen receptor that only T cells have; recognizes antigen as non-self, and MHC markers as self. **598**

taiga *See* boreal forest.

taproot system In plants, an enlarged primary root together with all of the lateral roots that branch from it. Typical of eudicots. **432**

taste receptor Chemoreceptor involved in the sense of taste. **524**

taxon (TAX-on) Plural **taxa**. A ranked group of organisms that share a unique set of traits; a species, genus, family, order, class, phylum, kingdom, or domain. **11**

taxonomy (tax-ON-oh-me) Naming and classifying species in a systematic way. **10**

telomere (TELL-a-meer) Region of noncoding DNA at the end of a chromosome; provides a buffer against loss of coding sequences. **183**

telophase (TELL-uh-faze) Stage of mitosis during which chromosomes arrive at opposite spindle poles, decondense, and become enclosed by a new nuclear envelope. **181**

temperate deciduous forest Northern Hemisphere biome in which the predominant plants are broadleaf trees that lose their leaves in fall and become dormant during cold winters. **772**

temperate grassland Biome in the interior of continents; perennial grasses and other non-woody plants adapted to grazing and fire predominate. **773**

temperature Measure of molecular motion. **31**

tendon Strap of dense connective tissue that connects a skeletal muscle to bone. **562**

teratogen (teh-RAT-uh-jenn) Environmental factor that disrupts normal development and causes birth defects. **691**

territory Area from which an animal or group of animals actively excludes others. **704**

testcross Method of determining the genotype of an individual with a dominant phenotype: a cross between the individual and another individual known to be homozygous recessive. **208**

testis Sperm-producing animal gonad. **670**

testosterone Hormone secreted by testes; functions in sperm formation and development of male secondary sex characteristics. **547, 680**

tetrapod (TEH-trah-pod) A vertebrate with four bony limbs, or a descendant of a four-legged ancestor. **403**

TH *See* thyroid hormone.

thalamus (THAL-ah-mus) Forebrain region that relays signals to the cerebrum; affects sleep–wake cycles. **513**

thermal radiation An object emits heat into the space around it. **660**

thermoreceptor Temperature-sensitive sensory receptor. **522**

thermoregulation Maintaining body temperature within the limited range that cells can tolerate. **660**

thigmotropism (thig-MA-truh-pizm) In plants, growth in a direction influenced by contact. **471**

threatened species Species likely to become endangered in the near future. **786**

threshold potential Neuron membrane potential at which voltage-gated sodium channels open, causing an action potential to occur. **504**

thylakoid membrane Inner membrane system that carries out light-dependent reactions in chloroplasts and cyanobacteria. **101**

thymus Hormone-producing organ in which T cells mature. **586**

thyroid gland Endocrine gland in the base of the neck that secretes thyroid hormone and calcitonin. **544**

thyroid hormone (**TH**) Iodine-containing hormones that collectively increase metabolic rate and play a role in development. **544**

thyroid-stimulating hormone (**TSH**) Anterior pituitary hormone; stimulates secretion of thyroid hormone by the thyroid. **543**

tight junction Cell junction that fastens together the plasma membrane of adjacent epithelial cells; collectively prevent fluids from leaking between the cells. Composed of adhesion proteins. **69, 484**

tissue In multicelled organisms, specialized cells organized in a pattern that allows them to perform a collective function. **4, 482**

tissue culture propagation Laboratory method in which individual plant cells (typically from meristem) are induced to form embryos. **463**

T lymphocyte *See* T cell.

topsoil Uppermost layer of soil; contains the most organic matter and plant nutrients. **443**

total fertility rate Expected number of children a woman will bear over the course of a lifetime. **725**

tracer A substance with a detectable component that can be followed in a biological system. **25**

trachea (TRAY-key-uh) Airway to the lungs; windpipe. **621**

tracheal system (TRAY-key-ul) Of insects, a system of air-filled tubes that convey gases between openings at the body surface (spiracles) and internal tissues. **617**

tracheids (TRAY-key-idz) Component of xylem. Tapered cells that die when mature; pitted walls that remain interconnect with other tracheids to form water-conducting tubes. **446**

tract Bundle of axons in the central nervous system. **510**

transcription Process in which enzymes use DNA as a template to assemble RNA. Part of gene expression. **148**

transcription factor Regulatory protein that influences transcription by binding directly to DNA; e.g., an activator or repressor. **164**

transfer RNA (**tRNA**) RNA that delivers amino acids to a ribosome during translation. **148**

transgenic (trans-JEN-ick) Refers to a genetically modified organism that carries a gene from a different species. **246**

translation Process in which a polypeptide chain is assembled from amino acids in the order specified by an mRNA. **148**

translocation In genetics, structural rearrangement in which a broken piece of chromosome has become reattached in the wrong location. **228** In plants, movement of organic compounds through phloem. **448**

transpiration (trans-purr-A-shun) Evaporation of water from aboveground plant parts. **447**

transport protein Protein that moves specific ions or molecules across a cell membrane. **57**

transposon *See* transposable element.

transposable element (**transposon**) Segment of DNA that can move spontaneously ("jump") to a new location in a chromosome or between chromosomes. **228**

triglyceride (tri-GLISS-a-ride) Molecule with three fatty acid tails bonded to a glycerol; main component of fats. Unsaturated fat is the common name for a triglyceride with three unsaturated fatty acid tails; saturated fats have one or more saturated fatty acid tails. **42**

tRNA *See* transfer RNA.

trophic level (TROE-fick) Position of an organism in a food chain. **751**

tropical rain forest Highly productive and species-rich biome in which year-round rains and warmth support continuous growth of evergreen broadleaf trees. **771**

tropism (TROPE-izm) A directional movement toward or away from a stimulus that triggers it. **470**

trypanosome (trip-AN-oh-sohm) Parasitic flagellated protist in the supergroup Excavata; has a single mitochondrion and a flagellum that is enclosed within a membrane. **335**

TSH *See* thyroid-stimulating hormone.

tubular reabsorption Water and solutes move from the filtrate inside a kidney tubule into the peritubular capillaries. **656**

tubular secretion Ions and breakdown products of organic molecules move out of peritubular capillaries and into filtrate. **656**

tumor (TOO-murr) A neoplasm that forms a lump in the body. **184**

tunicate (TOON-ih-kit) Invertebrate chordate that loses most of its defining chordate traits during the transition to adulthood. **402**

turgor (TRR-grr) Pressure that a fluid exerts against a structure that contains it. **89**

ultimate cause Of a behavior, the evolutionary factors that shaped the behavior. **698**

unsaturated fat *See* triglyceride.

unsaturated fatty acid Fatty acid with one or more carbon–carbon double bonds in its tail. **42**

urea (yuh-REE-uh) Nitrogen-containing waste compound derived from ammonia; water is required to excrete it. **652**

ureter (YUR-et-er) One of two tubes that carries urine from a kidney to the bladder. **654**

urethra (yur-EETH-rah) Tube through which urine from the bladder exits the body. **654**

uric acid Nitrogen-containing waste compound derived from ammonia; it can be excreted as crystals. **652**

urinary bladder Hollow, muscular organ that stores urine. **654**

urine Fluid that consists of water and soluble wastes; formed and excreted by the vertebrate kidneys. **653**

uterus Muscular chamber where offspring develop; womb. **676**

vaccine (vax-SEEN) A preparation introduced into the body in order to elicit immunity to a specific antigen. **610**

vacuole (VAK-you-ole) Large, fluid-filled vesicle that isolates or breaks down waste, debris, toxins, or food. *See also* central vacuole. **62**

vagina Female organ of intercourse and the birth canal. **676**

variable (VAIR-ee-uh-bull) In an experiment, a characteristic or event that differs among individuals or over time. **13**

vascular bundle Multistranded bundle of xylem, phloem, and sclerenchyma fibers running through a stem or leaf. **428**

vascular cambium Of plants, lateral meristem that produces secondary xylem and phloem. **435**

vascular cylinder Of a plant root, central column that consists of vascular tissue, pericycle, and supporting cells. **433**

vascular plant Plant that has xylem and phloem. **348**

vascular tissue system Tissues that carry water and nutrients through the plant body. *See* xylem, phloem. **424**

vas deferens (vas DEF-er-ens) One of a pair of long ducts that carry mature sperm to the ejaculatory duct. **680**

vasoconstriction Narrowing of a blood vessel when smooth muscle that rings it contracts. **580**

vasodilation Widening of a blood vessel when smooth muscle that rings it relaxes. **580**

vector In biotechnology, a plasmid or other DNA molecule that can accept foreign DNA and be replicated inside a host cell. **238**
In epidemiology, an animal that carries a pathogen from one host to another. **318**

vegetative reproduction Form of asexual reproduction in plants; new roots and shoots grow from extensions or fragments of a parent plant. **462**

vein Large-diameter vessel that returns blood to the heart. **574**

ventricle (VEN-trick-uhl) Heart chamber that receives blood from an atrium and pumps it out of the heart. **573**

venule (VEEN-yule) Blood vessel that conveys blood from capillaries to a vein. **583**

vernalization (vurr-null-iz-A-shun) Stimulation of flowering in spring by long exposure to low temperature in winter. **473**

vertebrae Bones of the vertebral column (backbone). **559**

vertebral column Backbone. **559**

vertebrate Animal with a backbone. **402**

vesicle (VESS-ih-cull) Saclike, membrane-enclosed organelle; different kinds store, transport, or break down their contents. **62**

vessel elements Of xylem, cells that form in stacks and die when mature; their pitted walls remain to form water-conducting tubes. Each tube consists of a stack of vessel elements that meet end to end at perforation plates. **446**

vestibular apparatus System of fluid-filled sacs and canals in the inner ear; contains organs of equilibrium. **532**

villi Multicelled projections from the lining of the small intestine. **637**

viral envelope Of some viral particles, an outer membrane derived from the host cell in which the particle formed. **316**

viral particle Virus that is not inside a host cell. **316**

viral reassortment Multiple strains of virus infect a host simultaneously and swap genes. **319**

virus Noncellular, infectious particle of protein and nucleic acid; replicates only in a host cell. **316**

visceral sensations Sensations that arise when sensory neurons associated with organs inside body cavities are activated. **523**

vision Detection of light in a way that provides a mental image of objects in the environment. **526**

visual accommodation Process of making adjustments to lens shape so light from an object falls on the retina. **527**

vital capacity Maximum amount of air moved in and out of lungs with forced inhalation and exhalation. **622**

vitamin Organic substance required in the diet in small amounts for normal metabolism. **642**

vomeronasal organ Pheromone-detecting organ of vertebrates. **524**

water cycle Movement of water among Earth's oceans, atmosphere, and the freshwater reservoirs on land. **754**

water molds Heterotrophic protists (stramenopiles) that grow as a mesh of nutrient-absorbing filaments. **336**

water–vascular system Of echinoderms, a system of fluid-filled tubes and tube feet that function in locomotion. **396**

wavelength Distance between the crests of two successive waves. **102**

wax Water-repellent substance that is a complex, varying mixture of lipids with long fatty acid tails bonded to long-chain alcohols. **43**

white blood cell Leukocyte; a blood cell with a role in housekeeping and defense. Lymphocytes are a special category of white blood cells that carry out adaptive immune responses. **579**

white matter Central nervous system tissue consisting primarily of myelin-wrapped axons. **510**

wood Accumulated secondary xylem. Forms inside the cylinder of vascular cambium in an older plant stem or root. **435**

X chromosome inactivation Mechanism of dosage compensation: developmental shutdown of one of the two X chromosomes in the cells of female mammals. *See also* Barr body. **168**

xylem (ZY-lum) Complex vascular tissue of plants; cell walls of its dead tracheids and vessel elements form tubes that distribute water and mineral ions through the plant body. **348, 427**

yeast Single-celled fungus. **366**

yellow marrow Bone marrow that is mostly fat; fills the interior of most adult long bones. **560**

yolk Nutritious material in many animal eggs. **669**

zero population growth Interval during which the number of births equals the number of deaths. **716**

zooplankton (zoe-eh-PLANK-ton) Community of tiny drifting or swimming heterotrophic organisms. **337**

zygote (ZYE-goat) Diploid cell that forms when two gametes fuse; the first cell of a new individual. **195**

zygote fungi (ZY-goat) Fungi that form a thick-walled zygospore during sexual reproduction. **367**

Index